Contributors

Christy C. Andrade
Department of Biology
Willamette University
Salem, OR
USA

andradec@willamette.edu

Nicole C. Arrigo
Center for Tropical Research
Institute of the Environment and Sustainability
University of California, Los Angeles
Los Angeles, CA
USA

ncarrigo@gmail.com

Barry J. Beaty
Arthropod-Borne and Infectious Diseases Laboratory
Department of Microbiology, Immunology, and Pathology
Colorado State University
Fort Collins, CO
USA

barry.beaty@colostate.edu

bbeaty@colostate.edu

Bethany G. Bolling
Arbovirus-Entomology Laboratory
Texas Department of State Health Services
Austin, TX
USA

bethany.bolling@dshs.state.tx.us

Aaron C. Brault
Division of Vector-Borne Diseases
National Center for Emerging, Zoonotic Infectious Diseases
Centers for Disease Control and Prevention
Fort Collins, CO
USA

zlu5@cdc.gov

Charles H. Calisher
Arthropod-borne and Infectious Diseases Laboratory
Department of Microbiology, Immunology and Pathology
College of Veterinary Medicine and Biomedical Sciences
Colorado State University
Fort Collins, CO
USA

calisher@cybersafe.net

Alexander T. Ciota
Arbovirus Laboratories
Wadsworth Center
New York State Department of Health
Slingerlands, NY
USA

alexander.ciota@health.ny.gov

Lijia Cui
School of Medicine
Tsinghua University
Beijing
China

celiaclj@163.com

Gregor Devine
Mosquito Control Laboratory
QIMR-Berghofer Institute of Medical Research
Brisbane, QLD
Australia

greg.devine@qimrberghofer.edu.au

George Dimopoulos
W. Harry Feinstone Department of Molecular Microbiology and Immunology
Bloomberg School of Public Health
Johns Hopkins University
Baltimore, MD
USA

gdimopou@jhsph.edu

Gregory D. Ebel
Arthropod-borne and Infectious Diseases Laboratory
Department of Microbiology, Immunology and Pathology
Colorado State University
Fort Collins, CO
USA

gregory.ebel@colostate.edu

Adam Fitch
Division of Pulmonary, Allergy and Critical Care Medicine
University of Pittsburgh School of Medicine
Pittsburgh, PA
USA

fitchad@pitt.edu

Alexander W.E. Franz
Department of Veterinary Pathobiology
College of Veterinary Medicine
University of Missouri
Columbia, MO
USA

franza@missouri.edu

Brian Friedrich
Department of Microbiology and Immunology
University of Texas Medical Branch
Galveston, TX
USA

bmfriedr@utmb.edu

Elodie Ghedin
Center for Genomics and Systems Biology
Department of Biology
New York University
New York, NY
USA

elodie.ghedin@nyu.edu

eg121@nyu.edu

Duane J. Gubler
Program on Emerging Infectious Diseases
Duke-NUS Graduate Medical School
Singapore

duane.gubler@duke-nus.edu.sg

Roy A. Hall
Australian Infectious Diseases Research Centre
School of Chemistry and Molecular Biosciences
The University of Queensland
St Lucia, QLD
Australia

roy.hall@uq.edu.au

Sonja Hall-Mendelin
Public Health Virology
Queensland Health Forensic and Scientific Services
Brisbane, QLD
Australia

sonja.hall-mendelin@health.qld.gov.au

Scott B. Halstead
Dengue Vaccine Initiative
International Vaccine Institute
Seoul
Republic of Korea

halsteads@erols.com

Kathryn A. Hanley
Department of Biology
New Mexico State University
Las Cruces, NM
USA

khanley@nmsu.edu

Jody Hobson-Peters
Australian Infectious Diseases Research Centre
School of Chemistry and Molecular Biosciences
The University of Queensland
St Lucia, QLD
Australia

j.peters2@uq.edu.au

Natapong Jupatanakul
W. Harry Feinstone Department of Molecular Microbiology and Immunology
Bloomberg School of Public Health
Johns Hopkins University
Baltimore, MD
USA

njupata1@jhu.edu

Birte Kalveram
Department of Experimental Pathology
University of Texas Medical Branch
Galveston, TX
USA

bkkalver@utmb.edu

William B. Klimstra
Department of Microbiology and Molecular Genetics and Center for Vaccine Research
University of Pittsburgh
Pittsburgh, PA
USA

klimstra@cvr.pitt.edu

Richard J. Kuhn
Markey Center for Structural Biology
Department of Biological Sciences; and
Bindley Bioscience Center
Purdue University
W. Lafayette, IN
USA

kuhnr@purdue.edu

Goro Kuno
Fort Collins, CO
USA

gykuno@gmail.com

Ivan V. Kuzmin
University of Texas Medical Branch
Galveston, TX
USA

ivkuzmin@utmb.edu

Amy J. Lambert
Division of Vector-Borne Diseases
National Center for Emerging, Zoonotic Infectious Diseases
Centers for Disease Control and Prevention
Fort Collins, CO
USA

ahk7@cdc.gov

Robert S. Lanciotti
Division of Vector-Borne Diseases
National Center for Emerging, Zoonotic Infectious Diseases
Centers for Disease Control and Prevention
Fort Collins, CO
USA

rsl2@cdc.gov

John S. Mackenzie
Faculty of Health Sciences
Curtin University
Perth, WA
Australia

J.Mackenzie@curtin.edu.au

N. James MacLachlan
Department of Pathology, Microbiology and Immunology
University of California, Davis
Davis, CA
USA

njmaclachlan@ucdavis.edu

Gonzalo Moratorio
Viral Populations and Pathogenesis Unit
Institut Pasteur
Paris
France

gonzalo.moratorio@pasteur.fr

Ken E. Olson
Arthropod-borne and Infectious Diseases Laboratory
Deptartment of Microbiology, Immunology and Pathology
Colorado State University
Fort Collins, CO
USA

kenneth.olson@colostate.edu

Rushika Perera
Arthropod-borne and Infectious Diseases Laboratory
Department of Microbiology, Immunology and Pathology
Colorado State University
Fort Collins, CO
USA

rushika.perera@colostate.edu

Natalie A. Prow
Australian Infectious Diseases Research Centre
School of Chemistry and Molecular Biosciences
The University of Queensland
St Lucia, QLD; and
QIMR Berghofer Medical Research Institute
Brisbane, QLD
Australia

natalie.prow@qimrberghofer.edu.au

Scott Ritchie
College of Public Health, Medical and Veterinary Sciences
James Cook University
Cairns, QLD
Australia

scott.ritchie@jcu.edu.au

Matthew B. Rogers
Department of Surgery
University of Pittsburgh Children's Hospital
Pittsburgh, PA
USA

rogersm@pitt.edu

Shannan L. Rossi
Department of Experimental Pathology
University of Texas Medical Branch
Galveston, TX
USA

slrossi@utmb.edu

†Kate D. Ryman
Department of Microbiology and Molecular Genetics and Center for Vaccine Research
University of Pittsburgh
Pittsburgh, PA
USA

ryman@cvr.pitt.edu

Pei-Yong Shi
Novartis Institute for Tropical Diseases
Chromos
Singapore

pei_yong.shi@novartis.com

Steven P. Sinkins
Biomedical and Life Sciences
Lancaster University
Lancaster
UK

s.sinkins@lancaster.ac.uk

Kenneth A. Stapleford
Viral Populations and Pathogenesis Unit
Institut Pasteur
Paris
France

kenneth.stapleford@pasteur.fr

† Deceased

Robert B. Tesh
Department of Pathology
Center for Biodefense and Emerging Infectious Diseases
University of Texas Medical Branch
Galveston, TX
USA

rtesh@utmb.edu

Nikos Vasilakis
Department of Pathology
Center for Biodefense and Emerging Infectious Diseases
Center of Tropical Diseases and Institute of Human Infections and Immunity
University of Texas Medical Branch
Galveston, TX
USA

nivasila@utmb.edu

Marco Vignuzzi
Viral Populations and Pathogenesis Unit
Institut Pasteur
Paris
France

marco.vignuzzi@pasteur.fr

Peter J. Walker
CSIRO Animal, Food and Health Sciences
Australian Animal Health Laboratory
Geelong, VIC
Australia

peter.walker@csiro.au

Thomas Walker
Department of Disease Control
London School of Hygiene and Tropical Medicine
London
UK

thomas.walker@lshtm.ac.uk

Qing-Yin Wang
Novartis Institute for Tropical Diseases
Chromos
Singapore

qing_yin.wang@novartis.com

Scott C. Weaver
Institute for Human Infections and Immunity
Center for Tropical Diseases and Department of Pathology
University of Texas Medical Branch
Galveston, TX
USA

sweaver@utmb.edu

Foreword

More than 100 years since the demonstration of *Aedes (Stegomyia) aegypti* transmission of yellow fever virus by Walter Reed and colleagues, arboviruses continue to be the causes of major public health challenges, the subjects of many new viral discoveries, and the etiologic agents of new and repeated disease emergence. Following peaks of arbovirus discovery during the 1930s, due to technological advances in virus isolation and identification, and the during 1950–1960s largely thanks to the efforts of the Rockefeller Foundation and their international virus discovery programmes, arbovirology underwent a dramatic transformation beginning in the 1980s with the advent of modern molecular virology. Although field ecology and epidemiology studies have in many cases declined during this molecular biology era, advances in the study of arboviral genetics and the molecular basis of arbovirus–host and –vector interactions have led to exciting new opportunities for vaccine development and promising new targets for antiviral development, offering hope for the control in the near future of some of the most important arboviral diseases. Increasingly efficient amplification and sequencing of viral genomes and the ability to manipulate RNA viral genomes via cDNA rescue systems has revolutionized the study of arbovirus evolution and pathogenesis. Mechanisms of emergence into new transmission cycles that either directly impact spillover from enzootic cycles or mediate the development of urban, human–human transmission mediated by anthropophilic vectors have begun to be elucidated. Improved understanding of the effects of anthropogenic change on arbovirus transmission cycles and exposure of humans and domesticated animals has also revealed the increasing challenges that human activities such as international travel and commerce, deforestation, and climate change will place on our ability to control arboviral diseases.

In this comprehensive book on arboviruses, the first in several decades, all of the major arbovirus groups are reviewed with emphasis on taxonomy and discovery, including recently identified and characterized 'insect-specific' viruses that are revolutionizing our understanding of the evolution of their related arbovirus taxa, and advances in the understanding of virus–host interactions from the organismal to molecular levels. The advent of deep sequencing has revealed details of arbovirus population structure throughout the transmission cycle, which influences transmission and evolution, and the recent discovery of major effects of small RNAs on infection and replication in both arthropod vectors and vertebrate hosts is already having major impacts on the field. Molecular genetics is also beginning to impact directly our strategies for controlling arboviral diseases, with novel transgenic mosquito approaches and *Wolbachia* bacterial infections of mosquitoes already being tested in field settings for dengue control.

Despite the exciting advances in science and technology of the past three decades and their promise for applications aimed at controlling arboviruses, the challenges remain sobering: dengue viruses continue to expand and are now estimated to infect about 400 million persons annually, and chikungunya virus has recently spread to near pandemic proportions, affecting both tropical and temperate regions. Bluetongue virus is spreading northward into Europe as increasing temperatures permit northward expansion of the *Culicoides* spp. vectors, and West Nile virus has undergone a dramatic resurgence in the United States since 2012, without a clear explanation or any major improvement in our ability to predict outbreaks in time or space, let alone to mitigate them. Insecticide resistance continues to limit our ability to reduce exposure to arboviruses through vector control, and the anthropophilic behaviour of mosquito vectors like *A. aegypti* and *A. (Stegomyia) albopictus*, which continues its invasive spread on four continents, challenges control even when vectors remain susceptible. These and many other continuing challenges will require new, interdisciplinary approaches to better understand the determinants of enzootic and epidemic circulation, and both scientific and public policy advances to accelerate product development and clinical trials to bring antivirals and vaccines to the markets where they are desperately needed. This book, written by experts in their respective fields, provides a comprehensive treatment of the many topics that will need to be considered and included in these interdisciplinary efforts to improve in our ability to predict, prevent, and control arboviral diseases in the 21st century.

Scott C. Weaver

Current Books of Interest

Preface

In the waning years of the 20th century, arboviruses (viruses transmitted to humans and other animals by arthropod vectors) re-emerged as major global public health problems. Today, this diverse array of viruses is among the most important cause of emerging epidemic infectious diseases worldwide. It has been almost 30 years since the last comprehensive coverage of the taxonomy, epidemiology, ecology, evolution and biology of the arboviruses. Since then, there has been a tremendous amount of new knowledge published on these viruses. This book revisits all of the above aspects of arbovirus research plus new information on the molecular biology, risk factors underlying the re-emergence of arboviral disease epidemics in human and animal populations. We are fortunate to have recruited the expertise of a diverse set of contributors, whose effort undoubtedly required sacrifices of both time and energy in the face of many other commitments.

The recent adoption of new detection technologies [next-generation sequencing (NGS)] has resulted in a cataclysm of discovery of newly recognized viruses with restricted host range in arthropods whose detection was not possible with the classic serology and isolation methods of the past. As result the question of how arboviruses should be defined has been brought to the forefront. An important question is should the definition be broadened to include arthropod-associated viruses that are in the same families as known arboviruses such as Togaviridae, Flaviviridae, Bunyaviridae, Rhabdoviridae, Reoviridae, etc.?

Our intention with this book is to provide a forum in which members of the arbovirus community could highlight new approaches, concepts or concerns without any preconceptions, thus providing a stimulating baseline where new investigators from a wide variety of disciplines might consider joining in arbovirus research providing a set of 'fresh eyes' and perspectives. This book is divided into four sections that cover: (i) molecular biology, (ii) viral diversity and evolution, (iii) arbovirus diagnosis and control and (iv) future trends.

We owe a large amount of gratitude to the many people who made this book a reality. To the fellow contributors and authors of each of the chapters, whose dedication, patience and collaboration made this endeavour a delightful experience. We would like to acknowledge Shannan Rossi, Nicholas Bergren and Ashley Rhame for assistance with proofreading the chapters and preparing the index. We must also thank Annette Griffin, Acquisitions Editor at Horizon Press, and Melanie Woodward, Project Leader at Prepress Projects Ltd, for their invaluable help.

Finally, we dedicate this book to our spouses, whose patience, dedication, understanding and everlasting support for our scientific journeys made it all worth it.

It is with great sadness that the editors note the passing of one of our authors, Dr Kate Ryman, in November 2015 just before this book went to press. Dr Ryman was an outstanding virologist who made seminal contributions to our understanding of the pathogenesis of many arboviruses in the alphavirus and flavivirus genera, and she will be sorely missed by us and her many friends in the arbovirology community.

Nikos Vasilakis
Duane J. Gubler

The Arboviruses: Quo Vadis?

1

Duane J. Gubler and Nikos Vasilakis

Abstract

Arthropod-borne viruses (arboviruses) are the causative agents of significant morbidity and mortality among humans and domestic animals globally. They are maintained in complex biological life cycles, involving a primary vertebrate host and a primary arthropod vector. These cycles exist in natural sylvatic or urban foci of transmission. Arboviruses may emerge from their ecologically distinct nidus when humans or domestic animals unknowingly encroach on their environment, leading to local, regional or, in some instances, global epidemics. The principal drivers of epidemic arboviral disease emergence are anthropogenic, fuelled by uncontrolled human population growth, economic development and globalization, combined with societal and technological changes. Other factors contributing to arboviral disease emergence are environmental changes, including urbanization, changes in land and water use, agricultural and animal husbandry practices, new irrigation systems and deforestation. Unless these trends are controlled, and eventually reversed, the future will probably see more widespread and larger epidemics of arboviral disease.

Introduction

Arboviruses, long considered relatively unimportant as causes of human disease, have emerged in the past 40 years as important public health and biosecurity threats (Gubler, 2002). The term arbovirus is a contraction of arthropod-borne virus. It has no taxonomic significance, but is an ecological term used to define viruses that are transmitted among vertebrate hosts by blood sucking arthropods when they take a bloodmeal (WHO, 1985). All known arboviruses are natural parasites of animals other than humans (zoonotic pathogens). They include a diverse group of viruses belonging to nine viral families (Table 1.1) (Karabatsos, 1985, 2001 update). Mosquitoes, flies and ticks are the most important vectors while rodents and birds are the most important vertebrate reservoir hosts of those viruses affecting humans and their domestic animals.

Arboviruses have at least two natural hosts, a vertebrate and an arthropod, that are required for maintenance in nature. They are usually maintained in complex biological life cycles involving a primary vertebrate host and a primary arthropod vector (Fig. 1.1) (Gubler, 2001). These cycles exist in natural, usually focal foci that are unknown until humans or their domestic animals encroach on the natural nidus, causing an

Table 1.1 Taxonomic status of known, probable or possible arboviruses

Family	Number of genera	Number of viruses
Bunyaviridae	5*	248
Flaviviridae	1	53
Reoviridae	2	77
Rhabdoviridae	12#	68
Togaviridae	1	28
Orthomyxoviridae	1	3
Arenaviridae	1	1
Poxviridae	1	1
Unclassified	1	13

*The numbers of genera and viruses depicted are based on the ninth report of the International Committee on Taxonomy of viruses (ICTV) (Plyusnin *et al*., 2012). These numbers are currently revised by the ICTV.
#The number of genera has been increased to reflect the diversity of the family *Rhabdoviridae* (see Chapter 3), and ratification is currently under consideration by the ICTV (Walker *et al*., 2015).

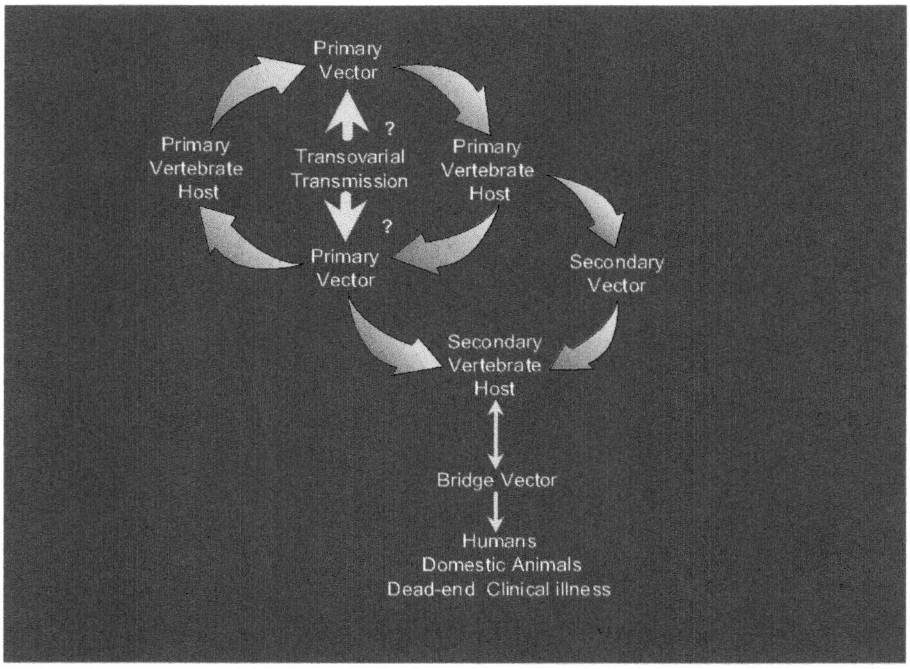

Figure 1.1 Hypothetical arbovirus life cycle.

epidemic. In some instances, feral animals or the arthropod vectors encroach on the human ecosystem causing epidemics. Once the virus has escaped the primary cycle, it may establish a secondary cycle in new foci with different vectors and vertebrate hosts. In most cases, the secondary vertebrate hosts such as humans are incidental and do not contribute to the transmission cycle. Exceptions to this include a number of viruses that have adapted to humans, including dengue (DENV), yellow fever (YFV), zika (ZIKV), Chikungunya (CHIKV) and Ross River (RRV). Moreover, with the recent advent of new detection technologies [next generation sequencing (NGS)], a number of viruses with restricted host range in arthropods have been discovered (Nasar *et al.*, 2012; Marklewitz *et al.*, 2013; Vasilakis *et al.*, 2013a,b, 2014; Zirkel *et al.*, 2013; Kallies *et al.*, 2014), begging the question of how arboviruses should be defined (Vasilakis *et al.*, 2013a, 2014). Should the definition be broadened to include arthropod-associated viruses that are in the same families as known arboviruses such as *Togaviridae*, *Flaviviridae*, *Bunyaviridae*, *Rhabdoviridae*, *Reoviridae*, etc.? Historically, the term 'arbovirus' has been based on biological and ecological characteristics. While modern virus taxonomy is based principally on morphological and molecular characteristics, more recently biological and ecological relationships have again become important, adding another layer of complexity in understanding the relationships of diverse virus families (Kuno, 2007). Arboviruses must be defined biologically as well as molecularly. This would allow the arthropod-specific viruses to be defined as arboviruses (Charles Calisher, 2014, personal communication).

There are approximately 135 arboviruses known to infect humans. They cause a broad spectrum of illness ranging from asymptomatic infection to severe and fatal disease. In general, arboviruses cause three basic clinical syndromes in humans: (i) systemic febrile illness; (ii) haemorrhagic fever; and (iii) invasive neurological disease (Gubler, 2001). Most arbovirus infections cause a non-specific febrile illness in the acute phase, followed by full recovery. In those patients who develop severe disease, the illness is often biphasic, with a non-specific acute febrile phase often going unrecognized, followed by arthritic, haemorrhagic or neurological disease. Arthritic sequelae are usually transient. Haemorrhagic manifestation can vary from mild skin and mucous membrane bleeding to severe viscerotropic disease with frank haemorrhage and death. Neurological disease usually presents as meningoencephalitis with permanent neurological sequelae and sometimes death. The same virus may cause all three syndromes in different individuals. As noted above, humans are generally secondary hosts for these viruses and are thus dead-end hosts, although there are exceptions, such as DENV, YFV, ZIKV, CHIKV, RRV and other viruses.

Current status of arboviral diseases

Diseases thought to have been caused by arboviruses have been known for centuries. However, the first arbovirus isolated was yellow fever in 1927. The isolation of other arboviruses causing human disease followed, with Japanese encephalitis in 1924 (retrospectively), St. Louis encephalitis in 1933, West Nile in 1937, Chikungunya in 1938 and dengue in 1943 (Karabatsos, 1985). The Rockefeller Foundation, founded in 1914, was instrumental in the discovery of most known arboviruses. With laboratories in Trinidad, Colombia and Brazil in South America, Egypt, Nigeria, South Africa and Uganda in Africa, and India in Asia, Rockefeller scientists isolated and characterized many viruses between 1935 and 1970. Because the Rockefeller laboratories were mostly located in South America and Africa, that is where most arboviruses were discovered. Africa accounts for 25% and North and South America for 42% of all known and potential arboviruses (Fig. 1.2). After 1970, support for this kind of discovery field research waned and only four new viruses were described during the last 30 years of the 20th century (Gubler, 2001).

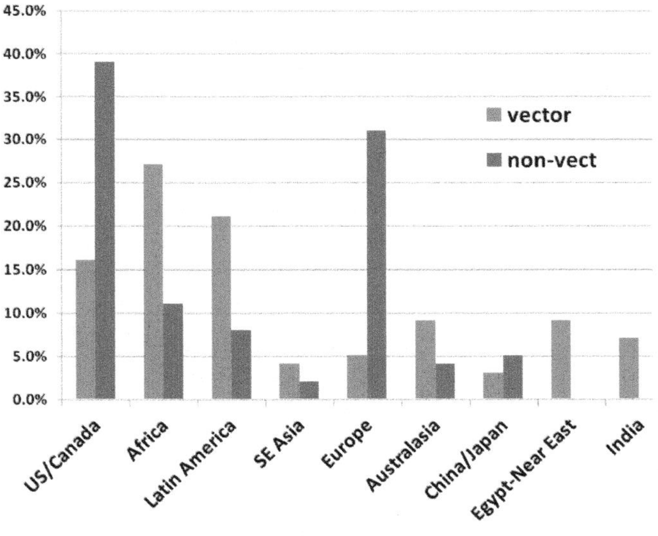

Figure 1.2 Human virus sites of discovery.

Table 1.2 Resurgent and/or emergent arboviruses causing major global epidemics the last 30 years

Family	Disease
Flaviviridae	Dengue Yellow fever Japanese encephalitis West Nile Zika Kyasanur Forest disease
Togaviridae	Chikungunya Venezuelan equine encephalitis Ross River Barmah Forest
Bunyaviridae	Rift Valley fever Oropouche California encephalitis Crimean–Congo haemorrhagic fever Severe febrile thrombocytopenia syndrome
Reoviridae	Bluetongue

In recent years, there has been a dramatic emergence of epidemic arboviral diseases, including several new viruses that have been identified (Fig. 1.3 and Table 1.2). The majority of those emergent viruses causing epidemics belonged to three families: *Flaviviridae* (six viruses), *Bunyaviridae* (five viruses) and *Togaviridae* (four viruses) (Table 1.2). Only one (bluetongue, family *Reoviridae*) was not infective for humans. Major geographic spread accompanied by increased frequency and magnitude of epidemics have occurred in the past 30 years with dengue, West Nile, Zika, Kyasanur Forest disease, Alkhurma, Japanese encephalitis, Chikungunya, Ross River, Barmah Forest, Venezuelan equine encephalitis, Rift Valley fever, Oropouche, and Crimean–Congo haemorrhagic fever being the most important (Table 1.2). Some are newly described viruses [severe febrile thrombocytopenic syndrome and Bourbon (Yu *et al.*, 2011; Kosoy, *et al.*, 2015)], but most are viruses that have been known for many years, only recently emerging to cause major epidemics. The factors responsible for the dramatic emergence of epidemic arboviral diseases in the past 40 years are complex and involve global trends (demographic, economic, environmental and societal) that have emerged since the Second World War (see below and Fig. 1.3).

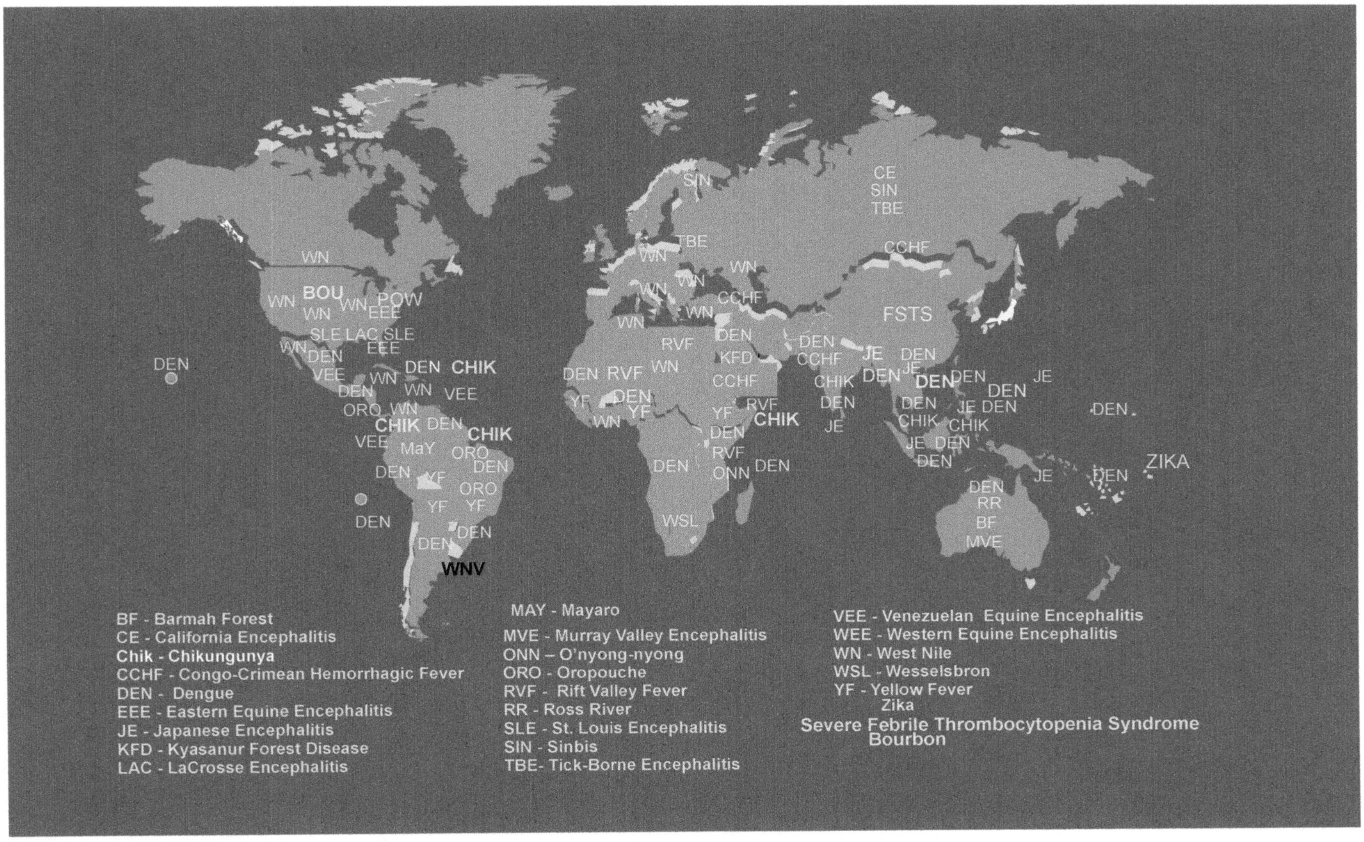

Figure 1.3 Global resurgence of epidemic arboviral disease.

Yellow fever virus is an enigma. It has been maintained for decades in sylvatic cycles in Africa and South America, with only periodic small epidemics in humans. In urban settings, YFV has the same epidemiology as DENV, CHIKV, and ZIKV, which have all spread dramatically in recent years. Of special interest is South America, where YFV has been contained in the rain forests of the Amazon Basin for over 70 years. During that time, the urban growth in the American region has been unprecedented and many of the urban areas located in the YFV enzootic area have been re-infested with the principal urban mosquito vector, *Aedes aegypti*, putting the risk of urban epidemics of YF at its highest level in history (Gubler, 2004). Despite the increased risk, major urban YF epidemics have not occurred in the region.

Arboviruses currently have a worldwide distribution (Fig. 1.3) (Gubler, 2001). As noted, however, each virus will generally have a focal distribution because of the specific vertebrate host and vector ecological requirements to maintain the primary cycle. Thus most epidemics of arboviral disease are short lived. With the increased ease with which the viruses, vectors and vertebrate hosts are transported to new geographic areas via globalization and modern air transportation, however, that may be changing (Gubler, 2002). A quick glance at Fig. 1.3 suggests that persons will be at risk of arboviral infection anywhere in the world with the exception of the Arctic and Antarctica.

Reasons for global emergence

The principal reasons for the dramatic emergence of epidemic arboviral diseases are summarized in Fig. 1.4. Human population growth and the resulting economic development, societal and technological changes have been the major drivers of environmental change, which have included unprecedented urbanization, changes in land and water use, changes in agricultural practices, new irrigation systems and deforestation (Gubler, 2002, 2011). This resulted in habitat alteration and encroachment from the natural ecosystem to the urban ecosystems and vice versa. A circular human rural to urban migration drove uncontrolled urbanization and brought exotic pathogens into more frequent contact with humans in an urban environment, where there was greater probability of secondary transmission and spread by aeroplane to new geographic areas. New animal husbandry methods were developed to increase food production to feed the increased human population. Finally, globalization provided the ideal mechanism for rapid spread of pathogens to new geographic locations where public health infrastructure for vector-borne infectious diseases had deteriorated in the wake of the successful control of malaria and YF in the 1950s and 1960s. Collectively, this complex web of interrelated factors resulted in the emergence of newly recognized arboviruses and the re-emergence of viruses that had been successfully controlled after the Second World War (Gubler, 2002, 2004, 2011).

It is intuitive that climate change would play a role in this emergence and spread because the arthropod vectors of arboviruses are cold-blooded animals whose biology is greatly influenced by temperature and other environmental factors. Climate change, however, has not been a major factor in determining emergence and spread of these or other infectious diseases, compared with the demographic and societal drivers noted above.

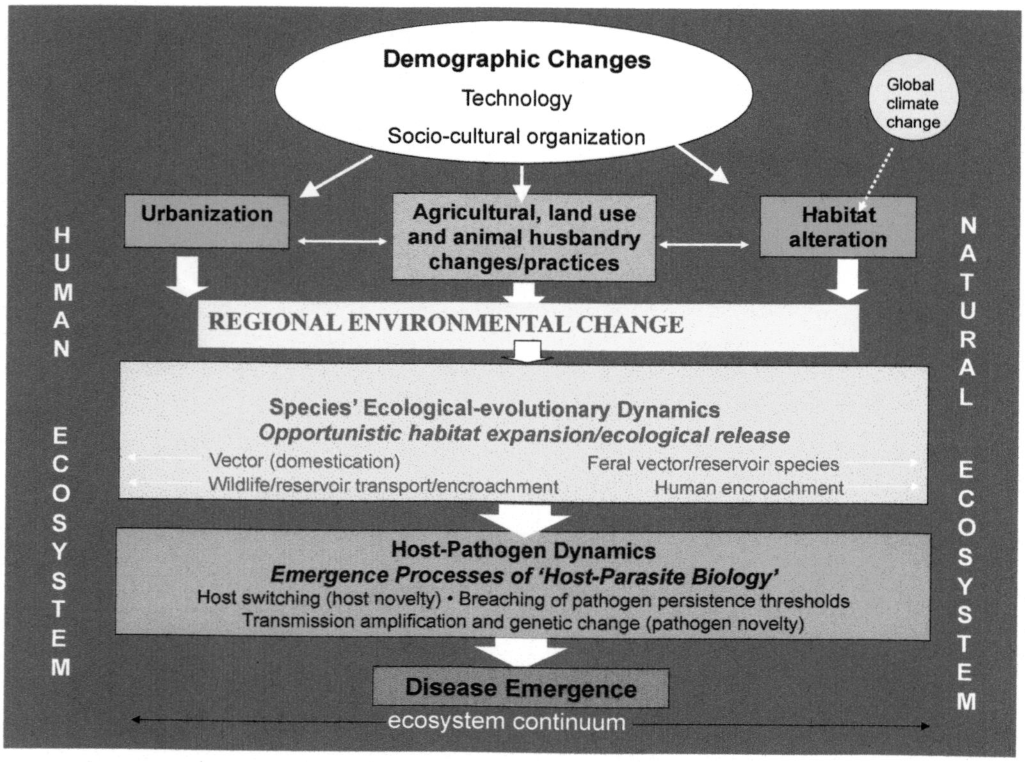

Figure 1.4 Factors responsible for the emergence of arboviral diseases.

Quo vadis

What does the future hold for arboviruses? It is difficult and perhaps foolish to try to predict the future. However, the global trends that have been primarily responsible for the increased transmission and epidemic activity of the recent past, are projected to continue. Thus the global human population is expected to reach 10 billion by 2050 (UN Department of Economic and Social Affairs, 2011). That trend alone will drive major environmental and societal changes, including urban growth, changes in land use, water use, agriculture, animal husbandry, deforestation and other environmental parameters. It will also drive human migration and movement, which in turn will increase human encroachment on natural habitats where many microbial populations exist in harmony with the natural ecosystems, unknown to humans until the systems are disturbed by the encroachment of man or domestic animals. As urban centres expand, there will likely also be increased encroachment of feral animals from the natural ecosystems into the urban ecosystems, putting humans at further risk (Fig. 1.4). Anthropogenic environmental change driven by human population growth and all of its indirect ramifications, will probably have a major impact on many arboviral diseases.

Add to that, globalization and deteriorating public health infrastructure, and the result is increased arboviral epidemics and disease. An estimated 3.3 billion people travelled by air in 2014 (IATA Annual Review, 2014), and the numbers and diversity of animals being transported by air is far in excess of that number. The Global Air Network (Fig. 1.5) consists of over 50,000 routes (IATA Annual Review, 2014), and provides the ideal mechanism to facilitate the rapid geographic spread of both the viruses and the arthropod vectors (Gubler, 2002, 2011). Unless the global trends of population growth, urbanization and globalization can be reversed, the future will probably see more widespread and larger epidemics of arboviral disease.

Fortunately, the future also holds major advancements in technology that can be harnessed to help contain, control and prevent epidemic arborviral diseases. It is anticipated that new technology and understanding of the molecular biology and genetics of arboviruses will allow more effective early warning surveillance systems to be put in place, and that new antiviral drugs, vaccines, therapeutic antibodies and vector control tools will be developed as the above scenario continues to unfold, thus allowing public health officials to prevent and control these diseases before they become established in new geographical regions (Gubler, 2015). The success of such prevention programmes, however, will require political will and sustained economic support for a broad research agenda to develop new and innovative tools to detect, contain and control these viral diseases.

Acknowledgements

This work was supported in part by the Duke-NUS Signature Research Program funded by the Ministry of Health, Singapore, and by NIH contract HHSN272201000040I/HHSN27200004/D04 (NV).

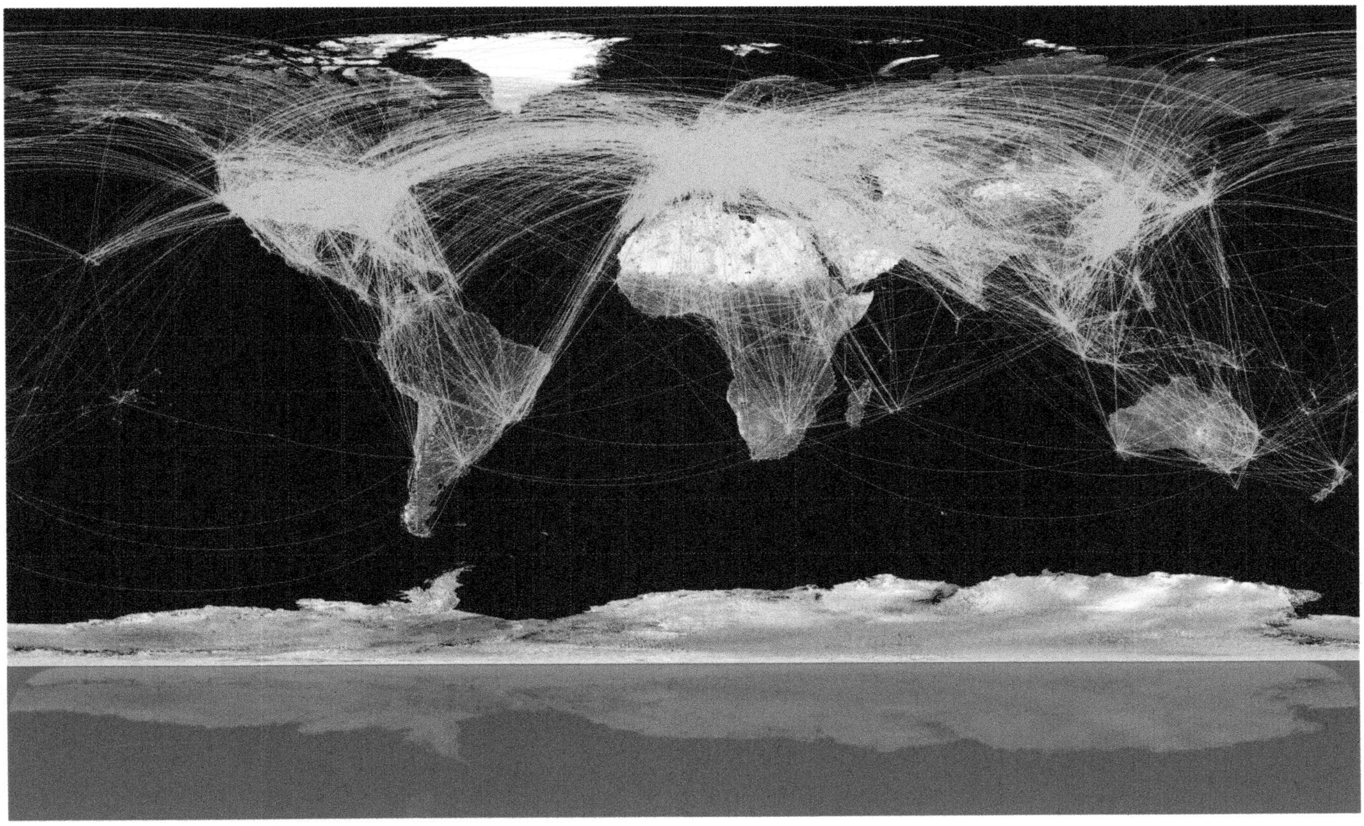

Figure 1.5 The Global Air Network.

References

Gubler, D.J. (2001). Human arbovirus infections worldwide. In West Nile Virus. Detection, Surveillance, and Control, vol. 951, D.J. White and D.L. Morse, eds. (Annals of the New York Academy of Sciences, New York), pp. 13–25.

Gubler, D.J. (2002). The global emergence/resurgence of arboviral diseases as public health problems. Arch. Med. Res. 33, 330–342.

Gubler, D.J. (2004). The changing epidemiology of yellow fever and dengue, 1900 to 2003: full circle? Comp. Immunol., Microbiol. Infect. Dis. 27, 319–330.

Gubler, D.J. (2011). Dengue, urbanization and globalization: the unholy trinity of the 21st century. Trop. Med. Health (Suppl.), *39*, 1–9.

Gubler, D.J. (2015). The partnership for dengue control – a new global alliance for the prevention and control of dengue. Vaccine 33, 1233.

IATA (2014). Annual review. Available online: http://www.iata.org. Accessed spring 2015.

Kallies, R., Kopp, A., Zirkel, F., Estrada, A., Gillespie, T.R., Drosten, C., and Junglen, S. (2014). Genetic characterization of goutanap virus, a novel virus related to negeviruses, cileviruses and higreviruses. Viruses *6*, 4346–4357.

Karabatsos, N. (1985). International Catalogue of Arboviruses, Including Certain Other Viruses of Vertebrates, 2001 Update (American Society of Tropical Medicine and Hygiene, San Antonio, TX).

Kosoy, O.I., Lambert, A.J., Hawkinson, D.J., Pastula, D.M., Goldsmith, C.S., and Hunt, D.C. (2015). Novel thogotovirus associated with febrile illness and death, United States, 2014. Emerg. Infect. Dis. *21*, 760–776.

Kuno, G. (2007). Host range specificity of flaviviruses: correlation with in vitro replication. J. Med. Entomol. *44*, 93–101.

Marklewitz, M., Zirkel, F., Rwego, I.B., Heidemann, H., Trippner, P., Kurth, A., Kallies, R., Briese, T., Lipkin, W.I., Drosten, C., Gillespie, T.R., and Junglen, S. (2013). Discovery of a unique novel clade of mosquito-associated bunyaviruses. J. Virol. *87*, 12850–12865.

Nasar, F., Palacios, G., Gorchakov, R.V., Guzman, H., Da Rosa, A.P., Savji, N., Popov, V.L., Sherman, M.B., Lipkin, W.I., Tesh, R.B., and Weaver, S.C. (2012). Eilat virus, a unique alphavirus with host range restricted to insects by RNA replication. Proc. Natl. Acad. Sci. U.S.A. *109*, 14622–14627.

Plyusnin, A., Beaty, B.J., Elliott, R.M., Goldbach, R., Kormelink, R., Lundkvist, A., Schmaljohn, C.S., and Tesh, R.B. (2012). Bunyaviridae. In Virus Taxonomy, Classification and Nomenclature of Viruses. Ninth Report of the International Committee on Taxonomy of Viruses, A.M.Q. King, M.J. Adams, E.B. Carstens, and E.J. Lefkowitz, eds. (Elsevier Academic Press, San Diego, CA, USA), pp. 725–741.

UN Department of Economic and Social Affairs, Population Division (2011). World Population Prospects: The 2010 Revision (UN Department of Economic and Social Affairs, New York, NY, USA).

Vasilakis, N., Widen, S., Mayer, S.V., Seymour, R., Wood, T.G., Popov, V., Guzman, H., Travassos da Rosa, A.P.A., Ghedin, E., Holmes, E.C., *et al.* (2013a). Niakha virus: a novel member of the *Rhabdoviridae* family isolated from phlebotomine sandflies in Senegal. Virology *444*, 80–89.

Vasilakis, N., Forrester, N.L., Palacios, G., Nasar, F., Rossi, S.L., Wood, T.G., Popov, P., Haddow, A.D., Gorchakov, R., Watts, D.M., *et al.* (2013b). Negev – a new insect-specific taxon with a dispersed geographic distribution. J. Virol. *87*, 2475–2488.

Vasilakis, N., Guzman, H., Forrester, N.L., Firth, C., Widen, S., Wood, T.G., Rossi, S.L., Ghedin, E., Popov, V., Blasdell, K., *et al.* (2014). Mesoniviruses are insect-specific viruses with extensive geographic distribution. Virol. J. *11*, 97.

Walker, P.J., Firth, C., Widen, S.G., Blasdell, K.R., Guzman, H., Wood, T.G., Paradkar, P.N., Holmes, E.C., Tesh, R.B., and Vasilakis, N. (2015). Evolution of genome size and complexity in the rhabdoviridae. PLoS Pathogens *11*, e1004664.

WHO. (1985). Arthropod-borne and rodent-borne viral diseases. World Health Organization Technical Report Series, No. 719 (World Health Organization, Geneva).

Yu, X.J., Liang, M.F., Zhang, S.Y., Liu, Y., Li, J.D., Sun, Y.L., Zhang, L., Zhang, Q.F., Popov, V.L., Li, C., *et al.* (2011). Fever with thrombocytopenia associated with a novel bunyavirus in China. N. Engl. J. Med. *364*, 1523–1532.

Zirkel, F., Roth, H., Kurth, A., Drosten, C., Ziebuhr, J., and Junglen, S. (2013). Identification and characterization of genetically divergent members of the newly established family Mesoniviridae. J. Virol. *87*, 6346–6358.

Part I

Molecular Biology

The Taxonomy of Arboviruses

Nicole C. Arrigo, Scott C. Weaver and Charles H. Calisher

2

Abstract

The purpose of viral taxonomy, as with all taxonomy, is to categorize viruses in a way that reflects their evolutionary relatedness, and the taxonomy of arboviruses is based on the same principles as that of all other viruses. Arboviruses are considered together only because of their unique biological characteristics and complex natural cycles, and are not collectively a rational taxonomic grouping based on ancestry. Historically, the definition of an arbovirus is a virus transmitted between vertebrate hosts by haematophagus (blood-feeding) arthropods, such as mosquitoes, ticks, sandflies, and culicoids. However, recent findings brought about by remarkable advances and applications of molecular techniques and viral genetics have revealed the presence of 'arthropod viruses' (sometimes called 'insect-specific viruses') within taxa traditionally including arboviruses, causing arbovirologists to reconsider the traditional definition of an arbovirus and to determine how best to organize them into a logical and useful taxonomic framework. Here we present a historical and contemporary perspective of arbovirus taxonomy and provide a current listing of conventionally defined arboviruses, as well as those 'arthropod viruses' currently challenging convention.

Introduction

The taxonomy of arthropod-borne viruses ('arboviruses') is based on the same principles as that of other viruses (King *et al.*, 2012). Arboviruses are considered together because of their unique biological characteristics and complex life cycles. Prior to formulation of an official, evolution-oriented taxonomic scheme, viruses were grouped together based on the diseases they cause or the physical and biological properties they share, e.g. encephalitis, hepatitis, or arthropod-borne viruses. As technology advanced, other defining features of viruses, such as antigenic relationships, were characterized and incorporated into a universal classification system, creating some conflicts with prior categorizations based on disease or transmission mode. Nonetheless, the custom of considering arboviruses as a cohesive grouping has persisted because of the history, collegial community, and scientific methodology of 'arbovirology.' More recently, findings brought about by remarkable advances and applications of molecular techniques and viral genetics have challenged traditional dogma, causing arbovirologists to reconsider the definition of an arbovirus and to reconsider how best to organize them into a logical and useful taxonomic framework.

Virus taxonomy

The overall purpose of taxonomy is to help the scientific community better understand the relationships among taxa, or classes, and evolution. Therefore, virus taxonomy refers to the systematic classification of viral taxa in ways that reflect their relatedness. Viral taxa are not actual replicating viruses; rather, they are categories or lists of names of viruses and groups of viruses and thus are abstract constructs (Van Regenmortel, 2007). Viruses are infectious agents subject to detection or manipulation by scientists; viral taxa are the emblematic representatives of viruses that are the subjects of virus taxonomy. Therefore, the systematics of viral taxa is a theoretical framework that provides an overview of the commonalities and possible evolutionary histories of their concrete virus counterparts. Although the scientific community often misunderstands this distinction, it is important to understand when considering some of the fundamental issues in virus taxonomy, including taxa delineation and virus nomenclature.

The International Committee for Taxonomy of Viruses (ICTV), a Division of the International Union of Microbiological Societies, is responsible for maintaining a dynamic and exponentially growing catalogue of all viral taxa. Within the ICTV, the Executive Committee, Subcommittees, and Study Groups comprising scientists actively involved in corresponding areas of virology officially assign recognized viruses to viral taxa and organize them into a descending classification scheme of class, order, family, genus, and species. The lowest level for virus classification by the ICTV is the species taxon and the ICTV takes no responsibility for classifying or assigning names to anything below this level, i.e. viruses, strains, isolates, subtypes, varieties, or genotypes. Although most virus species are classified in genera and genera into families, it is not obligatory that all levels of the taxonomic hierarchy be used, and those not assigned to higher orders are given the status 'unassigned'.

Because the primary mode of scientific communication is in writing, italicization is used to distinguish viral taxa from viruses that constitute the taxa. That is, names of orders, families, genera, and species are italicized, whereas the names of viruses are not italicized. For example, Kunjin virus belongs to the family *Flaviviridae*, genus *Flavivirus*, species *West Nile virus*. Note also that virus species are always capitalized, while viruses are not unless they begin with a formal name such as Kunjin. Confusing to some, a species taxon often shares the same name as its member virus, further emphasizing the need for italicization, e.g. the virus eastern equine encephalitis virus is a member of the family *Togaviridae*, genus *Alphavirus*, species *Eastern equine encephalitis virus*. A binomial system has been proposed in which the genus name (ending in *-virus*) is used as the second word of a virus species name (Van Regenmortel *et al.*, 2010). For example, the species name *West Nile virus* would become *West Nile flavivirus*. While this system has been discussed for many years and would clarify some misunderstandings of shared species and virus names, it has not been universally adopted by the virology community or the ICTV.

The species taxon is not only the most fundamental, commonly used, and typographically confused taxonomic category, as well as the subject of numerous contentious topics in virus taxonomy, but it is also the most influenced by technological advances in science. Official usage of the species taxon in virus classification was first accepted by the ICTV in 1991 with the definition: 'A virus species is a polythetic class of viruses that constitute a replicating lineage and occupy a particular ecological niche' (Van Regenmortel *et al.*, 1991; Van Regenmortel and Mahy, 2004).

In this definition, a polythetic class is one whose members have several properties in common, but which do not necessarily all share a single, common defining property. Also by this definition, members of a virus species are defined collectively by a consensus group of properties, including but not limited to genome sequence relatedness, natural host range, cell and tissue tropism, pathogenicity, mode of transmission, and antigenic properties (Van Regenmortel *et al.*, 1997). The species taxon thus differs from the higher viral taxa (order, family, and genus), which are 'universal' classes and as such are defined by a few properties that are both necessary and sufficient for membership, e.g. morphology, physiochemical properties, and genome organization.

First published in the 7th ICTV report (Van Regenmortel *et al.*, 2000) and remaining in the 8th ICTV Report (Fauquet, 2005), this concept of a polythetic class made placement of a virus into a particular species and the task of species delineation somewhat subjective and often challenging. However, this definition has also provided the necessary flexibility to accommodate the continually and rapidly evolving nature of viruses, their often complex modes of evolution, including nucleotide sequence divergence, reassortment of genetic segments, and recombination, and the technology used to identify and characterize them. Despite its effective use for over two decades, the ICTV definition of a virus species according to the 9th ICTV Report (King *et al.*, 2012) was changed to: 'A species is a monophyletic group of viruses whose properties can be distinguished from those of other species by multiple criteria.' In addition to this official change, the ICTV further commented that the criteria by which different species within a genus can be distinguished may include, but are not limited to, natural and experimental host range, cell and tissue tropism, pathogenicity, vector specificity, antigenicity, and the degree of relatedness of their genomes or genes. These distinctions fall under the purview of each Study Group, although the ICTV Executive Committee retains the right to change or overrule the expert Study Groups and has done so.

While the definition of a virus species has always generated controversy, this change in definition from a polythetic class to a monophyletic group incited a feverish debate amongst virologists (Van Regenmortel *et al.*, 2013). In a polythetic class, no defining property is necessarily present in all of its members, whereas a monothetic class is simply a universal class in which all members share one or more defining properties (synapomorphies). Furthermore, the latest definition recognizes a species as a monophyletic group of viruses, emphasizing a common and exclusive phylogenetic relatedness and the contemporary emphasis of genetic similarity as a sole species-defining property. The inclusion of 'monophyletic' in the new species definition therefore implies a greater emphasis on genetic relationships, which provide the most direct evidence of monophyly. It also precludes other types of groups, such as paraphyletic, in which selected (e.g. with certain phenotypes) but not all members of a clade descended from a common ancestor.

Without the need for any other property to be simultaneously different amongst all members of a species, the task of species demarcation risks becoming an arbitrary line of genetic similarity, sure to need continual modification as new genetic information is accumulated and more viruses are discovered. Furthermore, reassortment and recombination can confuse simple phylogenetic relationships based on genome sequences, confounding traditional phylogenetically based classifications. Appendix 2.1 in this chapter combines polythetic and monothetic philosophies and repairs obvious errors and omissions in the current ICTV taxonomic system, those of which are noted in explanatory footnotes. Note: Review of the literature ended 1 July 2014 upon submission of this chapter to the editors. Therefore, post hoc changes were not considered unless a significant and discernible virus was discovered or an ICTV report legitimized the taxonomy of the virus.

Taxonomy: from classical virology to molecular biology

Prior to acceptance of the modern classification scheme and the common use of the species taxon, classification of most viruses was based primarily on their morphology as demonstrated by electron microscopy and on their antigenic properties and serological cross-reactivities. Different viruses were identified by a reciprocal, fourfold or greater difference in antibody–antigen cross-reactivity, i.e. the heterologous versus homologous titres of antibodies in reactions against two viruses. A fourfold or greater difference in only one direction (non-reciprocal) signified an antigenic subtype, while antigenic varieties were distinguishable only with special serological tests, such as kinetic haemagglutination inhibition (Casals, 1964).

In the past few decades, advances in molecular biology have had a profound influence on virologists' approach to and perspectives on virus taxonomy. Technological innovations in genetics, genomics, and bioinformatics have considerably improved our ability to detect the presence of virus genomes in a wide array of biological samples, including insects, vertebrate host tissues, and vertebrate body fluids and excreta, and have enhanced our use of laboratory-based systems, such as tissue cultures and experimental models of natural transmission cycles. Furthermore, the ability to visualize, bank, compare, and analyse genetic information has revolutionized our contextual interpretation of viruses and their taxonomic management, particularly sequencing and phylogenetic inferences.

The discovery of reverse transcriptases (RTs) in 1970 (Temin and Mizutani, 1970) and the development of the polymerase chain reaction (PCR) by Kary Mullis in the early 1980s (Mullis *et al.*, 1986) transformed virology by providing the molecular tools to detect and genetically characterize viruses in an efficient manner. The genetic sequences of viral nucleic acid products, amplified through either RT-PCR for viral RNA or PCR for viral DNA, were recognized through various early sequencing technologies, including Maxam–Gilbert (chemical modification) (Maxam and Gilbert, 1977) and Sanger (chain-termination) (Sanger and Coulson, 1975) methods. While technical variations of early chain-termination sequencing methods are still widely used, 'next generation' or 'high-throughput' metagenomic sequencing advancements are currently providing innovative approaches for detecting the presence of known and hitherto unrecognized sequences in samples that contain little genetic material, material in which host nucleic acids far outnumber their viral counterparts, or samples that are presently unculturable, contain viral sequences insufficiently defined to be amplified by PCR, or contain multiple viral agents.

While limitations with these next generation sequencing technologies (namely cost, technical and informatic capabilities, and sample manipulation) presently restrict their routine universal use, more conventional and targeted sequencing of DNA products has become the approach of choice of many virologists, for both detection and characterization. Publicly available genetic databases continue to source the development of singleplex, multiplex, degenerate, consensus, and sequence-specific PCR assays to detect the presence of viral genomes and target characteristic genetic motifs at the species, genus, and family levels. Although some higher level taxa-consensus assays are more difficult to develop for diverse virus groups, such as the bunyaviruses, those that target well-characterized taxa and virus groups are capable of detecting known and previously unrecognized viruses with sensitivities comparable to or more so than virus isolation, with the benefit of efficiently providing additional genetic information for evolutionary analyses.

With the benefits of molecular biology so vast and widely accessible, many virologists now feel that genetic consensus sequences and the proteins they encode provide enough information to sufficiently classify viruses (Gibbs, 2013). However, the advantages of working with a viable and culturable agent cannot be overstated, not only because it allows a multifaceted virus population characterization that provides more useful biological and taxonomic information, but because it also retains the ability to derive additional genetic material for experimental manipulation, assay development, banking for future use, and validation of genetic studies, which remain cornerstones of the scientific process. In some cases, complete genomic sequences derived from a sample without virus isolation can be used to rescue virus containing the master sequence using synthetic or PCR-amplified cDNA clones (Nasar *et al.*, 2012), but the use of only partial genome sequences from single RT-PCR amplicons for virus detection and characterization, without virus isolation, precludes most further characterizations.

A taxonomic ideal

An ideal taxonomic scenario is one in which classical virology and biology are complemented by more modern molecular techniques to provide a multifaceted approach to virus classification. The recent characterization of a 'mosquito-only' alphavirus, Eilat virus (EILV), can be used as an example of a polythetically based classification that is far more useful for informing long-term public health and basic research needs than a simple genetic analysis for taxonomic purposes (Nasar *et al.*, 2012). Eilat virus was detected in a mosquito pool, but remained difficult to identify because it did not cause cytopathic effects (CPE) in vertebrate cells or disease in vertebrates, including infant mice. Only the presence of CPE noted in mosquito cell cultures inoculated with a triturated mosquito pool suggested the presence of a virus. Following the visualization of virus-like particles in these cells by electron microscopy (although most were a second, coinfecting virus; see below), next generation sequencing revealed the presence of RNA sequences homologous to those of alphaviruses. Thus, a combination of classical and modern techniques was critical for the identification of EILV.

The characterization of EILV was also challenging due to the presence of another RNA virus in the same mosquito pool, a negevirus (family unassigned) never before identified (Vasilakis *et al.*, 2013). Because this negevirus replicated more rapidly in mosquito cell cultures than did EILV, the latter could not be isolated using the traditional virological methods of plaque assay or limiting dilutions. Molecular methods were used to generate a full-length, infectious cDNA clone after determination of the EILV genomic sequence. This clone permitted rescue of EILV in mosquito cells uninfected by the negevirus, followed by detailed antigenic characterization and high-resolution imaging using cryoelectron microscopy. Combined with host range studies performed using cell cultures through electroporations of transcribed RNA, EILV was determined to have a fundamental restriction of RNA replication in vertebrate cells, to be most closely related to the western equine encephalitis complex of alphaviruses, and to be a distinct species based not only on its sequence divergence but also on multiple phenotypic characteristics that indicated its uniqueness in the genus *Alphavirus* (family *Togaviridae*). Sequence comparisons alone would not have provided a clear rationale for its species distinction, emphasizing the benefit of a polythetic and multifaceted approach to virus taxonomy.

Arbovirus taxonomy

The taxonomic approach to arboviruses mirrors that of all other viruses; however, a contemporary perspective of arbovirus taxonomy as a unified group requires a re-evaluation of what has traditionally been considered an arbovirus. The original definition of an arbovirus is restricted to a virus transmitted between vertebrate hosts by haematophagus (blood-feeding) arthropods, such as mosquitoes, ticks, sandflies, and culicoids. With molecular advances and more recent discoveries, the traditional definition of an arbovirus is no longer sufficiently inclusive or functional, and has not been so for many years. A number of examples demonstrate the limitations of the classical definition and the need for a broader perspective of these viruses, particularly when considering their taxonomic classifications.

Historically, arboviruses have been considered collectively, based on their unique biological characteristics and complex natural cycles; however, these commonalities are not taxonomically defining, such that arboviruses are a grouping of viruses, not a rational taxon or a series of taxa. Even prior to the widespread use of molecular tools and the emphasis on genomic characterization, arboviruses have principally belonged to a limited number of virus families: *Togaviridae* (genus *Alphavirus*), *Flaviviridae* (genus *Flavivirus*), *Bunyaviridae* (genera *Orthobunyavirus*, *Phlebovirus*, and *Nairovirus*, and viruses not assigned to a genus), *Reoviridae* (genera *Coltivirus*, *Orbivirus*, and *Seadornavirus*), *Rhabdoviridae* (genera *Vesiculovirus*, *Ephemerovirus*, and *Tibrovirus*, and viruses not assigned to a genus), *Orthomyxoviridae* (genera *Thogotovirus* and *Quaranjavirus*) and *Asfarviridae*.

Despite this pattern, taxonomic inclusion in these families has not been and is not currently sufficient to define an arbovirus. In fact, recognition of the membership of arboviruses in multiple families substantiates their wide variety biologically, geographically, and evolutionarily. A number of well-defined arbovirus species are members of the families *Flaviviridae* (e.g. *Yellow fever virus*, *Dengue virus*, and *West Nile virus*) and *Togaviridae* (e.g. *Sindbis virus* and *Venezuelan equine encephalitis virus*), yet many non-arboviral flaviviruses have been associated only with small mammals, such as rodents and bats, and some alphaviruses only with fish, and not with arthropod-borne transmission. At the same time, hantaviruses (family *Bunyaviridae*, genus *Hantavirus*), which are thought to be restricted to direct mammal-to-mammal transmission, have been placed in the family *Bunyaviridae* because of their molecular similarities to arthropod-associated members of the family [the orthobunyaviruses (genus *Orthobunyavirus*; primarily mosquito-transmitted), nairoviruses (genus *Nairovirus*; primarily tick-transmitted), phleboviruses (genus *Phlebovirus*; primarily biting fly transmitted), and the plant-infecting tospoviruses (genus *Tospovirus*; primarily thrip-transmitted)], even though hantaviruses are not known to be transmitted by arthropods. These examples demonstrate that arbovirologists have historically chosen to ignore these dichotomies, principally to avoid splitting arboviruses from taxonomic groups, rather than to expand or reconsider the dogmatic view of an arbovirus.

Further emphasizing the limitations of the traditional view of arboviruses are those viruses, e.g. Gamboa virus and vesicular stomatitis Indiana virus, that replicate in and are transmitted by arthropods, and which replicate in vertebrates, but not as an obligatory part of their natural cycles. These incidental infections produce a transient low-level or undetectable viraemia in vertebrate hosts, which results in an evolutionarily dead-end infection that does not influence persistence in nature. Arboviruses are also generally considered to be cytolytic for their vertebrate hosts, but cause little or no cytopathological changes in their arthropod hosts. However, eastern and western equine encephalitis viruses and West Nile virus have long been considered arboviruses despite their apparent pathogenicity for their enzootic vectors (Girard *et al.*, 2005; Weaver *et al.*, 1988, 1992). In addition, some arboviruses have single or multiple genome segments, providing the opportunity for reassortment to rapidly generate new variants that may have altered host ranges or virulence for either the vertebrate host or the arthropod vector (Briese *et al.*, 2013).

The traditional view of arboviruses has met its most challenging issue with the recent discovery of viruses that are genetically closely related to traditional arboviruses, yet appear restricted to infection only of arthropods and altogether incapable of infecting vertebrates. Through the application of PCR and new high-throughput sequencing technologies, discovery efforts have revealed the existence of nucleotide sequences of previously unrecognized flavivirus, alphavirus, and other virus sequences. As described earlier with the vignette of the 'mosquito-only' alphavirus, Eilat virus, these newly recognized viruses have been shown to be incapable of replication in vertebrate cell cultures or in laboratory rodents, but do replicate in cell cultures of arthropod origin, and can be detected in the saliva of arthropods. Furthermore, these viruses may influence vector susceptibility to oral infection by traditional arboviruses (Bolling *et al.*, 2012). Therefore, despite their genetic and biological relatedness to arboviruses, should these apparently arthropod-specific viruses be omitted from lists of arboviruses because, to our current knowledge, they do not infect vertebrates and are not associated with haematophagus transmission? This is a difficult question to answer both biologically and philosophically, particularly in light of the numerous aforementioned inconsistencies with traditional arbovirus dogma.

With the use of molecular tools, the discovery of arthropod viruses within traditional arbovirus taxa, such as the flaviviruses and alphaviruses, appears to be accelerating. This suggests that these viruses will eventually dominate several if not all of the taxonomic families that include traditional arboviruses, as well as virus taxa from other arthropods in the order Diptera, or even other members of the phylum *Arthropoda*. It is also possible that the discovery of additional arthropod viruses, within taxa of traditional arboviruses, will eventually lead to the inference that the ancestors of many traditional arboviruses gained the ability to infect vertebrates during the course of evolution. Assuming that most arthropod viruses are maintained by vertical transmission (Bolling *et al.*, 2012), this also implies that many traditional arboviruses have lost the ability to be efficiently transmitted vertically, leaving them dependent to varying degrees on vertebrates for horizontal transmission. However, the historically longer and more extensive search for viruses of vertebrates but not of arthropods suggests that there are probably many arthropod virus taxa that remain undiscovered and have not yet made this adaptation to infect vertebrates. It is our view that arthropod-borne, arthropod-specific, and arthropod-associated viruses continue to warrant collective consideration and study based on their complex biological involvement with arthropods. While current researchers have focused on 'insect-specific viruses' and have therefore adopted this terminology, it is not yet proven that these viruses are strictly insect-specific. Therefore, we feel a more comprehensive and inclusive term of 'arthropod virus' better describes this rapidly evolving and expanding group of viruses.

Appendix 2.1 lists all traditionally considered arboviruses according to listings maintained by the ICTV, as well as other arboviruses that we felt were important to recognize because of their epidemiological and historic significance. Appendix 2.2 lists those arthropod viruses that have been detected only in arthropods, and which, thus far, appear to be arthropod-specific. It is clear that the surface has only been scratched insofar as detecting such viruses. Therefore, Appendix 2.2 is limited to those arthropod viruses that are members of families comprising traditional arboviruses (*Flaviviridae, Togaviridae, Bunyaviridae, Reoviridae, Rhabdoviridae,* and unassigned viruses of the genus *Negivirus* and of the family *Mesoniviridae*). Both tables also provide the taxa to which these viruses belong [order (when so placed), family, genus, and species, all written in italics], and the ICTV traditional or informal abbreviations for virus names.

Conclusions

Prior to the widespread use of molecular technologies, the descriptions of the biological characteristics and natural cycle of a virus were typically more comprehensive and its acceptance as an 'arbovirus' was clearer. Molecular tools and the shift towards genomic characterization have provided the world of virology with immense technical advantages, despite the taxonomic challenges they pose to traditional dogmas. The taxonomy of arthropod viruses will continue to evolve, not only to provide accurate and meaningful indications of the evolutionary relationships that inform the most natural classifications, but also to incorporate the phenotypic diversity within these fascinating and important groups of viruses.

Acknowledgements

We would like ot thank the following individuals for their contributions to this chapter. Ernest Gould, Emergence des Pathologies Virales, Aix-Marseille Univ – Institut de Recherche pour le Développement – Ecole des Hautes Etudes en Santé Publique, Unité des Virus Emergents, Faculté de Médecine de Marseille, Marseille, France; Jens Kuhn, Tunnell Government Services Contractor, Lead Virologist, Integrated Research Facility at Fort Detrick, National Institute of Allergy and Infectious Diseases, National Institutes of Health, Fort Detrick, Frederick, Maryland, USA; Dimitry K. Lvov, D.I. Ivanovsky Institute of Virology RAMS, Gamaleya 16, Moscow, Russia; Eric C. Mossel, US Centers for Disease Control and Prevention, Division of Vector-Borne Infectious Diseases, Fort Collins, Colorado, USA; Hideki Ebihara, Laboratory of Virology, National Institute of Allergy and Infectious Diseases, National Institutes of Health, Missoula, Montana, USA; Peter Simmonds, Infection and Immunity Division, Roslin Institute, University of Edinburgh, Easter Bush, Edinburgh, Scotland; Robert B. Tesh, Department of Pathology, University of Texas Medical Branch, Galveston, Texas, USA; Amelia P.A. Travassos da Rosa, Department of Pathology, University of Texas Medical Branch, Galveston, Texas, USA.

References

Bolling, B.G., Olea-Popelka, F.J., Eisen, L., Moore, C.G., and Blair, C.D. (2012). Transmission dynamics of an insect-specific flavivirus in a naturally infected *Culex pipiens* laboratory colony and effects of co-infection on vector competence for West Nile virus. Virology *427*, 90–97.

Briese, T., Calisher, C.H., and Higgs, S. (2013). Viruses of the family *Bunyaviridae*: are all available isolates reassortants? Virology *446*, 207–216.

Casals, J. (1964). Antigenic variants of eastern equine encephalitis virus. J. Exp. Med. *119*, 547–565.

Fauquet, C.M. (2005). Virus Taxonomy: VIIIth Report of the International Committee on Taxonomy of Viruses (Burlington, MA, Elsevier Academic Press).

Gibbs, A.J. (2013). Viral taxonomy needs a spring clean; its exploration era is over. Virol. J. *10*, 254.

Girard, Y.A., Popov, V., Wen, J., Han, V., and Higgs, S. (2005). Ultrastructural study of West Nile virus pathogenesis in *Culex pipiens quinquefasciatus* (Diptera: Culicidae). J. Med. Entomol. *42*, 429–444.

King, A.M., Adams, M.J., Carstens, E.B., and Lefkowitz, E.J., eds. (2012). Virus taxonomy: classification and nomenclature of viruses: Ninth Report of the International Committee on Taxonomy of Viruses (San Diego, CA: Elsevier Academic Press).

Maxam, A.M., and Gilbert, W. (1977). A new method for sequencing DNA. Proc. Natl. Acad. Sci. U.S.A. *74*, 560–564.

Mullis, K., Faloona, F., Scharf, S., Saiki, R., Horn, G., and Erlich, H. (1986). Specific enzymatic amplification of DNA in vitro: the polymerase chain reaction. Cold Spring Harbor symposia on quantitative biology *51*, 263–273.

Nasar, F., Palacios, G., Gorchakov, R.V., Guzman, H., Da Rosa, A.P., Savji, N., Popov, V.L., Sherman, M.B., Lipkin, W.I., Tesh, R.B., *et al.* (2012). Eilat virus, a unique alphavirus with host range restricted to insects by RNA replication. Proc. Natl. Acad. Sci. U.S.A. *109*, 14622–14627.

Sanger, F., and Coulson, A.R. (1975). A rapid method for determining sequences in DNA by primed synthesis with DNA polymerase. J. Mol. Biol. *94*, 441–448.

Temin, H.M., and Mizutani, S. (1970). RNA-dependent DNA polymerase in virions of Rous sarcoma virus. Nature *226*, 1211–1213.

Van Regenmortel, M.H. (2007). Virus species and virus identification: past and current controversies. Infect. Genet. Evol. *7*, 133–144.

Van Regenmortel, M.H., and Mahy, B.W. (2004). Emerging issues in virus taxonomy. Emerg. Infect. Dis. *10*, 8–13.

Van Regenmortel, M.H., Maniloff, J., and Calisher, C. (1991). The concept of virus species. Arch. Virol. *120*, 313–314.

Van Regenmortel, M.H., Bishop, D.H., Fauquet, C.M., Mayo, M.A., Maniloff, J., and Calisher, C.H. (1997). Guidelines to the demarcation of virus species. Arch. Virol. *142*, 1505–1518.

Van Regenmortel, M.H., Fauquet, C.M., Bishop, D.H.L., Carstens, E.B., Estes, M.K., Lemon, S.M., Maniloff, J., Mayo, M.A., McGeogh, D.J., Pringle, C.R., *et al.*, eds. (2000). Virus Taxonomy. Classification and Nomenclature of Viruses. Seventh Report of the International Committee on Taxonomy of Viruses (San Diego, CA: Academic Press).

Van Regenmortel, M.H., Burke, D.S., Calisher, C.H., Dietzgen, R.G., Fauquet, C.M., Ghabrial, S.A., Jahrling, P.B., Johnson, K.M., Holbrook, M.R., Horzinek, M.C., *et al.* (2010). A proposal to change existing virus species names to non-Latinized binomials. Arch. Virol. *155*, 1909–1919.

Van Regenmortel, M.H., Ackermann, H.W., Calisher, C.H., Dietzgen, R.G., Horzinek, M.C., Keil, G.M., Mahy, B.W., Martelli, G.P., Murphy, F.A., Pringle, C., *et al.* (2013). Virus species polemics: 14 senior virologists oppose a proposed change to the ICTV definition of virus species. Arch. Virol. *158*, 1115–1119.

Vasilakis, N., Forrester, N.L., Palacios, G., Nasar, F., Savji, N., Rossi, S.L., Guzman, H., Wood, T.G., Popov, V., Gorchakov, R., *et al.* (2013). *Negevirus*: a proposed new taxon of insect-specific viruses with wide geographic distribution. J. Virol. *87*, 2475–2488.

Weaver, S.C., Scott, T.W., Lorenz, L.H., Lerdthusnee, K., and Romoser, W.S. (1988). Togavirus-associated pathologic changes in the midgut of a natural mosquito vector. J. Virol. *62*, 2083–2090.

Weaver, S.C., Lorenz, L.H., and Scott, T.W. (1992). Pathologic changes in the midgut of *Culex tarsalis* following infection with western equine encephalomyelitis virus. Am. J. Trop. Med. Hyg. *47*, 691–701.

Appendix 2.1 Taxonomy of arboviruses[a,b]

Viruses of the family *Togaviridae*, genus *Alphavirus*

Species	Virus	Abbreviation
Aura virus	Aura virus	AURAV
Barmah Forest virus	Barmah Forest virus	BFV
Bebaru virus	Bebaru virus	BEBV
Cabassou virus	Cabassou virus	CABV
Chikungunya virus	chikungunya virus	CHIKV
Eastern equine encephalitis virus	eastern equine encephalitis virus	EEEV
Everglades virus	Everglades virus	EVEV
Fort Morgan virus	Fort Morgan virus	FMV
	Buggy Creek virus	BCV
Getah virus	Getah virus	GETV
Highlands J virus	Highlands J virus	HJV
Mayaro virus	Mayaro virus	MAYV
Middelburg virus	Middelburg virus	MIDV
Mosso das Pedras virus	Mosso das Pedras virus	MDPV
Mucambo virus	Mucambo virus	MUCV
Ndumu virus	Ndumu virus	NDUV
O'nyong-nyong virus	o'nyong-nyong virus	ONNV
Pixuna virus	Pixuna virus	PIXV
Rio Negro virus	Rio Negro virus	RNV
Ross River virus	Ross River virus	RRV
	Sagiyama virus	SAGV
Salmon pancreas disease virus	salmon pancreas disease virus	SPDV
	sleeping disease virus	SDV
Semliki Forest virus	Semliki Forest virus	SFV
Sindbis virus	Sindbis virus	SINV
	Babanki virus	BBKV
	Kyzylagach virus	KYZV
	Ockelbo virus	OCKV
Southern elephant seal virus	southern elephant seal virus	SESV
Tonate virus	Tonate virus	TONV
Trocara virus	Trocara virus	TROV
Una virus	Una virus	UNAV
Venezuelan equine encephalitis virus	Venezuelan equine encephalitis virus	VEEV
Western equine encephalitis virus	western equine encephalitis virus	WEEV
Whataroa virus	Whataroa virus	WHAV

Viruses of the family *Flaviviridae*, genus *Flavivirus*

Tick-borne flaviviruses

Mammalian tick-borne virus group

Species	Virus	Abbreviation
Kyasanur Forest disease virus	Kyasanur Forest disease virus	KFDV
	Alkhumra haemorrhagic fever virus	AHFV
Langat virus	Langat virus	LGTV
Omsk haemorrhagic fever virus	Omsk haemorrhagic fever virus	OHFV
Powassan virus	Powassan virus	POWV
Royal Farm virus	Royal Farm virus	RFV
Tick-borne encephalitis virus[c]	Central European encephalitis virus	CEEV

Species	Virus	Abbreviation
	Russian Spring–Summer encephalitis virus	RSSEV
	louping ill virus[d]	LIV

Seabird tick-borne virus group, i.e. Tyuleniy virus group

Species	Virus	Abbreviation
Gadgets Gully virus	Gadgets Gully virus	GGYV
Kama virus	Kama virus	KAMAV
Meaban virus	Meaban virus	MEAV
Saumarez Reef virus	Saumarez Reef virus	SREV
Tyuleniy virus	Tyuleniy virus	TYUV

Kadam virus group (probably tick-borne)

Species	Virus	Abbreviation
Kadam virus	Kadam virus	KADV

Mosquito-borne flaviviruses

Species	Virus	Abbreviation
Aroa virus group		
Aroa virus	Aroa virus	AROAV
	Bussuquara virus	BSQV
	Iguape virus	IGUV
	Naranjal virus	NJLV
Dengue virus group		
Dengue virus	dengue virus 1	DENV-1
	dengue virus 2	DENV-2
	dengue virus 3	DENV-3
	dengue virus 4	DENV-4
Japanese encephalitis virus group		
Cacipacore virus	Cacipacore virus	CPCV
Japanese encephalitis virus	Japanese encephalitis virus	JEV
Koutango virus	Koutango virus	KOUV
Murray Valley encephalitis virus	Alfuy virus	ALFV
	Murray Valley encephalitis virus	MVEV
St. Louis encephalitis virus	St. Louis encephalitis virus	SLEV
Usutu virus	Usutu virus	USUV
West Nile virus	Kunjin virus	KUNV
	West Nile virus	WNV
Yaounde virus	Yaounde virus	YAOV
Kokobera virus group		
Kokobera virus	Kokobera virus	KOKV
	Stratford virus	STRV
Ntaya virus group		
Bagaza virus	Bagaza virus	BAGV
Ilheus virus	Ilheus virus	ILHV
	Rocio virus	ROCV
Israel turkey meningoencephalitis virus	Israel turkey meningoencephalitis virus	ITV
Ntaya virus	Ntaya virus	NTAV
Tembusu virus	Tembusu virus	TMUV

Species	Virus	Abbreviation
Zika virus	Zika virus	ZIKAV
Yellow fever virus group		
Sepik virus	Sepik virus	SEPV
Wesselsbron virus	Wesselsbron virus	WSLV
Yellow fever virus	yellow fever virus	YFV

Probable mosquito-borne flaviviruses

Species	Virus	Abbreviation
Kedougou virus group		
Kedougou virus	Kedougou virus	KEDV
Edge Hill virus group		
Banzi virus	Banzi virus	BANV
Bouboui virus	Bouboui virus	BOUV
Edge Hill virus	Edge Hill virus	EHV
Jugra virus	Jugra virus	JUGV
Saboya virus	Saboya virus	SABV
	Potiskum virus	POTV
Uganda S virus	Uganda S virus	UGSV

Flaviviruses with no known arthropod vector

Species	Virus	Abbreviation
Entebbe bat virus group		
Entebbe bat virus	Entebbe bat virus	ENTV
	Sokuluk virus	SOKV
Yokose virus	Yokose virus	YOKV
Modoc virus group		
Apoi virus	Apoi virus	APOIV
Cowbone Ridge virus	Cowbone Ridge virus	CRV
Jutiapa virus	Jutiapa virus	JUTV
Modoc virus	Modoc virus	MODV
Sal Vieja virus	Sal Vieja virus	SVV
San Perlita virus	San Perlita virus	SPV
Rio Bravo virus group		
Bukalasa bat virus	Bukalasa bat virus	BBV
Carey Island virus	Carey Island virus	CIV
Dakar bat virus	Dakar bat virus	DBV
Montana myotis leukoencephalitis virus	Montana myotis leukoencephalitis virus	MMLV
Phnom Penh bat virus	Phnom Penh bat virus	PPBV
	Batu Cave virus	BCV
Rio Bravo virus	Rio Bravo virus	RBV

Other related viruses that may be members of the genus *Flavivirus* but have not been approved as species

Vector	Virus	Abbreviation
Mammalian tick	Karshi virus	KSIV
Mosquito	Quang Binh virus	QBV
	Spondweni virus	SPOV
No known arthropod vector	Tamana bat virus	TABV

Viruses of the family *Bunyaviridae*, genus *Orthobunyavirus*

Species	Virus	Abbreviation
Acara virus	Acara virus	ACAV
	Moriche virus	MORV
Akabane virus	Akabane virus	AKAV
	Sabo virus	SABOV
	Tinaroo virus	TINV
	Yaba-7 virus	Y7V
Alajuela virus	Alajuela virus	ALJV
	San Juan virus	SJV
Anopheles A virus	Anopheles A virus	ANAV
	Arumateua virus	ARTV
	Caraipé virus	CPEV
	Las Maloyas virus	LMV
	Lukuni virus	LUKV
	Trombetas virus	TRMV
	Tucuruí virus	TUCRV
Anopheles B virus	Anopheles B virus	ANBV
	Boraceia virus	BORV
Bakau virus	Bakau virus	BAKV
	Ketapang virus	KETV
	Nola virus	NOLAV
	Tanjong Rabok virus	TRV
	Telok Forest virus	TFV
Batama virus	Batama virus	BMAV
Benevides virus	Benevides virus	BVSV
Bertioga virus	Bertioga virus	BERV
	Cananeia virus	CNAV
	Guaratuba virus	GTBV
	Itimirim virus	ITIV
	Mirim virus	MIRV
Bimiti virus	Bimiti virus	BIMV
Botambi virus	Botambi virus	BOTV
Bunyamwera virus	Bunyamwera virus	BUNV
	Batai (Chittoor, Čalovo) virus	BATV
	Birao virus	BIRV
	Bozo virus	BOZOV
	Cache Valley virus	CVV
	Fort Sherman virus	FSV
	Germiston virus	GERV
	Iaco virus	IACOV
	Ilesha virus	ILEV
	Lokern virus	LOKV
	Maguari virus	MAGV
	Mboke virus	MBOV
	Ngari virus[a]	NRIV
	Northway virus	NORV
	Playas virus	PLAV
	Potosi virus	POTV
	Santa Rosa virus	SARV
	Shokwe virus	SHOV
	Tensaw virus	TENV
	Tlacotalpan virus	TLAV
	Xingu virus	XINV

Species	Virus	Abbreviation
Bushbush virus	Bushbush virus	BSBV
	Juan Diaz virus	JDV
Bwamba virus	Bwamba virus	BWAV
	Pongola virus	PGAV
California encephalitis virus	California encephalitis virus	CEV
	Inkoo virus	INKV
	Morro Bay virus	MBV
	Keystone virus	KEYV
	Khatanga virus	CHATV
	Melao virus	MELV
	Jamestown Canyon virus	JCV
	South River virus	SoRV
	La Crosse virus	LACV
	San Angelo virus	SAV
	Serra do Navio virus	SDNV
	snowshoe hare virus	SSHV
Capim virus	Capim virus	CAPV
Caraparu virus	Caraparú virus	CARV
	Apeú virus	APEUV
	Bruconha virus	BRUV
	Ossa virus	OSSAV
	Vinces virus	VINV
	Zungarococha virus	ZUNV
Catu virus	Catu virus	CATUV
Gamboa virus	Gamboa virus	GAMV
	Pueblo Viejo	PVV
Guajara virus	Guajara virus	GJAV
Guama virus	Guama virus	GMAV
	Ananindeua virus	ANUV
	Mahogany Hammock virus	MHV
	Moju virus	MOJUV
Guaroa virus	Guaroa virus	GROV
Kaeng Khoi virus	Kaeng Khoi virus	KKV
Kairi virus	Kairi virus	KRIV
Koongol virus	Koongol virus	KOOV
	Wongal virus	WONV
Madrid virus	Madrid virus	MADV
Main Drain virus	Main Drain virus	MDV
Manzanilla virus	Manzanilla virus	MANV
	Buttonwillow virus	BUTV
	Ingwavuma virus	INGV
	Inini virus	INIV
	Mermet virus	MERV
Marituba virus	Marituba virus	MTBV
	Gumbo Limbo virus	GLV
	Murutucú virus	MURV
	Nepuyo virus	NEPV
	Restan virus	RESV
Minatitlan virus	Minatitlan virus	MNTV
	Palestina virus	PLSV
M'Poko virus	M'Poko virus	MPOV
	Yaba-1 virus	Y1V
Nyando virus	Nyando virus	NDOV
	Eretmapodites virus	ERETV

Species	Virus	Abbreviation
Olifantsvlei virus	Olifantsvlei virus	OLIV
	Bobia virus	BIAV
	Dabakala virus	DABV
	Oubi virus	OUBIV
Oriboca virus	Oriboca virus	ORIV
	Itaqui virus	ITQV
Oropouche virus	Oropouche virus	OROV
	Facey's Paddock virus	FPV
Patois virus	Patois virus	PATV
	Abras virus	ABRV
	Babahoyo virus	BABV
	Pahayokee virus	PAHV
	Shark River virus	SRV
Sathuperi virus	Sathuperi virus	SATV
	Douglas virus	DOUV
Sedlec virus	Sedlec virus	SEDV
	Oyo virus	OYOV
	I 612045	–
Shamonda virus	Shamonda virus	SHAV
	Peaton virus	PEAV
	Sango virus	SANV
Shuni virus	Shuni virus	SHUV
	Aino virus	AINOV
	Kaikalur virus	KAIV
Simbu virus	Simbu virus	SIMV
	Pintupo virus	PINTV
	Utinga virus	UTIV
	Utive virus	UTVEV
Tacaiuma virus	Tacaiuma virus	TCMV
	CoAr 1071 virus	CA1071V
	CoAr 3627 virus	CA3627V
	Virgin River virus	VRV
Tahyña virus	Tahyña virus	TAHC
	(Lumbo virus)	
Tete virus	Tete virus	TETEV
	Bahig virus	BAHV
	Matruh virus	MTRV
	Tsuruse virus	TSUV
	Weldona virus	WELV
Thimiri virus	Thimiri virus	THIV
Timboteua virus	Timboteua virus	TBTV
Trivitattus virus	trivitattus virus	TVTV
Turlock virus	Turlock virus	TURV
	Lednice virus	LEDV
	Umbre virus	UMBV
Wyeomyia virus	Wyeomyia virus	WYOV
	Anhembi virus	AMBV
	Cachoeira Porteira virus	CPOV
	Iaco virus	IACOV
	Macaua virus	MCAV
	Sororoca virus	SORV
	Taiassui virus	TAIAV
	Tucunduba virus	TUCV
Zegla virus	Zegla virus	ZEGV

In addition, the relatively recently detected Schmallenberg virus, Iquitos virus and Jatobal virus are RNA reassortants of previously recognized viruses (see Briese, T. Calisher, C.H., and Higgs, S. (2013) Viruses of the family *Bunyaviridae*: Are all available isolates reassortants? Virology. 446:207–216).

Other related viruses that are or may be members of the genus *Orthobunyavirus* but have not been approved as species include Brazoran virus (BRAZV), Khurdun virus (KHURV), Leanyer virus (LEAV), Gouléako virus (GOUV), Murrumbidgee virus (MURBV), Mojui dos Campos virus (MDCV), Salt Ash virus (SASHV), Sedlec virus (SEDV) and Termeil virus (TERV).

Viruses of the family *Bunyaviridae*, genus *Nairovirus*

Species	Virus	Abbreviation
Caspiy virus	Caspiy virus	CASV
Crimean–Congo haemorrhagic fever virus	Crimean–Congo haemorrhagic fever virus	CCHFV
	Hazara virus	HAZV
	Khasan virus	KHAV
Dera Ghazi Khan virus	Dera Ghazi Khan virus	DGKV
	Abu Hammad virus	AHV
	Abu Mina virus	AMV
	Kao Shuan virus	KSV
	Pathum Thani virus	PTHV
	Pretoria virus	PREV
Dugbe virus	Dugbe virus	DUGV
	Kupe virus	KUPEV
	Nairobi sheep disease virus (Ganjam virus)	NSDV
Estero Real virus	Estero Real virus	ERV
Hughes virus	Hughes virus	HUGV
	Farallon virus	FARV
	Fraser Point virus	FPV
	Great Saltee virus	GRSV
	Puffin Island virus	PIV
	Punta Salinas virus	PSV
	Raza virus	RAZAV
	Sapphire II virus	SAPV
	Soldado virus	SOLV
	Zirqa virus	ZIRV
Qalyub virus	Qalyub virus	QYBV
	Bakel virus	BAKLV
	Bandia virus	BDAV
	Omo virus	OMOV
Sakhalin virus	Sakhalin virus (Avalon virus)	SAKV
	Clo Mor virus	CLMV
	Finch Creek virus	FINCV
	Kachemak Bay virus	KBV
	Taggert virus	TAGV
	Tillamook virus	TILLV
Thiafora virus	Thiafora virus	TFAV
	Erve virus	ERVEV

Other related viruses that may be members of the genus *Nairovirus* but have not been approved as species include Gossas virus (GOSV), Issyk-Kul virus (ISKV), Tamdy virus (TDYV), Keterah (KTRV), Kasokero (KASOV) and Yogue virus (YOGV).

Viruses of the family *Bunyaviridae*, genus *Phlebovirus*

Species	Virus	Abbreviation
Aguacate virus	Aguacate virus	AGUV
	Armero virus	ARMV

Species	Virus	Abbreviation
	Durania virus	DURNV
	Ixcanal virus	IXCV
Bhanja virus	Bhanja virus	BHAV
	(Palma virus)	
	Forécariah virus	FORV
	Heartland virus	HRTV
	Hunter Island Group virus	HIGV
	Kismayo virus	KISV
	Lone Star virus	LSV
	severe fever with thrombocytopenia syndrome virus	SFTSV
Bujaru virus	Bujaru virus	BUJV
	Munguba virus	MUNV
Candiru virus	Candiru virus	CDUV
	Alenquer	ALEV
	Ariquemes virus	ARQV
	Escharate virus	ESCV
	Itaituba virus	ITAIV
	Jacunda virus	JCNV
	Maldonado virus	MLOV
	Morumbi virus	MRBV
	Mucura virus	MRAV
	Nique virus	NIQV
	Oriximina virus	ORXV
	Serra Norte virus	SRNV
	Turuna virus	TUAV
Chilibre virus	Chilibre virus	CHIV
	Cacao virus	CACV
Frijoles virus	Frijoles virus	FRIV
	Joa virus	JOAV
Grand Arbaud virus	Grand Arbaud virus	GAV
Icoaraci virus	Icoaraci virus	ICOV
	Belterra virus	BELTV
	Salobo virus	SLBOV
Kaisodi virus	Kaisodi virus	KSOV
	Khasan virus	KHAV
	Silverwater virus	SILV
	Lanjan virus	LJNV
Karimabad virus	Karimabad virus	KARV
	Gabek Forest virus	GFV
Manawa virus	Manawa virus	MWAV
	Komandory virus	KOMV
Murre virus	Murre virus	MURRV
	RML 105355 virus	RMLV
	Sunday Canyon virus	SCAV
Precarious Point virus	Precarious Point virus	PPV
Punta Toro virus	Punta Toro virus	PTV
	Buenaventura virus	BUEV
Rift Valley fever virus	Rift Valley fever virus	RVFV
Salehabad virus	Salehabad virus	SALV
	Adria virus	ADRV
	Arbia virus	ARBV
	Arumowot virus	AMTV
	Odrenisrou	ODRV

Species	Virus	Abbreviation
Sandfly fever Naples virus	sandfly fever Naples virus	SFNV
	Gordil virus	GORV
	Granada virus	GRAV
	Massilia virus	MASLV
	Punique virus	PUNV
	Saint-Floris virus	SAFV
	Tehran virus	TEHV
	Toscana virus	TOSV
Sandfly fever Sicilian virus	sandfly fever Sicilian virus	SFSV
	Chagres virus	CHGV
	Corfou virus	CFUV
Uukuniemi virus	Uukuniemi virus	UUKV
	Catch-me-cave virus	CMCV
	Chizé virus	CHZV
	EgAn 1825–61 virus	EGAV
	Fin V707 virus	FINV
	Oceanside virus	OCEV
	Ponteves virus	PTVV
	St. Abb's Head virus	SAHV
	Tunis virus	TUNV
	Zaliv Terpeniya virus	ZTV

Other related viruses that may be members of the genus *Phlebovirus* but have not been approved as species include Ambe virus (AMBEV), American dog tick phlebovirus (ADTV), Anhanga virus (ANHV), Arboledas virus (ADSV), Bradypus 4 virus (BDPS4 virus), Caimito virus (CAIV), Itaporanga virus (ITPV), Kala Iris virus (KAIRV), Leticia virus (LTCV), Mariquita virus (MRQV), Morolillo virus (MOLV), Olbia virus (OLBV), Pacui virus (PACV), Provencia virus (PROVV), Rio Grande virus (RGV), Salanga virus (SALAV), Salobo virus (SBOV), Shibuyunji virus (SHIBV), Tapara virus (TAPV), Uriurana virus (URIV) and Urucuri virus (URUV).

There are many viruses that have not been assigned to a recognized genus in the family *Bunyaviridae*. For most, no biochemical characterization, useful for determining taxonomic status, has been reported.

Genus-unassigned viruses of the family *Bunyaviridae*

Species	Virus	Abbreviation
	Antequera virus	ANTV
	Barranqueras virus	BQSV
	Bangui virus	BGIV
	Belem virus	BLMV
	Belmont virus	BELV
	Bobaya virus	BOBV
	Caddo Canyon virus	CDCV
	Chim virus	CHIMV
	Enseada virus	ENSV
	Gan Gan virus	GGV
	Kaisodi virus	KSOV
	Kowanyama virus	KOWV
	Lanjan virus	LJNV
	Mapputta virus	MAPV
	Maprik virus	MPKV
	Okola virus	OKOV
	Pacora virus	PCAV
	Para virus	PARAV
	Resistencia virus	RTAV
	Santarem virus	STMV
	Silverwater virus	SILV

Species	Virus	Abbreviation
	Tanga virus	TANV
	Tataguine virus	TATV
	Trubanaman virus	TRUV
	Wanowrie virus	WANV
	Witwatersrand virus	WITV
	Yacaaba virus	YACV

Viruses of the family *Reoviridae*, subfamily *Spinareovirinae*, genus *Coltivirus*

Species	Virus	Abbreviation
Colorado tick fever virus	Colorado tick fever virus	CTFV
Eyach virus	Eyach virus	EYAV

Another related virus that may be a member of the genus *Coltivirus* but has not been approved as a species is Salmon River virus (SaRV).

Viruses of the family *Reoviridae*, subfamily *Sedoreovirinae*, genus *Orbivirus*

Species	Virus	Abbreviation
African horse sickness virus	African horse sickness virus 1	AHSV-1
	African horse sickness virus 2	AHSV-2
	African horse sickness virus 3	AHSV-3
	African horse sickness virus 4	AHSV-4
	African horse sickness virus 5	AHSV-5
	African horse sickness virus 6	AHSV-6
	African horse sickness virus 7	AHSV-7
	African horse sickness virus 8	AHSV-8
	African horse sickness virus 9	AHSV-9
Bluetongue virus	bluetongue virus 1	BTV-1
	bluetongue virus 2	BTV-2
	bluetongue virus 3	BTV-3
	bluetongue virus 4	BTV-4
	bluetongue virus 5	BTV-5
	bluetongue virus 6	BTV-6
	bluetongue virus 7	BTV-7
	bluetongue virus 8	BTV-8
	bluetongue virus 9	BTV-9
	bluetongue virus 10	BTV-10
	bluetongue virus 11	BTV-11
	bluetongue virus 12	BTV-12
	bluetongue virus 13	BTV-13
	bluetongue virus 14	BTV-14
	bluetongue virus 15	BTV-15
	bluetongue virus 16	BTV-16
	bluetongue virus 17	BTV-17
	bluetongue virus 18	BTV-18
	bluetongue virus 19	BTV-19
	bluetongue virus 20	BTV-20
	bluetongue virus 21	BTV-21
	bluetongue virus 22	BTV-22
	bluetongue virus 23	BTV-23
	bluetongue virus 24	BTV-24
	bluetongue virus 25	BTV-25
	bluetongue virus 26	BTV-26
Changuinola virus	Changuinola virus	CGLV

Species	Virus	Abbreviation
	Almeirim virus	ALMV
	Altamira virus	ALTV
	Caninde virus	CANV
	Gurupi virus	GURV
	Irituia virus	IRIV
	Jamanxi virus	JAMV
	Jari virus	JARIV
	Monte Dourado virus	MDOV
	Ourem virus	OURV
	Purus virus	PURV
	Saraca virus	SRAV
Chenuda virus	Chenuda virus	CNUV
	Baku virus	BAKUV
	Essaouira virus	ESSV
	Huacho virus	HUAV
	Kala Iris virus	KIRV
	Mono Lake virus	MLV
	Sixgun City virus	SCV
Chobar Gorge virus	Chobar Gorge virus	CGV
	Fomede virus	FOMV
Corriparta virus	Corriparta virus	CORV
	Acado virus	ACDV
	California mosquito pool virus	CMPV
	Jacareacanga virus	JACV
Epizootic haemorrhagic disease virus	epizootic haemorrhagic disease virus 1	EHDV-1
	epizootic haemorrhagic disease virus 2 (Ibaraki virus)	EHDV-2
	epizootic haemorrhagic disease virus 3	EHDV-3
	epizootic haemorrhagic disease virus 4	EHDV-4
	epizootic haemorrhagic disease virus 5	EHDV-5
	epizootic haemorrhagic disease virus 6	EHDV-6
	epizootic haemorrhagic disease virus 7	EHDV-7
	epizootic haemorrhagic disease virus 8	EHDV-8
Equine encephalosis virus	equine encephalosis virus 1	EEV-1
	equine encephalosis virus 2	EEV-2
	equine encephalosis virus 3	EEV-3
	equine encephalosis virus 4	EEV-4
	equine encephalosis virus 5	EEV-5
	equine encephalosis virus 6	EEV-6
	equine encephalosis virus 7	EEV-7
Eubenangee virus	Eubenangee virus	EUBV
	Ngoupe virus	NGOV
	Pata virus	PATAV
	Tilligerry virus	TILV
Great Island virus	Great Island virus	GIV
	Above Maiden virus	ABMV
	Arbroath virus	ABRV
	Bauline virus	BAUV
	Broadhaven virus	BRDV
	Cape Wrath virus	CWV
	Colony virus	COYV
	Colony B North virus	CBNV
	Ellidaey virus	ELLV

Species	Virus	Abbreviation
	Foula virus	FOUV
	Great Saltee Island virus	GSIV
	Grimsey virus	GSYV
	Inner Farne virus	INFV
	Kemerovo virus	KEMV
	Kenai virus	KENV
	Kharagysh virus	KHAV
	Lipovnik virus	LIPV
	Lundy virus	LUNV
	Maiden virus	MDNV
	Mill Door virus	MDRV
	Mykines virus	MYKV
	North Clett virus	NCLV
	North End virus	NEDV
	Nugget virus	NUGV
	Okhotskiy virus	OKHV
	Poovoot virus	POOV
	Rost Island virus	RSTV
	Shiant Islands virus	SHIV
	Thormodseyjarlettur virus	THRV
	Tillamook virus	TLMV
	Tindholmur virus	TDMV
	Tribec virus	TRBV
	Vearoy virus	VAEV
	Wexford virus	WEXV
	Yaquina Head virus	YHV
Ieri virus	Ieri virus	IERIV
	Gomoka virus	GOMV
	Arkonam virus	ARKV
Lebombo virus	Lebombo virus 1	LEBV-1
Orungo virus	Orungo virus 1	ORUV-1
	Orungo virus 2	ORUV-2
	Orungo virus 3	ORUV-3
	Orungo virus 4	ORUV-4
Palyam virus	Palyam virus	PALV
	(Kasba virus/Chuzan virus)	
	Abadina virus	ABAV
	Bunyip Creek virus	BCV
	CSIRO village virus	CVGV
	D'Aguilar virus	DAGV
	Gweru virus	GWV
	Kindia virus	KINV
	Marrakai virus	MARV
	Marondera virus	MRDV
	Nyabira virus	NYAV
	Petevo virus	PETV
	Vellore virus	VELV
Peruvian horse sickness virus	Peruvian horse sickness virus-1	PHSV-1
	Elsey virus	ELSV
Umatilla virus	Umatilla virus	UMAV
	Llano Seco virus	LLSV
	Minnal virus	MINV
	Netivot virus	NETV

Species	Virus	Abbreviation
Wad Medani virus	Wad Medani virus	WMV
	Seletar virus	SELV
Wallal virus	Wallal virus	WALV
	Mudjinbarry virus	MUDV
	Wallal K virus	WALKV
Warrego virus	Warrego virus	WARV
	Mitchell river virus	MRV
	Warrego K virus	WARKV
Wongorr virus	Wongorr virus	WGRV
	Paroo river virus	PRV
	Picola virus	PIAV
	Wongorr virus CS131	WGRV
Yunnan orbivirus	Yunnan orbivirus 1	YUOV-1
	Rioja virus	RIOV
	Yunnan orbivirus 2	YOUV-2
	Middle Point orbivirus	

Other related viruses that may be members of the genus *Orbivirus* but have not been approved as species include Andasibe virus (ANDV), Aniva virus (ANIV), Codajas virus (COV), Ife virus (IFEV), Itupiranga virus (ITUV), Japanaut virus (JAPV), Kammavanpettai virus (KMPV), Lake Clarendon virus (LCV), Matucare virus (MATV), Stretch Lagoon virus (SLOV), Tembe virus (TMEV) and Tracambe virus (TRAV).

Viruses of the family *Reoviridae*, subfamily *Sedoreovirinae*, genus *Seadornavirus*

Species	Virus	Abbreviation
Banna virus	Banna virus	BAV
Kadipiro virus	Kadipiro virus-Java-7075	KDV-Ja7075
Liao ning virus	Liao ning virus (NE97–12)	LNV-NE97–12
Balatone virus	Balatone virus	BALV

Viruses of the order *Mononegavirales*, family *Rhabdoviridae*, genus *Vesiculovirus*

Species	Virus	Abbreviation
Carajas virus	Carajas virus	CJSV
Chandipura virus	Chandipura virus	CHPV
Cocal virus	Cocal virus	COCV
Isfahan virus	Isfahan virus	ISFV
Maraba virus	Maraba virus	MARAV
Piry virus	Piry virus	PIRYV
Vesicular stomatitis Alagoas virus	vesicular stomatitis Alagoas virus	VSAV
Vesicular stomatitis Indiana virus	vesicular stomatitis Indiana virus	VSIV
Vesicular stomatitis New Jersey virus	vesicular stomatitis New Jersey virus	VSNJV

Other related viruses that may be members of the genus *Vesiculovirus* but have not been approved as species include BeAn 157575 virus (BeAnV-157575), Boteke virus (BTKV), Calchaqui virus (CQIV), Gray Lodge virus (GLOV), Jurona virus (JURV), Klamath virus (KLAV), Kwatta virus (KWAV), La Joya virus (LJV), Malpais Spring virus (MSPV), Perinet virus (PERV), Porton virus (PORV), Radi virus (RADIV) and Yug Bogdanovac virus (YBV).

Viruses of the order *Mononegavirales*, family *Rhabdoviridae*, genus *Ephemerovirus*

Species	Virus	Abbreviation
Adelaide River virus	Adelaide River virus	ARV
Berrimah virus	Berrimah virus	BRMV
Bovine ephemeral fever virus	bovine ephemeral fever virus	BEFV
Kotonkan virus	kotonkan virus	KOTV
Obodhiang virus	Obodhiang virus	OBOV

Other related viruses that may be members of the genus *Ephemerovirus* but have not been approved as species are Kimberley virus (KIMV), Malakal virus (MALV) and Puchong virus (PUCV).

Viruses of the order *Mononegavirales*, family *Rhabdoviridae*, genus *Tibrovirus*

Species	Virus	Abbreviation
Tibrogargan virus	Tibrogargan virus	TIBV
	Bas Congo virus	BASV
	Bivens Arm virus	BAV
	Coastal Plains virus	CPV

Unassigned species in the family *Rhabdoviridae*

Species	Virus	Abbreviation
Flanders virus	Flanders virus	FLAV
	Wongabel virus	WGBLV
Ngaingan virus	Ngaingan virus	NGAV
Sigma virus	Sigma virus HAP 23	SIGMAV-HAP23
	Sigma virus AP30	SIGMAV-AP30
Tupaia virus	Tupaia virus	TUPV

Other related viruses that may be members of the family *Rhabdoviridae* but have not been approved as species include Almpiwar virus (ALMV), Aruac virus (ARUV), Bahia Grande virus (BGV), Bangoran virus (BGNV), Barur virus (BARV), Beaumont virus (BEAUV), Bimbo virus (BBOV), Blue crab virus (BCV), Chaco virus (CHOV), Charleville virus (CHVV), Connecticut virus (CNTV), Cuiaba virus (CUIV), Curionopolis virus (CURV), DakArK 7292 virus (DAKV-7292), Durham virus (DURV), Entamoeba virus (ENTV), Farmington virus (FARV), Fukuoka virus (FUKAV), Garba virus (GARV), Harlingen virus (HARV), Hart Park virus (HPV), Humpty Doo virus (HDOOV), Iriri virus (IRIV, duplicate abbreviation), Inhangapi virus (INHV), Itacaiunas virus (ITAV), Joinjakaka virus (JOIV), Kamese virus (KAMV), Kannamangalam virus (KANV), Kern Canyon virus (KCV), Keuraliba virus (KEUV), Kolongo virus (KOLV), Koolpinyah virus (KOOLV), Landjia virus (LJAV), Le Dantec virus (LDV), Manitoba virus (MNTBV), Marco virus (MCOV), Morreton virus (MRTV), Mosqueiro virus (MQOV), Mossuril virus (MOSV), Mount Elgon bat virus (MEBV), Muir Springs virus (MSV), Nasoule virus (NASV), Navarro virus (NAVV), New Minto virus (NMV), Nkolbisson virus (NKOV), North Creek virus (NORCV), Oak Vale virus (OVRV), Oita virus (OITAV), Ouango virus (OUAV), Parry Creek virus (PCRV), Reed Ranch virus (RERAV), Rochambeau virus (RBUV), Sandjimba virus (SJAV), Sawgrass virus (SAWV), Sena Madureira virus (SMV), Sripur virus (SRIV), Sweetwater Branch virus (SWBV), Timbo virus (TIMV), Xiburema virus (XIBV) and Yata virus (YATAV).

Viruses of the family *Orthomyxoviridae*, genus *Thogotovirus*

Species	Virus	Abbreviation
Thogoto virus	Thogoto virus	THOV
Dhori virus	Dhori virus	DHOV
	Batken virus	BKNV

In addition to these, Upolu virus (UPOV) and Aransas Bay virus (ABV) (related) and Jos virus (JOSV) have been placed in the genus *Thogotovirus*, but are not members of either recognized species.

Viruses of the family *Orthomyxoviridae*, genus *Quaranjavirus*

Species	Virus	Abbreviation
Quaranfil virus	Quaranfil virus	QRFV
	Johnston Atoll virus	JAV
	Lake Chad virus	LKCV
	Tyulek virus	TLKV

Araguari virus, Wellfleet Bay virus and Cygnet River virus have been shown to be orthomyxoviruses but have not been placed in a genus or species.

Viruses of the family *Asfarviridae*, genus *Asfavirus* (DNA virus)

Species	Virus	Abbreviation
African swine fever virus	African swine fever virus	ASFV

Viruses of undetermined taxonomy

Species	Virus	Abbreviation
	Jingman tick virus	JMTV

[a]All genera traditionally considered arboviruses are shown in the tables below. There are other genera within these virus families but, as they do not contain arboviruses, suspect arboviruses, or otherwise are only considered possible arboviruses, they are not included in this table.
[b]Species names are in italics; names of virus isolates are in Roman script. Assigned abbreviations for virus names also are listed; there are no abbreviations for species names. Note that certain abbreviations are duplicative of abbreviations for other viruses. Until ICTV corrects these duplications, we have used the official ICTV abbreviations.
[c]The ICTV lists subtypes of tick-borne encephalitis virus [European subtype (TBEV-Eur), Far Eastern subtype (TBEV-FE), and Siberian subtype (TBEV-Sib)], and provides abbreviations for those subtypes but does not list the virus itself. As ICTV does not involve itself with taxa below species, we have chosen to list tick-borne encephalitis viruses using their classical names.
[d]The ICTV lists subtypes of louping ill virus [British subtype (LIV-Brit), Irish subtype (LIV-Ir), Spanish subtype (LIV-Spain), Turkish sheep encephalitis virus subtype (TSEV), and Greek goat encephalitis virus subtype (GGEV)], and provides abbreviations for those subtypes but does not list the virus itself. As ICTV does not involve itself with taxa below species, we have chosen to list louping ill virus. Negishi virus, formerly considered a distinct virus, now is considered a subtype of prototype louping ill virus.
[e]Ngari (a.k.a. Garissa) virus is a reassortant of Bunyamwera (S and L segments) and Batai (M segment) viruses. Many other viruses on this list and some that are not listed are reassortants or may be reassortants. For further details, see Briese, T. Calisher, C.H., and Higgs, S. (2013) Viruses of the family *Bunyaviridae*: Are all available isolates reassortants? Virology. 446:207–216.

Appendix 2.2 Taxonomy of arthropod viruses[a]

Species	Virus	Abbreviation
Arthropod viruses of the family *Flaviviridae*[b]		
	Aedes cinereus flavivirus	–[c]
	Aedes galloisi flavivirus	AGFV
	Aedes flavivirus	AEFV
	Aedes vexans flavivirus	AeveFV
	Aripo virus	ARIPV
	Barkedji virus	BJV
	Calbertado virus	CBTV
	Cell fusing agent virus	CFAV
	Chaoyang virus	CHAOV
	Culex flavivirus	CXFV
	Culex theileri flavivirus	CTFV[d]
	Czech *Aedes vexans* flavivirus	Czech AeveFV
	Donggang virus	–
	Hanko virus	HANKV
	Kamiti River virus	KRV
	Lammi virus	LAMV
	La Tina virus	–
	Marisma mosquito virus	MMV
	Nakiwogo virus	NAKV
	Nanay virus	NANV
	Ngoye virus	NGOV
	Nounané virus	NOUV
	Ochlerotatus flavivirus	OcFV
	Ochlerotatus caspius flavivirus	OCFVPT
	Palm Creek virus	PCrV
	Spanish *Culex* flavivirus	SCxFV
	Spanish Ochlerotatus flavivirus	SOcFV
Arthropod viruses of the family *Togaviridae*		
	Eilat virus	EILV

Species	Virus	Abbreviation
Arthropod viruses of the family *Bunyaviridae*		
	Cumuto virus	CUMV
	Herbert virus	HEBV
	Kibale virus	KIBV
	Taï virus	TAIV
Arthropod viruses of the family *Mesoniviridae*		
	Bontag Baru virus	BotV
	Cavally virus	CAVV
	Dak Nong virus	DKNV
	Hana virus	HanaV
	Kamphaeng Phet virus	KphV
	Méno virus	MénoV
	Moumo virus	MoumoV
	Nam Dinh virus	NDiV
	Nsé virus	NséV
Arthropod viruses of the family *Reoviridae*		
	Aedes pseudoscutellaris reovirus	APRV
	St Croix River virus	SCRV
Arthropod viruses of the family *Rhabdoviridae*		
	Arboretum virus	ABTV
	Moussa virus	MOUV
	Long Island tick rhabdovirus	LITRV
	Puerto Almendras virus	PTAMV
Arthropod viruses for which the family is unassigned, genus *Negivirus*		
	Dezidougou virus	DEZV
	Loreto virus	LORV
	Negev virus	NEGV
	Ngewotan virus	NWTV
	Piura virus	PIUV
	Santana virus	SANV[e]
	Wallerfield virus	WALV[f]

[a]This is a provisional list of viruses detected in haematophagus arthropods ('arthropod viruses') that, thus far, appear to be arthropod-specific and detected in field materials principally by polymerase chain reactions. Many other viruses have been isolated from or detected in arthropods, but the viruses listed here are in virus families that also contain known arboviruses. We consider this a provisional list because it is clear that the surface has only been scratched insofar as detecting viruses in arthropods. Among the arthropods that have not been adequately surveyed for viruses or not surveyed at all are: chelicerates (spiders, mites, and scorpions), myriapods (millipedes, centipedes), crustaceans (lobsters, crabs, barnacles, crayfish, shrimp), and hexapods.
[b]These flaviviruses may be separated into insect-specific viruses and mosquito-borne flaviviruses as the former are restricted to a single genetic group clearly distinct from the latter.
[c]– Indicates there is no official or suggested abbreviation for this virus.
[d]The abbreviation CTFV is the established abbreviation for Colorado tick fever virus.
[e]The abbreviation SANV is the established abbreviation for Sango virus.
[f]The abbreviation WALV is the established abbreviation for Wallal virus.

Genomic Organization of Arboviral Families

3

Nikos Vasilakis, Amy Lambert, N. James MacLachlan and Aaron C. Brault

Abstract

Arboviruses (arthropod-borne viruses) are represented almost exclusively by viruses comprising RNA genomes for which a number of genomic organization and replication strategies have been observed. Herein, we have reviewed six viral families that constitute the majority of arboviruses transmissible to humans and livestock including: alphaviruses (*Togaviridae*), flaviviruses (*Flaviviridae*), rhabdoviruses (*Rhabdoviridae*), bunyaviruses (*Bunyaviridae*), reoviruses (*Reoviridae*) and orthomyxoviruses (*Orthomyxoviridae*). Descriptions of the overall genomic architecture, viral proteins and replicative strategies have been included and serve to compare and contrast the breadth of variation that can be observed among these diverse groups of viruses that are transmitted by arthropod vectors. This diversity indicates the likelihood of convergence of disparate viral groups for arthropod transmission yet such a comparison can serve to highlight conserved genetic strategies between dipartite viruses for elucidation of the evolutionary pressures imposed by the necessity for replication in both vertebrate and invertebrate hosts.

Introduction

Arboviruses are viruses that are biologically transmitted by arthropod vectors, mosquitoes, ticks, sandflies, hemipterans and midges. Commensurate with the diversity of arthropod vectors that transmit them, a plethora of diverse arboviral taxa have evolved with various genomic organization structures and associated replication strategies (Table 3.1). We have included herein a brief description of the genome organization of the four principal viral families [*Togaviridae* (alphaviruses), *Flaviviridae* (flaviviruses), *Reoviridae* (orbiviruses), *Rhabdoviridae* (rhabdoviruses), *Bunyaviridae* (bunyaviruses) and *Orthomyxoviridae* (orthomyxoviruses)] that comprise the vast majority of arboviruses known to be biologically transmitted to humans and livestock. All viral families described herein comprise RNA genomes of varying lengths and polarities with segmented and non-segmented genomes. Genome length of viruses coupled with the low fidelity of their viral encoded RNA dependent RNA polymerases allows for a high degree of genetic heterogeneity of all of these RNA viruses that affords the capacity for rapid adaptation to new ecological niches. This chapter seeks to survey the genomic structures of the various arboviral families, with special emphasis on description of the unique features of arboviral agents within those taxonomic designations.

Alphaviruses

Alphaviruses, like all other members of the family *Togaviridae*, comprise a single-stranded, positive polarity (message-sense) RNA genome of approximately 11–11.7 kilobases in length. Viral RNAs comprise both a genomic 42S RNA and a subgenomic 26S RNA that encodes the terminal 3′ region of the viral genome and encodes the viral structural proteins. The alphaviral RNA genome and subgenomic RNAs are very similar to that of a host mRNA with the 5′ terminus capped by a 7-methylguanosine moiety and the 3′ terminus of the genomic and subgenomic RNA containing poly-A tracts. The positive

Table 3.1 Description of the families of arboviruses with genome structure, RNA polarity, genome segmentation, length of genome and cap structure

Virus family	Genome structure	Polarity	Strandedness	Genome length (kb)
Togaviridae	RNA/unsegmented/poly-A	Positive	Single-stranded	11–11.7
Flaviviridae	RNA/unsegmented	Positive	Single-stranded	11–11.2
Bunyaviridae	RNA/segmented (3)	Negative/ambisense	Single-stranded	11–17
Reoviridae	RNA/segmented (10–12)	Negative	Double-stranded	Variable
Rhabdoviridae	RNA/unsegmented	Negative	Single-stranded	11–15
Orthomyxoviridae	RNA/segmented (6–8)	Negative	Single-stranded	Variable

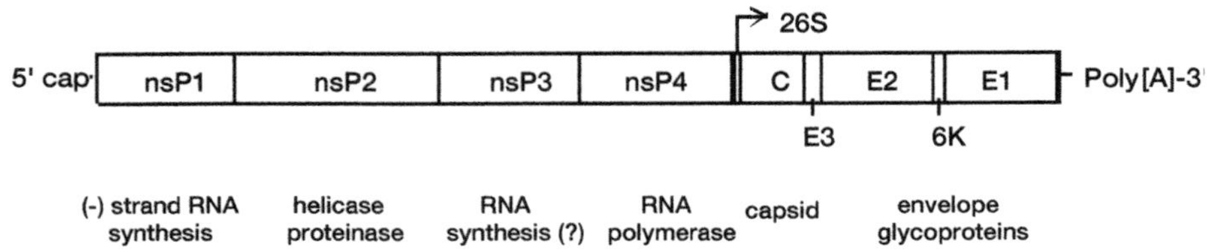

Figure 3.1 Alphaviral genome organization. The 5′ two-thirds of the genome encodes four non-structural proteins (nsP1–4). The 3′ one-third of the genome contains a 26S subgenomic RNA that encodes three structural proteins (capsid, E2 and E1 glycoproteins). The proposed genetic function of each viral protein is designated below the gene region. Beyond its role in negative strand RNA synthesis, the function of the nsP3 gene is not well understood (Strauss and Strauss, 1994).

polarity RNA genome is functionally divided into two separate open reading frames (ORFs) encoding four non-structural proteins (nsP1–4) in the 5′ two-thirds of the viral genome and the structural proteins that are expressed by a second ORF in the remaining one-third of the genome. In between the non-structural and structural coding regions lies a small intergenic region encoding a 26S subgenomic promoter that drives transcription of the three major structural proteins (capsid, E1 and E2 glycoproteins; Fig. 3.1) (Schlesinger and Schlesinger, 1996). Two additional proteins of undetermined function, 6K and E3, are present in the structural protein gene region (Strauss and Strauss, 1994).

Direct translation of the 42S and 26S subgenomic RNAs result in the production of the alphaviral replicase proteins and structural proteins, respectively. Differential translation of the non-structural polyprotein consisting of nsP1–3 (P123) or nsP1–4 (P1234) results from translational termination at a stop codon between nsP3–4 or read-through, respectively (Jose *et al.*, 2009; Myles *et al.*, 2006).

The original genomic alphaviral RNA, in addition to being translated by cellular components to produce a non-structural polyprotein, serves as the template for minus-strand RNA synthesis involving non-structural proteins. Initial viral RNA replication occurs by the generation of negative strand synthesis from the 3′ terminus of the 42S genomic message sense-RNA required by P123 or P23 and nsP4. The negative strands that are generated then serve to produce additional positive-sense, genome length RNAs that are packaged during virion assembly. Switch-over from minus- to plus-strand RNA synthesis is believed to occur as a result of cleavage products produced between the nsP1–2 and nsP2–3 proteins likely resulting from a change in the conformation of the replication complex with altered strand-specific replication affinities (Kuhn, 2007; Strauss and Strauss, 1994). The subgenomic 26S mRNA generated from the internal RNA promoter is also translated as a polyprotein; the capsid protein is cleaved in the cytoplasm and the remaining polyprotein is processed in the secretory pathway to yield the E1 and E2 glycoproteins which are inserted into the plasma membrane as a heterodimer (Strauss and Strauss, 1994).

The non-structural protein 1 (nsP1) is a membrane encoded viral protein for which guanyltransferase activity for RNA capping of synthesized genomic viral RNAs has been ascribed (Mi and Stollar, 1990, 1991). The nsP2 protein is a multifunctional protein that houses motifs responsible for and helicase, nucleoside triphosphatase (NTPase) and protease activities (Gomez de Cedron *et al.*, 1999; Rikkonen *et al.*, 1994; Vasiljeva *et al.*, 2001). Nuclear localization signals are also present within multiple domains of the protein that target the protein to the nucleus where the protein is associated with host proteins (Rikkonen *et al.*, 1992). This viral protein has also been reported to serve as an innate immune antagonist by reducing type I and type II IFN stimulated JAK-STAT signalling (Fros *et al.*, 2010). Of all of the alphaviral proteins, the role of nsP3 is the least characterized. The protein is intrinsically required for both positive and negative strand RNA replication in the P123 polyprotein as well as in its monomeric cleaved form (Kuhn, 2007). The amino terminus of the protein has been associated with effectors for intracellular signalling and apoptotic pathways (Malet *et al.*, 2009; Shin *et al.*, 2012). The carboxy terminus of the nsP3 is highly variable among alphaviral taxa and has been associated with insertion and deletions (Foy *et al.*, 2012) and has been shown to be hyper-phosphorylated (Li *et al.*, 1990) and differential mosquito infectivity phenotypes (Saxton-Shaw *et al.*, 2013). The final linear non-structural protein, nsP4, serves as the viral RNA-dependent RNA polymerase that is translated as part of the NS polyprotein as a product of read-through of an opal stop codon between nsP3 and the start of nsP4 (Strauss *et al.*, 1983). The carboxyl terminus of the protein has demonstrated RdRP domains determined by homology with other RdRPs as well as predicted secondary structures responsible for positive and negative sense vRNA replication as well as for generation of the 26S RNA (Kuhn, 2007). The RdRP activity of the nsP4 has also been associated with maintenance and repair of the poly-A tail (Tomar *et al.*, 2006).

Alphaviral structural proteins are translated directly from the 26S RNA which is transcribed from the 26S subgenomic promoter located −19 to +5 relative to the site of mRNA initiation. Translation of the 26S message occurs at a rate approximately 3-fold higher than that of the 42S genomic RNA as a result of the high transcription rate of the subgenomic RNA due to enhanced promoter activity (Raju and Huang, 1991). The structural polyprotein is cleaved by host and viral proteases. The alphaviral capsid protein, located at the 5′ end of the polyprotein, contains tracts of basic amino acid residues at the N-terminus associated with direct binding of genomic RNA sequences that have been mapped to genomic nucleotide positions 945 to 1076 for Sindbis viral 42 S sequences (Linger *et al.*, 2004); however, alternative encapsidation sequences could exist for other alphaviruses such as Aura virus, which has been found to encapsidate 26S RNA as well (Rumenapf *et al.*, 1995). Additional domains in the capsid allow for capsid dimerization (Frolova *et al.*, 1997) as well as a serine protease that dictates autoproteolytic cleavage from the structural polyprotein (Choi *et al.*, 1996). The E3 protein serves as a signal sequence for transport of the polyprotein through the ER as well as interaction with the E2 protein to prevent premature fusion with the E1 under basic conditions in the Golgi (Garoff *et al.*, 1990; Li *et al.*, 2010). The E2 protein serves as the predominant structural protein exposed at the surface of the mature virion and subsequently is involved in viral attachment and entry as well as the most prominent target for the development of neutralizing immune responses. Between the E2 and E1 proteins lies the 6K transmembrane protein that serves to facilitate E1 protein trafficking in the ER (Melton *et al.*, 2002) as well as a domain that interacts with the E2 protein during the maturation process (Loewy *et al.*, 1995). A frame-shifting product, the transframe protein (TF) has been identified in a number of Old World alphaviruses (chikungunya, Sindbis and Semliki Forest viruses) to result from ribosomal translational slippage as at a conserved tract within the 6K protein. This ribosomal frameshifting occurs in less than 20% of translated polyproteins and results in the translation of an additional approximately 70 amino acid protein (Firth *et al.*, 2008; Snyder *et al.*, 2013) that, although not directly required for RNA replication and/or viral infectivity, has been associated with altered virion release and reduced neuroinvasive properties in murine models (Snyder *et al.*, 2013). The final structural protein, E1, serves to mediate fusion of the viral envelope with the host endosome. This fusion event is dictated by the exposure of E1 fusion domains during the maturation process as well as potentially through the *de novo* ion-permeable pore formation activity of the E1 protein that can actively reduce the pH of endosome, thus facilitating low-pH-mediated endosomal fusion (Wengler *et al.*, 2003).

Flaviviruses

The family *Flaviviridae* currently consists of four genera: *Flavivirus*, *Pestivirus*, *Hepacivirus*, and the newly proposed *Pegivirus* genus (Lindenbach *et al.*, 2013; Stapleton *et al.*, 2011). Although members of the family have a large host range that includes both vertebrates and invertebrates, only members of the genus *Flavivirus* are known as arboviruses, vectored either by mosquitos or ticks.

Flaviviral genomes are approximately 11 kb in length with a single open reading frame of approximately 3400 codons flanked by a type I cap (Cleaves and Dubin, 1979) and lacking a 3′ polyadenylate (pAn) tail (Wengler *et al.*, 1978). The cap (m^7GpppAmN) functions as a stabilizing structure for the viral RNA, as well as to initiate translation and quench the innate antiviral defence (Daffis *et al.*, 2010). As the viral polypeptide is synthesized, proteins are co- and post-translationally processed into three structural [capsid (C), membrane (prM/M) and envelope (E)] and seven non-structural (NS1, NS2A, NS2B, NS3, NS4A, NS4B and NS5) proteins. All of the non-structural proteins have been implicated in viral RNA synthesis or protein processing (Fig. 3.2) (Lindenbach *et al.*, 2013). The genomic RNA includes 5′ (ca 100 nt) and 3′ (ca. 400–800 nt) untranslated regions (UTRs) (Fig. 3.2). The 5′-UTR does not appear to be well conserved among different flaviviruses, albeit the presence of common structural elements, such as a bifurcating 5′-stem loop (5′-SL) (Fig. 3.2). The function of the 5′-SL is twofold: (i) initiation of viral genome translation; and (ii) initiation of RNA replication by association with the viral polymerase (Dong *et al.*, 2008; Filomatori *et al.*, 2011; Filomatori *et al.*, 2006). The 3′-UTR includes several conserved

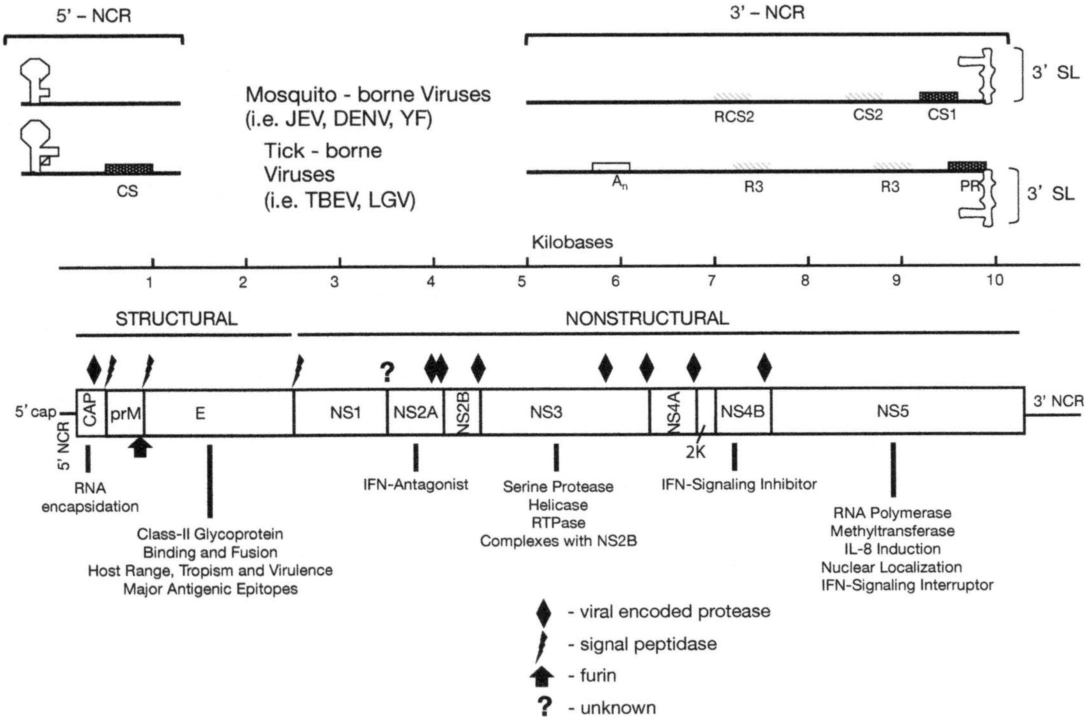

Figure 3.2 Flavivirus genome organization. Three structural proteins (C-prM-E) and seven non-structural proteins (NS1/NS1′, 2A, 2B, 3, 4A, 4B and 5) with an additional NS1′ as a frameshift extension of the NS1 are translated as a single polyprotein directly from the 11 kb positive-sense RNA genome. The single polyprotein is post translationally cleaved by viral and host-specific proteases to generate the aforementioned described viral proteins.

regions, structures and sequence duplications that play important roles in viral replication and translation and are shared between tick-borne and mosquito-borne flaviviruses, but their organization is not (Fig. 3.2). These include (i) a long (90–120 nt) nucleotide sequence that forms the stem–loop (3′-SL), which is the most common shared structural similarity shared among all flaviviruses (Brinton *et al.*, 1986; Hahn *et al.*, 1987; Lindenbach *et al.*, 2013). Within the 3′-SL are several required functional domains that influence virus–host interactions including virus replication (Elghonemy *et al.*, 2005; Yu and Markoff, 2005; Zeng *et al.*, 1998), translation (Chiu *et al.*, 2005; Holden and Harris, 2004; Holden *et al.*, 2006) and regulation of RNA synthesis (Tilgner *et al.*, 2005). Additionally, the 3′-SL is involved in interactions with viral, such as NS2A, NS3 and NS5 (Chen *et al.*, 1997; Cui *et al.*, 1998; Mackenzie *et al.*, 1998) and cellular proteins of functional importance, such as translation elongation factor 1A (Blackwell and Brinton, 1997; Davis *et al.*, 2007; De Nova-Ocampo *et al.*, 2002), La autoantigen (De Nova-Ocampo *et al.*, 2002; Garcia-Montalvo *et al.*, 2004; Vashist *et al.*, 2009; Vashist *et al.*, 2011; Yocupicio-Monroy *et al.*, 2007), polypyrimidine track binding protein (Anwar *et al.*, 2009; De Nova-Ocampo *et al.*, 2002; Kim and Jeong, 2006), Y box-binding protein 1 (Paranjape and Harris, 2007) as well as Mov34 protein (Ta and Vrati, 2000); and (ii) the well-conserved motifs CS1, CS2, RCS2 (conserved sequence 1, 2 and repeated conserved sequence, respectively), pseudoknot structures and a few tandem sequence repeats (R) that are found among mosquito-borne flaviviruses, which are located upstream of the 3′-SL (Gritsun and Gould, 2007; Hahn *et al.*, 1987; Proutski *et al.*, 1997). In tick-borne flaviviruses the PR, R3 and a polyadenylate sequence that is found in some tick-borne viruses, constitute the common conserved motifs of the 3′-NCR (Lindenbach *et al.*, 2013; Mandl *et al.*, 1993; Wallner *et al.*, 1995). Lastly, the 5′- and 3′-flaviviral termini include untranslated sequences important for viral RNA replication and translation, and interact with cellular factors involved in these functions. Specifically, genomes are circularized through base pairing between conserved sequence elements such as the terminal two plus-strand 3′-UTR nucleotides (UC-3′), which are complementary to the terminal 5′-UTR nucleotides (5′-AG) and assisted by other conserved sequence elements (Alvarez *et al.*, 2008; Khromykh *et al.*, 2001a; Kofler *et al.*, 2006; Markoff, 2003; Rice *et al.*, 1985; Wengler and Wengler, 1981). Circularization of the flavivirus genome facilitates initiation of replication by bringing into close proximity the polymerase complex to the 5′-SL and the 3′ site of initiation of replication (Filomatori *et al.*, 2006, 2011; Friebe *et al.*, 2011; Iglesias and Gamarnik, 2011; Khromykh *et al.*, 2001a; Kofler *et al.*, 2006).

Translation of the single ORF results in a single polyprotein that is cleaved by host and virus-derived proteases to produce the structural (capsid-premembrane/membrane-envelope; C-prM/M-E) and non-structural (NS1-NS2A-NS2B-NS3-NS4A-2K-NS4B-NS5) proteins (Fig. 3.2). The capsid protein (C), an 11 kDa homodimer protein with an unusually high net charge (Trent, 1977), is essential for RNA genome binding and encapsidation (Chang *et al.*, 2001; Khromykh and Westaway, 1996). Flaviviral C proteins demonstrate a large degree of flexibility where deletions are tolerated without loss of function (Kofler *et al.*, 2002, 2003; Patkar *et al.*, 2007; Schlick *et al.*, 2009). The membrane protein (prM/M), a 27–31 kDa N-glycosylated protein cleaved in the trans-Golgi network (TGN) by a cell-encoded furin-like protease during the late stages of virus assembly, to release mature virions (Kuhn *et al.*, 2002; Li *et al.*, 2008; Yu *et al.*, 2008) and assists in the proper folding of the envelope protein (Heinz *et al.*, 1994; Konishi and Mason, 1993; Kostyuchenko *et al.*, 2013; Lorenz *et al.*, 2002; Zhang *et al.*, 2003; Zhang *et al.*, 2013). The envelope (E) protein is a 53 kDa class II N-glycosylated dimeric membrane fusion protein that mediates virus binding and fusion to host cell membrane (Allison *et al.*, 1995; Bressanelli *et al.*, 2004; Johnson *et al.*, 1994; Modis *et al.*, 2004; Rey *et al.*, 1995), as well as confers protective immune responses by eliciting neutralizing, antifusion, and replication-enhancing antibodies (Chen *et al.*, 1996; Liao and Kielian, 2005; Nybakken *et al.*, 2005; Roehrig *et al.*, 1998; Rouvinski *et al.*, 2015). Each of the monomeric subunits of the E is composed of three distinct domains: (i) domain I, an eight-stranded β-barrel structure oriented parallel to the viral membrane; (ii) domain II, a finger-like structure composed of a pair of discontinuous loops, one of which is highly conserved among all flaviviruses, functioning as an internal fusion peptide and is stabilized by three disulfide bridges; and (iii) domain III, located in the outer lateral surface of the dimer, which interacts with cellular receptors for target cell entry (Beasley and Barrett, 2002; Chu *et al.*, 2005, 2007; Crill and Roehrig, 2001; Hung *et al.*, 1999), contains amino acid residues that are responsible for the determination of host range, tropism and virulence among flaviviruses (Beltramello *et al.*, 2010; Cecilia and Gould, 1991; McElroy *et al.*, 2006; Rey *et al.*, 1995; Wu *et al.*, 1997), as well as the most important epitopes that elicit a strong neutralizing immune response (Beltramello *et al.*, 2010; Chu *et al.*, 2005, 2007; Deng *et al.*, 2014; Edeling *et al.*, 2014; Lai *et al.*, 2007; Oliphant *et al.*, 2005; Shrestha *et al.*, 2010; Sukupolvi-Petty *et al.*, 2007; and reviewed in Pierson *et al.*, 2008).

Among the non-structural proteins, NS1 is a ~46 kDa dimeric N-glycosylated glycosyl-phosphatidylinositol (GPI) anchored protein that exists in both intra- and extracellular forms (Jacobs *et al.*, 2000; Lee *et al.*, 1989; Smith *et al.*, 1970; Wengler *et al.*, 1990; Winkler *et al.*, 1988, 1989). The intracellular form plays an important role in the replication of the genome (Mackenzie *et al.*, 1996; Muylaert *et al.*, 1996; Westaway *et al.*, 1997), probably through an interaction with NS4A (Lindenbach and Rice, 1999), whereas the extracellular form is highly antigenic and induces a strong humoral immune response (Amorim *et al.*, 2012; Chung *et al.*, 2006; Henriques *et al.*, 2013; Lin *et al.*, 1998). NS2A, is a ~22 kDa hydrophobic protein involved in the coordination of change between RNA packaging and replication (Khromykh *et al.*, 2001b) and inhibition of interferon (IFN) signalling (Jones *et al.*, 2005; Liu *et al.*, 2004, 2006; Munoz-Jordan *et al.*, 2003; Rodriguez-Madoz *et al.*, 2010). NS2A is also under weak selection pressures, a process that influences virus evolution during natural transmission (Bennett *et al.*, 2003; Vasilakis *et al.*, 2007; Zhang *et al.*, 2005). NS2B, is a membrane-associated ~14 kDa protein which associates with NS3 to form the viral protease complex and serves as a cofactor in the structural activation of the NS3 serine protease (Assenberg *et al.*, 2009; Clum *et al.*, 1997; Erbel *et al.*, 2006; Falgout *et al.*, 1991; Leung *et al.*, 2001). NS3, is a ~70 kDa multifunctional protein with trypsin-like serine protease (Bazan and Fletterick, 1989; Chambers *et al.*, 1990; Gorbalenya *et al.*, 1989a), helicase (Gorbalenya *et al.*, 1989a,b; Luo *et al.*, 2008;

Warrener *et al.*, 1993; Wu *et al.*, 2005; Xu *et al.*, 2005), and RNA triphosphatase (RTPase) enzyme activities (Bartelma and Padmanabhan, 2002; Benarroch *et al.*, 2004; Warrener *et al.*, 1993; Wengler and Wengler, 1993) and is involved in the processing of the viral polyprotein, as well as RNA replication. NS3 also may also play a role inducing apoptosis through activation of caspace-8 (Prikhod'ko *et al.*, 2002; Ramanathan *et al.*, 2006; Shafee and AbuBakar, 2003). NS4A and NS4B, are small hydrophobic proteins of ~16 and ~27 kDa respectively. NS4A plays a role in RNA replication through its cis-acting interactions with the replication complex and NS1 (Lindenbach and Rice, 1999; Mackenzie *et al.*, 1998; and reviewed in Lindenbach *et al.*, 2013). NS4B functions as an interferon (IFN)-signalling inhibitor (Ambrose and Mackenzie, 2011; Jones *et al.*, 2005; Munoz-Jordan *et al.*, 2003, 2005). Lastly, NS5 is a large multifunctional and well-conserved protein of 103 kDa with RNA cap-processing (encoded by its N-terminal region) (Egloff *et al.*, 2002; Issur *et al.*, 2009; Ray *et al.*, 2006), RNA-dependent RNA polymerase (RdRp) (encoded by its C-terminal region) (Ackermann and Padmanabhan, 2001; Guyatt *et al.*, 2001; Rice *et al.*, 1985; Tan *et al.*, 1996), interleukin 8 induction (IL-8) (Medin *et al.*, 2005) and nuclear localization (encoded by amino acid residues 387–404) (Kapoor *et al.*, 1995; Pryor *et al.*, 2007; Uchil *et al.*, 2006) and interruption of IFN-signalling (Best *et al.*, 2005) activities.

Rhabdoviruses

The family *Rhabdoviridae* currently consists of 11 genera: *Cytorhabdovirus*, *Ephemerovirus*, *Lyssavirus*, *Novirhabdovirus*, *Nucleorhabdovirus*, *Perhabdovirus*, *Sigmavirus*, *Sprivivirus*, *Tibrovirus*, *Tupavirus* and *Vesiculovirus* (Walker *et al.*, 2015). Recently a landmark study extended the range of the known genera to 17, including the *Almendravirus*, *Bahiavirus*, *Curiovirus*, *Hapavirus*, *Ledantevirus*, *Sawgravirus* and *Sripuvirus*, pending ratification by the International Committee on Taxonomy of Viruses (ICTV) (Walker *et al.*, 2015). Like members of the family *Flaviviridae*, members of this family have a large host range that include both vertebrates (terrestrial and aquatic), invertebrates and plants, and vector-borne rhabdoviruses are present in 12 of the 17 groups. Vector-borne rhabdoviruses are absent from genera associated with bats (*Lyssavirus*), flies (*Sigmavirus*) and fish (*Perhabdovirus*, *Sprivivirus*). The exception to this trend are the genera *Tupavirus* and *Almendravirus*, which comprise viruses that have not yet been associated with a vector species and vertebrate host, respectively, and for which little is known about their ecology or distribution (Vasilakis *et al.*, 2014; Walker *et al.*, 2015).

The negative polarity, single-stranded RNA rhabdovirus genome comprises of five structural proteins: the nucleoprotein (N), polymerase-associated phosphoprotein (P), matrix protein (M), glycoprotein (G) and the RNA-dependent RNA polymerase (L) (Dietzgen *et al.*, 2012) (Fig. 3.3). The rhabdoviral genomes feature partially complementary, untranslated leader (l) and trailer (t) sequences and the five ORFs are arranged in the descending order of 3′-N-P-M-G-L-5′ (Dietzgen *et al.*, 2012). Each of the ORFs is flanked by a conserved transcription initiation (TI) and transcription termination/polyadenylation (TTP) sequence, whose main functions are to coordinate and arrange the expression of the five corresponding capped and polyadenylated mRNAs (Dietzgen *et al.*, 2012). The rhabdovirus genomes may also contain additional ORFs encoding putative proteins mostly of unknown function (Fig. 3.3). These alternative or overlapping ORFs are located mostly

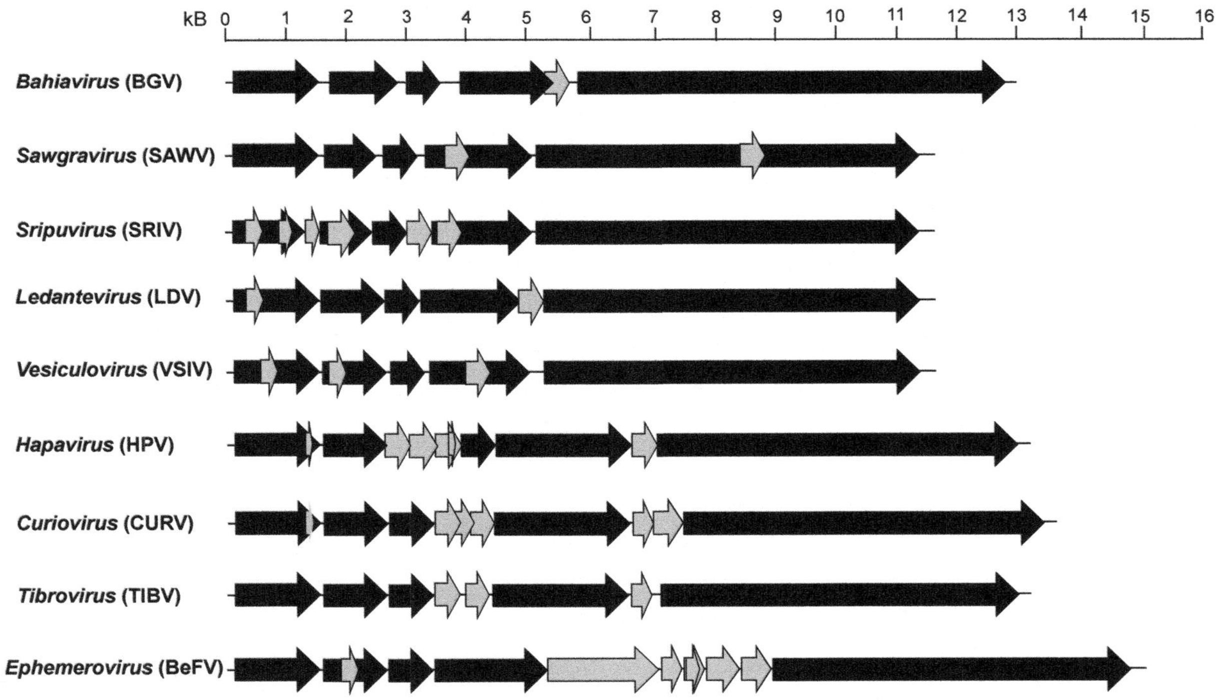

Figure 3.3 Rhabdoviral genome organization. Five structural genes 3′-N-P-M-G-L-5′ are transcribed from the viral template as a 5′-capped, 3′-polyadenylated and monocistronic mRNAs. Prototypic viruses (abbreviation) of each genus are grouped according to established and proposed genera. Common genes (N-P-M-G-L) are shaded in black and alternative or overlapping genes are shaded in light grey.

within the major structural genes or as independent ORFs flanked by TI or TTP sequences in the regions between the structural protein genes, some of which appear to have arisen by gene duplication (Allison *et al.*, 2014; Simon-Loriere and Holmes, 2013; Walker *et al.*, 1992; Wang and Walker, 1993 and reviewed in Walker *et al.*, 2011, 2015). Most of these ORFs are ≥ 180 nt in length, an arbitrary cut-off value selected on the basis that two small basic proteins of 55 and 65 amino acids (C and C¢) have been shown to be expressed from an alternative ORF within the VSIV P gene (Peluso *et al.*, 1996), and share no detectable protein sequence similarity with any other protein with those in public databases. Additionally, these ORFs are assigned names according to the following rules: (i) each has the designated description 'U' (unknown), followed by a number as the ORF appears in order in the genome (i.e. U1, U2, U3, etc); (ii) the first ORF within each transcriptional unit is assigned the same designation as the transcriptional unit itself; and (iii) each subsequent ORF within any transcriptional unit (e.g. alternative, overlapping or consecutive) is designated by letter (i.e. U1x, U1y, U1z). Alternative ORFs occur in a different frame within another longer ORF; overlapping ORFs are alternative ORFs which extend beyond the end of the primary ORF; and consecutive ORFs are those which do not overlap the primary ORF but follow consecutively within the same transcriptional unit (Walker *et al.*, 2011).

Bunyaviruses

The family *Bunyaviridae* includes more than 300 distinct members, with over 60 viruses associated with human illness, predominantly organized into four genera: *Hantavirus*, *Orthobunyavirus*, *Phlebovirus* and *Nairovirus*, according to structural, genetic and antigenic characteristics (Barrett and Shope, 1995; Nichol *et al.*, 2005). Exemplifying the great diversity of species classified within the family, the fifth genus of the family *Bunyaviridae*, the genus *Tospovirus*, contains viruses that are known to cause disease in plants, not animals (Nichol *et al.*, 2005). Further illustrating this extraordinary diversity, the taxonomical hierarchy of *Bunyaviridae* is complex, with members of each genus additionally classified into individual groups, subtypes and complexes according to serological characteristics (Barrett and Shope, 1995).

The bunyavirus particle contains a tripartite genome of mostly negative polarity. The three genomic segments, L, M and S, can occur in end-hydrogen bonded circularized forms and generally encode an RNA dependent RNA polymerase (L), envelope glycoproteins (Gn and Gc) and a nucleocapsid (N) protein, respectively. Each genomic segment is complexed within a ribonucleocapsid that contains both an abundance of N and a minority of L-proteins. A unique property of the bunyavirus genome is the highly conserved, complementary nature of the 5′ and 3′ termini of each genomic segment, which facilitates circularization within the ribonucleocapsid. The organization of the coding regions within each genomic segment varies by genus, demonstrating a molecular basis for the extraordinary diversity of species described within the family (Fig. 3.4) (Elliott, 1990).

The critical functions of the key proteins encoded by the bunyavirus genome are well illustrated within the generalized replication strategy of *Bunyaviridae*, which is entirely cytoplasmic and begins with the interaction of virus surface glycoproteins with host cell receptors. Both Gn and Gc proteins have been implicated in host cell attachment (Arikawa *et al.*, 1989; Keegan and Collett, 1986). However, Gc appears to be the primary attachment protein for members of the genus *Orthobunyavirus* (Plassmeyer *et al.*, 2005). After attachment, uptake into the cell is driven by receptor-mediated endocytosis. Acidification of endocytic vesicles is thought to cause conformational changes in Gn and/or Gc, allowing fusion of viral and cellular membranes (Gonzalez-Scarano *et al.*, 1984),

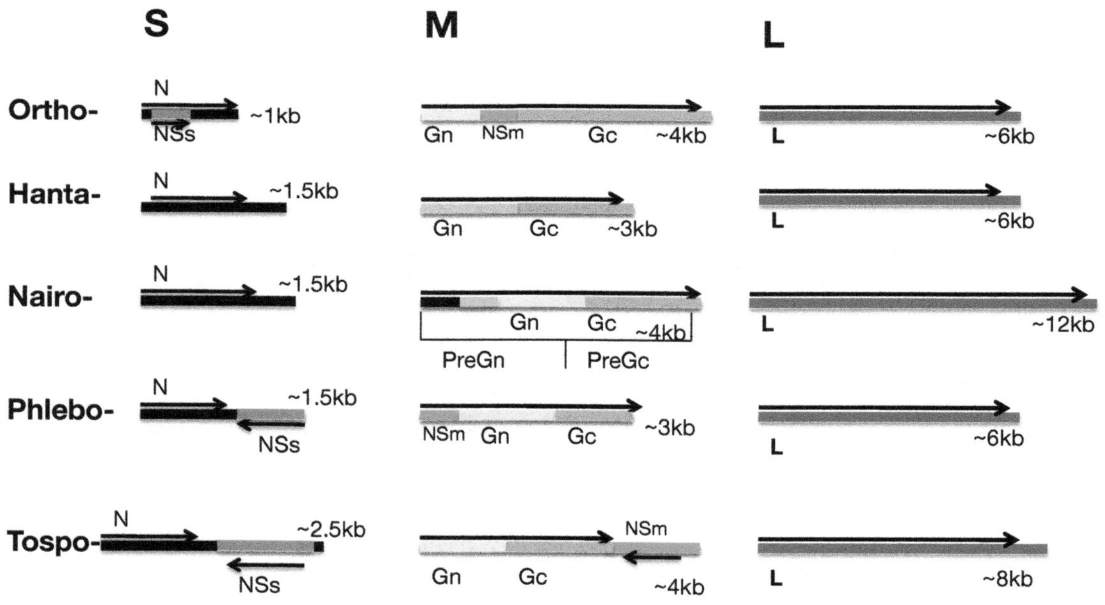

Figure 3.4 Generalized coding strategies of the family *Bunyaviridae*. Orthobunyavirus, hantavirus, nairovirus, phlebovirus and tospovirus vcRNAs are depicted 5′–3′. ORF directionality is indicated by arrows and the relative sizes of ORFs and RNA segments are not drawn to scale. The N and NSs overlapping ORFs of the S segment of the genus *Orthobunyavirus* and the ambisense nature of the phlebovirus and tospovirus S and M segments clearly represent the molecular diversity of the family *Bunyaviridae*.

thus facilitating the release of the ribonucleocapsid into the cytoplasm. Primary transcription of mRNA is primed by 'cap snatching' from cytoplasmic host cellular mRNAs through endonucleolytic activity of the L-protein (Patterson *et al.*, 1984) and is facilitated by the N protein (Mir and Panganiban, 2006). The L-protein is responsible for both primary transcription of mRNAs and genomic replication via a positive sense intermediate (Jin and Elliott, 1993). At some point in the replication cycle, L-protein activity switches from primed, primary transcription to unprimed genomic replication through an unknown mechanism. For the M segment, translation of polypeptides occurs via ER bound ribosomes and the nascent polypeptide is co-translationally cleaved to generate Gn and Gc, which are dimerized within the ER (Schmaljohn and Nichol, 2007). Translation of L and S segment encoded proteins occurs on free ribosomes in the cytoplasm (Schmaljohn and Nichol, 2007). Virus assembly and maturation occurs in the ER and Golgi membranes (Lyons and Heyduk, 1973; Novoa *et al.*, 2005) and release of mature virions occurs by either transport through migration of Golgi vesicles and fusion of vesicular and plasma membranes (Bishop, 1996; Smith and Pifat, 1982) or alternatively, via direct budding at the plasma membrane (Ravkov *et al.*, 1997).

While not directly implicated within the generalized replication strategy, non-structural proteins, NSs and NSm that are encoded within the *Orthobunyavirus*, *Phlebovirus* and *Tospovirus* genomes (Fig. 3.4) are of growing research interest. Of these proteins, NSs functions have been most well characterized in orthobunyaviruses as contributing to the shutdown of mammalian, but not mosquito host cell protein synthesis (Hart *et al.*, 2009; Thomas *et al.*, 2004; Weber *et al.*, 2002). In addition, ortho- and phlebovirus NSs proteins have well documented anti-IFN activity in mammalian systems (Blakqori and Weber, 2005; Habjan *et al.*, 2009; Weber *et al.*, 2002). As such, NSs is thought to have a role in potentiating the zoonotic capacity of orthobunyaviruses (Hart *et al.*, 2009) and phleboviruses by allowing these viruses to overcome vertebrate host innate immune responses.

The extraordinary diversity of the family *Bunyaviridae* harkens of both an ancient origin and an entirely unique evolutionary potential among all other known virus families. Indeed, for the arthropod-borne animal viruses of the family, the demanding ecology of their transmission cycle requires enormous plasticity for virus survival and propagation in nature. Presumably potentiating this capability, in addition to evolution through genetic drift, the segmented nature of the bunyavirus genome allows for the possibility of evolution through the reassortment of genomic segments between heterologous viruses. In fact, co-infection with more than one bunyavirus resulting in the reassortment of genomic segments has been well established in both laboratory and natural settings (Beaty *et al.*, 1985, 1997; Bishop and Beaty, 1988; Borucki *et al.*, 1999; Briese *et al.*, 2007; Cheng *et al.*, 1999; Li *et al.*, 1995; Nunes *et al.*, 2005; Yanase *et al.*, 2006). Of particular importance to analyses of the arthropod-borne genera, mosquitoes have been shown to be effective reservoirs for the reassortment of genomic segments between heterologous bunyaviruses of shared serological character (Beaty *et al.*, 1985, 1997; Borucki *et al.*, 1999). In addition, bunyavirus segment reassortment has been associated with outbreaks of human disease (Bowen *et al.*, 2001; Briese *et al.*, 2006; Gerrard *et al.*, 2004). Despite this significant documentation, the role of segment reassortment in bunyavirus evolution and pathogenicity is largely unknown owing to the previously described lack of comprehensive sequence data for members of the family *Bunyaviridae* (Gerrard *et al.*, 2004).

Reoviruses

The family *Reoviridae* is complex in that it includes a large number of viruses assigned to 15 genera (*Orthoreovirus*, *Aquareovirus*, *Oryzavirus*, *Fijivirus*, *Mycoreovirus*, *Cypovirus*, *Idnoreovirus*, *Dinovernavirus*, *Coltivirus*, *Orbivirus*, *Rotavirus*, *Seadornavirus*, *Phytovirus*, *Cardoreovirus*, *Mimoreovirus*), which are further subdivided among two subfamilies (*Spinareovirinae* and *Sedoreovirinae*) depending on virion morphology; specifically, the presence or absence of spikes (turrets) on the virion surface and the number (either 2 or 3) of protein shells (King *et al.*, 2012). All members of the family have virus particles with icosahedral symmetry and segmented genomes of linear dsRNA, however, the number of genome segments and their coding assignments differ markedly among the viruses included within the various genera of the family.

Individual viruses within the family *Reoviridae* infect a wide variety of hosts including mammals, birds, reptiles, amphibians, fish, molluscs, crustaceans, insects, plants and fungi (King *et al.*, 2012; MacLachlan and Dubovi, 2011). The biological properties of the viruses contained within the family vary remarkably between, and even within, genera. Viruses included within the genera *Orbivirus*, *Coltivirus* and *Seadornavirus* replicate in both vertebrates as well as insect vectors, specifically ticks, mosquitoes, sandflies, and *Culicoides* biting midges depending on the specific virus. Recent findings indicate that co-evolution of orbiviruses with their respective arthropod vectors results in the unique features of individual virus species (Mohd Jaafar *et al.*, 2014).

This review will focus on the replication of orbiviruses as this has been especially well characterized with bluetongue virus (BTV), which is the type species of the genus (reviewed: (Patel and Roy, 2014)). Orbiviruses are included in the subfamily *Sedoreovirinae* because their virions have a relatively featureless outer capsid. Individual orbiviruses are transmitted to their mammalian hosts by different haematophagous insects, but BTV and other pathogenic orbiviruses such as African horse sickness virus, epizootic haemorrhagic disease virus and equine encephalosis virus are all transmitted by *Culicoides* biting midges.

The orbivirus virion is non-enveloped and includes seven structural viral proteins (VP1–7), of which four (VP2, VP3, VP5, VP7) are major structural determinants whereas the three others (VP1, VP4, VP6) constitute the transcriptase complex and are included in relatively low concentrations in virions. The BTV virion consists of an outer capsid of VP2 and VP5 and a double-layered core particle of VP3 and 7 that encapsidates the transcriptase complex and the 10 dsRNA genome segments. In ruminants, BTV is tropic to vascular endothelial cells and to various mononuclear inflammatory cells, notably dendritic cells, monocyte/macrophages, and lymphocytes (Maclachlan *et al.*, 2009). Virus replication has been characterized in BTV-infected cell cultures grown *in vitro* (Patel and Roy, 2014). Virus replication follows the typical process of attachment and

Table 3.2 Coding assignments of the individual dsRNA genome segments of BTV (from Patel and Roy, 2014)

Gene segment (length)	Protein(s)	Functions
1 (3944 nt)	VP1	RNA-dependent RNA polymerase
2 (2926 nt)	VP2	Outer capsid protein, receptor binding, serotype determination and induction of neutralizing antibodies (Maclachlan *et al.*, 2014)
3 (2772 nt)	VP3	Core particle structural protein
4 (2011 nt)	VP4	Capping enzyme
5 (1769 nt)	NS1	Cytoplasmic tubules that promote translation
6 (1638 nt)	VP5	Outer capsid protein that conformationally interacts with VP2, membrane permeabilization
7 (1156 nt)	VP7	Core particle structural protein, group specific antigen
8 (1124 nt)	NS2	Cytoplasmic inclusion bodies
9 (1046 nt)	VP6 NS4	Helicase Host–virus interactions, possible interferon modulation (Ratinier *et al.*, 2011)
10 (822 nt)	NS3/3A	Virus egress from cells, antagonist of type I interferon production (Vitour *et al.*, 2014)

entry, replication, assembly and egress. Translation of the 10 genome segments leads to the production of the seven structural VPs as well as five non-structural (NS) proteins (NS1, NS2, NS3/3A, NS4). Transcription of the genome dsRNA leads to the production from the replicase complexes included in virus core particles of 10 positive-sense ssRNAs that range in size from 833 (gene segment 10) to 3944 (gene segment 1) base pairs (Table 3.2). These ssRNA are released from the core particle into the cell cytoplasm where they serve as templates both for translation as well as transcription to form new copies of genomic dsRNA. Eight of the ssRNAs are monocistronic whereas the two smallest are bi-cistronic. Functions of the individual proteins are listed in Table 3.2.

The genus *Coltivirus* includes Colorado tick fever and Eyach viruses that are transmitted by ticks to humans. Although the virion organization of coltiviruses is generally similar to that of BTV, coltiviruses are included in the subfamily *Spinareovirinae* because of the morphological appearance of their virions. Furthermore, the genome of coltiviruses includes 12 segments of dsRNA that range in size from 675 to 4350 nt.

Orthomyxoviruses

The family *Orthomyxoviridae* includes viruses that possess negative-sense, single-stranded, segmented RNA genomes of a varied number from six to eight segments. There are six genera within *Orthomyxoviridae*, including the influenzaviruses A, B, and C in addition to the *Isavirus*, *Thogotovirus* and *Quaranjavirus* genera. Much of what is known about the genomics and molecular biology of orthomyxoviruses has been determined through the relatively thorough investigation of influenzaviruses. Here, we briefly introduce what is known of the genomic organization of the *Thogotovirus* and *Quaranjavirus* genera, two comparatively small and less well-described tick-associated genera of *Orthomyxoviridae*.

Named after the type species, the genus *Thogotovirus* comprises at least six distinct viruses that have been associated with both hard and soft ticks and are global in their distribution. Historically, only two members of the genus, Thogoto and Dhori viruses have been associated with human illness. Thogotoviruses possess six genomic segments that encode seven structural proteins (Hagmaier *et al.*, 2004; Weber *et al.*, 1998) and have a generalized genome size of approximately 12 kb. The three largest genomic segments, RNA segments 1, 2, and 3 encode the canonical polymerase subunits PB2, PB1 and PA, respectively that are common to all viruses of the *Orthomyxoviridae*. Together, these proteins interact to form the RNA-dependent RNA polymerase complex (Weber *et al.*, 1998) that is foundational to replication. Unlike the influenzaviruses, thogotoviruses possess only a single surface glycoprotein, G that is encoded within RNA segment 4 and is distantly similar to the baculovirus glycoprotein GP64 (Morse *et al.*, 1992). RNA segment 5 encodes the nucleoprotein, NP (Weber *et al.*, 1996). Lastly, RNA segment 6 encodes a matrix protein (M) of 29.8 kDa and an interferon antagonist (ML) of 34.4 kDa (Hagmaier *et al.*, 2003) that is virion associated. The smaller of the two proteins (M) is generated from an alternatively spliced mRNA transcript (Kochs, 2000).

As part of more comprehensive investigations, genomic sequencing of three viruses, Lake Chad, Johnston Atoll and Quaranfil viruses led to the proposed establishment of the genus *Quaranjavirus*, a tick-associated genus of distant relation to the thogotoviruses (Presti *et al.*, 2009). Viruses of the genus are distributed throughout regions of Africa, the Middle East and the Pacific. Quaranfil virus has been associated with a mild, febrile illness in children and possesses a genome that is approximately 11.5 kb in size and comprises six known RNA segments (Presti *et al.*, 2009). As with thogotoviruses, Quaranfil virus RNA segments 1, 2 and 3 encode open reading frames that possess homology with influenza polymerase subunit proteins PB2, PA and PB1, respectively (Presti *et al.*, 2009). Segments 4 and 6 encode proteins of unknown function with no significant homology with other orthomyxoviral proteins (Presti *et al.*, 2009). Segment 5 encodes a putative glycoprotein with distant homology to thogotovirus and baculovirus glycoproteins (Presti *et al.*, 2009).

Acknowledgement

This work was supported in part by NIH contract HHSN272201000040I/HHSN27200004/D04 (NV).

References

Ackermann, M., and Padmanabhan, R. (2001). De novo synthesis of RNA by the dengue virus RNA-dependent RNA polymerase exhibits temperature dependence at the initiation but not elongation phase. J. Biol. Chem. 276, 39926–39937.

Allison, A.B., Mead, D.G., Palacios, G.F., Tesh, R.B., and Holmes, E.C. (2014). Gene duplication and phylogeography of North American members of the Hart Park serogroup of avian rhabdoviruses. Virology *448*, 284–292.

Allison, S.L., Schalich, J., Stiasny, K., Mandl, C.W., Kunz, C., and Heinz, F.X. (1995). Oligomeric rearrangement of tick-borne encephalitis virus envelope proteins induced by an acidic pH. J. Virol. *69*, 695–700.

Alvarez, D.E., Filomatori, C.V., and Gamarnik, A.V. (2008). Functional analysis of dengue virus cyclization sequences located at the 5′ and 3′ UTRs. Virology *375*, 223–235.

Ambrose, R.L., and Mackenzie, J.M. (2011). West Nile virus differentially modulates the unfolded protein response to facilitate replication and immune evasion. J. Virol. *85*, 2723–2732.

Amorim, J.H., Diniz, M.O., Cariri, F.A., Rodrigues, J.F., Bizerra, R.S., Goncalves, A.J., de Barcelos Alves, A.M., and de Souza Ferreira, L.C. (2012). Protective immunity to DENV2 after immunization with a recombinant NS1 protein using a genetically detoxified heat-labile toxin as an adjuvant. Vaccine *30*, 837–845.

Anwar, A., Leong, K.M., Ng, M.L., Chu, J.J., and Garcia-Blanco, M.A. (2009). The polypyrimidine tract-binding protein is required for efficient dengue virus propagation and associates with the viral replication machinery. J. Biol. Chem. *284*, 17021–17029.

Arikawa, J., Schmaljohn, A.L., Dalrymple, J.M., and Schmaljohn, C.S. (1989). Characterization of Hantaan virus envelope glycoprotein antigenic determinants defined by monoclonal antibodies. J. Gen. Virol. *70*, 615–624.

Assenberg, R., Mastrangelo, E., Walter, T.S., Verma, A., Milani, M., Owens, R.J., Stuart, D.I., Grimes, J.M., and Mancini, E.J. (2009). Crystal structure of a novel conformational state of the flavivirus NS3 protein: implications for polyprotein processing and viral replication. J. Virol. *83*, 12895–12906.

Attoui, H., Mertens, P.P.C., Becnel, J., Belaganahalli, S., Bergoin, M., Brussaard, C.P., Chappell, J.D., Ciarlet, M., del Vas, M., Dermody, T.S., *et al.* (2012). The Reoviridae. In Virus Taxonomy, Classification and Nomenclature of Viruses: 9th Report of the International Committee on Taxonomy of Viruses, King, A.M., Adams, M.J., Carstens, E.B., and Lefkowitz, E.J., eds. (Elsevier Academic Press, London, UK), pp. 541–637.

Barrett, A.D., and Shope, R.E. (1995). Bunyaviridae. In Topley and Wilson's Microbiology and Microbial Infections, B.W.J. Mahy, and J.H. Ter Meulen, eds. (ASM Press, Washington, DC), pp. 1025–1058.

Bartelma, G., and Padmanabhan, R. (2002). Expression, purification, and characterization of the RNA 5′-triphosphatase activity of dengue virus type 2 non-structural protein 3. Virology *299*, 122–132.

Bazan, J.F., and Fletterick, R.J. (1989). Detection of a trypsin-like serine protease domain in flaviviruses and pestiviruses. Virology *171*, 637–639.

Beasley, D.W., and Barrett, A.D. (2002). Identification of neutralizing epitopes within structural domain III of the West Nile virus envelope protein. J. Virol. *76*, 13097–13100.

Beaty, B.J., Sundin, D.R., Chandler, L.J., and Bishop, D.H. (1985). Evolution of bunyaviruses by genome reassortment in dually infected mosquitoes (*Aedes triseriatus*). Science *230*, 548–550.

Beaty, B.J., Borucki, M.K., Farfan Ale, J.A., and White, D. (1997). Arbovirus–vector interactions: determinants of arbovirus evolution. In Factors in the Emergence of Arbovirus Diseases, J.E. Saluzzo, and B. Dodet, eds. (Elsevier, Paris), pp. 23–35.

Beltramello, M., Williams, K.L., Simmons, C.P., Macagno, A., Simonelli, L., Quyen, N.T., Sukupolvi-Petty, S., Navarro-Sanchez, E., Young, P.R., de Silva, A.M., *et al.* (2010). The human immune response to Dengue virus is dominated by highly cross-reactive antibodies endowed with neutralizing and enhancing activity. Cell Host Microbe *8*, 271–283.

Benarroch, D., Selisko, B., Locatelli, G.A., Maga, G., Romette, J.L., and Canard, B. (2004). The RNA helicase, nucleotide 5′-triphosphatase, and RNA 5′-triphosphatase activities of Dengue virus protein NS3 are Mg2+-dependent and require a functional Walker B motif in the helicase catalytic core. Virology *328*, 208–218.

Bennett, S.N., Holmes, E.C., Chirivella, M., Rodriguez, D.M., Beltran, M., Vorndam, V., Gubler, D.J., and McMillan, W.O. (2003). Selection-driven evolution of emergent dengue virus. Mol. Biol. Evol. *20*, 1650–1658.

Best, S.M., Morris, K.L., Shannon, J.G., Robertson, S.J., Mitzel, D.N., Park, G.S., Boer, E., Wolfinbarger, J.B., and Bloom, M.E. (2005). Inhibition of interferon-stimulated JAK-STAT signaling by a tick-borne flavivirus and identification of NS5 as an interferon antagonist. J. Virol. *79*, 12828–12839.

Bishop, D.H. (1996). Biology and molecular biology of bunyaviruses. In The Bunyaviridae, R.H. Elliot, ed. (Plenum Press, London), pp. 19–53.

Bishop, D.H., and Beaty, B.J. (1988). Molecular and biochemical studies of the evolution, infection and transmission of insect bunyaviruses. Phil. Trans. R. Soc. B *321*, 463–483.

Blackwell, J.L., and Brinton, M.A. (1997). Translation elongation factor-1 alpha interacts with the 3′ stem–loop region of West Nile virus genomic RNA. J. Virol. *71*, 6433–6444.

Blakqori, G., and Weber, F. (2005). Efficient cDNA-based rescue of La Crosse bunyaviruses expressing or lacking the non-structural protein NSs. J. Virol. *79*, 10420–10428.

Borucki, M.K., Chandler, L.J., Parker, B.M., Blair, C.D., and Beaty, B.J. (1999). Bunyavirus superinfection and segment reassortment in transovarially infected mosquitoes. J. Gen. Virol. *80*, 3173–3179.

Bowen, M.D., Trappier, S.G., Sanchez, A.J., Meyer, R.F., Goldsmith, C.S., Zaki, S.R., Dunster, L.M., Peters, C.J., Ksiazek, T.G., and Nichol, S.T. (2001). A reassortant bunyavirus isolated from acute hemorrhagic fever cases in Kenya and Somalia. Virology *291*, 185–190.

Bressanelli, S., Stiasny, K., Allison, S.L., Stura, E.A., Duquerroy, S., Lescar, J., Heinz, F.X., and Rey, F.A. (2004). Structure of a flavivirus envelope glycoprotein in its low-pH-induced membrane fusion conformation. EMBO J. *23*, 728–738.

Briese, T., Bird, B., Kapoor, V., Nichol, S.T., and Lipkin, W.I. (2006). Batai and Ngari viruses: M segment reassortment and association with severe febrile disease outbreaks in East Africa. J. Virol. *80*, 5627–5630.

Briese, T., Kapoor, V., and Lipkin, W.I. (2007). Natural M-segment reassortment in Potosi and Main Drain viruses: implications for the evolution of orthobunyaviruses. Arch. Virol. *152*, 2237–2247.

Brinton, M.A., Fernandez, A.V., and Dispoto, J.H. (1986). The 3′-nucleotides of flavivirus genomic RNA form a conserved secondary structure. Virology *153*, 113–121.

Cecilia, D., and Gould, E.A. (1991). Nucleotide changes responsible for loss of neuroinvasiveness in Japanese encephalitis virus neutralization-resistant mutants. Virology *181*, 70–77.

Chambers, T.J., Weir, R.C., Grakoui, A., McCourt, D.W., Bazan, J.F., Fletterick, R.J., and Rice, C.M. (1990). Evidence that the N-terminal domain of non-structural protein NS3 from yellow fever virus is a serine protease responsible for site-specific cleavages in the viral polyprotein. Proc. Natl. Acad. Sci. U.S.A. *87*, 8898–8902.

Chang, C.J., Luh, H.W., Wang, S.H., Lin, H.J., Lee, S.C., and Hu, S.T. (2001). The heterogeneous nuclear ribonucleoprotein K (hnRNP K) interacts with dengue virus core protein. DNA Cell Biol. *20*, 569–577.

Chen, C.J., Kuo, M.D., Chien, L.J., Hsu, S.L., Wang, Y.M., and Lin, J.H. (1997). RNA–protein interactions: involvement of NS3, NS5, and 3′ noncoding regions of Japanese encephalitis virus genomic RNA. J. Virol. *71*, 3466–3473.

Chen, Y., Maguire, T., and Marks, R.M. (1996). Demonstration of binding of dengue virus envelope protein to target cells. J. Virol. *70*, 8765–8772.

Cheng, L.L., Rodas, J.D., Schultz, K.T., Christensen, B.M., Yuill, T.M., and Israel, B.A. (1999). Potential for evolution of California serogroup bunyaviruses by genome reassortment in *Aedes albopictus*. Am. J. Trop. Med. Hyg. *60*, 430–438.

Chiu, W.W., Kinney, R.M., and Dreher, T.W. (2005). Control of translation by the 5′- and 3′-terminal regions of the dengue virus genome. J. Virol. *79*, 8303–8315.

Choi, H.K., Lee, S., Zhang, Y.P., McKinney, B.R., Wengler, G., Rossmann, M.G., and Kuhn, R.J. (1996). Structural analysis of Sindbis virus capsid mutants involving assembly and catalysis. J. Mol. Biol. *262*, 151–167.

Chu, J.H., Chiang, C.C., and Ng, M.L. (2007). Immunization of flavivirus West Nile recombinant envelope domain III protein induced specific immune response and protection against West Nile virus infection. J. Immunol. *178*, 2699–2705.

Chu, J.J., Rajamanonmani, R., Li, J., Bhuvanakantham, R., Lescar, J., and Ng, M.L. (2005). Inhibition of West Nile virus entry by using a recombinant domain III from the envelope glycoprotein. J. Gen. Virol. *86*, 405–412.

Chung, K.M., Nybakken, G.E., Thompson, B.S., Engle, M.J., Marri, A., Fremont, D.H., and Diamond, M.S. (2006). Antibodies against West Nile Virus non-structural protein NS1 prevent lethal infection through Fc gamma receptor-dependent and -independent mechanisms. J. Virol. *80*, 1340–1351.

Cleaves, G.R., and Dubin, D.T. (1979). Methylation status of intracellular dengue type 2 40 S RNA. Virology 96, 159–165.

Clum, S., Ebner, K.E., and Padmanabhan, R. (1997). Cotranslational membrane insertion of the serine proteinase precursor NS2B-NS3(Pro) of dengue virus type 2 is required for efficient in vitro processing and is mediated through the hydrophobic regions of NS2B. J. Biol. Chem. 272, 30715–30723.

Crill, W.D., and Roehrig, J.T. (2001). Monoclonal antibodies that bind to domain III of dengue virus E glycoprotein are the most efficient blockers of virus adsorption to Vero cells. J. Virology 75, 7769–7773.

Cui, T., Sugrue, R.J., Xu, Q., Lee, A.K., Chan, Y.C., and Fu, J. (1998). Recombinant dengue virus type 1 NS3 protein exhibits specific viral RNA binding and NTPase activity regulated by the NS5 protein. Virology 246, 409–417.

Daffis, S., Szretter, K.J., Schriewer, J., Li, J., Youn, S., Errett, J., Lin, T.Y., Schneller, S., Zust, R., Dong, H., *et al.* (2010). 2′-O methylation of the viral mRNA cap evades host restriction by IFIT family members. Nature 468, 452–456.

Davis, W.G., Blackwell, J.L., Shi, P.Y., and Brinton, M.A. (2007). Interaction between the cellular protein eEF1A and the 3′-terminal stem–loop of West Nile virus genomic RNA facilitates viral minus-strand RNA synthesis. J. Virology 81, 10172–10187.

De Nova-Ocampo, M., Villegas-Sepulveda, N., and del Angel, R.M. (2002). Translation elongation factor-1alpha, La, and PTB interact with the 3′ untranslated region of dengue 4 virus RNA. Virology 295, 337–347.

Deng, W.L., Guan, C.Y., Liu, K., Zhang, X.M., Feng, X.L., Zhou, B., Su, X.D., and Chen, P.Y. (2014). Fine mapping of a linear epitope on EDIII of Japanese encephalitis virus using a novel neutralizing monoclonal antibody. Virus Res. 179, 133–139.

Dietzgen, R.G., Calisher, C.H., Kurath, G., Kuzman, I.V., and Rodriquez, L.L. (2012). Rhabdoviridae. In Virus Taxonomy, Ninth Report of the International Committee on Taxonomy of Viruses, A.M.Q. King, M.J. Adams, E.B. Carstens, E.J. Lefkowitz, D.M. Stone, R.B. Tesh, N. Tordo, P.J. Walker, T. Wetzel, and A.E. Whitfield, eds. (Elsevier, San Diego).

Dong, H., Zhang, B., and Shi, P.Y. (2008). Terminal structures of West Nile virus genomic RNA and their interactions with viral NS5 protein. Virology 381, 123–135.

Edeling, M.A., Austin, S.K., Shrestha, B., Dowd, K.A., Mukherjee, S., Nelson, C.A., Johnson, S., Mabila, M.N., Christian, E.A., Rucker, J., *et al.* (2014). Potent dengue virus neutralization by a therapeutic antibody with low monovalent affinity requires bivalent engagement. PLoS Pathogens 10, e1004072.

Egloff, M.P., Benarroch, D., Selisko, B., Romette, J.L., and Canard, B. (2002). An RNA cap (nucleoside-2′-O-)-methyltransferase in the flavivirus RNA polymerase NS5: crystal structure and functional characterization. EMBO J. 21, 2757–2768.

Elghonemy, S., Davis, W.G., and Brinton, M.A. (2005). The majority of the nucleotides in the top loop of the genomic 3′ terminal stem loop structure are cis-acting in a West Nile virus infectious clone. Virology 331, 238–246.

Elliott, R.M. (1990). Molecular biology of the Bunyaviridae. J. Gen. Virol. 71, 501–522.

Erbel, P., Schiering, N., D'Arcy, A., Renatus, M., Kroemer, M., Lim, S.P., Yin, Z., Keller, T.H., Vasudevan, S.G., and Hommel, U. (2006). Structural basis for the activation of flaviviral NS3 proteases from dengue and West Nile virus. Nature Struct. Mol. Biol. 13, 372–373.

Falgout, B., Pethel, M., Zhang, Y.M., and Lai, C.J. (1991). Both non-structural proteins NS2B and NS3 are required for the proteolytic processing of dengue virus non-structural proteins. J. Virol. 65, 2467–2475.

Filomatori, C.V., Lodeiro, M.F., Alvarez, D.E., Samsa, M.M., Pietrasanta, L., and Gamarnik, A.V. (2006). A 5′ RNA element promotes dengue virus RNA synthesis on a circular genome. Genes Dev. 20, 2238–2249.

Filomatori, C.V., Iglesias, N.G., Villordo, S.M., Alvarez, D.E., and Gamarnik, A.V. (2011). RNA sequences and structures required for the recruitment and activity of the dengue virus polymerase. J. Biol. Chem. 286, 6929–6939.

Firth, A.E., Chung, B.Y., Fleeton, M.N., and Atkins, J.F. (2008). Discovery of frameshifting in Alphavirus 6K resolves a 20-year enigma. Virology J. 5, 108.

Foy, N.J., Akhrymuk, M., Akhrymuk, I., Atasheva, S., Bopda-Waffo, A., Frolov, I., and Frolova, E.I. (2012). Hypervariable domains of nsP3 proteins of the New World and the Old World alphaviruses mediate formation of distinct, virus-specific protein complexes. J. Virol. 87, 1997–2010.

Friebe, P., Shi, P.Y., and Harris, E. (2011). The 5′ and 3′ downstream AUG region elements are required for mosquito-borne flavivirus RNA replication. J. Virol. 85, 1900–1905.

Frolova, E., Frolov, I., and Schlesinger, S. (1997). Packaging signals in alphaviruses. J. Virol. 71, 248–258.

Fros, J.J., Liu, W.J., Prow, N.A., Geertsema, C., Ligtenberg, M., Vanlandingham, D.L., Schnettler, E., Vlak, J.M., Suhrbier, A., Khromykh, A.A., *et al.* (2010). Chikungunya virus non-structural protein 2 inhibits type I/II interferon-stimulated JAK-STAT signaling. J. Virol. 84, 10877–10887.

Garcia-Montalvo, B.M., Medina, F., and del Angel, R.M. (2004). La protein binds to NS5 and NS3 and to the 5′ and 3′ ends of Dengue 4 virus RNA. Virus Res. 102, 141–150.

Garoff, H., Huylebroeck, D., Robinson, A., Tillman, U., and Liljestrom, P. (1990). The signal sequence of the p62 protein of Semliki Forest virus is involved in initiation but not in completing chain translocation. J. Cell Biol. 111, 867–876.

Gerrard, S.R., Li, L., Barrett, A.D., and Nichol, S.T. (2004). Ngari virus is a *Bunyamwera* virus reassortant that can be associated with large outbreaks of hemorrhagic fever in Africa. J. Virol. 78, 8922–8926.

Gomez de Cedron, M., Ehsani, N., Mikkola, M.L., Garcia, J.A., and Kaariainen, L. (1999). RNA helicase activity of Semliki Forest virus replicase protein NSP2. FEBS Lett. 448, 19–22.

Gonzalez-Scarano, F., Pobjecky, N., and Nathanson, N. (1984). La Crosse bunyavirus can mediate pH-dependent fusion from without. Virology 132, 222–225.

Gorbalenya, A.E., Donchenko, A.P., Koonin, E.V., and Blinov, V.M. (1989a). N-terminal domains of putative helicases of flavi- and pestiviruses may be serine proteases. Nucleic Acids Res. 17, 3889–3897.

Gorbalenya, A.E., Koonin, E.V., Donchenko, A.P., and Blinov, V.M. (1989b). Two related superfamilies of putative helicases involved in replication, recombination, repair and expression of DNA and RNA genomes. Nucleic Acids Res. 17, 4713–4730.

Gritsun, T.S., and Gould, E.A. (2007). Origin and evolution of 3′UTR of flaviviruses: long direct repeats as a basis for the formation of secondary structures and their significance for virus transmission. Adv. Virus Res. 69, 203–248.

Guyatt, K.J., Westaway, E.G., and Khromykh, A.A. (2001). Expression and purification of enzymatically active recombinant RNA-dependent RNA polymerase (NS5) of the flavivirus Kunjin. J. Virol. Methods 92, 37–44.

Habjan, M., Pichlmair, A., Elliott, R.M., Overby, A.K., Glatter, T., Gstaiger, M., Superti-Furga, G., Unger, H., and Weber, F. (2009). NSs protein of rift valley fever virus induces the specific degradation of the double-stranded RNA-dependent protein kinase. J. Virol. 83, 4365–4375.

Hahn, C.S., Hahn, Y.S., Rice, C.M., Lee, E., Dalgarno, L., Strauss, E.G., and Strauss, J.H. (1987). Conserved elements in the 3′ untranslated region of flavivirus RNAs and potential cyclization sequences. J. Mol. Biol. 198, 33–41.

Hart, T.J., Kohl, A., and Elliott, R.M. (2009). Role of the NSs protein in the zoonotic capacity of Orthobunyaviruses. Zoonoses Public Health 56, 285–296.

Heinz, F.X., Stiasny, K., Puschner-Auer, G., Holzmann, H., Allison, S.L., Mandl, C.W., and Kunz, C. (1994). Structural changes and functional control of the tick-borne encephalitis virus glycoprotein E by the heterodimeric association with protein prM. Virology 198, 109–117.

Henriques, H.R., Rampazo, E.V., Goncalves, A.J., Vicentin, E.C., Amorim, J.H., Panatieri, R.H., Amorim, K.N., Yamamoto, M.M., Ferreira, L.C., Alves, A.M., *et al.* (2013). Targeting the non-structural protein 1 from dengue virus to a dendritic cell population confers protective immunity to lethal virus challenge. PLoS Neglect. Trop. Dis. 7, e2330.

Holden, K.L., and Harris, E. (2004). Enhancement of dengue virus translation: role of the 3′ untranslated region and the terminal 3′ stem–loop domain. Virology 329, 119–133.

Holden, K.L., Stein, D.A., Pierson, T.C., Ahmed, A.A., Clyde, K., Iversen, P.L., and Harris, E. (2006). Inhibition of dengue virus translation and RNA synthesis by a morpholino oligomer targeted to the top of the terminal 3′ stem–loop structure. Virology 344, 439–452.

Hung, S.L., Lee, P.L., Chen, H.W., Chen, L.K., Kao, C.L., and King, C.C. (1999). Analysis of the steps involved in Dengue virus entry into host cells. Virology 257, 156–167.

Iglesias, N.G., and Gamarnik, A.V. (2011). Dynamic RNA structures in the dengue virus genome. RNA Biol. *8*, 249–257.

Issur, M., Geiss, B.J., Bougie, I., Picard-Jean, F., Despins, S., Mayette, J., Hobdey, S.E., and Bisaillon, M. (2009). The flavivirus NS5 protein is a true RNA guanylyltransferase that catalyzes a two-step reaction to form the RNA cap structure. RNA *15*, 2340–2350.

Jacobs, M.G., Robinson, P.J., Bletchly, C., Mackenzie, J.M., and Young, P.R. (2000). Dengue virus non-structural protein 1 is expressed in a glycosyl-phosphatidylinositol-linked form that is capable of signal transduction. FASEB J. *14*, 1603–1610.

Jin, H., and Elliott, R.M. (1993). Characterization of Bunyamwera virus S RNA that is transcribed and replicated by the L-protein expressed from recombinant vaccinia virus. J. Virol. *67*, 1396–1404.

Johnson, A.J., Guirakhoo, F., and Roehrig, J.T. (1994). The envelope glycoproteins of dengue 1 and dengue 2 viruses grown in mosquito cells differ in their utilization of potential glycosylation sites. Virology *203*, 241–249.

Jones, M., Davidson, A., Hibbert, L., Gruenwald, P., Schlaak, J., Ball, S., Foster, G.R., and Jacobs, M. (2005). Dengue virus inhibits alpha interferon signaling by reducing STAT2 expression. J. Virol. *79*, 5414–5420.

Jose, J., Snyder, J.E., and Kuhn, R.J. (2009). A structural and functional perspective of alphavirus replication and assembly. Future Microbiol. *4*, 837–856.

Kapoor, M., Zhang, L., Ramachandra, M., Kusukawa, J., Ebner, K.E., and Padmanabhan, R. (1995). Association between NS3 and NS5 proteins of dengue virus type 2 in the putative RNA replicase is linked to differential phosphorylation of NS5. J. Biol. Chem. *270*, 19100–19106.

Keegan, K., and Collett, M.S. (1986). Use of bacterial expression cloning to define the amino acid sequences of antigenic determinants on the G2 glycoprotein of Rift Valley fever virus. J. Virol. *58*, 263–270.

Khromykh, A.A., and Westaway, E.G. (1996). RNA binding properties of core protein of the flavivirus Kunjin. Arch. Virol. *141*, 685–699.

Khromykh, A.A., Meka, H., Guyatt, K.J., and Westaway, E.G. (2001a). Essential role of cyclization sequences in flavivirus RNA replication. J. Virol. *75*, 6719–6728.

Khromykh, A.A., Varnavski, A.N., Sedlak, P.L., and Westaway, E.G. (2001b). Coupling between replication and packaging of flavivirus RNA: evidence derived from the use of DNA-based full-length cDNA clones of Kunjin virus. J. Virol. *75*, 4633–4640.

Kim, S.M., and Jeong, Y.S. (2006). Polypyrimidine tract-binding protein interacts with the 3′ stem–loop region of Japanese encephalitis virus negative-strand RNA. Virus Res. *115*, 131–140.

Kofler, R.M., Heinz, F.X., and Mandl, C.W. (2002). Capsid protein C of tick-borne encephalitis virus tolerates large internal deletions and is a favorable target for attenuation of virulence. J. Virol. *76*, 3534–3543.

Kofler, R.M., Leitner, A., O'Riordain, G., Heinz, F.X., and Mandl, C.W. (2003). Spontaneous mutations restore the viability of tick-borne encephalitis virus mutants with large deletions in protein C. J. Virol. *77*, 443–451.

Kofler, R.M., Hoenninger, V.M., Thurner, C., and Mandl, C.W. (2006). Functional analysis of the tick-borne encephalitis virus cyclization elements indicates major differences between mosquito-borne and tick-borne flaviviruses. J. Virol. *80*, 4099–4113.

Konishi, E., and Mason, P.W. (1993). Proper maturation of the Japanese encephalitis virus envelope glycoprotein requires cosynthesis with the premembrane protein. J. Virol. *67*, 1672–1675.

Kostyuchenko, V.A., Zhang, Q., Tan, J.L., Ng, T.S., and Lok, S.M. (2013). Immature and mature dengue serotype 1 virus structures provide insight into the maturation process. J. Virol. *87*, 7700–7707.

Kuhn, R.J. (2007). Togaviridae: The viruses and their replication. In Virology, Fields, B.N., Knipe, D.M., and Howley, P.M., eds. (Lippincott-Raven, New York, NY, USA), pp. 1001–1022.

Kuhn, R.J., Zhang, W., Rossmann, M.G., Pletnev, S.V., Corver, J., Lenches, E., Jones, C.T., Mukhopadhyay, S., Chipman, P.R., Strauss, E.G., *et al.* (2002). Structure of dengue virus: implications for flavivirus organization, maturation, and fusion. Cell *108*, 717–725.

Lai, C.J., Goncalvez, A.P., Men, R., Wernly, C., Donau, O., Engle, R.E., and Purcell, R.H. (2007). Epitope determinants of a chimpanzee dengue virus type 4 (DENV-4)-neutralizing antibody and protection against DENV-4 challenge in mice and rhesus monkeys by passively transferred humanized antibody. J. Virol. *81*, 12766–12774.

Lee, J.M., Crooks, A.J., and Stephenson, J.R. (1989). The synthesis and maturation of a non-structural extracellular antigen from tick-borne encephalitis virus and its relationship to the intracellular NS1 protein. J. Gen. Virol. *70*, 335–343.

Leung, D., Schroder, K., White, H., Fang, N.X., Stoermer, M.J., Abbenante, G., Martin, J.L., Young, P.R., and Fairlie, D.P. (2001). Activity of recombinant dengue 2 virus NS3 protease in the presence of a truncated NS2B co-factor, small peptide substrates, and inhibitors. J. Biol. Chem. *276*, 45762–45771.

Li, D., Schmaljohn, A.L., Anderson, K., and Schmaljohn, C.S. (1995). Complete nucleotide sequences of the M and S segments of two hantavirus isolates from California: evidence for reassortment in nature among viruses related to hantavirus pulmonary syndrome. Virology *206*, 973–983.

Li, G.P., La Starza, M.W., Hardy, W.R., Strauss, J.H., and Rice, C.M. (1990). Phosphorylation of Sindbis virus nsP3 in vivo and in vitro. Virology *179*, 416–427.

Li, L., Lok, S.M., Yu, I.M., Zhang, Y., Kuhn, R.J., Chen, J., and Rossmann, M.G. (2008). The flavivirus precursor membrane-envelope protein complex: structure and maturation. Science *319*, 1830–1834.

Li, L., Jose, J., Xiang, Y., Kuhn, R.J., and Rossmann, M.G. (2010). Structural changes of envelope proteins during alphavirus fusion. Nature *468*, 705–708.

Liao, M., and Kielian, M. (2005). Domain III from class II fusion proteins functions as a dominant-negative inhibitor of virus membrane fusion. J. Cell Biol. *171*, 111–120.

Lin, Y.L., Chen, L.K., Liao, C.L., Yeh, C.T., Ma, S.H., Chen, J.L., Huang, Y.L., Chen, S.S., and Chiang, H.Y. (1998). DNA immunization with Japanese encephalitis virus non-structural protein NS1 elicits protective immunity in mice. J. Virol. *72*, 191–200.

Lindenbach, B.D., and Rice, C.M. (1999). Genetic interaction of flavivirus non-structural proteins NS1 and NS4A as a determinant of replicase function. J. Virol. *73*, 4611–4621.

Lindenbach, B.D., Murray, C.L., Thiel, H.-J., and Rice, C.M. (2013). Flaviviridae. In Fields Virology, D.M. Knipe, and P.M. Howley, eds. (Lippincott Williams & Wilkins, Philadelphia), pp. 712–746.

Linger, B.R., Kunovska, L., Kuhn, R.J., and Golden, B.L. (2004). Sindbis virus nucleocapsid assembly: RNA folding promotes capsid protein dimerization. RNA *10*, 128–138.

Liu, W.J., Chen, H.B., Wang, X.J., Huang, H., and Khromykh, A.A. (2004). Analysis of adaptive mutations in Kunjin virus replicon RNA reveals a novel role for the flavivirus non-structural protein NS2A in inhibition of beta interferon promoter-driven transcription. J. Virol. *78*, 12225–12235.

Liu, W.J., Wang, X.J., Clark, D.C., Lobigs, M., Hall, R.A., and Khromykh, A.A. (2006). A single amino acid substitution in the West Nile virus non-structural protein NS2A disables its ability to inhibit alpha/beta interferon induction and attenuates virus virulence in mice. J. Virol. *80*, 2396–2404.

Loewy, A., Smyth, J., von Bonsdorff, C.H., Liljestrom, P., and Schlesinger, M.J. (1995). The 6-kilodalton membrane protein of Semliki Forest virus is involved in the budding process. J. Virol. *69*, 469–475.

Lorenz, I.C., Allison, S.L., Heinz, F.X., and Helenius, A. (2002). Folding and dimerization of tick-borne encephalitis virus envelope proteins prM and E in the endoplasmic reticulum. J. Virol. *76*, 5480–5491.

Luo, D., Xu, T., Watson, R.P., Scherer-Becker, D., Sampath, A., Jahnke, W., Yeong, S.S., Wang, C.H., Lim, S.P., Strongin, A., *et al.* (2008). Insights into RNA unwinding and ATP hydrolysis by the flavivirus NS3 protein. EMBO J. *27*, 3209–3219.

Lyons, M.J., and Heyduk, J. (1973). Aspects of the developmental morphology of California encephalitis virus in cultured vertebrae and arthropod cells and in mouse brain. Virology *54*, 37–52.

McElroy, K.L., Tsetsarkin, K.A., Vanlandingham, D.L., and Higgs, S. (2006). Role of the yellow fever virus structural protein genes in viral dissemination from the *Aedes aegypti* mosquito midgut. J. Gen. Virol. *87*, 2993–3001.

Mackenzie, J.M., Jones, M.K., and Young, P.R. (1996). Immunolocalization of the dengue virus non-structural glycoprotein NS1 suggests a role in viral RNA replication. Virology *220*, 232–240.

Mackenzie, J.M., Khromykh, A.A., Jones, M.K., and Westaway, E.G. (1998). Subcellular localization and some biochemical properties of the flavivirus Kunjin non-structural proteins NS2A and NS4A. Virology *245*, 203–215.

MacLachlan, N.J., and Dubovi, E.J. (2011). Reoviridae. In Fenner's Veterinary Virology, MacLachlan, N.J., and Dubovi, E.J., eds. (Academic Press, London), pp. 275–291.

Maclachlan, N.J., Drew, C.P., Darpel, K.E., and Worwa, G. (2009). The pathology and pathogenesis of bluetongue. J. Comp. Pathol. *141*, 1–16.

Maclachlan, N.J., Henderson, C., Schwartz-Cornil, I., and Zientara, S. (2014). The immune response of ruminant livestock to bluetongue virus: from type I interferon to antibody. Virus Res. *182*, 71–77.

Malet, H., Coutard, B., Jamal, S., Dutartre, H., Papageorgiou, N., Neuvonen, M., Ahola, T., Forrester, N., Gould, E.A., Lafitte, D., *et al.* (2009). The crystal structures of chikungunya and Venezuelan equine encephalitis virus nsP3 macro domains define a conserved adenosine binding pocket. J. Virol. *83*, 6534–6545.

Mandl, C.W., Holzmann, H., Kunz, C., and Heinz, F.X. (1993). Complete genomic sequence of Powassan virus: evaluation of genetic elements in tick-borne versus mosquito-borne flaviviruses. Virology *194*, 173–184.

Markoff, L. (2003). 5′- and 3′-noncoding regions in flavivirus RNA. In The Flaviviruses: Structure, Replication and Evolution, T.P. Monath, and T.J. Chambers, eds. (Elsevier Academic Press, San Diego), pp. 177–228.

Medin, C.L., Fitzgerald, K.A., and Rothman, A.L. (2005). Dengue virus non-structural protein NS5 induces interleukin-8 transcription and secretion. J. Virol. *79*, 11053–11061.

Melton, J.V., Ewart, G.D., Weir, R.C., Board, P.G., Lee, E., and Gage, P.W. (2002). Alphavirus 6K proteins form ion channels. J. Biol. Chem. *277*, 46923–46931.

Mi, S., and Stollar, V. (1990). Both amino acid changes in nsP1 of Sindbis virusLM21 contribute to and are required for efficient expression of the mutant phenotype. Virology *178*, 429–434.

Mi, S., and Stollar, V. (1991). Expression of Sindbis virus nsP1 and methyltransferase activity in *Escherichia coli*. Virology *184*, 423–427.

Mir, M.A., and Panganiban, A.T. (2006). The bunyavirus nucleocapsid protein is an RNA chaperone: possible roles in viral RNA panhandle formation and genome replication. RNA *12*, 272–282.

Modis, Y., Ogata, S., Clements, D., and Harrison, S.C. (2004). Structure of the dengue virus envelope protein after membrane fusion. Nature *427*, 313–319.

Mohd Jaafar, F., Belhouchet, M., Belaganahalli, M., Tesh, R.B., Mertens, P.P., and Attoui, H. (2014). Full-genome characterisation of Orungo, Lebombo and Changuinola viruses provides evidence for co-evolution of orbiviruses with their arthropod vectors. PloS One *9*, e86392.

Munoz-Jordan, J.L., Sanchez-Burgos, G.G., Laurent-Rolle, M., and Garcia-Sastre, A. (2003). Inhibition of interferon signaling by dengue virus. Proc. Natl. Acad. Sci. U.S.A. *100*, 14333–14338.

Munoz-Jordan, J.L., Laurent-Rolle, M., Ashour, J., Martinez-Sobrido, L., Ashok, M., Lipkin, W.I., and Garcia-Sastre, A. (2005). Inhibition of alpha/beta interferon signaling by the NS4B protein of flaviviruses. J. Virol. *79*, 8004–8013.

Muylaert, I.R., Chambers, T.J., Galler, R., and Rice, C.M. (1996). Mutagenesis of the N-linked glycosylation sites of the yellow fever virus NS1 protein: effects on virus replication and mouse neurovirulence. Virology *222*, 159–168.

Myles, K.M., Kelly, C.L., Ledermann, J.P., and Powers, A.M. (2006). Effects of an opal termination codon preceding the nsP4 gene sequence in the O'Nyong-Nyong virus genome on *Anopheles gambiae* infectivity. J. Virol. *80*, 4992–4997.

Nichol, S.T., Beaty, B.J., Elliot, R.H., Goldbach, A., Plyusnin, A., Schmaljohn, C.S., and Tesh, R.B. (2005). Bunyaviridae. In Virus Taxonomy: VIIIth Report of the International Committee on Taxonomy of Viruses, C.M. Fauquet, M.A. Mayo, J. Maniloff, U. Desselberger, and L.A. Ball, eds. (Academic Press, Amsterdam), pp. 695–716.

Novoa, R.R., Calderita, G., Cabezas, P., Elliott, R.M., and Risco, C. (2005). Key Golgi factors for structural and functional maturation of bunyamwera virus. J. Virol. *79*, 10852–10863.

Nunes, M.R., Travassos da Rosa, A.P., Weaver, S.C., Tesh, R.B., and Vasconcelos, P.F. (2005). Molecular epidemiology of group C viruses (Bunyaviridae, Orthobunyavirus) isolated in the Americas. J. Virol. *79*, 10561–10570.

Nybakken, G.E., Oliphant, T., Johnson, S., Burke, S., Diamond, M.S., and Fremont, D.H. (2005). Structural basis of West Nile virus neutralization by a therapeutic antibody. Nature *437*, 764–769.

Oliphant, T., Engle, M., Nybakken, G.E., Doane, C., Johnson, S., Huang, L., Gorlatov, S., Mehlhop, E., Marri, A., Chung, K.M., *et al.* (2005). Development of a humanized monoclonal antibody with therapeutic potential against West Nile virus. Nature Med. *11*, 522–530.

Paranjape, S.M., and Harris, E. (2007). Y box-binding protein-1 binds to the dengue virus 3′-untranslated region and mediates antiviral effects. J. Biol. Chem. *282*, 30497–30508.

Patel, A., and Roy, P. (2014). The molecular biology of Bluetongue virus replication. Virus Res. *182*, 5–20.

Patkar, C.G., Jones, C.T., Chang, Y.H., Warrier, R., and Kuhn, R.J. (2007). Functional requirements of the yellow fever virus capsid protein. J. Virol. *81*, 6471–6481.

Patterson, J.L., Holloway, B., and Kolakofsky, D. (1984). La Crosse virions contain a primer-stimulated RNA polymerase and a methylated cap-dependent endonuclease. J. Virol. *52*, 215–222.

Peluso, R.W., Richardson, J.C., Talon, J., and Lock, M. (1996). Identification of a set of proteins (C′ and C) encoded by the bicistronic P gene of the Indiana serotype of vesicular stomatitis virus and analysis of their effect on transcription by the viral RNA polymerase. Virology *218*, 335–342.

Pierson, T.C., Fremont, D.H., Kuhn, R.J., and Diamond, M.S. (2008). Structural insights into the mechanisms of antibody-mediated neutralization of flavivirus infection: implications for vaccine development. Cell Host Microbe *4*, 229–238.

Plassmeyer, M.L., Soldan, S.S., Stachelek, K.M., Martin-Garcia, J., and Gonzalez-Scarano, F. (2005). California serogroup Gc (G1) glycoprotein is the principal determinant of pH-dependent cell fusion and entry. Virology *338*, 121–132.

Prikhod'ko, G.G., Prikhod'ko, E.A., Pletnev, A.G., and Cohen, J.I. (2002). Langat flavivirus protease NS3 binds caspase-8 and induces apoptosis. J. Virol. *76*, 5701–5710.

Proutski, V., Gould, E.A., and Holmes, E.C. (1997). Secondary structure of the 3′ untranslated region of flaviviruses: similarities and differences. Nucleic Acids Res. *25*, 1194–1202.

Pryor, M.J., Rawlinson, S.M., Butcher, R.E., Barton, C.L., Waterhouse, T.A., Vasudevan, S.G., Bardin, P.G., Wright, P.J., Jans, D.A., and Davidson, A.D. (2007). Nuclear localization of dengue virus non-structural protein 5 through its importin alpha/beta-recognized nuclear localization sequences is integral to viral infection. Traffic *8*, 795–807.

Raju, R., and Huang, H.V. (1991). Analysis of Sindbis virus promoter recognition in vivo, using novel vectors with two subgenomic mRNA promoters. J. Virol. *65*, 2501–2510.

Ramanathan, M.P., Chambers, J.A., Pankhong, P., Chattergoon, M., Attatippaholkun, W., Dang, K., Shah, N., and Weiner, D.B. (2006). Host cell killing by the West Nile Virus NS2B-NS3 proteolytic complex: NS3 alone is sufficient to recruit caspase-8-based apoptotic pathway. Virology *345*, 56–72.

Ratinier, M., Caporale, M., Golder, M., Franzoni, G., Allan, K., Nunes, S.F., Armezzani, A., Bayoumy, A., Rixon, F., Shaw, A., *et al.* (2011). Identification and characterization of a novel non-structural protein of bluetongue virus. PLoS Pathogens *7*, e1002477.

Ravkov, E.V., Nichol, S.T., and Compans, R.W. (1997). Polarized entry and release in epithelial cells of Black Creek Canal virus, a New World hantavirus. J. Virol. *71*, 1147–1154.

Ray, D., Shah, A., Tilgner, M., Guo, Y., Zhao, Y., Dong, H., Deas, T.S., Zhou, Y., Li, H., and Shi, P.Y. (2006). West Nile virus 5′-cap structure is formed by sequential guanine N-7 and ribose 2′-O methylations by non-structural protein 5. J. Virol. *80*, 8362–8370.

Rey, F.A., Heinz, F.X., Mandl, C., Kunz, C., and Harrison, S.C. (1995). The envelope glycoprotein from tick-borne encephalitis virus at 2 A resolution. Nature *375*, 291–298.

Rice, C.M., Lenches, E.M., Eddy, S.R., Shin, S.J., Sheets, R.L., and Strauss, J.H. (1985). Nucleotide sequence of yellow fever virus: implications for flavivirus gene expression and evolution. Science *229*, 726–733.

Rikkonen, M., Peranen, J., and Kaariainen, L. (1992). Nuclear and nucleolar targeting signals of Semliki Forest virus non-structural protein nsP2. Virology *189*, 462–473.

Rikkonen, M., Peranen, J., and Kaariainen, L. (1994). ATPase and GTPase activities associated with Semliki Forest virus non-structural protein nsP2. J. Virol. *68*, 5804–5810.

Rodriguez-Madoz, J.R., Belicha-Villanueva, A., Bernal-Rubio, D., Ashour, J., Ayllon, J., and Fernandez-Sesma, A. (2010). Inhibition of the type I interferon response in human dendritic cells by dengue virus infection requires a catalytically active NS2B3 complex. J. Virol. *84*, 9760–9774.

Roehrig, J.T., Bolin, R.A., and Kelly, R.G. (1998). Monoclonal antibody mapping of the envelope glycoprotein of the dengue 2 virus, Jamaica. Virology *246*, 317–328.

Rouvinski, A., Guardado-Calvo, P., Barba-Spaeth, G., Duquerroy, S., Vaney, M.C., Kikuti, C.M., Navarro Sanchez, M.E., Dejnirattisai, W., Wongwiwat, W., Haouz, A., *et al.* (2015). Recognition determinants of broadly neutralizing human antibodies against dengue viruses. Nature *520*, 109–113.

Rumenapf, T., Brown, D.T., Strauss, E.G., Konig, M., Rameriz-Mitchel, R., and Strauss, J.H. (1995). Aura alphavirus subgenomic RNA is packaged into virions of two sizes. J. Virol. *69*, 1741–1746.

Saxton-Shaw, K.D., Ledermann, J.P., Borland, E.M., Stovall, J.L., Mossel, E.C., Singh, A.J., Wilusz, J., and Powers, A.M. (2013). O'nyong nyong virus molecular determinants of unique vector specificity reside in non-structural protein 3. PLoS Neglected Trop. Dis. *7*, e1931.

Schlick, P., Taucher, C., Schittl, B., Tran, J.L., Kofler, R.M., Schueler, W., von Gabain, A., Meinke, A., and Mandl, C.W. (2009). Helices alpha2 and alpha3 of West Nile virus capsid protein are dispensable for assembly of infectious virions. J. Virol. *83*, 5581–5591.

Schmaljohn, C.S., and Nichol, S.N. (2007). Bunyaviridae. In Field's Virology, B.N. Fields, D.M. Knipe, P.M. Howley, and D.E. Griffin, eds. (Lippincott Williams & Wilkins, Philadelphia), pp. 1741–1778.

Shafee, N., and AbuBakar, S. (2003). Dengue virus type 2 NS3 protease and NS2B-NS3 protease precursor induce apoptosis. J. Gen. Virol. *84*, 2191–2195.

Shin, G., Yost, S.A., Miller, M.T., Elrod, E.J., Grakoui, A., and Marcotrigiano, J. (2012). Structural and functional insights into alphavirus polyprotein processing and pathogenesis. Proc. Natl. Acad. Sci. U.S.A. *109*, 16534–16539.

Shrestha, B., Brien, J.D., Sukupolvi-Petty, S., Austin, S.K., Edeling, M.A., Kim, T., O'Brien, K.M., Nelson, C.A., Johnson, S., Fremont, D.H., *et al.* (2010). The development of therapeutic antibodies that neutralize homologous and heterologous genotypes of dengue virus type 1. PLoS Pathogens *6*, e1000823.

Simon-Loriere, E., and Holmes, E.C. (2013). Gene duplication is infrequent in the recent evolutionary history of RNA viruses. Mol. Biol. Evol. *30*, 1263–1269.

Smith, J.F., and Pifat, D.Y. (1982). Morphogenesis of sandfly viruses (Bunyaviridae family). Virology *121*, 61–81.

Smith, T.J., Brandt, W.E., Swanson, J.L., McCown, J.M., and Buescher, E.L. (1970). Physical and biological properties of dengue-2 virus and associated antigens. J. Virol. *5*, 524–532.

Snyder, J.E., Kulcsar, K.A., Schultz, K.L., Riley, C.P., Neary, J.T., Marr, S., Jose, J., Griffin, D.E., and Kuhn, R.J. (2013). Functional characterization of the alphavirus TF protein. J. Virol. *87*, 8511–8523.

Stapleton, J.T., Foung, S., Muerhoff, A.S., Bukh, J., and Simmonds, P. (2011). The GB viruses: a review and proposed classification of GBV-A, GBV-C (HGV), and GBV-D in genus Pegivirus within the family Flaviviridae. J. Gen. Virol. *92*, 233–246.

Strauss, E.G., Rice, C.M., and Strauss, J.H. (1983). Sequence coding for the alphavirus non-structural proteins is interrupted by an opal termination codon. Proc. Natl. Acad. USA. *80*, 5271–5275.

Strauss, J.H., and Strauss, E.G. (1994). The alphaviruses: gene expression, replication, and evolution. Microbiol. Rev. *58*, 491–562.

Sukupolvi-Petty, S., Austin, S.K., Purtha, W.E., Oliphant, T., Nybakken, G.E., Schlesinger, J.J., Roehrig, J.T., Gromowski, G.D., Barrett, A.D., Fremont, D.H., *et al.* (2007). Type- and subcomplex-specific neutralizing antibodies against domain III of dengue virus type 2 envelope protein recognize adjacent epitopes. J. Virol. *81*, 12816–12826.

Ta, M., and Vrati, S. (2000). Mov34 protein from mouse brain interacts with the 3′ noncoding region of Japanese encephalitis virus. J. Virol. *74*, 5108–5115.

Tan, B.H., Fu, J., Sugrue, R.J., Yap, E.H., Chan, Y.C., and Tan, Y.H. (1996). Recombinant dengue type 1 virus NS5 protein expressed in *Escherichia coli* exhibits RNA-dependent RNA polymerase activity. Virology *216*, 317–325.

Thomas, D., Blakqori, G., Wagner, V., Banholzer, M., Kessler, N., Elliott, R.M., Haller, O., and Weber, F. (2004). Inhibition of RNA polymerase II phosphorylation by a viral interferon antagonist. J. Biol. Chem. *279*, 31471–31477.

Tilgner, M., Deas, T.S., and Shi, P.Y. (2005). The flavivirus-conserved penta-nucleotide in the 3′ stem–loop of the West Nile virus genome requires a specific sequence and structure for RNA synthesis, but not for viral translation. Virology *331*, 375–386.

Tomar, S., Hardy, R.W., Smith, J.L., and Kuhn, R.J. (2006). Catalytic core of alphavirus non-structural protein nsP4 possesses terminal adenylyltransferase activity. J. Virol. *80*, 9962–9969.

Trent, D.W. (1977). Antigenic characterization of flavivirus structural proteins separated by isoelectric focusing. J. Virol. *22*, 608–618.

Uchil, P.D., Kumar, A.V., and Satchidanandam, V. (2006). Nuclear localization of flavivirus RNA synthesis in infected cells. J. Virol. *80*, 5451–5464.

Vashist, S., Anantpadma, M., Sharma, H., and Vrati, S. (2009). La protein binds the predicted loop structures in the 3′ non-coding region of Japanese encephalitis virus genome: role in virus replication. J. Gen. Virol. *90*, 1343–1352.

Vashist, S., Bhullar, D., and Vrati, S. (2011). La protein can simultaneously bind to both 3′- and 5′-noncoding regions of Japanese encephalitis virus genome. DNA Cell Biol. *30*, 339–346.

Vasilakis, N., Holmes, E.C., Fokam, E.B., Faye, O., Diallo, M., Sall, A.A., and Weaver, S.C. (2007). Evolutionary processes among sylvatic dengue type 2 viruses. J. Virol. *81*, 9591–9595.

Vasilakis, N., Castro-Llanos, F., Widen, S.G., Aguilar, P.V., Guzman, H., Guevara, C., Fernandez, R., Auguste, A.J., Wood, T.G., Popov, V., *et al.* (2014). Arboretum and Puerto Almendras viruses: two novel rhabdoviruses isolated from mosquitoes in Peru. J. Gen. Virol. *95*, 787–792.

Vasiljeva, L., Valmu, L., Kaariainen, L., and Merits, A. (2001). Site-specific protease activity of the carboxyl-terminal domain of Semliki Forest virus replicase protein nsP2. J. Biol. Chem. *276*, 30786–30793.

Vitour, D., Doceul, V., Ruscanu, S., Chauveau, E., Schwartz-Cornil, I., and Zientara, S. (2014). Induction and control of the type I interferon pathway by Bluetongue virus. Virus Res. *182*, 59–70.

Walker, P.J., Byrne, K.A., Riding, G.A., Cowley, J.A., Wang, Y., and McWilliam, S. (1992). The genome of bovine ephemeral fever rhabdovirus contains two related glycoprotein genes. Virology *191*, 49–61.

Walker, P.J., Dietzgen, R.G., Joubert, D.A., and Blasdell, K.R. (2011). Rhabdovirus accessory genes. Virus Res. *162*, 110–125.

Walker, P.J., Firth, C., Widen, S.G., Blasdell, K.R., Guzman, H., Wood, T.G., Paradkar, P.N., Holmes, E.C., Tesh, R.B., and Vasilakis, N. (2015). Evolution of genome size and complexity in the rhabdoviridae. PLoS Pathogens *11*, e1004664.

Wallner, G., Mandl, C.W., Kunz, C., and Heinz, F.X. (1995). The flavivirus 3′-noncoding region: extensive size heterogeneity independent of evolutionary relationships among strains of tick-borne encephalitis virus. Virology *213*, 169–178.

Wang, Y., and Walker, P.J. (1993). Adelaide river rhabdovirus expresses consecutive glycoprotein genes as polycistronic mRNAs: new evidence of gene duplication as an evolutionary process. Virology *195*, 719–731.

Warrener, P., Tamura, J.K., and Collett, M.S. (1993). RNA-stimulated NTPase activity associated with yellow fever virus NS3 protein expressed in bacteria. J. Virol. *67*, 989–996.

Weber, F., Jambrina, E., González, S., Dessens, J.T., Leahy, M., Kochs, G., Portela, A., Nuttall, P.A., Haller, O., Ortín, J., and Zürcher, T. (1998). *In vivo* reconstitution of active Thogoto virus polymerase: assays for the compatibility with other orthomyxovirus core proteins and template RNAs. Virus Res. *58*, 13–20.

Weber, F., Bridgen, A., Fazakerley, J.K., Streitenfeld, H., Kessler, N., Randall, R.E., and Elliott, R.M. (2002). Bunyamwera bunyavirus non-structural protein NSs counteracts the induction of alpha/beta interferon. J. Virol. *76*, 7949–7955.

Wengler, G., and Wengler, G. (1981). Terminal sequences of the genome and replicative-from RNA of the flavivirus West Nile virus: absence of poly(A) and possible role in RNA replication. Virology *113*, 544–555.

Wengler, G., and Wengler, G. (1993). The NS 3 non-structural protein of flaviviruses contains an RNA triphosphatase activity. Virology *197*, 265–273.

Wengler, G., Wengler, G., and Gross, H.J. (1978). Studies on virus-specific nucleic acids synthesized in vertebrate and mosquito cells infected with flaviviruses. Virology *89*, 423–437.

Wengler, G., Wengler, G., Nowak, T., and Castle, E. (1990). Description of a procedure which allows isolation of viral non-structural proteins from BHK vertebrate cells infected with the West Nile flavivirus in a state which allows their direct chemical characterization. Virology *177*, 795–801.

Wengler, G., Koschinski, A., Wengler, G., and Dreyer, F. (2003). Entry of alphaviruses at the plasma membrane converts the viral surface proteins into an ion-permeable pore that can be detected by electrophysiological analyses of whole-cell membrane currents. J. Gen. Virol. *84*, 173–181.

Westaway, E.G., Mackenzie, J.M., Kenney, M.T., Jones, M.K., and Khromykh, A.A. (1997). Ultrastructure of Kunjin virus-infected cells: colocalization of NS1 and NS3 with double-stranded RNA, and of NS2B with NS3, in virus-induced membrane structures. J. Virol. *71*, 6650–6661.

Winkler, G., Randolph, V.B., Cleaves, G.R., Ryan, T.E., and Stollar, V. (1988). Evidence that the mature form of the flavivirus non-structural protein NS1 is a dimer. Virology *162*, 187–196.

Winkler, G., Maxwell, S.E., Ruemmler, C., and Stollar, V. (1989). Newly synthesized dengue-2 virus non-structural protein NS1 is a soluble protein but becomes partially hydrophobic and membrane-associated after dimerization. Virology *171*, 302–305.

Wu, J., Bera, A.K., Kuhn, R.J., and Smith, J.L. (2005). Structure of the Flavivirus helicase: implications for catalytic activity, protein interactions, and proteolytic processing. J. Virol. *79*, 10268–10277.

Wu, S.C., Lian, W.C., Hsu, L.C., and Liau, M.Y. (1997). Japanese encephalitis virus antigenic variants with characteristic differences in neutralization resistance and mouse virulence. Virus Res. *51*, 173–181.

Xu, T., Sampath, A., Chao, A., Wen, D., Nanao, M., Chene, P., Vasudevan, S.G., and Lescar, J. (2005). Structure of the Dengue virus helicase/nucleoside triphosphatase catalytic domain at a resolution of 2.4 A. J. Virol. *79*, 10278–10288.

Yanase, T., Kato, T., Yamakawa, M., Takayoshi, K., Nakamura, K., Kokuba, T., and Tsuda, T. (2006). Genetic characterization of Batai virus indicates a genomic reassortment between orthobunyaviruses in nature. Arch. Virol. *151*, 2253–2260.

Yocupicio-Monroy, M., Padmanabhan, R., Medina, F., and del Angel, R.M. (2007). Mosquito La protein binds to the 3′ untranslated region of the positive and negative polarity dengue virus RNAs and relocates to the cytoplasm of infected cells. Virology 357, 29–40.

Yu, I.M., Zhang, W., Holdaway, H.A., Li, L., Kostyuchenko, V.A., Chipman, P.R., Kuhn, R.J., Rossmann, M.G., and Chen, J. (2008). Structure of the immature dengue virus at low pH primes proteolytic maturation. Science *319*, 1834–1837.

Yu, L., and Markoff, L. (2005). The topology of bulges in the long stem of the flavivirus 3′ stem–loop is a major determinant of RNA replication competence. J. Virol. 79, 2309–2324.

Zeng, L., Falgout, B., and Markoff, L. (1998). Identification of specific nucleotide sequences within the conserved 3′-SL in the dengue type 2 virus genome required for replication. J. Virol. 72, 7510–7522.

Zhang, C., Mammen, M.P., Jr., Chinnawirotpisan, P., Klungthong, C., Rodpradit, P., Monkongdee, P., Nimmannitya, S., Kalayanarooj, S., and Holmes, E.C. (2005). Clade replacements in dengue virus serotypes 1 and 3 are associated with changing serotype prevalence. J. Virol. *79*, 15123–15130.

Zhang, W., Chipman, P.R., Corver, J., Johnson, P.R., Zhang, Y., Mukhopadhyay, S., Baker, T.S., Strauss, J.H., Rossmann, M.G., and Kuhn, R.J. (2003). Visualization of membrane protein domains by cryo-electron microscopy of dengue virus. Nature Struct. Biol. *10*, 907–912.

Zhang, X., Ge, P., Yu, X., Brannan, J.M., Bi, G., Zhang, Q., Schein, S., and Zhou, Z.H. (2013). Cryo-EM structure of the mature dengue virus at 3.5-A resolution. Nature Struct. Mol. Biol. *20*, 105–110.

Host Metabolism and its Contribution in Flavivirus Biogenesis

4

Rushika Perera and Richard J. Kuhn

Abstract

Intracellular communication is key to cellular homeostasis. During infection of their hosts, viruses require a subversion of normal cellular communication pathways such that attention is focused towards efficient viral replication and virion biogenesis. Thus, the cellular environment is converted to a viral replication factory. Many viruses achieve this by significantly rearranging the intracellular membrane environment to form 'viral cobwebs' that extend throughout the cytoplasm. These cobwebs increase membrane contact sites within the cell and enhance cellular communication through signalling and transport of raw materials required for viral replication. Any form of intracellular conversion is directly linked to the metabolic pathways that drive cellular homeostasis. This chapter will therefore focus on the specific cellular metabolic pathways that are hijacked by viruses to achieve the formation of these viral replication factories. Specifically, we discuss how flaviviruses achieve an optimal cellular environment for replication and how these mechanisms seem conserved across species barriers as they replicate in both the human host and mosquito vector.

Introduction

The flaviviruses are a genus within the *Flaviviridae* family and are arthropod-borne viruses that infect a wide variety of animal species. They consist of approximately 70 members and include important human pathogens such as dengue (DENV), yellow fever (YFV), West Nile (WNV), and Japanese encephalitis virus (JEV) (Lindenbach *et al.*, 2001; Kuhn 2011). The flavivirus genome is a positive sense single stranded RNA of ~11 kb and encodes three structural proteins and seven non-structural proteins (Fig. 4.1). The three structural proteins include capsid (C), pre-membrane (prM) and envelope (E) that form the virus particle. The seven non-structural proteins include NS1, NS2A, NS2B, NS3, NS4A, NS4B and NS5 and are involved in genome replication. The genome is translated into a single polyprotein and individual viral proteins are processed co-translationally by viral and cellular proteases. Many of the viral non-structural and some structural proteins are also heavily invested in interfering with and/or recruiting host factors for virion biogenesis (Khadka *et al.*, 2011).

Every aspect of the flaviviruses life cycle is intimately connected to cellular membranes. This may be primarily driven by the fact that all flaviviral proteins are either directly or indirectly associated with cellular membranes. During their life cycle, flaviviruses infect cells through receptor-mediated endocytosis. Following entry, their genome is translated and processed in the cytoplasm to produce the individual viral proteins. These viral proteins then function to replicate the genome and assemble new viral particles. While early events in viral RNA translation and replication are unclear, once a critical level of the viral proteins is achieved, intracellular membranes are rearranged to form microenvironments that are committed to further viral protein translation and processing and viral RNA replication. These microenvironments also protect replicating virus from detection by the host defence mechanisms. Newly replicated viral RNA is packaged into virions that bud into the endoplasmic reticulum (ER) and travel through the secretory pathway to be released from the infected cell (Fig. 4.2) (Perera *et al.*, 2008; Chatel-Chaix *et al.*, 2014).

During these events, the newly translated viral proteins also function to subvert host cell metabolism and antiviral pathways such that the intracellular environment is primed for viral replication. Given the intimacy of the viral life cycle with intracellular membranes, many of the metabolic pathways that are perturbed either directly or indirectly impact membrane homeostasis. Specifically, there is a delicate interconnectivity between altered metabolic pathways and a new physical membrane architecture that provides an optimal environment for cellular communication during virus replication. This chapter will discuss perturbations in cellular metabolism observed during viral infection and how they might relate to the membrane requirements for replication and virion biogenesis.

The players

The structural proteins

The C protein assists in packaging the viral RNA into the virus particles. It is a dimer in solution and has a signal sequence at its C-terminus that anchors it in the ER membrane. The solution structure for the DENV C protein has been solved by nuclear

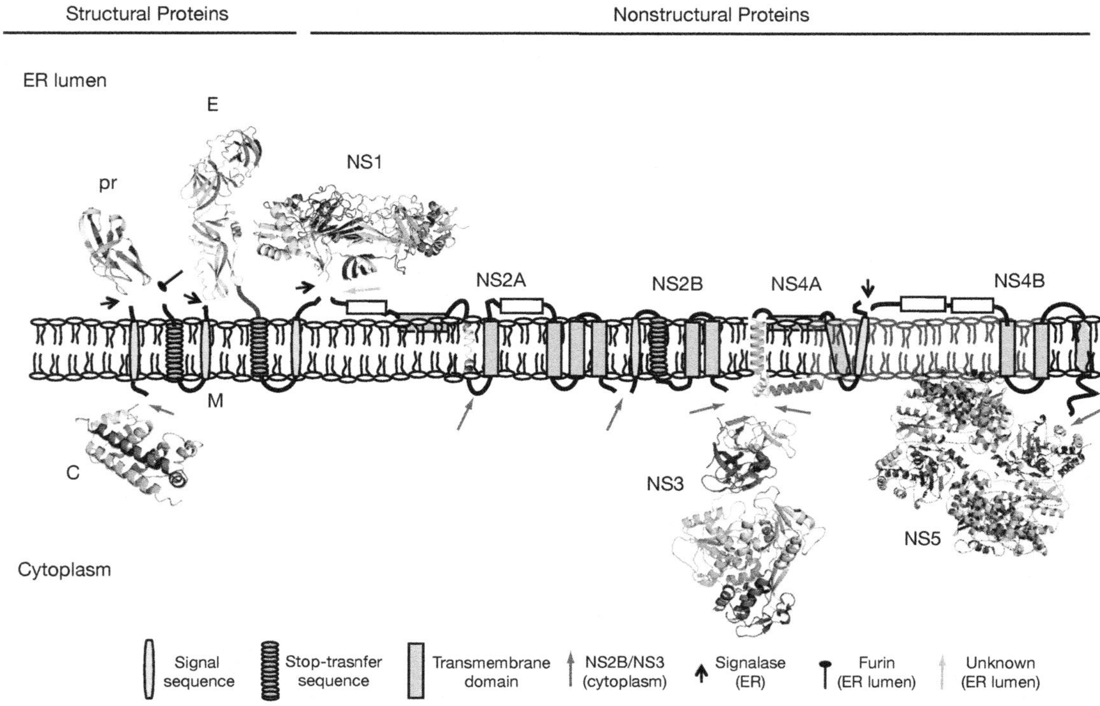

Figure 4.1 Schematic diagram of the flavivirus polyprotein. The three structural proteins include capsid (C), pre-membrane (prM) and envelope (E). The seven non-structural proteins include NS1, NS2A, NS2B, NS3, NS4A, NS4B and NS5. The atomic structures for all soluble domains of the flaviviral proteins have been solved with the exception of the integral membrane proteins NS2A, parts of NS2B, NS4A and NS4B. PDB identifiers; capsid NMR structure, 1R6R; pr peptide (as prM-E structure), 3C6E; E, 1OKE; NS1, 4O6B; NS3, 2VBC; NS5, 4VOR. The NMR structure of the third transmembrane domain (pTMS3), of NS2A (2MOS) and the N-terminal membrane associated regions of NS4A have been solved. The latter is shown as a schematic since the PDB identifier is unavailable. The polyprotein is cleaved on the cytoplasmic side by viral protein NS2B/3, and on the ER luminal side by signalase and an unknown protease. In the trans Golgi, the pr peptide is cleaved from the M protein by furin. These cleavage sites are shown.

magnetic resonance spectroscopy and shows that the monomeric protein consists of four alpha helices with one alpha helix forming the dimer interface (Ma *et al.*, 2004). Mutagenesis studies have demonstrated that the dimeric conformation is required for function and yet the protein displays remarkable functional plasticity. Almost one-third of the protein can be deleted without affecting its RNA-binding capacity (Jones *et al.*, 2003; Patkar *et al.*, 2007). It is yet unclear how much of the population of C protein is membrane bound versus free in solution during infection. Interestingly, while its role in viral RNA packaging and assembly is required in the cytoplasm, the C protein is primarily observed in speckled regions in the nucleus. In the nucleus, it has been implicated in inhibiting nucleosome formation and induction of apoptosis (Sangiambut *et al.*, 2008; Bhuvanakantham *et al.*, 2009; Netsawang *et al.*, 2010; Colpitts *et al.*, 2011).

The pre-membrane (prM) protein is an integral membrane glycoprotein embedded in the ER. It exists as a heterodimer together with the E protein and forms the surface glycoprotein shell of immature flavivirus particles (Kuhn *et al.*, 2002). Specifically, prM forms a cap-like structure and primarily functions to protect the fusion peptide on the E protein to prevent premature fusion of the virus with host cell membranes (Guirakhoo *et al.*, 1992). Flavivirus fusion occurs at an approximate pH threshold of 6.6–6.8 (Ueba *et al.*, 1977; Gollins *et al.*, 1986; Summers *et al.*, 1989; Randolph *et al.*, 1990; Guirakhoo *et al.*, 1991, 1993; Despres *et al.*, 1993; McMinn *et al.*, 1996; Corver *et al.*, 2000; Stiasny *et al.*, 2003). The structure of the ectodomain of prM has been solved at low and neutral pH and shows that pH does not impact its tertiary fold (Li *et al.*, 2008). However, due to several conserved histidine residues that line the interface between the prM and E proteins within the heterodimer, the interaction of prM with the E protein can be disrupted by changes in pH. This forms the basis for some of the structural rearrangements that occur in the glycoproteins shell (discussed below) during maturation of flaviviral particles. During the flavivirus life cycle, the prM protein undergoes furin-mediated cleavage in the trans-Golgi network (TGN) converting immature virus particles into mature virus particles.

The E protein is the first point of contact between the virus and the host cell and is also the primary effector of structural heterogeneity in the flavivirus particles (Kuhn *et al.*, 2002; Mukhopadhyay *et al.*, 2003; Dowd *et al.*, 2014; Plevka *et al.*, 2014). Together with the prM protein, it forms the glycoprotein shell of the virus. Specifically, the E protein can form a heterodimer with the prM protein as well as a homodimer with other E protein molecules. Within these interactions, it can assume either a trimeric or a dimeric conformation. The detailed structural transitions of the E protein in the glycoprotein shell during viral maturation are discussed below. The atomic structure of the E protein from several flaviviruses has been solved (Rey *et al.*, 1995; Modis *et al.*, 2003, 2004, 2005; Zhang *et al.*, 2004; Kanai *et al.*, 2006; Nybakken *et al.*, 2006). It is a class II fusion protein and consists of three β-barrel domains; domain I forms the N-terminus but is centrally located in the molecule, domain II is elongated and mediates dimerization of E protein and also includes the hydrophobic, well-conserved fusion peptide at its distal end (Allison *et al.*, 2001), domain III (DIII) is an

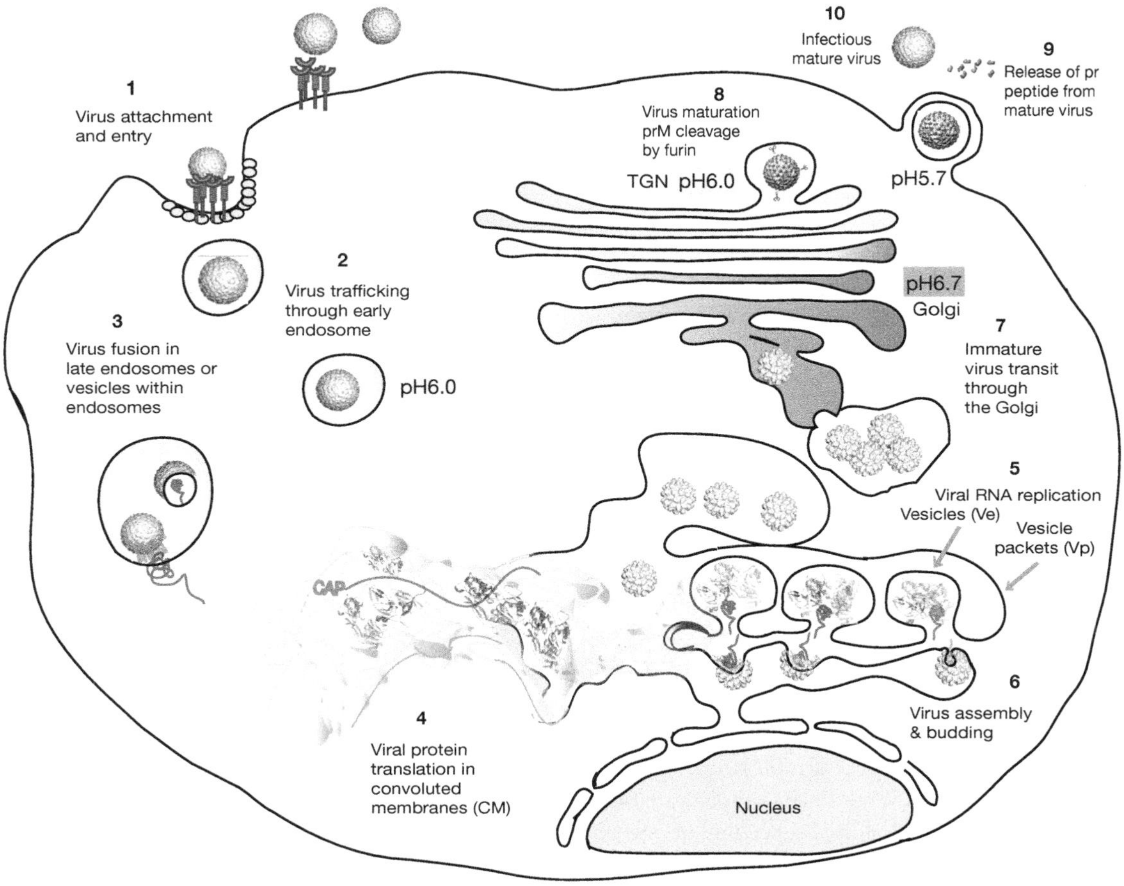

Figure 4.2 Life cycle of flaviviruses showing the altered virus-induced membrane architecture. 1. Virions attach to cell surface attachment molecules and receptors. B. Following attachment, they are internalized through endocytosis. In the low-pH environment of the early endosomes, the viral glycoproteins rearrange to form the fusion intermediates. 3. The lipid repertoire in the membranes of the multivesicular bodies of the late endosome allow fusion to occur between the viral and cellular membranes allowing disassembly of the virus and release of the viral RNA into the cytoplasm. 4. Translation and polyprotein processing occur in virus-induced convoluted membranes and paracrystalline arrays (CM/PC) that are derived from the ER membranes. 5. Mature viral proteins are transported to vesicles (Ve) within vesicle packets (Vp). These structures are also induced through viral protein expression. The Vp/Ve are sites where viral RNA replication complexes are assembled. There is a presumed interconnectivity between the CM/PC structures and Vp/Ve structures on both the luminal and cytoplasmic sides of the ER. 6. Following RNA replication, newly synthesized viral RNA is transported to sites of virus assembly through association with the capsid protein. Here, the capsid protein–RNA complex associates with host membranes and glycoproteins and buds into the ER lumen. 7. Newly assembled viral particles are immature (spiky). They transit through the secretory pathway to the trans Golgi (TGN) compartments. 8. In the low pH of the TGN, structural rearrangements in the glycoprotein layer of the virus particle exposes the furin cleavage site on the prM proteins and allows furin to cleave the pr peptide from the M protein. The pr peptide, however, remains associated with the M protein to prevent premature fusion of the virus within the infected cell. 9. Upon release into the extracellular milieu, the pr peptide is released. 10. The new mature virus becomes available to infect a new cell.

immunoglobulin (Ig)-like domain that is involved in receptor binding (Rey *et al.*, 1995; Bhardwaj *et al.*, 2001) and antibody neutralization (Crill *et al.*, 2001; Beasley *et al.*, 2002; Halstead *et al.*, 2005; Stiasny *et al.*, 2006; Pierson *et al.*, 2007, 2008; Sukupolvi-Petty *et al.*, 2007).

The non-structural proteins

NS1 is a glycosylated transmembrane protein that functions in its dimeric form, as a cofactor in viral RNA replication possibly through interactions with other viral non-structural transmembrane proteins such as NS4A and NS4B (Mackenzie *et al.*, 1998; Lindenbach *et al.*, 1999; Youn *et al.*, 2012). It is also secreted as a hexamer functioning as a viral antigen that circulates in the sera of dengue-infected patients in the form of a proteolipid particle (Muller *et al.*, 2013). Both the cell-associated and secreted forms are highly immunogenic and are involved in immune evasion. Several structures for the secreted form of NS1 were solved by cryo-electron microscopy and indicate that it forms a symmetrical barrel-shaped hexamer carrying a cargo of approximately 70 lipid molecules (Gutsche *et al.*, 2011; Muller *et al.*, 2012). Recently a breakthrough in the structural analysis of this protein was achieved resulting in the atomic structure for full-length WNV and DENV NS1 in both its dimer and hexameric forms. This structure has revealed great insight into the functional domains involved in membrane binding and immune complex interactions (Akey *et al.*, 2014; Edeling *et al.*, 2014).

NS2A is a transmembrane protein that functions in both viral RNA replication and assembly. While the atomic structure of NS2A is not available, the topology of the protein has recently being determined through extensive biochemical analyses. It has five ER membrane-spanning regions and two-luminal membrane associated regions with the N-terminus residing in the ER lumen and the C-terminus in the cytosol (Xie *et al.*,

2013, 2015). NS2A displays multifunctionality mediating both viral RNA replication and virus assembly. Interestingly, two distinct populations of NS2A seem to be involved in these two processes (Xie *et al.*, 2015).

NS3 is a serine protease and requires a co-factor, NS2B for activation. It functions in processing of the viral polyprotein (Bera *et al.*, 2007). It also has helicase and nucleotide triphosphatase activities important for viral RNA replication (Wengler *et al.*, 1991; Warrener *et al.*, 1995; Wu *et al.*, 2005). Additionally, similar to NS2A, it is also involved in virion assembly (Kummerer *et al.*, 2002; Patkar *et al.*, 2008). While NS3 does not have membrane-spanning regions, its functionality within the viral RNA replication complex requires that it is intimately surrounded by and positioned on viral induced membranes (Welsch *et al.*, 2009; Gillespie *et al.*, 2010). Although many atomic structures of individual domains have been solved through the years, the structure for the full-length NS3 protein was also recently solved (Luo *et al.*, 2008, 2010). The structure reveals that the protein is an elongated molecule with a linker region connecting the protease and helicase domains that allows for the correct level of flexibility between the domains to ensure optimal functionality. In addition to its role in viral RNA processing, replication and assembly, NS3 has a multifunctional role in interfering with or controlling host pathways (Khadka *et al.*, 2011). Recently, it has been demonstrated that flaviviruses significantly perturb the lipid repertoire of infected cells and that the primary enzyme in lipid biosynthesis, fatty acid synthase (FASN) is recruited to specific, virally induced membranes in the cell (Perera *et al.*, 2012). It was determined that NS3 was the primary viral protein responsible for recruiting FASN and modulating its activity (Heaton *et al.*, 2010a). It is now known that other lipid biosynthetic enzymes may also be recruited by NS3 to similar sites producing localized lipid factories in association with viral RNA replication (R.C. Gullberg, N. Chotiwan, R.J. Kuhn and R. Perera, unpublished). The NS2B/NS3 protease has also been implicated in suppressing the interferon production through selectively cleaving human STING, a protein that interacts with RIG-1 to stimulate IFN production (Aguirre *et al.*, 2012).

NS4A and NS4B are non-enzymatic, transmembrane proteins and the primary effectors of intracellular membrane rearrangements that occur during flavivirus infection. The atomic structures of these proteins are not known, but extensive biochemical analyses have been carried out to determine the topology of these proteins and their roles in viral RNA replication (Miller *et al.*, 2006, 2007; Roosendaal *et al.*, 2006). The proposed topology of NS4A suggests that it has four hydrophobic regions that span the ER membrane. Its N-terminal one-third localizes to the cytoplasm and the C-terminal end to the ER lumen. The C-terminal region of NS4A (referred to as 2K) is a signal sequence that assists in the translocation of NS4B into the ER lumen (Roosendaal *et al.*, 2006; Miller *et al.*, 2007; Ambrose and Mackenzie, 2015). Both Roosendaal *et al.* and Miller *et al.* observed that NS4A localized to viral RNA replication complexes and that its expression was required for inducing the altered viral membrane architecture. However, some disagreement exists as to the requirement of the 2K- peptide for induction of these membranes. Recently, the solution NMR structure of the N-terminal region of NS4A was solved and indicated that residues 17–80 from NS4A formed two amphipathic α helices, where helix α1 included residues 17–32 and helix α2 included residues 40–47. Both α1 and α2 associated with the cytosolic side of ER membrane. A third helix, α3 (residues 52–75) observed in the structure was proposed to transverse the ER membrane (Zou *et al.*, 2015b). Detailed interaction analyses of NS4A and NS4B carried out in the same study suggested that the first transmembrane domain of NS4A (residues 40–76) interacted with the first transmembrane domain of NS4B (80–146). Mutational analyses also indicated that this interaction between the two proteins was crucial for viral RNA replication.

NS4B has been shown to be a dimer in solution and in infected cells (Zou *et al.*, 2014) and it has also been implicated in suppressing the innate antiviral response (Munoz-Jordan *et al.*, 2003, 2005). Its topology suggests that two helices in the N-terminal region are in the ER lumen possibly associated with the luminal leaflet and interacting with NS4A. These helices are followed by three transmembrane regions that span the ER membrane. The C-terminal region is in the cytoplasm (Miller *et al.*, 2006). NS4B has been shown to interact with NS3. Specifically, subdomains 2 and 3 of the NS3 helicase interact with 12 amino acids from the cytoplasmic tail of NS4B. NMR analyses of this C-terminal region of NS4B has suggested it forms a three-turn α-helix and two short β-strands. Four conserved residues within this structure are essential for DENV replication (Zou *et al.*, 2015b). Both NS4A and NS4B together with the other viral proteins (NS2A, NS3 and NS5) are integral to the formation and function of viral RNA replication complexes and are also primarily responsible for rearranging the internal membrane architecture of infected cells required for the assembly of these complexes (discussed below) (Miller *et al.*, 2006, 2007; Roosendaal *et al.*, 2006; Welsch *et al.*, 2009; Gillespie *et al.*, 2010; Zou *et al.*, 2015a,b).

NS5 functions as a methyltransferase and an RNA-dependent RNA polymerase. Its methyltransferase domain is responsible for capping the viral genome and internal RNA methylation. This domain forms the N-terminal one-third of the protein. The C-terminal two-thirds of the protein is the RNA-dependent RNA polymerase required for viral genome replication. The two domains are connected through an alpha helical linker region that contains a bipartite nuclear localization signal (NLS). This NLS interacts with both importin α/β as well as NS3 (Johansson *et al.*, 2001). Similar to the NS3 protein, several atomic structures of the individual domains of NS5 have been solved (Yap *et al.*, 2010; Lim *et al.*, 2013; Brecher *et al.*, 2015). However, recently, the full-length structure has also been solved (Zhao *et al.*, 2015). While NS5 is functionally required in the cytoplasm for viral RNA replication, in some flaviviruses a majority of this viral protein translocates to the nucleus via its NLS (Rawlinson *et al.*, 2009; Fraser *et al.*, 2014). Therefore, it is hypothesized that two populations of NS5 molecules may exist in infected cells. One population is retained in the cytoplasm by its interaction with NS3 and the other population is a hyperphosphorylated form of NS5 that immediately translocates following synthesis to the nucleus for host modulation (Bhattacharya *et al.*, 2009). However, to date, its specific role in the nucleus has not been determined. One study has suggested that nuclear NS5 controls IL-8 gene expression and subsequently assists in immune evasion (Medin *et al.*, 2005; Rawlinson *et al.*, 2009). Another attractive hypothesis is that NS5 might be controlling the transcription of genes involved in modulation of cellular metabolism. It has also been shown that nuclear NS5 could increase NF-kB binding to the RANTES promoter and activates RANTES activity in an antiviral response (Khunchai *et al.*, 2012, 2015). In the cytoplasm, NS5 assists in evading the immune response by phosphorylating STAT1 and preventing its nuclear localization (Best *et al.*, 2005; Park *et al.*, 2007). It also marks STAT2 for degradation (Ashour *et al.*, 2009; Laurent-Rolle *et al.*, 2014). Therefore, NS5 while functioning mostly in a proviral capacity might also be exploited by the host to function in an antiviral capacity. Apart from NS3, NS5 is the only other soluble protein. However, recently, it has been demonstrated that NS5, similar to all other flaviviral proteins has a high affinity for membranes and that this membrane affinity is required for its role in viral RNA replication (R. Sengupta, R. Perera and R.J. Kuhn, unpublished).

The membranes

Translation and replication

The membrane architecture of flavivirus infected cells is conserved across species and optimized for intracellular signalling. During infection of both its human and mosquito hosts, flaviviruses cause significant rearrangement and expansion of intracellular membranes. These membrane rearrangements are induced following the expression of viral proteins NS4A and NS4B (Roosendaal *et al.*, 2006; Miller *et al.*, 2007). Studies have demonstrated that for these viruses, the membranes are derived from intermediate compartments (IC) or the rough endoplasmic reticulum (RER) and have defined morphologies that are conserved across species. Three distinct types of morphologies have been observed

1. *Convoluted membranes (CM).* These are highly curved complex reticular networks of membranes that are best described as a 'tangled ball of twine.' They are presumed to be continuous with the ER lumen to allow for dynamic transport of nutrients required for function.
2. *Paracrystaline arrays (PC).* These structures are only seen in WNV-infected cells and take the form of individual vesicle clusters that are immersed within and presumably connected to the CM.
3. *Vesicles (Ve) and vesicle packets (Vp).* These are luminal invaginations of the RER, where the Ve are small vesicles that are internally confined or 'packeted' within a surrounding membrane referred to as the Vp. Multiple Ve can be found within a single Vp. The Ve have neck-like structures that are directly open to the cytoplasm or connected to adjacent Ve that may be open to the cytoplasm to facilitate reagent exchange.

Immunofluorescence and immunoelectron microscopy studies have suggested that the CM may be sites of viral protein translation and processing due to the presence of the viral protein NS3 and its co-factor NS2B, that function in proteolytic processing of the viral genome. The same studies have also identified double stranded RNA (dsRNA) and the RNA-dependent RNA polymerase, NS5 in the Vp/Ve structures suggesting these membranes may function as platforms for viral RNA synthesis. Interestingly, ssRNA has not been observed within the Vp/Ve structures, suggesting that ssRNA might be rapidly transported away from these membranes for packaging into virus particles, or to the CM for translation of viral proteins. The current hypothesis is that many of these membranous structures are continuous with each other and with the ER to optimize transport of components required for viral protein translation and replication (Welsch *et al.*, 2009; Gillespie *et al.*, 2010).

In addition to the CM/PC and Vp/Ve structures, tube-like structures (T) have also been observed, but their function still remains illusive. A most striking observation it that the CM/PC, Vp/Ve and T structures described above seems to be conserved in human and mosquito cells as well as mosquitoes infected with mosquito-borne flaviviruses (Junjhon *et al.*, 2014). They are also observed in mammalian and tick cells infected with tick-borne flaviviruses (TBFs) (Offerdahl *et al.*, 2012). However, unlike the mosquito-borne viruses, TBFs cause an acute as well as persistent infection in tick cells. Comparison of the membrane architecture in acute versus persistently infected tick cells indicated that the primary difference was in the T structures. While in the acute phase, the T structures were observed as single tubules; in the persistent phase the tubular structures occurred in fascile-like bundles of multiple tubules wrapped in a single membranous sheath. The relevance of this observation to persistence of infection is yet unknown.

Virus assembly and budding

In the flaviviruses, only actively replicated viral RNA is packaged into virions (Kromykh *et al.*, 2001). Viral assembly and budding occur in close proximity to the ER-derived Vp/Ve structures. Specifically, structural protein complexes have been observed in invaginated ER cisternae that are directly apposed to the neck-like structure of the Ve. It is proposed that newly replicated ssRNA is transferred through these neck-like pores to the vicinity of the structural proteins to facilitate assembly. This transfer may involve interactions between the non-structural proteins such as NS2A or NS4A/4B and the capsid protein. The capsid protein facilitates viral RNA packaging into the virus particles through interaction of the capsid–RNA complex on the cytoplasmic side of the ER with the prM and E glycoproteins that are embedded in the ER membrane and forms a particle that buds into the ER lumen. This process allows the virus particle to acquire an ER-derived lipid membrane that resides between the capsid-RNA layer and the prM/E glycoprotein shell. Packets of these newly assembled viral particles are observed in infected cells and are released from the cells following maturation and transit through the TGN.

Flavivirus particles are icosahedrally symmetrical structures that display approximately $T=3$ symmetry (Kuhn *et al.*, 2002; Mukhopadhyay *et al.*, 2003; Zhang *et al.*, 2003, 2007a, 2013a; Kostyuchenko *et al.*, 2014). As described above, the viral particles consist of three structural proteins C, prM and E. Unlike in the alphaviruses that demonstrate an icosahedrally symmetric nucleocapsid, all structural studies carried out to date have not been able to identify a core or capsid in the flaviviruses that conforms to icosahedral symmetry (Zhang *et al.*, 2007b). Therefore, it is hypothesized that the capsid protein binds the genomic RNA similar to histones binding DNA and has a role assisting in packaging of the genome. The capsid–RNA complex is surrounded by the ER-derived membrane acquired through viral budding into the lumen of ER. The prM and E glycoproteins that are embedded in the ER membrane form a glycoprotein shell that hides the host-derived membrane from surface exposure. The icosahedral symmetry observed in the particle primarily arises from the arrangement of the proteins in this glycoprotein layer.

Significant structural plasticity is observed in the structure of flavivirus particles that assemble and bud into the ER (Perera *et al.*, 2008; Yu *et al.*, 2008). Specifically, the glycoproteins undergo pH dependent structural transitions during the maturation process as they transit through the TGN producing immature, partially mature and mature viruses that display pleomorphic structures. Newly assembled virus particles that bud into the ER are referred to as 'immature' particles. They have a diameter of ~60 nm and consist of 90 prM-E heterodimers arranged as 60 trimeric spikes extending outward from the surface of the particle. Maturation of the viral particles occurs as they transit through the TGN. The low-pH environment of the TGN drives dissociation of the prM-E heterodimers and drives structural transitions in the glycoprotein shell that results in the formation of 90 E protein homodimers. These homodimers lay flat against the viral membrane giving the viral particles a smooth morphology. The transition from 'spiky' to 'smooth' is a reversible process and is dependent on pH. This new smooth virus morphology allows furin in the TGN to cleave the pr-M protein resulting in the cleavage of the pr peptide from the M protein. The cleaved pr peptide however remains associated with the viral particles until it encounters a neutral pH environment upon release of the virus into the extracellular milieu. It is hypothesized that this is to ensure that the fusion loop that exists within the E protein is protected by the pr peptide to prevent premature fusion of the viral particles within the infected cell. The viral particles with a cleaved pr-M protein are considered 'mature.' Both immature and pr-associated mature viruses are non-infectious (or have reduced infectivity). Following release of the pr peptide the mature virus is infectious and can undergo structural transitions under low-pH conditions necessary to facilitate fusion within a new host cell.

Interestingly, not all flaviviral particles follow the immature to mature structural transitions to completion. Cryo-electron microscopy images of flavivirus particles often show pleomorphic maturation intermediates. Structural analysis of these intermediates have shown that maturation is nucleated at a single point on the particle and that as it progresses through the particle, there is a transient loss of ordered icosahedral organization in the glycoprotein shell (Plevka *et al.*, 2014). This is primarily facilitated by the fact that the glycoproteins while capable of transitioning between heterodimer and homodimer complexes are still tethered to the membrane through their transmembrane domains thus forcing their interactions to also be defined by their relationship with the host membrane. It has been postulated that this transient flexibility in the structural organization of the virus may be crucial for assembly, maturation and subsequent fusion with a new host cell (Stiasny *et al.*, 2006; Plevka *et al.*, 2014). This hypothesis is further supported by the observation that temperature differences between the human host and mosquito vector may alter or introduce new pleomorphisms between viral particles. Specifically, in some flavivirus strains, particles exposed to 37°C show a 'bumpy' larger (~550 Å) appearance with some membrane exposed on the surface while virus exposed to 28°C, representative of the body temperature of a mosquito vector, show little surface exposure of the membrane and smaller (~500 Å), smoother architecture (Zhang *et al.*, 2013b). These structural studies have made a crucial observation; that the human host or mosquito vector environment could have a significant impact on the structural organization and thus the resulting infectivity of the virus particles released from infected cells. This impact has to be significantly influenced by the host intracellular membrane environment, which becomes an intimate component of the structure of virus particle. Therefore, virus–host interactions, specifically those that change the intracellular membrane environment could be crucial to defining the outcome of infection.

Virus entry

Successful fusion of flaviviruses is dependent on the host membrane lipid composition. Prior to entry however, flavivirus particles must first attach to host cells via interactions with attachment molecules and/or receptors. While a specific receptor has not yet been identified for these viruses, several C-type lectins such as DC-SIGN in dendritic cells (Tassaneetrithep *et al.*, 2003; Lozach *et al.*, 2005; Fernandez-Garcia *et al.*, 2009), the mannose receptor in macrophages (Miller *et al.*, 2008) and L-SIGN in live endothelial cells (Tassaneetrithep *et al.*, 2003) have been suggested as attachment molecules facilitating entry into host cells (Chu *et al.*, 2004; Barba-Spaeth *et al.*, 2005) (Jindadamrongwech *et al.*, 2004; Navarro-Sanchez *et al.*, 2003; Lozach *et al.*, 2005; Pokidysheva *et al.*, 2006; Krishnan *et al.*, 2007). In addition to these attachment molecules, recent work by Morizono and Chen have suggested that a phosphatidylserine (PS)-mediated mechanism might also be at play during enveloped virus attachment (Morizono *et al.*, 2011, 2014). Specifically, they showed that a bifunctional serum protein, Gas6, is capable of linking phosphatidylserine (PS) found in the envelope of many viruses to Axl, a receptor tyrosine kinase found on target cells. Subsequently, several other molecules have also been discovered that facilitate this PS-mediated recognition between virus and host. For DENV and WNV, Meertens *et al.* (2012) demonstrated that the T cell immunoglobulin domain and mucin domain family of proteins (TIM 1 and 4) binds directly to PS in the envelope of DENV particles, while TAM proteins utilize their natural ligands Gas6 and ProS as bridging molecules to bind indirectly to PS exposed on the viral envelope.

Following attachment, flaviviruses primarily enter through receptor-mediated endocytosis. This is favoured by the low-pH environment of the endosome that is required to drive the structural transitions of the outer glycoproteins into their fusogenic form (Allison *et al.*, 1995; Acosta *et al.*, 2008; Harrison 2008; van der Schaar *et al.*, 2008; Smit *et al.*, 2011; Stiasny *et al.*, 2011). However, in addition to low pH, flaviviruses also require a specific lipid repertoire in the endosomal membrane for complete fusion to occur. Work by Zaitseva *et al.* demonstrated that even though the viral envelope glycoproteins could transition into the fusogenic structure in a low pH environment, completion of fusion between viral and host membranes required an anionic lipid repertoire, specifically bis(monoacyl glycero) phosphate, BMP in the host membrane. This is a lipid that is enriched in the inner leaflet of endosomes and based on the pH of the compartment will form multivesicular bodies within the endosomes (Zaitseva *et al.*, 2010). Additionally, Khadkha *et al.* (2011) have shown that several proteins in the ESCRT-III pathway that stabilize multivesicular body formation interact with DENV proteins. These data support the hypothesis that flavivirus fusion occurs in these late endosomal compartments (van der Schaar *et al.*, 2008; Zaitseva *et al.*, 2010). However, it should be noted that other studies have hinted that the coupling of a specific membrane lipid repertoire to flaviviral fusion and genome release may be specific to the viral and host membrane pair being studied (Krishnan *et al.*, 2007; van der Schaar *et al.*, 2008). Interestingly, in mosquito cells, the plasma membrane lipid composition is analogous to late endosomal membranes in mammalian cells in that it is enriched in anionic lipids. Therefore, under low-pH conditions, flaviviruses can fuse at the plasma membrane of mosquito cells. The mechanism of how these flavivirus class II fusion proteins might utilize or interact with the anionic phospholipids to facilitate fusion is yet unclear since not all class II fusion proteins require such interactions (i.e. alphaviruses). It is probable that the high negative curvature of anionic lipid containing membranes may play a significant role. It should also be noted here that other lipids such as phosphatidylcholine (PC), phosphatidylethanolamine (PE), cholesterol and sphingolipids also promote flavivirus fusion and may contribute to the overall physico-chemical properties required in the fusogenic membranes (Gollins *et al.*, 1986; Corver *et al.*, 2000; Stiasny *et al.*, 2003; Umashankar *et al.*, 2008; Poh *et al.*, 2009; Yu *et al.*, 2009; Moesker *et al.*, 2010; Schmidt *et al.*, 2010). Given that flaviviruses do not possess a structured nucleocapsid like the alphaviruses (Zhang *et al.*, 2007b), viral fusion with late endosomal membranes releases the RNA genome (and associated capsid proteins) into the cytoplasm for immediate translation, processing and replication.

The master metabolic switches

The requirement for highly specialized membranes to support every aspect of the flavivirus life cycle imposes a heavy burden on the viral proteins to expand their multifunctional capacity and reprogram the cellular metabolome such that the energy and biosynthetic needs for membrane biogenesis are met. This reprogramming is observed in key metabolic pathways including central carbon metabolism, lipid biosynthesis and lipolysis with the last two processes being impacted by activation of autophagy and the unfolded protein response (UPR) (Fig. 4.3). In this section we will discuss our current knowledge on how flaviviruses reprogram these metabolic pathways and how this reprogramming might impact flavivirus biogenesis.

Central carbon metabolism

Glycolysis and the tricarboxylic acid cycle (TCA) are the gatekeepers of cellular energy and biosynthetic needs. Cells utilize glucose and glutamine as the central carbon source to produce ATP through oxidative phosphorylation. This process can however be subverted by using these carbon sources for other macromolecular biosynthetic needs such as lipid biosynthesis. While the importance of central carbon metabolism is appreciated, only one study thus far has investigated its activation during flavivirus replication. Fontaine *et al.* (2015) demonstrated that glucose uptake was increased during a time course of DENV infection and that the glucose transporter (GLUT1) and the first enzyme in glycolysis, hexokinase 2 (HK2) was upregulated. Pharmacologically inhibiting glycolysis was detrimental to viral replication reducing both viral RNA synthesis and virion production. Metabolic profiling studies indicated that levels of early glycolytic intermediates, glucose-6-phosphate and fructose-6-phosphate increased with time of infection (at 10, 24 and 48 h post infection) but levels of late glycolytic intermediates such as 3-phosphoglycerate and phosphoenolpyruvate started at a higher level in infected cells compared to mock infected cells but decreased over time. An attractive hypothesis to explain these observations is that the later glycolytic intermediates are being actively utilized as infection progresses for the biosynthesis of lipids and that a differential flux through the individual steps in the glycolytic pathway drives faster utilization of the later versus early intermediates. Flux analysis through metabolic pathways is the next frontier in mass spectrometry based metabolomics and is currently being carried out to decipher the impact (in real time) of DENV infection on central carbon metabolism (J. Steel, R.C. Gullberg, J. Kirkwood, C. Broeckling and R. Perera, unpublished).

Fontaine *et al.* (2015) also showed that glutaminolysis was impacted during DENV infection. They proposed that glutamine could function as an anaplerotic substrate for the TCA cycle. Alternatively, it could also be a nitrogen source and drive purine and pyrimidine nucleotide biosynthesis. Interestingly, the intermediates for nucleotide biosynthesis were elevated during DENV infection, possibly supporting the latter hypothesis. Other studies (see 'Autophagy and lipolysis', below) have also demonstrated the increased requirement for ATP and NADH during the DENV life cycle (Heaton *et al.*, 2010a). Once again, metabolic flux analysis through these pathways would be critical to determine the requirement of the different carbon sources during flavivirus infection. Specifically, developing the ability to link metabolic flux with the kinetics of viral replication and virion biogenesis would lend great insight into the changing landscape of the cellular metabolome during infection.

De novo lipid biosynthesis

Initially, when electron microscopy studies revealed that flavivirus infected cells had a significantly rearranged membrane architecture, it was unknown whether existing membranes were reconfigured to form the specialized membranes, or whether new membrane biogenesis was activated. Based on recent studies investigating lipid biosynthesis in these infected cells, it is hypothesized now that new membrane biogenesis may be at play (Perera *et al.*, 2012; Aktepe *et al.*, 2015; A. Mikulasova, R.L. Ambrose, J.M. Makenzie *et al.*, unpublished). Following a primary lead from yeast two hybrid studies that identified an interaction between viral protein NS3 and fatty acid synthase (FASN), the primary enzyme in de novo phospholipid biosynthesis, it has been demonstrated that in DENV infected cells, NS3 relocalizes and activates FASN to initiate de novo lipid biosynthesis (Heaton *et al.*, 2010b; Khadka *et al.*, 2011). However, while in DENV-infected cells FASN relocates to sites of active viral RNA synthesis (within Vp/Ve structures), in WNV-infected cells FASN has been found to relocate to sites of viral RNA translation (CM/PC) yet unidentified sites (presumably

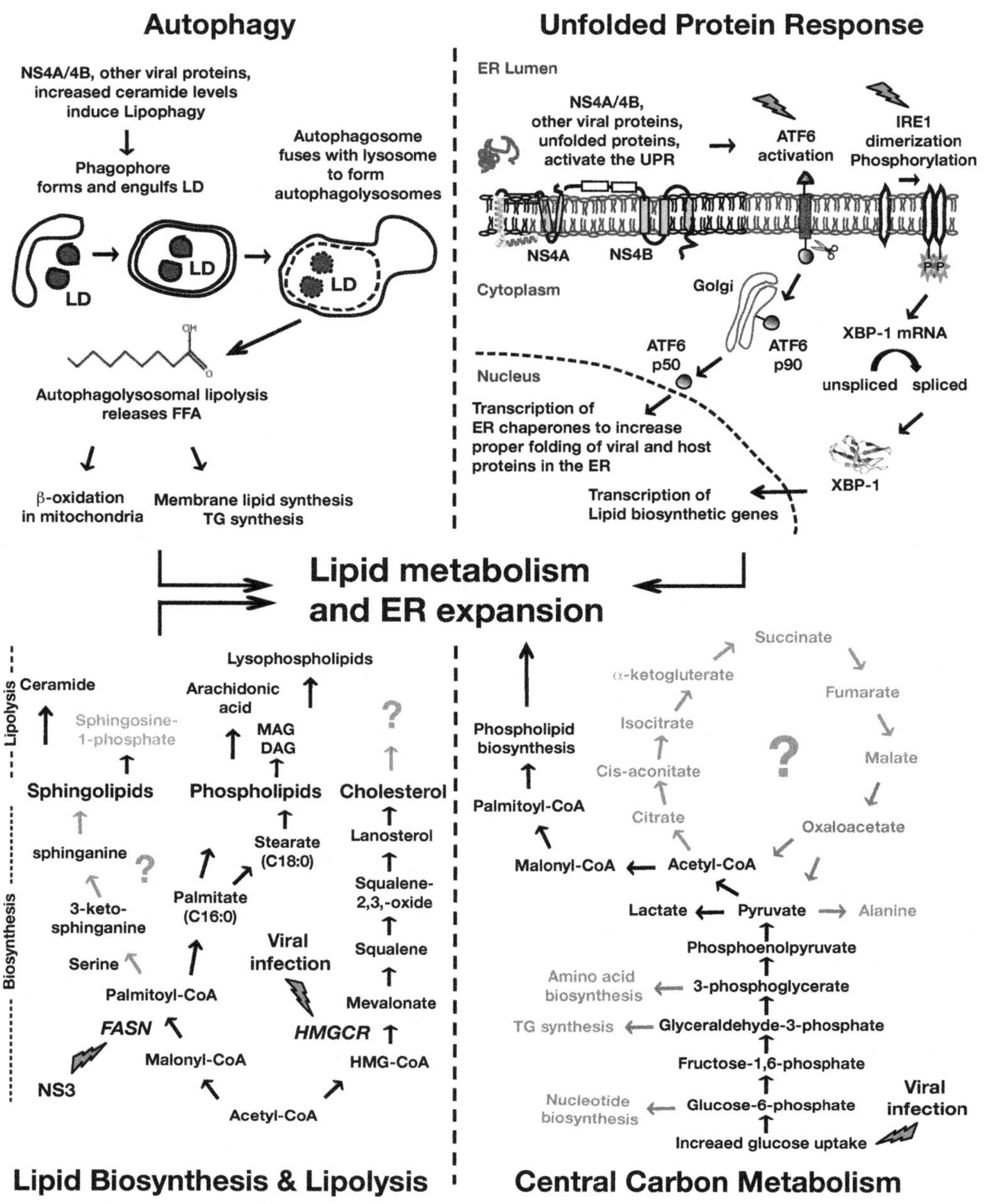

Figure 4.3 Flavivirus-induced ER membrane expansion is intimately linked to the metabolic pathways activated during infection. All metabolic pathways that are activated during flavivirus infection have the capacity to support the process of ER expansion. Top left; *Autophagy:* NS4A and NS4B (and possibly other viral proteins, depending on the flavivirus) or elevated ceramide (CER) levels (observed in DENV infected cells), induce autophagy. Activation of autophagy has been shown to be important for the degradation of lipid droplets, LD (a process called 'lipophagy'), which in turn releases free fatty acids (FFA) for b-oxidation and/or membrane lipid synthesis. These two processes could support ER expansion through stimulating energy homeostasis or providing raw materials for membrane synthesis. Top right; *Unfolded protein response:* The expression of membrane bound viral proteins such as NS4A and NS4B, or the increased accumulation of unfolded proteins in the ER lumen activates the UPR. Specifically, flavivirus infection activates ATF6 and IRE1. Upon activation, the cytoplasmic domain of ATF6 translocates to the nucleus to enhance transcription of ER chaperones that assist in the proper folding of viral and host proteins. Activation of IRE1 involves dimerization and phosphorylation leading to the splicing of XBP-1 mRNA and expression of the XBP-1 protein. Similar to ATF6, XBP-1 translocates to the nucleus and activates transcription of both ER chaperones for proper folding of viral and host proteins and lipid biosynthetic genes required for enhancing the lipid and membrane repertoire in infected cells. Bottom left; *Lipid biosynthesis and lipolysis:* Flaviviruses induce *de novo* phospholipid biosynthesis through the activation of fatty acid synthase (FASN). Specifically, FASN is activated by NS3 and catalyses the synthesis of fatty acids from acetyl-CoA and malonyl-CoA precursors. There is a preferential increase in the repertoire of specific phospholipids such as PC and PE. Additionally, elevated levels of lipid intermediates such as mono- and di-acylglycerols (MAG and DAG), arachidonic acid and lysophospholipids suggest that lipolysis of phospholipids are also activated during infection. While it is unclear if *de novo* synthesis of sphingolipids from palmitoyl-CoA and serine occur (grey arrows), the presence of elevated levels of CER suggests that lipolysis of complex sphingolipids could be occurring during infection. Cholesterol synthesis is also a key pathway perturbed during flavivirus infection. Specifically, HMG-CoA reductase (HMGCR), the rate-limiting enzyme in cholesterol biosynthesis is activated by flaviviruses. Cholesterol plays a crucial role in forming lipid rafts with sphingolipids (also elevated during infection) and provides platforms for the assembly of viral and host protein complexes. The role of cholesterol degradation in virus infection is unknown (grey). Overall, the activation of de novo lipid biosynthesis and lipolysis provides a specialized lipid repertoire important for virus-induced ER expansion. Bottom right; *Central carbon metabolism:* It has been demonstrated that glucoses uptake is enhanced during flavivirus infection (specifically during DENV infection). While flux analysis through this pathway is still ongoing, it is known that it is partially responsible for activating membrane lipid biosynthesis. However, its role in activating nucleotide and amino acid biosynthesis, or if flux through the TCA cycle is altered during infection is currently unknown (grey arrows).

other domains in the ER) and does not co-localize with viral proteins or viral RNA (Heaton *et al.*, 2010a; A. Mikulasova, R.L. Ambrose, J.M. Mackenzie *et al.*, unpublished). Pharmacological inhibition of FASN impacts both DENV and WNV replication indicating that while the activity of this enzyme is necessary for overall flavivirus replication, its activity may be required at different microenvironments in cells infected with DENV versus WNV. While these two viruses share many of the membrane morphological changes observed in infected cells (Vp/Ve and CM), PC structures are only observed in WNV infected cells. Therefore, these differences in FASN localization and activity may contribute to the differences in membrane architecture observed between WNV and DENV (A. Mikulasova, R.L. Ambrose, J.M. Mackenzie *et al.*, unpublished).

With the advances in liquid chromatography–mass spectrometry techniques (LC-MS), changes in lipid metabolism mediated through FASN activity or other pathways have been profiled in flavivirus-infected cells. Unfortunately, much of the work in mammalian cells is still in progress (A. Mikulasova, R.L. Ambrose, J.M. Mackenzie *et al.*, and R.C. Gullberg, N. Chotiwan, S. Khadka, D.J. LaCaount, R.J. Kuhn and R. Perera, unpublished). However, a comprehensive lipid profile in mosquito cells has been carried out (Perera *et al.*, 2012). This study suggested that there was an overall increase in the lipid content in DENV infected cells. Specifically, cellular membranes were enriched in lipids that can physically alter the curvature and permeability of the membranes as well as the recruitment and assembly of protein complexes. Among the phospholipids, unsaturated phosphotidylcholine (PC) showed the highest levels of expression. Generally PCs are planar, rigid bilayer-forming lipids. However, upon incorporation of unsaturated acyl chains, they can induce bending and increase the flexibility of the membrane. Therefore, incorporation of unsaturated PCs into virus-induced membranes could achieve the highly curved and fluid membrane architecture characteristic of the Vp/Ve and CM structures. Bioactive lipids such as sphingomyelin (SM) and ceramide (CER) were also upregulated in infected cells. While these lipids are important for signalling cascades, CER was enriched in replication complex membranes. CER is a cone-shaped lipid that can induce negative curvature (inward budding) in membranes and could assist in the inward budding of the Ve into the ER lumen (Hannun *et al.*, 2011). Therefore, viral infection could be upregulating CER to promote its required membrane architecture.

Interestingly, lysophospholipids (LPL) were also enriched in DENV infected cells. These lipids are inverted cone-shaped lipids that upon asymmetric incorporation into lipid bilayers could promote positive curvature of membranes (Cullis *et al.*, 1979; Janmey *et al.*, 2006; Rivera *et al.*, 2008). The neck-like structures in the Ve require positive curvature of the membrane. Additionally, positive curvature is utilized during vesicle fission and budding events such as those that occur during virion assembly. LPLs also function as water-soluble surfactants that increase the permeability of membranes thus increasing the 'leakiness' and increased transport of molecules across the bilayer. Leaky membranes are assets in virus-infected cells, as they increase communication with the cytosol for exchange of reagents necessary for genome replication and virion assembly, specifically when these activities are confined to the membranous microenvironments as those in the Vp/Ve and CM structures. It should be noted here that CER and LPLs are also upregulated as a stress response in the cell. Therefore, elevated levels of these lipids may not only be utilized for their physical attributes in membranes, but may also be a cellular signalling response to virus infection. The assembly of viral protein and host protein complexes in these highly modified ER membranes requires hydrophobic interactions between the proteins and lipids in the membrane. Several lipid intermediates that may assist in this process were also identified. Specifically, intermediates with small head groups such as mono- and diacylglycerol (MG and DG) and phosphatidic acid (PA) were upregulated during DENV infection. Transient accumulation of these lipids in membranes can induce bilayer instability and promote negative curvature. For instance, they can incorporate into membranes and locally expose non-polar regions in neighbouring lipids inducing strong hydrophobic interactions between the proteins and lipids.

Once again, the effort to determine lipid profiles in mammalian cells during flavivirus infection is still in progress. However, Mackenzie *et al.* (2007) have shown that, during WNV infection, cholesterol biosynthesis is upregulated and there is a redistribution of cellular cholesterol to viral replication membranes with a concomitant reduction in cholesterol at the plasma membrane. They also demonstrated that this requirement for cholesterol impacted both viral RNA replication and virus production, and that it was tightly linked to the activity of a cellular protein, HMG-CoA reductase (HMGCR) responsible for geranylgeranylation of cellular proteins involved in cholesterol biosynthesis. Essentially, HMGCR-induced membrane proliferation was required for WNV replication. Cholesterol is also an important lipid for DENV replication and the total cellular cholesterol and cholesterol-containing lipid raft formation as well as HMGCR activity are increased early during DENV infection. Modulation of cholesterol biosynthesis through genetic and pharmacological means reduced DENV RNA replication and virion biogenesis (Rothwell *et al.*, 2009; Martinez-Gutierrez *et al.*, 2011; Soto-Acosta *et al.*, 2013). It has been shown previously that cholesterol and SM containing lipid rafts are important for flavivirus fusion (Medigeshi *et al.*, 2008; Umashankar *et al.*, 2008; Moesker *et al.*, 2010; Carro *et al.*, 2013). Clearly, these lipid platforms are also required for the assembly of viral RNA replication complexes and the assembly of infectious viral particles (possibly impacting the conformational organization of the glycoprotein shell). It could be envisaged that the combination of rigidity induced by the presence of lipid rafts and fluidity introduced through the phospholipid composition (unsaturated PCs, cone-shaped or inverted cone-shaped lipids) may prime the viral induced membranes for optimal functionality.

Oxidative stress and membrane structure

Recently, the importance of oxidative stress has been investigated in DENV-infected cells. Several studies have shown that flaviviruses induce oxidative stress during infection of both mammalian and mosquito cells (Liao *et al.*, 2002; Kumar *et al.*, 2009; Lin *et al.*, 2000, 2004; Raung *et al.*, 2001; Verma *et al.*, 2008; Yang *et al.*, 2010). These studies also showed that antioxidants reduced viral RNA replication and virion biogenesis (Nazmi *et al.*, 2010; Chen *et al.*, 2011, 2012; Pan *et al.*, 2012). Historically, oxidation of molecules resulting in a cellular stress response was considered to merely be a response to infection. However, Gullberg *et al.* (2015) have recently demonstrated that flaviviruses require oxidative stress for viral RNA replication. Specifically, they showed that antioxidants reduced levels of viral RNA specifically inhibiting the capping of these RNAs by the viral protein NS5. They further determined that this effect was primarily due to the enhancement of its guanylyltransferase activity (Gullberg *et al.*, 2015). Oxidative stress can, however, have a detrimental effect on the structure of biological membranes (Borza *et al.*, 2013). It can lead to changes in acyl chain composition of phospholipids, changes in the ratio of polyunsaturated phospholipids and other fatty acids and subsequent reduction in membrane fluidity. It can also impose changes in the porocity of membranes and active ion transport. Therefore, given the extremely membranous microenvironments required for flaviviral replication, there must be a fine balance in these environments between the positive and negative influences of oxidative stress pathways. This warrants more detailed investigations of this pathway possibly linking the kinetics of viral RNA replication, virion biogenesis and changes in membrane lipid composition to the activation of oxidative stress. More detailed chemical measurements of the downstream metabolites of lipid peroxidation (aldehydes such as malondialdehyde or polymerized carbonyl compounds such as lipofuscin) will also lend great insight to how oxidative stress might impact the composition and function of the viral induced microenvironements.

Autophagy and lipolysis

Autophagy is a physiological response that controls cellular energy balance and nutritional status (Singh *et al.*, 2009). Cytosolic components including organelles are sequestered in autophagosomes (double membrane vesicles) that eventually fuse with lysosomes to degrade cellular components and provide raw material for biosynthesis of macromolecules or prevent accumulation of toxic components or microbes in the cell. It also functions as a process that liberates fatty acids from storage lipids such as triglycerides, to enhance β-oxidation or provide raw material for membrane biogenesis. During viral infection autophagy can function as an antiviral response (VSV, CHIKV, WNV) or be actively diverted to provide support for viral replication (DENV, HCV, PV).

During infection with flaviviruses, WNV and DENV seem to have opposing interactions with the autophagy pathway. In WNV infected cells, an increase in autophagic flux has been observed as evidenced by the increased availability of LC3-II, a lipidated form of LC3 that heralds the activation of autophagy. However, inhibition of this pathway through gene knockout experiments increased WNV RNA replication suggesting that autophagy was activated as an antiviral response by the host during WNV infection (Beatman *et al.*, 2012; Kobayashi *et al.*, 2014).

In contrast, several studies have suggested that DENV activates this pathway as a means of supporting viral RNA replication, virus assembly and maturation. During DENV infection, an increase in autophagosomes is observed, and these structures are found to co-localize with markers of lysosomes (LAMP1) forming autolysosomes or markers of late endosomes (M6P-R) forming amphisomes (Khakpoor *et al.*, 2009; Heaton *et al.*, 2010c; Panyasrivanit *et al.*, 2011). In the latter case, dsRNA was observed in the amphisomes, suggesting that the entry of these viruses through late endosomes may be directly linked to early events in viral RNA replication. One of the most interesting observations, however, is that DENV could be up-regulating autophagy to take advantage of its ability to control lipid and energy homeostasis in the cell (Singh *et al.*, 2009; Heaton *et al.*, 2010b). Work by Heaton and Randall has shown that DENV infection induces autophagosome formation at 24–48 h post infection and that these structures co-localize with lipid droplets (LD) in the cell. They also observed a negative correlation between the number of autophagosomes and the LD area per cell and demonstrated that the LD diameter was decreased by ~ 35%, leading to an overall reduction in LD volume by ~ 70% upon autophagy activation. Furthermore, through co-localization studies of markers for LDs, autophagosomes and lysosomes they showed that the LDs were delivered to autolysosomes for degradation during infection. Specifically, triglycerides, which are storage lipids in LDs, were depleted in an autophagy-dependent manner. The lipolysis of triglycerides is generally activated when there is an increased need for energy in the cell leading to the release of free fatty acids that are then transported to the mitochondria for β-oxidation. This study demonstrated that during DENV infection, the kinetics of LD depletion (triglyceride lipolysis) coincided with the increase in β-oxidation and these metabolic changes were required for both RNA replication and infectious virus release (Heaton and Randall, 2010). McLean *et al.* (2011) have shown that this process of autophagy activation may be coordinated by viral protein expression in a cell type specific manner. Specifically, they showed that the expression of DENV NS4A activated autophagy and PI3K-dependent pro-survival signalling in epithelial cells (and not in macrophages or neuronal cells). Pro-survival signalling is a common consequence of autophagy (Broker *et al.*, 2005; Kroemer *et al.*, 2005; Ogata *et al.*, 2006).

Subsequently, Mateo *et al.* (2013) showed that autophagy had a different impact on virus infection. In their studies, they stimulated autophagy using nicardipine or rapamycin and showed that infectious virus production was increased in both cells and mice. They also observed an increase in the pathology caused by virus infection. Furthermore, stimulation of autophagy increased the specific infectivity of virus particles as determined by the particle–PFU ratio of viruses released from cells treated with inducers. Inhibition of autophagy using an inhibitor with enhanced specificity towards autophagy, Spautin-1, however, decreased the production of infectious virions, and produced an abundance of non-infectious virions with maturation defects. Specifically, they observed an array of non-infectious particles of varying sizes and shapes (as determined by sedimentation experiments) and altered physicochemical properties possibly resembling assembly or disassembly intermediates. Analysis of the physical composition of the virions indicated that in the presence or absence of inhibitor, there was a similar ratio of envelope protein (E) to membrane protein (M) in the virions released from the cell, suggesting that inhibition of autophagy did not interfere with the normal maturation cleavage of the prM protein. However, the peptide, pr, that is cleaved during maturation was still associated with the virions released from inhibitor treated cells. During a normal infection (in the absence of inhibitor), this pr peptide is released from the virus particles upon exit from the cell due to the neutral pH of the extracellular environment (Liu *et al.*, 2011; Mateo *et al.*, 2013). The biochemical details leading to the observations in the current study are unclear. It has been shown that Spautiin-1 impacts the PI3K pathway and reduces the levels of PI3P lipids in the cell (Liu *et al.*, 2011). While previous studies have shown that PI lipids are dispensable for DENV replication, these observations raise the question of whether changes in cellular lipid composition might influence the flavivirus maturation process and impact the production of infectious particles.

The studies discussed above suggest that there is a complex interaction between autophagy and the flaviviral life cycle. Two specific themes arise. The first is where viral protein expression-mediated control of autophagy impacts cellular metabolic homeostasis by stimulating energy production for viral RNA replication and possibly pro-survival cascades in the cell. The second is the influence of autophagy on the physicochemical environment influencing virus assembly. It would be an attractive hypothesis to suggest that autophagy directed lipolysis and resulting free fatty acids have a direct influence on the lipid and protein composition of the ER membranes which are incorporated into the virion envelopes upon assembly and budding. This envelope composition would then influence the structural protein transitions necessary for virion maturation and conversion into infectious particles.

Unfolded protein response and lipid biosynthesis

The UPR pathway is triggered during ER stress, when there is an increase in the levels of misfolded or unfolded proteins. It is also triggered by nutrient deprivation, altered lipid metabolism and calcium homeostasis, over expression of abnormal proteins and virus infection (Volmer *et al.*, 2013, 2014). There are three known ER stress transducers: inositol-requiring enzyme 1 (IRE1), activating transcription factor 6 (ATF6) and PKR-like ER kinase (PERK) that initiate the UPR. These are ER transmembrane proteins and have luminal stress-sensing domains and cytosolic effector domains that control downstream gene expression cascades to respond to and alleviate ER stress. They function by increasing chaperone expression to assist in protein folding, increasing protein degradation by activating the ER-associated degradation pathway (ERAD), increasing ER volume by stimulating lipid biosynthesis and reducing protein expression by inhibiting translation.

There are several aspects of the UPR that are activated or manipulated during viral infection. This is specifically evident during flavivirus infection since viral protein translation, modification and virion assembly all occur in the ER placing significant stress on the organelle. The antiviral response of the UPR is controlled by PERK and is generally activated upon detection of increased viral protein expression leading to the inhibition of protein translation and apoptosis. Additionally, the other two arms of the UPR, ATF6 and IRE-1, drive the expression of full-length X box binding protein 1 (Xbp-1), a transcription factor that upregulates ERAD resulting in the enhanced degradation of viral proteins. Ambrose and Mackenzie have shown that during WNV infection of Vero cells the expression of downstream effectors of ERAD are not increased suggesting that WNV manipulation of the UPR is skewed towards a proviral influence. However, Medigeshi *et al.* observed that in neuronal cells infected with WNV, eIF2alpha was transiently phosphorylated to inhibit protein translation, and there was an induction of the proapoptotic cyclic AMP response element-binding transcription factor homologous protein. Additionally, ATF6 was rapidly degraded by proteasomes suggesting an up-regulation of ER stress (Medigeshi *et al.*, 2007).

The most interesting observation, however, is that the UPR is hijacked by these viruses to modulate lipid metabolism (Ambrose *et al.*, 2011). Through the activation of ATF6 and IRE-1, increased expression of Xbp-1 has a direct impact on lipid metabolism related genes and can be activated to drive ER expansion (Sriburi *et al.*, 2004; Volmer *et al.*, 2014). Given that flavivirus-infected cells demonstrate significant expansion of ER-derived membranes, this particular function of Xbp-1 is advantageous to the virus. Ambrose and Mackenzie have shown that WNV infected cells demonstrate a selective activation of ER sensors that modulate lipid metabolism. Specifically, in WNV infected Vero cells peak viral protein translation and RNA replication (~12–36 h post infection) coincided with an increase in Xbp-1 transcription and splicing controlled by ATF6 and IRE-1 activation. Additionally, they observed that the expression of two ER-resident transmembrane viral proteins NS4A and NS4B stimulated this effect specifically through the hydrophobic regions of the proteins (Ambrose *et al.*, 2011, 2013). It has been previously shown that hydrophobic (transmembrane) regions of ATF6 and IRE-1 are sufficient for the required oligomerization that activates these biosensors for signal transduction leading to increased lipid biosynthesis (Volmer and Ron, 2014). Thus, cross-talk between the hydrophobic regions of the ER-resident viral proteins, NS4A and NS4B and UPR biosensor could be triggering the increase in lipid biosynthesis. Interestingly, both these viral proteins are also the primary effectors of the membrane biogenesis observed in WNV infected cells. Therefore, it is an attractive hypothesis that NS4A and NS4B function by directly manipulating the UPR to drive membrane biogenesis. Interestingly, the work by Medigeshi *et al.* (2007) in neuronal cells has shown that Xbp-1 is non-essential for WNV replication. This poses the question whether lipid biosynthesis and membrane expansion may be different in neuronal versus non-neuronal cells or whether alternate mechanism other than the UPR are at play to control these processes.

Studies by Yu *et al.* (2006) have shown that in DENV- and JEV infected cells several ER resident viral proteins (prM-E, NS1, NS2A, NS2B-3 and 2K-NS4B), trigger splicing of Xbp-1 and increased expression of down stream molecules such as EDEM-1, ERdj4 and p58(IPK). However, it seems that this impact on Xbp-1 expression is not directly required to support viral replication, but rather is beneficial to combat the cytopathic effects induced by viral infection. Early dengue virus protein synthesis induces extensive rearrangement of the endoplasmic reticulum independent of the UPR and SREBP-2 pathway (Pena and Harris, 2012). These are significant observations as they suggest that WNV and DENV may have different utilities for the UPR and may employ different control mechanisms for lipid biosynthesis and membrane biogenesis. It would be important to conduct parallel studies in the same cell types for instance to directly compare the utility of the UPR in DENV and WNV paying close attention to the kinetics of virus replication.

Calcium signalling

Calcium (Ca^{2+}) is a universal signalling molecule that plays an important role in many cellular processes. Ca^{2+} also significantly influences several aspects of virus replication including viral entry, viral protein expression and processing, virion assembly, maturation and budding (Zhou *et al.*, 2009). The influx of Ca^{2+} into the cytosol can occur through the opening of Ca^{2+} channels on the surface of cells, through the acidification of endosomes or through the release of Ca^{2+} from intracellular stores in the ER. Unfortunately, only one study has investigated the importance of calcium signalling in flavivirus replication. Scherbik and Brington showed that WNV attachment and/or entry triggers measurable calcium influx into infected cells through calcium channels on the plasma membrane. While Ca^{2+} influx is not required for WNV entry, it seems to play a significant role early (within the first 2 h) during virus replication (Scherbik *et al.*, 2010). Specifically, they observed that Ca^{2+} signalling was required to activate the focal adhesion kinase (FAK) that in turn activated several cell survival pathways such as ERK1/2 and PI3K/Akt early during infection. For instance, It still remains to be seen if intracellular calcium stores are also utilized through the newly formed ER-derived membrane contact sites in flavivirus infected cells to carry out signalling required for aspects of the replication cycle.

Conclusions and perspectives

Mimicking cellular membrane contact sites

The extensive membrane rearrangements in flavivirus infected cells are like a membrane 'cobweb' that extends across the cytoplasm linking the ER, the nuclear membrane and other organelles in the cell. In essence, they mimic cellular 'membrane contact sites' (MCSs). In an uninfected cell, tethering of membranes from compartments of the same organelle or membranes from two different organelles occur to exchange material or transmit signals. These tethered morphologies are known as MCS and are characterized by specific criteria: (i) membranes from two cellular compartments are in close apposition (~30 nm) to each other; (ii) the membranes do not fuse but may transiently show hemi-fusion; (iii) specific proteins and/or lipids are enriched at the MCSs; and (iv) the formation of the MCS impacts the function or cellular lipid and protein composition of at least one of the two organelles. Functionally, these MCSs are critical for the exchange of lipids, specifically cholesterol, for the control of calcium and other intracellular signalling cascades, organelle trafficking, inheritance and most importantly, the assembly and function of regulatory protein complexes in the cell (Prinz, 2014; Lahiri *et al.*, 2015).

The flavivirus induced membranes have a distinct similarity to the MCS architecture and fulfil the criteria described above. The extensive expansion of the ER membranes into highly curved and convoluted CM/PC structures or the Vp/Ve structures bring the ER luminal compartments into close apposition without showing signs of membrane fusion. Additionally, the increased membrane surface area expands the platforms available for the assembly of viral-host protein regulatory complexes facilitating a local enrichment of both viral and host proteins in these membranes. Recent evidence has also shown that the recruitment of specific lipid biosynthetic enzymes into the CM/PC and Vp/Ve structures function to increase the local synthesis of specific lipids required for the highly curved membrane architecture, causing changes in or increasing the functionality of these specialized ER compartment. Recent evidence has also shown that the composition of these viral induced membranes may be unique with respect to both protein and lipid content compared with their originating ER membranes. The favoured hypothesis is that the recruitment of specific lipid biosynthetic enzymes into the CM/PC and Vp/Ve structures by viral proteins function to increase the local synthesis of specific lipids required for the specialized membrane architecture. The newly formed MCSs may assist in these recruitment efforts. The ER performs many essential functions including protein folding, secretion, and calcium homeostasis, the unfolded protein response and lipid metabolism. It is possible that by modifying the ER, flaviviruses also hijack some of these essential functions of the organelle specifically enhancing intracellular signalling through these newly formed membrane contact sites within the CM/PC and Vp/Ve structures. These new membrane contacts may also assist in modulating other cellular metabolic pathways such as autophagy and oxidative stress.

The unknown

Much of the work on cellular metabolism has been carried out in mammalian cells. Little is know about the comparative requirements of these metabolic networks in the invertebrate hosts (mosquitoes and ticks). The only study in the mosquito vector was carried out on mosquito cells and investigated the overall changes in lipid profile during DENV infection. Cellular metabolism is a critical aspect of virus–host interactions in the flaviviruses since their transmission cycle requires successful replication within both vertebrate and invertebrate hosts. Specifically, since flaviviral replication is so intimately linked to membrane biogenesis in both hosts, cellular metabolic alterations must also be critical in both environments. For instance, are the same metabolic pathways altered to impact viral RNA replication and virion biogenesis in the arthropod vector compared the human host? Does the metabolic repertoire play a role in defining viral escape from the midgut and effective dissemination through the arthropod vector? In essence does the cellular metabolic environment influence the midgut escape barrier? Does the infectivity of virus particles that are released from the invertebrate hosts directly into humans depend on the membrane composition available to the virus upon budding from different tissues in the arthropod vector (i.e. midguts and salivary glands)? Do we see the same pleomorphisms in mature, partially mature and immature particles released from the arthropod vector? If so, what metabolic repertoire might influence this process? How does autophagy, the UPR pathway or oxidative stress play a role in each of these hosts to impact virus replication and dissemination? Finally, since cellular metabolism is the key to understanding the efficacy of antiviral or anti-vector interventions can it be exploited to identify novel targets for intervention? Although much has recently been described for flavivirus–host interactions, there remain many fundamental questions to be answered.

References

Acosta, E.G., Castilla, V., and Damonte, E.B. (2008). Functional entry of dengue virus into *Aedes albopictus* mosquito cells is dependent on clathrin-mediated endocytosis. J. Gen. Virol. *89*, 474–484.

Aguirre, S., Maestre, A.M., Pagni, S., Patel, J.R., Savage, T., Gutman, D., Maringer, K., Bernal-Rubio, D., Shabman, R.S., Simon, V., *et al.* (2012). DENV inhibits type I IFN production in infected cells by cleaving human STING. PLoS Pathog. *8*, e1002934.

Akey, D.L., Brown, W.C., Dutta, S., Konwerski, J., Jose, J., Jurkiw, T.J., DelProposto, J., Ogata, C.M., Skiniotis, G., Kuhn, R.J., and Smith, J.L. (2014). Flavivirus NS1 structures reveal surfaces for associations with membranes and the immune system. Science *343*, 881–885.

Allison, S.L., Schalich, J., Stiasny, K., Mandl, C.W., Kunz, C., and Heinz, F.X. (1995). Oligomeric rearrangement of tick-borne encephalitis virus envelope proteins induced by an acidic pH. J. Virol. *69*, 695–700.

Allison, S.L., Schalich, J., Stiasny, K., Mandl, C.W., and Heinz, F.X. (2001). Mutational evidence for an internal fusion peptide in flavivirus envelope protein E. J. Virol. *75*, 4268–4275.

Ambrose, R.L., and Mackenzie, J.M. (2011). West Nile virus differentially modulates the unfolded protein response to facilitate replication and immune evasion. J. Virol. *85*, 2723–2732.

Ambrose, R.L., and Mackenzie, J.M. (2013). ATF6 signaling is required for efficient West Nile virus replication by promoting cell survival and inhibition of innate immune responses. J. Virol. *87*, 2206–2214.

Ashour, J., Laurent-Rolle, M., Shi, P.Y., and Garcia-Sastre, A. (2009). NS5 of dengue virus mediates STAT2 binding and degradation. J. Virol. *83*, 5408–5418.

Barba-Spaeth, G., Longman, R.S., Albert, M.L., and Rice, C.M. (2005). Live attenuated yellow fever 17D infects human DCs and allows for presentation of endogenous and recombinant T cell epitopes. J. Exp. Med. *202*, 1179–1184.

Bhattacharya, D., Mayuri, Best, S.M., Perera, R., Kuhn, R.J., and Striker, R. (2009). Protein kinase G phosphorylates mosquito-borne flavivirus NS5. J. Virol. *83*, 9195–9205.

Beasley, D.W., and Barrett, A.D. (2002). Identification of neutralizing epitopes within structural domain III of the West Nile virus envelope protein. J. Virol. *76*, 13097–13100.

Beatman, E., Oyer, R., Shives, K.D., Hedman, K., Brault, A.C., Tyler, K.L., and Beckham, J.D. (2012). West Nile virus growth is independent of autophagy activation. Virology *433*, 262–272.

Bera, A.K., Kuhn, R.J., and Smith, J.L. (2007). Functional characterization of cis and trans activity of the Flavivirus NS2B-NS3 protease. J. Biol. Chem. *282*, 12883–12892.

Best, S.M., Morris, K.L., Shannon, J.G., Robertson, S.J., Mitzel, D.N., Park, G.S., Boer, E., Wolfinbarger, J.B., and Bloom, M.E. (2005). Inhibition of interferon-stimulated JAK-STAT signaling by a tick-borne flavivirus and identification of NS5 as an interferon antagonist. J. Virol. *79*, 12828–12839.

Bhardwaj, S., Holbrook, M., Shope, R.E., Barrett, A.D., and Watowich, S.J. (2001). Biophysical characterization and vector-specific antagonist activity of domain III of the tick-borne flavivirus envelope protein. J. Virol. *75*, 4002–4007.

Bhuvanakantham, R., Chong, M.K., and Ng, M.L. (2009). Specific interaction of capsid protein and importin-alpha/beta influences West Nile virus production. Biochem. Biophys. Res. Comm. *389*, 63–69.

Borza, C., Muntean, D., Dehelean, C., Săvoiu, G., Şerban, C., Simu, G., Andoni, M., Butur, M., and Drăgan, S. (2013). Oxidative stress and lipid peroxidation – a lipid metabolism dysfunction. In Lipid Metabolism, R.V. Baez, ed. (InTech), pp. 23–38.

Brecher, M.B., Li, Z., Zhang, J., Chen, H., Lin, Q., Liu, B., and Li, H. (2015). Refolding of a fully functional flavivirus methyltransferase revealed that S-adenosyl methionine but not S-adenosyl homocysteine is copurified with flavivirus methyltransferase. Protein Sci. *24*, 117–128.

Broker, L.E., Kruyt, F.A., and Giaccone, G. (2005). Cell death independent of caspases: a review. Clin. Cancer Res. *11*, 3155–3162.

Carro, A.C., and Damonte, E.B. (2013). Requirement of cholesterol in the viral envelope for dengue virus infection. Virus Res. *174*, 78–87.

Chatel-Chaix, L., and Bartenschlager, R. (2014). Dengue virus- and hepatitis C virus-induced replication and assembly compartments: the enemy inside – caught in the web. J. Virol. *88*, 5907–5911.

Chen, T.H., Tang, P., Yang, C.F., Kao, L.H., Lo, Y.P., Chuang, C.K., Shih, Y.T., and Chen, W.J. (2011). Antioxidant defense is one of the mechanisms by which mosquito cells survive dengue 2 viral infection. Virology *410*, 410–417.

Chen, T.H., Lo, Y.P., Yang, C.F., and Chen, W.J. (2012). Additive protection by antioxidant and apoptosis-inhibiting effects on mosquito cells with dengue 2 virus infection. PLoS Negl. Trop. Dis. *6*, e1613.

Chu, J.J., and Ng, M.L. (2004). Interaction of West Nile virus with alpha v beta 3 integrin mediates virus entry into cells. J. Biol. Chem. *279*, 54533–54541.

Colpitts, T.M., Barthel, S., Wang, P., and Fikrig, E. (2011). Dengue virus capsid protein binds core histones and inhibits nucleosome formation in human liver cells. PLoS One *6*, e24365.

Corver, J., Ortiz, A., Allison, S.L., Schalich, J., Heinz, F.X., and Wilschut, J. (2000). Membrane fusion activity of tick-borne encephalitis virus and recombinant subviral particles in a liposomal model system. Virology *269*, 37–46.

Crill, W.D., and Roehrig, J.T. (2001). Monoclonal antibodies that bind to domain III of dengue virus E glycoprotein are the most efficient blockers of virus adsorption to Vero cells. J. Virol. *75*, 7769–7773.

Cullis, P.R., and de Kruijff, B. (1979). Lipid polymorphism and the functional roles of lipids in biological membranes. Biochim. Biophys. Acta *559*, 399–420.

Despres, P., Frenkiel, M.P., and Deubel, V. (1993). Differences between cell membrane fusion activities of two dengue type-1 isolates reflect modifications of viral structure. Virology *196*, 209–219.

Dowd, K.A., Mukherjee, S., Kuhn, R.J., and Pierson, T.C. (2014). Combined effects of the structural heterogeneity and dynamics of flaviviruses on antibody recognition. J. Virol. *88*, 11726–11737.

Edeling, M.A., Diamond, M.S., and Fremont, D.H. (2014). Structural basis of Flavivirus NS1 assembly and antibody recognition. Proc. Natl. Acad. Sci. U.S.A. *111*, 4285–4290.

Fernandez-Garcia, M.D., Mazzon, M., Jacobs, M., and Amara, A. (2009). Pathogenesis of flavivirus infections: using and abusing the host cell. Cell Host Microbe *5*, 318–328.

Fontaine, K.A., Sanchez, E.L., Camarda, R., and Lagunoff, M. (2015). Dengue virus induces and requires glycolysis for optimal replication. J. Virol. *89*, 2358–2366.

Fraser, J.E., Rawlinson, S.M., Wang, C., Jans, D.A., and Wagstaff, K.M. (2014). Investigating dengue virus non-structural protein 5 (NS5) nuclear import. Methods Mol. Biol. *1138*, 301–328.

Gillespie, L.K., Hoenen, A., Morgan, G., and Mackenzie, J.M. (2010). The endoplasmic reticulum provides the membrane platform for biogenesis of the flavivirus replication complex. J. Virol. *84*, 10438–10447.

Gollins, S.W., and Porterfield, J.S. (1986). pH-dependent fusion between the flavivirus West Nile and liposomal model membranes. J. Gen. Virol. *67*, 157–166.

Guirakhoo, F., Heinz, F.X., Mandl, C.W., Holzmann, H., and Kunz, C. (1991). Fusion activity of flaviviruses: comparison of mature and immature (prM-containing) tick-borne encephalitis virions. J. Gen. Virol. 72, 1323–1329.

Guirakhoo, F., Bolin, R.A., and Roehrig, J.T. (1992). The Murray Valley encephalitis virus prM protein confers acid resistance to virus particles and alters the expression of epitopes within the R2 domain of E glycoprotein. Virology *191*, 921–931.

Guirakhoo, F., Hunt, A.R., Lewis, J.G., and Roehrig, J.T. (1993). Selection and partial characterization of dengue 2 virus mutants that induce fusion at elevated pH. Virology *194*, 219–223.

Gullberg, R.C., Jordan Steel, J., Moon, S.L., Soltani, E., and Geiss, B.J. (2015). Oxidative stress influences positive strand RNA virus genome synthesis and capping. Virology *475*, 219–229.

Gutsche, I., Coulibaly, F., Voss, J.E., Salmon, J., d'Alayer, J., Ermonval, M., Larquet, E., Charneau, P., Krey, T., Megret, F., *et al.* (2011). Secreted dengue virus non-structural protein NS1 is an atypical barrel-shaped high-density lipoprotein. Proc. Natl. Acad. Sci. U.S.A. *108*, 8003–8008.

Halstead, S.B., Heinz, F.X., Barrett, A.D., and Roehrig, J.T. (2005). Dengue virus: molecular basis of cell entry and pathogenesis, 25–27 June 2003, Vienna, Austria. Vaccine *23*, 849–856.

Hannun, Y.A., and Obeid, L.M. (2011). Many ceramides. J. Biol. Chem. *286*, 27855–27862.

Harrison, S.C. (2008). The pH sensor for flavivirus membrane fusion. J. Cell Biol. *183*, 177–179.

Heaton, N.S., and Randall, G. (2010). Dengue virus-induced autophagy regulates lipid metabolism. Cell Host Microbe *8*, 422–432.

Heaton, N.S., Perera, R., Berger, K.L., Khadka, S., LaCount, D.J., Kuhn, R.J., and Randall, G. (2010a). Dengue virus non-structural protein 3 redistributes fatty acid synthase to sites of viral replication and increases cellular fatty acid synthesis. Proc. Natl. Acad. Sci. U.S.A. *107*, 17345–17350.

Heaton, N.S., and Randall, G. (2010b). Dengue virus-induced autophagy regulates lipid metabolism. Cell Host Microbe *8*, 422–432.

Janmey, P.A., and Kinnunen, P.K. (2006). Biophysical properties of lipids and dynamic membranes. Trends Cell Biol. *16*, 538–546.

Jindadamrongwech, S., Thepparit, C., and Smith, D.R. (2004). Identification of GRP 78 (BiP) as a liver cell expressed receptor element for dengue virus serotype 2. Arch. Virol. *149*, 915–927.

Johansson, M., Brooks, A.J., Jans, D.A., and Vasudevan, S.G. (2001). A small region of the dengue virus-encoded RNA-dependent RNA polymerase, NS5, confers interaction with both the nuclear transport receptor importin-beta and the viral helicase, NS3. J. Gen. Virol. *82*, 735–745.

Jones, C.T., Ma, L., Burgner, J.W., Groesch, T.D., Post, C.B., and Kuhn, R.J. (2003). Flavivirus capsid is a dimeric alpha-helical protein. J. Virol. *77*, 7143–7149.

Junjhon, J., Pennington, J.G., Edwards, T.J., Perera, R., Lanman, J., and Kuhn, R.J. (2014). Ultrastructural characterization and three-dimensional architecture of replication sites in dengue virus-infected mosquito cells. J. Virol. *88*, 4687–4697.

Kanai, R., Kar, K., Anthony, K., Gould, L.H., Ledizet, M., Fikrig, E., Marasco, W.A., Koski, R.A., and Modis, Y. (2006). Crystal structure of West Nile virus envelope glycoprotein reveals viral surface epitopes. J. Virol. *80*, 11000–11008.

Khadka, S., Vangeloff, A.D., Zhang, C., Siddavatam, P., Heaton, N.S., Wang, L., Sengupta, R., Sahasrabudhe, S., Randall, G., Gribskov, M., Kuhn, R.J., Perera, R., and LaCount, D.J. (2011). A physical interaction network of dengue virus and human proteins. Mol. Cell Proteomics *10*, M111.012187.

Khakpoor, A., Panyasrivanit, M., Wikan, N., and Smith, D.R. (2009). A role for autophagolysosomes in dengue virus 3 production in HepG2 cells. J. Gen. Virol. *90*, 1093–1103.

Kromykh, A.A., Varnavski, A.N., Sedlak, P.L., and Westaway, E.G. (2001). Coupling between replication and packaging of flavivirus RNA: evidence derived from the use of DNA-based full-length cDNA clones of Kunjin virus. J. Virol. *75*, 4633–4640.

Khunchai, S., Junking, M., Suttitheptumrong, A., Yasamut, U., Sawasdee, N., Netsawang, J., Morchang, A., Chaowalit, P., Noisakran, S., Yenchitsomanus, P.T., and Limjindaporn, T. (2012). Interaction of dengue virus non-structural protein 5 with Daxx modulates RANTES production. Biochem. Biophys. Res. Commun. *423*, 398–403.

Khunchai, S., Junking, M., Suttitheptumrong, A., Kooptiwut, S., Haegeman, G., Limjindaporn, T., and Yenchitsomanus, P.T. (2015). NF-kappaB is required for dengue virus NS5-induced RANTES expression. Virus Res. *197*, 92–100.

Kobayashi, S., Orba, Y., Yamaguchi, H., Takahashi, K., Sasaki, M., Hasebe, R., Kimura, T., and Sawa, H. (2014). Autophagy inhibits viral genome replication and gene expression stages in West Nile virus infection. Virus Res. *191*, 83–91.

Kostyuchenko, V.A., Chew, P.L., Ng, T.S., and Lok, S.M. (2014). Near-atomic resolution cryo-electron microscopic structure of dengue serotype 4 virus. J. Virol. *88*, 477–482.

Krishnan, M.N., Sukumaran, B., Pal, U., Agaisse, H., Murray, J.L., Hodge, T.W., and Fikrig, E. (2007). Rab 5 is required for the cellular entry of dengue and West Nile viruses. J. Virol. *81*, 4881–4885.

Kroemer, G., and Jaattela, M. (2005). Lysosomes and autophagy in cell death control. Nat Rev Cancer *5*, 886–897.

Kuhn, R.J. (2011). Flaviviruses. In Fundamentals of Molecular Virology, 2nd edn, Acheson, N.H., ed. (John Wiley & Sons, Hoboken, NJ, USA), pp. 137–147.

Kuhn, R.J., Zhang, W., Rossmann, M.G., Pletnev, S.V., Corver, J., Lenches, E., Jones, C.T., Mukhopadhyay, S., Chipman, P.R., Strauss, E.G., Baker, T.S., and Strauss, J.H. (2002a). Structure of dengue virus: implications for flavivirus organization, maturation, and fusion. Cell *108*, 717–725.

Kumar, S., Misra, U.K., Kalita, J., Khanna, V.K., and Khan, M.Y. (2009). Imbalance in oxidant/antioxidant system in different brain regions of rat after the infection of Japanese encephalitis virus. Neurochem. Int. *55*, 648–654.

Kummerer, B.M., and Rice, C.M. (2002). Mutations in the yellow fever virus non-structural protein NS2A selectively block production of infectious particles. J. Virol. *76*, 4773–4784.

Lahiri, S., Toulmay, A., and Prinz, W.A. (2015). Membrane contact sites, gateways for lipid homeostasis. Curr. Opin. Cell Biol. *33C*, 82–87.

Laurent-Rolle, M., Morrison, J., Rajsbaum, R., Macleod, J.M., Pisanelli, G., Pham, A., Ayllon, J., Miorin, L., Martinez-Romero, C., tenOever, B.R., and Garcia-Sastre, A. (2014). The interferon signaling antagonist function of yellow fever virus NS5 protein is activated by type I interferon. Cell Host Microbe *16*, 314–327.

Li, L., Lok, S.M., Yu, I.M., Zhang, Y., Kuhn, R.J., Chen, J., and Rossmann, M.G. (2008). The flavivirus precursor membrane-envelope protein complex: structure and maturation. Science *319*, 1830–1834.

Liao, S.L., Raung, S.L., and Chen, C.J. (2002). Japanese encephalitis virus stimulates superoxide dismutase activity in rat glial cultures. Neurosci. Lett. *324*, 133–136.

Lim, S.P., Koh, J.H., She, C.C., Liew, C.W., Davidson, A.D., Chua, L.S., Chandrasekaran, R., Cornvik, T.C., Shi, P.Y., and Lescar, J. (2013). A crystal structure of the dengue virus non-structural protein 5 (NS5) polymerase delineates interdomain amino acid residues that enhance its thermostability and de novo initiation activities. J. Biol. Chem. *288*, 31105–31114.

Lin, Y.L., Liu, C.C., Chuang, J.I., Lei, H.Y., Yeh, T.M., Lin, Y.S., Huang, Y.H., and Liu, H.S. (2000). Involvement of oxidative stress, NF-IL-6, and RANTES expression in dengue-2-virus-infected human liver cells. Virology *276*, 114–126.

Lin, R.J., Liao, C.L., and Lin, Y.L. (2004). Replication-incompetent virions of Japanese encephalitis virus trigger neuronal cell death by oxidative stress in a culture system. J. Gen. Virol. *85*, 521–533.

Lindenbach, B.D., and Rice, C.M. (1999). Genetic interaction of flavivirus non-structural proteins NS1 and NS4A as a determinant of replicase function. J. Virol. *73*, 4611–4621.

Lindenbach, B.D., and Rice, C.M. (2001). *Flaviviridae*: the viruses and their replication. In Fields Virology, D.M. Knipe and P.M. Howley, eds. (Lippincott Williams & Wilkins, Philadelphia), pp. 991–1041.

Liu, J., Xia, H., Kim, M., Xu, L., Li, Y., Zhang, L., Cai, Y., Norberg, H.V., Zhang, T., Furuya, T., *et al.* (2011). Beclin1 controls the levels of p53 by regulating the deubiquitination activity of USP10 and USP13. Cell *147*, 223–234.

Lozach, P.Y., Burleigh, L., Staropoli, I., Navarro-Sanchez, E., Harriague, J., Virelizier, J.L., Rey, F.A., Despres, P., Arenzana-Seisdedos, F., and Amara, A. (2005). Dendritic cell-specific intercellular adhesion molecule 3-grabbing non-integrin (DC-SIGN)-mediated enhancement of dengue virus infection is independent of DC-SIGN internalization signals. J. Biol. Chem. *280*, 23698–23708.

Luo, D., Xu, T., Hunke, C., Gruber, G., Vasudevan, S.G., and Lescar, J. (2008). Crystal structure of the NS3 protease-helicase from dengue virus. J. Virol. *82*, 173–183.

Luo, D., Wei, N., Doan, D.N., Paradkar, P.N., Chong, Y., Davidson, A.D., Kotaka, M., Lescar, J., and Vasudevan, S.G. (2010). Flexibility between the protease and helicase domains of the dengue virus NS3 protein conferred by the linker region and its functional implications. J. Biol. Chem. *285*, 18817–18827.

Ma, L., Jones, C.T., Groesch, T.D., Kuhn, R.J., and Post, C.B. (2004). Solution structure of dengue virus capsid protein reveals another fold. Proc. Natl. Acad. Sci. U.S.A. *101*, 3414–3419.

Mackenzie, J.M., Khromykh, A.A., Jones, M.K., and Westaway, E.G. (1998). Subcellular localization and some biochemical properties of the flavivirus Kunjin non-structural proteins NS2A and NS4A. Virology *245*, 203–215.

Mackenzie, J.M., Khromykh, A.A., and Parton, R.G. (2007). Cholesterol manipulation by West Nile virus perturbs the cellular immune response. Cell Host Microbe *2*, 229–239.

McLean, J.E., Wudzinska, A., Datan, E., Quaglino, D., and Zakeri, Z. (2011). Flavivirus NS4A-induced autophagy protects cells against death and enhances virus replication. J. Biol. Chem. *286*, 22147–22159.

McMinn, P.C., Weir, R.C., and Dalgarno, L. (1996). A mouse-attenuated envelope protein variant of Murray Valley encephalitis virus with altered fusion activity. J. Gen. Virol. *77*, 2085–2088.

Martinez-Gutierrez, M., Castellanos, J.E., and Gallego-Gomez, J.C. (2011). Statins reduce dengue virus production via decreased virion assembly. Intervirology *54*, 202–216.

Mateo, R., Nagamine, C.M., Spagnolo, J., Mendez, E., Rahe, M., Gale, M., Jr., Yuan, J., and Kirkegaard, K. (2013). Inhibition of cellular autophagy deranges dengue virion maturation. J. Virol. *87*, 1312–1321.

Medigeshi, G.R., Lancaster, A.M., Hirsch, A.J., Briese, T., Lipkin, W.I., Defilippis, V., Fruh, K., Mason, P.W., Nikolich-Zugich, J., and Nelson, J.A. (2007). West Nile virus infection activates the unfolded protein response, leading to CHOP induction and apoptosis. J. Virol. *81*, 10849–10860.

Medigeshi, G.R., Hirsch, A.J., Streblow, D.N., Nikolich-Zugich, J., and Nelson, J.A. (2008). West Nile virus entry requires cholesterol-rich membrane microdomains and is independent of alphavbeta3 integrin. J. Virol. *82*, 5212–5219.

Medin, C.L., Fitzgerald, K.A., and Rothman, A.L. (2005). Dengue virus non-structural protein NS5 induces interleukin-8 transcription and secretion. J. Virol. *79*, 11053–11061.

Meertens, L., Carnec, X., Lecoin, M.P., Ramdasi, R., Guivel-Benhassine, F., Lew, E., Lemke, G., Schwartz, O., and Amara, A. (2012). The TIM and TAM families of phosphatidylserine receptors mediate dengue virus entry. Cell Host Microbe *12*, 544–557.

Miller, J.L., de Wet, B.J., Martinez-Pomares, L., Radcliffe, C.M., Dwek, R.A., Rudd, P.M., and Gordon, S. (2008). The mannose receptor mediates dengue virus infection of macrophages. PLoS Pathog. *4*, e17.

Miller, S., Sparacio, S., and Bartenschlager, R. (2006). Subcellular localization and membrane topology of the Dengue virus type 2 Non-structural protein 4B. J. Biol. Chem. *281*, 8854–8863.

Miller, S., Kastner, S., Krijnse-Locker, J., Buhler, S., and Bartenschlager, R. (2007). The non-structural protein 4A of dengue virus is an integral membrane protein inducing membrane alterations in a 2K-regulated manner. J. Biol. Chem. *282*, 8873–8882.

Modis, Y., Ogata, S., Clements, D., and Harrison, S.C. (2003). A ligand-binding pocket in the dengue virus envelope glycoprotein. Proc. Natl. Acad. Sci. U.S.A. *100*, 6986–6991.

Modis, Y., Ogata, S., Clements, D., and Harrison, S.C. (2004). Structure of the Dengue virus envelope protein after fusion. Nature *427*, 313–319.

Modis, Y., Ogata, S., Clements, D., and Harrison, S.C. (2005). Variable surface epitopes in the crystal structure of dengue virus type 3 envelope glycoprotein. J. Virol. *79*, 1223–1231.

Moesker, B., Rodenhuis-Zybert, I.A., Meijerhof, T., Wilschut, J., and Smit, J.M. (2010). Characterization of the functional requirements of West Nile virus membrane fusion. J. Gen. Virol. *91*, 389–393.

Morizono, K., and Chen, I.S. (2014). Role of phosphatidylserine receptors in enveloped virus infection. J. Virol. *88*, 4275–4290.

Morizono, K., Xie, Y., Olafsen, T., Lee, B., Dasgupta, A., Wu, A.M., and Chen, I.S. (2011). The soluble serum protein Gas6 bridges virion envelope phosphatidylserine to the TAM receptor tyrosine kinase Axl to mediate viral entry. Cell Host Microbe *9*, 286–298.

Mukhopadhyay, S., Kim, B.-S., Chipman, P.R., Rossmann, M.G., and Kuhn, R.J. (2003a). Structure of West Nile virus. Science *303*, 248.

Muller, D.A., and Young, P.R. (2013). The flavivirus NS1 protein: molecular and structural biology, immunology, role in pathogenesis and application as a diagnostic biomarker. Antiviral Res. *98*, 192–208.

Muller, D.A., Landsberg, M.J., Bletchly, C., Rothnagel, R., Waddington, L., Hankamer, B., and Young, P.R. (2012). Structure of the dengue virus glycoprotein non-structural protein 1 by electron microscopy and single-particle analysis. J. Gen. Virol. *93*, 771–779.

Munoz-Jordan, J.L., Sanchez-Burgos, G.G., Laurent-Rolle, M., and Garcia-Sastre, A. (2003). Inhibition of interferon signaling by dengue virus. Proc. Natl. Acad. Sci. U.S.A. *100*, 14333–14338.

Munoz-Jordan, J.L., Laurent-Rolle, M., Ashour, J., Martinez-Sobrido, L., Ashok, M., Lipkin, W.I., and Garcia-Sastre, A. (2005). Inhibition of alpha/beta interferon signaling by the NS4B protein of flaviviruses. J. Virol. *79*, 8004–8013.

Navarro-Sanchez, E., Altmeyer, R., Amara, A., Schwartz, O., Fieschi, F., Virelizier, J.L., Arenzana-Seisdedos, F., and Despres, P. (2003). Dendritic-cell-specific ICAM3-grabbing non-integrin is essential for the productive infection of human dendritic cells by mosquito-cell-derived dengue viruses. EMBO Rep. *4*, 723–728.

Nazmi, A., Dutta, K., and Basu, A. (2010). Antiviral and neuroprotective role of octaguanidinium dendrimer-conjugated morpholino oligomers in Japanese encephalitis. PLoS Negl. Trop. Dis. *4*, e892.

Netsawang, J., Noisakran, S., Puttikhunt, C., Kasinrerk, W., Wongwiwat, W., Malasit, P., Yenchitsomanus, P.T., and Limjindaporn, T. (2010). Nuclear localization of dengue virus capsid protein is required for DAXX interaction and apoptosis. Virus Res. *147*, 275–283.

Nybakken, G.E., Nelson, C.A., Chen, B.R., Diamond, M.S., and Fremont, D.H. (2006). Crystal structure of the West Nile virus envelope glycoprotein. J. Virol. *80*, 11467–11474.

Offerdahl, D.K., Dorward, D.W., Hansen, B.T., and Bloom, M.E. (2012). A three-dimensional comparison of tick-borne flavivirus infection in mammalian and tick cell lines. PLoS One *7*, e47912.

Ogata, M., Hino, S., Saito, A., Morikawa, K., Kondo, S., Kanemoto, S., Murakami, T., Taniguchi, M., Tanii, I., Yoshinaga, K., *et al.* (2006). Autophagy is activated for cell survival after endoplasmic reticulum stress. Mol. Cell Biol. *26*, 9220–9231.

Pan, X., Zhou, G., Wu, J., Bian, G., Lu, P., Raikhel, A.S., and Xi, Z. (2012). Wolbachia induces reactive oxygen species (ROS)-dependent activation of the Toll pathway to control dengue virus in the mosquito *Aedes aegypti*. Proc. Natl Acad. Sci. U.S.A. *109*, E23–31.

Panyasrivanit, M., Greenwood, M.P., Murphy, D., Isidoro, C., Auewarakul, P., and Smith, D.R. (2011). Induced autophagy reduces virus output in dengue infected monocytic cells. Virology *418*, 74–84.

Park, G.S., Morris, K.L., Hallett, R.G., Bloom, M.E., and Best, S.M. (2007). Identification of residues critical for the interferon antagonist function of Langat virus NS5 reveals a role for the RNA-dependent RNA polymerase domain. J. Virol. *81*, 6936–6946.

Patkar, C.G., and Kuhn, R.J. (2008). Yellow Fever virus NS3 plays an essential role in virus assembly independent of its known enzymatic functions. J. Virol. *82*, 3342–3352.

Patkar, C.G., Jones, C.T., Chang, Y.H., Warrier, R., and Kuhn, R.J. (2007). Functional requirements of the yellow fever virus capsid protein. J. Virol. *81*, 6471–6481.

Pena, J., and Harris, E. (2012). Early dengue virus protein synthesis induces extensive rearrangement of the endoplasmic reticulum independent of the UPR and SREBP-2 pathway. PLoS One *7*, e38202.

Perera, R., and Kuhn, R.J. (2008). Structural proteomics of dengue virus. Curr. Opin. Microbiol. *11*, 369–377.

Perera, R., Riley, C., Isaac, G., Hopf-Jannasch, A.S., Moore, R.J., Weitz, K.W., Pasa-Tolic, L., Metz, T.O., Adamec, J., and Kuhn, R.J. (2012). Dengue virus infection perturbs lipid homeostasis in infected mosquito cells. PLoS Pathog. *8*, e1002584.

Pierson, T.C., Xu, Q., Nelson, S., Oliphant, T., Nybakken, G.E., Fremont, D.H., and Diamond, M.S. (2007). The stoichiometry of antibody-mediated neutralization and enhancement of West Nile virus infection. Cell Host Microbe *1*, 135–145.

Pierson, T.C., Fremont, D.H., Kuhn, R.J., and Diamond, M.S. (2008). Structural insights into the mechanisms of antibody-mediated neutralization of flavivirus infection: implications for vaccine development. Cell Host Microbe *4*, 229–238.

Plevka, P., Battisti, A.J., Sheng, J., and Rossmann, M.G. (2014). Mechanism for maturation-related reorganization of flavivirus glycoproteins. J. Struct. Biol. *185*, 27–31.

Poh, M.K., Yip, A., Zhang, S., Priestle, J.P., Ma, N.L., Smit, J.M., Wilschut, J., Shi, P.Y., Wenk, M.R., and Schul, W. (2009). A small molecule fusion inhibitor of dengue virus. Antiviral Res. *84*, 260–266.

Pokidysheva, E., Zhang, Y., Battisti, A.J., Bator-Kelly, C.M., Chipman, P.R., Xiao, C., Gregorio, G.G., Hendrickson, W.A., Kuhn, R.J., and Rossmann, M.G. (2006). Cryo-EM reconstruction of dengue virus in complex with the carbohydrate recognition domain of DC-SIGN. Cell *124*, 485–493.

Prinz, W.A. (2014). Bridging the gap: membrane contact sites in signaling, metabolism, and organelle dynamics. J. Cell Biol. *205*, 759–769.

Randolph, V.B., and Stollar, V. (1990). Low pH-induced cell fusion in flavivirus-infected *Aedes albopictus* cell cultures. J. Gen. Virol. *71*, 1845–1850.

Raung, S.L., Kuo, M.D., Wang, Y.M., and Chen, C.J. (2001). Role of reactive oxygen intermediates in Japanese encephalitis virus infection in murine neuroblastoma cells. Neurosci. Lett. *315*, 9–12.

Rawlinson, S.M., Pryor, M.J., Wright, P.J., and Jans, D.A. (2009). CRM1-mediated nuclear export of dengue virus RNA polymerase NS5 modulates interleukin-8 induction and virus production. J. Biol. Chem. *284*, 15589–15597.

Rey, F.A., Heinz, F.X., Mandl, C., Kunz, C., and Harrison, S.C. (1995). The envelope glycoprotein from tick-borne encephalitis virus at 2 A resolution. Nature *375*, 291–298.

Rivera, R., and Chun, J. (2008). Biological effects of lysophospholipids. Rev. Physiol. Biochem. Pharmacol. *160*, 25–46.

Roosendaal, J., Westaway, E.G., Khromykh, A., and Mackenzie, J.M. (2006). Regulated cleavages at the West Nile virus NS4A-2K-NS4B junctions play a major role in rearranging cytoplasmic membranes and Golgi trafficking of the NS4A protein. J. Virol. *80*, 4623–4632.

Rothwell, C., Lebreton, A., Young Ng, C., Lim, J.Y., Liu, W., Vasudevan, S., Labow, M., Gu, F., and Gaither, L.A. (2009). Cholesterol biosynthesis modulation regulates dengue viral replication. Virology *389*, 8–19.

Sangiambut, S., Keelapang, P., Aaskov, J., Puttikhunt, C., Kasinrerk, W., Malasit, P., and Sittisombut, N. (2008). Multiple regions in dengue virus capsid protein contribute to nuclear localization during virus infection. J. Gen. Virol. *89*, 1254–1264.

van der Schaar, H.M., Rust, M.J., Chen, C., van der Ende-Metselaar, H., Wilschut, J., Zhuang, X., and Smit, J.M. (2008). Dissecting the cell entry pathway of dengue virus by single-particle tracking in living cells. PLoS Pathog. *4*, e1000244.

Scherbik, S.V., and Brinton, M.A. (2010). Virus-induced Ca2+ influx extends survival of west nile virus-infected cells. J. Virol. *84*, 8721–8731.

Schmidt, A.G., Yang, P.L., and Harrison, S.C. (2010). Peptide inhibitors of flavivirus entry derived from the E protein stem. J. Virol. *84*, 12549–12554.

Singh, R., Kaushik, S., Wang, Y., Xiang, Y., Novak, I., Komatsu, M., Tanaka, K., Cuervo, A.M., and Czaja, M.J. (2009). Autophagy regulates lipid metabolism. Nature *458*, 1131–1135.

Smit, J.M., Moesker, B., Rodenhuis-Zybert, I., and Wilschut, J. (2011). Flavivirus cell entry and membrane fusion. Viruses 3, 160–171.

Soto-Acosta, R., Mosso, C., Cervantes-Salazar, M., Puerta-Guardo, H., Medina, F., Favari, L., Ludert, J.E., and del Angel, R.M. (2013). The increase in cholesterol levels at early stages after dengue virus infection correlates with an augment in LDL particle uptake and HMG-CoA reductase activity. Virology *442*, 132–147.

Sriburi, R., Jackowski, S., Mori, K., and Brewer, J.W. (2004). XBP1: a link between the unfolded protein response, lipid biosynthesis, and biogenesis of the endoplasmic reticulum. J. Cell. Biol. *167*, 35–41.

Stiasny, K., Koessl, C., and Heinz, F.X. (2003). Involvement of lipids in different steps of the Flavivirus fusion mechanism. J. Virol. *77*, 7856–7862.

Stiasny, K., Kiermayr, S., and Heinz, F.X. (2006). Entry functions and antigenic structure of flavivirus envelope proteins. Novartis Found Symp. *277*, 57–65; discussion 65–73, 251–253.

Stiasny, K., Fritz, R., Pangerl, K., and Heinz, F.X. (2011). Molecular mechanisms of flavivirus membrane fusion. Amino Acids *41*, 1159–1163.

Sukupolvi-Petty, S., Austin, S.K., Purtha, W.E., Oliphant, T., Nybakken, G.E., Schlesinger, J.J., Roehrig, J.T., Gromowski, G.D., Barrett, A.D., Fremont, D.H., and Diamond, M.S. (2007). Type- and subcomplex-specific neutralizing antibodies against domain III of dengue virus type 2 envelope protein recognize adjacent epitopes. J. Virol. *81*, 12816–12826.

Summers, P.L., Cohen, W.H., Ruiz, M.M., Hase, T., and Eckels, K.H. (1989). Flaviviruses can mediate fusion from without in *Aedes albopictus* mosquito cell cultures. Virus Res. *12*, 383–392.

Tassaneetrithep, B., Burgess, T.H., Granelli-Piperno, A., Trumpfheller, C., Finke, J., Sun, W., Eller, M.A., Pattanapanyasat, K., Sarasombath, S., Birx, D.L., *et al.* (2003). DC-SIGN (CD209) mediates dengue virus infection of human dendritic cells. J. Exp. Med. *197*, 823–829.

Ueba, N., and Kimura, T. (1977). Polykaryocytosis induced by certain arboviruses in monolayers of BHK-21-528 cells. J. Gen. Virol. *34*, 369–373.

Umashankar, M., Sanchez-San Martin, C., Liao, M., Reilly, B., Guo, A., Taylor, G., and Kielian, M. (2008). Differential cholesterol binding by class II fusion proteins determines membrane fusion properties. J. Virol. *82*, 9245–9253.

Verma, S., Molina, Y., Lo, Y.Y., Cropp, B., Nakano, C., Yanagihara, R., and Nerurkar, V.R. (2008). In vitro effects of selenium deficiency on West Nile virus replication and cytopathogenicity. Virol. J. *5*, 66.

Volmer, R., and Ron, D. (2014). Lipid-dependent regulation of the unfolded protein response. Curr. Opin. Cell Biol. *33C*, 67–73.

Volmer, R., van der Ploeg, K., and Ron, D. (2013). Membrane lipid saturation activates endoplasmic reticulum unfolded protein response transducers through their transmembrane domains. Proc. Natl. Acad. Sci. U.S.A. *110*, 4628–4633.

Warrener, P., and Collett, M.S. (1995). Pestivirus NS3 (p80) protein possesses RNA helicase activity. J. Virol. *69*, 1720–1726.

Welsch, S., Miller, S., Romero-Brey, I., Merz, A., Bleck, C.K., Walther, P., Fuller, S.D., Antony, C., Krijnse-Locker, J., and Bartenschlager, R. (2009). Composition and three-dimensional architecture of the dengue virus replication and assembly sites. Cell Host Microbe *5*, 365–375.

Wengler, G., and Wengler, G. (1991). The carboxy-terminal part of the NS 3 protein of the West Nile flavivirus can be isolated as a soluble protein after proteolytic cleavage and represents an RNA-stimulated NTPase. Virology *184*, 707–715.

Wu, J., Bera, A.K., Kuhn, R.J., and Smith, J.L. (2005). Structure of the Flavivirus helicase: implications for catalytic activity, protein interactions, and proteolytic processing. J. Virol. 79, 10268–10277.

Xie, X., Gayen, S., Kang, C., Yuan, Z., and Shi, P.Y. (2013). Membrane topology and function of dengue virus NS2A protein. J. Virol. *87*, 4609–4622.

Xie, X., Zou, J., Puttikhunt, C., Yuan, Z., and Shi, P.Y. (2015). Two distinct sets of NS2A molecules are responsible for dengue virus RNA synthesis and virion assembly. J. Virol. *89*, 1298–1313.

Yang, T.C., Lai, C.C., Shiu, S.L., Chuang, P.H., Tzou, B.C., Lin, Y.Y., Tsai, F.J., and Lin, C.W. (2010). Japanese encephalitis virus down-regulates thioredoxin and induces ROS-mediated ASK1-ERK/p38 MAPK activation in human promonocyte cells. Microbes Infect. *12*, 643–651.

Yap, L.J., Luo, D., Chung, K.Y., Lim, S.P., Bodenreider, C., Noble, C., Shi, P.Y., and Lescar, J. (2010). Crystal structure of the dengue virus methyltransferase bound to a 5′-capped octameric RNA. PLoS One *5*(9), e12836.

Youn, S., Li, T., McCune, B.T., Edeling, M.A., Fremont, D.H., Cristea, I.M., and Diamond, M.S. (2012). Evidence for a genetic and physical interaction between non-structural proteins NS1 and NS4B that modulates replication of West Nile virus. J. Virol. *86*, 7360–7371.

Yu, C.Y., Hsu, Y.W., Liao, C.L., and Lin, Y.L. (2006). Flavivirus infection activates the XBP1 pathway of the unfolded protein response to cope with endoplasmic reticulum stress. J. Virol. *80*, 11868–11880.

Yu, I.M., Zhang, W., Holdaway, H.A., Li, L., Kostyuchenko, V.A., Chipman, P.R., Kuhn, R.J., Rossmann, M.G., and Chen, J. (2008). Structure of the immature dengue virus at low pH primes proteolytic maturation. Science *319*, 1834–1837.

Yu, I.M., Holdaway, H.A., Chipman, P.R., Kuhn, R.J., Rossmann, M.G., and Chen, J. (2009). Association of the pr peptides with dengue virus at acidic pH blocks membrane fusion. J. Virol. *83*, 12101–12107.

Zaitseva, E., Yang, S.T., Melikov, K., Pourmal, S., and Chernomordik, L.V. (2010). Dengue virus ensures its fusion in late endosomes using compartment-specific lipids. PLoS Pathog. *6*, e1001131.

Zhang, W., Chipman, P.R., Corver, J., Johnson, P.R., Zhang, Y., Mukhopadhyay, S., Baker, T.S., Strauss, J.H., Rossmann, M.G., and Kuhn, R.J. (2003). Visualization of membrane protein domains by cryo-electron microscopy of dengue virus. Nat. Struct. Biol. *10*, 907–912.

Zhang, X., Ge, P., Yu, X., Brannan, J.M., Bi, G., Zhang, Q., Schein, S., and Zhou, Z.H. (2013a). Cryo-EM structure of the mature dengue virus at 3.5-A resolution. Nat. Struct. Mol. Biol. *20*, 105–110.

Zhang, X., Sheng, J., Plevka, P., Kuhn, R.J., Diamond, M.S., and Rossmann, M.G. (2013b). Dengue structure differs at the temperatures of its human and mosquito hosts. Proc. Natl. Acad. Sci. U.S.A. *110*, 6795–6799.

Zhang, Y., Zhang, W., Ogata, S., Clements, D., Strauss, J.H., Baker, T.S., Kuhn, R.J., and Rossmann, M.G. (2004). Conformational changes of the flavivirus E glycoprotein. Structure *12*, 1607–1618.

Zhang, Y., Kaufmann, B., Chipman, P.R., Kuhn, R.J., and Rossmann, M.G. (2007a). Structure of immature West Nile virus. J. Virol. *81*, 6141–6145.

Zhang, Y., Kostyuchenko, V.A., and Rossmann, M.G. (2007b). Structural analysis of viral nucleocapsids by subtraction of partial projections. J. Struct. Biol. *157*, 356–364.

Zhao, Y., Soh, T.S., Zheng, J., Chan, K.W., Phoo, W.W., Lee, C.C., Tay, M.Y., Swaminathan, K., Cornvik, T.C., Lim, S.P., et al. (2015). A crystal structure of the Dengue virus NS5 protein reveals a novel inter-domain interface essential for protein flexibility and virus replication. PLoS Pathog. *11*, e1004682.

Zhou, Y., Frey, T.K., and Yang, J.J. (2009). Viral calciomics: interplays between Ca2+ and virus. Cell Calcium *46*, 1–17.

Zou, J., Xie, X., Lee le, T., Chandrasekaran, R., Reynaud, A., Yap, L., Wang, Q.Y., Dong, H., Kang, C., Yuan, Z., Lescar, J., and Shi, P.Y. (2014). Dimerization of flavivirus NS4B protein. J. Virol. *88*, 3379–3391.

Zou, J., Lee, L.T., Wang, Q.Y., Xie, X., Lu, S., Yau, Y.H., Yuan, Z., Geifman Shochat, S., Kang, C., Lescar, J., and Shi, P.Y. (2015a). Mapping the interactions between the NS4B and NS3 proteins of dengue virus. J Virol. *89*, 3471–3483.

Zou, J., Xie, X., Wang, Q.Y., Dong, H., Lee, M.Y., Kang, C., Yuan, Z., and Shi, P.Y. (2015b). Characterization of dengue virus NS4A and NS4B protein interaction. J Virol. *89*, 3455–3470.

Vector-borne Bunyavirus Pathogenesis and Innate Immune Evasion

5

Brian Friedrich, Birte Kalveram and Shannan L. Rossi

Abstract

The family *Bunyaviridae* comprises viruses that include several human and veterinary pathogens of note with signs and symptoms ranging from haemorrhagic fever to severe encephalitis in humans and fetal malformations and abortion storms in livestock. These viruses are found on every continent except Antarctica and use a variety of methods for transmission ranging from direct spread between vertebrates, use of haemophilic insects, sexual transmission among insects and even thrips that infect plants. Two non-structural proteins, NSs and NSm, serve as major virulence factors which act to subvert the host anti-viral response by interrupting signalling cascades leading to interferon induction, suppressing transcription of interferon and other host-response genes, inhibiting interferon-stimulated genes, and modulating the regulation of cell death. This chapter will focus on virus–host interactions of the arthropod-borne genera *Orthobunyavirus*, *Phlebovirus* and *Nairovirus*.

Introduction

The family *Bunyaviridae* is one of the most diverse groupings of viruses and comprises (at least) five genera: *Orthobunyavirus*, *Hantavirus*, *Nairovirus*, *Phlebovirus* and *Tospovirus*. It is worth noting the recent discovery of three novel bunyaviruses from Côte d'Ivoire, Ghana and Uganda, which were shown to replicate only in insect cells and are phylogenetically distinct from the other genera, suggests the potential existence of another, insect-restricted genus (Marklewitz *et al.*, 2013). These genera combined contain more than 350 characterized viruses that infect both vertebrates and use many types of invertebrates as vectors. Most of these genera contain species known to cause severe disease in humans and/or livestock. The only viruses that are not vector-borne are in the genus *Hantavirus*, which are transmitted directly through aerosolized rodent excreta. Hantaviruses are of critical human importance and include Andes, Dobrava-Belgrade, Hantaan, Sin Nombre and Seoul viruses. Interestingly, tospoviruses are transmitted between plants by at least 13 different species of thrips, causing disease in peanut, tomato, watermelon and courgette plants. *Hantavirus* and *Tospovirus* genera will not be discussed here since they are not vertebrate arboviruses. Instead, this chapter will focus predominantly on the arthropod-borne genera *Orthobunyavirus*, *Phlebovirus* and *Nairovirus*.

Transmission

In nature, insect-vectored bunyaviruses are transmitted between a variety of arthropods and an equally diverse group of vertebrates. Vertebrates become exposed when the infected arthropod takes a bloodmeal. In most cases, humans are not considered an essential part of the virus' maintenance in nature and infections are incidental. As such, the majority of these cases are seen during summer months when the insects are active, especially in areas with cold winters. If the viraemia, or amount of virus present in the blood, is of a sufficient amplitude, a naive feeding arthropod may become infected, and thus the cycle continues. In some cases, the vertebrate intermediate is not required if the virus can be passed directly from mother to offspring via transovarial transmission. Here, the eggs are infected and when the vector matures and takes its first bloodmeal, virus transmission can occur. Orthobunyaviruses are mainly vectored by mosquitoes and flies belonging to the genus *Culicoides*, while nairoviruses are vectored by ixodid ticks. Phleboviruses primarily by phlebotomine sandflies, although two important human pathogens in this genus, Rift Valley fever virus (RVFV) and severe fever with thrombocytopenia syndrome virus (SFTSV) are exceptions to this rule and are vectored by mosquitoes and ticks, respectively.

Orthobunyaviruses

Orthobunyaviruses account for the largest number of viral species in the *Bunyavirus* family. These viruses are divided into three main groups: the California, the Bunyamwera and the Simbu serogroups.

California serogroup

In the USA, the most medically important bunyaviruses are in the California serogroup, consisting of viruses including La Crosse (LACV), snowshoe hare (SSHV) Jamestown Canyon (JCV) viruses. LACV is primarily found in the eastern part of the US along the Mississippi River valley. The principal hosts are rodents such as chipmunks and tree squirrels (Moulton and Thompson, 1971). The main vector is *Aedes triseriatus*. In addition to mammal–mosquito transmission, LACV may also be maintained directly in the mosquitoes by transovarial transmission (Baldridge *et al.*, 1989; Tesh and Gubler, 1975). SSHV is very closely related to La Crosse virus. In nature, SSHV circulates primarily between *Aedes* spp. of mosquito and rodents such as hares, voles and Arctic lemmings (Mitchell *et al.*, 1993). JCV circulates in the north-east and upper midwest areas of the USA and neighbouring Canada between white-tailed deer and various *Aedes* spp. mosquitoes (Armstrong and Andreadis, 2007; Campbell *et al.*, 1989).

Bunyamwera serogroup

Bunyamwera virus (BUNV) is present in sub-Saharan Africa and a high prevalence of antibody has been documented in in human populations. Primarily transmitted by *Aedes* mosquitos, it infects a variety of vertebrates. While anti-BUNV antibodies have been found in birds, rodents, bats, and primates, the vertebrate host that contributes most heavily to maintenance in nature is unknown (Tauro *et al.*, 2009). Cache Valley virus (CCV) is present in the US and also uses a variety of mosquitoes as vectors, including *Aedes* spp. (*Ae. taeniorhynchus*, *Ae. canadensis*, *Ae. vexans*), *Culex tarsalis* and, interestingly, *Anopheles quadrimaculatus* (Hubalek *et al.*, 2014). Transovarial transmission has been observed in *Culiseta inornata*, from which the virus was originally isolated, but the role of this maintenance method is unclear (Hayles and Lversen, 1980; Holden and Hess, 1959; Hubalek *et al.*, 2014). The vertebrate host is believed to be deer as evidenced by a high antibody prevalence (Blackmore and Grimstad, 1998).

Simbu serogroup

Simbu serogroup viruses such as Oropouche (OROV) and Akabane (AKAV) viruses are vectored by *Culicoides* midges. Large outbreaks of OROV in South America and across the southern Caribbean, as well as sufficiently high viraemia in patients, point to humans as part of the amplification cycle (Pinheiro *et al.*, 1981). Antibody prevalence data in sloths and nonhuman primates suggest that they might also be part of the virus' maintenance cycle (Pinheiro *et al.*, 1981). AKAV is a livestock pathogen and its vector depends on the geographic location, using *Culicoides* spp. across Asia, *C. brevitarsis* in Australia and *C. imicola* in in the Middle East (Moulton and Thompson, 1971).

Phleboviruses

As the name phlebovirus suggests, most viruses of this genus are transmitted by phlebotomine sandflies. However, sandflies are not the only arthropod vectors used by this genus, as mosquitos and ticks also act as vectors in this very diverse group.

Sandfly fever Sicilian virus (SFSV) and sandfly fever Naples virus (SFNV) have been found over a wide geographical region extending from the Mediterranean basin, through the Middle East, and as far east as India. These viruses are primarily vectored by *Phlebotomus papatasi* (Eitrem *et al.*, 1991; Schmidt *et al.*, 1971). Toscana virus (TOSV) is closely related to the sandfly fever viruses, is found mainly in southern and western Europe, and is vectored by *P. perniciosus* (Eitrem *et al.*, 1991). Transovarial and venereal transmission have both been observed with TOSV in *P. perniciosus* (Ciufolini *et al.*, 1989; Tesh *et al.*, 1992; Tesh and Modi, 1987). While the wildlife host has not been identified, antibodies present against TOSV have been found in many animal species.

Rift Valley fever virus is responsible for large, sporadic outbreaks that have occurred in regions ranging from much of sub-Saharan Africa to the Arabian Peninsula. RVFV is readily transmitted by a number of mosquito species (at least six genera, and greater than 30 species) as well as sandflies and ticks (Bird *et al.*, 2009). Evidence of transovarial transmission has been shown with *Ae. lineatopennis* mosquitoes (Linthicum *et al.*, 1985). The significance of this type of transmission for virus maintenance in nature still requires further investigation. There is much debate on which vertebrate hosts play a role in the maintenance of RVFV. There is evidence that African buffalo and other wild ruminants may play a role, but the contribution of bats or rodents cannot be ruled out yet either (Olive *et al.*, 2012).

Nairoviruses

The Crimean–Congo haemorrhagic fever group consists of Crimean–Congo haemorrhagic fever (CCHFV) and Hazara viruses. CCHFV has a wide geographic relevance, with cases ranging from western China and southern Asia, to parts of Europe, the Middle East, and most of Africa (Bente *et al.*, 2013). Ticks belonging to the genus *Hyalomma* are the primary vectors of CCHFV, but the species depends upon the geographic location. The species most utilized by CCHFV are *H. marginatum*, *H. lusitanicum*, and *H. marginatum rufipes*. Additionally, it has been shown that *Rhipicephalus* spp. and *Dermacenter* spp. ticks are able to transmit the virus (Bente *et al.*, 2013). The vertebrate hosts for CCHFV have not been well studied, but most *Hyalomma* spp. are multi-host and studies suggest that both small vertebrates (such as hares) and large vertebrates (such as sheep) can play a role in the maintenance of CCHFV (Gonzalez *et al.*, 1998; Shepherd *et al.*, 1989).

Nairoviruses in general are almost all vectored by ticks. Of the existing serogroups in this genus, the two most important are the Nairobi sheep disease group and the Crimean–Congo haemorrhagic fever group. The Nairobi sheep disease group includes Nairobi sheep disease virus (NSDV) and Dugbe virus (DUGV). Nairobi sheep disease virus is a disease of ruminants in Africa and India (Ganjam virus) that circulates between these vertebrate and ixodid ticks (*Rhipicephalus appendiculatus*, *Haemaphysalis wellingtoni*) with the potential for the rat (*Arvicanthis abyssinicus*) to serve as a reservoir (Hubalek *et al.*, 2014). Dugbe virus is commonly found circulating throughout western Africa circulating between cattle and *Amblyomma variegatum* ticks (Booth *et al.*, 1991).

Pathogenesis

Bunyaviruses, as a group, have the potential to cause significant morbidity and mortality in both humans and livestock, and thus are important for both health and economic reasons. Human disease caused by bunyaviruses can range from mild, asymptomatic infections to more severe infections which can include pulmonary disease, encephalitis, haemorrhagic fever, and even death. Replication of the different bunyaviruses is very similar. They enter the host cell through receptor-mediated endocytosis and replicate in the cytoplasm of the cell. Following replication, assembly, budding, and maturation of virions takes place in the Golgi (Fontana *et al.*, 2008; Lozach *et al.*, 2010).

Orthobunyaviruses

At least 30 orthobunyaviruses have been associated with human disease, with a wide range of symptoms/signs and severity. Bunyamwera virus is the namesake of the family and is considered the prototypic bunyavirus. Orthobunyaviruses usually have an incubation period of 3–7 days and can cause a wide range of diseases, including non-specific, acute febrile illness, encephalitis, haemorrhagic fever, or have abortive/teratogenic effects in animals (Eldridge *et al.*, 2001; Gerrard *et al.*, 2004; Hart *et al.*, 2009; Hoffmann *et al.*, 2012; Konno *et al.*, 1982; Konno and Nakagawa, 1982; McJunkin *et al.*, 2001; Pinheiro *et al.*, 1981; Rust *et al.*, 1999).

Oropouche virus causes an acute febrile illness that generally lasts 2–7 days. Symptoms include fever, chills, headache, myalgia, and arthralgia (Pinheiro *et al.*, 1981). Ngari virus is the causative agent of a recent outbreak of haemorrhagic fever in East Africa. Symptoms include acute onset fever, headache, mucosal bleeding, and gastrointestinal haemorrhage. Prior to this outbreak, Ngari virus was not associated with haemorrhagic disease, which is very similar to the discovery of RVFV (Bowen *et al.*, 2001; Gerrard *et al.*, 2004). Other orthobunyavirus infections, such as AKAV, CVV, and Schmallenberg virus (SBV), can cause abortion and teratogenesis in livestock (Edwards *et al.*, 1989; Hoffmann *et al.*, 2012; Konno *et al.*, 1982).

La Crosse virus is a major cause of paediatric encephalitis, with children under 15 years of age being the most susceptible, and a case fatality of about 1% (Gaensbauer *et al.*, 2014; McJunkin *et al.*, 2001). In most cases, LACV is a non-specific febrile illness; however, a small proportion of cases develop into acute encephalitis. Following infection, the incubation period is typically 3–8 days. Symptoms can include vomiting, headache, fever, joint/muscle pain, stiff neck, and in severe cases, seizures or coma. LACV primarily infects the neurons in the brain (Kallfass *et al.*, 2012). Most patients completely recover, but a small number of survivors of severe encephalitis can have long-term or permanent deficits, including cognitive disabilities or persistent paresis (McJunkin *et al.*, 2001; Rust *et al.*, 1999; Soldan *et al.*, 2010). Jamestown Canyon virus is also a causative agent of encephalitis, and is similar to LACV, but usually causes a milder meningitis in adults rather than children (Srihongse *et al.*, 1984).

Orthobunyavirus infection in humans is initiated after a mosquito bite, followed by extensive replication in the striated muscles near the site of infection. The virus then spreads to most of the organs and can cross the blood–brain barrier to infect the brain (Taylor and Peterson, 2014). Infection of the CNS could possibly also happen via the olfactory neurons (Bennett *et al.*, 2008). Neuronal necrosis and damage continues to happen after viral clearance, suggesting that pro-apoptotic factors and cytokines, at least in part, play an important role in the pathogenesis of disease (Soldan *et al.*, 2010).

Phleboviruses

Of the nearly 70 different phleboviruses, only a small fraction have been linked to disease in humans, including TOSV, SFNV, SFSV, Candiru virus, Chagres virus, Punta Toro virus (PTV), Heartland virus, severe fever with thrombocytopenia syndrome virus (SFTSV), and the most pathogenic member, RVFV (McMullan *et al.*, 2012; Muehlenbachs *et al.*, 2014; Palacios *et al.*, 2011; Perrone *et al.*, 2007; Srihongse and Johnson, 1974; Yu *et al.*, 2011). Most human phlebovirus infections have an incubation period of 1–6 days and result in asymptomatic or self-limiting, non-specific febrile illness. However, some infections do proceed to more severe disease. For example, while over 60% of TOSV infections are mild, about 30% will progress to more severe aseptic encephalitis or meningitis (Braito *et al.*, 1997, 1998). More recently discovered phleboviruses, SFTSV and Heartland virus, have been shown to cause human disease. SFTSV infections can have symptoms such as leucopenia, moderate neutropenia, thrombocytopenia, and elevated serum enzyme levels (Lei *et al.*, 2015; Park *et al.*, 2014; Yu *et al.*, 2011). Additionally, SFTSV can cause multi-organ failure and has a case-fatality rate of 6–30% (Park *et al.*, 2014). Heartland virus also causes a febrile illness with leucopenia and thrombocytopenia, and has resulted in at least one death (McMullan *et al.*, 2012; Yu *et al.*, 2011).

RVFV can infect both humans and livestock and has caused widespread outbreaks of severe disease in Africa and the Arabian Peninsula (Abdo-Salem *et al.*, 2006; Aradaib *et al.*, 2013; Bird *et al.*, 2009; Himeidan *et al.*, 2014; Madani *et al.*, 2003; McIntosh *et al.*, 1980; Meegan *et al.*, 1979; Morvan *et al.*, 1992; Woods *et al.*, 2002). In most human infections, RVFV causes an acute, non-specific febrile illness 2–6 days after initial infection. However, a small percentage of patients will develop a more severe disease, which may including symptoms such as acute liver disease, delayed-onset encephalitis, retinitis, blindness, or haemorrhagic symptoms (Madani *et al.*, 2003). RVFV is primarily a hepatotropic disease during severe infection, and this hepatic damage may account for many early clinical symptoms. Livestock, particularly ruminants such as sheep, cattle, and goats, are particularly susceptible to RVFV infection. The fatality among neonates (< 1 month old) sheep can approach 100%. Older animals are less susceptible, but still case fatality ranges from 5% to 30% (Bird *et al.*, 2009). Signs in livestock include loss of appetite, nasal discharge, diarrhoea, hepatic necrosis, splenomegaly and gastrointestinal haemorrhage. Additionally, RVFV induces spontaneous abortion or 'abortion storms' in pregnant animals (Balkhy and Memish, 2003; Bird *et al.*, 2009; Coetzer, 1982).

Phlebovirus infection is initiated after the bite of an infected vector and spreads from the site of infection to the regional lymph node. The virus then spreads through the blood stream to most of the organs, including the liver, spleen, adrenal glands, lungs, kidney, and sometimes the brain. RVFV is able to cross the blood–brain barrier, but there is evidence that neuroinvasion can also happen either via the olfactory neurones or by ascending infection up the cranial nerves (Smith *et al.*, 2010).

Nairoviruses

There are at least 34 species of nairovirus, the most important of which is CCHFV. Other important members include NSDV, DUGV and Kupe virus (Lasecka and Baron, 2014). Most nairoviruses have an incubation of 3–7 days, followed by symptoms or disease ranging from asymptomatic infections to acute, self-limiting febrile illness to more severe nausea, diarrhoea, myalgia, chills, mucosal bleeding, and ecchymosis (Bente *et al.*, 2013; Bin Tarif *et al.*, 2012; Boyd *et al.*, 2006; Cevik *et al.*, 2008;

Crabtree *et al.*, 2009; Davies, 1978; Lasecka and Baron, 2014; Peyrefitte *et al.*, 2010). NSDV is highly pathogenic in sheep and goats, with fatality as high as 90%; however, infection in humans is more limited and can cause a brief, febrile illness. NSDV causes a fever, followed by profound leucopenia and haemorrhagic symptoms such as gastrointestinal and mucosal bleeding (Bin Tarif *et al.*, 2012). DUGV does not often cause disease in humans, but on the rare occasions it does infect humans, it usually causes a benign febrile illness. One extreme case was associated with thrombocytopenia and meningitis (Burt *et al.*, 1996). Kupe virus was recently discovered and is closely related to DUGV; however, despite being present in ticks feeding on livestock, it is currently unknown if it causes disease in mammals (Crabtree *et al.*, 2009; Lutomiah *et al.*, 2014).

CCHFV infects humans and is known to cause severe disease over a widespread geographical area. It is known to have a wide range of severity, ranging from a non-specific febrile illness to severe haemorrhagic symptoms (Bente *et al.*, 2013). Disease is usually divided into four phases: incubation, pre-haemorrhagic, haemorrhagic, and convalescent (Ergonul, 2006). Following an incubation period of 3–7 days, the most common symptoms reported in the pre-haemorrhagic phase are non-specific signs and symptoms such as fever, headache, myalgia, nausea, vomiting and diarrhoea. Patients usually develop haemorrhagic signs 3–5 days after onset of disease. These include petechial rash, mucosal bleeding, and large ecchymosis (Bente *et al.*, 2013; Cevik *et al.*, 2008; Ergonul, 2006). The case fatality for CCHFV ranges from 5–50%, and death usually occurs between days 5 and 15 after onset of symptoms (Bente *et al.*, 2013; Cevik *et al.*, 2008; Coates and Sweet, 1990; Ergonul, 2006).

Nairovirus infection is initiated after the bite of a tick and spreads from the site of infection to the lymph and blood stream where it spreads to most of the organs. Disseminated intravascular coagulation (DIC), which results in formation of blood clots throughout the vessels in the body, is usually evident early in infection. Disease progression also is helped by an early increase in the prothrombin time ratio and activated partial thromboplastin time which reduces the blood's ability to coagulate and clot (Cevik *et al.*, 2008).

Genome

The bunyavirus genome is organized into three single-stranded negative or ambisense segments designated as large (L), medium (M) and small (S). The four structural proteins are encoded in a negative-sense, with the polymerase (L) encoded in the L-segment, the glycoproteins Gn and Gc encoded in the M-segment and the nucleoprotein (N) in the S-segment.

In addition to the structural proteins, a number of bunyaviruses encode one or two non-structural proteins in their S- and M-segments which are generally dispensable for viral replication, but play an important role in viral pathogenesis through their interactions with components of the host cell innate immunity. NSm proteins are found in orthobunyaviruses, phleboviruses and nairoviruses and are derived from an M-segment encoded polyprotein precursor, along with the Gn and Gc glycoproteins through cleavage by host proteases. In the orthobunyaviruses and nairoviruses, NSm lies between Gn and Gc, whereas in the phleboviruses, NSm is encoded in the N-terminus of Gn. S-segment derived NSs proteins are found in both orthobunyaviruses and phleboviruses, although each genus employs a distinct coding strategy for NSs. In orthobunyaviruses, N and NSs are translated from overlapping reading frames in a single S-segment mRNA. In phleboviruses, on the other hand, N and NSs are translated from two distinct subgenomic mRNAs, with N – like all structural proteins – being encoded in a negative-sense orientation and NSs being encoded in an ambisense manner. Interestingly, even though bunyavirus NSs proteins share little to no sequence similarity and vary greatly in length, they display remarkable convergence in their ability to suppress the host innate immune response.

Innate immunity

As is common with other viruses and often a critical aspect of pathogenesis, bunyaviruses have developed mechanisms to inhibit the host's innate immune response. The specific ways this is achieved may vary between families or even viruses. The inhibition mechanisms employed by bunyaviruses function in both the insect and vertebrate portions of the transmission cycle. Immunity evasion techniques in the insect will be examined in another chapter, and therefore will not be discussed here. The signalling pathways involved in initiating an innate immune response in vertebrates are specific yet slightly redundant. Even though our understanding is detailed, new discoveries are made every year expanding upon how cells work to raise the alarm and respond to an outside threat of infection. Despite this, the main components have been well studied. Briefly, this response can be divided into two intertwined and equally important halves: interferon production and interferon signalling.

Type I interferon (IFN) α/β production is initiated when danger signals called pathogen-associated molecular patterns (PAMPs), like double-stranded (ds)RNA or 5′ triphosphate single-stranded (ss)RNA produced during infection are detected by proteins specifically attuned to recognize these signals. These detector proteins are called pathogen recognition receptors (PRRs). The markers of viral infection are detected by host intracellular molecules protein kinase R (PKR), RNA helicases MDA-5 and RIG-I, and toll-like receptors (TLR) 3 and 7. RIG-I binds to MAVS (mitochondria anti-viral signalling protein), a protein located on mitochondria that activates IRF3 and NFκB as well as caspase-dependent apoptosis (Lei *et al.*, 2009; Seth *et al.*, 2005). RIG-I, MDA-5 and TLR3 can lead to the activation of IRF3. Upon activation, IRF3 dimerizes and translocates to the nucleus, where it serves as an activation factor on the IFNβ promoter.

After IFNα/β is secreted from the original cell, receptors on neighbouring cells can bind the interferon and, through a signalling cascade, trigger the production of interferon stimulated genes (ISGs). This process involves STAT proteins and IRF 9, which form a complex called ISFG-3, which translocates into the nucleus to drive the transcription of multiple ISGs. These ISGs perform a variety of functions that ultimately establish an antiviral state in the cell. MxA protein is part of the GTPase family and accumulates in the cytoplasm of interferon-treated cells, inhibiting the multiplication of several RNA viruses by preventing transport of vRNPs into the nucleus (Weber *et al.*, 2000). ISG15 is an ubiquitin-like molecule that helps protect the cell from virus-related death in several ways. First, ISG15 can conjugate to both viral and host proteins, which is important for proper regulation of the antiviral response. Secondly, unconjugated ISG15 has been shown to modulate cytokine production in response to infection (Morales and Lenschow, 2013). PKR is a dsRNA and 5′ppp ssRNA-detecting protein that dimerizes when activated and performs two key roles: the phosphorylation of eIF2α resulting in the shutdown of cellular translation and the activation of NFκB (Garcia *et al.*, 2006; Nallagatla *et al.*, 2007). 2′–5′-Oligoadenylate synthetase (OAS) also detects dsRNA, but instead leads to the activation of RNase L, which degrades RNA. The exact role of viperin (virus inhibitory protein, endoplasmic reticulum-associated interferon inducible) has not been fully elucidated, but it has been implicated in a number of antiviral activities. Studies have shown that viperin can have direct antiviral activity and also modulate innate immune signalling. Additionally, viperin has been shown to be able to bind to the cytosolic surface of the endoplasmic reticulum (ER) and inhibit the secretion of soluble proteins as well as reduce the rate of ER to Golgi traffic, which potentially could affect viral entry, assembly, and/or budding (Helbig and Beard, 2014). The variety of functions performed by these ISGs serves to make the cell an inhospitable atmosphere for virus replication.

Bunyaviruses have developed the ability to evade both IFN induction and innate immune signalling pathways. In general, NSs is the main protein responsible for these activities. Studies using viruses genetically engineered to lack a functional NSs have been instrumental in flushing out the contribution of NSs protein in IFN antagonism and pathogenesis. In general, viruses lacking NSs function are strong inducers of the antiviral response *in vitro* and are attenuated *in vivo*. In addition to blocking the innate immune response directly, global mechanisms including protein synthesis shut-off may be effective in dampening an immune response. Bunyaviruses have developed means to block host cell transcription either by phosphorylating the RNA polymerase II (RNAPII) or targeting specific subunits for proteasome-mediated degradation.

Orthobunyaviruses

Most of the studies in the genus *Orthobunyavirus* have been completed with the prototype virus Bunyamwera virus. NSs ablation mutant viruses (BUNdelNSs) are still functional, indicating that NSs is non-essential for virus propagation (Bridgen *et al.*, 2001). Instead, BUNdelNSs strongly induces IFNβ production *in vitro* (Bridgen *et al.*, 2001; Varela *et al.*, 2013; Weber *et al.*, 2002) and is attenuated in mice with a competent IFN response (Weber *et al.*, 2002). *In vitro* studies implicated dsRNA as the PAMP that initiates the IFN induction pathway. IRF3 is activated (Weber *et al.*, 2002) as evidenced by its dimerization and translocation to the nucleus (Kohl *et al.*, 2003) in wild type BUNV, indicating that BUNV inhibits IFN production downstream of IRF3. BUNV also blocks the functionality of NFκB (Weber *et al.*, 2002). Priming cells with type I IFN establishes a cellular antiviral state that prevents BUNV infection. PKR is both activated by (Streitenfeld *et al.*, 2003) and inhibits BUNV infection (Carlton-Smith and Elliott, 2012). Other ISGs that were found to have anti-bunyavirus effects include MTAP44 and viperin (Carlton-Smith and Elliott, 2012).

In addition to directly targeting IFN induction/signalling, host cellular transcription is also affected by BUNV. NSs inhibits the functionality of host cell RNA polymerase II (RNAPII) by blocking its phosphorylation at serine 2 in the C-terminal domain (Thomas *et al.*, 2004). This occurs though the binding of NSs to MED8 (Leonard *et al.*, 2006), a mediator protein that binds to RNAPII to regulate transcription (Leonard *et al.*, 2006). This interaction leads to the degradation of RNAPII (Leonard *et al.*, 2006). In this way, BUNV blocks the transcription of all host cell messages, including those required for IFN production.

Other prominent members of the genus orthobunyavirus, including LACV and SBV, have been studied more extensively in recent years. These viruses inhibit IFN production as well via NSs *in vitro* (Blakqori *et al.*, 2007; Kraatz *et al.*, 2015). In LACV-infected mice, NSs primarily antagonizes IFN production from astrocytes, but not microglial cells, even though the neurons are the primary cell infected (Kallfass *et al.*, 2012). LACV engages the RIG-I response, as well as downstream IRF3 dimerization and nuclear translocation (Verbruggen *et al.*, 2011) and MAVS activation (Mukherjee *et al.*, 2013). Despite this, interferon is still not produced due to host cell transcriptional ablation. These viruses affect the RNAPII, but using a different mechanism than BUNV. Instead, events similar to those seen during DNA damage response are engaged. NSs targets the RPB1 subunit of the transcriptionally active RNAPII, ultimately leading to its degradation by a proteasome-dependent mechanism (Verbruggen *et al.*, 2011). DNA damage response proteins pak6 and H2A.X were activated by LACV NSs (Verbruggen *et al.*, 2011). Likewise, SBV blocks universal cellular transcription by targeting the RPB1 subunit for degradation (Barry *et al.*, 2014). However, the ISG MxA is able to combat LACV infection by sequestering N protein, thereby reducing viral genome replication (Kochs *et al.*, 2002).

Interestingly, there are naturally occurring orthobunyaviruses that lack NSs. Viruses in the Anopheles A, B and Tete serotypes lack functional NSs proteins (Mohamed *et al.*, 2009). Only Tete among these has been shown to cause disease in vertebrates (Calisher *et al.*, 1990). Tacaiuma virus, a member of the Anopheles A serotype, can still inhibit IFNα production *in vitro* (Mohamed *et al.*, 2009). Additionally, the Wyeomyia group may also lack a functional NSs based upon phylogenetic and genetic comparisons with other orthobunyaviruses, but experiments have not been performed to back up this hypothesis (Chowdhary *et al.*, 2012). These viruses point to other mechanisms that need to be elucidated through further study.

Phleboviruses

There are three main ways RVFV combats the host cell's response to infection: degradation of PKR, specific shut-off of IFNβ transcription, and global transcription shut-off. These events are initiated quickly after infection and work in concert to prevent the immune response while still allowing virus protein translation, and perhaps help to answer why RVFV is so highly pathogenic. In addition, RVFV NSs protein can directly target PKR for degradation (Habjan *et al.*, 2009), thereby keeping the translational machinery intact. This is done solely by NSs since expressing the protein alone leads to a reduction in cellular PKR (Ikegami *et al.*, 2009). RVFV lacking NSs induced PKR activation and eIF2α phosphorylation in a caspase-independent manner (Ikegami *et al.*, 2009) and were attenuated *in vivo* (Habjan *et al.*, 2009).

Concurrently, cellular transcription is shutdown. Early during infection, NSs forms 'filamentous' structures in the nucleus (Le May *et al.*, 2004). Here, RVFV NSs directly interacts with a variety of protein to disrupt SAP30, YYI and TFIIH function. NSs specifically blocks the transcription of IFNβ by binding to two of the gene's transcription factors: SAP30 and YY1. A complex containing SAP3 along with NCoR, HDAC3 and Sin3A binds to the YYI protein at the −90 position in the promoter, acting as a repressor. During RVFV infection, YYI directly binds at the −90 position, but NSs filaments bind to both YYI and SAP30 thereby keeping the repressor complex in place. This interaction prevents the recruitment of the CBP activator protein from binding (along with other enhancer elements like IRF3) to drive IFNβ transcription. In this model, dimerized IRF3 can still bind to the IFNβ promoter but has no effect (Le May *et al.*, 2008). TFIIH comprises 10 subunit proteins, namely XPD, XPB, MAT1, cyclin H, cdk7, p8, p34, p44, p52 and p62, and is essential for RNA polymerases I and II (Hampsey, 1998; Orphanides *et al.*, 1996). NSs is able to inhibit cellular transcription by outcompeting XPD and binding to p44 in its place (Le May *et al.*, 2004). The filamentous structures also contain p44 and XPB, thereby reducing the amount available to form TFIIH complexes (Le May *et al.*, 2004). Additionally, p62 appears to be targeted for proteosomal degradation by NSs (Kalveram *et al.*, 2011).

Other phleboviruses interfere with IFN using other pathways. For example, TOSV NSs inhibits IFNβ production by directly targeting IRF3 for proteasome degradation (Gori-Savellini *et al.*, 2011, 2013). Moreover, TOSV inhibits PKR activation (Kalveram and Ikegami, 2013). Unlike RVFV, TOSV does not produce a global cellular reduction in transcription (Kalveram and Ikegami, 2013). The current area of research is investigating the innate immune response evasion of phleboviruses in the Uukuniemi (UUKV)-like tick-borne group. Viruses like SFTSV and Heartland virus have pushed scientists to more closely examine these viruses and those closely related to them (Matsuno *et al.*, 2015). As such, much will be elucidated in the future when reverse genetic systems like that just described for SFTSV are used (Brennan *et al.*, 2014). However, NSs expression studies have shown NSs forms not filamentous structures like the NSs of RVFV, but instead 'viroplasm-like structures' in the cytoplasm which colocalize with dsRNA (Wu *et al.*, 2014a). Wu *et al.* (2014a) hypothesize NSs here may act as a scaffold for virion assembly, which would be a novel function of bunyaviral NSs proteins. Another hypothesis for these 'viroplasm-like structures' is the sequestration of innate immune proteins such as RIG-I, IRF3, TRIM25 and TBK1 (Brennan *et al.*, 2014; Ning *et al.*, 2014; Santiago *et al.*, 2014; Wu *et al.*, 2014b). Indeed, NSs is able to inhibit IFNβ and NFκB (Ning *et al.*, 2014; Qu *et al.*, 2012; Santiago *et al.*, 2014). It is worth noting that UUKV has been previously used as a prototypic member of this group. UUKV NSs is a weak IFN antagonist, which may help to explain why this virus is a relatively mild pathogen in humans (Rezelj *et al.*, 2015).

Nairoviruses

The innate immune evasion techniques for CCHFV are a little different from those mentioned above. This virus is hypothesized to avoid detection by PRRs rather than seek to actively inhibit them. For example, RIG-I is not activated since 5′-triphosphate ss-RNA produced during infection is removed (Habjan *et al.*, 2008). CCHFV also inhibits the innate immune response, perhaps by an IRF3-dependent mechanism (Andersson *et al.*, 2008). Indeed, neither CCHFV nor NSDV appears to stimulate IFNβ transcription *in vitro* (Habjan *et al.*, 2008; Holzer *et al.*, 2011). Interestingly, part of the viral RNA-dependent RNA-polymerase (L-protein) contains an ovarian tumour domain, which has the unique property of cleaving the ubiquitin-like protein ISG15 (Frias-Staheli *et al.*, 2007).

Once exogenous IFN is recognized, however, the antiviral state of the cell is primed to prevent CCHFV infection; *in vivo* studies have shown mice lacking STAT1 or IFNα receptor were susceptible to CCHFV infection whereas mice with intact IFN signalling pathways were not (Bente *et al.*, 2010; Bereczky *et al.*, 2010). Human MxA protein, an ISG, interacts with N protein, and it is hypothesized this interaction sequesters N protein from participating in viral replication (Andersson *et al.*, 2004).

Cellular death and apoptosis

When the cell has exhausted all other survival mechanisms or when the damage is too great, the programmed cell death response apoptosis may occur. This death, as opposed to the energy-independent necrosis, is the preferential way to limit the damage done to host tissues. Many viruses either avoid or induce apoptosis, depending on their life cycle. Bunyaviruses do both, as summarized below.

It is worth noting that the innate immune and apoptosis signalling pathways converge as multiple points. For example, it has recently been shown that IRF3 activation directly contributes to apoptosis in the following ways: IRF3 dimers serve as transcription activators for the Noxa and PUMA proapoptotic proteins and Bax can interact directly with IRF3 to promote cytochrome *c* release from mitochondria (Chattopadhyay *et al.*, 2010, 2013). Cytochrome *c* release ultimately results in the caspase cleavage cascade and the induction of irreversible apoptotic events. PKR can be cleaved, leaving a catalytically active byproduct capable of eIF2α phosphorylation, by activated caspases 3, 7 and 8 (Saelens *et al.*, 2001).

Orthobunyaviruses

Viruses in the genus *Orthobunyavirus* either induce or suppress apoptosis in infected cells through their NSs protein. Similar to the cellular DNA damage response, the NSs protein of both LACV and SBV specifically targets the RPB1 subunit of elongating RNA polymerase II (Barry *et al.*, 2014; Verbruggen *et al.*, 2011), which ultimately leads to the induction of caspase-dependent apoptosis (Sancar *et al.*, 2004). Furthermore, the C-terminus of the LACV NSs protein contains a putative reaper-like region (RLR). Like its namesake Reaper, the NSs of another California serogroup virus, San Angelo virus, appears to induce apoptosis through interaction with the apoptotic regulator Scythe (Colon-Ramos *et al.*, 2003). Recently, a protein found in neurons called SARM1, has been implicated in contributing to cell death during LACV infection and blocking this protein increases cell survival (Mukherjee *et al.*, 2013). Likewise, OROV virus induces apoptosis as characterized by cytochrome c release and caspase 3 and 9 activity (Acrani *et al.*, 2010).

BUNV NSs, on the other hand, appears to have anti-apoptotic activity, as a NSs deletion mutant induces apoptosis earlier during infection than wild type virus. By interfering with the downstream signalling of IRF-3, BUNV NSs not only suppresses RIG-I and MDA5 mediated induction of interferons, but also IRF-3 dependent apoptosis (Kohl *et al.*, 2003). The exact mechanism of this signalling interference, however, remains to be characterized in full. As with the induction of apoptosis by LACV and SBV through activation of the DNA damage response, it is tempting to speculate that this suppression of IRF-3-dependent apoptosis may have evolved as a side effect of NSs-mediated suppression of the interferon response. However, it cannot be ruled out that the modulation of apoptotic pathways in itself is also beneficial for viral replication.

Phleboviruses

Like no other bunyavirus, RVFV has proteins which specifically exhibit both pro- and anti-apoptotic functions. While most replicating viruses trigger apoptosis non-specifically as a cellular defence mechanism (Everett and McFadden, 1999), RVFV induces apoptosis via its NSs protein while simultaneously inhibiting apoptosis via its NSm protein. NSs induces apoptosis by promoting Ser-15 and Ser-46 phosphorylation of the central regulator p53 (Austin *et al.*, 2012) and activates the classical DNA damage signalling proteins, ataxia telangiectasia mutated (ATM), Chk.2 and H2A.X (Baer *et al.*, 2012), although the exact mechanisms for these activations remain to be elucidated. Strikingly, viral replication is impaired in cells lacking the p53 gene (Austin *et al.*, 2012) or treated with checkpoint inhibitors (Baer *et al.*, 2012), which strongly suggests that this proapoptotic state creates an environment favourable for viral replication. On the other hand, the NSm protein of RVFV has been well characterized as an anti-apoptotic protein. NSm localizes to the mitochondrial outer membrane via its C-terminal region (Terasaki *et al.*, 2013) and is able to suppress caspase-8- and caspase-9-mediated apoptosis (Won *et al.*, 2007).

Nairoviruses

CCHFV also causes apoptosis *in vitro*, characterized by caspase 3 activation, and the cleavage of the viral nucleocapsid protein (Karlberg *et al.*, 2011). Infected cells also show evidence of PUMA, Noxa and pro-apoptotic ER-stress protein CHOP (Rodrigues *et al.*, 2012). In contrast, DUGV, which shows less cytopathic effect in tissue culture, did not show similar apoptosis markers (Rodrigues *et al.*, 2012). The difference between CCHFV and DUGV highlight the differences that pathogenesis at the cellular level can direct the course of disease in the host.

Acknowledgement

We would like to thank Thomas Ksiazek for his critical insights and review of this chapter.

References

Abdo-Salem, S., Gerbier, G., Bonnet, P., Al-Qadasi, M., Tran, A., Thiry, E., Al-Eryni, G., and Roger, F. (2006). Descriptive and spatial epidemiology of Rift valley fever outbreak in Yemen 2000–2001. Ann. N. Y. Acad. Sci. *1081*, 240–242.

Acrani, G.O., Gomes, R., Proenca-Modena, J.L., da Silva, A.F., Carminati, P.O., Silva, M.L., Santos, R.I., and Arruda, E. (2010). Apoptosis induced by Oropouche virus infection in HeLa cells is dependent on virus protein expression. Virus Res. *149*, 56–63.

Andersson, I., Bladh, L., Mousavi-Jazi, M., Magnusson, K.E., Lundkvist, A., Haller, O., and Mirazimi, A. (2004). Human MxA protein inhibits the replication of Crimean–Congo hemorrhagic fever virus. J. Virol. *78*, 4323–4329.

Andersson, I., Karlberg, H., Mousavi-Jazi, M., Martinez-Sobrido, L., Weber, F., and Mirazimi, A. (2008). Crimean–Congo hemorrhagic fever virus delays activation of the innate immune response. J. Med. Virol. *80*, 1397–1404.

Aradaib, I.E., Erickson, B.R., Elageb, R.M., Khristova, M.L., Carroll, S.A., Elkhidir, I.M., Karsany, M.E., Karrar, A.E., Elbashir, M.I., and Nichol, S.T. (2013). Rift Valley fever, Sudan, 2007 and 2010. Emerg. Infect. Dis. *19*, 246–253.

Armstrong, P.M., and Andreadis, T.G. (2007). Genetic relationships of Jamestown Canyon virus strains infecting mosquitoes collected in Connecticut. Am. J. Trop. Med. Hyg. *77*, 1157–1162.

Austin, D., Baer, A., Lundberg, L., Shafagati, N., Schoonmaker, A., Narayanan, A., Popova, T., Panthier, J.J., Kashanchi, F., Bailey, C., *et al.* (2012). p53 activation following Rift Valley fever virus infection contributes to cell death and viral production. PloS One 7, e36327.

Baer, A., Austin, D., Narayanan, A., Popova, T., Kainulainen, M., Bailey, C., Kashanchi, F., Weber, F., and Kehn-Hall, K. (2012). Induction of DNA damage signaling upon Rift Valley fever virus infection results in cell cycle arrest and increased viral replication. J. Biol. Chem. *287*, 7399–7410.

Baldridge, G.D., Beaty, B.J., and Hewlett, M.J. (1989). Genomic stability of La Crosse virus during vertical and horizontal transmission. Arch. Virol. *108*, 89–99.

Balkhy, H.H., and Memish, Z.A. (2003). Rift Valley fever: an uninvited zoonosis in the Arabian peninsula. Int. J. Antimicrobial Agents *21*, 153–157.

Barry, G., Varela, M., Ratinier, M., Blomstrom, A.L., Caporale, M., Seehusen, F., Hahn, K., Schnettler, E., Baumgartner, W., Kohl, A., *et al.* (2014). NSs protein of Schmallenberg virus counteracts the antiviral response of the cell by inhibiting its transcriptional machinery. J. Gen. Virol. *95*, 1640–1646.

Bennett, R.S., Cress, C.M., Ward, J.M., Firestone, C.Y., Murphy, B.R., and Whitehead, S.S. (2008). La Crosse virus infectivity, pathogenesis, and immunogenicity in mice and monkeys. Virol. J. *5*, 25.

Bente, D.A., Alimonti, J.B., Shieh, W.J., Camus, G., Stroher, U., Zaki, S., and Jones, S.M. (2010). Pathogenesis and immune response of Crimean–Congo hemorrhagic fever virus in a STAT-1 knockout mouse model. J. Virol. *84*, 11089–11100.

Bente, D.A., Forrester, N.L., Watts, D.M., McAuley, A.J., Whitehouse, C.A., and Bray, M. (2013). Crimean–Congo hemorrhagic fever: history, epidemiology, pathogenesis, clinical syndrome and genetic diversity. Antiviral Res. *100*, 159–189.

Bereczky, S., Lindegren, G., Karlberg, H., Akerstrom, S., Klingstrom, J., and Mirazimi, A. (2010). Crimean–Congo hemorrhagic fever virus infection is lethal for adult type I interferon receptor-knockout mice. J. Gen. Virol. *91*, 1473–1477.

Bin Tarif, A., Lasecka, L., Holzer, B., and Baron, M.D. (2012). Ganjam virus/Nairobi sheep disease virus induces a pro-inflammatory response in infected sheep. Vet. Res. *43*, 71.

Bird, B.H., Ksiazek, T.G., Nichol, S.T., and Maclachlan, N.J. (2009). Rift Valley fever virus. J. Am. Vet. Med. Assoc. *234*, 883–893.

Blackmore, C.G., and Grimstad, P.R. (1998). Cache Valley and Potosi viruses (Bunyaviridae) in white-tailed deer (*Odocoileus virginianus*): experimental infections and antibody prevalence in natural populations. Am. J. Trop. Med. Hyg. *59*, 704–709.

Blakqori, G., Delhaye, S., Habjan, M., Blair, C.D., Sanchez-Vargas, I., Olson, K.E., Attarzadeh-Yazdi, G., Fragkoudis, R., Kohl, A., Kalinke, U., *et al.* (2007). La Crosse bunyavirus non-structural protein NSs serves to suppress the type I interferon system of mammalian hosts. J. Virol. *81*, 4991–4999.

Booth, T.F., Steele, G.M., Marriott, A.C., and Nuttall, P.A. (1991). Dissemination, replication, and trans-stadial persistence of Dugbe virus (Nairovirus, Bunyaviridae) in the tick vector *Amblyomma variegatum*. Am. J. Trop. Med. Hyg. *45*, 146–157.

Bowen, M.D., Trappier, S.G., Sanchez, A.J., Meyer, R.F., Goldsmith, C.S., Zaki, S.R., Dunster, L.M., Peters, C.J., Ksiazek, T.G., Nichol, S.T., *et al.* (2001). A reassortant bunyavirus isolated from acute hemorrhagic fever cases in Kenya and Somalia. Virology *291*, 185–190.

Boyd, A., Fazakerley, J.K., and Bridgen, A. (2006). Pathogenesis of Dugbe virus infection in wild-type and interferon-deficient mice. J. Gen. Virol. *87*, 2005–2009.

Braito, A., Corbisiero, R., Corradini, S., Marchi, B., Sancasciani, N., Fiorentini, C., and Ciufolini, M.G. (1997). Evidence of Toscana virus infections without central nervous system involvement: a serological study. Eur. J. Epidemiol. *13*, 761–764.

Braito, A., Ciufolini, M.G., Pippi, L., Corbisiero, R., Fiorentini, C., Gistri, A., and Toscano, L. (1998). Phlebotomus-transmitted toscana virus infections of the central nervous system: a seven-year experience in Tuscany. Scand. J. Infect. Dis. *30*, 505–508.

Brennan, B., Li, P., Zhang, S., Li, A., Liang, M., Li, D., and Elliott, R.M. (2014). A reverse genetic system for severe fever with thrombocytopenia syndrome virus. J. Virol. *89*, 3026–3037.

Bridgen, A., Weber, F., Fazakerley, J.K., and Elliott, R.M. (2001). Bunyamwera bunyavirus non-structural protein NSs is a nonessential gene product that contributes to viral pathogenesis. Proc. Natl. Acad. Sci. U.S.A. *98*, 664–669.

Burt, F.J., Spencer, D.C., Leman, P.A., Patterson, B., and Swanepoel, R. (1996). Investigation of tick-borne viruses as pathogens of humans in South Africa and evidence of Dugbe virus infection in a patient with prolonged thrombocytopenia. Epidemiol. Infect. *116*, 353–361.

Calisher, C.H., McLean, R.G., Zeller, H.G., Francy, D.B., Karabatsos, N., and Bowen, R.A. (1990). Isolation of Tete serogroup bunyaviruses from Ceratopogonidae collected in Colorado. Am. J. Trop. Med. Hyg. *43*, 314–318.

Campbell, G.L., Eldridge, B.F., Hardy, J.L., Reeves, W.C., Jessup, D.A., and Presser, S.B. (1989). Prevalence of neutralizing antibodies against California and Bunyamwera serogroup viruses in deer from mountainous areas of California. Am. J. Trop. Med. Hyg. *40*, 428–437.

Carlton-Smith, C., and Elliott, R.M. (2012). Viperin, MTAP44, and protein kinase R contribute to the interferon-induced inhibition of Bunyamwera Orthobunyavirus replication. J. Virol. *86*, 11548–11557.

Cevik, M.A., Erbay, A., Bodur, H., Gulderen, E., Bastug, A., Kubar, A., and Akinci, E. (2008). Clinical and laboratory features of Crimean-Congo hemorrhagic fever: predictors of fatality. Int. J. Infect. Dis. *12*, 374–379.

Chattopadhyay, S., Marques, J.T., Yamashita, M., Peters, K.L., Smith, K., Desai, A., Williams, B.R., and Sen, G.C. (2010). Viral apoptosis is induced by IRF-3-mediated activation of Bax. EMBO J. *29*, 1762–1773.

Chattopadhyay, S., Fensterl, V., Zhang, Y., Veleeparambil, M., Yamashita, M., and Sen, G.C. (2013). Role of interferon regulatory factor 3-mediated apoptosis in the establishment and maintenance of persistent infection by Sendai virus. J. Virol. *87*, 16–24.

Chowdhary, R., Street, C., Travassos da Rosa, A., Nunes, M.R., Tee, K.K., Hutchison, S.K., Vasconcelos, P.F., Tesh, R.B., Lipkin, W.I., and Briese, T. (2012). Genetic characterization of the Wyeomyia group of orthobunyaviruses and their phylogenetic relationships. J. Gen. Virol. *93*, 1023–1034.

Ciufolini, M.G., Maroli, M., Guandalini, E., Marchi, A., and Verani, P. (1989). Experimental studies on the maintenance of Toscana and Arbia viruses (Bunyaviridae: Phlebovirus). Am. J. Trop. Med. Hyg. *40*, 669–675.

Coates, D.M., and Sweet, C. (1990). Studies on the pathogenicity of a nairovirus, Dugbe virus, in normal and immunosuppressed mice. J. Gen. Virol. *71*, 325–332.

Coetzer, J.A. (1982). The pathology of Rift Valley fever. II. Lesions occurring in field cases in adult cattle, calves and aborted foetuses. Onderstepoort J. Vet. Res. *49*, 11–17.

Colon-Ramos, D.A., Irusta, P.M., Gan, E.C., Olson, M.R., Song, J., Morimoto, R.I., Elliott, R.M., Lombard, M., Hollingsworth, R., Hardwick, J.M., *et al.* (2003). Inhibition of translation and induction of apoptosis by Bunyaviral non-structural proteins bearing sequence similarity to reaper. Mol. Biol. Cell *14*, 4162–4172.

Crabtree, M.B., Sang, R., and Miller, B.R. (2009). Kupe virus, a new virus in the family *Bunyaviridae*, genus *Nairovirus*, Kenya. Emerg. Infect. Dis. *15*, 147–154.

Davies, F.G. (1978). Nairobi sheep disease in Kenya. The isolation of virus from sheep and goats, ticks and possible maintenance hosts. J. Hyg. *81*, 259–265.

Edwards, J.F., Livingston, C.W., Chung, S.I., and Collisson, E.C. (1989). Ovine arthrogryposis and central nervous system malformations associated with in utero Cache Valley virus infection: spontaneous disease. Vet. Pathol. *26*, 33–39.

Eitrem, R., Stylianou, M., and Niklasson, B. (1991). High prevalence rates of antibody to three sandfly fever viruses (Sicilian, Naples, and Toscana) among Cypriots. Epidemiol. Infect. *107*, 685–691.

Eldridge, B.F., Glaser, C., Pedrin, R.E., and Chiles, R.E. (2001). The first reported case of California encephalitis in more than 50 years. Emerg. Infect. Dis. *7*, 451–452.

Ergonul, O. (2006). Crimean–Congo haemorrhagic fever. Lancet Infect. Dis. *6*, 203–214.

Everett, H., and McFadden, G. (1999). Apoptosis: an innate immune response to virus infection. Trends Microbiol. *7*, 160–165.

Fontana, J., Lopez-Montero, N., Elliott, R.M., Fernandez, J.J., and Risco, C. (2008). The unique architecture of Bunyamwera virus factories around the Golgi complex. Cell Microbiol. *10*, 2012–2028.

Frias-Staheli, N., Giannakopoulos, N.V., Kikkert, M., Taylor, S.L., Bridgen, A., Paragas, J., Richt, J.A., Rowland, R.R., Schmaljohn, C.S., Lenschow, D.J., *et al.* (2007). Ovarian tumor domain-containing viral proteases evade ubiquitin- and ISG15-dependent innate immune responses. Cell Host Microbe *2*, 404–416.

Gaensbauer, J.T., Lindsey, N.P., Messacar, K., Staples, J.E., and Fischer, M. (2014). Neuroinvasive arboviral disease in the United States: 2003 to 2012. Pediatrics *134*, e642–650.

Garcia, M.A., Gil, J., Ventoso, I., Guerra, S., Domingo, E., Rivas, C., and Esteban, M. (2006). Impact of protein kinase PKR in cell biology: from antiviral to antiproliferative action. Microbiol. Mol. Biol. Rev. *70*, 1032–1060.

Gerrard, S.R., Li, L., Barrett, A.D., and Nichol, S.T. (2004). Ngari virus is a Bunyamwera virus reassortant that can be associated with large outbreaks of hemorrhagic fever in Africa. J. Virol. *78*, 8922–8926.

Gonzalez, J.P., Camicas, J.L., Cornet, J.P., and Wilson, M.L. (1998). Biological and clinical responses of west African sheep to Crimean–Congo haemorrhagic fever virus experimental infection. Res. Virol. *149*, 445–455.

Gori Savellini, G., Weber, F., Terrosi, C., Habjan, M., Martorelli, B., and Cusi, M.G. (2011). Toscana virus induces interferon although its NSs protein reveals antagonistic activity. J. Gen. Virol. *92*, 71–79.

Gori-Savellini, G., Valentini, M., and Cusi, M.G. (2013). Toscana virus NSs protein inhibits the induction of type I interferon by interacting with RIG-I. J. Virol. *87*, 6660–6667.

Habjan, M., Andersson, I., Klingstrom, J., Schumann, M., Martin, A., Zimmermann, P., Wagner, V., Pichlmair, A., Schneider, U., Muhlberger, E., *et al.* (2008). Processing of genome 5′ termini as a strategy of negative-strand RNA viruses to avoid RIG-I-dependent interferon induction. PloS One *3*, e2032.

Habjan, M., Pichlmair, A., Elliott, R.M., Overby, A.K., Glatter, T., Gstaiger, M., Superti-Furga, G., Unger, H., and Weber, F. (2009). NSs protein of rift valley fever virus induces the specific degradation of the double-stranded RNA-dependent protein kinase. J. Virol. *83*, 4365–4375.

Hampsey, M. (1998). Molecular genetics of the RNA polymerase II general transcriptional machinery. Microbiol. Mol. Biol. Rev. *62*, 465–503.

Hart, T.J., Kohl, A., and Elliott, R.M. (2009). Role of the NSs protein in the zoonotic capacity of Orthobunyaviruses. Zoonoses Public Health *56*, 285–296.

Hayles, L.B., and Lversen, J.O. (1980). Cache Valley virus: experimental infection in Culiseta inornata. Can. J. Microbiol. *26*, 287–290.

Helbig, K.J., and Beard, M.R. (2014). The role of viperin in the innate antiviral response. J. Mol. Biol. *426*, 1210–1219.

Himeidan, Y.E., Kweka, E.J., Mahgoub, M.M., El Rayah el, A., and Ouma, J.O. (2014). Recent outbreaks of Rift Valley fever in East Africa and the Middle East. Frontiers Public Health *2*, 169.

Hoffmann, B., Scheuch, M., Hoper, D., Jungblut, R., Holsteg, M., Schirrmeier, H., Eschbaumer, M., Goller, K.V., Wernike, K., Fischer, M., *et al.* (2012). Novel orthobunyavirus in Cattle, Europe, 2011. Emerg. Infect. Dis. *18*, 469–472.

Holden, P., and Hess, A.D. (1959). Cache Valley virus, a previously undescribed mosquito-borne agent. Science *130*, 1187–1188.

Holzer, B., Bakshi, S., Bridgen, A., and Baron, M.D. (2011). Inhibition of interferon induction and action by the nairovirus Nairobi sheep disease virus/Ganjam virus. PloS One *6*, e28594.

Hubalek, Z., Rudolf, I., and Nowotny, N. (2014). Arboviruses pathogenic for domestic and wild animals. Adv. Virus Res. *89*, 201–275.

Ikegami, T., Narayanan, K., Won, S., Kamitani, W., Peters, C.J., and Makino, S. (2009). Rift Valley fever virus NSs protein promotes post-transcriptional downregulation of protein kinase PKR and inhibits eIF2alpha phosphorylation. PLoS Pathogens *5*, e1000287.

Kallfass, C., Ackerman, A., Lienenklaus, S., Weiss, S., Heimrich, B., and Staeheli, P. (2012). Visualizing production of beta interferon by astrocytes and microglia in brain of La Crosse virus-infected mice. J. Virol. *86*, 11223–11230.

Kalveram, B., and Ikegami, T. (2013). Toscana Virus NSs Protein promotes degradation of double-stranded RNA-dependent protein kinase. J. Virol. *87*, 3710–3718.

Kalveram, B., Lihoradova, O., and Ikegami, T. (2011). NSs protein of rift valley fever virus promotes posttranslational downregulation of the TFIIH subunit p62. J. Virol. *85*, 6234–6243.

Karlberg, H., Tan, Y.J., and Mirazimi, A. (2011). Induction of caspase activation and cleavage of the viral nucleocapsid protein in different cell types during Crimean–Congo hemorrhagic fever virus infection. J. Biol. Chem. *286*, 3227–3234.

Kochs, G., Janzen, C., Hohenberg, H., and Haller, O. (2002). Antivirally active MxA protein sequesters La Crosse virus nucleocapsid protein into perinuclear complexes. Proc. Natl. Acad. Sci. U.S.A. *99*, 3153–3158.

Kohl, A., Clayton, R.F., Weber, F., Bridgen, A., Randall, R.E., and Elliott, R.M. (2003). Bunyamwera virus non-structural protein NSs counteracts interferon regulatory factor 3-mediated induction of early cell death. J. Virol. *77*, 7999–8008.

Konno, S., and Nakagawa, M. (1982). Akabane disease in cattle: congenital abnormalities caused by viral infection. Experimental disease. Vet. Pathol. *19*, 267–279.

Konno, S., Moriwaki, M., and Nakagawa, M. (1982). Akabane disease in cattle: congenital abnormalities caused by viral infection. Spontaneous disease. Vet. Pathol. *19*, 246–266.

Kraatz, F., Wernike, K., Hechinger, S., Konig, P., Granzow, H., Reimann, I., and Beer, M. (2015). Deletion mutants of schmallenberg virus are avirulent and protect from virus challenge. J. Virol. *89*, 1825–1837.

Lasecka, L., and Baron, M.D. (2014). The molecular biology of nairoviruses, an emerging group of tick-borne arboviruses. Arch. Virol. *159*, 1249–1265.

Le May, N., Dubaele, S., Proietti De Santis, L., Billecocq, A., Bouloy, M., and Egly, J.M. (2004). TFIIH transcription factor, a target for the Rift Valley hemorrhagic fever virus. Cell *116*, 541–550.

Le May, N., Mansuroglu, Z., Leger, P., Josse, T., Blot, G., Billecocq, A., Flick, R., Jacob, Y., Bonnefoy, E., and Bouloy, M. (2008). A SAP30 complex inhibits IFN-beta expression in Rift Valley fever virus infected cells. PLoS Pathogens *4*, e13.

Lei, X.Y., Liu, M.M., and Yu, X.J. (2015). Severe fever with thrombocytopenia syndrome and its pathogen SFTSV. Microbes and infection/Institut Pasteur *17*, 149–154.

Lei, Y., Moore, C.B., Liesman, R.M., O'Connor, B.P., Bergstralh, D.T., Chen, Z.J., Pickles, R.J., and Ting, J.P. (2009). MAVS-mediated apoptosis and its inhibition by viral proteins. PloS One *4*, e5466.

Leonard, V.H., Kohl, A., Hart, T.J., and Elliott, R.M. (2006). Interaction of Bunyamwera Orthobunyavirus NSs protein with mediator protein MED8: a mechanism for inhibiting the interferon response. J. Virol. *80*, 9667–9675.

Linthicum, K.J., Davies, F.G., Kairo, A., and Bailey, C.L. (1985). Rift Valley fever virus (family Bunyaviridae, genus Phlebovirus). Isolations from Diptera collected during an inter-epizootic period in Kenya. J. Hyg. *95*, 197–209.

Lozach, P.Y., Mancini, R., Bitto, D., Meier, R., Oestereich, L., Overby, A.K., Pettersson, R.F., and Helenius, A. (2010). Entry of bunyaviruses into mammalian cells. Cell Host Microbe *7*, 488–499.

Lutomiah, J., Musila, L., Makio, A., Ochieng, C., Koka, H., Chepkorir, E., Mutisya, J., Mulwa, F., Khamadi, S., Miller, B.R., *et al.* (2014). Ticks and tick-borne viruses from livestock hosts in arid and semiarid regions of the eastern and northeastern parts of Kenya. J. Med. Entomol. *51*, 269–277.

McIntosh, B.M., Russell, D., dos Santos, I., and Gear, J.H. (1980). Rift Valley fever in humans in South Africa. South African Med. J. *58*, 803–806.

McJunkin, J.E., de los Reyes, E.C., Irazuzta, J.E., Caceres, M.J., Khan, R.R., Minnich, L.L., Fu, K.D., Lovett, G.D., Tsai, T., and Thompson, A. (2001). La Crosse encephalitis in children. New Engl. J. Med. *344*, 801–807.

McMullan, L.K., Folk, S.M., Kelly, A.J., MacNeil, A., Goldsmith, C.S., Metcalfe, M.G., Batten, B.C., Albarino, C.G., Zaki, S.R., Rollin, P.E., *et al.* (2012). A new phlebovirus associated with severe febrile illness in Missouri. N. Engl. J. Med. *367*, 834–841.

Madani, T.A., Al-Mazrou, Y.Y., Al-Jeffri, M.H., Mishkhas, A.A., Al-Rabeah, A.M., Turkistani, A.M., Al-Sayed, M.O., Abodahish, A.A., Khan, A.S., Ksiazek, T.G., *et al.* (2003). Rift Valley fever epidemic in Saudi Arabia: epidemiological, clinical, and laboratory characteristics. Clin. Infect. Dis. *37*, 1084–1092.

Marklewitz, M., Zirkel, F., Rwego, I.B., Heidemann, H., Trippner, P., Kurth, A., Kallies, R., Briese, T., Lipkin, W.I., Drosten, C., *et al.* (2013). Discovery of a unique novel clade of mosquito-associated bunyaviruses. J. Virol. *87*, 12850–12865.

Matsuno, K., Weisend, C., Kajihara, M., Matysiak, C., Williamson, B.N., Simuunza, M., Mweene, A.S., Takada, A., Tesh, R.B., and Ebihara, H. (2015). Comprehensive molecular detection of tick-borne phleboviruses leads to the retrospective identification of taxonomically unassigned bunyaviruses and the discovery of a novel member of the genus phlebovirus. J. Virol. *89*, 594–604.

Meegan, J.M., Hoogstraal, H., and Moussa, M.I. (1979). An epizootic of Rift Valley fever in Egypt in 1977. Vet. Record *105*, 124–125.

Mitchell, C.J., Lvov, S.D., Savage, H.M., Calisher, C.H., Smith, G.C., Lvov, D.K., and Gubler, D.J. (1993). Vector and host relationships of California serogroup viruses in western Siberia. Am. J. Trop. Med. Hyg. *49*, 53–62.

Mohamed, M., McLees, A., and Elliott, R.M. (2009). Viruses in the Anopheles A, Anopheles B, and Tete serogroups in the Orthobunyavirus genus (family Bunyaviridae) do not encode an NSs protein. J. Virol. *83*, 7612–7618.

Morales, D.J., and Lenschow, D.J. (2013). The antiviral activities of ISG15. J. Mol. Biol. *425*, 4995–5008.

Morvan, J., Rollin, P.E., Laventure, S., Rakotoarivony, I., and Roux, J. (1992). Rift Valley fever epizootic in the central highlands of Madagascar. Res. Virol. *143*, 407–415.

Moulton, D.W., and Thompson, W.H. (1971). California group virus infections in small, forest-dwelling mammals of Wisconsin. Some ecological considerations. Am. J. Trop. Med. Hyg. *20*, 474–482.

Muehlenbachs, A., Fata, C.R., Lambert, A.J., Paddock, C.D., Velez, J.O., Blau, D.M., Staples, J.E., Karlekar, M.B., Bhatnagar, J., Nasci, R.S., *et al.* (2014). Heartland virus-associated death in Tennessee. Clin. Infect. Dis. *59*, 845–850.

Mukherjee, P., Woods, T.A., Moore, R.A., and Peterson, K.E. (2013). Activation of the innate signaling molecule MAVS by bunyavirus infection upregulates the adaptor protein SARM1, leading to neuronal death. Immunity *38*, 705–716.

Nallagatla, S.R., Hwang, J., Toroney, R., Zheng, X., Cameron, C.E., and Bevilacqua, P.C. (2007). 5′-triphosphate-dependent activation of PKR by RNAs with short stem–loops. Science *318*, 1455–1458.

Ning, Y.J., Wang, M., Deng, M., Shen, S., Liu, W., Cao, W.C., Deng, F., Wang, Y.Y., Hu, Z., and Wang, H. (2014). Viral suppression of innate immunity via spatial isolation of TBK1/IKKepsilon from mitochondrial antiviral platform. J. Mol. Cell Biol. *6*, 324–337.

Olive, M.M., Goodman, S.M., and Reynes, J.M. (2012). The role of wild mammals in the maintenance of Rift Valley fever virus. J. Wildlife Dis. *48*, 241–266.

Orphanides, G., Lagrange, T., and Reinberg, D. (1996). The general transcription factors of RNA polymerase II. Genes Dev. *10*, 2657–2683.

Palacios, G., Tesh, R., Travassos da Rosa, A., Savji, N., Sze, W., Jain, K., Serge, R., Guzman, H., Guevara, C., Nunes, M.R., *et al.* (2011). Characterization of the Candiru antigenic complex (Bunyaviridae: Phlebovirus), a highly diverse and reassorting group of viruses affecting humans in tropical America. J. Virol. *85*, 3811–3820.

Park, S.W., Han, M.G., Yun, S.M., Park, C., Lee, W.J., and Ryou, J. (2014). Severe fever with thrombocytopenia syndrome virus, South Korea, 2013. Emerg. Infect. Dis. *20*, 1880–1882.

Perrone, L.A., Narayanan, K., Worthy, M., and Peters, C.J. (2007). The S segment of Punta Toro virus (Bunyaviridae, Phlebovirus) is a major determinant of lethality in the Syrian hamster and codes for a type I interferon antagonist. J. Virol. *81*, 884–892.

Peyrefitte, C.N., Perret, M., Garcia, S., Rodrigues, R., Bagnaud, A., Lacote, S., Crance, J.M., Vernet, G., Garin, D., Bouloy, M., *et al.* (2010). Differential activation profiles of Crimean–Congo hemorrhagic fever virus- and Dugbe virus-infected antigen-presenting cells. J. Gen. Virol. *91*, 189–198.

Pinheiro, F.P., Travassos da Rosa, A.P., Travassos da Rosa, J.F., Ishak, R., Freitas, R.B., Gomes, M.L., LeDuc, J.W., and Oliva, O.F. (1981). Oropouche virus. I. A review of clinical, epidemiological, and ecological findings. Am. J. Trop. Med. Hyg. *30*, 149–160.

Qu, B., Qi, X., Wu, X., Liang, M., Li, C., Cardona, C.J., Xu, W., Tang, F., Li, Z., Wu, B., *et al.* (2012). Suppression of the interferon and NF-kappaB responses by severe fever with thrombocytopenia syndrome virus. J. Virol. *86*, 8388–8401.

Rezelj, V.V., Overby, A.K., and Elliott, R.M. (2015). Generation of mutant Uukuniemi viruses lacking the non-structural protein NSs by reverse genetics indicates that NSs is a weak interferon antagonist. J. Virol. *89*, 4849–4856.

Rodrigues, R., Paranhos-Baccala, G., Vernet, G., and Peyrefitte, C.N. (2012). Crimean–Congo hemorrhagic fever virus-infected hepatocytes induce ER-stress and apoptosis cross-talk. PloS One *7*, e29712.

Rust, R.S., Thompson, W.H., Matthews, C.G., Beaty, B.J., and Chun, R.W. (1999). La Crosse and other forms of California encephalitis. J. Child Neurol. *14*, 1–14.

Saelens, X., Kalai, M., and Vandenabeele, P. (2001). Translation inhibition in apoptosis: caspase-dependent PKR activation and eIF2-alpha phosphorylation. J. Biol. Chem. *276*, 41620–41628.

Sancar, A., Lindsey-Boltz, L.A., Unsal-Kacmaz, K., and Linn, S. (2004). Molecular mechanisms of mammalian DNA repair and the DNA damage checkpoints. Ann. Rev. Biochem. *73*, 39–85.

Santiago, F.W., Covaleda, L.M., Sanchez-Aparicio, M.T., Silvas, J.A., Diaz-Vizarreta, A.C., Patel, J.R., Popov, V., Yu, X.J., Garcia-Sastre, A., and Aguilar, P.V. (2014). Hijacking of RIG-I signaling proteins into virus-induced cytoplasmic structures correlates with the inhibition of type I interferon responses. J. Virol. *88*, 4572–4585.

Schmidt, J.R., Schmidt, M.L., and Said, M.I. (1971). Phlebotomus fever in Egypt. Isolation of phlebotomus fever viruses from Phlebotomus papatasi. Am. J. Trop. Med. Hyg. *20*, 483–490.

Seth, R.B., Sun, L., Ea, C.K., and Chen, Z.J. (2005). Identification and characterization of MAVS, a mitochondrial antiviral signaling protein that activates NF-kappaB and IRF 3. Cell *122*, 669–682.

Shepherd, A.J., Leman, P.A., and Swanepoel, R. (1989). Viremia and antibody response of small African and laboratory animals to Crimean–Congo hemorrhagic fever virus infection. Am. J. Trop. Med. Hyg. *40*, 541–547.

Smith, D.R., Steele, K.E., Shamblin, J., Honko, A., Johnson, J., Reed, C., Kennedy, M., Chapman, J.L., and Hensley, L.E. (2010). The pathogenesis of Rift Valley fever virus in the mouse model. Virology *407*, 256–267.

Soldan, S.S., Hollidge, B.S., Wagner, V., Weber, F., and Gonzalez-Scarano, F. (2010). La Crosse virus (LACV) Gc fusion peptide mutants have impaired growth and fusion phenotypes, but remain neurotoxic. Virology *404*, 139–147.

Srihongse, S., and Johnson, C.M. (1974). Human infections with Chagres virus in Panama. Am. J. Trop. Med. Hyg. *23*, 690–693.

Srihongse, S., Grayson, M.A., and Deibel, R. (1984). California serogroup viruses in New York State: the role of subtypes in human infections. Am. J. Trop. Med. Hyg. 33, 1218–1227.

Streitenfeld, H., Boyd, A., Fazakerley, J.K., Bridgen, A., Elliott, R.M., and Weber, F. (2003). Activation of PKR by Bunyamwera virus is independent of the viral interferon antagonist NSs. J. Virol. *77*, 5507–5511.

Tauro, L.B., Diaz, L.A., Almiron, W.R., and Contigiani, M.S. (2009). Infection by Bunyamwera virus (Orthobunyavirus) in free ranging birds of Cordoba city (Argentina). Vet. Microbiol. *139*, 153–155.

Taylor, K.G., and Peterson, K.E. (2014). Innate immune response to La Crosse virus infection. J. Neurovirol. *20*, 150–156.

Terasaki, K., Won, S., and Makino, S. (2013). The C-terminal region of Rift Valley fever virus NSm protein targets the protein to the mitochondrial outer membrane and exerts antiapoptotic function. J. Virol. *87*, 676–682.

Tesh, R.B., and Gubler, D.J. (1975). Laboratory studies of transovarial transmission of La Crosse and other arboviruses by *Aedes albopictus* and *Culex fatigans*. Am. J. Trop. Med. Hyg. *24*, 876–880.

Tesh, R.B., and Modi, G.B. (1987). Maintenance of Toscana virus in Phlebotomus perniciosus by vertical transmission. Am. J. Trop. Med. Hyg. *36*, 189–193.

Tesh, R.B., Lubroth, J., and Guzman, H. (1992). Simulation of arbovirus overwintering: survival of Toscana virus (Bunyaviridae:Phlebovirus) in its natural sand fly vector Phlebotomus perniciosus. Am. J. Trop. Med. Hyg. *47*, 574–581.

Thomas, D., Blakqori, G., Wagner, V., Banholzer, M., Kessler, N., Elliott, R.M., Haller, O., and Weber, F. (2004). Inhibition of RNA polymerase II phosphorylation by a viral interferon antagonist. J. Biol. Chem. *279*, 31471–31477.

Varela, M., Schnettler, E., Caporale, M., Murgia, C., Barry, G., McFarlane, M., McGregor, E., Piras, I.M., Shaw, A., Lamm, C., *et al.* (2013). Schmallenberg virus pathogenesis, tropism and interaction with the innate immune system of the host. PLoS Pathogens *9*, e1003133.

Verbruggen, P., Ruf, M., Blakqori, G., Overby, A.K., Heidemann, M., Eick, D., and Weber, F. (2011). Interferon antagonist NSs of La Crosse virus triggers a DNA damage response-like degradation of transcribing RNA polymerase II. J. Biol. Chem. *286*, 3681–3692.

Weber, F., Haller, O., and Kochs, G. (2000). MxA GTPase blocks reporter gene expression of reconstituted Thogoto virus ribonucleoprotein complexes. J. Virol. *74*, 560–563.

Weber, F., Bridgen, A., Fazakerley, J.K., Streitenfeld, H., Kessler, N., Randall, R.E., and Elliott, R.M. (2002). Bunyamwera bunyavirus non-structural protein NSs counteracts the induction of alpha/beta interferon. J. Virol. *76*, 7949–7955.

Won, S., Ikegami, T., Peters, C.J., and Makino, S. (2007). NSm protein of Rift Valley fever virus suppresses virus-induced apoptosis. J. Virol. *81*, 13335–13345.

Woods, C.W., Karpati, A.M., Grein, T., McCarthy, N., Gaturuku, P., Muchiri, E., Dunster, L., Henderson, A., Khan, A.S., Swanepoel, R., *et al.* (2002). An outbreak of Rift Valley fever in Northeastern Kenya, 1997–98. Emerg. Infect. Dis. *8*, 138–144.

Wu, X., Qi, X., Liang, M., Li, C., Cardona, C.J., Li, D., and Xing, Z. (2014a). Roles of viroplasm-like structures formed by non-structural protein NSs in infection with severe fever with thrombocytopenia syndrome virus. FASEB J. *28*, 2504–2516.

Wu, X., Qi, X., Qu, B., Zhang, Z., Liang, M., Li, C., Cardona, C.J., Li, D., and Xing, Z. (2014b). Evasion of antiviral immunity through sequestering of TBK1/IKKepsilon/IRF3 into viral inclusion bodies. J. Virol. *88*, 3067–3076.

Yu, X.J., Liang, M.F., Zhang, S.Y., Liu, Y., Li, J.D., Sun, Y.L., Zhang, L., Zhang, Q.F., Popov, V.L., Li, C., *et al.* (2011). Fever with thrombocytopenia associated with a novel bunyavirus in China. N. Engl. J. Med. *364*, 1523–1532.

Vector-borne Rhabdoviruses

Ivan V. Kuzmin and Peter J. Walker

6

Abstract

Rhabdoviruses constitute a large group of single-stranded negative-sense RNA viruses distinguished by their morphology and common phylogenetic origin. They infect a wide range of organisms including placental mammals, marsupials, birds, reptiles, fish, insects and plants. The majority of rhabdoviruses are arthropod-borne although some of them have adapted to circulation in vertebrate hosts without participation of arthropod vectors (such as lyssaviruses, novirhabdoviruses, spriviviruses and perhabdoviruses) or to circulation within insect populations (such as sigmaviruses). Rhabdoviruses have relatively simple genomes consisting of the five canonical genes (N, P, M, G and L) which may be overprinted, overlapped and interspersed with accessory genes. Despite the fact that the vesicular stomatitis virus has served as a model of negative-sense RNA viruses in various studies, many aspects of virus–host interactions, transmission and circulation patterns are still poorly understood. As appears from rhabdovirus pathobiology, even the canonical genes are multifunctional and serve distinct functions in the diverse family representatives. Functions of the accessory genes are largely unknown. Rhabdoviruses pose global threat for public health, agri- and aquaculture, and to economy. Conversely, some of them have been successfully used for the development of novel recombinant biologics. The progress in molecular techniques facilitates virus characterization and discovery of novel pathogens but more studies in the laboratory and clinical setting are needed to understand pathobiology of rhabdoviruses and to design feasible mechanisms for prevention and control of rhabdoviral diseases.

Introduction: general characteristics of rhabdoviruses

Rhabdoviruses (family *Rhabdoviridae*) constitute a large diverse group of enveloped single-stranded negative-sense RNA [(–) ssRNA] viruses, with more than 200 representatives documented to date (Dietzgen *et al.*, 2011). In addition to monophyletic origin revealed from phylogenetic reconstructions, rhabdovirus virions have specific morphology that distinguish them from other taxa in the order *Mononegavirales*, the *Bornaviridae*, the *Filoviridae*, and the *Paramyxoviridae*. Those rhabdoviruses that infect vertebrates are bullet- or cone-shaped, while plant-adapted rhabdoviruses may appear bacilliform, bullet-shaped or pleomorphic. The virions (Fig. 6.1) are large, 100–430 nm in length and 45–100 nm in diameter. Their outer surface is covered with glycoprotein spikes, protruding through the virion lipid envelope, which are 5–10 nm long and about 3 nm in diameter. Internally, the nucleocapsid core, about 30–70 nm in diameter, exhibits helical symmetry and can be seen as cross-striations (spacing about 5 nm) in negatively stained and thin-sectioned virions. The nucleocapsid consists of a ribonucleoprotein (RNP) complex comprising the genomic RNA tightly bound to the nucleoprotein (N), together with an RNA-dependent RNA polymerase (L) and phosphoprotein (P). The nucleocapsid is active for transcription and replication: the N-RNA template is processed by the L-protein, which contains most enzymatic activities, and its cofactor the P-protein. The nucleocapsid has a helical symmetry, is uncoiled and filamentous, about 700 × 20 nm in size. In the virion, the matrix protein (M) condenses the nucleocapsid; it interacts

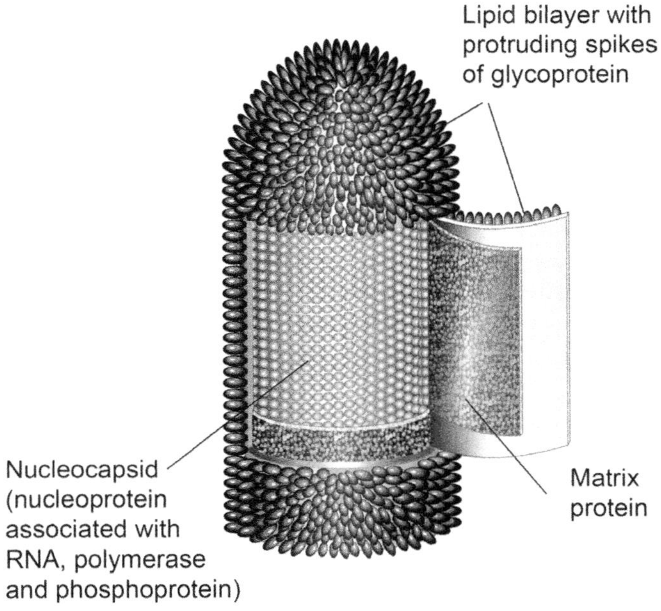

Figure 6.1 Schematic illustration of rhabdovirus virion.

with the N–RNA complex and associates with the host-derived lipid bilayer containing the transmembrane G-protein (Dietzgen, 2012).

Rhabdoviruses are adapted to diverse ecological niches infecting a variety of animals (invertebrate and vertebrate) and plants. Their ecology includes marine, freshwater and terrestrial environments. The majority of family members have been identified in insects, and even the viruses infecting plants are usually transmitted by insects. Some rhabdoviruses (as shown for members of genus *Sigmavirus*) are adapted to circulation among insects only, being transmitted vertically through infected eggs or sperm (Longdon *et al.*, 2012). Such vertebrate pathogens as ephemeroviruses, tibroviruses and vesiculoviruses are transmitted via insect vectors in which they replicate and therefore are true arboviruses. In addition to insects, several rhabdoviruses have been identified in ticks (Ghedin *et al.*, 2013) although transmission mechanisms and host spectrum for these are unknown, as well as for many other rhabdoviruses isolated from invertebrates and vertebrates on a limited number of occasions (Dietzgen *et al.*, 2011). The members of genera *Lyssavirus, Novirhabdovirus, Sprivivirus* and *Perhabdovirus* are adapted to circulation among vertebrates without participation of arthropod vectors.

It cannot be inferred either from genome structure or phylogeny whether rhabdoviruses originated initially from plants or arthropods. Putative rhabdovirus gene fragments have been identified in the genomes of plants (Chiba *et al.*, 2011) and arthropods (Fort *et al.*, 2011; Li *et al.*, 2015). For the latter, fragments (putatively originating from rhabdoviral N, G and L genes) were found in mosquitoes (*Aedes aegypti, Culex quinquefasciatus* and *Anopheles* spp.), and ticks (*Ixodes scapularis* and *Tetranychus urticae*), as well as other terrestrial and marine arthropods. Moreover, the fragments in *A. aegypti* are likely subjected to significant purifying selection, which supports their functional role in the host's biology and suggests that mutual adaptation between rhabdoviruses and their invertebrate hosts has been ongoing during several million years (Fort *et al.*, 2011). Li *et al.* (2015) have also reported the detection of complete or partial genomes of a large number of novel rhabdoviruses and other mononegaviruses in the RNA transcriptomes of 70 diverse arthropod species. Phylogenetic analysis using RdRp sequences showed that many of these are basal to rhabdoviruses isolated from vertebrates and plants, suggesting that arthropods have had an important role in their deep evolution.

Organization of the 11–16 kb rhabdovirus genome (Fig. 6.2) is similar to that of all mononegaviruses. It includes five major genes that are arranged in the conserved linear order 3′-N-P-M-G-L-5′. At each end of the genome, 3′ leader and 5′ trailer sequences of approximately 50–100 nt contain the promoter sequences that initiate replication of the genome and anti-genome, respectively (Dietzgen, 2012). One or more additional interposed genes have been identified in many rhabdoviruses (Walker *et al.*, 2011). The accessory genes may be present either in the same or an overlapping open reading frame (ORF) within an existing gene or in a novel ORF between the structural protein genes. Several studies have inferred that their roles may be accessory to infection in that they are not essential but may enhance replication and transcription, alter host innate immune responses, modulate apoptotic pathways and other cellular functions, and facilitate cell-to-cell movement (Walker *et al.*, 2011). Wongabel virus U3 protein has been shown to interact with the SWI/SNF chromatin remodelling complex to inhibit expression of innate immune response genes (Joubert *et al.*, 2015). Some of the additional proteins have characteristics resembling viroporins, or carrier proteins, which are small, dispensable proteins that may have roles in the formation of pores in membranes to facilitate passage of ions and small molecules during infection. Viroporins are known to interact with the cell vesicle system and to be involved in glycoprotein trafficking (Gonzalez and Carrasco, 2003). The small hydrophobic α1 proteins of ephemeroviruses have characteristics of viroporins (McWilliam *et al.*, 1997; Joubert *et al.*, 2014). However, most accessory genes encode proteins of unknown function that are unrelated to any other known proteins. The challenges in detecting the accessory proteins in cell culture suggest that they are probably of greater relevance in the animal host (Gubala, 2012). The unique feature of plant-adapted rhabdoviruses is that they encode a viral movement protein encoded between P and M genes which facilitates cell-to cell transport (Jackson *et al.*, 2005). One of the most interesting additional genes is a second non-structural glycoprotein (G_{NS}) gene that immediately follows the G gene in ephemeroviruses (Walker *et al.*, 1992; Wang and Walker, 1993; Blasdell *et al.*, 2012a; Blasdell *et al.*, 2012b) and Ngaingan virus (Gubala *et al.*, 2010). The function of G_{NS} proteins is currently unknown. Although G_{NS} shares sequence similarity with the structural G, it does not have important antigenic and neutralizing sites of the latter. Moreover, unlike the G-protein, ephemerovirus G_{NS} does not induce spontaneous cell fusion at low pH (Johal *et al.*, 2008), suggesting that it functions neither in cell attachment nor fusion in vertebrates or invertebrates.

The evolution of rhabdovirus accessory genes may have occurred via gene duplication facilitated by a copy-choice mechanism involving polymerase jumping (Wang and Walker, 1993). Other possible mechanisms for the acquisition of accessory genes include homologous genetic recombination (which is believed to occur rarely in mononegaviruses) or lateral gene transfer. Life-long persistent infections occur commonly in insects and mixed infections with different viruses have been reported (Chen *et al.*, 2004; Mourya *et al.*, 2001; Thavara *et al.*, 2006). Unlike gene duplication, in which the new gene must be

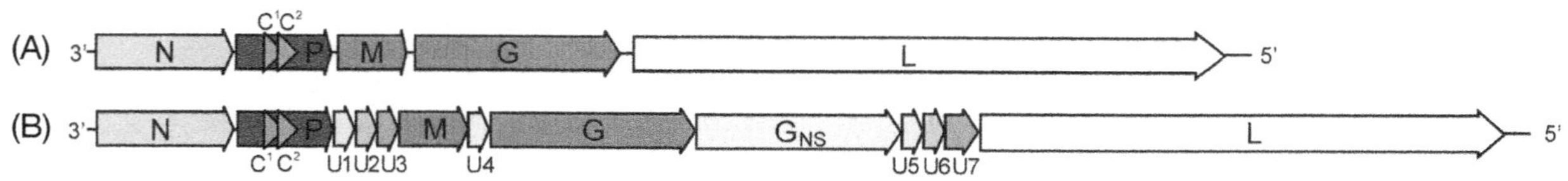

Figure 6.2 Schematic illustration of the most simple rhabdovirus genome present in vesuculoviruses (A) and the complex genome of Ngaingan virus (B).

preserved while evolving a new function, recombination could offer immediate selective advantage in larger more complex genomes by increasing tissue tropism, host range or replication efficiency or more effective blocking of host defensive mechanisms (Walker *et al.*, 2011).

Arthropod-transmitted rhabdoviruses have either the most simple or the most complex genomes. For example, vesiculoviruses that circulate in mammals and insects have as simple genomes as lyssaviruses that circulate only in mammals and as spririviruses and perhabdoviruses that circulate in fish without participation of arthropod vectors. In contrast, novirhabdoviruses which are also transmitted directly between fish have an additional NV gene. Conversely, many insect-borne rhabdoviruses (ephemeroviruses and such unclassified rhabdoviruses as Wongabel and Ngaingan) have the most complex genomes with multiple accessory genes. This observation is opposite to the general suggestion that vector-borne RNA viruses are more constrained than non-vector-borne (Woelk and Holmes, 2002; Jenkins *et al.*, 2002) and suggests that host adaptations and evolutionary pathways of different viruses are distinct, and genome conservation itself does not imply directly whether a virus is adapted to circulation in a single or alternate hosts.

Replication cycle

Most details of rhabdovirus replication in cell culture and laboratory animals were obtained from the model of such typical insect-borne pathogens as vesicular stomatitis virus (VSV), one of the best studied RNA viruses to date.

As other rhabdoviruses, VSV appears to use a variety of different receptors for attachment to different types of host cells that appear to be of low affinity and often are not easily saturable. Cell attachment is mediated by G-protein protruding through virion surface. Negatively charged lipids and phosphatidyl serine have been proposed as cellular receptors. In addition, nonspecific electrostatic and hydrophobic interactions mediate attachment of VSV to cells (Lyles *et al.*, 2013). Following attachment, virions migrate to clathrin-coated pits where they undergo endocytosis into coated vesicles. These vesicles lose their clathrin coats to become early endosomes. The contents of early endosomes are transported to late endosomes and lysosomes for degradation. As virions progress through the endocytic pathway, they are exposed to progressively lower pH, because of acidification of endosomal contents by the adenosine triphosphate (ATP)-dependent endosomal proton pump. At low pH (< 6.5), the G-protein mediates fusion of the viral envelope with the endosome membrane. Many of the virions that attach to the cell surface are not internalized, and many of the internalized virions appear to be degraded by proteases and other enzymes in lysosomes, presumably because they have failed to release their nucleocapsids into the cytoplasm at earlier stages in the endocytic pathway. Likely, the inefficiency of these early events in virus infection is largely responsible for the relatively high virus particle to plaque-forming unit (PFU) ratios of rhabdoviruses (and other virus types), which are typically 10 to 100 virus particles per a plaque forming unit (Matlin *et al.*, 1982).

A general principle by which viral envelope proteins promote fusion is that they must insert into the target membrane through a region of their sequence referred to as the fusion peptide. The fusion peptide of rhabdovirus G-proteins consists of an internal region of the protein sequence, and proteolysis is not involved in activating fusion (Lyles *et al.*, 2013). This is similar to alphaviruses and flaviviruses in which the fusion peptide appears to be an internal region of the protein sequence. A region of G-protein that inserts into target membranes at low pH has been mapped to amino acids 59–221 of the VSV G-protein (Durrer *et al.*, 1995). A second region of the G-protein sequence functionally involved in fusion is the membrane-proximal ectodomain sequence immediately N-terminal to the membrane anchor sequence (Jeetendra, 2002). The cooperation of the fusion peptide and the membrane proximal sequence may be analogous to similarly separate sequences in other viral fusion proteins in bringing the viral and host membranes together for fusion (Lyles *et al.*, 2013). Fusion releases the internal virion components into the cytoplasm of the host cell.

The stable N–RNA complex is the template for both transcription and replication mediated by the viral RNA-dependent RNA polymerase (L–P) complex. The N–RNA association is highly specific since in infected cells N protein does not associate with any other RNAs, including viral mRNAs. Based on the atomic crystal structure of the N–RNA complex, two domains of the N protein form a positively charged RNA binding groove into which the RNA binds by interactions with the sugar-phosphate backbone (Albertini *et al.*, 2008; Ruigrok *et al.*, 2011). Every N protein molecule binds to 9 or 10 nucleotides and completely enwraps the genomic RNA to form a long helical N–RNA complex. The viral polymerase complex consists of two subunits, the enzymatically active L-protein and the non-catalytic cofactor P. The L-protein has multiple domains and performs the functions required for genome transcription and replication, including RNA-dependent RNA polymerase, mRNA 5′ capping enzyme, cap methyltransferase, 3′ poly (A) polymerase and protein kinase activities (Dietzgen, 2012). The L–P complex recognizes the polymerase entry site at the RNA 3′ end. L is positioned on the N–RNA complex through P. The P-protein also acts as a chaperone of nascent RNA-free N by forming N–P complexes that prevent N from binding to cellular RNAs and from polymerizing in the absence of RNA. The viral polymerase complex attaches to this RNP complex. The viral polymerase is fully competent to synthesize all of the viral mRNA without synthesis of viral proteins and without additional host proteins (Ruigrok *et al.*, 2011).

Once the nucleocapsid is released into the cytoplasm, the RNA genome is repetitively transcribed. The transcription is generally considered to be a sequential stop–start mechanism, in which *cis*-acting signals in the template RNA govern the activities of the transcriptase complex at each gene junction. With the exception of the junction between the leader and N genes, each of the VSV gene junctions contains a gene end sequence for the upstream gene (3′AUACUUUUUUU5′), an intergenic dinucleotide (G/CA), which is not transcribed, and a gene start sequence for the downstream gene (3′UUGUC5′). Mutagenesis studies have clearly established that this sequence at each gene junction functions as a signal for polyadenylation and termination of the upstream mRNA, and also as a signal for the initiation, capping, and methylation of the downstream mRNA (Barr *et al.*, 1997, 2001; Hinzman *et al.*, 2002; Hwang *et al.*, 1998). In other rhabdoviruses, variations occur in sequences of these transcription control sequences and the lengths of the intergenic sequences vary but the requirement for a stretch of seven U residues in the transcription termination/polyadenylation signal appears to be invariant, at least amongst the animal rhabdoviruses.

The viral mRNAs are capped by guanosine in a 5′–5′ triphosphate linkage, as are host mRNA. The capping reaction differs, however, from that of host capping enzymes in that both the a and b phosphates are derived from the guanosine triphosphate (GTP) donor, whereas for host capping enzymes, only the a phosphate is derived from the GTP donor (Abraham and Banerjee, 1976). A small amount of the host translation factor EF-1 is associated with VSV virions, and it has been proposed that the α subunit of EF-1 plays a role in the capping reaction through its guanine nucleotide-binding activity (Das *et al.*, 1998). Viral mRNA cap methylation uses S-adenosyl methionine as a methyl donor, as with host enzymes, but is mediated by the viral L-protein (Ogino and Banerjee, 2011). Following elongation of viral mRNA, the transcriptase complex encounters a termination signal at the end of each gene consisting of the sequence 3′AUACUUUUUUU. This signals the polymerase to 'stutter' over the seven Us in the template, resulting in polyadenylation of the viral mRNA (Barr *et al.*, 1997, 2001). In certain cases the 5′-end of an mRNA may overlap the 3′-end of the preceding gene (Dietzgen *et al.*, 2011; Walker *et al.*, 2011).

Following the polyadenylation reaction, two possible fates exist for the transcriptase complex at each gene junction. The most common outcome is that the transcriptase complex traverses the two intergenic nucleotides and resumes transcription at the initiation signal of the downstream gene. Approximately 20% to 30% of transcriptase complexes fail to resume transcription of the downstream gene, however, and presumably dissociate from the template, leading to a 20% to 30% attenuation of expression of the downstream gene at each gene junction (Iverson and Rose, 1981; Wertz *et al.*, 1998). This transcription attenuation results in a gradient of mRNA and protein expression, such that the abundance of each gene product depends on its distance from the 3′ end of the genome (i.e. N > P > M > G > L). The G–L gene junction is unusual in that the level of attenuation is much higher than that at the other gene junctions, resulting in much lower levels of L-protein relative to the other viral proteins (Ball and White, 1976; Wertz *et al.*, 1998).

The initiation and termination of transcription of one element differs from that of all the others, namely that of the leader RNA which is transcribed from a short 3′-terminal sequence at the end of the genome. Transcription of the leader RNA is independently initiated by polymerase entry at the 3′ end of the template and it differs from the other transcribed genes both in terms of the *cis*-acting signals in the template that initiate transcription, and the nature of the product leader RNA, which is phosphorylated at the 5′ end and lacks a cap structure (Lyles *et al.*, 2013).

Encapsidation of nascent RNA appears to constitute a signal for the viral RNA polymerase to ignore the sequences in the genome template at each gene junction that govern the stop–start mechanism for transcription, thereby generating full-length, encapsidated RNA that are complementary to the genome (i.e. antigenomes). Use of antigenomes as templates results in synthesis of progeny genomes. The mechanism of RNA replication appears to be the same regardless of whether genomes or antigenomes are used as templates. The critical cis-acting RNA sequences that govern replication are located at the 3′ ends of the genome and antigenome. These sequences serve as promoters to initiate RNA synthesis, and their complementary sequences at the 5′ end of the product RNA serve as encapsidation signals, with the resulting encapsidation permitting elongation of leader and trailer RNA into full-length products. The 3′ termini of the genome and antigenome of VSV are identical at 15 of 18 positions. These 18 nucleotides are essential elements of both the genomic and antigenomic promoters. However, the genomic and antigenomic promoters of VSV differ substantially at positions 19–29 and 34–46. These sequences in the genomic promoter are required for mRNA synthesis, but not for replication (Li and Pattnaik, 1999; Whelan and Wertz, 1999). In contrast, these sequences in the antigenomic promoter serve as an enhancer of replication. As a result of this enhancer activity in the antigenomic promoter, replication of genomes versus antigenomes is asymmetric: many more genomes than antigenomes are synthesized in virus-infected cells (Lyles *et al.*, 2013).

Once nucleocapsids containing progeny genomes begin to accumulate in infected cells, they are used as templates for secondary transcription, and are assembled into progeny virions. The viral mRNAs are generally translated in cytoplasmic polysomes, except for the G mRNA which is translated on membrane-bound polysomes. Cellular chaperones are engaged to transport the M protein to the plasma membrane and G-protein from the endoplasmic reticulum to the Golgi apparatus and then to the plasma membrane. Post-translational modification of G-protein involves removal of the amino-proximal signal peptide and stepwise glycosylation in compartments of the Golgi apparatus. Viral nucleocapsid structures are assembled in association with M protein (Dietzgen, 2012).

The final step in virus assembly is release of the budding virion. The budding process is mediated by interaction of M protein with host proteins involved in the formation of multivesicular bodies (Harty *et al.*, 2001; Irie *et al.*, 2004; Jayakar *et al.*, 2000). At the plasma membrane of infected cells, G-protein is organized into clusters or microdomains. These G-protein-containing microdomains are formed independently of other viral components (Brown and Lyles, 2003) and appear to be similar to cholesterol- and sphingolipid-rich lipid rafts that serve as sites of assembly for other viruses, such as influenza viruses (Luan *et al.*, 1995; Pickl *et al.*, 2001). An interesting feature of the budding process is that envelope glycoproteins from many unrelated viruses, as well as some host integral membrane proteins, can be incorporated into the viral envelopes in a process referred to as pseudotype formation or phenotypic mixing (Zavada, 1982; Brown and Lyles, 2005).

Virus–host interactions

The majority of insights described above were obtained from cell culture studies. Indeed, in real life the viruses face many challenges, dealing with the complexity of different tissues and systems present in a living organism (in the case of arboviruses, several hosts from very diverse animal taxa may be involved), and even with more complex variables at the population level to facilitate their sustained transmission and perpetuation. The same viruses (for example, VSV) may cause persistent infection in insect hosts but acute disease in vertebrate hosts. Even rhabdoviruses with the most simple and similar genome organization (such as vesiculoviruses, spriviviruses, perhabdoviruses) utilize a diversity of ecological niches infecting insects, mammals and fish in a variety of different manners. Because of the very limited genome size of rhabdoviruses, individual genomic regions may be responsible for multiple functions. Therefore, a single mutation may have multiple antagonistic effects. This may explain the significant purifying selection constraints observed in rhabdovirus genes (Woelk and Holmes, 2002). Other observations (obtained from *in vitro* experiments with VSV) suggest that virus populations may include several quasi-species, or phenotypes, with each phenotype adapted to circulation in a specific host. Depending on host environment, only one of these phenotypes will dominate in virus population (and be determined in the consensus nucleotide sequence) whereas other phenotypes will be under-represented. When the virus is transmitted to a different host species, other phenotypes will gain advantage and become dominant in the virus population (Novella and Presloid, 2012). We may expect a much greater repertoire of phenotype changes from viruses with multiple accessory genes, such as ephemeroviruses and various non-classified rhabdoviruses. We certainly do not understand many important aspects of rhabdovirus–host interactions, but several available observations and experimental findings are useful for conducting further research in this direction. In this book we focus on arboviruses and omit the knowledge accumulated for other important members of the *Rhabdoviridae* unless it can be reasonably generalized.

As discussed above, the diversity and phylogeny of rhabdoviruses strongly suggest that they evolved in invertebrates (most likely in insects) with secondary colonization of vertebrates. An interesting assessment of rhabdovirus–insect interactions was prepared by Hogenhout *et al.* (2003). The first step of the virus transmission cycle is acquisition of the pathogen from the vertebrate or plant host by the insect's piercing and sucking mouthparts, and subsequent infection of the epithelial cell layers in the insect midgut. In addition to interaction with cellular receptors, insect-transmitted rhabdoviruses must be able to withstand degradation by potent proteases in the insect's saliva and midgut lumen, and to cross physical barriers, such as the multilayer peritrophic-like membranes separating the microvilli from the midgut lumen (Ammar and Nault, 1985). After escape from gut cells, rhabdoviruses spread to other organs and tissues in the vector through the nervous system and/or haemolymph. Rhabdoviruses may establish long-term persistence in insects, and at least sigmaviruses can be transmitted vertically via infected eggs and sperm (Longdon *et al.*, 2012). However, for transmission to vertebrate hosts they must be excreted with saliva and both infection of and escape from the salivary gland can be barriers for rhabdovirus transmission. For example, sowthistle yellow vein virus is found in various tissues of non-vectors but is only found in the salivary gland of competent vectors (Sylvester and Richardson, 1992). To increase the efficiency of transmission, viruses can utilize components of arthropod saliva to increase local tissue penetration or virus replication in the inoculation site. For example, tick salivary gland extracts inhibit the antiviral activity of interferon (IFN), and increase replication of vesiculoviruses (Hajnicka *et al.*, 2000). In cattle, bites by VSV-infected insects resulted in lesion formation at much lower virus doses than necessary to cause lesions via needle injection or scarification (Reis *et al.*, 2011). In addition, by mechanisms similar to those utilized by lyssaviruses to increase excretion of infectious saliva and biting activity in rabid animals, arthropod-infecting rhabdoviruses may change physiology and behaviour of infected arthropods. Other arthropod-borne viruses of animals and plants have been shown to modify insect feeding behaviour to enhance transmission efficiency (Jackson *et al.*, 2012; Moreno-Delafuente *et al.*, 2013) and several arboviruses have been shown to infect nervous tissues in insects (Girard *et al.*, 2004; McElroy *et al.*, 2008).

Insects, vertebrates and plants all have an innate and adaptive immune response that aids in protection against pathogens. Although most of the facets of these defensive mechanisms in arthropods remain to be elucidated, a significant body of knowledge exists on host responses to rhabdovirus infection in vertebrates as described below using the example of vesiculoviruses.

Vertebrate response to vesiculovirus infection

The release of viral nucleoprotein because of degradation of virions in the endocytic pathway during early infection, and its detection by the pattern recognition receptor TLR7, has been proposed as a mechanism of IFN induction in dendritic cells infected with VSV (Lund *et al.*, 2004). Most cell types, however, lack TLR7, and activation of antiviral responses usually requires transcription and translation of viral gene products (TenOever *et al.*, 2004), indicating that most cells probably detect viral products by intracellular receptors. An RNA helicase, RIG-I, is required for activation of IFN gene expression in response to VSV infection of fibroblasts and hepatocytes (Jayakar and Whitt, 2002; Sumpter *et al.*, 2005), and a related helicase, MDA5, can mediate an antiviral response against VSV in transfected cells (Yoneyama *et al.*, 2005), suggesting that these helicases may be intracellular pattern recognition receptors that detect the presence of viral gene products. The major source of IFN in VSV-infected mice appears to be a subclass of plasmacytoid dendritic cells residing in the marginal zone of the spleen (Barchet *et al.*, 2002). Other cell types involved in innate immunity include chemokine-secreting marginal zone macrophages which eliminate the virus and infected cells from circulation (Ciavarra *et al.*, 2005; Oehen *et al.*, 2002). Fish-infecting rhabdoviruses induce production of Mx protein, *vig-1* and *vig-2* proteins which are analogous to the mammalian IFN regulatory factors. These proteins elicit protection against infection *in vitro* and *in vivo* (Kim *et al.*, 2000; Yabu *et al.*, 1998; Boudinot *et al.*, 2001). Virus induced fish interferons – named IFNφ – are structurally and functionally similar to type I IFN and induce many conserved effector genes (Verrier *et al.*, 2011), but they differ from mammalian type I interferons by the presence of introns in their genes and the structure of their receptor (Altmann *et al.*, 2003; Levraud *et al.*, 2007; Aggad *et al.*, 2009). Additionally, IFNφ expression can be induced through the toll-like receptor (TLR) pathways (Oshiumi *et al.*, 2008). In fish, TLR3 has been reported in endoplasmic reticulum where it binds small size ssRNAs while TLR22 is expressed at the cell membrane and recognizes long dsRNAs. These TLRs recruit TIR containing adaptors (TICAMs) and trigger IRF-dependent IFNφ production. Finally, IFNφ binding to their cognate receptors results in the activation of JAK/STAT canonical pathways and subsequent induction of many interferon stimulated genes (ISG), some of which have a known anti-viral activity like Mx, PKR, ISG15 or *Vig-1*/viperin (Verrier *et al.*, 2012).

Apoptosis is another defensive mechanism against infection. In fact, most of the cytopathic effects of VSV result from the induction of apoptosis, caused by inhibition of host gene expression and translation (Lyles *et al.*, 2013). Host gene expression is inhibited at three levels: (a) transcription of host mRNA, (b) transport of host mRNA from the nucleus to the cytoplasm, and (c) translation of host mRNA into proteins. Two viral gene products have been implicated in the shut-off of host gene expression by VSV: leader RNA and M protein. However, the significance of leader RNA was likely over-estimated (Dunigan *et al.*, 1986) whereas the role of M protein has been demonstrated repeatedly in several studies. Expression of M protein in transfected cells inhibits expression of co-transfected genes driven by a wide variety of promoters (Ahmed *et al.*, 2003; Ahmed and Lyles, 1998; Black and Lyles, 1992; Ferran and Lucas-Lenard, 1997; Lyles *et al.*, 1996) whereas mutated M protein lacks this ability (Ahmed and Lyles, 1997; Desforges *et al.*, 2001; Jayakar and Whitt, 2002; Stojdl *et al.*, 2003). The inhibition of host translation appears to result from the inhibition of initiation of protein synthesis. The likely target of this action is the eIF4F translation initiation complex, which is responsible for recruiting mRNA to the ribosome (Connor and Lyles, 2002). The inhibition of both host transcription and translation generally occur in parallel and are usually 80% to 90% complete by 4 to 6 h post infection (Ahmed *et al.*, 2003). In general, viruses such as VSV that replicate rapidly and target general aspects of host gene expression and translation are likely to induce host responses that result in death of host cells via apoptosis, as occurs in the vertebrate cells. In many insect cells, however, VSV replication is attenuated and a persistent infection is established with little, if any, cytopathic effect (Wyers *et al.*, 1980).

In addition to innate immune responses, adaptive immune responses are critical for virus clearance. The significant representation of G-protein on virion envelopes is able to induce a T-cell-independent IgM response (Bachmann *et al.*, 1995). When G is expressed on the surfaces of infected cells, T-cells are required for production of IgG (Maloy *et al.*, 1998). VSV-infected dendritic cells are responsible for presentation of viral antigens to T-cells and B-cells (Ciavarra *et al.*, 2000, 2005; Ludewig *et al.*, 2000). The CD4+ T-helper cell (Th) response to VSV infection includes elicitation of both Th1 and Th2 cells. The response is predominantly of the Th1 type, resulting in secretion of IFN-γ and isotype switching in B cells to produce predominantly IgG2a antibodies. Isotype switching to IgG2a is also mediated by IFN-γ-producing γ-Δ T cells. This polarization of the T-cell response presumably reflects secretion of IL-12 by dendritic cells and other antigen-presenting cells. The infection also elicits CD8+ cytolytic T cells (Tc) which recognize peptides containing conserved sequences derived from G-protein or N-protein (and perhaps other viral proteins) presented in the context of class I major histocompatibility complex (MHC) molecules on virus-infected cells. In addition to CD8+ Tc, VSV also elicits CD4+ cytotoxic T cells, which recognize epitopes derived from G-protein presented in the context of class II MHC molecules (reviewed by Lyles *et al.*, 2013). Such complex of the innate and adaptive immune responses is able to clear VSV (with lysis of infected cells) from vertebrate hosts after several days of acute infection.

Pathobiology of vesiculovirus infections

The pathobiology of VSV in vertebrates is quite complex and still poorly understood. The virus is unable to penetrate intact skin or mucosa. For successful transmission, it needs to be introduced beneath the skin and mucous membranes via wounds and abrasions (Jubb, 1985). The virus can also be transmitted by the bite of insect vectors such as mosquitoes (*Aedes* spp.), sand flies (*Lutzomyia* spp.), blackflies (*Simulium* spp.), and other diptera (Comer *et al.*, 1990; Cupp *et al.*, 1992; Webb and Holbrook, 1989; Smith *et al.*, 2009). In addition, although inconsistently, it can be transmitted between cattle via direct contact (Mead *et al.*, 2004). Disease signs include only localized skin lesions. Remarkably, such lesions appear only in the specific regions, such as the snout, mouth, lip, coronary band of the hoof, and teats (Rodrigues and Pauszek, 2012). Vesicular lesions are not observed after inoculation of the skin of the neck or abdomen, nor by the intranasal or intravenous routes (Scherer *et al.*, 2007). Viraemia is not observed in VSV-infected cattle. It appears that virus replicates only locally in the inoculation area, and is rapidly cleared by the immune system.

The pathobiology of VSV in a laboratory rodent model is totally different. Adult mice are relatively resistant to VSV inoculated intravenously or intraperitoneally, although they are highly susceptible to intranasal and intracranial inoculation, with the development of encephalitis resulting in nearly 50% fatality rate (Lyles *et al.*, 2013). The virus spreads primarily via neuronal pathways. Following intranasal inoculation, the virus first replicates in olfactory receptor neurons (Poirot *et al.*, 1985), from where it quickly spreads to the glomerular cells of the olfactory bulb and the anterior olfactory nuclei, and further to other parts of the central nervous system (CNS). In addition to neuronal spread, VSV infects cells lining the ventricular system, where it can be released into the cerebrospinal fluid and can spread to other parts of the brain and spinal cord (Huneycutt *et al.*, 1994; Plakhov *et al.*, 1995). The morbidity and mortality associated with VSV infection in mice is attributed to virus-induced apoptosis of infected cells in the CNS. Intranasal inoculation can lead to the infection of the respiratory epithelium and virus spread through the respiratory tract to the lungs; however, little, if any, pathology is observed in these tissues (Forger *et al.*, 1991).

The majority of human exposures to vesicular stomatitis viruses (for example, in cattle-handlers and veterinarians) remain asymptomatic and lead to the development of an antibody response. Skin lesions are uncommon. More frequently humans develop a mild flu-like disease (Rodriguez and Pauszek, 2012) and only occasionally encephalitides (Quiroz *et al.*, 1988). Similarly, a flu-like febrile disease was documented in humans infected by another vesiculovirus, Isfahan virus (Tesh *et al.*, 1977). In contrast, human infection with such phylogenetically and ecologically related pathogen as Chandipura virus (CHPV) usually manifests as severe acute encephalitis with nearly 50% fatality rate (Rao *et al.*, 2004; Narasimha Rao *et al.*, 2008; Tandale *et al.*, 2008; Gurav *et al.*, 2010).

The most complex are interactions of vesiculoviruses with insects. Infection in insects is usually persistent and noncytolytic. As discussed above, it remains to be established whether the viruses alter insect physiology and behaviour to increase transmissibility. Vertical and venereal transmission of CHPV was demonstrated in the mosquito *A. aegypti*. The minimum filial infection rate among the progeny was 0.9–1.4%. The venereal infection rate of CHPV among inseminated females was 32.7% (Mavale *et al.*, 2005). Similarly, VSV and Alagoas virus were detected in a small proportion of progeny of infected sand flies *Lutzomyia* spp. (Comer *et al.*, 1990; Tesh *et al.*, 1987). In other settings, VSV can be transmitted between insects via feeding on the infected mammals. Since the mammals do not develop viraemia, feeding on the vesicular skin lesions is essential (Smith *et al.*, 2009). However, co-feeding of infected and non-infected sand flies on the same mammal can also result in horizontal transmission of VSV (Mead *et al.*, 2000). Viral replication occurs in insects in epithelial, neural, and haemolymph cells. Tissues of the alimentary canal are infected in a temporal pattern that parallels the route of digestion/absorption: foregut and midgut by day 1, surrounding hemolymph and Malpighian tubules by day 3, and finally the midgut/hindgut junction, hindgut, and rectal region by day 5. Circulation of virus in the hemolymph coincided with infection of the dermis and fat bodies, the salivary glands, eyes, cerebral and subthoracic ganglia, and the ovaries. Neural infections were seen in the subabdominal ganglia innervating the midgut in 33% of insects (Drolet *et al.*, 2005).

Fish rhabdoviruses that were previously classified in the genus *Vesiculovirus* but recently separated in two independent genera *Sprivivirus* and *Perhabdovirus* demonstrate significant genome similarity to vesiculoviruses but their ecology and virus–host interactions are very distinct. Although participation of arthropod vectors (such as the carp louse, *Argulus foliaceus*, and leeches, *Piscicola geometra*) may potentially occur, the most common mechanism of their transmission is water-borne (Ahne *et al.*, 2002). The viruses primarily replicate in gills and further disseminate to internal organs via blood causing severe panvasculitis. They are extensively excreted via faeces, urine and skin lesions to ensure transmission to other susceptible hosts. The disease patterns are influenced by water temperature, age and condition of the fish, population density and stress factors. The immune status of the fish is also an important factor with both innate and adaptive. Clinical disease is usually observed at water temperatures between 5–18°C and is most severe at temperatures below 10°C, when it is believed the host immune response is suppressed or delayed (Rodriguez and Pauszek, 2012).

Pathobiology of bovine ephemeral fever

Bovine ephemeral fever virus (BEFV) is one of the most important vector-borne viruses of livestock. It causes an acute febrile illness in cattle and water buffaloes throughout most tropical and sub-tropical regions of Africa, the Middle-East, Asia and Australia, extending seasonally into temperate zones (Walker, 2005; Walker *et al.*, 2012). It does not occur in the Americas or Europe (other than in western regions of Turkey) where it is considered a significant exotic disease. Clinical expression of bovine ephemeral fever (BEF) can vary from unapparent to very severe. In mild cases, the disease is of short duration with clinical signs including fever, ocular and nasal discharge, loss of appetite and temporary lameness; in the most severe cases, there is profuse salivation, ruminal stasis, joint swelling and limb paralysis, sometimes leading to death (Mackerras *et al.*, 1940; Basson *et al.*, 1970). Prolonged paralysis can also occur, preventing affected animals from standing for weeks or months after other signs have subsided (Hill and Schultz, 1977). Morbidity rates in affected herds may often be very high in non-immune animals and, although mortality rates are typically quite low (< 1%), there have been various reports in recent years of severe epizootics with mortality rates exceeding 5–10%. In Taiwan, a severe BEF epizootic in 1996 resulted in a case fatality rate of 11.3% due to the combined effects of disease and culling (Liao *et al.*, 1998). An epizootic in the Jordan Valley of Israel in 1999 was reported to have affected 97.7% of dairy herds with a 38.6% morbidity rate and a case fatality rate of 8.6% (Yeruham *et al.*, 2010). In China, where BEF can occur over a vast area and infect tens of millions of cattle during the summer and early autumn, low mortality rates (<1%) were common prior to the year 2000 (Bai, 1993; Zheng and Qui, 2012). However, during an epizootic in Henan Province in central China in 2004, it has been reported that 12,200 dairy cattle were affected with a case fatality rate of 18.0%; in 2005, the mortality rate was similar, and in 2011 5360 of 32,051 infected cattle (16.7%) died (Zheng and Qui, 2012). In Turkey, a widespread BEF epizootic in 2012 that affected most parts of the country extending to the European land mass caused mortality rates estimated at 10–20% (Tonbak *et al.*, 2013). Phylogenetic analyses have indicated that, while some 2012 BEFV isolates from this epizootic clustered with isolates from Turkey obtained during 2008, other were most closely related to BEFV isolates collected in China during the 2011 epizootic. Although China is known to export cattle in large numbers to the Middle-East (Aziz-Boaron *et al.*, 2012), the reasons for this apparent change and/or variation in BEFV pathogenicity have not been fully investigated.

The clinical signs of BEF in cattle are consistent with pathology caused by increased vascular permeability, primarily affecting endothelial cells of the small blood vessels, and an associated inflammatory response (Young and Spradbrow, 1990a; St George, 1993). Indeed, clinical signs can be completely prevented by treatment with anti-inflammatory drugs (Uren *et al.*, 1989). In experimentally infected cattle, the incubation period is usually 2–4 days and viraemia persists for up to 5 days (Mackerras *et al.*, 1940; Young and Spradbrow, 1990b; Uren *et al.*, 1992). Peak viraemia usually precedes the onset of fever which occurs in two or more phases at intervals of 12–24 h. Fever is accompanied by a marked neutrophilic leucocytosis, lymphopenia, eosinopenia and elevated levels of blood calcium, fibrinogen and cytokines including INF-a, IL-I and TNF, which are potent mediators of inflammation (Uren *et al.*, 1992; Uren and Zakrzewski, 1989). Although the initial site of BEFV replication is unknown, the virus is present in leucocytes and blood plasma during viraemia. Virus has also been recovered from synovial, pericardial, thoracic and abdominal fluids (Young and Spradbrow, 1985). BEFV antigens can be detected by immunofluorescence in neutrophils obtained from synovial, abdominal, thoracic, and pericardial fluids, and in neutrophils in affected synovial membranes and epicardium (Young and Spradbrow, 1980, 1985). However, it has not been determined whether the detected antigen represents replicating virus or has simply been phagocytosed. As there is no evidence of extensive tissue damage other than vasculitis, it has been suggested that interferon toxicity may be the cause of the general inflammatory response and other clinical signs of BEFV infection (St George, 1993; St George *et al.*, 1995).

The neutralizing antibody response to BEFV infection follows the cessation of viraemia by several days (Young and Spradbrow, 1990b). Neutralizing antibody is protective and, although some exceptions have been recorded, a solid body of evidence from epidemiological studies and field observations indicates that BEFV infection usually leads to life-long immunity (St George, 1985; St George *et al.*, 1977; Uren *et al.*, 1994). The brief viraemia and associated transmission window that precedes the onset of adaptive immunity can explain the existence of a single BEFV serotype worldwide and the limited antigenic variation detected in the major neutralization sites in the G-protein. Indeed, as discussed below, cross-reactive antibody to Kimberley virus (KIMV) and other ephemeroviruses in cattle that are subsequently infected with BEFV appears to be driving evolution of the virus, at least in the Australian episystem (Trinidad *et al.*, 2014).

Although BEFV has been isolated from biting midges (*Culicoides* spp.) in Kenya and Zimbabwe (Davies and Walker, 1974; Blackburn *et al.*, 1985), and from mosquitoes and biting midges on several occasions in Australia (Cybinski and Muller, 1990; Standfast *et al.*, 1976; Muller and Standfast, 1986; St George *et al.*, 1976), various lines of evidence indicate that mosquitoes are likely to be the major vectors. Experimental studies in cattle have established that BEFV infection is possible only by intravenous but not by intradermal, intramuscular or subcutaneous injection, suggesting that capillary feeders (mosquitoes) rather than pool feeders (midges) would be required for efficient transmission (St George, 1993). Furthermore, as observed by St. George (1993, 2009), pooled blood in intradermal lacerations caused by feeding midges would be drained via the lymphatic system and the available evidence suggests that BEFV does not appear in lymphatic fluid or the cellular fraction during the febrile phase of illness. There is also experimental evidence that BEFV is secreted in the saliva of experimentally infected mosquitoes but not midges (St George, 2004), and that the most abundant livestock-associated culicoides species in South Africa is refractory to oral infection with BEFV (Venter *et al.*, 2003). It has also been argued that the epidemiology of the disease differs from that of known midge-transmitted bluetongue and Akabane diseases and is inconsistent with the distribution and seasonal abundance of *Culicoides* species (St George, 1993; Kirkland, 1993; Murray, 1997). Interestingly, it has been reported that BEFV-infected mosquitoes, but not midges, display the CO_2 sensitivity that is characteristic of sigmaviruses replicating in drosophila (Muller and Srtandfast, 1993).

Other ephemeroviruses

Other recognized ephemeroviruses include Berrimah virus (BRMV), Kimberley virus, Adelaide River virus (ARV), Obodhiang virus (OBOV) and kotonkan virus (KOTV). BRMV was isolated on only a single occasion from a healthy sentinel steer in the Northern Territory of Australia (Gard *et al.*, 1983). It is most closely related antigenically to BEFV but it is distant phylogenetically from all available BEFV isolates. BRMV has not been associated with disease and little is known of its geographic distribution or ecology. KIMV has been isolated in Australia from healthy sentinel cattle as well as from mosquitoes *Cx. annulirostris* and biting midges *C. brevitarsis* (Liehne *et al.*, 1981; Cybinski and Zakrzewski, 1983; Zakrzewski and Cybinski, 1984). KIMV antibodies have been detected in cattle, water buffalo, horses and goats in Indonesia, Papua New Guinea, China and Australia, but the virus has not been associated with disease (Cybinski and Zakrzewski, 1983; Jiang and Yan, 1989; Soleha *et al.*, 1993a,b). Although KIMV and BEFV can be clearly distinguished antigenically and are distinct phylogenetically, they have a similar geographic distribution in Australia. Indeed, the evolution of BEFV in Australia, which is under strong positive selection, appears to be driven in part by cross-reactive neutralizing antibody to KIMV which shares a single neutralizing epitope with the BEFV G-protein (Cybinski *et al.*, 1990; Trinidad *et al.*, 2014). Comparative serological and genetic studies have shown that Malakal virus which was isolated from mosquitoes *Mansonia uniformis* in Sudan (Schmidt *et al.*, 1965) can also be considered a geographic variant of KIMV, suggesting that BEFV and KIMV may have a similar geographic distribution globally (Blasdell *et al.*, 2012b).

ARV was also isolated from a healthy sentinel steer in northern Australia (Gard *et al.*, 1984) and is most closely related to OBOV which had been isolated previously from *Mansonia uniformis* mosquitoes in Sudan (Schmidt *et al.*, 1965). Nevertheless, ARV and OBOV are distinct antigenically and phylogenetically and have been assigned as separate species in the genus *Ephemerovirus* (Blasdell *et al.*, 2012a). Antibodies to ARV have been detected in cattle in Australia, China, Papua New Guinea and New Caledonia and there is evidence of infection in water buffalo and pigs in Australia (Bai, 1993; Gard *et al.*, 1984; Daniels *et al.*, 1995).

KOTV is the only other ephemerovirus known to be associated with clinical bovine ephemeral fever (Blasdell *et al.*, 2012a). KOTV was isolated in Nigeria from a mixed pool of biting midges (Kemp *et al.*, 1973). It was subsequently shown that an ephemeral fever-like illness in local and imported cattle, characterized by anorexia, nasal discharge, lameness, recumbency and low level mortalities, was associated with seroconversion to KOTV (Tomori *et al.*, 1974). The disease has also been reproduced experimentally in cattle. A high prevalence of KOTV-neutralizing antibody has been reported in cattle, hedgehog (*Atelerix albiventris*) and rodents (*Cricetomys gambianis*) in Nigeria and there is also evidence of infection in horses, cattle egrets (*Bubulcus ibis*) and humans (2 of 39 sera collected from Jos in central Nigeria) (Kemp *et al.*, 1973; Tomori *et al.*, 1974). Koolpinyah virus (KOOLV), which is closely related antigenically and phylogenetically to KOTV, was isolated on two occasions from healthy sentinel cattle in the Northern Territory of Australia (Gard *et al.*, 1992). Our analysis of the complete genome sequences indicate that KOOLV and KOTV are very closely related viruses with very similar genome organizations and predicted transcription strategies (Walker *et al.*, 2015). The potential role of KOOLV in the epidemiology of ephemeral fever in Australia requires further investigation.

Other potential ephemeroviruses, each of which was isolated from *Mansonia uniformis* mosquitoes, are Puchong virus (PUCV) from Malaysia and Yata virus (YATV) from the Central African Republic (Karabastos, 1985). Little is known about their ecology or pathobiology but a new ephemerovirus most closely related to PUCV has recently been identified in cattle from northern Australia (K.R. Blasdell, personal communication). As *Mansonia uniformis* is distributed in Africa, Asia and Australia (that reflects the known distribution of ephemeroviruses), has a feeding preference for cattle and has been a common source of ephemerovirus isolation, its role in the epizootiology of BEFV and other ephemeroviruses is also worthy of further investigation.

Tibroviruses

The genus *Tibrovirus* currently comprises two species, *Tibrogargan virus* (TIBV) and *Coastal Plains virus* (CPV). The viruses are distinct phylogenetically and antigenically but they have very similar and unusual genome organizations with three additional ORFs located between the M and G genes (U1 and U2) and between the G and L genes (U3) (Gubala *et al.*, 2011). Each virus has been isolated from healthy sentinel cattle in Australia (Cybinski and Gard, 1986; Gubala *et al.*, 2011). TIBV has also been isolated from biting midges *Culicoides brevitarsis* and serosurveys of cattle indicated that each of these viruses has a similar distribution across northern, central and eastern Australia and Papua New Guinea that roughly parallels the distribution of *C. brevitarsis* (Cybinski *et al.*, 1980; Cybinski and Gard, 1986). TIBV-neutralizing antibodies have been detected in a high proportion of water buffalo tested but not in camels, dogs, goats, horses, pigs, sheep, marsupials or humans (Cybinski *et al.*, 1980). Neutralizing antibody to CPV has been detected in water buffalo, dogs and a horse but not in other species tested including 86 humans from regions where cattle sera were positive (Cybinski and Gard, 1986). Although experimental infections have not been reported for either virus, monitoring of cattle in sentinel herds in Australia during periods of recorded seroconversion did not detect any evidence of an association with disease (Cybinski and Gard, 1986).

Bivens Arm virus (BAV) and Sweetwater Branch virus (SBV) are serologically related rhabdoviruses that were isolated in Florida from *Culicoides insignis* which were feeding on, or in the vicinity of, healthy water buffalo imported from Trinidad (Gibbs *et al.*, 1989). Each virus is related antigenically to TIBV and CPV and recent studies have indicated that they have the same characteristic genome organization and cluster with them phylogenetically (Walker *et al.*, 2015). Serosurveys conducted in Florida, Puerto Rico and St. Croix indicated a high prevalence of BAV-neutralizing antibody in cattle suggesting that the virus was present in Florida prior to the importation of water buffalo from the Caribbean (Gibbs *et al.*, 1989; Tuekam *et al.*, 1991). Antibody to BAV (or a closely related virus) was also detected in a horse and a white-tailed deer but not in sheep or wildebeest in Florida (Gibbs *et al.*, 1989). Neither BAV nor SBV has been associated with disease in cattle or water buffalo but their close relationship to TIBV and CPV suggests that culicoides-borne tibroviruses may result in primarily asymptomatic infection of bovines over a wide geographic range.

Tibroviruses have also been detected in humans in Africa. In May and June 2009, an outbreak of haemorrhagic fever occurred in Mangala village in the Democratic Republic of Congo (DRC), affecting three villagers, two of whom died within 3 days of the onset of high fever and mucosal haemorrhage (Grard *et al.*, 2012). PCR screening of an acute serum sample from the single survivor revealed no evidence of infection with known agents of viral haemorrhagic fevers. However, although no virus was isolated, deep sequencing of the sample revealed the near complete genome sequence of a novel rhabdovirus that was named Bas Congo virus (BASV). Phylogenetic analysis based on L-protein sequences indicated that BASV is most closely related to TIBV and CPV, and that these viruses share the same genome organization, including each of the accessory genes U1, U2 and U3. BASV RNA was detected in the serum ($\sim 10^6$ copies/ml) of the patient and virus-neutralization tests conducted with a VSV pseudotype bearing the *in-vitro* synthesized BASV G-protein detected neutralizing antibodies in convalescent serum from the surviving patient and in one of five asymptomatic health care workers from Mangala that were identified as close contacts. A serosurvey of undiagnosed haemorrhagic fever patients and healthy individuals from DRC failed to detect any other positive records (Grard *et al.*, 2012).

Further evidence of tibrovirus infection in humans was obtained from the analysis of blood samples from Irrua in southern Nigeria (Stremlau *et al.*, 2015). Screening by deep sequencing of plasma from patients with unexplained acute febrile illnesses and healthy individuals from the same community revealed the complete genomes of two novel rhabdoviruses. Ekpoma virus-1 (EKV-1) was detected in an apparently healthy female who, upon subsequent interview, could not recall any episode of febrile illness in period following collection of the blood sample. Ekpoma virus-2 (EKV-2) was detected in an apparently healthy female from the same village who was admitted to hospital two weeks after blood collection with a febrile illness that, based on clinical symptoms, was diagnosed as malaria. EKV-1 and EKV-2 share the typical genome organization of tibroviruses, including accessory genes U1, U2 and U3, and cluster phylogenetically with the TIBV, CPV and BASV. EKV-1 is more closely related to TIBV and CPV; EKV-2 is more closely related to BASV. Analysis of viral genetic loads in blood plasma revealed 4.5×10^6 copies EKV-1 RNA/ml in one individual and 4.6×10^3 copies EKV-2 RNA/ml in the other but attempts to culture the viruses in a range of mammalian and insect cell lines were unsuccessful (Stremlau *et al.*, 2015).

Thus, although various tibroviruses have been detected in cattle or humans, it is not clear if any are capable of causing clinical disease. After many years of monitoring and a relatively high prevalence of infection in some regions, there is certainly no evidence that tibroviruses are associated with disease outbreaks in cattle. In humans, the evidence linking BASV to haemorrhagic fever is very tenuous and the detection of EKV-1 and EKV-2 in apparently healthy individuals in Africa supports the view that BASV may simply have been present in the blood sample as a concurrent asymptomatic infection. It is also unclear if human tibrovirus infections are due to spill-over from an infection cycle in which bovines are the natural host or if these viruses, of possible bovine origin, have adapted to infection of humans. Serosurveys of human, cattle and other livestock populations in western Africa would help illuminate aspects of the host range and distribution of tibroviruses and the risk of human exposure. There is also a need for further studies to determine the role of insects in the transmission of human tibroviruses. Although there is good direct and indirect evidence that tibroviruses infecting cattle are transmitted by biting midges, nothing is yet known about the transmission cycle of BASV, EKV-1 or EKV-2.

Other rhabdoviruses with established or inferred vector transmission

At present only rhabdoviruses in the genus *Sigmavirus* are considered to be true insect-specific viruses which circulate exclusively among *Drosophila* flies (Longdon *et al.*, 2012). For other rhabdoviruses isolated from insects, an involvement of vertebrate hosts is either confirmed or proposed. For example, if a virus is detected in a blood-sucking insect or a tick, it is reasonable to expect that it may also have a vertebrate host. Recent advances in the next-generation sequencing have led to genetic characterization of many novel pathogens, including rhabdoviruses (Gubala, 2012). Phylogenetic relationships of such recently characterized viruses may provide additional inferences to their circulation patterns even if no direct ecological and epidemiological observations are yet available.

The Hart Park group was established on the basis of two serologically related viruses, Hart Park virus (HPV) and Flanders virus (FLAV) (Whitney, 1964; Johnson, 1965; Tesh *et al.*, 1983). Both viruses have been isolated numerous times from mosquitoes and birds throughout the USA, they are capable of causing fatal encephalitis in suckling mice, and FLAV causes haemorrhage in chick embryos (Jenson *et al.*, 1967; Karabatsos, 1985). Kamese virus (KAMV), Mossuril virus (MOSV) and Mosqueiro virus (MQOV) are proposed members of the Hart Park group based on their antigenic similarity to HPV and FLAV (Karabatsos *et al.*, 1973; Calisher *et al.*, 1989). KAMV was isolated from *Culex annulioris* in Uganda and MOSV was isolated from *Culex sitiens* collected in Mozambique (Kokernot *et al.*, 1962). Numerous isolations of MOSV have also been made in central Africa from various species of mosquito (Karabatsos, 1985). Evidence suggests that both viruses are capable of infecting humans, and MOSV antibodies have also been found in a baboon. MQOV was isolated in Brazil from *Culex portesi* and from mosquitoes of other species but no genetic information exists for this virus (Karabatsos, 1985; Calisher *et al.*, 1989). Comparisons of short L-protein sequences suggest that Bangoran virus (BGNV), isolated from *Culex perfuscus* and from the brain of a bird in the Central African Republic, and Porton virus (PORV), isolated from *Mansonia uniformis* in Malaysia, may also belong to the Hart Park group (Dacheux *et al.*, 2010), but there is no serological or other support for this suggestion.

Wongabel virus (WONV) was isolated from biting midges (*Culicoides austropalpalis*) collected in Queensland, Australia. Midges of this species have been observed to have a feeding preference for birds, and neutralizing antibodies to this virus have been detected in sea birds collected off the Great Barrier Reef, suggesting that WONV may have an avian reservoir host (Muller *et al.*, 1981; Humphery-Smith *et al.*, 1991). No neutralizing antibodies have been detected in human sera from the island residents. A survey of over 1600 animal sera (including cattle, buffalo, pigs, wild and domestic birds, wallabies, kangaroos, wallaroos, horses, rats and pythons) collected between 1997 and 2007 in the Northern Territory, Australia, did not conclusively identify any significant seroprevalence to WONV (Gubala, 2012). Low level neutralization was observed in cattle and wallabies, possibly indicative of infection with another serologically related virus. Ngaingan virus (NGAV) was isolated from biting midges (*Culicoides brevitarsis* and *C. actoni*) that were collected in northern Queensland, Australia (Doherty *et al.*, 1973; Kay *et al.*, 1978). Two independent serological surveys of Australian livestock and wildlife identified macropods (wallabies, kangaroos, and wallaroos) and possibly cattle as primary hosts for the virus (Doherty *et al.*, 1973; Gubala *et al.*, 2010). At the antigenic level NGAV shares no obvious relationship to the ephemeroviruses or any of the tentative Hart Park viruses (Calisher *et al.*, 1989; Gibbs *et al.*, 1989). Genomic sequencing of Holmes Jungle virus (HOJV) and Ord River virus (ORRV) suggests that they have a very close relationship to WONV. These three viruses are similar phylogenetically and contain almost identical genome structures (Gubala, 2012). The available short fragment of the L-protein of Parry Creek virus (PCRV) similarly suggests a close relationship to WONV (Bourhy *et al.*, 2005), but this needs confirmation by additional sequencing. PCRV was isolated from the same species of mosquito as HOJV and ORRV (*Culex annulirostris*) (Liehne *et al.*, 1976, 1981). ORRV and PCRV have unknown reservoir hosts but HOJV, or a closely related virus, may be present in cattle (Gubala, 2012).

The genomes of tupaia virus (TUPV), Oak Vale virus (OVRV), Durham virus (DURV) and Harrison Dam virus (HARDV) suggest that these viruses have a monophyletic origin. In addition, a distinguishing feature found in these four rhabdoviruses is the presence of a gene between the M and G genes, which putatively codes for a small hydrophobic protein (SH) (Springfeld *et al.*, 2005; Allison *et al.*, 2011; Quan *et al.*, 2011; Gubala, 2012). TUPV was isolated from the hepatocellular carcinoma cells of a moribund tree shrew (*Tupaia belangeri*) which was imported from Thailand (Kurz *et al.*, 1986; Springfeld *et al.*, 2005) and to date was propagated in cell culture of this host species only. DURV was isolated from the brain of an ataxic American coot in North Carolina, USA, and no other hosts for this virus have been established as of yet (Allison *et al.*, 2011). OVRV was isolated several times from *Culex edwardsi* mosquitoes collected in Queensland, Australia (Cybinski and Muller, 1990). In addition, OVRV was isolated from a pool of *Aedes* (*Ochlerotatus*) *vigilax* mosquitoes which feed on a broad range of animals including humans and birds (Calisher *et al.*, 1989; Weir *et al.*, 1996) and from *Anopheles annulipes* mosquitoes collected in in Western Australia (Quan *et al.*, 2011). Although OVRV was isolated in association with an outbreak of BEF in cattle, the serological and insect data collectively suggest that OVRV is not a cattle virus (*Cx. edwardsi* rarely feeds on cattle). However, the ability of OVRV to propagate well through mammalian cell culture, insect cell culture and live mosquitoes suggests that likely it is an arbovirus (Gubala, 2012). OVRV was speculated to circulate between mosquitoes and an avian host, but attempts to demonstrate compatibility with birds by performing replication studies in cattle egrets suggest this is unlikely the case (Cybinski and Muller, 1990).

Moussa virus (MOUV) was isolated several times from mosquitoes (*Culex decens* and other species) collected in Ivory Coast (Quan *et al.*, 2010). No ecological or serological surveys in animal and human populations have been performed to date. Based on distinguished genomic and phylogenetic properties MOUV is recognized as a species within *Rhabdoviridae*, without particular relatedness to any of the established viral genera (Dietzgen *et al.*, 2011).

Several phylogenetically related rhabdoviruses were isolated from insectivorous bats and blood-suckling arthropods in different continents. Kern Canyon virus (KCV) was isolated from a bat *Myotis yumanensis* in California, USA (Murphy and Fields, 1967). Mount Elgon bat virus (MEBV) was isolated from a bat *Rhinolophus* sp. in Kenya (Metselaar *et al.*, 1969). Oita virus (OITAV) was isolated from a bat *Rhinolophus cornutus* in Japan (Iwasaki *et al.*, 2004). Fikirini rhabdovirus (FKRV) was isolated from a bat *Hipposideros vittatus* in Kenya (Kading *et al.*, 2013). Kolente virus (KOLEV) was isolated from a bat *Hipposideros jonesi* and from a pool of *Amblyomma* (*Theileriella*) *variegatum* ticks in Guinea (Ghedin *et al.*, 2013). The detection of these monophyletic viruses in bats around the globe suggested that they may be ecologically associated and restricted to bat hosts (Kuzmin *et al.*, 2006). However, several arthropod-borne viruses were also identified as member of this group when characterized genetically or serologically, such as Barur virus (BARV), Fukuoka virus (FUKAV), Nkolbisson virus (NKOV), Le Dantec virus (LDV) and Keuraliba virus (KEUV) (Gubala, 2012). BARV was originally isolated from rats (*Rattus rattus*) in India (Ghosh and Rajagopalan, 1973). The virus was subsequently isolated from ticks in India, and from mosquitoes and ticks in Africa (Johnson *et al.*, 1977; Butenko *et al.*, 1981). NKOV was isolated from mosquitoes (*Eretmapodites leucopus*, *Aedes* spp. and *Culex* spp. in Cameroon, and has subsequently been isolated from *Culicidea* spp. in the Ivory Coast and from humans in the Central African Republic (Salaun *et al.*, 1969; Bourhy *et al.*, 2008). FUKAV was first isolated from *Culicoides punctalis* biting midges, *Culex tritaeniorhynchus* mosquitoes, and from calves with fever and leucopoenia in Japan (Noda *et al.*, 1992). LDV was isolated from a human with febrile illness in Senegal (Cropp *et al.*, 1985). The related KEUV was isolated from several rodents, also in Senegal (Tesh *et al.*, 1983; Cropp *et al.*, 1985).

Another monophyletic clade based on limited gene sequences available for comparisons constitute Kolongo virus (KOLV), Sanjimba virus (SJAV), Nasoule virus (NASV), Garba virus (GARV), Bimbo virus (BBOV), Ouango virus (OUAV) and Boteke virus (BTKV) (Bourhy *et al.*, 2005; Kuzmin *et al.*, 2006; Dacheux *et al.*, 2010; Gubala *et al.*, 2012). All these were isolated from birds in Africa, except BTKV, which was isolated from mosquitoes (Karabastos, 1985).

Bahia Grande virus (BGV), Muir Springs virus (MSV) and Reed Ranch virus (RRV) constitute a serologically related group of rhabdoviruses isolated from mosquitoes in the USA. Short L gene fragments available for BGV and MSV suggest their phylogenetic relatedness (Allison *et al.*, 2011). A serological survey demonstrated that humans, cattle, sheep, reptiles and wild mammals from Texas had neutralizing antibodies to BGV (Kerschner *et al.*, 1986).

A group of phylogenetically related viruses associated with reptiles in Australia includes Almpiwar virus (ALMV), Charleville virus (CHVV), and Humpty Doo virus (HDOOV) (Bourhy *et al.*, 2005). Although ALMV has never been isolated from arthropods, it is an assumed arbovirus on its ability to multiply in experimentally infected mosquitoes (Carley *et al.*, 1973). CHVV was isolated several times from sandflies (*Phlebotomus* spp.), biting midges (*Lasiohelea* spp.) and a lizard *Gehyra australis* (Doherty *et al.*, 1973; Karabastos, 1985). HDOOV was isolated from biting midges *Culicoides marksi* and *Lasiohelea* spp. (Standfast *et al.*, 1984).

Rochambeau virus (RBUV), isolated from mosquitoes (*Coquillettidia albicosta*) in French Guiana, was initially suggested to be a lyssavirus based on limited serological cross-reactivity (Calisher *et al.*, 1989). However, the available partial N gene sequence of RBUV is more similar to that of ephemeroviruses and Hart Park viruses but in general does not allow inclusion of RBUV in any of the established rhabdovirus genera (Kuzmin *et al.*, 2006).

There are still many viruses in *Rhabdoviridae* for which not only ecological but also genetic information is lacking. Their preliminary characterization was based on morphology, in particular cases on serological cross-reactivity. An example is the Sawgrass serogroup (Calisher *et al.*, 1989) consisting of Sawgrass virus (SAWV), Connecticut virus (CNTV) and New Minto virus (NMV). All three viruses were isolated from ticks collected on mammals in the USA (Sather *et al.*, 1970; Wellings *et al.*, 1972; Ritter *et al.*, 1978; Main and Carey, 1980). Neutralizing antibodies to NMV were detected in local population of cottontail rabbits, suggesting that NMV is likely an arbovirus (Main and Carey, 1980).

Given the increasing development of sequencing techniques (Coffey *et al.*, 2014) it is reasonable to anticipate generation of a great number of novel rhabdovirus genomes in the near future. This information will certainly help to explain phylogenetic and evolutionary relationships within the family. However, extensive research on virus ecology, including host interactions, circulation patterns and role in clinical diseases are of paramount significance. For example, the recently described Bas Congo virus was implicated as a potential causative agent of a severe haemorrhagic fever in humans (Grard *et al.*, 2012) but much more studies are needed to corroborate this suggestion and to elucidate ecology of the virus. Such studies are quite laborious and expensive, and usually are limited to pathogens of public health and economic significance. Nevertheless, this is the major direction of further work, not only because many rhabdoviruses are the significant pathogens but also because they have been successfully used as recombinant vaccine vectors (Schnell *et al.*, 1993; Papaneri *et al.*, 2012; Mire *et al.*, 2013). This role may be significantly extended for prevention of various infectious diseases in humans, domestic animals and wildlife once we better understand biology of various rhabdoviruses circulating in the world.

References

Abraham, G., and Banerjee, A.K. (1976). Sequential transcription of the genes of vesicular stomatitis virus. Proc. Natl. Acad. Sci. U.S.A. 73, 1504–1508.

Aggad, D., Mazel, M., Boudinot, P., Mogensen, K.E., Hamming, O.J., Hartmann, R., Kotenko, S., Herbomel, P., Lutfalla, G., and Levraud, J.P. (2009). The two groups of zebrafish virus-induced interferons signal via distinct receptors with specific and shared chains. J. Immunol. *183*, 3924–3931.

Ahmed, M., and Lyles, D.S. (1997). Identification of a consensus mutation in M protein of vesicular stomatitis virus from persistently infected cells that affects inhibition of host-directed gene expression. Virology *237*, 378–388.

Ahmed, M., and Lyles, D.S. (1998). Effect of vesicular stomatitis virus matrix protein on transcription directed by host RNA polymerases I, II, and III. J. Virol. *72*, 8413–8419.

Ahmed, M., McKenzie, M.O., Puckett, S., Hojnacki, M., Poliquin, L., and Lyles, D.S. (2003). Ability of the matrix protein of vesicular stomatitis virus to suppress beta interferon gene expression is genetically correlated with the inhibition of host RNA and protein synthesis. J. Virol. *77*, 4646–4657.

Ahne, W., Bjorklund, H.V., Essbauer, S., Fijan, N., Kurath, G., and Winton, J.R. (2002). Spring viremia of carp (SVC). Dis. Aquat. Organ. *52*, 261–272.

Albertini, A.A.V., Schoehn, G., Weissenhorn, W., and Ruigrok, R.W.H. (2008). Structural aspects of rabies virus replication. Cell. Mol. Life Sci. *65*, 282–294.

Allison, A.B., Palacios, G., Travassos da Rosa, A., Popov, V.L., Lu, L., Xiao, S.Y., DeToy, K., Briese, T., Lipkin, W.I., Keel, M.K., *et al.* (2011). Characterization of Durham virus, a novel rhabdovirus that encodes both a C and SH protein. Virus Res. *155*, 112–122.

Altmann, S.M., Mellon, M.T., Distel, D.L., and Kim, C.H. (2003). Molecular and functional analysis of an interferon gene from the zebrafish, *Danio rerio*. J. Virol. *77*, 1992–2002.

Ammar, E.-D., and Nault, L.R. (1985). Assembly and accumulation sites of maize mosaic virus in its planthopper vector. Intervirology *24*, 33–41.

Aziz-Boaron, O., Klausner, Z., Shenkar, J., Gafni, O., Gelman, B., David, D., and Klement, E. (2012). Circulation of bovine ephemeral fever in the Middle-East – strong evidence for transmission by winds and animal transport. Vet. Microbiol. *158*, 300–307.

Bachmann, M.F., Hengartner, H., and Zinkernagel, R.M. (1995). T-helper cell-independent neutralizing B-cell response against vesicular stomatitis virus: role of antigen patterns in B-cell induction? Eur. J. Immunol. *25*, 3445–3451.

Bai, W.B. (1993). Epidemiology and control of bovine ephemeral fever in China. In Bovine Ephemeral Fever and Related Rhabdoviruses, T.D. St George, M.F. Uren, P.L. Young, and D. Hoffmann, eds. (ACIAR, Canberra, Australia), pp. 20–22.

Ball, L.A., and White, C.N. (1976). Order of transcription of genes of vesicular stomatitis virus. Proc. Natl. Acad. Sci. U.S.A. *73*, 442–446.

Barchet, W., Cella, M., Odermatt, B., Asselin-Paturel, C., Colonna, M., and Kalinke, U. (2002). Virus-induced interferon alpha production by a dendritic cell subset in the absence of feedback signaling in vivo. J. Exp. Med. *195*, 507–516.

Barr, J.N., and Wertz, G.W. (2001). Polymerase slippage at vesicular stomatitis virus gene junctions to generate poly(A) is regulated by the upstream 3′-AUAC-5′ tetranucleotide: implications for the mechanism of transcription termination. J. Virol. *75*, 6901–6913.

Barr, J.N., Whelan, S.P., and Wertz, G.W. (1997). Role of the intergenic dinucleotide in vesicular stomatitis virus RNA transcription. J. Virol. *71*, 1794–1801.

Basson, P.A., Pienaar, J.G., and van der Westhuizen, B. (1970). The pathology of ephemeral fever: a study of the experimental disease in cattle. J. S. Afr. Vet. Med. Ass. *40*, 385–397.

Black, B.L., and Lyles, D.S. (1992). Vesicular stomatitis virus matrix protein inhibits host cell-directed transcription of target genes in vivo. J. Virol. *66*, 4058–4064.

Blackburn, N.K., Searle, L., and Phleps, R.J. (1985). Viruses isolated from *Culicoides* (Diptera: *Ceratopogonidae*) caught at a veterinary research farm, Mazowe, Zimbabwe. J. Ent. Soc. S. Afr. *48*, 331–336.

Blasdell, K.R., Voysey, R., Bulach, D., Joubert, D.A., Tesh, R.B., *et al.* (2012a) Kotonkan and Obodhiang viruses: African ephemeroviruses with large and complex genomes. Virology *425*, 143–153.

Blasdell, K.R., Voysey, R., Bulach, D.M., Trinidad, L., Tesh, R.B., Boyle, D.B., and Walker, P.J. (2012b). Malakal virus from Africa and Kimberley virus from Australia are geographic variants of a widely distributed ephemerovirus. Virology *433*, 236–244.

Boudinot, P., Salhi, S., Blanco, M., and Benmansour, A. (2001). Viral haemorrhagic septicaemia virus induces *vig-2*, a new interferon-responsive gene in rainbow trout. Fish Shellfish Immunol. *11*, 383–397.

Bourhy, H., Cowley, J.A., Larrous, F., Holmes, E.C., and Walker, P.J. (2005). Phylogenetic relationships among rhabdoviruses inferred using the L polymerase gene. J. Gen. Virol. *86*, 2849–2858.

Bourhy, H., Gubala, A., Weir, R., and Boyle, D. (2008). Animal rhabdoviruses. In Encyclopedia of virology, B.W.J. Mahy, and M.H.V. van Regenmortel, eds. (Elsevier Academic Press, Oxford, UK), pp. 111–121.

Brown, E.L., and Lyles, D.S. (2003). Organization of the vesicular stomatitis virus glycoprotein into membrane microdomains occurs independently of intracellular viral components. J. Virol. *77*, 3985–3992.

Brown, E.L., and Lyles, D.S. (2005). Pseudotypes of vesicular stomatitis virus with CD4 formed by clustering of membrane microdomains during budding. J. Virol. *79*, 7077–7086.

Butenko, A.M., Gromashevsky, V.L., L'Vov D.K., and Popov, V.F. (1981). First isolations of Barur virus (*Rhabdoviridae*) from ticks (Acari: Ixodidae) in Africa. J. Med. Entomol. *18*, 232–234.

Calisher, C.H., Karabatsos, N., Zeller, H., Digoutte, J.P., Tesh, R.B., Shope, R.E., Travassos da Rosa, A.P., and St George, T.D. (1989). Antigenic relationships among rhabdoviruses from vertebrates and hematophagous arthropods. Intervirology *30*, 241–257.

Carley, J.G., Standfast, H.A., and Kay, B.H. (1973). Multiplication of viruses isolated from arthopods and vertebrates in Australia in experimentally infected mosquitoes. J. Med. Entomol. *10*, 244–249.

Chen, Y., Zhao, Y., Hammond, J., Hsu, H., Evans, J., and Feldlaufer, M. (2004). Multiple virus infections in the honey bee and genome divergence of honey bee viruses. J. Invert. Pathol. *87*, 84–93.

Chiba, S., Kondo, H., Tani, A., Saisho, D., Sakamoto, W., Kanematsu, S., and Suzuki, N. (2011). Widespread endogenization of genome sequences of non-retroviral RNA viruses into plant genomes. PLoS Path. 7, e1002146.

Ciavarra, R.P., Greene, A.R., Horeth, D.R., Buhrer, K., van Rooijen, N., and Tedeschi, B. (2000). Antigen processing of vesicular stomatitis virus in situ. Interdigitating dendritic cells present viral antigens independent of marginal dendritic cells but fail to prime CD4(+) and CD8(+) T cells. Immunology. *101*, 512–520.

Ciavarra, R.P., Taylor, L., Greene, A.R., Yousefieh, N., Horeth, D., van Rooijen, N., Steel, C., Gregory, B., Birkenbach, M., and Sekellick, M. (2005). Impact of macrophage and dendritic cell subset elimination on antiviral immunity, viral clearance and production of type 1 interferon. Virology *342*, 177–189.

Coffey, L.L., Page, B.L., Greninger, A.L., Herring, B.L., Russell, R.C., Doggett, S.L., Haniotis, J., Wang, C., Deng, X., and Delwart, E.L. (1914). Enhanced arbovirus surveillance with deep sequencing: Identification of novel rhabdoviruses and bunyaviruses in Australian mosquitoes. Virology *448*, 146–158.

Comer, J.A., Tesh, R.B., Modi, G.B., Corn, J.L., and Nettles, V.F. (1990). Vesicular stomatitis virus, New Jersey serotype: replication in and transmission by *Lutzomyia shannoni* (Diptera: Psychodidae). Am. J. Trop. Med. Hyg. *42*, 483–490.

Comer, J.A., Stallknecht, D.E., Corn, J.L., and Nettles, V.F. (1991). *Lutzomyia shannoni* (Diptera: Psychodidae): a biological vector of the New Jersey serotype of vesicular stomatitis virus on Ossabaw Island, Georgia. Parassitologia 33 (Suppl.), 151–158.

Connor, J.H., and Lyles, D.S. (2002). Vesicular stomatitis virus infection alters the eIF4F translation initiation complex and causes dephosphorylation of the eIF4E binding protein 4E-BP1. J. Virol. *76*, 10177–10187.

Cropp, C.B., Prange, W.C., and Monath, T.P. (1985). LeDantec virus: identification as a rhabdovirus associated with human infection and formation of a new serogroup. J. Gen. Virol. *66*, 2749–2754.

Cupp, E.W., Mare, C.J., Cupp, M.S., and Ramberg, F.B. (1992). Biological transmission of vesicular stomatitis virus (New Jersey) by *Simulium vittatum* (Diptera: Simuliidae). J. Med. Entomol. *29*, 137–140.

Cybinski, D.H., and Gard, G.P. (1986). Isolation of a new rhabdovirus in Australia related to Tibrogargan virus. Aust. J. Biol. Sci. *39*, 225–232.

Cybinski, D.H., and Muller, M.J. (1990). Isolation of arboviruses from cattle and insects at two sentinel sites in Queensland, Australia, 1979–85. Aust. J. Zool. *38*, 25–32.

Cybinski, D.H., and Zakrzewski, H. (1983). The isolation and preliminary characterization of a rhabdovirus in Australia related to bovine ephemeral fever virus. Vet. Microbiol. *8*, 221–235.

Cybinski, D.H., St George, T.D., Standfast, H.A., and McGregor, A. (1980). Isolation of Tibrogargan virus, a new Australian rhabdovirus, from *Culicoides brevitarsis*. Vet. Microbiol. *5*, 301–308.

Cybinski, D.H., Walker, P.J., Byrne, K.A., and Zakrzewski, H. (1990). Mapping of antigenic sites on the bovine ephemeral fever virus glycoprotein using monoclonal antibodies. J. Gen. Virol. *71*, 2065–2072.

Dacheux, L., Berthet, N., Dissard, G., Holmes, E.C., Delmas, O., Larrous, F., Guigon, G., Dickinson, P., Faye, O., Sall, A.A., *et al.* (2010). Application of broad-spectrum resequencing microarray for genotyping rhabdoviruses. J. Virol. *84*, 9557–9574.

Daniels, P.W., Sendow, I., Soleha, E., Hunt, S.N.T., and Bahri, S. (1995). Australian-Indonesian collaboration in veterinary arbovirology – a review. Vet. Microbiol. *46*, 151–174.

Das, T., Mathur, M., Gupta, A.K., Janssen, G.M., and Banerjee, A.K. (1998). RNA polymerase of vesicular stomatitis virus specifically associates with translation elongation factor-1 alphabetagamma for its activity. Proc. Natl. Acad. Sci. U.S.A. *95*,1449–1454.

Davies, F.G., and Walker, A.R. (1974). The isolation of ephemeral fever virus from cattle and *Culicoides* midges in Kenya. Vet. Rec. *95*, 63–64.

Desforges, M., Charron, J., Bérard, S., Beausoleil, S., Stojdl, D.F., Despars, G., Laverdière, B., Bell, J.C., Talbot, P.J., Stanners, C.P., and Poliquin, L. (2001). Different host-cell shut-off strategies related to the matrix protein lead to persistence of vesicular stomatitis virus mutants on fibroblast cells. Virus Res. *76*, 87–102.

Dietzgen, R.G. (2012). Morphology, genome organization, transcription and replication of rhabdoviruses. In Rhabdoviruses: Molecular Taxonomy, Evolution, Genomics, Ecology, Host–Vector Interactions, Cytopathology and Control, R.G. Dietzgen, and I.V. Kuzmin, eds. (Caister Academic Press, Norfolk, UK), pp. 5–12.

Dietzgen, R.G., Calisher, C.H., Kurath, G., Kuzmin, I.V., Rodriguez, L.L., Stone, D.M., Tesh, R.B., Tordo, N., Walker, P.J., Wetzel, T., and Whitfield, A.E. (2011). Family *Rhabdoviridae*. In Virus Taxonomy: Ninth Report of the International Committee on Taxonomy of Viruses, A.M.Q. King, M.J. Adams, E.B. Carstens, and E.J. Lefkowitz, eds. (Elsevier, Oxford, UK), pp. 686–714.

Doherty, R.L., Carley, J.G., Standfast, H.A., Dyce, A.L., Kay, B.H., and Snowdon, W.A. (1973). Isolation of arboviruses from mosquitoes, biting midges, sandflies and vertebrates collected in Queensland, 1969 and 1970. Trans. R. Soc. Trop. Med. Hyg. *67*, 536–543.

Drolet, B.S., Stuart, M.A., and Derner, J.D. (2009). Infection of Melanoplus sanguinipes grasshoppers following ingestion of rangeland plant species harboring vesicular stomatitis virus. Appl. Environ. Microbiol. *75*, 3029–3033.

Dunigan, D.D., Baird, S., and Lucas-Lenard, J. (1986). Lack of correlation between the accumulation of plus-strand leader RNA and the inhibition of protein and RNA synthesis in vesicular stomatitis virus infected mouse L cells. Virology *150*, 231–246.

Durrer, P., Gaudin, Y., Ruigrok, R.W., Graf, R., and Brunner, J. (1995). Photolabeling identifies a putative fusion domain in the envelope glycoprotein of rabies and vesicular stomatitis viruses. J. Biol. Chem. *270*, 17575–17581.

Ferran, M.C., and Lucas-Lenard, J.M. (1997). The vesicular stomatitis virus matrix protein inhibits transcription from the human beta interferon promoter. J. Virol. *71*, 371–377.

Forger, J.M. 3rd, Bronson, R.T., Huang, A.S., and Reiss, C.S. (1991). Murine infection by vesicular stomatitis virus: initial characterization of the H-2d system. J. Virol. *65*, 4950–4958.

Fort, P., Albertini, A., Van-Hua, A., Berthomieu, A., Roche, S., Delsuc, F., Pasteur, N., Capy, P., Gaudin, Y., and Weill, M. (2012). Fossil rhabdoviral sequences integrated into arthropod genomes: ontogeny, evolution, and potential functionality. Mol. Biol. Evol. *29*, 381–390.

Gard, G.P., Cybinski, D.H., and St George, T.D. (1983). The isolation in Australia of a new virus related to bovine ephemeral fever virus. Aust. Vet. J. *60*, 89–90.

Gard, G.P., Cybinski, D.H., and Zakrzewski, H. (1984). The isolation of a fourth bovine ephemeral fever group virus. Aust. Vet. J. *61*, 332.

Gard, G.P., Melville, L.F., Calisher, C.H., and Karabatsos, N. (1992). Koolpinyah: a virus related to kotonkan from cattle in northern Australia. Intervirology *34*, 142–145.

Ghedin, E., Rogers, M.B., Widen, S.G., Guzman, H., Travassos da Rosa, A.P., Wood, T.G., Fitch, A., Popov, V., Holmes, E.C., Walker, P.J., Vasilakis, N., and Tesh, R.B. (2013). Kolente virus, a rhabdovirus species isolated from ticks and bats in the Republic of Guinea. J. Gen. Virol. *94*, 2609–2615.

Ghosh, S.N., and Rajagopalan, P.K. (1973). Encephalomyocarditis virus activity in *Mus booduga* (Gray) in Barur Village (1961–1962), Sagar KFD area, Mysore State, India. Indian J. Med. Res. *61*, 989–991.

Gibbs, E.P., Calisher, C.H., Tesh, R.B., Lazuick, J.S., Bowen, R., and Greiner, E.C. (1989). Bivens arm virus: a new rhabdovirus isolated from *Culicoides insignis* in Florida and related to Tibrogargan virus of Australia. Vet. Microbiol. *19*, 141–150.

Girard, Y.A., Klingler, K.A., and Higgs, S. (2004). West Nile virus dissemination and tissue tropisms in orally infected *Culex pipiens quinquefasciatus*. Vector-Borne Zoonotic Dis. *4*, 109–122.

Gonzalez, M.E., and Carrasco, L. (2003). Viroporins. FEBS Lett. *552*, 28–34.

Grard, G., Fair, J.N., Lee, D., Slikas, E., Steffen, I., Muyembe, J.J., Sittler, T., Veeraraghavan, N., Ruby, J.G., Wang, C., Makuwa, M., Mulembakani, P., Tesh, R.B., Mazet, J., Rimoin, A.W., Taylor, T., Schneider, B.S., Simmons, G., Delwart, E., Wolfe, N.D., Chiu, C.Y., and Leroy, E.M. (2012). A novel rhabdovirus associated with acute hemorrhagic fever in central Africa. PLoS Pathog. *8*, e1002924.

Gulaba, A. (2012). Recent advances in the characterization of animal rhabdoviruses. In Rhabdoviruses: Molecular Taxonomy, Evolution, Genomics, Ecology, Host–Vector Interactions, Cytopathology and Control, R.G. Dietzgen, and I.V. Kuzmin, eds. (Caister Academic Press, Norfolk, UK), pp. 165–204.

Gubala, A., Davis, S., Weir, R., Melville, L., Cowled, C., Walker, P., and Boyle, D. (2010). Ngaingan virus, a macropod-associated rhabdovirus, contains a second glycoprotein gene and seven novel open reading frames. Virology *399*, 98–108.

Gubala, A., Davis, S., Weir, R., Melville, L., Cowled, C., and Boyle, D. (2011). Tibrogargan and Coastal Plains rhabdoviruses: genomic characterisation, evolution of novel genes and seroprevalence in Australian livestock. J. Gen. Virol. *92*, 2160–2170.

Gurav, Y.K., Tandale, B.V., Jadi, R.S., Gunjikar, R.S., Tikute, S.S., Jamgaonkar, A.V., Khadse, R.K., Jalgaonkar, S.V., Arankalle, V.A., and Mishra, A.C. (2010). Chandipura virus encephalitis outbreak among children in Nagpur division, Maharashtra, 2007. Ind. J. Med. Res. *132*, 395–399.

Hajnická, V., Kocakova, P., Slovak, M., Labuda, M., Fuchsberger, N., and Nuttall, P.A. (2000). Inhibition of the antiviral action of interferon by tick salivary gland extract. Parasite Immunol. *22*, 201–206.

Harty, R.N., Brown, M.E., McGettigan, J.P., Wang, G., Jayakar, H.R., Huibregtse, J.M., Whitt, M.A., and Schnell, M.J. (2001). Rhabdoviruses and the cellular ubiquitin-proteasome system: a budding interaction. J. Virol. *75*, 10623–10629.

Hill, M.W.M., and Schultz, K. (1977). Ataxia and paralysis associated with bovine ephemeral fever infection. Aust. Vet. J. *53*, 217–221.

Hinzman, E.E., Barr, J.N., and Wertz, G.W. (2002). Identification of an upstream sequence element required for vesicular stomatitis virus mRNA transcription. J. Virol. *76*, 7632–7641.

Hogenhout, S.A., Redinbaugh, M.G., and Ammar, el-D. (2003). Plant and animal rhabdovirus host range: a bug's view. Trends Microbiol. *11*, 264–271.

Humphery-Smith, I., Cybinski, D.H., Byrnes, K.A., and St George, T.D. (1991). Seroepidemiology of arboviruses among seabirds and island residents of the Great Barrier Reef and Coral Sea. Epidemiol. Infect. *107*, 435–440.

Huneycutt, B.S., Bi, Z., Aoki, C.J., and Reiss, C.S. (1993). Central neuropathogenesis of vesicular stomatitis virus infection of immunodeficient mice. J. Virol. *67*, 6698–6706.

Hwang, L.N., Englund, N., and Pattnaik, A.K. (1998). Polyadenylation of vesicular stomatitis virus mRNA dictates efficient transcription termination at the intercistronic gene junctions. J. Virol. *72*, 1805–1813.

Irie, T., Licata, J.M., Jayakar, H.R., Whitt, M.A., Bell, P., and Harty, R.N. (2004). Functional analysis of late-budding domain activity associated with the PSAP motif within the vesicular stomatitis virus M protein. J. Virol. *78*, 7823–7827.

Iverson, L.E., and Rose, J.K. (1981). Localized attenuation and discontinuous synthesis during vesicular stomatitis virus transcription. Cell 23, 477–484.

Iwasaki, T., Inoue, S., Tanaka, K., Sato, Y., Morikawa, S., Hayasaka, D., Moriyama, M., Ono, T., Kanai, S., Yamada, A., and Kurata, T. (2004). Characterization of Oita virus 296/1972 of *Rhabdoviridae* isolated from a horseshoe bat bearing characteristics of both lyssavirus and vesiculovirus. Arch. Virol. *149*, 1139–1154.

Jackson, B.T., Brewster, C.C. & Paulson, S.L. (2012). La Crosse virus infection alters blood feeding behavior in *Aedes triseriatus* and *Aedes albopictus* (Diptera: Culicidae). J. Med. Entomol. *49*, 1424–1429.

Jayakar, H.R., and Whitt, M.A. (2002). Identification of two additional translation products from the matrix (M) gene that contribute to vesicular stomatitis virus cytopathology. J. Virol. *76*, 8011–8018.

Jayakar, H.R., Murti, K.G., and Whitt, M.A. (2000). Mutations in the PPPY motif of vesicular stomatitis virus matrix protein reduce virus budding by inhibiting a late step in virion release. J. Virol. *74*, 9818–9827.

Jeetendra, E., Robison, C.S., Albritton, L.M., and Whitt, M.A. (2002). The membrane-proximal domain of vesicular stomatitis virus G-protein functions as a membrane fusion potentiator and can induce hemifusion. J. Virol. *76*, 12300–12311.

Jenkins, G.M., Rambaut, A., Pybus, O.G., and Holmes, E.C. (2002). Rates of molecular evolution in RNA viruses: a quantitative phylogenetic analysis. J. Mol. Evol. *54*, 156–165.

Jenson, A.B., Rabin, E.R., Wende, R.D., and Melnick, J.L. (1967). A comparative light and electron microscopic study of rabies and Hart Park virus encephalitis. Exp. Mol. Pathol. *7*, 1–10.

Jiang, C.L., and Yan, J.D. (1989). Evidence of Kimberley virus infection of cattle in China. Trop. Anim. Health Prod. *21*, 85–86.

Johal, J., Gresty, K., Kongsuwan, K., and Walker, P.J. (2008). Antigenic characterization of bovine ephemeral fever rhabdovirus G and GNS glycoproteins expressed from recombinant baculoviruses. Arch. Virol. *153*, 1657–1665.

Johnson, B.K., Shockley, P., Chanas, A.C., Squires, E.J., Gardner, P., Wallace, C., Simpson, D.I., Bowen, E.T., Platt, G.S., Way, H., Chandler, J.A., Highton, R.B., and Hill, M.N. (1977). Arbovirus isolations from mosquitoes: Kano Plain, Kenya. Trans. R. Soc. Trop. Med. Hyg. *71*, 518–521.

Johnson, H.N. (1965). Diseases derived from wildlife. Calif. Health *23*, 35–39.

Joubert, D.A., Blasdell, K.R., Audsley, M.D., Trinidad, L., Monaghan, P., Dave, K.A., Lieu, K., Amos-Ritchie, R., Jans, D.A., Moseley, G.W., Gorman, J.J., and Walker, P.J. (2014). Bovine ephemeral fever rhabdovirus α1 protein has viroporin-like properties and binds importin β1 and importin 7. J. Virol. *88*, 1591–1603.

Joubert, D.A., Rodriguez-Andres, J., Monaghan, P., Cummins, M., McKinstry, W.J., Paradkar, P.N., Moseley, G.W., and Walker, P.J. (2015). Wongabel rhabdovirus accessory protein U3 targets the SWI/SNF chromatin remodelling complex. J. Virol. *89*, 1377–1388.

Jubb, K.V.F. (1985). Vesicular stomatitis virus. In Pathology of Domestic Animals, 3rd ed. K.V.F. Jubb, P.C. Kennedy, and N. Palmer, eds., vol. 3. (Academic Press, New York, USA), pp. 92–93.

Kading, R.C., Gilbert, A.T., Mossel, E.C., Crabtree, M.B., Kuzmin, I.V., Niezgoda, M., Agwanda, B., Markotter, W., Weil, M.R., Montgomery, J.M., Rupprecht, C.E., and Miller, B.R. (2013). Isolation and molecular characterization of Fikirini rhabdovirus, a novel virus from a Kenyan bat. J. Gen. Virol. *94*, 2393–2398.

Karabatsos, N., ed. (1985). The American Committee on Arthropod-borne Viruses: International Catalog of Arboviruses Including Certain Other Viruses of Vertebrates, 3rd edn (American Society for Tropical Medicine and Hygiene, San Antonio, Texas).

Karabatsos, N., Lipman, M.B., Garrison, M.S., and Mongillo, C.A. (1973). The morphology, morphogenesis, and serological characterization of the rhabdoviruses Navarro, Kwatta, and Mossuril. J. Gen. Virol. *21*, 429–433.

Kay, B.H., Boreham, P.F.L., Dyce, A.L., and Standfast, H.A. (1978). Blood feeding of biting midges (Diptera:Ceratopogonidae) at Kowanyama, Cape York Peninsula, North Queensland. J. Aust. Ent. Soc. *17*, 145–149.

Kemp, G.E., Lee, V.H., Moore, D.L., Shope, R.E., Causey, O.R., and Murphey, F.A. (1973). Kotonkan, a new rhabdovirus related to Mokola virus of the rabies serogroup. Am. J. Epidemiol. *98*, 43–49.

Kerschner, J.H., Calisher, C.H., Vorndam, A.V., and Francy, D.B. (1986). Identification and characterization of Bahia Grande, Reed Ranch and Muir Springs viruses, related members of the family *Rhabdoviridae* with widespread distribution in the United States. J. Gen. Virol. *67*, 1081–1089.

Kim, C.H., Johnson, M.C., Drennan, J.D., Simon, B.E., Thomann, E., and Leong, J.A. (2000). DNA vaccines encoding viral glycoproteins induce nonspecific immunity and Mx protein synthesis in fish. J. Virol. *74*, 7048–7054.

Kirkland, P.D. (1993). The epidemiology of bovine ephemeral fever in south-eastern Australia: evidence for a mosquito vector. In Bovine Ephemeral Fever and Related Rhabdoviruses, T.D. St George, M.F. Uren, P.L. Young, and D. Hoffmann, eds. (Canberra, Australia: ACIAR), 33–37.

Kokernot, R.H., Mc, I.B., Worth, C.B., and De Sousa, J. (1962). Isolation of viruses from mosquitoes collected at Lumbo, Mozambique. II. Mossuril virus, a new virus isolated from the *Culex* (*culex*) *sitiens* Wiedemann group. Am. J. Trop. Med. Hyg. *11*, 683–684.

Kurz, W., Gelderblom, H., Flugel, R.M., and Darai, G. (1986). Isolation and characterization of a tupaia rhabdovirus. Intervirology *25*, 88–96.

Kuzmin, I.V., Hughes, G.J., and Rupprecht, C.E. (2006). Phylogenetic relationships of seven previously unclassified viruses within the family *Rhabdoviridae* using partial nucleoprotein gene sequences. J. Gen. Virol. *87*, 2323–2331.

Levraud, J.P., Boudinot, P., Colin, I., Benmansour, A., Peyrieras, N., Herbomel, P., and Lutfalla, G. (2007). Identification of the zebrafish IFN receptor: implications for the origin of the vertebrate IFN system. J. Immunol. *178*, 4385–4394.

Li, C.X., Shi, M., Tian, J.H., Lin, X.D., Kang, Y.J., Chen, L.J., Qin, X.C., Xu, J., Holmes, E.C., and Zhang, Y.Z. (2015). Unprecedented genomic diversity of RNA viruses in arthropods reveals the ancestry of negative-sense RNA viruses. eLife *4*, e05378.

Li, T., and Pattnaik, A.K. (1999). Overlapping signals for transcription and replication at the 3′ terminus of the vesicular stomatitis virus genome. J. Virol. *73*, 444–452.

Liao, Y.K., Inaba, Y., Li, N.J., Chain, C.Y., Lee, S.L., and Liou, P.P. (1998). Epidemiology of bovine ephemeral fever virus in Taiwan. Microbiol. Res. *153*, 289–295.

Liehne, P.F., Stanley, N.F., Alpers, M.P., and Liehne, C.G. (1976). Ord River arboviruses – the study site and mosquitoes. Aust. J. Exp. Biol. Med. Sci. *54*, 487–497.

Liehne, P.F.S., Anderson, S., Stanley, N.F., Liehne, C.G., Wright, A.E., Chan, K.H., Leivers, S., Britten, D.K., and Hamilton, N.P. (1981). Isolation of Murray Valley encephalitis virus and other arboviruses in the Ord River Valley, 1972–1976. Aust. J. Exp. Biol. Med. Sci. *59*, 347–356.

Longdon, B., Wilfert, L., and Jiggins, M. (2012). The Sigma viruses from *Drosophila*. In Rhabdoviruses: Molecular Taxonomy, Evolution, Genomics, Ecology, Host–Vector Interactions, Cytopathology and Control, R.G. Dietzgen, and I.V. Kuzmin, eds. (Caister Academic Press, Norfolk, UK), pp. 117–132.

Luan, P., Yang, L., and Glaser, M. (1995). Formation of membrane domains created during the budding of vesicular stomatitis virus. A model for selective lipid and protein sorting in biological membranes. Biochemistry *34*, 9874–9883.

Ludewig, B., Maloy, K.J., Lopez-Macias, C., Odermatt, B., Hengartner, H., and Zinkernagel, R.M. (2000). Induction of optimal anti-viral neutralizing B cell responses by dendritic cells requires transport and release of virus particles in secondary lymphoid organs. Eur. J. Immunol. *30*, 185–196.

Lund, J.M., Alexopoulou, L., Sato, A., Karow, M., Adams, N.C., Gale, N.W., Iwasaki, A., and Flavell, R.A. (2004). Recognition of single-stranded RNA viruses by Toll-like receptor 7. Proc. Natl. Acad. Sci. U. S. A. *101*, 5598–5603.

Lyles, D.S., McKenzie, M.O., Kaptur, P.E., Grant, K.W., and Jerome, W.G. (1996). Complementation of M gene mutants of vesicular stomatitis virus by plasmid-derived M protein converts spherical extracellular particles into native bullet shapes. Virology *217*, 76–87.

Lyles, D.S., Kuzmin, I.V., and Rupprecht, C.E. (2013). Rhabdoviridae. In Fields Virology, 6th edition. D.M. Knipe, and P.M. Howley, eds. (Lippincott Williams and Wilkins, Philadelphia), pp. 885–922.

McElroy, K.L., Girard, Y.A., McGee, C.E., Tsetsarkin, K.A., Vanlandingham, D.L., and Higgs, S. (2008). Characterization of the antigen distribution and tissue tropisms of three phenotypically distinct yellow fever virus variants in orally infected *Aedes aegypti* mosquitoes. Vector-Borne Zoonotic Dis. *8*, 675–687.

Mackerras, I.M., Mackerras, M.J., and Burnet, F.M. (1940). Experimental studies of ephemeral fever in Australian cattle. CSIRO Bull. *136*, 1–116.

McWilliam, S.M., Kongsuwan, K., Cowley, J.A., Byrne, K.A., and Walker, P.J. (1997). Genome organization and transcription strategy in the complex GNS-L intergenic region of bovine ephemeral fever rhabdovirus. J. Gen. Virol. *78*, 1309–1317.

Main, A.J., and Carey, A.B. (1980). Connecticut virus: a new Sawgrass group virus from *Ixodes dentatus* (Acari: Ixodidae). J. Med. Entomol. *17*, 473–476.

Maloy, K.J., Odermatt, B., Hengartner, H., and Zinkernagel, R.M. (1998). Interferon gamma-producing gammadelta T cell-dependent antibody isotype switching in the absence of germinal center formation during virus infection. Proc. Natl. Acad. Sci. U.S.A. *95*, 1160–1165.

Matlin, K.S., Reggio, H., Helenius, A., and Simons, K. (1982). Pathway of vesicular stomatitis virus entry leading to infection. J. Mol. Biol. *156*, 609–631.

Mavale, M.S., Geevarghese, G., Ghodke, Y.S., Fulmali, P.V., Singh, A., and Mishra, A.C. (2005). Vertical and venereal transmission of Chandipura virus (Rhabdoviridae) by *Aedes aegypti* (Diptera: Culicidae). J. Med. Entomol. *42*, 909–911.

Mead, D.G., Ramberg, F.B., Besselsen, D.G., and Mare, C.J. (2000). Transmission of vesicular stomatitis virus from infected to noninfected black flies co-feeding on nonviremic deer mice. Science *287*, 485–487.

Mead, D.G., Gray, E.W., Noblet, R., Murphy, M.D., Howerth, E.W., and Stallknecht, D.E. (2004). Biological transmission of vesicular stomatitis virus (New Jersey serotype) by *Simulium vittatum* (Diptera: Simuliidae) to domestic swine (*Sus scrofa*). J. Med. Entomol. *41*, 78–82.

Metselaar, D., Williams, M.C., Simpson, D.I., West, R., and Mutere, F.A. (1969). Mount Elgon bat virus: a hitherto undescribed virus from *Rhinolophus hildebrandtii eloquens* K. Anderson. Arch. Virol. *26*, 183–193.

Mire, C.E., Geisbert, J.B., Marzi, A., Agans, K.N., Feldmann, H., and Geisbert, T.W. (2013). Vesicular Stomatitis Virus-Based Vaccines Protect Nonhuman Primates against Bundibugyo ebolavirus. PLoS Negl. Trop. Dis. *7*, e2600.

Moreno-Delafuente, A., Garzo, E., Moreno, A., and Fereres, A. (2013). A plant virus manipulates the behavior of its whitefly vector to enhance its transmission efficiency and spread. PLoS One *8*, e61543.

Mourya, D.T., Thakare, J.R., Gokhale, M.D., Powers, A.M., Hundekar, S.L., Jayakumar, P.C., Bondre, V.P., Shouche, Y.S., and Padbirdi, V.S. (2001). Isolation of chikungunya virus from *Aedes aegypti* mosquitoes collected in the town of Yawat, Pune District, Maharashtra State, India. Acta Virol. *45*, 305–309.

Muller, M.J., and Standfast, H.A. (1986). Vectors of ephemeral fever group viruses. In Arbovirus Research in Australia: Proceedings of the Fourth Symposium. T.D. St George, B.H. Kay, and J. Blok, eds. (CSIRO/QIMR, Brisbane, Australia), pp. 295–300.

Muller, M.J., and Standfast, H.A. (1993). Investigation of the vectors of bovine ephemeral fever virus in Australia. In Bovine Ephemeral Fever and Related Rhabdoviruses, T.D. St George, M.F. Uren, P.L. Young, and D. Hoffmann, eds. (ACIAR, Canberra, Australia), 29–32.

Muller, M.J., Murray, M.D., and Edwards, J.A. (1981). Blood-sucking midges and mosquitoes feeding on mammals at Beatrice Hill, N.T. Aust. J. Zool. *29*, 573–588.

Murphy, F.A., and Fields, B.N. (1967). Kern Canyon virus: electron microscopic and immunological studies. Virology *33*, 625–637.

Murray, M.D. (1997). Possible vectors of bovine ephemeral fever in the 1967/68 epizootic in northern Victoria. Aus. Vet. J. *75*, 220–220.

Narasimha Rao, S., Wairagkar, N.S., Murali Mohan, V., Khetan, M., and Somarathi, S. (2008). Brain stem encephalitis associated with Chandipura in Andhra Pradesh outbreak. J. Trop. Pediatr. *54*, 25–30.

Noda, M., Inaba, Y., Banjo, M., and Kubo, M. (1992). Isolation of Fukuoka virus, a member of the Kern Canyon serogroup viruses of the family *Rhabdoviridae*, from cattle. Vet. Microbiol. *32*, 267–271.

Novella, I.S., and Presloid, J.B. (2012). In Rhabdoviruses: Molecular Taxonomy, Evolution, Genomics, Ecology, Host–Vector Interactions, Cytopathology and Control, R.G. Dietzgen, and I.V. Kuzmin, eds. (Caister Academic Press, Norfolk, UK), pp. 205–214.

Oehen, S., Odermatt, B., Karrer, U., Hengartner, H., Zinkernagel, R., and López-Macías, C. (2002). Marginal zone macrophages and immune responses against viruses. J. Immunol. *169*, 1453–1458.

Ogino, T., and Banerjee, A.K. (2011). An unconventional pathway of mRNA cap formation by vesiculoviruses. Virus Res. *162*, 100–109.

Oshiumi, H., Matsuo, A., Matsumoto, M., and Seya, T. (2008). Pan-vertebrate toll-like receptors during evolution. Curr. Genomics *9*, 488–493.

Papaneri, A.B., Wirblich, C., Cann, J.A., Cooper, K., Jahrling, P.B., Schnell, M.J., and Blaney, J.E. (2012). A replication-deficient rabies virus vaccine expressing Ebola virus glycoprotein is highly attenuated for neurovirulence. Virology. *434*, 18–26.

Pickl, W.F., Pimentel-Muinos, F.X., and Seed, B. (2001). Lipid rafts and pseudotyping. J. Virol. *75*, 7175–7183.

Plakhov, I.V., Arlund, E.E., Aoki, C., and Reiss, C.S. (1995). The earliest events in vesicular stomatitis virus infection of the murine olfactory neuroepithelium and entry of the central nervous system. Virology *209*, 257–262.

Poirot, M.K., Schnitzlein, W.M., and Reichmann, M.E. (1985). The requirement of protein synthesis for VSV inhibition of host cell RNA synthesis. Virology *140*, 91–101.

Quan, P.L., Junglen, S., Tashmukhamedova, A., Conlan, S., Hutchison, S.K., Kurth, A., Ellerbrok, H., Egholm, M., Briese, T., Leendertz, F.H., and Lipkin, W.I. (2010). Moussa virus: a new member of the *Rhabdoviridae* family isolated from *Culex decens* mosquitoes in Cote d'Ivoire. Virus Res. *147*, 17–24.

Quan, P.L., Williams, D.T., Johansen, C.A., Jain, K., Petrosov, A., Diviney, S.M., Tashmukhamedova, A., Hutchison, S.K., Tesh, R.B., Mackenzie, J.S., Briese, T., and Lipkin, W.I. (2011). Genetic characterization of K13965, a strain of Oak Vale virus from Western Australia. Virus Res. *160*, 206–213.

Quiroz, E., Moreno, N., Peralta, P.H., and Tesh, R.B. (1988). A human case of encephalitis associated with vesicular stomatitis virus (Indiana serotype) infection. Am. J. Trop. Med. Hyg. *39*, 312–314.

Rao, B.L., Basu, A., Wairagkar, N.S., Gore, M.M., Arankalle, V.A., Thakare, J.P., Jadi, R.S., Rao, K.A., and Mishra, A.C. (2004). A large outbreak of acute encephalitis with high fatality rate in children in Andhra Pradesh, India, in 2003, associated with Chandipura virus. Lancet *364*, 869–874.

Reis, J.L. Jr, Rodriguez, L.L., Mead, D.G., Smoliga, G., and Brown, C.C. (2011). Lesion development and replication kinetics during early infection in cattle inoculated with Vesicular stomatitis New Jersey virus via scarification and black fly (*Simulium vittatum*) bite. Vet. Pathol. *48*, 547–557.

Ritter, D.G., Calisher, C.H., Muth, D.J., Shope, R.E., Murphy, F.A., and Whitfield, S.G. (1978). New Minto virus: a new rhabdovirus from ticks in Alaska. Can. J. Microbiol. *24*, 422–426.

Rodriguez, L.L., and Pauszek, S.J. (2012). Genus *Vesiculovirus*. In Rhabdoviruses: Molecular Taxonomy, Evolution, Genomics, Ecology, Host–Vector Interactions, Cytopathology and Control, R.G. Dietzgen, and I.V. Kuzmin, eds. (Caister Academic Press, Norfolk, UK), pp. 23–36.

Ruigrok, R.W.H., Crepin, T., and Kolakofsky, D. (2011). Nucleoproteins and nucleocapsids of negative-strand RNA viruses. Curr. Opin. Microbiol. *14*, 504–510.

Salaun, J.J., Rickenbach, A., Bres, P., Brottes, H., Germain, M., Eouzan, J.P., and Ferrara, L. (1969). The Nkolbisson virus (YM 31–65), a new prototype of arbovirus isolated in Cameroun. Ann. Inst. Pasteur. Paris *116*, 254–260.

Sather, G.E., Lewis, A.L., Jennings, W., Bond, J.O., and Hammon, W.M. (1970). Sawgrass virus: a newly described arbovirus in Florida. Am. J. Trop. Med. Hyg. *19*, 319–326.

Scherer, C.F., O'Donnell, V., Golde, W.T., Gregg, D., Estes, D.M., and Rodriguez, L.L. (2007). Vesicular stomatitis New Jersey virus (VSNJV) infects keratinocytes and is restricted to lesion sites and local lymph nodes in the bovine, a natural host. Vet. Res. *38*, 375–390.

Schmidt, J.R., Williams, M.C., Lulu, M., Mivule, A., and Mujombe, E. (1965). Viruses isolated from mosquitoes collected in the Southern Sudan and Western Ethiopia. East Afr. Virus Res. Inst. Rep. *15*, 24.

Schnell, M.J., Buonocore, L., Kretzschmar, E., Johnson, E., and Rose, J.K. (1996). Foreign glycoproteins expressed from recombinant vesicular stomatitis viruses are incorporated efficiently into virus particles. Proc. Natl. Acad. Sci. U.S.A. *93*, 11359–11365.

Smith, P.F., Howerth, E.W., Carter, D., Gray, E.W., Noblet, R., and Mead, D.G. (2009). Mechanical transmission of vesicular stomatitis New Jersey virus by *Simulium vittatum* (Diptera: Simuliidae) to domestic swine (*Sus scrofa*). J. Med. Entomol. *46*, 1537–1540.

Soleha, E., Daniels, P.W., Sendow, I., and Sukarsih (1993a). Seroepidemiological studies of bovine ephemeral fever group viral infections in Indonesia. In Arbovirus Research in Australia: Proceedings of the Sixth Symposium. M.F. Uren, and B.H. Kay, eds. (CSIRO/QIMR, Brisbane, Australia), pp. 179–183.

Soleha, E., Daniels, P.W., Sukarsih, and Sendow, I. (1993b). A study of bovine ephemeral fever group rhabdoviral infections in West Java, Indonesia. In Bovine Ephemeral Fever and Related Rhabdoviruses, T.D. St. George, M.F. Uren, P.L. Young, and D. Hoffmann, eds. (ACIAR, Canberra), pp. 45–50.

Springfeld, C., Darai, G., and Cattaneo, R. (2005). Characterization of the Tupaia rhabdovirus genome reveals a long open reading frame overlapping with P and a novel gene encoding a small hydrophobic protein. J. Virol. *79*, 6781–6790.

St George, T.D. (1985). Studies on the pathogenesis of bovine ephemeral fever in sentinel cattle. I. Virology and serology. Vet. Microbiol. *10*, 493–504.

St George, T.D. (1993). The natural history of ephemeral fever of cattle. In Bovine Ephemeral Fever and Related Rhabdoviruses, T.D. St George, M.F. Uren, P.L. Young, and D. Hoffmann, eds. (ACIAR, Canberra), pp. 13–19.

St George, T.D. (2004). Bovine ephemeral fever. In Infectious diseases of Livestock. Second ed. J.A.W. Coetzer, and R.C. Tustin, eds. (Oxford University Press, Cape Town, South Africa), pp. 1183–1193.

St George, T.D. (2009). Evidence that mosquitoes are the vectors of bovine epmemeral fever virus. In Arbovirus Research in Australia. P. Ryan, J. Aaskov, and R. Russell, eds. (QIMR, Coffs Harbour, Australia), pp. 161–164.

St George, T.D., Standfast, H.A., and Dyce, A.L. (1976). The isolation of ephemeral fever virus from mosquitoes in Australia. Aust. Vet. J. *52*, 242.

St George, T.D., Standfast, H.A., Christie, D.G., Knott, S.G., and Morgan, I.R. (1977). The epizootiology of bovine ephemeral fever in Australia and Papua – New Guinea. Aust. Vet. J. *53*, 17–28.

St George, T.D., Murphy, G.M., Burren, B., and Uren, M.F. (1995). Studies on the pathogenesis of bovine ephemeral fever. IV: A comparison with the inflammatory events in milk fever of cattle. Vet. Microbiol. *46*, 131–142.

Standfast, H.A., St George, T.D., and Dyce, A.L. (1976). The isolation of ephemeral fever virus from mosquitoes in Australia. Aust. Vet. J. *52*, 242.

Stojdl, D.F., Lichty, B.D., tenOever, B.R., Paterson, J.M., Power, A.T., Knowles, S., Marius, R., Reynard, J., Poliquin, L., Atkins, H., Brown, E.G., Durbin, R.K., Durbin, J.E., Hiscott, J., and Bell, J.C. (2003). VSV strains with defects in their ability to shutdown innate immunity are potent systemic anti-cancer agents. Cancer Cell *4*, 263–275.

Stremlau, M.H., Andersen, K.G., Folarin, O.A., Odia, I., Ehiane, P.E., Omoniwa, O., Omoregie, O., Jiang, P.P., Yozwiak, N.L., Matranga, C.B., *et al.* (2015) Discovery of novel rhabdoviruses in the blood of healthy individuals from West Africa. PLoS Negl. Trop. Dis. *9*, e0003631.

Sumpter, R. Jr., Loo, Y.M., Foy, E., Li, K., Yoneyama, M., Fujita, T., Lemon, S.M., and Gale, M. Jr. (2005). Regulating intracellular antiviral defense and permissiveness to hepatitis C virus RNA replication through a cellular RNA helicase, RIG-I. J. Virol. *79*, 2689–2699.

Sylvester, E.S., and Richardson, J. (1992), Aphid-borne rhabdoviruses relationship with their aphid vectors. In Advances in Disease Vector Research, Harris, K.F., ed. (Springer-Verlag, Berlin), pp. 313–341.

Tandale, B.V., Tikute, S.S., Arankalle, V.A., Sathe, P.S., Joshi, M.V., Ranadive, S.N., Kanojia, P.C., Eshwarachary, D., Kumarswamy, M., and Mishra, A.C. (2008). Chandipura virus: a major cause of acute encephalitis in children in North Telangana, Andhra Pradesh, India. J. Med. Virol. *80*, 118–124.

TenOever, B.R., Sharma, S., Zou, W., Sun, Q., Grandvaux, N., Julkunen, I., Hemmi, H., Yamamoto, M., Akira, S., Yeh, W.C., Lin, R., and Hiscott, J. (2004). Activation of TBK1 and IKKvarepsilon kinases by vesicular stomatitis virus infection and the role of viral ribonucleoprotein in the development of interferon antiviral immunity. J. Virol. *78*, 10636–10649.

Tesh, R., Saidi, S., Javadian, E., Loh, P., and Nadim, A. (1977). Isfahan virus, a new vesiculovirus infecting humans, gerbils, and sandflies in Iran. Am. J. Trop. Med. Hyg. *26*, 299–306.

Tesh, R.B., Travassos Da Rosa, A.P., and Travassos Da Rosa, J.S. (1983). Antigenic relationship among rhabdoviruses infecting terrestrial vertebrates. J. Gen. Virol. *64*, 169–176.

Tesh, R.B., Boshell, J., Modi, G.B., Morales, A., Young, D.G., Corredor, A., Ferro de Carrasquilla, C., de Rodriguez, C., Walters, L.L., and Gaitan, M.O. (1987). Natural infection of humans, animals, and phlebotomine sand flies with the Alagoas serotype of vesicular stomatitis virus in Colombia. Am. J. Trop. Med. Hyg. *36*, 653–661.

Thavara, U., Siriyasatien, P., Tawatsin, A., Asavadachanukorn, P., Anantapreecha, S., Wongwanich, R., and Mulla, M.S. (2006). Double infection of heteroserotypes of dengue viruses in field populations of *Aedes aegypti* and *Aedes albopictus* (Diptera: Culicidae) and serological features of dengue viruses found in patients in southern Thailand. S. A. J. Trop. Med. Pub. Health *37*, 468–476.

Tomori, O., Fagbami, A., and Kemp, G. (1974). Kotonkan virus: experimental infection of white Fulani calves. Bull. Epiz. Dis. Afr. *22*, 195–200.

Tonbak, S., Berber, E., Yoruk, M.D., Azkur, A.K., Pestil, Z., and Bulut, H. (2013). A large-scale outbreak of bovine ephemeral fever in Turkey, 2012. J. Vet. Med. Sci. *75*, 1511–1514.

Trinidad, L., Blasdell, K.R., Joubert, D.A., Davis, S.S., Melville, L., Kirkland, P.D., Coulibaly, F., Holmes, E.C., and Walker, P.J. (2014). Evolution of bovine ephemeral fever virus in the Australian episystem. J. Virol. *88*, 1525–1535.

Tuekam, T., Greiner, E.C., and Gibbs, E.P.J. (1991). Seroepidemiology of Bivens Arm virus infections of cattle in Florida, St Croix and Puerto Rico. Vet. Microbiol. *28*, 121–127.

Uren, M.F., and Zakrzewski, H. (1989). Mechanisms of immunity to BEF. In Arbovirus Research in Australia: Proceedings of the Fifth Symposium. M.F. Uren, J. Blok, and L.H Manderson, eds. (CSIRO/QIMR, Brisbane, Australia), pp. 274–276.

Uren, M.F., St George, T.D., and Zakrzewski, H. (1989). The effect of anti-inflammatory agents on the clinical expression of bovine ephemeral fever. Vet. Microbiol. *19*, 99–111.

Uren, M.F., St George, T.D., and Murphy, G.M. (1992). Studies on the pathogenesis of bovine ephemeral fever in experimental cattle. III. Virological and biochemical data. Vet. Microbiol. *30*, 297–307.

Uren, M.F., Walker, P.J., Zakrzewski, H., St George, T.D., and Byrne, K.A. (1994). Effective vaccination of cattle using the virion G-protein of bovine ephemeral fever virus as an antigen. Vaccine *12*, 845–850.

Venter, G.J., Hamblin, C., and Paweska, J.T. (2003). Determination of the oral susceptibility of South African livestock-associated biting midges, *Culicoides* species, to bovine ephemeral fever virus. Med. Vet. Entomol. *17*, 133–137.

Verrier, E.R., Langevin, C., Benmansour, A., and Boudinot, P. (2011). Early antiviral response and virus-induced genes in fish. Dev. Comp. Immunol. *35*, 1204–1214.

Verrier, E.R., Langevin, C., Tohry, C., Houel, A., Ducrocq, V., Benmansour, A., Quillet, E., and Boudinot, P. (2012). Genetic resistance to rhabdovirus infection in teleost fish is paralleled to the derived cell resistance status. PLoS One *7*, e33935.

Walker, P.J. (2005). Bovine ephemeral fever in Australia and the world. Curr. Topics Microbiol. Immunol. *292*, 57–80.

Walker, P.J., Byrne, K.A., Riding, G.A., Cowley, J.A., Wang, Y., and McWilliam, S. (1992). The genome of bovine ephemeral fever rhabdovirus contains two related glycoprotein genes. Virology *191*, 49–61.

Walker, P.J., Dietzgen, R.G., Joubert, D.A., and Blasdell, K.R. (2011). Rhabdovirus accessory genes. Virus Res. *162*, 110–125.

Walker, P.J., Blasdell, K.R., and Joubert, D.A. (2012). Ephemeroviruses: arthropod-borne rhabdoviruses of ruminants, with large and complex genomes. In Rhabdoviruses: Molecular Taxonomy, Evolution, Genomics, Ecology, Cytopathology and Control. R.G. Dietzgen, and I.V. Kuzmin, eds. (Horizon Scientific Press, Norwich, UK), pp. 59–88.

Walker, P.J., Firth, C., Widen, S.G., Blasdell, K.R., Guzman, H., Wood, T.G., Paradkar, P.N., Holmes, E.C., Tesh, R.B., and Vasilakis, N. (2015). Evolution of genome size and complexity in the rhabdoviridae. PLoS Pathog. *11*, e1004664.

Wang, Y., and Walker, P.J. (1993). Adelaide river rhabdovirus expresses consecutive glycoprotein genes as polycistronic mRNAs: new evidence of gene duplication as an evolutionary process. Virology *195*, 719–731.

Webb, P.A., and Holbrook, F.R. (1989). Vesicular stomatitis. In The Arboviruses, T.P. Monath, ed. Vol. 5. (CRC Press, Boca Raton, Florida), pp. 1–29.

Weir, R.P., Hyatt, A.D., Calisher, C.H., and Whelan, P.I. (1996). New records of arboviruses isolated from mosquitoes in the Northern Territory 1982–1992. In Arbovirus research in Australia. Proceedings 7th Symposium 25–29 November, B.H. Kay, M.D. Brown, and J.G. Aaskov, eds. (CSIRO/QIMR, Brisbane, Australia), pp. 311–321.

Wellings, F.M., Lewis, A.L., and Pierce, L.V. (1972). Agents encountered during arboviral ecological studies: Tampa Bay area, Florida, 1963 to 1970. Am. J. Trop. Med. Hyg. *21*, 201–213.

Wertz, G.W., Perepelitsa, V.P., and Ball, L.A. (1998). Gene rearrangement attenuates expression and lethality of a nonsegmented negative strand RNA virus. Proc. Natl. Acad. Sci. U.S.A. *95*, 3501–3506.

Whelan, S.P., and Wertz, G.W. (1999). Regulation of RNA synthesis by the genomic termini of vesicular stomatitis virus: identification of distinct sequences essential for transcription but not replication. J. Virol. *73*, 297–306.

Whitney, E. (1964). Flanders strain, an arbovirus newly isolated from mosquitoes and birds of New York State. Am. J. Trop. Med. Hyg. *13*, 123–131.

Woelk, C.H., and Holmes, E.C. (2002). Reduced positive selection in vector-borne RNA viruses. Mol. Biol. Evol. *19*, 2333–2336.

Wyers, F., Richard-Molard, C., Blondel, D., and Dezelee, S. (1980). Vesicular stomatitis virus growth in Drosophila melanogaster cells: G-protein deficiency. J. Virol. 33, 411–422.

Yabu, T., Hirose, H., Hirono, I., Katagiri, T., Aoki, T., and Yamamoto, E. (1998). Molecular cloning of a novel interferon regulatory factor in Japanese flounder, *Paralichthys olivaceus*. Mol. Mar. Biol. Biotechnol. 7, 138–144.

Yeruham, I., Van Ham, M., Stram, Y., Friedgut, O., Yadin, H., Mumcuoglu, K.Y., and Braverman, Y. (2010). Epidemiological investigation of bovine ephemeral fever outbreaks in Israel. Vet. Med. Int. 2010, e290541.

Yoneyama, M., Kikuchi, M., Matsumoto, K., Imaizumi, T., Miyagishi, M., Taira, K., Foy, E., Loo, Y.M., Gale, M. Jr., Akira, S., Yonehara, S., Kato, A., and Fujita, T. (2005). Shared and unique functions of the DExD/H-box helicases RIG-I, MDA5, and LGP2 in antiviral innate immunity. J. Immunol. *175*, 2851–2858.

Young, P.L., and Spradbrow, P.B. (1980). The role of neutrophils in bovine ephemeral fever virus infection of cattle. J. Infec. Dis. *142*, 50–55.

Young, P.L., and Spradbrow, P.B. (1985). Transmission of virus from serosal fluids and demonstration of antigen in neutrophils and mesothelial cells of cattle infected with bovine ephemeral fever virus. Vet. Microbiol. *10*, 199–207.

Young, P.L., and Spradbrow, P.B. (1990a). Demonstration of vascular permeability changes in cattle infected with bovine ephemeral fever virus. J. Comp. Pathol. *102*, 55–62.

Young, P.L., and Spradbrow. P.B. (1990b). Clinical response of cattle to experimental infection with bovine ephemeral fever virus. Vet. Rec. *126*, 86–88.

Zakrzewski, H., and Cybinski, D.H. (1984). Isolation of Kimberley virus, a rhabdovirus, from *Culicoides brevitarsis*. Aust. J. Exp. Biol. Med. Sci. *62*, 779–780.

Zavada, J. (1982). The pseudotypic paradox. J. Gen. Virol. *63*, 15–24.

Zheng, F.Y., and Qui, C.Q. (2012), Phylogenetic relationships of the glycoprotein gene of bovine ephemeral fever virus isolated from mainland China, Taiwan, Japan, Turkey, Israel and Australia. Virol. J. *9*, e268.

Alphavirus–Host Interactions

Kate D. Ryman and William B. Klimstra

7

Abstract

Members of the genus *Alphavirus*, family *Togaviridae*, are arthropod-borne viruses, mostly vectored in nature by specific mosquito species between various vertebrate reservoir hosts including birds, rodents, equids and non-human primates (NHPs) in classical arbovirus transmission cycles. The exceptions are the so-called salmonid alphaviruses, salmon pancreatic disease virus and its subtype, sleeping disease virus, which infect fish, causing mortality in farmed salmon and trout (McLoughlin and Graham, 2007), and the southern elephant seal virus (SESV; La Linn *et al.*, 2001). The presence of SESV within lice (La Linn *et al.*, 2001) suggests an arthropod-borne cycle, but the vector–host relationships have yet to be confirmed (Forrester *et al.*, 2012). The second genus in the family is *Rubivirus*, containing only the rubella virus, which shares genomic and structural features with the alphaviruses (Strauss *et al.*, 1984) but is not an arbovirus and so not discussed further. Within the *Alphavirus* genus, viruses are grouped into four major subgroups by antibody cross-reactivity: the Semliki Forest (SF), Venezuelan equine encephalitis (VEE), eastern equine encephalitis (EEE) and western equine encephalitis (WEE) serocomplexes. Fascinatingly, recent phylogenetic evidence suggests that the salmonid alphaviruses may be the progenitors of the entire genus (Forrester *et al.*, 2012; Nasar *et al.*, 2012).

As a consequence of their transmission cycle, the alphaviruses must be replication-competent in their mosquito vectors and reservoir hosts, which vary between viruses. It has become increasingly clear that these highly divergent hosts each exert strongly selective pressures on the viruses that constrain viral genetic variation (Coffey *et al.*, 2013) but may also drive particular aspects of host interaction that result in disease (Coffey *et al.*, 2013; Trobaugh *et al.*, 2014a; Ventoso *et al.*, 2006). In this chapter, we will describe recent advances in our understanding of the characteristics of alphavirus interaction with vertebrate hosts, focusing upon the relationship of particular infection/replication characteristics to disease manifestations and immune responses.

Alphaviruses of medical importance

Of the approximately 30 individual alphaviruses that have been identified to-date, a large subset cause medically and economically important disease in humans, and in some cases, in domesticated animals (e.g. horses, mules, emus, ostriches, and farmed fish). In general, humans represent 'dead-end' hosts for the alphaviruses, meaning that they are not part of the natural transmission cycle, the major exception being chikungunya virus (CHIKV), discussed further below. Based upon the symptoms of the diseases that they cause, the alphaviruses can be loosely divided into two groups: agents of encephalitic and febrile/arthritogenic disease, although these categories can overlap significantly and appear to represent different points on a disease continuum. These categories, and major representatives from each, are discussed below.

The encephalitic alphaviruses

The encephalitic alphaviruses are found in the VEE, EEE and WEE serocomplexes and are currently confined geographically to the Americas. These viruses are named literally for the severe, and often fatal, neuroinflammatory disease they cause in equids. In humans, while all viruses in this category are capable of causing encephalitic disease, the frequency and severity of neurological symptoms varies depending upon the virus (Calisher *et al.*, 1983; Deresiewicz *et al.*, 1997; Rozdilsky *et al.*, 1968; Watts *et al.*, 1998; Zacks and Paessler, 2010), as will be described below. In addition, as for many arboviral diseases, only a minority of infected individuals develops symptomatic disease. Disease rates and severity are strongly dependent on age, with children and the elderly much more susceptible. In the last few years, research studies have begun to reveal that differential disease symptoms, host range and epidemiology are underpinned by divergent virus–host interactions at the molecular, cellular and organismal level, discussed in later sections.

Venezuelan equine encephalitis

There are six virus subtypes (I–VI) found within the VEE serocomplex: a collection of viruses defined by antibody cross-reactivity and therefore representing similarity of surface glycoprotein topology (Aguilar *et al.*, 2011). Although in the past endemic and epidemic cases have occurred in the southeastern USA (Aguilar *et al.*, 2011; Weaver and Reisen, 2010), these viruses are currently geographically restricted to Central and South America with the exception of the Everglades virus (EVEV), which circulates in Florida (Bigler *et al.*, 1974; Coffey *et al.*, 2006) and occasionally infects humans (Monath *et al.*, 1980). VEE subtype I has six variants (A-F) of VEEV, the prototypic virus. Subtypes II–VI include lesser-known viruses such as Mucambo (III-A), Tonate (III-B), Pixuna (IV) and Rio Negro (VI). Subtype I varieties D, E, and F and subtypes II–VI are enzootic, equine-avirulent strains (Weaver *et al.*, 2004), not associated with major epizootic or epidemic outbreaks. However, they can cause isolated incidents of potentially fatal human illness and represent a threat for emergence (Adams *et al.*, 2012).

Currently, it is the repeated emergence of VEEV variants I-AB and I-C from enzootic Subtype I strains (Powers *et al.*, 1997) that is responsible for major equine epizootics and epidemic outbreaks, coincident with adaptation of the viruses to efficient infection of equines and/or transmission by equiphilic mosquitoes (Anishchenko *et al.*, 2004) and amino acid changes in the E2 glycoprotein (Brault *et al.*, 2002). The I-AB and I-C variants are highly neuropathogenic in equines, causing up to 80% lethality and there is evidence for vector transmission between equines and from equine to human during epizootic outbreaks. Even in epidemics, however, humans seem to remain a dead-end host for the virus, perhaps due to insufficient viraemic titres for infection of the feeding mosquito. The most recent large epidemic occurred in 1995 in Venezuela and Colombia with an estimated 100,000 symptomatic cases caused by a I-C strain (Brault *et al.*, 2001; Rico-Hesse *et al.*, 1995; Weaver *et al.*, 1996), but there are likely many sporadic cases and small outbreaks that are misdiagnosed as DENV or another arbovirus (Aguilar *et al.*, 2011). Many patients experience no more than a relatively mild, febrile illness (Weaver *et al.*, 2004). Neurological signs, including disorientation, depression, ataxia and convulsions, and sequelae are relatively rare (< 15% of symptomatic cases) and fatality rates are less than 0.5%, although both are more common in children (Aguilar *et al.*, 2011). Prevalent histopathological findings in these fatal human infections include destruction of lymphoid tissues, pharyngitis, pneumonia, and meningoencephalitis with oedema, haemorrhage and vasculitis in the brain. Although sporadic, human infections with other VEE serocomplex viruses result in similar symptoms and mortality rates.

Eastern equine encephalitis

EEEV strains are found throughout the Americas. North American EEEV strains (NA-EEEV) are genetically highly related with little temporal or geographic diversity (Young *et al.*, 2008). The circulating virus is highly neurovirulent in equines. Occurring primarily along the eastern seaboard of North America from Canada to Florida, NA-EEEV is also far more neurovirulent than VEEV or WEEV in humans, causing mortality in around half of symptomatic cases, and long-term neurological *sequelae* in one third of the survivors (Steele and Twenhafel, 2010). Fortunately human infections are rare but the incidence has been increasing in recent years. In the last 10 years there have been over 100 EEE cases confirmed by the Centers for Disease Control and Prevention (CDC), while only 200 probable and confirmed cases occurred in the four decades prior (US Department of Health and Human Services, CDC).

NA-EEEV is primarily maintained in a cycle by *Culiseta* mosquito species that selectively feed on passerine (perching) birds; the most typical reservoir host. The frequency with which virus is found in nature increases throughout the summer as more birds and mosquitoes become infected. Transmission of the virus from its avian-mosquito cycle in swampy woodland areas to humans and equines occurs sporadically via ‘bridging’ vectors; mosquitoes that are less selective feeders and bite both avian and mammalian hosts. No adaptation of the emerging EEEV is evident thus far and, although the virus is highly virulent in equines (Zacks and Paessler, 2010), exotic birds (e.g. ostriches and emus (Chenier *et al.*, 2010; Verwoerd, 2000) and humans (Silverman *et al.*, 2013), mammals are dead-end hosts so epizootic/epidemic outbreaks do not occur. Also unlike VEEV infections, prodromal febrile illness is short or absent during NA-EEEV infection of humans and symptoms often begin suddenly with myalgia and increasingly severe headache, followed by vomiting, shortness of breath (dyspnoea), lack of energy and strength (asthenia), seizures and coma. The absence of a prodrome is prognostic for more severe disease, and more common in children (Silverman *et al.*, 2013). Clinical imaging studies have revealed changes in the basal ganglia and thalami, with circulatory shock and brain oedema in the early stages of disease. Histopathological examination of fatal cases indicates brain oedema with necrosis, cerebral haemorrhage, vasculitis and encephalomyelitis (Deresiewicz *et al.*, 1997).

Interestingly, South American strains of the virus (SA-EEEV) are genetically much more diverse than NA-EEEV strains (Carrera *et al.*, 2013) are primarily reservoired in rodents and cause little disease in humans or equines relative to the high seroprevalence and virus circulation. Recently increasing human cases of EEE in Latin America (Carrera *et al.*, 2013) are likely associated with ecological changes that have increased human contact with enzootic transmission cycles, genetic changes in the viruses that resulted in increased equine and human virulence, and/or an altered host range (Carrera *et al.*, 2013), making it more likely that the SA-EEEV strains may emerge as significant causative agents of human disease in the future.

Western equine encephalitis

Detailed phylogenetic analyses have revealed that WEEV strains emerged by recombination of EEEV-like non-coding sequences, non-structural and capsid protein genes with Sindbis virus (SINV)-like envelope glycoprotein genes (Hahn *et al.*, 1988; Weaver *et al.*, 1997). Consequently the WEE serocomplex contains the encephalitic WEEV and a number of other viruses, classified as SIN-group, that fall in the febrile/arthritogenic category. Thus, if WEEV were classified based upon non-coding and nsP gene sequences, it would cluster closely with EEEV (Bergren *et al.*, 2014).

WEEV is endemic to the western regions of North America. Between 1964 and 2003, there were over 600 confirmed human cases of WEE in the US (US Department of Health and Human Services, CDC) but the annual incidence declined over this period, likely due to changes in agricultural irrigation practices and the introduction of mosquito abatement programmes since the virulence of the circulating virus does not seem to have diminished (Forrester *et al.*, 2008), and there have been no confirmed cases in the last decade (US. Department of Health and Human Services, CDC). The severity of WEEV infection is very strongly age-dependent with young children much more likely to develop symptomatic and severe disease, whereas infections in adults tend to be asymptomatic or mild with non-specific 'flu-like' symptoms, e.g. sudden onset of fever, headache, nausea, vomiting, anorexia and malaise. In some cases, additional symptoms of altered mental state, asthenia and signs of meningitis occur. In a minority of infected individuals, encephalitis or encephalomyelitis occurs with neck stiffness, confusion, seizures, somnolence, coma and death. The overall symptomatic case-fatality rate in humans is estimated to be in the range 3–7% in encephalitic cases with 30% of the survivors experiencing neurological sequelae. WEEV encephalitis in humans is associated with vasculitis and focal haemorrhage in the basal ganglia and thalamus.

The arthritogenic alphaviruses

A number of alphaviruses have been identified that together are responsible for millions of cases of serious, though primarily not life-threatening, febrile illness with arthritis/arthralgia in humans (Suhrbier *et al.*, 2012; Toivanen, 2008). Many of these viruses are classified in the SF serocomplex viruses including Semliki Forest (SFV), chikungunya (CHIKV), o'nyong nyong (ONNV), Mayaro (MAYV), Ross River (RRV) and Barmah Forest (BFV) viruses, while others are SIN-group viruses in the WEE serocomplex (see below). Until recently these were often termed the 'Old World' alphaviruses because of their general geographic restriction to these continents; however, this classification needs to be reevaluated with the recent global spread and reestablishment of CHIKV in the Americas, in addition to the historic presence of MAYV and AURAV. The global distribution of these viruses, increased international travel, economic development, changes in mosquito vectors and the potentially explosive nature of epidemics are all likely to contribute to more frequent incidence of these diseases in the future (Suhrbier *et al.*, 2012).

Sindbis group viruses

The original Sindbis virus (SINV), strain AR339, was isolated in 1952 near Cairo, Egypt from mosquitoes and is the prototypic alphavirus (Taylor *et al.*, 1955), discovered as one of hundreds of arboviruses in the Rockefeller Foundation Virus Program (Rosenberg *et al.*, 2013). Other closely related viruses in the SIN-group are Girdwood, Okelbo, Karelian fever, Whataroa and Pogosta. The SIN-group viruses are transmitted between avian reservoir hosts by *Culex* spp. mosquitoes. SIN-group strains are endemic in large parts of Europe, Africa, Asia, and Australia, but human disease has been mostly reported from northern Europe and South Africa (Bergqvist *et al.*, 2015; Lundstrom and Pfeffer, 2010; Niklasson and Espmark, 1986; Strauss *et al.*, 1984). SINV causes Sindbis fever in humans with arthralgia/arthritis (sometimes persisting), rash and malaise, most commonly in South-East Africa, Egypt, Israel, the Philippines and Australia. In Northern Europe and Russia, SIN-group viruses are also associated with Pogosta, Okelbo and Karelian fever, all of which have similar symptoms. Intermittent outbreaks are usually observed in late summer or early fall.

Ross River virus

RRV is endemic and enzootic in Australia, Papua New Guinea, and many islands in the South Pacific, with most cases occurring in northern Australia during the rainy season when mosquito vectors are at their most abundant (Suhrbier *et al.*, 2012). However, epidemiological analyses have suggested that changing land practices and global warming will likely increase the range and activity of this and related alphaviruses such as BFV in the future. Reservoir hosts are native mammals such as kangaroos and wallabies, as well as rodents, bats and horses. Currently, around 5000 Australians are infected annually. The infection, sometimes called 'epidemic polyarthritis,' is commonly associated with fever, flu-like symptoms, rash, severe muscle and joint pains, and sometimes swelling of the joints (especially in the fingers, wrists and feet). The lethargy and fatigue experienced by patients can often be debilitating and, although symptoms typically resolve within several weeks to a few months, roughly 10% of people have ongoing joint pains, depression and fatigue for many months.

Chikungunya virus

Currently, CHIKV is by far the most medically prevalent arthritogenic alphavirus worldwide. CHIKV was first isolated in 1952 in Tanzania (Robinson, 1955); however, at least three epidemics most probably of this virus were recorded as early as the late 1700s in Indonesia but confused with DENV (Carey, 1971; Halstead, 2015). Regular pandemics have occurred since then, primarily arising in Africa and spreading into Asia but also probably in the Americas, with early outbreaks called *'kidinga pepo'* and thus confused with epidemic dengue fever which eventually took this name (Halstead, 2015). CHIKV strains form three distinct clades: West African; East/Central/South African (ECSA); and Asian (Caglioti *et al.*, 2013). Primates are the reservoirs hosts for CHIKV and the virus shares a mosquito vector, *Ae. aegypti*, with the flaviviruses, yellow fever and dengue.

Historically CHIKV has been geographically confined by its requirement for certain NHP reservoir hosts, but a single amino acid mutation arose in the E1 glycoprotein of an ECSA strain, which facilitated transmission from human to human by Ae. albopictus mosquitoes (Tsetsarkin *et al.*, 2007), a much more aggressively biting species. The ongoing pandemic of CHIK fever began in Kenya (Lo Presti *et al.*, 2012) and spread to La Réunion Island, neighbouring Indian Ocean islands and then to the surrounding mainland. Estimates as high as 6.5 million cases have occurred. Cases in travellers returning from these regions have been diagnosed in Japan, parts of Europe and the Americas (Centers for Disease and Prevention, 2006; Gibney *et al.*, 2011; Johansson *et al.*, 2014; Khan *et al.*, 2014; Lindsey *et al.*, 2015; Pfeffer and Loscher, 2006; Seyler *et al.*, 2009); at least 40 countries have experienced such cases. Without the requirement for the NHP reservoir host, autochthonous cases have inevitably occurred in Southern Europe (Bonilauri *et al.*, 2008; Fischer *et al.*, 2013; Gould and Higgs, 2009; Tilston *et al.*, 2009; Vazeille *et al.*, 2008). CHIKV has also recently emerged (or reemerged (Carey, 1971; Halstead, 2015)) in the Americas, arriving first in the Caribbean Islands (December, 2013) before spreading to the Central, South and North American mainland (Fischer *et al.*, 2014; Johansson *et al.*, 2014; Kendrick *et al.*, 2014; Staples and Fischer, 2014; Weaver and Forrester, 2015).

Interestingly, this virus is from the Asian clade (Diaz *et al.*, 2015), not derived from the La Réunion epidemic, and so we may have temporarily avoided a full-blown epidemic outbreak in the US (Weaver, 2014).

Although CHIKV infection can be asymptomatic, attack rates are often extremely high within local populations. In symptomatic cases the acute phase of disease presents with fever, chills, headache, eye pain, diarrhoea, vomiting, rash, generalized myalgia and polyarthralgia (reviewed in Caglioti *et al.*, 2013). Roughly two thirds of these cases progress to significant chronic and/or flaring arthralgia/arthritis in humans, persisting for months to years and reinforcing the seriousness of the current global pandemic for which there is still no licensed therapeutic or vaccine (Rashad *et al.*, 2014; Weaver *et al.*, 2012).

Alphavirus molecular biology related to host interaction

The alphaviruses are small, enveloped viruses with a single-stranded, positive-sense RNA genome that encodes only seven mature proteins (Strauss *et al.*, 1984). Each viral component, both nucleic acid and protein, must therefore be multifunctional and highly efficient for the virus to survive in its battle against the host's antiviral response at the molecule, cell and organism level. Virus replication and protein expression is regulated tightly, such that small changes may dramatically alter susceptibility to the host's innate and adaptive immune responses. Here, we briefly introduce aspects of alphavirus infection and replication pertinent to virus virulence/attenuation, pathogenesis and host immune responses, assuming that the cell is highly permissive to alphavirus infection. Very early molecular level virus–host interactions essentially determine everything (Frolov *et al.*, 2012) and it should be remembered that efficiency alters with virus genotype, cell type and antiviral state.

The alphavirus virion

Alphavirus structural components

All alphavirus virions comprise an icosahedral nucleocapsid of capsid (C) protein monomers, enveloped in a host cell-derived lipid membrane that is studded with membrane-anchored glycoprotein spikes: trimer-heterodimers of proteins E1 and E2 (Mukhopadhyay *et al.*, 2006; Paredes *et al.*, 1993, 2001; Voss *et al.*, 2010). Both envelope proteins have transmembrane domains reaching through the lipid envelope to interact via their C-termini with the nucleocapsid. Each trimer-heterodimer interacts with one capsid monomer, forming a relatively rigid and highly organized, double-layered $T=4$ icosahedron. Unlike many, less structured, enveloped viruses, there is no evidence that any host proteins are present in the lipid membrane or nucleocapsid of the virion. However, the specific lipid composition of the membrane can affect virus infectivity (Chatterjee *et al.*, 2002; Hafer *et al.*, 2009). The E2 glycoprotein interacts with the cell-surface attachment receptors (Roehrig *et al.*, 1988) and is the major target for antibody reactivity (e.g. (Brehin *et al.*, 2008; Chua *et al.*, 2014; Hunt *et al.*, 2010; Roehrig *et al.*, 1980; Schmaljohn *et al.*, 1982; Westarp *et al.*, 1988) while E1 mediates pH-dependent fusion with endosomal membranes once the virus particle is internalized (Kielian, 2010; Pletnev *et al.*, 2001; Sanchez-San Martin *et al.*, 2009).

The alphavirus genome

The encapsulated alphavirus genome is a single-stranded, positive-sense RNA molecule of approximately 11 kB with a 5′ cap and 3′ polyadenylation (Strauss *et al.*, 1984), meaning that it can be translated immediately upon release into the cytoplasm, somewhat resembling a cellular mRNA but with some significant differences. The 5′-termini of alphavirus genomic RNAs have a type-0 cap structure, methylated at the N-7 position of the 5′ nucleotide (nt1) by the nsP1 protein (Ahola and Kaariainen, 1995), but lacking methylation at the 2′-O positions on nt2 or nt3 that form the type-1 and -2 cap structures typically found on the mRNAs of higher eukaryotes. This is an interesting difference from viruses in the most closely related Flaviviridae family, which have type-1 caps (Hyde *et al.*, 2014). The genome has 5′ and 3′ terminal non-translated regions (NTRs) and can be thought of as being divided into two open reading frames (ORFs) separated by a third NTR. All three NTRs have extensive, highly conserved, secondary and tertiary structure, serving a multitude of functions (Hyde *et al.*, 2015) discussed further below. The upstream and downstream ORFs encode the four non-structural (nsP1–4), and three structural [C, PE2 (E3/E2) and 6K/E1] proteins, respectively, although the structural protein-encoding ORF is only translated from a sub-genomic mRNA synthesized after replication from the complementary or 'negative-sense' strand, not directly from the genomic RNA (Strauss *et al.*, 1984).

Virion binding, entry and uncoating of the genome

The relationship between alphaviruses and their receptors is a complex one with the identity of the receptor(s) utilized by wild-type isolates mostly elusive. Multiple attachment/entry receptors on various cell types have been identified for alphaviruses over five decades. However, demonstrations since the late 1990s that receptor usage changes rapidly with only a few passages in cultured cells (Bernard *et al.*, 2000; Gardner *et al.*, 2014; Heil *et al.*, 2001; Klimstra *et al.*, 1998; Smit *et al.*, 2002) rendered the validity of many earlier studies, all performed with cell culture-adapted strains, difficult to determine. Studies using genotypically wild-type viruses, derived using unpassaged genome sequences stabilized and resurrected using cDNA clone technologies are discussed below. Sadly, however, receptor identification and most other areas of alphavirus biology continue to be muddied in the literature by the careless use of strains that are adapted to growth *in vitro* in cultured cells.

Arthritogenic alphavirus interactions with cell-surface receptors

SINV particles derived from a cDNA-derived, consensus wild-type AR339 strain called TR339 (Klimstra *et al.*, 1998; McKnight *et al.*, 1996) bind very poorly to cells in general and have orders of magnitude lower infectivity per particle for cultured cells than do cell culture-adapted laboratory strains (Klimstra *et al.*, 1998). Importantly, the cell attachment phenotype and genomic sequence of TR339 are indistinguishable from that of an ancestral SINV AR339 isolate preserved prior to cell culture passage (Klimstra and Ryman, unpublished observations). When the natural genotype is carefully maintained, similar findings have subsequently been made for other arthritogenic alphaviruses including Girdwood (Klimstra and Heise, unpublished), SFV (Smit *et al.*, 2002), RRV (Heil *et al.*, 2001) and representatives of all three CHIKV clades including the ECSA La Réunion isolate (Gardner *et al.*, 2012, 2014; Gorchakov *et al.*). Furthermore, SINV isolated from the sera of mice infected with a cell culture-adapted strain has restored wild-type binding phenotype (Byrnes and Griffin, 2000). Poor binding is presumably beneficial to the virus preventing non-productive sequestration and extending the duration and level of serum viraemia for transmission back to the mosquito vector. However, how the virus binds and enters remains a mystery except that the virus particles can rearrange their glycoproteins to reveal 'transitional epitopes' related to improved infectivity under certain conditions (Flynn *et al.*, 1990; Meyer *et al.*, 1992; Meyer and Johnston, 1993; Olmsted *et al.*, 1986; Klimstra, Ryman and Johnston, unpublished). Importantly, the improved infectivity of cell culture-adapted arthritogenic alphaviruses *in vitro* is typically highly attenuating *in vivo*, and may alter the course of disease (Gardner *et al.*, 2014; Ryman *et al.*, 2007a). This is, of course, central to the development of live-attenuated vaccines by serial cell culture passage (Bernard *et al.*, 2000; Gardner *et al.*, 2014; Gorchakov *et al.*, 2012).

Encephalitic alphavirus interactions with cell-surface receptors

In contrast to the arthritogenic alphaviruses, wild-type strains of VEEV, WEEV and particularly EEEV have a significantly higher infectivity per particle (Bernard *et al.*, 2000; Gardner *et al.*, 2011, 2013). Whereas cell culture adaptation of the arthritogenic alphaviruses results in the acquisition of positively charged amino acid substitutions in the E2 glycoprotein that facilitate binding and infection via cell surface heparan sulfate proteoglycans (HSPGs) (Gardner *et al.*, 2014; Heil *et al.*, 2001; Klimstra *et al.*, 1998; Smit *et al.*, 2002), naturally circulating wild-type strains of the encephalitic alphaviruses exhibit degrees of HS-dependent infectivity. EEEV in particular utilizes HS efficiently as an attachment receptor and is attenuated by the deliberate ablation of this ability (Gardner *et al.*, 2011), but virulent wild-type VEEV and WEEV strains also exhibit partial dependence upon HS for infection (Bernard *et al.*, 2000; Gardner *et al.*, 2011). It has also been shown that selected neurovirulence of SINV strains requires HS binding coupled with a second mutation (Ryman *et al.*, 2007a), pointing to HSPG receptor usage as a neurovirulence factor for alphaviruses generally. However, this is a fine balance as more strongly HS-binding, attenuated variants of VEEV can be selected by passage of the wild-type virus (Bernard *et al.*, 2000). Clearly there is more to this phenomenon than we yet understand.

C-type lectins DC-SIGN and SIGN-R

Notably, mosquito cells do not process N-linked carbohydrate modifications of proteins to the complex carbohydrate but instead end processing after addition and trimming of high mannose carbohydrates. Growth of SINV either in mosquito cells or in mammalian cells under conditions that limited complex carbohydrate content of the glycoproteins, greatly increases binding to and infection via C-type lectin molecules (Klimstra *et al.*, 2003). C-type lectins, which bind mannose-enriched carbohydrate modifications of host and pathogen proteins, have been shown to bind glycoproteins of several viruses and facilitate either *cis* or *trans* infection of cells in addition to their many other functions (Garcia-Vallejo and van Kooyk). DC-SIGN and SIGN-R are expressed on several early targets of alphavirus infection, including dendritic cells (DCs), macrophages and cells of the reticuloendothelial system. Both function as attachment/entry receptors for the mosquito cell-derived TR339 strain of SINV on DC-SIGN/SIGN-R transfected human monocytic Raji cells or untreated primary human DCs (Klimstra *et al.*, 2003).

Preliminary studies to test the interaction of mosquito- and mammalian-derived EEEV, WEEV, RRV and VEEV virions with DC-SIGN/L-SIGN molecules have indicated that each of these alphaviruses can use C-type lectins as attachment receptors when complex carbohydrate processing is limited (Ryman and Klimstra, unpublished observations). It has also been proposed that alphavirus interaction with this receptor family alters the response of myeloid cells to virus infection, perhaps reducing type I IFN induction and other antiviral responses, while enhancing virus replication (Shabman *et al.*, 2007; Silva *et al.*, 2014). C-type lectin molecules appear to be common attachment and entry receptors for arthropod-produced viruses, influencing the infection efficiency and potentially the immune response early after introduction into a vertebrate host (Davis *et al.*, 2006).

Entry and uncoating

Infection commences when the virus binds to its cell-surface receptor and is internalized within an endosome possibly by a clathrin-dependent mechanism (Kielian and Jungerwirth, 1990). As the pH of the endosome drops, a *'fusion loop'* is exposed in the E1 protein by a structural rearrangement of E2 (Kielian, 2010). E2 releases E1 from each heterodimer in the trimer, and the three E1 proteins form an extended E1 homotrimer that inserts into the endosome/lysosome membrane. The E1 proteins then refold to a hairpin-like structure, enabling membrane fusion (Fields and Kielian, 2013; Kielian *et al.*, 2010; Liu and Kielian, 2009; Roman-Sosa and Kielian, 2011; Zheng *et al.*, 2011). This essentially turns the virus envelope inside out, exposing the nucleocapsid to the cellular cytoplasm. There is some evidence that the C proteins of incoming nucleocapsids associate with ribosomes (Singh and Helenius, 1992; Wengler and Wengler, 1984; Wengler *et al.*, 1992). While the ribosome-bound capsid proteins become distributed throughout the cytoplasm, the viral RNA remains associated with membrane-bound ribosomes.

Viral protein synthesis and genome replication

Specificity, and subtle temporal regulation, of alphavirus RNA replication are achieved via recognition of *cis*-acting conserved sequence elements (CSEs) present in the termini of the viral genome and the negative-sense genome template, the secondary structures of which are highly conserved. The four nsPs are translated by host cell translation initiation factors (eIFs) and ribosomes as a single polyprotein from the upstream ORF of the genome as either P123 or P1234 depending upon read-through of an opal stop codon found in some, but not all, alphaviruses between the sequences encoding nsP3 and nsP4 (Takkinen, 1986). This opal stop codon likely serves to optimize the quantity of the nsP4 RNA-dependent RNA polymerase (RdRp) synthesized.

The formation of virus replication complexes (RCs) in infected cells determines virus replication and ultimately pathogenesis on the molecular and cellular levels. Alphavirus replication leads to formation of 'cytopathic vacuoles' (CPVs), modified endosomes, lysosomes and plasma membranes, with numerous membrane invaginations (spherules) in which the viral RCs accumulate. These CPVs have been proposed as the sites of viral RNA replication (Frolova *et al.*, 2010; Froshauer *et al.*, 1988; Grimley *et al.*, 1972), sequester multiple host proteins (Varjak *et al.*, 2013), and may hide the viral pathogen-associated molecular patterns (PAMPs) from immediate detection by the cell.

Early in infection, partially processed P123/nsP4 RCs transcribe the genome to produce genome-length, complementary RNAs, creating partially double-stranded (ds) RNA replicative intermediates (Lemm and Rice, 1993a,b). A 19-nt CSE in the 3′ terminus of the genome, immediately upstream of the poly(A) tail, is the core promoter for synthesis of the complementary strand (Ou *et al.*, 1981), and seems to interact with the 5′ NTR via translation initiation factors to initiate replication and/or translation (Frolov *et al.*, 2001). A 51-nt CSE with two-stem–loop secondary structures is also found at the 5′ end of the genomic RNA in the nsP1-encoding gene (Niesters and Strauss, 1990), serving as a replication enhancer (Frolov *et al.*, 2001). After a short time, during which the replicase is further processed, there is a switch to preferential synthesis of progeny positive-sense genomes by transcription of the negative-sense RNA from a promoter found in its 3′ terminus (sequence complementary to the 5′ terminus of the genome; Lemm *et al.*, 1994; Shirako and Strauss, 1994).

Sub-genome length, positive-sense RNAs, identical in sequence to the 3′ one-third of the genome and encoding the viral structural proteins, are transcribed from the internal NTR '26S' promoter at a 5- to 20-fold molar excess over genomic RNA (Strauss *et al.*, 1984). Finally, a 24-nt CSE comprising the internal NTR and overlapping the sub-genomic RNA transcript in the negative-strand forms the sub-genomic promoter and a 'translation enhancer element' in the subgenomic RNA (Frolov and Schlesinger, 1994; McInerney *et al.*, 2005; Ventoso *et al.*, 2006), discussed further below.

Virion assembly and egress

As the nascent structural polyprotein is produced, capsid autoproteolytically cleaves on the cytoplasmic side of the endoplasmic reticulum (ER) and assembles around progeny genomes by interaction with a highly conserved 'packaging signal' within the nsP1-encoding sequence to form the nucleocapsid structure (Kim *et al.*, 2011). The remaining polyprotein is processed and cleaved in the secretory pathway by host enzymes to yield the mature E1 and E2 glycoproteins anchored in the ER by transmembrane domains in preparation for envelopment of the nucleocapsid and budding from the plasma membrane. Final cleavage of the pE2 proteins into E2 and E3 occurs late in the cycle to protect against premature fusion with cellular membranes while exiting the cell (Martinez *et al.*, 2014; Uchime *et al.*, 2013).

Cellular detection of and response to alphavirus infection

Often, murine embryo fibroblasts (MEFs) are used to assess the roles of particular proteins in pathogen sensing and antiviral induction cascades because they can be readily manipulated. However, their use is problematic because control MEF cultures vary widely in their permissiveness and response to alphavirus infection (Burke *et al.*, 2009; Clavarino *et al.*, 2012), and are susceptible to clonal variation during passage and/or immortalization (e.g. Balachandran *et al.*, 2000; Wang *et al.*, 2009). In addition, when MEFs or other non-myeloid cultures relevant to pathogenesis (e.g. neurones or osteoblasts) are efficiently infected with a particular alphavirus they often fail to respond or release detectable type I IFN into supernatants, due to host macromolecular synthesis shut-off by the virus (discussed above). While this may reflect some of what occurs locally in the host organism, the cells are being studied in the absence of a network of other cell-types that may be essential to the overall *in vivo* phenotype, for example by low level type I IFN priming. Consequently, the findings of these studies should be interpreted with caution.

Detection of the invading alphavirus

Cells possess pathogen recognition receptors (PRRs) upstream of complex signalling pathway networks. When pathogen-associated molecular patterns (PAMPs) are detected, virus stress-inducible gene (VSIG) expression is induced (Sarkar and Sen, 2004) to defend against the invading virus and to sound the alarm to surrounding cells. In non-myeloid cells alphavirus PAMPs are detected by cytoplasmic PRRs including MDA5, PKR and possibly RIG-I (Barry *et al.*, 2009; Burke *et al.*, 2009). To-date, there is no clear evidence that cell-surface or endoplasmic toll-like receptors (TLRs) are involved. MDA5 and PKR are known to detect cytoplasmic double-stranded or highly structured RNAs, while RIG-I detects uncapped, 5′-triphosphorylated, single-stranded RNAs (Gitlin *et al.*, 2006; Hornung *et al.*, 2006; Pichlmair *et al.*, 2006). The early transcription of VISGs (including type I IFN genes) in response to alphavirus infection is partially dependent upon MDA5 and/or PKR (Barry *et al.*, 2009; Burke *et al.*, 2009; Ryman *et al.*, 2002, 2005; Schulz *et al.*, 2010). The presence of PKR augments type I IFN synthesis, at least in some circumstances, most likely via its role in maintaining IFN-β mRNA integrity (Schulz *et al.*, 2010), and/or its effects on levels of the NF-κB inhibitor, IκB-α (McAllister *et al.*, 2012). In contrast, detection of alphavirus infection by RIG-I significantly decreases type I IFN induction while eliciting antiviral activity against alphaviruses by an IFN-α/β induction-independent mechanism (Burke *et al.*, 2009; Schoggins and Rice, 2011).

Canonical cytosolic RNA virus-sensing pathways in non-myeloid cells converge via recruitment of IPS-1/CARDIF, activation of TBK1, TBK1-mediated phosphorylation, dimerization and nuclear translocation of constitutively expressed transcription factor, IFN regulatory factor (IRF)-3. Infections of various non-myeloid cell-types from mice and primates by wild-type alphaviruses result in IRF3 activation and nuclear accumulation with minimal interference by the virus (Breakwell *et al.*, 2007; Burke *et al.*, 2009; Peltier *et al.*, 2013; White *et al.*, 2011), although IRF3 activation may be delayed by the sequestration of alphavirus PAMPs within CPV structures thus avoiding detection early on. This activates transcription of VSIGs including IFN-β and IRF7 (Burke *et al.*, 2009; Grandvaux *et al.*, 2002; Schafer *et al.*, 1998; Schilte *et al.*, 2010; White *et al.*, 2011).

Downstream of IRF3 activation and translocation, findings diverge depending upon cell-type and virus. In SINV-infected cells, little/no transcription or translation of the IFN-β gene or other VSIGs are detected (Burke *et al.*, 2009; Garmashova *et al.*, 2006; Yin *et al.*, 2009) but this inhibition occurs coincident with global host macromolecular synthesis arrest, and thus specific inhibition of the IRF3 pathway, if present, cannot be distinguished (see below). However, IRF3-dependent genes including IFN-β are transcribed in CHIKV-, VEEV- and WEEV-infected cells in the early stages but the mRNAs are not translated (Peltier *et al.*, 2013; White *et al.*, 2011; Yin *et al.*, 2009), again in an environment of generalized protein synthesis arrest. Later in infection, VSIG transcription is also curtailed (White *et al.*, 2011). A non-cytopathic SINV strain with a mutation in nsP2 that abrogates the virus' ability to arrest host macromolecular synthesis (Frolov *et al.*, 1999) induces a robust type I IFN response in MEFs (Burke *et al.*, 2009).

Importantly, however, findings for non-myeloid cell infections are not consistent with type I IFN induction profiles for myeloid cell-tropic alphaviruses in cultured DCs or macrophages *in vitro* or *in vivo*. In particular, IFN-β gene expression in fibroblasts and neurons in response to alphavirus infection is extremely delayed compared with myeloid cell responses or detection of type I IFN in sera of infected animals (Ryman and Klimstra, 2008). Cells have complex networks of pathways that respond to a multitude of stressors by modulating the translation of its own mRNAs once they are synthesized. In some cell-types (e.g. DCs and macrophages), this provides a rapid response capability by releasing pre-synthesized mRNAs for translation. Extensive further studies are needed to unravel these networks and interactions in order to understand the *in vivo* situation.

Antagonism of host gene expression in infected cells

Alphaviruses are well known to interfere with cellular gene transcription (Aguilar *et al.*, 2007; Frolova *et al.*, 2002) and mRNA translation (Gorchakov *et al.*, 2005; Peranen *et al.*, 1990; Yin *et al.*, 2009) to varying degrees offering the viruses several advantages: (i) increased availability of host cell factors for translation of the viral RNA; (ii) inhibition/suppression of VSIG responses by the infected cell; and (iii) retardation of infected cell's ability to communicate its infected state to neighbouring cells [e.g. via the release of interferon (IFN)]. Of course this requires that the virus achieve translation of its own RNA in a translation-arrested environment, which will be discussed in the following section. Interestingly, these abilities seem to have been acquired coincident with replication in mammalian cells (Ventoso, 2012) as similar processes do not occur in infected mosquito viruses (Atasheva *et al.*, 2008) and the recently identified mosquito-restricted Eilat virus cannot achieve these effects in mammalian cells (Nasar *et al.*, 2015).

Inhibition of host gene transcription

The majority of alphaviruses studied so far appear to have the ability to inhibit host gene transcription in permissive cells. The importance of this ability is emphasized by the fact that the Old and New World alphaviruses appear to have independently evolved mechanisms to achieve the same goal. Because the alphavirus replication cycle does not require DNA-dependent RNA polymerase (DdRp)-mediated nuclear transcription of RNAs, the virus is unlikely to be adversely affected by suppression of this process. However, the timing/efficiency with which host transcription declines seems relatively late to have tremendous effect under physiological conditions.

For some of the arthritogenic alphaviruses (SFV, SINV), a large proportion of the nsP2 protein translocates to the nucleus during infection (Atasheva *et al.*, 2007; Breakwell *et al.*, 2007; Garmashova *et al.*, 2006; Peranen *et al.*; Rikkonen, 1996; Tamm *et al.*, 2008), often (although not always) associated with host cell macromolecular synthesis shut-off and induction of cytopathic effect (CPE) in highly permissive cell lines (Burke *et al.*, 2009; Garmashova *et al.*, 2006, 2007a,b). Old World SINV and CHIKV alphavirus nsP2 mutants that were selected for non-cytopathogenicity (Frolova *et al.*, 2002; Fros *et al.*, 2013), or an SFV mutant with a defective nsP2 nuclear localization signal (Tamm *et al.*, 2008) do not inhibit host cell transcription or translation activities, replicate poorly compared to their wild-type parents and induce type I IFN release (Burke *et al.*, 2009; Frolov *et al.*, 2002, 2009; Garmashova *et al.*, 2006). Inhibition of cell gene transcription by SINV, SFV and CHIKV nsP2 proteins may be associated with rapid ubiquitinylation and degradation of Rpb1, a catalytic subunit of the RNA polymerase (Pol) II complex (Akhrymuk *et al.*, 2012). At high multiplicity of infection (MOI) and in highly permissive fibroblast cell lines this occurs within six hours post-infection; however it remains to be determined whether or not this is critical for subversion of antiviral responses in more differentiated cells *in vitro* or *in vivo*, and to what extent the arrest of host translation, also mediated by nsP2, dominates host macromolecular shut-off.

In the New World alphaviruses, the capsid proteins are major mediators of host cell transcriptional shut-off *in vitro* (Aguilar *et al.*, 2007; Garmashova *et al.*, 2006, 2007a,b). The VEEV capsid protein is partially localized to the nuclear membrane (Garmashova *et al.*, 2007a), specifically to nuclear pore structures (Atasheva *et al.*, 2008). A tetrameric complex of capsid protein, importin-α/β, and the CRM1 nuclear export receptor appears to block nucleocytoplasmic trafficking of capsid and cellular proteins through the nuclear pore. This unusual complex accumulates in the centre of the nuclear pores and alters nuclear import of capsid (Lundberg *et al.*, 2013), coincident with the development of transcriptional shut-off (Atasheva *et al.*, 2008, 2010a,b). The effects of the capsid protein map to a positively charged amino terminal fragment in VEEV and EEEV, and viruses with small deletions in this domain lack the activity described above (Aguilar *et al.*, 2008; Garmashova *et al.*, 2007a,b). Although it has been demonstrated that the EEEV and VEEV capsid mutant viruses lacking transcription arrest capabilities are attenuated *in vivo* (Aguilar *et al.*, 2008; Reynaud *et al.*, 2015), it has not been confirmed that attenuation coincides with greater type I IFN induction or reduced transcription shut-off as posited.

Arrest of host mRNA translation

Infection of permissive cells by alphaviruses typically leads to the rapid arrest of host protein synthesis independently of gene transcription (Frolova *et al.*, 2002; Yin *et al.*, 2009). These are difficult to distinguish when studying nsP2 from the arthritogenic viruses, as similar domains within the protein impact both functions, and the mutation of nsP2 has strong secondary effects upon the kinetics of virus replication and accumulation of replicative intermediates (Burke *et al.*, 2009; Frolov *et al.*, 2009; Fros *et al.*, 2010). However, it has recently been shown that nsP2-mediated translation arrest occurs in the absence of other viral proteins or virus replication (Fros *et al.*, 2010, 2013). Additionally, data from CHIKV and SINV (Gorchakov *et al.*, 2005; White *et al.*, 2011; Yin *et al.*, 2009) indicate that translation shut-off occurs before transcription declines during infection, and these activities are mediated through distinct mechanisms. The viral protein involved in host translation arrest during New World alphavirus infection has not been determined. Interestingly, VEEV replicon particles that do not express the capsid or glycoproteins potently shut off translation in primary cortical neurone cultures in the absence of dramatic host transcription inhibition (Yin *et al.*, 2009) indicating a separate translation-inhibiting activity associated with replication and/or nsP expression of the encephalitic viruses which may be cell type dependent.

In vitro, particularly in commonly used, highly permissive, immortalized cell lines, the ability to arrest translation of host mRNAs is common to many viruses. Of the three major steps in host protein synthesis (initiation, elongation and termination), viruses typically interfere with translation initiation by commandeering or modifying important eukaryotic initiation factors (eIFs) to ensure an efficient translation of viral mRNAs and the simultaneous decline of host translation, in addition to blocking the synthesis of antiviral response proteins. As yet the alphaviruses have not been shown to target components of the cap-binding complex (eIF4F) that are required for the recruitment of ribosomes to mRNAs. Another important point of translation control in infected cells relies upon eIF2α-mediated recruitment of the initiator Met-tRNA to the 40S ribosomal subunit, which is blocked when eIF2α is phosphorylated by one of four stress-responsive kinases: PKR, PERK, GCN2 and/or HRI. Detection of dsRNA during virus infection by constitutively expressed, cytoplasmic PKR causes its activation, dimerization and phosphorylation eIF2α, blocking general translation in the cell. Many viruses prevent PKR activation (reviewed in Domingo-Gil *et al.*, 2011) but in cells infected with SINV or SFV viral mRNAs continue to be translated efficiently despite almost complete phosphorylation of eIF2α (Clavarino *et al.*, 2012; McInerney *et al.*, 2005). A hairpin loop located downstream from the AUG in viral subgenomic transcripts, allows these mRNAs to be translated by an eIF2α-independent mechanism in infected cells (Garcia-Moreno *et al.*, 2013; Ventoso *et al.*, 2006). It is not known whether the encephalitic alphaviruses use a similar mechanism.

This area of alphavirus–host interaction remains highly controversial as other groups have shown that, although the phosphorylation of eIF2α is delayed in immortalized PKR$^{-/-}$ cells compared with PKR+/+ counterparts, translation is arrested early (possibly earlier due to increased virus replication) and throughout infection via primarily eIF2α/PKR-independent mechanisms (Gorchakov *et al.*, 2004; White *et al.*, 2011). Furthermore, cell lines and primary cells lacking PKR are dramatically more permissive to virus infection and PKR$^{-/-}$ mice show increased susceptibility in the earliest stages of infection, although overall the absence of PKR has little effect on disease outcome (Barry *et al.*, 2009; Gorchakov *et al.*, 2004; Ryman *et al.*, 2005; Ryman *et al.*, 2002; Tesfay *et al.*, 2008; White *et al.*, 2011; Zhang *et al.*, 2007). In cells expressing only a non-phosphorylatable form of eIF2α, little translation arrest was observed (McInerney *et al.*, 2005), but these cells are highly refractory to infection. We conclude that the phosphorylation of eIF2α actually enhances SINV replication (and translation of the subgenomic RNA) and consequently the delayed phosphorylation of eIF2α in PKR$^{-/-}$ cells is not the mechanism of increased permissivity. Thus, it appears that PKR possesses alternative antiviral activity one of which has been shown to be a role in IFN induction by alphaviruses (Barry *et al.*, 2009; Burke *et al.*, 2009). It is tempting to speculate that this mechanism is related to the interaction of the eIF4E cap-binding protein with eIF4G as seen for another positive-sense RNA arbovirus, vesicular stomatitis virus (Connor and Lyles, 2002, 2005; Whitlow *et al.*, 2008), particularly as translation of the alphavirus subgenome does not require eIF4G (Castello *et al.*, 2006).

Relevance of host macromolecular synthesis arrest to pathogenesis *in vivo*

So far, the phenomena of transcription and translation arrest have been extensively studied in cultured cells, but not often in primary cells or in infected animals, so that evidence for virus-induced shut-off of host macromolecular synthesis arrest *in vivo* is lacking. *In vitro* experiments described above often require somewhat artifactual conditions, such as the use of immortalized cells lines highly susceptible to the virus and high multiplicities of infection; conditions that are difficult to find during virus replication in animals. Where the effects of certain pathways such as the eIF2α/PKR (Barry *et al.*, 2009; Ryman *et al.*, 2002, 2005) or eIF2α/GCN2 (Berlanga *et al.*, 2006), effects have been relatively minor. Overall, the intimate balance between virus replication and host stress responses that are curtailed by shut-off confounds these studies (Frolov *et al.*, 2012).

Although it remains to be determined if the IFN induction inhibitory capacity of the alphavirus nsP or capsid proteins in cultured cells is an indirect effect of a global block to transcription and/or translation or if more specific effects are involved, it appears that global arrest of translation is typically evident *before* substantial effects on gene transcription, with the kinetics varying between cell-types. Further characterization of the mechanism through which nsP2 and capsid proteins affect gene expression and the careful dissection of the apparently multiple antagonistic activities of nsP2 and possibly capsid (Frolov *et al.*, 2012; Peltier *et al.*, 2013) will be extremely informative regarding this issue.

Disruption/avoidance of host cell signaling pathways

Both arthritogenic and encephalitic alphaviruses can interfere with the type I IFN signalling through its receptor and the STAT1-dependent pathway in infected cells (Aguilar *et* al., 2008; Fros et al., 2010; Simmons et al., 2009; Yin et al., 2009). Most studies have concluded that type I IFN response antagonism and general disruption of host cell macromolecular synthesis are overlapping activities. However, for the SFV nsP2 mutant with minimal nuclear translocation effects upon host transcription early after infection were similar to those elicited by the wild-type virus (Peranen et al., 1990), and a reduced rate of viral RNA and protein accumulation (Breakwell et al., 2007). Notably, wild-type SFV also blocked accumulation of the TNF-α protein in cultured cell supernatants (but the mutant did not), implying an effect upon host responses in addition to the IFN pathways, but did not interfere with nuclear translocation of IRF3 or NF-κB transcription factors (Breakwell et al., 2007). It is important to note that similar domains of nsP2 have been implicated in alphavirus infection-mediated blockade of STAT1-dependent signal transduction of autocrine type I IFN responses (Fros et al., 2010; Simmons et al., 2009; Yin et al., 2009).

IFN-inducible and other antiviral effectors

Protection of cells by type I IFN priming

Exposure of cells to IFN-α/β prior to virus infection (known as priming) triggers an antiviral response *in vivo* that can be mimicked *in vitro*. It has emerged that the effectors of the refractory state are many and complex (Schoggins et al.; Zhang et al., 2007). Alphaviruses display differential sensitivity to the pre-existing antiviral state. We have perhaps learned the most about anti-alphaviral effectors from the highly type I IFN-sensitive SINV (Chan et al., 2008; Lenschow et al., 2005; MacDonald et al., 2007; Ryman et al., 2002, 2005; Tesfay et al., 2008; Zhang et al., 2007) as this virus clearly exhibits so little ability to withstand/antagonize IFN responses, particularly *in vivo* (Ryman et al., 2000). SINV infection of mice lacking the type I IFN receptor (IFNAR1$^{-/-}$) is fatal within 2–3 days (Ryman et al., 2000).

In the last ten years a number of IFN-stimulated gene (ISG) products have been shown to have antiviral activity against one or more alphaviruses *in vitro* and/or *in vivo* (Bick et al., 2003; Hyde et al., 2014; Lenschow et al., 2005; Schoggins et al., 2011; Zhang et al., 2007) and are discussed further below.

PKR and eIF2α pathways

As discussed above, constitutively expressed and modestly IFN-inducible PKR is activated and/or eIF2α is phosphorylated in SINV (Gorchakov et al., 2004; Ryman et al., 2002, 2005; Sanz et al., 2009; Ventoso et al., 2006), SFV (Barry et al., 2009; McInerney et al., 2005; Ventoso et al., 2006), CHIKV (Clavarino et al., 2012; White et al., 2011) and EEEV (Aguilar et al., 2008) infected cells. The PKR/eIF2α pathway of translation control has primarily been studied as a contributor to cell protein arrest, which is generally considered pro-viral, freeing host translation machinery for synthesis of alphavirus proteins and suppressing translation of IFNs, VSIGs and ISGs. However, this pathway appears to play some antiviral role early in infection both *in vitro* and *in vivo* (Barry et al., 2009; Ryman et al., 2002), and as described above augments type I IFN production from infected cells even in the absence of global translation shut-off. Thus, the complex and often contradictory interactions between PKR/eIF2α and alphaviruses remain controversial and incompletely understood. Overall, potent IFN-α/β-inducible inhibition of alphavirus replication occurs in the absence of PKR or other eIF2α kinases suggesting limited importance (Gorchakov et al., 2004; Ryman et al., 2002, 2005; Tesfay et al., 2008; White et al., 2011).

The IFIT protein family

Although the IFIT family proteins are highly induced in response to type I IFN treatment and by certain viral PAMPs via IRF3 activation (reviewed by (Hyde et al., 2014; Zhou et al., 2013), only the IFIT1 protein (also known as p56 or ISG56) confers significant antiviral activity against SINV and the effects of overexpression in untreated cells, siRNA-mediated knockdown in IFN-α/β-primed cells or IFIT1 deficiency (IFIT1$^{-/-}$ cells) *in vitro* are modest (Zhang et al., 2007), whereas wild-type strains of VEEV and CHIKV are essentially resistant to this antiviral effector. As discussed above, alphavirus RNA genomes lack the 2′ O methylation of the 5′ cap structure used by some viruses to avoid IFIT1-mediated inhibition and masquerade as host mRNAs (Daffis et al., 2010; Hyde et al., 2014; Szretter et al., 2012), but have instead acquired secondary structure motifs in the 5′ NTR that attenuate IFIT1 binding and antiviral function (Hyde et al., 2014). The introduction of point mutations into the 5′ NTRs of wild-type SINV and VEEV, predicted to weaken the first stem–loop structure in the genome (SL1) and known to attenuate these viruses *in vivo* (McKnight et al., 1996; White et al., 2001), dramatically increased IFIT1 sensitivity, at least in part via inhibition of genome translation (Hyde et al., 2014). IFIT1 is also known to sequester single-stranded viral RNAs with 5′ triphosphates in a complex with other IFIT family members (Pichlmair et al., 2011), a process for which alphaviral negative-strand replicative intermediates could be a target, and is involved in regulation of eIF4F translational activity through interaction with eIF3 (Guo et al., 2000; Hui et al., 2003, 2005). Whether or not IFIT1 also mediates antiviral activity via these mechanisms remains to be tested.

Ubiquitin-like ISG15

ISG15 encodes a ubiquitin-like protein that functions as a modifier of protein function (Zhao et al., 2013). ISG15 conjugates to target proteins via isopeptide bonding with a series of other IFN-α/β-inducible enzymes: E1-activating enzyme, UbE1L; E2-conjugating enzyme UbcH8/UbcM8; E3 ligase enzymes Herc5, HHARI, and Efp; and deconjugating enzyme, UBP43 (reviewed by Zhao et al., 2013). There is accumulating evidence implicating ISG15 as a highly IFN-inducible protein with broad-spectrum antiviral activity, inhibiting by conjugation (or 'ISGylation') and also as a free, unconjugated protein (reviewed by (Lenschow, 2010). While ISG15 has substantial antiviral activity against SINV (Giannakopoulos et al., 2009; Lenschow et al., 2005, 2007; Zhang et al., 2007) and CHIKV (Werneke et al., 2011) *in vivo*, it has little to no effect *in vitro* (Zhang et al., 2007). ISG15 does not directly inhibit SINV or CHIKV replication *in vivo*, but instead seems to perform an immunomodulatory role in keeping with the lack of efficacy *in vitro*. Surprisingly, the mechanism of action by which ISG15 limits SINV and CHIKV infection differ substantially (Giannakopoulos et al., 2009; Ritchie et al., 2004; Werneke et al., 2011).

ISG20 3′–5′ exoribonuclease

Strongly IFN-α/β-inducible ISG20 is an antiviral effector with an apparent preference for RNA viruses (Espert *et al.*, 2003, 2005; Jiang *et al.*, 2008; Zhang *et al.*, 2007). ISG20 is a member of the DEDDh subgroup of the DEDD exonuclease superfamily (Espert *et al.*, 2003; Horio *et al.*, 2004) with 3′–5′ exonuclease activity, and a preference for single-stranded RNA substrates. The mechanism by which ISG20 interferes with virus replication is not known but, as little/no antiviral effect was observed in cells overexpressing a mutated ISG20 protein defective in exonuclease activity, it is tempting to speculate that ISG20 controls virus replication by degrading viral RNA via its 3′–5′ exoribonuclease activity. However, 3′-terminal stem–loop structures in the RNA substrate (perhaps such as those in the alphavirus genomic RNAs) strongly reduce its enzymatic activity (Espert *et al.*, 2003). Moreover, we have shown indirectly that viral RNA extracted from un-primed or IFN-primed (ISG20-expressing) cells and introduced into unprimed cells was equally infectious and therefore intact (Tesfay *et al.*, 2008) and others have suggested that global mRNA is not altered (Espert *et al.*, 2003; Zhou *et al.*, 2011). Identification of ISG20 RNA targets in virus-infected cells thus remains a major challenge for the understanding and exploitation of its mechanism of antiviral action and for the elucidation of ways in which relatively IFN-α/β-resistant viruses might circumvent its activity.

Zinc finger antiviral protein (ZAP)

The zinc finger antiviral protein (ZAP) is a highly inducible ISG and VISG-protein (Ryman *et al.*, 2005; Wang *et al.*, 2010), with strong antiviral activity against multiple alphaviruses including SINV, SFV, RRV and VEEV (Bick *et al.*, 2003; Kerns *et al.*, 2008; Zhang *et al.*, 2007). ZAP is binds to viral RNA (Chen *et al.*, 2012; Guo *et al.*, 2004) leading to its exosome-mediated degradation (Guo *et al.*, 2007; Zhu and Gao, 2008; Zhu *et al.*, 2012). Via this mechanism, ZAP blocks an early step after entry and at or before the synthesis of the viral polyproteins (Bick *et al.*, 2003). Two distinct isoforms of ZAP are expressed by alternative splicing: the long isoform has the greater anti-alphaviral activity (Kerns *et al.*, 2008) and synergizes with other ISGs to confer maximal protection against alphavirus infections (Karki *et al.*, 2012). Thus, ZAP is pivotal in the cellular response to alphaviral infection at several levels, although its activity *in vivo* has yet to be tested.

Inhibition of alphavirus replication by host microRNAs

The 3′ NTR of the NA-EEEV RNA genome contains binding site sequences for the myeloid-specific microRNA (miRNA), miR-142-3p (Trobaugh *et al.*, 2014a). Binding of the miRNA represses viral translation and replication in myeloid cells altering the virus tropism. Removal of the microRNA binding sites rescues replication in miR-142-3p expressing cells both *in vitro* and *in vivo* (Trobaugh *et al.*, 2014a). Currently, it is unclear whether or not other alphaviruses utilize miRNA inhibition to determine tissue tropism *in vivo* but evidence has been presented that the 3′NTRs of other alphaviruses are involved in cell and host tropism and may be important virulence factors.

Host–alphavirus interactions in pathogenesis

In this final section, the pathogenesis sequences of the most studied alphaviruses are reviewed in the light of the information provided above.

Pathogenesis of encephalitic alphaviruses

Wild-type strains of VEEV, EEEV and WEEV cause fatal disease in adult immunocompetent mice following subcutaneous inoculation (Aguilar *et al.*, 2005; Anishchenko *et al.*, 2004; Davis *et al.*, 1994; Davis *et al.*, 1989; Gardner *et al.*, 2008; Steele and Twenhafel, 2010), modelling the most severe human disease state. VEEV pathogenesis has been studied the most extensively and it has been assumed, incorrectly in retrospect, that the pathogenesis of EEEV and WEEV would be similar. We will review first the pathogenesis of VEEV and then the ways in which the pathogenesis of EEEV in particular has been shown to differ.

VEEV pathogenesis

For wild-type strains of VEEV (both epizootic I-A/B and enzootic I-D) infection by mosquito bite is mimicked in mice by deposition of virus in the subcutaneous tissues. The virus replicates at the site of inoculation, and disseminates very quickly to the lymph nodes that drain the inoculation site (DLNs). In modern studies, mice are inoculated subcutaneously, usually in the hind-leg footpad, and the earliest infected cells are found primarily in the footpad area, in the popliteal (PLN) behind the knee and in the inguinal LN (Aronson *et al.*, 2000; MacDonald and Johnston, 2000). Studies demonstrating this first step in virus spread using traditional virus titration/IHC approaches (Gardner *et al.*, 2008; Grieder and Nguyen, 1996; MacDonald and Johnston, 2000) can now be confirmed using *in vivo* imaging technologies (Gardner *et al.*, 2008). VEEV infects LN cells incredibly efficiently and causes the LN to greatly increase in size suggesting infiltrates but major questions remain: (i) exactly what cell-type(s) are infected in the DLN (and later in other lymphoid tissues); and (ii) do these cells become infected at the site of inoculation and subsequently migrate to the DLN and/or does free virus travel via the lymphatics and infect LN-resident cells that are designed to filter out foreign materials. MacDonald *et al.* first described infection by VEEV of DCs and/or Langerhans' cells (LCs) (MacDonald and Johnston, 2000). VEEV amplifies further in the DLN and at the site of inoculation, seeding a primary serum viraemia presumably by release of virus into the efferent lymph, which drains into the circulatory system via the thoracic duct. By this mechanism, high levels of cell-free, infectious virus disseminate systemically and infect tissues distant from the site of inoculation.

During this early phase of infection, high levels of proinflammatory cytokines including type I and type II IFN and IL-12 p70, and chemokines CCL-2 [monocyte chemotactic protein 1 (MCP-1)], CXCL9 [monokine-induced by IFN-γ (MIG)] and CXCL10 [IFN-inducible protein 10 (IP-10)] are released into the bloodstream (Gardner *et al.*, 2011). Both LN and spleen have been implicated as sites of production for various cytokines (Grieder *et al.*, 1997) and lymphoid tissues are destroyed by the infection (Grieder *et al.*, 1995). Innate immune cytokine and chemokine induction within infected lymphoid tissue cells has the potential to influence permissivity of cells distant from the site of infection (e.g. by type IIFN priming) or access of virus to certain tissues (e.g. by permeabilization of blood vessels and the blood–brain barrier (Gardner *et al.*, 2009; Konopka *et al.*, 2009). Extremely high levels of type I IFN are induced within a few hours of VEEV infection but have minimal effect (Gardner *et al.*, 2008; White *et al.*, 2001).

The BBB is formed by BMECs joined by extremely tight junctions via their interaction with perivascular feet of astrocytes. This tight layer partitions the brain from the blood, regulates blood–brain exchange of substances and cells, maintains CNS homeostasis, and helps protect the brain from pathological insults such as neuroinvasion by blood-borne, neurotropic viruses. Pivotal to virulence, VEEV can 'cross' the BBB, invade the CNS, and induce brain pathologies. Less than two days following subcutaneous infection of mice with VEEV, virus infects olfactory sensory neurons in the neuroepithelium and rapidly invades the brain at multiple sites, effectively circumventing the BBB (Charles *et al.*, 1995; Schafer *et al.*, 2011). Virions are released into the nasal passages, perhaps from infected neuroepithelial or cervical LN-resident macrophages and DCs. It has recently been posited that initial infection of a few olfactory sensory neurons indirectly increases BBB permeability via induction of MMP-9, facilitating neuroinvasion by a massive second wave of viraemic virus (Schafer *et al.*, 2011). Both chemical destruction of the neuroepithelium (Charles *et al.*, 1995) and non-invasive drug treatment to reduce MMP-9 levels (Schafer *et al.*, 2011) delay VEEV neuroinvasion and improve prognosis.

EEEV pathogenesis

Similar to human infections, wild-type NA-EEEV causes extremely severe neurological disease in subcutaneously inoculated adult mice (Aguilar *et al.*, 2005; Gardner *et al.*, 2008; Vogel *et al.*, 2005). NA-EEEV-infected mice tend not to develop clinical signs of febrile illness early after inoculation, but progress directly to the neurological stages of disease, similar to severe human cases. Although viral antigen is detectable in the DLN, infection of DLN cells is rare (Gardner *et al.*, 2008; Vogel *et al.*, 2005). As mentioned above, this virus fails to produce a dramatic febrile response in infected humans or mice as a result of the limited replication of this virus in cells associated with febrile responses (i.e. DCs and macrophages; (Gardner *et al.*, 2008; Vogel *et al.*, 2005), due to the ability of myeloid cell-specific miR142–5p to suppress EEEV replication (Trobaugh *et al.*, 2014a). NA-EEEV amplifies primarily in osteoblast-lineage cells of the bone endosteum and periosteum, and in muscle, while sparing myeloid cell-types such as DCs and macrophages, and lymphoid tissues in general (Gardner *et al.*, 2008; Vogel *et al.*, 2005). This tropism is readily reproduced *in vitro* in primary cell cultures (Gardner *et al.*, 2008) and appears to be a strategy by which this virus avoids antiviral and proinflammatory immune induction (Gardner *et al.*, 2008, 2011; Trobaugh *et al.*, 2014a). A relatively low-level serum viraemia is seeded, but nevertheless neuroinvasion occurs and the virus replicates and spreads extremely efficiently within CNS tissues, causing extensive damage (Gardner *et al.*, 2008, 2013; Vogel *et al.*, 2005). In contrast to VEEV, the neuroepithelium and olfactory sensory neurons are spared and infected cells appear simultaneously in the brain parenchyma with widespread distribution (Gardner *et al.*, 2013; Vogel *et al.*, 2005), suggestive of direct invasion across the BBB by increasing tight junction permeability and/or by infecting BMECs. Importantly, the difference in neuroinvasion routes may be due to failure of EEEV to infect lymphoid tissue/myeloid-lineage cells due to HS binding (Gardner *et al.*, 2011) and host microRNA-dependent growth suppression (Trobaugh *et al.*, 2014a) that result in limited proinflammatory cytokine/chemokine induction (Gardner *et al.*, 2011).

Pathogenesis of arthritogenic alphaviruses

The arthritogenic alphaviruses tend to be attenuated or even avirulent in adult mice, causing mild or subclinical infection and eliciting a protective immune response. However, the severity of SINV, SFV, RRV, CHIKV and ONNV infection increases dramatically in younger animals or in animals deficient in type I IFN-mediated innate immune responses.

SINV pathogenesis

In adult mice, joint-associated tissues become largely refractory to SINV replication, but in the absence of an IFN-α/β response the virus replicates to similar levels as in neonates, causing extensive pathology (Ryman *et al.*, 2000). These data are indicative that the joint involvement in human infections may be the result of IFN-α/β response evasion/antagonism by the arthritogenic alphaviruses, which does not occur in mice. Wild-type SINV infection in neonatal mice results in a rapidly fatal, systemic inflammatory response syndrome (SIRS; Klimstra *et al.*, 1999). A similar disease was observed in adult IFNAR1$^{-/-}$ mice, indicating the critical role of this antiviral response in protection of mature animals (Ryman *et al.*, 2000). In adult mice lacking both the type I and type II (IFN-γ) IFN receptors, we have observed evidence of fibrin deposition and haemorrhage in tissue samples suggesting vascular disease is also a component of virulent, uncontrolled SINV infection (Ryman *et al.*, 2007b). Furthermore, we have demonstrated that SINV-induced SIRS diminishes with development of the host or cell culture-adaptation of the virus, coincident with attenuation of virus virulence.

The acquired ability to bind HS and efficiently infect cultured cells due to the presence of positively charged mutations in E2 is inversely correlated with SINV virulence *in vivo* following subcutaneous inoculation (Klimstra *et al.*, 1998, 1999; McKnight *et al.*, 1996; Byrnes and Griffin, 2000). While a systemic inflammatory disease is a major feature of neonatal mouse pathogenesis induced by TR339, encephalitis is more likely to be observed when the systemic pathology is ameliorated by cell culture-adaptive mutations that confer HS binding (Klimstra *et al.*, 1999). The acquisition of HS-binding capability appears to promote non-productive binding and sequestration by HS structures exposed to virus in serum, limiting viraemic potential and dissemination (Klimstra *et al.*, 1999; Klimstra *et al.*, 1998) and can be rapidly selected against *in vivo* in viruses isolated from the blood (Byrnes and Griffin, 2000).

CHIKV pathogenesis

Adult wild-type mice develop local symptoms when inoculated subcutaneously with wild type CHIKV in the hind footpad (Gardner *et al.*, 2010, 2012;). Younger mice (Morrison *et al.*, 2011) or mice with a partial or complete defect in the type I IFN pathway develop more severe musculoskeletal disease (MSD) (Gardner *et al.*, 2012). Susceptible mice develop viraemia in the first few days after infection. Skeletal muscles and joints have high viral load and pathological changes are also observed in joint-associated connective tissues adjacent to affected muscles. CHIKV RNA can persist in joint-associated tissues for at least four months (Hawman *et al.*, 2013). Interestingly, CHIKV is detected largely in mesenchymal not myeloid lineage cells (Couderc *et al.*, 2008), in keeping with *in vitro* studies (Gardner *et al.*, 2012). Infiltration of monocyte-macrophages into the joint, attracted by MCP-1, contributes to bone erosion and swelling (Chen *et al.*, 2014; Rulli *et al.*, 2011). In contradiction, however, CHIKV-infection of mice lacking the MCP-1 receptor (CCR2) was more severe due to an excessive recruitment of neutrophils to the inflamed joint (Poo *et al.*, 2014a). CHIKV-specific CD4+ (but not CD8+) T cells appear to contribute to MSD and joint swelling but do not control virus replication and dissemination (Hawman *et al.*, 2013; Teo *et al.*, 2013). Mice that lack B cells are unable to clear CHIKV viraemia with increased disease severity (Lum *et al.*, 2013; Poo *et al.*, 2014a; Poo *et al.*, 2014b). As yet, although there appear to be many parallels to human infection, the chronic, flaring polyarthritis seen in humans is not observed in mice.

Future perspectives

Our knowledge of the alphavirus–host interaction continues to expand and deepen, aided by knew molecular and imaging techniques. However, much remains to be resolved and we still lack any licensed antivirals, therapeutics or vaccines. The last few years have revealed exciting new areas of investigation such as the direct impact of host microRNAs on alphavirus pathogenesis and identification of specific antiviral effectors and their mechanisms of action; knowledge that can be readily converted into methods for rational virus attenuation (Trobaugh *et al.*, 2014a), and vulnerable points in the virus life-cycle that can be targeted with drugs. In particular, we are learning ever more about the differences between individual alphaviruses and the underpinning molecular reasons for them, as well as the importance of studying wild type viruses, and *in vivo* systems as much as possible.

References

Adams, A.P., Navarro-Lopez, R., Ramirez-Aguilar, F.J., Lopez-Gonzalez, I., Leal, G., Flores-Mayorga, J.M., Travassos da Rosa, A.P., Saxton-Shaw, K.D., Singh, A.J., Borland, E.M., *et al.* (2012). Venezuelan equine encephalitis virus activity in the Gulf Coast region of Mexico, 2003–2010. PLoS Neglect. Trop. Dis. *6*, e1875.

Aguilar, P.V., Paessler, S., Carrara, A.S., Baron, S., Poast, J., Wang, E., Moncayo, A.C., Anishchenko, M., Watts, D., Tesh, R.B., *et al.* (2005). Variation in interferon sensitivity and induction among strains of eastern equine encephalitis virus. J. Virol. *79*, 11300–11310.

Aguilar, P.V., Weaver, S.C., and Basler, C.F. (2007). Capsid protein of eastern equine encephalitis virus inhibits host cell gene expression. J. Virol. *81*, 3866–3876.

Aguilar, P.V., Leung, L.W., Wang, E., Weaver, S.C., and Basler, C.F. (2008). A five-amino-acid deletion of the eastern equine encephalitis virus capsid protein attenuates replication in mammalian systems but not in mosquito cells. J. Virol. *82*, 6972–6983.

Aguilar, P.V., Estrada-Franco, J.G., Navarro-Lopez, R., Ferro, C., Haddow, A.D., and Weaver, S.C. (2011). Endemic Venezuelan equine encephalitis in the Americas: hidden under the dengue umbrella. Future Virol. *6*, 721–740.

Ahola, T., and Kaariainen, L. (1995). Reaction in alphavirus mRNA capping: formation of a covalent complex of non-structural protein nsP1 with 7-methyl-GMP. Proc. Natl. Acad. Sci. U.S.A. *92*, 507–511.

Akhrymuk, I., Kulemzin, S.V., and Frolova, E.I. (2012). Evasion of the innate immune response: the Old World alphavirus nsP2 protein induces rapid degradation of Rpb1, a catalytic subunit of RNA polymerase II. J. Virol. *86*, 7180–7191.

Anishchenko, M., Paessler, S., Greene, I.P., Aguilar, P.V., Carrara, A.S., and Weaver, S.C. (2004). Generation and characterization of closely related epizootic and enzootic infectious cDNA clones for studying interferon sensitivity and emergence mechanisms of Venezuelan equine encephalitis virus. J. Virol. *78*, 1–8.

Aronson, J.F., Grieder, F.B., Davis, N.L., Charles, P.C., Knott, T., Brown, K., and Johnston, R.E. (2000). A single-site mutant and revertants arising in vivo define early steps in the pathogenesis of Venezuelan equine encephalitis virus. Virology *270*, 111–123.

Atasheva, S., Gorchakov, R., English, R., Frolov, I., and Frolova, E. (2007). Development of Sindbis viruses encoding nsP2/GFP chimeric proteins and their application for studying nsP2 functioning. J. Virol. *81*, 5046–5057.

Atasheva, S., Garmashova, N., Frolov, I., and Frolova, E. (2008). Venezuelan equine encephalitis virus capsid protein inhibits nuclear import in Mammalian but not in mosquito cells. J. Virol. *82*, 4028–4041.

Atasheva, S., Fish, A., Fornerod, M., and Frolova, E.I. (2010a). Venezuelan equine Encephalitis virus capsid protein forms a tetrameric complex with CRM1 and importin alpha/beta that obstructs nuclear pore complex function. J. Virol. *84*, 4158–4171.

Atasheva, S., Krendelchtchikova, V., Liopo, A., Frolova, E., and Frolov, I. (2010b). Interplay of acute and persistent infections caused by Venezuelan equine encephalitis virus encoding mutated capsid protein. J. Virol. *84*, 10004–10015.

Balachandran, S., Roberts, P.C., Kipperman, T., Bhalla, K.N., Compans, R.W., Archer, D.R., and Barber, G.N. (2000). Alpha/beta interferons potentiate virus-induced apoptosis through activation of the FADD/Caspase-8 death signaling pathway. J. Virol. *74*, 1513–1523.

Barry, G., Breakwell, L., Fragkoudis, R., Attarzadeh-Yazdi, G., Rodriguez-Andres, J., Kohl, A., and Fazakerley, J.K. (2009). PKR acts early in infection to suppress Semliki Forest virus production and strongly enhances the type I interferon response. J. Gen. Virol. *90*, 1382–1391.

Bergqvist, J., Forsman, O., Larsson, P., Naslund, J., Lilja, T., Engdahl, C., Lindstrom, A., Gylfe, A., Ahlm, C., Evander, M., *et al.* (2015). Detection and isolation of Sindbis virus from mosquitoes captured during an outbreak in Sweden, 2013. Vector Borne Zoonotic Dis. *15*, 133–140.

Bergren, N.A., Auguste, A.J., Forrester, N.L., Negi, S.S., Braun, W.A., and Weaver, S.C. (2014). Western equine encephalitis virus: evolutionary analysis of a declining alphavirus based on complete genome sequences. J. Virol. *88*, 9260–9267.

Berlanga, J.J., Ventoso, I., Harding, H.P., Deng, J., Ron, D., Sonenberg, N., Carrasco, L., and de Haro, C. (2006). Antiviral effect of the mammalian translation initiation factor 2alpha kinase GCN2 against RNA viruses. EMBO J. *25*, 1730–1740.

Bernard, K.A., Klimstra, W.B., and Johnston, R.E. (2000). Mutations in the E2 glycoprotein of Venezuelan equine encephalitis virus confer heparan sulfate interaction, low morbidity, and rapid clearance from blood of mice. Virology *276*, 93–103.

Bick, M.J., Carroll, J.W., Gao, G., Goff, S.P., Rice, C.M., and MacDonald, M.R. (2003). Expression of the zinc-finger antiviral protein inhibits alphavirus replication. J. Virol. *77*, 11555–11562.

Bigler, W.J., Ventura, A.K., Lewis, A.L., Wellings, F.M., and Ehrenkranz, N.J. (1974). Venezuelan equine encephalomyelitis in Florida: endemic virus circulation in native rodent populations of Everglades hammocks. Am. J. Trop. Med. Hyg. *23*, 513–521.

Bonilauri, P., Bellini, R., Calzolari, M., Angelini, R., Venturi, L., Fallacara, F., Cordioli, P., Angelini, P., Venturelli, C., Merialdi, G., *et al.* (2008). Chikungunya virus in *Aedes albopictus*, Italy. Emerg. Infect. Dis. *14*, 852–854.

Brault, A.C., Powers, A.M., Medina, G., Wang, E., Kang, W., Salas, R.A., De Siger, J., and Weaver, S.C. (2001). Potential sources of the 1995 Venezuelan equine encephalitis subtype IC epidemic. J. Virol. *75*, 5823–5832.

Brault, A.C., Powers, A.M., Holmes, E.C., Woelk, C.H., and Weaver, S.C. (2002). Positively charged amino acid substitutions in the e2 envelope glycoprotein are associated with the emergence of Venezuelan equine encephalitis virus. J. Virol. *76*, 1718–1730.

Breakwell, L., Dosenovic, P., Karlsson Hedestam, G.B., D'Amato, M., Liljestrom, P., Fazakerley, J., and McInerney, G.M. (2007). Semliki

Forest virus non-structural protein 2 is involved in suppression of the type I interferon response. J. Virol. *81*, 8677–8684.
Brehin, A.C., Rubrecht, L., Navarro-Sanchez, M.E., Marechal, V., Frenkiel, M.P., Lapalud, P., Laune, D., Sall, A.A., and Despres, P. (2008). Production and characterization of mouse monoclonal antibodies reactive to Chikungunya envelope E2 glycoprotein. Virology *371*, 185–195.
Burke, C.W., Gardner, C.L., Steffan, J.J., Ryman, K.D., and Klimstra, W.B. (2009). Characteristics of alpha/beta interferon induction after infection of murine fibroblasts with wild-type and mutant alphaviruses. Virology *395*, 121–132.
Byrnes, A.P., and Griffin, D.E. (2000). Large-plaque mutants of Sindbis virus show reduced binding to heparan sulfate, heightened viremia, and slower clearance from the circulation. J. Virol. *74*, 644–651.
Caglioti, C., Lalle, E., Castilletti, C., Carletti, F., Capobianchi, M.R., and Bordi, L. (2013). Chikungunya virus infection: an overview. N. Microbiologica *36*, 211–227.
Calisher, C.H., Emerson, J.K., Muth, D.J., Lazuick, J.S., and Monath, T.P. (1983). Serodiagnosis of western equine encephalitis virus infections: relationships of antibody titer and test to observed onset of clinical illness. J. Am. Vet. Med. Assoc. *183*, 438–440.
Carey, D.E. (1971). Chikungunya and dengue: a case of mistaken identity? J. Hist. Med. Allied Sci. *26*, 243–262.
Carrera, J.P., Forrester, N., Wang, E., Vittor, A.Y., Haddow, A.D., Lopez-Verges, S., Abadia, I., Castano, E., Sosa, N., Baez, C., *et al.* (2013). Eastern equine encephalitis in Latin America. N. Engl. J. Med. *369*, 732–744.
Castello, A., Sanz, M.A., Molina, S., and Carrasco, L. (2006). Translation of Sindbis virus 26S mRNA does not require intact eukariotic initiation factor 4G. J. Mol. Biol. *355*, 942–956.
Centers for Disease Control and Prevention (2006). Chikungunya fever diagnosed among international travelers –United States, 2005–2006. MMWR *55*, 1040–1042.
Chan, Y.L., Chang, T.H., Liao, C.L., and Lin, Y.L. (2008). The cellular antiviral protein viperin is attenuated by proteasome-mediated protein degradation in Japanese encephalitis virus-infected cells. J. Virol. *82*, 10455–10464.
Charles, P.C., Walters, E., Margolis, F., and Johnston, R.E. (1995). Mechanism of neuroinvasion of Venezuelan equine encephalitis virus in the mouse. Virology *208*, 662–671.
Chatterjee, P.K., Eng, C.H., and Kielian, M. (2002). Novel mutations that control the sphingolipid and cholesterol dependence of the Semliki Forest virus fusion protein. J. Virol. *76*, 12712–12722.
Chen, S., Xu, Y., Zhang, K., Wang, X., Sun, J., Gao, G., and Liu, Y. (2012). Structure of N-terminal domain of ZAP indicates how a zinc-finger protein recognizes complex RNA. Nature Struct. Mol. Biol. *19*, 430–435.
Chen, W., Foo, S.S., Rulli, N.E., Taylor, A., Sheng, K.C., Herrero, L.J., Herring, B.L., Lidbury, B.A., Li, R.W., Walsh, N.C., *et al.* (2014). Arthritogenic alphaviral infection perturbs osteoblast function and triggers pathologic bone loss. Proc. Natl. Acad. Sci. U.S.A. *111*, 6040–6045.
Chenier, S., Cote, G., Vanderstock, J., Macieira, S., Laperle, A., and Helie, P. (2010). An eastern equine encephalomyelitis (EEE) outbreak in Quebec in the fall of 2008. Can. Vet. J. *51*, 1011–1015.
Chua, C.L., Chan, Y.F., and Sam, I.C. (2014). Characterisation of mouse monoclonal antibodies targeting linear epitopes on Chikungunya virus E2 glycoprotein. J. Virol. Methods *195*, 126–133.
Clavarino, G., Claudio, N., Couderc, T., Dalet, A., Judith, D., Camosseto, V., Schmidt, E.K., Wenger, T., Lecuit, M., Gatti, E., *et al.* (2012). Induction of GADD34 is necessary for dsRNA-dependent interferon-beta production and participates in the control of Chikungunya virus infection. PLoS Path. *8*, e1002708.
Coffey, L.L., Crawford, C., Dee, J., Miller, R., Freier, J., and Weaver, S.C. (2006). Serologic evidence of widespread everglades virus activity in dogs, Florida. Emerg. Infect. Dis. *12*, 1873–1879.
Coffey, L.L., Forrester, N., Tsetsarkin, K., Vasilakis, N., and Weaver, S.C. (2013). Factors shaping the adaptive landscape for arboviruses: implications for the emergence of disease. Future Microbiol. *8*, 155–176.
Connor, J.H., and Lyles, D.S. (2002). Vesicular stomatitis virus infection alters the eIF4F translation initiation complex and causes dephosphorylation of the eIF4E binding protein 4E-BP1. J. Virol. *76*, 10177–10187.
Connor, J.H., and Lyles, D.S. (2005). Inhibition of host and viral translation during vesicular stomatitis virus infection. eIF2 is responsible for the inhibition of viral but not host translation. J. Biol. Chem. *280*, 13512–13519.
Couderc, T., Chretien, F., Schilte, C., Disson, O., Brigitte, M., Guivel-Benhassine, F., Touret, Y., Barau, G., Cayet, N., Schuffenecker, I., *et al.* (2008). A mouse model for Chikungunya: young age and inefficient type-I interferon signaling are risk factors for severe disease. PLoS Path. *4*, e29.
Daffis, S., Szretter, K.J., Schriewer, J., Li, J., Youn, S., Errett, J., Lin, T.Y., Schneller, S., Zust, R., Dong, H., *et al.* (2010). 2′-O methylation of the viral mRNA cap evades host restriction by IFIT family members. Nature *468*, 452–456.
Davis, C.W., Nguyen, H.Y., Hanna, S.L., Sanchez, M.D., Doms, R.W., and Pierson, T.C. (2006). West Nile virus discriminates between DC-SIGN and DC-SIGNR for cellular attachment and infection. J. Virol. *80*, 1290–1301.
Davis, N.L., Willis, L.V., Smith, J.F., and Johnston, R.E. (1989). In vitro synthesis of infectious venezuelan equine encephalitis virus RNA from a cDNA clone: analysis of a viable deletion mutant. Virology *171*, 189–204.
Davis, N.L., Grieder, F.B., Smith, J.F., Greenwald, G.F., Valenski, M.L., Sellon, D.C., Charles, P.C., and Johnston, R.E. (1994). A molecular genetic approach to the study of Venezuelan equine encephalitis virus pathogenesis. Arch. Virol. Suppl. *9*, 99–109.
Deresiewicz, R.L., Thaler, S.J., Hsu, L., and Zamani, A.A. (1997). Clinical and neuroradiographic manifestations of eastern equine encephalitis. N. Engl. J. Med. *336*, 1867–1874.
Diaz, Y., Carrera, J.P., Cerezo, L., Arauz, D., Guerra, I., Cisneros, J., Armien, B., Botello, A.M., Arauz, A.B., Gonzalez, V., *et al.* (2015). Chikungunya virus infection: first detection of imported and autochthonous cases in Panama. Am. J. Trop. Med. Hyg. *92*, 482–485.
Domingo-Gil, E., Toribio, R., Najera, J.L., Esteban, M., and Ventoso, I. (2011). Diversity in viral anti-PKR mechanisms: a remarkable case of evolutionary convergence. PloS One *6*, e16711.
Espert, L., Degols, G., Gongora, C., Blondel, D., Williams, B.R., Silverman, R.H., and Mechti, N. (2003). ISG20, a new interferon-induced RNase specific for single-stranded RNA, defines an alternative antiviral pathway against RNA genomic viruses. The J. Biol. Chem. *278*, 16151–16158.
Espert, L., Degols, G., Lin, Y.L., Vincent, T., Benkirane, M., and Mechti, N. (2005). Interferon-induced exonuclease ISG20 exhibits an antiviral activity against human immunodeficiency virus type 1. The J. Gen. Virol. *86*, 2221–2229.
Fields, W., and Kielian, M. (2013). A key interaction between the alphavirus envelope proteins responsible for initial dimer dissociation during fusion. J. Virol. *87*, 3774–3781.
Fischer, D., Thomas, S.M., Suk, J.E., Sudre, B., Hess, A., Tjaden, N.B., Beierkuhnlein, C., and Semenza, J.C. (2013). Climate change effects on Chikungunya transmission in Europe: geospatial analysis of vector's climatic suitability and virus' temperature requirements. Int. J. Health Geographics *12*, 51.
Fischer, M., Staples, J.E., Arboviral Diseases Branch, N.C.f.E., and Zoonotic Infectious Diseases, C.D.C. (2014). Notes from the field: chikungunya virus spreads in the Americas – Caribbean and South America, 2013–2014. MMWR *63*, 500–501.
Flynn, D.C., Meyer, W.J., Mackenzie, J.M. Jr., and Johnston, R.E. (1990). A conformational change in Sindbis virus glycoproteins E1 and E2 is detected at the plasma membrane as a consequence of early virus–cell interaction. J. Virol. *64*, 3643–3653.
Forrester, N.L., Kenney, J.L., Deardorff, E., Wang, E., and Weaver, S.C. (2008). Western Equine Encephalitis submergence: lack of evidence for a decline in virus virulence. Virology *380*, 170–172.
Forrester, N.L., Palacios, G., Tesh, R.B., Savji, N., Guzman, H., Sherman, M., Weaver, S.C., and Lipkin, W.I. (2012). Genome-scale phylogeny of the alphavirus genus suggests a marine origin. J. Virol. *86*, 2729–2738.
Frolov, I., and Schlesinger, S. (1994). Translation of Sindbis virus mRNA: effects of sequences downstream of the initiating codon. J. Virol. *68*, 8111–8117.
Frolov, I., Agapov, E., Hoffman, T.A. Jr., Pragai, B.M., Lippa, M., Schlesinger, S., and Rice, C.M. (1999). Selection of RNA replicons capable of persistent noncytopathic replication in mammalian cells. J. Virol. 73, 3854–3865.

Frolov, I., Hardy, R., and Rice, C.M. (2001). Cis-acting RNA elements at the 5′ end of Sindbis virus genome RNA regulate minus- and plus-strand RNA synthesis. RNA *7*, 1638–1651.

Frolov, I., Garmashova, N., Atasheva, S., and Frolova, E.I. (2009). Random insertion mutagenesis of sindbis virus non-structural protein 2 and selection of variants incapable of downregulating cellular transcription. J. Virol. *83*, 9031–9044.

Frolov, I., Akhrymuk, M., Akhrymuk, I., Atasheva, S., and Frolova, E.I. (2012). Early events in alphavirus replication determine the outcome of infection. J. Virol. *86*, 5055–5066.

Frolova, E.I., Fayzulin, R.Z., Cook, S.H., Griffin, D.E., Rice, C.M., and Frolov, I. (2002). Roles of non-structural protein nsP2 and Alpha/Beta interferons in determining the outcome of Sindbis virus infection. J. Virol. *76*, 11254–11264.

Frolova, E.I., Gorchakov, R., Pereboeva, L., Atasheva, S., and Frolov, I. (2010). Functional Sindbis virus replicative complexes are formed at the plasma membrane. J. Virol. *84*, 11679–11695.

Fros, J.J., Liu, W.J., Prow, N.A., Geertsema, C., Ligtenberg, M., Vanlandingham, D.L., Schnettler, E., Vlak, J.M., Suhrbier, A., Khromykh, A.A., *et al.* (2010). Chikungunya virus non-structural protein 2 inhibits type I/II interferon-stimulated JAK-STAT signaling. J. Virol. *84*, 10877–10887.

Fros, J.J., van der Maten, E., Vlak, J.M., and Pijlman, G.P. (2013). The C-terminal domain of chikungunya virus nsP2 independently governs viral RNA replication, cytopathicity, and inhibition of interferon signaling. J. Virol. *87*, 10394–10400.

Froshauer, S., Kartenbeck, J., and Helenius, A. (1988). Alphavirus RNA replicase is located on the cytoplasmic surface of endosomes and lysosomes. J. Cell Biol. *107*, 2075–2086.

Garcia-Moreno, M., Sanz, M.A., Pelletier, J., and Carrasco, L. (2013). Requirements for eIF4A and eIF2 during translation of Sindbis virus subgenomic mRNA in vertebrate and invertebrate host cells. Cell. Microbiol. *15*, 823–840.

Garcia-Vallejo, J.J., and van Kooyk, Y. (2013). The physiological role of DC-SIGN: a tale of mice and men. Trends Immunol. *34*, 482–486.

Gardner, C.L., Burke, C.W., Tesfay, M.Z., Glass, P.J., Klimstra, W.B., and Ryman, K.D. (2008). Eastern and Venezuelan equine encephalitis viruses differ in their ability to infect dendritic cells and macrophages: impact of altered cell tropism on pathogenesis. J. Virol. *82*, 10634–10646.

Gardner, C.L., Yin, J., Burke, C.W., Klimstra, W.B., and Ryman, K.D. (2009). Type I interferon induction is correlated with attenuation of a South American eastern equine encephalitis virus strain in mice. Virology *390*, 338–347.

Gardner, C.L., Ebel, G.D., Ryman, K.D., and Klimstra, W.B. (2011). Heparan sulfate binding by natural eastern equine encephalitis viruses promotes neurovirulence. Proc. Natl. Acad. Sci. U.S.A. *108*, 16026–16031.

Gardner, C.L., Burke, C.W., Higgs, S.T., Klimstra, W.B., and Ryman, K.D. (2012). Interferon-alpha/beta deficiency greatly exacerbates arthritogenic disease in mice infected with wild-type chikungunya virus but not with the cell culture-adapted live-attenuated 181/25 vaccine candidate. Virology *425*, 103–112.

Gardner, C.L., Choi-Nurvitadhi, J., Sun, C., Bayer, A., Hritz, J., Ryman, K.D., and Klimstra, W.B. (2013). Natural variation in the heparan sulfate binding domain of the eastern equine encephalitis virus E2 glycoprotein alters interactions with cell surfaces and virulence in mice. J. Virol. *87*, 8582–8590.

Gardner, C.L., Hritz, J., Sun, C., Vanlandingham, D.L., Song, T.Y., Ghedin, E., Higgs, S., Klimstra, W.B., and Ryman, K.D. (2014). Deliberate attenuation of chikungunya virus by adaptation to heparan sulfate-dependent infectivity: a model for rational arboviral vaccine design. PLoS Neglect. Trop. Dis. *8*, e2719.

Gardner, J., Anraku, I., Le, T.T., Larcher, T., Major, L., Roques, P., Schroder, W.A., Higgs, S., and Suhrbier, A. (2010). Chikungunya virus arthritis in adult wild-type mice. J. Virol. *84*, 8021–8032.

Garmashova, N., Gorchakov, R., Frolova, E., and Frolov, I. (2006). Sindbis virus non-structural protein nsP2 is cytotoxic and inhibits cellular transcription. J. Virol. *80*, 5686–5696.

Garmashova, N., Atasheva, S., Kang, W., Weaver, S.C., Frolova, E., and Frolov, I. (2007a). Analysis of Venezuelan equine encephalitis virus capsid protein function in the inhibition of cellular transcription. J. Virol. *81*, 13552–13565.

Garmashova, N., Gorchakov, R., Volkova, E., Paessler, S., Frolova, E., and Frolov, I. (2007b). The Old World and New World alphaviruses use different virus-specific proteins for induction of transcriptional shut-off. J. Virol. *81*, 2472–2484.

Giannakopoulos, N.V., Arutyunova, E., Lai, C., Lenschow, D.J., Haas, A.L., and Virgin, H.W. (2009). ISG15 Arg151 and the ISG15-conjugating enzyme UbE1L are important for innate immune control of Sindbis virus. J. Virol. *83*, 1602–1610.

Gibney, K.B., Fischer, M., Prince, H.E., Kramer, L.D., St George, K., Kosoy, O.L., Laven, J.J., and Staples, J.E. (2011). Chikungunya fever in the United States: a fifteen year review of cases. Clin. Infect. Dis. *52*, e121–126.

Gitlin, L., Barchet, W., Gilfillan, S., Cella, M., Beutler, B., Flavell, R.A., Diamond, M.S., and Colonna, M. (2006). Essential role of mda-5 in type I IFN responses to polyriboinosinic:polyribocytidylic acid and encephalomyocarditis picornavirus. Proc. Natl. Acad. Sci. U.S.A. *103*, 8459–8464.

Gorchakov, R., Frolova, E., Williams, B.R., Rice, C.M., and Frolov, I. (2004). PKR-dependent and -independent mechanisms are involved in translational shut-off during Sindbis virus infection. J. Virol. *78*, 8455–8467.

Gorchakov, R., Frolova, E., and Frolov, I. (2005). Inhibition of transcription and translation in Sindbis virus-infected cells. J. Virol. *79*, 9397–9409.

Gorchakov, R., Wang, E., Leal, G., Forrester, N.L., Plante, K., Rossi, S.L., Partidos, C.D., Adams, A.P., Seymour, R.L., Weger, J., *et al.* (2012). Attenuation of Chikungunya virus vaccine strain 181/clone 25 is determined by two amino acid substitutions in the E2 envelope glycoprotein. J. Virol. *86*, 6084–6096.

Gould, E.A., and Higgs, S. (2009). Impact of climate change and other factors on emerging arbovirus diseases. Trans. R. Soc. Trop. Med. Hyg. *103*, 109–121.

Grandvaux, N., Servant, M.J., tenOever, B., Sen, G.C., Balachandran, S., Barber, G.N., Lin, R., and Hiscott, J. (2002). Transcriptional profiling of interferon regulatory factor 3 target genes: direct involvement in the regulation of interferon-stimulated genes. J. Virol. *76*, 5532–5539.

Grieder, F.B., and Nguyen, H.T. (1996). Virulent and attenuated mutant Venezuelan equine encephalitis virus show marked differences in replication in infection in murine macrophages. Microbial Pathogen. *21*, 85–95.

Grieder, F.B., Davis, N.L., Aronson, J.F., Charles, P.C., Sellon, D.C., Suzuki, K., and Johnston, R.E. (1995). Specific restrictions in the progression of Venezuelan equine encephalitis virus-induced disease resulting from single amino acid changes in the glycoproteins. Virology *206*, 994–1006.

Grieder, F.B., Davis, B.K., Zhou, X.D., Chen, S.J., Finkelman, F.D., and Gause, W.C. (1997). Kinetics of cytokine expression and regulation of host protection following infection with molecularly cloned Venezuelan equine encephalitis virus. Virology *233*, 302–312.

Grimley, P.M., Levin, J.G., Berezesky, I.K., and Friedman, R.M. (1972). Specific membranous structures associated with the replication of group A arboviruses. J. Virol. *10*, 492–503.

Guo, J., Hui, D.J., Merrick, W.C., and Sen, G.C. (2000). A new pathway of translational regulation mediated by eukaryotic initiation factor 3. EMBO J. *19*, 6891–6899.

Guo, X., Carroll, J.W., Macdonald, M.R., Goff, S.P., and Gao, G. (2004). The zinc finger antiviral protein directly binds to specific viral mRNAs through the CCCH zinc finger motifs. J. Virol. *78*, 12781–12787.

Guo, X., Ma, J., Sun, J., and Gao, G. (2007). The zinc-finger antiviral protein recruits the RNA processing exosome to degrade the target mRNA. Proc. Natl. Acad. Sci. U.S.A. *104*, 151–156.

Hafer, A., Whittlesey, R., Brown, D.T., and Hernandez, R. (2009). Differential incorporation of cholesterol by Sindbis virus grown in mammalian or insect cells. J. Virol. *83*, 9113–9121.

Hahn, C.S., Lustig, S., Strauss, E.G., and Strauss, J.H. (1988). Western equine encephalitis virus is a recombinant virus. Proc. Natl. Acad. Sci. U.S.A. *85*, 5997–6001.

Halstead, S.B. (2015). Reappearance of chikungunya, formerly called dengue, in the Americas. Emerg. Infect. Dis. *21*, 557–561.

Hawman, D.W., Stoermer, K.A., Montgomery, S.A., Pal, P., Oko, L., Diamond, M.S., and Morrison, T.E. (2013). Chronic joint disease caused by persistent Chikungunya virus infection is controlled by the adaptive immune response. J. Virol. *87*, 13878–13888.

Heil, M.L., Albee, A., Strauss, J.H., and Kuhn, R.J. (2001). An amino acid substitution in the coding region of the E2 glycoprotein adapts Ross River virus to utilize heparan sulfate as an attachment moiety. J. Virol. *75*, 6303–6309.

Horio, T., Murai, M., Inoue, T., Hamasaki, T., Tanaka, T., and Ohgi, T. (2004). Crystal structure of human ISG20, an interferon-induced antiviral ribonuclease. FEBS Lett. *577*, 111–116.

Hornung, V., Ellegast, J., Kim, S., Brzozka, K., Jung, A., Kato, H., Poeck, H., Akira, S., Conzelmann, K.K., Schlee, M., *et al.* (2006). 5′-Triphosphate RNA is the ligand for RIG-I. Science *314*, 994–997.

Hui, D.J., Bhasker, C.R., Merrick, W.C., and Sen, G.C. (2003). Viral stress-inducible protein p56 inhibits translation by blocking the interaction of eIF3 with the ternary complex eIF2.GTP.Met-tRNAi. J. Biol. Chem. *278*, 39477–39482.

Hui, D.J., Terenzi, F., Merrick, W.C., and Sen, G.C. (2005). Mouse p56 blocks a distinct function of eukaryotic initiation factor 3 in translation initiation. J. Biol. Chem. *280*, 3433–3440.

Hunt, A.R., Frederickson, S., Maruyama, T., Roehrig, J.T., and Blair, C.D. (2010). The first human epitope map of the alphaviral E1 and E2 proteins reveals a new E2 epitope with significant virus neutralizing activity. PLoS Neglect. Trop. Dis. *4*, e739.

Hyde, J.L., Gardner, C.L., Kimura, T., White, J.P., Liu, G., Trobaugh, D.W., Huang, C., Tonelli, M., Paessler, S., Takeda, K., *et al.* (2014). A viral RNA structural element alters host recognition of nonself RNA. Science *343*, 783–787.

Hyde, J.L., Chen, R., Trobaugh, D.W., Diamond, M.S., Weaver, S.C., Klimstra, W.B., and Wilusz, J. (2015). The 5′ and 3′ ends of alphavirus RNAs – Non-coding is not non-functional. Virus Res. *206*, 99–107.

Jiang, D., Guo, H., Xu, C., Chang, J., Gu, B., Wang, L., Block, T.M., and Guo, J.T. (2008). Identification of three interferon-inducible cellular enzymes that inhibit the replication of hepatitis C virus. J. Virol. *82*, 1665–1678.

Johansson, M.A., Powers, A.M., Pesik, N., Cohen, N.J., and Staples, J.E. (2014). Nowcasting the spread of chikungunya virus in the Americas. PLoS One *9*, e104915.

Karki, S., Li, M.M., Schoggins, J.W., Tian, S., Rice, C.M., and MacDonald, M.R. (2012). Multiple interferon stimulated genes synergize with the zinc finger antiviral protein to mediate anti-alphavirus activity. PLoS One 7, e37398.

Kendrick, K., Stanek, D., Blackmore, C., Centers for Disease, C., and Prevention (2014). Notes from the field: Transmission of chikungunya virus in the continental United States--Florida, 2014. MMWR *63*, 1137.

Kerns, J.A., Emerman, M., and Malik, H.S. (2008). Positive selection and increased antiviral activity associated with the PARP-containing isoform of human zinc-finger antiviral protein. PLoS Genet. *4*, e21.

Khan, K., Bogoch, I., Brownstein, J.S., Miniota, J., Nicolucci, A., Hu, W., Nsoesie, E.O., Cetron, M., Creatore, M.I., German, M., *et al.* (2014). Assessing the origin of and potential for international spread of chikungunya virus from the Caribbean. PLoS Curr. *6*. doi: 10.1371/currents.outbreaks.2134a0a7bf37fd8d388181539fea2da5

Kielian, M. (2010). Structural biology: An alphavirus puzzle solved. Nature *468*, 645–646.

Kielian, M., and Jungerwirth, S. (1990). Mechanisms of enveloped virus entry into cells. Mol. Biol. Med. *7*, 17–31.

Kielian, M., Chanel-Vos, C., and Liao, M. (2010). Alphavirus entry and membrane fusion. Viruses *2*, 796–825.

Kim, D.Y., Firth, A.E., Atasheva, S., Frolova, E.I., and Frolov, I. (2011). Conservation of a packaging signal and the viral genome RNA packaging mechanism in alphavirus evolution. J. Virol. *85*, 8022–8036.

Klimstra, W.B., Ryman, K.D., and Johnston, R.E. (1998). Adaptation of Sindbis virus to BHK cells selects for use of heparan sulfate as an attachment receptor. J. Virol. *72*, 7357–7366.

Klimstra, W.B., Ryman, K.D., Bernard, K.A., Nguyen, K.B., Biron, C.A., and Johnston, R.E. (1999). Infection of neonatal mice with sindbis virus results in a systemic inflammatory response syndrome. J. Virol. *73*, 10387–10398.

Klimstra, W.B., Nangle, E.M., Smith, M.S., Yurochko, A.D., and Ryman, K.D. (2003). DC-SIGN and L-SIGN can act as attachment receptors for alphaviruses and distinguish between mosquito cell- and mammalian cell-derived viruses. J. Virol. *77*, 12022–12032.

Konopka, J.L., Thompson, J.M., Whitmore, A.C., Webb, D.L., and Johnston, R.E. (2009). Acute infection with venezuelan equine encephalitis virus replicon particles catalyzes a systemic antiviral state and protects from lethal virus challenge. J. Virol. *83*, 12432–12442.

La Linn, M., Gardner, J., Warrilow, D., Darnell, G.A., McMahon, C.R., Field, I., Hyatt, A.D., Slade, R.W., and Suhrbier, A. (2001). Arbovirus of marine mammals: a new alphavirus isolated from the elephant seal louse, Lepidophthirus macrorhini. J. Virol. *75*, 4103–4109.

Lemm, J.A., and Rice, C.M. (1993a). Assembly of functional Sindbis virus RNA replication complexes: requirement for coexpression of P123 and P34. J. Virol. *67*, 1905–1915.

Lemm, J.A., and Rice, C.M. (1993b). Roles of non-structural polyproteins and cleavage products in regulating Sindbis virus RNA replication and transcription. J. Virol. *67*, 1916–1926.

Lemm, J.A., Rumenapf, T., Strauss, E.G., Strauss, J.H., and Rice, C.M. (1994). Polypeptide requirements for assembly of functional Sindbis virus replication complexes: a model for the temporal regulation of minus- and plus-strand RNA synthesis. EMBO J. *13*, 2925–2934.

Lenschow, D.J. (2010). Antiviral Properties of ISG15. Viruses *2*, 2154–2168.

Lenschow, D.J., Giannakopoulos, N.V., Gunn, L.J., Johnston, C., O'Guin, A.K., Schmidt, R.E., Levine, B., and Virgin, H.W.t. (2005). Identification of interferon-stimulated gene 15 as an antiviral molecule during Sindbis virus infection in vivo. J. Virol. *79*, 13974–13983.

Lenschow, D.J., Lai, C., Frias-Staheli, N., Giannakopoulos, N.V., Lutz, A., Wolff, T., Osiak, A., Levine, B., Schmidt, R.E., Garcia-Sastre, A., *et al.* (2007). IFN-stimulated gene 15 functions as a critical antiviral molecule against influenza, herpes, and Sindbis viruses. Proc. Natl. Acad. Sci. U.S.A. *104*, 1371–1376.

Lindsey, N.P., Prince, H.E., Kosoy, O., Laven, J., Messenger, S., Staples, J.E., and Fischer, M. (2015). Chikungunya virus infections among travelers-United States, 2010–2013. Am. J. Trop. Med. Hyg. *92*, 82–87.

Liu, C.Y., and Kielian, M. (2009). E1 mutants identify a critical region in the trimer interface of the Semliki forest virus fusion protein. J. Virol. *83*, 11298–11306.

Lo Presti, A., Ciccozzi, M., Cella, E., Lai, A., Simonetti, F.R., Galli, M., Zehender, G., and Rezza, G. (2012). Origin, evolution, and phylogeography of recent epidemic CHIKV strains. Infect. Genet. Evol. *12*, 392–398.

Lum, F.M., Teo, T.H., Lee, W.W., Kam, Y.W., Renia, L., and Ng, L.F. (2013). An essential role of antibodies in the control of Chikungunya virus infection. J. Immunol. *190*, 6295–6302.

Lundberg, L., Pinkham, C., Baer, A., Amaya, M., Narayanan, A., Wagstaff, K.M., Jans, D.A., and Kehn-Hall, K. (2013). Nuclear import and export inhibitors alter capsid protein distribution in mammalian cells and reduce Venezuelan Equine Encephalitis Virus replication. Antiviral Res. *100*, 662–672.

Lundstrom, J.O., and Pfeffer, M. (2010). Phylogeographic structure and evolutionary history of Sindbis virus. Vector Borne Zoonotic Dis. *10*, 889–907.

McAllister, C.S., Taghavi, N., and Samuel, C.E. (2012). Protein kinase PKR amplification of interferon beta induction occurs through initiation factor eIF-2alpha-mediated translational control. J. Biol. Chem. *287*, 36384–36392.

MacDonald, G.H., and Johnston, R.E. (2000). Role of dendritic cell targeting in Venezuelan equine encephalitis virus pathogenesis. J. Virol. *74*, 914–922.

MacDonald, M.R., Machlin, E.S., Albin, O.R., and Levy, D.E. (2007). The zinc finger antiviral protein acts synergistically with an interferon-induced factor for maximal activity against alphaviruses. J. Virol. *81*, 13509–13518.

McInerney, G.M., Kedersha, N.L., Kaufman, R.J., Anderson, P., and Liljestrom, P. (2005). Importance of eIF2alpha phosphorylation and stress granule assembly in alphavirus translation regulation. Mol. Biol. Cell *16*, 3753–3763.

McKnight, K.L., Simpson, D.A., Lin, S.C., Knott, T.A., Polo, J.M., Pence, D.F., Johannsen, D.B., Heidner, H.W., Davis, N.L., and Johnston, R.E. (1996). Deduced consensus sequence of Sindbis virus strain AR339: mutations contained in laboratory strains which affect cell culture and in vivo phenotypes. J. Virol. *70*, 1981–1989.

McLoughlin, M.F., and Graham, D.A. (2007). Alphavirus infections in salmonids – a review. J Fish Dis *30*, 511–531.

Martinez, M.G., Snapp, E.L., Perumal, G.S., Macaluso, F.P., and Kielian, M. (2014). Imaging the alphavirus exit pathway. J. Virol. *88*, 6922–6933.

Meyer, W.J., and Johnston, R.E. (1993). Structural rearrangement of infecting Sindbis virions at the cell surface: mapping of newly accessible epitopes. J. Virol. *67*, 5117–5125.

Meyer, W.J., Gidwitz, S., Ayers, V.K., Schoepp, R.J., and Johnston, R.E. (1992). Conformational alteration of Sindbis virion glycoproteins induced by heat, reducing agents, or low pH. J. Virol. *66*, 3504–3513.

Monath, T.P., Craven, R.B., Muth, D.J., Trautt, C.J., Calisher, C.H., and Fitzgerald, S.A. (1980). Limitations of the complement-fixation test for distinguishing naturally acquired from vaccine-induced yellow fever infection in flavivirus-hyperendemic areas. Am. J. Trop. Med. Hyg. *29*, 624–634.

Morrison, T.E., Oko, L., Montgomery, S.A., Whitmore, A.C., Lotstein, A.R., Gunn, B.M., Elmore, S.A., and Heise, M.T. (2011). A mouse model of chikungunya virus-induced musculoskeletal inflammatory disease: evidence of arthritis, tenosynovitis, myositis, and persistence. Am. J. Pathol. *178*, 32–40.

Mukhopadhyay, S., Zhang, W., Gabler, S., Chipman, P.R., Strauss, E.G., Strauss, J.H., Baker, T.S., Kuhn, R.J., and Rossmann, M.G. (2006). Mapping the structure and function of the E1 and E2 glycoproteins in alphaviruses. Structure *14*, 63–73.

Nasar, F., Palacios, G., Gorchakov, R.V., Guzman, H., Da Rosa, A.P., Savji, N., Popov, V.L., Sherman, M.B., Lipkin, W.I., Tesh, R.B., *et al.* (2012). Eilat virus, a unique alphavirus with host range restricted to insects by RNA replication. Proc. Natl. Acad. Sci. U.S.A. *109*, 14622–14627.

Nasar, F., Gorchakov, R.V., Tesh, R.B., and Weaver, S.C. (2015). Eilat virus host range restriction is present at multiple levels of the virus life cycle. J. Virol. *89*, 1404–1418.

Niesters, H.G., and Strauss, J.H. (1990). Mutagenesis of the conserved 51-nucleotide region of Sindbis virus. J. Virol. *64*, 1639–1647.

Niklasson, B., and Espmark, A. (1986). Ockelbo disease: arthralgia 3–4 years after infection with a Sindbis virus related agent. Lancet *1*, 1039–1040.

Olmsted, R.A., Meyer, W.J., and Johnston, R.E. (1986). Characterization of Sindbis virus epitopes important for penetration in cell culture and pathogenesis in animals. Virology *148*, 245–254.

Ou, J.H., Strauss, E.G., and Strauss, J.H. (1981). Comparative studies of the 3′-terminal sequences of several alpha virus RNAs. Virology *109*, 281–289.

Paredes, A., Alwell-Warda, K., Weaver, S.C., Chiu, W., and Watowich, S.J. (2001). Venezuelan equine encephalomyelitis virus structure and its divergence from old world alphaviruses. J. Virol. *75*, 9532–9537.

Paredes, A.M., Brown, D.T., Rothnagel, R., Chiu, W., Schoepp, R.J., Johnston, R.E., and Prasad, B.V. (1993). Three-dimensional structure of a membrane-containing virus. Proc. Natl. Acad. Sci. U.S.A. *90*, 9095–9099.

Peltier, D.C., Lazear, H.M., Farmer, J.R., Diamond, M.S., and Miller, D.J. (2013). Neurotropic arboviruses induce interferon regulatory factor 3-mediated neuronal responses that are cytoprotective, interferon independent, and inhibited by Western equine encephalitis virus capsid. J. Virol. *87*, 1821–1833.

Peranen, J., Rikkonen, M., Liljestrom, P., and Kaariainen, L. (1990). Nuclear localization of Semliki Forest virus-specific non-structural protein nsP2. J. Virol. *64*, 1888–1896.

Pfeffer, M., and Loscher, T. (2006). Cases of chikungunya imported into Europe. Euro surveillance: bulletin Europeen sur les maladies transmissibles = European communicable disease bulletin *11*, E060316 060312.

Pichlmair, A., Schulz, O., Tan, C.P., Naslund, T.I., Liljestrom, P., Weber, F., and Reis e Sousa, C. (2006). RIG-I-mediated antiviral responses to single-stranded RNA bearing 5′-phosphates. Science *314*, 997–1001.

Pichlmair, A., Lassnig, C., Eberle, C.A., Gorna, M.W., Baumann, C.L., Burkard, T.R., Burckstummer, T., Stefanovic, A., Krieger, S., Bennett, K.L., *et al.* (2011). IFIT1 is an antiviral protein that recognizes 5′-triphosphate RNA. Nature Immunol. *12*, 624–630.

Pletnev, S.V., Zhang, W., Mukhopadhyay, S., Fisher, B.R., Hernandez, R., Brown, D.T., Baker, T.S., Rossmann, M.G., and Kuhn, R.J. (2001). Locations of carbohydrate sites on alphavirus glycoproteins show that E1 forms an icosahedral scaffold. Cell *105*, 127–136.

Poo, Y.S., Nakaya, H., Gardner, J., Larcher, T., Schroder, W.A., Le, T.T., Major, L.D., and Suhrbier, A. (2014a). CCR2 deficiency promotes exacerbated chronic erosive neutrophil-dominated chikungunya virus arthritis. J. Virol. *88*, 6862–6872.

Poo, Y.S., Rudd, P.A., Gardner, J., Wilson, J.A., Larcher, T., Colle, M.A., Le, T.T., Nakaya, H.I., Warrilow, D., Allcock, R., *et al.* (2014b). Multiple immune factors are involved in controlling acute and chronic chikungunya virus infection. PLoS Neglect. Trop. Dis. *8*, e3354.

Powers, A.M., Oberste, M.S., Brault, A.C., Rico-Hesse, R., Schmura, S.M., Smith, J.F., Kang, W., Sweeney, W.P., and Weaver, S.C. (1997). Repeated emergence of epidemic/epizootic Venezuelan equine encephalitis from a single genotype of enzootic subtype ID virus. J. Virol. *71*, 6697–6705.

Rashad, A.A., Mahalingam, S., and Keller, P.A. (2014). Chikungunya virus: emerging targets and new opportunities for medicinal chemistry. J. Med. Chem. *57*, 1147–1166.

Reynaud, J.M., Kim, D.Y., Atasheva, S., Rasalouskaya, A., White, J.P., Diamond, M.S., Weaver, S.C., Frolova, E.I., and Frolov, I. (2015). IFIT1 differentially interferes with translation and replication of alphavirus genomes and promotes induction of type I interferon. PLoS Path. *11*, e1004863.

Rico-Hesse, R., Weaver, S.C., de Siger, J., Medina, G., and Salas, R.A. (1995). Emergence of a new epidemic/epizootic Venezuelan equine encephalitis virus in South America. Proc. Natl. Acad. Sci. U.S.A. *92*, 5278–5281.

Rikkonen, M. (1996). Functional significance of the nuclear-targeting and NTP-binding motifs of Semliki Forest virus non-structural protein nsP2. Virology *218*, 352–361.

Ritchie, K.J., Hahn, C.S., Kim, K.I., Yan, M., Rosario, D., Li, L., de la Torre, J.C., and Zhang, D.E. (2004). Role of ISG15 protease UBP43 (USP18) in innate immunity to viral infection. Nature Med. *10*, 1374–1378.

Robinson, M.C. (1955). An epidemic of virus disease in Southern Province, Tanganyika Territory, in 1952–53. I. Clinical features. Trans. R. Soc. Trop. Med. Hyg. *49*, 28–32.

Roehrig, J.T., Corser, J.A., and Schlesinger, M.J. (1980). Isolation and characterization of hybrid cell lines producing monoclonal antibodies directed against the structural proteins of Sindbis virus. Virology *101*, 41–49.

Roehrig, J.T., Hunt, A.R., Kinney, R.M., and Mathews, J.H. (1988). In vitro mechanisms of monoclonal antibody neutralization of alphaviruses. Virology *165*, 66–73.

Roman-Sosa, G., and Kielian, M. (2011). The interaction of alphavirus E1 protein with exogenous domain III defines stages in virus–membrane fusion. J. Virol. *85*, 12271–12279.

Rosenberg, R., Johansson, M.A., Powers, A.M., and Miller, B.R. (2013). Search strategy has influenced the discovery rate of human viruses. Proc. Natl. Acad. Sci. U.S.A. *110*, 13961–13964.

Rozdilsky, B., Robertson, H.E., and Chorney, J. (1968). Western encephalitis: report of eight fatal cases. Saskatchewan epidemic, 1965. Can. Med. Assoc. J. *98*, 79–86.

Rulli, N.E., Rolph, M.S., Srikiatkhachorn, A., Anantapreecha, S., Guglielmotti, A., and Mahalingam, S. (2011). Protection from arthritis and myositis in a mouse model of acute chikungunya virus disease by bindarit, an inhibitor of monocyte chemotactic protein-1 synthesis. J. Infectious Dis. *204*, 1026–1030.

Ryman, K.D., and Klimstra, W.B. (2008). Host responses to alphavirus infection. Immunol. Rev. *225*, 27–45.

Ryman, K.D., Klimstra, W.B., Nguyen, K.B., Biron, C.A., and Johnston, R.E. (2000). Alpha/beta interferon protects adult mice from fatal Sindbis virus infection and is an important determinant of cell and tissue tropism. J. Virol. *74*, 3366–3378.

Ryman, K.D., White, L.J., Johnston, R.E., and Klimstra, W.B. (2002). Effects of PKR/RNase L-dependent and alternative antiviral pathways on alphavirus replication and pathogenesis. Viral Immunol. *15*, 53–76.

Ryman, K.D., Meier, K.C., Nangle, E.M., Ragsdale, S.L., Korneeva, N.L., Rhoads, R.E., MacDonald, M.R., and Klimstra, W.B. (2005). Sindbis virus translation is inhibited by a PKR/RNase L-independent effector induced by alpha/beta interferon priming of dendritic cells. J. Virol. *79*, 1487–1499.

Ryman, K.D., Gardner, C.L., Burke, C.W., Meier, K.C., Thompson, J.M., and Klimstra, W.B. (2007a). Heparan sulfate binding can contribute to the neurovirulence of neuroadapted and nonneuroadapted Sindbis viruses. J. Virol. *81*, 3563–3573.

Ryman, K.D., Meier, K.C., Gardner, C.L., Adegboyega, P.A., and Klimstra, W.B. (2007b). Non-pathogenic Sindbis virus causes hemorrhagic fever in the absence of alpha/beta and gamma interferons. Virology *368*, 273–285.

Sanchez-San Martin, C., Liu, C.Y., and Kielian, M. (2009). Dealing with low pH: entry and exit of alphaviruses and flaviviruses. Trends Microbiol. *17*, 514–521.

Sanz, M.A., Castello, A., Ventoso, I., Berlanga, J.J., and Carrasco, L. (2009). Dual mechanism for the translation of subgenomic mRNA from Sindbis virus in infected and uninfected cells. PloS One *4*, e4772.

Sarkar, S.N., and Sen, G.C. (2004). Novel functions of proteins encoded by viral stress-inducible genes. Pharmacol. Ther. *103*, 245–259.

Schafer, A., Brooke, C.B., Whitmore, A.C., and Johnston, R.E. (2011). The role of the blood–brain barrier during Venezuelan equine encephalitis virus infection. J. Virol. *85*, 10682–10690.

Schafer, S.L., Lin, R., Moore, P.A., Hiscott, J., and Pitha, P.M. (1998). Regulation of type I interferon gene expression by interferon regulatory factor-3. J. Biol. Chem. *273*, 2714–2720.

Schilte, C., Couderc, T., Chretien, F., Sourisseau, M., Gangneux, N., Guivel-Benhassine, F., Kraxner, A., Tschopp, J., Higgs, S., Michault, A., *et al.* (2010). Type I IFN controls chikungunya virus via its action on nonhematopoietic cells. J. Exp. Med. *207*, 429–442.

Schmaljohn, A.L., Johnson, E.D., Dalrymple, J.M., and Cole, G.A. (1982). Non-neutralizing monoclonal antibodies can prevent lethal alphavirus encephalitis. Nature *297*, 70–72.

Schoggins, J.W., and Rice, C.M. (2011). Interferon-stimulated genes and their antiviral effector functions. Curr. Opin. Virol. *1*, 519–525.

Schoggins, J.W., Wilson, S.J., Panis, M., Murphy, M.Y., Jones, C.T., Bieniasz, P., and Rice, C.M. (2011). A diverse range of gene products are effectors of the type I interferon antiviral response. Nature *472*, 481–485.

Schulz, O., Pichlmair, A., Rehwinkel, J., Rogers, N.C., Scheuner, D., Kato, H., Takeuchi, O., Akira, S., Kaufman, R.J., and Reis e Sousa, C. (2010). Protein kinase R contributes to immunity against specific viruses by regulating interferon mRNA integrity. Cell Host Microbe 7, 354–361.

Seyler, T., Grandesso, F., Le Strat, Y., Tarantola, A., and Depoortere, E. (2009). Assessing the risk of importing dengue and chikungunya viruses to the European Union. Epidemics *1*, 175–184.

Shabman, R.S., Morrison, T.E., Moore, C., White, L., Suthar, M.S., Hueston, L., Rulli, N., Lidbury, B., Ting, J.P., Mahalingam, S., *et al.* (2007). Differential induction of type I interferon responses in myeloid dendritic cells by mosquito and mammalian-cell-derived alphaviruses. J. Virol. *81*, 237–247.

Shirako, Y., and Strauss, J.H. (1994). Regulation of Sindbis virus RNA replication: uncleaved P123 and nsP4 function in minus-strand RNA synthesis, whereas cleaved products from P123 are required for efficient plus-strand RNA synthesis. J. Virol. *68*, 1874–1885.

Silva, L.A., Khomandiak, S., Ashbrook, A.W., Weller, R., Heise, M.T., Morrison, T.E., and Dermody, T.S. (2014). A single-amino-acid polymorphism in Chikungunya virus E2 glycoprotein influences glycosaminoglycan utilization. J. Virol. *88*, 2385–2397.

Silverman, M.A., Misasi, J., Smole, S., Feldman, H.A., Cohen, A.B., Santagata, S., McManus, M., and Ahmed, A.A. (2013). Eastern equine encephalitis in children, Massachusetts and New Hampshire,USA, 1970–2010. Emerg. Infect. Dis. *19*, 194–201; quiz 352.

Simmons, J.D., White, L.J., Morrison, T.E., Montgomery, S.A., Whitmore, A.C., Johnston, R.E., and Heise, M.T. (2009). Venezuelan equine encephalitis virus disrupts STAT1 signaling by distinct mechanisms independent of host shut-off. J. Virol. *83*, 10571–10581.

Singh, I., and Helenius, A. (1992). Role of ribosomes in Semliki Forest virus nucleocapsid uncoating. J. Virol. *66*, 7049–7058.

Smit, J.M., Waarts, B.L., Kimata, K., Klimstra, W.B., Bittman, R., and Wilschut, J. (2002). Adaptation of alphaviruses to heparan sulfate: interaction of Sindbis and Semliki forest viruses with liposomes containing lipid-conjugated heparin. J. Virol. *76*, 10128–10137.

Staples, J.E., and Fischer, M. (2014). Chikungunya virus in the Americas – what a vectorborne pathogen can do. N. Engl. J. Med. *371*, 887–889.

Steele, K.E., and Twenhafel, N.A. (2010). Review paper: pathology of animal models of alphavirus encephalitis. Vet. Pathol. *47*, 790–805.

Strauss, E.G., Rice, C.M., and Strauss, J.H. (1984). Complete nucleotide sequence of the genomic RNA of Sindbis virus. Virology *133*, 92–110.

Suhrbier, A., Jaffar-Bandjee, M.C., and Gasque, P. (2012). Arthritogenic alphaviruses – an overview. Nature Rev. Rheumatol. *8*, 420–429.

Szretter, K.J., Daniels, B.P., Cho, H., Gainey, M.D., Yokoyama, W.M., Gale, M., Jr., Virgin, H.W., Klein, R.S., Sen, G.C., and Diamond, M.S. (2012). 2′-O methylation of the viral mRNA cap by West Nile virus evades ifit1-dependent and -independent mechanisms of host restriction in vivo. PLoS Path. *8*, e1002698.

Takkinen, K. (1986). Complete nucleotide sequence of the non-structural protein genes of Semliki Forest virus. Nucleic Acids Res. *14*, 5667–5682.

Tamm, K., Merits, A., and Sarand, I. (2008). Mutations in the nuclear localization signal of nsP2 influencing RNA synthesis, protein expression and cytotoxicity of Semliki Forest virus. J. Gen. Virol. *89*, 676–686.

Taylor, R.M., Hurlbut, H.S., Work, T.H., Kingston, J.R., and Frothingham, T.E. (1955). Sindbis virus: a newly recognized arthropod transmitted virus. Am. J. Trop. Med. Hyg. *4*, 844–862.

Teo, T.H., Lum, F.M., Claser, C., Lulla, V., Lulla, A., Merits, A., Renia, L., and Ng, L.F. (2013). A pathogenic role for CD4+ T cells during Chikungunya virus infection in mice. J. Immunol. *190*, 259–269.

Tesfay, M.Z., Yin, J., Gardner, C.L., Khoretonenko, M.V., Korneeva, N.L., Rhoads, R.E., Ryman, K.D., and Klimstra, W.B. (2008). Alpha/beta interferon inhibits cap-dependent translation of viral but not cellular mRNA by a PKR-independent mechanism. J. Virol. *82*, 2620–2630.

Tilston, N., Skelly, C., and Weinstein, P. (2009). Pan-European Chikungunya surveillance: designing risk stratified surveillance zones. Int. J. Health Geographics *8*, 61.

Toivanen, A. (2008). Alphaviruses: an emerging cause of arthritis? Curr. Opin. Rheumatol. *20*, 486–490.

Trobaugh, D.W., Gardner, C.L., Sun, C., Haddow, A.D., Wang, E., Chapnik, E., Mildner, A., Weaver, S.C., Ryman, K.D., and Klimstra, W.B. (2014a). RNA viruses can hijack vertebrate microRNAs to suppress innate immunity. Nature *506*, 245–248.

Trobaugh, D.W., Ryman, K.D., and Klimstra, W.B. (2014b). Can understanding the virulence mechanisms of RNA viruses lead us to a vaccine against eastern equine encephalitis virus and other alphaviruses? Exp. Rev. Vaccines *13*, 1423–1425.

Tsetsarkin, K.A., Vanlandingham, D.L., McGee, C.E., and Higgs, S. (2007). A single mutation in chikungunya virus affects vector specificity and epidemic potential. PLoS Path. *3*, e201.

Uchime, O., Fields, W., and Kielian, M. (2013). The role of E3 in pH protection during alphavirus assembly and exit. J. Virol. *87*, 10255–10262.

Varjak, M., Saul, S., Arike, L., Lulla, A., Peil, L., and Merits, A. (2013). Magnetic fractionation and proteomic dissection of cellular organelles occupied by the late replication complexes of semliki forest virus. J. Virol. *87*, 10295–10312.

Vazeille, M., Jeannin, C., Martin, E., Schaffner, F., and Failloux, A.B. (2008). Chikungunya: a risk for Mediterranean countries? Acta Tropica *105*, 200–202.

Ventoso, I. (2012). Adaptive changes in alphavirus mRNA translation allowed colonization of vertebrate hosts. J. Virol. *86*, 9484–9494.

Ventoso, I., Sanz, M.A., Molina, S., Berlanga, J.J., Carrasco, L., and Esteban, M. (2006). Translational resistance of late alphavirus mRNA to eIF2alpha phosphorylation: a strategy to overcome the antiviral effect of protein kinase PKR. Genes Dev. *20*, 87–100.

Verwoerd, D.J. (2000). Ostrich diseases. Rev. Sci. Tech. *19*, 638–661.

Vogel, P., Kell, W.M., Fritz, D.L., Parker, M.D., and Schoepp, R.J. (2005). Early events in the pathogenesis of eastern equine encephalitis virus in mice. Am. J. Pathol. *166*, 159–171.

Voss, J.E., Vaney, M.C., Duquerroy, S., Vonrhein, C., Girard-Blanc, C., Crublet, E., Thompson, A., Bricogne, G., and Rey, F.A. (2010). Glycoprotein organization of Chikungunya virus particles revealed by X-ray crystallography. Nature *468*, 709–712.

Wang, F., Barrett, J.W., Ma, Y., Dekaban, G.A., and McFadden, G. (2009). Induction of alpha/beta interferon by myxoma virus is selectively abrogated when primary mouse embryo fibroblasts become immortalized. J. Virol. *83*, 5928–5932.

Wang, X., Lv, F., and Gao, G. (2010). Mutagenesis analysis of the zinc-finger antiviral protein. Retrovirology *7*, 19.

Watts, D.M., Callahan, J., Rossi, C., Oberste, M.S., Roehrig, J.T., Wooster, M.T., Smith, J.F., Cropp, C.B., Gentrau, E.M., Karabatsos, N., *et al.* (1998). Venezuelan equine encephalitis febrile cases among humans in the Peruvian Amazon River region. Am. J. Trop. Med. Hyg. *58*, 35–40.

Weaver, S.C. (2014). Arrival of chikungunya virus in the new world: prospects for spread and impact on public health. PLoS Neglect. Trop. Dis. *8*, e2921.

Weaver, S.C., and Forrester, N.L. (2015). Chikungunya: Evolutionary history and recent epidemic spread. Antiviral Res. *120*, 32–39.

Weaver, S.C., and Reisen, W.K. (2010). Present and future arboviral threats. Antiviral Res. *85*, 328–345.

Weaver, S.C., Salas, R., Rico-Hesse, R., Ludwig, G.V., Oberste, M.S., Boshell, J., and Tesh, R.B. (1996). Re-emergence of epidemic Venezuelan equine encephalomyelitis in South America. VEE Study Group. Lancet *348*, 436–440.

Weaver, S.C., Kang, W., Shirako, Y., Rumenapf, T., Strauss, E.G., and Strauss, J.H. (1997). Recombinational history and molecular evolution of western equine encephalomyelitis complex alphaviruses. J. Virol. *71*, 613–623.

Weaver, S.C., Ferro, C., Barrera, R., Boshell, J., and Navarro, J.C. (2004). Venezuelan equine encephalitis. Ann. Rev. Entomol. *49*, 141–174.

Weaver, S.C., Osorio, J.E., Livengood, J.A., Chen, R., and Stinchcomb, D.T. (2012). Chikungunya virus and prospects for a vaccine. Exp. Rev. Vaccines *11*, 1087–1101.

Wengler, G., and Wengler, G. (1984). Identification of a transfer of viral core protein to cellular ribosomes during the early stages of alphavirus infection. Virology *134*, 435–442.

Wengler, G., Wurkner, D., and Wengler, G. (1992). Identification of a sequence element in the alphavirus core protein which mediates interaction of cores with ribosomes and the disassembly of cores. Virology *191*, 880–888.

Werneke, S.W., Schilte, C., Rohatgi, A., Monte, K.J., Michault, A., Arenzana-Seisdedos, F., Vanlandingham, D.L., Higgs, S., Fontanet, A., Albert, M.L., *et al.* (2011). ISG15 is critical in the control of Chikungunya virus infection independent of UbE1L mediated conjugation. PLoS Path. *7*, e1002322.

Westarp, M.E., Stanley, J., and Griffin, D.E. (1988). Sindbis virus neutralization. Ann. N. Y. Acad. Sci. *540*, 566–567.

White, L.J., Wang, J.G., Davis, N.L., and Johnston, R.E. (2001). Role of alpha/beta interferon in Venezuelan equine encephalitis virus pathogenesis: effect of an attenuating mutation in the 5′ untranslated region. J. Virol. *75*, 3706–3718.

White, L.K., Sali, T., Alvarado, D., Gatti, E., Pierre, P., Streblow, D., and Defilippis, V.R. (2011). Chikungunya virus induces IPS-1-dependent innate immune activation and protein kinase R-independent translational shut-off. J. Virol. *85*, 606–620.

Whitlow, Z.W., Connor, J.H., and Lyles, D.S. (2008). New mRNAs are preferentially translated during vesicular stomatitis virus infection. J. Virol. *82*, 2286–2294.

Yin, J., Gardner, C.L., Burke, C.W., Ryman, K.D., and Klimstra, W.B. (2009). Similarities and differences in antagonism of neuron alpha/beta interferon responses by Venezuelan equine encephalitis and Sindbis alphaviruses. J. Virol. *83*, 10036–10047.

Young, D.S., Kramer, L.D., Maffei, J.G., Dusek, R.J., Backenson, P.B., Mores, C.N., Bernard, K.A., and Ebel, G.D. (2008). Molecular epidemiology of eastern equine encephalitis virus, New York. Emerg. Infect. Dis. *14*, 454–460.

Zacks, M.A., and Paessler, S. (2010). Encephalitic alphaviruses. Vet. Microbiol. *140*, 281–286.

Zhang, Y., Burke, C.W., Ryman, K.D., and Klimstra, W.B. (2007). Identification and characterization of interferon-induced proteins that inhibit alphavirus replication. J. Virol. *81*, 11246–11255.

Zhao, C., Collins, M.N., Hsiang, T.Y., and Krug, R.M. (2013). Interferon-induced ISG15 pathway: an ongoing virus–host battle. Trends Microbiol. *21*, 181–186.

Zheng, Y., Sanchez-San Martin, C., Qin, Z.L., and Kielian, M. (2011). The domain I-domain III linker plays an important role in the fusogenic conformational change of the alphavirus membrane fusion protein. J. Virol. *85*, 6334–6342.

Zhou, X., Michal, J.J., Zhang, L., Ding, B., Lunney, J.K., Liu, B., and Jiang, Z. (2013). Interferon induced IFIT family genes in host antiviral defense. Int. J. Biol. Sci. *9*, 200–208.

Zhou, Z., Wang, N., Woodson, S.E., Dong, Q., Wang, J., Liang, Y., Rijnbrand, R., Wei, L., Nichols, J.E., Guo, J.T., *et al.* (2011). Antiviral activities of ISG20 in positive-strand RNA virus infections. Virology *409*, 175–188.

Zhu, Y., and Gao, G. (2008). ZAP-mediated mRNA degradation. RNA Biol. *5*, 65–67.

Zhu, Y., Wang, X., Goff, S.P., and Gao, G. (2012). Translational repression precedes and is required for ZAP-mediated mRNA decay. EMBO J. *31*, 4236–4246.

Molecular Interactions Between Arboviruses and Insect Vectors: Insects' Immune Responses to Virus Infection

8

Natapong Jupatanakul and George Dimopoulos

Abstract

Despite the efficient spread of arboviral diseases, the mosquito represents a bottleneck in the transmission of pathogenic viruses. Within the mosquito, arboviruses have to overcome several barriers, imposed by the mosquito innate immune system, microbiota and other factors, to reach the salivary glands from where they can be transmitted to a new host. Here we will address the molecular interactions between viruses and mosquito vectors with emphasis on the innate immune system and microbiota, and discuss their potential for the development of disease control strategies.

Introduction

Arboviruses are insect-transmitted viruses, and mosquitoes, ticks, sandflies, and midges are vectors for arboviral diseases (the list of arboviruses and insect vectors is thoroughly reviewed in Cleton *et al.*, 2012). Arboviruses have been recognized as emerging medically important viral pathogens. Their increasing incidence in recent decades is the result of a variety of factors: urbanization, the rapid growth of the human population, the increasingly widespread distribution of insect vectors, an increase in travel activity from endemic to non-endemic regions, and global warming (Cleton *et al.*, 2012; Reiter, 2010; Soverow *et al.*, 2009; Weaver, 2013; Weaver and Reisen, 2010). There is no current estimate of the overall incidence of all arbovirus infections; however, a recent study put the estimate of dengue alone at 390 million infections annually (Bhatt *et al.*, 2013).

Because of the lack of licensed drugs and vaccines against several of the arboviruses, the elimination and control of insect vectors remains the only option to reduce the global disease burden. However, conventional mosquito elimination programmes depend on the use of insecticides, which carries with it ecological, environmental, and effectiveness concerns (Ault, 1994; Dong, 2007; Gubler, 1998; Rivero *et al.*, 2010). Given these concerns, recent research has focused on using molecular entomological approaches to develop alternative and sustainable arbovirus and vector control strategies. This chapter will review recent advances in the field of molecular interactions between insect vectors and arboviruses. The novel insect vector control approaches derived from these findings will be discussed in Chapter 19.

Molecular biology of the arbovirus–insect interaction

Although arboviruses can cause serious diseases in humans, they rarely cause mosquito pathology, and they can persistently infect the mosquito vector (for life) (Salazar *et al.*, 2007). Among the arboviruses, the interactions between the *Aedes* mosquitoes and dengue virus and between *Culex* mosquitoes and West Nile virus (WNV) are the best studied because of the high public health significance and disease burden associated with these insects.

Arboviruses are generally maintained in human populations by horizontal transmission between insect vectors and vertebrate hosts or reservoirs, meaning that the viruses have to be able to complete their life cycle in at least two very distinct species. The insect arbovirus life cycle has been most extensively studied in dengue virus (DENV). In order to complete its transmission cycle, DENV has to overcome infection barriers in the mosquito at the midgut infection, midgut escape, and transmission stages (Black *et al.*, 2002).

After female mosquitoes feed on infectious blood, the virus infects and propagates to establish infection in the mosquito midgut epithelium (the midgut infection barrier) (Black *et al.*, 2002). Next, the viruses have to be able to escape from the midgut tissue, then enter and replicate in other organs (midgut escape barrier). Finally, in order to complete the transmission cycle, the virus has to infect and propagate in the salivary glands, from which it can be transmitted to naive individuals through an infectious bite (transmission barrier). DENV level in the midgut generally peaks at 7–10 days, and the virus is then disseminated to other parts of the body through the trachea. It typically takes about 10–14 days after the original infectious bloodmeal for DENV to reach the salivary glands, from which it is transmitted to naive individuals (Salazar *et al.*, 2007). The incubation period; however, can change depending on temperature, mosquito and virus strains, and other factors (Hardy *et al.*, 1983; Sim *et al.*, 2013; Tjaden *et al.*, 2013; Watts *et al.*, 1987). Novel arbovirus control approaches aim to manipulate mosquito's immune responses so that the virus cannot establish efficient infection in the insect vector, or to shorten the mosquito's lifespan so that it will not be able to live long enough to complete the transmission cycle. The insect's immune responses largely determine the viral load, extrinsic incubation period, and mortality of the insect vector after viral infection, all of which directly affect the final outcome, disease transmission (Cirimotich *et al.*, 2009; Ocampo *et al.*, 2013; Sim *et al.*, 2013).

Knowledge of the mosquito's immune responses has been largely based on research in the insect model organism *Drosophila melanogaster* (reviewed in Lemaitre and Hoffmann, 2007) and the *Anopheles gambiae mosquito* (Barillas-Mury and Kumar, 2005; Levashina, 2004; Whitten *et al.*, 2006). In contrast to vertebrates, insects lack adaptive immunity and rely mainly on their innate immune system to fight infection (Vilmos and Kurucz, 1998). The insect innate immune system is complex and has various mechanisms to allow different immune responses to specific groups of pathogen. The classical insect immune response comprises both humoral and cellular responses (Dimopoulos, 2003; Lavine and Strand, 2002). The humoral response relies on soluble effectors such as antimicrobial peptides (AMPs), melanization by prophenoloxidase activation, and a complement like-system (Blandin *et al.*, 2004; Cerenius and Söderhäll, 2004; Cerenius *et al.*, 2008; Dimarcq *et al.*, 1994). The cellular response involves haemocyte-mediated mechanisms such as phagocytosis and encapsulation (Blandin and Levashina, 2007; Irving *et al.*, 2005). The activation of these cellular and humoral immune responses is regulated by a variety of immune signalling pathways, including the Toll, Immune deficiency (Imd), and Janus kinase-signal transduction and activation of transcription (JAK/STAT) pathways (Agaisse and Perrimon, 2004; Georgel *et al.*, 2001; Lemaitre *et al.*, 1996; Zambon *et al.*, 2005).

Sensing and combating pathogens: immune signaling pathways and antiviral mechanisms

The field of mosquito molecular immunity has recently expanded as a result of the availability of the complete sequence of the *Ae. aegypti* genome (Nene *et al.*, 2007) and the *C. quinquefasciatus* genome (Arensburger *et al.*, 2010), together with web-based bioinformatics tools on the Vectorbase website (www.vectorbase.org) (Lawson *et al.*, 2009; Megy *et al.*, 2011). Recent studies have shown that the classical immune pathways that were previously shown to be involved in mosquito responses to infection with bacteria and fungi are also important for the insect's antiviral response. Thanks to genome-wide transcriptomic analyses using microarrays, together with RNA interference (RNAi)-mediated gene silencing, the Toll, Imd, and JAK-STAT pathways have been identified as key players in the control of mosquito immune responses to DENV and WNV infection (Dong *et al.*, 2012; Luplertlop *et al.*, 2011; Paradkar *et al.*, 2012; Ramirez and Dimopoulos, 2010; Souza-Neto *et al.*, 2009; Xi *et al.*, 2008). A recent study in a panel of field-derived *Ae. aegypti* revealed that the basal level of immune pathway activity contributes to the mosquitoes' susceptibility to DENV infection (Sim *et al.*, 2013). In addition to the classical immune pathways, insects also use other specialized responses, such as the RNAi mechanism, prophenoloxidase activation cascade, and apoptosis, to combat arbovirus infection (Adelman *et al.*, 2001; Franz *et al.*, 2006; Ocampo *et al.*, 2013).

Innate immune signalling is triggered by specific pathogen recognition receptors (PRRs) that recognize molecular patterns conserved among microbes, termed pathogen-associated molecular patterns (PAMPs). These PAMPs include lipopolysaccharides, peptidoglycans, mannans, and dsRNA (Blair, 2011; Shi *et al.*, 2012; Tsakas and Marmaras, 2010; Welchman *et al.*, 2009). Once pathogens are recognized, PRRs activate different pathogen-specific signalling cascades, which regulate the transcription of effector molecules (Blair, 2011; De Gregorio *et al.*, 2001; Irving *et al.*, 2001). The innate immune response to virus infection in both vertebrates and insects is activated by dsRNA, a PAMP for viruses. In vertebrates, the Toll-like receptors (TLRs) recognize viral RNA (Alexopoulou *et al.*, 2001; Diebold *et al.*, 2004) and are important in the immune responses to virus infection (Alexopoulou *et al.*, 2001; Diebold, 2008). In addition to TLRs, PRRs containing a DExD/H-Box RNA helicase domain, the retinoic acid-inducible gene I (RIG-I), and the melanoma-associated gene 5 can also recognize cytoplasmic viral RNA (Baum and García-Sastre, 2010). In mammals, virus recognition by PRRs induces the expression of type I interferon (IFN), which triggers the activation of the JAK/STAT pathway (Shuai and Liu, 2003). JAK/STAT activation results in the expression of the anti-viral gene, mRNA degradation, and translational arrest. However, although mammals and insects recognize similar PAMPs, the mosquito's immune response to virus infection is very different from the vertebrate system. A summary of the immune signalling pathways is presented in Fig. 8.1.

The Toll pathway

The Toll pathway is an NF-κB signalling pathway that was first characterized in terms of its role in *Drosophila* development and subsequently shown to play a role in insect immune responses against Gram-positive bacteria, fungi, and viruses (Belvin and Anderson, 1996; Kim and Kim, 2005; Lemaitre *et al.*, 1996; Rutschmann *et al.*, 2002; Valanne *et al.*, 2011; Zambon *et al.*, 2005). The vertebrate TLR signalling pathway was later shown to participate in vertebrate immune activation by recognizing the PAMPs and mounting an immune response associated with a specific pathogen (Aderem and Ulevitch, 2000; Medzhitov *et al.*, 1997; Takeda and Akira, 2001). Unlike the vertebrate TLR signalling pathway, the insect Toll pathway does not directly interact with PAMPs. Instead, the recognition of pathogens by peptidoglycan recognition proteins (PGRPs) or Gram-negative binding proteins (GNBPs) triggers a proteolytic cascade that cleaves Spätzle (Spz), a cytokine ligand for the Toll receptor (DeLotto and DeLotto, 1998; Lemaitre *et al.*, 1996). Activated Spz binds to the Toll receptor and triggers Toll pathway signalling through phosphorylation of the cytoplasmic adaptor proteins, MyD88, Tube, and the kinase Pelle (Lemaitre *et al.*, 1996). The activation of the Toll signalling pathway leads to the degradation of the negative regulator, Cactus, which binds to the NF-kB transcription factor Dorsal (Rel1 in mosquitoes) (Nicolas *et al.*, 1998). After being released from Cactus, Dorsal is translocated to the nucleus and binds to *cis*-acting elements of the promoters of antimicrobial peptides and other immune effector genes (Lemaitre *et al.*, 1996; Lemaitre and Hoffmann, 2007; Shin *et al.*, 2005).

The role of the Toll pathway in the insect anti-viral response was first described in terms of the *Drosophila* response to

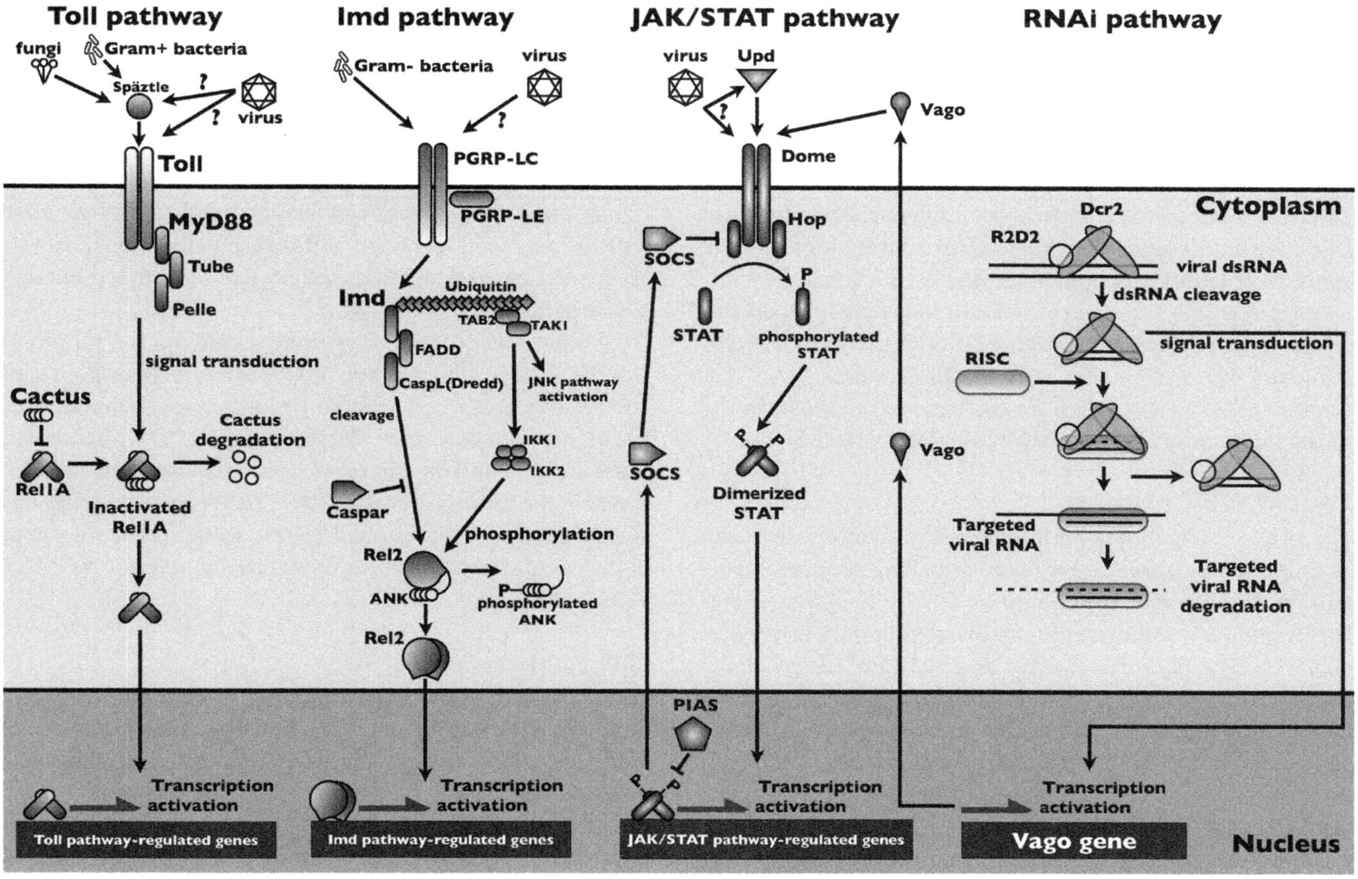

Figure 8.1 Summary of the insect Toll, Imd, JAK/STAT, and RNAi pathways (based on *Drosophila* and mosquito model). Three classical immune pathways are specifically activated upon recognition of PAMPs by receptors on the cell surface. The recognition trigger signalling cascades that results in the activation of transcription factors that control expression of AMPs and other immune factors. The RNAi pathway serves as an intracellular PRR and immune defence mechanism. The RNAi pathway can also activate the JAK/STAT pathway and provide systemic protection against arbovirus infection.

Drosophila X virus (DXV) infection (Zambon *et al.*, 2005). *Drosophila* mutants deficient in Toll pathway activation had higher virus titres and mortality upon DXV infection when compared to wild-type flies. A genome-wide transcriptomic comparison between DENV-infected and naive *Ae. aegypti* revealed that Toll pathway-associated genes (Spz, Toll, Rel1A, and several AMPs) were activated in response to DENV infection (Xi *et al.*, 2008). Activation of the Toll pathway through the RNAi-mediated gene silencing of Cactus resulted in a reduction of midgut DENV titres, while inactivation of Toll pathway signalling by silencing the adaptor protein MyD88 resulted in higher midgut DENV titres (Xi *et al.*, 2008). Subsequent experiments revealed that the role of the Toll pathway in controlling DENV is conserved in field-derived *Ae. aegypti* and in various DENV serotypes (Ramirez and Dimopoulos, 2010). In a study using *Ae. aegypti* strains with different degrees of susceptibility to DENV infection (Sim *et al.*, 2013), activation of the Toll pathway resulted in decreased DENV titres in susceptible *Ae. aegypti* strains, while silencing of the Toll pathway rendered the DENV-refractory *Ae. aegypti* strains more susceptible to infection. These results demonstrate the importance of the Toll pathway in controlling DENV infection. In addition to its link to DENV infection in *Ae. aegpti*, the Toll pathway has also been linked to the immediate response of *Ae. aegypti* to Sindbis virus (SINV) infection (Sanders *et al.*, 2005) and the response of *C. quinquefasciatus* to West Nile virus (WNV) infection (Chelsea T Smartt, 2009).

The Imd pathway

The immune deficiency (Imd) pathway is another major immune signalling pathway in insects that regulates an NF-kB family transcription factor (the detailed regulation of this pathway is reviewed in Aggarwal and Silverman, 2008). The pathway has been linked to the insect responses to bacterial infection, and its activation results in the production of AMPs and the induction of apoptosis (Georgel *et al.*, 2001). In contrast to the Toll pathway, which requires Spz as a ligand, the Imd pathway is activated by pathogens PAMPs through the peptidoglycan recognition proteins (PGRPs) PGRP-LC and PGRP-LE (Choe *et al.*, 2002; Kaneko and Silverman, 2005; Takehana *et al.*, 2002). The recognition of pathogen by PGRPs leads to an activation of the adaptor protein Imd and results in an activation of either the NF-kB signalling pathway or the c-Jun NH_2terminal kinase (JNK) pathway (Aggarwal and Silverman, 2008; Cirimotich *et al.*, 2010). It has been suggested that the recognition of bacteria activates the NF-kB branch by causing the degradation of the negative regulator, Caspar. The activation of the Imd pathway finally results in an activation of the Relish (Rel2 in mosquitoes) transcription factor by enzymatic cleavage of the IkB domain of Relish (Meister *et al.*, 2005). Activated Relish then promotes the transcription of AMPs and other immune defence mechanisms to fight against microbes (Cirimotich *et al.*, 2010; Meister *et al.*, 2005).

Up-regulation of the Imd modulators and Imd-regulated genes in response to arbovirus infection in mosquitoes has also been observed in SINV and DENV infection (Luplertlop *et al.*, 2011; Sanders *et al.*, 2005). However, direct activation of the Imd pathway by RNAi-mediated gene silencing of Caspar had no effect on mosquito midgut DENV titres (Xi *et al.*, 2008). A more recent comparison of different laboratory and field-derived *Ae. aegypti* has shown that activation of the Imd pathway in DENV-susceptible mosquito strains has no effect on midgut DENV titres, but an inactivation of the Imd pathway results in an increase in the midgut DENV titres of the DENV-refractory strains (Sim *et al.*, 2013). This result suggests that the Imd pathway is required for the anti-viral response in the mosquito but may not be sufficient to inhibit virus infection.

The JAK/STAT pathway

The JAK/STAT pathway was discovered in a vertebrate model as an interferon (cytokine)-induced signalling pathway important for development (Macchi *et al.*, 1995). The pathway was later found to be important for anti-viral immunity (Dupuis *et al.*, 2003; Fu *et al.*, 1992; Karst *et al.*, 2003). In *Drosophila*, the JAK/STAT pathway plays a crucial role as a signalling pathway in insect development and in the immune response against pathogenic bacteria and viruses (Arbouzova and Zeidler, 2006; Bina and Zeidler, 2009; Cronin *et al.*, 2009; Wang and Ligoxygakis, 2006).

Drosophila JAK/STAT signalling pathway is initiated by the binding of the activated cytokine-like Unpaired ligand (Upd) to the extracellular domain of the Domeless receptor (Dome) (Harrison *et al.*, 1998). The binding of Upd to the receptor triggers a conformational change and dimerization of the Dome receptor (Brown *et al.*, 2003). This dimerization then triggers the Janus kinase Hopscotch (Hop) to phosphorylate the cytosolic tail of the Dome receptor, which in turn activates STAT (Bina and Zeidler, 2009). The activated STAT is dimerized and translocated to the nucleus and triggers the transcription of JAK/STAT pathway-regulated genes (Wang and Ligoxygakis, 2006). The JAK/STAT pathway is negatively regulated by the protein inhibitors of activated STAT (PIAS), and suppressors of cytokine signalling (SOCS) repressor proteins to prevent its over-activation (Shuai and Liu, 2003).

A study of the Drosophila C virus (DCV) in the *Drosophila* model has revealed that DCV infection and virus replication result in an up-regulation of hundreds of JAK/STAT pathway-regulated genes, more than half of which do not overlap with genes induced by bacterial or fungal infection (Dostert *et al.*, 2005). Flies with the *hop* mutation have higher mortality after DCV infection than do wild-type flies, emphasizing the role of the JAK/STAT pathway in controlling virus infection (Dostert *et al.*, 2005).

An analysis of the *Ae. aegypti* mosquito genome has identified several homologues of the fly's JAK/STAT components, and the pathway has also been shown to play a role in the mosquito's responses to DENV infection (Dong *et al.*, 2012; Souza-Neto *et al.*, 2009). Transient activation of the JAK/STAT pathway by silencing of PIAS has been found to inhibit DENV replication in mosquito midguts, and transient blocking of the signal transduction by silencing Dome or Hop resulted in an increase in DENV titres (Souza-Neto *et al.*, 2009). This study also identified two JAK/STAT pathway-regulated DENV restriction factors, DVRF1 and DVRF2. DVRF1 contains a putative transmembrane domain and is suspected to be a pathway receptor. DVRF2 is a putative secreted protein thought to contain allergen C domains, which have been found in *Anopheles gambiae* PRRs that recognize the *Plasmodium* parasite, suggesting that DVRF2 might play a role as a DENV PRR. (Dong *et al.*, 2006; Souza-Neto *et al.*, 2009). Although these DVRFs have been identified and their putative functions was proposed, their real functions still need to be analysed through biochemical assays.

Another recent study suggesting a role for the mosquito JAK/STAT signalling pathway in the anti-DENV responses of DENV-susceptible and -resistant mosquitoes (Behura *et al.*, 2011) has demonstrated that several JAK/STAT-associated genes are differentially expressed in different mosquito strains; however, the function of these JAK/STAT-associated genes was not experimentally confirmed in this study. These mosquito strains are powerful tools for dissecting the roles of the JAK/STAT pathway in *Ae. aegypti*.

RNA interference

The identified immune signalling pathways discussed earlier are responsible for sensing virus infection and activating anti-viral mechanisms. However, the mechanisms that mosquitoes use to control arbovirus infection are not very well understood. In addition to classical immune signalling pathways, recent research suggests that RNA interference is a main mechanism in mosquito antiviral responses (Adelman *et al.*, 2001; Franz *et al.*, 2006). Other anti-viral mechanisms might also be involved but they are not as well characterized.

RNAi is a mechanism that can target foreign RNA for degradation, and it has long been recognized to be a key player among the mechanisms of anti-viral immunity in insects. This process relies on the Dicer2 (Dcr2) enzyme, which contains the DExD/H-Box RNA helicase domain and acts as a pattern recognition receptor in RNAi's recognition of exogenous dsRNAs (Deddouche *et al.*, 2008). Once they are recognized, Dcr2 cleaves long exogenous dsRNAs to generate 21–22 basepairs small-interfering RNAs (siRNAs). The siRNAs, together with Dcr2, can be loaded onto the RNA-induced silencing complex (RISC). During the effector stage of the pathway, RISC unwinds the siRNAs, degrades one of the RNA strands, and then guides it to the complementary RNA. Argonaut 2 (Ago2), a protein in the RISC complex that contains endonuclease activity, then degrades the target RNA strand (Ding and Voinnet, 2007). RNAi was previously characterized as an antiviral mechanism, but recent studies have shown that it can also function as a PRR for immune signalling pathways. In the *Drosophila* system, in addition to degrading target RNA, RNAi can also induce the expression of antiviral effectors, in a manner similar to RIG-I in mammals (Yoneyama *et al.*, 2004); for example, recognition of DCV by the DExD/H-Box RNA helicase domain of Dcr2 can induce the expression of the anti-viral effector Vago (Deddouche *et al.*, 2008).

Recent studies have demonstrated that the RNAi pathway also serves as an anti-DENV mechanism in *Ae. aegypti*. The very first evidence of a role for RNAi in modulating DENV infection was obtained with the transformation of the plasmid expressing inverted repeat DENV RNA (irRNA) in mosquito cells (Adelman *et al.*, 2001). Later, transgenic mosquitoes expressing inducible irRNA were also used to confirm the importance of RNAi (Franz *et al.*, 2006). These transgenic mosquitoes had lower DENV titres when compared with wild-type mosquitoes, suggesting a role for siRNA in anti-DENV responses. Knockdown of Ago2 in the transgenic mosquitoes negated the protective effect of the irRNA, confirming the importance of the RNAi mechanism (Franz *et al.*, 2006). However, the role of the RNAi mechanism in the anti-DENV defence in wildtype mosquitoes was not confirmed until 2009 (Sánchez-Vargas *et al.*, 2009). DENV infection in mosquito cell lines and adult female mosquitoes resulted in the production of siRNAs that could inhibit virus replication (Sánchez-Vargas *et al.*, 2009). On the other hand, transient silencing of the RNAi pathway components (Dcr2, R2D2, and Ago2) resulted in an increase in DENV titres and a reduction in the DENV extrinsic incubation period in mosquitoes (Sánchez-Vargas *et al.*, 2009).

Although siRNA has long been shown to be able to inhibit DENV in the mosquito, the biogenesis of the siRNA was still unclear until deep sequencing of DENV-specific small RNAs in mosquito cells was performed (Scott *et al.*, 2010). It has been speculated that the DENV-siRNA could be derived from the replicative intermediates of the virus or from secondary structure within the same RNA strand. The small RNA profile in DENV-infected mosquito cells has revealed an equal ratio of positive and negative strands, suggesting that the siRNAs were derived from the dsRNA replicative intermediates (Scott *et al.*, 2010). A study by Hess *et al.* (2011) identified 24–30-bp small RNAs during the early stage of DENV infection, suggesting that these RNAs are derived from the PIWI-RNAi pathway, which normally regulates germline gene expression, stem cell maintenance, and suppression of retrotransposons (Carmell *et al.*, 2002; Kalmykova *et al.*, 2005). The PIWI-mediated RNAi pathway was also shown to be sufficient to inhibit Rift Valley fever virus in the mosquito C6/36 cells (Leger *et al.*, 2013). It is therefore possible that alternative RNAi pathways also play a role in anti-viral mechanisms, and the link between the canonical and alternative RNAi pathways should be pursued.

Characterization of the role of RNAi in the systemic immune response is also important for our understanding of how the mosquito systemically controls virus infection. Previous studies of DCV in *Drosophila* have found a systemic spread of RNAi through the uptake of dsRNA from the cellular environment (Saleh *et al.*, 2009). However, a similar mechanism has not yet been identified in the mosquito's anti-DENV response. It is complicated to confirm this phenomenon in mosquitoes because systemic RNAi was originally discovered in a *Drosophila* mutant that is deficient in the dsRNA uptake pathway, and no such mutant is available in the mosquito system.

Cross-talk between immune signaling pathways

A genome-wide transcriptomic comparison between susceptible and refractory *Ae. aegypti* strains revealed that genes in several immune signalling pathways are similarly expressed among strains with similar vector competence (Behura *et al.*, 2011; Sim *et al.*, 2013). Transcriptomic comparison between DENV infected and uninfected mosquitoes showed that genes from different immune signalling pathways were simultaneously activated upon DENV infection (Sim *et al.*, 2012; Xi *et al.*, 2008). These studies suggested that the mosquito's immune response to arbovirus infection is more complex than expected and that there might be cross-talk between multiple signalling pathways. Recent studies have suggested the importance of the Toll and JAK/STAT pathways in controlling DENV infection in the mosquito (Souza-Neto *et al.*, 2009; Xi *et al.*, 2008). However, it is still unclear whether these two pathways work independently or synergistically. A study in the *Drosophila* has suggested that there is cross-talk between immune signalling pathways during the insect anti-viral response, since activation of the JAK/STAT pathway alone failed to induce the expression of anti-viral effectors (Dostert *et al.*, 2005). In the vertebrate system, cross-talk has been demonstrated between the Toll and JAK/STAT pathways in the defence against virus infection. Both Toll and JAK/STAT pathway activation result in the over-expression of genes of the negative regulator family, Suppressor of cytokine signalling (SOCS), which in turn inhibits the signal transduction of both pathways to prevent over-activation of the immune response (Shuai and Liu, 2003). However, the role of SOCS genes in the mosquito JAK/STAT pathway is not well characterized, and their role in the mosquito Toll pathway has not been studied. Functional assays of these genes using RNAi-mediated gene silencing can help us understand how these genes regulate the immune signalling pathways and also confirm whether the cross-talk actually exists.

Another important missing link is the relationship between the classical immune pathways and the RNAi mechanism in the mosquito's responses to arbovirus infection. Thus far, the only identified PRR for DENV is Dcr2 in the RNAi pathway, but whether Dcr2 serves as a PRR for the Toll and JAK/STAT pathways in the mosquito is still unclear. The evidence for cross-talk between the JAK/STAT and RNAi pathways comes from the observation that induction of the *Culex* Vago gene by Dcr2 can inhibit West Nile virus (WNV) through JAK/STAT pathway activation (Paradkar *et al.*, 2012). This molecule and mechanism is limited to insects, and the *Ae. aegypti* homologue of the Vago gene may also act through an anti-viral mechanism similar to that of *Culex* Vago. Understanding the cross-talk between pathways can help our understanding of the mechanisms that underlie systemic anti-viral immune responses. This knowledge is also useful for designing Vago-overexpressing transgenic mosquitoes as a novel vector control measure. One benefit of Vago-overexpressing transgenic mosquitoes is that the JAK/STAT activation by Vago is more virus-specific than an over-expression of the JAK/STAT components.

Other anti-viral mechanisms

Several innate immune mechanisms such as apoptosis and melanization have been shown to be used as insect defence mechanisms against virus infection (Liu *et al.*, 2011, 2013; Rodriguez-Andres *et al.*, 2012). Transcriptomic comparisons between refractory and susceptible mosquito strains have suggested that programmed cell death or apoptosis is involved in the control of DENV in mosquitoes (Ocampo *et al.*, 2013); this study found that an inhibition of apoptosis through silencing of caspase 16 resulted in higher DENV titres, but an activation of apoptosis by transient silencing of an Inhibitor of the Apoptosis I (IAP-1) gene resulted in lower DENV titres. These findings suggest that apoptosis is important as an anti-DENV mechanism in *Ae. aegypti*. Michelob_x (mx), an antagonist of the IAP gene, was also up-regulated after DENV infection of refractory mosquitoes, suggesting that it also plays a role in DENV infection (Liu *et al.*, 2013). However, one study has shown that induction of apoptosis resulted in an increase in midgut Sindbis virus (SINV) titres, while the suppression of apoptosis reduced SINV titres (Wang *et al.*, 2012). This study suggests that arbovirus can use apoptosis as a way to facilitate dissemination.

Melanization is an innate immune response triggered by the prophenoloxidase (PPO) enzymatic cascade (reviewed in (Cerenius and Söderhäll, 2004)) that results in phagocytosis and encapsulation (Cerenius and Söderhäll, 2004; Cerenius *et al.*, 2008). A role for the prophenoloxidase system in antiviral response was shown in a study of Semliki Forest virus (SFV) infection in *Ae. aegypti* (Rodriguez-Andres *et al.*, 2012). Infection with a recombinant SFV that expressed the PPO cascade inhibitor, Egf1.0, resulted in higher virus titres than the wild-type virus, suggesting the implication of the PPO cascade in immune responses to arbovirus infection.

Arboviruses influence the vector immune system to facilitate infections

Arboviruses and insect vectors have co-evolved together for a long time, and have both evolved ways to counteract each other. Insect vectors developed an immune system to control virus infection, while viruses suppress insect immune responses to establish an infection.

DENV has been reported to be able to antagonize immune activation in the *Ae. aegypti* Aag2 cell line (Sim and Dimopoulos, 2010); this study found that the induction of Toll and IMD pathway-responsive genes in response to bacterial challenge is inhibited in DENV-infected cells. However, the mechanism(s) the virus used to suppress the insect immune response was not identified. A more recent study identified lipid-binding molecules in the myeloid differentiation 2-related lipid recognition protein (ML) gene family as potential negative immune regulators; it found that DENV might induce these genes in order to suppress mosquito immune pathways (Jupatanakul *et al.*, 2014). In addition to DENV, SFV has also been found to be able to suppress immune signalling pathways in the *Ae. albopictus* U4.4 cell line (Fragkoudis *et al.*, 2008).

In order to deal with the effects of the RNAi mechanism, the genome of arboviruses contains non-coding regions that can suppress RNAi in both mammalian and insect cells (Schnettler *et al.*, 2012). Schnettler and coworkers used a model of flaviviral RNA replication-suppressed siRNA-induced gene silencing in WNV and DENV replicon-expressing S2 *Drosophila* cells. WNV and DENV 3′ untranslated region-derived RNA molecules were found to be able to suppress RNAi activity, and thus promote the replication of the WNV and DENV constructs. However, this study was only done *in vitro*, and whether such events occur *in vitro* needs to be confirmed.

Beyond insects and arboviruses: the impact of symbionts and the microbiome on arbovirus vectorial capacity

Mosquitoes, like other organisms, are exposed to a wide range of microbes from their environments, as well as during feeding. Therefore, microorganisms in the environment to which the mosquito is exposed are likely to play a role in shaping the insect's immune response and finally determine the outcome of the disease transmission. As in other organisms, the symbiotic relationship between the host (mosquitoes) and microorganisms plays a crucial role in mosquito digestion, metabolism, and immune maturation (Charroux and Royet, 2010; Dillon and Dillon, 2004; Ryu *et al.*, 2008). Studies in the *Drosophila* model have suggested that the Toll pathway is involved in responses to a broad spectrum of microorganisms, and the presence of microorganisms in contact with insect vectors is likely to influence the arboviral vectorial capacity. In other words, exposure to one microorganism can provide cross-protection against another microorganism.

The study of these tripartite interactions is currently a growing field that is aimed at developing novel disease control strategies. Several microorganisms and insect models have been shown to be able to reduce vector competence for pathogens such as *Plasmodium* and arboviruses (Cirimotich *et al.*, 2011a; Xi *et al.*, 2008). This area of study of the tripartite interaction in insects can be divided into two main research focuses, the mosquito microbiome and *Wolbachia* endosymbionts.

The microbiome and mosquitoes: a complex tripartite relationship

Several studies suggest an importance of gut microbiome on insects immune responses. The immune interactions between insects and their symbionts have been well characterized in tsetse flies. Weiss *et al.* (2011) have shown that when the tsetse fly lacks its mutualist *Wigglesworthia glossinidia* during development, its humoral and cellular immune responses failed to be activated. Commensal gut bacteria in *Drosophila* can activate Relish, thereby triggering translocation of Relish and activating the Imd pathway in intestinal tissue (Ryu *et al.*, 2008). Even though the gut bacteria activate the Imd pathway, the expression of AMPs was not induced in this study. It was found that Caudal is a negative regulator that controls the expression of AMPs in the presence of commensal bacteria, thus preventing the elimination of the microflora from the insect gut (Ryu *et al.*, 2008).

Several studies have also described the effect of *Serratia* and *Enterobacter* bacteria in *Anopheline* mosquito guts on the inhibition of *Plasmodium* development (Cirimotich *et al.*, 2011a; Gonzalez-Ceron *et al.*, 2003). However, very few studies have investigated the influences of the mosquito microbiota on arbovirus infection and disease transmission. The mosquito's gut microbiome is of particular interest to the field because the gut is the first line of pathogen exposure, and it is also relatively easy to manipulate microbes in the mosquito gut.

Studies have tried to identify the composition of the *Aedes* mosquito gut microbiome by using either classical cultivation methods or 16s rRNA sequencing (Chouaia *et al.*, 2010; Terenius *et al.*, 2012; Valiente Moro *et al.*, 2013; Zouache *et al.*, 2011). These studies have identified bacteria of the phylum Firmicutes, such as *Bacillus*; the phylum Actinobacteria, such as *Streptomyces*, *Microbacterium*, and *Terrabacterium*; and the phylum Proteobacteria, such as *Asaia*, *Pantoea*, *Acenetobacter*, *Enterobacter* and *Pseudomonas*. It has been suggested that the composition of the gut bacteria can differ depending on the environment and source of the insect's diet, since male and female *Aedes* mosquitoes have been shown to vary in bacterial composition (Valiente Moro *et al.*, 2013). For example, Proteobacteria are predominant in the female mosquito gut, perhaps because these microorganisms can tolerate the redox stress that occurs during bloodmeal ingestion better than other groups of bacteria. In contrast, Actinobacteria, which are commonly associated with soil and water, are predominant in males, and are potentially acquired from the sugar meal or the environment (Valiente Moro *et al.*, 2013). Many studies have shown that the microbiome can either promote or inhibit arbovirus infection in insects through various mechanisms (Fig. 8.2).

It has been shown that the presence of certain bacterial isolates in the mosquito midgut provides a basal level of immune activity that protects the mosquito from arbovirus infection. The first evidence showing that mosquito gut bacteria can influence DENV susceptibility was provided by Xi *et al.* (2008). The study showed that the presence of mosquito gut microbes provides mosquitoes with basal immune activation, resulting in lower DENV titres than in aseptic mosquitoes (Xi *et al.*, 2008). The field mosquito-derived bacterial isolates *Proteus* sp. and *Paenibacillus* sp. were also shown to prime mosquito gut immune responses against DENV infection when the bacteria were ingested via bloodmeal, and *Proteus* sp. also affected DENV infection when ingested via a sugar meal (Ramirez *et al.*, 2012). Conversely, this study also showed that DENV infection resulted in a decrease in the number of midgut bacteria as a result of immune activation and AMP production (Ramirez *et al.*, 2012). In the entomopathogenic fungus *Beauveria bassiana* has also been shown to activate the Toll and JAK/STAT pathways, and mediate suppression of DENV infection (Dong *et al.*, 2012). The mosquito microbiome could also serve as a physical barrier and thereby directly influence pathogen infection by blocking mosquito gut epithelial cells from pathogen exposure (Cirimotich *et al.*, 2011b). Alternatively, these microorganisms could produce secondary metabolites that inhibit arboviruses. *Enterobacter ludwigii*, *Pseudomonas rhodesiae* and *Vagococcus salmoninarium* inhibit La Crosse virus when co-incubated with the virus prior to infection of Vero cell (Joyce *et al.*, 2011).

Most of the studies on the mosquito microbiome have focused almost exclusively on bacteria. However, studies have also suggested a role for fungi and yeast in mosquito biology. Several studies have identified *Candida* and *Pichia* yeast in the mosquito gut, but the impact of these yeasts on mosquito immunity was not documented (Gusmão *et al.*, 2010; Ignatova *et al.*, 1996). More recent studies have identified *Wickerhamomyces anomalus* (*Pichia anomala*) in the midguts and reproductive organs of both malaria and dengue mosquito vectors (Ricci *et al.*, 2011). One interesting aspect of this yeast is that it has already been used as a biocontrol agent against other fungi in agriculture, because it can produce and secrete toxins. However, the effect of this yeast on arbovirus vectorial capacity has yet to be studied. Another advantage of this yeast is that molecular tools such as established transformation plasmids, coupled with a thorough knowledge of the secretion system of *Pichia*, also make it a good candidate for constructing recombinant yeast expressing anti-DENV effectors such as Cecropin C or DVRF 1 and 2. In term of field applications, this yeast is classified as a biosafety level 1 organism, meaning that it has no known hazards to healthy individuals. Given its lack of restrictions on handling, it is more likely to be used in the field than are bacteria (De Hoog, 1996; Fredlund *et al.*, 2002; Ricci *et al.*, 2012).

Application of microbiome in the filed also carries the risk

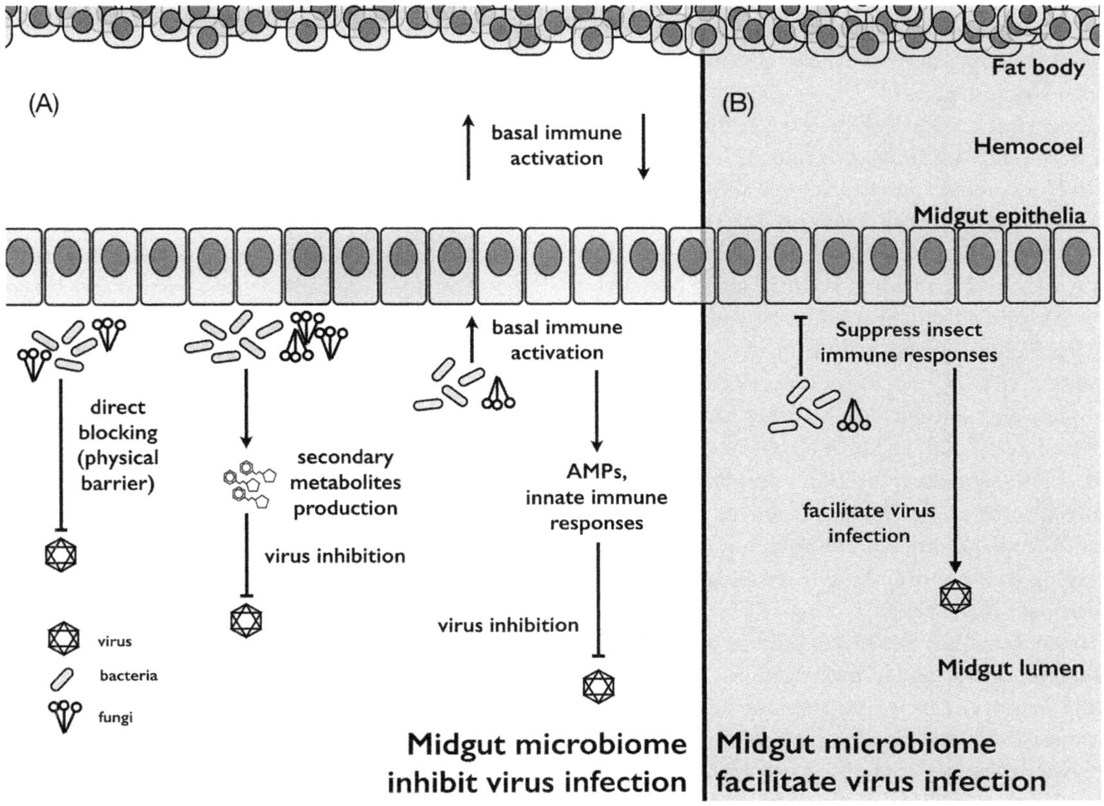

Figure 8.2 Tripartite interactions between the insect gut microbiome and arboviruses. (A) The gut microbiome can inhibit arbovirus infection by a direct mechanism involving physical blocking, secondary metabolites, or an indirect mechanism involving the activation of immune signalling pathways. (B) Some of the microbiota can facilitate arbovirus infection through suppression of insects' immune responses.

of increasing arbovirus vector competence. Insect gut microbiome can also facilitate arbovirus infection as demonstrated in a study between midgut bacteria, *Serratia odorifera*, and DENV (Apte-Deshpande *et al.*, 2012). The study found that the bacteria bound to prohibitin, a protein that has been linked to immune responses including the Toll pathway, thus block DENV recognition and immune signalling.

Endosymbionts: the effect of *Wolbachia* symbiosis on mosquito immune responses and arboviral vectorial capacity

Wolbachia is a maternally inherited insect endosymbiont estimated to infect 66% of insects in the world (Hilgenboecker *et al.*, 2008). *Wolbachia* was first identified in the ovaries of *Culex* mosquitoes almost a century ago (Hertig and Wolbach, 1924), but its interaction with mosquitoes has only been dissected in the past decade. *Wolbachia* is able to spread among insects rapidly because of its ability to change mosquito reproduction (feminization, parthenogenesis, and cytoplasmic incompatibility) to increase the reproductive success of the infected female (Iturbe-Ormaetxe *et al.*, 2011). One of the important contributing factors to the widespread success of *Wolbachia* in insects is its ability to generate cytoplasmic incompatibility (CI), meaning that uninfected eggs fertilized with sperm from infected males will not survive (Yen and Barr, 1971; Dobson *et al.*, 2002). This phenotypic advantage has proved to be useful for driving and maintaining the *Wolbachia* population in mosquitoes in recent field release tests (Hoffmann *et al.*, 2011). More recently, *Wolbachia* strains from *Drosophila* have been shown to provide the insect host with resistance to infection by viral pathogens (Teixeira *et al.*, 2008).

Despite the fact that several mosquito species, including *Ae. Albopictus*, are naturally infected with *Wolbachia*, the main DENV vector *Ae. aegypti* does not have a native *Wolbachia* endosymbiont (Iturbe-Ormaetxe *et al.*, 2011; Kittayapong *et al.*, 2000; Ricci *et al.*, 2002; Sinkins *et al.*, 1995). *Wolbachia* from other insects such as *Drosophila* and *Ae. albopictus* has been successfully introduced into *Ae. aegypti* mosquitoes, and the pathogen protection property is inherited by the *Ae. aegypti* (Bian *et al.*, 2010; Moreira *et al.*, 2009; Walker *et al.*, 2011). *Wolbachia* is a powerful biological control agent because it can interfere with a broad spectrum of pathogens. In addition to DENV, *Wolbachia* has also been shown to be able to inhibit yellow fever virus (YFV), Chikugunya virus (CHIKV), WNV, filarial nematodes, and *Plasmodium* parasites (Cirimotich *et al.*, 2011b; Hussain *et al.*, 2013; Iturbe-Ormaetxe *et al.*, 2011; Kambris *et al.*, 2010; Moreira *et al.*, 2009; van den Hurk *et al.*, 2012). Moreover, some of the *Wolbachia* strains also shorten the lifespan of the mosquitoes, thereby reducing the chance of DENV developing in the mosquito and reaching the salivary glands (McMeniman *et al.*, 2009).

The mechanism by which *Wolbachia* inhibits a broad spectrum of pathogens is still under investigation. Studies have shown that the activation of innate immune responses and the production of AMPs and effector molecules, for example, cecropin, defensin, thioester-containing proteins (TEP), and C-type lectins, both contribute to the pathogen's inhibition (Kambris *et al.*, 2010; Moreira *et al.*, 2009). A study by Pan *et al.* has demonstrated that *Wolbachia* can activate the Toll pathway through the induction of reactive oxygen species (Pan *et al.*, 2012). Melanization, a cellular immune response in insects, is induced by *Wolbachia* infection in both *Drosophila* and *Ae. aegypti*, suggesting that pre-activation of the innate immune response might be a contributing factor to pathogen inhibition (Thomas *et al.*, 2011).

Immune activation is not likely to be the only mechanism of *Wolbachia*-mediated pathogen inhibition, since one of the *Wolbachia* strains has been shown to inhibit *Drosophila* C virus without immune activation (Bourtzis *et al.*, 2000; Osborne *et al.*, 2009). One possible alternative cause of pathogen inhibition is competition for the cellular resources involved in replication. *Wolbachia* can grow in mosquito cells to very high densities, and *Wolbachia*-mediated pathogen inhibition seems to correlate with the bacterial load in insect cells (Iturbe-Ormaetxe *et al.*, 2011; Moreira *et al.*, 2009; Osborne *et al.*, 2009). A more recent study has revealed that *Wolbachia* can also influence mosquito DNA methylation through the expression of microRNA aae-miR-2940, which promotes *Wolbachia* replication but inhibits DENV replication (Zhang *et al.*, 2013). These findings suggest that different phenotypic changes in *Wolbachia*-infected mosquitoes result in a synergistic inhibition of infection with pathogens.

Wolbachia-infected mosquitoes have now been tested as a DENV biocontrol agent in the field setting (Hoffmann *et al.*, 2011; Rasgon, 2011; Walker *et al.*, 2011). A *Wolbachia*-infected mosquito wMel strain was approved to be released in two sites in Queensland, Australia. The mosquitoes were released at regular intervals over 2½ months, and it was found that they replaced natural mosquito inhabitants almost completely in one site and to about 80% in the other site (Hoffmann *et al.*, 2011; Rasgon, 2011). Although the DENV protection efficiency has not yet been assessed, this study is a proof-of-principle for the feasibility of using *Wolbachia* as a novel tool for arthropod-borne disease intervention.

Conclusion

The global situation with regard to arboviral diseases is worsening over time, and the control of these diseases requires integrated measures. Studies of molecular interactions between insect vectors and arboviruses, particularly those interactions involving the insect immune system, provide us alternative approaches for novel insect and disease control strategies. It has been shown that the molecular interactions are far more complex than just a simple insect–virus relationship; instead, other microorganisms, both prokaryotic and eukaryotic, also affect arboviral vector competence. The dissection of these complex relationships may lead to powerful tools that will decrease the burden of arboviral diseases.

References

Adelman, Z.N., Blair, C.D., Carlson, J.O., Beaty, B.J., and Olson, K.E. (2001). Sindbis virus-induced silencing of dengue viruses in mosquitoes. Insect Mol. Biol. *10*, 265–273.

Aderem, A., and Ulevitch, R.J. (2000). Toll-like receptors in the induction of the innate immune response. Nature *406*, 782–787.

Agaisse, H., and Perrimon, N. (2004). The roles of JAK/STAT signaling in *Drosophila* immune responses. Immunol. Rev. *198*, 72–82.

Aggarwal, K., and Silverman, N. (2008). Positive and negative regulation of the *Drosophila* immune response. BMB Rep. *41*, 267–277.

Alexopoulou, L., Holt, A.C., Medzhitov, R., and Flavell, R.A. (2001). Recognition of double-stranded RNA and activation of NF-kappaB by Toll-like receptor 3. Nature *413*, 732–738.

Apte-Deshpande, A., Paingankar, M., Gokhale, M.D., and Deobagkar, D.N. (2012). *Serratia odorifera* a Midgut Inhabitant of *Aedes aegypti* Mosquito Enhances Its Susceptibility to Dengue-2 Virus. PLoS ONE *7*, e40401.

Arbouzova, N.I., and Zeidler, M.P. (2006). JAK/STAT signalling in *Drosophila*: insights into conserved regulatory and cellular functions. Development *133*, 2605–2616.

Arensburger, P., Megy, K., Waterhouse, R.M., Abrudan, J., Amedeo, P., Antelo, B., Bartholomay, L., Bidwell, S., Caler, E., Camara, F., *et al.* (2010). Sequencing of *Culex quinquefasciatus* establishes a platform for mosquito comparative genomics. Science *330*, 86–88.

Ault, S.K. (1994). Environmental management: a re-emerging vector control strategy. Am. J. Trop. Med. Hyg. *50*, 35–49.

Barillas-Mury, C.C., and Kumar, S.S. (2005). *Plasmodium*–mosquito interactions: a tale of dangerous liaisons. Cell. Microbiol. *7*, 1539–1545.

Baum, A., and García-Sastre, A. (2010). Induction of type I interferon by RNA viruses: cellular receptors and their substrates. Amino Acids *38*, 1283–1299.

Behura, S.K., Gomez-Machorro, C., Harker, B.W., deBruyn, B., Lovin, D.D., Hemme, R.R., Mori, A., Romero-Severson, J., and Severson, D.W. (2011). Global cross-talk of genes of the mosquito *Aedes aegypti* in response to dengue virus infection. PLoS Neglect. Trop. Dis. *5*, e1385.

Belvin, M.P., and Anderson, K.V. (1996). A conserved signaling pathway: the *Drosophila* toll-dorsal pathway. Annu. Rev. Cell Dev. Biol. *12*, 393–416.

Bhatt, S., Gething, P.W., Brady, O.J., Messina, J.P., Farlow, A.W., Moyes, C.L., Drake, J.M., Brownstein, J.S., Hoen, A.G., Sankoh, O., *et al.* (2013). The global distribution and burden of dengue. Nature *496*, 504–507.

Bian, G., Xu, Y., Lu, P., Xie, Y., and Xi, Z. (2010). The endosymbiotic bacterium *Wolbachia* induces resistance to dengue virus in *Aedes aegypti*. PLoS Pathog. *6*, e1000833.

Bina, S., and Zeidler, M. (2009). JAK/STAT pathway signalling in *Drosophila melanogaster*. In JAK-STAT Pathway in Disease, A. Stephanou, ed. (Landes Bioscience, Austin, TX, USA), pp. 24–42.

Black, W.C., Bennett, K.E., Gorrochótegui-Escalante, N., Barillas-Mury, C.V., Fernández-Salas, I., de Lourdes Muñoz, M., Farfán-Alé, J.A., Olson, K.E., and Beaty, B.J. (2002). Flavivirus susceptibility in *Aedes aegypti*. Arch. Med. Res. *33*, 379–388.

Blair, C.D. (2011). Mosquito RNAi is the major innate immune pathway controlling arbovirus infection and transmission. Future Microbiol. *6*, 265–277.

Blandin, S.A., and Levashina, E.A. (2007). Phagocytosis in mosquito immune responses. Immunol. Rev. *219*, 8–16.

Blandin, S., Shiao, S.-H., Moita, L.F., Janse, C.J., Waters, A.P., Kafatos, F.C., and Levashina, E.A. (2004). Complement-like protein TEP1 is a determinant of vectorial capacity in the malaria vector *Anopheles gambiae*. Cell *116*, 661–670.

Bourtzis, K., Pettigrew, M.M., and O'Neill, S.L. (2000). *Wolbachia* neither induces nor suppresses transcripts encoding antimicrobial peptides. Insect Mol. Biol. *9*, 635–639.

Brown, S., Hu, N., and Hombría, J.C.-G. (2003). Novel level of signalling control in the JAK/STAT pathway revealed by *in situ* visualisation of protein–protein interaction during *Drosophila* development. Development *130*, 3077–3084.

Carmell, M.A., Xuan, Z., Zhang, M.Q., and Hannon, G.J. (2002). The Argonaute family: tentacles that reach into RNAi, developmental control, stem cell maintenance, and tumorigenesis. Genes Dev. *16*, 2733–2742.

Cerenius, L., and Söderhäll, K. (2004). The prophenoloxidase-activating system in invertebrates. Immunol. Rev. *198*, 116–126.

Cerenius, L., Lee, B.L., and Söderhäll, K. (2008). The proPO-system: pros and cons for its role in invertebrate immunity. Trends Immunol. *29*, 263–271.

Charroux, B., and Royet, J. (2010). *Drosophila* immune response: From systemic antimicrobial peptide production in fat body cells to local defense in the intestinal tract. Fly (Austin) *4*, 40–47.

Choe, K.-M., Werner, T., Stöven, S., Hultmark, D., and Anderson, K.V. (2002). Requirement for a peptidoglycan recognition protein (PGRP) in Relish activation and antibacterial immune responses in *Drosophila*. Science *296*, 359–362.

Chouaia, B., Rossi, P., Montagna, M., Ricci, I., Crotti, E., Damiani, C., Epis, S., Faye, I., Sagnon, N., Alma, A., *et al.* (2010). Molecular evidence for multiple infections as revealed by typing of *Asaia* bacterial symbionts of four mosquito species. Appl. Environ. Microbiol. *76*, 7444–7450.

Cirimotich, C.M., Scott, J.C., Phillips, A.T., Geiss, B.J., and Olson, K.E. (2009). Suppression of RNA interference increases alphavirus replication and virus-associated mortality in *Aedes aegypti* mosquitoes. BMC Microbiol. *9*, 49.

Cirimotich, C.M., Dong, Y., Garver, L.S., Sim, S., and Dimopoulos, G. (2010). Mosquito immune defenses against *Plasmodium* infection. Dev. Comparative Immunol. *34*, 387–395.

Cirimotich, C.M., Dong, Y., Clayton, A.M., Sandiford, S.L., Souza-Neto, J.A., Mulenga, M., and Dimopoulos, G. (2011a). Natural microbe-mediated refractoriness to *Plasmodium* infection in *Anopheles gambiae*. Science *332*, 855–858.

Cirimotich, C.M., Ramirez, J.L., and Dimopoulos, G. (2011b). Native Microbiota Shape Insect Vector Competence for Human Pathogens. Cell Host Microbe *10*, 307–310.

Cleton, N., Koopmans, M., Reimerink, J., Godeke, G.-J., and Reusken, C. (2012). Come fly with me: review of clinically important arboviruses for global travelers. J. Clin. Virol. *55*, 191–203.

Cronin, S.J.F., Nehme, N.T., Limmer, S., Liegeois, S., Pospisilik, J.A., Schramek, D., Leibbrandt, A., Simoes, R. de M., Gruber, S., Puc, U., *et al.* (2009). Genome-wide RNAi screen identifies genes involved in intestinal pathogenic bacterial infection. Science *325*, 340–343.

De Gregorio, E., Spellman, P.T., Rubin, G.M., and Lemaitre, B. (2001). Genome-wide analysis of the *Drosophila* immune response by using oligonucleotide microarrays. Proc. Natl. Acad. Sci. U.S.A. *98*, 12590–12595.

De Hoog, G.S. (1996). Risk assessment of fungi reported from humans and animals. Mycoses *39*, 407–417.

Deddouche, S., Matt, N., Budd, A., Mueller, S., Kemp, C., Galiana-Arnoux, D., Dostert, C., Antoniewski, C., Hoffmann, J.A., and Imler, J.-L. (2008). The DExD/H-box helicase Dicer-2 mediates the induction of antiviral activity in *Drosophila*. Nat. Immunol. *9*, 1425–1432.

DeLotto, Y.Y., and DeLotto, R.R. (1998). Proteolytic processing of the *Drosophila* Spätzle protein by easter generates a dimeric NGF-like molecule with ventralising activity. Mech. Dev. *72*, 141–148.

Diebold, S.S. (2008). Recognition of viral single-stranded RNA by Toll-like receptors. Adv. Drug Deliv. Rev. *60*, 813–823.

Diebold, S.S., Kaisho, T., Hemmi, H., Akira, S., and Reis e Sousa, C. (2004). Innate antiviral responses by means of TLR7-mediated recognition of single-stranded RNA. Science *303*, 1529–1531.

Dillon, R.J., and Dillon, V.M. (2004). The gut bacteria of insects: nonpathogenic interactions. Annu. Rev. Entomol. *49*, 71–92.

Dimarcq, J.L., Hoffmann, D., Meister, M., Bulet, P., Lanot, R., Reichhart, J.M., and Hoffmann, J.A. (1994). Characterization and transcriptional profiles of a *Drosophila* gene encoding an insect defensin. Eur. J. Biochem. *221*, 201–209.

Dimopoulos, G. (2003). Insect immunity and its implication in mosquito–malaria interactions. Cell. Microbiol. *5*, 3–14.

Ding, S.-W., and Voinnet, O. (2007). Antiviral Immunity Directed by Small RNAs. Cell *130*, 413–426.

Dobson, S.L., Fox, C.W., and Jiggins, F.M. (2002). The effect of *Wolbachia*-induced cytoplasmic incompatibility on host population size in natural and manipulated systems. Proc. Biol. Sci. *269*, 437–445.

Dong, K. (2007). Insect sodium channels and insecticide resistance. Invert. Neurosci. *7*, 17–30.

Dong, Y., Aguilar, R., Xi, Z., Warr, E., Mongin, E., and Dimopoulos, G. (2006). *Anopheles gambiae* immune responses to human and rodent *Plasmodium* parasite species. PLoS Pathog. *2*, e52–e525.

Dong, Y., Morton, J.C., Ramirez, J.L., Souza-Neto, J.A., and Dimopoulos, G. (2012). The entomopathogenic fungus *Beauveria bassiana* activate toll and JAK-STAT pathway-controlled effector genes and anti-dengue activity in *Aedes aegypti*. Insect Biochem. Mol. Biol. *42*, 126–132.

Dostert, C., Jouanguy, E., Irving, P., Troxler, L., Galiana-Arnoux, D., Hetru, C., Hoffmann, J.A., and Imler, J.-L. (2005). The Jak-STAT signaling pathway is required but not sufficient for the antiviral response of *Drosophila*. Nat. Immunol. *6*, 946–953.

Dupuis, S., Jouanguy, E., Al-Hajjar, S., Fieschi, C., Al-Mohsen, I.Z., Al-Jumaah, S., Yang, K., Chapgier, A., Eidenschenk, C., Eid, P., *et al.* (2003). Impaired response to interferon-α/β and lethal viral disease in human STAT1 deficiency. Nat. Genet. *33*, 388–391.

Fragkoudis, R., Chi, Y., Siu, R.W.C., Barry, G., Attarzadeh-Yazdi, G., Merits, A., Nash, A.A., Fazakerley, J.K., and Kohl, A. (2008). Semliki Forest virus strongly reduces mosquito host defence signaling. Insect Mol. Biol. *17*, 647–656.

Franz, A.W.E., Sánchez-Vargas, I., Adelman, Z.N., Blair, C.D., Beaty, B.J., James, A.A., and Olson, K.E. (2006). Engineering RNA interference-based resistance to dengue virus type 2 in genetically modified *Aedes aegypti*. Proc. Natl. Acad. Sci. U.S.A. *103*, 4198–4203.

Fredlund, E., Druvefors, U., Boysen, M.E., Lingsten, K.-J., and Schnürer, J. (2002). Physiological characteristics of the biocontrol yeast *Pichia anomala* J121. FEMS Yeast Res. *2*, 395–402.

Fu, X.Y., Schindler, C., Improta, T., Aebersold, R., and Darnell, J.E. (1992). The proteins of ISGF-3, the interferon alpha-induced transcriptional activator, define a gene family involved in signal transduction. Proc. Natl. Acad. Sci. U.S.A. *89*, 7840–7843.

Georgel, P., Naitza, S., Kappler, C., Ferrandon, D., Zachary, D., Swimmer, C., Kopczynski, C., Duyk, G., Reichhart, J.M., and Hoffmann, J.A. (2001). *Drosophila* immune deficiency (IMD) is a death domain protein that activates antibacterial defense and can promote apoptosis. Dev. Cell *1*, 503–514.

Gonzalez-Ceron, L., Santillan, F., Rodriguez, M.H., Mendez, D., and Hernandez-Avila, J.E. (2003). Bacteria in midguts of field-collected *Anopheles albimanus* block *Plasmodium vivax* sporogonic development. J. Med. Entomol. *40*, 371–374.

Gubler, D.J. (1998). Dengue and dengue hemorrhagic fever. Clin. Microbiol. Rev. *11*, 480–496.

Gusmão, D.S., Santos, A.V., Marini, D.C., Bacci, M., Jr., Berbert-Molina, M.A., and Lemos, F.J.A. (2010). Culture-dependent and culture-independent characterization of microorganisms associated with *Aedes aegypti* (Diptera: Culicidae) (L.) and dynamics of bacterial colonization in the midgut. Acta Trop. *115*, 275–281.

Hardy, J.L., Houk, E.J., Kramer, L.D., and Reeves, W.C. (1983). Intrinsic factors affecting vector competence of mosquitoes for arboviruses. Annu. Rev. Entomol. *28*, 229–262.

Harrison, D.A., McCoon, P.E., Binari, R., Gilman, M., and Perrimon, N. (1998). *Drosophila* unpaired encodes a secreted protein that activates the JAK signaling pathway. Genes Dev. *12*, 3252–3263.

Hertig, M., and Wolbach, S.B. (1924). Studies on *Rickettsia*-Like Micro-Organisms in Insects. J. Med. Res. *44*, 329–374.7.

Hess, A.M., Prasad, A.N., Ptitsyn, A., Ebel, G.D., Olson, K.E., Barbacioru, C., Monighetti, C., and Campbell, C.L. (2011). Small RNA profiling of Dengue virus–mosquito interactions implicates the PIWI RNA pathway in anti-viral defense. BMC Microbiol. *11*, 45.

Hilgenboecker, K., Hammerstein, P., Schlattmann, P., Telschow, A., and Werren, J.H. (2008). How many species are infected with *Wolbachia*? – A statistical analysis of current data. FEMS Microbiol. Lett. *281*, 215–220.

Hoffmann, A.A., Montgomery, B.L., Popovici, J., Iturbe-Ormaetxe, I., Johnson, P.H., Muzzi, F., Greenfield, M., Durkan, M., Leong, Y.S., Dong, Y., *et al.* (2011). Successful establishment of *Wolbachia* in *Aedes* populations to suppress dengue transmission. Nature *476*, 454–457.

van den Hurk, A.F., Hall-Mendelin, S., Pyke, A.T., Frentiu, F.D., McElroy, K., Day, A., Higgs, S., and O'Neill, S.L. (2012). Impact of *Wolbachia* on infection with chikungunya and yellow fever viruses in the mosquito vector *Aedes aegypti*. PLoS Neglect. Trop. Dis. *6*, e1892.

Hussain, M., Lu, G., Torres, S., Edmonds, J.H., Kay, B.H., Khromykh, A.A., and Asgari, S. (2013). Effect of *Wolbachia* on replication of West Nile virus in a mosquito cell line and adult mosquitoes. J. Virol. *87*, 851–858.

Ignatova, E.A., Nagornaia, S.S., Povazhnaia, T.N., and Ianishevskaia, G.S. (1996). [The yeast flora of blood-sucking mosquitoes]. Mikrobiol. Z. *58*, 12–15.

Irving, P., Troxler, L., Heuer, T.S., Belvin, M., Kopczynski, C., Reichhart, J.M., Hoffmann, J.A., and Hetru, C. (2001). A genome-wide analysis of immune responses in *Drosophila*. Proc. Natl. Acad. Sci. U.S.A. *98*, 15119–15124.

Irving, P., Ubeda, J.-M., Doucet, D., Troxler, L., Lagueux, M., Zachary, D., Hoffmann, J.A., Hetru, C., and Meister, M. (2005). New insights into *Drosophila* larval haemocyte functions through genome-wide analysis. Cell. Microbiol. *7*, 335–350.

Iturbe-Ormaetxe, I., Walker, T., and O'Neill, S.L. (2011). *Wolbachia* and the biological control of mosquito-borne disease. EMBO Rep. *12*, 508–518.

Joyce, J.D., Nogueira, J.R., Bales, A.A., Pittman, K.E., and Anderson, J.R. (2011). Interactions between La Crosse virus and bacteria isolated from the digestive tract of *Aedes albopictus* (Diptera: Culicidae). J. Med. Entomol. *48*, 389–394.

Jupatanakul, N., Sim, S., and Dimopoulos, G. (2013). *Aedes aegypti* ML and Niemann-Pick type C family members are agonists of dengue virus infection. Dev. Comparative Immunol. *43*, 1–9.

Kalmykova, A.I., Klenov, M.S., and Gvozdev, V.A. (2005). Argonaute protein PIWI controls mobilization of retrotransposons in the *Drosophila* male germline. Nucleic Acids Res. *33*, 2052–2059.

Kambris, Z., Blagborough, A.M., Pinto, S.B., Blagrove, M.S.C., Godfray, H.C.J., Sinden, R.E., and Sinkins, S.P. (2010). *Wolbachia* stimulates immune gene expression and inhibits *Plasmodium* development in *Anopheles gambiae*. PLoS Pathog. *6*, e1001143.

Kaneko, T., and Silverman, N. (2005). Bacterial recognition and signalling by the *Drosophila* IMD pathway. Cell. Microbiol. *7*, 461–469.

Karst, S.M., Wobus, C.E., Lay, M., Davidson, J., and Virgin, H.W. (2003). STAT1-dependent innate immunity to a Norwalk-like virus. Science *299*, 1575–1578.

Kim, T., and Kim, Y.-J. (2005). Overview of innate immunity in *Drosophila*. J. Biochem. Mol. Biol. *38*, 121–127.

Kittayapong, P., Baisley, K.J., Baimai, V., and O'Neill, S.L. (2000). Distribution and diversity of *Wolbachia* infections in Southeast Asian mosquitoes (Diptera: Culicidae). J. Med. Entomol. *37*, 340–345.

Lavine, M.D., and Strand, M.R. (2002). Insect hemocytes and their role in immunity. Insect Biochem. Mol. Biol. *32*, 1295–1309.

Lawson, D., Arensburger, P., Atkinson, P., Besansky, N.J., Bruggner, R.V., Butler, R., Campbell, K.S., Christophides, G.K., Christley, S., Dialynas, E., *et al.* (2009). VectorBase: a data resource for invertebrate vector genomics. Nucleic Acids Res. *37*, D583–D587.

Leger, P., Lara, E., Jagla, B., Sismeiro, O., Mansuroglu, Z., Coppee, J.Y., Bonnefoy, E., and Bouloy, M. (2013). Dicer-2- and Piwi-mediated RNA interference in Rift Valley fever virus-infected mosquito cells. J. Virol. *87*, 1631–1648.

Lemaitre, B., and Hoffmann, J. (2007). The host defense of *Drosophila melanogaster*. Annu. Rev. Immunol. *25*, 697–743.

Lemaitre, B., Nicolas, E., Michaut, L., Reichhart, J.M., and Hoffmann, J.A. (1996). The dorsoventral regulatory gene cassette spätzle/Toll/cactus controls the potent antifungal response in *Drosophila* adults. Cell *86*, 973–983.

Levashina, E.A. (2004). Immune responses in *Anopheles gambiae*. Insect Biochem. Mol. Biol. *34*, 673–678.

Liu, B., Becnel, J.J., Zhang, Y., and Zhou, L. (2011). Induction of reaper ortholog mx in mosquito midgut cells following Baculovirus infection. Cell Death Differentiation *18*, 1337–1345.

Liu, B., Behura, S.K., Clem, R.J., Schneemann, A., Becnel, J., Severson, D.W., and Zhou, L. (2013). P53-Mediated Rapid Induction of Apoptosis Conveys Resistance to Viral Infection in *Drosophila melanogaster*. PLoS Pathog. *9*, e1003137.

Luplertlop, N., Surasombatpattana, P., Patramool, S., Dumas, E., Wasinpiyamongkol, L., Saune, L., Hamel, R., Bernard, E., Sereno, D., Thomas, F., *et al.* (2011). Induction of a peptide with activity against a broad spectrum of pathogens in the *Aedes aegypti* salivary gland, following infection with dengue virus. PLoS Pathog. *7*, e1001252.

Macchi, P., Villa, A., Giliani, S., Sacco, M.G., Frattini, A., Porta, F., Ugazio, A.G., Johnston, J.A., Candotti, F., and O'Shea, J.J. (1995). Mutations of Jak-3 gene in patients with autosomal severe combined immune deficiency (SCID). Nature *377*, 65–68.

McMeniman, C.J., Lane, R.V., Cass, B.N., Fong, A.W.C., Sidhu, M., Wang, Y.F., and O'Neill, S.L. (2009). Stable introduction of a life-shortening *Wolbachia* infection into the mosquito *Aedes aegypti*. Science *323*, 141–144.

Medzhitov, R.R., Preston-Hurlburt, P.P., and Janeway, C.A.C. (1997). A human homologue of the *Drosophila* Toll protein signals activation of adaptive immunity. Nature *388*, 394–397.

Megy, K., Emrich, S.J., Lawson, D., Campbell, D., Dialynas, E., Hughes, D.S.T., Koscielny, G., Louis, C., MacCallum, R.M., Redmond, S.N., *et al.* (2011). VectorBase: improvements to a bioinformatics resource for invertebrate vector genomics. Nucleic Acids Res. *40*, D729–D734.

Meister, S., Kanzok, S.M., Zheng, X.-L., Luna, C., Li, T.-R., Hoa, N.T., Clayton, J.R., White, K.P., Kafatos, F.C., Christophides, G.K., *et al.* (2005). Immune signaling pathways regulating bacterial and malaria parasite infection of the mosquito *Anopheles gambiae*. Proc. Natl. Acad. Sci. U.S.A. *102*, 11420–11425.

Moreira, L.A., Iturbe-Ormaetxe, I., Jeffery, J.A., Lu, G., Pyke, A.T., Hedges, L.M., Rocha, B.C., Hall-Mendelin, S., Day, A., Riegler, M., *et al.* (2009). A *Wolbachia* symbiont in *Aedes aegypti* limits infection with dengue, Chikungunya, and *Plasmodium*. Cell *139*, 1268–1278.

Nene, V., Wortman, J.R., Lawson, D., Haas, B., Kodira, C., Tu, Z.J., Loftus, B., Xi, Z., Megy, K., Grabherr, M., *et al.* (2007). Genome sequence of *Aedes aegypti*, a major arbovirus vector. Science *316*, 1718–1723.

Nicolas, E.E., Reichhart, J.M.J., Hoffmann, J.A.J., and Lemaitre, B.B. (1998). In vivo regulation of the IkappaB homologue cactus during the immune response of *Drosophila*. J. Biol. Chem. *273*, 10463–10469.

Ocampo, C.B., Caicedo, P.A., Jaramillo, G., Ursic Bedoya, R., Baron, O., Serrato, I.M., Cooper, D.M., and Lowenberger, C. (2013). Differential expression of apoptosis related genes in selected strains of *Aedes aegypti* with different susceptibilities to dengue virus. PLoS ONE *8*, e61187.

Osborne, S.E., Leong, Y.S., O'Neill, S.L., and Johnson, K.N. (2009). Variation in Antiviral Protection Mediated by Different *Wolbachia* Strains in *Drosophila simulans*. PLoS Pathog. *5*, e1000656.

Pan, X., Zhou, G., Wu, J., Bian, G., Lu, P., Raikhel, A.S., and Xi, Z. (2012). *Wolbachia* induces reactive oxygen species (ROS)-dependent activation of the Toll pathway to control dengue virus in the mosquito *Aedes aegypti*. Proc. Natl. Acad. Sci. U.S.A. *109*, E23–E31.

Paradkar, P.N., Trinidad, L., Voysey, R., Duchemin, J.-B., and Walker, P.J. (2012). Secreted Vago restricts West Nile virus infection in *Culex* mosquito cells by activating the Jak-STAT pathway. Proc. Natl. Acad. Sci. U.S.A. *109*, 18915–18920.

Ramirez, J.L., and Dimopoulos, G. (2010). The Toll immune signaling pathway control conserved anti-dengue defenses across diverse *Ae. aegypti* strains and against multiple dengue virus serotypes. Dev. Comparative Immunol. *34*, 625–629.

Ramirez, J.L., Souza-Neto, J., Torres Cosme, R., Rovira, J., Ortiz, A., Pascale, J.M., and Dimopoulos, G. (2012). Reciprocal tripartite interactions between the *Aedes aegypti* midgut microbiota, innate immune system and dengue virus influences vector competence. PLoS Neglect. Trop. Dis. *6*, e1561.

Rasgon, J.L. (2011). Dengue fever: Mosquitoes attacked from within. Nature *476*, 407–408.

Reiter, P. (2010). Yellow fever and dengue: a threat to Europe? Eur. Surveill. *15*, 19509.

Ricci, I., Cancrini, G., Gabrielli, S., D'Amelio, S., and Favi, G. (2002). Searching for *Wolbachia* (Rickettsiales: Rickettsiaceae) in mosquitoes (Diptera: Culicidae): large polymerase chain reaction survey and new identifications. J. Med. Entomol. *39*, 562–567.

Ricci, I., Mosca, M., Valzano, M., Damiani, C., Scuppa, P., Rossi, P., Crotti, E., Cappelli, A., Ulissi, U., Capone, A., *et al.* (2011). Different mosquito species host *Wickerhamomyces anomalus* (*Pichia anomala*): perspectives on vector-borne diseases symbiotic control. Antonie Van Leeuwenhoek *99*, 43–50.

Ricci, I., Damiani, C., Capone, A., DeFreece C., Rossi, P., *et al.* (2012). Mosquito/microbiota interactions: from complex relationships to biotechnological perspectives. Curr. Opin. Microbiol. *15*, 278–284.

Rivero, A., Vézilier, J., Weill, M., Read, A.F., and Gandon, S. (2010). Insecticide control of vector-borne diseases: when is insecticide resistance a problem? PLoS Pathog. *6*, e1001000.

Rodriguez-Andres, J., Rani, S., Varjak, M., Chase-Topping, M.E., Beck, M.H., Ferguson, M.C., Schnettler, E., Fragkoudis, R., Barry, G., Merits, A., *et al.* (2012). Phenoloxidase activity acts as a mosquito innate immune response against infection with Semliki Forest virus. PLoS Pathog. *8*, e1002977.

Rutschmann, S., Kilinc, A., and Ferrandon, D. (2002). Cutting edge: the toll pathway is required for resistance to gram-positive bacterial infections in *Drosophila*. J. Immunol. *168*, 1542–1546.

Ryu, J.-H., Kim, S.-H., Lee, H.-Y., Bai, J.Y., Nam, Y.-D., Bae, J.-W., Lee, D.G., Shin, S.C., Ha, E.-M., and Lee, W.-J. (2008). Innate immune homeostasis by the homeobox gene caudal and commensal-gut mutualism in *Drosophila*. Science *319*, 777–782.

Salazar, M.I., Richardson, J.H., Sánchez-Vargas, I., Olson, K.E., and Beaty, B.J. (2007). Dengue virus type 2 replication and tropisms in orally infected. BMC Microbiol. *7*, 9–13.

Saleh, M.-C., Tassetto, M., van Rij, R.P., Goic, B., Gausson, V., Berry, B., Jacquier, C., Antoniewski, C., and Andino, R. (2009). Antiviral immunity in *Drosophila* requires systemic RNA interference spread. Nature *458*, 346–350.

Sanders, H.R., Foy, B.D., Evans, A.M., Ross, L.S., Beaty, B.J., Olson, K.E., and Gill, S.S. (2005). Sindbis virus induces transport processes and alters expression of innate immunity pathway genes in the midgut of the disease vector, *Aedes aegypti*. Insect Biochem. Mol. Biol. *35*, 1293–1307.

Sánchez-Vargas, I., Scott, J.C., Poole-Smith, B.K., Franz, A.W.E., Barbosa-Solomieu, V., Wilusz, J., Olson, K.E., and Blair, C.D. (2009). Dengue virus type 2 infections of *Aedes aegypti* are modulated by the mosquito's RNA interference pathway. PLoS Pathog. *5*, e1000299.

Schnettler, E., Sterken, M.G., Leung, J.Y., Metz, S.W., Geertsema, C., Goldbach, R.W., Vlak, J.M., Kohl, A., Khromykh, A.A., and Pijlman, G.P. (2012). Noncoding flavivirus RNA displays RNA interference suppressor activity in insect and Mammalian cells. J. Virol. *86*, 13486–13500.

Scott, J.C., Brackney, D.E., Campbell, C.L., Bondu-Hawkins, V., Hjelle, B., Ebel, G.D., Olson, K.E., and Blair, C.D. (2010). Comparison of dengue virus type 2-specific small RNAs from RNA interference-competent and -incompetent mosquito cells. PLoS Neglect. Trop. Dis. *4*, e848.

Shi, X.-Z., Zhong, X., and Yu, X.-Q. (2012). *Drosophila melanogaster* NPC2 proteins bind bacterial cell wall components and may function in immune signal pathways. Insect Biochem. Mol. Biol. *42*, 545–556.

Shin, S.W., Kokoza, V., Bian, G., Cheon, H.-M., Kim, Y.J., and Raikhel, A.S. (2005). REL1, a homologue of *Drosophila* dorsal, regulates toll antifungal immune pathway in the female mosquito *Aedes aegypti*. J. Biol. Chem. *280*, 16499–16507.

Shuai, K., and Liu, B. (2003). Regulation of JAK–STAT signalling in the immune system. Nat. Rev. Immunol. *3*, 900–911.

Sim, S., and Dimopoulos, G. (2010). Dengue virus inhibits immune responses in *Aedes aegypti* cells. PLoS ONE *5*, e10678.

Sim, S., Ramirez, J.L., and Dimopoulos, G. (2012). Dengue virus infection of the *Aedes aegypti* salivary gland and chemosensory apparatus induces genes that modulate infection and blood-feeding behavior. PLoS Pathog. *8*, e1002631.

Sim, S., Jupatanakul, N., Ramirez, J.L., Kang, S., Romero-Vivas, C.M., Mohammed, H., and Dimopoulos, G. (2013). Transcriptomic profiling of diverse *Aedes aegypti* strains reveals increased basal-level immune activation in dengue virus-refractory populations and identifies novel virus-vector molecular interactions. PLoS Neglect. Trop. Dis. *7*, e2295.

Sinkins, S.P., Braig, H.R., and O'Neill, S.L. (1995). *Wolbachia pipientis*: bacterial density and unidirectional cytoplasmic incompatibility between infected populations of *Aedes albopictus*. Exp. Parasitol. *81*, 284–291.

Smartt, C.T., Richards, S.L., Anderson, S.L., and Erickson, J.S. (2009). West Nile Virus Infection Alters Midgut Gene Expression in *Culex pipiens quinquefasciatus* Say (Diptera: Culicidae). Am. J. Trop. Med. Hyg. *81*, 258. Souza-Neto, J.A., Sim, S., and Dimopoulos, G. (2009). An evolutionary conserved function of the JAK-STAT pathway in anti-dengue defense. Proc. Natl. Acad. Sci. U.S.A. *106*, 17841–17846.

Soverow, J.E., Wellenius, G.A., Fisman, D.N., and Mittleman, M.A. (2009). Infectious disease in a warming world: how weather influenced West Nile virus in the United States (2001–2005). Environ. Health Perspect. *117*, 1049–1052.

Takeda, K., and Akira, S. (2001). Roles of Toll-like receptors in innate immune responses. Genes Cells *6*, 733–742.

Takehana, A., Katsuyama, T., Yano, T., Oshima, Y., Takada, H., Aigaki, T., and Kurata, S. (2002). Overexpression of a pattern-recognition receptor, peptidoglycan-recognition protein-LE, activates imd/relish-mediated antibacterial defense and the prophenoloxidase cascade in *Drosophila* larvae. Proc. Natl. Acad. Sci. U.S.A. *99*, 13705–13710.

Teixeira, L., Ferreira, A., and Ashburner, M. (2008). The bacterial symbiont *Wolbachia* induces resistance to RNA viral infections in *Drosophila melanogaster*. PLoS Biol. *6*, e2.

Terenius, O., Lindh, J.M., Eriksson-Gonzales, K., Bussière, L., Laugen, A.T., Bergquist, H., Titanji, K., and Faye, I. (2012). Midgut bacterial dynamics in *Aedes aegypti*. FEMS Microbiol. Ecol. *80*, 556–565.

Thomas, P., Kenny, N., Eyles, D., Moreira, L.A., O'Neill, S.L., and Asgari, S. (2011). Infection with the wMel and wMelPop strains of *Wolbachia* leads to higher levels of melanization in the hemolymph of *Drosophila melanogaster*, *Drosophila simulans* and *Aedes aegypti*. Dev. Comparative Immunol. *35*, 360–365.

Tjaden, N.B., Thomas, S.M., Fischer, D., and Beierkuhnlein, C. (2013). Extrinsic Incubation Period of Dengue: Knowledge, Backlog, and Applications of Temperature Dependence. PLoS Neglect. Trop. Dis. *7*, e2207.

Tsakas, S., and Marmaras, V. (2010). Insect immunity and its signalling: an overview. Invertebrate Surv. J. *7*, 228–238.

Valanne, S., Wang, J.-H., and Rämet, M. (2011). The *Drosophila* Toll signaling pathway. J. Immunol. *186*, 649–656.

Valiente Moro, C., Tran, F.H., Raharimalala, F.N., Ravelonandro, P., and Mavingui, P. (2013). Diversity of culturable bacteria including *Pantoea* in wild mosquito *Aedes albopictus*. BMC Microbiol *13*, 70.

Vilmos, P., and Kurucz, E. (1998). Insect immunity: evolutionary roots of the mammalian innate immune system. Immunol. Lett. *62*, 59–66.

Walker, T., Johnson, P.H., Moreira, L.A., Iturbe-Ormaetxe, I., Frentiu, F.D., McMeniman, C.J., Leong, Y.S., Dong, Y., Axford, J., Kriesner, P., *et al.* (2011). The *w*Mel *Wolbachia* strain blocks dengue and invades caged *Aedes aegypti* populations. Nature *476*, 450–453.

Wang, H., Gort, T., Boyle, D.L., and Clem, R.J. (2012). Effects of manipulating apoptosis on Sindbis virus infection of *Aedes aegypti* mosquitoes. J. Virol. *86*, 6546–6554.

Wang, L., and Ligoxygakis, P. (2006). Pathogen recognition and signalling in the *Drosophila* innate immune response. Immunobiology *211*, 251–261.

Watts, D.M., Burke, D.S., Harrison, B.A., Whitmire, R.E., and Nisalak, A. (1987). Effect of temperature on the vector efficiency of *Aedes aegypti* for dengue 2 virus. Am. J. Trop. Med. Hyg. *36*, 143–152.

Weaver, S.C. (2013). Urbanization and geographic expansion of zoonotic arboviral diseases: mechanisms and potential strategies for prevention. Trends Microbiol. *21*, 360–363.

Weaver, S.C., and Reisen, W.K. (2010). Present and future arboviral threats. Antiviral Res. *85*, 328–345.

Weiss, B.L., Wang, J.-H., and Aksoy, S. (2011). Tsetse immune system maturation requires the presence of obligate symbionts in larvae. PLoS Biol. *9*, e1000619.

Welchman, D.P., Aksoy, S., Jiggins, F., and Lemaitre, B. (2009). Insect immunity: from pattern recognition to symbiont-mediated host defense. Cell Host Microbe *6*, 107–114.

Whitten, M.M.A., Shiao, J., and Levashina, E.A. (2006). Mosquito midguts and malaria: cell biology, compartmentalization and immunology. Parasite Immunol. *28*, 121–130.

Xi, Z., Ramirez, J.L., and Dimopoulos, G. (2008). The *Aedes aegypti* toll pathway controls dengue virus infection. PLoS Pathog. *4*, e1000098.

Yen, J.H., and Barr, A.R. (1971). New hypothesis of the cause of cytoplasmic incompatibility in *Culex pipiens* L. Nature *232*, 657–658.

Yoneyama, M., Kikuchi, M., Natsukawa, T., Shinobu, N., Imaizumi, T., Miyagishi, M., Taira, K., Akira, S., and Fujita, T. (2004). The RNA helicase RIG-I has an essential function in double-stranded RNA-induced innate antiviral responses. Nat. Immunol. *5*, 730–737.

Zambon, R.A., Nandakumar, M., Vakharia, V.N., and Wu, L.P. (2005). The Toll pathway is important for an antiviral response in *Drosophila*. Proc. Natl. Acad. Sci. U.S.A. *102*, 7257–7262.

Zhang, G., Hussain, M., O'Neill, S.L., and Asgari, S. (2013). *Wolbachia* uses a host microRNA to regulate transcripts of a methyltransferase, contributing to dengue virus inhibition in *Aedes aegypti*. Proc. Natl. Acad. Sci. U.S.A. *110*, 10276–10281.

Zouache, K., Raharimalala, F.N., Raquin, V., Tran-Van, V., Raveloson, L.H.R., Ravelonandro, P., and Mavingui, P. (2011). Bacterial diversity of field-caught mosquitoes, *Aedes albopictus* and *Aedes aegypti*, from different geographic regions of Madagascar. FEMS Microbiol. Ecol. *75*, 377–389.

Part II

Viral Diversity and Evolution

Genetic Diversity of Arboviruses

9

Kenneth A. Stapleford, Gonzalo Moratorio and Marco Vignuzzi

Abstract

Arthropod-borne viruses (arboviruses) encompass a wide range of genetically distinct and diverse viral species, with many of these viral pathogens capable of causing severe disease in mammals and plants. Arboviruses are unique in that they must be able to infect and replicate in both invertebrate (mosquitoes, ticks, sand flies, midges, dipteran, and thrips) as well as vertebrate (humans, primates, birds, rodents) and plant hosts in order to maintain a successful viral lifecycle. To facilitate these processes, arboviruses have taken advantage of a wide range of genome compositions, structures, and organizations, as well as the ability to evolve and generate diverse populations within viral species. This unique genetic diversity, both at the genome structure level as well as the nucleotide level, plays essential roles in viral genome replication, transcription, translation, transmission, vector tropism, and pathogenesis. This chapter will introduce the role of genetic diversity in specific arboviruses and the implications of levels of diversity in the viral life cycle.

Introduction

Arboviruses make up a long list of pathogens of clinical, veterinary and agricultural importance with a world-wide distribution (Weaver and Reisen, 2010; Hubalek *et al.*, 2014). Currently, there are few vaccines and antiviral therapies targeted towards these viruses, in part due to the vast genetic diversity not only between families of arboviruses, but within the individual viral species themselves. It is this fundamental diversity of genome composition, structure, and organization that gives way to unique molecular and genetic mechanisms employed by arboviruses for replication, transmission, pathogenesis, and evolution (Gray and Banerjee, 1999). For instance, with the majority of arboviruses employing a RNA genome and thus, an error-prone polymerase lacking proof-reading mechanisms, these viruses undergo rapid evolution and may quickly adapt to their environment, which in many cases leads to significant genetic diversity at the nucleotide level within specific viral species themselves. Subsequently, upon selection for individual viruses through host, vector, climate or other external environmental pressures, these nucleotide level changes can lead to the emergence of novel viral strains and in the worst case scenario, a new epidemic. In addition to error-prone replication, these viruses utilize other mechanisms to generate genetic diversity, such as genetic recombination in the case of the alphaviruses and flaviviruses (Allison and Stallknecht, 2009; Hahn *et al.*, 1988; Weaver *et al.*, 1997), and genetic reassortment for the segmented viruses, such as members of the *Bunyaviridae*, *Reoviridae* and *Orthomyxoviridae* families (Cheng *et al.*, 1999; Reese *et al.*, 2008).

In this chapter, we provide an overview of the genetic diversity within and between arbovirus-containing virus families, the molecular mechanisms responsible for this diversity and some of the viral, host and environmental factors affecting it, that are explored in detail in other chapters.

Diversity in genome structure and organization between and within viral families

Because arboviruses are a categorical, rather than taxonomic grouping, its members belong to a variety of viral families that include the *Togaviridae, Flaviviridae, Bunyaviridae, Rhabdoviridae, Reoviridae, Orthomxyoviridae* and *Asfarviridae.* Consequently, the genetic diversity of arboviruses taken as an ensemble is extreme, as nearly every type of genome composition and structure is represented (positive-sense, single-stranded RNA viruses; negative-sense, single-stranded RNA viruses; segmented, negative-strand RNA viruses, double-stranded RNA viruses; and DNA viruses). Below, we highlight some of the shared and unique features in genome organization among these different families.

Togaviridae

The *Togaviridae* family consists of two genera, *Alphavirus* and *Rubivirus*, of which the genus *Rubivirus* consists of one member, rubella virus, which will not be discussed here. The alphaviruses, however, are composed of roughly 30 viral species, of which the majority are confirmed arboviruses (Fig. 9.1). Alphaviruses are 50–70 nm spherical viral particles encapsidating a single-stranded, positive-sense RNA genome of 11–12 kb (Coffey *et al.*, 2013; Weaver *et al.*, 2012). The alphavirus genome contains a 5′ 7-methylguanosine cap and is polyadenylated at the 3′ end (Fig. 9.2A), characteristics distinct from the single-stranded,

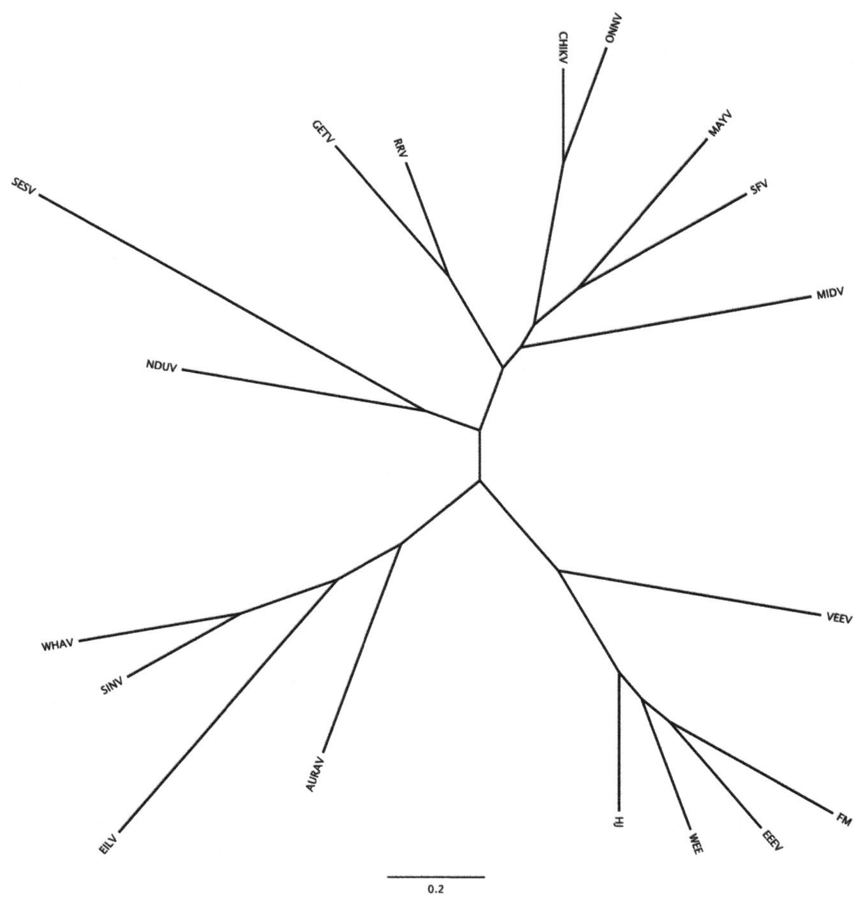

Figure 9.1 Phylogenetic relationships of the main arboviral members within the *Togaviridae* family. Phylogenetic relationships of the main arboviral members within the *Togaviridae* family. A maximum likelihood phylogenetic tree based on the alignment of the RdRp gene is shown. Branch lengths are proportional to the evolutionary distance (scale bar) between the taxa. Different members within the family are shown. AEFV (Aedes Flavivirus); ALFV (Alfuy virus); APOIV (Apoi virus); AROAV (Aroa virus); BAGV (Bagaza virus); BANV (Banzi virus); BOUV (Bouboi virus); BUSV (Bussuquara virus); CFAV (Cell Fusing Agent virus); CFV (*Culex* Flavivirus); CHAV (Chaoyang virus); DEDSV (Duck Egg Drop Syndrome virus); DENV-1 (Dengue vírus 1); DENV-2 (Dengue vírus 2); DENV-3 (Dengue vírus 3); DENV-4 (Dengue vírus 4); DGV (Donggang vírus); EHV (Edge Hill vírus); ENTV (Entebbe Bat vírus); GGEV (Greek Goat Encephalitis vírus); GGYV (Gadgets Gully vírus); IGV (Iguape vírus); ILHV (Ilheus vírus); JEV (Japanese Encephalitis vírus); JUGV (Jugra vírus); KADV (Kadam vírus); KEDV (Kedougou vírus); KOKV (Kokobera vírus); KOUV (Koutango vírus); KRV (Kamiti River vírus); KSIV (Karshi vírus); KYDV (Kyasanur forest Disease vírus); LGTV (Langat vírus); LIV (Louping Ill vírus); MEAV (Meaban vírus); MMLV (Montana Myotis Leukoencephalitis vírus); MODV (Modoc vírus); MVEV (Murray Valley Encephalitis vírus); NOUV (Nounane vírus); NTAV (Ntaya vírus); OHFU (Omsk Haemorrhagic Fever vírus); PCV (Palm Creek vírus);POWV (Powassan vírus); QBV (Quang Binh vírus); RBV (Rio Bravo vírus); RFV (Royal Farm vírus); ROCV (Rocio vírus); SABV (Saboya virus); SLEV (Saint Louis Encephalitis vírus); SREV (Saumarez Reef virus); SSEV (Spanish Sheep Encephalitis vírus); TBEV (Tick Borne Encephalitis vírus); TYUV (Tyuleniy virus); UGSV (Uganda S virus); UV (Usutu virus); WESSV (Wesselsbron virus); WNV (West Nile virus);YAOV (Yaounde virus); YFV (Yellow Fever virus); YOKV (Yokose vírus); ZV (Zika virus).

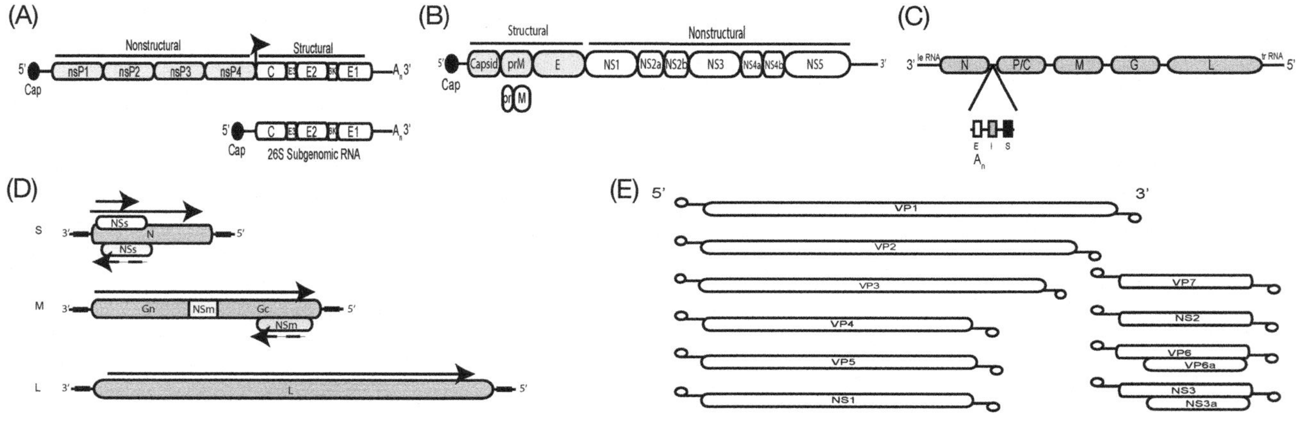

Figure 9.2 Generalized genomic organization of arbovirus containing families and genera. A. *Togaviridae*, Alphavirus B. *Flaviviridae*, Flavivirus C. *Rhabdoviridae* D. *Bunyaviridae*. E. *Reoviridae.* SINV (Sindbis virus); WHAV (Whataroa virus); AURAV (Aura virus); EILV (Eilat virus); WEEV (Western equine encephalitis virus); HJ (Highlands J virus); EEEV (Eastern equine encephalitis virus); FM (Fort Morgan virus); VEEV (Venezuelan equine encephalitis virus); SESV (Southern Elephant Seal virus); NDUV (Ndumu virus); MIDV (Middelburg virus); GETV (Getah virus); RRV (Ross River virus); SFV (Semliki Forest virus); MAYV (Mayaro virus); ONNV (O'yong-nyong virus); CHIKV (Chikungunya virus).

positive-sense flaviviruses that will be discussed later. Unlike the single-stranded, negative-sense RNA viruses, the alphavirus genomes resemble a cellular mRNA and are thus rapidly and efficiently translated once inside the cell, a feature unique to the positive-strand RNA viruses (Strauss and Strauss, 1994). Alphavirus genomes encode four non-structural proteins at the 5′ end required for positive and negative-strand RNA synthesis and five structural proteins at the 3′ end needed for maturation and assembly. The non-structural proteins consist of nsP1, required for RNA capping; nsP2, a multifunctional helicase/protease necessary for polyprotein processing, negative-strand RNA synthesis, host transcriptional shut-off and evasion of innate immunity; nsP3, an accessory protein for RNA synthesis; and nsP4, the RNA-dependent RNA polymerase. The structural proteins include the capsid protein, which forms the nucleocapsid with the viral RNA; E3, a leader protein for E2; the E2 glycoprotein, required for cellular attachment and receptor binding; 6k, a leader protein for E1; and the E1 fusogenic glycoprotein required for viral entry and cellular membrane fusion. These viral proteins are synthesized as a polyprotein and subsequently cleaved by cellular and viral proteases into their mature and functioning forms, a strategy used by many positive sense RNA viruses. Alphaviruses have acquired a number of unique genetic elements to regulate their replication cycle. For example, the genome contains an Opal stop codon following the nsP3 gene, which allows for early synthesis of the non-structural proteins nsP1-nsP2-nsP3 as one polyprotein, and in lesser amounts, the non-structural proteins nsP1-nsP2-nsP3-nsP4 as another polypeptide by read-through translation (Kato *et al.*, 2009; Kim *et al.*, 2004). This Opal codon thus creates a temporal stoichiometry of the non-structural proteins, which has been shown to regulate negative- and positive-strand genome synthesis and in the case of some viruses, increase infectivity (Myles *et al.*, 2006). In addition, to decouple the production of non structural proteins from structural ones (that are required in larger amounts for virus assembly), the alphaviruses have evolved a unique genetic organization. These two regions of the genome are separated by a subgenomic promoter, which during replication gives rise to a capped and polyadenylated 26S subgenomic RNA from which the structural proteins are translated. Finally, flanking the genome, as well as in the subgenomic promoter region, are conserved sequence elements (Birket *et al.*, 2013) required for efficient RNA genome synthesis (Fig. 9.2A).

Flaviviridae

The *Flaviviridae* family is made of three genera: *Flavivirus*, *Pestivirus* and *Hepacivirus*. Of these genera, the flaviviruses contain roughly 70 species that include numerous arboviruses and human pathogens, such as yellow fever virus, dengue virus, West Nile virus, tick-borne encephalitis virus, Japanese encephalitis virus and St. Louis encephalitis virus (Fig. 9.3). Among the flaviviruses, we can also distinguish between the mosquito- or tick-borne viruses and even further, between the encephalitic flaviviruses transmitted by *Culex* mosquitoes and the haemorrhagic viruses transmitted by *Aedes* mosquito species (Gould and Solomon, 2008).

Similar to the alphaviruses, flaviviruses are also small, enveloped viruses of 50 nm diameter that contain a single-stranded, positive-sense RNA genome of approximately 11 kb (Harris *et al.*, 2006) (Fig. 9.2B). The genome contains a 5′ cap and, with the exception of some tick-borne viruses, it is not polyadenylated at the 3′ end. As another distinction from the alphaviruses, the flavivirus genome is organized with the structural genes at the 5′ end of the genome, followed by the 3′ non-structural genes, without the decoupling strategy of a subgenomic promoter. Rather, the flavivirus genome is initially translated as a single polyprotein, and is subsequently cleaved by cellular and viral proteases into the mature forms of the viral proteins. The genome encodes three structural proteins: the capsid protein necessary for nucleocapsid production; the prM protein, which is subsequently cleaved by host furin protease into pr and M during viral maturation; and the E glycoprotein facilitating receptor binding and fusion and providing a major antigenic determinant for flaviviruses. The non-structural proteins include NS1, required for RNA replication; NS2A, necessary for RNA replication, assembly and an innate immune response antagonist; NSAB cofactor for NS3 protease activity; NS3 serine-protease, which cleaves the polyprotein and is the RNA helicase; NS4A and NS4B, required for RNA replication and inhibiting innate immune response; and NS5, the viral RNA-dependent RNA polymerase. Although flaviviruses do not contain a conventional subgenomic RNA as in the alphaviruses, they do encode a small RNA that corresponds to the 3′ portion of the genome, termed the subgenomic flavivirus RNA (sfRNA) (Pijlman, 2008; Chapman, 2014). The sfRNA is a product of incomplete RNA degradation due to its highly structured sequence. This small RNA plays distinct roles in viral replication and pathogenesis and has been shown to increase virulence in mice (Roby *et al.*, 2014). Once again, as for many RNA viruses, the 5′ and 3′ untranslated regions (UTR) of the flavivirus genome contain several conserved genetic elements that regulate RNA replication. However, these regions also contain significant variability among the viruses in this genus. The most dramatic differences are vector-related, between mosquito and tick borne flaviviruses, where most notably, some tick-borne flaviviruses do encode a polyA signal that is not the typical characteristic of the *Flaviviridae* (Wallner *et al.*, 1995).

Rhabdoviridae

The *Rhabdoviridae* family consists of 190 members divided into six genera (Fig. 9.4) (Kuzmin *et al.*, 2009). The *Lyssavirus* genus, which includes the well-studied rabies virus, and the *Novirhabdovirus* genus that infects fish are not known to be transmitted by arthoropod vectors. In contrast, arthropods can transmit the *Vesiculoviruses* (including the prototypical vesticular stomatitis virus) and the *Ephemeroviruses* [consisting of only bovine ephermeral fever virus (BEFV), Adelaide river virus, and Berrimah virus (BRMV))] to mammals and humans; while the *Cytorhabdoviruses* and *Nucleorhabdoviruses* are transmitted to plants (Jackson *et al.*, 2005). The members of the *Rhabdoviridae* family are composed of a single-strand, negative-sense RNA genome encapsidated within an enveloped 100- to 430-nm-long rod-shaped viral particle. The genome ranges from 10 kb for the vesiculoviruses to 12–15 kb for the plant viruses, and is uncapped and non-polyadenylated at the 3′ and 5′ ends, respectively. The rhabdoviruses are nonsegmented viruses and the viral genome encodes five genes in the order 3′-N-P-M-G-L-5′ (Fig. 9.2C). The N gene, encodes the viral nucleocapsid; P, a highly phosphorylated

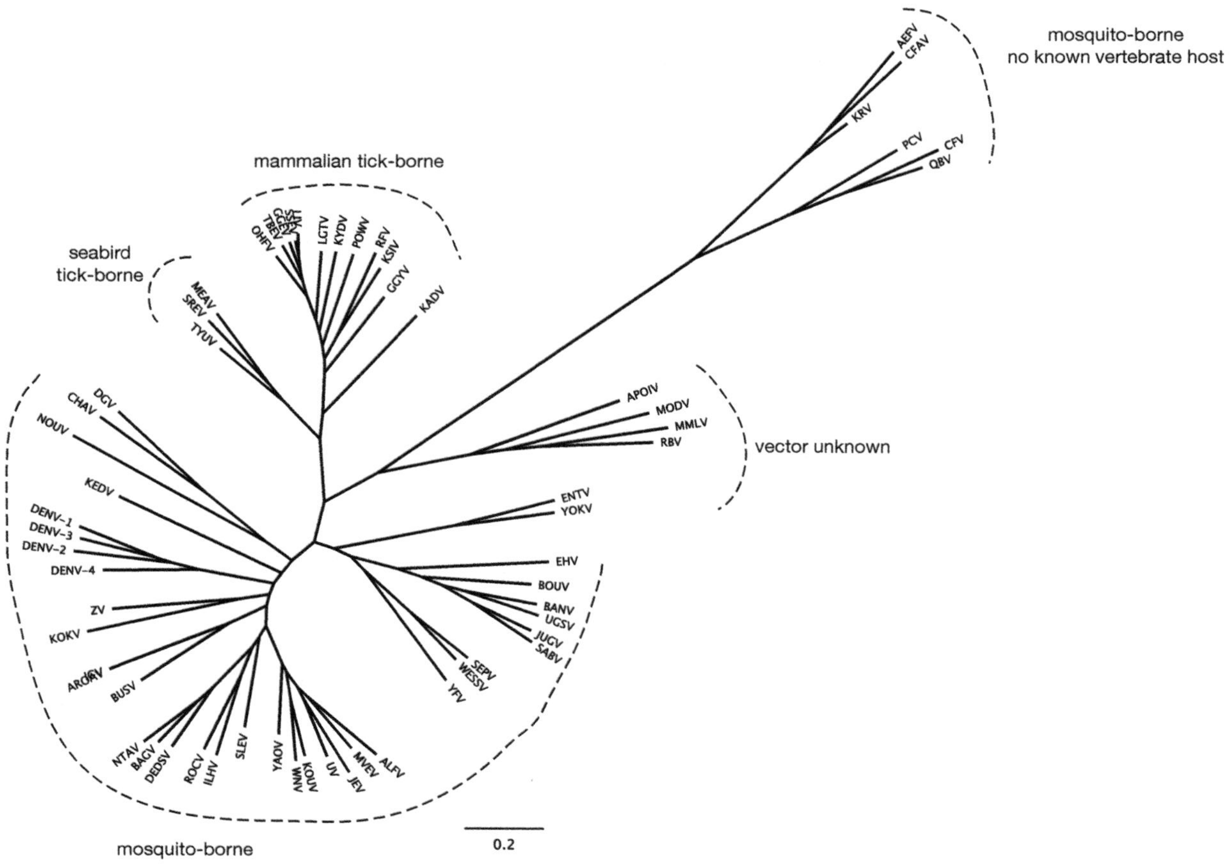

Figure 9.3 Phylogenetic relationships of the main arboviral members within the *Flaviviridae* family. A maximum likelihood phylogenetic tree based on the alignment of the NS5 region is shown. Branch lengths are proportional to the evolutionary distance (scale bar) between the taxa. The transmission modes, i.e. mosquito-borne, tick-borne and no known vector are indicated. AEFV (Aedes Flavivirus); ALFV (Alfuy virus); APOIV (Apoi virus); AROAV (Aroa virus); BAGV (Bagaza virus); BANV (Banzi virus); BOUV (Bouboi virus); BUSV (Bussuquara virus); CFAV (Cell Fusing Agent virus); CFV (*Culex* Flavivirus); CHAV (Chaoyang virus); DEDSV (Duck Egg Drop Syndrome virus); DENV-1 (Dengue vírus 1); DENV-2 (Dengue vírus 2); DENV-3 (Dengue vírus 3); DENV-4 (Dengue vírus 4); DGV (Donggang vírus); EHV (Edge Hill vírus); ENTV (Entebbe Bat vírus); GGEV (Greek Goat Encephalitis vírus); GGYV (Gadgets Gully vírus); IGV (Iguape vírus); ILHV (Ilheus vírus); JEV (Japanese Encephalitis vírus); JUGV (Jugra vírus); KADV (Kadam vírus); KEDV (Kedougou vírus); KOKV (Kokobera vírus); KOUV (Koutango vírus); KRV (Kamiti River vírus); KSIV (Karshi vírus); KYDV (Kyasanur forest Disease vírus); LGTV (Langat vírus); LIV (Louping Ill vírus); MEAV (Meaban vírus); MMLV (Montana Myotis Leukoencephalitis vírus); MODV (Modoc vírus); MVEV (Murray Valley Encephalitis vírus); NOUV (Nounane vírus); NTAV (Ntaya vírus); OHFU (Omsk Haemorrhagic Fever vírus); PCV (Palm Creek vírus);POWV (Powassan vírus); QBV (Quang Binh vírus); RBV (Rio Bravo vírus); RFV (Royal Farm vírus); ROCV (Rocio vírus); SABV (Saboya virus); SLEV (Saint Louis Encephalitis vírus); SREV (Saumarez Reef virus); SSEV (Spanish Sheep Encephalitis vírus); TBEV (Tick Borne Encephalitis vírus); TYUV (Tyuleniy virus); UGSV (Uganda S virus); UV (Usutu virus); WESSV (Wesselsbron virus); WNV (West Nile virus);YAOV (Yaounde virus); YFV (Yellow Fever virus); YOKV (Yokose vírus); ZV (Zika virus).

phosphoprotein; M, the matrix protein; G, the viral glycoprotein; and L, the RNA-dependent RNA polymerase. In addition to this set of proteins, different rhabdoviruses also encode a variety of accessory proteins that account for the large discrepancies found in genome size between genera (Kuzmin *et al.*, 2009; Walker *et al.*, 2011).

Bunyaviridae

Bunyaviridae is the largest virus family consisting of over 300 species (Soldan and Gonzalez-Scarano, 2005) (Kormelink *et al.*, 2011) (Fig. 9.5). The *Bunyaviridae* family contains five genera – *Orthobunyavirus, Hantavirus, Nairovirus, Phlebovirus* and *Tospovirus* – and represent a large group of human, animal and plant pathogens, including the medically relevant Rift Valley fever virus. The majority of bunyaviruses, with the exception of the *Hantavirus* genus, are transmitted by arthropods including mosquitoes, ticks, phlebotomine sandflies, midges and, in the case of plants, thrips. *Bunyaviridae* members are enveloped viruses of 80 to 120 nm in diameter containing a segmented, single-stranded, negative-sense RNA genome. The tripartite genome consists of large (L), medium (M), and small (S) segments, which encode individual viral proteins in individual open reading frames (Eifan *et al.*, 2013) (Fig. 9.2C). The L segment encodes the viral RNA-dependent RNA polymerase, required for initial transcription of viral messages and subsequent RNA synthesis. The M segment encodes the glycoproteins Gn and Gc, necessary for virion assembly. The S segment encodes the viral nucleocapsid protein, N. At the 3′ and 5′ ends of the genome, each genus of the *Bunyaviridae* family contains a specific and conserved sequence that can be used to distinguish among the genera. These sequences of 10–15 nucleotides are complementary and thought to basepair to form circular RNA molecules required for primary transcription and RNA replication. In addition, a certain degree of diversity exists with respect to additional accessory proteins encoded by the genome. For instance, within the *Tospovirus* genus, viruses encode two additional non-structural proteins, NSs and NSm, that function as an RNAi inhibitor (Demmers

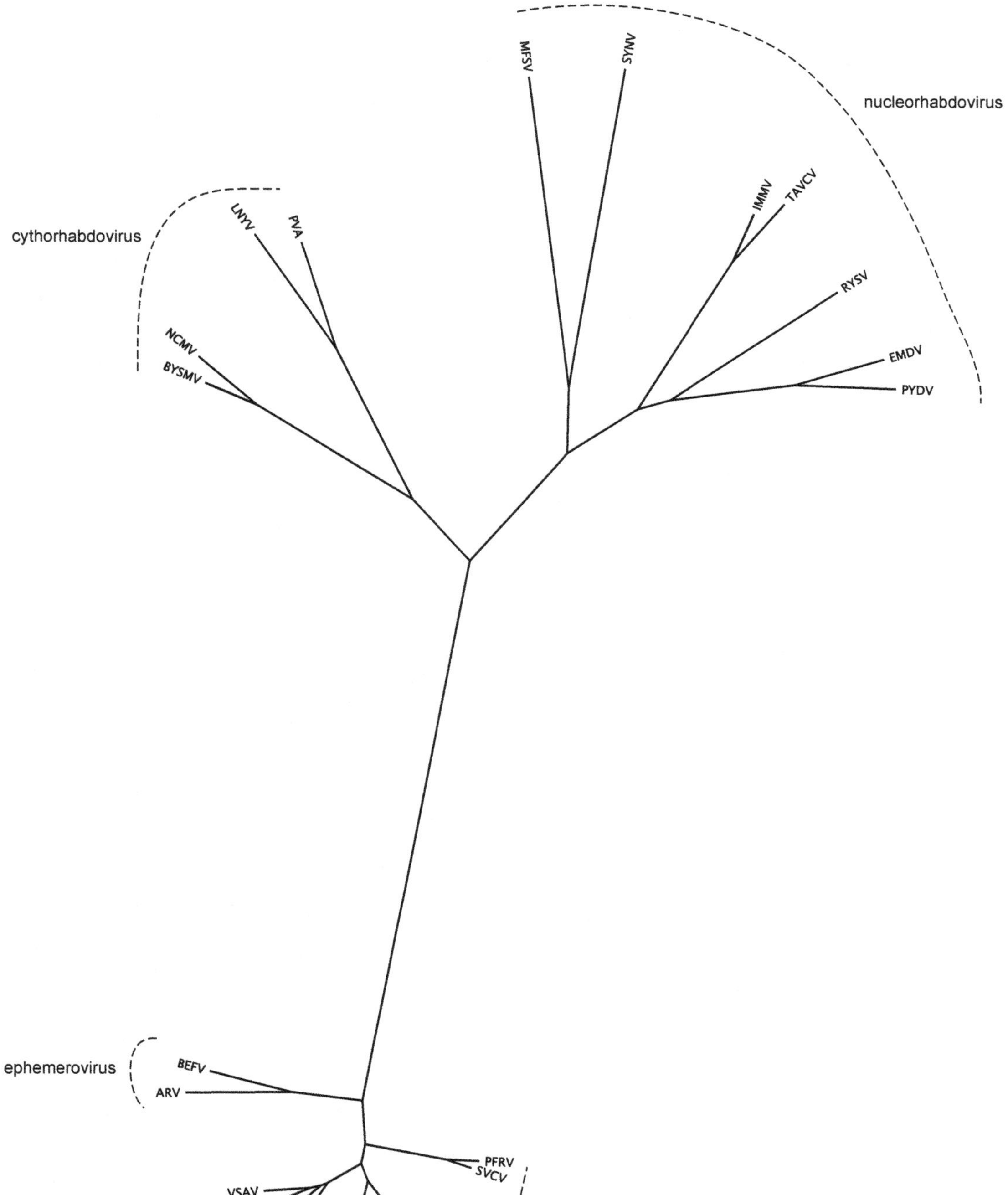

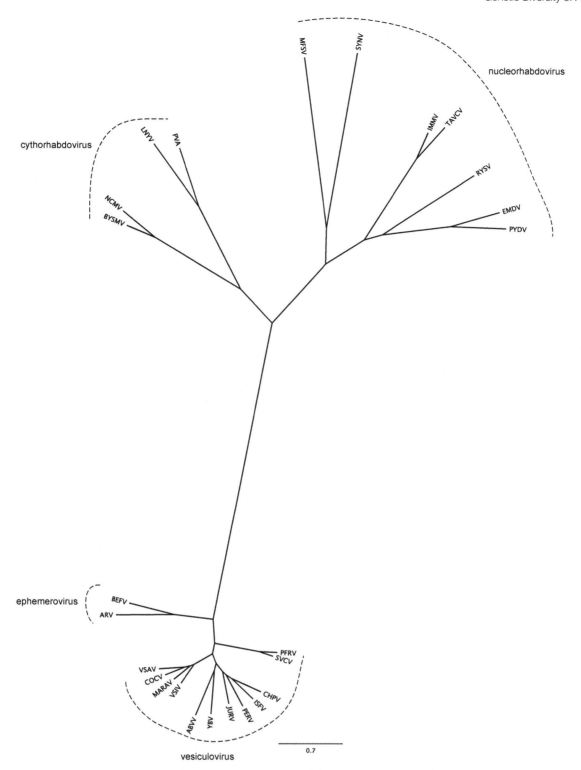

Figure 9.4 Phylogenetic relationships of the main arboviral members within the *Rhabdoviridae* family. A maximum likelihood phylogenetic tree based on the alignment of the L (RdRp) region is shown. Branch lengths are proportional to the evolutionary distance (scale bar) between the taxa. Different genera within the *Rhabdoviridae* family are observed. ABVV (American Bat Vesiculovirus); ARV (Adelaide River virus); BEFV (Bovine Ephemeral Fever virus); BYSMV (Barley Yellow Striate Mosaic virus; CHPV (Chandipura virus); COCV (Cocal virus); EMDV (Aubergine Mottled Dwarf virus); IMMV (Iranian Maize Mosaic virus); ISFV (Isfahan virus); JURV (Jurona virus); LNYV (Lettuce Necrotic Yellows virus; MARAV (Maraba virus); MFSV (Maize Fine Streak virus); NCMV (Northern Cereal Mosaic virus); PERV (Perinet virus); PFRV (Pike Fry Rhabdovirus); PVA (Persimmon virus A); PYDV (Potato Yellow Dwarf virus); RYSV (Rice Yellow Stunt Virus); SVCV (Spring Viraemia of Carp virus; SYNV (Sonchus Yellow Net virus); TAVCV (Taro Vein Chlorosis virus); VSAV (Vesicular Stomatitis Alagoas virus); VSIV (Vesicular Stomatitis Indiana virus); YBV (Yug Bogdanovac virus).

et al., 2014) and a movement protein (NSm) (Lewandowski and Adkins, 2005) (Hallwass *et al.*, 2014) respectively, functions found in many plant viruses and necessary for viral spread within the plant and vector host. Furthermore, the orthobunyaviruses and phleboviruses also encode a NSs protein, which has been shown to inhibit various cellular processes including cellular transcription and the innate immune response (Varela *et al.*, 2013).

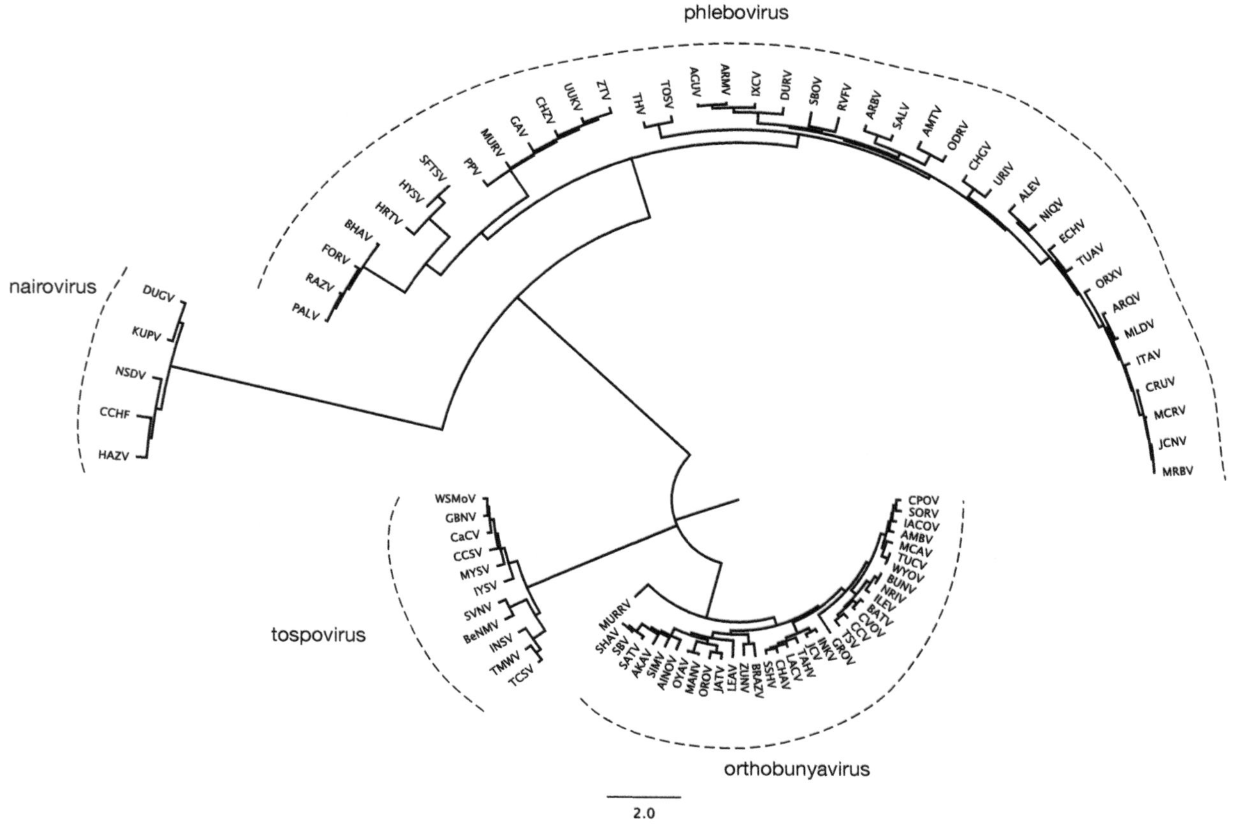

Figure 9.5 Phylogenetic relationships among the main arboviral members of the *Bunyaviridae* family. A maximum likelihood phylogenetic tree based on the alignment of the RPRD is shown. Branch lengths are proportional to the evolutionary distance (scale bar) between the taxa. Different genera within the family are observed. Genus *Tospovirus*: BeNMV (Bean Necrotic Mosaic virus); CaCV (Capsicum Chlorosis virus); CCSV (Calla lily Chlorotic Spot virus); GBNV (Groundnut Bud Necrosis virus); INSV (Impatiens Necrotic Spot virus); IYSV (Iris Yellow Spot virus); MYSV (Melon Yellow Spot virus); SVNV (Soybean Vein Necrosis virus); TCSV (Tomato Chlorotic Spot virus); TSWV (Tomato Spotted Wilt virus); WSMoV (Watermelon Silver Mottle virus). Genus *Nairovirus*: CCHF (Crimean–Congo haemorrhagic fever); DUGV (Dugbe virus); HAZV (Hazara virus); KUPV (Kupe virus); NSDV (Nairobi Sheep Disease virus). Genus Orthobunyavirus: AINOV (Aino virus); AKAV (Akabane virus); AMBV (Anhembi virus); BATV (Batai virus); BRAZV (Brazoran virus); BUNV (Bunyamwera virus); CCV (Cache Valley virus); CHAV (Chatanga virus); CPOV (Cachoeira Porteira virus); CVOV (Calovo virus); GROV (Guaroa virus); IACOV (Iaco virus); ILEV (Ilesha virus); INKV (Inkoo virus); JATV (Jatobal virus); JCV (Jamestown Canyon virus); LACV (La Crosse Encephalitis virus); LEAV (Leanyer virus); MANV (Manzanilla virus); MCAV (Macaua virus); MURRV (Murrumbidgee virus); NRIV (Ngari virus); OROV (Oropouche virus); OYAV (Oya virus); SATV (Sathuperi virus); SBV (Schmallenberg virus); SHAV (Shamonda virus); SIMV (Simbu virus); SORV (Sororoca virus); SSHV (Snowshoe Hare virus); TAHV (Tahyna virus); TSV (Tensaw virus); TUCV (Tucunduba virus); WYOV (Wyeomyia virus); ZUNV (Zungarococha virus). Genus *Phlebovirus*: AGUV (Aguacate virus); ALEV (Alenquer virus); ARBV (Arbia virus); ARQV (Ariquemes virus); ARMV (Armero virus); AMTV (Arumowot virus); BHAV (Bhanja virus); CRUV (Candiru virus); CHGV (Chagres virus); CHZV (Chize virus); DURV (Durania virus); ECHV (Echarte virus); FORV (Forecariah virus); GAV (Grand Arbaud virus); HRTV (Heartland virus); HYSV (Huaiyangshan virus); ITAV (Itaituba virus); IXCV (Ixcanal virus); JCNV (Jacunda virus); MLDV (Maldonado virus); MRBV (Morumbi virus); MCRV (Mucura virus); MURV (Murre virus); NIQV (Nique virus); ODRV (Odrenisrou virus); ORXV (Oriximina virus); PALV (Palma virus); PPV (Precarious Point virus); RAZV (Razdan virus); SALV (Salehabad virus); SBOV (Salobo virus); SFTSV (Severe Fever with Thrombocytopenia Syndrome virus); THV (Tehran virus); TOSV (Toscana virus); TUAV (Turuna virus); URIV (Uriurana virus); UUKV (Uukuniemi virus); ZTV (Zaliv Terpenia virus).

Reoviridae

The *Reoviridae* family consists of 12 genera, six of which contain arboviruses: *Orbivirus, Coltivirus, Seadornavirus, Phytoreovirus, Fijivirus* and *Oryzaviruses* (Fig. 9.6). Three of these genera, *Orbivirus, Coltivirus* and *Seadornavirus,* infect mammals, whereas members of the genera *Phytoreovirus, Fijivirus* and *Oryzavirus* infect plants. Reoviruses are unique among the arboviruses in that they encapsidate a segmented, double-stranded RNA genome, which ranges from 10 to 12 distinct segments depending on the genus. The segments of the reovirus genome are classified by size: large (L), medium (M), and small (S), with each segment encoding for individual or in some cases multiple viral proteins (Fig. 9.2E) (Silverstein *et al.*, 1976). Similar to other segmented viruses such as the *Bunyaviridae* family, RNA segment reassortment has considerable impact on their genetic diversity (Cheng *et al.*, 1999; Reese *et al.*, 2008). Of all the segments, the S1 segment exhibits the most diversity between viruses (Bird *et al.*, 2007).

Orthomyxoviridae

The *Orthomyxoviridae* family is made up of five genera: influenza viruses A, B, and C, *Isavirus* and *Thogotovirus*. Only the *Thogotovirus* genus consists of two viruses in the arbovirus category, Thogoto virus (THOV) and Dhori virus (DHOV), which are transmitted by ticks and infect mammals and humans. These orthomyxoviruses are segmented, single-stranded, negative-sense RNA viruses with six to eight genomic segments (Neumann *et al.*, 2004). The *Thogotovirus* genome consists of six (THOV) or seven (DHOV) segments encoding seven viral proteins. These segments (1–6) encode PB1, PB2, and PA,

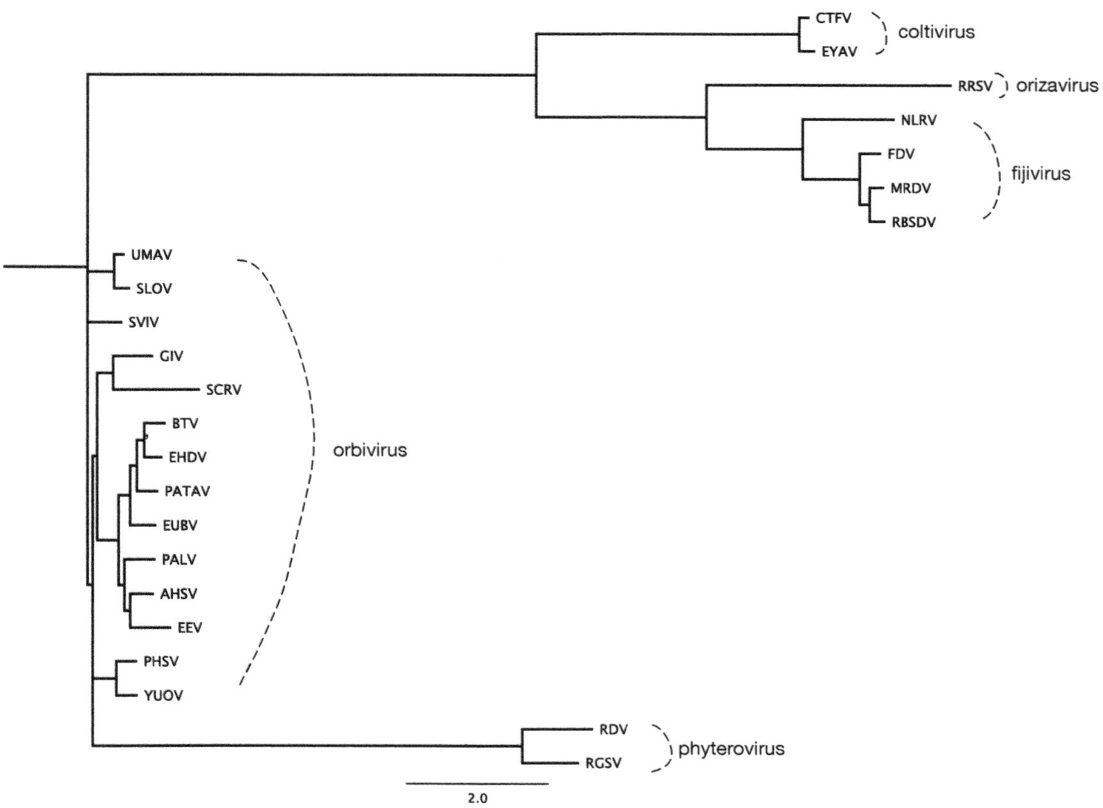

Figure 9.6 Phylogenetic relationships of the main arboviral members within the *Reoviridae* family. A maximum likelihood phylogenetic tree based on the alignment of the RPRD is shown. Branch lengths are proportional to the evolutionary distance (scale bar) between the taxa. Different genera within the family are observed. RBSDV (Rice black streaked dwarf virus); MRDV (Maize rough dwarf virus); FDV (Fiji disease virus); NLRV (Nilaparvata lugens virus); RRSV (Rice ragged stunt virus); EYAV (Eyach virus); CTFV (Colorado tick fever virus); RGDV (Rice gall dwarf virus); RDV (Rice dwarf virus); SCRV (St. croix river virus); GSIV (Great saltee island virus); SLOV (Stretch Lagoon virus); UMAV (Umatilla virus); SVIV (Sathuvachari virus); YUOV (Yunnan orbivirus); PHSV (Peruvian horse sickness virus); AHSV (African horse sickness virus); PALV (Palyam virus); EEV (Equine encephalosis virus); EUBV (Eubenangee virus); PATAV (Pata virus); EHDV (Epizootic haemorrhagic disease virus); BTV (Bluetongue virus).

which together form the RNA-dependent RNA polymerase complex, the viral glycoprotein GP, the viral nucleocapsid NP, and viral matrix proteins M and M1. Similar to the *Rhabdoviridae* family and distinct from *Bunyaviridae*, the genome is uncapped and nonpolyadenylated. However, unlike other negative-strand arboviruses, the orthomxyoviruses replicate in the nucleus of infected cells providing access to distinct cellular factors for replication and unique transcription/replication strategies such as gene splicing to generate two proteins from a single segment, as is the case for the thogotovirus genome segment 6, which produces the ML and M protein from a spliced transcript (Kochs *et al.*, 2000). These proteins are thought to function as interferon antagonists (Pichlmair *et al.*, 2004) (Buettner *et al.*, 2010) (Vogt *et al.*, 2008). Unlike the influenza A and B viruses, and more analogous to influenza C, thogotoviruses express a single surface glycoprotein rather than separate haemagglutinins and neuraminidases. However, it is genetically more similar to the glycoprotein of Baculoviruses, an exclusively insect virus. Finally, similar to other segmented negative-strand RNA viruses, genetic reassortment during infection has been demonstrated and plays signification roles in producing and maintaining genetic diversity among these viruses.

Asfarviridae

African swine fever virus (ASFV) is the only known DNA arbovirus and the sole member of the *Asfarviridae* family (Dixon *et al.*, 2013; Michaud *et al.*, 2013). ASFV is part of the nucleocytoplasmic large DNA virus group (NCLDV), which includes members such as the *Poxviridae* and the recently discovered *Mimiviridae*. The virus is composed of a large icosahedral, enveloped particle containing a 170–200 kb linear double-stranded DNA genome. As such, ASFV is drastically different from RNA arboviruses, not only in genomic material, but also in that it encodes 150 open reading frames, many of which function as multigene families (MGF) for viral replication and host immune evasion.

Molecular mechanisms responsible for genetic diversity

Genetic drift and point mutation

A primary source of genetic diversity for the RNA arboviruses is the low fidelity of the RNA dependent RNA polymerases (RdRp) that replicate their genomes. Because RdRps lack the classic proof-reading and repair mechanisms of DNA polymerases, this error prone replication is a continual source of random point mutations (insertions, deletions, substitutions). It is estimated that on average, one to two mutations are introduced into every newly synthesized genome (Drake and Holland, 1999). As a result, RNA virus populations comprise swarms of variants, which have been extensively studied in the framework of viral quasispecies and the genetic diversity occurring within a strain, rather than between strains (Domingo et al., 2012). Many arboviruses including vesicular stomatitis virus, and several alpha- and flavi-viruses, have been characterized in this manner, particularly in terms of fitness trade-offs occurring during host-alternation between vertebrates and invertebrates (Combe and Sanjuan, 2014; Coffey and Vignuzzi, 2011).

Over the long term, as virus populations are transmitted between hosts, this relatively stochastic generation of point mutations is counterbalanced by selective pressures (adaptation to new environments, escape from host immunity, etc.) and further compounded by factors affecting population dynamics (such as changes in multiplicities of infection and effective population sizes, genetic bottlenecks). In result, related genotypes may neutrally drift apart, may diverge or may converge over time. The gradual process of point mutation, countered by selection, over the long-term is perhaps best described for influenza A viruses, for which yearly drifts of 0.6% have been observed and attributed to positive selection of antigenic escape variants in the human population (Klein et al., 2014; McDonald et al., 2007). Examples of genetic drift within the arboviruses can be found in every family. For instance, phylogenetic analysis of YFV isolates collected over a 65-year period in Brazil revealed the temporal evolution of South American genotype I, during which genetic variability reached up to 7.4% (4.5% amino acid changes) between strains (Vasconcelos et al., 2004). In the case of CCHFV, one of the most genetically diverse arboviruses, the S, L and M segments diverge at the nucleotide level by 20%, 22% and 31% between isolates (Deyde et al., 2006). It is important to highlight that although much of this can be attributed to genetic drift, phylogenetic analyses have identified other events such as reassortment (Armstrong and Andreadis, 2007; Reese et al., 2008) and recombination (Hahn et al., 1988; Twiddy and Holmes, 2003) that have accelerated these differences throughout the years.

Genetic recombination

Whereas point mutations permit incremental generation of genetic diversity and step-wise adaptation, recombination between genotypes of the same or related virus species results in more dramatic jumps, as larger stretches of genomic sequence are exchanged between partners. Recombination occurs most frequently between members of the same species with high sequence homology, and particularly among the positive-sense RNA viruses, although less frequent recombination has been documented in negative-sense viruses as well (Chare et al., 2003). The likelihood of recombination increases with multiplicity of infection, as the parental genomes must be in close proximity within the same cell while the replication complex is synthesizing nascent genomes. In cell culture, these conditions are easily achieved, while in nature, it is unclear to what extent viruses co-infect the same cell for a given host tissue. Nevertheless, these conditions certainly do occur, as studies using fluorescently marked viruses revealed co-infection of cells in the midguts of infected mosquitoes (Smith et al., 2008).

One of the best examples of recombination having occurred in nature is found within the alphaviruses, in the Western Equine Encephalitis virus (WEEV) complex. WEEV is the product of genetic recombination between Eastern Equine Encephalitis virus (EEEV) and a Sindbis-like virus. Specifically, the WEEV genome is a combination of the non-structural proteins, capsid, and 3′ UTR from EEEV and the structural glycoproteins from a Sindbis-like virus (Hahn et al., 1988; Weaver et al., 1997). Since this recombination event, the WEE complex has continued to evolve with related viruses such as Fort Morgan virus (FMV), Highlands J virus (HJV), and Buggy creek virus (BCV) (Twiddy and Holmes, 2003). For chikungunya virus (CHIKV), the potential for recombination, at least within the same species, was demonstrated *in vitro* using pairs of genomes containing intra-genic deletions, found to occur in 1 out of 10^5 co-infections (McGee et al., 2011; Cui et al., 2011).

Similar to the CHIKV studies, *in vitro* experiments using flavivirus constructs designed to detect recombination events between tick-borne encephalitis virus (TBEV), West Nile virus (WNV), and Japanese encephalitis virus revealed that recombination can in principle occur, but with very low frequency (Twiddy and Holmes, 2003; Taucher et al., 2010). However, genetic recombination between two different flavivirus species has not yet been confirmed in nature. On the other hand, various reports suggest that recombination within the same species has indeed occurred, such as between subtypes of dengue virus (Perez-Ramirez et al., 2009; Worobey et al., 1999), and between related strains of Japanese encephalitis virus or St. Louis encephalitis virus (May et al., 2008; Twiddy and Holmes, 2003). Indeed, a strain of St. Louis encephalitis virus in Guatemala was found to be a recombinant of strains from Argentina and the USA (Baillie et al., 2008).

For CCHFV, phylogenetic analyses provide evidence for recombination having occurred in the S segment of some strains, and possibly in the L segments of group IV viruses (Chare et al., 2003; Lukashev, 2005; Deyde et al., 2006). Similarly, mosaic patterns in segments of the orbivirus BTV are also genetic signatures of the role that recombination may play in the diversity of this virus family (He et al., 2010).

Genetic reassortment

For viruses with segmented genomes (*Bunyaviridae*, *Reoviridae* and *Orthomxyoviridae*), reassortment can occur during co-infection where segments are exchanged like mini-chromosomes to generate new combinations. By such mechanisms, viruses expand their genetic diversity by acquiring entirely new genetic combinations. The process is perhaps best known for the non-arbovirus member of orthomyxoviruses, influenza A, where reassortment was at the origin of the 1957, 1968 and 2009 (swine flu) pandemic strains (Westgeest et al., 2014; Tao et al., 2014). Among the arboviruses in this family, the reassortment of Thogoto virus was shown to occur *in vivo* in both vertebrate (Jones et al., 1987) and invertebrate (Davies et al., 1987) hosts.

For the orbivirus, Bluetongue virus (BTV) with 26 known serotypes, genetic variation is primarily attributed to reassortment between local, circulating strains (Maclachlan and Guthrie, 2010). Phylogenetic evidence of reassortment has been found in numerous studies (Batten et al., 2008; Maan et al., 2012; Shafiq et al., 2013). As a result, these genetic differences have led to significant phenotypic differences, even within the same genotype. For example, genotype BTV-4 from South Africa is more virulent then BTV-4 from South America. Experimental co-infection reveals that reassorted virus can represent 100% of the progeny after only four passages in cell culture (Shaw *et al.*, 2013), up to 70% of progeny in *Culicoides* flies (el Hussein *et al.*, 1989; Samal *et al.*, 1987a), and 5% in sheep (Samal *et al.*, 1987b) and cows (Oberst *et al.*, 1987). In the closely related African horse sickness virus (AHSV), reassortment between the attenuated live vaccine and circulating strains was also observed (von Teichman and Smit, 2008).

The extent of reassortment within the orthobunyaviruses is considerable, such that researchers question whether most, if not all, members of this group are reassortants (Briese *et al.*, 2013). In the recently emerging tick-borne Severe Fever with Thrombocytopenia Syndrome virus (SFTSV), isolates in Central China indicate separate origins for the S versus M and L segments (Ding *et al.*, 2013). Another recent example is the mosquito-borne Ngari virus, whose origin lies in the reassortment between a Bunyamwera virus (BUNV) and a Batai virus (BATV) (Groseth *et al.*, 2012). Schmallenberg virus, currently causing significant damage in ruminary populations in northern Europe, also seems to be a reassortment between RNA segments of Sathuperi and Shamonda viruses (Yanase *et al.*, 2012). Up to 25% of mosquitoes caught in the wild, carried La Crosse virus reassortants (Reese *et al.*, 2008) and experimental co-infection of mosquitoes with La Crosse and snowshoe hare viruses generated reassortants (Borucki *et al.*, 1999). Finally, reassortment has also been documented in bunyaviruses infecting plants, where a new virus affecting tomatoes is a reassortant between Groundnut ringspot virus (GRSV) and Tomato chlorotic spot virus (TCSV) (Webster *et al.*, 2011).

Heterogeneity of genome size and protein repertoire within viral families

Although viruses within the same family, and especially within the same genera, tend to present the same genome organization and similar size, some heterogeneity and noteworthy exceptions can be mentioned, particularly for families with larger size genomes.

The single-stranded, positive-sense *Togaviridae* and *Flaviviridae* families have members with highly similar genetic structure, yet whose genomes vary between 11 and 12 kb and been 10 and 11 kb, respectively, owing in the most part to varying lengths of the 5′ and 3′ UTR regions (Wallner *et al.*, 1995). Studies comparing the untranslated regions of flaviviruses reveal that size heterogeneity, resulting from deletions or duplications of short sequences, can be found within the genus, and even within subgroups, such as among members of the Yellow Fever group (Mutebi *et al.*, 2004; Wang *et al.*, 1996).

A significant diversity is observed in terms of size of each genome segment between the *Bunyaviridae* genera, leading to diverse protein expression profiles during viral replication. The *Orthobunyavirus* genus contains the smallest viral genome (~12.4 kb), while the *Tospovirus* genus encapsidates the largest (~16.5 kb). For example, nairoviruses are genetically distinct in the *Bunyaviridae* family as they do not encode a NSs protein as other genera, and their L segment is nearly twice the size (~12 kb) of the other genera, suggesting that the RdRp may carry multiple functions not present in other bunyaviruses(Eifan *et al.*, 2013). Furthermore, *Tospovirus* is the only genus of the *Bunyaviridae* family infecting plants. Most notable within this genus is tomato spotted wilt virus (TSWV) that has been used extensively to study plant virus replication and molecular biology. TSWV has a wide host range, able to infect over 900 types of plants. The tospovirus genome is distinct from other bunyaviruses in that it encodes a much larger L segment then phlebovirus counterparts (~8.8 kb compared with ~6.4 kb). Additionally, ambisense transcription from both the M and S segment generates products essential for the survival of tospoviruses by evasion of the RNAi machinery of plants and insects, and movement of virus within plants (Eifan *et al.*, 2013; Turina *et al.*, 2012; Kormelink *et al.*, 2011).

As compared to the prototypical vesiculovirus counterparts in the *Rhabdoviridae* family which encode five genes and has a genome of roughly 11 kb, the ephemeroviruses contain a slightly different genome structure encoding additional proteins between the G and L genes, resulting in an increase in genome size of 5 kb. In particular, BEFV encodes 10 genes including two isoforms of the matrix protein (M1 and M2), a non-structural glycoprotein (Gns) of unknown function, and several additional proteins α1, α2, β, and γ with unknown function (Walker *et al.*, 2011). The closely related ARV, contains nine genes lacking the g gene. The cyto- and nucleorhabdoviruses are the two genera of this family that infect plants and are transmitted by leafhoppers and aphids. Similar to the ephemeroviruses, plant rhabdoviruses contain additional genes and the number, position, and function of these genes vary between species. For instance, the cytorhabdovirus, LNYV, contains the genome 3′-N-4a-4b-M-G-L-5′ where 4a and 4b encode phosphoproteins, and SYNV contains the genome organization, 3′-N-P-sc4-M-G-L-5′, where sc4 encodes a potential movement protein (Huang *et al.*, 2005; Jackson *et al.*, 2005). Furthermore, the plant rhabdoviruses 3′ and 5′ leader and trailer RNA sequences are much longer than other genera (Huang *et al.*, 2005; Jackson *et al.*, 2005). Finally, at least in the case of the nucleorhabdovirus SYNV, the leader RNA is polyadenylated, although its function is unknown.

Within the segmented *Reoviridae* family, heterogeneity in number of genome segments is observed (Silverstein *et al.*, 1976). Unlike the orbiviruses, coltiviruses and seadornaviruses comprise 12 genome segments instead of the typical ten. Members of this genus include Colorado tick fever virus (CTFV), California hare coltivirus (CTFV-Ca), Eyach virus (EYAV), and Salmon River virus (SRV). Viruses composing the three genera Phytoreovirus, Fijivirus, Oryzavirus are plant pathogens transmitted by leafhoppers and diptera. As for the mammalian arboviruses above, the members of each genera have the same number of segments, but there are differences between genera. The fijiviruses and oryzaviruses each contain 10 segments encoding 12 proteins, whereas the phytoreoviruses encapsidate 12 segments encoding 15 proteins. As with the plant bunyaviruses, these plant viruses also encode proteins not found in the mammalian reovirus counterparts, such as movement proteins (Ji *et al.*, 2011) that facilitate cell-to-cell spread within the plant host.

As can be expected for large DNA viruses encoding many more proteins than any RNA virus, significant differences can exist between ASFV genotypes. The majority of genetic differences between genotypes have mapped to gain and loss of function of multigene families (MGF). For example, MGF 360 and MGF 110 contains 22 and 14 paralogous genes, respectively (Dixon *et al.*, 2013). The gene copy numbers between ASFV genotypes can play essential roles in virulence, as a low virulence Portuguese isolate contains 11 copies of MGF 360 whereas an isolate form Kenya contains 18 (Michaud *et al.*, 2013).

Genetic diversity and the host: influence of the invertebrate vector and pathogenesis in the vertebrate host

The invertebrate host is recognized as playing a key role, and potentially more so than the vertebrate host, in arbovirus evolution and genetic diversity (Kuno and Chang, 2005; Weaver, 2006). This may principally be due to the persistent nature of arbovirus infection in invertebrates, relative to acute infections in vertebrate hosts, which result in longer periods of virus–host interaction and adaptation. In result, a virus that becomes particularly specialized to a specific host, could become constrained by that host. A prime example of this can be since among the flavivirus genus where, although these viruses contain similar genome structures they can be readily segregated into mosquito- and tick-borne viruses suggesting that various level of diversity lead to this viral specialization (Fig. 9.3). This can also be seen with distinct viruses within the other arbovirus containing families which have become specialized to infect a variety of different invertebrates. In addition, not only do these viruses become specialized to only one insect vector they can be confined to only one species of a particular insect as well. For example, Ross River virus (RRV) an alphavirus and member of the Semliki Forest antigenic complex is predominantly localized to Australia. To date, three genetically distinct subgroups of RRV circulating throughout the continent are recognized: the northeastern, southeastern, and western isolates. Unlike other alphaviruses that have successfully spread world-wide, it is thought that the overall diversity of RRV may be restrained because of its vector (Sammels *et al.*, 1995), which has a range limited to the Oceania region (Harley *et al.*, 2001). Another example among the alphaviruses, is O'nyong'yong virus (ONNV), the closest relative to CHIKV with the key distinction of being the only arbovirus known to infect and be transmitted by the *Anopheles* mosquito (Powers *et al.*, 2000). The genetic differences permitting infection of *Anopheles* mosquitoes are currently unknown; however, recent studies implicate differences in the nsP3 protein as a key determinant for vector specificity (Saxton-Shaw *et al.*, 2013). In specializing to specific vector species, these viruses may have lost the opportunity to readily adapt to other mosquito species.

If, however, an arbovirus does manage to jump into a different, less competent host, it will likely diversify from the original strain as it adapts to the new host environment. One of the best recent examples of vector-specific influences on arbovirus evolution involves chikungunya virus diversity, which divides into three major genotypes; the West African strain, the East-Central-South Africa strain (ECSA), and the Asian strain. In 2005, a large outbreak of chikungunya occurred on the small French island of La Réunion, which introduced a novel lineage of the ECSA strain into circulation that resulted in the spread of the Indian Ocean lineage (IOL) throughout Asia. Genetic comparisons revealed that adaptive mutations in the E1 and E2 glycoproteins allowed CHIKV to more readily infect and be transmitted by the wide-spread asian tiger mosquito, *Aedes albopictus* (Tsetsarkin *et al.*, 2007, 2011). These studies suggest that during its current expansion, CHIKV will continue to evolve and generate more diverse strains with increased fitness for certain mosquito vectors.

Finally, the genetic differences between viruses even within the same genus can lead to drastic differences in pathogenesis and disease. For example, the alphaviruses can be divided into encephalitic(Zacks and Paessler, 2010) (Hollidge *et al.*, 2010) and arthralgic alphaviruses (Suhrbier *et al.*, 2012) and specific adaptive mutations have been linked to be necessary for these distinctions. Adaptive mutations in the Sindbis virus E2 protein has been linked to neurovirulence where as glycosylation plays key roles in Ross River virus arthritis and inflammation. In addition, within the flavivirus genus there exists encephalitic and haemorrhagic viruses. The encephalitic flavivirus, West Nile virus diversity branches into eight lineages, with lineages 1 and 2 being the two major lineages, and lineage 1 being the more virulent of the two. Within lineage 1, the NY99 strain that swept across North America is highly virulent, compared to the Kunjin strains of WNV in Australia. Interestingly, genetic differences affecting the glycosylation states of the surface glycoproteins seem to be responsible. Nonetheless, these differences in diseases by arboviruses further indicates that genetic differences influence many aspects of arbovirus biology.

Genetic diversity and geographic distribution: compounding effects of host, climate and globalization?

When examining genetic differences between genotypes and lineages within the same species, diversity often segregates according to geographic distribution. As one can expect, this is clearly influenced by the presence and relative mobility of both invertebrate vector and primary host; but geographic barriers and environmental forces can further impact to what extent genotypes segregate and diversify, or mix and co-circulate. Phylogenetic analysis of any arbovirus reveals the geographical distribution of subtypes. The diversity of YFV, for example, comprises seven genotypes that can be further divided based on geographic location. Five genotypes localize to Africa (West Africa I and II, East Africa, East/Central Africa, and Angola genotypes), whereas two are found in South America (South America I and II). As another example, the diversity of TBEV is distributed across three subtypes based on geographical location: the European, Far East, and Siberian subtypes (Weidmann *et al.*, 2013). Phylogenetic analysis of the ephemerovirus BEFV G-protein identifies three distinct genotypes based on geographic location (Middle Eastern, East Asian and Australian).

Based on their geographic location, alphaviruses can be split to two classes (Luers *et al.*, 2005; Suhrbier *et al.*, 2012; Weaver *et al.*, 2012). The Old World alphaviruses, including chikungunya, O'nyong-nyong, Sindbis, Ross River, and Semliki Forest viruses, are found throughout Europe, Africa, Asia, and Australia (Sammels *et al.*, 1995). In contrast, the new world alphaviruses, including the eastern equine encephalitis, venezuelan equine encephalitis and western equine encephalitis viruses, localize to the Americas (Zacks and Paessler, 2010). Each of these viruses can be further divided by geographical localization. EEEV, for example, circulates primarily between mosquitoes and birds and consists of four major lineages (I-IV), with distinct geographical distribution. Lineage I is the North

American lineage (NA EEEV), whereas EEEV found in Central and South American encompasses lineages II to IV (SA EEEV). The relatively high genetic diversity among the South American strains is thought to reflect geographical features such as host and vector species differences (Arrigo *et al.*, 2010); while the strains of North America are relatively homogenous (Davis *et al.*, 2008). VEEV comprises six distinct subtypes (I–VI), and subtype I, for example, is further divided into five strains (IAB, IC, ID, IE and IF). The vast genetic diversity of VEEV is attributed to the fact that VEEV is primarily amplified in small rodents, which allow for local evolution of the virus because of their limited geographical mobility (Weaver and Barrett, 2004). Finally, even though ASFV is a DNA virus, there has been a large genetic difference found between circulating strains. Thus far, 22 genotypes (genotype I–XXII) of ASFV are divided into four distinct lineages (L1–4). The greatest amount of genetic diversity between ASFV genotypes is observed within eastern and southern Africa where the sylvatic cycle is most prevalent, suggesting that viral evolution may have been under pressure in this area (Dixon *et al.*, 2013; Michaud *et al.*, 2013).

Although these viruses may seem to remain constrained to specific geographic regions of the world for the moment, the growing levels of world trade, travel, and climate change can drastically change this landscape in very little time. For instance, there are currently five genotypes (I-V) of JEV circulating throughout Asia and it has been shown that monsoon rain and winds (Upadhyayula *et al.*, 2012) as well as climate (Schuh *et al.*, 2013) have been responsible for the distribution of infected mosquitoes in these areas as well as segregation of phylogenetic clades of JEV throughout Asia respectively. In addition, as temperatures rise, tropical vectors may of the chance to move into new regions bringing arboviruses along with them. Finally, the combination world travel and trade, coupled with the spread of a highly invasive mosquito, *Aedes albopictus*, have been instrumental in the spread of chikungunya virus from Africa, throughout Asia, and recently to the Americas, moving chikungunya from an Old World alphavirus to the new world (Fischer and Staples, 2014; Charrel *et al.*, 2014).

Concluding remarks and future directions

Arboviruses consist of a plethora of genetically diverse viral species capable of causing devastating disease in not only humans but animals and plants as well. These viruses are made of a variety of genomic compositions and organizations which in turn allow individual virus families and species to employ unique genetic and molecular mechanism for viral replication and pathogenesis. These mechanisms give rise to diverse populations of viruses which infect a variety of organisms and cause distinct diseases. Similar features, including rapid replication rates and error-prone replication strategies allow for arboviruses to adapt rapidly to their invertebrate and vertebrate hosts and in certain situations leading to novel viral pathogens with increase fitness advantages in their hosts.

The diversity of arboviruses, not only at the genomic structure level but also at the nucleotide level has been studied in detail. The development of nucleotide sequencing has allowed for detailed phylogenetic analyses of many arboviruses to be completed, which in turn can be efficiently linked to viral pathogenesis and disease outcome. In addition, these phylogenic and genetic studies are being combined with reverse genetic systems to study the function of viral variants present in nature at the molecular level, further adding to our knowledge of arbovirus diversity. Unfortunately sequence data for many arboviruses is not currently available, incomplete, or insufficient indicating that more work needs to be done to complete studies of arbovirus diversity. To help with these limitations, the employment of next generation deep-sequencing (Coffey *et al.*, 2014) has proven a novel and important method to further characterize large populations of viruses, as well as identify novel arboviruses in insects, animals, and plants. Additionally, the emerging field of phylogeography will help place the ongoing evolution of arboviruses in a real world context. Taken together, these techniques, along with advances in bioinformatic tools will prove essential to continue to investigate and understand the complex and genetically diverse world of arboviruses.

References

Allison, A.B., and Stallknecht, D.E. (2009). Genomic sequencing of Highlands J virus: a comparison to western and eastern equine encephalitis viruses. Virus Res. *145*, 334–340.

Armstrong, P.M., and Andreadis, T.G. (2007). Genetic relationships of Jamestown Canyon virus strains infecting mosquitoes collected in Connecticut. Am. J. Trop. Med. Hyg. *77*, 1157–1162.

Arrigo, N.C., Adams, A.P., and Weaver, S.C. (2010). Evolutionary patterns of eastern equine encephalitis virus in North versus South America suggest ecological differences and taxonomic revision. J. Virol. *84*, 1014–1025.

Baillie, G.J., Kolokotronis, S.O., Waltari, E., Maffei, J.G., Kramer, L.D., and Perkins, S.L. (2008). Phylogenetic and evolutionary analyses of St. Louis encephalitis virus genomes. Mol. Phylogenet. Evol. *47*, 717–728.

Batten, C.A., Maan, S., Shaw, A.E., Maan, N.S., and Mertens, P.P. (2008). A European field strain of bluetongue virus derived from two parental vaccine strains by genome segment reassortment. Virus Res. *137*, 56–63.

Bird, B.H., Khristova, M.L., Rollin, P.E., Ksiazek, T.G., and Nichol, S.T. (2007). Complete genome analysis of 33 ecologically and biologically diverse Rift Valley fever virus strains reveals widespread virus movement and low genetic diversity due to recent common ancestry. J. Virol. *81*, 2805–2816.

Birket, M.J., Casini, S., Kosmidis, G., Elliott, D.A., Gerencser, A.A., Baartscheer, A., Schumacher, C., Mastroberardino, P.G., Elefanty, A.G., Stanley, E.G., *et al.* (2013). PGC-1alpha and Reactive Oxygen Species Regulate Human Embryonic Stem Cell-Derived Cardiomyocyte Function. Stem Cell Reports *1*, 560–574.

Borucki, M.K., Chandler, L.J., Parker, B.M., Blair, C.D., and Beaty, B.J. (1999). Bunyavirus superinfection and segment reassortment in transovarially infected mosquitoes. J. Gen. Virol. *80*, 3173–3179.

Briese, T., Calisher, C.H., and Higgs, S. (2013). Viruses of the family Bunyaviridae: are all available isolates reassortants? Virology *446*, 207–216.

Buettner, N., Vogt, C., Martinez-Sobrido, L., Weber, F., Waibler, Z., and Kochs, G. (2010). Thogoto virus ML-protein is a potent inhibitor of the interferon regulatory factor-7 transcription factor. J. Gen. Virol. *91*, 220–227.

Chapman, E.G., Constantino, D.A., Rabe, J.L., Moon, S.L., Wilusz, J., Nix, J.C., and Kleft, J.S. (2014). The structural basis of pathogenic subgenomic flaviviruses (sfRNA) production. Science *344*, 307–310.

Chare, E.R., Gould, E.A., and Holmes, E.C. (2003). Phylogenetic analysis reveals a low rate of homologous recombination in negative-sense RNA viruses. J. Gen. Virol. *84*, 2691–2703.

Charrel, R., Leparc-Goffart, I., Gallian, P., and de Lamballerie, X. (2014). Globalization of Chikungunya: 10 years to invade the world. Clin. Microbiol. Infect. *20*, 662–663.

Cheng, L.L., Rodas, J.D., Schultz, K.T., Christensen, B.M., Yuill, T.M., and Israel, B.A. (1999). Potential for evolution of California serogroup bunyaviruses by genome reassortment in *Aedes albopictus*. Am. J. Trop. Med. Hyg. *60*, 430–438.

Coffey, L.L., and Vignuzzi, M. (2011). Host alternation of chikungunya virus increases fitness while restricting population diversity and adaptability to novel selective pressures. J. Virol. *85*, 1025–1035.

Coffey, L.L., Forrester, N., Tsetsarkin, K., Vasilakis, N., and Weaver, S.C. (2013). Factors shaping the adaptive landscape for arboviruses: implications for the emergence of disease. Future Microbiol. *8*, 155–176.

Coffey, L.L., Page, B.L., Greninger, A.L., Herring, B.L., Russell, R.C., Doggett, S.L., Haniotis, J., Wang, C., Deng, X., and Delwart, E.L. (2014). Enhanced arbovirus surveillance with deep sequencing: identification of novel rhabdoviruses and bunyaviruses in Australian mosquitoes. Virology *448*, 146–158.

Combe, M., and Sanjuan, R. (2014). Variation in RNA virus mutation rates across host cells. PLoS Pathog. *10*, e1003855.

Cui, J., Gao, M., and Ren, X. (2011). Phylogeny and homologous recombination in Chikungunya viruses. Infect. Genet. Evol. *11*, 1957–1963.

Davies, C.R., Jones, L.D., Green, B.M., and Nuttall, P.A. (1987). In vivo reassortment of Thogoto virus (a tick-borne influenza-like virus) following oral infection of Rhipicephalus appendiculatus ticks. J. Gen. Virol. *68*, 2331–2338.

Davis, L.E., Beckham, J.D., and Tyler, K.L. (2008). North American encephalitic arboviruses. Neurol. Clin. *26*, 727–757, ix.

Demmers, M.W., Baan, C.C., Janssen, M., Litjens, N.H., Ijzermans, J.N., Betjes, M.G., Weimar, W., and Rowshani, A.T. (2014). Substantial proliferation of human renal tubular epithelial cell-reactive CD4+CD28null memory T cells, which is resistant to tacrolimus and everolimus. Transplantation *97*, 47–55.

Deyde, V.M., Khristova, M.L., Rollin, P.E., Ksiazek, T.G., and Nichol, S.T. (2006). Crimean–Congo hemorrhagic fever virus genomics and global diversity. J. Virol. *80*, 8834–8842.

Ding, N.Z., Luo, Z.F., Niu, D.D., Ji, W., Kang, X.H., Cai, S.S., Xu, D.S., Wang, Q.W., and He, C.Q. (2013). Identification of two severe fever with thrombocytopenia syndrome virus strains originating from reassortment. Virus Res. *178*, 543–546.

Dixon, L.K., Chapman, D.A., Netherton, C.L., and Upton, C. (2013). African swine fever virus replication and genomics. Virus Res. *173*, 3–14.

Domingo, E., Sheldon, J., and Perales, C. (2012). Viral quasispecies evolution. Microbiol. Mol. Biol. Rev. *76*, 159–216.

Drake, J.W., and Holland, J.J. (1999). Mutation rates among RNA viruses. Proc. Natl. Acad. Sci. U.S.A. *96*, 13910–13913.

Eifan, S., Schnettler, E., Dietrich, I., Kohl, A., and Blomstrom, A.L. (2013). Non-structural proteins of arthropod-borne bunyaviruses: roles and functions. Viruses *5*, 2447–2468.

Fischer, M., and Staples, J.E. (2014). Notes from the field: chikungunya virus spreads in the Americas – Caribbean and South america, 2013–2014. MMWR Morb. Mortal. Wkly Rep. *63*, 500–501.

Gould, E.A., and Solomon, T. (2008). Pathogenic flaviviruses. Lancet *371*, 500–509.

Gray, S.M., and Banerjee, N. (1999). Mechanisms of arthropod transmission of plant and animal viruses. Microbiol. Mol. Biol. Rev. *63*, 128–148.

Groseth, A., Matsuno, K., Dahlstrom, E., Anzick, S.L., Porcella, S.F., and Ebihara, H. (2012). Complete genome sequencing of four geographically diverse strains of Batai virus. J. Virol. *86*, 13844–13845.

Hahn, C.S., Lustig, S., Strauss, E.G., and Strauss, J.H. (1988). Western equine encephalitis virus is a recombinant virus. Proc. Natl. Acad. Sci. U.S.A. *85*, 5997–6001.

Hallwass, M., de Oliveira, A.S., de Campos Dianese, E., Lohuis, D., Boiteux, L.S., Inoue-Nagata, A.K., Resende, R.O., and Kormelink, R. (2014). The Tomato spotted wilt virus cell-to-cell movement protein (NS) triggers a hypersensitive response in Sw-5-containing resistant tomato lines and in *Nicotiana benthamiana* transformed with the functional Sw-5b resistance gene copy. Mol. Plant Pathol. *15*, 871–880.

Harley, D., Sleigh, A., and Ritchie, S. (2001). Ross River virus transmission, infection, and disease: a cross-disciplinary review. Clin. Microbiol. Rev. *14*, 909–932, table of contents.

Harris, E., Holden, K.L., Edgil, D., Polacek, C., and Clyde, K. (2006). Molecular biology of flaviviruses. Novartis Found. Symp. *277*, 23–39; discussion *40*, 71–23, 251–253.

He, C.Q., Ding, N.Z., He, M., Li, S.N., Wang, X.M., He, H.B., Liu, X.F., and Guo, H.S. (2010). Intragenic recombination as a mechanism of genetic diversity in bluetongue virus. J. Virol. *84*, 11487–11495.

Hollidge, B.S., Gonzalez-Scarano, F., and Soldan, S.S. (2010). Arboviral encephalitides: transmission, emergence, and pathogenesis. J. Neuroimmun. Pharmacol. *5*, 428–442.

Huang, Y.W., Geng, Y.F., Ying, X.B., Chen, X.Y., and Fang, R.X. (2005). Identification of a movement protein of rice yellow stunt rhabdovirus. J. Virol. *79*, 2108–2114.

Hubalek, Z., Rudolf, I., and Nowotny, N. (2014). Arboviruses pathogenic for domestic and wild animals. Adv. Virus Res. *89*, 201–275.

el Hussein, A., Ramig, R.F., Holbrook, F.R., and Beaty, B.J. (1989). Asynchronous mixed infection of Culicoides variipennis with bluetongue virus serotypes 10 and 17. J. Gen. Virol. *70*, 3355–3362.

Jackson, A.O., Dietzgen, R.G., Goodin, M.M., Bragg, J.N., and Deng, M. (2005). Biology of plant rhabdoviruses. Annu. Rev. Phytopathol. *43*, 623–660.

Ji, X., Qian, D., Wei, C., Ye, G., Zhang, Z., Wu, Z., Xie, L., and Li, Y. (2011). Movement protein Pns6 of rice dwarf phytoreovirus has both ATPase and RNA binding activities. PLoS One *6*, e24986.

Jones, L.D., Davies, C.R., Green, B.M., and Nuttall, P.A. (1987). Reassortment of Thogoto virus (a tick-borne influenza-like virus) in a vertebrate host. J. Gen. Virol. *68*, 1299–1306.

Kato, T., Aizawa, M., Takayoshi, K., Kokuba, T., Yanase, T., Shirafuji, H., Tsuda, T., and Yamakawa, M. (2009). Phylogenetic relationships of the G gene sequence of bovine ephemeral fever virus isolated in Japan, Taiwan and Australia. Vet. Microbiol. *137*, 217–223.

Kim, K.H., Rumenapf, T., Strauss, E.G., and Strauss, J.H. (2004). Regulation of Semliki Forest virus RNA replication: a model for the control of alphavirus pathogenesis in invertebrate hosts. Virology *323*, 153–163.

Klein, E.Y., Serohijos, A.W., Choi, J.M., Shakhnovich, E.I., and Pekosz, A. (2014). Influenza A H1N1 pandemic strain evolution--divergence and the potential for antigenic drift variants. PLoS One *9*, e93632.

Kochs, G., Weber, F., Gruber, S., Delvendahl, A., Leitz, C., and Haller, O. (2000). Thogoto virus matrix protein is encoded by a spliced mRNA. J. Virol. *74*, 10785–10789.

Kormelink, R., Garcia, M.L., Goodin, M., Sasaya, T., and Haenni, A.L. (2011). Negative-strand RNA viruses: the plant-infecting counterparts. Virus Res. *162*, 184–202.

Kuno, G., and Chang, G.J. (2005). Biological transmission of arboviruses: reexamination of and new insights into components, mechanisms, and unique traits as well as their evolutionary trends. Clin. Microbiol. Rev. *18*, 608–637.

Kuzmin, I.V., Novella, I.S., Dietzgen, R.G., Padhi, A., and Rupprecht, C.E. (2009). The rhabdoviruses: biodiversity, phylogenetics, and evolution. Infect. Genet. Evol. *9*, 541–553.

Lewandowski, D.J., and Adkins, S. (2005). The tubule-forming NSm protein from Tomato spotted wilt virus complements cell-to-cell and long-distance movement of Tobacco mosaic virus hybrids. Virology *342*, 26–37.

Luers, A.J., Adams, S.D., Smalley, J.V., and Campanella, J.J. (2005). A phylogenomic study of the genus Alphavirus employing whole genome comparison. Comp. Funct. Genomics *6*, 217–227.

Lukashev, A.N. (2005). Evidence for recombination in Crimean–Congo hemorrhagic fever virus. J. Gen. Virol. *86*, 2333–2338.

Maan, N.S., Maan, S., Nomikou, K., Guimera, M., Pullinger, G., Singh, K.P., Belaganahalli, M.N., and Mertens, P.P. (2012). The genome sequence of bluetongue virus type 2 from India: evidence for reassortment between eastern and western topotype field strains. J. Virol. *86*, 5967–5968.

McDonald, N.J., Smith, C.B., and Cox, N.J. (2007). Antigenic drift in the evolution of H1N1 influenza A viruses resulting from deletion of a single amino acid in the haemagglutinin gene. J. Gen. Virol. *88*, 3209–3213.

McGee, C.E., Tsetsarkin, K.A., Guy, B., Lang, J., Plante, K., Vanlandingham, D.L., and Higgs, S. (2011). Stability of yellow fever virus under recombinatory pressure as compared with chikungunya virus. PLoS One *6*, e23247.

Maclachlan, N.J., and Guthrie, A.J. (2010). Re-emergence of bluetongue, African horse sickness, and other orbivirus diseases. Vet. Res. *41*, 35.

May, F.J., Li, L., Zhang, S., Guzman, H., Beasley, D.W., Tesh, R.B., Higgs, S., Raj, P., Bueno, R., Jr., Randle, Y., *et al.* (2008). Genetic variation of St. Louis encephalitis virus. J. Gen. Virol. *89*, 1901–1910.

Michaud, V., Randriamparany, T., and Albina, E. (2013). Comprehensive phylogenetic reconstructions of African swine fever virus: proposal for a new classification and molecular dating of the virus. PLoS One *8*, e69662.

Mutebi, J.P., Rijnbrand, R.C., Wang, H., Ryman, K.D., Wang, E., Fulop, L.D., Titball, R., and Barrett, A.D. (2004). Genetic relationships and evolution of genotypes of yellow fever virus and other members of the yellow fever virus group within the Flavivirus genus based on the 3′ noncoding region. J. Virol. *78*, 9652–9665.

Myles, K.M., Kelly, C.L., Ledermann, J.P., and Powers, A.M. (2006). Effects of an opal termination codon preceding the nsP4 gene sequence in the O'Nyong-Nyong virus genome on Anopheles gambiae infectivity. J. Virol. *80*, 4992–4997.

Neumann, G., Brownlee, G.G., Fodor, E., and Kawaoka, Y. (2004). Orthomyxovirus replication, transcription, and polyadenylation. Curr. Top. Microbiol. Immunol. *283*, 121–143.

Oberst, R.D., Stott, J.L., Blanchard-Channell, M., and Osburn, B.I. (1987). Genetic reassortment of bluetongue virus serotype 11 strains in the bovine. Vet. Microbiol. *15*, 11–18.

Perez-Ramirez, G., Diaz-Badillo, A., Camacho-Nuez, M., Cisneros, A., and Munoz Mde, L. (2009). Multiple recombinants in two dengue virus, serotype-2 isolates from patients from Oaxaca, Mexico. BMC Microbiol. *9*, 260.

Pichlmair, A., Buse, J., Jennings, S., Haller, O., Kochs, G., and Staeheli, P. (2004). Thogoto virus lacking interferon-antagonistic protein ML is strongly attenuated in newborn Mx1-positive but not Mx1-negative mice. J. Virol. *78*, 11422–11424.

Piljman, G.P., Funk, A., Kondratieva, N., Leung, J., Torres, S., van der Aa, L., Liu, W.J., Palmenberg, A.C., Shi, P.Y., Hall, R.A., and Khromykh, A.A. (2008) A highly structured, nuclease-resistant, noncoding RNA produced by flaviviruses is required for pathogenesis. Cell Host Microbe *6*, 579–591.

Powers, A.M., Brault, A.C., Tesh, R.B., and Weaver, S.C. (2000). Re-emergence of Chikungunya and O'nyong-nyong viruses: evidence for distinct geographical lineages and distant evolutionary relationships. J. Gen. Virol. *81*, 471–479.

Reese, S.M., Blitvich, B.J., Blair, C.D., Geske, D., Beaty, B.J., and Black, W.C.t. (2008). Potential for La Crosse virus segment reassortment in nature. Virol. J. *5*, 164.

Roby, J.A., Pijlman, G.P., Wilusz, J., and Khromykh, A.A. (2014). Noncoding subgenomic flavivirus RNA: multiple functions in West Nile virus pathogenesis and modulation of host responses. Viruses *6*, 404–427.

Samal, S.K., el-Hussein, A., Holbrook, F.R., Beaty, B.J., and Ramig, R.F. (1987a). Mixed infection of Culicoides variipennis with bluetongue virus serotypes 10 and 17: evidence for high frequency reassortment in the vector. J. Gen. Virol. *68*, 2319–2329.

Samal, S.K., Livingston, C.W., Jr., McConnell, S., and Ramig, R.F. (1987b). Analysis of mixed infection of sheep with bluetongue virus serotypes 10 and 17: evidence for genetic reassortment in the vertebrate host. J. Virol. *61*, 1086–1091.

Sammels, L.M., Coelen, R.J., Lindsay, M.D., and Mackenzie, J.S. (1995). Geographic distribution and evolution of Ross River virus in Australia and the Pacific Islands. Virology *212*, 20–29.

Saxton-Shaw, K.D., Ledermann, J.P., Borland, E.M., Stovall, J.L., Mossel, E.C., Singh, A.J., Wilusz, J., and Powers, A.M. (2013). O'nyong nyong virus molecular determinants of unique vector specificity reside in non-structural protein 3. PLoS Negl. Trop. Dis. *7*, e1931.

Schuh, A.J., Ward, M.J., Brown, A.J., and Barrett, A.D. (2013). Phylogeography of Japanese encephalitis virus: genotype is associated with climate. PLoS Negl. Trop. Dis. *7*, e2411.

Shafiq, M., Minakshi, P., Bhateja, A., Ranjan, K., and Prasad, G. (2013). Evidence of genetic reassortment between Indian isolate of bluetongue virus serotype 21 (BTV-21) and bluetongue virus serotype 16 (BTV-16). Virus Res. *173*, 336–343.

Shaw, A.E., Ratinier, M., Nunes, S.F., Nomikou, K., Caporale, M., Golder, M., Allan, K., Hamers, C., Hudelet, P., Zientara, S., *et al.* (2013). Reassortment between two serologically unrelated bluetongue virus strains is flexible and can involve any genome segment. J. Virol. *87*, 543–557.

Silverstein, S.C., Christman, J.K., and Acs, G. (1976). The reovirus replicative cycle. Annu. Rev. Biochem. *45*, 375–408.

Smith, D.R., Adams, A.P., Kenney, J.L., Wang, E., and Weaver, S.C. (2008). Venezuelan equine encephalitis virus in the mosquito vector *Aedes taeniorhynchus*: infection initiated by a small number of susceptible epithelial cells and a population bottleneck. Virology *372*, 176–186.

Soldan, S.S., and Gonzalez-Scarano, F. (2005). Emerging infectious diseases: the Bunyaviridae. J. Neurovirol. *11*, 412–423.

Strauss, J.H., and Strauss, E.G. (1994). The alphaviruses: gene expression, replication, and evolution. Microbiol. Rev. *58*, 806.

Suhrbier, A., Jaffar-Bandjee, M.C., and Gasque, P. (2012). Arthritogenic alphaviruses--an overview. Nat. Rev. Rheumatol. *8*, 420–429.

Tao, H., Steel, J., and Lowen, A.C. (2014). Intrahost dynamics of influenza virus reassortment. J. Virol. *88*, 7485–7492.

Taucher, C., Berger, A., and Mandl, C.W. (2010). A trans-complementing recombination trap demonstrates a low propensity of flaviviruses for intermolecular recombination. J. Virol. *84*, 599–611.

von Teichman, B.F., and Smit, T.K. (2008). Evaluation of the pathogenicity of African Horsesickness (AHS) isolates in vaccinated animals. Vaccine *26*, 5014–5021.

Tsetsarkin, K.A., Vanlandingham, D.L., McGee, C.E., and Higgs, S. (2007). A single mutation in chikungunya virus affects vector specificity and epidemic potential. PLoS Pathog. *3*, e201.

Tsetsarkin, K.A., Chen, R., Sherman, M.B., and Weaver, S.C. (2011). Chikungunya virus: evolution and genetic determinants of emergence. Curr. Opin. Virol. *1*, 310–317.

Turina, M., Tavella, L., and Ciuffo, M. (2012). Tospoviruses in the Mediterranean area. Adv. Virus Res. *84*, 403–437.

Twiddy, S.S., and Holmes, E.C. (2003). The extent of homologous recombination in members of the genus Flavivirus. J. Gen. Virol. *84*, 429–440.

Upadhyayula, S.M., Rao Mutheneni, S., Nayanoori, H.K., Natarajan, A., and Goswami, P. (2012). Impact of weather variables on mosquitoes infected with Japanese encephalitis virus in Kurnool district, Andhra Pradesh. Asian Pac. J. Trop. Med. *5*, 337–341.

Varela, M., Schnettler, E., Caporale, M., Murgia, C., Barry, G., McFarlane, M., McGregor, E., Piras, I.M., Shaw, A., Lamm, C., *et al.* (2013). Schmallenberg virus pathogenesis, tropism and interaction with the innate immune system of the host. PLoS Pathog. *9*, e1003133.

Vasconcelos, P.F., Bryant, J.E., da Rosa, T.P., Tesh, R.B., Rodrigues, S.G., and Barrett, A.D. (2004). Genetic divergence and dispersal of yellow fever virus, Brazil. Emerg. Infect. Dis. *10*, 1578–1584.

Vogt, C., Preuss, E., Mayer, D., Weber, F., Schwemmle, M., and Kochs, G. (2008). The interferon antagonist ML-protein of thogoto virus targets general transcription factor IIB. J. Virol. *82*, 11446–11453.

Walker, P.J., Dietzgen, R.G., Joubert, D.A., and Blasdell, K.R. (2011). Rhabdovirus accessory genes. Virus Res. *162*, 110–125.

Wallner, G., Mandl, C.W., Kunz, C., and Heinz, F.X. (1995). The flavivirus 3′-noncoding region: extensive size heterogeneity independent of evolutionary relationships among strains of tick-borne encephalitis virus. Virology *213*, 169–178.

Wang, E., Weaver, S.C., Shope, R.E., Tesh, R.B., Watts, D.M., and Barrett, A.D. (1996). Genetic variation in yellow fever virus: duplication in the 3′ noncoding region of strains from Africa. Virology *225*, 274–281.

Weaver, S.C. (2006). Evolutionary influences in arboviral disease. Curr. Top. Microbiol. Immunol. *299*, 285–314.

Weaver, S.C., and Barrett, A.D. (2004). Transmission cycles, host range, evolution and emergence of arboviral disease. Nat. Rev. Microbiol. *2*, 789–801.

Weaver, S.C., and Reisen, W.K. (2010). Present and future arboviral threats. Antiviral Res. *85*, 328–345.

Weaver, S.C., Kang, W., Shirako, Y., Rumenapf, T., Strauss, E.G., and Strauss, J.H. (1997). Recombinational history and molecular evolution of western equine encephalomyelitis complex alphaviruses. J. Virol. *71*, 613–623.

Weaver, S.C., Winegar, R., Manger, I.D., and Forrester, N.L. (2012). Alphaviruses: population genetics and determinants of emergence. Antiviral Res. *94*, 242–257.

Webster, C.G., Reitz, S.R., Perry, K.L., and Adkins, S. (2011). A natural M RNA reassortant arising from two species of plant- and insect-infecting bunyaviruses and comparison of its sequence and biological properties to parental species. Virology *413*, 216–225.

Weidmann, M., Frey, S., Freire, C.C., Essbauer, S., Ruzek, D., Klempa, B., Zubrikova, D., Vogerl, M., Pfeffer, M., Hufert, F.T., *et al.* (2013). Molecular phylogeography of tick-borne encephalitis virus in central Europe. J. Gen. Virol. *94*, 2129–2139.

Westgeest, K.B., Russell, C.A., Lin, X., Spronken, M.I., Bestebroer, T.M., Bahl, J., van Beek, R., Skepner, E., Halpin, R.A., de Jong, J.C., *et al.* (2014). Genomewide analysis of reassortment and evolution of human influenza A(H3N2) viruses circulating between 1968 and 2011. J. Virol. *88*, 2844–2857.

Worobey, M., Rambaut, A., and Holmes, E.C. (1999). Widespread intra-serotype recombination in natural populations of dengue virus. Proc. Natl. Acad. Sci. U.S.A. *96*, 7352–7357.

Yanase, T., Kato, T., Aizawa, M., Shuto, Y., Shirafuji, H., Yamakawa, M., and Tsuda, T. (2012). Genetic reassortment between Sathuperi and Shamonda viruses of the genus Orthobunyavirus in nature: implications for their genetic relationship to Schmallenberg virus. Arch. Virol. *157*, 1611–1616.

Zacks, M.A., and Paessler, S. (2010). Encephalitic alphaviruses. Vet. Microbiol. *140*, 281–286.

Ecological and Epidemiological Factors Influencing Arbovirus Diversity, Evolution and Spread

10

Roy A. Hall, Sonja Hall-Mendelin, Jody Hobson-Peters, Natalie A. Prow and John S. Mackenzie

Abstract

The requirement for arboviruses to be transmitted between arthropod vectors and vertebrate hosts provides enormous complexity in their strategies to persist in populations and spread into new areas. In this chapter, we discuss ecological, environmental and anthropological factors that influence the spread, establishment and evolution of vector-borne viruses and the emergence of novel arboviral diseases. These topics include the introduction of viruses and vectors into new geographical regions, the adaption of viruses to new vectors and hosts and how these events influence the evolution of new viral strains, new vector–host relationships and the emergence of viral strains with enhanced virulence. The potential effects of climate change, deforestation, and the encroachment of human habitation and agriculture on the emergence and distribution of arboviral diseases are also discussed. We conclude that enhanced arboviral surveillance will be crucial in the management and prevention of vector-borne viral disease in the future, and that new viral detection strategies that are independent of viral sequence or antigenic structure and can be integrated into high-throughput, cost-effective real-time surveillance systems are essential for monitoring the spread of arboviruses and identifying new and emerging vector-borne viruses.

Introduction

The requirement for arboviruses to be transmitted between arthropod vectors and vertebrate hosts and to propagate in both, places stringent control on the replication methods of these viruses, and provides enormous complexity in their strategies to persist and spread through populations and into new areas. The impact of arboviruses spreading into new regions of the world is perhaps best exemplified by the invasion of the Americas by the mosquito-borne West Nile virus (WNV) in 1999, the spread of Chikungunya virus (CHIKV) into southern Europe, and to island nations in the Pacific and Caribbean, and the spread of bluetongue virus (BTV) into Europe. The introduction of a virulent WNV strain into virgin territory containing highly competent vectors and susceptible hosts provided very efficient modes of transmission resulting in unprecedented levels of human and animal disease. Similarly, the introduction of arthropod vectors such as *Aedes albopictus* and *Culicoides* species to new areas in Europe has been responsible for the first evidence of autochthonous transmission leading to outbreaks of disease due to CHIKV and dengue viruses (DENV) in humans and BTV in livestock respectively. The emergence of a new global strain of CHIKV has also been associated with selective adaption to the *Ae. albopictus* mosquito, while the recent appearance of novel viruses such as Schmallenberg and a possible new serotype of dengue remind us that new arboviruses are continually evolving. However our understanding of the processes involved and the origins of these viruses is still relatively limited.

In this chapter, we briefly describe the transmission cycles of arboviruses, important components in their ecology, and then discuss the role of environmental, anthropological and ecological factors that influence the spread, establishment and evolution of vector-borne viruses and the emergence of novel arboviral diseases. We refer to the introduction of viruses and vectors into new geographical regions, the adaptation of viruses to new vectors and hosts and how these events influence the evolution of new viral strains, new vector–host relationships and the emergence of viral strains with enhanced virulence. We also discuss the effects of climate change, deforestation, and the encroachment of human habitation and agriculture on the evolution of arboviruses and the emergence and distribution of the diseases they cause. Finally, we examine the effect of viral co-infection of vectors and hosts on arboviral evolution and of transmission with particular reference to the recent discovery of many insect-specific viruses and an upsurge of interest in investigating the insect virome by new generation sequencing platforms.

Arbovirus transmission cycles and persistence

Nearly all arboviruses are maintained in nature by zoonotic transmission cycles involving an arthropod and a non-human vertebrate host, most commonly birds, rodents, bats or primates. Importantly, the virus replicates in both the arthropod and the vertebrate host. Epidemics occur when transmission spills over to incidental hosts or the pathogen jumps into a new vector species. A few viruses are maintained in cycles involving arthropod vectors and human hosts, such as urban transmission cycles of DENV, yellow fever virus (YFV) and CHIKV. Most arboviruses are vectored by mosquitoes, ticks (Nuttall *et al.*, 1994; Labuda and Nuttall, 2004), *Culicoides* (Mellor *et al.*, 2000), and phlebotomine sandflies (Depaquit *et al.*, 2010; Alkan *et al.*, 2013). Other arthropod species may be involved occasionally, such as some members of the family *Cimicidae* (reviewed by Dogett *et al.*, 2012; Adelman *et al.*, 2013), lice (La Linn *et al.*, 2001), louse fly (Farajollahi *et al.*, 2005a; Grancz *et al.*, 2004), and possibly mites (Mumcuoglu *et al.*, 2005). Mites have also been shown to transmit Venezuelan equine encephalitis virus (VEEV) mechanically, with no replication in the mite (Durden *et al.*, 1992). Although some arboviruses may be restricted to certain mosquito vector species, such as with DENV and YFV, others may be transmitted by a wide variety of mosquito vectors, such as is encountered with WNV and other members of the Japanese encephalitis serological complex of flaviviruses, and by Sindbis (SINV), Ross River and other alphaviruses. Ticks may also be infected with tick-borne encephalitis virus (TBEV) by non-viraemic transmission, whereby leucocytes migrate between tick feeding sites on the same host, bearing infectious virus, and so providing a transport route for the virus between co-feeding ticks independent of systemic viraemia (Labuda *et al.*, 1993a,b; Randolph, 2011). *Culicoides* or biting midges are major transmitters of arboviruses causing diseases in animals, and with the single exception of the orthobunyavirus, Oropouche virus (Pinheiro *et al.*, 1981; Carpenter *et al.*, 2013), do not transmit arboviruses to humans. Some mosquito-borne arboviruses may be transmitted by both mosquitoes and ticks, as demonstrated by WNV (Schmidt and Mufeed, 1964; Lvov *et al.*, 2004; Lwande *et al.*, 2013), by mosquitoes and *Culicoides* as found for bovine ephemeral fever (Walker, 2005), and by mosquitoes, ticks and sandflies as found for Chandipura virus (Menghani *et al.*, 2012). The effect of multiple vectors, which may or may not overlap in space and time, may greatly assist in virus persistence and virus abundance and thus epidemic peaks (Lord, 2010), and may lead to greater diversity. The transmission of arboviruses to vertebrates by blood-sucking arthropod vectors is commonly known as biological transmission (Kuno and Chang, 2005).

There are many complex interactions concerned with the virus, vector and environment inherent in the transmission of both mosquito-borne (Kenney and Brault, 2014; Ciota and Kramer, 2013) and tick-borne viruses (Estrada-Peña *et al.*, 2013) which impact on the fitness and the capability of arboviruses to evolve and spread. These interactions are shown in Fig. 10.1. The ability of viruses to survive over winter or through periods of drought is an essential property for virus persistence. Drought may even increase virus outbreaks under certain conditions (Epstein and Defilippo, 2001; Shaman *et al.*, 2005), but can also prevent outbreaks through preventing vector development (Komar and Spielman, 1994). Overwintering can be via vertical transmission, by survival of adults in hibernacula, by persistent infections in vertebrate hosts, or in a few cases, by dormancy in hibernating vertebrate hosts (Rosen, 1987; Kuno, 2001a,b). Vertical transmission, in the mosquito egg, larvae or pupa, has been recorded for a number of viruses in a wide range of different mosquito, tick and phlebotomine species (Rosen, 1987; Turell, 1988; Nelms *et al.*, 2013). Overwintering of infected adult mosquitoes in hibernacula has been reported in several studies, as exemplified by the detection of WNV (Nasci *et al.*, 2001; Bugbee and Forte, 2004; Farajollahi *et al.*, 2005b; Andreadis *et al.*, 2010), and in adult *Culicoides*, as exemplified by BTV (Gerry and Mullins, 2000). Persistence in vertebrate hosts has been described for arboviruses belonging

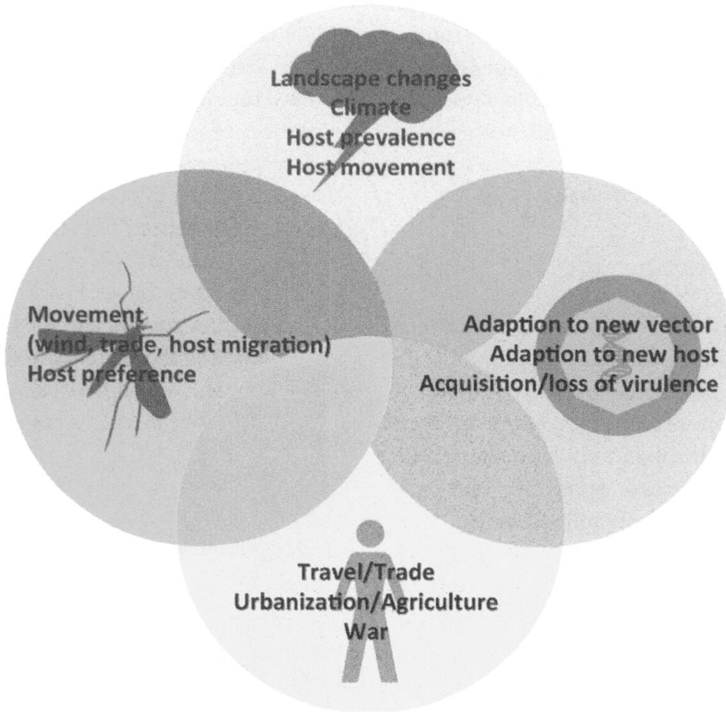

Figure 10.1 Venn diagram illustrating the complex interactions between ecological, environmental and anthroplogical factors associated with driving arboviral diversity and evolution and the emergence and spread of disease.

to most virus families (reviewed by Kuno, 2001b). These infections are often asymptomatic, cryptic, and prolonged, and may provide an effective method for virus dispersal. Thus long-term persistence lasting several months to years has been described in bats with Japanese encephalitis virus (JEV), St. Louis encephalitis virus (SLEV) and Rio Bravo virus; in birds with Eastern equine encephalitis virus (EEEV) and Western equine encephalitis virus (WEEV); in snakes with EEEV and WEEV; in monkeys with Kyasanur Forest disease virus (KFDV), WNV, YFV and TBEV; and in cattle, sheep and elk with BTV. Long-term persistence of infectious virus has also been reported occasionally in some human infections with JEV and TBEV (Kuno, 2001b).

It is also important to recognize that transmission of arboviruses can occur occasionally by non-biological methods, that is, without the intermediary of a blood-sucking arthropod. Aerosol, mucocutaneous, percutaneous and oral transmission have been described for various arboviruses (Kuno, 2001b; Pedrosa and Cardoso, 2011). Thus aerosol transmission has been reported for VEEV (Seymour *et al.*, 1978; Laboratory Safety for Arboviruses and Certain Other Viruses, 1980; Pratt *et al.*, 1998); percutaneous needle-stick transmission has been described for DENV (Nemes *et al.*, 2004; Wagner *et al.*, 2004) and for Crimean–Congo haemorrhagic fever virus (CCHFV) (Celikbas *et al.*, 2014); mucocutaneous transmission has been reported for DENV (Chen and Wilson, 2004) and for CCHFV (Naderi *et al.*, 2013); probable breast milk transmission has been reported for DENV (Barthel *et al.*, 2013), WNV (CDC, 2002) and the live attenuated vaccine strain of YFV (Kuhn *et al.*, 2011; Traiber *et al.*, 2011); transmission through infected blood donations has been reported for various arboviruses (Petersen and Bausch, 2010) including DENV (Tambyah *et al.*, 2008) and WNV (Pealer *et al.*, 2003; Busch *et al.*, 2005); transmission through organ transplantation of WNV (Iwamoto *et al.*, 2003; Yango *et al.*, 2014) and DENV (Prasad *et al.*, 2012); outbreaks of TBEV infection have been reported from ingestion of unpasteurized dairy products from viraemic herds (Gresíkova *et al.*, 1975; Kohl *et al.*, 1996; Holzmann *et al.*, 2009; Caini *et al.*, 2012); and one case of sexual transmission has been described for Zika virus (Foy *et al.*, 2011). Rift Valley fever virus (RVFV) can also be transmitted by both mosquito transmission and by direct transmission. Indeed the vast majority of human cases result from direct or indirect contact with the blood or organs of infected animals. The virus can be transmitted to humans through the handling of animal tissue during slaughtering or butchering, or from infected carcasses. Human infections arise from virus entry through broken skin or by aerosol, and possibly by ingestion of unpasteurized milk from infected animals. Certain occupational groups such as herders, farmers, slaughterhouse workers and veterinarians are therefore at higher risk of infection (World Health Organization, 2008). WNV has been detected in the urine of patients (Tonry *et al.*, 2005), and in one study, virus was detected in the urine of patients up to 6 years after initial infection (Murray *et al.*, 2010), although this could not be confirmed in subsequent studies (Gibney *et al.*, 2011; Baty *et al.*, 2012). These few examples all demonstrate that non-biological or direct transmission can readily occur, most often through accidents to health care workers and laboratory workers, or via blood donations in epidemic areas, but occasionally as part of the normal ecology of the virus.

Arbovirus spread and establishment

Arbovirus distribution, spread and emergence is inextricably linked to a number of factors, including vector competence, the presence of susceptible hosts, the genetics of the virus, human activities and societal changes, transportation and global trade, landscape change, animal movements and bird migration, and environmental factors such as weather and climate (Sellers, 1980; Institute of Medicine, 1992; Gubler, 2001; Mackenzie *et al.*, 2004; Weaver and Reisen, 2010). As most arboviruses are zoonoses, the presence of a suitable vertebrate host is important to sustain virus emergence in a new ecosystem. Some viruses utilize humans or domestic animals as their primary vertebrate hosts, such as DENV, urban YFV, and CHIKV in humans or Rift Valley fever (RVFV) and BTV viruses in domestic animal transmission cycles. There are many examples of arbovirus emergence and spread into new habitats, with the earliest known example being the spread of YFV from West Africa to the Americas during the 15th to 17th centuries (Marr, 1982). Over recent years, it is best demonstrated by the global spread of dengue throughout tropical parts of the world (Figs. 10.2 and 10.3), and particularly the emergence of dengue haemorrhagic fever in the Americas following the arrival of a south-east Asian genotype of DENV-2 into Cuba in 1981 (Rico-Hesse *et al.*, 1997; Gubler, 2007), and by the emergence of WNV into the Americas (Lanciotti *et al.*, 1999; Jia *et al.*, 1999) (Fig. 10.4), but other recent examples include DENV into Nepal (Pandey *et al.*, 2004), Easter Island (Perret *et al.*, 2003) and France (La Ruche *et al.*, 2010); Zika virus into Yap (Duffy *et al.*, 2009) and French Polynesia (CDC, 2013) (Fig. 10.5); CHIKV into northern Italy (Beltrame *et al.*, 2007) and elsewhere in southern Europe (Tomasello and Schlagenhauf, 2013), and in late 2013 into the Caribbean where it is spreading rapidly (CDC, 2014) (Fig. 10.6); JEV in South-East Asia (Solomon *et al.*, 2003) and Australasia (Hanna *et al.*, 1996; Hanna *et al.*, 1999; Mackenzie *et al.*, 2002; Mackenzie *et al.*, 2007); and RVFV into Saudi Arabia and Yemen (World Heath Organization, 2000; Balkhy and Memish, 2003), most recently, to the Comores and Mayotte (Sissoko *et al.*, 2009; Roger *et al.*, 2011) (Fig. 10.7).

Spread of competent vectors

The presence of competent vectors is perhaps the most crucial component, and integral for transmission between reservoir hosts. Some viruses, such as those that utilize *Culex* sp. mosquitoes primarily for transmission, can be transmitted by a large variety of mosquito genera and species, and thus have a greater potential for finding an appropriate vector, whereas those that utilize *Aedes* sp. of mosquitoes tend to be more restricted. Examples of the former are WNV (Komar, 2003; Kramer *et al.*, 2008; Monini *et al.*, 2010; Mackenzie *et al.*, 2004) and JEV (van den Hurk *et al.*, 2009; Le Flohic *et al.*, 2013), whereas examples of the latter are the dengue viruses (Rodhain and Rosen, 1997), especially in urban environments. The same specificity can be observed for bluetongue virus which is also limited globally by its *Culicoides* sp. vectors (Mellor, 2004; Tabachnick, 2004; Purse *et al.*, 2005).

The earliest example of arboviral spread and emergence is believed to be the spread of YFV from West Africa to the New World during the slave trade in the 15th to 17th centuries (Christophers, 1960; Marr, 1982; Lounibos, 2002). The only known urban mosquito vector of YFV, *Ae. aegypti*, also

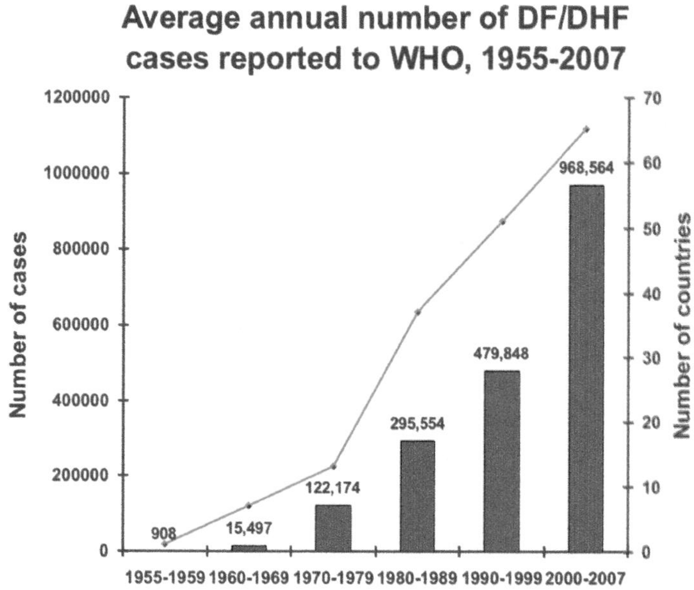

Figure 10.2 Impact of DENV showing the increase in case numbers for dengue fever (DF) and dengue haemmorrhagic fever (DHF) by decade and by number of countries reporting DENV, Figure courtesy of the World Health Organization.

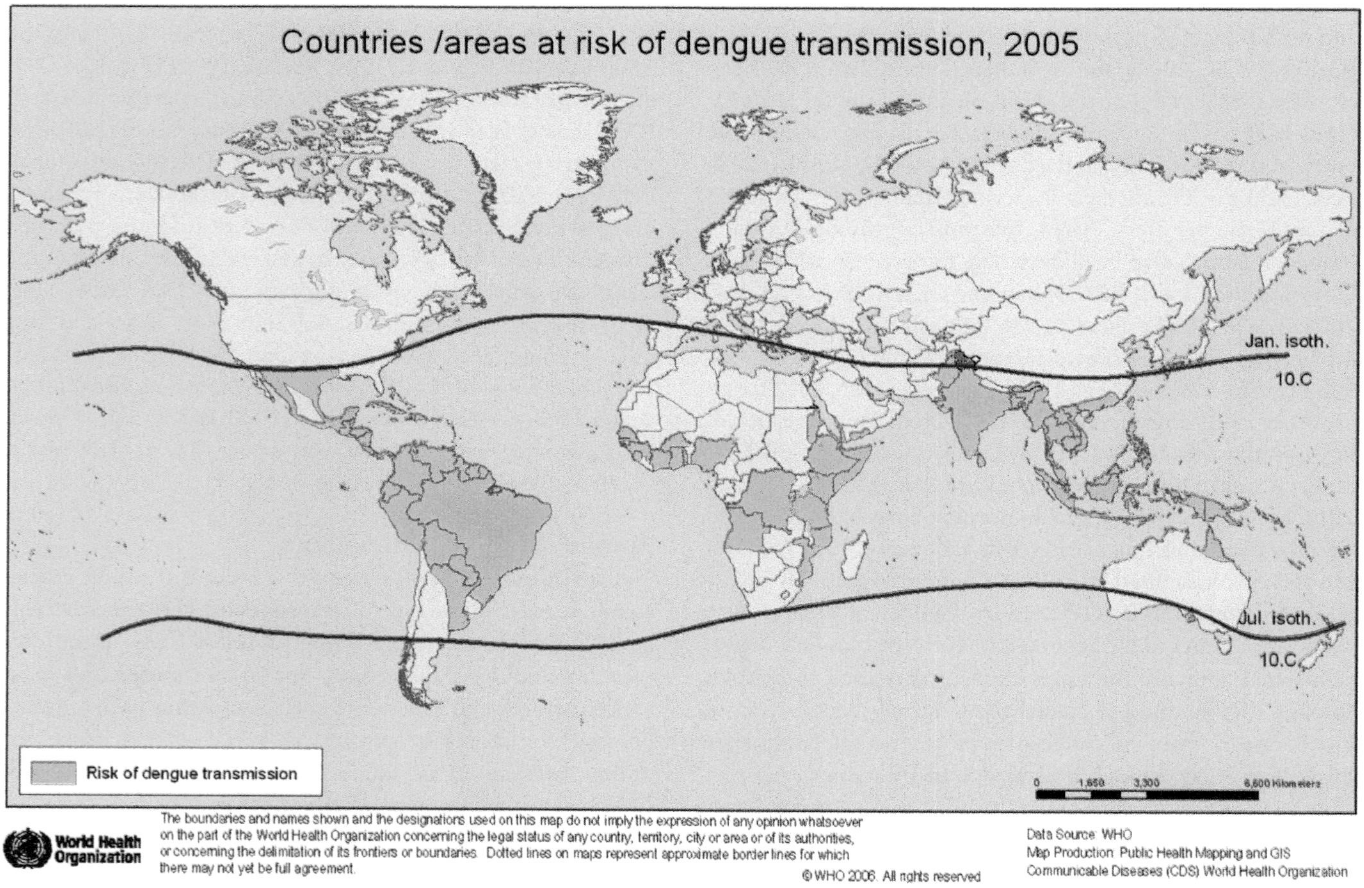

Figure 10.3 Map showing the countries and areas at risk of DENV transmission. The contour lines of the January and July isotherms indicate areas at risk by the geographical limits of the northern and southern hemispheres for year-round survival of *Aedes aegypti*, the principal mosquito vector of DENV. Map courtesy of the World Health Organization.

migrated from West Africa, and is believed to have preceded YFV as a passenger on the slave vessels, probably breeding in water casks kept on deck, but it wasn't until the mosquito had become established that YFV could itself gain a foothold in the New World. This led to the first known yellow fever outbreak in the Americas in Yucatan in 1648, with many subsequent outbreaks, largely in port cities, as far north as Boston and New York (Patterson, 1992).

The introduction of exotic mosquito species and other insect vectors of arboviruses to novel, susceptible environments along

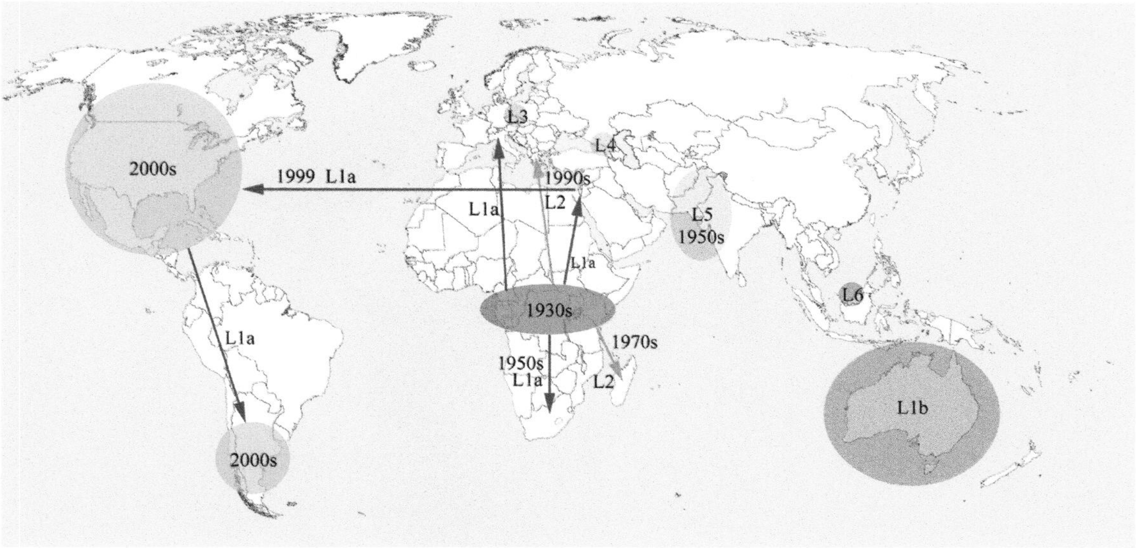

Figure 10.4 Map depicting the spread of WNV, and the geographic distribution of lineages 1–6.

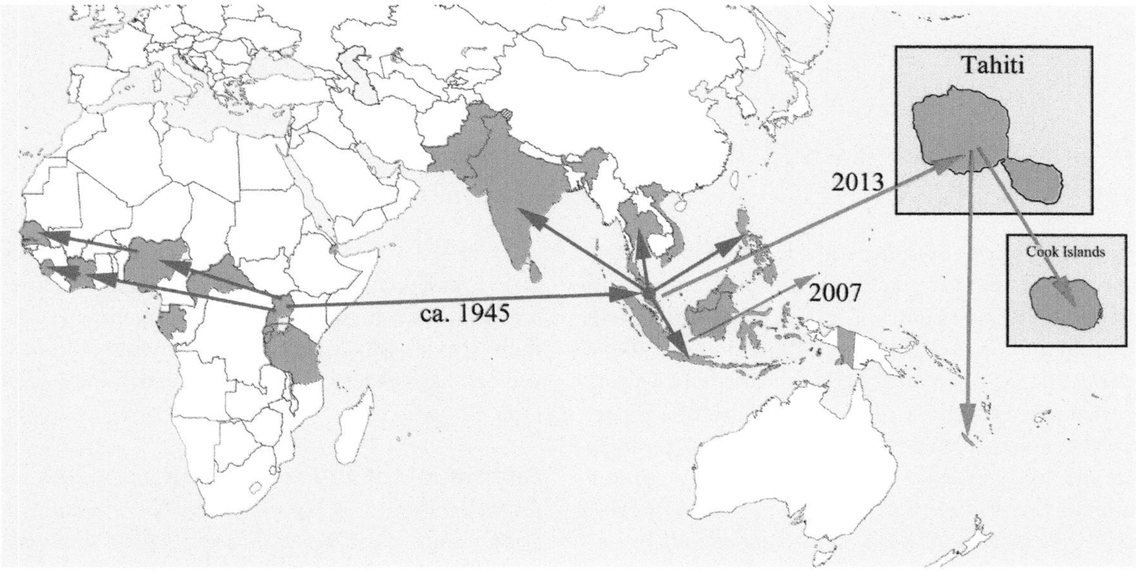

Figure 10.5 Map depicting the spread of Zika virus. The two lineages are the African and Asian lineages, with the Asian lineage evolving from an African virus transported to South-east Asia about 1945.

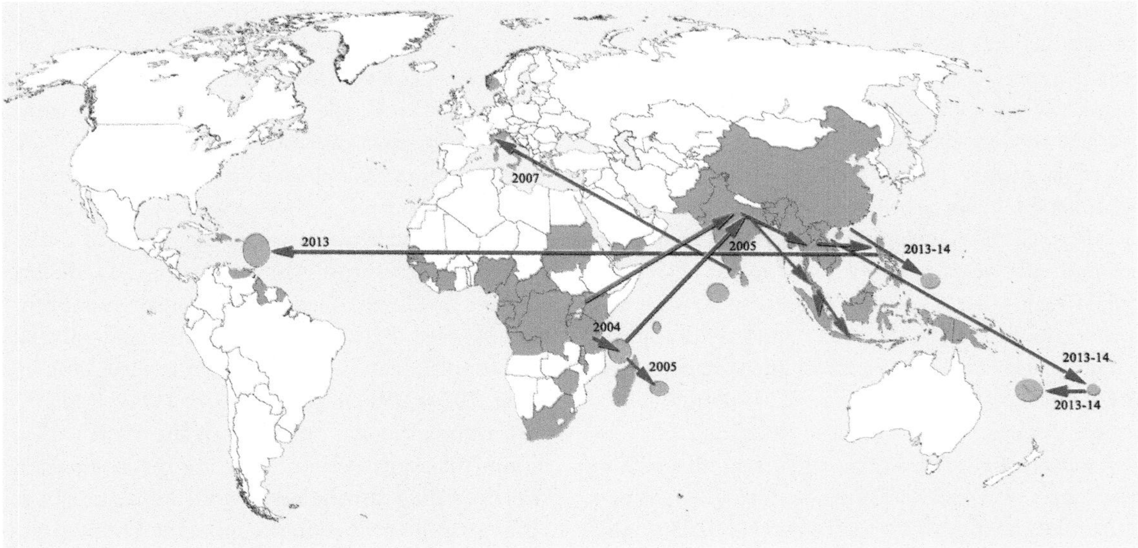

Figure 10.6 Map depicting the spread of CHIKV. Most recent spread if that of the Asian lineage to the Caribbean in 2013.

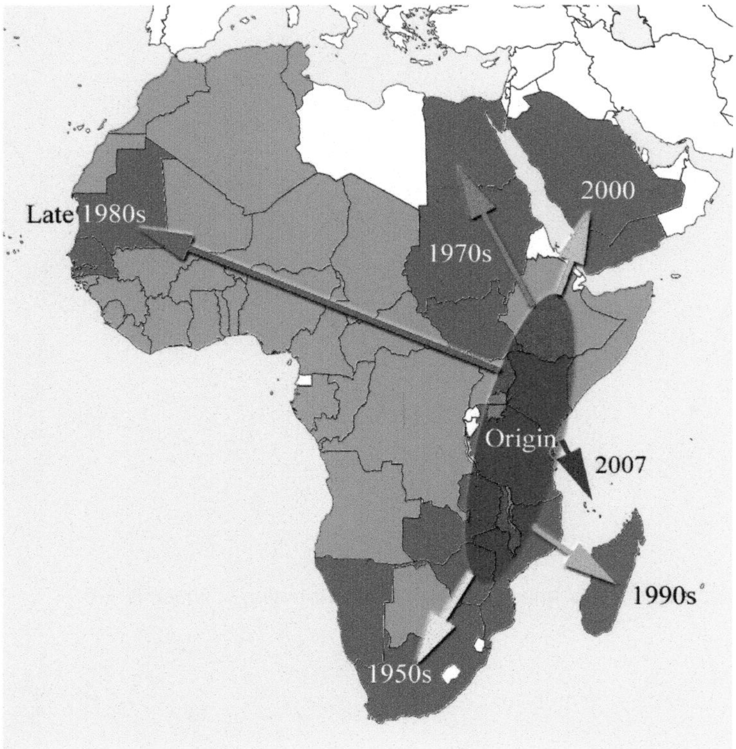

Figure 10.7 Map depicting the spread of RVFV from its origins in the Rift Valley of eastern Africa.

seaborne trade routes and, more recently, by global air travel, has overcome the previous geographic barriers and led to many examples of biological invasion and allowed the increasing spread of human and animal pathogens (Tatem *et al.*, 2006; Lounibos, 2002, 2011; Rogers *et al.*, 2006), of which *Ae. aegypti* and *Ae. albopictus* are the two most important vector species. While air travel has been an important mode of spread through transporting viraemic passengers, it has been less important than sea trade for vector dispersal, probably in part to fumigation or aircraft disinsection (Russell, 1989; Russell and Paton, 1989), although air transport of mosquitoes infected with JEV has been suggested as the cause of outbreaks of JE in Guam (Hammon *et al.*, 1958) and Saipan (Mitchell *et al.*, 1993), and of Zika virus in Yap (Duffy *et al.*, 2009). Indeed, exotic strains of *Ae. aegypti* and *Ae. albopictus* have been imported at Australian international airports in Melbourne, Adelaide and Perth over a three month period in 2014, probably in large passenger baggage bins (A. Sly, personal communication).

Ae. aegypti has become established throughout most of the tropical and subtropical world through sea trade over very many years, and is the major urban vector of DENV and YFV, and in some areas, CHIKV, including the current outbreak in the Caribbean. *Ae. albopictus*, is now a major vector of CHIKV and a secondary major vector of DENV. It is also a vector of a number of other arboviruses (Gratz, 2004), and has been spread to new geographic areas through sea trade, and particularly by the transport of used car tyres (Fig. 10.8a and b). Consequently, it is now widely dispersed in Europe, the Americas, Asia and Africa (Hawley *et al.*, 1987; Laird *et al.*, 1994; Lounibos, 2002; Gratz, 2004; Rogers *et al.*, 2006; Benedict *et al.*, 2007; Paupy *et al.*, 2009; Medlock *et al.*, 2012; Bonizzoni *et al.*, 2013), and is currently threatening Australia (Beebe *et al.*, 2013). It has been responsible for autochthonous transmission of DENV and CHIKV in southern Europe (Tomasello and Schlagenhauf, 2013), and made many areas elsewhere receptive to future arboviral transmission (Fig. 10.8b). A number of other mosquito species which are known arbovirus vectors, or potential vectors, have also spread widely through the trade in car tyres (Lounibos, 2011).

Adaptation of viruses to vector species

Genetic changes in the virus can have a major effect, especially with respect to vector competence, transmissibility and pathogenesis, and provide the mechanism for cross-species adaptation. Recent examples of this have been found for CHIKV where a mutation in the E1 gene, resulting in the A226V substitution, has resulted in a significant increase in the ability of the virus to be effectively transmitted by *Ae. albopictus*, leading to an increased pathogenicity of the virus (Tsetsarkin *et al.*, 2007). A second novel substitution subsequently arose in the virus in Kerala, India, the E2-L210Q, which led to significantly greater circulation, dissemination, and persistence in *Ae. albopictus*, and increased the risk of more severe and expanded epidemics. These results provided the first evidence supporting the hypothesis that adaptation of CHIKV (and possibly other arboviruses) to new niches is a sequential multistep process involving at least two adaptive mutations (Tsetsarkin and Weaver, 2011). A single amino acid substitution affecting vector infection has also been reported for VEEV (Brault *et al.*, 2004a). The rapid spread of WNV in the United States also resulted in the evolution of the virus with a new single dominant clade, WN02, replacing the original NY99 lineage, and becoming distributed across America (Davis *et al.*, 2005b). The extrinsic incubation period of the WN02 genotype was up to four days shorter than the NY99 lineage after peroral inection in *Cx. pipiens* and *Cx. tarsalis* mosquitoes, probably due

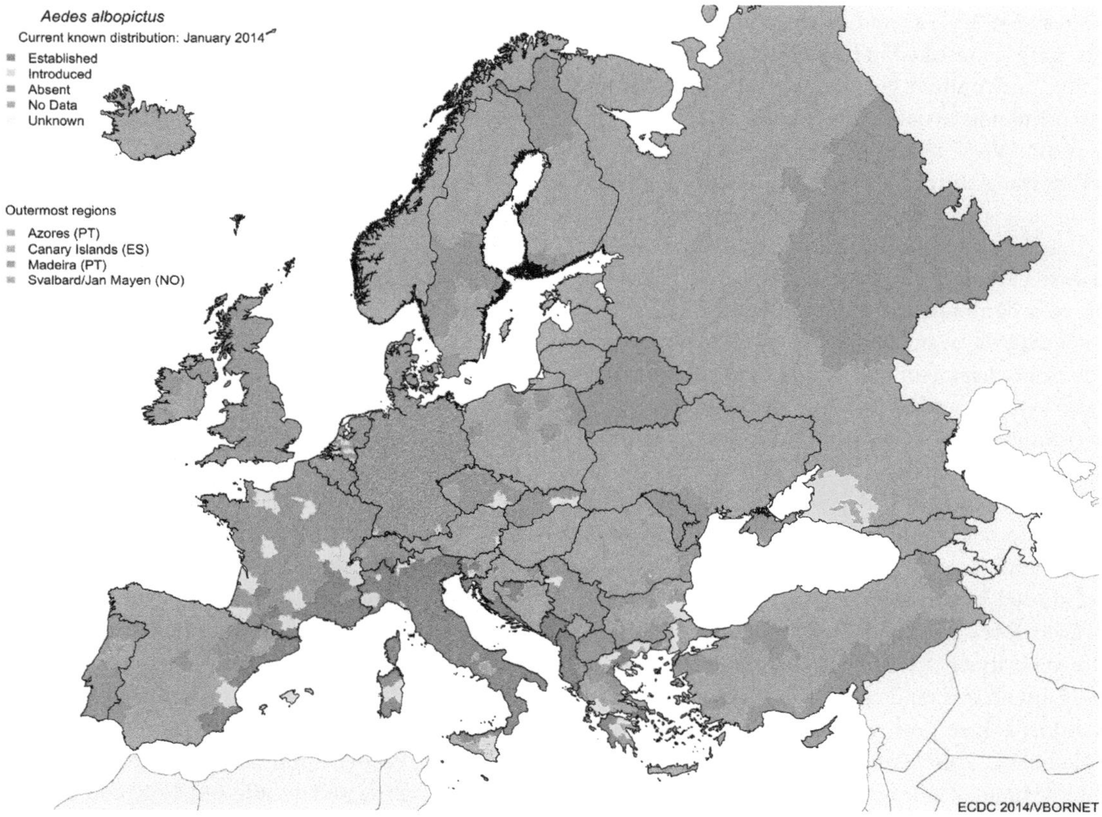

Figure 10.8 (a) Map depicting the spread of *Aedes albopictus*; the areas in blue represent the original geographic distribution of *Ae. albopictus*, and those in green show areas where the mosquito has spread and become established. (b) Map depicting the establishment of *Aedes albopictus* in Europe. Areas in red represent those areas in which the mosquito has established; those in orange are areas where the mosquito has been introduced but eradicated; those in green are areas which are free of the mosquito; and those in grey are areas where no data are available. Map is courtesy of European Centre for Disease Control ECDC201/VBORNET.

to differences in the interaction of the virus with the mosquito midgut (Moudy *et al.*, 2007), but this could not be confirmed by Anderson *et al.* (2012), who found no significant difference between transmission of WN02 and NY99 in *Cx. tarsalis* mosquitoes.

Adaptation of viruses to host species

The introduction of WNV into the USA was initially heralded by the unusual deaths of several bird species in the Bronx Zoo (Ludwig *et al.*, 2002). Indeed the particularly virulent phenotype of the North American strain for avian species, especially the American Crow, provided a unique monitoring system for the spread of the virus through the US, based on the reporting of dead or moribund crows (Komar, 2001). While the initial reason for pronounced WNV bird virulence in the US was attributed to a lack of local host resistance to a newly introduced virus, further investigations with other WNV strains, including avian infections studies, complete genome sequence comparisons and reverse genetics revealed that a single amino acid substitution (A249P) in the NS3 viral protein was responsible (Brault *et al.*, 2004b). This motif enhanced virus replication in the birds with significantly higher levels of virus in the blood providing more efficient transmission to feeding mosquitoes. Contrary to the dogma that enhanced virulence is a disadvantage to viral persistence and survival, Brault *et al.* (2007) proposed that the high level of mortality in American crows was associated with a complete lack of WNV-immunity developing in that species ensuring continued susceptibility to infection. Furthermore, owing to the severe symptoms that manifest in infected crows, the birds are usually roosting on low branches or lying moribund on the ground, providing ground-dwelling mosquito species easier access to feed (Brault *et al.*, 2007). Furthermore, the extremely high viral titres reached in the blood and tissues of WNV-infected American crows (> 10 billion infectious units/g) resulted in high levels of virus released during excretion and mutual grooming (feathers), allowing bird-to-bird transmission to occur (Brault *et al.*, 2004b).

However, most WNV strains from other regions of the world, including the Australian WNV Kunjin lineage strain (WNV_{KUN}), do not carry this virulence determinant and infection of American crows with these viruses produced no significant disease (Brault *et al.*, 2004b). It is also worth noting that infection of a common Australian crow, the little raven (*Corvus mellori*) with the avian-virulent North American strain produced little or no disease in this species, and produced significantly lower (> 1000-fold) viraemic responses, suggesting that the American crows were particularly susceptible to this virus (Bingham *et al.*, 2010).

An unprecedented outbreak of WNV_{KUNV} encephalitis in horses in south-eastern Australia in early 2011, which involved more than 1000 animals and more than one hundred fatalities (Roche *et al.*, 2013), was shown to be associated with the emergence of a virulent strain of WNV_{KUN} in Australia (Frost *et al.*, 2012). The majority of WNV_{KUN} isolates examined to date are significantly attenuated, and reports of disease in infected humans and animals have been rare and the symptoms relatively mild (Hall *et al.*, 2002; Audsley *et al.*, 2011). However, an isolate from the brain of a diseased horse infected during the 2011 outbreak showed significantly enhanced neuroinvasion in mice (Frost *et al.*, 2012). While the motif responsible for enhanced equine disease has yet to be identified, the new strain lacked virulence markers present in the highly pathogenic North American WNV strains. These included the avian virulence determinant (NS3 249Pro) discussed above, consistent with a lack of bird morbidity during the outbreak, and specific residues in prM shown to enhance mouse neuroinvasion (Setoh *et al.*, 2012). Studies on the 2011 WNV_{KUN} strain are currently under way to understand the basis for its enhanced pathogenesis in equines and why there were no human cases during the outbreak.

The distribution of equine cases in coastal areas of NSW during 2011, where no activity has previously been detected for WNV_{KUN}, also pointed to a shift in the ecology and epidemiology of the virus (Hawkes *et al.*, 1985; Frost *et al.*, 2012). This may be due to a novel mechanism of virus introduction to the area and/or a change in vector–host dynamics allowing efficient transmission to occur in coastal regions. One theory involves the adaptation of the virus to a new host (different avian species or terrestrial animal, such as kangaroos or rabbits) in high prevalence at the time (Prow, 2013). Indeed several recent examples of emerging viral diseases have been attributed to adaptation of a virus to a new host species (Taubenberger and Kash, 2010; Longdon *et al.*, 2011). Another theory implicates mosquito vectors and their possible modifications in transmission patterns (van den Hurk *et al.*, 2014a). In the context of the 2011 WNV_{KUN} outbreak, these hypotheses warrant further investigation.

Demographic changes, societal changes, urbanization and travel

The importance of societal changes and urbanization are evident for a number of viruses, but especially for dengue. Increased urbanization has occurred progressively since the end of the Second World War, following unprecedented population growth, especially in tropical developing countries, which led to population movement from rural communities to city fringes where uncontrolled and unplanned urbanization led to the development of shanty towns without running water or adequate sanitation. The lack of sanitation and rubbish removal in these shanty towns, and the increased use of non-biodegradable plastic containers, provided ideal larval habitats for *Ae. aegypti*, and resulted in a massive increase in the number of dengue and dengue haemorrhagic fever cases (Gubler, 1997; Gubler, 1998; Gubler, 2001; Mackenzie *et al.*, 2004). This increase is clearly depicted in Fig. 10.9. In addition, the increased use of motor vehicles, resulting in discarded motor and truck tyres, also provided ideal larval habitats (Gubler, 1998). Overall, urbanization has undoubtedly been the single most important factor in the amplification of dengue in individual countries.

Travel and transportation, both people and trade, have been major avenues for arbovirus spread between continents or over long distances. The role of viraemic travellers in carrying DENV and CHIKV on aircraft has been an important method of spreading these viruses from country to country, and across the globe (Wilder-Smith and Gubler, 2008; Burt *et al.*, 2012). Indeed, over the past decade, nearly 40 countries have imported cases of CHIK in viraemic travellers (Suhrbier *et al.*, 2012). The movement of CHIKV is shown in Fig. 10.6. Cargo and container vessels have been the means of dispersal of both mosquitoes and viruses. As mentioned above, car tyres have long been known to be a perfect mode for transporting mosquitoes around the globe, with a wide range of species recorded in tyres (Reiter and Sprenger, 1987; Yee, 2008; Schaffner *et al.*, 2009; Lounibos, 2011) and other receptacles, such as lucky bamboo plants, *Dracaena sanderiana* [Sparagalus: Dracaenaceae (Agavaceae)] (Madon *et al.*, 2002; Scholte *et al.*, 2008; Qualls *et al.*, 2013).

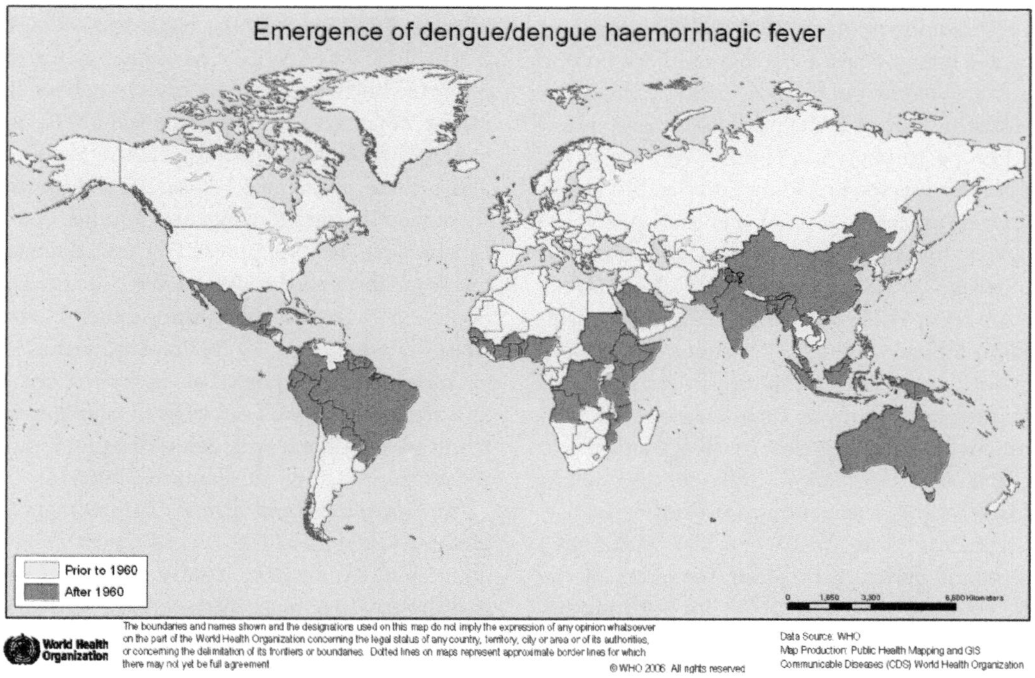

Figure 10.9 Map depicting the spread of DENV. Countries in yellow reported cases of DF/DHF prior to 1960, and the countries in red are those which have reported DF/DHF over the past 50 years. Map courtesy of the World Health Organization.

Viral dispersal by host migration and livestock movement

Bird migration has long been recognized as a mechanism for long distance dissemination and the wide geographic distribution and spread of certain arboviruses, including flaviviruses of the Japanese encephalitis and tick-borne encephalitis sero-complexes, and alphaviruses of the Eastern equine encephalitis and Western equine encephalitis antigenic complexes (reviewed by Hubálek, 2004; Gould and Solomon, 2008).

JEV utilizes various ardeid wading birds as maintenance hosts, especially the Black-crowned night heron (*Nycticorax nycticorax*), Plumed egrets (*Egretta intermedia*), Little egrets (*Egretta garzetta*) and other ardeid birds, that spread virus through vagrant movement and migration (Buescher *et al.*, 1959; Innis, 1995; Soman *et al.*, 1977; Endy and Nisalak, 2002; Gubler, 2007; Mackenzie *et al.*, 2007; van den Hurk *et al.*, 2009; Huang *et al.*, 2010). Indeed ardeid birds have been implicated as the source and reservoir of virus causing outbreaks where pigs are uncommon (Soman *et al.*, 1977; Rosen, 1986).

Murray Valley encephalitis virus (MVEV) also utilizes ardeid birds as vertebrate hosts, especially the Nankeen night heron (*Nycticorax caledonicus*) (Boyle *et al.*, 1983; Marshall, 1988; Carver *et al.*, 2009). It is probable that MVEV is spread by the movement of vagrant birds following water sources in arid areas (Broom *et al.*, 2003). Nankeen night herons are found throughout Australia and on off-shore islands, including the Cocos Keeling Islands, and in Indonesia, and Papua New Guinea.

Various species of birds are the major reservoirs and long distance disseminators of WNV in both the Old and New Worlds (Hayes, 1989; Tsai, 1997; Komar, 2000; Malkinson and Banet, 2002; McLean *et al.*, 2002; Gubler, 2007; Artsob *et al.*, 2009; Kilpatrick, 2011). In the Old World, various migratory and resident species have been implicated in WNV transmission cycles, including passerines and non-passerines (Marka *et al.*, 2013). Many wild bird species have been shown to be infected and potentially able to be involved in the dispersal of WNV, including various raptors (Bakonyi *et al.*, 2013), members of the family *Corvidae*, especially the European magpie (*Pica pica*) and the Hooded crow (*Corvus cornix*) (Work *et al.*, 1953; Work *et al.*, 1955; Jourdain *et al.*, 2007; Angelini *et al.*, 2010; Valiakos *et al.*, 2012; Calzolari *et al.*, 2013). One of the major trans-equatorial migrants transporting WNV from Africa to the Middle East and Europe is the white stork (*Ciconia ciconia*) which has yielded virus isolates and been shown to have a high incidence of antibodies (Malkinson and Banet, 2002; Malkinson *et al.*, 2002; Linke *et al.*, 2007; Seidowski *et al.*, 2010). As with MVEV, the Nankeen night heron is the major host and disseminator of WNV_{KUN} (Marshall, 1988).

In the New World, WNV infection has been demonstrated in over 300 species of birds and over 30 species of non-avian hosts (Gubler, 2007). The principal reservoir of WNV in North America is the American robin (*Turdus migratorius*) (Kilpatrick *et al.*, 2006), and it, together with the common grackle (*Quiscalus quiscula*), is believed to be responsible for the rapid dispersal and epidemic proliferation of WNV across North America (Artsob *et al.*, 2009). A low prevalence of WNV was found in a number of migratory species sampled in the Atlantic and Mississippi flyways, with evidence indicating that two species, northern cardinals (*Cardinalis cardinalis*) and grey catbirds (*Dumetella carolinensis*), may play a role in virus dispersal (Dusek *et al.*, 2009). Many avian species are more adversely affected by WNV infection in North America compared to European species (Artsob *et al.*, 2009), especially passerines, with declines in bird populations reported (Koenig *et al.*, 2007; LeDeau *et al.*, 2007).

Usutu virus (USUV) emerged in Central Europe from Africa in 2001, probably introduced by migratory birds, and was the cause of death of a number of Eurasian blackbirds (*Turdus merula*) and other species of owls and swallows around

Vienna, Austria (Weissenböck *et al.*, 2002), although recent studies have indicated it might have arrived in southern Europe several years earlier (Weissenböck *et al.*, 2013). The virus overwintered in Austria, and over the past decade has spread across much of Europe through wild birds (Vazquez *et al.*, 2011), infecting various species, including Eurasian blackbirds (*T. merula*), European magpies (*P. pica*), Eurasian jays (*Garrulus glandarius*), song thrushes (*T. philomelos*), and European starlings (*Sturnus vulgaris*) (Tamba *et al.*, 2011; Calzolari *et al.*, 2013; Bakonyi *et al.*, 2007; Höfle *et al.*, 2013). It is now considered to be a resident pathogen in Europe (Vazquez *et al.*, 2011). USUV isolates from *Cx. pipiens* mosquitoes in Spain appear to be genetically distinct from those circulating elsewhere in Central and southern Europe, and closer phylogenetically to African isolates (Bakonyi *et al.*, 2014). They are also of low virulence for birds, resulting in covert transmission cycles.

St. Louis encephalitis virus (SLEV) is also maintained in local mosquito–bird cycles, particularly members of the passeriformes and columbiformes. There has not been any evidence to indicate migratory bird involvement between SLEV genotypes from South Central and North America, but recent studies have indicated that the gene flow is from south to north, with the Gulf of Mexico region being the major source of virus for North American SLEV populations. This suggests that the virus is acquired by migratory birds while wintering in the Gulf region, which carry it back to southern US in the spring along the Central, Mississippi and Atlantic migratory flyways (Auguste *et al.*, 2009).

Many passerine and non-passerine bird species are parasitized by ticks, which may be infected with tick-borne pathogens (Hasle, 2013), including various flaviviruses and bunyaviruses. Tick-borne flaviviruses fall into two major groups, the seabird tick-borne virus group and the mammalian tick-borne virus group. The seabird tick-borne flaviruses are believed to be widely dispersed by seabirds, and have been isolated from ticks collected in seabird nests or from dead or captured birds. They include Saumarez Reef virus from *Ixodes eudyptidis* and *Ornithodoros capensis* ticks (St George *et al.*, 1977), Tyuleniy virus from *I. uriae* ticks (D.K. Lvov and V.D. Ilichev, referred to by Hubalek, 2004), and Meaban virus from *O. maritimus* ticks collected in France and elsewhere in the western Mediterranean basin (Chastel *et al.*, 1985; Arnal *et al.*, 2014). Although the major vertebrate hosts of mammalian tick-borne viruses are various mammalian species, especially rodents, migratory birds have long been implicated in their dispersal (Hubalek, 2004). Thus, recent studies of migratory birds in the Baltic region of Russia, Siberia and Sweden found TBEV in *Ixodes* ticks, particularly *I. ricinus*, from a variety of migratory passerines (Mikryukova *et al.*, 2013; Movila *et al.*, 2013; Waldenström *et al.*, 2007).

Birds have also been implicated in transmission cycles of other tick-borne flaviviruses such as louping ill disease (Hudson *et al.*, 1995) and KFDV (Mehla *et al.*, 2009), but as local hosts rather than in dispersal. It has been hypothesized that KFDV might be a variant of Omsk haemorrhagic fever virus that was transported to the Indian subcontinent by migratory birds (Work, 1958), but this explanation has been questioned (Pattnaik, 2006).

A number of alphaviruses employ wild birds as their reservoir hosts including EEEV (Morris, 1988), WEEV (Reisen and Monath, 1988), and SINV (Niklasson, 1988; Lundström *et al.*, 2001). EEEV and WEEV have been isolated from migrating passerine birds, and in some cases have been thought responsible for initiating outbreaks (reviewed by Hubálek, 2004). SINV is widely distributed in Europe, Africa, Asia and Australasia, and occurs as distinct Paleoarctic/Ethiopian and Oriental/Australian lineages. Phylogenetic studies of a number of virus isolates have clearly supported the concept that migratory birds move and disperse the virus in north–south directions between South Africa and northern Europe and between Asia and Australia (Sammels *et al.*, 1999; Lundström and Pfeffer, 2010).

Various other arboviruses, including certain bunyaviruses and orbiviruses, have been isolated from migratory birds in different parts of the world, demonstrating their importance as a means of virus dispersal (Hubalek, 2004).

The movement and spread of arboviruses by human travellers, especially YFV, DENV and CHIKV, has been alluded to above, but of the other potential mammalian hosts in virus dispersal, bats are potentially the most effective, whereas rodents, primates and other mammalian hosts are more sedentary and unlikely to play a significant role in virus dispersal except in local spread. Over the past decade, bats have been shown to harbour a variety of different viruses (Calisher *et al.*, 2006; Smith and Wang, 2013; Hayman *et al.*, 2013; Chen *et al.*, 2014; Dacheux *et al.*, 2014). This is largely due to their evolutionary diversity, long life spans, abundance and social behaviour, their ability to fly, and their sympatry (Calisher *et al.*, 2006; Mackenzie *et al.*, 2008; Luis *et al.*, 2013). Bats are known to be reservoirs of a number of flaviviruses, particularly the groups comprising viruses with no known vector (Entebbe bat virus, Modoc virus and Rio Bravo virus groups), many of which have been observed only in bats, but other medically important flaviviruses have also been reported from bats, either by isolation or from seroepidemiological studies, including JEV (Sulkin and Allen, 1974; Banerjee *et al.*, 1988; Wang *et al.*, 2009; Liu et al., 2013), SLEV (Sulkin *et al.*, 1966; Allen *et al.*, 1970; Thompson *et al.*, 2014), and WNV (Davis *et al.*, 2005a; Thompson *et al.*, 2014). In addition, various other arboviruses have been described in bats including the alphavirus VEEV (Correa-Giron *et al.*, 1972; Seymour *et al.*, 1978; Calisher *et al.*, 1982); the Orthobunyavirus Kaeng Khoi virus (Williams *et al.*, 1976; Osborne *et al.*, 2003); and the Phleboviruses Malsoor virus, a novel bat virus related to Heartland and severe fever with thrombocytopenia viruses (see below) (Mourya *et al.*, 2014), RVFV (Olive *et al.*, 2012) and Toscana virus (Verani *et al.*, 1988). However, the role of bats as possible vertebrate hosts in transmission cycles of these viruses remains to be determined, and thus any role in virus spread.

Livestock movement has also assisted the spread of arbovirus diseases. It has long been known that BTV and RVFV can be spread through livestock movement, as has been seen in recent years in Europe for BTV (Sellers and Taylor, 1980; Wilson and Mellor, 2009) and in Africa for RVFV (Abdo-Salem *et al.*, 2011; Davies, 2006; Xue *et al.*, 2013). BTV types 1, 9 and 16 outbreaks in south-eastern Europe were attributable at least in part to animal movements along the 'Eurasian Ruminant Street', a contiguous region with high densities of ruminants stretching from India and Pakistan through Turkey and Iran to south-eastern Europe (Slingenbergh *et al.*, 2004; Wilson and Mellor, 2009). The epidemic of BTV type 8 in 2006 also provided examples of the role of livestock movements as a major contributing factor to virus spread, and the importance of preventing animal movements in outbreak control strategies (Mintiens *et al.*, 2008). Since RVFV was first reported from the greater Rift Valley in Kenya in the 1930s, the geographic distribution has grown significantly, and now includes most countries of Africa as well as Madagascar (Pepin *et al.*, 2010). This spread has been largely due to movement of infected livestock. RVFV moved outside the African continent for the first time in 2000 with outbreaks in Saudi Arabia and Yemen (Fig. 10.7), raising concerns of emergence in other continents (Turell and Kay, 1998; Moutailler *et al.*, 2008; Turell *et al.*, 2008; Chevalier *et al.*, 2010; Weaver and Reisen, 2010). Most recently, the virus has spread to Mayotte and Grande Camore in the Comores Archipelago (Sissoko *et al.*, 2009). In both these instances, the spread of RVFV to the Middle East and Mayotte, is believed to have been due to the livestock trade and the movement of infected small ruminants. The movement of RVFV is depicted in Fig. 10.7.

Vertebrate hosts of some medically important arboviruses, and their transmission cycles, remain to be determined. This is particularly so for the sandfly borne phleboviruses (Alkan *et al.*, 2013; Maroli *et al.*, 2013). Sandfly fever (or Pappataci fever or phlebotomus fever) is caused by viruses belonging to the *Phlebovirus* genus of the *Bunyaviridae* family comprising the Sicilian-like and Corfu viruses (Sandfly fever Sicilian serocomplex), Naples-like and Toscana virus (Sandfly fever Naples serocomplex) and viruses of the Salehabad serocomplex. These viruses cover a large area from the Mediterranean region, Africa, Middle East, and India to central Asia. The vectors on the African and the Eurasian continents belong to the genera *Phlebotomus* and *Sergentomyia*, in the New World to the genus *Lutzomyia* (Alkan *et al.*, 2013). In endemic areas, seroprevalence in humans and animals is high, posing a public health concern for immunologically naive visitors. Sandfly fever has been reported from travellers returning from endemic regions, where infection was acquired, to non-endemic areas, often posing difficulty in diagnosis (Eitrem *et al.*, 1990; Ellis *et al.*, 2008). While most of the sandfly fever viruses cause mild symptoms, others such as Toscana (Jaijakul *et al.*, 2012; Schwarz *et al.*, 1995) and Sandfly fever Turkey virus (Ergunay *et al.*, 2012; Becker *et al.*, 1997) can involve the central nervous system, presenting with more severe clinical features. Furthermore, new serotypes have recently emerged in Europe, Africa and east and central Asia (Papa *et al.*, 2011; Navarro-Marí *et al.*, 2013; Charrel *et al.*, 2009; Zhioua *et al.*, 2010; Es-Sette *et al.*, 2012; Amaro *et al.*, 2011; Punda-Polic *et al.*, 2012, Bhatt *et al.*, 1967), highlighting the potential for this disease to spread into new areas.

Landscape changes

The increasing world population is driving an intensification of agriculture as it spreads into areas once covered with forests. These changes to land use are seen most commonly in tropical and subtropical areas, and result in fragmentation of habitats and alterations in vector–host–parasite relationships (Sutherst, 2004). Landscape changes include deforestation, environmental damage from mining, changes to agriculture including paddy field expansion, and dam building, all of which can alter the prevalence, spatial variations and species of vertebrate hosts and mosquito vectors, and assist virus spread (Gratz, 1999; Vasconcelos *et al.*, 2001; Lambin *et al.*, 2010). JEV is an excellent example of the spread due to landscape changes (Tsai, 1997; Le Flohic *et al.*, 2013); it has shown a propensity to spread and establish in new areas (Mackenzie *et al.*, 2007) due largely to increased areas of irrigated agriculture, often following deforestation, and water impoundment in dams for irrigation and human consumption (Tsai, 1997; Le Flohic *et al.*, 2013).

Deforestation has been associated with the emergence and spread of KFDV in Karnataka State of India (Banerjee, 1988; Walsh *et al.*, 1993). Forest clearance for agricultural purposes, grazing cattle, and collecting firewood as villages encroached on forested areas provided ideal opportunities for virus transmission, especially through the introduction of cattle which amplified the major tick vector species, *Haemaphysalis spinigera*. Deforestation for cattle ranching was also a major factor in the re-emergence and increase in reported yellow fever cases in Brazil (Vasconcelos *et al.*, 2001), and may have contributed to the spread of Oropouche and Mayaro viruses. Indeed epidemics of Oropouche virus were closely associated with population movement into new dwellings in rural and semi-urban areas close to forests, where agricultural practices and rotten organic material such as cacao husks and banana tree stumps provided ideal environmental conditions for vector breeding (LeDuc and Pinheiro, 1989). Indeed the rapid colonization of the Amazon region and the consequent increase in human populations in contact with forest areas, together with the environmental changes, has raised the concern that new pathogenic arboviruses may emerge (Travassos da Rosa *et al.*, 1989). This has been exacerbated by the building of an extensive network of highways across vast areas of virgin tropical forests which, together with the environmental changes caused by mining and extraction industries, has also led to the emergence and spread of many novel arboviruses (Vasconcelos *et al.*, 2001).

Landscape fragmentation has been a major contributing factor to the recent emergence of tick-borne viral diseases. This is often associated with arable land that has been abandoned allowing secondary vegetation growth and repopulation with potential vertebrate hosts such as hares, rodents and deer (Estrada-Pena *et al.*, 2013). Indeed this has been the likely cause of the sudden emergence of Crimean–Congo haemorrhagic fever (CCHF) in eastern Turkey (Estrada-Peña *et al.*, 2010). Crimean–Congo haemorrhagic fever virus (CCHFV) is transmitted by Ixodid (hard) ticks of the *Hyalomma* genus and is arguably the most important tick-borne viral pathogen on a global scale (Bente *et al.*, 2013; Mertens *et al.*, 2013). It occurs in south-eastern Europe, the Middle East, Asia and Africa. Virus infection in humans causes a potentially fatal form of haemorrhagic fever, with about 10–20% of infections producing clinical disease and a case fatality rate of about 5% (Mertens *et al.*, 2013). For a more comprehensive discussion on the virus, the disease and its transmission, readers are directed to recent reviews by Bente *et al.* (2013) and Mertens *et al.* (2013). Prior to 2002, CCHFV was unheard of in Turkey. By 2009 however nearly 5000 cases had been reported (Karti *et al.*, 2004, Albayrak *et al.*, 2012; Gunes *et al.*, 2009). This appears to be associated with farmers being exposed to infected ticks while reclaiming lands for agricultural purposes after years of abandonment due to civil unrest. A similar scenario has been reported in other parts of south-eastern Europe (Estrada-Pena *et al.*, 2012).

The building of dams has been associated with virus emergence and spread in many parts of the world. One of the best examples has been in Brazil, where the Amazonian region appears to be one of the richest reservoirs of arboviruses (Dégallier *et al.*, 1992). Dams built in Para, Amazonas and Rondonia States have caused profound environmental changes and enhanced arthropod breeding, and resulted in an increased circulation and epizootic activity of a number of arboviruses including Oropouche virus and SLEV, and yielded a large number of novel viruses not previously reported (Dégallier *et al.*, 1992; Vasconcelos *et al.*, 2001). Dam construction on the Senegal River in Mauritania is another excellent example, and led to a major epidemic of RVF in 1987 with 200 deaths (Digoutte and Peters, 1989). This was the first outbreak of RVF in Mauritania, and was followed by further smaller outbreaks (Zeller *et al.*, 1995; Nabeth *et al.*, 2001). The construction of two dams on the Ord River and the development of a large irrigation area in the Kimberley region of north-west Australia resulted in profound ecological changes including greatly enhanced condition for mosquito breeding and in attracting large numbers of water birds, the primary vertebrate hosts for MVEV. These changes are believed to have contributed to an increase in the incidence of MVEV activity and human cases in Western Australia and the Northern Territory (Mackenzie and Broom, 1999).

Virus dispersal via wind-borne vectors

The wind-borne movement of vectors has been a major mode of spread for both mosquito-borne and *Culicoides*-borne viral diseases (Sellers, 1980; Mackenzie *et al.*, 2000). Flights of up to 175 km have been recorded for *Ae. taeniorhynchus* and *Ae. sollicitans* mosquitoes, vectors of VEEV and EEEV (Sellers, 1980), and as much as 500 km for *Cx. tritaeniorhynchus* mosquitoes, the major vector of JEV (Asahina, 1970). *Culicoides* sp. have also been carried very long distances at heights of up to 1500 metres. Wind has been suggested as a way for JEV genotypes to disperse in the eastern and south-eastern Asian region (Nga *et al.*, 2004), and for JEV to move from Papua New Guinea into Australia (Ritchie and Rochester, 2001). Indeed, the regular introduction of JEV into the Torres Strait or the Australian mainland since 1995 was proposed to have been mediated via wind-borne *Culex* sp. mosquitoes from the Western Province of Papua New Guinea, associated with cyclonic winds, however a role for infected vertebrates (e.g. water birds) could not be ruled out (Ritchie and Rochester, 2001). The recent discovery of genetically homogenous strains of a new insect-specific bunyavirus (Badu virus) from *Cx. annulirostris* in the Western Province of Papua New Guinea (PNG) and on Badu island in the Torres Strait, that does not infect mammalian or avian cells, provides strong evidence for viral dispersal in wind-borne mosquitoes (J. Hobson-Peters, D. Warrilow, B.J. McLean, D. Watterson, A.M.G. Colmant, A.F. van den Hurk, S. Hall-Mendelin, M. Hastie, J. Gorman, J. Harrison, N.A. Prow, R.T. Bernard, and R.A. Hall, unpublished data). Wind-borne dispersal of *Culicoides*-borne viruses has long been an accepted mode of spread for diseases such as bluetongue and Schmallenberg viruses. Thus, bluetongue viruses have spread to and in Europe (Wilson and Mellor, 2009; Burgin *et al.*, 2013) and to Australia through wind-borne *Culicoides* sp. (St George, 1986; Eagles *et al.*, 2012; Eagles *et al.*, 2013), and the recently emerged Schmallenberg virus has spread within Europe through wind-borne vectors (Sedda and Rogers, 2013). The spread of the *Culicoides*-borne epizootic haemorrhagic disease virus in Israel has also been associated with wind dispersal (Kedmi *et al.*, 2010).

The effect of climate change on arboviruses

There has been much speculation about the potential impact of climate change on the spread and incidence of mosquito-borne viral diseases. The effect of global warming, with higher rainfall and humidity and rising sea levels, could factor into an increased prevalence and distribution of arboviral diseases. These changes can influence several components of arboviral ecology including increased breeding and survival of vectors, expansion of their geographical boundaries, shorter extrinsic incubation periods and enhanced virus transmission, all of which impact on human vulnerability to vector-borne diseases (Sutherst, 2004; Gould and Higgs, 2009). Recent examples have been the northward expansion of *Culicoides imicola* and bluetongue into northern Europe due to warmer temperatures in this part of the world (Purse *et al.*, 2005), and the potential continued global expansion of *Ae. albopictus* (Fischer *et al.*, 2013). Indeed, the efficient transmission of bluetongue virus by a Palearctic species of *Culicoides* which is known to occur across much of the upper portions of the northern hemisphere, including North America, is of considerable concern (Maclachlan, 2010). In addition, recent weather extremes have also been shown to impact on vector-borne disease outbreaks, as described by Anyamba *et al.* (2014) for outbreaks of DEN in East Africa and WN in the USA following droughts, and outbreaks of RVF in southern Africa and MVE in Australia following above average rainfall events.

The ecology of RVFV is intimately associated with rainfall and flooding of mosquito breeding habitats, with all outbreaks occurring after heavy rainfall of mosquito breeding habitats known as 'dambos' (Linthicum *et al.*, 1985; Logan *et al.*, 1992), in which the virus persists during dry season or inter-epizootic periods by vertical transmission in *Aedes* mosquito eggs, and after substantial rainfall events, infected mosquitoes hatch and initiate outbreaks in domestic animal populations. If the mosquito habitats remain flooded, this then drives epizootic or epidemic activity through various *Culex* mosquito populations as secondary vectors. The epizootics have been closely linked to the El Ninõ/Southern Oscillation (ENSO) phenomenon and elevated Indian Ocean temperatures that lead to heavy rainfall (Linthicum *et al.*, 1999), and thus may be particularly sensitive to climate change. The use of satellite measurements for various parameters including sea surface temperatures, outgoing long-wave radiation, rainfall, and landscape ecology using the normalized difference vegetation index, have provided the development of methods to successfully predict epizootic RVFV activity in space and time with a lead time of 2–4 months (Anyamba *et al.*, 2010). Concern has also been raised that the incidence of MVEV in Australia may be susceptible to climate change. As with RVFV, ENSO is a major indicator of potential transmission and spread of virus activity from endemic regions in northern Australia to south-eastern Australia (Nicholls,

1993; Russell, 1998). It presumes that increased heavy rainfall events and flooding in northern and central Australia will provide the mechanism for the virus to move from the north to the populous south-east through water bird and mosquito transmission cycles as the birds follow the flood waters (Mackenzie *et al.*, 1993; Forbes, 1998).

The transmission of tick-borne viruses and the geographical range of their vectors may also be effected by climate change. Survival of ticks is dependent on sufficiently warm periods during the year to allow ticks to moult into adults so they can successfully overwinter, even under relative harsh winter temperatures. Thus changes in climate, particularly associated with warmer weather in spring and autumn, may facilitate enhanced tick survival and spread (Estrada-Pena *et al.*, 2012). Since tick-borne viruses are sustained in the vector populations by transstadial, transovarial and venereal transmission (Matser *et al.*, 2009), as well as horizontally to naive ticks via some vertebrate hosts that become viraemic (Gunes *et al.*, 2011), enhanced tick survival and spread will have a significant effect on viral transmission.

However, several expert reviews on the effect of climate change on the incidence and spread of arboviral diseases caution against ignoring the complex and dynamic interplay between the many factors associated with arboviral transmission (Russell, 1998; Reiter, 2008; Russell *et al.*, 2009; Maclachlan and Guthrie, 2010). Indeed, it is likely that trends and predictions will vary greatly from region to region around the globe. As pointed out by Russell (Russell, 1998; Russell *et al.*, 2009), the complexity of arbovirus ecology and the uncertainty of climate change models provide little confidence in predicting the effect of climate change on incidence, spread or emergence of arboviral diseases. For example, predictions of warmer winters may extend the period of arbovirus activity in cooler regions by allowing for earlier seasonal activity of some vector species (Russell *et al.*, 2009). However, reduced rainfall during winter and spring may provide fewer sites to initiate vector breeding and may also result in reduced numbers of wildlife reservoirs of arboviruses in rural areas. More frequent flooding events may result in extensive breeding of some vectors, however these events may also flush out the breeding sites of others and temporarily reduce their numbers (Russell, 1998). Conversely, longer periods of drought and lower humidity in some regions will likely reduce the survival of vector populations, such as overwintering adult mosquitoes or their desiccant-resistant eggs. It has also been predicted that global warming will impact on sea levels (Ramasamy and Surendran, 2011) increasing the potential for salt marsh inundation and extending the breeding areas for insect vectors that inhabit these sites. However, this will depend on the topology and drainage of inundated areas and whether larval predator populations have become established (Russell, 1998). Collectively, these scenarios make it difficult to predict the outcomes of climate change on the transmission of various arboviruses and the incidence of disease even with the most confident estimates of each environmental parameter. While it is likely that changes in disease incidence will occur, it will be unpredictable, with some regions experiencing more cases and others less.

Climate change may also affect the diversity and evolution of arboviruses. As mentioned above, increased winter temperatures with lower rainfall and humidity may reduce mosquito survival and the chance of persistence of viruses that rely on vertical transmission and diapause (Russell, 1998). Could this provide selective pressure for viruses to evolve more efficient mechanisms for infecting the egg in the female mosquito? In this context the recent discovery of mosquito-specific viruses representing most of the common arbovirus families (see following section), and evidence that they persist in mosquitoes predominantly by highly efficient vertical transmission, may provide interesting models to explore the molecular determinants that may facilitate such an adaption for arboviruses (Cook *et al.*, 2006, 2012; Warrilow *et al.*, 2014; Hobson-Peters *et al.*, 2013). The shorter life spans of mosquitoes at higher temperatures may also select for viruses that have shorter extrinsic incubation periods, such as that described for the WNV02 strain that emerged in the USA (Moudy *et al.*, 2007).

Genetic factors underlying diversity and spread

Genetic diversity in alternating hosts

All arboviruses are RNA viruses, except for African swine fever, the only DNA arbovirus, and typically occur as genetically diverse populations or quasispecies because of their error-prone replication process which accumulates mutations (reviewed by Lauring and Andino, 2010; Bordería *et al.*, 2011). Quasispecies have been described for mosquito-borne arboviruses, as exemplified by WNV (Beasley *et al.*, 2003; Jerzak *et al.*, 2005; Ciota *et al.*, 2007; Brackney *et al.*, 2011) and DENV (Wang *et al.*, 2002a; Lin *et al.*, 2004; Kurosu, 2011). This genetic plasticity is important in RNA virus evolution and spread by providing viruses with the genetic capability to adapt and prosper in different and changing environments and in different hosts and vectors, and thus providing the mechanism for establishing in novel ecosystems and in generating epidemic activity. Minor genetic changes may be sufficient to make dramatic changes in viral phenotype, as described above for WNV, when the New York 99 strain was displaced by the WN02 strain (Davis *et al.*, 2003; Beasley *et al.*, 2003; Ebel *et al.*, 2004; Moudy *et al.*, 2007) due to a fitness advantage derived from a shorter extrinsic incubation period (Moudy *et al.*, 2007), and found with CHIKV where mutations in the E1 (E1-A226V) and E2 (E2-L210Q) proteins increased the infectivity and dissemination in *Ae. albopictus* (Tsetsarkin *et al.*, 2007, 2009; Tsetsarkin and Weaver, 2011).

The role of genetic diversity within natural arbovirus transmission cycles and its importance to host range changes, virus perpetuation, replicative capability, and virus evolution remain poorly understood, and are subject to a number of constraints (reviewed by Coffey *et al.*, 2013). The alternating replication in arthropod and vertebrate hosts required by most arboviruses may impose constraints on their diversity through purifying selection and/or population bottlenecks in one or both of the vector and host, although most commonly in the vector (Smith *et al.*, 2008; Coffey *et al.*, 2008, 2013; Brackney *et al.*, 2010; Ciota *et al.*, 2012; Forrester *et al.*, 2012) through fitness trade-offs in which viral traits favoured in one host are purified in the other (Holmes, 2003). The trade-off proposes that viruses maintain adequate replicative fitness in two disparate hosts in exchange for superior fitness in one host. Experimental infection studies with VEEV suggested that only certain midgut epithelial cells became infected (Smith *et al.*, 2008) thus providing an anatomical barrier, but this appeared to be specific for an epidemic strain of VEEV as an endemic strain of virus was able to infect all the midgut cells, indicating that a possible long-term vector–virus relationship may lead to higher midgut infectivity (Kenney *et al.*, 2012). However, despite possible anatomical barriers or bottlenecks, genetic variation of WNV is maintained in mosquitoes (Jerzak *et al.*, 2005, 2007; Brackney *et al.*, 2011), and is significantly greater than that found in naturally infected birds (Jerzak *et al.*, 2005) or from passage of virus in chickens (Jerzak *et al.*, 2007). Indeed the intrahost genetic diversity of WNV that arose from alternating passage was significantly greater than the diversity generated during chicken-only passage, and was similar to mosquito-only passage (Jerzak *et al.*, 2008). Alternate host cell passage with CHIKV was shown to increase fitness without drastic changes in diversity, whereas the greatest increase in genetic diversity was observed after serial passage, suggesting an evolutionary trade-off between fitness and diversity (Coffey and Vignuzzi, 2011).

Experiments using an *in vitro* model for DENV infection of mosquito cells and human cells, either as serial passages or alternating passages supported the hypothesis that releasing DENV from host alternation facilitates adaptation, but does not fully support the hypothesis that alternation necessitates a fitness trade-off (Vasilakis *et al.*, 2009). Intrahost diversity of DENV has been investigated in both mosquitoes and in human patients (Wang *et al.*, 2002a,b; Lin *et al.*, 2004; Descloux *et al.*, 2009; Thai *et al.*, 2012) with conflicting results. Earlier studies described low but measurable levels of diversity, and in one case, this was greater in humans than in mosquitoes (Lin *et al.*, 2004), but most recently, in a more rigorous process, intrahost sequence diversity was found to be very low (Thai *et al.*, 2012), and there was clear evidence for mixed infection with multiple phylogenetically distinct lineages. Furthermore and unexpectedly, a study of French Polynesian strains of DEN-1 found a significant correlation between intrahost genetic diversity and clinical outcome, in which severe cases were characterized by more homogeneous viral populations with lower diversity (Descloux *et al.*, 2009).

Interestingly, Powassan virus is more homogeneous in ticks, and less intrahost genetic diversity is observed, whereas interhost diversity was similar to WNV, suggesting that purifying selection and/or population bottlenecks constrained quasispecies diversity in ticks (Brackney *et al.*, 2010).

This clearly shows that arboviruses, like all RNA viruses, generate considerable diversity or quasispecies during the replicative process, but unlike single host viruses, the requirement of arboviruses to replicate alternately in vertebrates and invertebrates places considerable constraints on the degree of diversity through purifying selection and/or population bottlenecks, especially in the vector, through fitness trade-offs. These trade-offs require the maintenance of adequate replicative fitness in both hosts. The bottlenecks in the vector are most often anatomical, for instance with few midgut cells capable of being infected, but despite the various constraints, genetic variation is appreciably greater in the vector than in the vertebrate host.

Genetic recombination and reassortment

Diversity can also be generated by genetic recombination and reassortment. Recombination has been shown to occur in DENV isolates obtained from human patients (Tolou *et al.*, 2001; Uzcategui *et al.*, 2001; Aaskov *et al.*, 2007; Chen *et al.*, 2008; Perez-Ramirez *et al.*, 2009) or from published sequences (Holmes *et al.*, 1999; Worobey *et al.*, 1999; Twiddy and Holmes, 2003). Results from published sequence studies also suggested that recombination might occur with JEV and SLEV (Twiddy and Holmes, 2003), but no evidence was found to suggest recombination in WNV, YFV or TBEV. Subsequently, sequence studies have provided support for recombination with JEV (Chuang and Chen, 2009; Carney *et al.*, 2012) and WNV (Pickett and Lefkowitz, 2009). TBEV, however, was believed to have evolved in a clonal manner, with early studies being unable to demonstrate any evidence of recombination, but more recent observations have described recombination for louping ill virus (Bertrand *et al.*, 2012) and in various clades of TBEV (Norberg *et al.*, 2013). The latter study also suggested that geographic barriers might restrict recombination to involve only closely related viruses.

The *Bunyavirus* superfamily comprises approximately 300 viruses, all of which possess three genetic segments, and reassortment of segments between members within the different genera is well established, and for some serogroups in the Orthobunyaviruses such as the Simbu and Bunyamwere serogroups, is very common (reviewed by Briese *et al.*, 2013). Indeed, there is increasing evidence that novel viruses arise by reassortment, sometimes with one donor unknown, as demonstrated by Iquitos virus in Peru (Aguilar *et al.*, 2011), Cholul virus in Mexico (Blitvich *et al.*, 2012), and Schmallenberg virus in Europe (Yanase *et al.*, 2012).

Genetic reassortment is also well-established in the *Orbivirus* genus of the family *Reoviridae*, with frequent evidence for reassortment between bluetongue virus strains and types (Shaw *et al.*, 2013; Batten *et al.*, 2008), epizootic haemorrhagic disease virus (Allison *et al.*, 2012), and African horse sickness virus (Meiring *et al.*, 2009).

Genetic lineages and molecular phylogeography in epidemiology and virus spread

Molecular epidemiology using genomic sequence information has provided the means to determine the directions and timescales of virus dispersal and spread – where a virus strain causing an outbreak originated from, and in some cases, the stepping stones en route. Thus, the diversity of RNA viruses has led to the development of specific genotypes and/or lineages and clades of individual virus species in different geographic areas that are then used to trace the movements of the virus. The different genotypes or lineages are usually indistinguishable serologically, but with molecular data, it has been possible to demonstrate multiple introductions and re-introductions. Genetic diversity was clearly demonstrated for a number of arboviruses in the early 1980s using RNAse oligonucleotide fingerprinting and antigen signature analysis, particularly for DENV-2 (Trent *et al.*, 1983, 1989; Monath *et al.*, 1986). Subsequently, nucleotide sequence studies of DENV strains isolated from different regions and countries demonstrated the sensitivity that this provided in deciphering phylogenetic and evolutionary relationships (Rico-Hesse, 1990, 1997; Chungue *et al.*, 1993; Lewis *et al.*, 1993; Lanciotti *et al.*, 1994; Chungue *et al.*, 1995), with five genotypes for DEN-1 and DENV-2, four genotypes for DENV-3 and three genotypes for DENV-4. These genotypes gave ample evidence of the dispersal of these viruses over wide areas in space and time (reviewed by Thomas *et al.*, 2003; Rico-Hesse, 2003). This was particularly exemplified by the genotypes of DENV-1 and DENV-2. DENV-1 genotype 1 was shown to comprise viruses from the Americas, Africa and SE Asia; genotype 2 comprised a single isolate from Sri Lanka; genotype 3 also comprised a single isolate from Jamaica; genotype 4 comprised viruses from SE Asia, the South Pacific, Australia and Mexico; and genotype 5 comprised viruses from China Taiwan and Thailand (Rico-Hesse, 1990). Similarly, for DENV-2, genotype 1 comprised viruses from the Caribbean and the South Pacific; genotype 2 comprised the prototype PNG virus and viruses from China Taiwan, the Philippines, and an early Thai virus; genotype 3 comprised strains from Vietnam, Jamaica and Thailand; genotype 4 comprised viruses from Indonesia, the Seychelles, Burkina Faso, and Sri Lanka; and interestingly, the fifth genotype comprised isolates from rural Africa, and represented sylvatic strains of virus (Rico-Hesse, 1990).

JEV also evolved into five distinct genotypes or genetic lineages, probably in or around the Indo-Malaysian region of South-East Asia, which subsequently spread through eastern and southern Asia (Chen *et al.*, 1990, 1992; Uchil and Satchidanandam, 2001; Solomon *et al.*, 2003; Mackenzie and Williams, 2009; Li *et al.*, 2011; Mohammed *et al.*, 2011) and, more recently, into Australasia (Hanna *et al.*, 1996; Pyke *et al.*, 2001; Mackenzie and Williams, 2009). In many areas of Asia, the previously dominant genotype, genotype 3, is being replaced by genotype 1 (Schuh *et al.*, 2014), including in northern Vietnam (Nga *et al.*, 2004), China (Wang *et al.*, 2007), Korea (Yun *et al.*, 2010), Japan (Saito *et al.*, 2007), and Thailand (Nitatpattana *et al.*, 2008). In addition, JEV genotype V, which had only been isolated once previously in 1952 from human brain tissue (Solomon *et al.*, 2003), has been increasingly detected in China (Li *et al.*, 2011), Korea (Takhampunya *et al.*, 2011), and most recently from Tibet, China (Li *et al.*, 2014). Conversely, genotype 4 appears restricted to Indonesia (Schuh *et al.*, 2013a). The emergence of JEV to Australasia in 1995 was due to a genotype 2 virus which could be shown to be related to the genotype 2 viruses from Java in Indonesia (Mackenzie *et al.*, 2002; Schuh *et al.*, 2013a) Molecular epidemiological studies carried out in Japan have demonstrated that the viruses circulating in Japan comprise both local and introduced strains of JEV, and that introductions have come from several parts of the Asian mainland (Nabeshima *et al.*, 2009). Interestingly, a molecular phylogeographical analysis of 487 JEV isolates collected from 12 countries over 75 years has linked molecular evolution of the isolates with climate. Thus two genotypes, Ib and III, were shown to be temperate genotypes with epidemic transmission and most cases in summer months and which were probably maintained throughout the year in northern latitudes, whereas genotypes Ia and II were tropical genotypes with endemic transmission resulting in fewer cases but throughout the year, and which were probably maintained primarily in mosquito-avian and mosquito-swine transmission cycles (Schuh *et al.*, 2013b).

WNV also exists in at least seven distinct lineages (Mackenzie and Williams, 2009). Lineages 1 and 2 are the most important, with the majority of pathogenic strains found in lineage 1 and the prototype strain in lineage 2. Lineage 1 is found on all continents, and as lineage 1b or Kunjin virus in Australia (WNV_{KUN}) (Lanciotti *et al.*, 1999; Scherret *et al.*, 2001; Scherret *et al.*, 2002); lineage 2 was thought to be largely confined to Africa, but in recent years has become more frequent in Europe (Ciccozzi *et al.*, 2013; McMullen *et al.*, 2013); lineage 3 has only been found in central Europe (Bakonyi *et al.*, 2005); lineage 4 has only been found in a tick from the Caucasus region on the border of Europe and Asia (Lvov *et al.*, 2004); lineage 5 has been reported from India (Bondre *et al.*, 2007); lineage 6 has only been described from Malaysian Sarawak (Mackenzie and Williams, 2009); and a possible lineage 7 is an African flavivirus, Koutango virus (Mackenzie and Williams, 2009). An eighth lineage has recently been described from Senegal (Fall *et al.*, 2014), and an additional possible lineage has been described (Vazquez *et al.*, 2010). Lineages 1–6 are depicted in Fig. 10.4. Of other flaviviruses, four distinct genetic lineages have been found for MVEV, two in Papua New Guinea and two in Australia (Mackenzie and Williams, 2009); seven genetic lineages have been described for SLEV with a number of clades (Kramer and Chandler, 2001; May *et al.*, 2008). The data were consistent with multiple introductions from South America into North America.

Zika virus has been known to circulate in parts of sub-Saharan Africa for many years since its first isolation in Uganda in 1947–48 (Dick *et al.*, 1952), and it probably spread to south-east Asia in about 1945–50 (Faye *et al.*, 2014). Two major lineages then evolved, one in Africa and one in Asia (Haddow *et al.*, 2012), where it caused a relatively minor disease of fever with a rash and arthralgia. The virus spread from Asia to Yap in 2007 causing the largest known outbreak of human disease (Lanciotti *et al.*, 2008; Duffy *et al.*, 2009), and from Asia to the South Pacific in 2013–14, including French Polynesia where it caused an explosive outbreak with over 28,000 cases (Musso *et al.*, 2014; Cao-Lormeau *et al.*, 2014), and subsequently spread to Cook Islands, Easter Island, and New Caledonia (Fig. 10.5). Phylogenetic studies using isolates obtained from both outbreaks, Yap and the South Pacific, found that the outbreaks were caused by closely related viruses belonging to the Asian lineage (Haddow *et al.*, 2012; Cao-Lormeau *et al.*, 2014). Viruses circulating in Africa indicated that the African lineage comprises two tightly bound clades, one containing isolates from Nigeria and Senegal, and the other comprising isolates from Uganda, Senegal and Central African Republic, but both clades contained viruses from Côte d'Ivoire, and there was evidence consistent with at least two introductions into West Africa (Faye *et al.*, 2014).

A good example of the sensitivity of molecular epidemiological studies is well demonstrated by TBEV. TBEV exists in at least three genetic lineages – a European lineage, a Far Eastern lineage and a Siberian lineage (Ecker *et al.*, 1999). Recently, isolates of the Far Eastern lineage were found in the European region of the former Soviet Union thousands of kilometres west of their endemic region, but on further analysis, this anomaly was found to be due to the introduction of game animals from the Far Eastern endemic region which had been brought west for economic purposes, and they had brought the virus with their ticks (Kovalev *et al.*, 2010).

Most, if not all, mosquito-borne and tick-borne flaviviruses comprise distinct genetic lineages associated with geographic movements of viruses and their vectors. It is probable that this is also the case with all the mosquito-borne alphaviruses. Thus the alphavirus, CHIKV, exists in three distinct lineages, the West African lineage (the most distant), the Asian lineage, and the Indian Ocean/East, Central and South African lineage, although the latter lineage is divided into two clades, the Indian Ocean clade and the East, Central and South African clade. The acquisition of two new mutations in this latter lineage resulted in improved infection, dissemination and transmission of the virus in *Ae. albopictus* mosquitoes, described above, that saw the virus move initially from East Africa, to cause huge outbreaks in Indian Ocean island nations, and then in India. From India, the virus moved west to Italy and east into Malaysia and Singapore. These movements were clearly shown by phylogenetic analyses (Lo Presti *et al.*, 2012; Tsetsarkin *et al.*, 2011). CHIKV moved again in late 2013, this time to emerge in the Caribbean on the island of Saint Martin (Van Bortel *et al.*, 2014), but, unlike the major spread over the past decade, the incursion into the Caribbean was due to the Asian lineage (Fig. 10.6).

Considerable information has been generated through many studies on the movement of arboviruses by exploring their phylogeographic relationships. Almost all viruses have yielded valuable information on mechanisms, source and speed of spread, the epidemic waves, and the pathogenetic variables related to their spread. Some movements are complex, particularly for dengue (Chen and Vasilakis, 2011; Villa-Arenas and Zanotto, 2013; Messer *et al.*, 2003), whereas information on some others, notably YFV, may be limited due to the lack of data; although five genotypes of YFV have been recognized in Africa and two in the Americas (Barrett and Monath, 2003; de Souza *et al.*, 2010; Stock *et al.*, 2013), not many isolates have been fully sequenced, and thus detailed knowledge of the phylogeography is incomplete.

Emergence of insect-specific viruses and their effect on arboviral transmission

The recent isolation of a novel group of flaviviruses that do not replicate in vertebrate cells (termed 'insect-specific flaviviruses' – ISFs) has created intense interest in their modes of transmission and their evolutionary origins (Cook *et al.*, 2012). These viruses, including Cell fusing agent virus (CFA), Kamiti River virus (KRV) and *Culex* Flavivirus (CxFV), have been isolated from mosquitoes in many parts of the world (Crabtree *et al.*, 2003; Cook *et al.*, 2006; Hoshino *et al.*, 2007). The worldwide prevalence of ISFs has also led to the discovery of insect-specific viruses (ISVs) of other arboviral families in natural mosquito populations, including the *Togaviridae* and *Bunyaviridae*. To date, only one alphavirus with an insect-specific host range has been discovered. Eilat virus was isolated from *Anopheles* mosquitoes collected in Israel and represents a novel complex within the alphavirus genus (Nasar *et al.*, 2012). Recently, two new groups of insect-specific bunyaviruses have been discovered in mosquitoes of north-western Africa, each likely to represent a new genus in the *Bunyaviridae* family (Marklewitz *et al.*, 2011, 2013; Auguste *et al.*, 2014). Three related viruses (named Herbert, Tai and Kibale) were isolated predominantly from *Culex* mosquitoes and determined to form a sister clade to the *Orthobunyavirus* genus (Markelwitz *et al.*, 2013), while Gouleako and Cumuto viruses were shown to distantly cluster with the *Phlebovirus* genus. A new insect-specific bunyavirus in *Culex* mosquito populations of the Torres Strait and Papua New Guinea within the Australasian region has also been identified, which groups distantly with Gouleako and Cumuto viruses (J. Hobson-Peters, D. Warrilow, B.J. McLean, D. Watterson, A.M.G. Colmant, A.F. van den Hurk, S. Hall-Mendelin, M. Hastie, J. Gorman, J. Harrison, N.A. Prow, R.T. Bernard, and R.A. Hall, unpublished data). Together, the insect cell-specific tropism, along with differences in genome length and the absence of cross-reactivity with other bunyaviruses supports that these new viruses are highly divergent from known vertebrate-infecting bunyaviruses and may provide valuable insight into the evolution of these viruses.

The apparent lack of vertebrate hosts in the transmission cycle suggests ISVs have evolved other modes of transmission between mosquitoes. The most likely being vertical transmission whereby the virus is passed from the female mosquito to her offspring through the ova and involves virus entry into the fully formed egg during oviposition (Tesh and Cornet, 1981). While this mode of transmission occurs quite frequently with some bunyaviruses and may be a mechanism of short term virus survival under certain conditions, only a very small percentage of the progeny of most flavivirus-infected mosquitoes acquire infection by this route (Kuno and Chang, 2005). However, recent laboratory experiments with *Culex pipiens* mosquitoes naturally infected with the ISF, CxFV, indicate that up to >80% of their progeny were infected via this mode of transmission (Bolling *et al.*, 2011; Bolling *et al.*, 2012). These findings support the theory that ISFs, and probably other insect-specific viruses, are maintained in nature by this route, and may have evolved a more efficient mechanism for vertical transmission. Research to elucidate the viral determinants and exact mechanism responsible for enhanced vertical transmission by ISFs and other ISVs is a priority, as similar adaptive strategies for persistence in arthropod populations may also be employed by arboviruses of medical and veterinary significance. Furthermore, assessing the level of co-evolution of ISFs and other ISVs with their host mosquito species, and the possible trade-offs they have made in adapting to efficient vertical transmission (e.g. inability to grow in vertebrate cells and/or lack of vertical transmission in other mosquito species), is another intriguing question for research.

Although insect-specific viruses are not directly associated with human disease, recent reports indicate that persistent infection of mosquito vectors with ISFs may regulate the transmission of arboviruses of medical significance (Kent *et al.*, 2010; Newman *et al.*, 2011; Bolling *et al.*, 2012; Hobson-Peters *et al.*, 2013). Our own observations that an Australian ISF, Palm Creek virus, restricts replication of the local pathogenic flaviviruses MVEV and WNV_{KUN} in mosquito cells (Hobson-Peters *et al.*, 2013) and those of Bolling *et al.* (2012) that show delayed transmission of WNV in *Culex* species co-infected with CxFV, has created intense interest in how ISFs may naturally regulate the transmission of flaviviruses of medical and veterinary significance in mosquito populations. Given the relatively high efficiency of vertical transmission by ISFs and their ability to persist in mosquitoes in nature, it is feasible that a significant reduction in vector competence for medically significant flaviviruses may occur in some populations of mosquitoes persistently infected with ISFs.

Newly found arboviruses of humans and animals

In addition to the appearance of genetic variants of known arboviruses with enhanced virulence or altered host or vector range, new viral species have recently been described that are associated with disease, or that have the potential to cause disease. Novel viruses have also been recognized during epidemiological investigations that were not associated with disease, but may have the potential to be pathogenic.

Bunyaviridae – genus *Orthobunyavirus*

In 2011, Schmallenberg virus (SBV), a new member of the Simbu group of orthobunyaviruses, was isolated in Germany and shown to be responsible for disease in cattle, sheep and goats (Hoffmann *et al.*, 2012). Transmitted by various *Culicoides* species (Rasmussen *et al.*, 2012; De Regge *et al.*, 2012; Balenghien *et al.*, 2014), it causes high fever, a decrease in milk production and severe diarrhoea, and is also responsible for abortions, stillbirths and congenital malformations (Hoffmann *et al.*, 2012; Garigliany *et al.*, 2012). The virus continues to spread rapidly across Europe (Afonso *et al.*, 2014), with evidence of infection ranging from Scandinavia in the north to southern Spain in the south, and from the United Kingdom in the west across to Turkey in the east. The genomic segments of the virus have been sequenced and shown to be similar to Shamonda (S segment), Sathuperi (M segment) and Shamonda (L segment) viruses, thus it appears to be a reassortment virus between a Shamonda-like virus and a Sathuperi-like virus (Hoffmann *et al.*, 2012; Yanase *et al.*, 2012). These two viruses belong to the Simbu serogroup of the genus *Orthobunyavirus*, with Douglas, Tinaroo and Peaton viruses, all suspected to cause congenital deformities in domestic animals, mainly in Asia, Africa and Oceania (Charles, 1994). Shamonda virus was first isolated from cattle in Nigeria in 1965 (Causey *et al.*, 1972) and recently re-emerged in Japan (Yanase *et al.*, 2005). It has also been suggested, however, that Shamonda virus should be reclassified as a species of Sathuperi virus, and consequently SBV may not have arisen by reassortment, but may be a species of Sathuperi virus and a possible ancestor of Shamonda virus (Goller *et al.*, 2012). The origin of SBV and the reason for its sudden emergence are unknown, but it is suspected to have been circulating in parts of Europe for some time prior to its discovery.

A novel orthobunyavirus, Oya virus, was isolated from the lungs of a pig during the Nipah virus outbreak in Sarawak, Malaysia, in 2000 and shown to be a member of the Simbu serocomplex (Kono *et al.*, 2002). Serological studies of pigs from a number of farms in six pig-breeding states in Malaysia indicated that Oya virus was widespread. Although no direct evidence was found to implicate Oya virus as the cause of disease in pigs, the possibility exists that the virus is a pig pathogen.

Bunyaviridae – genus *Phlebovirus*

Several isolates of a new virus were obtained in 2009 and 2010 from patients suffering fever, fatigue, conjunctival congestion, diarrhoea, abdominal pain, leucocytopenia, thrombocytopenia, proteinuria and haematuria in several provinces of China (Yu *et al.*, 2011). The new virus, named severe fever with thrombocytopenia syndrome virus (SFTSV) or Huaiyangshan virus (Yu *et al.*, 2011; Zhang *et al.*, 2012), was shown to be a new species in the *Phlebovirus* genus. A history of tick bites for many of the patients and isolation of the virus from *Haemaphysalis longicornis* and *Rhipicephalus microplus* species implicated ticks as likely vectors (Zhang *et al.*, 2012). Cases of SFTS have also been reported from South Korea (Kim *et al.*, 2013) and Japan (Takahashi *et al.*, 2014). Two closely related viruses, Heartland and Malsoor viruses, have recently been isolated (McMullan *et al.*, 2012; Mourya *et al.*, 2014). Heartland virus has also been associated with severe febrile illness, thrombocytopenia and tick bites in humans in Missouri, USA, in 2009 (McMullan *et al.*, 2012). Six additional cases of Heartland virus disease in humans were subsequently confirmed between 2012–2013 (Pastula *et al.*, 2014), while serological assays have identified the presence of antibodies to Heartland virus (or a close relative) in serum from farmed animals within the same region, although the identification of the virus requires substantiation (Xing *et al.*, 2013; Nasci *et al.*, 2014). Two related viruses have recently been described, Malsoor virus and Hunter Island Group virus, but although neither virus has yet been associated with human disease, their genetic similarity to known human pathogens makes their discovery of more than passing interest. Malsoor virus was isolated from *Rousettus leschenaultia* fruit bats in western India (Mourya *et al.*, 2014), but has not been associated with any human or animal disease at this time. Hunter Island Group virus was isolated from ticks collected from sick and healthy birds during a severe, fatal disease of shy albatrosses (*Thalassarche cauta*) on an island in the Hunter Island Group in north-western Tasmania, Australia (Wang *et al.*, 2014). As the virus was obtained from ticks collected from both sick and healthy birds, and none of the birds were shown to have antibodies to the virus, the authors concluded that the virus may not have been the causative agent of the disease. These recent novel virus isolations clearly suggest that other genetically related and potentially pathogenic phleboviruses may be widely distributed in different parts of the world.

A possible novel phlebovirus was recently obtained from *Phlebotomus perfiliewi* sandflies in Italy and named Fermo virus. It has not been associated with human disease, and is a member of the Sandfly fever Naples virus serocomplex (Remoli *et al.*, 2014).

Lone Star virus, a previously unclassified *Bunyavirus* from Lone Star ticks (*Amblyomma americanum*) collected in the Ohio–Mississippi basin (Kokernot *et al.*, 1969), was shown phylogenetically to be highly divergent but part of a clade comprising members of the Bhanji virus group and closely related to Heartland and SFTS viruses (Swei *et al.*, 2013).

A number of other bunyaviruses have been isolated from mosquitoes or ticks but have not been associated with human or animal disease (Major *et al.*, 2009; Coffey *et al.*, 2014; Lanciotti *et al.*, 2013), and it is probable that, as further arthropod collections are made, especially from mosquitoes, ticks and *Culicoides*, additional members of this large virus family will be found. In addition, as bunyaviruses have segmented genomes, novel reassortants will also continue to arise.

Rhabdoviridae

A novel virus was detected in 2009 in clinical samples from patients experiencing acute haemorrhagic fever in the Democratic Republic of Congo (Grard *et al.*, 2012). Deep sequencing of the samples enabled the assembly of a complete genome of a new rhabdovirus, named Bas Congo virus, the first rhabdovirus to be associated with haemorrhagic disease. Phylogenetic studies demonstrated that Bas Congo virus was most closely related to Australian rhabdoviruses in the Tibrogargan group (Gubala *et al.*, 2011) and to members of the *Ephemerovirus* genus, which are transmitted to cattle and buffalo by *Culicoides* biting midges and various mosquito species (Walker, 2005). Thus based on the similarity to other vector-borne rhabdoviruses and on preliminary epidemiological evidence, Bas Congo virus is probably an arthropod-borne virus, and if this is later shown to be the case, it would be the first medically significant arbovirus in the *Rhabdoviridae* family. Indeed, the isolation of Moussa virus, a novel rhabdovirus isolated from *Culex* mosquitoes in Cote d'Ivoire, demonstrates the presence of arthropod vectors of rhabdoviruses in Africa (Quan *et al.*, 2010).

Reoviridae

Recently a new seadornavirus (*Family Reoviridae*), Liao Ning virus (LNV), was isolated from mosquitoes in China (Attoui *et al.*, 2006). Genetic variants have been subsequently isolated in Xinjiang province of China and Australia (Lv *et al.*, 2012; Lu *et al.*, 2011; Coffey *et al.*, 2014; B.J. McLean, A.M.G. Colmant, D. Warrilow, A.F. van den Hurk, N.A. Prow, C.E. Webb, R.A. Hall and J. Hobson-Peters, unpublished data). While only isolated from arthropods to date (mosquitoes and *Culicoides*), it has been shown to replicate in vertebrate cells, and after two injections of the Chinese strain into mice, induced fatal haemorrhagic symptoms (Attoui *et al.*, 2006). Until further studies are carried out to confirm the initial mouse studies and assess the virulence of the additional LNV lineages in vertebrate species, they pose a potential medical and veterinary risk in the Australasian region. Indeed, LNV has been detected in common mosquito species such as *Ae. vigilax*, at very high frequencies, often in conjunction with the common arbovirus pathogen RRV (Coffey *et al.*, 2014; B.J. McLean, A.M.G. Colmant, D. Warrilow, A.F. van den Hurk, N.A. Prow, C.E. Webb, R.A. Hall, and J. Hobson-Peters, unpublished data). Considering that there is a high incidence of RRV infection in humans, horses and some marsupial species in Australia and RRV is predominantly transmitted by *Ae. vigilax* in coastal regions of Australia, this suggests that humans and horses and other animals are also at high risk of infection with LNV, providing that it can be demonstrated that this virus can disseminate to the salivary gland of infected mosquitoes.

Flaviviridae

Occasionally novel flaviviruses emerge, as evidenced by Goose/Jiangsu/808/2010 virus, a Tembusu-related virus that causes egg-drop and neurological disease of geese in China (Huang *et al.*, 2013).

Alphaviridae

Novel alphaviruses have been reported, including Trocara virus from *Ae. serratus* mosquitoes in the Amazon Basin, the first member of a newly recognized antigenic complex in the *Alphavirus* genus (Travassos da Rosa *et al.*, 2001), and new subtypes of VEEV (Rio Negro virus in Argentina; Pisano *et al.*, 2012). Two unusual viruses which have not been associated with animal or human disease are a novel lineage of SINV in Western Australia (Saleh *et al.*, 2003), and southern elephant seal virus (La Linn *et al.*, 2001) from an elephant seal louse (*Mirounga leonine*) on Macquarie Island.

The appearance of several new arboviruses of potential medical and veterinary significance emphasizes the need for continued real-time surveillance utilizing 'open-minded' approaches that allow the detection of previously unknown viruses. These include sequence- and antigen- independent methods discussed in the next section.

Protocols to detect, discover and monitor newly emerged arboviruses

As discussed above, the emergence of new arboviruses is a complex and dynamic process, influenced not only by the ecology of the virus, vector and host but also by environmental conditions that affect vector-borne transmission. Furthermore, increased international trade and travel, often to exotic locations in the tropics and subtropics, enhances the chance of arboviruses being transported to new geographical areas. Together, these variables make the emergence of arboviral diseases and the frequency of outbreaks unpredictable. Therefore, arbovirus surveillance programmes, which accurately monitor the activity of endemic, new and emerging viruses in real time, are essential public health priorities. In this section we briefly discuss how surveillance systems have advanced to allow high throughput monitoring of arbovirus activity in remote areas, and enable the detection and isolation of newly emerged, vector-borne viruses.

Traditionally, virus isolation from trapped arthropods (mosquitoes and biting midges) or wild and domestic animals has been performed to monitor the prevalence and distribution of arboviruses in specific regions (van den Hurk *et al.*, 2012). Isolation of viruses by inoculation of samples onto cell cultures allows for the detection of known viruses using virus-specific antibodies (Broom *et al.*, 1998). New viruses can also be discovered in this way, but usually require serial passage of inoculated cultures onto multiple cells lines until cytopathic effects are observed due to viral infection. While this approach has been useful for the detection of many novel arboviruses, it is labour intensive, time consuming and logistically expensive and thus not suited to high-throughput, real-time monitoring of arbovirus activity. Furthermore, isolation of arboviruses from mosquitoes does not necessarily demonstrate transmission. This approach also does not enable the identification of ISVs in most cases due to the absence of distinct cytopathic effect.

Sentinel animals have also been employed to monitor the activity of medically important arboviruses such as chickens for MVEV and WNV (Spencer *et al.*, 2001; Komar, 2001) and sentinel pigs for JEV (Hanna *et al.*, 1999), as well as cattle for the veterinary relevant viruses bluetongue, akabane and bovine ephemeral fever viruses (Gard *et al.*, 1988; Animal Health Australia, 2001). Regular sampling of these animals and detection of virus-specific antibodies in their serum provides useful information on arboviral transmission at a particular place and time, and in some cases, especially for veterinary viruses from sentinel cattle, can be used for virus isolation. While these approaches can be effective when targeting known arbovirus threats, they are not usually suitable for detecting viruses that have newly emerged or have recently been introduced to the region, except in some cases from sentinel cattle. They are also logistically difficult and expensive to maintain.

In response to the JEV incursions into northern Australia and the Torres Strait in the late 1990s, it became clear that traditional methods of mosquito trapping and virus isolation and sampling of sentinel pigs was not practical in remote locations. Mechanical failure of the mosquito traps, the need to collect trap catches within 24 h, the processing of large numbers of mosquitoes (approximately one JEV detection in over 48,000 mosquitoes; van den Hurk *et al.*, 2001; van den Hurk *et al.*, 2006) and the logistics of maintaining a cold chain between the field and the laboratory all contributed to the difficulty and expense of the operation. Furthermore, mosquitoes that were caught in traps set in remote areas for one week were contaminated with fungus due to the high levels of humidity. To overcome these obstacles, a series of modifications were made to the traps, and a novel viral detection system was developed.

To simplify mosquito trapping in remote areas, a non-powered, passive box trap was developed (Ritchie *et al.*, 2013), still baited with CO_2 but without moving parts or battery power. This was a major improvement, and eliminated the mechanical components that were subject to failure due to damage during transport or as a result of the high humidity in tropical environments.

To complement the design of these traps, a novel viral detection method was developed to suit deployment in remote tropical locations. This was based on observations that when infected mosquitoes feed on cotton wads soaked in a sugar solution, they salivate and expectorate virus onto the cotton wad (Doggett *et al.*, 2001, van den Hurk *et al.*, 2007). To take advantage of this, commercially available filter paper cards (FTA) impregnated with chemicals to preserve nucleic acids, inhibit fungal and bacteria growth and to inactivate viruses were used (Hall-Mendelin *et al.*, 2010). The FTA cards were coated with honey and strategically placed in the passive box traps to encourage wild mosquitoes to feed and expectorate any viruses contained in their saliva. The traps were run for 7 days and cards removed and sent to the laboratory via post. Although no infectious virus remained on the cards, viral RNA was eluted, detected and identified using virus-specific primers in real-time RT-PCR (van den Hurk *et al.*, 2014b). In preliminary field trials near Perth, Western Australia, and in Cairns, North Queensland, the alphaviruses RRV and BFV were successfully detected on FTA cards placed in traps for up to 7 days (Hall-Mendelin *et al.*, 2010). The new strategy showed similar sensitivity when compared with traditional virus isolation from the mosquitoes caught in the same traps. Additional field trials conducted in Northern Australia were run in parallel with a sentinel chicken programme. While WNV_{KUN} was detected on FTA cards in several traps, none of the sentinel chickens in the same locations seroconverted to the virus at this time (van den Hurk *et al.*, 2014b). These results suggest that the use of FTA cards in passive box traps is a logistically simple and highly sensitive early warning system for arboviruses. This strategy is currently being integrated into arbovirus surveillance programmes by several agencies and laboratories in Australia and the US.

The use of virus-specific primers and RT-PCR to amplify viral RNA deposited on FTA cards has to date enabled the detection of common arboviruses such as RRV, BFV, WNV and MVEV and the differentiation between viral strains through sequencing the amplicons (van den Hurk *et al.*, 2014b). This system also has the potential to detect novel introduced viruses and the discovery of new and emerging viruses by using cDNA from randomly primed RT-PCR amplification of the viral RNA eluted form FTA cards as the template for deep sequencing using next generation sequencing (NGS) technology (see Hall-Mendelin *et al.*, 2013, and below).

Recent advances in NGS platforms have enabled the discovery of unprecedented numbers of new viral genomes from arthropod samples or clinical specimens in a sequence independent manner (Zirkel *et al.*, 2011; Hang *et al.*, 2012; Cook *et al.*, 2013; Coffey *et al.*, 2014; Maruyama *et al.*, 2014; Warrilow *et al.*, 2014). This technology facilitates the sequencing of entire viral genomes, providing valuable information on the evolution of individual viral species and insights into genetic divergence, as was the case in the sequencing of field samples from the SBV epidemic in Europe (Rosseel *et al.*, 2012). A likely advantage of NGS will be its capacity to identify un-culturable viruses directly from the original sample.

NGS will likely play an important role in the future of arbovirus surveillance, particularly for the rapid identification and genetic characterization of novel viruses in mosquitoes. Recent studies by two groups revealed the successful detection by NGS of dengue, JEV and YFV sequences in experimentally infected mosquitoes (Bishop-Lilly *et al.*, 2010; Hall-Mendelin *et al.*, 2013). While the approach has been effective for detecting a range of other microorganisms directly in mosquito and tick samples (Hall-Mendelin *et al.*, 2013; Vayssier-Taussat *et al.*, 2013; Bonnet *et al.*, 2014), RNA viruses have proven difficult to sequence by NGS directly from samples, with the low abundance of viral RNA in comparison to host RNA a recurrent problem (Bishop-Lilley *et al.*, 2010; Liu *et al.*, 2011; Radford *et al.*, 2012; Marston *et al.*, 2013). However, recent reports using deep sequencing to identify unknown viruses isolated from mosquitoes in culture have been very promising. When Coffey *et al.* (2014) tested several culture supernatants from cells showing CPE after inoculation with mosquito homogenates, they obtained near full length sequences for several new viruses including bunyaviruses, rhabdoviruses and reoviruses. Of particular interest was the detection of up to three different viruses in a single culture. These finding indicate that NGS has great potential in arbovirus surveillance.

While NGS technology is obviously a leap forward in the rapid genetic analysis and taxonomic classification of new arboviruses, particularly when coupled with cell culture isolation, it remains too expensive for use in routine testing of samples in arbovirus surveillance. However, an inexpensive strategy that provides high throughput screening of mosquito samples for new and known viruses would allow a more selective and cost effective use of NGS for genome sequencing and characterization of novel viruses. As part of an investigation to explore the mosquito virome and define the biodiversity of mosquito-borne viruses in Australia, our lab developed a new approach for rapid, sensitive and high throughput discovery of new mosquito borne viruses (O'Brien *et al.*, 2015). This approach, known as MAVRIC (monoclonal antibody against viral RNA intermediates in cells), specifically detects the double-stranded RNA (dsRNA) intermediate, which is only present in significant amounts in cells during the replication cycle of positive strand and double stranded RNA viruses (Weber *et al.*, 2006). Staining for viral dsRNA in mosquito cell cultures previously inoculated with mosquito samples using MAVRIC in ELISA or IFA has been highly successful in identifying cultures infected with medically and agriculturally significant viruses from several families (including viruses not previously seen before in Australia or other parts of the world). These include flaviviruses, alphaviruses, reoviruses (orbiviruses and seadornaviruses) and nidoviruses (mesoniviruses). This unique reagent provides a simple, yet extremely powerful tool to rapidly detect and isolate a wide range of indigenous, exotic and newly discovered viruses. When linked directly to new generation sequencing platforms, this approach enables acquisition of near full genome sequences and the rapid taxonomic classification of new viruses (Warrilow *et al.*, 2014).

On a final note, the computer analysis of NGS can be significant barrier for most researchers, requiring specialized computer software and skilled interpretation by experienced personnel. However, an approach to address this problem has been recently developed. A new computational pipeline named 'virus hunter' has been made freely available for the analysis of NGS data specifically for the detection of novel and known viral sequences (Zhao *et al.*, 2013).

Conclusions

In this chapter we have discussed the many factors that influence arbovirus spread, persistence, transmission and pathogenesis, and hence drive the evolution of new viral strains and the emergence of new arboviral diseases. From these discussions it is clear that the complex interaction of these variables make it difficult to predict the prevalence of arboviral disease in specific regions of the world. While there is little doubt that significant changes in climate, landscape, culture, travel and trade will occur on the planet in the next 50 years, how these will collectively influence the frequency of arboviral transmission and disease will vary between regions and between viruses.

Although the precise pattern of change is impossible to predict, several recent examples of arboviruses adapting to new vector and host species to enhance transmission or to survive in a new habitat suggests we will see more changes to the ecology of known viruses in the future. The emergence of many previously unknown arboviruses in the last few decades also indicates that more new viruses will appear as a result of a change in their ecology.

A strong message to emerge from these conclusions is that enhanced, real-time arboviral surveillance systems will be crucial in the management and prevention of arboviral disease in the future. New detection strategies that are independent of viral sequence or antigenic structure and can be integrated into a high throughput, cost-effective strategy will not only monitor the activity of known arboviral pathogens, but will have the capability to detect new and emerging arboviral diseases.

Acknowledgement

The authors gratefully acknowledge the help of Tobias Hall in preparing the figures used in this chapter.

References

Aaskov, J., Buzacott, K., Field, E., Lowry, K., Berlioz-Arthaud, A., and Holmes, E.C. (2007). Multiple recombinant dengue type 1 viruses in an isolate from a dengue patient. J. Gen. Virol. *88*, 3334–3340.

Abdo-Salem, S., Waret-Szkuta, A., Roger, F., Olive, M.M., Saeed, K., and Chevalier, V. (2011). Risk assessment of the introduction of Rift Valley fever from the Horn of Africa to Yemen via legal trade of small ruminants. Trop. Anim. Health Prod. *43*, 471–480.

Adelman, Z.N., Miller, D.M., and Myles, K.M. (2013). Bed bugs and infectious disease: a case for the arboviruses. PLoS Pathog. *9(8)*, e1003462 [Epub ahead of print]

Afonso, A., Abrahantes, J.C., Conraths, F., Veldhuis, A., Elbers, A., Roberts, H., Van der Stede, Y., Méroc, E., Gache, K., and Richardson, J. (2014). The Schmallenberg virus epidemic in Europe – 2011–2013. Prev. Vet. Med. Mar 11. pii: S0167-5877(14)00082-8 [Epub ahead of print]

Aguilar, P.V., Barrett, A.D., Saeed, M.F., Watts, D.M., Russell, K., Guevara, C., Ampuero, J.S., Suarez, L., Cespedes, M., Montgomery, J.M., Halsey, E.S., and Kochel, T.J. (2011). Iquitos virus: a novel reassortant orthobunyavirus associated with human illness in Peru. PLoS Negl. Trop. Dis. *5*, e1315.

Albayrak, H., Ozan, E., and Kurt, M. (2012). Serosurvey and molecular detection of Crimean–Congo haemorrhagic fever virus (CCHFV) in northern Turkey. Trop. Anim. Health Prod. *44*, 1667–1671.

Alkan, C., Bichaud, L., de Lamballerie, X., Alten, B., Gould, E.A., and Charrel, R.N. (2013). Sandfly-borne phleboviruses of Eurasia and Africa: epidemiology, genetic diversity, geographic range, control measures. Antiviral Res. *100*, 54–74.

Allen, R., Taylor, S.K., and Sulkin, S.E. (1970). Studies of arthropod-borne virus infections in Chiroptera. 8. Evidence of natural St. Louis encephalitis virus infections in bats. Am. J. Trop. Med. Hyg. *19*, 851–859.

Allison, A.B., Holmes, E.C., Potgieter, A.C., Wright, I.M., Sailleau, C., Breard, E., Ruder, M.G., and Stallknecht, D.E. (2012). Segmental configuration and putative origin of the reassortant orbivirus, epizootic haemorrhagic disease virus serotype 6, strain Indiana. Virology *424*, 67–75.

Arnal, A., Gómez-Díaz, E., Cerdà-Cuéllar, M., Lecollinet, S., Pearce-Duvet, J., Busquets, N., García-Bocanegra, I., Pagès, N., Vittecoq, M., Hammouda, A., *et al.* (2014). Circulation of a Meaban-like virus in yellow-legged gulls and seabird ticks in the western Mediterranean basin. PLoS One *9*, e89601.

Amaro, F., Luz, T., Parreira, P., Ciufolini, M.G., Marchi, A., Janeiro, N., Zagalo, A., Proença, P., Ramos, M.I., and Alves, M.J. (2011). Toscana virus in the Portuguese population: serosurvey and clinical cases. Acta Med. Port. *24*, 503–508.

Anderson, J.F., Main, A.J., Cheng, G., Ferrandino, F.J., and Fikrig, E. (2012). Horizontal and vertical transmission of West Nile virus genotype NY99 by *Culex salinarius* and genotypes NY99 and WN02 by *Culex tarsalis*. Am. J. Trop. Med. Hyg. *86*, 134–139.

Andreadis, T.G., Armstrong, P.M., and Bajwa, W.I. (2010). Studies on hibernating populations of *Culex pipiens* from a West Nile virus endemic focus in New York City: parity rates and isolation of West Nile virus. J. Am. Mosq. Control Assoc. *26*, 257–264.

Angelini, P., Tamba, M., Finarelli, A.C., Bellini, R., Albieri, A., Bonilauri, P., Cavrini, F., Dottori, M., Gaibani, P., Martini, E., *et al.* (2010). West Nile virus circulation in Emilia-Romagna, Italy: the integrated surveillance system 2009. Eur. Surveill. *15*, 19547.

Animal Health Australia (2001). The history of bluetongue, Akabane, and ephemeral fever viruses and their vectors in Australia. Animal Health Australia, Deakin, ACT, Australia. Available online: http://www.animalhealthaustralia.com.au/programs/disease-surveillance/national-arbovirus-monitoring-program/namp-archived-reports/. Accessed 20 June 2014.

Anyamba, A., Linthicum, K.J., Small, J., Britch, S.C., Pak, E., de La Rocque, S., Formenty, P., Hightower, A.W., Breiman, R.F., Chretien, J.P., *et al.* (2010). Prediction, assessment of the Rift Valley fever activity in East and Southern Africa 2006–2008 and possible vector control strategies. Am. J. Trop. Med. Hyg. *83(Suppl.)*, 43–51.

Anyamba, A., Small, J.L., Britch, S.C., Tucker, C.J., Pak, E.W., Reynolds, C.A., Crutchfield, J., and Linthicum, K.J. (2014). Recent weather extremes and impacts on agricultural production and vector-borne disease outbreak patterns. PLoS One *9*, e92538.

Artsob, H., Gubler, D.J., Enria, D.A., Morales, M.A., Pupo, M., Bunning, M.L., and Dudley, J.P. (2009). West Nile Virus in the New World:

trends in the spread and proliferation of West Nile Virus in the Western Hemisphere. Zoonoses Publ. Health *56*, 357–369.
Asahina, S. (1970). Transoceanic flight of mosquitoes on the Northwest Pacific. Jpn. J. Med. Sci. Biol. *23*, 255–258.
Attoui, H., Mohd Jaafar, F., Belhouchet, M., Tao, S., Chen, B., Liang, G., Tesh, R.B., de Micco, P., and de Lamballerie, X. (2006). Liao Ning virus, a new Chinese seadornavirus that replicates in transformed and embryonic mammalian cells. J. Gen. Virol. *87*, 199–208.
Audsley, M., Edmonds, J., Liu, W., Mokhonov, V., Mokhonova, E., Melian, E.B., Prow, N., Hall, R.A., and Khromykh, A.A. (2011). Virulence determinants between New York 99 and Kunjin strains of West Nile virus. Virology *414*, 63–73.
Auguste, A.J., Pybus, O.G., and Carrington, C.V. (2009). Evolution and dispersal of St. Louis encephalitis virus in the Americas. Infect. Genet. Evol. *9*, 709–715.
Auguste, A.J., Carrington, C.V., Forrester, N.L., Popov, V.L., Guzman, H., Widen, S.G., Wood, T.G., Weaver, S.C., and Tesh, R.B. (2014). Characterization of a novel Negevirus and a novel Bunyavirus isolated from *Culex (Culex) declarator* mosquitoes in Trinidad. J. Gen. Virol. *95*, 481–485.
Bakonyi, T., Hubálek, Z., Rudolf, I., and Nowotny, N. (2005). Novel flavivirus or new lineage of West Nile virus, Central Europe. Emerg. Infect. Dis. *11*, 225–231.
Bakonyi, T., Ferenczi, E., Erdélyi, K., Kutasi, O., Csörgő, T., Seidel, B., Weissenböck, H., Brugger, K., Bán, E., and Nowotny, N. (2013). Explosive spread of a neuroinvasive lineage 2 West Nile virus in Central Europe, 2008/2009. Vet. Microbiol. *165*, 61–70.
Bakonyi, T., Busquets, N., and Nowotny, N. (2014). Comparison of complete genome sequences of Usutu virus strains detected in Spain, Central Europe and Africa. Vector Borne Zoonotic Dis. *14*, 324–329.
Balenghien, T., Pagès, N., Goffredo, M., Carpenter, S., Augot, D., Jacquier, E., Talavera, S., Monaco, F., Depaquit, J., Grillet, C., *et al.* (2014). The emergence of Schmallenberg virus across Culicoides communities and ecosystems in Europe. Prev. Vet. Med. Mar 18. pii: S0167-5877(14)00102-0. [Epub ahead of print]
Balkhy, H.H., and Memish, Z.A. (2003). Rift Valley fever: an uninvited zoonosis in the Arabian peninsula. Int. J. Antimicrob. Agents *21*, 153–157.
Banerjee, K. (1988). Kyasanur Forest disease. In The Arboviruses: Epidemiology and Ecology, Vol III, T.P. Monath, ed. (CRC Press Inc., Boca Raton, Florida), pp. 93–116.
Banerjee, K., Bhat, H.R., Geevarghese, G., Jacob, P.G., and Malunjkar, A.S. (1988). Antibodies against Japanese encephalitis virus in insectivorous bats from Karnataka. Indian J. Med. Res. *87*, 527–530.
Barrett, A.D., and Monath, T.P. (2003). Epidemiology and ecology of yellow fever virus. Adv. Virus Res. *61*, 291–315.
Barthel, A., Gourinat, A.C., Cazorla, C., Joubert, C., Dupont-Rouzeyrol, M., and Descloux, E. (2013). Breast milk as a possible route of vertical transmission of dengue virus? Clin. Infect. Dis. *57*, 415–417.
Batten, C.A., Maan, S., Shaw, A.E., Maan, N.S., and Mertens, P.P. (2008). A European field strain of bluetongue virus derived from two parental vaccine strains by genome segment reassortment. Virus Res. *137*, 56–63.
Baty, S.A., Gibney, K.B., Staples, J.E., Patterson, A.B., Levy, C., Lehman, J., Wadleigh, T., Feld, J., Lanciotti, R., Nugent, C.T., and Fischer, M. (2012). Evaluation for West Nile Virus (WNV), RNA in urine of patients within 5 months of WNV infection. J. Infect. Dis. *205*, 1476–1477.
Beasley, D.W., Davis, C.T., Guzman, H., Vanlandingham, D.L., Travassos da Rosa, A.P., Parsons, R.E., Higgs, S., Tesh, R.B., and Barrett, A.D. (2003). Limited evolution of West Nile virus has occurred during its southwesterly spread in the United States. Virology *309*, 190–195.
Becker, M., Zielen, S., Schwarz, T.F., Linde, R., and Hofmann, D. (1997). Pappataci fever. Klin. Padiatr. *209*, 377–379.
Beebe, N.W., Ambrose, L., Hill, L.A., Davis, J.B., Hapgood, G., Cooper, R.D., Russell, R.C., Ritchie, S.A., Reimer, L.J., Lobo, N.F., *et al.* (2013). Tracing the tiger: population genetics provides valuable insights into the *Aedes* (Stegomyia) albopictus invasion of the Australasian Region. PLoS Negl. Trop. Dis. *7*, e2361.
Beltrame, A., Angheben, A., Bisoffi, Z., Monteiro, G., Marocco, S., Calleri, G., Lipani, F., Gobbi, F., Canta, F., Castelli, F., *et al.* (2007). Imported Chikungunya infection, Italy. Emerg. Infect. Dis. *13*, 1264–1266.
Benedict, M.Q., Levine, R.S., Hawley, W.A., and Lounibos, L.P. (2007). Spread of the tiger: global risk of invasion by the mosquito *Aedes albopictus*. Vector Borne Zoonotic Dis. *7*, 76–85.
Bente, D.A., Forrester, N.L., Watts, D.M., McAuley, A.J., Whitehouse, C.A., and Bray, M. (2013). Crimean–Congo hemorrhagic fever: history, epidemiology, pathogenesis, clinical syndrome and genetic diversity. Antiviral Res. *100*, 159–189.
Bertrand, Y., Töpel, M., Elväng, A., Melik, W., and Johansson, M. (2012). First dating of a recombination event in mammalian tick-borne flaviviruses. PLoS One *7*, e31981.
Bhatt, P.N., and Rodrigues, F.M. (1967). Chandipura: a new Arbovirus isolated in India from patients with febrile illness. Indian J. Med. Res. *55*, 1295–1305.
Bingham, J., Lunt, R.A., Green, D.J., Davies, K.R., Stevens, V., and Wong, F.Y. (2010). Experimental studies of the role of the little raven (*Corvus mellori*) in surveillance for West Nile virus in Australia. Aust. Vet. J. *88*, 204–210.
Bishop-Lilly, K.A., Turell, M.J., Willner, K.M., Butani, A., Nolan, N.M., Lentz, S.M., Akmal, A., Mateczun, A., Brahmbhatt, T.N., Sozhamannan, S., Whitehouse, C.A., and Read, T.D. (2010). Arbovirus detection in insect vectors by rapid, high-throughput pyrosequencing. PLoS Negl. Trop. Dis. *4*, e878.
Blitvich, B.J., Saiyasombat, R., Dorman, K.S., Garcia-Rejon, J.E., Farfan-Ale, J.A., and Loroño-Pino, M.A. (2012). Sequence and phylogenetic data indicate that an orthobunyavirus recently detected in the Yucatan Peninsula of Mexico is a novel reassortant of Potosi and Cache Valley viruses. Arch. Virol. *157*, 1199–1204.
Bolling, B.G., Eisen, L., Moore, C.G., and Blair, C.D. (2011). Insect-specific flaviviruses from *Culex* mosquitoes in Colorado, with evidence of vertical transmission. Am. J. Trop. Med. Hyg. *85*, 169–177.
Bolling, B.G., Olea-Popelka, F.J., Eisen, L., Moore, C.G., and Blair, C.D. (2012). Transmission dynamics of an insect-specific flavivirus in a naturally infected *Culex pipiens* laboratory colony and effects of co-infection on vector competence for West Nile virus. Virology *427*, 90–97.
Bondre, V.P., Jadi, R.S., Mishra, A.C., Yergolkar, P.N., and Arankalle, V.A. (2007). West Nile virus isolates from India: evidence for a distinct genetic lineage. J. Gen. Virol. *88*, 875–884.
Bonizzoni, M., Gasperi, G., Chen, X., and James, A.A. (2013). The invasive mosquito species *Aedes albopictus*: current knowledge and future perspectives. Trends Parasitol. *29*, 460–468.
Bonnet, S., Michelet, L., Moutailler, S., Cheval, J., Hebert, C., Vayssier-Taussat, M., and Eloit, M. (2014). Identification of parasitic communities within European ticks using next-generation sequencing. PLoS Negl. Trop. Dis. *8*(3), e2753.
Bordería, A.V., Stapleford, K.A., and Vignuzzi, M. (2011). RNA virus population diversity: implications for inter-species transmission. Curr. Opin. Virol. *1*, 643–648.
Boyle, D.B., Dickerman, R.W., and Marshall, I.D. (1983). Primary viraemia responses of herons to experimental infection with Murray Valley encephalitis, Kunjin and Japanese encephalitis viruses. Aust. J. Exp. Biol. Med. Sci. *61*, 655–664.
Brackney, D.E., Brown, I.K., Nofchissey, R.A., Fitzpatrick, K.A., and Ebel, G.D. (2010). Homogeneity of Powassan virus populations in naturally infected *Ixodes scapularis*. Virology *402*, 366–371.
Brackney, D.E., Pesko, K.N., Brown, I.K., Deardorff, E.R., Kawatachi, J., and Ebel, G.D. (2011). West Nile virus genetic diversity is maintained during transmission by *Culex pipiens quinquefasciatus* mosquitoes. PLoS One *6*(9), e24466.
Brault, A.C., Powers, A.M., Ortiz, D., Estrada-Franco, J.G., Navarro-Lopez, R., and Weaver, S.C. (2004a). Venezuelan equine encephalitis emergence: enhanced vector infection from a single amino acid substitution in the envelope glycoprotein. Proc. Natl. Acad. Sci. U.S.A. *101*, 11344–11349.
Brault, A.C., Langevin, S.A., Bowen, R.A., Panella, N.A., Biggerstaff, B.J., Miller, B.R., and Komar, N. (2004b). Differential virulence of West Nile strains for American crows. Emerg. Infect. Dis. *10*, 2161–2168.
Brault, A.C., Huang, C.Y., Langevin, S.A., Kinney, R.M., Bowen, R.A., Ramey, W.N., Panella, N.A., Holmes, E.C., Powers, A.M., and Miller, B.R. (2007). A single positively selected West Nile viral mutation confers increased virogenesis in American crows. Nat. Genet. *39*, 1162–1166.
Briese, T., Calisher, C.H., and Higgs, S. (2013). Viruses of the family Bunyaviridae: are all available isolates reassortants? Virology *446*, 207–216.
Broom, A.K., Hall, R.A., Johansen, C.A., Oliveira, N., Howard, M.A., Lindsay, M.D., Kay, B.H., and Mackenzie, J.S. (1998). Identification

of Australian arboviruses in inoculated cell cultures using monoclonal antibodies in ELISA. Pathology 30, 286–288.

Broom, A.K., Lindsay, M.D., Wright, A.E., Smith, D.W., and Mackenzie, J.S. (2003). Epizootic activity of Murray Valley encephalitis and Kunjin viruses in an Aboriginal community in the southeast Kimberley region of Western Australia: results of mosquito fauna and virus isolation studies. Am. J. Trop. Med. Hyg. 69, 277–283.

Buescher, E.L., Scherer, W.F., McClure, H.E., Moyer, J.T., Rosenberg, M.Z., Yoshii, M., and Okada, Y. (1959). Ecologic studies of Japanese encephalitis virus in Japan. IV. Avian infection. Am. J. Trop. Med. Hyg. 8, 678–688.

Bugbee, L., and Forte, L.R. (2004). The discovery of West Nile virus in overwintering *Culex pipiens* (Diptera: Culicidae) mosquitoes in Lehigh County, Pennsylvania. J. Am. Mosq. Control Assoc. 20, 326–327.

Burgin, L.E., Gloster, J., Sanders, C., Mellor, P.S., Gubbins, S., and Carpenter, S. (2013). Investigating incursions of bluetongue virus using a model of long-distance Culicoides biting midge dispersal. Transbound. Emerg. Dis. 60, 263–272.

Burt, F.J., Rolph, M.S., Rulli, N.E., Mahalingam, S., and Heise, M.T. (2012). Chikungunya: a re-emerging virus. Lancet 379, 662–671.

Busch, M.P., Caglioti, S., Robertson, E.F., McAuley, J.D., Tobler, L.H., Kamel, H., Linnen, J.M., Shyamala, V., Tomasulo, P., and Kleinman, S.H. (2005). Screening the blood supply for West Nile virus RNA by nucleic acid amplification testing. N. Engl. J. Med. 353, 460–467.

Caini, S., Szomor, K., Ferenczi, E., Szekelyne Gasper, A., Csohan, A., Krisztalovics, K., Molnar, Z., and Horvath, J. (2012). Tick-borne encephalitis transmitted by unpasteurised cow milk in western Hungary, September to October 2011. Eur. Surveill. 17(12), pii 20128.

Calisher, C.H., Kinney, R.M., de Sousa Lopes, O., Trent, D.W., Monath, T.P., and Francy, D.B. (1982). Identification of a new Venezuelan equine encephalitis virus from Brazil. Am. J. Trop. Med. Hyg. 31, 1260–1272.

Calisher, C.H., Childs, J.E., Field, H.E., Holmes, K.V., and Schountz, T. (2006). Bats: important reservoir hosts of emerging viruses. Clin. Microbiol. Rev. 19, 531–545.

Calzolari, M., Bonilauri, P., Bellini, R., Albieri, A., Defilippo, F., Tamba, M., Tassinari, M., Gelati, A., Cordioli, P., Angelini, P., and Dottori, M. (2013). Usutu virus persistence and West Nile virus inactivity in the Emilia-Romagna region (Italy) in 2011. PLoS One 8(5), e63978.

Cao-Lormeau, V.M., Roche, C., Teissier, A., Robin, E., Berry, A.L., Mallet, H.P., Sall, A.A., and Musso, D. (2014). Zika virus, French Polynesia, South Pacific, 2013. Emerg. Infect. Dis. 20, 1084–1086.

Carney, J., Daly, J.M., Nisalak, A., and Solomon, T. (2012). Recombination and possible selection identified in complete genome sequences of Japanese encephalitis virus. Arch. Virol. 157, 75–83.

Carpenter, S., Groschup, M.H., Garros, C., Felippe-Bauer, M.L., and Purse, B.V. (2013). *Culicoides* biting midges, arboviruses and public health in Europe. Antiviral Res. 100, 102–113.

Carver, S., Bestall, A., Jardine, A., and Ostfeld, R.S. (2009). Influence of hosts on the ecology of arboviral transmission: potential mechanisms influencing dengue, Murray Valley encephalitis, and Ross River virus in Australia. Vector-Borne Zoonotic Dis. 9, 51–64.

Causey, O.R., Kemp, G.E., Causey, C.E., and Lee, V.H. (1972). Isolations of Simbu-group viruses in Ibadan, Nigeria 1964–69, including the new types Sango, Shamonda, Sabo and Shuni. Ann. Trop. Med. Parasitol. 66, 357–362.

CDC (2002). Possible West Nile virus transmission to an infant through breast-feeding – Michigan, 2002. MMWR Morb. Mortal Wkly. Rep. 51, 877–878.

CDC (2013). Zika Fever in French Polynesia (Tahiti). Available online: http://wwwnc.cdc.gov/travel/notices/watch/zika-fever-french-polynesia-tahiti. Accessed 15 January 2014.

CDC (2014). Chikungunya in the Caribbean. Available online: http://wwwnc.cdc.gov/travel/notices/watch/chikungunya-saint-martin. Accessed 16 January 2014.

Celikbas, A.K., Dokuzoğuz, B., Baykam, N., Gok, S.E., Eroğlu, M.N., Midilli, K., Zeller, H., and Ergonul, O. (2014). Crimean–Congo hemorrhagic fever among health care workers, Turkey. Emerg. Infect. Dis. 20, 477–479.

Charles, J.A. (1994). Akabane virus. Vet. Clin. North Am. Food Anim. Pract. 10, 525–546.

Charrel, R.N., Moureau, G., Temmam, S., Izri, A., Marty, P., Parola, P., da Rosa, A.T., Tesh, R.B., and de Lamballerie, X. (2009). Massilia virus, a novel Phlebovirus (Bunyaviridae) isolated from sandflies in the Mediterranean. Vector Borne Zoonotic Dis. 9, 519–530.

Chastel, C., Main, A.J., Guiguen, C., le Lay, G., Quillien, M.C., Monnat, J.Y., and Beaucournu, J.C. (1985). The isolation of Meaban virus, a new Flavivirus from the seabird tick Ornithodoros (Alectorobius) maritimus in France. Arch. Virol. 83, 129–140.

Chen, L.H., and Wilson, M.E. (2004). Transmission of dengue virus without a mosquito vector: nosocomial mucocutaneous transmission and other routes of transmission. Clin. Infect. Dis. 39, e56–60.

Chen, L., Liu, B., Yang, J., and Jin, Q. (2014). DBatVir: the database of bat-associated viruses. Database (Oxford) March 18:bau021.

Chen, R., and Vasilakis, N. (2011). Dengue – quo tu et quo vadis? Viruses 3, 1562–1608.

Chen, S.P., Yu, M., Jiang, T., Deng, Y.Q., Qin, C.F., Han, J.F., and Qin, E.D. (2008). Identification of a recombinant dengue virus type 1 with 3 recombination regions in natural populations in Guandong province, China. Arch. Virol. 153, 1175–1179.

Chen, W.R., Tesh, R.B., and Rico-Hesse, R. (1990). Genetic variation of Japanese encephalitis virus in nature. J. Gen. Virol. 71, 2915–2922.

Chen, W.R., Rico-Hesse, R., and Tesh, R.B. (1992). A new genotype of Japanese encephalitis virus from Indonesia. Am. J. Trop. Med. Hyg. 47, 61–69.

Chevalier, V., Pépin, M., Plée, L., and Lancelot, R. (2010). Rift Valley fever--a threat for Europe? Euro Surveill. 15(10), 19506.

Christophers, R.C. (1960). *Aedes aegypti*. The Yellow Fever Mosquito: Its Life History, Bionomics, and Structure. (Cambridge University Press, Cambridge, UK).

Chuang, C.K., and Chen, W.J. (2009). Experimental evidence that RNA recombination occurs in the Japanese encephalitis virus. Virology 394, 286–297.

Chungue, E., Deubel, V., Cassar, O., Laille, M., and Martin, P.M. (1993). Molecular epidemiology of dengue3 viruses and genetic relatedness among dengue 3 strains isolated from patients with mild or severe form of dengue fever in French Polynesia. J. Gen. Virol. 74, 2765–2770.

Chungue, E., Cassar, O., Drouet, M.T., Guzman, M.G., Laille, M., Rosen, L., and Deubel, V. (1995). Molecular epidemiology of dengue-1 and dengue-4 viruses. J. Gen. Virol. 76, 1877–1884.

Ciccozzi, M., Peletto, S., Cella, E., Giovanetti, M., Lai, A., Gabanelli, E., Acutis, P.L., Modesto, P., Rezza, G., Platonov, A.E., Lo Presti, A., and Zehender, G. (2013). Epidemiological history and phylogeography of West Nile virus lineage 2. Infect. Genet. Evol. 17, 46–50.

Ciota, A.T., and Kramer, L.D. (2013). Vector–virus interactions and transmission dynamics of West Nile virus. Viruses 5, 3021–3947.

Ciota, A.T., Ngo, K.A., Lovelace, A.O., Payne, A.F., Zhou, Y., Shi, P.Y., and Kramer, L.D. (2007). Role of the mutant spectrum in adaptation and replication of West Nile virus. J. Gen. Virol. 88, 865–874.

Ciota, A.T., Ehrbar, D.J., Van Slyke, G.A., Payne, A.F., Willsey, G.G., Viscio, R.E., and Kramer, L.D. (2012). Quantification of intrahost bottlenecks of West Nile virus in *Culex pipiens* mosquitoes using an artificial mutant swarm. Infect. Genet. Evol. 12, 557–564.

Coffey, L.L., and Vignuzzi, M. (2011). Host alternation of chikungunya virus increases fitness while restricting population diversity and adaptability to novel selective pressures. J. Virol. 85, 1025–1035.

Coffey, L.L., Vasilakis, N., Brault, A.C., Powers, A.M., Tripet, F., and Weaver, S.C. (2008). Arbovirus evolution in vivo is constrained by host alternation. Proc. Natl. Acad. Sci. U.S.A. 105, 6970–6975.

Coffey, L.L., Forrester, N., Tsetsarkin, K., Vasilakis, N., and Weaver, S.C. (2013). Factors shaping the adaptive landscape for arboviruses: implications for the emergence of disease. Future Microbiol. 8, 155–178.

Coffey, L.L., Page, B.L., Greninger, A.L., Herring, B.L., Russell, R.C., Doggett, S.L., Haniotis, J., Wang, C., Deng, X., and Delwart, E.L. (2014). Enhanced arbovirus surveillance with deep sequencing: Identification of novel rhabdoviruses and bunyaviruses in Australian mosquitoes. Virology 448, 146–158.

Cook, S., Bennett, S.N., Holmes, E.C., De Chesse, R., Moureau, G., and de Lamballerie, X. (2006). Isolation of a new strain of the flavivirus cell fusing agent virus in a natural mosquito population from Puerto Rico. J. Gen. Virol. 87, 735–748.

Cook, S., Moureau, G., Kitchen, A., Gould, E.A., de Lamballerie, X., Holmes, E.C., and Harbach, R.E. (2012). Molecular evolution of the insect-specific flaviviruses. J. Gen. Virol. 93, 223–234.

Cook, S., Chung, B.Y., Bass, D., Moureau, G., Tang, S., McAlister, E., Culverwell, C.L., Glücksman, E., Wang, H., Brown, T.D., *et al.* (2013). Novel virus discovery and genome reconstruction from field RNA

samples reveals highly divergent viruses in dipteran hosts. PLoS One *8*(11), e80720.
Correa-Giron, P., Calisher, C.H., and Baer, G.M. (1972). Epidemic strain of Venezuelan equine encephalomyelitis virus from a vampire bat captured in Oaxaca, Mexico, 1970. Science *175*, 546–547.
Crabtree, M.B., Sang, R.C., Stollar, V., Dunster, L.M., and Miller, B.R. (2003). Genetic and phenotypic characterization of the newly described insect flavivirus, Kamiti River virus. Arch. Virol. *148*, 1095–1118.
Dacheux, L., Cervantes-Gonzalez, M., Guigon, G., Thiberge, J.M., Vandenbogaert, M., Maufrais, C., Caro, V., and Bourhy, H. (2014). A preliminary study of viral metagenomics of French bat species in contact with humans: identification of new mammalian viruses. PLoS One *9*(1), e87194.
Davies, F.G. (2006). Risk of Rift Valley fever epidemic at the haj in Mecca, Sausi Arabia. Rev. Sci. Tech. *25*, 137–147.
Davis, A., Bunning, M., Gordy, P., Panella, N., Blitvich, B., and Bowen, R. (2005). Experimental and natural infection of North American bats with West Nile virus. Am. J. Trop. Med. Hyg. *73*, 467–469.
Davis, C.T., Beasley, D.W., Guzman, H., Raj, R., D'Anton, M., Novak, R.J., Unnasch, T.R., Tesh, R.B., and Barrett, A.D. (2003). Genetic variation among temporally and geographically distinct West Nile virus isolates, United States, 2001, 2002. Emerg. Infect. Dis. *9*, 1423–1429.
Davis, C.T., Ebel, G.D., Lanciotti, R.S., Brault, A.C., Guzman, H., Siirin, M., Lambert, A., Parsons, R.E., Beasley, D.W., Novak, R.J., *et al.* (2005). Phylogenetic analysis of North American West Nile virus isolates, 2001–2004: evidence for the emergence of a dominant genotype. Virology *342*, 252–265.
Dégallier, N., Travassos da Rosa, A.P.A., Vasconcelos, P.F.C., Herve, J.P., Sá Filho, G.C., Travassos da Rosa, J.F.S., Travassos da Rosa, E.S., and Rodrigues, S.G. (1992). Modifications of arbovirus transmission in relation to construction of dams in Brazilian Amazonia. J. Brazilian Assoc. Advance. Sci. *44*, 124–135.
Depaquit, J., Grandadam, M., Fouque, F., Andry, P.E., and Peyrefitte, C. (2010). Arthropod-borne viruses transmitted by Phlebotomine sandflies in Europe: A review. Eur. Surveill., *15*, 19507.
De Regge, N., Deblauwe, I., De Deken, R., Vantieghem, P., Madder, M., Geysen, D., Smeets, F., Losson, B., van den Berg, T., and Cay, A.B. (2012). Detection of Schmallenberg virus in different Culicoides spp. by real-time RT-PCR. Transbound. Emerg. Dis. *59*, 471–475.
Descloux, E., Cao-Lormeau, V.M., Roche, C., and de Lamballerie, X. (2009). Dengue 1 diversity and microevolution, French Polynesia 2001–2006: connection with epidemiology an clinics. PLoS Negl. Trop. Dis. *3*(8), e493.
de Souza, R.P., Foster, P.G., Sallum, M.A., Coimbra, T.L., Maeda, A.Y., Silveira, V.R., Moreno, E.S., da Silva, F.G., Rocco, I.M., Ferreira, I.B., *et al.* (2010). Detection of a new yellow fever virus lineage within the South American genotype I in Brazil. J. Med. Virol. *82*, 175–185.
Dick, G.W., Kitchen, S.F., and Haddow, A.J. (1952). Zika virus. 1. Isolation and serological specificity. Trans. R. Soc. Trop. Med. Hyg. *46*, 509–520.
Digoutte, J.P., and Peters, C.J. (1989). General aspects of the 1987 Rift Valley fever epidemic in Mauritania. Res. Virol. *140*, 27–30.
Doggett, S.L., Klowden, M.J., and Russell, R.C. (2001). Are vector competence experiments competent vector experiments? Arbovirus Res. Austral. *8*, 126–130.
Doggett, S.L., Dwyer, D.E., Peñas, P.F., and Russell, R.C. (2012). Bed bugs: clinical relevance and control options. Clin. Microbiol. Rev. *25*, 164–192.
Duffy, M.R., Chen, T.H., Hancock, W.T., Powers, A.M., Kool, J.L., Lanciotti, R.S., Pretrick, M., Marfel, M., Holzbauer, S., Dubray, C., *et al.* (2009). Zika virus outbreak on Yap Island, Federated States of Micronesia. N. Engl. J. Med. *360*, 2536–2543.
Durden, L.A., Linthicum, K.J., and Turell, M.J. (1992). Mechanical transmission of Venezuelan equine encephalomyelitis virus by hematophagous mites (Acari). J. Med. Entomol. *29*, 118–121.
Dusek, R.J., McLean, R.G., Kramer, L.D., Ubico, S.R., Dupuis, A.P. 2nd, Ebel, G.D., and Guptill, S.C. (2009). Prevalence of West Nile virus in migratory birds during spring and fall migration. Am. J. Trop. Med. Hyg. *81*, 1151–1158.
Eagles, D., Deveson, T., Walker, P.J., Zalucki, M.P., and Durr, P. (2012). Evaluation of long-distance dispersal of *Culicoides* midges into northern Australia using a migration model. Med. Vet. Entomol. *26*, 334–340.
Eagles, D., Walker, P.J., Zalucki, M.P., and Durr, P.A. (2013). Modelling spatio-temporal patterns of long distance *Culicoides* dispersal into northern Australia. Prev. Vet. Med. *110*, 312–322.
Ebel, G.D., Carricaburu, J., Young, D., Bernard, K.A., and Kramer, L.D. (2004). Genetic and phenotypic variation of West Nile virus in New York, 2000–2003. Am. J. Trop. Med. Hyg. *71*, 493–500.
Ecker, M., Allison, S.L., Meixner, T., and Heinz, F.X. (1999). Sequence analysis and genetic classification of tick-borne encephalitis viruses from Europe and Asia. J. Gen. Virol. *80*, 179–185.
Eitrem, R., Vene, S., and Niklasson, B. (1990). Incidence of sand fly fever among Swedish United Nations soldiers on Cyprus during 1985. Am. J. Trop. Med. Hyg. *43*, 207–211.
Ellis, S.B., Appenzeller, G., Lee, H., Mullen, K., Swenness, R., Pimentel, G., Mohareb, E., and Warner, C. (2008). Outbreak of sandfly fever in central Iraq. Mil. Med. *173*, 949–953.
Endy, T.P., and Nisalak, A. (2002). Japanese encephalitis virus: ecology and epidemiology. Curr. Top. Microbiol. Immunol. *267*, 12–48.
Epstein, P.R., and Defilippo, C. (2001). West Nile virus and drought. Global Change Human Health *2*, 105–107.
Ergunay, K., Ismayilova, V., Colpak, I.A., Kansu, T., and Us, D. (2012). A case of central nervous system infection due to a novel Sandfly Fever Virus (SFV) variant: Sandfly Fever Turkey Virus (SFTV). J. Clin. Virol. *54*, 79–82.
Es-Sette, N., Nourlil, J., Hamdi, S., Mellouki, F., and Lemrani, M. (2012). First detection of Toscana virus RNA from sand flies in the genus *Phlebotomus* (Diptera: Phlebotomidae) naturally infected in Morocco. J. Med. Entomol. *49*, 1507–1509.
Estrada-Peña, A., Vatansever, Z., Gargili, A., and Ergönul, O. (2010). The trend towards habitat fragmentation is the key factor driving the spread of Crimean–Congo haemorrhagic fever. Epidemiol. Infect. *138*, 1194–1203.
Estrada-Peña, A., Ayllón, N., and de la Fuente, J. (2012). Impact on climate trends on tick-borne pathogen transmission. Front. Physiol. *3*, 64.
Estrada-Peña, A., Gray, J.S., Kahl, O., Lane, R.S., and Nijhof, A.M. (2013). Research on the ecology of ticks and tick-borne pathogens--methodological principles and caveats. Front. Cell. Infect. Microbiol. *3*, 29.
Evangelista, J., Cruz, C., Guevara, C., Astete, H., Carey, C., Kochel, T.J., Morrison, A.C., Williams, M., Halsey, E.S., and Forshey, B.M. (2013). Characterization of a novel flavivirus isolated from *Culex* (*Melanoconion*) *ocossa* mosquitoes from Iquitos, Peru. J. Gen. Virol. *94*, 1266–1272.
Fall, G., Diallo, M., Loucoubar, C., Faye, O., and Sall, A.A. (2014). Vector competence of *Culex neavei* and *Culex quinquefasciatus* (Diptera:Culicidae) from Senegal for lineages 1, 2, Koutango and a putative new lineage of West Nile virus. Am. J. Trop. Med. Hyg. *90*, 747–754.
Farajollahi, A., Crans, W.J., Nickerson, D., Bryant, P., Wolf, B., Glaser, A., and Andreadis, T.G. (2005a). Detection of West Nile virus RNA from the louse fly *Icosta americana* (Diptera: Hippoboscidae). J. Am Mosq. Control Assoc. *21*, 474–476.
Farajollahi, A., Crans, W.J., Bryant, P., Wolf, B., Burkhalter, K.L., Godsey, M.S., Aspen, S.E., and Nasci, R.S. (2005b). Detection of West Nile viral RNA from an overwintering pool of *Culex pipiens pipiens* (Diptera: Culicidae) in New Jersey, 2003. J. Med. Entomol. *42*, 490–494.
Faye, O., Freire, C.C.M., Iamarino, A., Faye, O., de Oliveira, J.V.C., Diallo, M., Zanotto, P.M.A., and Sall, A.A. (2014). Molecular evolution of Zika virus during its emergence in the 20th Century. PLoS Negl. Trop. Dis. *8*, e2636.
Fischer, D., Thomas, S.M., Suk, J.E., Sudre, B., Hess, A., Tjaden, N.B., Beierkuhnlein, C., and Semenza, J.C. (2013). Climate change effects on Chikungunya transmission in Europe: geospatial analysis of vector's climatic suitability and virus' temperature requirements. Int. J. Health Geogr. *12*, 51.
Forbes, J.A. (1978). Murray Valley encephalitis 1974. also The epidemic variance since 1914 and predisposing rainfall patterns. (Australasian Medical Publishing Co Ltd, Sydney, Australia).
Forrester, N.L., Guerbois, M., Seymour, R.L., Spratt, H., and Weaver, S.C. (2012). Vector-borne transmission imposes a severe bottleneck on an RNA virus population. PLoS Pathog. *8*, e1002897.
Foy, B.D., Kobylinski, K.C., Chilson Foy, J.L., Blitvich, B.J., Travassos da Rosa, A., Haddow, A.D., Lanciotti, R.S., and Tesh, R.B. (2011). Probable non-vector-borne transmission of Zika virus, Colorado, USA. Emerg. Infect. Dis. *17*, 880–882.

Frost, M.J., Zhang, J., Edmonds, J.H., Prow, N.A., Gu, X., Davis, R., Hornitzky, C., Arzey, K.E., Finlaison, D., Hick, P., *et al.* (2012). Characterization of virulent West Nile virus Kunjin strain, Australia, 2011. Emerg. Infect. Dis. *18*, 792–800.

Gard, G.P., Weir, R.P., and Walsh, S.J. (1988). Arboviruses recovered from sentinel cattle using several virus isolation methods. Vet. Microbiol. *18*, 119–125.

Garigliany, M.M., Bayrou, C., Kleijnen, D., Cassart, D., Jolly, S., Linden, A., and Desmecht, D. (2012). Schmallenberg virus: a new Shamonda/Sathuperi-like virus on the rise in Europe. Antiviral Res. *95*, 82–87.

Gerry, A.C., and Mullens, B.A. (2000). Seasonal abundance and survivorship of *Culicoides sonorensis* (Diptera: Ceratopogonidae) at a southern California dairy, with reference to potential bluetongue virus transmission and persistence. J. Med. Entomol. *37*, 675–688.

Gibney, K.B., Lanciotti, R.S., Sejvar, J.J., Nugent, C.T., Linnen, J.M., Delorey, M.J., Lehman, J.A., Boswell, E.N., Staples, J.E., and Fischer, M. (2011). West Nile virus RNA not detected in urine of 40 people tested 6 years after acute West Nile virus disease. J. Infect. Dis. *203*, 344–347.

Goller, K.V., Höper, D., Schirrmeier, H., Mettenleiter, T.C., and Beer, M. (2012). Schmallenberg virus as a possible ancestor of Shamonda virus. Emerg. Infect. Dis. *18*, 1644–1646.

Gould, E.A., and Higgs, S. (2009). Impact of climate change and other factors on emerging arbovirus diseases. Trans. R. Soc. Trop. Med. Hyg. *103*, 109–121.

Gould, E.A., and Solomon, T. (2008). Pathogenic flaviviruses. Lancet *371*, 500–509.

Grancz, A.Y., Barker, I.K., Lindsay, R., Dibernardo, A., McKeever, K., and Hunter, B. (2004). West Nile virus outbreak in North American owls, Ontario, 2002. Emerg. Infect. Dis. *10*, 2135–2142.

Grard, G., Fair, J.N., Lee, D., Slikas, E., Steffen, I., Muyembe, J.J., Sittler, T., Veeraraghavan, N., Ruby, J.G., Wang, C., *et al.* (2012). A novel rhabdovirus associated with acute haemorrhagic fever in central Africa. PLoS Pathog. *8*, e1002924.

Gratz, N.G. (1999). Emerging and resurging vector-borne diseases. Annu. Rev. Entomol. *44*, 51–75.

Gratz, N.G. (2004). Critical review of the vector status of *Aedes albopictus*. Med. Vet. Entomol. *18*, 215–227.

Gresíková, M., Sekeyová, M., Stúpalová, S., and Necas, S. (1975). Sheep milk-borne epidemic of tick-borne encephalitis in Slovakia. Intervirology *5*, 57–61.

Gubala, A., Davis, S., Weir, R., Melville, L., Cowled, C., and Boyle, D. (2011). Tibrogargan and Coastal Plains rhabdoviruses: genomic characterization, evolution of novel genes and seroprevalence in Australian livestock. J. Gen. Virol. *92*, 2160–2170.

Gubler, D.J. (1997). Dengue and dengue hemorrhagic fever: its history and resurgence as a global public health problem. In Dengue and Dengue Hemorrhagic Fever, D.J. Gubler and G. Kono, eds. (CAB International, London), pp. 1–22.

Gubler, D.J. (1998). Resurgent vector-borne diseases as a global health problem. Emerg. Infect. Dis. *4*, 442–450.

Gubler, D.J. (2001). Human arbovirus infections worldwide. Ann. N. Y. Acad. Sci. *951*, 13–24.

Gubler, D.J. (2007). The continuing spread of West Nile virus in the western hemisphere. Clin. Infect. Dis. *45*, 1039–1046.

Gunes, T., Engin, A., Poyraz, O., Elaldi, N., Kaya, S., Dokmetas, I., Bakir, M., and Cinar, Z. (2009). Crimean–Congo hemorrhagic fever virus in high-risk population, Turkey. Emerg. Infect. Dis. *15*, 461–464.

Gunes, T., Poyraz, O., and Vatansever, Z. (2011). Crimean–Congo hemorrhagic fever virus in ticks collected from humans, livestock, and picnic sites in the hyperendemic region of Turkey. Vector Borne Zoonotic Dis. *11*, 1411–1416.

Haddow, A.D., Schuh, A.J., Yasuda, C.Y., Kasper, M.R., Heang, V., Huy, R., Guzman, H., Tesh, R.B., and Weaver, S.C. (2012). Genetic characterisation of Zika virus strains: geographic expansion of the Asian lineage. PLoS Negl. Trop. Dis. *6*, e1477.

Hall, R.A., Broom, A.K., Smith, D.W., and Mackenzie, J.S. (2002). The ecology and epidemiology of Kunjin virus. Curr. Top. Microbiol. Immunol. *267*, 253–269.

Hall-Mendelin, S., Ritchie, S.A., Johansen, C.A., Zborowski, P., Cortis, G., Dandridge, S., Hall, R.A., van den Hurk, A.F. (2010). Exploiting mosquito sugar feeding to detect mosquito-borne pathogens. Proc. Natl. Acad. Sci. U.S.A. *107*, 11255–11259.

Hall-Mendelin, S., Allcock, R., Kresoje, N., van den Hurk, A.F., and Warrilow, D. (2013). Detection of arboviruses and other micro-organisms in experimentally infected mosquitoes using massively parallel sequencing. PLoS One *8*, e58026.

Hammon, W.M., Tigertt, W.D., Sather, G.E., Berge, T.O., and Meiklejohn, G. (1958). Epidemiologic studies of concurrent virgin epidemics of Japanese B encephalitis and of mumps on Guam, 1947–1948, with subsequent observations including dengue, through 1957. Am. J. Trop. Med. Hyg. *7*, 441–467.

Hang, J., Forshey, B.M., Kochel, T.J., Li, T., Solórzano, V.F., Halsey, E.S., and Kuschner, R.A. (2012). Random amplification and pyrosequencing for identification of novel viral genome sequences. J. Biomol. Tech. *23*, 4–10.

Hanna, J.N., Ritchie, S.A., Phillips, D.A., Shield, J., Bailey, M.C., Mackenzie, J.S., Poidinger, M., McCall, B.J., and Mills, P.J. (1996). An outbreak of Japanese encephalitis in the Torres Strait, Australia, 1995. Med. J. Aust. *165*, 256–260.

Hanna, J.N., Ritchie, S.A., Phillips, D.A., Lee, J.M., Hills, S.L., van den Hurk, A.F., Pyke, A.T., Johansen, C.A., and Mackenzie, J.S. (1999). Japanese encephalitis in north Queensland, Australia, 1998. Med. J. Aust. *170*, 533–536.

Hasle, G. (2013). Transport of ixodid ticks and tick-borne pathogens by migratory birds. Front. Cell. Infect. Microbiol. *3*, 48.

Hawkes, R.A., Boughton, C.R., Naim, H.M., Wild, J., and Chapman, B. (1985). Arbovirus infections of humans in New South Wales. Seroepidemiology of the flavivirus group of togaviruses. Med. J. Aust. *143*, 555–561.

Hawley, W.A., Reiter, P., Copeland, R.S., Pumpuni, C.B., and Craig, G.B. Jr. (1987). *Aedes albopictus* in North America: probable introduction in used tires from northern Asia. Science *236*, 1114–1115.

Hayes, C.G. (1989). West Nile fever. In The Arboviruses: Epidemiology and Ecology, T.P Monath, ed. (CRC Press Inc., Boca Raton, Florida), pp. 60–88.

Hayman, D.T., Bowen, R.A., Cryan, P.M., McCracken, G.F., O'Shea, T.J., Peel, A.J., Gilbert, A., Webb, C.T., and Wood, J.L. (2013). Ecology of zoonotic infectious diseases in bats: current knowledge and future directions. Zoonoses Publ. Health *60*, 2–21.

Hobson-Peters, J., Yam, A.W., Lu, J.W., Setoh, Y.X., May, F.J., Kurucz, N., Walsh, S., Prow, N.A., Davis, S.S., Weir, R., *et al.* (2013). A new insect-specific flavivirus from northern Australia suppresses replication of West Nile virus and Murray Valley encephalitis virus in co-infected mosquito cells. PLoS One *8*, e56534.

Hodneland, K., Bratland, A., Christie, K.E., Endresen, C., and Nylund, A. (2005). New subtype of salmonid alphavirus (SAV), Togaviridae, from Atlantic salmon Salmo salar and rainbow trout *Oncorhynchus mykiss* in Norway. Dis. Aquat. Organ. *66*, 113–120.

Hoffmann, B., Scheuch, M., Höper, D., Jungblut, R., Holsteg, M., Schirrmeier, H., Eschbaumer, M., Goller, K.V., Wernike, K., Fischer, M., *et al.* (2012). Novel orthobunyavirus in Cattle, Europe, 2011. Emerg. Infect. Dis. *18*, 469–472.

Höfle, U., Gamino, V., de Mera, I.G., Mangold, A.J., Ortíz, J.A., and de la Fuente, J. (2013). Usutu virus in migratory song thrushes, Spain. Emerg. Infect. Dis. *19*, 1173–1175.

Holmes, E.C. (2003). Patterns of intra- and interhost nonsynonymous variation reveal strong purifying selection in dengue virus. J. Virol. *77*, 11296–11298.

Holmes, E.C., Worobey, M., and Rambaut, A. (1999). Phylogenetic evidence for recombination in dengue virus. Mol. Biol. Evol. *16*, 405–409.

Holzmann, H., Aberle, S.W., Stiasny, K., Werner, P., Mischak, A., Zainer, B., Netzer, M., Koppi, S., Bechter, E., and Heinz, F.X. (2009). Tick-borne encephalitis from eating goat cheese in a mountain region of Austria. Emerg. Infect. Dis. *15*, 1671–1673.

Hoshino, K., Isawa, H., Tsuda, Y., Yano, K., Sasaki, T., Yuda, M., Takasaki, T., Kobayashi, M., and Sawabe, K. (2007). Genetic characterization of a new insect flavivirus isolated from *Culex pipiens* mosquito in Japan. Virology *359*, 405–414.

Huang, J.H., Lin, T.H., Teng, H.J., Su, C.L., Tsai, K.H., Lu, L.C., Yang, C.F., Chang, S.F., Liao, T.L., Yu, S.K., *et al.* (2010). Molecular epidemiology of Japanese encephalitis virus, Taiwan. Emerg. Infect. Dis. *16*, 876–878.

Huang, X., Han, K., Zhao, D., Liu, Y., Zhang, J., Niu, H., Zhang, K., Zhu, J., Wu, D., Gao, L., and Li, Y. (2013). Identification and molecular characterization of a novel flavivirus isolated from geese in China. Res. Vet. Sci. *94*, 774–780.

Hubálek, Z. (2004). An annotated checklist of pathogenic microorganisms associated with migratory birds. J. Wildl. Dis. *40*, 639–659.

Hudson, P.J., Norman, R., Laurenson, M.K., Newborn, D., Gaunt, M., Jones, L., Reid, H., Gould, E., Bowers, R., and Dobson, A. (1995). Persistence and transmission of tick-borne viruses: *Ixodes ricinus* and louping-ill virus in red grouse populations. Parasitology *111*, S49-S58.

Huhtamo, E., Putkuri, N., Kurkela, S., Manni, T., Vaheri, A., Vapalahti, O., and Uzcátegui, N.Y. (2009). Characterization of a novel flavivirus from mosquitoes in northern Europe that is related to mosquito-borne flaviviruses of the tropics. J. Virol. *83*, 9532–9540.

van den Hurk, A.F., Johansen, C.A., Zborowski, P., Phillips, D.A., Pyke, A.T., Mackenzie, J.S., and Ritchie, S.A. (2001). Flaviviruses isolated from mosquitoes collected during the first recorded outbreak of Japanese encephalitis virus on Cape York Peninsula, Australia. Am. J. Trop. Med. Hyg. *64*, 125–130.

van den Hurk, A.F., Montgomery, B.L., Northill, J.A., Smith, I.L., Zborowski, P., Ritchie, S.A., Mackenzie, J.S., and Smith, G.A. (2006). Short Report: the first isolation of Japanese encephalitis virus from mosquitoes collected from mainland Australia. Am. J. Trop. Med. Hyg. *75*, 21–25.

van den Hurk, A.F., Johnson, P.H., Hall-Mendelin, S., Northill, J.A., Simmons, R.J., Jansen, C.C., Frances, S.P., Smith, G.A., and Ritchie, S.A. (2007). Expectoration of Flaviviruses during sugar feeding by mosquitoes (Diptera: Culicidae). J. Med. Entomol. *44*, 845–850.

van den Hurk, A.F., Ritchie, S.A., and Mackenzie, J.S. (2009). Ecology and geographical expansion of Japanese encephalitis virus. Annu. Rev. Entomol. *54*, 17–35.

van den Hurk, A.F., Hall-Mendelin, S., Johansen, C.A., Warrilow, D., and Ritchie, S.A. (2012). Evolution of mosquito-based arbovirus surveillance systems in Australia. J. Biomed. Biotechnol. *2012*, 325659.

van den Hurk, A.F., Hall-Mendelin, S., Webb, C.E., Tan, C.S., Frentiu, F.D., Prow, N.A., and Hall, R.A. (2014a). Role of enhanced vector transmission of a new West Nile virus strain in an outbreak of equine disease in Australia in 2011. Parasit. Vectors *12*, 586.

van den Hurk, A.F., Hall-Mendelin, S., Townsend, M., Kurucz, N., Edwards, J., Ehlers, G., Rodwell, C., Moore, F.A., McMahon, J.L., Northill, J.A., *et al.* (2014b). Applications of a sugar-based surveillance system to track arboviruses in wild mosquito populations. Vector Borne Zoonotic Dis. *14*, 66–73.

Innis, B.L. (1995). Japanese encephalitis. In Exotic Viral Infections, J.S. Porterfield, ed. (Chapman and Hall, London), pp. 147–174.

Institute of Medicine (1992). Emerging infections. In Microbial Threats to the United States. J. Lederberg, R.E. Shope, S.C. Oaks, eds. (National Academy Press, Washington DC).

Iwamoto, M., Jernigan, D.B., Guasch, A., Trepka, M.J., Blackmore, C.G., Hellinger, W.C., Pham, S.M., Zaki, S., Lanciotti, R.S., Lance-Parker, S.E., *et al.* (2003). Transmission of West Nile virus from an organ donor to four transplant recipients. N. Engl. J. Med. *348*, 2196–22203.

Jaijakul, S., Arias, C.A., Hossain, M., Arduino, R.C., Wootton, S.H., and Hasbun, R. (2012). Toscana meningoencephalitis: a comparison to other viral central nervous system infections. J. Clin. Virol. *55*, 204–208.

Jerzak, G., Bernard, K.A., Kramer, L.D., and Ebel, G.D. (2005). Genetic variation in West Nile virus from naturally infected mosquitoes and birds suggests quasispecies structure and strong purifying selection. J. Gen. Virol. *86*, 2175–2183.

Jerzak, G.V., Bernard, K., Kramer, L.D., Shi, P.Y., and Ebel, G.D. (2007). The West Nile virus mutant spectrum is host-dependent and a determinant of mortality in mice. Virology *360*, 469–476.

Jerzak, G.V., Brown, I., Shi, P.Y., Kramer, L.D., and Ebel, G.D. (2008). Genetic diversity and purifying selection in West Nile virus populations are maintained during host switching. Virology *374*, 256–260.

Jia, X.Y., Briese, T., Jordan, I., Rambaut, A., Chi, H.C., Mackenzie, J.S., Hall, R.A., Scherret, J., and Lipkin, W.I. (1999). Genetic analysis of West Nile New York 1999 encephalitis virus. Lancet *354*, 1971–1972.

Jourdain, E., Schuffenecker, I., Korimbocus, J., Reynard, S., Murri, S., Kayser, Y., Gauthier-Clerc, M., Sabatier, P., and Zeller, H.G. (2007). West Nile virus in wild resident birds, Southern France, 2004. Vector Borne Zoonotic Dis. *7*, 448–452.

Junglen, S., Kopp, A., Kurth, A., Pauli, G., Ellerbrok, H., and Leendertz, F.H. (2009). A new flavivirus and a new vector: characterization of a novel flavivirus isolated from uranotaenia mosquitoes. J. Virol. *83*, 4462–4468.

Karti, S.S., Odabasi, Z., Korten, V., Yilmaz, M., Sonmez, M., Caylan, R., Akdogan, E., Eren, N., Koksal, I., Ovali, E., *et al.* (2004). Crimean–Congo hemorrhagic fever in Turkey. Emerg. Infect. Dis. *10*, 1379–1384.

Kedmi, M., Herziger, Y., Galon, N., Cohen, R.M., Perel, M., Batten, C., Braverman, Y., Gottlieb, Y., Shpigel, N., and Klement, E. (2010). The association of winds with the spread of EHDV in dairy cattle in Israel during an outbreak in 2006. Prev. Vet. Med. *96*, 152–160.

Kent, R.J., Crabtree, M.B., and Miller, B.R. (2010). Transmission of West Nile virus by *Culex quinquefasciatus* Say infected with *Culex* Flavivirus Izabal. PLoS Negl. Trop. Dis. *4*, e671.

Kenney, J.L., and Brault, A.C. (2014). The role of environmental, virological and vector interactions in dictating biological transmission of arthropod-borne viruses by mosquitoes. Adv. Virus Res. *89*, 39–83.

Kenney, J.L., Adams, A.P., Gorchakov, R., Leal, G., and Weaver, S.C. (2012). Genetic and anatomic determinants of enzootic Venezuelan equine encephalitis virus infection of *Culex* (Malanoconion) *taeniopus*. PLoS Negl. Trop. Dis. *6*, e1606.

Kilpatrick, A.M. (2011). Gobalization, land use, and the invasion of West Nile virus. Science *334*, 323–327.

Kilpatrick, A.M., Daszak, P., Jones, M.J., Marra, P.P., and Kramer, L.D. (2006). Host heterogeneity dominates West Nile virus transmission. Proc. Biol. Sci. *273*, 2327–2333.

Kim, K.H., Yi, J., Kim, G., Choi, S.J., Jun, K.I., Kim, N.H., Choe, P.G., Kim, N.J., Lee, J.K., and Oh, M.D. (2013). Severe fever with thrombocytopenia syndrome, South Korea, 2012. Emerg. Infect. Dis. *19*, 1892–1894.

Koenig, D.L., Marcus, L., Scott, T.W., and Dickinson, L. (2007). West Nile virus and California breeding bird declines. Eco. Health *4*, 18–24.

Kohl, I., Kozuch, O., Elecková, E., Labuda, M., and Zaludko, J. (1996). Family outbreak of alimentary tick-borne encephalitis in Slovakia associated with a natural focus of infection. Eur. J. Epidemiol. *12*, 373–375.

Kokernot, R.H., Calisher, C.H., Stannard, L.J., and Hayes, J. (1969). Arbovirus studies in the Ohio-Mississippi Basin, 1964–1967. VII. Lone Star virus, a hitherto unknown agent isolated from the tick *Amblyomma americanum* (Linn). Am. J. Trop. Med. Hyg. *18*, 789–795.

Kolodziejek, J., Pachler, K., Bin, H., Mendelson, E., Shulman, L., Orshan, L., and Nowotny, N. (2013). Barkedji virus, a novel mosquito-borne flavivirus identified in *Culex perexiguus* mosquitoes, Israel, 2011. J. Gen. Virol. *94*, 2449–2457.

Komar, N. (2000). West Nile viral encephalitis. Rev. Sci. Tech. *19*, 166–176.

Komar, N. (2001). West Nile virus surveillance using sentinel birds. Ann. N. Y. Acad. Sci. *951*, 58–73.

Komar, N. (2003). West Nile virus: epidemiology and ecology in North America. Adv. Virus Res. *61*, 185–234.

Komar, N., and Spielman, A. (1994). Emergence of eastern encephalitis in Massachusetts. Ann. N. Y. Acad. Sci. *740*, 157–168.

Kono, Y., Yusnita, Y., Mohd Ali, A.R., Maizan, M., Sharifah, S.H., Fauzia, O., Kubo, M., and Aziz, A.J. (2002). Characterization and identification of Oya virus, a Simbu serogroup virus of the genus Bunyavirus, isolated from a pig suspected of Nipah virus infection. Arch. Virol. *147*, 1623–1630.

Kovalev, S.Y., Kokorev, V.S., and Belyaeva, I.V. (2010). Distribution of Far-Eastern tick-borne encephalitis virus subtype strains in the former Soviet Union. J. Gen. Virol. *91*, 2941–2946.

Kramer, L.D., and Chandler, L.J. (2001). Phylogenetic analysis of the envelope gene of St. Louis encephalitis virus. Arch. Virol. *146*, 2341–2355.

Kramer, L.D., Styer, L.M., and Ebel, G.D. (2008). A global perspective on the epidemiology of West Nile virus. Annu. Rev. Entomol. *53*, 61–81.

Kuhn, S., Twele-Montecinos, L., MacDonald, J., Webster, P., and Law, B. (2011). Case report: probable transmission of vaccine strain of yellow fever virus to an infant via breast milk. CMAJ *183(4)*, E243–E245.

Kuno, G. (2001a). Transmission of arboviruses without involvement of arthropod vectors. Acta Virol. *45*, 139–150.

Kuno, G. (2001b). Persistence of arboviruses and antiviral antibodies in vertebrate hosts: its occurrence and impacts. Rev. Med. Virol. *11*, 165–190.

Kuno, G., and Chang, G.J. (2005). Biological transmission of arboviruses: reexamination of and new insights into components, mechanisms, and unique traits as well as evolutionary trends. Clin. Microbiol. Rev. *18*, 608–637.

Kurosu, T. (2011). Quasispecies of dengue virus. Trop. Med. Health *39*(Suppl.), 29–36.

Laboratory safety for arboviruses and certain other viruses of vertebrates. The Subcommittee on Arbovirus Laboratory Safety of the American Committee on Arthropod-Borne Viruses. (1980). Am. J. Trop. Med. Hyg. *29*, 1359–1381.

Labuda, M., and Nuttall, P.A. (2004). Tick-borne viruses. Parasitology *129*, S221–S245.

Labuda, M., Danielova, V., Jones, L.D., and Nuttall, P.A. (1993a). Amplification of tick-borne encephalitis virus infection during co-feeding of ticks. Med. Vet. Entomol. *7*, 339–342.

Labuda, M., Jones, L.D., Williams, T., Danielova, V., and Nuttall, P.A. (1993b). Efficient transmission of tick-borne encephalitis virus between cofeeding ticks. J. Med. Entomol. *30*, 295–299.

Laird, M., Calder, L., Thornton, R.C., Syme, R., Holder, P.W., and Mogi, M. (1994). Japanese *Aedes albopictus* among four mosquito species reaching New Zealand in used tires. J. Am. Mosq. Control Assoc. *10*, 14–23.

La Linn, M., Gardner, J., Warrilow, D., Darnell, G.A., McMahon, C.R., Field, I., Hyatt, A.D., Slade, R.W., and Suhrbier, A. (2001). Arbovirus of marine mammals: a new alphavirus isolated from the elephant seal louse, *Lepidophthirus macrorhini*. J. Virol. *75*, 4103–4109.

Lambin, E.F., Tran, A., Vanwambeke, S.O., and Soti, V. (2010). Pathogenic landscapes: interactions between land, people, disease vectors, and their animal hosts. Int. J. Health Geogr. *9*, 54.

Lanciotti, R.S., Lewis, J.G., Gubler, D.J., and Trent, D.W. (1994). Molecular evolution and epidemiology of dengue-3 viruses. J. Gen. Virol. *75*, 65–75.

Lanciotti, R.S., Roehrig, J.T., Deubel, V., Smith, J., Parker, M., Steele, K., Crise, B., Volpe, K.E., Crabtree, M.B., Scherret, J.H., *et al.* (1999). Origin of the West Nile virus responsible for an outbreak of encephalitis in the northeastern United States. Science *286*, 2333–2337.

Lanciotti, R.S., Kosoy, O.L., Laven, J.J., Velez, J.O., Lambert, A.J., and Johnson, A.J. (2008). Genetic and serologic properties of Zika virus associated with an epidemic, Yap State, Micronesia, 2007. Emerg. Infect. Dis. *14*, 1232–1239.

Lanciotti, R.S., Kosoy, O.I., Bosco-Lauth, A.M., Pohl, J., Stuchlik, O., Reed, M., and Lambert, A.J. (2013). Isolation of a novel orthobunyavirus (Brazoran virus) with a 1.7kb S segment that encodes a unique nucleocapsid protein possessing two putative functional domains. Virology *444*, 55–63.

La Ruche, G., Souarès, Y., Armengaud, A., Peloux-Petiot, F., Delaunay, P., Desprès, P., Lenglet, A., Jourdain, F., Leparc-Goffart, I., Charlet, F., *et al.* (2010). First two autochthonous dengue virus infections in metropolitan France, September 2010. Eur. Surveill. *15(39)*, 19676.

Lauring, A.S., and Andino, R. (2010). Quasispecies theory and the behavior of RNA viruses. PLoS Pathog. *6*, e1001005.

LeDeau, S.L., Kilpatrick, A.M., and Marra, P.P. (2007). West Nile virus emergence and large-scale declines of North American bird populations. Nature *447*, 710–713.

LeDuc, J.W., and Pinheiro, F.P. (1989). Oropouche fever. In The Arboviruses: Epidemiology and Ecology, Vol. IV, T.P. Monath, ed. (CRC Press Inc., Boca Raton, Florida). pp. 2–14.

Le Flohic, G., Porphyre, V., Barbazan, P., and Gonzalez, J.P. (2013). Review of climate, landscape, and viral genetics as drivers of the Japanese encephalitis virus ecology. PLoS Negl. Trop. Dis. *7*, e2208.

Lewis, J.A., Chang, G.J., Lanciotti, R.S., Kinney, R.M., Mayer, L.W., and Trent, D.W. (1993). Phylogenetic relationships of dengue-2 viruses. Virology *197*, 216–224.

Li, M.H., Fu, S.H., Chen, W.X., Wang, H.Y., Guo, Y.H., Liu, Q.Y., Li, Y.X., Luo, H.M., Da, W., Duo, Ji, D.Z., *et al.* (2011). Genotype v Japanese encephalitis virus is emerging. PLoS Negl. Trop. Dis. *5*, e1231.

Li, M.H., Fu, S.H., Chen, W.X., Cao, Y.X., and Liang, G.D. (2014). Molecular characterization of full-length genome of Japanese encephalitis virus genotype V isolated from Tibet, China. Biomed. Environ. Sci. *27*, 231–239.

Lin, S.R., Hsieh, S.C., Yueh, Y.Y., Lin, T.H., Chao, D.Y., Chen, W.J., King, C.C., and Wang, W.K. (2004). Study of sequence variation of dengue type 3 virus in naturally infected mosquitoes and human hosts: implications for transmission and evolution. J. Virol. *78*, 12717–12721.

Linke, S., Niedrig, M., Kaiser, A., Ellerbrok, H., Müller, K., Müller, T., Conraths, F.J., Mühle, R.U., Schmidt, D., Köppen, U., Bairlein, F., Berthold, P., and Pauli, G. (2007). Serologic evidence of West Nile virus infections in wild birds captured in Germany. Am. J. Trop. Med. Hyg. *77*, 358–364.

Linthicum, K.J., Davies, F.G., Kairo, A., and Bailey, C.L. (1985). Rift Valley fever virus (family Bunyaviridae, genus Phlebovirus) isolations from diptera collected during an interepizootic period in Kenya. J. Hyg. (Camb.) *95*, 197–209.

Linthicum, K.J., Anyamba, A., Tucker, C.J., Kelley, P.W., Myers, M.F., and Peters, C.J. (1999). Climate and satellite indicators to forecast Rift Valley fever epidemics in Kenya. Science *285*, 397–400.

Liu, S., Vijayendran, D., and Bonning, B.C. (2011). Next generation sequencing technologies for insect virus discovery. Viruses *3*, 1849–1869.

Liu, S., Li, X., Chen, Z., Chen, Y., Zhang, Q., Liao, Y., Zhou, J., Ke, X., Ma, L., Xiao, J., *et al.* (2013). Comparison of genomic and amino acid sequences of eight Japanese encephalitis virus isolates from bats. Arch. Virol. *158*, 2543–2552.

Lo Presti, A., Ciccozzi, M., Cella, E., Lai, A., Simonetti, F.R., Galli, M., Zehender, G., and Rezza, G. (2012). Origin, evolution, and phylogeography of recent epidemic CHIKV strains. Infect. Genet. Evol. *12*, 392–398.

Logan, T.M., Linthicum, K.J., and Ksiazek, T.G. (1992). Isolation of Rift Valley fever virus from mosquitoes collected during an outbreak in domestic animals in Kenya. J. Med. Entomol. *28*, 293–295.

Longdon, B., Hadfield, J.D., Webster, C.L., Obbard, D.J., and Jiggins, F.M. (2011). Host phylogeny determines viral persistence and replication in novel hosts. PLoS Pathog. *7*, e1002260.

Lord, C.C. (2010). The effect of multiple vectors on arbovirus transmission. Isr. J. Ecol. Evol. *56*, 371–392.

Lounibos, L.P. (2002). Invasions by insect vectors of human disease. Annu. Rev. Entomol. *47*, 233–266.

Lounibos, L.P. (2011). Mosquitoes. In Encyclopedia of Biological Invasions, D. Simberloff and M. Rejmánek, eds. (University of California Press, Berkeley, California), pp. 462–466.

Lu, Z., Liu, H., Fu, S., Lu, X., Dong, Q., Zhang, S., Tong, S., Li, M., Li, W., Tang, Q., and Liang, G. (2011). Liao ning virus in China. Virol. J. *8*, 282.

Ludwig, G.V., Calle, P.P., Mangiafico, J.A., Raphael, B.L., Danner, D.K., Hile, J.A., Clippinger, T.L., Smith, J.F., Cook, R.A., and McNamara, T. (2002). An outbreak of West Nile virus in a New York City captive wildlife population. Am. J. Trop. Med. Hyg. *67*, 67–75.

Luis, A.D., Hayman, D.T., O'Shea, T.J., Cryan, P.M., Gilbert, A.T., Pulliam, J.R., Mills, J.N., Timonin, M.E., Wills, C.K., Cunningham, A.A., *et al.* (2013). A comparison of bats and rodents as reservoirs of zoonotic viruses: are bats special? Proc. Biol. Sci. *280*, 201222753.

Lundström, J.O., and Pfeffer, M. (2010). Phylogeographic structure and evolutionary history of Sindbis virus. Vector Borne Zoonotic Dis. *10*, 889–907.

Lundström, J.O., Lindström, K.M., Olsen, B., Dufva, R., and Krakower, D.S. (2001). Prevalence of sindbis virus neutralizing antibodies among Swedish passerines indicates that thrushes are the main amplifying hosts. J. Med. Entomol. *38*, 289–297.

Lv, X., Mohd Jaafar, F., Sun, X., Belhouchet, M., Fu, S., Zhang, S., Tong, S.X., Lv, Z., Mertens, P.P., Liang, G., and Attoui, H. (2012). Isolates of Liao ning virus from wild-caught mosquitoes in the Xinjiang province of China in 2005. PLoS One 7: e37732.

Lvov, D.K., Butenko, A.M., Gromashevsky, V.L., Kovtunov, A.I., Prilipov, A.G., Kinney, R., Aristova, V.A., Dzharkenov, A.F., Samokhvalov, E.I., Savage, H.M., *et al.* (2004). West Nile virus and other zoonotic viruses in Russia: examples of emerging-reemerging situations. Arch. Virol. Suppl. *18*, 85–96.

Lwande, O.W., Lutomiah, J., Obanda, V., Gakuya, F., Mutisya, J., Mulwa, F., Michuki, G., Chepkorir, E., Fischer, A., Venter, M., and Sang, R. (2013). Isolation of tick and mosquito-borne arboviruses from ticks sampled from livestock and wild animal hosts in Ijara District, Kenya. Vector Borne Zoonotic Dis. *13*, 637–642.

Mackenzie, J.S., and Broom, A.K. (1999). Old river irrigation area: the effect of dam construction and irrigation on the incidence of Murray Valley encephalitis virus. In Water Resources – Health, Environment and Development, BH Kay, ed. (Spon Press, London), pp. 108–122.

Mackenzie, J.S., and Williams, D.T. (2009). The zoonotic flaviviruses of southern, south-eastern and eastern Asia, and Australasia: the potential for emergent viruses. Zoonoses Publ. Health, *56*, 338–356.

Mackenzie, J.S., Lindsay, M.D., and Broom, A.K. (1993). Climate changes and vector-borne diseases: potential consequences for human health. In Health in the Greenhouse. The Medical and Environmental Health Effects of Global Climate Change, C.E. Ewan, E.A. Bryant, G.D. Calvert, J.A. Garrick, eds. (Australian Government Publishing Service, Canberra, Australia), pp. 229–234.

Mackenzie, J.S., Lindsay, M.D., and Daniels, P.W. (2000). The effect of climate on the incidence of vector-borne viral diseases: the potential value of seasonal forecasting. In Applications of Seasonal Climate Forecasting in Agriculture and Natural Ecosystems – The Australian Experience, G. Hammer, N. Nicholls, and C. Mitchell, eds. (Kluwer Academic Publishers, Dordrecht, The Netherlands), pp. 429–452.

Mackenzie, J.S., Johansen, C.A., Ritchie, S.A., van den Hurk, A.F., and Hall, R.A. (2002). Japanese encephalitis as an emerging virus: the emergence and spread of Japanese encephalitis virus in Australia. Curr. Top. Microbiol. Immunol. *267*, 49–73.

Mackenzie, J.S., Gubler, D.J., and Petersen, L.R. (2004). Emerging flaviviruses: the spread and resurgence of Japanese encephalitis, West Nile and dengue viruses. Nat. Med. *10(12 Suppl)*, S98-S109.

Mackenzie, J.S., Williams, D.T., and Smith, D.W. (2007). Japanese encephalitis virus: an example of a virus with a propensity to emerge and spread in new geographic areas. In Emerging Viruses in Human Populations, Perspective in Medical Virology 16, E. Tabor, ed. (Elsevier, Amsterdam, The Netherlands), pp. 201–268.

Mackenzie, J.S., Childs, J.E., Field, H.E., Wang, L.F., and Breed, A.C. (2008). The role of bats as reservoir hosts of emerging neurological viruses. In Neurotropic virus infections, C.S. Reiss, ed. (Cambridge University Press, Cambridge, UK), pp. 382–406.

Maclachlan, N.J. (2010). Global implications of the recent emergence of bluetongue virus in Europe. Vet. Clin. North Am. Food Anim. Pract. *26*, 163–171.

Maclachlan, N.J., and Guthrie, A.J. (2010). Re-emergence of bluetongue, African horse sickness, and other orbivirus diseases. Vet. Res. *41*, 35.

Madon, M.B., Mulla, M.S., Shaw, M.W., Kluh, S., and Hazelrigg, J.E. (2002). Introduction of *Aedes albopictus* (Skuse) in southern California and potential for its establishment. J. Vector Ecol. *27*, 149–154.

Major, L., Linn, M.L., Slade, R.W., Schroder, W.A., Hyatt, A.D., Gardner, J., Cowley, J., and Suhrbier, A. (2009). Ticks associated with Macquarie Island penguins carry arboviruses from four genera. PLoS One *4*, e4375.

Malkinson, M., and Banet, C. (2002). The role of birds in the ecology of West Nile virus in Europe and Africa. Curr. Top. Microbiol. Immunol. *267*, 309–322.

Malkinson, M., Banet, C., Weisman, Y., Pokamunski, S., King, R., Drouet, M.T., and Deubel, V. (2002). Introduction of West Nile virus in the Middle East by migrating white storks. Emerg. Infect. Dis. *8*, 392–397.

Marka, A., Diamantidis, A., Papa, A., Valiakos, G., Chaintoutis, S.C., Doukas, D., Tserkezou, P., Giannakopoulos, A., Papaspyropoulos, K., Patsoula, E., et al. (2013). West Nile virus state of the art report of the MALWEST project. Int. J. Environ. Res. Public Health *10*, 6534–6610.

Marklewitz, M., Handrick, S., Grasse, W., Kurth, A., Lukashev, A., Drosten, C., Ellerbrok, H., Leendertz, F.H., Pauli, G., and Junglen, S. (2011). Gouleako virus isolated from West African mosquitoes constitutes a proposed novel genus in the family Bunyaviridae. J. Virol. *85*, 9227–9234.

Marklewitz, M., Zirkel, F., Rwego, I.B., Heidemann, H., Trippner, P., Kurth, A., Kallies, R., Briese, T., Lipkin, W.I., Drosten, C., Gillespie, T.R., and Junglen, S. (2013). Discovery of a unique novel clade of mosquito-associated bunyaviruses. J. Virol. *87*, 12850–12865.

Maroli, M., Feliciangeli, M.D., Bichaud, L., Charrel, R.N., and Gradoni, L. (2013). Phlebotomine sandflies and the spreading of leishmaniasis and other diseases of public health concern. Med. Vet. Entomol. *27*, 123–147.

Marr, J.J. (1982). Merchants of death: the role of the slave trade in the transmission of disease from Africa to the Americas. Pharos Alpha Omega Alpha Honor Med. Soc. *45*, 31–35.

Marshall, I.D. (1988). Murray Valley and Kunjin encephalitis. In the Arboviruses: Epidemiology and Ecology, Vol III, T.P. Monath, ed. (CRC Press, Boca Raton, Florida), pp. 152–189.

Marston, D.A., McElhinney, L.M., Ellis, R.J., Horton, D.L., Wise, E.L., Leech, S.L., David, D., de Lamballerie, X., and Fooks, A.R. (2013). Next generation sequencing of viral RNA genomes. BMC Genomics *14*, 444.

Maruyama, S.R., Castro-Jorge, L.A., Ribeiro, J.M., Gardinassi, L.G., Garcia, G.R., Brandão, L.G., Rodrigues, A.R., Okada, M.I., Abrão, E.P., Ferreira, B.R., *et al.* (2014). Characterisation of divergent flavivirus NS3 and NS5 protein sequences detected in *Rhipicephalus microplus* ticks from Brazil. Mem. Inst. Oswaldo Cruz *109*, 38–50.

Matser, A., Hartemink, N., Heesterbeek, H., Galvani, A., and Davis, S. (2009). Elasticity analysis in epidemiology: an application to tick-borne infections. Ecol. Lett. *12*, 1298–1305.

May, F.J., Li, L., Zhang, S., Guzman, H., Beasley, D.W., Tesh, R.B., Higgs, S., Raj, P., Bueno, R. Jr., Randle, Y., Chandler, L., and Barrett, A.D. (2008). Genetic variation of St. Louis encephalitis virus. J. Gen. Virol. *89*, 1901–1910.

McLean, R.G., Ubico, S.R., Bourne, D., and Komar, N. (2002). West Nile virus in livestock and wildlife. Curr. Top. Microbiol. Immunol. *267*, 271–308.

McMullan, L.K., Folk, S.M., Kelly, A.J., MacNeil, A., Goldsmith, C.S., Metcalfe, M.G., Batten, B.C., Albariño, C.G., Zaki, S.R., Rollin, P.E., Nicholson, W.L., and Nichol, S.T. (2012). A new phlebovirus associated with severe febrile illness in Missouri. N. Engl. J. Med. *367*, 834–841.

McMullen, A.R., Albayrak, H., May, F.J., Davis, C.T., Beasley, D.W., and Barrett, A.D. (2013). Molecular evolution of lineage 2 West Nile virus. J. Gen.Virol. *94*, 18–25.

Medlock, J.M., Hansford, K.M., Schaffner, F., Versteirt, V., Hendrickx, G., Zeller, H., and Van Bortel, W. (2012). A review of the invasive mosquitoes in Europe: ecology, public health risks, and control options. Vector Borne Zoonotic Dis. *12*, 435–447.

Mehla, R., Kumar, S.R., Yadav, P., Barde, P.V., Yergolkar, P.N., Erickson, B.R., Carroll, S.A., Mishra, A.C., Nichol, S.T., and Mourya, D.T. (2009). Recent ancestry of Kyasanur Forest disease virus. Emerg. Infect. Dis. *15*, 1431–1437.

Meiring, T.L., Huismans, H., and van Staden, V. (2009). Genome segment reassortment identifies non-structural protein NS3 as a key protein in African horsesickness virus release and alteration of membrane permeability. Arch. Virol. *154*, 263–271.

Mellor, P.S. (2004). Infection of the vectors and bluetongue epidemiology in Europe. Vet. Ital. *40*, 167–174.

Mellor, P.S., Boorman, J., and Baylis, M. (2000). *Culicoides* biting midges: their role as arbovirus vectors. Annu. Rev. Entomol. *45*, 307–340.

Menghani, S., Chikhale, R., Raval, A., Wadibhasme, P., and Khedekar, P. (2012). Chandipura Virus: an emerging tropical pathogen. Acta Trop. *124*, 1–14.

Mertens, M., Schmidt, K., Ozkul, A., and Groschup, M.H. (2013). The impact of Crimean–Congo haemorrhagic fever virus on public health. Antiviral Res. *98*, 248–260.

Messer, W.B., Gubler, D.J., Harris, E., Sivananthan, K., De Silva, A.M. (2003). Emergence and global spread of a dengue serotype 3, subtype III virus. Emerg. Infect. Dis. *9*, 800–809.

Mikryukova, T.P., Moskvitina, N.S., Kononova, Y.V., Korobitsyn, I.G., Kartashov, M.Y., Tyuten Kov, O.Y., Protopopova, E.V., Romanenko, V.N., Chausov, E.V., Gashkov, S.I., *et al.* (2013). Surveillance of tick-borne encephalitis virus in wild birds and ticks in Tomsk city and its suburbs (Western Siberia). Ticks Tick Borne Dis. *5*, 145–151.

Mintiens, K., Méroc, E., Faes, C., Abrahantes, J.C., Hendrickx, G., Staubach, C., Gerbier, G., Elbers, A.R., Aerts, M., and De Clercq, K. (2008). Impact of human interventions on the spread of bluetongue virus serotype 8 during the 2006 epidemic in north-western Europe. Prev. Vet. Med. *87*, 145–161.

Mitchell, C.J., Savage, H.M., Smith, G.C., Flood, S.P., Castro, L.T., and Roppul, M. (1993). Japanese encephalitis on Saipan: a survey of suspected mosquito vectors. Am. J. Trop. Med. Hyg. *48*, 585–590.

Mohammed, M.A., Galbraith, S.E., Radford, A.D., Dove, W., Takasaki, T., Kurane, I., and Solomon, T. (2011). Molecular phylogenetic and evolutionary analyses of Muar strain of Japanese encephalitis virus reveal it is the missing fifth genotype. Infect. Genet. Evol. *11*, 855–862.

Monath, T.P., Wands, J.R., Hill, L.J., Brown, N.V., Marciniak, R.A., Wong, M.A., Gentry, M.K., Burke, D.S., Grant, J.A., and Trent, D.W. (1986). Geographic classification of dengue-2 virus strains by antigen signature analysis. Virology *154*, 313–324.

Monini, M., Falcone, E., Busani, L., Romi, R., and Ruggeri, F.M. (2010). West Nile virus: characteristics of an African virus adapting to the third Millennium world. Open Virol. J. *4*, 42–51.

Morris, C.D. (1988). Eastern equine encephalomyelitis. In The Arboviruses: Epidemiology and Ecology, Volume III, T.P. Monath, ed. (CRC Press Inc., Boca Raton, Florida), pp. 1–20.

Moudy, R.M., Meola, M.A., Morin, L.L., Ebel, G.D., and Kramer, L.D. (2007). A newly emergent genotype of West Nile virus is transmitted earlier and more efficiently by *Culex* mosquitoes. Am. J. Trop. Med. Hyg. *77*, 365–370.

Mourya, D.T., Yadav, P.D., Basu, A., Shete, A., Patil, D.Y., Zawar, D., Majumdar, T.D., Kokate, P., Sarkale, P., Raut, C.G., and Jadhav, S.M. (2014). Malsoor virus, a novel bat phlebovirus, is closely related to

severe fever with thrombocytopenia syndrome virus and heartland virus. J. Virol. *88*, 3605–3609.
Moutailler, S., Krida, G., Schaffner, F., Vazeille, M., and Failloux, A.B. (2008). Potential vectors of Rift Valley fever virus in the Mediterranean region. Vector Borne Zoonotic Dis. *8*, 749–754.
Movila, A., Alekseev, A.N., Dubinina, H.V., and Toderas, I. (2013). Detection of tick-borne pathogens in ticks from migratory birds in the Baltic region of Russia. Med. Vet. Entomol. *27*, 113–117.
Mumcuoglu, K.Y., Banet-Noach, C., Malkinson, M., Shalom, U., and Galun, R. (2005). Argasid ticks as possible vectors of West Nile virus in Israel. Vector Borne Zoonotic Dis. *5*, 65–71.
Murray, K., Walker, C., Herrington, E., Lewis, J.A., McCormick, J., Beasley, D.W., Tesh, R.B., and Fisher-Hoch, S. (2010). Persistent infection with West Nile virus years after initial infection. J. Infect. Dis. *201*, 2–4.
Musso, D., Nilles, E.J., and Cao-Lormeau, V.M. (2014). Rapid spreading of emerging Zika virus in the Pacific area. Clin. Microbiol. Infect. [Epub ahead of print].
Nabeshima, T., Loan, H.T., Inoue, S., Sumiyoshi, M., Haruta, Y., Nga, P.T., Huoung, V.T., del Carmen Parquet, M., Hasebe, F., and Morita, K. (2009). Evidence of frequent introductions of Japanese encephalitis virus from south-east and continental east Asia to Japan. J. Gen. Virol. *90*, 827–832.
Nabeth, P., Kane, Y., Abdalahi, M.O., Diallo, M., Ndiaye, K., Ba, K., Schneegans, F., Sall, A.A., and Mathiot, C. (2001). Rift Valley fever outbreak, Mauritania, 1998: seroepidemiologic, virologic, entomologic, and zoologic investigations. Emerg. Infect. Dis. *7*, 1052–1054.
Naderi, H., Sheybani, F., Bojdi, A., Khosravi, N., and Mostafavi, I. (2013). Fatal nosocomial spread of Crimean–Congo haemorrhagic fever with very short incubation period. Am. J. Trop. Med. Hyg. *88*, 469–471.
Nasar, F., Palacios, G., Gorchakov, R.V., Guzman, H., Da Rosa, A.P., Savji, N., Popov, V.L., Sherman, M.B., Lipkin, W.I., Tesh, R.B., and Weaver, S.C. (2012). Eilat virus, a unique alphavirus with host range restricted to insects by RNA replication. Proc. Natl. Acad. Sci. U.S.A. *109*, 14622–14627.
Nasci, R.S., Savage, H.M., White, D.J., Miller, J.R., Cropp, B.C., Godsey, M.S., Kerst, A.J., Bennett, P., Gottfried, K., and Lanciotti, R.S. (2001). West Nile virus in overwintering *Culex* mosquitoes, New York City, 2000. Emerg. Infect. Dis. *7*, 742–744.
Nasci, R.S., Lambert, A.J., and Savage, H.M. (2014). Novel bunyavirus in domestic and captive farmed animals, Minnesota, USA. Emerg. Infect. Dis. *20*, 336.
Navarro-Marí, J.M., Gómez-Camarasa, C., Pérez-Ruiz, M., Sanbonmatsu-Gámez, S., Pedrosa-Corral, I., and Jiménez-Valera, M. (2013). Clinic-epidemiologic study of human infection by Granada virus, a new phlebovirus within the sandfly fever Naples serocomplex. Am. J. Trop. Med. Hyg. *88*, 1003–1006.
Nelms, B.M., Fechter-Leggett, E., Carroll, B.D., Macedo, P., Kluh, S., and Reisen, W.K. (2013). Experimental and natural vertical transmission of West Nile virus by California *Culex* mosquitoes (Diptera: Culicidae). J. Med. Entomol. *50*, 371–378.
Nemes, Z., Kiss, G., Madarassi, E.P., Peterfi, Z., Ferenczi, E., Bakonyi, T., and Ternak, G. (2004). Nosocomial transmission of dengue. Emerg. Infect. Dis. *10*, 1880–1881.
Newman, C.M., Cerutti, F., Anderson, T.K., Hamer, G.L., Walker, E.D., Kitron, U.D., Ruiz, M.O., Brawn, J.D., and Goldberg, T.L. (2011). *Culex* flavivirus and West Nile virus mosquito coinfection and positive ecological association in Chicago, United States. Vector Borne Zoonotic Dis. *11*, 1099–1105.
Nga, P.T., del Carmen Parquet, M., Cuong, V.D., Ma, S.P., Hasebe, F., Inoue, S., Makino, Y., Takagi, M., Nam, V.S., and Morita, K. (2004). Shift in Japanese encephalitis virus (JEV) genotype circulating in northern Vietnam: implications for frequent introductions of JEV from Southeast Asia to East Asia. J. Gen. Virol. *85*, 1625–1631.
Nicholls, N. (1993). El ninõ-southern oscillation and vector-borne disease. Lancet *342*, 1284–1285.
Niklasson, B. (1988). Sindbis and Sindbis-like viruses. In The Arboviruses: Epidemiology and Ecology, Volume IV, T.P. Monath, ed. (CRC Press Inc., Boca Raton, Florida), pp. 168–176.
Nitatpattana, N., Dubot-Pérès, A., Gouilh, M.A., Souris, M., Barbazan, P., Yoksan, S., de Laballerie, X., and Gonzalez, J.O. (2008). Change in Japanese encephalitis virus distribution, Thailand. Emerg. Infect. Dis. *14*, 1762–1765.
Norberg, P., Roth, A., and Bergström, T. (2013). Genetic recombination of tick-borne flaviviruses among wild-type strains. Virology *440*, 105–116.
Nuttall, P.A., Jones, L.D., Labuda, M., and Kaufman, W.R. (1994). Adaptations of arboviruses to ticks. J. Med. Entomol. *31*, 1–9.
O'Brien, C.A., Hobson-Peters, J., Yam, A.W., Colmant, A.M., McLean, B.J., Prow, N.A., Watterson, D., Hall-Mendelin, S., Warrilow, D., Ng, M.L., *et al.* (2015). Viral RNA intermediates as targets for detection and discovery of novel and emerging mosquito-borne viruses. PLoS Negl. Trop. Dis. *9*, e0003629.
Olive, M.M., Goodman, S.M., and Reynes, J.M. (2012). The role of wild mammals in the maintenance of Rift Valley fever virus. J. Wildl. Dis. *48*, 241–266.
Osborne, J.C., Rupprecht, C.E., Olson, J.G., Ksiazek, T.G., Rollin, P.E., Niezgoda, M., Goldsmith, C.S., An, U.S., and Nichol, S.T. (2003). Isolation of Kaeng Khoi virus from dead *Chaerephon plicata* bats in Cambodia. J. Gen. Virol. *84*, 2685–2689.
Pandey, B.D., Rai, S.K., Morita, K., and Kurane, I. (2004). First case of Dengue virus infection in Nepal. Nepal Med. Coll. J. *6*, 157–159.
Papa, A., Velo, E., and Bino, S. (2011). A novel phlebovirus in Albanian sandflies. Clin. Microbiol. Infect. *17*, 585–587.
Pastula, D.M., Turabelidze, G., Yates, K.F., Jones, T.F., Lambert, A.J., Panella, A.J., Kosoy, O.I., Velez, J.O., Fisher, M., and Staples, E. (2014). Notes from the field: Heartland virus disease – United States, 2012–2013. MMWR Morb. Mortal. Wkly. Rep. *63*, 270–271.
Patterson, K.D. (1992). Yellow fever epidemics and mortality in the United States, 1693–1905. Soc. Sci. Med. *34*, 855–865.
Pattnaik, P. (2006). Kyasanur forest disease: an epidemiological view in India. Rev. Med. Virol. *16*, 151–165.
Paupy, C., Delatte, H., Bagny, L., Corbel, V., and Fontenille, D. (2009). *Aedes albopictus*, an arbovirus vector: from the darkness to the light. Microbes Infect. *11*, 1177–1185.
Pealer, L.N., Marfin, A.A., Petersen, L.R., Lanciotti, R.S., Page, P.L., Stramer, S.L., Stobierski, M.G., Signs, K., Newman, B., Kapoor, H., Goodman, J.L., and Chamberland, M.E. (2003). Transmission of West Nile virus through blood transfusion in the United States in 2002. N. Engl. J. Med. *349*, 1236–1245.
Pedrosa, P.B., and Cardoso, T.A. (2011). Viral infections in workers in hospital and research laboratory settings: a comparative review of infection modes and respective biosafety aspects. Int. J. Infect. Dis. *15*, e366–376.
Pepin, M., Bouloy, M., Bird, B.H., Kemp, A., and Paweska, J. (2010). Rift Valley fever virus (Bunyaviridae: Phlebovirus): an update on pathogenesis, molecular epidemiology, vectors, diagnostics and prevention. Vet. Res. *41*, 61.
Perez-Ramirez, G., Diaz-Badillo, A., Camacho-Nuez, M., Cisneros, A., and Munoz Mde, L. (2009). Multiple recombinants in two dengue virus, serotype 2 isolates from patients from Oaxaca, Mexico. BMC Microbiol. *9*, 260.
Perret, C., Abarca, K., Ovalle, J., Ferrer, P., Godoy, P., Olea, A., Aguilera, X., and Ferrés, M. (2003). Dengue-1 virus isolation during first dengue fever outbreak on Easter Island, Chile. Emerg. Infect. Dis. *9*, 1465–1467.
Petersen, L.R., and Busch, M.P. (2010). Transfusion-transmitted arboviruses. Vox Sang. *98*, 495–503.
Pickett, B.E., and Lefkowitz, E.J. (2009). Recombination in West Nile Virus: minimal contribution to genomic diversity. Virol. J. *6*, 165.
Pinheiro, F.P., Travassos da Rosa, A.P., Travassos da Rosa, J.F., Ishak, R., Freitas, R.B., Gomes, M.L., LeDuc, J.W., and Oliva, O.F. (1981). Oropouche virus. I. A review of clinical, epidemiological, and ecological findings. Am. J. Trop. Med. Hyg. *30*, 149–160.
Pisano, M.B., Spinsanti, L.I., Díaz, L.A., Farías, A.A., Almirón, W.R., Ré, V.E., and Contigiani, M.S. (2012). First detection of Rio Negro virus (Venezuelan equine encephalitis complex subtype VI) in Córdoba, Argentina. Mem. Inst. Oswaldo Cruz *107*, 125–128.
Prasad, N., Bhadauria, D., Sharma, R.K., Gupta, A., Kaul, A., and Srivastava, A. (2012). Dengue virus infection in renal allograft recipients: a case series during 2010 outbreak. Transpl. Infect. Dis. *14*, 163–168.
Pratt, W.D., Gibbs, P., Pitt, M.L., and Schmaljohn, A.L. (1998). Use of telemetry to assess vaccine-induced protection against parenteral and aerosol infections with Venezuelan equine encephalitis virus in non-human primates. Vaccine *16*, 1056–1064.
Prow, N.A. (2013). The changing epidemiology of Kunjin virus in Australia. Int. J. Environ. Res. Public Health *10*, 6255–6272.

Punda-Polić, V., Mohar, B., Duh, D., Bradarić, N., Korva, M., Fajs, L., Saksida, A., and Avšič-Županc, T. (2012). Evidence of an autochthonous Toscana virus strain in Croatia. J. Clin. Virol. *55*, 4–7.

Purse, B.V., Mellor, P.S., Rogers, D.J., Samuel, A.R., Mertens, P.P., and Baylis, M. (2005). Climate change and the recent emergence of bluetongue in Europe. Nature Rev. Microbiol. 3, 171–181.

Pyke, A.T., Williams, D.T., Nisbet, D.J., van den Hurk, A.F., Taylor, C.T., Johansen, C.A., Macdonald, J., Hall, R.A., Simmons, R.J., Mason, R.J.V., Lee, J.M., Ritchie, S.A., Smith, G.A., and Mackenzie, J.S. (2001). The appearance of a second genotype of Japanese encephalitis virus in the Australasian region. Am. J. Trop. Med. Hyg. *65*, 747–753.

Qualls, W.A., Xue, R.D., Beier, J.C., and Müller, G.C. (2013). Survivorship of adult *Aedes albopictus* (Diptera: Culicidae) feeding on indoor ornamental plants with no inflorescence. Parasitol. Res. *112*, 2313–2318.

Quan, P.L., Junglen, S., Tashmukhamedova, A., Conlan, S., Hutchison, S.K., Kurth, A., Ellerbrok, H., Egholm, M., Briese, T., Leendertz, F.H., and Lipkin, W.I. (2010). Moussa virus: a new member of the *Rhabdoviridae* family isolated from *Culex decens* mosquitoes in Côte d'Ivoire. Virus Res. *147*, 17–24.

Radford, A.D., Chapman, D., Dixon, L., Chantrey, J., Darby, A.C., and Hall, N. (2012). Application of next-generation sequencing technologies in virology. J. Gen. Virol. *93*, 1853–1868.

Ramasamy, R., and Surendran, S.N. (2011). Possible impact of rising sea levels on vector-borne infectious diseases. BMC Infect. Dis. *11*, 18.

Randolph, S.E. (2011). Transmission off tick-borne pathogens between co-feeding ticks: Milan Labuda's enduring paradigm. Ticks Tick Borne Dis. *2*, 179–182.

Rasmussen, L.D., Kristensen, B., Kirkeby, C., Rasmussen, T.B., Belsham, G.J., Bødker, R., and Bøtner, A. (2012). Culicoids as vectors of Schmallenberg virus. Emerg. Infect. Dis. *18*, 1204–1206.

Reisen, W.K., and Monath, T.P. (1988). Western equine encephalomyelitis. In The Arboviruses: Epidemiology and Ecology, Volume V, T.P. Monath, ed. (CRC Press Inc., Boca Raton, Florida), pp. 89–137.

Reiter, P. (2008). Climate change and mosquito-borne disease: knowing the horse before hitching the cart. Rev. Sci. Tech. *27*, 383–398.

Reiter, P., and Sprenger, D. (1987).The used tire trade: a mechanism for the worldwide dispersal of container breeding mosquitoes. J. Am. Mosq. Control Assoc. *3*, 494–501.

Remoli, M.E., Fortuna, C., Marchi, A., Bucci, P., Argentini, C., Bongiorno, G., Maroli, M., Gradoni, L.,Gramiccia, M., and Ciufolini, M.G. (2014). Viral isolates of a novel putative phlebovirus in the Marche Region of Italy. Am. J. Trop. Med. Hyg. *90*, 760–763.

Rico-Hesse, R. (1990). Molecular evolution and distribution of dengue viruses types 1 and 2 in nature. Virology *174*, 479–493.

Rico-Hesse, R. (2003). Microevolution and virulence of dengue viruses. Adv. Virus Res. *59*, 315–341.

Rico-Hesse, R., Harrison, L.M., Salas, R.A., Tovar, D., Nisalak, A., Ramos, C., Boshell, J., de Mesa, M.T., Nogueira, R.M., and de Rosa, A.T. (1997). Origins of dengue type 2 viruses associated with increased pathogenicity in the Americas. Virology *230*, 244–251.

Ritchie, S.A., and Rochester, W. (2001). Wind-blown mosquitoes and introduction of Japanese encephalitis into Australia. Emerg. Infect. Dis. *7*, 900–903.

Ritchie, S.A., Cortis, G., Paton, C., Townsend, M., Shroyer, D., Zborowski, P., Hall-Mendelin, S., and van den Hurk, A.F. (2013). A simple non-powered passive trap for the collection of mosquitoes for arbovirus surveillance. J. Med. Entomol. *50*, 185–194.

Roche, S.E., Wicks, R., Garner, M.G., East, I.J., Paskin, R., Moloney, B.J., Carr, M., and Kirkland, P. (2013). Descriptive overview of the 2011 epidemic of arboviral disease in horses in Australia. Aust. Vet. J. *91*, 5–13.

Rodhain, F., and Rosen, L. (1997). Mosquito vectors and dengue virus–vector relationships. In Dengue and Dengue Hemorrhagic Fever, D.J. Gubler and G. Kono, eds. (CAB International, Wallingford, UK).

Roger, M., Girard, S., Faharoudine, A., Halifa, M., Bouloy, M., Cetre0Sossah, C., and Cardinale, E. (2011). Rift Valley fever in ruminants, Republic of Comoros, 2009. Emerg. Infect. Dis. *17*, 1319–1320.

Rogers, D.J., Wilson, A.J., Hay, S.I., and Graham, A.J. (2006). The global distribution of yellow fever and dengue. Adv. Parasitol. *62*, 181–220.

Rosen, L. (1986). The natural history of Japanese encephalitis virus. Annu. Rev. Microbiol. *40*, 395–414.

Rosen, L. (1987). Overwintering mechanisms of mosquito-borne arboviruses in temperate climates. Am. J. Trop. Med. Hyg. *37(Suppl)*, 69S-76S.

Rosseel, T., Scheuch, M., Höper, D., De Regge, N., Caij, A.B., Vandenbussche, F., and Van Borm, S. (2012). DNase SISPA-next generation sequencing confirms Schmallenberg virus in Belgian field samples and identifies genetic variation in Europe. PLoS One *7*, e41967.

Russell, R.C. (1989). Transport of insects of public health importance on international aircraft. Trav. Med. Intern. *7*, 26–31.

Russell, R.C. (1998). Mosquito-borne arboviruses in Australia: the current scene and implications of climate change for human health. Int. J. Parasitol. *28*, 955–969.

Russell, R.C., and Paton, R. (1989). In-flight disinsection as an efficacious procedure for preventing international transport of insects of public health importance. Bull. World Health Organ. *67*, 543–547.

Russell, R.C., Currie, B.J., Lindsay, M.D., Mackenzie, J.S., Ritchie, S.A., and Whelan, P.I. (2009). Dengue and climate change in Australia: predictions for the future should incorporate knowledge from the past. Med. J. Aust. *190*, 265–268.

St. George, T.D. (1986). Arboviruses infecting livestock in the Australian region. In Arbovirus Research in Australia, Proceedings of the 4th Symposium, T.D. St George, B.H. Kay, J. Blok, eds. (Queensland Institute of Medical Research and Commonwealth Scientific and Industrial Research Organization, Brisbane, Australia), pp. 23–25.

St. George, T.D., Standfast, H.A., Doherty, R.L., Carley, J.G., Fillipich, C., and Brandsma, J. (1977). The isolation of Saumarez Reef virus, a new flavivirus, from bird ticks *Ornithodoros capensis* and *Ixodes eudyptidis* in Australia. Aust. J. Exp. Biol. Med. Sci. *55*, 493–499.

Saito, M., Taira, K., Itokazu, K., and Mori, N. (2007). Recent change of the antigenicity and genotype of Japanese encephalitis viruses distributed on Okinawa Island, Japan. Am. J. Trop. Med. Hyg. *77*, 737–746.

Saleh, S.M., Poidinger, M., Mackenzie, J.S., Broom, A.K., Lindsay, M.D., and Hall, R.A. (2003). Complete genomic sequence of the Australian south-west genotype of Sindbis virus: comparisons with other Sindbis strains and identification of a unique deletion in the 3′-untranslated region. Virus Genes *26*, 317–327.

Sammels, L.M., Lindsay, M.D., Poidinger, M., Coelen, R.J., and Mackenzie, J.S. (1999). Geographic distribution and evolution of Sindbis virus in Australia. J. Gen. Virol. *80*, 739–748.

Schaffner, F., Kaufmann, C., Hegglin, D., and Mathis, A. (2009). The invasive mosquito *Aedes japonicus* in central Europe. Med. Vet. Entomol. *23*, 448–451.

Scherret, J., Mackenzie, J.S., Hall, R.A., Deubel, V., and Gould, E.A. (2002). Phylogeny and molecular epidemiology of West Nile and Kunjin viruses. Curr. Top. Microbiol. Immunol. *267*, 373–390.

Scherret, J.H., Poidinger, M., Mackenzie, J.S., Broom, A.K., Deubel, V., Lipkin, W.I., Briese, T., and Hall, R.A. (2001). The relationships between West Nile and Kunjin viruses. Emerg. Infect. Dis. *7*, 697–705.

Schmidt, J.R., and Said, M.I. (1964). Isolation of West Nile virus from the African bird argasid, *Argas reflexus hermanni*, in Egypt. J. Med. Entomol. *1*, 83–86.

Scholte, E.J., Dijkstra, E., Blok, H., De Vries, A., Takken, W., Hofhuis, A., Koopmans, M., De Boer, A., and Reusken, C.B. (2008). Accidental importation of the mosquito *Aedes albopictus* into the Netherlands: a survey of mosquito distribution and the presence of dengue virus. Med. Vet. Entomol. *22*, 352–358.

Schuh, A.J., Guzman, H., Tesh, R.B., and Barrett, A.D. (2013a). Genetic diversity of Japanese encephalitis virus isolates obtained from the Indonesian archipelago between 1974 and 1987. Vector Borne Zoonotic Dis. *13*, 479–488.

Schuh, A.J., Ward, M.J., Brown, A.J., and Barrett, A.D. (2013b). Phylogeography of Japanese encephalitis virus: genotype is associated with climate. PLoS Negl. Trop. Dis. *7*, e2411.

Schuh, A.J., Ward, M.J., Leigh Brown, A.J., and Barrett, A.D. (2014). Dynamics of the emergence and establishment of a newly dominant genotype of Japanese encephalitis virus throughout Asia. J. Virol. *88*, 4522–4532.

Schwarz, T.F., Gilch, S., and Jäger, G. (1995). Aseptic meningitis caused by sandfly fever virus, serotype Toscana. Clin. Infect. Dis. *21*, 669–671.

Sedda, L., and Rogers, D.J. (2013). The influence of the wind in the Schmallenberg virus outbreak in Europe. Sci. Rep. *3*, 3361.

Seidowski, D., Ziegler, U., von Rönn, J.A., Müller, K., Hüppop, K., Müller, T., Freuling, C., Mühle, R.U., Nowotny, N., Ulrich, R.G., Niedrig, M., and Groschup, M.H. (2010). West Nile virus monitoring of migratory

and resident birds in Germany. Vector Borne Zoonotic Dis. *10*, 639–647.
Sellers, R.F. (1980). Weather, host and vector – their interplay in the spread of insect-borne animal virus diseases. J. Hyg. (Lond.) *85*, 65–102.
Sellers, R.F., and Taylor, W.P. (1980). Epidemiology of bluetongue and the import and export of livestock, semen and embryos. Bull. l'Office Int. Epizoot. *92*, 587–592.
Setoh, Y.X., Prow, N.A., Hobson-Peters, J., Lobigs, M., Young, P.R., Khromykh, A.A., and Hall, R.A. (2012). Identification of residues in West Nile virus pre-membrane protein that influence viral particle secretion and virulence. J. Gen. Virol. *93*, 1965–1975.
Seymour, C., Dickerman, R.W., and Martin, M.S. (1978). Venezuelan encephalitis virus infection in neotropical bats. I. Natural infection in a Guatemalan enzootic focus. Am. J. Trop. Med. Hyg. *27*, 290–296.
Shaman, J., Day, J.F., and Stieglitz, M. (2005). Drought-induced amplification and epidemic transmission of West Nile virus in southern Florida. J. Med. Entomol. *42*, 134–141.
Shaw, A.E., Ratinier, M., Nunes, S.F., Nomikou, K., Caporale, M., Golder, M., Allan, K., Hamers, C., Hudelet, P., Zientara, S., Breard, E., Mertens, P., and Palmarini, M. (2013). Reassortment between two serologically unrelated bluetongue virus strains is flexible and can involve any genome segment. J. Virol. *87*, 543–557.
Sissoko, D., Giry, C., Gabrie, P., Tarantola, A., Pettinelli, F., Collet, L., D'Ortenzio, E., Renault, P., and Pierre, V. (2009). Rift Valley fever, Mayotte, 2007–2008. Emerg. Infect. Dis. *15*, 568–570.
Slingenbergh, J.I., Gilbert, M., de Balogh, K.I., and Wint, W. (2004). Ecological sources of zoonotic diseases. Rev. Sci. Tech. *23*, 467–484.
Smith, D.R., Adams, A.P., Kenney, J.L., Wang, E., and Weaver, S.C. (2008). Venezuelan equine encephalitis virus in the mosquito vector *Aedes taeniorhynchus*: infection initiated by a small number of susceptible epithelial cells and a population bottleneck. Virology *372*, 176–186.
Smith, I., and Wang, L.F. (2013). Bats and their virome: an important source of emerging viruses capable of infecting humans. Curr. Opin. Virol. *3*, 84–91.
Solomon, T., Ni, H., Beasley, D.W., Ekkelenkamp, M., Cardosa, M.J., and Barrett, A.D. (2003). Origin and evolution of Japanese encephalitis virus in Southeast Asia. J. Virol. *77*, 3091–3098.
Soman, R.S., Rodrigues, F.M., Guttikar, S.N., and Guru, P.Y. (1977). Experimental viraemia and transmission of Japanese encephalitis virus by mosquitoes in ardeid birds. Indian J. Med. Res. *66*, 709–718.
Spencer, J.D., Azoulas, J., Broom, A.K., Buick, T.D., Currie, B., Daniels, P.W., Doggett, S.L., Hapgood, G.D., Jarrett, P.J., Lindsay, M.D., *et al.* (2001). Murray Valley encephalitis virus surveillance and control initiatives in Australia. National Arbovirus Advisory Committee of the Communicable Diseases Network Australia. Commun. Dis. Intell. Q. Rep. *25*, 33–47. Erratum in: Commun. Dis. Intell. (2001) *25*, 155.
Stock, N.K., Laraway, H., Faye, O., Dialo, M., Niedrig, M., and Sall, A.A. (2013). Biological and phylogenetic characteristics of yellow fever virus lineages from West Africa. J. Virol. *87*, 2895–2907.
Sulkin, S.E., and Allen, R. (1974). Virus infections in bats. Monogr. Virol. *8*, 1–103.
Sulkin, S.E., Sims, R.A., and Allen, R. (1966). Isolation of St. Louis encephalitis virus from bats (*Tadarida b. mexicana*) in Texas. Science *152*, 223–225.
Suhrbier, A., Jaffar-Bandjee, M.C., and Gasque, P. (2012). Arthriticogenic alphaviruses – an overview. Nat. Rev. Rheumatol. *8*, 420–429.
Sutherst, R.W. (2004). Global change and human vulnerability to vector-borne diseases. Clin. Microbiol. Rev. *17*, 136–173.
Swei, A., Russell, B.J., Naccache, S.N., Kabre, B., Veeraraghavan, N., Pilgard, M.A., Johnson, B.J., and Chiu, C.Y. (2013). The genome sequence of Lone Star virus, a highly divergent bunyavirus found in the *Amblyomma americanum* tick. PLoS One *8*, e62083.
Tabachnick, W.J. (2004). *Culicoides* and the global epidemiology of bluetongue virus infection. Vet. Ital. *40*, 144–150.
Takahashi, T., Maeda, K., Suzuki, T., Ishido, A., Shigeoka, T., Tominaga, T., Kamei, T., Honda, M., Ninomiya, D., Sakai, T., *et al.* (2014). The first identification and retrospective study of severe fever with thrombocytopenia syndrome in Japan. J. Infect. Dis. *209*, 816–827.
Takhampunya, R., Kim, H.C., Tippayachai, B., Kengluecha, A., Klein, T.A., Lee, W.J., Grieco, J., and Evans, B.P. (2011). Emergence of Japanese encephalitis virus genotype V in the Republic of Korea. Virol. J. *8*, 449.
Tamba, M., Bonilauri, P., Bellini, R., Calzolari, M., Albieri, A., Sambri, V., Dottori, M., and Angelini, P. (2011). Detection of Usutu virus within a West Nile virus surveillance program in Northern Italy. Vector Borne Zoonotic Dis. *11*, 551–557.
Tambyah, P.A., Koay, E.S., Poon, M.L., Lin, R.V., and Ong, B.K. (2008). Dengue hemorrhagic fever transmitted by blood transfusion. N. Engl. J. Med. *359*, 526–1527.
Tatem, A.J., Hay, S.I., and Rogers, D.J. (2006). Global traffic and disease vector dispersal. Proc. Natl. Acad. Sci. U.S.A. *103*, 6242–6247.
Taubenberger, J.K., and Kash, J.C. (2010). Influenza virus evolution, host adaptation, and pandemic formation. Cell Host Microbe *7*, 440–451.
Tesh, R.B., and Cornet, M. (1981). The location of San Angelo virus in developing ovaries of transovarially infected *Aedes albopictus* mosquitoes as revealed by fluorescent antibody technique. Am. J. Trop. Med. Hyg. *30*, 212–218.
Thai, K.T., Henn, M.R., Zody, M.C., Tricou, V., Nguyet, N.M., Charlebois, P., Green, L., de Vries, P.J., Hien, T.T., Farrar, J., van Doorn, H.R., de Jong, M.D., Birren, B.W., Holmes, E.C., and Simmons, C.P. (2012). High-resolution analysis of intrahost genetic diversity in dengue serotype 1 infection identifies mixed infections. J. Virol. *86*, 835–843.
Thomas, S.J., Strickman, D., and Vaughn, D.W. (2003). Dengue epidemiology: virus epidemiology, ecology and emergence. Adv. Virus Res. *61*, 235–289.
Thompson, N.N., Auguste, A.J., Travassos da Rosa, A.P., Carrington, C.V., Blitvich, B.J., Chadee, D.D., Tesh, R.B., Weaver, S.C., and Adesiyun, A.A. (2014). Seroepidemiology of Selected Alphaviruses and Flaviviruses in Bats in Trinidad. Zoonoses Public Health [Epub ahead of print]
Tolou, H.J., Couissinier-Paris, P., Durand, J.P., Mercier, V., de Pina, J.J., de Micco, P., Billoir, F., Charrel, R.N., de Lamballerie, X. (2001). Evidence for recombination in natural populations of dengue virus type 1 based on the analysis of complete genome sequences. J. Gen. Virol. *82*, 1283–1290.
Tomasello, D., and Schlagenhauf, P. (2013). Chikungunya and dengue autochthonous cases in Europe, 2007–2012. Travel Med. Infect. Dis. *11*, 274–284.
Tonry, J.H., Brown, C.B., Cropp, C.B., Co, J.K., Bennett, S.N., Nerurkar, V.R., Kuberski, T., and Gubler, D.J. (2005). West Nile virus detection in urine. Emerg. Infect. Dis. *11*, 1294–1296.
Traiber, C., Coelho-Amaral, P., Ritter, V.R., and Winge, A. (2011). Infant meningoencephalitis caused by yellow fever vaccine virus transmitted via breastmilk. J. Pediatr. (Rio J.) *87*, 269–272.
Travassos da Rosa, A.P.A., Shope, R.E., Pinheiro, F.P., Travassos da Rosa, J.F.S., Vasconcelos, P.F.C., Herve, J.P., and Degallier, N. (1989). Arbovirus research in the Brazilian Amazon. In Arbovirus Research in Australia, Proceedings of the Fifth Symposium, M.F. Uren, J. Blok, and L.H. Manderson, eds. (Commonwealth Scientific and Industrial Research Organization and the Queensland Institute of Medical Research, Brisbane, Australia), pp. 1–8.
Travassos da Rosa, A.P., Turell, M.J., Watts, D.M., Powers, A.M., Vasconcelos, P.F., Jones, J.W., Klein, T.A., Dohm, D.J., Shope, R.E., Degallier, N., *et al.* (2001). Trocara virus: a newly recognized Alphavirus (Togaviridae) isolated from mosquitoes in the Amazon Basin. Am. J. Trop. Med. Hyg. *64*, 93–97.
Trent, D.W., Grant, J.A., Rosen, L., and Monath, T.P. (1983). Genetic variation among dengue 2 viruses of different geographic origin. Virology *128*, 271–284.
Trent, D.W., Grant, J.A., Monath, T.P., Manske. C.L., Corina, M., and Fox, G.E. (1989). Genetic variation and microevolution of dengue 2 virus in Southeast Asia. Virology, *172*, 523–535.
Tsai, T.F. (1997). Factors in the changing epidemiology of Japanese encephalitis and West Nile fever. In Factors in the Emergence of Arbovirus Diseases, J.F. Saluzzo and B. Dodet, eds. (Elsevier, Paris), pp. 179–189.
Tsetsarkin, K.A., and Weaver, S.C. (2011). Sequential adaptive mutations enhance efficient vector switching by Chikungunya virus and its epidemic emergence. PLoS Pathog. *7*, e1002412.
Tsetsarkin, K.A., Vanlandingham, D.L., McGee, C.E., and Higgs, S. (2007). A single mutation in chikungunya virus affects vector specificity and epidemic potential. PLoS Pathog. *3*, e201.
Tsetsarkin, K.A., McGee, C.E., Volk, S.M., Vanlandingham, D.L., Weaver, S.C., and Higgs, S. (2009). Epistatic roles of E2 glycoprotein mutations in adaptation of chikungunya virus to *Aedes albopictus* and *Ae. aegypti* mosquitoes. PLoS One *4*, e6835.
Tsetsarkin, K.A., Chen, R., Sherman, M.B., and Weaver, S.C. (2011). Chikungunya virus: evolution and genetic determinants of emergence. Curr. Opin. Virol. *1*, 310–317.

Turell, M.J. (1988). Horizontal and vertical transmission of viruses by insect and tick vectors. In The Arboviruses: Epidemiology and Ecology, T.P. Monath, ed. (CRC Press Inc., Boca Raton, Florida), pp. 127–152.

Turell, M.J., and Kay, B.H. (1998). Susceptibility of selected strains of Australia mosquitoes (Diptera: Culicidae) to Rift Valley fever virus. J. Med. Entomol. *35*, 132–135.

Turell, M.J., Dohm, D.J., Mores, C.N., Terracina, L., Wallette, D.L. Jr., Hribar, L.J., Pecor, J.E., and Blow, J.A. (2008). Potential for North American mosquitoes to transmit Rift Valley fever virus. J. Am. Mosq. Control Assoc. *24*, 502–507.

Twiddy, S.S., and Holmes, E.C. (2003). The extent of homologous recombination in members of the genus Flavivirus. J. Gen. Virol. *84*, 429–440.

Uzcategui, N.Y., Camacho, D., Comach, G., Cuello de Uzcategui, R., Holmes, E.C., and Gould, E.A. (2001). Molecular epidemiology of dengue type 2 virus in Venezuela: evidence for in situ virus evolution and recombination. J. Gen. Virol. *82*, 2945–2953.

Uchil, P.D., and Satchidanandam, V. (2001). Phylogenetic analysis of Japanese encephalitis virus: envelope gene based analysis reveals a fifth genotype, geographic clustering, and multiple introductions of the virus into the Indian subcontinent. Am. J. Trop. Med. Hyg. *65*, 242–251.

Valiakos, G., Touloudi, A., Athanasiou, L.V., Giannakopoulos, A., Iacovakis, C., Birtsas, P., Spyrou, V., Dalabiras, Z., Petrovska, L., and Billinis, C. (2012). Serological and molecular investigation into the role of wild birds in the epidemiology of West Nile virus in Greece. Virol. J. *9*, 266.

Van Bortel, W., Dorleans, F., Rosine, J., Blateau, A., Rousset, D., Matheus, S., Leparc-Goffart, I., Flusin, O., Prat, C., Cesaire, R., *et al.* (2014). Chikungunya outbreak in the Caribbean region, December 2013 to March 2014, and the significance for Europe. Eur. Surveill. *19*, pii: 20759.

Vasconcelos, P.F., Travassos da Rosa, A.P., Rodrigues, C.G., Travassos da Rosa, E.S., Dégallier, N., and Travassos da Rosa, J.F. (2001). Inadequate management of natural ecosystem in the Brazilian Amazon region results in the emergence and reemergence of arboviruses. Cad. Saude Publica *17*, S156–S164.

Vasilakis, N., Deardorff, E.R., Kenney, J.L., Rossi, S.L., Hanley, K.A., and Weaver, S.C. (2009). Mosquitoes put the brake on arbovirus evolution: experimental evolution reveals slower mutation accumulation in mosquito than vertebrate cells. PLoS Pathog. *5*, e1000467.

Vayssier-Taussat, M., Moutailler, S., Michelet, L., Devillers, E., Bonnet, S., Cheval, J., Hébert, C., and Eloit, M. (2013). Next generation sequencing uncovers unexpected bacterial pathogens in ticks in Western Europe. PLoS One *8*, e81439.

Vazquez, A., Sanchez-Seco, M.P., Ruiz, S., Molero, F., Hernandez, L., Moreno, J., Magallanes, A., Tejedor, C.G., and Tenorio, A. (2010). Putative new lineage of West Nile virus, Spain. Emerg. Infect. Dis. *16*, 549–552.

Vazquez, A., Jimenez-Clavero, M., Franco, L., Donoso-Mantke, O., Sambri, V., Niedrig, M., Zeller, H., and Tenorio, A. (2011). Usutu virus: a potential risk of human disease in Europe. Eur. Surveill. *16*, 19935.

Verani, P., Ciufolini, M.G., Caciolli, S., Renzi. A., Nicoletti, L., Sabatinelli, G., Bartolozzi, D., Volpi, G., Amaducci, L., and Coluzzi, M. (1988). Ecology of viruses isolated from sand flies in Italy and characterized of a new Phlebovirus (Arabia virus). Am. J. Trop. Med. Hyg. *38*, 433–439.

Villabona-Arenas, C.J., and Zanotto, P.M. (2013). Worldwide spread of Dengue virus type 1. PLoS One *8*, e62649.

Wagner, D., de With, K., Huzly, D., Hufert, F., Weidmann, M., Breisinger, S., Eppinger, S., Kern, W.V., and Bauer, T.M. (2004). Nosocomial acquisition of dengue. Emerg. Infect. Dis. *10*, 1872–1873.

Waldenström, J., Lundkvist, A., Falk, K.I., Garpmo, U., Bergström, S., Lindegren, G., Sjöstedt, A., Mejlon, H., Fransson, T., Haemig, P.D., and Olsen, B. (2007). Migrating birds and tickborne encephalitis virus. Emerg. Infect. Dis. *13*, 1215–1218.

Walker, P.J. (2005). Bovine ephemeral fever in Australia and the world. Curr. Top. Microbiol. Immunol. *292*, 57–80.

Walsh, J.F., Molyneux, D.H., and Birley, M.H. (1993). Deforestation: effects on vector-borne disease. Parasitology *106*, S55–S75.

Wang, H.Y., Takasaki, T., Fu, S.H., Sun, X.H., Zhang, H.L., Wang, Z.X., Hao, Z.Y., Zhang, J.K., Tang, Q., Kotaki, A., Tajima, S., Liang, X.F., Yang, W.Z., Kuane, I., and Liang, G.D. (2007). Molecular epidemiological analysis of Japanese encephalitis virus in China. J. Gen. Virol. *88*, 885–894.

Wang, J., Selleck, P., Yu, M., Ha, W., Rootes, C., Gales, R., Wise, T., Crameri, S., Chen, H., Broz, I., *et al.* (2014). Novel phlebovirus with zoonotic potential isolated from ticks, Australia. Emerg. Infect. Dis. *20*, 1040–1043.

Wang, J.L., Pan, X.L., Zhang, H.L., Fu, S.H., Wang, H.Y., Tang, Q., Wang, L.F., and Liang, G.D. (2009). Japanese encephalitis viruses from bats in Yunnan, China. Emerg. Infect. Dis. *15*, 939–942.

Wang, W.K., Lin, S.R., Lee, C.M., King, C.C., and Chang, S.C. (2002a). Dengue type 3 virus in plasma is a population of closely related genomes: quasispecies. J. Virol. *76*, 4662–4665.

Wang, W.K., Sung, T.L., Lee, C.N., Lin, T.Y., and King, C.C. (2002b). Sequece diversity of the capsid gene and the non-structural gene NS2B of dengue-3 virus *in vivo*. Virology *303*, 181–191.

Warrilow, D., Watterson, D., Hall, R.A., Davis, S.S., Weir, R., Kurucz, N., Whelan, P., Allcock, R., Hall-Mendelin, S., O'Brien, C.A., and Hobson-Peters, J. (2014). A new species of mesonivirus from the Northern Territory, Australia. PLoS One *9*, e91103.

Weaver, S.C., and Reisen, W.K. (2010). Present and future arboviral threats. Antiviral Res. *85*, 328–345.

Weber, F., Wagner, V., Rasmussen, S.B., Hartmann, R., and Paludan, S.R. (2006). Double-stranded RNA is produced by positive-strand RNA viruses and DNA viruses but not in detectable amounts by negative-strand RNA viruses. J. Virol. *80*, 5059–5064.

Weissenböck, H., Kolodziejek, J., Url, A., Lussy, H., Rebel-Bauder, B., and Nowotny, N. (2002). Emergence of Usutu virus, an African mosquito-borne flavivirus of the Japanese encephalitis virus group, Central Europe. Emerg. Infect. Dis. *8*, 652–656.

Weissenböck, H., Bakonyi, T., Rossi, G., Mani, P., and Nowotny, N. (2013). Usutu virus, Italy, 1996. Emerg. Infect. Dis. *19*, 274–277.

Wilder-Smith, A., and Gubler, D.J. (2008). Geographic expansion of dengue: the impact of international travel. Med. Clin. North Am. *92*, 1377–1390.

Wilson, A.J., and Mellor, P.S. (2009). Bluetongue in Europe: past, present and future. Philos. Trans. R. Soc. Lond. B. Biol. Sci. *364*, 2669–2681.

Williams, J.E., Imlarp, S., Top, F.H. Jr., Cavanaugh, D.C., and Russell, P.K. (1976). Kaeng Khoi virus from naturally infected bed bugs (*Cimicidae*) and immature free-tailed bats. Bull. World Health Organ. *53*, 365–369.

Work, T.H. (1958). Russian spring-summer virus in India: Kyasanur Forest disease. Prog. Med. Virol. *1*, 248–279.

Work, T.H., Hurlbut, H.S., and Taylor, R.M. (1953). Isolation of West Nile virus from hooded crow and rock pigeon in the Nile delta. Proc. Soc. Exp. Biol. Med. *84*, 719–722.

Work, T.H., Hurlbut, H.S., and Taylor, R.M. (1955). Indigenous wild birds of the Nile Delta as potential West Nile virus circulating reservoirs. Am. J. Trop. Med. Hyg. *4*, 872–888.

World Health Organization (2000). Rift Valley fever in Saudi Arabia. Available online: http://www.who.int/csr/don/2000_09_29/en/. Accessed 10 June 2014.

World Health Organization (2008). Rift Valley fever fact sheet. Weekly Epidemiol. Rec. *83*, 17–22.

Worobey, M., Rambaut, A., and Holmes, E.C. (1999). Widespread intra-serotype recombination in natural populations of dengue virus. Proc. Natl. Acad. Sci. U.S.A. *96*, 7352–7357.

Xing, Z., Schefers, J., Schwabenlander, M., Jiao, Y., Liang, M., Qi, X., Li, C., Goyal, S., Cardona, C.J., Wu, X., *et al.* (2013). Novel bunyavirus in domestic and captive farmed animals, Minnesota, USA. Emerg. Infect. Dis. *19*, 1487–1489.

Xue, L., Cohnstaedt, L.W., Scott, H.M., and Scoglio, C. (2013). A hierarchical network approach for modelling Rift Valley fever epidemics with applications in North America. PLoS One *8*, e62049.

Yanase, T., Maeda, K., Kato, T., Nyuta, S., Kamata, H., Yamakawa, M., and Tsuda, T. (2005). The resurgence of Shamonda virus, an African Simbu group virus of the genus Orthobunyavirus, in Japan. Arch. Virol. *150*, 361–369.

Yanase, T., Kato, T., Aizawa, M., Shuto, Y., Shirafuji, H., Yamakawa, M., and Tsuda, T. (2012). Genetic reassortment between Sathuperi and Shamonda viruses of the genus Orthabunyavirus in nature: implications for their genetic relationship to Schmallenberg virus. Arch. Virol. *157*, 1611–1616.

Yango, A.F., Fischbach, B.V., Levy, M., Chandrakantan, A., Tan, V., Spak, C., Melton, L., Rice, K., Barri, Y., Rajagopal, A., and Klintmalm, G. (2014). West Nile virus infection in kidney and pancreas transplant recipients in the Dallas-Fort worth metroplex during the 2012 Texas epidemic. Transplantation *97*, 953–957.

Yee, D.A. (2008). Tires as habitats for mosquitoes: a review of studies within the eastern United States. J. Med. Entomol. *45*, 581–593.

Yu, X.J., Liang, M.F., Zhang, S.Y., Liu, Y., Li, J.D., Sun, Y.L., Zhang, L., Zhang, Q.F., Popov, V.L., Li, C., *et al.* (2011). Fever with thrombocytopenia associated with a novel bunyavirus in China. N. Engl. J. Med. *364*, 1523–1532.

Yun, S.M., Cho, J.E., Ju, Y.R., Kim, S.Y., Tyou, J., Han, M.G., Choi, W.Y., and Jeong, Y.E. (2010). Molecular epidemiology of Japanese encephalitis virus circulating in South Korea, 1983–2005. Virol. J. *7*, 127.

Zeller, H.G., Akakpo, A.J., and Ba, M.M. (1995). Rift Valley fever epizootic in small ruminants in southern Mauritania (October 1993): risk of extensive outbreaks. Ann. Soc. Belg. Med. Trop. *75*, 135–140.

Zhang, Y.Z., Zhou, D.J., Qin, X.C., Tian, J.H., Xiong, Y., Wang, J.B., Chen, X.P., Gao, D.Y., He, Y.W., Jin, D., *et al.* (2012). The ecology, genetic diversity, and phylogeny of Huaiyangshan virus in China. J. Virol. *86*, 2864–2868.

Zhao, G., Krishnamurthy, S., Cai, Z., Popov, V.L., Travassos da Rosa, A.P., Guzman, H., Cao, S., Virgin, H.W., Tesh, R.B., and Wang, D. (2013). Identification of novel viruses using VirusHunter – an automated data analysis pipeline. PLoS One *8*, e78470.

Zhioua, E., Moureau, G., Chelbi, I., Ninove, L., Bichaud, L., Derbali, M., Champs, M., Cherni, S., Salez, N., Cook, S., de Lamballerie, X., and Charrel, R.N. (2010). Punique virus, a novel phlebovirus, related to sandfly fever Naples virus, isolated from sandflies collected in Tunisia. J. Gen. Virol. *91*, 1275–1283.

Zirkel, F., Kurth, A., Quan, P.L., Briese, T., Ellerbrok, H., Pauli, G., Leendertz, F.H., Lipkin, W.I., Ziebuhr, J., Drosten, C., and Junglen, S. (2011). An insect nidovirus emerging from a primary tropical rainforest. MBio *2*, e00077–00011.

Role of Inter- and Intra-host Genetics in Arbovirus Evolution

11

Alexander T. Ciota and Gregory D. Ebel

Abstract

The global expansion of arthropod-borne viruses (arboviruses) in recent decades has significantly increased the public health threat from these pathogens, but the inherently diverse nature of viral RNA genomes both within and between hosts and vectors makes accurately characterizing current and future threats from arboviral agents a challenge. Here, we review the contributions of studies of inter and intra-host arbovirus genetics to our understanding of the unique selective pressures shaping arbovirus evolution and the consequences of genetic change. In particular, we discuss how past studies have informed our knowledge of virus- and host-specific differences in shaping arboviral swarms and how this equates to the potential for phenotypic and/or epidemiological shifts; and how current and future studies utilizing next generation technologies and methods are beginning to provide the tools to gain a deeper understanding of arbovirus evolution in relevant systems.

Significance of genetic change in arboviruses

There are over 100 arthropod-borne viruses (arbovirus) that are known to be associated with human disease. The global public health risk from pathogens including Dengue virus (DENV; *Flaviviridae*: *Flavivirus*) West Nile virus (WNV; *Flaviviridae*: *Flavivirus*), and Chikungunya virus (CHIKV; *Togaviridae*: *Alphavirus*), among others, has rapidly expanded in recent years. Arboviruses are predominantly RNA viruses that persist in nature as dynamic and diverse strains possessing vast evolutionary potential. An understanding of the forces that shape the genetic diversity of arboviruses, as well as the phenotypic correlates of this diversity, is central to our capacity to predict the evolutionary and epidemiological trajectories of these pathogens. The genome diversity among and within arboviral strains is derived from the rapid error-prone replication associated with viral RNA-dependent RNA polymerases (RdRp), as well as characteristically high viral loads. Mutation rates, although likely both species and strain-dependent, are estimated to be 10^{-4}/site/round of replication (Acevedo *et al.*, 2014; Drake and Holland, 1999; Holland *et al.*, 1982), equating to approximately one substitution/genome for most arboviruses. Given the high viral loads often measured in competent vertebrate hosts and vectors, arbovirus amplification may at times result in the production of up to 1 billion new viral genomes, permitting exploration of vast areas of sequence space, including all possible single step mutations, within a single infection.

Although consensus-level substitution rates of arboviruses are generally low in relation to documented mutation rates (Duffy *et al.*, 2008), examples of adaptive evolution facilitating arbovirus emergence or spread have been noted. These include (a) the well-documented WNV strain displacement facilitated by decreased extrinsic incubation in *Culex* mosquitoes, particularly *Cx. tarsalis*, occurring from 2002 to 2005 in concert with the western expansion of WNV facilitated primarily by this species (Ebel *et al.*, 2004; Goldberg *et al.*, 2010; Moudy *et al.*, 2007); (b) increased vector competence for South American epidemic strains of Venezuelan equine encephalitis virus (VEEV; *Togaviridae*: *Alphavirus*) in *Ae. taeniorhyncus* (Brault *et al.*, 2004; Smith *et al.*, 2007), and, most recently, (c) adaptation of CHIKV to *Ae. albopictus*, enabling expansion out of Africa to the islands of the Indian Ocean and aiding in the subsequent global spread (Tsetsarkin *et al.*, 2007). Although adaptations to vertebrate hosts have also been noted (Anishchenko *et al.*, 2006; Brault *et al.*, 2007), these examples demonstrate how vector–virus interactions are often more specific and how genetic change resulting in alterations to vector competence therefore frequently govern the relationship between arbovirus evolution and epidemiological shifts. Remarkably, the majority of these adaptations are driven primarily by a limited number of base substitutions, a phenomenon that has also been noted with experimental evolution studies (Ciota *et al.*, 2010). The relative lack of evolutionary change identified with arboviruses is often noted and generally attributed to the evolutionary constraint created by divergent fitness landscapes of vertebrate and invertebrate environments (Holmes, 2003; Jenkins *et al.*, 2002; Twiddy *et al.*, 2002; Weaver *et al.*, 1992b; Zanotto *et al.*, 1995), yet experimental evidence that significant constraint is a fixed characteristic of host cycling is generally lacking (Ciota *et al.*, 2010). Additional explanations for evolutionary constraint include the need for epistatic interactions (Tsetsarkin *et al.*, 2011), intra and interhost bottlenecks (Ciota *et al.*, 2011, 2012a; Forrester *et al.*, 2012), virulence evolution

(Ciota *et al.*, 2013) and the possibility that many arboviruses already reside on or near maximum fitness peaks in relatively static host environments (Ciota *et al.*, 2009). In addition, this concept of evolutionary stasis refers only to changes in consensus sequences, whereas most arboviruses exist in equilibrium as complex quasispecies structures with a swarm of minority variants contributing to phenotypic outcomes including viral fitness, adaptability, and pathogenesis (Ciota *et al.*, 2007a,b; Coffey and Vignuzzi, 2011; Ebel *et al.*, 2011; Fitzpatrick *et al.*, 2010; Jerzak *et al.*, 2007; Novella *et al.*, 1995; Novella and Ebendick-Corpus, 2004; Vignuzzi *et al.*, 2006). In addition, the scale at which arbovirus adaptation and evolution are assessed may be too broad. Studies with DENV demonstrate that vector–virus genotype by genotype interactions are likely the rule, suggesting the possibility that local adaptation may occur more frequently (Lambrechts *et al.*, 2009), and that the documented examples of adaptive evolution may represent only the instances where generic species-level adaptation occurs.

Relationship between within and between-host mutational diversity

The capacity for RNA viruses to produce and maintain genetic diversity may provide an advantage in terms of phenotypic plasticity and therefore exploitation of new and changing environments. This characteristic, which has obvious epidemiological consequences, may indeed be a prerequisite for the success of arboviral pathogens given their inherent need for replication and transmission in taxonomically divergent hosts and diverse tissues. Although widespread evolutionary change is relatively slow among arboviruses, consensus level change is often noted and attributed to both stochastic and selective forces. Variable spatial and temporal evolutionary patterns are driven largely by unique transmission cycles, host and vector ecology, and mechanisms of maintenance among arboviral species and strains. For example, arboviruses such as VEEV which rely mainly on mammals as amplifying hosts maintain more geographically defined evolutionary clades then do avian dispersed pathogens such as WNV (McMullen *et al.*, 2011; Weaver *et al.*, 1992a). Genetic isolation can result in rapid diversification and defining new lineages and species can be challenging, particularly for RNA viruses with global distributions and large sequence distances. Molecular epidemiology studies of WNV, the most geographically dispersed arbovirus, demonstrate that classification could include as many five distinct lineages, with sequence divergences of over 25% for some isolates (Bondre *et al.*, 2007; Vazquez *et al.*, 2010). Similarly, North and South American strains of Eastern equine encephalitis virus (EEEV; *Togaviridae*: *Alphavirus*) often diverge by >25% (Arrigo *et al.*, 2010a). As is the case for both WNV and EEEV, putative new lineages of arboviruses are often phenotypically distinct and therefore unique in their epidemiological relevance (Aliota *et al.*, 2012; Arrigo *et al.*, 2010b), which is necessary to consider when defining appropriate arbovirus classification.

Interhost diversity results from differential stochastic or selective processes acting on individual intrahost populations. As these evolutionarily forces are both virus and host-dependent so too are levels of genetic diversity within and between hosts. Intrahost diversity of both WNV and its close relative St. Louis encephalitis virus (SLEV; *Flaviviridae*: *Flavivirus*) is generally higher in mosquitoes than avian hosts, likely due to relaxed purifying selection and/or balancing selection (Brackney *et al.*, 2009; Ciota *et al.*, 2009; Jerzak *et al.*, 2005, 2007). DENV strains derived from mosquitoes, on the other hand, were found to be less diverse than human isolates (Lin *et al.*, 2004). As shown with SLEV, significant temporal fluctuations in mutant swarm diversity also occur, possibly as a result of seasonal bottlenecks (Ciota *et al.*, 2011). The tick-borne Powassan virus (POWV; *Flaviviridae*: *Flavivirus*) was found to be relatively homogeneous, possibly due to selective pressures and/or bottlenecks unique to tick maintenance (Brackney *et al.*, 2010). Well-documented anatomical barriers also can result in population bottlenecks within mosquitoes (Ciota *et al.*, 2012a; Forrester *et al.*, 2012; Girard *et al.*, 2004; Hardy *et al.*, 1983) yet, as shown with WNV, are also likely species dependent (Brackney *et al.*, 2011).

Despite the advantage that may be associated with genetic diversity, the vast majority of new mutations produced by arboviruses are deleterious and the majority of nascent RNA genomes are in fact incapable of producing viable infections. Because of this accumulation of deleterious mutations, these pathogens may at times reside in close proximity to error catastrophe, a lethal spiral into sequence space in which purifying selection cannot outpace the accumulation of deleterious mutations (Domingo *et al.*, 2005; Eigen, 2002). Error catastrophe is essentially a progressive loss of information by the RNA genome. Although some would argue that low fidelity is simply a by-product of maximizing replication rates (Furio *et al.*, 2005) the capacity to uncouple mutation and replication rates suggests that there is indeed an evolutionary advantage to the production and maintenance of these diverse swarms (Coffey *et al.*, 2011). Recent studies suggesting that constraining RNA viral swarms through the creation of higher fidelity arbovirus variants is a viable strategy for virus attenuation support this idea (Vignuzzi *et al.*, 2008). Despite these findings, with notable exceptions (Brackney *et al.*, 2010; Ciota *et al.*, 2011; Jerzak *et al.*, 2005; Lin *et al.*, 2004), there remains a dearth of studies characterizing spatial and temporal variations of arboviral swarms of naturally circulating strains. Given that viral fitness and phenotypic potential do not require consensus change (Ciota *et al.*, 2007b; Ebel *et al.*, 2001, 2011; Novella *et al.*, 1995; Novella and Ebendick-Corpus, 2004), characterizing cryptic swarm evolution by utilizing deep sequencing technologies in future arbovirus sequence surveys would provide a needed glimpse into evolutionary pressures and latent potential of these pathogens.

Arboviral plasticity and robustness

Although the potential for phenotypic change is rooted in genetic diversity, the extent to which phenotypic plasticity correlates to genetic change is determined within the context of individual fitness landscapes. A robust strain or population is one that exists within a relatively flat fitness landscape and is therefore resistant to phenotypic change in the face of mutation. The distinction between robustness and plasticity is critical for understanding the evolutionary potential of arboviruses, as the genetic diversity within or between hosts may often represent phenotypically redundant strains. Such redundancy intuitively would be counter to adaptability, yet studies also demonstrate that robustness can at times promote adaptability, as more fragile mutant swarms may lack the capacity to navigate through sequence space (Goldhill *et al.*, 2014; van Nimwegen, 2006). Indeed, theory suggests that evolution at high mutation rates characteristic of RNA viruses will result in robust populations with broad fitness landscapes. This 'survival of the flattest' scenario is useful to buffer against the cost of frequent deleterious mutations, but could also act as a constraint against selection for strains with superior fitness when mutational neighbours are significantly less fit (Wilke *et al.*, 2001). A hallmark for robustness is relaxed purifying selection, i.e. an increased tolerance for mutation accumulation, as has often been noted for arboviruses (Holmes, 2003; Jerzak *et al.*, 2005). One way in which RNA viruses can overcome this need for robustness is by utilizing cooperative interactions. A study with mosquito-cell adapted WNV demonstrates that a population selected under conditions of frequent co-infection can evolve to cooperate, such that swarm fitness is higher than any individual and both genetic and phenotypic diversity are more readily retained (Ciota *et al.*, 2012b). Although the frequency of co-infection *in vivo* is not well studied, evidence that deleterious mutants can likely be maintained in mosquitoes through strain complementation has been shown for both WNV (Brackney *et al.*, 2011) and DENV (Aaskov *et al.*, 2006). Possessing a mechanism for the retention of diversity could be advantageous for host cycling pathogens, yet frequent co-infection could also lead to decreased diversity in the absence of complementation due to clonal interference or competition among haplotypes (Dennehy *et al.*, 2013). The challenge with arboviruses in particular is that ever-shifting fitness landscapes make experimentally determining swarm plasticity and/or robustness difficult. This is clear in studies that demonstrate that swarm size and strength of selection is host-dependent (Ciota *et al.*, 2009; Jerzak *et al.*, 2007; Lin *et al.*, 2004). The combination of deep sequencing with the identification of genetic markers or signatures associated with specific phenotypic changes (Lauring and Andino, 2011) would begin to allow us to assess the potential for adaptation or host shifts of arboviral swarms.

Multihost adaptation and host range expansion

Although traditional theory suggests an evolutionary bias for specialism, most pathogens are capable of utilizing multiple host environments (Woolhouse *et al.*, 2001). The success of arboviruses in particular demonstrates that exploiting highly divergent multihost systems for amplification and transmission is an evolutionarily viable strategy. Although there has been much written about the theoretical evolutionary and adaptive costs of generalism due to antagonistic pleiotropy and/or mutation accumulation (Turner and Elena, 2000), experimental evidence to support the ubiquitous nature of such costs both among arboviruses (Ciota *et al.*, 2010; Deardorff *et al.*, 2011) and species in general (Hereford, 2009) is lacking. As discussed, the capacity to maintain genetic diversity may be paramount to arboviruses avoiding significant cost from host cycling and therefore it is likely not coincidental, but rather by evolutionary design, that these pathogens are almost exclusively RNA viruses. The composition and breadth of viral swarms may be highly dynamic, and molecular memory, host-specific robustness, and perpetual influx of new mutations could provide a cryptic genetic and phenotypic fluidity in the absence of significant consensus genotypic change. In addition, the idea of tradeoffs holds an inherent assumption that host environments are exclusive in the way they interact with viruses, yet at times similar pathways, receptors and mechanisms of replication are shared and, as a result, fitness landscapes of seemingly divergent hosts may intersect more than is normally assumed (Baranowski *et al.*, 2001; Strauss *et al.*, 1994). Despite this, host-specific requirements, and therefore selective pressures, are certainly present to some extent for arboviruses (Ciota *et al.*, 2008, 2009; Coffey and Vignuzzi, 2011; Deardorff *et al.*, 2011; Fitzpatrick *et al.*, 2010; Turner and Elena, 2000; Vasilakis *et al.*, 2009) but the lesson from past experimental and epidemiological studies is that defining these requires consideration of the uniqueness of individual systems and viruses. Studies with VEEV demonstrate that host-specific fitness increases and modest costs in bypassed hosts are associated with specialization in mosquitoes or mice (Coffey *et al.*, 2008). Increased fitness of WNV in *Cx. pipiens* mosquitoes did not correlate to attenuated avian viraemia (Ciota *et al.*, 2008), yet similar studies demonstrate inconsistent and species-specific trade-offs, as well as the possibility that single host passage can at times result in co-adaptation (Deardorff *et al.*, 2011). Studies with SLEV demonstrate that further adaptation to *Cx. pipiens* mosquitoes may not be attainable and although other species may be capable of modest adaptations to primary vectors or hosts, what may be more epidemiologically important than further adaptation to current transmission cycles is the capacity for host range expansions or shifts. The generalist evolutionary history of arboviruses, together with the fact that they are likely to be frequently imbibed by a range of haematophagous vectors, makes them uniquely poised for adaptation to novel vector species (Turner *et al.*, 2010). Passage studies with SLEV in tick cell culture demonstrate that host shifts are achievable with modest genetic change and little cost (Ciota *et al.*, 2014). The simultaneous adaptation of CHIKV to *Ae. albopictus* in multiple locations, which is also not associated with a cost in *Ae. aegypti* or human hosts (Ng and Hapuarachchi, 2010), demonstrates how similar expansions in nature can have substantial epidemiological consequences.

Next generation studies of the relationship between mutational diversity and phenotype

Significant progress has been made towards understanding how mutational diversity impacts the phenotype of an RNA virus population. Groundbreaking studies of vesicular stomatitis virus (*Rhabdoviridae*: *Rhabdovirus*), foot and mouth disease virus (*Picornaviridae*: *Aphthovirus*) and others demonstrated that RNA viruses display several predicted properties of quasispecies populations, and made significant progress in attributing population diversity with measurable phenotypes such as replication efficiency and fitness (Domingo *et al.*, 1978; Ruiz-Jarabo *et al.*, 2000; Wilke and Novella, 2003). For example, seminal studies by John Holland and colleagues, among others, demonstrated that stochastic reductions in RNA virus population diversity, imposed by sequential passage at extremely low MOI, leads to fitness loss through a mechanism akin to 'Muller's ratchet' (Duarte *et al.*, 1992). Additional studies demonstrated that the average fitness of the individuals comprising any given virus population was lower than the fitness of the RNA virus population as a whole, consistent with classical quasispecies theoretical predictions (Duarte *et al.*, 1994). These studies further estimated that the fitness of virus populations can increase on average by 250% after an astonishingly short amount of time under selection, demonstrating the capacity for rapid evolution of these agents. Such fitness increases are extremely high, given that in population biology, a relative fitness (ω) increase of only 1% can have profound implications and readily become fixed in a given population. Finally, studies of RNA viruses and their population dynamics have supported several tenets of classic population biology. These include the competitive exclusion principle, which holds that complete competitors cannot coexist (Hardin, 1960), and the Red Queen hypothesis (Clarke *et al.*, 1994), which states that organisms must continually adapt in order to survive in a competitive or changing environment. These studies and many others have provided a critical foundation for the study of intrahost population biology as it relates to RNA viruses.

The vast majority of this foundational work, however, was conducted using tissue cultures. This presents obvious limitations in interpreting the results since in nature, viruses must infect intact organisms, circumvent or subvert innate and acquired immune responses and pass through physiological barriers in order to persist. The ultimate goal of much research into RNA viruses is to more deeply understand their biology *under the conditions they encounter in nature*. This, however, is easier said than done. Critical problems in studying RNA virus populations within natural hosts have been recently addressed, permitting the field to move into more realistic experimental systems and thus ask the questions that are currently most relevant.

Among these is the problem of sampling the virus population. Specifically, until recently it was not clear how one could sample a population containing 10^6–10^{10} members in a reliable and repeatable manner. The first efforts to study viral RNA sequences in the context of viral quasispecies used high-fidelity RT-PCR, cloning of amplicons and Sanger sequencing. In this way, a population could be sequenced at ~20–40′ coverage, but at great cost and though excessively labour-intensive procedures. While studies conducted using this approach revealed several important aspects of virus–host interactions (Jerzak *et al.*, 2005; Schneider and Roossinck, 2000, 2001), they were plagued by significant sampling issues. First, it was extremely difficult to sample entire genomes due to the inefficiency of whole-genome RT-PCR and cloning. As a result, most of these studies focused on only one or a few genomic regions. Second, the repeatability of any sampling effort was always questionable. Specifically, the likelihood of sampling any given member of an RNA virus population was a function of (a) its frequency in the population (b) the number of individuals sampled and (c) the overall size of the population. Moreover, low frequency variants cannot be sampled reliably. These problems have recently been overcome through the use of next-generation sequencing (NGS). The chief advantages of NGS over more traditional methods are the ability to sequence complete genomes at high coverage and to obtain specific information about the quality of individual reads. The disadvantage of NGS for estimation of viral population diversity is that most sequencing platforms [Illumina is currently dominant, but others (e.g. Roche 454 etc.) are frequently used] produce a relatively high error rate. Several computational approaches have been implemented in order to accurately identify which 'variants' represent existing RNA genomes and which are process-associated errors. These range from the fairly simple (imposing a quality-based cut-off to minimize the likelihood of poor quality bases contributing to the population profile) to the fairly sophisticated (using moving quality windows, phasing, etc.) (Macalalad *et al.*, 2012; Yang *et al.*, 2013). Additionally, several fairly common sense approaches have been proposed, such as deriving an expected error frequency from the distribution of quality scores at each position and using simple statistical tests to determine whether any putative variant exceeds that threshold. In any case, several options exist that vastly reduce the sequencing 'noise' that occurs at 50,000′ coverage. In other words, the problem has shifted from one of technologically imposed lack of sensitivity to the need for tools (currently largely computational) to reduce false-positive variant calls. In terms of deeply understanding the causes and consequences of RNA virus genetic diversity, the latter problem is, by far, the better one to have, and it has largely been solved.

Summary and future studies

Decades of studies have begun to shed light on the unique stochastic and selective forces shaping arbovirus genomes, as well as the phenotypically and epidemiological correlates of genetic change, but as with many biological systems, the more we uncover, the more complexity is revealed. Indeed, this complexity, as well as the specificity and dynamic nature of arbovirus–host systems, makes the prospect of predictive power with genotype-phenotype studies a daunting challenge. Despite this, the application of knowledge gained from these studies combined with experimental and technological advances has begun to pave the way for increasingly relevant investigations into arbovirus evolution.

One critical advance has been a move away from *in vitro* systems into those that more accurately mimic natural transmission cycles. Although this paradigm had been established for some years prior to the advent and widespread usage of NGS, and several important observations had been made on RNA virus population structure within hosts, recent studies have leveraged NGS and *in vivo* systems in impressive ways. Excellent examples of this have recently been published on CHIKV host interactions (Stapleford *et al.*, 2014). Notably, coupling experimental evolution studies with CHIKV in two important mosquito vectors with NGS has shown that (a) this approach can be used to accurately predict known fitness-enhancing mutations and (b) provide novel insights on additional mutations that could conceivably further enhance virus replication in these vectors and impact transmission. While the broad utility of this approach is not clear at present, these studies help demonstrate its possible application.

Another critical advancement has been combining NGS with Circle sequencing (Lou *et al.*, 2013) or duplex sequencing (Schmitt *et al.*, 2012), which utilize internal controls and redundancy to correct for NGS errors. The combination of these methods with traditional passage studies has demonstrated the capacity to effectively track mutant swarm fluctuations and fitness landscapes with high levels of resolution (Acevedo *et al.*, 2014). Utilizing similar methodology with *in vivo* studies will provide us with an unprecedented glimpse into evolving arboviral swarms.

Despite these advances, several challenges remain in this rapidly evolving field. First, it is currently unclear what the ability to completely characterize the population diversity of an RNA virus within a host adds to our understanding of phenotype. To be sure, examining mutation frequencies and selective pressures at the single nucleotide level provides unprecedented and elegant insights into virus–host interactions, but some would question the real impact of a single RNA genome present at 10^{-9} frequency. Several studies have demonstrated the ability of RNA virus populations to suppress the impact of minority variants that are of significantly higher fitness (De La Torre and Holland, 1990; Ebel *et al.*, 2011). What, then, becomes of these high-fitness, suppressed genomes? This, as well as many other questions pertaining to fitness landscapes, mutational robustness and multihost adaptation, can now be answered. However, the problem of attributing population genetic data to viral phenotype remains a challenge, particularly if virus genotype by host genotype outcomes, rather than generic interactions, dominate the phenotypic landscape. Achieving predictive power may therefore ultimately require advancements not just in our understanding of arbovirus evolution and fitness, but also of the genetic correlates of host and vector factors contributing to arbovirus fitness and transmission.

References

Aaskov, J., Buzacott, K., Thu, H.M., Lowry, K., and Holmes, E.C. (2006). Long-term transmission of defective RNA viruses in humans and *Aedes* mosquitoes. Science *311*, 236–238.

Acevedo, A., Brodsky, L., and Andino, R. (2014). Mutational and fitness landscapes of an RNA virus revealed through population sequencing. Nature *505*, 686–690.

Aliota, M.T., Jones, S.A., Dupuis, A.P., Ciota, A.T., Hubalek, Z., and Kramer, L.D. (2012). Characterization of Rabensburg virus, a flavivirus closely related to West Nile virus of the Japanese encephalitis antigenic group. PLoS One *7*, e39387.

Anishchenko, M., Bowen, R.A., Paessler, S., Austgen, L., Greene, I.P., and Weaver, S.C. (2006). Venezuelan encephalitis emergence mediated by a phylogenetically predicted viral mutation. Proc. Natl. Acad. Sci. U.S.A. *103*, 4994–4999.

Arrigo, N.C., Adams, A.P., and Weaver, S.C. (2010a). Evolutionary patterns of eastern equine encephalitis virus in North versus South America suggest ecological differences and taxonomic revision. J. Virol. *84*, 1014–1025.

Arrigo, N.C., Adams, A.P., Watts, D.M., Newman, P.C., and Weaver, S.C. (2010b). Cotton rats and house sparrows as hosts for North and South American strains of eastern equine encephalitis virus. Emerg. Infect. Dis. *16*, 1373–1380.

Baranowski, E., Ruiz-Jarabo, C.M., and Domingo, E. (2001). Evolution of cell recognition by viruses. Science *292*, 1102–1105.

Bondre, V.P., Jadi, R.S., Mishra, A.C., Yergolkar, P.N., and Arankalle, V.A. (2007). West Nile virus isolates from India: evidence for a distinct genetic lineage. J. Gen. Virol. *88*, 875–884.

Brackney, D.E., Beane, J.E., and Ebel, G.D. (2009). RNAi targeting of West Nile virus in mosquito midguts promotes virus diversification. PLoS Pathog. *5*, e1000502.

Brackney, D.E., Brown, I.K., Nofchissey, R.A., Fitzpatrick, K.A., and Ebel, G.D. (2010). Homogeneity of Powassan virus populations in naturally infected *Ixodes scapularis*. Virology *402*, 366–371.

Brackney, D.E., Pesko, K.N., Brown, I.K., Deardorff, E.R., Kawatachi, J., and Ebel, G.D. (2011). West Nile virus genetic diversity is maintained during transmission by *Culex pipiens* quinquefasciatus mosquitoes. PLoS One *6*, e24466.

Brault, A.C., Huang, C.Y., Langevin, S.A., Kinney, R.M., Bowen, R.A., Ramey, W.N., Panella, N.A., Holmes, E.C., Powers, A.M., and Miller, B.R. (2007).A single positively selected West Nile viral mutation confers increased virogenesis in American crows. Nat. Genet *39*, 1162–1166.

Brault, A.C., Powers, A.M., Ortiz, D., Estrada-Franco, J.G., Navarro-Lopez, R., and Weaver, S.C. (2004). Venezuelan equine encephalitis emergence: enhanced vector infection from a single amino acid substitution in the envelope glycoprotein. Proc. Natl. Acad. Sci. U.S.A. *101*, 11344–11349.

Ciota, A.T., and Kramer, L.D. (2010). Insights into arbovirus evolution and adaptation from experimental studies. Viruses *2*, 2594–2617.

Ciota, A.T., Lovelace, A.O., Jones, S.A., Payne, A., and Kramer, L.D. (2007a). Adaptation of two flaviviruses results in differences in genetic heterogeneity and virus adaptability. J. Gen. Virol. *88*, 2398–2406.

Ciota, A.T., Ngo, K.A., Lovelace, A.O., Payne, A.F., Zhou, Y., Shi, P.-Y., and Kramer, L.D. (2007b). Role of the mutant spectrum in adaptation and replication of West Nile virus. J. Gen. Virol. *88*, 865–874.

Ciota, A.T., Lovelace, A.O., Jia, Y., Davis, L.J., Young, D.S., and Kramer, L.D. (2008). Characterization of mosquito-adapted West Nile virus. J. Gen. Virol. *89*, 1633–1642.

Ciota, A.T., Jia, Y., Payne, A.F., Jerzak, G., Davis, L.J., Young, D.S., Ehrbar, D., and Kramer, L.D. (2009). Experimental passage of St. Louis encephalitis virus in vivo in mosquitoes and chickens reveals evolutionarily significant virus characteristics. PLoS One *4*, e7876.

Ciota, A.T., Koch, E.M., Willsey, G.G., Davis, L.J., Jerzak, G.V., Ehrbar, D.J., Wilke, C.O., and Kramer, L.D. (2011). Temporal and spatial alterations in mutant swarm size of St. Louis encephalitis virus in mosquito hosts. Infect. Genet. Evol. *11*, 460–468.

Ciota, A.T., Ehrbar, D.J., Van Slyke, G.A., Payne, A.F., Willsey, G.G., Viscio, R.E., and Kramer, L.D. (2012a). Quantification of intrahost bottlenecks of West Nile virus in *Culex pipiens* mosquitoes using an artificial mutant swarm. Infect. Genet. Evol. *12*, 557–564.

Ciota, A.T., Ehrbar, D.J., Van Slyke, G.A., Willsey, G.G., and Kramer, L.D. (2012b). Cooperative interactions in the West Nile virus mutant swarm. BMC Evol. Biol. *12*, 58.

Ciota, A.T., Ehrbar, D.J., Matacchiero, A.C., Van Slyke, G.A., and Kramer, L.D. (2013). The evolution of virulence of West Nile virus in a mosquito vector: implications for arbovirus adaptation and evolution. BMC Evol. Biol. *13*, 71.

Ciota, A.T., Payne, A.F., Ngo, K., and Kramer, L. (2014). Consequences of in vitro host shift for St. Louis encephalitis virus. J. Gen. Virol. *95*, 1281–1288.

Clarke, D.K., Duarte, E.A., Elena, S.F., Moya, A., Domingo, E., and Holland, J. (1994). The red queen reigns in the kingdom of RNA viruses. Proc. Natl. Acad. Sci. U.S.A. *91*, 4821–4824.

Coffey, L.L., and Vignuzzi, M. (2011). Host alternation of chikungunya virus increases fitness while restricting population diversity and adaptability to novel selective pressures. J. Virol. *85*, 1025–1035.

Coffey, L.L., Beeharry, Y., Borderia, A.V., Blanc, H., and Vignuzzi, M. (2011). Arbovirus high fidelity variant loses fitness in mosquitoes and mice. Proc. Natl. Acad. Sci. U.S.A. *108*, 16038–16043.

Coffey, L.L., Vasilakis, N., Brault, A.C., Powers, A.M., Tripet, F., and Weaver, S.C. (2008). Arbovirus evolution in vivo is constrained by host alternation. Proc. Natl. Acad. Sci. U.S.A. *105*, 6970–6975.

De La Torre, J.C., and Holland, J.J. (1990). RNA virus quasispecies populations can suppress vastly superior mutant progeny. J. Virol. *64*, 6278–6281.

Deardorff, E.R., Fitzpatrick, K.A., Jerzak, G.V., Shi, P.Y., Kramer, L.D., and Ebel, G.D. (2011). West Nile virus experimental evolution in vivo and the trade-off hypothesis. PLoS Pathog. *7*, e1002335.

Dennehy, J.J., Duffy, S., O'Keefe, K.J., Edwards, S.V., and Turner, P.E. (2013). Frequent coinfection reduces RNA virus population genetic diversity. J. Hered. *104*, 704–712.

Domingo, E., Sabo, D., Taniguchi, T., and Weissmann, C. (1978). Nucleotide sequence heterogeneity of an RNA phage population. Cell *13*, 735–744.

Domingo, E., Escarmis, C., Lazaro, E., and Manrubia, S.C. (2005). Quasispecies dynamics and RNA virus extinction. Virus Res. *107*, 129–139.

Drake, J.W., and Holland, J.J. (1999). Mutation rates among RNA viruses. Proc. Natl. Acad. Sci. U.S.A. *96*, 13910–13913.

Duarte, E., Clarke, D., Moya, A., Domingo, E., and Holland, J. (1992). Rapid fitness losses in mammalian RNA virus clones due to Mueller's ratchet. Proc. Natl. Acad. Sci. U.S.A. *89*, 6015–6019.

Duarte, E.A., Novella, I.S., Ledesma, S., Clarke, D.K., Moya, A., Elena, S.F., Domingo, E. & Holland, J.J. (1994). Subclonal components of consensus fitness in an RNA virus clone. J. Virol. *68*, 4295–4301.

Duffy, S., Shackelton, L.A., and Holmes, E.C. (2008). Rates of evolutionary change in viruses: patterns and determinants. Nat. Rev. Genet. *9*, 267–276.

Ebel, G.D., Dupuis, A.P. II, Ngo, K.A., Nicholas, D.C., Kauffman, E.B., Jones, S.A., Young, D.M., Maffei, J.G., Shi, P.-Y., Bernard, K.A., and Kramer, L.D. (2001). Partial genetic characterization of West Nile virus strains, New York State. Emerg. Infect. Dis. *7*, 650–653.

Ebel, G.D., Carricaburu, J., Young, D., Bernard, K.A., and Kramer, L.D. (2004). Genetic and phenotypic variation of West Nile virus in New York, 2000–2003. Am. J. Trop. Med. Hyg. *71*, 493–500.

Ebel, G.D., Fitzpatrick, K.A., Lim, P.Y., Bennett, C.J., Deardorff, E.R., Jerzak, G.V., Kramer, L.D., Zhou, Y., Shi, P.Y., and Bernard, K.A. (2011). Nonconsensus West Nile virus genomes arising during mosquito infection suppress pathogenesis and modulate virus fitness in vivo. J. Virol. *85*, 12605–12613.

Eigen, M. (2002). Error catastrophe and antiviral strategy. Proc. Natl. Acad. Sci. U.S.A. *99*, 13374–13376.

Fitzpatrick, K.A., Deardorff, E.R., Pesko, K., Brackney, D.E., Zhang, B., Bedrick, E., Shi, P.Y., and Ebel, G.D. (2010). Population variation of West Nile virus confers a host-specific fitness benefit in mosquitoes. Virology *404*, 89–95.

Forrester, N.L., Guerbois, M., Seymour, R.L., Spratt, H., and Weaver, S.C. (2012). Vector-borne transmission imposes a severe bottleneck on an RNA virus population. PLoS Pathog. *8*, e1002897.

Furio, V., Moya, A., and Sanjuan, R. (2005). The cost of replication fidelity in an RNA virus. Proc. Natl. Acad. Sci. U.S.A. *102*, 10233–10237.

Girard, Y.A., Klingler, K.A., and Higgs, S. (2004). West Nile virus dissemination and tissue tropisms in orally infected *Culex pipiens quinquefasciatus*. Vector Borne Zoonotic Dis. *4*, 109–122.

Goldberg, T.L., Anderson, T.K., and Hamer, G.L. (2010). West Nile virus may have hitched a ride across the Western United States on *Culex tarsalis* mosquitoes. Mol. Ecol. *19*, 1518–1519.

Goldhill, D., Lee, A., Williams, E.S., and Turner, P.E. (2014). Evolvability and robustness in populations of RNA virus Phi6. Front. Microbiol. *5*, 35.

Hardin, G. (1960). The competitive exclusion principle. Science *131*, 1292–1297.

Hardy, J.L., Houk, E.J., Kramer, L.D., and Reeves, W.C. (1983). Intrinsic factors affecting vector competence of mosquitoes for arboviruses. Annu. Rev. Entomol. *28*, 229–262.

Hereford, J. (2009). A quantitative survey of local adaptation and fitness trade-offs. Am. Nat. *173*, 579–588.

Holland, J.J., Spindler, K., Horodyski, F., Grabau, E., Nichol, S., and VandePol, S. (1982). Rapid evolution of RNA genomes. Science *215*, 1577–1585.

Holmes, E.C. (2003). Patterns of intra- and interhost nonsynonymous variation reveal strong purifying selection in dengue virus. J. Virol. *77*, 11296–11298.

Jenkins, G.M., Rambaut, A., Pybus, O.G., and Holmes, E.C. (2002). Rates of molecular evolution in RNA viruses: a quantitative phylogenetic analysis. J. Mol. Evol. *54*, 156–165.

Jerzak, G., Bernard, K.A., Kramer, L.D., and Ebel, G.D. (2005). Genetic variation in West Nile virus from naturally infected mosquitoes and birds suggests quasispecies structure and strong purifying selection. J. Gen. Virol. *86*, 2175–2183.

Jerzak, G.V., Bernard, K., Kramer, L.D., Shi, P.Y., and Ebel, G.D. (2007). The West Nile virus mutant spectrum is host-dependant and a determinant of mortality in mice. Virology *360*, 469–476.

Lambrechts, L., Chevillon, C., Albright, R.G., Thaisomboonsuk, B., Richardson, J.H., Jarman, R.G., and Scott, T.W. (2009). Genetic specificity and potential for local adaptation between dengue viruses and mosquito vectors. BMC Evol. Biol. *9*, 160.

Lauring, A.S., and Andino, R. (2011). Exploring the fitness landscape of an RNA virus by using a universal barcode microarray. J. Virol. *85*, 3780–3791.

Lin, S.R., Hsieh, S.C., Yueh, Y.Y., Lin, T.H., Chao, D.Y., Chen, W.J., King, C.C., and Wang, W.K. (2004). Study of sequence variation of dengue type 3 virus in naturally infected mosquitoes and human hosts: implications for transmission and evolution. J. Virol. *78*, 12717–12721.

Lou, D.I., Hussmann, J.A., McBee, R.M., Acevedo, A., Andino, R., Press, W.H., and Sawyer, S.L. (2013). High-throughput DNA sequencing errors are reduced by orders of magnitude using circle sequencing. Proc. Natl. Acad. Sci. U.S.A. *110*, 19872–19877.

Macalalad, A.R., Zody, M.C., Charlebois, P., Lennon, N.J., Newman, R.M., Malboeuf, C.M., Ryan, E.M., Boutwell, C.L., Power, K.A., Brackney, D.E., *et al.* (2012). Highly sensitive and specific detection of rare variants in mixed viral populations from massively parallel sequence data. PLoS Comput. Biol. *8*, e1002417.

McMullen, A.R., May, F.J., Li, L., Guzman, H., Bueno, R., Jr., Dennett, J.A., Tesh, R.B., and Barrett, A.D. (2011). Evolution of new genotype of West Nile virus in North America. Emerg. Infect. Dis. *17*, 785–793.

Moudy, R.M., Meola, M.A., Morin, L.L., Ebel, G.D., and Kramer, L.D. (2007). A newly emergent genotype of west nile virus is transmitted earlier and more efficiently by *Culex* mosquitoes. Am. J. Trop. Med. Hyg. *77*, 365–370.

Ng, L.C., and Hapuarachchi, H.C. (2010). Tracing the path of Chikungunya virus – evolution and adaptation. Infect. Genet. Evol. *10*, 876–885.

van Nimwegen, E. (2006). Epidemiology. Influenza escapes immunity along neutral networks. Science *314*, 1884–1886.

Novella, I.S., and Ebendick-Corpus, B.E. (2004). Molecular basis of fitness loss and fitness recovery in vesicular stomatitis virus. J. Mol. Biol. *342*, 1423–1430.

Novella, I.S., Clarke, D.K., Quer, J., Duarte, E.A., Lee, C.H., Weaver, S.C., Elena, S.F., Moya, A., Domingo, E., and Holland, J.J. (1995). Extreme fitness differences in mammalian and insect hosts after continuous replication of vesicular stomatitis virus in sandfly cells. J. Virol. *69*, 6805–6809.

Ruiz-Jarabo, C.M., Arias, A., Baranowski, E., Escarmis, C., and Domingo, E. (2000). Memory in viral quasispecies. J. Virol. *74*, 3543–3547.

Schmitt, M.W., Kennedy, S.R., Salk, J.J., Fox, E.J., Hiatt, J.B., and Loeb, L.A. (2012). Detection of ultra-rare mutations by next-generation sequencing. Proc. Natl. Acad. Sci. U.S.A. *109*, 14508–14513.

Schneider, W.L., and Roossinck, M.J. (2000). Evolutionarily related Sindbis-like plant viruses maintain different levels of population diversity in a common host. J. Virol. *74*, 3130–3134.

Schneider, W.L., and Roossinck, M.J. (2001). Genetic diversity in RNA virus quasispecies is controlled by host–virus interactions. J. Virol. *75*, 6566–6571.

Smith, D.R., Arrigo, N.C., Leal, G., Muehlberger, L.E., and Weaver, S.C. (2007). Infection and dissemination of Venezuelan equine encephalitis virus in the epidemic mosquito vector, *Aedes taeniorhynchus*. Am. J. Trop. Med. Hyg. *77*, 176–187.

Stapleford, K.A., Coffey, L.L., Lay, S., Borderia, A.V., Duong, V., Isakov, O., Rozen-Gagnon, K., Arias-Goeta, C., Blanc, H., Beaucourt, S., *et al.* (2014). Emergence and transmission of arbovirus evolutionary intermediates with epidemic potential. Cell Host Microbe *15*, 706–716.

Strauss, J.H., Wang, K.S., Schmaljohn, A.L., Kuhn, R.J., and Strauss, E.G. (1994). Host-cell receptors for Sindbis virus. Arch. Virol. *9*(Suppl.), 473–484.

Tsetsarkin, K.A., Vanlandingham, D.L., McGee, C.E., and Higgs, S. (2007). A single mutation in chikungunya virus affects vector specificity and epidemic potential. PLoS Pathog. *3*, e201.

Tsetsarkin, K.A., Chen, R., Leal, G., Forrester, N., Higgs, S., Huang, J., and Weaver, S.C. (2011). Chikungunya virus emergence is constrained in Asia by lineage-specific adaptive landscapes. Proc. Natl. Acad. Sci. U.S.A. *108*, 7872–7877.

Turner, P.E., and Elena, S.F. (2000). Cost of host radiation in an RNA virus. Genetics *156*, 1465–1470.

Turner, P.E., Morales, N.M., Alto, B.W., and Remold, S.K. (2010). Role of evolved host breadth in the initial emergence of an RNA virus. Evolution *64*, 3273–3286.

Twiddy, S.S., Farrar, J.J., Vinh Chau, N., Wills, B., Gould, E.A., Gritsun, T., Lloyd, G., and Holmes, E.C. (2002). Phylogenetic relationships and differential selection pressures among genotypes of dengue-2 virus. Virology *298*, 63–72.

Vasilakis, N., Deardorff, E.R., Kenney, J.L., Rossi, S.L., Hanley, K.A., and Weaver, S.C. (2009). Mosquitoes put the brake on arbovirus evolution: experimental evolution reveals slower mutation accumulation in mosquito than vertebrate cells. PLoS Pathog *5*, e1000467.

Vazquez, A., Sanchez-Seco, M.P., Ruiz, S., Molero, F., Hernandez, L., Moreno, J., Magallanes, A., Tejedor, C.G., and Tenorio, A. (2010). Putative new lineage of west nile virus, Spain. Emerg. Infect. Dis. *16*, 549–552.

Vignuzzi, M., Stone, J.K., Arnold, J.J., Cameron, C.E., and Andino, R. (2006). Quasispecies diversity determines pathogenesis through cooperative interactions in a viral population. Nature *439*, 344–348.

Vignuzzi, M., Wendt, E., and Andino, R. (2008). Engineering attenuated virus vaccines by controlling replication fidelity. Nat. Med. *14*, 154–161.

Weaver, S.C., Bellew, L.A., and Rico-Hesse, R. (1992a). Phylogenetic analysis of alphaviruses in the Venezuelan equine encephalitis complex and identification of the source of epizootic viruses. Virology *191*, 282–290.

Weaver, S.C., Rico-Hesse, R., and Scott, T.W. (1992b). Genetic diversity and slow rates of evolution in New World alphaviruses. Curr. Top. Microbiol. Immunol. *176*, 99–117.

Wilke, C.O., and Novella, I.S. (2003). Phenotypic mixing and hiding may contribute to memory in viral quasispecies. BMC Microbiol. 3, 11.

Wilke, C.O., Wang, J.L., Ofria, C., Lenski, R.E., and Adami, C. (2001). Evolution of digital organisms at high mutation rates leads to survival of the flattest. Nature *412*, 331–333.

Woolhouse, M.E., Taylor, L.H., and Haydon, D.T. (2001). Population biology of multihost pathogens. Science *292*, 1109–1112.

Yang, X., Charlebois, P., Macalalad, A., Henn, M.R., and Zody, M.C. (2013). V-Phaser 2: variant inference for viral populations. BMC Genomics *14*, 674.

Zanotto, P.M., Gao, G.F., Gritsun, T., Marin, M.S., Jiang, W.R., Venugopal, K., Reid, H.W., and Gould, E.A. (1995). An arbovirus cline across the northern hemisphere. Virology *210*, 152–159.

Arbovirus Genomics and Metagenomics

12

Adam Fitch, Matthew B. Rogers, Lijia Cui and Elodie Ghedin

Abstract

The arboviruses are a diverse group of arthropod transmitted viruses belonging to several viral families, almost all containing RNA genomes. New methodologies in viral genomics have enabled the probing of the different genomic structures observed in this group, including segmented and non-segmented single and double stranded RNA. We review sequence-dependent targeted gene approaches and sequence-independent metagenomic methods for the characterization of arbovirus genomes. The rapid expansion in the number of arboviral sequencing projects, focused on both known viruses for molecular epidemiology studies and unknown viruses for novel strain discovery, has been enabled by the development of new high throughput sequencing technologies. Outbreaks of both known and novel arboviruses mean that full genome characterization will continue to be salient to both research and healthcare organizations, informing vaccine development while fostering viral pathogenicity studies.

Introduction

This chapter aims to provide a practical overview on how to approach whole genome sequencing of arboviruses. The motivation for an arbovirus 'genome project' can be variable, from the characterization of a novel virus isolated in the field to the surveillance and comparative analysis of an epidemic strain. Arboviruses have a wide ranging pathology for domestic and wild animals (Hubalek *et al.*, 2014) and a high potential to emerge as human pathogens, as seen with the devastating outbreaks of West Nile virus (WNV) (Maxmen, 2012), chikungunya virus (CHKV) (Burt *et al.*, 2012), dengue virus (DENV) (Guzman *et al.*, 2010; Ritchie *et al.*, 2013) and the re-emergence of yellow fever virus (YFV) (Jentes *et al.*, 2011). Sequence characterization of this important group of viruses will continue to be vital as a measure to protect public health.

An arbovirus is generally defined as a virus transmitted to a vertebrate host by an arthropod vector, resulting in replication in the host. With such a broad meaning it is of no surprise that arboviruses are a diverse group with highly variable genome structures (Fig. 12.1). Alphaviruses (*Togaviridae*) such as CHKV, eastern equine encephalitis virus (EEEV), western equine encephalitis virus (WEEV) and Venezuelan equine encephalitis virus (VEEV) have single-stranded, positive-sense RNA genomes with a 5′7-methylguanosine cap ($m^7G5'ppp5'A$) and a 3′ polyA tail. Flaviviruses (*Flaviviridiae)* like YFV, DENV, and WNV also contain single-stranded, positive-sense RNA genomes with a $m^7G5'ppp5'A$ cap but lack a 3′ polyA tail. The bunyaviruses (*Bunyaviridae*), rhabdoviruses (*Rhabdoviridae*), and orthomyxoviruses (*Orthomyxoviridae*) are all negative sense ssRNA viruses with segmented genomes. Reoviruses also have segmented genomes, but consist of dsRNA (Murphy *et al.*, 1995). The lone DNA virus in the group, African swine fever virus (*Asfarviridae*) has a single dsDNA genomic segment. We can exploit some of these unique genomic features for more efficient sequencing.

Virus discovery in arthropods is still a fecund field. An estimate of the number of unknown viral species in mammals, a relatively small group of animals, indicated a minimum of 320,000 (Anthony *et al.*, 2013). It seems likely that this number is at least a log scale larger in arthropods, considering the number of undiscovered arthropod species is projected to be in the millions (Odegaard, 2000). As evidence of the discovery potential, a recent sequencing effort on the RNA of 70 arthropod species native to China uncovered 112 novel viruses (Li *et al.*, 2015). Viral genomics has become a significant area of ongoing research, precipitated by the advent of massively parallel sequencing technologies. These second and third-generation sequencing technological advances, commonly referred to as next-generation (NextGen) sequencing, coupled with increased interest in pandemic prediction and outbreak detection/prevention have pushed the field of viral genomics forward.

Approaches to whole genome sequencing

The choice of method (Fig. 12.2) for an arbovirus whole genome sequence project depends on several factors; most importantly, what is known about the viral isolate and the scale of the study, along with budgetary considerations and analysis goals. Well-characterized viruses are most efficiently sequenced

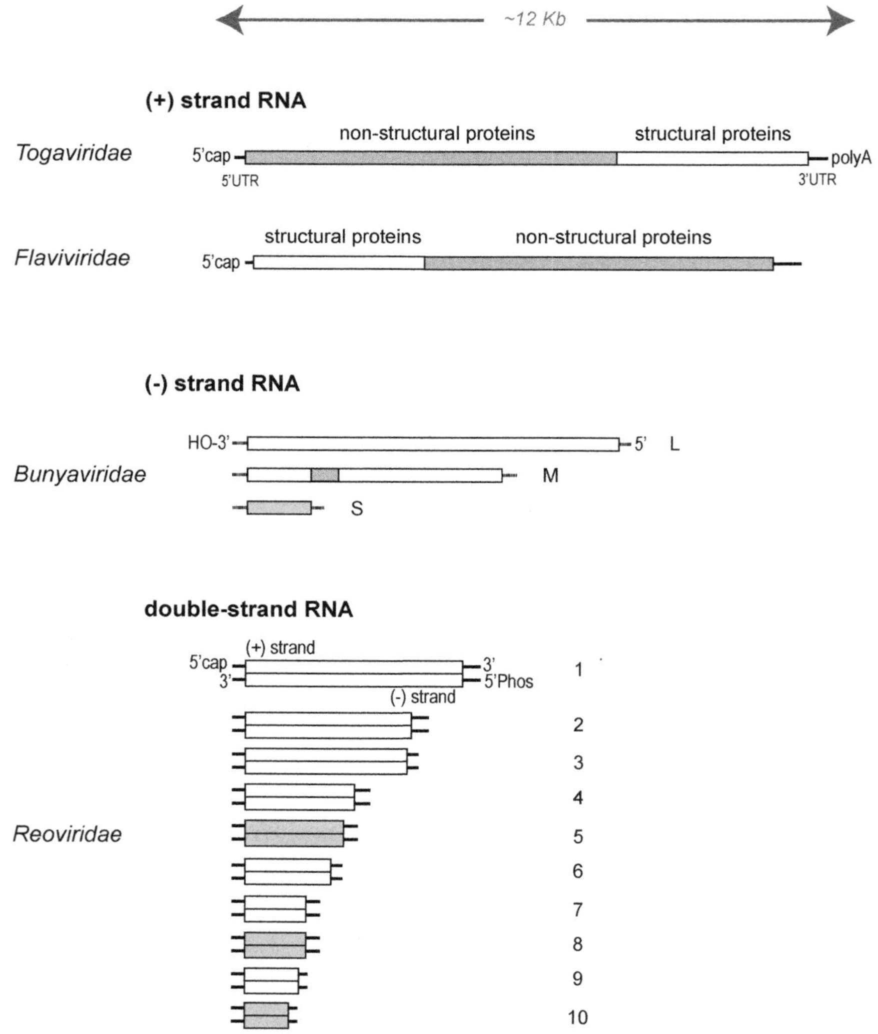

Figure 12.1 Examples of differing genome organizations and structures found in arboviruses. Regions that encode structural proteins are highlighted in grey. Representative species for each virus family listed here are Chikungunya (*Togaviridae*), Dengue (*Flaviviridae*), Rift Valley fever virus (*Bunyaviridae*), and Bluetongue virus (*Reoviridae*). UTR, untranslated region.

by a sequence-dependent approach, leveraging the already reported sequence data to rapidly generate the sequence of the new isolate(s). Poorly characterized viruses are a little more challenging, depending on what is known about the virus. If a partial assembly is available, or if there are sequences available from closely related viruses, it may be possible to design primer sets based on conserved regions and perform long PCRs, which can themselves be primer walked. Completely novel or poorly characterized viruses that fail sequence-dependent PCR can first be partially characterized in a sequence-independent manner with the likelihood that sequence dependent closure work will be necessary to finish the genome.

Analysis goals and project size also need to be considered when choosing an approach. Although small scale studies with the goal of generating a consensus sequence are likely best done with Sanger sequencing, the cost of NextGen sequencing has decreased so dramatically that it should be considered for nearly any size study. If variant calling is a goal of the analysis, then both sequencing depth (in order to produce significance) and replicates from the primary isolated RNA (to exclude PCR-generated diversity) are required. Any DNA created or used for a Sanger based sequencing approach can be fed into a NextGen sequencing pipeline in order to generate the proper depth; it is only a matter of producing the correct library for the desired NextGen platform.

Sequence-dependent and sequence-independent amplification of virus genomes

Both sequence-dependent and sequence-independent methods are commonly used in arbovirus genome sequencing projects. The sequence-dependent method requires prior knowledge of the pathogen to be sequenced. In this approach, multiple primer pairs are designed to 'walk' down the viral genome segment(s), generating overlapping amplified PCR products (referred to as amplicons). The primer pairs can be designed with conserved sequencing primers, such as the M13 sequencing tags, to facilitate throughput, or the products can be cloned into sequencing vectors. The advantage of the sequence-dependent approach is its specificity as it should only amplify the target sequence of the virus to the exclusion of 'contaminant' nucleic acids, such as host DNA and RNA that would have been extracted along with the virus. Although an efficient method, the disadvantage is that novel or poorly characterized arboviruses will not be captured in this manner. Additionally, it is possible that some sequence diversity is lost due to primer mismatch. To decrease the loss of possible interspecies variation and capture diversity,

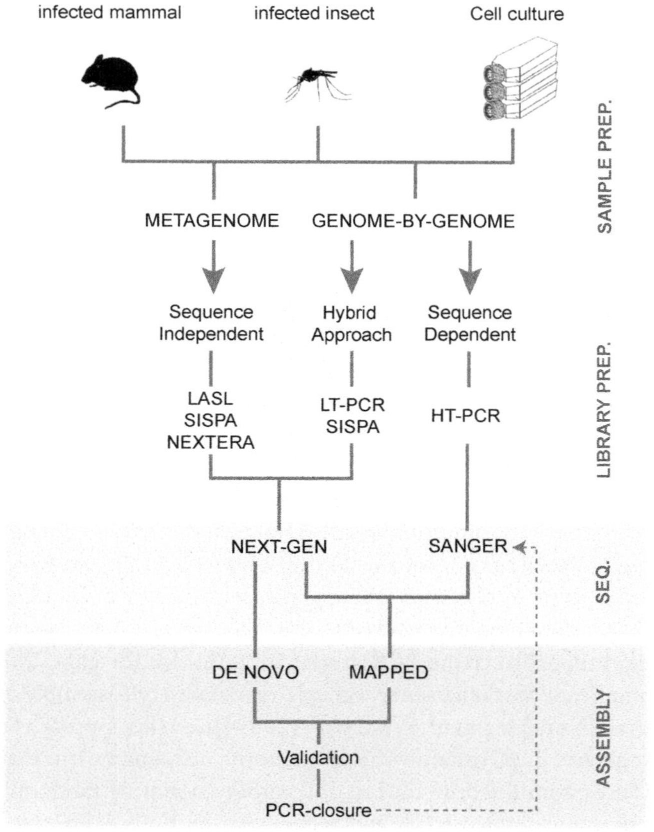

Figure 12.2 Overview of approaches for the sequencing of virus genomes. LASL, Linker Amplified Shotgun Library; SISPA, sequence-independent single primer amplification; LT-PCR, low-throughput PCR; HT-PCR, high-throughput PCR.

primers are designed with degeneracies and continuously updated throughout the length of the sequencing project.

The primer walking approach, although developed when Sanger sequencing was the only method available, is still used in many contemporary arbovirus projects because it remains the most direct and simple approach to characterize an RNA virus genome. A few examples of sequence-dependent genome projects for important arboviral pathogens include recent projects for WNV, CHKV and DENV. When WNV first emerged in the Western Hemisphere it was as an unknown agent causing human and bird encephalitis. A combination of epidemiology and microscopy indicated a flavivirus of arthropod origin as the cause. The Centers for Disease Control and Prevention (CDC) were able to serotype the isolate as the known and sequenced WNV, allowing for the use of long, overlapping amplicons to sequence the genomes of the first isolates of this virus from North America (Lanciotti *et al.*, 1999). The CHKV virus is a mosquito borne pathogen of significant public health concern that is rapidly expanding its geographic range. It quickly spread through the Caribbean, and in 2014 the first locally acquired cases emerged in the continental USA (Kendrick *et al.*, 2014). Sequencing projects targeting the virus use both primer walking of amplicons (Sreekumar *et al.*, 2010) and shearing of cDNA (Chen *et al.*, 2013). DENV is a major scourge of the tropics, infecting up to 390 million people annually according to a recent report (Bhatt *et al.*, 2013), resulting in significant morbidity and mortality on all populated continents except Europe, although the threat of a major outbreak there is looming. Because of its impact, DENV is also one of the best studied and sequence characterized arboviruses. Full genome sequencing pipelines with established primer sets are available to quickly primer walk all four subtypes (Chin-inmanu *et al.*, 2012; Christenbury *et al.*, 2010).

The sequence-independent approach requires no prior knowledge of the viral sequence and can take several forms. The choice of which sequence-independent approach to use is often informed by the type of sequencing project. One option is to simply shear extracted RNA and barcode it using a manufacturer's kit like the TruSeq RNA Sample kit from Illumina (Vasilakis *et al.*, 2013). Although a simpler approach, the method is expensive since each RNA sample will require its own library preparation (referred to as a library prep). A second option is to randomly amplify all nucleic acid present in a sample. Random amplification methods are generally much more cost effective than manufacturer's kits if a large group of viruses are being sequenced at the same time. Examples of this approach include Linker Amplified Shotgun Library (LASL) and sequence-independent single primer amplification (SISPA). LASL targets double-stranded DNA molecules, which are sheared, end polished and linker-adapted (Breitbart *et al.*, 2002). The introduced linker sequence is then used to prime a PCR reaction. This technique has been extended by Culley *et al.* (2006) to include RNA viruses, which are made into dsDNA by a process of reverse transcription and second strand synthesis. The SISPA method was originally developed by Reyes and Kim (1991) as a ligation-based technique, but was subsequently modified to include random PCR, filtering, and nuclease treatment (Allander *et al.*, 2001; Djikeng *et al.*, 2008; Froussard, 1992). SISPA compensates the prior knowledge limitation of the sequence-dependent approach by random priming all RNA molecules present in a sample. In SISPA, reverse transcription is performed on the arboviral RNA genome using a primer consisting of a fixed-tag (barcode) sequence followed by a random sequence. The random sequences used are of variable length (6N, 7N, 8N, 9N or 15N) depending on the study (Bishop-Lilly *et al.*, 2010; Hall-Mendelin *et al.*, 2013; Parameswaran *et al.*, 2012; Presti *et al.*, 2009; Victoria *et al.*, 2008). If the virus is known or suspected of having a poly-A tail, fixed-tag oligo-dT primers can also be added to increase coverage at the 3′ ends. Second strand synthesis is then performed using 3′–5′ Klenow exo-polymerase resulting in cDNA with the fixed-tag incorporated into both strands. A final PCR step is done to amplify cDNA sequences using the fixed-tag nucleotides.

The sequence-independent approach has often been used to characterize unknown isolates. Sierra Nevada, Kolente and Niakha viruses are examples of novel arboviruses sequenced with sequence independent methods as part of larger viral discovery and characterization projects (Ghedin *et al.*, 2013; Rogers *et al.*, 2014; Vasilakis *et al.*, 2013). RNA isolated from cell culture amplified and purified virions was either randomly amplified with SISPA (Sierra Nevada and Kolente) or randomly sheared with the Illumina TruSeq kit (Niakha). Although not on par with sequence-dependent methods, where the majority of the sequence data correspond to the virus genome, the percentage of non-contaminant reads attributed to each isolate was between 5% and 15% resulting in nearly full contiguous assemblies (contigs) for each virus. Eyach virus, a reovirus previously known to be distributed only in Europe, was found in mosquito homogenates from California by using sequence-independent priming on RNA isolated from inoculated mouse brain. Assemblies from the random sequencing methods produced enough sequence data to create significant contig matches in 8 of the 12 reported Eyach virus segments (Victoria *et al.*, 2008). An isolate of Highlands J virus, which is an alphavirus closely related to WEEV, was found by using SISPA on total RNA extracted from VERO cells that had been inoculated with filtered brain homogenate from a diseased Mississippi sand hill crane. Assembly produced a nearly complete (98.5%) single contig (Ip *et al.*, 2014). In addition to the detection of RNA arboviruses, Ng *et al.* (2011) reported that DNA viruses such as anelloviruses, circoviruses, herpesviruses, poxviruses, and papillomaviruses can be detected in mosquitoes using sequence independent methods, although these are likely not true arboviruses but transient residents picked up from a host during blood feeding.

Metagenomics for unbiased arbovirus discovery

The traditional methods of arbovirus discovery rely on isolating the virus from field-collected arthropods, amplifying virus in cell or animal culture, and isolating viral nucleic acids. These methods are time consuming, rely on culture methodology and require some prior knowledge of the causative virus (Bishop-Lilly *et al.*, 2010; Hall-Mendelin *et al.*, 2013). True viral metagenomics, on the other hand, is the amplification and sequencing of a viral community directly from primary samples. With the advent of NextGen sequencing and development of sequence-independent tools such as SISPA and LASL, it is now possible to sequence directly from the infected insect tissue.

Several groups have performed validation studies to test the suitability of this methodology for metagenomic samples by generating mock arbovirus metagenomes in arthropod vectors. For example, researchers used metagenomic approaches to track DENV, CHKV or YFV isolates that were experimentally introduced into mosquitoes (Bishop-Lilly *et al.*, 2010; Hall-Mendelin *et al.*, 2013). They reported not only the assembly of nearly the whole genome of those expected arboviruses, but also the discovery of some commensal mosquito bacteria, such as *Wolbachia*, *Pirellula* and *Asaia*. Until recently, the metagenomic approach had not been used in arbovirus discovery (Junglen and Drosten, 2013). In early 2014, a study investigated the DNA and RNA viral metagenome from the agricultural pest *Bemisia tabaci* and found RNA viruses from at least five different viral families as well as DNA from begomoviruses. Partial genome sequences generated in the metagenome analysis were used to produce primer pairs to fully sequence a novel carlavirus (Rosario *et al.*, 2014).

Notwithstanding the advantages of metagenomics, obtaining complete virus genomes from such data remains a challenge owing to the presence of non-viral nucleic acids. In the previously mentioned study on mosquitoes experimentally infected with DENV-1, the dengue virus sequences consisted of no more than 0.1% of the total number of reads (Bishop-Lilly *et al.*, 2010). However, these challenges are not insurmountable and the number of viral sequences generated will depend on many factors including viral titre, sample type, and the success of contaminant remediation methods.

Hybrid approach to whole genome capture

Sequence-dependent and -independent methods are not mutually exclusive and in fact are often used in combination. The hybrid approach is an efficient way to sequence a large number of known viruses in a high throughput, cost-effective manner. First, PCR with specific or partially degenerate primers is used to produce long, overlapping amplicons. These amplicons are combined in equimolar amounts (if feasible) and then fed into the SISPA protocol resulting in randomly tagged products that are ready for pooling and NextGen library preparation. This approach is one of the more popular methodologies for high throughput sequencing of the non-arboviral influenza virus and it can easily be adapted for arboviral sequencing (Ghedin *et al.*, 2012).

On the other hand, the results of a sequence-independent investigation of a sample can inform sequencing-dependent methods. Random priming and amplification from a complex sample, such as a pool of infected mosquitoes, will almost always create gaps in coverage. These gaps are closed by assembling the random sequence reads into larger contigs and then designing primers to create PCR amplicons to bridge the gaps. The amplicons are then primer-walked to generate a full assembly of the genomic segment. Presti *et al.* (2009) used this approach to sequence the Quaranfil virus, a tick borne orthomyxovirus that causes a mild febrile illness. The authors began by randomly priming extracted RNA and sequencing by both Sanger and 454, generating contigs from several segments of a virus that shared some amino acid similarity to the PB1 and HA of other orthomyxoviruses. Knowing that othromyxoviruses typically have conserved ends, they then used these contig sequences to generate primers for PCR and RACE (discussed in the next section), eventually fully sequencing five segments of the virus.

Sequencing of the terminal ends

One of the more difficult challenges in any viral sequencing project is coverage at the terminal ends of the genome or genomic segments. Random priming methods result in poor coverage and sequence dependent methods can completely miss terminal regions depending on where the primers are designed. Untranslated regions can be captured by rapid amplification of cDNA ends (RACE), but as arboviruses include a large number of viral types with variable genome structures, finishing work for unknown viruses will have to occur after the primary sequence data are analysed. Fortunately, once a virus is classified and the structure of its genome understood, it is possible to tailor an approach. For the alphaviruses, the 5′ cap and 3′ poly-A tail lend themselves well to a standard 5′ and 3′ RACE. The genome of the Highlands J virus was finished in this manner (Allison and Stallknecht, 2009). The flaviviruses also have a 5′ cap but lack a poly-A tail. In this case the best approach is to simply add a poly-A tail using poly-A polymerase (Lanciotti *et al.*, 1999) and then proceed to 5′ and 3′ RACE, as was done recently for the characterization of a novel duck flavivirus (Liu *et al.*, 2012). The reoviruses contain segmented dsRNA. They can be finished with standard 5′ RACE, along with poly-A tail addition plus gene-specific primers for each segment for 3′ RACE (Li *et al.*, 2014). The orthomyxoviruses have ssRNA segments with conserved ends. These ends can be targeted in RACE with a combination of gene-specific priming and homopolymeric tail addition. Briese *et al.* (2014) used this method to sequence the ends of the novel tick-borne Upolu and Arkansas Bay viruses.

Wet-lab techniques for preparation of samples for sequencing

Virus RNA isolation from cell culture and whole tissue

The methods used to isolate viral particles are diverse and the type of sequencing to be performed should be determined prior to deciding on a protocol. Approaches using sequence dependent priming to target a known virus can be quite a bit simpler as there is not much concern with amplifying contaminants. If extracting from cell culture supernatant, simply spinning the sample down at low speed to pellet cellular debris, and then extracting from the supernatant will produce good quality results. On the other hand, methods relying on sequence independent amplification using random priming will result in very little viral sequence data if clean-up methods are not utilized.

The specifics of culture based methods to amplify and purify virus will vary depending on the eccentricities of the individual virus. Classic methods are covered in the *Virology Methods Manual* (1996) and utilize mammalian or insect cell culture, polyethylene glycol (PEG) precipitation, and gradient purification. Filtering and nuclease treatment, if performed, should be done following purification (see 'Removal of host contaminant nucleic acids', below). RNA isolation can then be accomplished with one of the extraction methods listed in this chapter.

Isolating viral particles from whole tissues first involves homogenizing the tissue using any one of a variety of mechanical means. The most expensive option is to use a bead beating system such as the TissueLyser II (Qiagen) or FastPrep (MP Biomedicals). These systems are highly effective, but the cost for a complete system is significant. A much less expensive – but also less efficient – bead beater is a vortex mixer with a special tube holding rack, or with the tubes simply taped to the stage of the mixer. Handheld homogenizers can also produce acceptable results. The Omni TH Tissue Homogenizer (Omni International) utilizes disposable plastic tips to help prevent contamination and features tips tailored to different tissue types. The simplest homogenization method is to grind the tissue in buffer in a 1.5-ml microcentrifuge tube with a disposable plastic pestle. Protocols may also utilize dry ice freeze–thaw cycles to help disrupt the virus containing tissue. Heavy cellular debris is then pelleted away and supernatants may be passed through 0.22 or 0.45 μM filters to remove remaining bacterial and eukaryotic cells. As an example of a protocol using a handheld homogenizer to extract virions from tissue, Daly *et al.* (2011) extracted hepatitis C virus from clinical biopsy samples by using a tissue homogenizer and freeze–thaw methodology. Briefly:

1. A small section of liver tissue (2 mm^3) was placed into an ice-cold solution of buffered saline, pH 7.2 supplemented with 0.7% BSA. The tissue was then homogenized for 15 s on ice using an Omni TH hard tip tissue probe.
2. The homogenate was freeze–thawed three times by placing on dry ice until frozen, then at 37°C until thawed.
3. To pellet nuclei and other cellular materials, the samples were centrifuged at 600 × *g* for 10 min, 4°C.
4. Free floating nucleic acid contaminants were removed by digestion in a protocol similar to that indicated in the 'Removal of host contaminant nucleic acid' section of this chapter.
5. Finally, viral RNA was purified using Trizol LS.

Hall-Mendelin's (2013) mock metagenome project, discussed in detail in the introduction, used a simpler bead-based method to lyse insects artificially infected with CHKV, DENV-3 and YFV. They omitted nuclease treatment and filtering prior to lysis as there was an interest in detecting additional microorganisms, including *Wolbachia* bacteria.

1. Infected mosquitoes were placed into a 1.5-ml tube containing Opti-MEM media supplemented with 3% FBS.
2. A metal bead (Qiagen) was then added to the tube, and the tube placed into a TissueLyser II automated shaker. The type of bead and buffer used can vary with the type of tissue that is to be homogenized and it is recommended to follow manufacturer's instructions for your specific tissue type.
3. The tube was shaken for 2 min to completely lyse the mosquitoes, and nucleic acid extraction followed the QIAamp method listed in 'Virus genomic RNA extraction', below.

Removal of host contaminant nucleic acids

The expression 'garbage in–garbage out' is particularly appropriate when attempting a metagenome project. While sample preparation can enrich for viral RNA, there will remain enough RNA from the host tissue to dominate the sequencing data, resulting in less than 0.1% of reads from viral genomes. Extra clean-up steps are recommended unless the follow-up sequencing library preparation involves the use of specific primers. If random primers are to be used, we recommend at a minimum a low speed spin, filtering and nuclease treatment. A final ultracentrifugation step through a sucrose gradient or cushion can also be performed to concentrate viral particles (Ghedin *et al.*, 2013; Rogers *et al.*, 2014; Rosario *et al.*, 2014; Vasilakis *et al.*, 2013). Treatment of a virus containing solution has to occur before the addition of lysis media as it is dependent upon the viral capsid to protect viral nucleic acids from being digested. Cleanup of a supernatant generally starts by briefly pelleting a sample (1000–5000 rpm for 1–10 min) to remove host cells. Pre-cleared samples are then filtered through a 0.22 or 0.45 μm filter unit to remove any remaining large contaminants such as bacterial cells. It is worth noting that some groups avoid the use of filters out of concern that the viral particles will stick to the filter. Nuclease cocktails such as Turbo DNase (10–30 U), RNase One (20–30 U), and/or micrococcal nuclease is then added and the solution incubated at 37°C for 1–2 h. The quantity of nuclease used and the incubation time can be adjusted and should be determined empirically. After digestion of host contaminants, viral nucleic acids are extracted following one of the extraction protocols in this chapter.

Hall *et al.* (2014) sought to experimentally validate these clean-up methods against a mock sample containing human, bacterial and viral elements. They found that low-speed spinning, filtering, and nuclease treatment independently or in combination all decreased the amount of human and bacterial DNA as judged by qPCR; however, only a three step procedure using all of the treatments resulted in increased viral sequencing reads.

A newer, potentially useful approach to cut down on host contaminants is to specifically bind the methylated DNA of the host. The NEBNext Microbiome DNA Enrichment Kit does this by binding to the methyl-CpG binding domain of human MBD2-Fc protein and pelleting it out of solution. Unmethylated or partially methylated viral DNA remains in the supernatant and can then be recovered. As this approach only works to remove highly methylated DNA contaminants it would be best applied to arboviral discovery from isolates cultured on mammalian cells.

Virus genomic RNA extraction

Viral samples with or without host contaminant depletion can be RNA extracted through a variety of means, with two of the most popular being phenol–chloroform extraction and silica column purification.

The acid guanidinium thiocyanate–phenol–chloroform procedure is a standard method to extract RNA from both liquid and solid samples (Chomczynski and Sacchi, 1987). Commercial preparations of this chemistry for liquid samples such as TRIzol LS Reagent (Invitrogen) and TRI Reagent LS (Molecular Research Center, Inc.) are widely available. Both are used at a 3:1 ratio of reagent to liquid sample. For simplicity, adjust the volume of liquid sample to 250 µl and add 750 µl reagent. RNA extraction then follows the manufacturer's protocol. In brief, the following steps are performed:

1 Homogenize the sample by pipetting up and down several times through the pipette tip.
2 Incubate the homogenate for 5 min at room temperature in order to complete lysis, and then add 0.2 ml of chloroform. Shake vigorously for 15 s.
3 Incubate the phenolchloroform mix at 15–30°C for 2 to 15 min, then spin at 12,000 × *g*, 4°C, for 15 min in order to separate the phases.
4 Post spin, the solution will be in three distinct phases – a large red lower phase, a thin, cloudy interphase, and a clear upper aqueous phase. The total volume of the aqueous phase is about 70% of the initial Trizol/TriReagent LS volume, in this case approximately 525 µl. Transfer most of the aqueous phase to a new tube, avoiding the interphase. Generally, it is recommended to transfer no more than 450 µl of the aqueous phase.
5 The RNA can now be precipitated out of the aqueous phase. Optional, but recommended, is to add 5–10 µg of a glycogen carrier (e.g. Glycoblue) to the aqueous phase prior to adding isopropanol. Glycogen helps precipitate low concentration RNA and makes it easier to observe the pellet during processing. Add 0.5ml of isopropanol and incubate sample at room temperature for 10 min, then centrifuge at 12,000 × *g*, 4°C, for 10 min.
6 Remove the isopropanol with a pipette tip and wash the pellet (repeated twice) by adding 1 ml of 75% ethanol, vortexing briefly, then spinning at 7500 × *g*, 4°C, for 5 min.
7 Remove as much of the second ethanol wash as possible and air dry the pellet for 5–10 min at room temperature. Avoid overdrying pellet as it will make the sample very difficult to dissolve. Do not speedvac or heat sample during drying as this often results in overdrying.
8 Add 25–100 µl RNase-free water, pipette several times, and if necessary heat at 55°C for 10 min to help dissolve RNA. Expect some DNA carryover in the extracted RNA. If this is unacceptable, you can completely eliminate DNA contamination by performing two rounds of DNase I treatment followed each time by phenol:chloroform extraction. Make aliquots of the final purified RNA and freeze at −80°C immediately.

Commercial kits for RNA extraction, such as the QIAmp viral RNA mini extraction kit (QIAgen), may be used in place of phenol–choloroform. These kits generally promise much faster extraction times as well as the elimination of toxic phenol compounds. Most commercial kits follow the same methodology of silica column binding, and there is no selection specifically for RNA over DNA although each kit typically has a recommended protocol for eliminating DNA contamination. Additionally, there have been recent reports of these kits being contaminated with viral nucleic acids (Naccache *et al.*, 2013), requiring strict negative controls. Commercial kits also typically recommend using repetitive RNA as a carrier (polyA, polyAT, polyT). Since this may interfere with downstream random amplifications we recommend avoiding it. Follow the manufacturer's instructions for the specific kit used.

Amplification of viral nucleic acids

The specificity of the sequence-dependent approach is its greatest strength, and it will yield high-quality data with low levels of host contamination, allowing for greater multiplexing or the choice to perform lower-sequencing output providing cost savings.

Most well-known arboviruses have existing primer sets that can be utilized. The previously mentioned efforts directed at CHKV (Sreekumar *et al.*, 2010), DENV (Chin-inmanu *et al.*, 2012; Christenbury *et al.*, 2010), and WNV (Lanciotti *et al.*, 1999) are a few examples. Primer sets for novel or poorly characterized viruses will need to be created using available sequence data, if possible. Sequence from related viruses may prove useful to identify conserved regions within families that could then be targeted with degenerate primers. A general protocol for reverse transcription and amplification is as follows:

1 Perform reverse transcription with the Superscript III RT kit (Invitrogen) in triplicate, following the manufacturer's instructions, with 6 µl of extracted RNA and 1 µl of one of three primers – 50 ng/µl random hexamers, 50 µM oligo(dT)-20 or 10 µM gene specific primer.
2 Treat reaction with RNase H to remove RNA template.
3 Combine random hexamer and oligo-dT cDNA reactions.
4 Perform PCR on the cDNA using specific primers to create tiled, overlapping amplicons.
5 Purify with AMPure XP beads (Beckham-Coulter)
6 Sanger sequence from both ends of each amplicon with M13 sequencing primers.

Sequence-independent SISPA methodology is used to randomly amplify and tag a DNA or cDNA sample, and can be used to create large multiplexed pools. RNA samples will first have to be reverse-transcribed into cDNA. The end result of the SISPA protocol is a long smear of randomly amplified DNA containing multiplex tags. This DNA must be size selected for the appropriate platform, and then converted into a sequencing library (Djikeng and Spiro, 2009).

SISPA for RNA:

1. Convert RNA to cDNA using reverse transcriptase: add 1 µl of 20 µM SISPA-6N (and 1 µl 100 nM SISPA-oligo-dT, if applicable), 50–200 ng RNA template, 0.6 µl DMSO, then adjust volume to 12.3 µl with DEPC treated H_2O.
2. Mix, spin, and incubate at 65°C for 5 min, then place on ice for 5 min.
3. Spin briefly again, then add 4 µl First Strand Buffer, 2 µl DTT, 1 µl of 10 mM each dNTPs, 0.2 µl RNase OUT (80U), 0.5 µl Superscript III RT (100 U/µl).
4. Mix and spin, then place at 25°C for 10 min to allow the random primers to anneal. Incubate at 50°C for 50 min, 85°C for 10 min, 4°C hold.
5. Create a second strand of cDNA by adding 0.5 µl 3′–5′ exo- Klenow polymerase (0.5 U) and incubating at 37°C for 60 min, 75°C for 10 min, 4°C hold.
6. Continue with step 5 from the DNA section to degrade primers, and then amplify with Accuprime Taq Hifi polymerase.

SISPA for DNA:

1. Combine, on ice, 1 µl of 20 µM SISPA-6N Primer, 50–200 ng DNA, 0.8 µl DMSO, then adjust volume to 10 µl with nuclease-free water.
2. Mix, briefly spin, and then incubate for 5 min at 95°C followed by 5 min on ice.
3. Spin sample again and add 2 µl 10× NEB Buffer II, 1 µl 3′–5′ exo- Klenow polymerase (5 U), 2 µl of 10 mM each dNTP mix, and 5 µl H_2O.
4. Mix, spin, and incubate at 37°C for 60 min, 75°C for 10 min.
5. Exo-SAP treat cDNA to remove primers by adding 1 µl SAP (1 U), 0.2 µl Exonuclease I (4U), 1 µl SAP buffer, and 17.8 µl H_2O. Mix, spin and incubate at 37°C for 60 min, 72°C for 15 min.
6. Amplify tagged DNA in a PCR reaction. Combine 2.5 µl Accuprime Buffer I, 0.2 µl (1U) Accuprime Taq Hifi, 1 µl of 12.5 µM SISPA primer 2 µl Exo/Sap treated cDNA and 19.3 µl H_2O.
7. Cycle as follows – 95°C for 2 min, then 25–35 cycles of 95°C for 30 s, 55°C for 30 s, 68°C for 60 s, then a final extension of 68°C for 10 min, 4°C hold.

PCR amplification of the SISPA products will result in a large smear of amplified materials, each individual sample barcoded with its own tag. These materials should be multiplexed prior to initial size selection, and the multiplexing should occur on ice to prevent chimera formation due to the still active Taq polymerase. When multiplexing large numbers of samples, for example from a 96-well plate, it is best to first combine them in smaller batches of a single column or row.

1. Combine, on ice, equal volumes of each PCR reaction into pre measured PB1 buffer. Pipette mix after each sample is added and vortex after adding the final sample in the group.
2. To bind sample to silica membrane, add 700 µl of combined sample to a chilled Qiagen MinElute column and spin at maximum speed, 4°C, for 1 min.
3. Discard flowthrough and then repeat as necessary until the entire sample is bound.
4. Wash bound samples with 2 × 700 µl of ice-cold wash buffer, discarding wash buffer after each spin.
5. Place column back into recovery tube and spin at max speed for 3 min to remove residual wash buffer.
6. Place column into a clean recovery tube and open lid to air dry for 5 min.
7. Add 15 µl of elution buffer (EB) directly to the centre of the membrane and incubate for 5 min at room temperature.
8. Spin at max speed for 1 min to elute DNA.
9. Optionally, add another 10 µl of EB to the centre of the membrane and spin at max speed for another minute.
10. Next, size-select the combined columns/rows by using the Millipore Gel Extraction system. First, create a gel using the low EDTA TAE included with the kit.
11. Run out ~75% of the eluted material on the gel, reserving the remaining 25% in case of a mistake.
12. Cut out fragments appropriate for the platform to be used. For Ion Torrent the maximum recommended size is 420 bp for a 400 base library and for Illumina systems it is generally from 300–600 bp.
13. Add the cut gel piece to the Millipore column and spin at 5000 × *g* for 10 min.
14. Measure the concentration of the eluted DNA.
15. To make the final sequencing pool, combine equal amounts of sample from each pooled and size selected column/row into PB1 Buffer and repeat the Qiagen MinElute protocol.

Another sequence-independent way to amplify and tag viral nucleic acids is to use the Nextera kit from Illumina. This approach to library production is referred to as 'tagmentation' as it involves simultaneously fragmenting input cDNA while adding an adapter (tag) sequence. The adapter sequence is utilized in a subsequent limited cycle PCR to add sequencing primer binding sites and indices. Tagmentation is accomplished by using a hyperactive mutant of the transposase Tn5. The Nextera system has many advantages, notably its extreme speed – going from input DNA to tagged sequencing library can take as little as 90 min – and that it generates consistent coverage across the genome because there is no primer binding bias, which occurs with random amplification. Input DNA concentrations are markedly reduced from the recommendations of Illumina's TruSeq kit with Nextera only requiring 50 ng of DNA and the newer Nextera XT requiring as little as 1 ng. Nextera XT also includes library normalization beads, which are DNA binding beads with a low maximum binding capacity that allows for rapid normalization and multiplexing of library pool components. The main disadvantage to using Nextera kits is the high cost of the reagents, as a Nextera library prep including a requisite index kit costs ~$100 per sample and Nextera XT costs ~$40 per sample. Also, the nature of the transposon mediated DNA insertion means that the last 50 bases on the distal ends of the input DNA cannot be adapted and will not produce sequence data. Finally, Nextera sample preps are dependent on PCR to add sequencing binding sites and indices.

Library and template preparation for NextGen sequencing

The following section details a few popular manufacturer's protocols for library preparation, but as these tend to change over time please check for updates prior to starting library creation. Sample libraries intended for NextGen platforms and generated by long PCRs or whole genome amplification will need to be sheared, end repaired, and barcoded/adapter ligated prior to combining into a final sample pool. Shearing methodologies vary and can be either mechanical (sonication, nebulization, acoustic) or enzymatic. All methods can be used to produce acceptable quality DNA for library production and the method chosen is up to the individual researcher. Tagged sample pools produced from random amplifications like SISPA do not need to be sheared, however they do need to be end repaired and adapter ligated for the specific platform.

Here is a summarized enzymatic shearing protocol based on Life technologies' instructions for preparing DNA for the Ion Torrent. If starting with SISPA products, begin at step 7.

1 Shear 100 ng of input DNA with the Ion Xpress fragmentase kit. The length of the shearing reaction is dependent on the initial size of the input DNA and test reactions at several time points are recommended prior to moving on to experimental samples.
2 Add, in order, to a 1.5-ml LoBind tube – 100 ng DNA, 5 μl 10 × Ion Shear Reaction Buffer, nuclease-free water to total volume of 40 μl. Mix by vortexing for 5 s, then pulse spin. Do not make a master mix or scale up the reactions.
3 Add 10 μl Ion Shear Plus Enzyme mix. Immediately proceed to mixing the solution by pipetting up and down 8–10 times with a P100–P200 pipette set at 40 μl.
4 Incubate the mix in a water bath or heat block set at 37°C for the time frame established by your test samples from step 1.
5 After incubation, stop the shearing reaction by adding 5 μl of Ion Shear Stop Buffer and vortexing for 5 s. Store the stopped reaction on ice.
6 Purify the fragmented DNA with AMPure XP beads as indicated in the Ion Torrent manual.
7 If using SISPA product, begin at this step. For DNA sheared using the Ion Xpress kit this step is not necessary. Combine 100 ng of SISPA product, 20 μl 5′ End Repair Buffer, and 1 μl End Repair Enzyme in a total volume of 100 μl. Incubate at room temperature for 20 min, and then purify with AMPure XP beads as per Ion Torrent manual.
8 Next, ligate sequencing adaptors and barcodes. To the 25 μl of purified DNA add: 10 μl of 10 × ligase buffer, 2 μl of P1 adapter, 2 μl of an Ion Xpress Barcode, 2 μl dNTP mix, 49 μl nuclease-free water, 2 μl DNA Ligase, and 8 μl Nick Repair Polymerase. Mix well by pipetting up and down, then spin briefly.
9 Place in a thermal cycler and incubate at 25°C for 15 min, 72°C for 5 min, 4°C hold. Proceed immediately to the next steps.
10 Purify the reaction again with AMPure XP beads as directed in the Ion Torrent manual.
11 Size-select the libraries by using the E-Gel SizeSelect agarose gel system. Alternative size selection methods are also possible but can result in wider size ranges.
12 The E-Gel Size select system has eight wells with a ladder lane in the middle. Life Technologies recommends avoiding the middle two lanes (#4, #5) due to ladder contamination issues, the outer two lanes (#1, #8) due to retarded migration patterns, and to not run different libraries side by side in lanes #2, #3 and #5, #6 as mixing of the libraries can occur. These recommendations result in two useable lanes per well. We have found that using lanes #1 and #8 is completely acceptable as long as the run time for these specific lanes is increased by 30 s. Using lanes #1 and #8 in addition to lanes #3 and #6 will cut the cost and time required for size selection in half and is a recommended change to the Life Technologies protocol.
13 Without pre-running the gel, load all 20 μl of adapter ligated, AMPure purified DNA from step 19. Load up to four different libraries per gel, using lanes #1, #3, #6, #8 as previously mentioned. Avoid loading sample in all other lanes.
14 Dilute the 50 bp DNA ladder (Life Technologies cat# 10416–014) at a ratio of 1:40 in low-TE buffer. Mix well and add 10 μl of this dilution to the middle marker well, lane M. Do not substitute other ladders for this marker.
15 Add 25 μl of nuclease-free water to all empty wells on both the top and bottom rows, with the exception of the bottom well from lane M in which you add 10 μl.
16 Run the gel to separate the fragments for purification, and monitor the progress of the bands closely. The 350 bp band in the 50 bp ladder is two- to three-fold brighter than the other bands allowing for easy orientation. Remember to run the outer lanes (#1, #8) for an extra 30 s after purifying the inner (#3, #6) wells. Estimated run times for the gels are 12–14 min for a 200 base-read library and 16–20 min for a 400 base-read library. Stop the gel when the appropriate reference marker band (500 bp for 400 base-read, 350 bp for 200 base-read) reaches the reference line on the gel and add an additional ~10 μl Low TE to the collection wells. Do not overfill. Continue running the gel. For a 400 base-read library, run until the 500 bp marker band is at the top edge of the collection well. For a 200 base-read library, run until the 350-bp marker band has just completely entered the top edge of the collection well. Stop the gel and remove the sample from the collection wells, then add an additional 10 μl of low TE buffer and collect again. Continue running the gel for 30 s longer and repeat the collection process for lanes #1 and #8.
17 Amplify the library using Platinum PCR SuperMix High Fidelity by combing in a 1.5-ml LoBind tube – 100 μl Platinum PCR SuperMix High Fidelity, 5 μl Library Amplification Primer Mix, 25 μl unamplified, size-selected library. Mix and split the reaction into two 0.2-ml PCR tubes.

18 Run the following PCR cycle: 95°C for 5 min, then eight cycles of 95°C for 15 s, 58°C for 15 s, 70°C for 1 min. Hold at 4°C.
19 Combine the duplicate PCRs back into a LoBind tube, and purify once again using 195 µl AMPure XP beads (1.5X sample volume) following the Ion Torrent manual.
20 Analyse a 1:10 dilution of each barcoded sample on the Bioanalyzer with a high -ensitivity chip, and determine the pmol/L of each barcoded sample. Make an equimolar pool of barcoded samples at the highest possible concentration, and check several dilutions (1:5, 1:10, 1:20, 1:50) on a Bioanalyzer high sensitivity chip. Freeze the final undiluted pool in aliquots to minimize freeze–thaw cycles.
21 When you are ready to proceed to sequencing, make a final dilution in low-TE buffer of the equimolar pooled library using the following formula:

Dilution factor = Library concentration in pM/26 pM

Library Preparation for the Illumina platforms using the NEBNext Ultra kit for Illumina is summarized below.

1 Shear DNA using one of the methodologies mentioned above or start with SISPA randomly generated materials. Use 5 ng–1 µg of this fragmented DNA along with the NEBNext Ultra kit for Illumina to create a sequencing library.
2 Repair ends by combining in a PCR tube or plate: 3 µl End Repair Enzyme, 6.5 µl End Repair Reaction Buffer, and 55.5 µl of fragmented DNA. Pipette mix then quick spin.
3 Place in a thermocycler with the heated lid option on, and run a program of 30 min at 20°C, 30 min at 65°C, 4°C hold.
4 Ligate adapters by adding 15 µl of Blunt/TA Ligase Master Mix, 2.5 µl NEBNext Adapter for Illumina, and 1 µl Ligation Enhancer.
5 Incubate at 20°C for 15 min in a thermocycler.
6 Remove tubes from thermocycler and add 3 µl of USER enzyme to each ligation reaction.
7 Mix well, quick spin, and then incubate at 37°C for 15 min.
8 Clean up adapter ligated DNA by using AMPure XP beads as directed in the NEBNext manual.
9 In order to index samples and increase DNA concentration, set up a limited cycle PCR amplification by combining 25 µl NEBNext High Fidelity 2′ PCR Master Mix, 1 µl Index Primer, 1 µl Universal PCR Primer, and 23 µl Adaptor Ligated DNA Fragments.
10 Run a PCR cycle of 98°C for 30 s, then 6–15 cycles of 98°C for 10 s, 65°C for 30 s, 72°C for 30 s, followed by a final extension of 72°C for 5 min. Hold at 4°C. The number of cycles for this reaction is determined by the amount of initial input DNA, with more cycles necessary for lower concentration inputs.
11 Clean up the PCR reaction with AMPure XP bead as directed in the NEBNext manual.
12 Check a 1:5 dilution of the library(ies) on a Bioanalyzer high-sensitivity chip. If multiplexing libraries, combine into an equimolar pool and check the concentration using a reliable method such as Picogreen, Qubit, or qPCR.

Sequencing on NextGen platforms

Next generation platforms feature variable read lengths, types of sequence reads, run times and output sizes. The most practical systems for arboviral sequencing projects are made by Illumina and Ion Torrent, due to their significantly reduced cost per base and large output as compared to 454 and SOLID systems, two platforms that are now largely being phased out. Pacific Bioscience's platform has more limited utility because of its low output, however its massive read length can be used to link mutations if that is an analysis goal. The Illumina systems (HiSeq, MiSeq, NextSeq) run natively as a paired end platform, generating sequence reads from both ends of a given template. The systems can be made to perform single read chemistry; however the significant decline in sequence quality as the read length increases limits the utility of this approach in most cases. The numerous HiSeq systems (1500, 2000, 2500, 3000, 4000) feature different read lengths and run times, but all have the unifying characteristic of massive outputs up to, and in some configurations exceeding, the 1Tb range. Of these systems the HiSeq 2500 has the longest possible read lengths at 2 × 250 bp. It also features a rapid run mode that can decrease output and run times to 3–5 days, from the 11 days required by the older HiSeq 1500/2000 systems. The 3000/4000 systems also utilize a speeded-up chemistry with maximum read lengths of 2 × 150 bp and run times of 1–3.5 days. The huge output of these platforms may, however, be more than what is needed for most arbovirus sequencing projects where a small number of virus samples are being characterized. The NextSeq is a benchtop and mid-range output platform with a maximum read length of 2 × 150 bases, a speeded-up run time of 29 h, and maximum output in the 100–120 Gb range. The MiSeq, also a benchtop model, features the longest read lengths, up to 2 × 300 bases, but the smallest maximum output of 15 Gb. Run times are from 1–3 days depending on the length of the reads. Life Technologies' Ion Torrent and Ion Proton systems run as single read platforms. The Ion Torrent was the first model of this semiconductor based platform and has a smaller output of 1 Gb maximum, but generates longer reads of up to 400 bases. The Ion Proton is the newer, genome scale system that generates up to 10 Gb, but with read lengths currently limited to 200 bases. Run times are up to four hours for both the Torrent and Proton. Lastly, the Pacific Biosciences RS II platform features the greatest read length of any NextGen platform, up to several kilobases, but with limited output and a higher error rate compared to the Illumina and Life Technologies systems. Run times for the RS II are some of the shortest, on the order of 0.5 to 3 h. Ultimately, each platform has different strengths, weaknesses, and error profiles, which should be understood before beginning a genomics project.

Bioinformatics protocols for second generation sequencing platforms

As a result of the high volume of sequence data now being generated through current NextGen sequencing platforms, it has become feasible to assemble whole genome sequences for viruses without PCR and/or cloning. The scale of data generated is such that it is possible to assemble whole virus genomes from primary samples or cell culture isolates, sequencing through the host contaminants that, in some instances, can

Table 12.1 List of some useful bioinformatics tools used in virus genomics projects

Tool	Description	URL
CASAVA	Illumina's proprietary base caller	http://www.illumina.com/software/genome_analyzer_software.ilmn
Biopython Seqio library	converts QUAL and FASTA files to FASTQ	http://biopython.org/wiki/SeqIO
bam2fastq	convert BAM format alignments back into FASTQ read file(s)	http://www.hudsonalpha.org/gsl/information/software/bam2fastq
trim.pl	Script for quality trimming of sequence reads	http://wiki.bioinformatics.ucdavis.edu/index.php/Trim.pl
FastX_barcode_splitter.pl	demultiplexing single-end reads	http://hannonlab.cshl.edu/fastx_toolkit/
SMALT BOWTIE BWA	Sequence aligners	http://www.sanger.ac.uk/resources/software/smalt http://bowtie-bio.sourceforge.net/index.shtml http://bio-bwa.sourceforge.net/
Velvet VICUNA	De novo assemblers	https://www.ebi.ac.uk/~zerbino/velvet/ http://www.broadinstitute.org/scientific-community/science/projects/viral-genomics/vicuna
ABySS MIRA	De novo assemblers that work well with pyrosequencing data	http://www.bcgsc.ca/platform/bioinfo/software/abyss http://sourceforge.net/projects/mira-assembler/
amosCMP	Mapped assembler	http://sourceforge.net/projects/amos/ http://sourceforge.net/apps/mediawiki/amos/index.php?title=AMOScmp
ABACAS	Contig ordering	http://abacas.sourceforge.net/Manual.html

represent over 99% of the read data. The following paragraphs touch on bioinformatics protocols for pre-processing raw read data, library de-multiplexing, contaminant filtering, viral whole genome assembly, assembly finishing, and annotation of viral scaffolds. A list of suggested bioinformatics tools is provided in Table 12.1.

Data structure and handling

File formats

Libraries sequenced on current NextGen platforms are typically received in FASTQ format (Fig. 12.3). This is a modified FASTA format that includes not only a header name and sequence, but also an additional string of ASCII characters encoding PHRED (Ewing *et al.*, 1998) quality scores. Most current NextGen platforms output Sanger range PHRED scores, but some older Illumina platforms may still output Solexa, Illumina 1.3 and 1.5 PHRED range values. Current versions of Illumina's proprietary base caller CASAVA (as of 2010) generate quality scores comparable to original Sanger values (http://www.illumina.com/software/genome_analyzer_software.ilmn). As mappers and assemblers take advantage of these quality scores it is essential first to determine what range of PHRED scores are outputted with read data. This can often be done by inspecting the quality string by eye and determining the range of ASCII characters, but can also be accomplished through the use of programs such as SolexaQA, which will detect the quality range, and perform necessary conversions. Many aligners such as Bowtie (Langmead *et al.*, 2009) and BWA (Li and Durbin, 2009) will allow users to specify the range of PHRED scores, but others such as SMALT (http://www.sanger.ac.uk/resources/software/smalt/) expect Sanger style quality scores. For paired end libraries, a forward and reverse file, suffixed _1, and _2, respectively, are generated. Paired reads are related to each other by their respective positions in the files (e.g. a forward read at line position 40,000 in the forward read file has a mate at the same position in the reverse file).

Two frequently encountered exceptions to FASTQ format are the original base quality encoding of Sanger read data, which consists of individual FASTA and QUAL files. These are extracted by default from the standard flowgram format (SFF) archives generated by 454 platforms and older capillary sequencing platforms. In these, the qualities are encoded in a separate (.qual) file, with PHRED scores represented by space separated integers rather than ASCII characters. Multiple programs exist for converting QUAL and FASTA files to FASTQ as well as the popular Biopython Seqio library (http://biopython.org/wiki/SeqIO). Finally, it is also becoming common to receive sequenced reads in binary alignment format (BAM), particularly from Ion Torrent platforms. This occurs often when the sequenced reads have been aligned against a reference, though due to the efficient compression of BAM format it is becoming common even for unaligned reads. Programs such as *bam2fastq* (http://www.hudsonalpha.org/gsl/information/software/bam2fastq) can be used to convert the BAM format alignments back into FASTQ read file(s).

Deconvoluting multiplexed data

The sequencing capacity of NextGen sequencing platforms is so large that it is often feasible to combine multiple libraries into a single sequencing run (see 'Wet-lab techniques for preparation of samples for sequencing'). While this greatly lowers the cost of the experiment, it also poses the additional challenge of computationally splitting reads from each library back into separate pools by barcode sequence (de-multiplexing). This process inevitably results in the loss of some data where a barcode sequence cannot be identified. When both library preparation and sequencing is done by a sequencing facility, it is common for read data to be sent back already de-multiplexed by the facility's software; CASAVA, for example, can do this. When library preparation is done in-house, it is often the case that this is done with custom primers and barcodes not used by sequencing centres, or recognized by their de-multiplexing program. In these instances it is up to the end user to perform the de-multiplexing of the raw reads through open-source or custom de-multiplexing scripts. Demultiplexing by barcode is

Fastq Format

Instrument Name, Run ID, Flow Cell ID, Flow Cell Lane, Tile Number, Cluster X-coord, Cluster Y-coord, Forward/Reverse (1:N:0/2:N:0)

```
Header      @HWI-ST980:352:C1PTRACXX:5:2202:10457:19076 2:N:0
Sequence    TCCTGAAGTTGGTGGTAGGGTTAAAGTCTTCTATCCTCCTGGTAGTCACGCAGATCTGCTTGTTGATGTTCCACTTCCCTTTTGTGAATGGATACT
            +
Quality     DDDHFFHDHGGIEFGCBEEEHIIIHFEGIIEFDHHGEEHGIG??BD?FG6DGGIIGHIICGIGHC?;ACFHD;BFFBCEEEEE:A=;5@A::>;CC
Phred33
ASCII
```

Fasta + Qual Format

```
Header      >gnl|ti|1172642810 1047111480567
Sequence    CTGCTTATGCAGTGGAATCGTCCTACATTAAGTTAAAGG

Header      >gnl|ti|1172642810 1047111480567
Quality      35 35 14 13 10 10 12 13 13 13 13 13 13 13  9 13 13  9  6  6 9  9  9  9 13  9  9 10 11 10  9  9  9 15 14 21 26 34 44 40
Phred33
```

Csfasta + Qual Format (SOLID Colour Space)

```
Header      >1279_19_1175_F5-BC
Sequence    G22320303213222132222200102102132311

Header      >1279_19_1175_F5-BC
Quality     32 17 24 14 30 17 23 33 24 25 27 29 25 27 21 30 27 18 23 6 24 9 8 4 7 15 17 16 6 6 4 26 5 6 8
Phred33
```

Figure 12.3 Information contained in FASTA, QUAL and FASTQ formats.

relatively straightforward when dealing with single-end read data, but presents multiple challenges to paired-end data. Among the most commonly used applications for demultiplexing single-end reads is the *FastX_barcode_splitter.pl* (http://hannonlab.cshl.edu/fastx_toolkit/). This script works well at splitting reads into pools by barcode where the reads are single-ended, but will not use information from mate-paired reads in determining the sequencing pool, or detect chimeric inserts (i.e. instances where there are different barcodes on the forward and reverse reads). The challenge with paired-end reads is not only to fully take advantage of barcode information on both forward and reverse reads, but also to maintain the order of mate paired reads in both forward and reverse FASTQ files for downstream use. The Velvet assembler (Zerbino and Birney, 2008) accepts an interleaved format of forward and reverse reads where the forward and reverse sequences are staggered according to their order in the input files.

In house, we use custom barcodes and have generated python and perl scripts to handle barcode identification. The scripts first identify the barcode at the start of read 1 and then compare that to the barcode at the start of read 2. Reads are then binned into matching barcodes, single barcode, or mismatching barcode files before further processing.

Cleaning of reads and removal of host contamination

It is standard practice to quality trim read data prior to assembly. As with de-multiplexing, it is important to choose a quality trimming program that will conserve the paired-end relationship of both forward and reverse reads. For our applications we use the open source *trim.pl* (http://wiki.bioinformatics.ucdavis.edu/index.php/Trim.pl). The benefit of *trim.pl* is that it outputs both a forward and reverse trimmed read file, as well as a file of broken reads (where one of the pairs has been discarded). Most assemblers will accept a combination of paired end and single reads, so these files can be used in combination to generate the assembly (Rogers *et al.*, 2014).

Another practical recommendation is to first extract reads that map to a known contaminant – be it a host organism, a cell-culture type, or a contaminating microbe. For mapped assemblers such as amosCMP (Pop *et al.*, 2004), and consensus layout assemblers this step may be required in order to make the assembly tractable in a reasonable amount of time. Many read aligners currently exist, and vary according to ease of use, supported platforms and data formats accepted, and availability as open source. We have used Bowtie2 (Langmead and Salzberg, 2012) extensively in removing contaminating reads. Both Bowtie and Bowtie2 are very popular readers due to their relative ease of use, while also providing the user with a wide range of options for alignment type (local vs. global), control over mismatch scoring, gap and extension penalties and word seed matches. Bowtie2 also offers control over output files, including the option of dumping unaligned reads to a FASTQ file format. Control over gap extension and creation penalties becomes important when aligning reads from 454 or Ion torrent platforms where insertions and deletions (indels) due to homopolymer repeats become common. To remove contaminants using other aligners (e.g. BWA), the output SAM format file can be converted to BAM format, then unaligned pairs of reads can be filtered by bit-flag with Samtools to generate a BAM file of unmapped reads. This file can then be converted to forward and reverse FASTQ files of unmapped reads using bam2fastq, which can then be used in an assembly.

Types of assembly and choice of assembler

The choice of genome assembler should be informed by a combination of expected depth of sequencing, sequencing platform type, availability of a closely related reference and whether the data are paired- or single-end. In the case of depth of sequence data (or coverage), it is important to consider not only the size of the library, but the extent of non-viral reads contaminating the library. Most recently developed de-novo assemblers that operate by manipulating a de-brujin graph (Zerbino and Birney, 2008) work best with very large numbers of reads (typically in the millions). Overlap-layout-consensus assemblers such as the Celera Assembler can generate high-quality assemblies with considerably fewer reads but are time and computationally intensive, so are not commonly used with high depth second-generation platforms. One exception is the recently developed VICUNA algorithm (Yang *et al.*, 2012). VICUNA produces assemblies using an overlap-layout-consensus style algorithm like the Celera assembler and outputs multiple sequence alignments of the reads comprising each contig, making it applicable to viral communities where large variation is expected. VICUNA also allows for pre-filtering of reads against a contamination database, and joining of contigs using a reference genome. Recent comparison of VICUNA to Velvet found that the best Velvet assemblies were comparable to the best VICUNA assemblies (Wong *et al.*, 2013), though Velvet was far more sensitive to adjustments in parameters such as expected coverage and k-mer.

Guided assemblers can be used where a closely related viral genome to the target genome is known. These work well when host contamination is high, and can generate robust assemblies from a minimal depth of read data in a short amount of time. Use of guided assemblers come with the caveat that they will fail to assemble unique and highly variable regions between the target and reference genomes. Our experience has been principally with Velvet for de-novo assemblies, and amosCMP for guided assemblies, so the following sections deal with our use of these.

De novo assembly of viral genomes

We have had much success assembling arbovirus genomes with Velvet (Zerbino and Birney, 2008; Zerbino *et al.*, 2009), a de Brujin graph assembler. For the *de novo* assembly of virus genomes, Velvet proves to be a powerful assembler. It operates by manipulating de Brujin graphs of k-mers (short sequence words) that overlap by a k-mer size −1. Operationally, Velvet functions in two steps. First the read data are hashed into k-mers of user-specified size with *velveth*, and then a graph is constructed of these k-mers. This graph is then manipulated by *velvetg* to generate contigs. The choice of k-mer length influences the quality of the assembly in that very small k-mers may result in misassembled reads, while large ones risk wasting data resulting in gaps. We have found from our data that most viral genomes assemble best at high k-mers (typically greater than 49); we suspect at larger k-mers it is less likely that low coverage reads of host origin will be mistakenly assembled, resulting in a more complete assembly of the high copy viral genome. For uncontaminated samples, much lower k-mers (e.g. 19–33) are typically recommended. Nevertheless, the speed of the assembly process with Velvet allows for low-cost experimentation with a range of different k-mer and expected coverage sizes. Velvet also has the advantage of accepting multiple library types, such as paired end trimmed reads and broken reads (i.e. a read without a mate). Sanger reads, produced to span low-coverage areas, can also be incorporated at the sequence hashing step and used in assembly.

While Velvet works very well with Illumina data and Sanger data, it is not as well suited for assembly of Ion torrent or 454 reads due to issues with homopolymer repeats. For these data, de-brujin graph assemblers tailored towards pyrosequencing such as Assembly By Short Sequence (ABySS) (Simpson *et al.*, 2009) and MIRA (http://sourceforge.net/projects/mira-assembler/) should be used.

Mapped/guided assembly of viral genomes

It is often not feasible to assemble viral genomes using a de novo assembler due to low depth of viral reads. In these instances, if the sampled virus is known and is expected to be highly similar at the nucleotide level to a known public viral genome, it may be possible to assemble the viral genome using a mapped assembler. We have previously used AmosCMP to assemble strains of poxviruses (Kerr *et al.*, 2012). AmosCMP is a module of the popular Amos assembler project (Treangen *et al.*, 2011) that will generate mapped assemblies using Nucmer (Kurtz *et al.*, 2004) to map reads to the reference, then construct a consensus layout assembly based on these mapped reads (http://sourceforge.net/apps/mediawiki/amos/index.php?title=AMOScmp). Unlike the previous programs mentioned in this chapter, AmosCMP requires sequence and quality files encoded in separate files (.fasta and .qual). This conversion can be achieved with the biopython SeqIO library. The sequence and quality files then need to be converted into an Amos message file, which can be accepted as input by amosCMP.

Assembly finishing and annotation

Assembly quality can be assessed statistically. Size, maximum contig size and N50 values are outputted by Velvet. The size of the entire assembly is typically not a useful statistic as the final assembly inevitably includes contigs assembled from contaminating reads, but it is important nevertheless that the complete assembly size is not smaller than the expected viral genome size when known. N50 is a measure of how fragmented an assembly is. The N50 value of an assembly is the minimum size of contigs required to assemble half of the genome. They are also frequently reported along with an N90 value, and often plots of a range of N values can be informative in assessing the quality of an assembly (Gurevich *et al.*, 2013). Finally, maximum contig size can be a useful measure where a viral segment of known size is expected.

De novo assemblies invariably contain a mixture of contigs from both the target genome, and contaminants in the sample. To identify contigs specific to the target arbovirus genome, we perform BlastX queries on a local version of the nr database restricted to viral peptide GI numbers. This frequently allows for the identification of complete, or near complete segments, which can then be extracted from the assembly file.

Where segments are broken across multiple contigs we use the *abacas.pl* script (Assefa *et al.*, 2009) to align assembled contigs to a closely related reference using Promer (Kurtz *et al.*, 2004). This returns contigs ordered against reference segments by synteny with gaps introduced in the query sequence. With this 'ordered' assembly created, primers can then be designed for regions flanking the gaps either by manually examining the segment in Artemis (Carver *et al.*, 2008), or through the use of primer3 to computationally select good primer sites in the vicinity of the gaps. Following PCR closure of these gaps (see above), we annotate coding sequences (CDS) manually using Artemis, and then check the conceptual translations with BlastX searches against the Genbank nr database. As most arboviruses encode few protein coding genes, and CDS tend to occupy the length of entire segments, manual annotation is not a labour-intensive process for a few dozen samples. For larger DNA viruses with a closely related reference we have used the rapid annotation transfer script (Otto *et al.*, 2011), which will take EMBL format annotation from a closely related virus, and based on Nucmer (Kurtz *et al.*, 2004) matches attempt to predict related coding sequences in an unannotated genome.

Following a first round of annotation, genes with internal stop codons, or apparent frameshifts are checked by PCR, and a corrected version of the genome produced. Functional annotations of coding sequences are based on close matches to other viral peptides, or based on protein domains determined by searches against the on-line Interpro database (Hunter *et al.*, 2012).

Conclusion

Arboviral sequencing projects have become quite commonplace and the facilities necessary for such projects are accessible to any size laboratory. The goals of the sequencing projects may vary but the underlying techniques necessary to sequence any type of arbovirus have been well established. The advent of NextGen sequencing has significantly decreased the cost and time necessary to successfully sequence large numbers of arboviruses but has resulted in an increase in the reliance on computational methods for data storage and analysis. Bioinformatic analysis will remain the largest bottleneck as it is relatively easy for a single lab technician to generate large datasets that can then occupy months of a bioinformatic analyst's time.

Future trends

Arboviruses will continue to represent a major threat to human health. As the range of important disease-carrying arthropods, such as *Aedes* mosquitoes and hard ticks, expand so will the prevalence of arboviral disease. Large scale sequencing projects to track the outbreaks of disease as well as understand the evolution of viral pathogens will continue to follow suit. Indeed, major sequencing projects such as the Broad Institute's DENV (http://www.broadinstitute.org/annotation/viral/Dengue) and WNV (http://www.broadinstitute.org/annotation/viral/WNV/Home.html) projects and JCVI's EEEV project (http://gsc.jcvi.org/projects/gsc/arbovirus2/index.php) are generating data from thousands of field isolates.

Viral metagenomics studies from arthropod samples are in an expansion phase. Nearly all published arboviral studies involve a known target or culture dependent methodology. Increased funding for, and interest in, all types of 'biome' projects, as well as increased interest in pandemic prediction will lead to more metagenomic sequencing effort directed at arthropods and their unexplored viral metagenomes. NextGen platforms will continue to innovate and evolve, increasing output while decreasing sequencing run times. Inexpensive benchtop sequencers from both Illumina and Ion Torrent are capable of outputting gigabases worth of data at a fraction of the cost of generating the same data on the smaller output 454 or SOLID platforms from half a decade ago. These significant cost savings will continue the democratizing of sequencing allowing for nearly any size laboratory to conduct arboviral sequencing projects. Cloud based computing solutions and innovation in sequence analysis pipelines will allow greater access to NextGen data, fostering collaboration and hastening understanding. Technical innovations such as single virus genomics (Allen *et al.*, 2011), single molecule sequencing (Alquezar-Planas *et al.*, 2013) and circular sequencing library preparation (Acevedo *et al.*, 2014) will allow researchers to truly determine viral diversity and linked mutations within a population, cutting out platform and PCR errors. NextGen sequencers are beginning to move into the clinical lab and will become a significant tool in the detection of viral outbreaks, especially those of unknown aetiology (Firth and Lipkin, 2013). The use of sequence-independent protocols combined with rapid analysis pipelines such as SURPI (Naccache *et al.*, 2014) will allow clinicians to quickly determine the cause of an outbreak while scanning for important genomic characteristics such as drug resistance and virulence factors.

References

Acevedo, A., Brodsky, L., and Andino, R. (2014). Mutational and fitness landscapes of an RNA virus revealed through population sequencing. Nature *505*, 686–690.

Allander, T., Emerson, S.U., Engle, R.E., Purcell, R.H., and Bukh, J. (2001). A virus discovery method incorporating DNase treatment and its application to the identification of two bovine parvovirus species. Proc. Natl. Acad. Sci. U.S.A. *98*, 11609–11614.

Allen, L.Z., Ishoey, T., Novotny, M.A., McLean, J.S., Lasken, R.S., and Williamson, S.J. (2011). Single virus genomics: a new tool for virus discovery. PloS One *6*, e17722.

Allison, A.B., and Stallknecht, D.E. (2009). Genomic sequencing of Highlands J virus: A comparison to western and eastern equine encephalitis virus. Virus Res. *145*, 334–340.

Alquezar-Planas, D.E., Vitcetz, S.N., Sethuraman, A., Mourier, T., Mørk, S., Paxinos, E.E., Bruhn, C.A.W., Gorodkin, J., Shan, T., Nielsen, H.A., *et al.* (2013). Discovery of a divergent HPIV4 from respiratory secretions using second and third generation metagenomic sequencing. Scientific Reports 3, 2468.

Anthony, S.J., Epstein, J.H., Murray, K.A., Navarrete-Macias, I., Zambrana-Torrelio, C.M., Solovyov, A., Ojeda-Flores, R., Arrigo, N.C., Islam, A., Ali Khan, S., *et al.* (2013). A strategy to estimate unknown viral diversity in mammals. mBio *4*, e00598–00513.

Assefa, S., Keane, T.M., Otto, T.D., Newbold, C., and Berriman, M. (2009). ABACAS: algorithm-based automatic contiguation of assembled sequences. Bioinformatics *25*, 1968–1969.

Bhatt, S., Gething, P.W., Brady, O.J., Messina, J.P., Farlow, A.W., Moyes, C.L., Drake, J.M., Brownstein, J.S., Hoen, A.G., Sankoh, O., *et al.* (2013). The global distribution and burden of dengue. Nature *496*, 504–507.

Bishop-Lilly, K.A., Turell, M.J., Willner, K.M., Butani, A., Nolan, N.M., Lentz, S.M., Akmal, A., Mateczun, A., Brahmbhatt, T.N., and Sozhamannan, S. (2010). Arbovirus detection in insect vectors by rapid, high-throughput pyrosequencing. PLoS Neglect. Trop. Dis. *4*, e878.

Breitbart, M., Salamon, P., Andresen, B., Mahaffy, J.M., Segall, A.M., Mead, D., Azam, F., and Rohwer, F. (2002). Genomic analysis of uncultured marine viral communities. Proc. Natl. Acad. Sci. U.S.A. *99*, 14250–14255.

Briese, T., Chowdhary, R., Travassos da Rosa, A., Hutchison, S.K., Popov, V., Street, C., Tesh, R.B., and Lipkin, W.I. (2014). Upolu virus and Aransas Bay virus, two presumptive bunyaviruses, are novel members of the family Orthomyxoviridae. J. Virol. *88*, 5298–5309.

Burt, F.J., Rolph, M.S., Rulli, N.E., Mahalingam, S., and Heise, M.T. (2012). Chikungunya: a re-emerging virus. Lancet *379*, 662–671.

Carver, T., Berriman, M., Tivey, A., Patel, C., Böhme, U., Barrell, B.G., Parkhill, J., and Rajandream, M.-A. (2008). Artemis and ACT: viewing, annotating and comparing sequences stored in a relational database. Bioinformatics *24*, 2672–2676.

Chen, K.C., Kam, Y.W., Lin, R.T., Ng, M.M., Ng, L.F., and Chu, J.J. (2013). Comparative analysis of the genome sequences and replication profiles of chikungunya virus isolates within the East, Central and South African (ECSA) lineage. Virol. J. *10*, 169.

Chin-inmanu, K., Suttitheptumrong, A., Sangsrakru, D., Tangphatsornruang, S., Tragoonrung, S., Malasit, P., Tungpradabkul, S., and Suriyaphol, P. (2012). Feasibility of using 454 pyrosequencing for studying quasispecies of the whole dengue viral genome. BMC Genomics *13*, S7.

Chomczynski, P., and Sacchi, N. (1987). Single-step method of RNA isolation by acid guanidinium thiocyanate-phenol-chloroform extraction. Anal. Biochem. *162*, 156–159.

Christenbury, J.G., Aw, P.P., Ong, S.H., Schreiber, M.J., Chow, A., Gubler, D.J., Vasudevan, S.G., Ooi, E.E., and Hibberd, M.L. (2010). A method for full genome sequencing of all four serotypes of the dengue virus. J. Virol. Methods *169*, 202–206.

Culley, A.I., Lang, A.S., and Suttle, C.A. (2006). Metagenomic analysis of coastal RNA virus communities. Science *312*, 1795–1798.

Daly, G.M., Bexfield, N., Heaney, J., Stubbs, S., Mayer, A.P., Palser, A., Kellam, P., Drou, N., Caccamo, M., Tiley, L., *et al.* (2011). A viral discovery methodology for clinical biopsy samples utilising massively parallel next generation sequencing. PloS One *6*, e28879.

Djikeng, A., and Spiro, D. (2009). Advancing full length genome sequencing for human RNA viral pathogens. Future Virol. *4*, 47–53.

Djikeng, A., Halpin, R., Kuzmickas, R., DePasse, J., Feldblyum, J., Sengamalay, N., Afonso, C., Zhang, X., Anderson, N.G., and Ghedin, E. (2008). Viral genome sequencing by random priming methods. BMC Genomics *9*, 5.

Ewing, B., Hillier, L., Wendl, M.C., and Green, P. (1998). Base-calling of automated sequencer traces using Phred. I. Accuracy assessment. Genome Res. *8*, 175–185.

Firth, C., and Lipkin, W.I. (2013). The genomics of emerging pathogens. Annu. Rev. Genomics Hum. Genet. *14*, 281–300.

Froussard, P. (1992). A random-POR method (rPCR) to construct whole cDNA library from low amounts of RNA. Nucleic Acids Res. *20*, 2900.

Ghedin, E., Holmes, E.C., DePasse, J.V., Pinilla, L.T., Fitch, A., Hamelin, M.E., Papenburg, J., and Boivin, G. (2012). Presence of oseltamivir-resistant pandemic A/H1N1 minor variants before drug therapy with subsequent selection and transmission. J. Infect. Dis. *206*, 1504–1511.

Ghedin, E., Rogers, M.B., Widen, S.G., Guzman, H., da Rosa, A.P.T., Wood, T.G., Fitch, A., Popov, V., Holmes, E.C., and Walker, P.J. (2013). Kolente virus, a rhabdovirus species isolated from ticks and bats in the Republic of Guinea. J. Gen. Virol. *94*, 2609–2615.

Gurevich, A., Saveliev, V., Vyahhi, N., and Tesler, G. (2013). QUAST: quality assessment tool for genome assemblies. Bioinformatics *29*, 1072–1075.

Guzman, M.G., Halstead, S.B., Artsob, H., Buchy, P., Farrar, J., Gubler, D.J., Hunsperger, E., Kroeger, A., Margolis, H.S., Martinez, E., *et al.* (2010). Dengue: a continuing global threat. Nat. Rev. Microbiol. *8*, S7–S16.

Hall, R.J., Wang, J., Todd, A.K., Bissielo, A.B., Yen, S., Strydom, H., Moore, N.E., Ren, X., Huang, Q.S., Carter, P.E., *et al.* (2014). Evaluation of rapid and simple techniques for the enrichment of viruses prior to metagenomic virus discovery. J. Virol. Methods *195*, 194–204.

Hall-Mendelin, S., Allcock, R., Kresoje, N., van den Hurk, A.F., and Warrilow, D. (2013). Detection of arboviruses and other micro-organisms in experimentally infected mosquitoes using massively parallel sequencing. PloS One *8*, e58026.

Hubalek, Z., Rudolf, I., and Nowotny, N. (2014). Arboviruses pathogenic for domestic and wild animals. Advances in Virus Res. *89*, 201–275.

Hunter, S., Jones, P., Mitchell, A., Apweiler, R., Attwood, T.K., Bateman, A., Bernard, T., Binns, D., Bork, P., and Burge, S. (2012). InterPro in 2011: new developments in the family and domain prediction database. Nucleic Acids Res. *40*, D306–D312.

Ip, H.S., Wiley, M.R., Long, R., Palacios, G., Shearn-Bochsler, V., and Whitehouse, C.A. (2014). Identification and characterization of Highlands J virus from a Mississippi sandhill crane using unbiased next-generation sequencing. J. Virol. Methods *206*, 42–45.

Jentes, E.S., Poumerol, G., Gershman, M.D., Hill, D.R., Lemarchand, J., Lewis, R.F., Staples, J.E., Tomori, O., Wilder-Smith, A., and Monath, T.P. (2011). The revised global yellow fever risk map and recommendations for vaccination, 2010: consensus of the Informal WHO Working Group on Geographic Risk for Yellow Fever. Lancet Infect. Dis. *11*, 622–632.

Junglen, S., and Drosten, C. (2013). Virus discovery and recent insights into virus diversity in arthropods. Curr. Opin. Microbiol. *16*, 507–513.

Kendrick, K., Stanek, D., Blackmore, C., Centers for Disease Control and Prevention (2014). Notes from the field: Transmission of chikungunya virus in the continental United States – Florida, 2014. MMWR Morbid. Mortal. Weekly Rep. *63*, 1137.

Kerr, P.J., Ghedin, E., DePasse, J.V., Fitch, A., Cattadori, I.M., Hudson, P.J., Tscharke, D.C., Read, A.F., and Holmes, E.C. (2012). Evolutionary history and attenuation of myxoma virus on two continents. PLoS Path. *8*, e1002950.

Kurtz, S., Phillippy, A., Delcher, A.L., Smoot, M., Shumway, M., Antonescu, C., and Salzberg, S.L. (2004). Versatile and open software for comparing large genomes. Genome Biol. *5*, R12.

Lanciotti, R., Roehrig, J., Deubel, V., Smith, J., Parker, M., Steele, K., Crise, B., Volpe, K., Crabtree, M., and Scherret, J. (1999). Origin of the West Nile virus responsible for an outbreak of encephalitis in the northeastern United States. Science *286*, 2333–2337.

Langmead, B., and Salzberg, S.L. (2012). Fast gapped-read alignment with Bowtie 2. Nat. Methods *9*, 357–359.

Langmead, B., Trapnell, C., Pop, M., and Salzberg, S.L. (2009). Ultrafast and memory-efficient alignment of short DNA sequences to the human genome. Genome Biol. *10*, R25.

Li, C.X., Shi, M., Tian, J.H., Lin, X.D., Kang, Y.J., Chen, L.J., Qin, X.C., Xu, J., Holmes, E.C., and Zhang, Y.Z. (2015). Unprecedented genomic diversity of RNA viruses in arthropods reveals the ancestry of negative-sense RNA viruses. eLife *4*.

Li, H., and Durbin, R. (2009). Fast and accurate short read alignment with Burrows–Wheeler transform. Bioinformatics *25*, 1754–1760.

Li, M., Zheng, Y., Zhao, G., Fu, S., Wang, D., Wang, Z., and Liang, G. (2014). Tibet Orbivirus, a novel Orbivirus species isolated from *Anopheles maculatus* mosquitoes in Tibet, China. PloS One *9*, e88738.

Liu, M., Liu, C., Li, G., Li, X., Yin, X., Chen, Y., and Zhang, Y. (2012). Complete genome sequence of duck flavivirus from China. J. Virol. *86*, 3398–3399.

Mahy, B.W.J., and Kangro, H.O., eds. (1996). Virology Methods Manual (Academic Press Limited, London, UK).

Maxmen, A. (2012). The hidden threat of West Nile virus. Nature *489*, 349–350.

Murphy, F.A., Fauquet, C.M., Bishop, D.H.L., Ghabrial, S.A., Jarvis, A.W., Martelli, G.P., Mayo, M.A., and Summers, M.D. (1995). Virus taxonomy, Sixth Report of the International Committee on Taxonomy of Viruses (Springer, Vienna, Austria).

Naccache, S.N., Greninger, A.L., Lee, D., Coffey, L.L., Phan, T., Rein-Weston, A., Aronsohn, A., Hackett, J., Delwart, E.L., and Chiu, C.Y. (2013). The perils of pathogen discovery: origin of a novel parvovirus-like hybrid genome traced to nucleic acid extraction spin columns. J. Virol. *87*, 11966–11977.

Naccache, S.N., Federman, S., Veeraraghavan, N., Zaharia, M., Lee, D., Samayoa, E., Bouquet, J., Greninger, A.L., Luk, K.C., Enge, B., *et al.* (2014). A cloud-compatible bioinformatics pipeline for ultrarapid pathogen identification from next-generation sequencing of clinical samples. Genome Res. *24*, 1180–1192.

Ng, T.F.F., Willner, D.L., Lim, Y.W., Schmieder, R., Chau, B., Nilsson, C., Anthony, S., Ruan, Y., Rohwer, F., and Breitbart, M. (2011). Broad surveys of DNA viral diversity obtained through viral metagenomics of mosquitoes. PloS One *6*, e20579.

Odegaard, F. (2000). How many species of arthropods? Erwin's estimate revised. Biol. J. Linnean Soc. *71*, 15.

Otto, T.D., Dillon, G.P., Degrave, W.S., and Berriman, M. (2011). RATT: rapid annotation transfer tool. Nucleic Acids Res. *39*, e57–e57.

Parameswaran, P., Charlebois, P., Tellez, Y., Nunez, A., Ryan, E.M., Malboeuf, C.M., Levin, J.Z., Lennon, N.J., Balmaseda, A., and Harris, E. (2012). Genome-wide patterns of intrahuman dengue virus diversity reveal associations with viral phylogenetic clade and interhost diversity. J. Virol. *86*, 8546–8558.

Pop, M., Phillippy, A., Delcher, A.L., and Salzberg, S.L. (2004). Comparative genome assembly. Brief. Bioinform. *5*, 237–248.

Presti, R.M., Zhao, G., Beatty, W.L., Mihindukulasuriya, K.A., da Rosa, A.P.T., Popov, V.L., Tesh, R.B., Virgin, H.W., and Wang, D. (2009). Quaranfil, Johnston Atoll, and Lake Chad viruses are novel members of the family Orthomyxoviridae. J. Virol. *83*, 11599–11606.

Reyes, G., and Kim, J. (1991). Sequence-independent, single-primer amplification (SISPA) of complex DNA populations. Mol. Cell. Probes *5*, 473–481.

Ritchie, S.A., Pyke, A.T., Hall-Mendelin, S., Day, A., Mores, C.N., Christofferson, R.C., Gubler, D.J., Bennett, S.N., and van den Hurk, A.F. (2013). An explosive epidemic of DENV-3 in Cairns, Australia. PloS One *8*, e68137.

Rogers, M.B., Cui, L., Fitch, A., Popov, V., Travassos da Rosa, A.P., Vasilakis, N., Tesh, R.B., and Ghedin, E. (2014). Whole Genome Analysis of Sierra Nevada Virus, a Novel Mononegavirus in the Family Nyamiviridae. Am. J. Trop. Med. Hyg. *91*, 159–164.

Rosario, K., Capobianco, H., Ng, T.F.F., Breitbart, M., and Polston, J.E. (2014). RNA viral metagenome of whiteflies leads to the discovery and characterization of a whitefly-transmitted carlavirus in North America. PloS One *9*, e86748.

Simpson, J.T., Wong, K., Jackman, S.D., Schein, J.E., Jones, S.J., and Birol, I. (2009). ABySS: a parallel assembler for short read sequence data. Genome Res. *19*, 1117–1123.

Sreekumar, E., Issac, A., Nair, S., Hariharan, R., Janki, M.B., Arathy, D.S., Regu, R., Mathew, T., Anoop, M., Niyas, K.P., *et al.* (2010). Genetic characterization of 2006–2008 isolates of Chikungunya virus from Kerala, South India, by whole genome sequence analysis. Virus Genes *40*, 14–27.

Treangen, T.J., Sommer, D.D., Angly, F.E., Koren, S., and Pop, M. (2011). Next generation sequence assembly with AMOS. Curr. Protoc. Bioinform. 33, 11. 18. 11–11. 18. 18.

Vasilakis, N., Widen, S., Mayer, S.V., Seymour, R., Wood, T.G., Popov, V., Guzman, H., Travassos da Rosa, A., Ghedin, E., and Holmes, E.C. (2013). Niakha virus: A novel member of the family *Rhabdoviridae* isolated from phlebotomine sandflies in Senegal. Virology *444*, 80–89.

Victoria, J.G., Kapoor, A., Dupuis, K., Schnurr, D.P., and Delwart, E.L. (2008). Rapid identification of known and new RNA viruses from animal tissues. PLoS Path. *4*, e1000163.

Wong, T.H., Dearlove, B.L., Hedge, J., Giess, A.P., Piazza, P., Trebes, A., Paul, J., Smit, E., Smith, E.G., Sutton, J.K., *et al.* (2013). Whole genome sequencing and de novo assembly identifies Sydney-like variant noroviruses and recombinants during the winter 2012/2013 outbreak in England. Virol. J. *10*, 335.

Yang, X., Charlebois, P., Gnerre, S., Coole, M.G., Lennon, N.J., Levin, J.Z., Qu, J., Ryan, E.M., Zody, M.C., and Henn, M.R. (2012). De novo assembly of highly diverse viral populations. BMC Genomics *13*, 475.

Zerbino, D.R., and Birney, E. (2008). Velvet: algorithms for de novo short read assembly using de Bruijn graphs. Genome Res. *18*, 821–829.

Zerbino, D.R., McEwen, G.K., Margulies, E.H., and Birney, E. (2009). Pebble and rock band: heuristic resolution of repeats and scaffolding in the velvet short-read de novo assembler. PloS One *4*, e8407.

Role of Vertical Transmission in Mosquito-borne Arbovirus Maintenance and Evolution

13

Robert B. Tesh, Bethany G. Bolling and Barry J. Beaty

Abstract

Vertical transmission (VT) of arboviruses provides a unique mechanism for arbovirus maintenance during adverse environmental conditions and amplification in nature. The mechanisms by which most arboviruses are trans-seasonally maintained remain unclear. There is evidence for VT by either transovarial or transovum transmission of the virus from mother to progeny in each of the major arbovirus families. However, the VT rates observed in field and laboratory studies seem insufficient to maintain the viruses in nature. Stabilized infection may condition maintenance of arboviruses and insect-specific viruses in the arbovirus families in nature. Stabilized infection of *Drosophila melanogaster* maintains Sigma virus (family *Rhabdoviridae*) in nature in the absence of horizontal amplification of the virus in a vertebrate host. La Crosse virus may be maintained in *Aedes triseriatus* by stabilized infection. Stabilized infection could be a unifying mechanism for trans-seasonal maintenance of arboviruses and related viruses from different families. VT and long term persistent infections promote arbovirus evolution in vectors. Understanding the role of VT in maintenance, evolution, and emergence of arboviruses is critical to prevent and control these threats to public and animal health.

Introduction

Arthropod-borne viruses (arboviruses) are maintained in nature in cycles involving vertebrate hosts and haematophagous arthropod vectors. Thus, the genetic information of the virus is expressed in two phylogenetically and physiologically disparate systems, the poikilothermic invertebrate vector and the homeothermic vertebrate host, but with dramatically different outcomes. Arbovirus replication in vertebrate cells is typically cytolytic and infections in vertebrates (especially tangential hosts) can result in devastating disease outcomes, such as encephalitis and haemorrhagic fever. In contrast, arbovirus replication in arthropod cells is typically noncytolytic, and major untoward effects on natural vectors generally do not occur. Why arboviruses can cause devastating diseases in their vertebrate hosts, yet establish life-long persistent infections in their vectors remains a major mystery in virology. Similarly the mechanism(s) by which mosquito-borne arboviruses are maintained trans-seasonally and persist through adverse environmental conditions in temperate or tropical regions with dry seasons are poorly understood. During permissive environmental conditions, arboviruses are maintained horizontally by propagative biological transmission between vectors and susceptible vertebrate hosts, thereby amplifying the virus in nature. However, in temperate and tropical regions that have a dry season, adverse environmental conditions interrupt the transmission cycle. The vectors may be inactive or in diapause and thus the transmission cycle ceases. It is embarrassing that despite the fact that most of the important mosquito-borne arboviruses, such as St. Louis encephalitis virus (SLEV), West Nile virus (WNV), Japanese encephalitis virus (JEV), eastern equine encephalitis virus (EEEV) and western equine encephalitis virus (WEEV) were isolated decades ago and that we now have unparalleled molecular information concerning their genomes and molecular biology, we still do not completely understand the biology of trans-seasonal or overwintering mechanisms that maintain these viruses during adverse conditions. Even where we do have insight (e.g. La Crosse virus (LACV)), mathematical models suggest that the mechanisms are insufficiently robust to sustain the virus. Understanding the virus–vector interactions that maintain these pathogens trans-seasonally, which is likely a weak link in these arbovirus cycles, is a major knowledge gap in our understanding of arbovirus epidemiology. Such information is critical for developing more effective surveillance and control programmes for these important diseases.

Mosquito-borne arbovirus cycles are frequently quite specific, involving one or a few preferred vectors and a limited number of vertebrate hosts. Exceptions to the rule do occur; for example, WNV has been associated with multiple vector and vertebrate host species as it has trafficked through and become established in the Western Hemisphere (Turell *et al.*, 2005). Nonetheless the integrity of the arbovirus cycle (i.e. the preferred vector and vertebrate host relationships) promotes arbovirus cycle maintenance and amplification. This relationship may be conditioned by vertical transmission (VT) of the arbovirus, which is defined as the transmission of a pathogen from a parent to its progeny and which is most efficient in an evolved and preferred arbovirus–vector cycle.

Historically, arboviruses were not thought to be maintained by VT, and much effort was exerted to find other mechanisms by which arboviruses could persist in nature, including, persistent infection of vertebrates, reintroduction by infected migratory birds, and even plant reservoirs (Reeves 1974). However, following the landmark studies demonstrating VT of vesicular stomatitis Indiana virus (VSIV) by phlebotomine sandflies and then LACV by *Aedes triseriatus* mosquitoes(Tesh *et al.*, 1972, Watts *et al.*, 1973), laboratories around the world began investigating VT of arboviruses. Numerous arboviruses from different families have now been demonstrated to be vertically transmitted from mother to progeny.

VT can occur by two principal anatomic pathways: (1) transovarial transmission (TOT) of the virus, wherein the arbovirus infects ovarian follicles and thus the developing progeny of the mosquito or (2) transovum transmission (TST), wherein the progeny become infected via the micropyle during oviposition when the chorionated oocyte is in close contact with the infected ovarian calyx and oviducts. Some arboviruses from the families *Bunyaviridae* and *Rhabdoviridae* are considered to be maintained and amplified in nature via relatively efficient VT. In these systems, a large percentage of the infected female vectors VT the respective virus to a large percentage of their progeny. Arboviruses with high TOT and filial infection rates (FIR: the percentage of infected larvae from a single infected female) are theoretically more likely to be maintained trans-seasonally in nature. Efficient TOT not only protects the virus in the eggs during adverse environmental conditions, but can also amplify the prevalence of infected mosquitoes in nature in each gonadotrophic cycle because of the high FIRs. Generally speaking, arboviruses from other families (e.g. *Flaviviridae* and *Togaviridae*) have inefficient VT rates; these viruses are likely vertically transmitted by TST during oviposition and typically FIRs are quite low. Reports of VT of arboviruses, especially dengue virus (*Flaviviridae*), have increased dramatically with the advent of PCR techniques. Whether or not detection of viral nucleic acid in field collected larvae and adult males is evidence of productive VT of virus is not clear. For TST viruses, VT rates, even those determined using PCR, are significantly lower than those for viruses that are maintained by TOT. Interpretation of results is even further complicated by the fact that VT rates observed in laboratory studies are often much greater than those found in nature. Even for the arboviruses that exhibit efficient TOT, most models show that without significant horizontal amplification, the viruses cannot be maintained in nature for any appreciable time.

For example, LACV is transmitted transovarially by its natural vector, *Ae. triseriatus*, survives the winter in diapausing eggs of the vector, and initiates new transmission cycles upon the emergence of infected adult mosquitoes. There is little doubt that TOT is essential for LACV overwintering and amplification in nature. However, TOT and FIRs observed, even in laboratory studies, would seem to be insufficient to sustain the virus in nature. Similarly the low SLEV and WNV infection rates in overwintering *Culex pipiens* adult mosquitoes also seem to be insufficient to maintain these viruses in nature and too low to be epidemiologically significant, especially in cycles where herd immunity levels are high in the vertebrate amplifying hosts. Alternate mechanism(s) must be functioning to maintain these viruses. The recently discovered group of insect-specific viruses, including cell fusing agent virus, *Culex* flavivirus, and Kamiti River virus, which do not replicate in vertebrate hosts, must be maintained by VT. Understanding the VT mechanisms that promote maintenance of these insect-specific viruses in nature could provide important insight into arbovirus maintenance. Sigma virus (family *Rhabdoviridae*) is maintained in nature, without horizontal amplification in a vertebrate host, in *Drosophila melanogaster* by stabilized infection, which could be a model of maintenance of insect-specific viruses and bonafide arboviruses. Clearly there is much to learn about the relative roles of VT and stabilized infection in arbovirus maintenance in nature. Similarly, there is much to be learned concerning the effects of long term persistence in eggs or other stages of vectors following VT on arbovirus evolution. These are critical knowledge gaps in our understanding of the transmission cycles and evolutionary potential of mosquito-borne arboviruses.

In this chapter, we will review the evidence for VT in the major arbovirus families. We use the well-studied LACV-*Ae. triseriatus* VT model to illustrate anatomic and molecular determinants of TOT of viruses and to illustrate the importance of and issues involved in VT maintenance and amplification of arboviruses in nature. We will review the Sigma virus – *D. melanogaster* VT system as the best current model for stabilized infection of arboviruses and highlight the similarities with the LACV-*Ae. triseriatus* system. Understanding the mechanisms by which arboviruses survive adverse environmental conditions and emerge each season to initiate new transmission cycles is critical for instituting effective surveillance, prevention, and control strategies for arbovirus pathogens. Finally, LACV evolution in long term persistent infections of vectors will be reviewed, and we will speculate about the evolutionary consequences of VT and long-term persistent infection of vectors for other arboviruses from other families.

Evidence for vertical transmission of arboviruses

VT of arboviruses in the families *Rhabodoviridae*, *Bunyaviridae*, *Flaviviridae*, *Togaviridae*, *Reoviridae* and *Asfarviridae* have been reported following laboratory studies and/or by detecting the respective virus in field collected larvae or adult male mosquitoes (Table 13.1). Little is known about VT of the tick-borne Thogoto and related viruses (family *Orthomyxoviridae*). The pertinent VT literature will be very briefly reviewed for each of families. We will also include a brief review of insect-specific viruses, which have now been detected in most of the arbovirus families. Understanding the mechanisms that maintain these viruses in nature, in the absence of amplification in vertebrate hosts, may provide invaluable insight into overwintering mechanisms of the 'classical' arboviruses.

VT of asfarviruses

African swine fever virus (AFSV) is a member of the family *Asfarviridae* and is the only arbovirus that has a DNA genome. Its vectors are *Ornithodoros moubata* complex ticks, most commonly *O. moubata moubata* which is often found in warthog burrows in southern Africa. Filial infection rates from 1.2% to 35.5% have been reported in female *O. moubata* orally infected

Table 13.1 Examples of arboviruses with evidence of vertical transmission

Family and genus	Virus	Arthropod	Source(s)*	Reference(s)
Bunyaviridae				
Orthobunyavirus	Cache Valley	*Cs. inornata*	L	Corner *et al.*, 1980
	California encephalitis	*Ae. dorsalis*	L, FA	Crane *et al.*, 1977; Turell *et al.*,1982a,b; Kramer *et al.*, 1992
		Ae. melanimon	L, M, FA	Turell *et al.*, 1980, 1982b; Reisen *et al.*,1990
		Ae. squamigar	L, FA	Eldridge *et al.*, 1991; Kramer *et al.*, 1992
	Gamboa	*Aedeomyia squamipennis*	FA, FL	Karabatsos, 1985; Dutary *et al.*, 1989
	Jamestown Canyon	*Ae. cataphylla*	M, FA	Campbell *et al.*, 1991; Hardy *et al.*, 1993
		Ae. hexodontus	FA	Hardy *et al.*, 1993
		Ae. provacans	M	Boromisa and Grayson, 1990
		Ae. tahoensis	FA	Hardy *et al.*, 1993
		Ae. triseriatus	FA	Berry *et al.*, 1977
		Cs. inornata	L	Kramer *et al.*, 1993
	Keystone	*Ae. atlanticus*	M, FA, FL	LeDuc *et al.*, 1975a,b
		Ae. albopictus	L	Tesh, 1980b
	La Crosse	*Ae. aegypti*	L	Hughes *et al.*, 2006
		Ae. albopictus	L	Tesh and Gubler, 1975; Hughes *et al.*, 2006
		Ae. atropalpus	L	Freier and Beier, 1984
		Ae. hendersoni	L	Paulson and Grimstad, 1989
		Ae. triseriatus	L, FA, FL	Watts *et al.*, 1973; Pantuwatana *et al.*, 1974; Miller *et al.*, 1977; Schopen *et al.*, 1991
		Cs. inornata	L	Schopen *et al.*, 1991
	San Angelo	*Ae. albopictus*	L	Tesh, 1980b
	Snowshoe hare	*Ae. communis*	FA	Belloncik *et al.*, 1982
		Ae. implicatus	FA	McLintock *et al.*, 1976
		Ae. triseriatus	L	Schopen *et al.*, 1991
		Cs. inornata	L	Schopen *et al.*, 1991
	Tahyna	*Ae. aegypti*	L	Labuda *et al.*, 1983
		Ae. diantaeus	M	Traavik *et al.*, 1978
		Ae. vexans	L	Danielova and Ryba, 1979
		Cs. annulata	FL	Bardos *et al.*, 1975
Nairovirus	Crimean–Congo haemorrhagic fever	*Ambylomma variegatum*	L	Faye, 1999a
		Dermacentor marginatus	L	Kondratenko, 1976
		Hyalomma m. marginatum	L	Kondratenko,1976
		Hyalomma m. rufipes	L	Lee, 1970; Zeller, 1994; Faye, 1999a
		Hyalomma truncatum	L	Faye, 1999a; Gonzalez, 1992; Wilson, 1991
		Rhipicephalus evertsi	L	Faye, 1999b
		Rhipicephalus rossicus	L	Kondratenko, 1976
Phlebovirus	Aguacate	*Lutzomyia trapidoi*	M	Tesh and Chaniotis, 1975
	Arbia	*Phlebotomus pernicious*	L	Tesh and Modi, 1984
		Phlebotomus spp.	M	Ciufolini *et al.*, 1985
	Arboledas	*Lutzomyia* spp.	M	Test *et al.*, 1986
	Cacao	*Lutzomyia trapidoi*	M	Tesh and Chaniotis, 1975
	Chagres	*Lutzomyia* spp.	M	Tesh *et al.*, 1974
	Chilibre	*Lutzomyia* spp.	M	Tesh *et al.*, 1974
	Karimabad	*Phlebotomus papatasi*	L	Tesh and Modi, 1984
		Phlebotomus spp.	M	Tesh *et al.*, 1977
	Pacui	*Lutzomyia flaviscutellata*	L, M	Herve *et al.*, 1984; Aitken *et al.*, 1975
		Lutzomyia longipalpis	L	Tesh and Modi, 1984
		Phlebotomus papatasi	L	Tesh and Modi, 1984
	Punto Toro	*Lutzomyia trapidoi*	M	Tesh *et al.*, 1974
	Rift Valley fever	*Ae. aegypti*	FL	Seufi and Galal, 2010
		Ae. lineatopennis	FA	Linthicum *et al.*, 1985
		Ae. mcintoshi	L	Romoser *et al.*, 2011

Table 13.1 Continued

Family and genus	Virus	Arthropod	Source(s)*	Reference(s)
		An. coustani	FL	Seufi and Galal, 2010
		An. gambiae	M, FL	Seufi and Galal, 2010
		Cx. pipiens	FL	Seufi and Galal, 2010
		Hyaloma truncatum	L	Linthicum *et al*., 1989
	Rio Grande	*Lutzomyia anthophora*	L	Endris *et al*., 1983
	Saint-Floris	*Phlebotomus papatasi*	L	Tesh and Modi, 1984
	Sandlfy fever (Sicilian)	*Phlebotomus papatasi*	L, M	Tesh and Modi, 1984; Schmidt *et al*., 1971
	Toscana	*Phlebotomus pernicious*	L	Tesh and Modi, 1984
		Phlebotomus spp.	M	Ciufolini *et al*., 1985
Flaviviridae				
Flavivirus	Dengue-1	*Ae. aegypti*	L, M, FA, FL	Rosen *et al*.,1983; Serufo *et al*., 1993; Kow *et al*., 2001; Le Goff *et al*., 2011
		Ae. albopictus	L, M, FL	Rosen *et al*., 1983; Kow *et al*., 2001; Cecilio *et al*., 2009
		Ae. alcasidi	L	Freier and Rosen, 1987
		Ae. cooki	L	Freier and Rosen, 1987
		Ae. hebrideus	L	Freier and Rosen, 1987
		Ae. katherinensis	L	Freier and Rosen, 1987
		Ae. malayensis	L	Freier and Rosen, 1987
		Ae. medioviittatus	L	Freier and Rosen, 1988
		Ae. polynesiensis	L	Freier and Rosen, 1987
		Ae. pseudoscutellaris	L	Freier and Rosen, 1987
		Ae. tongae tabu	L	Freier and Rosen, 1987
	Dengue-2	*Ae. aegypti*	L, M, FA, FL	Jousset, 1982; Khin and Than, 1983; Kow *et al*., 2001; Le Goff *et al*., 2011; Martins *et al*., 2012
		Ae. albopictus	L, M, FA, FL	Rosen *et al*., 1983; Ibanez-Bernal *et al*.,1997; Cecilio *et al*., 2009; Kow *et al*., 2001; Martins *et al*., 2012
		Ae. alcasidi	L	Freier and Rosen 1987
		Ae. cooki	L	Freier and Rosen 1987
		Ae. fulcifer/taylori	M	Cordellier *et al.* 1983
		Ae. medioviittatus	L	Freier and Rosen, 1988
		Ae. polynesiensis	L	Freier and Rosen, 1987
		Ae. pseudoscutellaris	L	Freier and Rosen, 1987
		Ae. tongae tabu	L	Freier and Rosen, 1987
	Dengue-3	*Ae. aegypti*	L, M, FA, FL	Joshi *et al*., 1996; Kow *et al*., 2001; Joshi *et al*., 2002; Vilela *et al*., 2010
		Ae. albopictus	L, M, FA	Rosen *et al*., 1983, Ibanez-Bernal *et al*., 1997, Kow *et al*., 2001; Martins *et al*., 2012
		Ae. medioviittatus	L	Freier and Rosen, 1988
		Ae. polynesiensis	L	Freier and Rosen, 1987
	Dengue-4	*Ae. aegypti*	M, FA	Hull *et al*., 1984; Kow *et al*., 2001
		Ae. albopictus	L, M	Rosen *et al*., 1983; Kow *et al*., 2001
		Ae. medioviittatus	L	Freier and Rosen, 1988
		Ae. polynesiensis	L	Freier and Rosen, 1987
	Japanese encephalitis	*Ae. albopictus*	L	Rosen *et al*.,1978
		Ae. alcasidi	L	Rosen *et al*., 1989
		Ae. flavus	L	Rosen *et al*., 1989
		Ae. japonicus	L	Takashima and Rosen, 1989
		Ae. togoi	L	Rosen *et al*., 1978
		Ae. vexans	L	Rosen *et al*., 1989
		Armigeres subalbatus	L	Rosen *et al*., 1989
		Cx. annulus	L	Rosen *et al*., 1989
		Cx. fuscocephala	FA	Dhanda *et al*., 1989
		Cx. pipiens	L	Rosen *et al*., 1989

Table 13.1 Continued

Family and genus	Virus	Arthropod	Source(s)*	Reference(s)
		Cx. pseudovishnui	L, M, FA	Dhanda *et al*., 1989; Mourya *et al*., 1991
		Cx. quinquefasciatus	L	Rosen *et al*., 1989
		Cx. tritaeniorhynchus	L, FA, FL	Rosen *et al*.,1980, 1989; Dhanda *et al*., 1989; Mourya *et al*., 1991
		Cx. univittatus	FA	Dhanda *et al*., 1989
		Cx. whitmorei	M	Dhanda *et al*., 1989
		Ma. indiana	M	Arunachalam *et al*., 2004
	Kokobera	*Ae. albopictus*	L	Tesh, 1980
	Koutango	*Ae. aegypti*	M	Coz *et al*., 1976
	Kunjin	*Ae. albopictus*	L	Tesh, 1980
	Kyasanur Forest disease	*Haemaphysalis spinigera*	L	Singh *et al*., 1963
		Ixodes petauristae	L	Singh *et al*., 1968
	Murray Valley encephalitis	*Ae. aegypti*	L	Kay and Carley, 1980
		Ae. tremulus	M	Broom *et al*., 1995
	St. Louis encephalitis	*Ae. albopictus*	L	Rosen, 1988
		Ae. atropalpis	L	Hardy *et al*., 1980, 1984
		Ae. bahamensis	L	Shroyer, 1991
		Ae. epacticus	L	Hardy *et al*., 1980, 1984
		Ae. taeniorhynchus	L	Nayar *et al*.,1986
		An. albimanus	L	Nayar *et al*., 1986
		An. quadrimaculatus	L	Nayar *et al*., 1986
		Cx. nigripalpus	L	Nayar *et al*., 1986
		Cx. opisthopus	L	Nayar *et al*., 1986
		Cx. pipiens	L	Francy *et al*., 1981; Hardy *et al*., 1984
		Cx. quinquefasciatus	L	Nayar *et al*., 1986
		Cx. restuans	L	Nayar *et al*., 1986
		Cs. salinarius	L	Nayar *et al*., 1986
		Cx. tarsalis	L	Hardy *et al*., 1984
		Dermacentor variabilis	L	Blattner *et al*., 1944
	Tick-borne encephalitis	*Ixodes persulcatus*	FL	Il'enko *et al*., 1970
		Ixodes ricinus	L, FL	Rehacek, 1962; Il'enko *et al*., 1970; Danielova *et al*., 2002
	West Nile virus	*Ae. albopictus*	L, FA	Baqar *et al*., 1993; Unlu *et al*., 2010
		Ae. aegypti	L	Baqar *et al*., 1993
		Ae. triseriatus	M	Unlu *et al*., 2010
		Argas arboreus	L	Abbassy *et al*., 1993
		Cx. erythrothorax	FL	Phillips and Christensen, 2006
		Cx. pipiens	L, M, FL	Turell *et al*., 2001; Anderson *et al*., 2006, 2008; Nelms *et al*., 2013b
		Cx. quinquefasciatus	L, FA	Goddard *et al*., 2003; Reisen *et al*., 2006; Unlu *et al*., 2010
		Cx. salinarius	L, M	Unlu *et al*., 2010; Anderson *et al*., 2012
		Cx. tarsalis	L, FL	Goddard *et al*., 2003; Reisen *et al*., 2006; Anderson *et al*.,2012; Nelms *et al*., 2013a
		Cx. tritaeniorhynchus	M	Baqar *et al*., 1993
		Cx. univittatus	M	Miller *et al*., 2000
		Cx. vishnui	L	Mishra and Mourya, 2001
	Yellow fever	*Ae. aegypti*	L, M, FA	Aitken *et al*., 1979; Beaty *et al*., 1980, Fontenille *et al*., 1997
		Ae. fulcifer/taylori	M	Cornet *et al*., 1979
		Ae. mascarensis	L	Beaty *et al*., 1980
		Haemagogus equinus	L	Dutary and LeDuc, 1981
		Ambylomma variegatum	L	Saluzzo *et al*., 1980
Reoviridae				
Orbivirus	Bluetongue	*Culicoides sonorensis*	FA, FL	White *et al*., 2005
	Changuinola	*Lutzomyia* spp.	M	Tesh *et al*., 1974
	Orungo	*Ae.* spp.	M	Cordellier *et al*., 1982

Table 13.1 Continued

Family and genus	Virus	Arthropod	Source(s)*	Reference(s)
Rhabdoviridae				
	Hart Park	*Cx. tarsalis*	M	Reisen *et al.*, 1990
Vesiculovirus	Carajas	*Lutzomyia longipalpis*	L	Travassos da Rosa *et al.*, 1984
	Chandipura	*Ae. aegypti*	L	Mavale *et al.*, 2005
		Phlebobomus papatasi	L	Tesh and Modi, 1983
	Maraba	*Lutzomyia* spp.	M	Travassos da Rosa *et al.*, 1984
		Lutzomyia longipalpis	L	Travassos da Rosa *et al.*, 1984
	Morreton	*Lutzomyia longipalpis*	L	Tesh *et al.*, 1987
	Vesicular stomatitis-New Jersey	*Lutzomyia shannoni*	L, M	Comer *et al.*, 1990, Comer *et al.*, 1992
	Vesicular stomatitis-Indiana	*Lutzomyia trapidoi*	L	Tesh *et al.*, 1972
	Vesicular stomatitis-Indiana	*Lutzomyia ylephilator*	L	Tesh *et al.*, 1972
Togaviridae				
	Buggy Creek	*Oeciacus vicarius*	FE	Brown *et al.*, 2009b
Alphavirus	Chikungunya	*Ae. albopictus*	L, M, FA	Thavara *et al.*, 2009; Niyas *et al.*, 2010; Bellini *et al.*, 2012
		Ae. aegypti	M	Thavara *et al.*, 2009
	Eastern equine encephalitis	*Cs. melanura*	FL	Chamberlain and Sudia, 1961
	Ross River	*Ae. camptorhynchus*	FA	Dhileepan *et al.* 1996
		Ae. vigilax	L, M	Kay, 1982; Lindsay *et al.*, 1993
		Macleaya tremula	M	Lindsay *et al.*, 1993
	Sindbis	*Ae. camptorhynchus*	FA	Dhileepan *et al.*, 1996
	Western equine encephalitis	*Dermacentor andersoni*	L	Syverton and Berry, 1941
Asfarviridae				
Asfavirus	African swine fever	*Ornithodoros moubata*	L	Plowright *et al.*, 1970
		Ornithodoros puertoricensis	L	Hess *et al.*, 1987

*Virus isolated from or detected in: FA, adults reared from field-collected larvae; FL, field-collected larvae; FE, field-collected eggs; L, laboratory studies; M, field-collected males.
Adapted from: Turell MJ. 1988. Horizontal and vertical transmission of viruses by insect and tick vectors. *In The arboviruses: epidemiology and ecology*. TP Monath (ed). CRC, Boca Raton, FL.

with AFSV (Rennie *et al.*, 2001). Infected male *O. moubata* also can transmit AFSV to uninfected females during copulation (Plowright *et al.*, 1974).

VT of bunyaviruses

More is probably known about VT of viruses in the family *Bunyaviridae* than in any other family. The bunyaviruses infecting vertebrates are presently assigned to four genera (*Orthobunyavirus, Phlebovirus, Nairovirus* and *Hantavirus*); there is experimental and field evidence of VT of representative viruses in each of the four genera in their respective arthropod vectors (Table 13.1). Much of the information about VT has accrued because of the studies investigating VT of the California group of orthobunyaviruses, especially LACV–*Ae. triseriatus* interactions (Beaty *et al.*, 2000), but VT of other viruses in the group has also been reported (Tesh and Shroyer, 1980; Tesh and Cornet, 1981; Turell *et al.*, 1982b). Another relatively well-studied orthobunyavirus-mosquito VT maintenance cycle is Gamboa virus in its Neotropical vector *Aedeomyia squamipennis* (Dutary *et al.*, 1989).

The genus *Phlebovirus* consists of viruses transmitted by phlebotomine sandflies, mosquitoes and ticks. VT of Rift Valley fever virus (RVFV) with passage through the eggs of infected *Aedes lineatopennis* is the presumed mechanism by which this virus persists during interepidemic periods in East Africa. Experimental VT of RVFV has also been demonstrated in the tick *Hyalomma truncatum* (Linthicum *et al.*, 1985, 1989), although its importance in the long-term maintenance of RVFV is uncertain. VT of a number of other phleboviruses (Arboledas, Arbia, Sicilian, Saint Floris, Karimabad, Pacui, Rio Grande and Toscana) has also been demonstrated experimentally in their sandfly vectors (Tesh and Modi, 1984, 1987; Tesh *et al.*, 1986, 1992; Endris *et al.*, 1983). The TOT rates among the experimentally infected sandflies varied from 1.5% to >80%, depending upon the virus type and sandfly species tested. In general, TOT rates were higher when a natural virus–vector combination was used (Tesh and Modi, 1984, 1987; Tesh *et al.*, 1986, 1992; Endris *et al.*, 1983). In one study, Toscana virus was maintained by VT for 13 consecutive generations in a laboratory colony of *Phlebotomus perniciosus* (a natural virus–vector combination) without significant phenotypic or genetic changes in the virus (Tesh *et al.*, 1987; Bilsel *et al.*, 1988). The isolation of phleboviruses from pools of field-collected male sandflies is also quite common, supporting the results of laboratory studies (Tesh, 1988).

Crimean–Congo haemorrhagic fever virus (CCHFV) is the most important member of the genus *Nairovirus*, because of its very wide geographic distribution in the Old World and the severe disease that it produces in humans. VT of CCHFV has been demonstrated in a variety of tick species. This mechanism as well as trans-stadial transmission are believed to be the principal manner in which the virus is maintained from generation to generation and from season to season (Turell, 2007).

The consensus among virologists, until recently, has been that hantaviruses are not arthropod-borne; but instead that rodents serve as the natural reservoir of human-pathogenic hantaviruses, and that humans are infected with these viruses by inhalation of aerosols of infected rodent excreta. However, recent reports (Yu and Tesh, 2014) of the isolation and bite transmission of both Hantaan and Seoul viruses from parasitic mites in China, as well as demonstration of VT of Hantaan virus by infected adult female mites (*Leptotrombidium scutellare*) to their progeny suggests that mites could play an important role in the transmission and epidemiology of haemorrhagic fever with renal syndrome.

Recently, a probable new genus of bunyaviruses has been described in Artic midges (*Chaoborus* sp.) (Ballinger *et al.*, 2014). Available data suggest that these insect-only viruses are also maintained in their non-haematophagous midge hosts by VT.

VT of flaviviruses

Arboviruses in the genus flavivirus are assumed to be primarily maintained in nature through horizontal transmission (HT) between blood feeding arthropods and susceptible vertebrate hosts. VT appears to play only an accessory role, at least with some flaviviruses, and may be the mechanism by which transmission is maintained during adverse climatic conditions. Flaviviruses have been detected in field-collected immatures and male mosquitoes, indicating that VT does occur in nature (Table 13.1). It is thought that most flaviviruses are vertically transmitted by TST at the time of oviposition during fertilization via the micropyle as the fully developed egg passes through the oviduct. TST is much less efficient than 'true' TOT, where the virus infects the developing egg. FI rates seen with vertically transmitted flaviviruses in mosquitoes are usually low (less than 1%), compared to much higher rates (greater than 20%) seen with bunyaviruses (Tesh, 1984; Turell, 1988; Nasci *et al.*, 2001).

The first experimental studies reporting VT of a flavivirus were done by Marchoux and Simond (1905) during epidemiological investigations of yellow fever (YF) virus in Brazil in the early 1900s. Subsequent studies were unable to confirm this work until much later when Rosen and colleagues (Aitken *et al.*, 1979, Beaty *et al.*, 1980) demonstrated experimental VT of YF virus in *Aedes* species mosquitoes. Interestingly, infection rates were similar between surface-sterilized and non-sterilized eggs suggesting that VT occurs via TOT, but it was subsequently realized that progeny were infected by TST (Beaty *et al.*, 1980). YF virus was also detected in field-caught *Ae. aegypti* males and adults reared from field-collected larvae that were obtained from mosquito collections made during an outbreak of YF in Senegal in 1995, confirming that VT occurs naturally in the primary vector species (Fontenille *et al.*, 1997).

Experimental VT of the four dengue serotypes was demonstrated in *Ae. albopictus* and *Ae. aegypti*, with variable transmission rates observed among different virus strains (Rosen *et al.*, 1983). In addition, different mosquito species and even different geographic strains of the same species can have variable FI rates of dengue viruses (Rosen *et al.*, 1983; Freier and Rosen, 1987; Rodhain and Rosen, 1997). Further investigations into the mechanism of VT revealed that transmission of dengue viruses occurs via TST, as the mature eggs are fertilized during oviposition (Rosen, 1987b, 1988). This is reflected in the low FI rates that have been observed in experimental studies as well as in the field. Evidence supporting VT of dengue viruses in nature was first reported in 1983 with the isolation of dengue 2 from adult *Ae. aegypti* reared from field-collected larvae (Khin and Than, 1983) and has subsequently been reported for the other dengue viruses as well (Hull *et al.*, 1984; Serufo *et al.*, 1993; Martins *et al.*, 2012).

Numerous laboratory studies have also documented VT for the encephalitic flaviviruses, including JEV, SLEV and WNV (Rosen, 1988, Rosen *et al.*, 1989, Baqar *et al.*, 1993). VT potentially plays a role in maintenance of these viruses during periods of unsuitable environmental conditions, as the primary vector species are *Culex* mosquitoes, which overwinter as diapausing adult females in temperate climates. Vertically infected females can survive the winter in hibernaculae and reemerge in the spring to initiate viral transmission. Indeed, WNV and SLEV have been isolated from overwintering *Culex* adults in the field (Bailey *et al.*, 1978; Nasci *et al.*, 2001; Bugbee and Forte, 2004, Farajollahi *et al.*, 2005; Andreadis *et al.*, 2010). Identifying locations that harbour infected diapausing females could help vector control agencies implement more effective mosquito control strategies by targeting populations during periods of low abundance and limited activity (Nelms *et al.*, 2013a).

VT of alphaviruses

Compared with the bunyaviruses and flaviviruses, evidence to support VT of alphaviruses has been slower to accumulate (Table 13.1). Indeed, historically there was such a paucity of evidence for VT, that the New World alphaviruses were thought not to be maintained in this manner (Rosen, 1987a). Eastern equine encephalitis virus (EEEV) was detected in one pool of *Culiseta melanura* larvae during field studies in Massachusetts (Hayes *et al.*, 1962), but further attempts to substantiate this finding were unsuccessful. Laboratory studies investigating dissemination patterns of EEEV in *Cs. melanura* did not result in ovariole infection after feeding on viraemic chicks and furthermore, virus was not isolated from the progeny of infected females during TOT studies (Scott *et al.*, 1984). Based on these findings, it appears that TOT of EEEV does not occur in *Cs. melanura*, and alternative overwintering mechanisms have been suggested, including, for example, garter snakes serving as reservoir hosts (White *et al.*, 2011). WEEV has also been isolated from reptiles (Thomas *et al.*, 1980), suggesting a potential role for these poikilothermic animals in alphavirus trans-seasonality in the new world.

Field investigations provided evidence of VT of Buggy Creek

virus (BCRV), an ecologically unique alphavirus in the WEEV complex. BCRV is transmitted by swallow bugs (*Oeciacus vicarius* Horvath) and the nesting cliff swallow (*Petrochelidon pyrrhonota* Vieillot) is the primary reservoir host (Brown *et al.*, 2009a). BCRV was isolated in early spring from swallow bugs collected in North Dakota, before the return of the cliff swallows to the nest, suggesting overwintering of this virus in the vector (Brown *et al.*, 2009a). Additionally, BCRV was isolated from field-collected swallow bug eggs, indicating VT may play a role in the transmission cycle of this virus (Brown *et al.*, 2009b).

Ross River virus (RRV), an alphavirus that causes epidemic polyarthritis among humans in Australia, has been isolated from pools of male mosquitoes (Lindsay *et al.*, 1993) and also adult mosquitoes reared from field-collected larvae (Dhileepan *et al.*, 1996), both implicating VT as a means of viral persistence during adverse climatic conditions. Laboratory studies conducted to examine transmission of RRV by *Aedes vigilax* showed that at 4 dpi, virus was present in the ovaries of 50% of females, and 2 of 122 adult F1 progeny were infected (Kay, 1982), also providing supporting evidence for the role of VT in RRV survival.

There is evidence supporting VT of several other alphaviruses, including Chikungunya virus (CHIKV) isolated from pools of field-collected male *Aedes aegypti* and *Ae. albopictus* (Thavara *et al.*, 2009) and Sindbis virus (SINV) isolated from *Ae. camptorhynchus* adults reared from field-collected larvae (Dhileepan *et al.*, 1996). Interestingly, there appears to be a pattern with the data supporting VT of Old World alphaviruses, and less evidence supporting VT of the New World alphaviruses. Further studies are warranted to examine the genetic basis for these differences.

VT of rhabdoviruses

The family *Rhabdoviridae* is a large and extremely diverse group of viruses; there are now 15 established or proposed genera within the family, which includes viruses infecting vertebrates, invertebrates and plants (Walker *et al.*, 2015). Some of the rhabdoviruses are arthropod-borne (i.e. vesiculoviruses and ephemeroviruses) while others have no known arthropod association (i.e. lyssaviruses and novirhabdoviruses). Many of the recognized vesiculoviruses have been associated with phlebotomine sandflies (Karabatsos, 1985). There is considerable evidence supporting VT of arthropod-borne rhabdoviruses (Table 13.1). In 1971, vesicular stomatitis Indiana virus (VSIV) was shown to be vertically transmitted in two Panamanian sandfly species (*Lutzomyia trapidoi* and *L. ylephilator*) (Tesh *et al.*, 1971). The VT rates in F_1 progeny of orally infected adult females of the two species were 21 and 20%, respectively (Tesh *et al.*, 1972). VSIV was demonstrated in each developmental stage of the flies (egg, four larval instars, pupa and adult) with virus titres increasing in each successive developmental stage. F_1 adult female flies contained a mean of $10^{5.1}$ $TCID_{50}$ units of virus per insect upon emerging and transmitted the virus by bite to young hamsters (Tesh *et al.*, 1972). In contrast VT could not be demonstrated in two other anthropophylic sandfly species that occurred locally, indicating that VT of VSIV is not a generalized phenomenon and only occurs in a few species. Subsequent laboratory studies with two mosquito species (*A. albopictus* and *Cx. quinquefasciatus*) also failed to show VT in these insects (R. Tesh, unpublished).

Later laboratory studies demonstrated VT of four additional vesiculoviruses, Carajas, Chandipura, Morreton (incorrectly described as VS Alagoas virus) and vesicular stomatitis New Jersey, by several different *Lutzomyia* and *Phlebotomus* species (Tesh *et al.*, 1971, 1972, 1987; Tesh and Modi, 1983; Comer *et al.*, 1990; Travassos da Rosa *et al.*, 1984).

VT of reoviruses

The family *Reoviridae* is an extremely diverse group of double-stranded, segmented RNA viruses which are currently divided into two subfamilies and 15 genera (Attoui *et al.*, 2012). Representatives of this family infect vertebrates, invertebrates, plants and fungi, depending upon the genus. Likewise, their modes of transmission vary widely from the respiratory to oral-faecal to arthropod-borne routes. A number of the plant reoviruses are vector-borne, but only three of the 15 genera contain vector-borne viruses of vertebrates (*Orbivirus*, *Coltivirus* and *Seadornavirus*) and can be considered as true arboviruses. Little experimental work has been with viruses in these latter three genera to test for VT (Table 13.1); however, there are two reports of isolation of Changuinola group viruses from field-collected male sandflies (Tesh *et al.*, 1974) and of Orungo virus from male mosquitoes (Cordellier, *et al.*, 1982). Field studies conducted in northern Colorado to investigate overwintering strategies of bluetongue viruses (BTVs) in *Culicoides* species found BTV RNA in *Culicoides sonorensis* larvae, supporting VT as a mechanism for virus maintenance (White *et al.*, 2005). Nested PCR was used to determine the prevalence of specific BTV genome segments in the larvae. An interesting finding during this study was the differences seen between the expression of outer capsid-encoding sequences compared to core-encoding sequences. The authors postulated that during persistent infection, expression of outer capsid-encoding genes are down regulated to reduce the metabolic burden of viral infection in the larvae, since they are not required for infection in *Culicoides* cells.

VT of insect-specific viruses

Information is limited on the transmission dynamics of insect-specific viruses, a group of viruses found to replicate only in invertebrates and invertebrate cells. In contrast, classical arboviruses replicate in vertebrates as well as invertebrates. The inability to infect and replicate in vertebrate cells indicates that this group of viruses has a distinct transmission cycle from the classical arboviruses, which are maintained between arthropod vectors and vertebrate hosts. During the past decade, with advances in molecular tools for viral detection, there has been a dramatic increase in the isolation and characterization of these types of viruses. The insect-specific viruses now include a diverse group of viruses representing many different families and genera (*Togaviridae*, *Flaviviridae*, *Bunyaviridae*, *Rhabdoviridae*, *Reoviridae*, *Birnaviridae*, *Mesoniviridae*, Negevirus) and other still unclassified RNA viruses (Attoui *et al.*, 2005; Cook *et al.*, 2012; Kolodziejek *et al.*, 2013; Nasar *et al.*, 2012; Quan *et al.*, 2010; Vasilakis *et al.*, 2013; Zirkel *et al.*, 2013). These insect-specific viruses appear to be ubiquitous in nature, having been found in many parts of the world from a large variety of mosquito species. There is growing evidence that VT plays an important role in the transmission of insect-specific viruses (Table 13.2), although

Table 13.2 Examples of insect-specific viruses evaluated for evidence of vertical transmission

Family and genus	Virus	Arthropod	VT	Source	Reference(s)
Bunyaviridae					
Unassigned	Gouleako virus	*Anopheles* spp., *Culex* spp.	Yes	M	Marklewitz *et al.*, 2011
	Herbert virus	*Anopheles* spp., *Culex* spp.	No	M	Marklewitz *et al.*, 2013
	Kibale virus	*Anopheles* spp., *Culex* spp.	No	M	Marklewitz *et al.*, 2013
	Tai virus	*Anopheles* spp., *Culex* spp.	No	M	Marklewitz *et al.*, 2013
Flaviviridae					
Flavivirus	*Aedes* flavivirus	*Ae. albopictus*	Yes	L, M	Hoshino *et al.*, 2009; Haddow *et al.*, 2013; Bolling *et al.*, 2015
		Ae. flavopictus	Yes	M	Hoshino *et al.*, 2009
	Calbertado virus	*Cx. tarsalis*	Yes	M	Bolling *et al.*, 2011
	Cell fusing agent virus	*Ae. aegypti*	Yes	L, M	Cook *et al.*, 2006; Yamanaka *et al.*, 2013; Bolling *et al.*, 2014
		Ae. albopictus	Yes	M	Cook *et al.*, 2006
		Culex spp.	Yes	M	Cook *et al.*, 2006
	Culex flavivirus	*Cx. pipiens*	Yes	M	Hoshino *et al.*, 2007; Bolling *et al.*, 2011
		Cx. pipiens	No	L	Saiyasombat *et al.*, 2011
		Cx. quinquefasciatus	Yes	M	Farfan-Ale *et al.*, 2009
	Kamiti River virus	*Ae. aegypti*	Yes	L	Lutomiah *et al.*, 2007
		Ae. macintoshi	Yes	M	Crabtree *et al.*, 2003
	Spanish Ochlerotatus flavivirus	*Ae. caspius*	Yes	M	Vazquez *et al.*, 2012
Mesoniviridae					
Alphamesonivirus	Cavally virus	*Aedes* spp., *Anopheles* spp., *Culex* spp.	No	M	Zirkel *et al.*, 2011

VT, vertical transmission; L, laboratory studies; M, field-collected males tested for virus.

mechanisms of persistence have not been evaluated for many of these newly described viruses. Since it appears that these viruses do not use a vertebrate reservoir host as a part of their transmission cycle, it seems likely that VT is the primary mode of maintenance.

Cell fusing agent virus (CFAV) was the first insect-specific flavivirus described (Stollar and Thomas, 1975). It was isolated from an *Ae. aegypti* cell line and caused massive syncytia formation when co-cultivated with *Ae. albopictus* cells. More recently CFAV has been detected in male and female mosquito pools collected in Puerto Rico (Cook *et al.*, 2006) and also in immature forms of *Ae. aegypti* collected in Mexico (Martinez-Vega *et al.*, unpublished), indicating the potential role of VT in the natural transmission cycle of this virus. Additionally, examination of an *Ae. aegypti* laboratory colony, persistently infected with CFAV, showed high infection rates, with viral RNA detected in 100% (10/10) of females tested and 90% (9/10) of the males tested (Bolling *et al.*, 2015).

Culex flavivirus (CxFV) and *Aedes* flavivirus (AeFV), two insect-specific viruses, were both initially isolated from adult males and females collected during mosquito field surveys in Japan (Hoshino *et al.*, 2007, 2009). These viruses have subsequently been detected in mosquitoes in other parts of the world (Blitvich *et al.*, 2009; Bolling *et al.*, 2011; Cook *et al.*, 2009; Farfan-Ale *et al.*, 2009; Haddow *et al.*, 2013; Kim *et al.*, 2009; Morales-Betoulle *et al.*, 2008). Phylogenetic analysis comparing different geographic strains of CxFV revealed intriguing results, with isolates from Iowa and Texas grouping with the Asian strain, prompting questions on how insect-specific viruses have spread across the globe (Blitvich *et al.*, 2009). VT of CxFV can be very efficient; egg rafts from field-collected *Cx. pipiens* were hatched and of 26 females that oviposted viable egg rafts, 18 were CxFV RNA-positive. F1 adults from these positive females were tested for CxFV RNA; 526 of 540 progeny were infected, yielding a FIR of 97.4% and a TOT rate of 100% (Saiyasombat, *et al.*, 2011) In contrast, VT was not observed with *Cx. pipiens* intrathoracically (IT) inoculated with CxFV, despite the fact that CxFV RNA was detected in the ovaries at the time of oviposition (Saiyasombat *et al.*, 2011). Progeny were only tested after one ovarian cycle, however, and it is possible that for VT to occur, additional ovarian cycles are needed to ensure enough time for the follicular epithelium to become infected after IT inoculation. Detection of insect-specific flaviviruses in all life stages of *Cx. pipiens* from a CxFV-infected laboratory colony (Bolling *et al.*, 2012) and an AeFV-infected *Ae. albopictus* colony (Bolling *et al.*, 2015), including adult mosquitoes of both sexes, provides further evidence for VT as the primary mechanism of viral maintenance.

Another insect-specific flavivirus, Kamiti River virus, was first isolated from *Ae. macintoshi* larvae and pupae collected from flooded dambos in Kenya (Sang *et al.*, 2003) and laboratory experiments conducted with *Ae. aegypti* mosquitoes, orally exposed to KRV, demonstrated VT after the second and third ovarian cycles, with a transmission rate of 3.9% (Lutomiah *et al.*, 2007).). Interestingly, several recent field studies investigating new insect-specific viruses belonging to the families *Bunyaviridae* and *Mesoniviridae* failed to detect virus in pools of male mosquitoes (Zirkel *et al.*, 2011, Marklewitz *et al.*, 2013), highlighting the need for additional studies to better characterize the transmission cycles of these viruses in nature.

La Crosse virus and *Ae. triseriatus*: a most remarkable relationship

LACV is not only a significant human pathogen, but it is also a powerful model for maintenance, amplification, and trans-seasonal persistence of mosquito-borne arboviruses and other vector-borne pathogens. We will first present the LACV–*Ae. triseriatus* natural cycle, then the molecular and anatomic bases for and barriers to efficient TOT of LACV. We will provide evidence for stabilized infection in *Ae. triseriatus*, the molecular mechanisms involved, the similarities of this system with the well described and molecularly characterized stabilized infection of *D. melanogaster* with Sigma virus, and will discuss the potential roles of TOT in the evolutionary potential of LACV. We will also discuss whether or not horizontal amplification in a vertebrate host is a necessary component of the transmission and maintenance cycle in the presence of stabilized infection.

LACV transmission and maintenance cycle

In the upper Midwest, LACV is vectored by *Ae. triseriatus*, with chipmunks and tree squirrels serving as the principal vertebrate hosts (Grimstad, 1988). LACV is maintained in oak hickory climax forests with treeholes serving as the principal breeding site for the mosquito vectors (Beaty and Calisher, 1991). These forested areas can be small and are often surrounded by farms and suburban development. The virus is presumably maintained in these areas by HT between susceptible vertebrate hosts during the summer and early fall and overwinters following VT in diapausing eggs.

LACV virogenesis in *Ae. triseriatus*

LACV–*Ae. triseriatus* interactions have been intensively investigated. *Ae. triseriatus* mosquitoes transmit LACV orally, transovarially and venereally (Beaty and Calisher, 1991; Borucki *et al.*, 2002). Upon ingestion of an infected bloodmeal, LACV infects the epithelial cells lining the mosquito midgut (Beaty and Thompson, 1978). The efficiency of viral dissemination to other tissues depends on both the virus and the vector (Paulson *et al.*, 1989). If the virus is able to escape from the midgut, progeny particles then disseminate to the haemocoel and travel to and replicate in the heart, neural ganglia, fat body, ovaries, and salivary glands. The salivary glands are the last organ infected, at 7–16 days after ingestion of virus (Beaty and Thompson, 1978). Once in the salivary gland, the virus replicates and can be horizontally transmitted to the vertebrate host in saliva. Infection of the ovaries results in efficient VT of LACV to the progeny (Beaty and Thompson, 1976, 1978; Tesh and Beaty, 1983). In laboratory studies, there are no obvious deleterious effects on the developing oocyte and embryo and no differences in fecundity or teratology between infected and non-infected mosquitoes (Miller *et al.*, 1977), even during critical periods of embryogenesis, follicular resting stages, ovarian diapause and diapause (Beaty and Thompson, 1976; Tesh and Beaty, 1983; Borucki *et al.*, 2002). The gonadal tissues of both sexes of progeny, including both female and male accessory sex glands, are infected. In females, viral antigen and RNA is demonstrable in most ovarian tissues, including follicular, epithelium, oocytes, nurse cells and calyx (Beaty and Thompson, 1976; Chandler *et al.*, 1996, 1998). LACV antigen has been detected in the germarium of some infected mosquitoes, but this has not been systematically or rigorously investigated. LACV infection of *Ae. triseriatus* is virtually pantropic (with a notable exception of flight muscle) in the vector throughout its lifetime, regardless of whether the mosquito was infected horizontally, vertically, or venereally (Beaty and Thompson, 1975, 1976; Thompson and Beaty, 1977, 1978). LACV overwinters in diapaused eggs of *Ae. triseriatus*, and may persist for several years in eggs, which is further testimony to this remarkable host–parasite relationship (Grimstad, 1988; Beaty and Calisher, 1991; Turell 1988; Watts *et al.*, 1973).

VT is absolutely fundamental to the epidemiology of LACV and other bunyaviruses. The virus overwinters in the diapausing eggs of the vector, and upon emergence the following spring, infected female progeny can HT virus to new susceptible vertebrate hosts and TOT the virus to progeny. TOT infected male progeny can venereally transmit the virus to females during mating, especially to females that have ingested a bloodmeal prior to mating (Thompson and Beaty, 1977, 1978). The reason for increased susceptibility of blood fed females to venereal infection remains to be determined. Following venereal transmission, LACV can be found in all examined organs, including male accessory salivary glands, by 6 days post mating (Kramer and Thompson, 1983). VT of bunyaviruses in their vectors is not only critical for trans-seasonal maintenance but also is critical for the virus to elude herd immunity, which can exceed 90% in small forested areas (B. Beaty and W. Thompson, unpublished; Fig. 13.1). Thus most of the preferred vertebrate hosts are dead end hosts, limiting opportunities for horizontal amplification of the virus. This barrier to horizontal transmission is overcome by VT; a LACV-infected female could feed on either a seropositive or seronegative chipmunk or tree squirrel, potentially resulting in > 100 infected progeny from each gonadotrophic cycle, regardless of the immune status of the vertebrate host. Thus, VT is central to both the amplification and overwintering of LACV and is the key mechanism for maintenance of LACV in nature.

Ae. triseriatus mosquitoes differ genetically in their permissiveness to TOT. LACV is transmitted efficiently transovarially by *Ae. triseriatus* mosquitoes in the Northern states, where it overwinters through an obligatory facultative diapause in the egg stage. In Florida, *Ae. triseriatus* is active throughout the year, and the FIR but not TOT rates are lower than in mosquitoes from WI (Woodring, *et al.*, 1998). Three QTLs (quantitative trait loci) condition VT of LACV in *Ae. triseriatus* (Graham *et al.*, 2003). There is significant variability in VT potential in mosquitoes, even from the hyperendemic area. For example, the LACV infection rate in eggs collected from individual tree hole breeding sites, ranged from 0% in most tree holes to 16.7%. This suggests that females contributing to the eggs in the tree holes differed in their VT potential or that stably infected mosquitoes deposited their eggs in multiple tree hole breeding sites (skip oviposition) (Beaty and Thompson, 1975).

VT efficiency is a function of the per cent of infected females that TOT virus to their progeny (TOT rate) and the per cent of progeny infected per female (filial infection – FI rate). In laboratory studies with optimal conditions, the TOT and FI rates can each exceed 80% (Miller *et al.*, 1977). With these

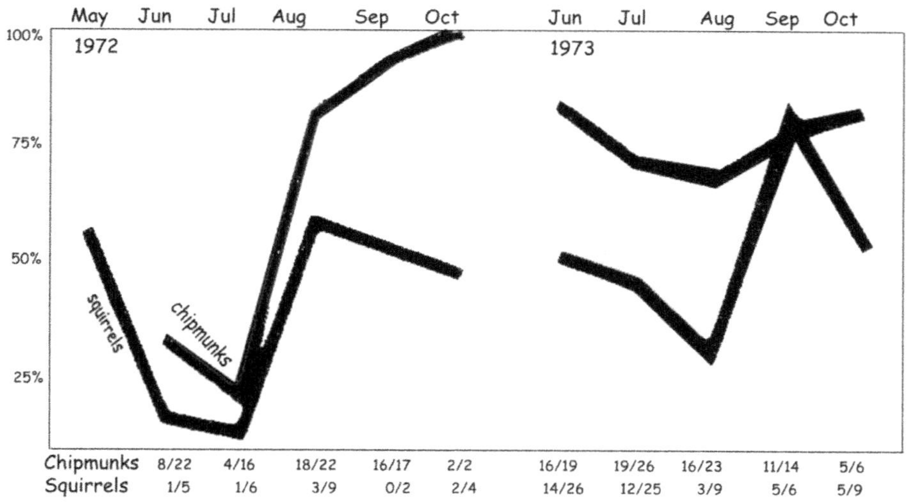

Figure 13.1 Monthly LaCrosse virus antibody prevalence rates in chipmunks and squirrels in two forested areas in suburban LaCrosse, Wisconsin.

rates, LACV could persist 4 years or longer in the absence of HT in vertebrate hosts (Miller *et al.*, 1977). However, studies of LACV overwintering in field collected eggs and larvae have yielded much lower infection rates, making it difficult to understand how LACV can survive in nature. In three studies of LACV overwintering in endemic regions, the infection rates in mosquitoes collected as eggs, larvae, or as adults, as determined by virus isolation, were 0.06%, 0.003% and 0.009% respectively (Lisitza *et al.*, 1977; Beaty and Thompson, 1975; Reese *et al.*, 2010) and are remarkably similar over the years. The reasons for the dramatic differences in TOT efficiency in laboratory and field studies remain to be determined.

In summary, VT amplifies LACV by TOT to progeny, by venereal transmission from TOT infected males to uninfected females, and by conferring a mating advantage to LACV infected females (see below). Despite all of these amplification mechanisms, quantitative models suggest that these mechanisms are insufficient to maintain LACV in nature.

Quantitative models of LACV transmission and maintenance

Quantitative models (DeFoliart, 1983; Fine, 1975, 1978) have been used to investigate the epidemiology of LACV and the closely related Keystone virus (KEYV) (*Bunyavirus*, California serogroup). The models were used to investigate the relative contributions of HT and VT to virus persistence in nature. With field determined infection rates, the models suggested that the cycles cannot be perpetuated by VT alone, and that HT is essential for maintenance of the virus in nature (Fine, 1975). Even when using laboratory determined TOT and FI rates, the models predicted that LACV would not be maintained more than a few generations in nature. Clearly, some mechanism maintains LACV in endemic foci. One possible mechanism for LACV persistence in nature could be stabilized infection of *Ae. triseriatus* (Tesh, 1980a; Tesh and Shroyer, 1980; Tesh and Beaty, 1983, Turell, 1988). If *Ae. triseriatus* females can become stably infected, LACV could theoretically be maintained indefinitely in nature. Stabilized infection of a dipteran was first observed with Sigma virus and *Drosophila melanogaster* (Seecof, 1968).

The heritable transmission model: stabilized SIGMAV infection of *Drosophila melanogaster*

Many insect pathogens and symbionts are maintained in their insect hosts by maternal and paternal transmission with no amplification in vertebrate hosts (Tesh, 1980a; Fleuriet, 1988; Longdon *et al.*, 2011; Longdon and Jiggins, 2012). Hereditary transmission is associated with stabilized infection wherein a very high proportion of progeny are infected by VT, thereby increasing the chance of survival of the pathogen in nature (Tesh 1980a; Tesh and Shroyer 1980; Turell *et al.*, 1982a). Stabilized infection requires that the pathogen must replicate and persist in virtually all of the insect tissues and cells and be noncytopathogenic. The pathogen must also not exert deleterious effects during oogenesis and follicular development, and embryogenesis, metamorphosis and pupation, adult emergence and fitness, and VT to the next generation (Tesh, 1980a; Fleuriet, 1988; Turell, 1988). Only truly co-evolved insect–pathogen interactions are likely to meet these criteria, which cross cut all heritably transmitted pathogens.

The SIGMAV (*Rhabdoviridae*)–*D. melanogaster* system is arguably the best understood example of co-evolution of virus and arthropod host. Stabilized SIGMAV infection of *D. melanogaster* permits the virus to be maintained at a constant level in nature by relatively few infected females. The prevalence, mechanisms, evolution, and consequences of SIGMAV stabilized and non-stabilized infections in laboratory and natural populations of *D. melanogaster* has been the subject of extensive research by Dr. Annie Fleuriet (Fleuriet, 1976, 1981, 1982, 1988, 1999). When *D. melanogaster* is infected with SIGMAV by inoculation, they develop a systemic infection and some of the developing oocytes become nonstably infected and the progeny of these eggs are infected (Fleuriet, 1982, 1988). However, if the germarium is infected, the virus stably infects the host and is transmitted to nearly 100% of the progeny. However, infection rates in the field are significantly lower: 8–30% (Fleuriet,1976). Maternal inheritance is complemented by venereal transmission of the virus by stably infected males (Fleuriet, 1982, 1988). SIGMAV infection can affect the fitness of flies, reduce fecundity, and increase mortality during overwintering, exposure to CO_2, and other environmental stressors

(Tesh and Shroyer, 1980; Turell *et al.*, 1982a; Fleuriet, 1981). The stabilized state counterbalances these selective pressures and maintains SIGMAV in natural populations even with a low frequency of infected females.

Genetic mapping studies have identified genetic loci that condition *D. melanogaster* resistance to SIGMAV (Longdon and Jiggins, 2012); two of these have been characterized at the molecular level – *CHKov* and *Ref(2)P*. The mechanism by which the *CHKov* genes condition resistance is unclear, but it may involve receptor events. The other molecularly characterized major determinant of resistance to SIGMAV infection in *D. melanogaster* is *ref(2)P* (Contamine *et al.*,1989). There are two alleles at the locus: ref(2)P^O (permissive to SIGMAV infection) and *ref(2)*P^P (resistant to SIGMAV). Two genetic haplotypes of SIGMAV have evolved in response to the resistance: Type I is sensitive to the gene product of the *ref(2)*P^P allele and Type II is resistant to the allele (Bangham *et al.*, 2007, 2008). Interactions between the SIGMAV types and the vector alleles condition the evolution and distribution of the viruses in natural populations of *D. melanogaster* (Bangham *et al.*, 2007, 2008; Dru *et al.*, 1993; Wayne *et al.*, 1996). The two *ref(2)P* gene common alleles, *ref(2)*P^P and *ref(2)*P^O, code for two mRNAS that differ by 100 nucleotides in size, due to different transcription initiation sites (Dezelee *et al.*, 1989; Contamine *et al.*, 1989). Only the larger mRNA is detected in ovaries where the gene product presumably permits SIGMAV to stably infect the germ line and to be transmitted in eggs. The transcriptional unit is divided into three exons and the mRNAs are heterogeneous in size differing only in the 5′ end of the first exon due to the use of different transcriptional initiation sites (Dezelee *et al.*, 1989; Contamine *et al.*, 1989). There is a long untranslated leader RNA region, and the translational product is 599 amino acids. The *ref(2)P* gene is highly variable, and the polymorphisms are hypothesized to be due to selective pressures to maintain a functional protein to modulate SIGMAV infection. Accumulation of polymorphisms in the gene suggests an ongoing innate immune arms race between the virus and host (Bangham *et al.*, 2007, 2008; Dru *et al.*, 1993; Wayne *et al.*, 1996; Longdon and Jiggins, 2012; Wilfert and Jiggins, 2013).

One or very few major gene effects commonly condition resistance in co-evolutionary relationships between pathogens and hosts (Bangham *et al.*, 2008). The *ref(2)P* gene – SIGMAV interplay is arguably the classic example of a gene–for-gene interaction in which a polymorphic gene in the parasite that confers pathogenicity, results in a corresponding polymorphism in the host gene that confers a response (Bangham *et al.*, 2007). The *ref(2)P* gene was shown to code for a signalling protein kinase that modulates the Toll pathway (Avila *et al.*, 2002). Thus perturbation of this innate immune pathway could condition SIGMAV infection and replication via a single gene effect. The specific mutation in the *ref(2)*P^P gene that conditions antiviral resistance is a CAG-AAT (glutamine-asparagine) change to a GGA (glycine) (Dru *et al.*, 1993; Wayne *et al.*, 1996). There are shared conformation-dependent epitopes between the ref(2)P-protein and the virus N (capsid) protein. The cell protein can associate with the viral polymerase protein, impairs the functionality of the polymerase (P) and N in the ribonucleoprotein (RNP) complex, and thereby reduces viral transcription and replication efficiency (Wyers *et al.*, 1993; Wyers *et al.*, 1995). This is provocative because LACV is also a negative strand arbovirus with a similar polymerase and RNP complex. Conversely, a mutant SIGMAV evolved an N protein epitope similar to the host protein, with virus trapping of the ref(2)P-protein and presumably down regulating Toll antiviral responses and promoting virus infection.

Are LAC and related viruses maintained by stabilized infection of vectors?

The SIGMAV–*D. melanogaster* system bears striking similarities to what has been observed with California serogroup viruses and their vectors. LACV TOT and FI rates in the laboratory can exceed 80% (Miller *et al.*, 1977; McGaw *et al.*, 1998). TOT$^+$ *Ae. triseriatus* that transmit LACV to progeny very efficiently can be readily selected (B. Beaty, unpublished). Both TOT and VT rates >80% can be readily achieved and maintained for multiple generations, with minimal selection, suggesting that stabilized infections of *Ae. triseriatus* do occur. However, these lines have not been rigorously characterized to determine if they represent stabilized infections, and very importantly, these rates have been attained in the laboratory. Stabilized infection of *Ae. dorsalis* mosquitoes with California encephalitis virus (CEV) has been demonstrated (Turell *et al.*, 1982a). These laboratory mosquitoes transmitted CEV vertically to over 90% of their progeny through five generations. However, field studies of LACV reveal that if stabilized infections do occur in nature, they occur at a much lower frequency (Lisitza *et al.*, 1977; Beaty and Thompson, 1975; Reese *et al.*, 2010). The fact that the infection rates detected in juveniles or male mosquitoes collected from the field are so low has contributed to the lack of acceptance of TOT as a mechanism for arbovirus maintenance (Tesh, 1980a). However if the females are stably infected, the virus could theoretically be maintained indefinitely in nature at a fairly constant level by a relatively small number of infected females in the absence of positive or negative selection, and the virus infection rate in the population could remain low as occurs with SIGMAV in *D. melanogaster* (Fleuriet, 1988).

In this regard, potentially stably infected mosquitoes in the La Crosse, WI region have been demonstrated (Reese *et al.*, 2010). Mosquito eggs were collected, returned to the Arthropod-borne and Infectious Diseases Laboratory (AIDL) Colorado State University, hatched and assayed individually for LACV infection by IFA, RT-PCR, and virus titration. Three infection phenotypes were identified in the field collected mosquitoes: superinfected (SI+) mosquitoes, with very large accumulations of virus antigen, LACV nucleic acids (NAs), and infectious virus; infected (I+) mosquitoes with dramatically lower levels of virus antigen and LACV NA, but lacking virus that could be isolated in cell culture, and I– mosquitoes with no detectable virus, antigen, or NAs (Fig. 13.2). SI+ mosquitoes were detected at a low frequency (0.08%), prevalence rates ranged from 0% to 0.12% (Reese *et al.*, 2010). These rates overlap those found for SIGMAV in some natural populations of *D. melanogaster* (Fleuriet, 1988). SI+ mosquitoes were widely distributed in the collection area, and in one WI site, SI+ mosquitoes were detected in both 2006 and 2007. The virus sequences were virtually identical for the two years, suggesting a stabilized infection in mosquitoes in that area (Reese *et al.*, 2010). The virus loads in these mosquitoes are essentially identical to those in selected TOT+ AIDL lines (B. Beaty, unpublished), which VT LACV very efficiently. This

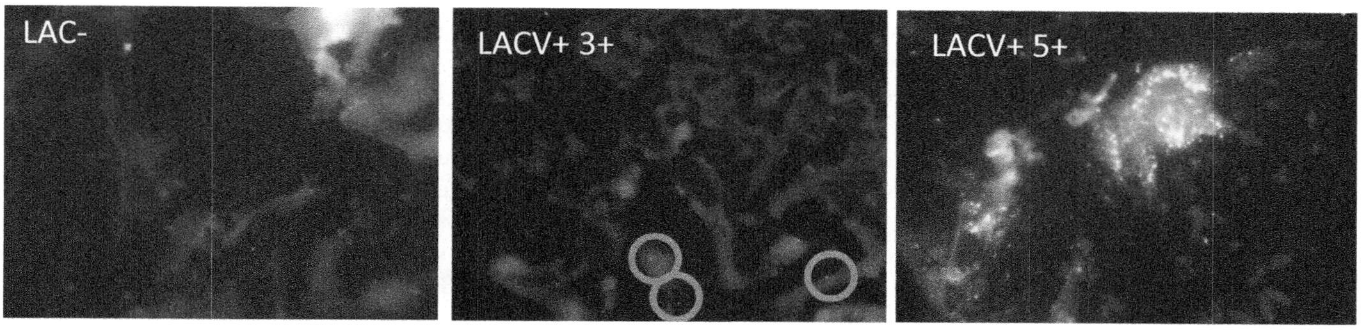

Figure 13.2 La Crosse virus antigen in infected, field collected *Aedes triseriatus* mosquitoes (with permission from *Virology Journal*, Reese *et al.*, 2010).

suggests that SI+ mosquitoes are stably infected. If so, this small proportion of SI+ mosquitoes (0.08%) could transmit LACV to most of their progeny and maintain LACV in nature at a fairly constant level (Fleuriet 1988) as occurs with SIGMAV in *D. melanogaster*. SIGMAV exerts an fitness effect on infected *D. melanogaster*, but a significant number of infected flies survive the adverse conditions to renew the transmission cycle. This is likely what happens with LACV in *Ae. triseriatus*.

Potential molecular mechanisms promoting LACV VT and stabilized infection in *Ae. triseriatus*

Multiple factors could promote or restrict establishment of stabilized infections in vectors. Because of the fundamental importance of hereditary transmission and the ovary in LACV transmission, amplification, persistence, and stabilized infection, the following will focus upon TOT and vector–virus interactions that potentially condition SI+, I+, and I- phenotypes. Clearly, non-ovarian anatomical barriers could preclude establishment of stabilized infections in horizontally infected vectors, but the focus will be upon virus virulence modulation and avoidance of innate immunity. All of these individually or synergistically could contribute to LACV TOT and maintenance in nature.

Co-regulation of vector and host transcription

Co-regulation of virus and vector transcription protects vulnerable life stages of the vector (e.g. the developing follicle and embryo) (Chandler *et al.*, 1996, 1998; Kempf *et al.*, 2006). Bunyaviruses scavenge host cell mRNA 5′ caps plus adjacent oligonucleotide sequences as primers for their own viral mRNA synthesis (Dobie *et al.*, 1997; Patterson *et al.*, 1984). LACV transcription is reduced during critical mosquito life stages (such as ovarian quiescence and embryo diapause, when host mRNA synthesis is restricted, thereby reducing the amount of capped mRNA available to prime virus transcription) and thus modulating viral replication and minimizing detrimental effects to the host (Chandler *et al.*, 1996). LACV replicative forms (RF) (virion complementary RNA and mRNA) were detected in metabolically active ovaries but not in resting stage ovaries. Ingestion of a bloodmeal induced ovarian metabolism and virus replication was reinitiated (Chandler *et al.*, 1996). Q-PCR was used to monitor viral transcription and replication in mosquito eggs during diapause (Kempf et al., 2006). Viral RF RNA was present in eggs, suggesting a low level of replication during diapause (McGaw *et al.*, 1998). Virus transcription was upregulated upon diapause cessation. To investigate productive TOT and overwintering efficiency in nature, a colony of *Ae. triseriatus* (TOT+) was selected that maintained TOT and FI rates > 80% for multiple generations in the laboratory (McGaw *et al.*, 1998). Eggs from the TOT+ and TOT− *Ae. triseriatus* colonies were induced into diapause, shipped to Wisconsin and maintained through the winter in natural environmental conditions. Eggs were returned to the laboratory at predetermined times for assay of diapause, mortality, filial infection, and hatching rates. Overwintering embryos from the TOT+ colony exhibited greater cumulative mortality (16.7%) than the TOT− eggs (7.3%). The increased mortality rate in TOT+ versus non-infected mosquitoes corresponded with emergence from diapause and the onset of host and virus transcription, suggesting that diapause modulated the adverse effects of virus infection (McGaw *et al.*, 1998). Thus, diapause and presumably ovarian quiescence restrict deleterious effects of viral infection (McGaw *et al.*, 1998; Chandler *et al.*, 1998).

Avoidance of innate immunity

LACV must avoid robust host antiviral innate immune responses (Kingsolver, *et al.*, 2013), especially programmed cell death (PCD) apoptotic and autophagic responses in ovarian tissues, for efficient TOT and stabilized infection in *Ae. triseriatus*. Apoptosis is induced by LACV infection in vertebrate but not in mosquito cell cultures (Pekosz *et al.*, 1996; B. Blitvich and B. Beaty, unpublished). Autophagy is a potent antiviral response in *D. melanogaster*. Autophagy has been demonstrated to be the key innate immune response of *D. melanogaster* pathogen resistance to VSV, a bonafide arbovirus (Shelley *et al.*, 2009; Cherry, 2009). Inhibition of autophagy by silencing *Atg* genes (e.g. *Atg18*) increased VSV titre in cells and resulted in lethal infections in flies. In contrast, activating autophagy resulted in decreased viral titres in cells and increased survival of infected flies. The VSV G-protein is a pathogen-associated molecular pattern that interacts with a Toll receptor and leads to a signalling cascade that activates autophagy (Nakamoto *et al.*, 2012). Clearly arboviruses must also avoid the autophagic and apoptotic responses in the vector host to be transmitted and particularly to establish stabilized infection. Investigations have revealed a number of mechanisms that could modulate PCD genes to enhance LACV TOT and to establish stabilized infections.

LACV could perturb host cell innate immune responses by targeted cap scavenging. Indeed, analysis of 5′ heterogeneous sequences on LACV mRNA revealed that cap scavenging did target specific mRNAs (Dobie *et al.*, 1997). One specific target was the *Ae. triseriatus* inhibitor of apoptosis protein 1 (ATIAP), which could perturb apoptotic and autophagic responses (Dobie *et al.*, 1997) and promote persistent infections in vectors (Beck *et al.*, 2007). Analysis of the 5′ untranslated region of the ATIAP-protein (Blitvich, *et al.*, 2002) revealed a presumed IRES structure, which could promote translation of ATIAP-protein even during times of mitosis, which would be critical during follicular development and embryogenesis (Beck *et al.*, 2007). The transcriptional unit of the ATIAP gene is very complex and results in multiple mRNA variants, which are differentially expressed in infected and noninfected mosquitoes (Blitvich, *et al.*, 2002; Beck *et al.*, 2007) There is also a long 5′ untranslated region (similar to *ref(2)P* mRNAs). Provocatively, mRNA variant 3, which is most abundantly produced in ovaries pre- and post-blood feeding, was down regulated in LACV infected mosquitoes (Fig. 13.3). mRNA variant abundance of *ATIAP* and *ref(2)P* in ovaries may be vector determinants of TOT and stabilized infection. The *ATIAP* gene from field collected mosquitoes is very polymorphic (see 'Evolutionary potential of LACV in endemic forested areas', below).

LACV could also evolve to perturb apoptosis and autophagy by interfering directly in the PCD pathways. Evidence supporting this hypothesis and identifying the potential molecular mechanisms involved has been accumulated in laboratory and field studies outlined below ('Evolutionary potential of LACV in endemic forested areas'). Laboratory studies comparing VT potential of WNV and LACV in *Ae. triseriatus* mosquitoes were most provocative in this regard. Following challenge with the two viruses, mosquito ovaries were harvested and examined for hallmarks of PCD. Evidence of an apoptotic response was detected in LACV infected follicular sheath cells but not in other ovarian cells. In contrast, ovarian follicles infected with WNV developed large autophagosomes, but LACV infections yielded no follicular pathology (Fig. 13.4; D. Dobie and B. Beaty, unpublished). Vacuoles with clear membrane-delimited morphology are a hallmark of autophagy and autophagic follicles do not yield viable embryos (McPhee and Baehrecke, 2009; Kourtis and Tavernarakis, 2009). Autophagy and apoptosis share a common pathway in Diptera. In *D. melanogaster*, a molecular pathway involving the effector caspase (Dcp-1) and an IAP regulate both apoptosis and autophagy (Hou *et al.*, 2008; Hou *et al.*, 2009). Importantly, ref(2)P-protein is also associated with regulation of autophagy in *Drosophila* (Nezis *et al.*, 2009). Whether similar mechanisms function in *Ae. triseriatus* remains to be determined. Identifying and characterizing the functions of the mosquito homologues of these genes would be a fruitful area of research. The *Ae. triseriatus* ovary is a unique system to assess the role of apoptosis and autophagy in the innate immune response to an invading pathogen (McPhee and Baehrecke, 2009; Deretic, 2009; Sir and Ou, 2010). Autophagy is a robust response in mosquito ovaries because of the necessity to resorb developing follicles in the absence of bloodmeals, fertilization, or nutrition (Hansen *et al.*, 2005; Nezis *et al.*, 2009). Clearly, avoidance of autophagy is critical for LACV persistence in the follicle.

LACV could also avoid the RNAi response to establish stabilized infections. RNAi, which is a robust determinant of vector competence for other arboviruses (Keene *et al.*, 2004; Franz *et al.*, 2006; Blair and Olson, 2014) does not seem to condition

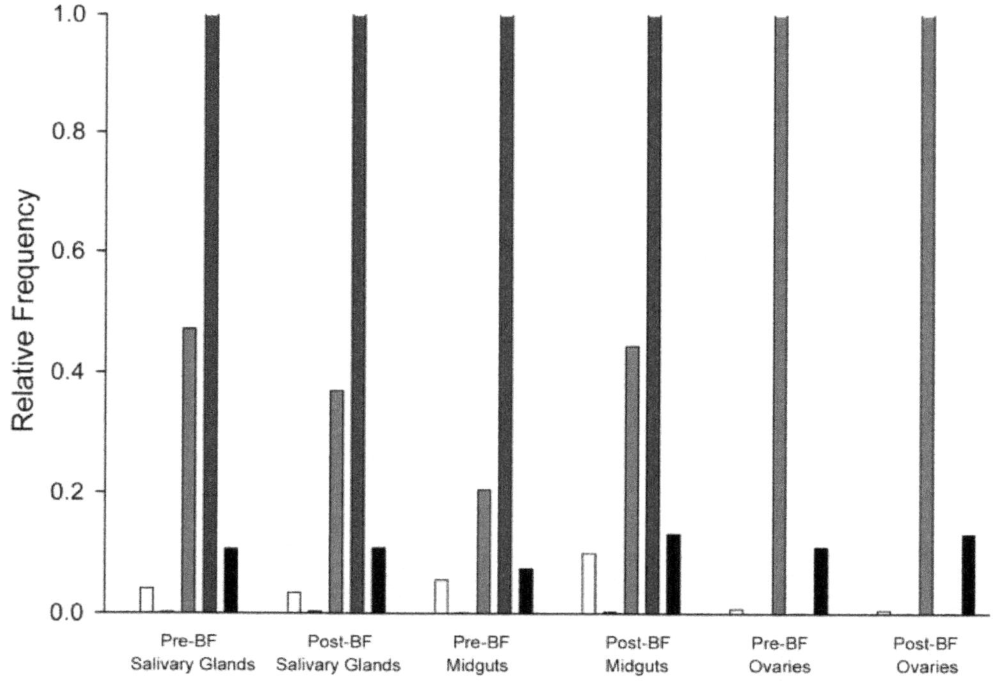

Figure 13.3 Tissue-specific expression of ATIAP1 mRNA variants in *Ae. triseriatus* tissues. Tissue-specific expression of each IAP1 mRNA variant from *Ae. triseriatus*, showing the relative amount of each AtIAP1 mRNA variant found in each life stage or tissue type. The most abundant mRNA variant for each stage or tissue was recorded as a relative frequency of one and the remaining AtIAP1 mRNA variants were recorded as a percentage of the most abundant variant based on their expression levels. Variant 1 is on the left of each group and proceeds to variant 5 on the right. (Quantitative RT-PCR amplifications were performed using mRNA variant-specific primers and total RNA extracted from various adult tissues, namely pre-bloodfed and 24 hour post-bloodfed female salivary glands, midguts, and ovaries (with permission from *Journal of Insect Biochemistry and Molecular Biology*, Beck *et al.*, 2007).

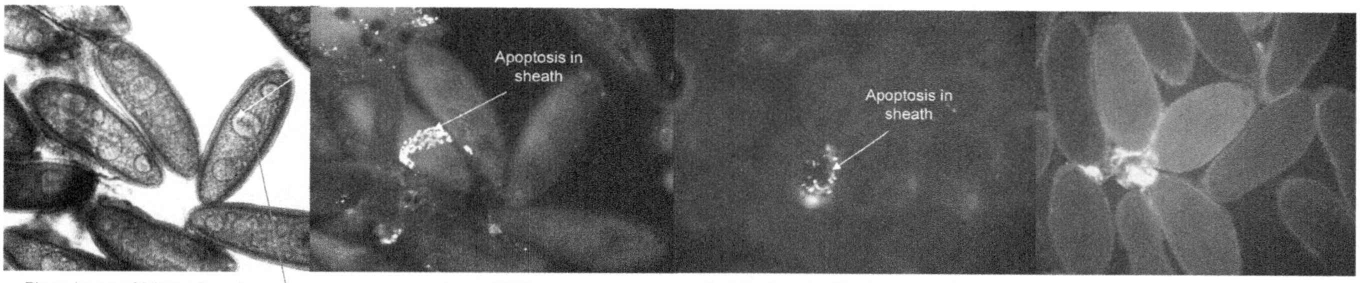

Phase image of follicles from *Ae. triseriatus* mosquito injected with WNV 3 days post injection (counterstained with Evans Blue)

Green: WNV
Orange: DNA fragmentation
Note: Evans Blue causes background fluorescence.

Follicles from *Ae triseriatus* mosquito 2 days post injection with LACV. There is slightly more apoptosis seen than in those injected with PBS

Follicles from *Ae. triseriatus* mosquito 3 days post injection with LACV. There is little to no apoptosis observed

Phase image of follicles from *Ae. triseriatus* mosquito injected with WNV 3 days post injection (counterstained with Evans Blue)

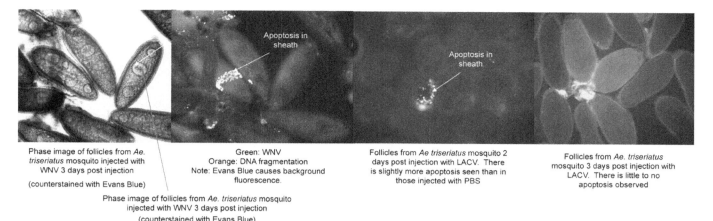

Figure 13.4 West Nile virus but not La Crosse virus induces a pathological response in *Aedes triseriatus* ovarian follicles.

SI+ and I+ phenotypes. The LACV NSs protein has been demonstrated to suppress type 1 interferon response and the RNAi response in vertebrate host cells (Soldan *et al.*, 2005; Blakqori *et al.*, 2007; Bouloy *et al.*, 2001; Bridgen *et al.*, 2001). LACV does induce specific siRNAs in infected mosquito cells, but the RNAi response does not seem to inhibit LACV in mosquito cells (B. Blivitch and B. Beaty, unpublished) or condition the SI+ phenotype in nature (see 'Evolutionary potential of LACV in endemic forested areas', below). LACV NSs sequences from SI+ and I+ mosquitoes were absolutely conserved, suggesting that NSs is not a candidate for conditioning the phenotypes. In contrast, extensive polymorphisms seen in NSm implicate perturbation of the PCD response in the SI+ phenotype (Reese *et al.*, 2010). Viruses can evolve viral inhibitors of RNAi (VIRs) to combat arthropod host RNAi responses (Chao *et al.*, 2005; Li *et al.*, 2002; Lingel *et al.*, 2005). It is feasible that LACV could have also evolved VIRs which could perturb the RNAi response in critical vector tissues or organs.

LACV could also perturb other innate immune signalling cascades. Microarray analysis has revealed that SINV infection of *Ae. aegypti* is associated with down regulation of the Toll pathway (Sanders *et al.*, 2005). Similar microarray analyses revealed that the Toll and JAK-STAT pathways are perturbed by DEN2 virus infection of mosquitoes (Xi *et al.*, 2008). Up-regulation of the Toll pathway led to a decrease in midgut virus titres while mosquitoes with the Toll pathway inactivated had higher midgut virus titres. Other effector systems of the Toll pathway or other pathways (e.g. Imd and JakStat) could contribute to the resistance to RNA viruses (Dostert *et al.*, 2005; Sanders *et al.*, 2005; Tsai *et al.*, 2008; Xi *et al.*, 2008). Considering the complexity and duration of LACV-*Ae. triseriatus* interactions, it is likely that multiple innate immune pathways may be involved in controlling virus infection. This could be a fruitful area for investigation.

Potential biological mechanisms promoting stabilized infection in *Ae. triseriatus* – mating advantages attributable to LACV infection

Although TOT is efficient in the LACV–*Ae. triseriatus* system, there is a fitness burden associated with virus infection (McGaw *et al.*, 1998). This fitness burden could be mitigated if stabilized infection conferred a fitness advantage to infected mosquitoes. In this regard, studies of LACV and *Ae. triseriatus* interactions revealed an unanticipated mating advantage for LACV infected mosquitoes. Initially, horizontally infected females were demonstrated to be more efficiently and rapidly inseminated than non-infected mosquitoes (Gabitszch, *et al.*, 2006). The fitness advantage also occurs in TOT+ laboratory colony mosquitoes. For example, in blood fed mosquitoes insemination rates were 68% for LACV TOT+ females vs. 39% for uninfected females (Reese *et al.*, 2009). Most importantly, we also demonstrated that TOT+, field-collected *Ae. triseriatus* mosquitoes also have a mating advantage. Twice as many TOT infected females (33%) were inseminated as uninfected females (15%) (Reese *et al.*, 2009; Fig. 13.5). Clearly, TOT+ mosquitoes have a mating fitness advantage over uninfected females. The reasons for this fitness advantage remain to be determined. The female accessory sex gland of infected mosquitoes contains large quantities of LACV (Gabitszch, *et al.*, 2006) and virus replication could perturb pheromone expression, thereby increasing mating efficiency. Whatever the mechanism, LACV infection in SI+ females could have could even be more pronounced in forested areas, where mate location would be more difficult. Persistent LACV infection may confer other unexpected fitness advantages that effect vector phenotype and could help maintain LACV in nature (Jackson, *et al.*, 2012).

Potential effects of TOT on LACV evolutionary potential

LACV and related bunyaviruses exhibit considerable evolutionary potential and substantial genetic diversity in nature. This diversity is attributable to both RNA segment reassortment (genetic shift) and intramolecular changes (genetic drift) (Beaty and Bishop, 1988; Beaty and Calisher, 1991, Borucki *et al.*, 2001; Beaty, *et al.*, 1997). LACV and related bunyaviruses have a tripartite, negative-sense RNA genome with the three segments designated large (L-coding for the RNA-dependent RNA polymerase), medium (M – coding for the G1 and G2 two glycoproteins and the non-structural protein NSm), and small (S – coding for the nucleocapsid protein and the non-structural protein NSs) (Beaty and Bishop, 1988). The negative-sense segmented genome clearly promotes bunyavirus evolutionary potential, both by segment reassortment in dually infected cells and by intramolecular genomic changes attributable to the infidelity of the RNA-dependent RNA polymerase and the lack of proof reading enzymes (Beaty and Bishop, 1988; Domingo

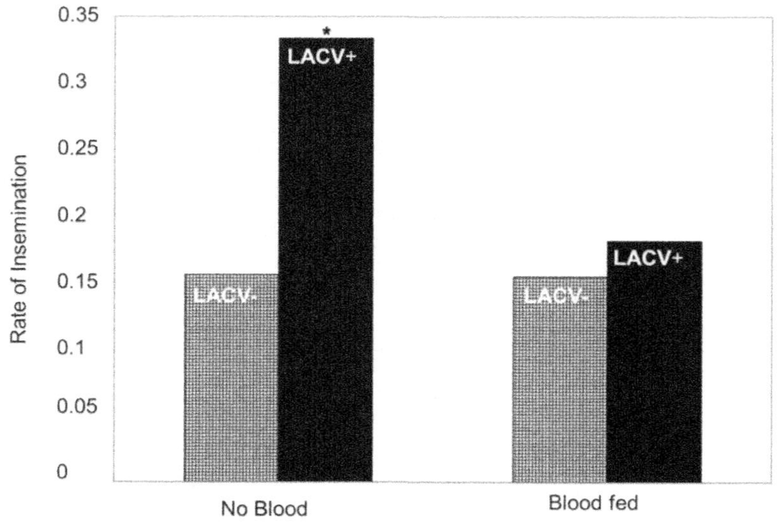

Figure 13.5 Insemination rates in field collected *Aedes triseriatus* mosquitoes. Field-collected LACV+ mosquitoes were more efficiently inseminated than FC LACV–female mosquitoes. The rate of insemination was higher for the FC LACV+ mosquitoes than the FC LACV– without a bloodmeal (*$P<0.05$). Insemination rates did not differ statistically when FC LACV+ and LACV– ingested a bloodmeal. The experiments were performed in triplicate (with permission from *Journal of Medical Entomology*, Reese *et al.*, 2009)

and Holland, 1997). LACV is genetically diverse in nature. It is likely that much of the genetic diversity in LACV is associated with vector passage, and TOT may be very important. Long-term vector infections could promote LACV evolution by both genetic drift and genetic shift.

Genetic drift

Considerable intramolecular evolution occurs with LACV (Klimas *et al.*, 1981; Borucki *et al.*, 2001) and the genetic diversity of LACV attributable to genetic drift is exceptional. Recent phylogenetic analysis of M segment sequences of LACV suggests there are at least three geographic lineages of the virus (Armstrong and Andreadis, 2006). Fatal human infections are associated with a narrow range of LACV genotypes (Huang *et al.*, 1997), and recent studies have revealed that nucleotide changes in the non-coding region and amino acid changes in the ORF of the RNA polymerase condition disease outcomes in mice (Bennet *et al.*, 2007). Interestingly, phylogenetic analyses have demonstrated that viruses from SI+ mosquitoes are very similar to viruses isolated from humans in the same areas (Reese *et al.*, 2010). Thus intramolecular evolution of SI+ LACVs could be very important epidemiologically, in terms of maintenance of both viruses in endemic foci and being more virulent in humans.

Genetic shift

Genome segment reassortment in mosquitoes is clearly a determinant of LACV evolution. Laboratory studies have demonstrated the occurrence of genetic shift (segment reassortment) in mosquitoes that have become dually infected by ingesting viruses of two different LACV genotypes, either simultaneously or within two days of each other (Beaty *et al.*, 1985). LACV reassortant viruses can be isolated from up to 25% of dually infected *Ae. triseriatus* and the newly generated viruses can be transmitted. High frequency reassortment also occurs when mosquitoes ingest heterologous California serogroup viruses simultaneously or via interrupted feeding (Chandler *et al.*, 1990, 1991). However, mosquitoes that ingest LACV become refractory to super infection with a second California serogroup virus (Sundin and Beaty, 1988). The level of interference increases over time from 24 h to 7 days, and clearly is an impediment to evolution in nature by preventing mosquitoes from being superinfected with a second virus (Sundin and Beaty, 1988). In nature, mosquitoes probably only rarely ingest two viruses simultaneously. A more likely scenario is that they would become superinfected when TOT-infected mosquitoes would ingest a viraemic bloodmeal. Indeed, the potential for segment reassortment increases when a transovarially infected mosquito takes a bloodmeal from a viraemic host (Borucki *et al.*, 1999). TOT+ mosquitoes do not exhibit complete interference to super infection (Borucki *et al.*, 1999). *Ae. triseriatus* mosquitoes TOT+ with LACV were challenged orally with Snowshoe Hare virus (SSHV). A significant proportion of the mosquitoes became infected when they ingested SSH virus in a bloodmeal. These mosquitoes were susceptible to super infection and served as a site for segment reassortment and the evolution of new virus genotypes (Borucki *et al.*, 1999). This was in dramatic contrast to orally infected mosquitoes which after a few days are resistant to oral super infection. The reason for the lack of complete interference to super infection in TOT+ mosquitoes remains to be determined. TOT+ mosquitoes have dramatically lower titres of virus present in their midguts than orally infected mosquitoes, which may condition superinfection when challenged with a second virus via blood feeding. Interestingly, although SSH virus infected a smaller proportion of previously non-infected mosquitoes (82% vs. 96% for LAC virus), the superinfection rate in TOT+ mosquitoes was equivalent to that of LAC virus. This suggests that there may be less interference to a heterologous superinfecting virus than to a homologous virus in infected mosquitoes, thereby promoting dual infection, RNA segment reassortment, and virus evolution.

The ovary is probably a major site of segment reassortment. Virus replication and reassortment is enhanced in the ovaries of female *Ae. triseriatus* that are provided with additional bloodmeals. Active transport of molecules into and rapid cell division in the ovarian follicles may overcome interference (Chandler *et al.*, 1990). Since TOT is a major maintenance and amplification mechanism for LACV in nature (Beaty and Thompson, 1975; Watts *et al.*, 1973), the ability of even a small percentage of TOT+ mosquitoes to become superinfected could be epidemiologically significant and result in generation of new reassortant viruses with new phenotypes, including altered vertebrate host and vector ranges, new tropisms or virulence (Beaty and Calisher, 1991).

Evolutionary potential of LACV in endemic forested areas

The amount of genetic shift and genetic drift occurring in a LACV hyperendemic area surrounding La Crosse, Wisconsin is truly extraordinary and testimony to the robust evolutionary potential of the *Bunyaviridae* viruses. The original landmark oligonucleotide finger printing studies of Klimas *et al.* (1981) revealed (1) extensive genetic variability in the LACV in a restricted area (no two isolates of LACV, even from the same tree hole, were identical), (2) the presence of multiple LACV genotypes in one area, and (3) the presence of reassortant viruses containing the S segment of one genotype and the L and M segments of the other genotype (Klimas *et al.*, 1981).

Modern molecular sequencing and analysis techniques were used to address in more depth the evolutionary potential of LACV in the Wisconsin/Minnesota hyperendemic area (Reese *et al.*, 2009; Reese *et al.*, 2010). *Ae. triseriatus* eggs were collected in oviposition traps by the La Crosse Public Health Department as part of ongoing surveillance programmes near premises or sites where LACV infections had occurred. Eggs were shipped to the AIDL at CSU, where they were hatched, reared to adults and assayed for LACV by immunofluorescence. LACV RNA was isolated from the abdomens of infected mosquitoes or viruses isolated in VERO cells, amplified, and portions of the L, M and S segments were sequenced. Sequence analysis revealed extensive nucleotide polymorphisms in the three segments, with the greatest genetic diversity in the M segment, which was twice that seen in the S segment and thrice that seen in the L segment (Reese *et al.*, 2009). Both phylogenetic and linkage disequilibrium analyses revealed that 24% of the mosquitoes contained reassortant LACVs. To put this into perspective, when *Ae. triseriatus* ingest two LACVs simultaneously, 100% of mosquitoes become dually infected and yield reassortant viruses (Beaty, *et al.*, 1985). When TOT infected mosquitoes ingest a bloodmeal containing an heterologous LACV, 19% become dually infected and yield reassortant viruses (Borucki *et al.*, 1999). This suggests that dual infection in the small forested areas may be attributable to superinfection of TOT+ mosquitoes. Regardless of the mechanism involved the evolutionary potential of LACV by segment reassortment is extraordinary.

Extensive genetic polymorphisms in LACVs were also detected in studies to identify stably infected mosquitoes in the hyperendemic area (Reese *et al.*, 2010). As described briefly above, three LACV infection phenotypes were detected in 17,825 field collected *Ae. triseriatus* mosquitoes collected as eggs in the hyperendemic area. Of these 17 (.09%) contained large amounts of viral antigen, viral RNA, and yielded LACV isolates and were designated superinfected (SI+) mosquitoes. I+ mosquitoes constituted 3.7% of the mosquitoes and contained detectable virus antigen and nucleic acid but no infectious virus. The remainder (96.2%) of the mosquitoes were designated I– and contained no detectable virus antigen, nucleic acid or infectious virus. To determine the molecular basis for the different virus phenotypes, sequences of the LACV S and M RNA segments from SI+ and I+ mosquitoes were compared. RNAi was not a determinant of the SI+ and I+ phenotypes; the NSs sequences of viruses from the two phenotypes were absolutely identical. Because NSs conditions RNAi in vertebrate cells (Soldan *et al.*, 2005), RNAi is not a likely determinant of the SI+ and I+ phenotypes (Reese *et al.*, 2010). In contrast, considerable gene polymorphisms were detected in LACV M sequences of viruses from the SI+ and I+ mosquitoes (Reese *et al.*, 2010). Apparently, the LACV M RNA is evolving through disruptive selection that maintains SI+ alleles at a higher frequency than the average mutation rate. Provocatively, a QTN was detected in the NSm gene of LACVs from SI+ but not I+ mosquitoes (Fig. 13.6). NSm conditions apoptosis in vertebrate cells (Bouloy *et al.*, 2001; Won *et al.*, 2007). The QTN that differentiated the SI+ and I+ is at position 246 of the NSm gene and corresponds to a U to A transversion in the third position of a CUN Leu codon. All CUA codons detected were in SI+ mosquitoes; none appeared in the I+ mosquitoes (Reese *et al.*, 2010). In one study site, both SI+ and I+ were detected. These differed in nucleotide sequences, and four amino acid changes in were detected in the NSm gene and may condition vector stabilized infection.

The genomic plasticity of LACV could easily result in evolution of viruses capable of inducing the SI+ phenotype. In this regard, the NSm virus mutations in SI+ mosquitoes are not monophyletic or phylogenetically distinct. Thus the mutations arise independently due to genetic drift and disruptive selection. The extraordinary mutation rates demonstrated in LACV in mosquitoes could account for the polyphyletic origins of the SI+ phenotypes. These mutations then could be subjected to disruptive selection in an innate immune arms race with vector host genes. Accumulating evidence suggests that interactions between the LACV NSm and the ATIAP gene may condition the SI+ phenotype. There is extensive gene flow in *Ae. triseriatus* in the hyperendemic area (Beck *et al.*, 2005). Phylogenetic analyses of the *Ae. triseriatus* ATIAP gene in mosquitoes from the hyperendemic area revealed extensive polymorphisms (approximately 3-fold greater than in typical mosquito genes) in the gene (Beck *et al.*, 2009). Despite the extensive genetic diversity, there was strong purifying selection in the baculovirus inhibitor of apoptosis repeat domain 1 (BIR1) (Fig. 13.7). The BIR functional domain is of course critical for preventing apoptosis, and the selection pressure may be exerted by LACV and associated with the NSm mutations discussed above. The ref(2)P gene of *D. melanogaster* is also highly polymorphic, presumably because of the host counteracting SIGMAV genetic changes (Dru et al., 1993; Longdon and Jiggins, 2012; Wilfert and Jiggins, 2013). Apoptosis and autophagy share common caspase regulatory pathway components, both function during degeneration of ovarian follicles, and autophagy is a key antiviral response in *D. melanogaster* (Shelley *et al.*, 2009; Cherry, 2009; Hou, 2008, 2009). Avoidance of these two innate immune responses would be absolutely critical for TOT and stabilized infection. Perhaps a major gene for gene interaction in an innate immune arms race conditions stabilized LACV infection in

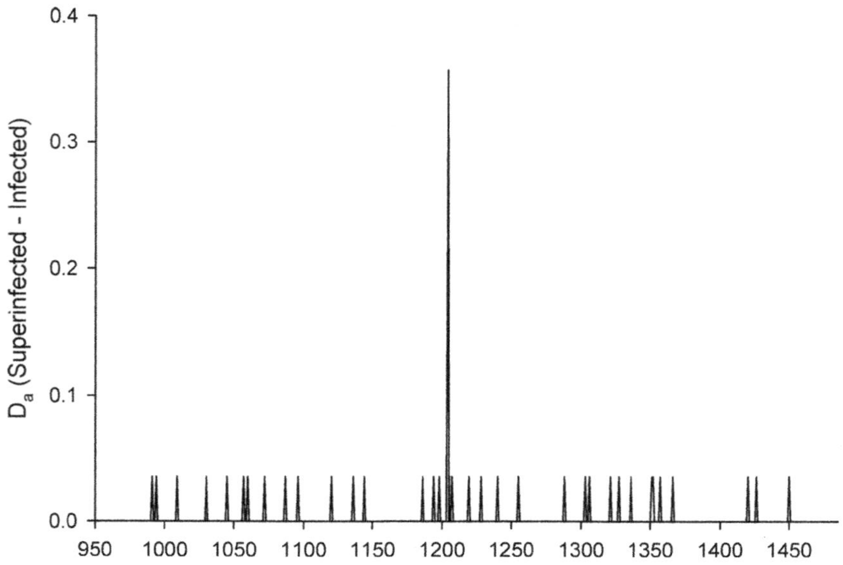

Figure 13.6 Nucleotide differences (D_a) between NSM sequences from SI+ and I+ mosquitoes (With permission from *Virology Journal*, Reese *et al.*, 2010).

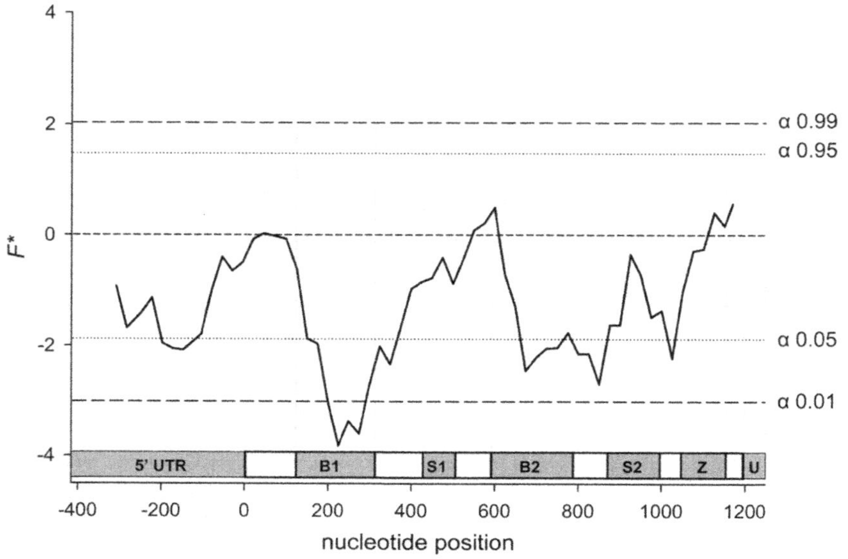

Figure 13.7 The BIR domain 1 of the ATIAP1 gene is under purifying selection. 5′ UTR, 5′ untranslated region; B1, 1st BIR domain; S1, 1st serine-rich domain; B2, 2nd BIR domain; S2, 2nd serine-rich domain; Z, zinc ring finger motif; U, 3′ untranslated region (with kind permission of Springer Science and Business Media; *Journal of Molecular Evolution*, Beck *et al.*, 2009).

Ae. triseriatus, similar to the SIGMAV–*D. melanogaster* system (Wilfert and Jiggins, 2013).

LACV SI+ *Ae. triseriatus* and SIGMAV-*D. melanogaster* – commonalities in VT mechanisms and potential importance of stabilized infections in arbovirology

To establish stabilized infection of a virus in its arthropod vector, there must be efficient TOT transmission, no cytopathology associated with PCD pathways, and no deleterious effect on the developing oocyte and embryo, even during critical periods such as embryogenesis, follicular resting stages, ovarian diapause and diapause. For both systems these criteria have been met, and while different molecular mechanisms are involved, in both systems a major gene-for-gene interaction seems to be a major determinant of conditioning stabilized infections. In the *D. melanogaster*-Rhabdovirus system, vector competence is conditioned in large part by the autophagic response. In addition, there is a *D. melanogaster*–SIGMAV gene-for gene interaction that conditions stabilized infection. The *Ref(2) P* gene conditions SIGMAV infection and persistence. Ref(2)P is a protein kinase in the Toll pathway. A Ref(2)P mutation CAG-AAT (glutamine-asparagine) to a GGA (glycine) restricts SIGMAV replication. *Ref(2)P* is very polymorphic, apparently due to an innate immune arms race between the virus and host. The gene is transcriptionally regulated, resulting in a short and long mRNA (and the long mRNA accumulates in the ovary. In the LACV-*Ae. triseriatus* cycle, the LACV NSm gene and the vector PCD pathway also seem to be in an innate immune arms race. The ATIAP1 protein is transcriptionally regulated, one mRNA variant accumulates in the ovary, regulation of autophagy is a major determinant of TOT. For both systems, unique virus–vector interactions have conditioned efficient TOT and stabilized infection. Both viruses are found in low frequency in infected arthropod hosts collected in nature, but seem to be able to maintain the virus in nature. If true, development of surveillance capacity for SI+ mosquitoes could provide a novel targeted and effective approach for control of LACV.

We have yet to prove that the LACV–*Ae. triseriatus* system is truly a stabilized infection. Conceptually, stabilized infection seems more likely in the SIGMAV–*D. melanogaster* system, where passage in vertebrate hosts is not part of the transmission cycle. Arboviruses need to infect and replicate in the arthropod vector and the vertebrate host, two dramatically different systems, which may constrain the ability of the virus to evolve to a stabilized infection genotype. However, the accumulating evidence with LACV and other arboviruses suggests that stabilized infection occurs. Theoretically stabilized infections in genotypically permissive lineages of a vector could maintain arboviruses and the newly discovered insect-specific viruses in nature. The molecular mechanisms may differ, but it is provocative to think that stabilized infections could be a potential unifying mechanism for VT maintenance and amplification of a number of arboviruses in nature.

Thus, if LACV and other arboviruses are maintained in stabilized infections, understanding the mechanisms involved is of fundamental importance to understanding the epidemiology of these important pathogens. Some questions and thoughts surrounding stabilized infection and its potential role in arbovirus and insect-specific virus cycles are presented below.

- *Does LACV even need a vertebrate host? Sigma virus does not.* This is a very important issue. If the vertebrate host is inconsequential for maintenance and amplification of the virus in nature, then interventions proposed to immunize vertebrate reservoir hosts in an attempt to control or eliminate LACV activity are doomed to failure. We feel that LACV probably stably infects *Ae. triseriatus* and may not require (or may only require a small level of horizontal amplification in vertebrates to be maintained in nature. Since the definition of an arbovirus includes biological transmission of the virus to vertebrate hosts, does that mean that LACV is not an arbovirus?
- *Do other negative strand arboviruses need a vertebrate host?* As noted, arboviruses in the family *Bunyaviridae* and *Rhabdoviridae* are the most efficiently transovarially transmitted viruses, and likely members of both families are maintained by stabilized infection. Both of the families have negative strand genomes, and thus unique polymerases. Association of the polymerase of Sigma virus with stabilized infection of *D. melanogaster* is provocative and needs to be investigated further as a determinant of stabilized infection.
- *Are the insect-specific viruses maintained by stabilized infections?* Discovery of these viruses, with representatives from each of the major arbovirus families, has caused great excitement in the field of arbovirology. VT and stabilized infection would seem to be the only feasible mechanism for maintenance of these viruses in nature. Understanding how these viruses are maintained will be of great interest and informative in our understanding of virus–vector interactions. This understanding may well provide invaluable information for understanding the overwintering and maintenance mechanisms of the true arboviruses in the different families; knowledge of which could be exploited for improved surveillance and control of these important human pathogens.
- *Are positive strand arboviruses maintained by stabilized infection?* Historically, TOT was not considered to be a part of the transmission cycle of positive-strand viruses. The detection of WNV in overwintered *Culex* mosquitoes, the amazing discovery of the *Culex* flaviviruses, the detection of DENV in field collected juvenile and male mosquitoes, the detection of Ross River virus in male mosquitoes, the isolation of Eilat virus, an insect-specific alphavirus, etc., indicates that stabilized infection needs to be revisited. For example, infection rates of flaviviruses detected in field collected larvae and males are in the range of those detected for SIGMAV and LACV in their field collected vectors.
- *Does stabilized infection and TOT condition the exceptional evolutionary potential of the* Bunyaviridae *to form new reassortant viruses?* There are more named Bunyaviruses than all of the other arboviruses combined (Beaty *et al.*, 1997). As more bunyaviruses are fully sequenced, it has become apparent that reassortment among closely related bunyaviruses is relatively common in nature (Briese *et al.*, 2013). Indeed, segment reassortment and emergence of new virus genotypes is a hallmark of the *Bunyaviridae*. The literature is replete with examples of newly detected reassortant bunyaviruses (Briese *et al.*, 2006, 2013; Blitvich *et al.*, 2012; Aguilar *et al.*, 2011; Garigliany, *et al.*, 2012; Hang, *et al.*, 2014; Goedhals, *et al.*, 2014;). Obviously segment reassortment promotes bunyavirus evolution and epidemic potential. The arthropod vector provides a unique site for dual infection and reassortment of new viruses with potentially new host range and virulence potential. Clearly, dual infection of vectors can lead to very efficient generation of reassortant viruses. This is most worrisome in terms of generation of new reassortant viruses with new virulence phenotypes and greater epidemic potential.

Summary

VT is an important component of the transmission and maintenance cycles of arboviruses in most of the arbovirus families, most notably the *Bunyaviridae* and the *Rhabdoviridae*. The role of VT in transmission and maintenance cycles of arboviruses in the *Flaviviridae* and *Togaviridae* is not as clear. VT may be more important for *Culex*-vectored flaviviruses than for *Aedes* vectored flaviviruses. Similarly VT may play more of a role in Old World than New World alphaviruses. Understanding the mechanisms and role of VT in these families is critical for surveillance and control of these important pathogens. The knowledge gap concerning overwintering remains for many arboviruses, and understanding the mechanisms conditioning trans-seasonality is critical for understanding arbovirus cycles in nature.

For mosquito-borne viruses in the family *Bunyaviridae*, VT is of fundamental importance for virus transmission, maintenance, and evolution. The LACV–*Ae. triseriatus* system illustrates potential VT mechanisms that condition virus survival and transmission in nature and promote virus evolution. The role of the vector in promoting segment reassortment is especially concerning. High frequency reassortment is likely minimized in the vertebrate hosts of segmented arboviruses due to the ephemeral nature of the infection and the small likelihood of dual infection of a host and cells in the host (Beaty *et al.*, 1997). In contrast, the vector with its long term persistent infections and its ability to become dually infected is a major source of virus evolution by segment reassortment for other arboviruses also. For example, following dual infection of *Culicoides variipennis* with two serotypes of bluetongue virus (10 genome segments), it is almost impossible to rescue the parental viruses (Samal *et al.*, 1987), illustrating the evolutionary potential of arboviruses in their vectors. Major genetic changes can lead rapidly to viruses with new host ranges, virulence potential, and vector host range, all of which pose threats to humans and domesticated animals.

Much remains to be learned about arbovirus–vector interactions including VT and stabilized infections. Understanding the role of these transmission mechanisms in maintenance, evolution, and emergence of arboviruses is critical to predict, prevent, and control these threats to public and animal health.

References

Abbassy, M.M., Osman, M., and Marzouk, A.S. (1993). West Nile virus (Flaviviridae:Flavivirus) in experimentally infected *Argas* ticks (Acari:Argasidae). Am. J. Trop. Med. Hyg. *48*, 726–737.

Aguilar, P.V., Barrett, A.D., Saeed, M.F., Watts, D.M., Russell, K., Guevara, C., Ampuero, J.S., Suarez, L., Cespedes, M., Montgomery, J.M., Halsey, E.S., and Kochel, T.J. (2011). Iquitos virus: a novel reassortant Orthobunyavirus associated with human illness in Peru. PLoS Negl. Trop. Dis. *5*, e1315.

Aitken, T.H.G., Woodall, J.P., De Andrade, A.H., Bensabath, G., and Shope, R.E. (1975). Pacui virus, phlebotomine flies, and small mammals in Brazil: an epidemiological study. Am. J. Trop. Med. Hyg. *24*, 358–368.

Aitken, T.H.G., Tesh, R.B., Beaty, B.J., and Rosen, L. (1979). Transovarial transmission of yellow fever virus by mosquitoes (*Aedes aegypti*). Am. J. Trop. Med. Hyg. *28*, 119–121.

Anderson, J.F., Andreadis, T.G., Main, A.J., Ferrandino, F.J., and Vossbrinck, C.R. (2006). West Nile virus from female and male mosquitoes (Diptera: Culicidae) in subterranean, ground, and canopy habitats in Connecticut. J. Med. Entomol. *43*, 1010–1019.

Anderson, J.F., Main, A.J., Delroux, K., and Fikrig, E. (2008). Extrinsic incubation periods for horizontal and vertical transmission of West Nile virus by *Culex pipiens pipiens* (Diptera: Culicidae). J. Med. Entomol. *45*, 445–451.

Anderson, J.F., Main, A.J., Cheng, G., Ferrandino, F.J., and Fikrig, E. (2012). Horizontal and vertical transmission of West Nile virus genotype NY99 by *Culex salinarius* and genotypes NY99 and WN02 by *Culex tarsalis*. Am. J. Trop. Med. Hyg. *86*, 134–139.

Andreadis, T.G., Armstrong, P.M., and Bajwa, W.I. (2010). Studies on hibernating populations of *Culex pipiens* from a West Nile virus endemic focus in New York City: parity rates and isolation of West Nile virus. J. Am. Mosq. Control Assoc. *26*, 257–264.

Armstrong, P.M., and Andreadis, T.G. (2006). A new genetic variant of La Crosse virus (Bunyaviridae) isolated from New England. Am. J. Trop. Med. Hyg. *75*, 491–496.

Arunachalam, N., Samuel, P.P., Hiriyan, J., Thenmozhi, V., and Gajanana, A. (2004). Japanese encephalitis in Kerala, south India: can *Mansonia* (Diptera: Culicidae) play a supplemental role in transmission? J. Med. Entomol. *41*, 456–461.

Attoui, H., Mohd Jaafar, F., Belhouchet, M., Biagini, P., Cantaloube, J.F., de Micco, P., de Lamballerie, X. (2005). Expansion of family Reoviridae to include nine-segmented dsRNA viruses: isolation and characterization of a new virus designated *Aedes pseudoscutellaris* reovirus assigned to a proposed genus (Dinovernavirus). Virology *343*, 212–223.

Attoui, H., Mertens, P.P.C., Becnel, J., Belaganahalli, S., Bergoin, M., Brussaard, C.P., Chappell, J.D., CiarletI M., Del Vas M., and Dermody, T.S. (2012). Family Rhabdoviridae. In Virus Taxonomy. Ninth Report of the International Committee on Taxonomy of Viruses, A.M.Q. King, M.J. Adams, E.B. Carstens and D.J. Lefkowitz, eds. (Elsevier Academic Press, Amsterdam), pp. 541–637.

Avila, A., Silverman, N., Diaz-Meco, M.T., and Moscat, J. (2002). The *Drosophila* atypical protein kinase C-ref(2)p complex constitutes a conserved module for signaling in the toll pathway. Mol. Cell. Biol. *22*, 8787–8795.

Bailey, C.L., Eldridge, B.F., Hayes, D.E., Watts, D.M., Tammariello, R.F., and Dalrymple, J.M. (1978). Isolation of St. Louis encephalitis virus from overwintering *Culex pipiens* mosquitoes. Science *199*, 1346–1349.

Ballinger, M.J., Bruenn, J.A., Hay, J., Czechowski, D., and Taylor, D.J. (2014). Discovery and evolution of Bunyavirids in arctic phantom midges and ancient Bunyavirid-like sequences in insect genomes. J. Virol. *88*, 8783–8794.

Bangham, J., Obbard, D.J., Kim, K.W., Haddrill, P.R., and Jiggins, F.M. (2007). The age and evolution of an antiviral resistance mutation in *Drosophila melanogaster*. Proc. Biol. Sci. *274*, 2027–2034.

Bangham, J., Kim, K.W., Webster, C.L., and Jiggins, F.M. (2008). Genetic variation affecting host–parasite interactions: different genes affect different aspects of sigma virus replication and transmission in *Drosophila melanogaster*. Genetics *178*, 2191–2199.

Baqar, S., Hayes, C.G., Murphy, J.R., and Watts, D.M. (1993). Vertical transmission of West Nile virus by *Culex* and *Aedes* species mosquitoes. Am. J. Trop. Med. Hyg. *48*, 757–762.

Bárdos, V., Ryba, J., and Hubálek, Z. (1975). Isolation of Tahyna virus from field collected *Culiseta annulata* (Schrk.) larvae. Acta. Virol. *19*, 446.

Beaty, B.J., and Bishop, D.H. (1988). Bunyavirus–vector interactions. Virus Res. *10*, 289–301.

Beaty, B.J., and Calisher, C.H. (1991). Bunyaviridae – natural history. Curr. Top. Microbiol. Immunol. *169*, 27–78.

Beaty, B.J., and Thompson, W.H. (1975). Emergence of La Crosse virus from endemic foci. Fluorescent antibody studies of overwintered *Aedes triseriatus*. Am. J. Trop. Med. Hyg. *24*, 685–691.

Beaty, B.J., and Thompson, W.H. (1976). Delineation of La Crosse virus in developmental stages of transovarially infected *Aedes triseriatus*. Am. J. Trop. Med. Hyg. *25*, 505–512.

Beaty, B.J., and Thompson, W.H. (1978). Tropisms of La Crosse virus in *Aedes triseriatus* (Diptera: Culicidae) following infective bloodmeals. J. Med. Entomol. *14*, 499–503.

Beaty, B.J., Tesh, R.B., and Aitken, T.H.G. (1980). Transovarial transmission of yellow fever virus in *Stegomyia* mosquitoes. Am. J. Trop. Med. Hyg. *29*, 125–132.

Beaty, B.J., Sundin, D.R., Chandler, L.J., and Bishop, D.H. (1985). Evolution of bunyaviruses by genome reassortment in dually infected mosquitoes (*Aedes triseriatus*). Science *230*, 548–550.

Beaty, B.J., Borucki, M., Farfan, J., and White, D. (1997). Arbovirus–vector interactions: determinants of arbovirus evolution. In Factors in the Emergence of Arbovirus Diseases, J.F. Saluzzo and B. Dodet, eds. (Elsevier, Paris), pp. 23–35.

Beaty, B.J., Rayms-Keller, F., Borucki, M.K., and Blair, C.D. (2000). La Crosse encephalitis virus and mosquitoes: a remarkable relationship. *ASM News 66*, 349–357.

Beck, E.T., Bosio, C.F., Geske, D.A., Blair, C.D., Beaty, B.J., Black 4th, W.C. (2005). An analysis of gene flow among midwestern populations of the mosquito *Ochlerotatus triseriatus*. Am. J. Trop. Med. Hyg. 73, 534–540.

Beck, E.T., Blair, C.D., Black 4th, W.C., Beaty, B.J., and Blitvich, B.J. (2007). Alternative splicing generates multiple transcripts of the inhibitor of apoptosis protein 1 in *Aedes* and *Culex* spp. mosquitoes. Insect Biochem. Mol. Biol. 37, 1222–1233.

Beck, E.T., Lozano Fuentes, S., Geske, D.A., Blair, C.D., Beaty, B.J., Black 4th, W.C. (2009). Patterns of variation in the inhibitor of apoptosis 1 gene of *Aedes triseriatus*, a transovarial vector of La Crosse virus. J. Mol. Evol. *68*, 403–413.

Bellini, R., Medici, A., Calzolari, M., Bonilauri, P., Cavrini, F., Sambri, V., Angelini, P., and Dottori, M. (2012). Impact of Chikungunya virus on *Aedes albopictus* females and possibility of vertical transmission using the actors of the 2007 outbreak in Italy. PLoS One 7, e28360.

Belloncik, S., Poulin, L., Maire, A., Aubin, A., Fauvel, M., and Jousset, F.X. (1982). Activity of California encephalitis group viruses in Entrelacs (province of Quebec, Canada). Can. J. Microbiol. *28*, 572–579.

Bennett, R.S., Ton, D.R., Hanson, C.T., Murphy, B.R., and Whitehead, S.S. (2007). Genome sequence analysis of La Crosse virus and in vitro and in vivo phenotypes. Virol. J. *4*, 41.

Berry, R.L., Lalonde-Weigert, B.J., Calisher, C.H., Parsons, M.A., and Bear, G.T. (1977). Evidence for transovarial transmission of Jamestown Canyon virus in Ohio. Mosq. News *37*, 494–496.

Bilsel, P.A., Tesh, R.B., and Nichol, S.T. (1988). RNA genome stability of Toscana virus during serial transovarial transmission in the sandfly *Phlebotomus perniciosus*. Virus Res. *11*, 87–94.

Blair, C.D., and Olson, K.E. (2014). Mosquito immune responses to arbovirus infections. Curr. Opin. Insect Sci. *3*, 22–29.

Blakqori, G., Delhaye, S., Habjan, M., Blair, C.D., Sánchez-Vargas, I., Olson, K.E., Attarzadeh-Yazdi, G., Fragkoudis, R., Kohl, A., Kalinke, U., et al. (2007). La Crosse bunyavirus non-structural protein NSs serves to suppress the type I interferon system of mammalian hosts. J. Virol. *81*, 4991–4999.

Blattner, R.J., and Heys, F.M. (1944). Blood-sucking vectors of encephalitis: experimental transmission of St. Louis encephalitis (Hubbard strain) to white Swiss mice by the American dog tick, *Dermacentor variabilis* Say. J. Exp. Med. *79*, 439–454.

Blitvich, B., Blair, C., Kempf, B., Hughes, M.T., Black 4th, W.C., Mackie, R.S., Meredith, C.T., Beaty, B.J., and Rayms-Keller, A. (2002). Developmental- and tissue-specific expression of an inhibitor of apoptosis protein 1 homologue from *Aedes triseriatus* mosquitoes. Insect Mol. Biol. *11*, 431–442.

Blitvich, B.J., Lin, M., Dorman, K.S., Soto, V., Hovav, E., Tucker, B.J., Staley, M., Platt, K.B., and Bartholomay, L.C. (2009). Genomic sequence and phylogenetic analysis of *Culex* flavivirus, an insect-specific flavivirus, isolated from *Culex pipiens* (Diptera: Culicidae) in Iowa. J. Med. Entomol. *46*, 934–941.

Blitvich, B.J., Saiyasombat, R., Dorman, K.S., Garcia-Rejon, J.E., Farfan-Ale, J.A., and Loroño-Pino, M.A. (2012). Sequence and phylogenetic data indicate that an orthobunyavirus recently detected in the Yucatan Peninsula of Mexico is a novel reassortant of Potosi and Cache Valley viruses. Arch. Virol. *157*, 1199–1204.

Bolling, B.G., Eisen, L., Moore, C.G., and Blair, C.D. (2011). Insect-specific flaviviruses from *Culex* mosquitoes in Colorado, with evidence of vertical transmission. Am. J. Trop. Med. Hyg. *85*, 169–177.

Bolling, B.G., Olea-Popelka, F.J., Eisen, L., Moore, C.G., and Blair, C.D. (2012). Transmission dynamics of an insect-specific flavivirus in a naturally infected *Culex pipiens* laboratory colony and effects of co-infection on vector competence for West Nile virus. Virology *427*, 90–97.

Bolling, B.G., Vasilakis, N., Guzman, H., Widen, S.G., Wood, T.G., Popov, V.L., Thangamani, S., and Tesh, R.B. (2015). Insect-specific viruses detected in laboratory mosquito colonies and potential implications for experiments evaluating arbovirus vector competence. Am. J. Trop. Med. Hyg. *92*, 422–428.

Boromisa, R.D., and Grayson, M.A. (1990). Incrimination of *Aedes provocans* as a vector of Jamestown Canyon virus in an enzootic focus of northeastern New York. J. Am. Mosq. Control Assoc. *6*, 504–509.

Borucki, M.K., Chandler, L.J., Parker, B.M., Blair, C.D., and Beaty, B.J. (1999). Bunyavirus superinfection and segment reassortment in transovarially infected mosquitoes. J. Gen. Virol. *80*, 3173–3179.

Borucki, M.K., Kempf, B.J., Blair, C.D., and Beaty, B.J. (2001). The effect of mosquito passage on the La Crosse virus genotype. J. Gen. Virol. *82*, 2919–2926.

Borucki, M.K., Kempf, B.J., Blitvich, B.J., Blair, C.D., and Beaty, B.J. (2002). La Crosse virus: replication in vertebrate and invertebrate hosts. Microbes Infect. *4*, 341–350.

Bouloy, M., Janzen, C., Vialat, P., Khun, H., Pavlovic, J., Huerre, M., and Haller, O. (2001). Genetic evidence for an interferon-antagonistic function of rift valley fever virus non-structural protein NSs. J. Virol. *75*, 1371–1377.

Briese, T., Bird, B., Kapoor, V., Nichol, S.T., and Lipkin, W.I. (2006). Batai and Ngari viruses: M segment reassortment and association with severe febrile disease outbreaks in East Africa. J. Virol. *80*, 5627–5630.

Briese, T., Calisher, C.H., and Higgs, S. (2013). Viruses of the family Bunyaviridae: are all available isolates reassortants? Virology *446*, 207–216.

Bridgen, A., Weber, F., Fazakerley, J.K., and Elliott, R.M. (2001). Bunyamwera bunyavirus non-structural protein NSs is a nonessential gene product that contributes to viral pathogenesis. Proc. Natl. Acad. Sci. U.S.A. *98*, 664–669.

Broom, A.K., Lindsay, M.D., Johansen, C.A., Wright, A.E., and Mackenzie, J.S. (1995). Two possible mechanisms for survival and initiation of Murray Valley encephalitis virus activity in the Kimberley region of Western Australia. Am. J. Trop. Med. Hyg. *53*, 95–99.

Brown, C.R., Moore, A.T., Knutie, S.A., and Komar, N. (2009a). Overwintering of infectious Buggy Creek virus (Togaviridae: Alphavirus) in *Oeciacus vicarius* (Hemiptera: Cimicidae) in North Dakota. J. Med. Entomol. *46*, 391–394.

Brown, C.R., Moore, A.T., Young, G.R., Padhi, A., and Komar, N. (2009b). Isolation of Buggy Creek virus (Togaviridae: Alphavirus) from field-collected eggs of *Oeciacus vicarius* (Hemiptera: Cimicidae). J. Med. Entomol. *46*, 375–379.

Bugbee, L.M., and Forte, L.R. (2004). The discovery of West Nile virus in overwintering *Culex pipiens* (Diptera: Culicidae) mosquitoes in Lehigh County, Pennsylvania. J. Am. Mosq. Control Assoc. *20*, 326–327.

Campbell, G.L., Eldridge, B.F., Reeves, W.C., and Hardy, J.L. (1991). Isolation of Jamestown Canyon virus from boreal *Aedes* mosquitoes from the Sierra Nevada of California. Am. J. Trop. Med. Hyg. *44*, 244–249.

Cecílio, A.B., Campanelli, E.S., Souza, K.P., Figueiredo, L.B., and Resende, M.C. (2009). Natural vertical transmission by *Stegomyia albopicta* as dengue vector in Brazil. Braz. J. Biol. *69*, 123–127.

Chamberlain, R.W., and Sudia, W.D. (1961). Mechanism of transmission of viruses by mosquitoes. Annu. Rev. Entomol. *6*, 371–390.

Chandler, L.J., Beaty, B.J., Baldridge, G.D., Bishop, D.H., and Hewlett, M.J. (1990). Heterologous reassortment of bunyaviruses in *Aedes triseriatus* mosquitoes and transovarial and oral transmission of newly evolved genotypes. J. Gen. Virol. *71*, 1045–1050.

Chandler, L.J., Hogge, G., Endres, M., Jacoby, D.R., Nathanson, N., and Beaty, B.J. (1991). Reassortment of La Crosse and Tahyna bunyaviruses in *Aedes triseriatus* mosquitoes. Virus Res. *20*, 181–191.

Chandler, L.J., Wasieloski, L.P., Blair, C.D., and Beaty, B.J. (1996). Analysis of La Crosse virus S-segment RNA and its positive-sense transcripts in persistently infected mosquito tissues. J. Virol. *70*, 8972–8976.

Chandler, L.J., Blair, C.D., and Beaty, B.J. (1998). La Crosse virus infection of *Aedes triseriatus* (Diptera: Culicidae) ovaries before dissemination of virus from the midgut. J. Med. Entomol. *35*, 567–572.

Chao, J.A., Lee, J.H., Chapados, B.R., Debler, E.W., Schneemann, A., and Williamson, J.R. (2005). Dual modes of RNA-silencing suppression by Flock House virus protein B2. Nat. Struct. Mol. Biol. *12*, 952–957.

Cherry, S. (2009). VSV infection is sensed by *Drosophila*, attenuates nutrient signaling, and thereby activates antiviral autophagy. Autophagy *5*, 1062–1063.

Ciufolini, M.G., Maroli, M., and Verani, P. (1985). Growth of two phleboviruses after experimental infection of their suspected sand fly vector, *Phlebotomus perniciosus* (Diptera: Psychodidae). Am. J. Trop. Med. Hyg. *34*, 174–179.

Comer, J.A., Tesh, R.B., Modi, G.B., Corn, J.L., and Nettles, V.F. (1990). Vesicular stomatitis virus, New Jersey serotype: replication in and transmission by *Lutzomyia shannoni* (Diptera: Psychodidae). Am. J. Trop. Med. Hyg. *42*, 483–490.

Comer, J.A., Corn, J.L., Stallknecht, D.E., Landgraf, J.G., and Nettles, V.F. (1992). Titers of vesicular stomatitis virus, New Jersey serotype, in naturally infected male and female *Lutzomyia shannoni* (Diptera: Psychodidae) in Georgia. J. Med. Entomol. *29*, 368–370.

Contamine, D., Petitjean, A.M., and Ashburner, M. (1989). Genetic resistance to viral infection: the molecular cloning of a *Drosophila* gene that restricts infection by the rhabdovirus sigma. Genetics *123*, 525–533.

Cook, S., Bennett, S.N., Holmes, E.C., De Chesse, R., Moureau, G., de Lamballerie, X. (2006). Isolation of a new strain of the flavivirus cell fusing agent virus in a natural mosquito population from Puerto Rico. J. Gen. Virol. *87*, 735–748.

Cook, S., Moureau, G., Harbach, R.E., Mukwaya, L., Goodger, K., Ssenfuka, F., Gould, E., Holmes, E.C., de Lamballerie, X. (2009). Isolation of a novel species of flavivirus and a new strain of *Culex* flavivirus (Flaviviridae) from a natural mosquito population in Uganda. J. Gen. Virol. *90*, 2669–2678.

Cook, S., Moureau, G., Kitchen, A., Gould, E.A., de Lamballerie, X., Holmes, E.C., and Harbach, R.E. (2012). Molecular evolution of the insect-specific flaviviruses. J. Gen. Virol. *93*, 223–234.

Cordellier, R., Chippaux, A., Monteny, N., Heme, G., Courtois, B., Germain, M., and Digoutte, J.P. (1982). Isolements du virus Orungo a partir de femelles et de males d' *Aedes selvatiques* catpures en Cote d'Ivoire. Cah. ORSTOM Ser. Entomol. Med. Parasitol. *20*, 265–267.

Cordellier, R., Bouchite, B., Roche, J.C., Monteny, N., Diaco, B., and Akoliba, P. (1983). The sylvatic distribution of dengue 2 virus in the sub-Sudanese savanna areas of the Ivory Coast in 1980. Entomological data and epidemiological study. Cah. ORSTOM Ser. Entomol. Med. Parasitol. *21*, 165–179.

Corner, L.C., Robertson, A.K., Hayles, L.B., and Iverson, J.O. (1980). Cache Valley virus: experimental infection in *Culiseta inornata*. Can. J. Microbiol. *26*, 287–290.

Cornet, M., Robin, Y., Heme, G., Adam, C., Renaudet, J., Valade, M., and Eyrand, M. (1979). Une poussée épizootique de fiévre jaune selvatique au Sénégal oriental. Isolement du virus de lots de moustiques adultes males et femelles. Medecine et Maladies Infectieuses *9*, 63–66.

Coz, J., Valade, M., Cornet, M., and Robin, Y. (1976). Transovarian transmission of a Flavivirus, the Koutango virus, in *Aedes aegypti* L. C.R. Acad. Sci. Hebd. Seances. Acad. Sci. D. *283*, 109–110.

Crabtree, M.B., Sang, R.C., Stollar, V., Dunster, L.M., and Miller, B.R. (2003). Genetic and phenotypic characterization of the newly described insect flavivirus, Kamiti River virus. Arch. Virol. *148*, 1095–1118.

Crane, G.T., Elbel, R.E., and Calisher, C.H. (1977). Transovarial transmission of California encephalitis virus in the mosquito *Aedes dorsalis* at Blue Lake, Utah. Mosq. News *37*, 479–482.

Danielova, V., and Ryba, J. (1979). Laboratory demonstration of transovarial transmission of Tahyna virus in *Aedes vexans* and the role of this mechanism in the overwintering of this arbovirus. Folia. Parasitol. *26*, 361–368.

Danielová, V., Holubová, J., Pejcoch, M., and Daniel, M. (2002). Potential significance of transovarial transmission in the circulation of tick-borne encephalitis virus. Folia. Parasitol. (Praha) *49*, 323–325.

DeFoliart, G.R. (1983). *Aedes triseriatus*: vector biology in relationship to the persistence of La Crosse virus in endemic foci. Prog. Clin. Biol. Res. *123*, 89–104.

Deretic, V. (2009). Autophagy in infection. Curr. Opin. Cell. Biol. *22*, 1–11.

Dezelee, S., Bras, F., Contamine, D., and Lopez-Ferber, M. (1989). Molecular analysis of ref(2)P, a *Drosophila* gene implicated in sigma rhabdovirus multiplication and necessary for male fertility. EMBO J. *8*, 3437–3446.

Dhanda, V., Mourya, D.T., Mishra, A.C., Ilkal, M.A., Pant, U., Jacob, P.G., and Bhat, H.R. (1989). Japanese encephalitis virus infection in mosquitoes reared from field-collected immatures and in wild-caught males. Am. J. Trop. Med. Hyg. *41*, 732–736.

Dhileepan, K., Azuolas, J.K., and Gibson, C.A. (1996). Evidence of vertical transmission of Ross River and Sindbis viruses (Togaviridae: Alphavirus) by mosquitoes (Diptera: Culicidae) in southeastern Australia. J. Med. Entomol. 33, 180–182.

Dobie, D.K., Blair, C.D., Chandler, L.J., Rayms-Keller, A., McGaw, M.M., Wasieloski, L.P., and Beaty, B.J. (1997). Analysis of La Crosse virus S mRNA 5′ termini in infected mosquito cells and *Aedes triseriatus* mosquitoes. J. Virol. *71*, 4395–4399.

Domingo, E., and Holland, J.J. (1997). RNA virus mutations and fitness for survival. Annu. Rev. Microbiol. *51*, 151–178.

Dostert, C., Jouanguy, E., Irving, P., Troxler, L., Galiana-Arnoux, D., Hetru, C., Hoffmann, J.A., and Imler, J.L. (2005). The Jak-STAT signaling pathway is required but not sufficient for the antiviral response of *Drosophila*. Nat. Immunol. *6*, 946–953.

Dru, P., Bras, F., Dezelee, S., Gay, P., Petitjean, A.M., Pierre-Deneubourg, A., Teninges, D., and Contamine, D. (1993). Unusual variability of the *Drosophila melanogaster* ref(2)P-protein which controls the multiplication of sigma rhabdovirus. Genetics *133*, 943–954.

Dutary, B.E., and LeDuc, J.W. (1981). Transovarial transmission of yellow fever virus by a sylvatic vector, *Haemagogus equinus*. Trans. R. Soc. Trop. Med. Hyg. *75*, 128.

Dutary, B.E., Petersen, J.L., Peralta, P.H., and Tesh, R.B. (1989). Transovarial transmission of Gamboa virus in a tropical mosquito, *Aedeomyia squamipennis*. Am. J. Trop. Med. Hyg. *40*, 108–113.

Eldridge, B.F., Lanzaro, G.C., Campbell, G.L., Reeves, W.C., and Hardy, J.L. (1991). Occurrence and evolutionary significance of a California encephalitis-like virus in *Aedes squamiger* (Diptera: Culicidae). J. Med. Entomol. *28*, 645–651.

Endris, R.G., Tesh, R.B., and Young, D.G. (1983). Transovarial transmission of Rio Grande virus (Bunyaviridae: Phlebovirus) by the sand fly, *Lutzomyia anthophora*. Am. J. Trop. Med. Hyg. 32, 862–864.

Farajollahi, A., Crans, W.J., Bryant, P., Wolf, B., Burkhalter, K.L., Godsey, M.S., Aspen, S.E., and Nasci, R.S. (2005). Detection of West Nile viral RNA from an overwintering pool of *Culex pipens pipiens* (Diptera: Culicidae) in New Jersey, 2003. J. Med. Entomol. *42*, 490–494.

Farfan-Ale, J.A., Loroño-Pino, M.A., Garcia-Rejon, J.E., Hovav, E., Powers, A.M., Lin, M., Dorman, K.S., Platt, K.B., Bartholomay, L.C., Soto, V., Beaty, B.J., Lanciotti, R.S., and Blitvich, B.J. (2009). Detection of RNA from a novel West Nile-like virus and high prevalence of an insect-specific flavivirus in mosquitoes in the Yucatan Peninsula of Mexico. Am. J. Trop. Med. Hyg. *80*, 85–95.

Faye, O., Cornet, J.R., Camicas, J.L., Fontenille, D., and Gonzalez, J.P. (1999a). Experimental transmission of Crimean–Congo hemorrhagic fever virus: role of 3 vector species in the maintenance and transmission cycles in Senegal. Parasite *6*, 27–32.

Faye, O., Fontenille, D., Thonnon, J., Gonzalez, J.E., Cornet, J.P., and Camicas, J.L. (1999b). Experimental transmission of Crimean–Congo hemorrhagic fever virus by *Rhipicephalus evertsi* evertsi (Acarina:Ixodidae). Bull. Soc. Pathol. Exot. *92*, 143–147.

Fine, P.E. (1975). Vectors and vertical transmission: an epidemiologic perspective. Ann. N. Y. Acad. Sci. *266*, 173–194.

Fine, P.E., and LeDuc, J.W. (1978). Towards a quantitative understanding of the epidemiology of Keystone virus in the eastern United States. Am. J. Trop. Med. Hyg. *27*, 322–338.

Fleuriet, A. (1976). Presence of the hereditary rhabdovirus sigma and polymorphism for a gene for resistance to this virus in natural populations of *Drosophila melanogaster*. Evolution *30*, 735–739.

Fleuriet, A. (1981). Effect of overwintering on the frequency of flies infected by the rhabdovirus sigma in experimental populations of *Drosophila melanogaster*. Arch. Virol. *69*, 253–260.

Fleuriet, A. (1982). Transmission efficiency of the sigma virus in natural populations of its host, *Drosophila melanogaster*. Arch. Virol. *71*, 155–167.

Fleuriet, A. (1988). Maintenance of a hereditary virus, the sigma virus, in populations of its host, *D. melanogaster*. In Evolutionary Biology, M Hecht and B Wallace, eds. (Plenum Press, New York), pp. 1–30.

Fleuriet, A. (1999). Evolution of the proportions of two sigma viral types in experimental populations of *Drosophila melanogaster* in the absence of the allele that is restrictive of viral multiplication. Genetics *153*, 1799–1808.

Fontenille, D., Diallo, M., Mondo, M., Ndiaye, M., and Thonnon, J. (1997). First evidence of natural vertical transmission of yellow fever virus in *Aedes aegypti*, its epidemic vector. Trans. R. Soc. Trop. Med. Hyg. *91*, 533–535.

Francy, D.B., Rush, W.A., Montoya, M., Inglish, D.S., and Bolin, R.A. (1981). Transovarial transmission of St. Louis encephalitis virus by *Culex pipiens* complex mosquitoes. Am. J. Trop. Med. Hyg. *30*, 699–705.

Franz, A.W., Sanchez-Vargas, I., Adelman, Z.N., Blair, C.D., Beaty, B.J., James, A.A., and Olson, K.E. (2006). Engineering RNA interference-based resistance to dengue virus type 2 in genetically modified *Aedes aegypti*. Proc. Natl. Acad. Sci. U.S.A. *103*, 4198–4203.

Freier, J.E., and Beier, J.C. (1984). Oral and transovarial transmission of La Crosse virus by *Aedes atropalpus*. Am. J. Trop. Med. Hyg. 33, 708–714.

Freier, J.E., and Rosen, L. (1987). Vertical transmission of dengue viruses by mosquitoes of the *Aedes scutellaris* group. Am. J. Trop. Med. Hyg. *37*, 640–647.

Freier, J.E., Rosen. L. (1988). Vertical transmission of dengue viruses by *Aedes mediovittatus*. Am. J. Trop. Med. Hyg. 39, 218–222.

Gabitzsch, E.S., Blair, C.D., and Beaty, B.J. (2006). Effect of La Crosse virus infection on insemination rates in female *Aedes triseriatus* (Diptera: Culicidae). J. Med. Entomol. *43*, 850–852.

Garigliany, M.M., Bayrou, C., Kleijnen, D., Cassart, D., Jolly, S., Linden, A., and Desmecht, D. (2012). Schmallenberg virus: a new Shamonda/Sathuperi-like virus on the rise in Europe. Antiviral Res. *95*, 82–87.

Goddard, L.B., Roth, A.E., Reisen, W.K., and Scott, T.W. (2003). Vertical transmission of West Nile Virus by three California *Culex* (Diptera: Culicidae) species. J. Med. Entomol. *40*, 743–746.

Goedhals, D., Bester, P.A., Paweska, J.T., Swanepoel, R., and Burt, F.J. (2014). Next-generation sequencing of southern African Crimean–Congo haemorrhagic fever virus isolates reveals a high frequency of M segment reassortment. Epidemiol. Infect. *1*, 1–11.

Gonzalez, J.P., Camicas, J.L., Cornet, J.P., Faye, O., and Wilson, M.L. (1992). Sexual and transovarian transmission of Crimean–Congo haemorrhagic fever virus in *Hyalomma truncatum* ticks. Res. Virol. *143*, 23–28.

Graham, D.H., Holmes, J.L., Beaty, B.J., Black 4th, W.C. (2003). Quantitative trait loci conditioning transovarial transmission of La Crosse virus in the eastern treehole mosquito, *Ochlerotatus triseriatus*. Insect. Mol. Biol. *12*, 307–318.

Grimstad, P.R. (1988). California group viruses. In The Arboviruses, TP Monath, ed. vol. 2. (CRC Press, Boca Raton, Florida), pp. 99–136.

Haddow, A.D., Guzman, H., Popov, V.L., Wood, T.G., Widen, S.G., Haddow, A.D., Tesh, R.B., and Weaver, S.C. (2013). First isolation of *Aedes* flavivirus in the Western Hemisphere and evidence of vertical transmission in the mosquito *Aedes* (Stegomyia) *albopictus* (Diptera: Culicidae). Virology *440*, 134–139.

Hang, J., Forshey, B.M., Yang, Y., Solórzano, V.F., Kuschner, R.A., Halsey, E.S., Jarman, R.G., and Kochel, T.J. (2014). Genomic characterization of group C Orthobunyavirus reference strains and recent South American clinical isolates. PLoS One *9*, e92114.

Hansen, I.A., Attardo, G.M., Roy, S.G., and Raikhel, A.S. (2005). Target of rapamycin-dependent activation of S6 kinase is a central step in the transduction of nutritional signals during egg development in a mosquito. J. Biol. Chem. *280*, 20565–20572.

Hardy, J.L., Rosen, L., Kramer, L.D., Presser, S.B., Shroyer, D.A., and Turell, M.J. (1980). Effect of rearing temperature on transovarial transmission of St. Louis encephalitis virus in mosquitoes. Am. J. Trop. Med. Hyg. *29*, 963–968.

Hardy, J.L., Rosen, L., Reeves, W.C., Scrivani, R.P., and Presser, S.B. (1984). Experimental transovarial transmission of St. Louis encephalitis virus by *Culex* and *Aedes* mosquitoes. Am. J. Trop. Med. Hyg. *33*, 166–175.

Hardy, J.L., Eldridge, B.F., Reeves, W.C., Schutz, S.J., and Presser, S.B. (1993). Isolations of Jamestown canyon virus (Bunyaviridae: California serogroup) from mosquitoes (Diptera: Culicidae) in the western United States, 1990–1992. J. Med. Entomol. *30*, 1053–1059.

Hayes, R.O., Beadle, L.D., Hess, A.D., Sussman, O., and Bonese, M.J. (1962). Entomological aspects of the 1959 outbreak of eastern encephalitis in New Jersey. Am. J. Trop. Med. Hyg. *11*, 115–121.

Herve, J.P., Travassos da Rosa, A.P.A., Sa-Filho, G.C., Travassos da Rosa, J.F., and Pinheiro, F.P. (1984). Demonstration de la transmission transovarienne due virus Pacui chez *Lutzomyia flaviscutellata* (Phlebotominae). Importance epidemiologique. Cah. ORSTOM Ser. Entomol. Med. Parasitol. *22*, 207–212.

Hess, W.R., Endris, R.G., Haslett, T.M., Monahan, M.J., and McCoy, J.P. (1987). Potential arthropod vectors of African swine fever virus in North America and the Caribbean basin. Vet. Parasitol. *26*, 145–155.

Hoshino, K., Isawa, H., Tsuda, Y., Yano, K., Sasaki, T., Yuda, M., Takasaki, T., Kobayashi, M., and Sawabe, K. (2007). Genetic characterization of a new insect flavivirus isolated from *Culex pipiens* mosquito in Japan. Virology *359*, 405–414.

Hoshino, K., Isawa, H., Tsuda, Y., Sawabe, K., and Kobayashi, M. (2009). Isolation and characterization of a new insect flavivirus from *Aedes albopictus* and *Aedes flavopictus* mosquitoes in Japan. Virology *391*, 119–129.

Hou, Y.C., Chittaranjan, S., Barbosa, S.G., McCall, K., and Gorski, S.M. (2008). Effector caspase Dcp-1 and IAP-protein Bruce regulate starvation-induced autophagy during *Drosophila melanogaster* oogenesis. J. Cell. Biol. *182*, 1127–1139.

Hou, Y.C., Hannigan, A.M., and Gorski, S.M. (2009). An executioner caspase regulates autophagy. Autophagy *5*, 530–533.

Huang, C., Thompson, W.H., Karabatsos, N., Grady, L., and Campbell, W.P. (1997). Evidence that fatal human infections with La Crosse virus may be associated with a narrow range of genotypes. Virus. Res. *48*, 143–148.

Hughes, M.T., Gonzalez, J.A., Reagan, K.L., Blair, C.D., and Beaty, B.J. (2006). Comparative potential of *Aedes triseriatus*, *Aedes albopictus*, and *Aedes aegypti* (Diptera: Culicidae) to transovarially transmit La Crosse virus. J. Med. Entomol. *43*, 757–761.

Hull, B., Tikasingh, E., de Souza, M., and Martinez, R. (1984). Natural transovarial transmission of dengue 4 virus in *Aedes aegypti* in Trinidad. Am. J. Trop. Med. Hyg. *33*, 1248–1250.

Ibáñez-Bernal, S., Briseño, B., Mutebi, J.P., Argot, E., Rodríguez, G., Martínez-Campos, C., Paz, R., de la Fuente-San Román, P., Tapia-Conyer, R., and Flisser, A. (1997). First record in America of *Aedes albopictus* naturally infected with dengue virus during the 1995 outbreak at Reynosa, Mexico. Med. Vet. Entomol. *11*, 305–309.

Il'enko, V.I., Gorozhankina, T.S., and Smorodintsev, A.A. (1970). Main patterns of transovarial transmission of tick-borne encephalitis virus by tick vectors. Med. Parazit. Moskova. *39*, 263–267.

Jackson, B.T., Brewster, C.C., and Paulson, S.L. (2012). LaCrosse virus infection alters blood feeding behaviour in *Aedes triseriatus* and *Aedes aegypti* mosquitoes. J. Insect Sci. *7*, 1–7.

Joshi, V., Singhi, M., and Chaudhary, R.C. (1996). Transovarial transmission of dengue 3 virus by *Aedes aegypti*. Trans. R. Soc. Trop. Med. Hyg. *90*, 643–644.

Joshi, V., Mourya, D.T., and Sharma, R.C. (2002). Persistence of dengue-3 virus through transovarial transmission passage in successive generations of *Aedes aegypti* mosquitoes. Am. J. Trop. Med. Hyg. *67*, 158–161.

Jousset, F.X. (1982). Geographic *Aedes aegypti* strains and dengue 2 virus susceptibility, ability to transmit to vertebrate and transovarial transmission. Ann. Virol. *132*, 357–370.

Karabatsos, N. (1985). International Catalogue of Arboviruses Including Certain Other Viruses of Vertebrates, 3rd ed. (American Society of Tropical Medicine and Hygiene, Washington, DC).

Kay, B.H. (1982). Three modes of transmission of Ross River virus by *Aedes vigilax* (Skuse). Aust. J. Exp. Biol. Med. Sci. *60*, 339–344.

Kay, B.H., and Carley, J.G. (1980). Transovarial transmission of Murray Valley encephalitis virus by *Aedes aegypti* (L). Aust. J. Exp. Biol. Med. Sci. *58*, 501–504.

Keene, K.M., Foy, B.D., Sanchez-Vargas, I., Beaty, B.J., Blair, C.D., and Olson, K.E. (2004). RNA interference acts as a natural antiviral response to O'nyong-nyong virus (Alphavirus; Togaviridae) infection of *Anopheles gambiae*. Proc. Natl. Acad. Sci. U.S.A. *101*, 17240–17245.

Kempf, B.J., Blair, C.D., and Beaty, B.J. (2006). Quantitative analysis of La Crosse virus transcription and replication in cell cultures and mosquitoes. Am. J. Trop. Med. Hyg. *74*, 224–232.

Khin, M.M., and Than, K.A. (1983). Transovarial transmission of dengue 2 virus by *Aedes aegypti* in nature. Am. J. Trop. Med. Hyg. *32*, 590–594.

Kim, D.Y., Guzman, H., Bueno Jr., R., Dennett, J.A., Auguste, A.J., Carrington, C.V., Popov, V.L., Weaver, S.C., Beasley, D.W., and Tesh, R.B. (2009). Characterization of *Culex* Flavivirus (Flaviviridae) strains isolated from mosquitoes in the United States and Trinidad. Virology *386*, 154–159.

Kingsolver, M.B., Huang, Z., and Hardy, R.W. (2013). Insect antiviral innate immunity: pathways, effectors, and connections. J. Mol. Biol. *425*, 4921–4936.

Klimas, R.A., Thompson, W.H., Calisher, C.H., Clark, G.G., Grimstad, P.R., and Bishop, D.H. (1981). Genotypic varieties of La Crosse virus isolated from different geographic regions of the continental United States and evidence for a naturally occurring intertypic recombinant La Crosse virus. Am. J. Epidemiol. *114*, 112–131.

Kolodziejek, J., Pachler, K., Bin, H., Mendelson, E., Shulman, L., Orshan, L., and Nowotny, N. (2013). Barkedji virus, a novel mosquito-borne flavivirus identified in *Culex perexiguus* mosquitoes, Israel, 2011. J. Gen. Virol. *94*, 2449–2457.

Kondratenko, V.F. (1976). Importance of ixodid ticks in the transmission and preservation of the causative agent of Crimean hemorrhagic fever in foci of the infection. Parazitologiia *10*, 297–302.

Kourtis, N., and Tavernarakis, N. (2009). Autophagy and cell death in model organisms. Cell Death Differ. *16*, 21–30.

Kow, C.Y., Koon, L.L., and Yin, P.F. (2001). Detection of dengue viruses in field caught male *Aedes aegypti* and *Aedes albopictus* (Diptera: Culicidae) in Singapore by type-specific PCR. J. Med. Entomol. *38*, 475–479.

Kramer, L.D., and Thompson, W.H. (1983). Quantitation of La Crosse virus in venereally infected *Aedes triseriatus*. Am. J. Trop. Med. Hyg. *32*, 1140–1146.

Kramer, L.D., Reeves, W.C., Hardy, J.L., Presser, S.B., Eldridge, B.F., and Bowen, M.D. (1992). Vector competence of California mosquitoes for California encephalitis and California encephalitis-like viruses. Am. J. Trop. Med. Hyg. *47*, 562–573.

Kramer, L.D., Bowen, M.D., Hardy, J.L., Reeves, W.C., Presser, S.B., and Eldridge, B.F. (1993). Vector competence of alpine, Central Valley, and coastal mosquitoes (Diptera: Culicidae) from California for Jamestown Canyon virus. J. Med. Entomol. *30*, 398–406.

Labuda, M., Ciampor, F., and Kozuch, O. (1983). Experimental model of transovarial transmission of Tahyna virus in *Aedes aegypti* mosquitoes. Acta. Virol. *27*, 245–250.

LeDuc, J.W., Burger, J.F., Eldridge, B.F., and Russell, P.K. (1975a). Ecology of Keystone virus, a transovarially maintained arbovirus. Ann. N. Y. Acad. Sci. *266*, 144–151.

LeDuc, J.W., Suyemoto, W., Eldridge, B.F., Russell, P.K., and Barr, A.R. (1975b). Ecology of California encephalitis viruses on the Del Mar Va Peninsula. II. Demonstration of transovarial transmission. Am. J. Trop. Med. Hyg. *24*, 124–126.

Lee, V.H., and Kemp, E. (1970). Congo virus: experimental infection of *Hyalomma rufipes* and transmission to a calf. Bull. Entomol. Soc. Nigeria 2, 133–135.

LeGoff, G., Revollo, J., Guerra, M., Cruz, M., Barja Simon, Z., Roca, Y., Vargas Florès, J., and Hervé, J.P. (2011). Natural vertical transmission of dengue viruses by *Aedes aegypti* in Bolivia. Parasite *18*, 277–280.

Li, H., Li, W.X., and Ding, S.W. (2002). Induction and suppression of RNA silencing by an animal virus. Science *296*, 1319–1321.

Lindsay, M.D., Broom, A.K., Wright, A.E., Johansen, C.A., and Mackenzie, J.S. (1993). Ross River virus isolations from mosquitoes in arid regions of Western Australia: implication of vertical transmission as a means of persistence of the virus. Am. J. Trop. Med. Hyg. *49*, 686–696.

Lingel, A., Simon, B., Izaurralde, E., and Sattler, M. (2005). The structure of the flock house virus B2 protein, a viral suppressor of RNA interference, shows a novel mode of double-stranded RNA recognition. EMBO Rep. *6*, 1149–1155.

Linthicum, K.J., Davies, F.G., Kairo, A., and Bailey, C.L. (1985). Rift Valley fever virus (family Bunyaviridae, genus Phlebovirus). Isolations from Diptera collected during an inter-epizootic period in Kenya. J. Hyg. (Lond). *95*, 197–209.

Linthicum, K.J., Logan, T.M., Bailey, C.L., Hohm, D.J., and Moulton, J.R. (1989). Transtadial and horizontal transmission of Rift Valley fever virus in *Hyaloma truncatum*. Am. J. Trop. Med. Hyg. *41*, 491–496.

Lisitza, M., DeFoliant, G., Yuill, T., and Karandemios, M. (1977). Prevalence rates of La Crosse virus (California encephalities group) in larvae from overwintered eggs of *Aedes triseriatus*. Mosq. News *37*, 745–750.

Longdon, B., and Jiggins, F.M. (2012). Vertically transmitted viral endosymbionts of insects: do sigma viruses walk alone? Proc. Biol. Sci. *279*, 3889–3898.

Longdon, B., Wilfert, L., Obbard, D.J., and Jiggins, F.M. (2011). Rhabdoviruses in two species of *Drosophila*: vertical transmission and a recent sweep. Genetics *188*, 141–150.

Lutomiah, J.J., Mwandawiro, C., Magambo, J., and Sang, R.C. (2007). Infection and vertical transmission of Kamiti river virus in laboratory bred *Aedes aegypti* mosquitoes. J. Insect Sci. *7*, 1–7.

McGaw, M.M., Chandler, L.J., Wasieloski, L.P., Blair, C.D., and Beaty, B.J. (1998). Effect of La Crosse virus infection on overwintering of *Aedes triseriatus*. Am. J. Trop. Med. Hyg. *58*, 168–175.

McLintock, J., Carry, P.S., Wagner, R.J., Leung, M.K., and Iverson, J.O. (1976). Isolation of snowshoe hare virus from *Aedes implicatus* larvae in Saskatchewan. Mosq. News *36*, 233–237.

McPhee, C.K., and Baehrecke, E.H. (2009). Autophagy in *Drosophila melanogaster*. Biochim. Biophys. Acta. *1793*, 1452–1460.

Marchoux E., Simond, P.-L. (1905). La transmission héréditaire du virus de la fiévre jaune chez le *Stegomyia faseiata*. C. R. Soc. Biol. (Paris) *59*, 259–260.

Marklewitz, M., Handrick, S., Grasse, W., Kurth, A., Lukashev, A., Drosten, C., Ellerbrok, H., Leendertz, F.H., Pauli, G., and Junglen, S. (2011). Gouleako virus isolated from West African mosquitoes constitutes a proposed novel genus in the family Bunyaviridae. J. Virol. *85*, 9227–9234.

Marklewitz, M., Zirkel, F., Rwego, I.B., Heidemann, H., Trippner, P., Kurth, A., Kallies, R., Briese, T., Lipkin, W.I., Drosten, C., Gillespie, T.R., and Junglen, S. (2013). Discovery of a unique novel clade of mosquito-associated bunyaviruses. J. Virol. *87*, 12850–12865.

Martins, V.E., Alencar, C.H., Kamimura, M.T., de Carvalho Araújo, F.M., De Simone, S.G., Dutra, R.F., and Guedes, M.I. (2012). Occurrence of natural vertical transmission of dengue-2 and dengue-3 viruses in *Aedes aegypti* and *Aedes albopictus* in Fortaleza, Ceará, Brazil. PLoS One *7*, e41386.

Mavale, M.S., Geevarghese, G., Ghodke, Y.S., Fulmali, P.V., Singh, A., and Mishra, A.C. (2005). Vertical and venereal transmission of Chandipura virus (Rhabdoviridae) by *Aedes aegypti* (Diptera: Culicidae). J. Med. Entomol. *42*, 909–911.

Miller, B.R., DeFoliart, G.R., and Yuill, T.M. (1977). Vertical transmission of La Crosse virus (California encephalitis group): transovarial and filial infection rates in *Aedes triseriatus* (Diptera: Culicidae). J. Med. Entomol. *14*, 437–440.

Miller, B.R., Nasci, R.S., Godsey, M.S., Savage, H.M., Lutwama, J.J., Lanciotti, R.S., and Peters, C.J. (2000). First field evidence for natural vertical transmission of West Nile virus in *Culex univittatus* complex mosquitoes from Rift Valley province, Kenya. Am. J. Trop. Med. Hyg. *62*, 240–246.

Mishra, A.C., and Mourya, D.T. (2001). Transovarial transmission of West Nile virus in *Culex vishnui* mosquito. Indian J. Med. Res. *114*, 212–214.

Morales-Betoulle, M.E., Monzón Pineda, M.L., Sosa, S.M., Panella, N., López, M.R., Cordón-Rosales, C., Komar, N., Powers, A., and Johnson, B.W. (2008). *Culex* flavivirus isolates from mosquitoes in Guatemala. J. Med. Entomol. *45*, 1187–1190.

Mourya, D.T., Mishra, A.C., and Soman, R.S. (1991). Transmission of Japanese encephalitis virus in *Culex pseudovishnui* & *C. tritaeniorhynchus* mosquitoes. Indian. J. Med. Res. *93*, 250–252.

Nakamoto, M., Moy, R.H., Xu, J., Bambina, S., Yasunaga, A., Shelly, S.S., Gold, B., and Cherry, S. (2012). Virus recognition by Toll-7 activates antiviral autophagy in *Drosophila*. Immunity *36*, 658–667.

Nasar, F., Palacios, G., Gorchakov, R.V., Guzman, H., Da Rosa, A.P., Savji, N., Popov, V.L., Sherman, M.B., Lipkin, W.I., Tesh, R.B., and Weaver, S.C. (2012). Eilat virus, a unique alphavirus with host range restricted to insects by RNA replication. Proc. Natl. Acad. Sci. U.S.A. *109*, 14622–14627.

Nasci, R.S., Savage, H.M., White, D.J., Miller, J.R., Cropp, B.C., Godsey, M.S., Kerst, A.J., Bennett, P., Gottfried, K., and Lanciotti, R.S. (2001). West Nile virus in overwintering *Culex* mosquitoes, New York City, 2000. Emerg. Infect. Dis. *7*, 742–744.

Nayar, J.K., Rosen, L., and Knight, J.W. (1986). Experimental vertical transmission of Saint Louis encephalitis virus by Florida mosquitoes. Am. J. Trop. Med. Hyg. *35*, 1296–1301.

Nelms, B.M., Fechter-Leggett, E., Carroll, B.D., Macedo, P., Kluh, S., and Reisen, W.K. (2013a). Experimental and natural vertical transmission of West Nile virus by California *Culex* (Diptera: Culicidae) mosquitoes. J. Med. Entomol. *50*, 371–378.

Nelms, B.M., Kothera, L., Thiemann, T., Macedo, P.A., Savage, H.M., and Reisen, W.K. (2013b). Phenotypic variation among *Culex pipiens* complex (Diptera: Culicidae) populations from the Sacramento Valley, California: horizontal and vertical transmission of West Nile virus, diapause potential, autogeny, and host selection. Am. J. Trop. Med. Hyg. *89*, 1168–1178.

Nezis, I.P., Simonsen, A., Sagona, A.P., Finley, K., Gaumer, S., Contamine, D., Rusten, T.E., Stenmark, H., and Brech, A. (2008). Ref(2)P, the *Drosophila melanogaster* homologue of mammalian p62, is required for the formation of protein aggregates in adult brain. J. Cell. Biol. *180*, 1065–1071.

Niyas, K.P., Abraham, R., Unnikrishnan, R.N., Mathew, T., Nair, S., Manakkadan, A., Issac, A., and Sreekumar, E. (2010). Molecular characterization of Chikungunya virus isolates from clinical samples and adult *Aedes albopictus* mosquitoes emerged from larvae from Kerala, South India. Virol. J. *7*, 189.

Pantuwatana, S., Thompson, W.H., Watts, D.M., Yuill, T.M., and Hanson, R.P. (1974). Isolation of La Crosse virus from field collected *Aedes triseriatus* larvae. Am. J. Trop. Med. Hyg. *23*, 246–250.

Patterson, J.L., Holloway, B., and Kolakofsky, D. (1984). La Crosse virions contain a primer-stimulated RNA polymerase and a methylated cap-dependent endonuclease. J. Virol. *52*, 215–222.

Paulson, S.L., and Grimstad, P.R. (1989). Replication and dissemination of La Crosse virus in the competent vector *Aedes triseriatus* and the incompetent vector *Aedes hendersoni* and evidence for transovarial transmission by *Aedes hendersoni* (Diptera: Culicidae). J. Med. Entomol. *26*, 602–609.

Paulson, S.L., Grimstad, P.R., Craig, G.B. Jr. (1989). Midgut and salivary gland barriers to La Crosse virus dissemination in mosquitoes of the *Aedes triseriatus* group. Med. Vet. Entomol. *3*, 113–123.

Pekosz, A., Phillips, J., Pleasure, D., Merry, D., and Gonzalez-Scarano, F. (1996). Induction of apoptosis by La Crosse virus infection and role of neuronal differentiation and human bcl-2 expression in its prevention. J. Virol. *70*, 5329–5335.

Phillips, R.A., and Christensen, K. (2006). Field-caught *Culex erythrothorax* larvae found naturally infected with West Nile virus in Grand County, Utah. J. Am. Mosq. Control Assoc. *22*, 561–562.

Plowright, W., Perry, C.T., and Peirce, M.A. (1970). Transovarial infection with African swine fever virus in the argasid tick, *Ornithodoros moubata porcinus*, Walton. Res. Vet. Sci. *11*, 582–584.

Quan, P.L., Junglen, S., Tashmukhamedova, A., Conlan, S., Hutchison, S.K., Kurth, A., Ellerbrok, H., Egholm, M., Briese, T., Leendertz, F.H., and Lipkin, W.I. (2010). Moussa virus: a new member of the Rhabdoviridae family isolated from *Culex decens* mosquitoes in Côte d'Ivoire. Virus Res. *147*, 17–24.

Reese, S.M., Blitvich, B.J., Blair, C.D., Geske, D., Beaty, B.J., and Black, W.C. 4th. (2008). Potential for La Crosse virus segment reassortment in nature. Virol. J. *5*, 164.

Reese, S., Beaty, M., Gabitzsch, E., Blair, C.D., and Beaty, B.J. (2009). *Aedes triseriatus* females transovarially infected with La Crosse virus mate more efficiently than uninfected mosquitoes. J. Med. Entomol. *46*, 1152–1158.

Reese, S., Mossel, E., Beaty, M., Beck, E.T., Geske, D., Blair, C.D., Beaty, B.J., and Black, W.C. 4th (2010). Identification of super-infected *Aedes triseriatus* mosquitoes collected as eggs from the field and partial characterization of the infecting La Crosse viruses. BMC Virology J. *7*, 76.

Rehacek, J. (1962). Transovarial transmission of tick-borne encephalitis virus by ticks. Acta. Virol. *6*, 220–226.

Reisen, W.K., Hardy, J.L., Reeves, W.C., Presser, S.B., Milby, M.M., and Meyer, R.P. (1990). Persistence of mosquito-borne viruses in Kern County, California, 1983–1988. Am. J. Trop. Med. Hyg. *43*, 419–437.

Reisen, W.K., Fang, Y., Lothrop, H.D., Martinez, V.M., Wilson, J., Oconnor, P., Carney, R., Cahoon-Young, B., Shafii, M., and Brault, A.C. (2006). Overwintering of West Nile virus in Southern California. J. Med. Entomol. *43*, 344–355.

Rennie, L., Wilkinson, P.J., and Mellor, P.S. (2001). Transovarial transmission of African swine fever virus in the argasid tick *Ornithodoros moubata*. Med. Vet. Entomol. *15*, 140–146.

Rodhain, F., and Rosen, L. (1997). Mosquito vectors and dengue virus–vector relationships. In Dengue and dengue hemorrhagic fever, D.J. Gubler and G. Kuno, eds. (CAB International, New York, NY, USA), pp. 45–60.

Romoser, W.S., Oviedo, M.N., Lerdthusnee, K., Patrican, L.A., Turell, M.J., Dohm, D.J., Linthicum, K.J., and Bailey, C.L. (2011). Rift Valley fever virus-infected mosquito ova and associated pathology: possible implications for endemic maintenance. Res. Rep. Trop. Med. *2*, 121–127.

Rosen, L. (1987a). Overwintering mechanisms of mosquito-borne arboviruses in temperate climates. Am. J. Trop. Med. Hyg. *37*, 69S-76S.

Rosen, L. (1987b). On the mechanism of vertical transmission of dengue virus in mosquitoes. C. R. Acad. Sc. Paris *13*, 347–350.

Rosen, L. (1988). Further observations on the mechanism of vertical transmission of flaviviruses by *Aedes* mosquitoes. Am. J. Trop. Med. Hyg. *39*, 123–126.

Rosen, L., Tesh, R.B., Lien, J.C., and Cross, J.H. (1978). Transovarial transmission of Japanese encephalitis virus by mosquitoes. Science *199*, 909–911.

Rosen, L., Shroyer, D.A., and Lien, J.C. (1980). Transovarial transmission of Japanese encephalitis virus by *Culex tritaeniorhynchus* mosquitoes. Am. J. Trop. Med. Hyg. *29*, 711–712.

Rosen, L., Shroyer, D.A., Tesh, R.B., Freier, J.E., and Lien, J.C. (1983). Transovarial transmission of dengue viruses by mosquitoes: *Aedes albopictus* and *Aedes aegypti*. Am. J. Trop. Med. Hyg. *32*, 1108–1119.

Rosen, L., Lien, J.C., Shroyer, D.A., Baker, R.H., and Lu, L.C. (1989). Experimental vertical transmission of Japanese encephalitis virus by *Culex tritaeniorhynchus* and other mosquitoes. Am. J. Trop. Med. Hyg. *40*, 548–556.

Saiyasombat, R., Bolling, B.G., Brault, A.C., Bartholomay, L.C., and Blitvich, B.J. (2011). Evidence of efficient transovarial transmission of *Culex* flavivirus by *Culex pipiens* (Diptera: Culicidae). J. Med. Entomol. *48*, 1031–1038.

Saluzzo J.F., Herve, J.P., Salaun, J.J., Germain, M., Comet, J.P., Camicas, J.L., Heine, G., and Robin, Y. (1980). Caractéristiques des souches du virus de la fiévre jaune isolées des oeufs et des larves d'une tique *Amblyomma variegatum*, recoltée sur le bétail a Bangui (Centrafrique). Ann. Virol. *131E*, 155–165.

Samal, S.K., el-Hussein, A., Holbrook, F.R., Beaty, B.J., and Ramig, R.F. (1987). Mixed infection of *Culicoides variipennis* with bluetongue virus serotypes 10 and 17: evidence for high frequency reassortment in the vector. J. Gen. Virol. *68*, 2319–2329.

Sanders, H.R., Foy, B.D., Evans, A.M., Ross, L.S., Beaty. B.J., Olson, K.E., and Gill, S.S. (2005). Sindbis virus induces transport processes and alters expression of innate immunity pathway genes in the midgut of the disease vector, *Aedes aegypti*. Insect Biochem. Mol. Biol. *35*, 1293–1307.

Sang, R.C., Gichogo, A., Gachoya, J., Dunster, M.D., Ofula, V., Hunt, A.R., Crabtree, M.B., Miller. B.R., and Dunster, L.M. (2003). Isolation of a new flavivirus related to cell fusing agent virus (CFAV) from field-collected flood-water *Aedes* mosquitoes sampled from a dambo in central Kenya. Arch. Virol. *148*, 1085–1093.

Schmidt, J.R., Schmidt, M.L., and Said, M.I. (1971). Phlebotomus fever in Egypt. Isolation of phlebotomus fever viruses from *Phlebotomus papatasi*. Am. J. Trop. Med. Hyg. *20*, 483–490.

Schopen, S., Labuda, M., and Beaty, B. (1991). Vertical and venereal transmission of California group viruses by *Aedes triseriatus* and *Culiseta inornata* mosquitoes. Acta. Virol. *35*, 373–382.

Scott, T.W., Hildreth, S.W., and Beaty, B.J. (1984). The distribution and development of eastern equine encephalitis virus in its enzootic mosquito vector, *Culiseta melanura*. Am. J. Trop. Med. Hyg. *33*, 300–310.

Seecof, R. (1968). The sigma virus infection of *Drosophila melanogaster*. Curr. Top. Microbiol. Immunol. *42*, 59–93.

Serufo, J.C., de Oca, H.M., Tavares, V.A., Souza, A.M., Rosa, R.V., Jamal, M.C., Lemos, J.R., Oliveira, M.A., Nogueira, R.M., and Schatzmayr, H.G. (1993). Isolation of dengue virus type 1 from larvae of *Aedes albopictus* in Campos Altos city, State of Minas Gerais, Brazil. Mem. Inst. Oswaldo. Cruz. *88*, 503–504.

Seufi, A.M., and Galal, F.H. (2010). Role of *Culex* and *Anopheles* mosquito species as potential vectors of rift valley fever virus in Sudan outbreak, 2007. BMC Infect. Dis. *10*, 65.

Shelly, S., Lukinova, N., Bambina, S., Berman, A., and Cherry, S. (2009). Autophagy is an essential component of *Drosophila* immunity against vesicular stomatitis virus. Immunity *30*, 588–598.

Shroyer, D.A. (1991). Preliminary studies of *Aedes bahamensis* as a host and potential vector of St. Louis encephalitis virus. J. Am. Mosq. Control Assoc. *7*, 63–65.

Singh, K.R., Pavri, K., and Anderson, C.R. (1963). Experimental transovarial transmission of Kyasanur Forest disease virus in *Haemaphysalis spinigera*. Nature *199*, 513.

Singh, K.R., Goverdhan, M.K., and Bhat, H.R. (1968). Transovarial transmission of Kyasanur Forest Disease virus by *Ixodes petauristae*. Indian J. Med. Res. *56*, 628–632.

Sir, D., and Ou, J.H. (2010). Autophagy in viral replication and pathogenesis. Mol. Cells *29*, 1–7.

Soldan, S.S., Plassmeyer, M.L., Matukonis, M.K., and Gonzalez-Scarano, F. (2005). La Crosse virus non-structural protein NSs counteracts the effects of short interfering RNA. J. Virol. *79*, 234–244.

Stollar, V., and Thomas, V.L. (1975). An agent in the *Aedes aegypti* cell line (Peleg) which causes fusion of *Aedes albopictus* cells. Virology *64*, 367–377.

Sundin, D.R., and Beaty, B.J. (1988). Interference to oral superinfection of *Aedes triseriatus* infected with La Crosse virus. Am. J. Trop. Med. Hyg. *38*, 428–432.

Syverton, J.T., and Berry, G.P. (1941). Hereditary transmission of the western type of equine encephalitis virus in the wood tick, *Dermacentor Andersoni* Stiles. J. Exp. Med. *73*, 507–530.

Takashima, I., and Rosen, L. (1989). Horizontal and vertical transmission of Japanese encephalitis virus by *Aedes japonicus* (Diptera: Culicidae). J. Med. Entomol. *26*, 454–458.

Tesh, R.B. (1980a). Vertical transmission of arthropod-borne viruses of vertebrates. In Vectors of Disease Agents, J.J. McKelvey and R.F. Harwood, eds. (Praeger Scientific Publishers, New York).

Tesh, R.B. (1980b). Experimental studies on the transovarial transmission of Kunjin and San Angelo viruses in mosquitoes. Am. J. Trop. Med. Hyg. *29*, 657–666.

Tesh, R.B. (1984). Transovarial transmission of arboviruses in their invertebrate vectors. Curr. Topics Vector Res. *2*, 57–76.

Tesh, R.B. (1988). The genus *Phlebovirus* and its vectors. Annu. Rev. Entomol. *33*, 169–181.

Tesh, R.B., and Beaty, B.J. (1983). Localization of California serogroup viruses in mosquitoes. Prog. Clin. Biol. Res. *123*, 67–75.

Tesh, R.B., and Chaniotis, B.N. (1975). Transovarial transmission of viruses by phlebotomine sandflies. Ann. N. Y. Acad. Sci. *266*, 125–134.

Tesh, R.B., and Cornet, M. (1981). The location of San Angelo virus in developing ovaries of transovarially infected *Aedes albopictus* mosquitoes as revealed by fluorescent antibody technique. Am. J. Trop. Med. Hyg. *30*, 212–218.

Tesh, R.B., and Gubler, D.J. (1975). Laboratory studies of transovarial transmission of La Crosse and other arboviruses by *Aedes albopictus* and *Culex fatigans*. Am. J. Trop. Med. Hyg. *24*, 876–880.

Tesh, R.B., and Modi, G.B. (1983). Growth and transovarial transmission of Chandipura virus (Rhabdoviridae: Vesiculovirus) in *phlebotomus papatasi*. Am. J. Trop. Med. Hyg. *32*, 621–623.

Tesh, R.B., and Modi, G.B. (1984). Studies on the biology of phleboviruses in sand flies (Diptera:Psychodidae). I. Experimental infection of the vector. Am. J. Trop. Med. Hyg. *33*, 1007–1016.

Tesh, R.B., and Modi, G.B. (1987). Maintenance of Toscana virus in *Phlebotomus perniciosus* by vertical transmission. Am. J. Trop. Med. Hyg. *36*, 189–193.

Tesh, R.B., and Shroyer, D.A. (1980). The mechanism of arbovirus transovarial transmission in mosquitoes: San Angelo virus in *Aedes albopictus*. Am. J. Trop. Med. Hyg. *29*, 1394–1404.

Tesh, R.B., Chaniotis, B.N., and Johnson, K.M. (1971). Vesicular stomatitis virus, Indiana serotype: Multiplication in and transmission by experimentally infected phlebotomine sandflies (*Lutzomyia trapidoi*). Am. J. Epidemiol. *93*, 491–495.

Tesh, R.B., Chaniotis, B.N., and Johnson, K.M. (1972). Vesicular stomatitis virus (Indiana serotype): transovarial transmission by phlebotomine sandflies. Science *175*, 1477–1479.

Tesh, R.B., Chaniotis, B.N., Peralta, P.H., and Johnson, K.M. (1974). Ecology of viruses isolated from Panamanian phlebotomine sandflies. Am. J. Trop. Med. Hyg. *23*, 258–269.

Tesh, R., Saidi, S., Javadian, E., and Nadim, A. (1977). Studies on the epidemiology of sandfly fever in Iran. I. Virus isolates obtained from *Phlebotomus*. Am. J. Trop. Med. Hyg. *26*, 282–287.

Tesh, R.B., Boshell, J., Young, D.G., Morales, A., Corredor, A., Modi, G.B., Ferro de Carrasquilla, C., de Rodriquez, C., and Gaitan, M.O. (1986). Biology of Arboledas virus, a new phlebotomus fever serogroup virus (Bunyaviridae: *Phlebovirus*) isolated from sand flies in Colombia. Am. J. Trop. Med. Hyg. *35*, 1310–1316.

Tesh, R.B., Boshell, J., Modi, G.B., Morales, A., Young, D.G., Corredor, A., Ferro de Carrasquilla, C., de Rodriguez, C., Walters, L.L., and Gaitan, M.O. (1987). Natural infection of humans, animals, and phlebotomine sand flies with the Alagoas serotype of vesicular stomatitis virus in Colombia. Am. J. Trop. Med. Hyg. *36*, 653–661.

Tesh, R.B., Lubroth, J., and Guzman, H. (1992). Simulation of arbovirus overwintering: survival of Toscana virus (Bunyaviridae: *Phlebovirus*) in its natural sandfly vector *Phlebotomus perniciosus*. Am. J. Trop. Med. Hyg. *47*, 574–581.

Thavara, U., Tawatsin, A., Pengsakul, T., Bhakdeenuan, P., Chanama, S., Anantapreecha, S., Molito, C., Chompoosri, J., Thammapalo, S., Sawanpanyalert, P., and Siriyasatien, P. (2009). Outbreak of chikungunya fever in Thailand and virus detection in field population of vector mosquitoes, *Aedes aegypti* (L.) and *Aedes albopictus* Skuse (Diptera: Culicidae). Southeast Asian J. Trop. Med. Public Health 40:951–962.

Thomas, L.A., Patzer, E.R., Cory, J.C., and Coe, J.E. (1980). Antibody development in garter snakes (*Thamnophis* spp.) experimentally infected with western equine encephalitis virus. Am. J. Trop. Med. Hyg. *29*, 112–117.

Thompson, W.H., and Beaty, B.J. (1977). Venereal transmission of La Crosse (California encephalitis) arbovirus in *Aedes triseriatus* mosquitoes. Science *196*, 530–531.

Thompson, W.H., and Beaty, B.J. (1978). Venereal transmission of La Crosse virus from male to female *Aedes triseriatus*. Am. J. Trop. Med. Hyg. *27*, 187–196.

Traavik, T., Mehl, R., and Wiger, R. (1978). California encephalitis group viruses isolated from mosquitoes collected in Southern and Arctic Norway. Acta. Pathol. Microbiol. Scand. B *86B*, 335–341.

Travassos da Rosa, A.P., Tesh, R.B., Travassos da Rosa, J.F., Herve, J.P., Main Jr., A.J. (1984). Carajas and Maraba viruses, two new vesiculoviruses isolated from phlebotomine sand flies in Brazil. Am. J. Trop. Med. Hyg. 33, 999–1006.

Tsai, C.W., McGraw, E.A., Ammar, E.D., Dietzgen, R.G., and Hogenhout, S.A. (2008). *Drosophila melanogaster* mounts a unique immune response to the Rhabdovirus sigma virus. Appl. Environ. Microbiol. *74*, 3251–3256.

Turell, M.J. (1988). Horizontal and vertical transmission of viruses by insect and tick vectors. In The Arboviruses: Epidemiology and Ecology, T.P. Monath, ed. (CRC Press, Boca Raton, Florida), 127–152.

Turell, M.J. (2007). Role of ticks in the transmission of Crimean–Congo hemorrhagic fever virus. In Crimean–Congo Hemorrhagic Fever, O. Ergonul and C.A. Whitehouse, eds. (Springer, Dordrecht, Netherlands), pp. 143–154.

Turell, M.J., Hardy, J.L., and Reeves, W.C. (1980). Demonstration of transovarial transmission of California encephalitis virus in experimentally infected *Aedes melanimon*. Proc. Calif. Mosq. Vector Control Assoc. *48*, 15–16.

Turell, M.J., Hardy, J.L., and Reeves, W.C. (1982a). Stabilized infection of California encephalitis virus in *Aedes dorsalis*, and its implications for viral maintenance in nature. Am. J. Trop. Med. Hyg. *31*, 1252–1259.

Turell, M.J., O'Guinn, M.L., Dohm, D.J., and Jones, J.W. (2001). Vector competence of North American mosquitoes (Diptera: Culicidae) for West Nile virus. J. Med. Entomol. *38*, 130–134.

Turell, M.J., Reeves, W.C., and Hardy, J.L. (1982b). Evaluation of the efficiency of transovarial transmission of California encephalitis viral strains in *Aedes dorsalis* and *Aedes melanimon*. Am. J. Trop. Med. Hyg. *31*, 382–388.

Unlu, I., Mackay, A.J., Roy, A., Yates, M.M., and Foil, L.D. (2010). Evidence of vertical transmission of West Nile virus in field-collected mosquitoes. J. Vector Ecol. *35*, 95–99.

Vasilakis, N., Forrester, N.L., Palacios, G., Nasar, F., Savji, N., Rossi, S.L., Guzman, H., Wood, T.G., Popov, V., Gorchakov, R., González, A.V., Haddow, A.D., Watts, D.M., Travassos da Rosa, A.P., Weaver, S.C., Lipkin, W.I., and Tesh, R.B. (2013). Negevirus: a proposed new taxon of insect-specific viruses with wide geographic distribution. J. Virol. *87*, 2475–2488.

Vázquez, A., Sánchez-Seco, M.P., Palacios, G., Molero, F., Reyes, N., Ruiz, S., Aranda, C., Marqués, E., Escosa, R., Moreno, J., Figuerola, J., and Tenorio, A. (2012). Novel flaviviruses detected in different species of mosquitoes in Spain. Vector Borne Zoonotic Dis. *12*, 223–229.

Vilela, A.P., Figueiredo, L.B., dos Santos, J.R., Eiras, A.E., Bonjardim, C.A., Ferreira, P.C., and Kroon, E.G. (2010). Dengue virus 3 genotype I in *Aedes aegypti* mosquitoes and eggs, Brazil, 2005–2006. Emerg. Infect. Dis. *16*, 989–992.

Walker, P.J., Firth, C., Widen, S.G., Blasdell, K.R., Guzman, H., Wood, T.G., Paradkar, P.N., Holmes, E.C., Tesh, R.B., and Vasilakis, N. (2015). Evolution of genome size and complexity in the *Rhabdoviridae*. PLoS Path. *11*, e1004664.

Watts, D.M., Pantuwatana, S., DeFoliart, G.R., Yuill, T.M., and Thompson, W.H. (1973). Transovarial transmission of La Crosse virus (California encephalitis group) in the mosquito, *Aedes triseriatus*. Science *182*, 1140–1141.

Wayne, M.L., Contamine, D., and Kreitman, M. (1996). Molecular population genetics of ref(2)P, a locus which confers viral resistance in *Drosophila*. Mol. Biol. Evol. *13*, 191–199.

White, D.M., Wilson, W.C., Blair, C.D., and Beaty, B.J. (2005). Studies on overwintering of bluetongue viruses in insects. J. Gen. Virol. *86*, 453–462.

White, G., Ottendorfer, C., Graham, S., and Unnasch, T.R. (2011). Competency of reptiles and amphibians for eastern equine encephalitis virus. Am. J. Trop. Med. Hyg. *85*, 421–425.

Wilfert, L., and Jiggins, F.M. (2013). The dynamics of reciprocal selective sweeps of host resistance and a parasite counter-adaptation in *Drosophila*. Evolution *67*, 761–773.

Wilson, M.L., Gonzalez, J.R., Cornet, J.R., and Camicas, J.L. (1991). Transmission of Crimean–Congo haemorrhagic fever virus from experimentally infected sheep to *Hyalomma truncatum* ticks. Res. Virol. *142*, 395–404.

Won, S., Ikegami, T., Peters, C.J., and Makino, S. (2007). NSm protein of Rift Valley fever virus suppresses virus-induced apoptosis. J. Virol. *81*, 13335–13345.

Woodring, J., Chandler, L.J., Oray, C.T., McGaw, M.M., Blair, C.D., and Beaty, B.J. (1998). Short report: Diapause, transovarial transmission, and filial infection rates in geographic strains of La Crosse virus-infected *Aedes triseriatus*. Am. J. Trop. Med. Hyg. *58*, 587–588.

Wyers, F., Dru, P., Simonet, B., and Contamine, D. (1993). Immunological cross-reactions and interactions between the *Drosophila melanogaster ref(2)P-protein* and sigma rhabdovirus proteins. J. Virol. *67*, 3208–3216.

Wyers, F., Petitjean, A.M., Dru, P., Gay, P., and Contamine, D. (1995). Localization of domains within the *Drosophila ref(2)P-protein* involved in the intracellular control of sigma rhabdovirus multiplication. J. Virol. *69*, 4463–4470.

Xi, Z., Ramirez, J.L., and Dimopoulos, G. (2008). The *Aedes aegypti* toll pathway controls dengue virus infection. PLoS Pathog. *4*, e1000098.

Yamanaka, A., Thongrungkiat, S., Ramasoota, P., and Konishi, E. (2013). Genetic and evolutionary analysis of cell-fusing agent virus based on Thai strains isolated in 2008 and 2012. Infect. Genet. Evol. *19*, 188–194.

Yu, X.-J., and Tesh, R.B. (2014). The role of mites in the transmission and maintenance of Hantaan virus (*Hantavirus: Bunyaviridae*). J. Infect. Dis. *210*, 1693–1699.

Zeller, H.G., Cornet, J.P., and Camicas, J.L. (1994). Experimental transmission of Crimean–Congo hemorrhagic fever virus by west African wild ground-feeding birds to *Hyalomma marginatum rufipes* ticks. Am. J. Trop. Med. Hyg. *50*, 676–681.

Zirkel, F., Kurth, A., Quan, P.L., Briese, T., Ellerbrok, H., Pauli, G., Leendertz, F.H., Lipkin, W.I., Ziebuhr, J., Drosten, C., and Junglen, S. (2011). An insect nidovirus emerging from a primary tropical rainforest. MBio 2, e00077–e00011.

Zirkel, F., Roth, H.I., Kurth, A., Drosten, C., Ziebuhr, J., and Junglen, S. (2013). Identification and characterization of genetically divergent members of newly established family Mesonviridae. J. Virol. *87*, 6346–6358.

The Boundaries of Arboviruses: Complexities Revealed in Their Host Ranges, Virus–Host Interactions and Evolutionary Relationships

14

Goro Kuno

Abstract

Reflecting the sharp increase in public health and veterinary problems caused by arboviruses in the past several decades, scientists worldwide have generated enormous amounts of data and publications about arboviruses and a variety of other related animal viruses currently not classified as arboviruses on the basis of the definition.

Upon a closer examination of the latter group of viruses, however, it has become clear that the definition of arbovirus, the tenets, and concepts on which arbovirology is established need to be re-evaluated for their ambiguity, absence of supporting evidence, inconsistency or difficulty of generalization, and an assortment of other problems in defining the boundaries between arboviruses and non-arboviruses.

To understand the origins of those problematic issues, first, it is necessary to trace the early histories of virus–host interactions conducive to the evolution of biological transmission. Accordingly, viral associations with arthropods, vertebrates and even plants in both terrestrial and/or aquatic environments are examined. The analyses are focused on the uniqueness of the biology of vectors, vertebrates, ecosystem, opportunity of virus–host contact, food chain, and the impacts of viral infection on hosts. The second foci of analyses concern the markers involved in the evolutionary process. Because virus–host interaction operating under particular ecological conditions is unique to each viral lineage, generalization of the markers among all arboviral lineages is difficult. Nevertheless, given rapidly accumulating useful information enriching the database, it is now possible to undertake such a study in at least a few arboviral lineages. In a study of flaviviruses, the history of host range shift, genome (or gene) length, and the motifs in the 3′ UTR were found to be useful markers for elucidating the evolutionary history of the lineage.

As by-products of this review process, many problematic issues surfaced. They include definition and measurement of virulence, existence of vertebrate reservoir, non-viraemic transmission, accuracy of molecular phylogeny without a support of empirical data, validity of *in vitro* experiments for the identification of genetic determinants of phenotypic shift, existence of the viruses which replicate in both arthropods and vertebrates (viruses with a biphylum host range) but without a need of biological transmission for survival, accuracy of host range of arboviruses and biological transmission in aquatic environments. Collectively, these issues and derived questions, coupled with the ambiguity of the boundaries of arboviruses, warrant an urgent re-assessment of the fundamentals of arbovirology.

Introduction

Given the fact that biological transmission, the unique characteristic of arboviruses, is not shared by all members in any virus family, clear segregation of arboviruses from non-arboviruses has presented a variety of problems. After reviewing the history of the nascent period of arbovirology, Jordi Casals, one of the distinguished early arbovirologists, concluded that although arboviruses could be segregated from non-arboviruses based on a set of specific criteria, nonetheless, the two groups were bonded together by other traits (Casals, 1971). The later establishment of two subcommittees (within the American Committee on Arthropod-Borne Viruses) dedicated to information exchange with respect to arbovirus status and inter-relationships of the viruses registered in the *International Catalogue of Arboviruses* was precisely designed to address this issue.

The difficulties of drawing clear boundaries become evident when one examines a variety of subjects concerning ambiguities, inconsistencies and puzzles presented in this chapter. Collectively, these subjects seriously question the practices of strict adherence only to the tenets of the traditional boundaries and of delegating the studies of ambiguous non-arboviruses to other branches of virology in the biomedical science that is rapidly becoming interdisciplinary and compel arbovirologists to re-evaluate the boundaries.

Furthermore, the ambiguities of the boundaries of arboviruses profoundly impact not only on the interests in and goals for professional career of researchers but also on adequacy of the missions of the institutions dedicated to arbovirus research as well as on disease control strategies in public health, veterinary medicine, agriculture, and wildlife conservation. The impact on biosafety level classification of the viruses also cannot be overlooked.

Accordingly, in a major departure from other chapters dedicated to updated reviews of arboviruses and their vector control, in the first part of this chapter, the boundaries between arboviruses and non-arboviruses are the major foci. The poorly defined boundaries in question are virus–host interface, ecological factors, host range, mode of transmission, transition of virus–host relationships, and phylogenetic relation. The viruses which do not satisfy all required traits of arboviruses for a variety of reasons but which are otherwise similar biologically or virologically or somehow connected either directly or indirectly constitute a large proportion of the viruses examined. For this discussion, the unique features of arthropods in the Animal Kingdom and the types of their associations with viruses are surveyed, featuring the kinds of opportunities of arthropods encountering viruses and vertebrates directly or indirectly in both terrestrial and aquatic environments.

Arthropod associations with plant viruses and vertebrate associations with viruses are similarly examined. Virus–host associations are then examined in the contexts of ecological determinants, establishment of persistent infection, pathologies in the hosts, modes of transmission, and food chain. After these analyses, foci are shifted to the evolutionary histories of selected arbovirus groups or lineages based on host range shift, phylogenetics, empirical data in support of phylogeny, and genetic determinants of host range. In the last section are discussed a variety of issues related to host range determination and re-examination of the tenets of arboviruses.

Readers are advised that in this chapter the word 'host' refers to both vectors and vertebrates, following the long tradition in biological sciences; and 'vector' refers only to invertebrates involved in the transmission of animal or plant viruses. Regarding abbreviation of virus names, the unfortunate lack of coordination for establishing a universal, species-specific system encompassing all families of viruses among more than 70 international study groups serving the ICTV (2012) has generated duplications of the same or similar acronyms for viruses belonging to different taxonomic groups and inevitable confusion when only acronyms of viruses belonging to multiple families are used in one report, typically for shortening text size. For arbovirologists, some examples of the problematic acronyms are as follows: ARAV (Araguari virus) vs. ARAV (Araraquara virus); ARV (Adelaide River virus) vs. ARV (Avian orthoreovirus) and ARV (Aravan virus); BAV (Banna virus) vs. BAV (Birens Arm virus) and BAV (Barramundi virus); BBV (Bukalasa bat virus) vs. BBV (Black beetle virus); BCV(Batu Cave virus) vs. BCV (Blue crab virus); DURV (Durania virus) vs. DURV (Durham virus); ENTV (Entebbe bat virus) vs. ENTV (Entamoeba virus); FMV (Fort Morgan virus) vs. FMV (Fig mosaic virus); GGV (Gan Gan virus) vs. GGV (Greasy grouper virus); GIV (Great Island virus) vs. GIV (Grouper iridovirus); JURV (Jurona virus) vs. JURV (Juruaca virus); KASV (Kasba virus) vs. KASV (Kasokero virus); KRV (Kamiti River virus) vs. KRV (Kilham rat virus); NDV (Nyando virus) vs. NDV (New Castle Disease virus); RBV (Rio Bravo virus) vs. RBV (Rotifer birnavirus); RRV (Ross River virus) vs. RRV (Reed Ranch virus); SABV (Saboya virus) vs. SABV (Sabía virus); SCRV (St Croix River virus) vs. SCRV (Siniperca chuatsi rhabdovirus) and SCRV (Santee-Cooper ranavirus). Similar problem is found in the literature for the viruses not yet officially approved by the ICTV: RabV (Rabensburg virus) vs. RABV (Rabies virus); HPV [Hapavirus (proposed)] vs. HPV (Human papilloma virus); SMV (Sena Madureira virus) vs. SMV (Smelt picornavirus).

Arthropod association with viruses

Currently, the number of species in the Phylum *Arthropoda* is at least 1.3 million and keeps rising, with the species of the Class *Insecta* accounting for nearly 83% of the known species of the Animal Kingdom and occupying nearly every corner of the terrestrial environments except for Antarctica. The high probability of arthropod-virus contact, which was abundantly revealed in recent metagenomic studies, reflects the fact that biosphere is awash with viruses, based on estimated quantity of roughly 10 virus particles per living cell (Dolja and Koonin, 2011). Naturally, over 6500 viruses reported (but only about 2300 officially accepted by the ICTV in 2012) today are thought to represent only a small fraction of the viruses that exist on earth.

With the enormous diversity in adaptability of arthropods in any ecosystem and biome, more than 14,000 species of haematophagous arthropods are found associated with practically all groups of vertebrates on earth; and many arthropods co-speciated as intimately associated parasites of mammals (Kim, 1985). Mosquitoes are found at >5000 m above sea level in the Himalayas, *Culex tarsalis* (a major vector of Western equine encephalomyelitis virus) occurs at an altitude of 3200 m in California, and anthropophilic *Aedes aegypti* mosquito has been found at an altitude of 2000–2150 m in Colombia and Mexico. As for disease transmission, Colorado tick fever infection occurs by tick bites at locations as high as 3300 m above sea level in North America; a dengue outbreak was recorded at an altitude of 1700 m in Mexico; autochthonous cases of Japanese encephalitis occurred at 1600 m above sea level in Nepal; Yellow fever virus (YFV) was isolated from sylvan mosquitoes at an altitude of 3000 m above sea level in Peru; and Tahyna virus was isolated from mosquitoes at an altitude of 2780 m in China. In other types of extreme conditions, arboviruses have been isolated from midges in the interior of the Sultanate of Oman where summer temperature exceeds 50°C even in the shade. Intimately anthropophilic *Ae. aegypti* follows human movement and is found breeding deep underground in gold mines in Australia (Eisler, 2003). In aquatic environments, ectoparasitic copepods and other arthropods (such as shrimps) are involved in viral transmission to aquatic vertebrates. Even in peri-Antarctic water, sea mammals or birds are found infested by lice or ticks carrying arboviruses (La Linn *et al.*, 2001; Major *et al.*, 2009). Arthropod-specific RNA viruses are found in association even with algae, as illustrated in the example of Kelp fly virus.

Most likely, the viruses of arthropods evolved over 350 million years ago in association with the emergence of their hosts. The long history (macroevolution) of close relationships between viruses and arthropods is illustrated in a report of cospeciation of dsDNA viruses (baculoviruses) and their insect hosts dating back to 310 million years ago (Thése *et al.*, 2011). In another study is reported the discovery of the viral inclusion bodies of the genus *Cypovirus* of the family *Reoviridae*, which can be distinguished from the similar inclusion bodies of the family *Baculoviridae*, in the sand flies embedded in Burmese ambers.

This confirmed the existence of the sand fly-associated cypovirus group and of the same virus–host association currently observed already in early Cretacean Period (circa 100 million years ago) (Poinar and Poinar, 2005). It has been proposed that the host range shift of primordial arthropod-associated bunyaviruses to vertebrates took place well over 100 million years ago (Plyusnin and Sironen, 2014). If this proposal is acceptable, it is deduced that primitive bunyavirus lineages with a host range in both arthropods and vertebrates evolved in about the same period. Similarly, some, who believe that arthropods could not become vectors in biological transmission until they developed antihemostatic mechanisms and antivasoconstricting factors, estimate the dates of the evolution of haematophagy at roughly 120 to 92 million years ago (Ribeiro, 1995).

On the other hand, when measured in the geological time scale, the progenitors of the extant lineages of RNA arboviruses were estimated to have originated only infinitely short periods of time before present (often much less than 5000 years for alphaviruses or <10,000 years for flaviviruses). However, the adequacy of constant rate of substitution as a basic assumption in the traditional molecular clock has been seriously questioned (Duchêne *et al.*, 2014). Although the early assumptions are still useful for microevolution, their application for viral lineages with a history of drastic shift in traits (i.e. host range) in the context of macroevolution is highly questionable. It is speculated that the extant lineages of flaviviruses evolved as early as 120,000 years ago (Pettersson and Fiz-Palacios, 2014). If so, one may question if the chromosomal integration of autosomal dominant gene responsible for resistance to flaviviral infection in certain breeds of rodents (Brinton, 1981) occurred so recently or these resistant genes had evolved in response to the ancient progenitors of the extant flaviviruses. In other study, in contrast, the inferred date of bifurcation of *Vesicular stomatitis virus* (VSV) into New Jersey and Indiana genotypes is estimated to be around 1.8 million years ago (Liang *et al.*, 2014). These reports clearly demonstrate that each major viral lineage emerged at a different geological period; and new lineages have been evolving continuously. It is, however, important to recognize that universal consensus on these estimates has not been obtained, as the acceptable concept and method used in dating have been constantly debated and are still evolving. As shown later in this chapter, many viruses investigated in arbovirology most likely evolved from the viruses of haematophagous arthropods through adaptation to replicate in vertebrates. This must be taken into consideration in dating the origins of those viruses.

Historical records of early recognition of viral diseases associated with arthropods

The historical accounts behind the discoveries of viral diseases associated with arthropods reveal fascinating records of the circumstances surrounding recognition of diseases, human curiosity about aetiology, and/or interest in documenting atypical conditions observed at a time when scientifically establishing an aetiological relationship was not possible or difficult. Listed below are the contributions to only selected viral diseases documented before 1900.

The oldest accurate record retrospectively identifiable as description of a viral disease of any insect was published in the form of poem ('De Bombyce') by Marco Girolamo Vida of Italy in 1527, in relation to jaundice (or 'grasserie') of the silkworm. At the time, silk industry was of enormous economic importance, and maintaining silkworm colony healthy critical. The observed yellow spots which developed on the abdomen of the caterpillars and emission of putrid odour are now recognized as a pathognomonic sign of a baculovirus infection. The chemical nature of the huge viral inclusion bodies (in which numerous virions are embedded) of this virus visible with a primitive microscope was determined to be a protein by J. Bolle as early as in 1894 (Smith, 1967). Involvement of leafhoppers in the development of a rice dwarf disease was determined by H. Hashimoto, not a scientist but a rice farmer, in 1894 (Corbett, 1964).

As for arboviral diseases, during an outbreak of a febrile illness in Zanzibar, Tanzania in 1871, which had been known among the natives at least since1823 and called 'kidinga pepo' in Swahili language, patients experienced prolonged and severe arthralgia (Christie, 1872). Retrospectively, it is highly probable that they were cases of chikungunya (Carey, 1971). According to an examination of an archival thesaurus written in Persian, a haemorrhagic disease in men revealing arthropods attached to the body was observed in present-day Tajikistan circa 1100 A.D. The disease was interpreted to be most compatible with Crimean–Congo haemorrhagic fever (Hoogstraal, 1979). This record is interesting because, according to a recent molecular clock dating, the common ancestor of this virus had existed approximately 1500–1100 BC (Carroll *et al.*, 2010). The contributions of Josiah Clark Nott, Louis Daniel Beauperthuy, Carlos Juan Finlay and Henry Rose Carter between 1848 and 1899 for elucidating the mosquito-borne transmission of yellow fever constitute a part of the celebrated early history of YF research and do not need a redundant elaboration of the significance again. On the other hand, little recognized in virology but which deserves attention is the hidden significance of the finding of Councilman body in the liver of YF patients in Cuba (Councilman, 1890). Nearly a century later, it was revealed that this pathology was the earliest scientifically proven evidence of apoptosis (Studzinski, 1999). The most reliable, earliest account of dengue haemorrhagic fever was described by Francis Everard Hare of England during his medical service in Queensland, Australia, where repeated outbreaks of dengue occurred, as increasing numbers of immigrants and merchants arrived by ships via dengue-endemic Asian ports (Hare, 1898).

The unmistakable blue colour of tongue, bluetongue, in diseased sheep was recognized in South Africa after the importation of Merino breed from Europe in late 18th century. The disease was first described by Francois de Vaillant of France in 1871–84 (Gorman, 1990) and in 1880 by Duncan Hutcheon (Verwoerd, 2012). The earliest record of bovine ephemeral fever (BEF) known by 'stiffsiekte' in Tanzania, Brundi, and Rwanda during German colonization was described by G. Schweinfurth in 1878. 'Cattle dengue' observed in Egypt in 1895 was also unmistakably BEF (Walker, 2005). The early recognition of sand fly fever caused by phleboviruses in the Mediterranean areas occurred in times of war through observations of the suffering of soldiers, first during the Napoleonic Wars beginning in 1799. In 1886, the illness was known in German as 'Hundskrankheit' (for dog sickness) but later as 'febbre da pappataci' in Italian ('pappataci' for gnat or sand fly). In the second recognition during WWII (1943–44), application of DDT for sand fly control in southern Italy by the Allied Forces proved the efficacy of the insecticide in the field for the first time.

The early outbreaks of a syndrome compatible with Japanese encephalitis were observed in repeated equine encephalitis outbreaks beginning in 1868 in Japan (Nakamura, 1967). The sudden emergence of the outbreak was most likely due to proliferation of pig farming as a consequence of relaxation of abstinence from meat consumption which had been religiously practised for centuries among Buddhists. Most likely, pigs played a key role as amplifying hosts in the outbreaks, but their role was unrecognized because of asymptomatic infection in swine. Louping ill ('louping' in Scottish language for leaping) in sheep was recognized in the Border districts between Scotland and England some time in the 18th century, which coincided with increased sheep population (Pool *et al.*, 1930) as the result of the Highland Clearances, a historically highly controversial forced displacement of the Gaels to convert the Highlands for an extensive sheep grazing. A seizure syndrome resembling tick-borne encephalitis was recorded in church registers in Åland of Finland in the 18th century. In 1889, Aleskei Kozhevnikov of Russia described a chronic seizure disorder in man ('Kozhevnikov epilepsy'), which is considered the first scientific document of TBE (Asher, 1979). The first of 10 major epizootics (1847–1930) of an equine disease compatible with equine encephalomyelitis (most likely by WEEV) was documented by veterinarians in the western territories of the US (Price, 1950) only 2 years before the 'Gold Rush' of 1849. At the time, migration of a large number of settlers and prospectors to the western territories across the continent was under way, which increased equine population considerably, since horses were essential in transportation, military, agriculture and many other human activities.

It has been long established that the Latin word 'virus,' which meant venomous substance, poison, slimy fluid and/or plant sap since late 14th century, was adopted by Martinus Beijerinck in 1898 in the context of filterable, infectious (and reproducible) substance ('contagium virum fluidum') for the agent of tobacco mosaic disease. It is of interest to note that the word had been applied to the cause of dengue in the context of communicable, infectious agent 3 years before the first proposal of filterability of the agent of tobacco mosaic disease by Adolf Mayer in 1886 (Hirsch, 1883). Earlier, 'hybridization' between communicable microbes based on Darwinian evolution had been proposed to explain a puzzlingly prevalent manifestation of arthralgia among atypical 'dengue' outbreaks (Christie, 1881), which was retrospectively suspected to be cases of chikungunya (Carey, 1971).

Viral genome integration in arthropod genome and commensalistic–mutualistic relationships

It is not surprising that the outcome of constant exposure of blood-feeding arthropods to exogenous genetic elements in blood over years facilitated integration of the foreign genetic elements in arthropods. Aside from ubiquitous integration of retrotransposons in the genomes of eukaryotic organisms including insects, other types of intimate virus–host relationships have been demonstrated in the discoveries of viral integration in the genome of arthropods, such as flaviviral genome in arthropods (Crochu *et al.*, 2004; Roiz *et al.*, 2009), nudivirus in leafhoppers (Cheng *et al.*, 2014), long segment of L-polymerase gene of rhabdoviruses in the genomes of ixodid ticks and mosquitoes (Katzourakis and Gifford, 2010; Fort *et al.*, 2012), long sequence of NS1-NS4A of flavivirus in the genome of *Aedes* mosquitoes, genome fragments of the genus *Quarajavirus* [proposed] (family *Orthomyxoviridae*) in ixodid ticks, and genome fragments of insect-specific flaviviruses and seadornaviruses in mosquitoes (Katzourakis and Gifford, 2010). Even more surprising was the discovery of endogenous bornavirus elements in insects and spiders (Horie *et al.*, 2013). As for chromosomal integration of viral genome in insect cells, it was experimentally demonstrated in an insect cell culture (Gundersen-Rindal and Lynn, 2003).

Also, intimate association of non-retroviruses and arthropods has also been revealed in frequent discoveries of 'indigenous' viruses of arthropods in arthropod cell lines established under sterile conditions, such as Cell fusing agent virus (a flavivirus), St. Croix River virus (an orbivirus), Aedes pseudoscutellaris reovirus (a dinovernavirus), Drosophila X virus (an entomobirnavirus) and many others. In the past, such examples of virus–host coexistence had been traditionally considered largely commensalistic in favour of viruses, but the functions beneficial to the arthropod hosts as a result of viral (including arboviral and plant viral) infection were recently recognized (Roossinck, 2011; Moreno-Delafuente *et al.*, 2013). Although arthropods do have innate mechanisms of self-defence against viral infection, the absence of antibody-based humoral immune system must have played an important role in serving as hosts for many virus groups.

Survey of arthropod groups regarding their associations with viruses

Many animal virus families have members which adopted insects as hosts, which reflects the enormous number of species of insects and the diversity of habitats they occupy. The magnitude of the diversity of arthropod associations with viruses was well recognized by early 1970s (Gibbs, 1973). Based on the latest virus classification (ICTV, 2012) and other new virus families added since but excluding plant viruses which replicate in arthropod vectors, at least 14 of 23 RNA animal virus families, two of four reverse-transcribing DNA/RNA virus families, and 12 of 18 DNA animal virus families have invertebrates as hosts. Including many unclassified viruses, most often the hosts of these invertebrate-associated viruses are insects. However, the current tallies are incomplete, since ICTV (2012) no longer lists many unclassified viruses of invertebrates; and we are constantly surprised to learn of new viruses discovered from arthropods, such as a new mimivirus (a giant dsDNA virus) from syrphid flies (Boughalmi *et al.*, 2013a). Thus, if splitting the family *Tetraviridae* and proposals of other new virus families [such as *Mesoniviridae*, tick-borne *Nyamiviridae*, *Chuviridae*, *Hytrosaviridae* of dipteran insects, as well as *Tenuivirus* (unclassified)] are approved, the number and proportion of the virus families replicating in arthropods will certainly increase in the future. Thus, it is not surprising that these facts, coupled with the role of arthropods as vectors of plant viruses and other functions and in interactions in the diversification of animal virus lineages to be introduced later, collectively left an impression on many virologists and biologists alike that arthropods, in particular insects, played the central role in the evolution of diverse lineages of animal and plant viruses (Andrewes, 1957;

Koblet, 1993; Lovisolo and Rösler, 2003; Fereres, 2015; Li *et al.*, 2015; Marklewitz *et al.*, 2015).

Among RNA virus families, the families *Flaviviridae* and *Togaviridae* stand out for their high proportions of arthropod-associated members. Eleven of 15 genera currently recognized in the family *Reoviridae*, 11 of 17 genera and four unclassified groups of the family *Rhabdoviridae* (Walker *et al.*, 2015), and four of five genera of the family *Bunyaviridae* are associated with arthropods.

Biological traits of arthropods with respect to association with viruses

One of the puzzles of arthropod–virus associations is scarcity of arboviruses transmitted by blood-sucking arthropods, such as fleas, lice and mites (Poinar and Poinar, 1998; Valiente Moro *et al.*, 2005; Adelman *et al.*, 2013), despite intimate contact with vertebrates, some of which are persistently infected with animal viruses (ranging from herpes viruses to rabies virus and retroviruses) and which present viraemia at least intermittently if not constantly. The other puzzle is why the blood-feeding dipteran flies in the suborders *Brachycera* and *Cyclorrhapha* (such as the families *Tabanidae* and *Muscidae*) are only mechanical vectors and have not evolved to become arbovirus vectors, despite larger volume of blood they imbibe, compared with mosquitoes, black flies, midges and sand flies (suborder *Nematocera*).

Mosquitoes

Currently, the total number of mosquito species is approximately 3400. They are distributed from the tropics to the Arctic region, practically ensuring contact with vertebrates in any types of terrestrial environments except for Antarctica. Excluding a large number of either pathogenic or commensalistic arthropod-specific viruses, nearly 50% of the viruses listed in the Arbovirus Catalogue (ASTMH, 1985) were isolated from mosquitoes. As of 2014, in addition to four DNA virus families, 11 RNA virus families and two unclassified RNA virus groups are known to replicate in mosquitoes. Five RNA virus families (*Bunyaviridae*, *Flaviviridae*, *Reoviridae*, *Rhabdoviridae*, and *Togaviridae*) contain non-arboviruses, viruses in tentative arbovirus status, and confirmed arboviruses; while in the *Nodaviridae* all members are non-arboviruses. In the remainder of the virus families [*Birnaviridae*, *Mesoniviridae* (proposed), *Roniviridae*, *Totiviridae*, *Tymoviridae*] and tentative groups [Picorna-like and *Negevirus* (proposed)], little is known about their vector-borne status.

Besides the diversity of ecosystems mosquitoes adapted to, the other factors contributing to the evolution of numerous mosquito-borne arboviruses are short life cycles, rapid gonotrophic cycle, multiple blood feeding of many species, and hence more frequent contact with vertebrates. Arboviral mutation accumulates more slowly in mosquitoes than in vertebrates (Vasilakis *et al.*, 2009). This is also compatible with the fact that arboviruses spend far longer time in mosquitoes than in vertebrates, given the fact that they are trans-stadially and transovarially transmitted in vectors.

Midges

They are a group of the smallest haematophagous flies. Among nearly 1400 species of the *Culicoides* midges, 96% of them are haematophagous on mammals or birds. They are distributed from the tropics to the subarctic regions and from sea level to the altitude of 4200 m above sea level. Females feed at dawn and twilight approximately every fourth day during their nearly 70-day lifespan. They transmit primarily three families of arboviruses (*Bunyaviridae*, *Reoviridae*, and *Rhabdoviridae*). In the *Flaviviridae*, *Israel turkey meningoencephalitis virus* is the only member transmitted by midges. Nearly 45% of the arboviruses isolated from the *Culicoides* midges have not been isolated from other arthropod groups. Although the flight ranges of most midges are short, some of them travel long distances, carried by air current and contributing to dissemination of arboviruses.

Sand flies

Nearly 700 species of sand flies exist, inhabiting primarily arid and savanna areas in tropical, subtropical, and temperate regions of the world. They are nocturnal-crepuscular and pool feeders. They feed on a wide range of domestic and wild mammals, birds, and poikilothermic vertebrates. The arboviruses transmitted by sand flies mostly belong to the families *Bunyaviridae*, *Reoviridae* and *Rhabdoviridae*. Currently, in the family *Flaviviridae* the only exception is *Saboya virus*; but a genome of an insect-specific flavivirus was recently detected in sand flies in the Mediterranean region. Many viruses are also transovarially transmitted in sand flies. The evidence of genome integration of ancient phlebovirus in daphnia revealed a close host relationship of an ancient phlebovirus lineage with freshwater crustaceans (Ballinger *et al.*, 2013).

The preponderance of phleboviruses in the New World is thought to reflect the greater number of species of sand flies in the New World compared with the number in the Old World (at about 3:1 ratio).

Ticks

Currently, nearly 900 species of ticks are recognized, of which 78% are hard ticks and the rest soft ticks. Ticks play a role in virus transmission in both terrestrial and aquatic environments. In fact, many ticks are ectoparasites of amphibians and mammals even in salt water. Human contacts with ticks occur mostly outdoors, but contacts with certain ticks (such as *Ornithodoros moubata*, a vector of African swine fever virus) occur in human dwellings. Generally, soft ticks are nest-dwelling (nidicolous), while hard ticks are free-living. The breadth of host range of ticks varies, depending on tick group. The types of association between ticks and vertebrates demonstrate an evolutionary gradient ranging from toxicity to the state of tolerance for the hosts, viral associations of ticks falling in the latter category. The host range diversity in ticks could be comprehensively analysed in the evolutionary context (Hoogstraal and Aeschlmann, 1982). While argasid ticks feed one bloodmeal per immature stage (larval and 2–8 nymphal) and multiple meals in adult, ixodid three-host ticks typically feed three times, each time on a different vertebrate at intervals of weeks, months or even years. Although most ticks feed only once per life stage, some soft ticks feed multiple times per life stage. On average, the volume of blood ticks imbibe is about 100 times those of mosquitoes. This mode of feeding favours viral transmission even when virus titre in blood is very low. In ticks, unlike in mosquitoes, bloodmeal is directly taken up by the midgut cells by means of receptor-mediated endocytosis into coated vesicles and digested in the lysosomal system in a process termed heterophagy. The molecular mechanism of tick's haematophagy was found similar to those of Platyhelminthes, nematodes and *Plasmodium* parasites (Sojka *et al.*, 2013).

At least six virus families [*Bunyaviridae*, *Flaviviridae*, *Orthomyxoviridae*, *Nyamiviridae* (proposed), *Reoviridae*, and *Rhabdoviridae*] have members transmitted by ticks. Some vertebrate-specific viruses not naturally transmitted but accidentally ingested by ticks (such as herpesviruses, Lymphocytic choriomeningitis virus (LCMV), Poliovirus, Rabies virus, and Foot-and-mouth disease virus) somehow remain infective for a long period in ticks, a unique trait among vectors. This may reflect the aforementioned direct bloodmeal ingestion unique to ticks. Puzzlingly, tick-borne viruses are conspicuously absent in the *Togaviridae*, although a few alphaviruses have been isolated from mites and ticks and replicate in tick cells *in vitro*.

Compared with mosquitoes, the life span of ticks is much longer, some living nearly 5–6 years and some soft ticks more than 15 years. Tick-borne viruses are trans-stadially transmitted through immature stages and vertically transmitted to eggs to ensure viral persistence for longer periods. For example, in an experiment, Nairobi sheep disease virus survived in larva for 245 days, in nymph for 359 days and in adult for as long as 871 days (Lewis, 1946), while Langat virus and Karshi virus infections in ticks persisted for more than 3 and 8 years, respectively (Turell *et al.*, 2004, 2015). It is estimated that TBEV spends more than 95% of its life cycle in Ixodid ticks, which explains the very low mutation rate of the virus (Nuttall *et al.*, 1991). Very long viral persistence is not limited to ticks. In fact, Buggy Creek virus (an alphavirus) persists in sedentary cimicid bugs in empty nests for at least 2 years after the migratory birds abandon the nests (Brown *et al.*, 2010a).

The long viral persistence in ticks facilitates geographical dispersal of tick-borne viruses through animal trade and vertebrate movement including bird migration.

Arthropod associations with plant viruses

Plants are a source of staple diet for a large number of animals. Even many haematophagous arthropods (including mosquitoes, black flies, sand flies, and midges) depend on plant juice at a particular stage of life for one or both sexes (Yuval, 1992; Braverman, 1994), providing contact opportunities between plant viruses and haematophagous insects. In fact, it was once hypothesized that Vesicular stomatitis virus (VSIV) was basically a plant virus, because the virus replicates in leafhoppers known as vectors of many plant viruses (Johnson *et al.*, 1969). Many phytophagous arthropods are, in turn, ingested by vertebrates. Thus, metagenomic studies on gut or stool contents of vertebrates typically reveal genomes of a large number of plant viruses transmitted by arthropods as well as animal viruses. In a study of stool samples, Pepper mild mottle virus was found in patients complaining of fever and abdominal pain who developed antibody to that plant virus (Colson *et al.*, 2010).

It has been well established that at least members of four virus families (*Bunyaviridae, Reoviridae, Rhabdoviridae, Tymoviridae*) and an unclassified group (*Tenuivirus*) infecting plants replicate in the arthropod vectors. Even honeybees important in pollination are systemically infected by a pollen-borne virus, tobacco ringspot virus (family *Secoviridae*) (Li, J.-L. *et al.*, 2014). Both Rhopalosiphum padi virus (family *Dicistroviridae*) of aphids and Nilaparvata lugens reovirus of leafhoppers, when injected into plant hosts by respective vector, circulate in the plants without replication. These viruses are nonetheless transmitted from infective to uninfected vectors cofeeding on the same plant, plants serving as passive reservoirs of the viruses.

It has been hypothesized that a plant virus (ssDNA nanovirus) switched hosts to infect a vertebrate, mediated by an unidentified retrovirus; and then it recombined with a vertebrate virus, to become a circovirus (Gibbs and Weiller, 1999). As revealed in a recent report, integration of an ancient plant virus gene homologous to movement gene in the genomes of insects further supports transfer of plant viral genes to animal genomes (Cui and Holmes, 2012).

The following reports of close relationships provide further support to the hypothesis of virus transfer between arthropod-associated viruses and plants: relation between negeviruses [of the *Negevirus* group (proposed)] isolated from mosquitoes or sand flies and a plant virus (Citrus leproses virus); replication in mosquito cells of a plant virus (of the family *Tymoviridae*) isolated from mosquitoes; replication of VSIV in leafhoppers which serve as vectors of plant-infecting rhabdoviruses in the Genus *Nucleorhabdovirus*; strong evidence of homology between nsP1 and nsP2 genes of Sindbis virus (SINV) and several plant viruses; and close relationships of Farmington virus (a rhabdovirus isolated from birds) with a varicosavirus (an unaffiliated negative ssRNA plant virus) (Lastra and Esparza, 1976; Ahlquist *et al.*, 1985; Selling *et al.*, 1990; Gordon *et al.*, 2001; Wang *et al.*, 2012; Palacios *et al.*, 2013a; Vasilakis *et al.*, 2013). Vector-borne plant viruses are thought to have derived originally from the viruses of the vectors through adaptation to the plants rather than in the reverse direction (Power, 2000; Hull, 2002).

Vertebrate associations with viruses

Integration of viral genome fragments in vertebrate genomes

Many vertebrates provide shelter and/or source of food to scavenging and haematophagous arthropods. Furthermore, for many vertebrates (including non-human primates) arthropods are an important source of food in both terrestrial and aquatic environments. All these forms of contact provide opportunities for viral infection from arthropods to vertebrates or in reverse direction as well as viral integration in host's genome over a long period of close association.

As an example, a genome study of DNA (such as the families *Baculoviridae* and *Polydnaviridae*) and RNA insect viruses (family *Reoviridae*) revealed molecular fossils of short sequences of these viruses integrated in the intergenic regions of the vertebrate genomes (Fan and Li, 2011). Another studies revealed a far more extensive integration of negative strand-RNA genomes of the families *Bornaviridae, Filoviridae, Orthomyxoviridae* and/or Midway virus (Belyi *et al.*, 2010a; Katzourakis and Gifford, 2010). Furthermore, integration of the fragments of ssDNA viruses (families *Parvoviridae* and *Circoviridae*), reverse-transcribing DNA viruses (family *Hepadnaviridae*), and dsDNA viruses (family *Adenoviridae*) was revealed in the vertebrate genomes (Belyi *et al.*, 2010b; Katzourakis and Gifford, 2010). It has been experimentally demonstrated that non-retroviral

RNA virus can persist in DNA form in a certain breed of rodent by means of a unique reverse transcriptase of the host (Klenerman *et al.*, 1997).

Survey of viral associations in selected vertebrate groups

Rodents

The Order *Rodentia* has 2500–3000 species and is the largest mammalian group, accounting for nearly 40% of mammals. In term of the total number of viruses transmitted as a mammalian group, rodents transmit more viruses than bats. Their gregarious behaviour favours direct viral transmission and maintenance among the members. As of 2013, nearly 180 viruses (including a new hepacivirus) have been known to infect rodents (Luis *et al.*, 2013). Rodents serve as hosts to more tick species than any other mammalian taxa, with more than 300 of approximately 600 three-host ixodid tick species feeding on rodents. For rodents, in turn, arthropods are a major source of food. Some RNA viruses are shed continuously in urine and faeces, which serve as sources of direct transmission. In the persistently infected rodent species moderate levels of pathology (measured by mortality) may be observed; but generally, mild or negligible severity of pathology permits chronic infection, as in the cases of arenaviruses, hantaviruses, and NKV flaviviruses. Coupled with high reproductive and turnover rates, milder pathologies favour rodents to serve as reservoirs of viruses.

Bats

In the wake of rapidly increased episodes of zoonotic viral disease outbreaks in the past few decades, bats have become the foci of attention as reservoirs of these agents (Calisher *et al.*, 2006); and the growing list of viruses in bats now includes Influenza A virus. Before these recent events, however, with an exception of Rabies virus, the most studied group of viruses in association with bats had been arboviruses (Sulkin, 1962; Sulkin and Allen, 1974). Some of the traits which favour bats to serve as reservoirs of non-arboviruses include (i) gregarious, nesting behaviour which facilitates direct transmission via contaminated urine, faeces, and bodily secretions; (ii) high intra- and inter-specific contact rates at roosting sites; (iii) long flight ranges of many species which enable dissemination of viruses over long distances; and (iv) long lifespans which are advantageous for perpetuation of viral infection in roosting sites (Luis *et al.*, 2013).

Additionally, still poorly understood mechanisms of immune evasion (including ephemeral or undetectable antibody response and virus sequestration in adipose tissue) despite endowment of adaptive immunity and cell-mediated immunity, facilitate establishment of chronic viral infection (Baker *et al.*, 2012). In fact, in infections by such viruses as Rio Bravo virus (a flavivirus), neutralizing antibody is not even elicited, and viraemia is absent (Baer and Woodall, 1966). Bats are also excellent carriers or amplifying hosts (but not reservoirs) of arboviruses, as demonstrated by a long list of arboviruses isolated from bats (Sulkin and Allen, 1974; Geevarghese and Banerjee, 1990; Calisher *et al.*, 2006).

Bats are found in all continents (except in Antarctica) and comprise the second largest group of mammals, with approximately 1150 species, making up nearly 25% of mammals. At least 140 viruses have been known to infect bats, and the total number of viruses is climbing. Thus, on the basis of number of viruses per host species, bats carry more virus species than rodents (Luis *et al.*, 2013). Insectivorous bats are ideal for acquiring arthropod-associated viruses by ingestion. In a virome study of bats in North America, among many families of viruses discovered, insect-specific viruses (such as members of the families *Iflaviridae* and *Dicistroviridae*) were dominant. More recently, bats were found to be the hosts of newly discovered hepaciviruses, vesiculovirus, phlebovirus, hantavirus, and nairovirus (Quan *et al.*, 2013; Ng *et al.*, 2013; Mourya *et al.*, 2014; Gu *et al.*, 2014; Ishii *et al.*, 2014).

Birds

Currently, arboviruses comprise none of the DNA but 7 of the 12 RNA virus families of birds. It has also been recognized that the mosquitoes transmitting arboviruses to birds mostly belong to the genus *Culex* rather than to the genus *Anopheles* or *Aedes*. The importance of birds in arboviral transmission was recognized early in relation to the transmission of alphaviruses and flaviviruses, such as Eastern equine encephalomyelitis virus (EEEV) and Western equine encephalomyelitis virus (WEEV), St. Louis encephalitis virus (SLEV) and West Nile virus (WNV). For example, WNV infection has been detected in more than 350 species of wild and domesticated birds worldwide. Avian mobility, in particular in long-range migration, is critically important in analysing the geographic distribution of arboviruses associated with birds. This is particularly important for tick-borne viruses associated with long-distance migratory birds because of longer viral persistence in ticks. It was early recognized in North America that for the transmission of indigenous encephalitic arboviruses to humans and domesticated animals to occur, imported birds (such as ring-necked pheasants, chukar, house sparrows, and house finches) rather than indigenous birds play a more important role as bridge vertebrate hosts, since indigenous birds are relatively refractory to infection by the indigenous viruses (Stamm, 1966).

Arboviral persistence in vertebrates

Viral persistence in vertebrates provides answers to the following two questions: (i) do vertebrates serve as reservoirs of arboviruses? and (ii) have the vertebrate viruses chronically infecting the hosts evolved directly to become arboviruses through adaptation to haematophagous arthropods? Because the first question is discussed in the section of arbovirus-related issues later, only the second question is commented on.

There are many RNA vertebrate viruses that persistently infect the hosts, such as arteriviruses, lyssaviruses, lentiviruses, pestiviruses, hepaciviruses, pegiviruses, rodent-borne bunya- or arenaviruses, many bat-borne viruses, and members of the NKV subgroup of genus *Flavivirus*. Because viraemia is observed in some of the chronically infected vertebrates which are constantly exposed to the bite of haematophagous arthropods, it would be valuable to find the viruses which persistently infect vertebrates and which are biologically transmitted by these arthropods. Thus far, no such an RNA virus has been found. In the early part of history, the prevalent concept on biological transmission developed by the biologists was that it evolved when vector-associated viruses secondarily adopted vertebrates as another hosts but not in the reverse direction

(Huff, 1938). As presented later in this chapter, at least in the family *Flaviviridae* the available data support two possibilities. First, arboviruses evolved as a result of adoption of vertebrates as secondary hosts by the viruses of haematophagous arthropods. In fact, 5 virus families containing traditional arboviruses (*Bunyaviridae*, *Flaviviridae*, *Reoviridae*, *Rhabdoviridae*, and *Togaviridae*) have a group of vector-specific viruses which do not replicate in vertebrates; and, these viruses are located at or near the root of phylogenetic tree, when family tree is available. Second, if vertebrate viruses were the origins, initially they evolved to become viruses of haematophagous arthropods, which, in turn, adopted vertebrates as second phylum of hosts.

A list of selected reports presenting ambiguous or intriguing data regarding the boundaries between arboviruses and non-arboviruses

In the first half of the 20th century, when isolation of aetiological agents of infectious diseases was gaining enormous importance, in the absence of the knowledge about the modes of transmission, experiments on insect-borne transmission were attempted either to confirm or rule out such a mode of transmission for a variety of agents ranging from Poliovirus and Rabies virus to LCMV and Classical swine fever virus. When more arboviruses began to be isolated following the first isolation in 1927 of vector-borne virus, Yellow fever virus (YFV), miscellaneous problems associated with clear segregation of arboviruses from non-arboviruses were recognized by the early researchers. These problems were illustrated in the difficulty of classifying the flaviviruses without a known vector and in the occasional isolation of rodent-borne bunyaviruses and arenaviruses from arthropods. Also, lack of information about arbovirus status of multiple virus groups, such as the family *Arteriviridae* and pestiviruses (family *Flaviviridae*), has been a constant challenge to arbovirologists (Brinton, 1980; Horzinek, 1981). Table 14.1 lists selected reports illustrating ambiguities of the boundaries separating arboviruses from non-arboviruses, phylogenetic relationships between arboviruses and non-arboviruses, possible transfer of virus lineages between terrestrial and aquatic environments, unusual host range of animal viruses, hypotheses, and even some provocative questions.

The diverse range of the puzzling or intriguing observations presented in Table 14.1 provides an important set of data for elucidating the mechanisms leading to the evolution of arboviruses. Among multiple hypotheses reviewed (Kuno and Chang, 2005), the 'blind alley' theory (Andrewes, 1957) captures best the essential denominator, a set of environmental factors providing the most ideal (but infinitely small) opportunity of the meeting of virus, vector, and vertebrate conducive to establishment of sustainable biological transmission of virus between two phyla of hosts.

Virological attributes of selected environments and hosts

For understanding the diversity of the host associations of arboviruses including host ranges, discussion of the ecological conditions conducive to the evolution of biological transmission is essential. The underlying mechanisms entail complex interactions among vectors, vertebrates, and other types of hosts, such as plants, in the contexts of fauna, flora, virome, population density, seasonality of host population size, dispersal of host, landscape, and climatological factors. Accordingly, the mechanism which has led to the evolution of arboviruses replicating in two disparate phyla of hosts (arthropods and vertebrates, separated in date of emergence on earth at least by several million years) was 'ecological routes mapped by available transmission mechanisms' (Reanney, 1982).

Development of metagenomics in the past decade dramatically changed our understanding of the depth of virus–host interactions, diversity of viruses, and the impacts of contact opportunities between viruses and hosts. Since ecological factors of arboviruses have been covered extensively in the past (Monath, 1988; Clements, 2012) and in other chapters of this book, only two environments, aquatic and peridomestic-domestic, are described in this chapter, the first for the importance in virus lineage transfer between terrestrial and aquatic environments by means of food chain and the second for the importance in arboviral transfer to humans.

Aquatic environments

Traditionally, the principles of arbovirology have been almost exclusively focused on terrestrial environments. However, many insects are known to inhabit marine environments (Cheng, 1976). A consideration of aquatic environments is indispensable, in particular, for studying the evolution of arboviruses. The intriguing relationships between aquatic and terrestrial viruses were made clearer in light of the genome detection of numerous insect-associated viruses and of even some arboviruses in the genomes of aquatic organisms (Djikeng *et al.*, 2009; Belyi *et al.*, 2010a; Ballinger *et al.*, 2013).

In the current virus taxonomy, out of 9 RNA animal virus families and picorna-like group infecting fishes, with only exceptions of the families *Caliciviridae* and *Paramyxoviridae*, all have members infecting insects or ticks. Similarly, aquatic arthropods, such as larvae of dipteran insects, shrimps, copepods, and other crustaceans, are infected by at least seven RNA animal virus families (*Bunyaviridae, Dicistroviridae, Nodaviridae, Reoviridae, Rhabdoviridae, Roniviridae, Totiviridae*) and five DNA virus families (*Baculoviridae, Iridoviridae, Nimaviridae, Parvoviridae* and *Poxviridae*). Currently, with the exceptions of the *Nimaviridae* and *Roniviridae*, all these aquatic virus families have members infecting arthropods in the terrestrial environments. The possible evidence of interbiome transfer of arboviral lineage between terrestrial and aquatic environments is strongest in the genus *Vesiculovirus* of the *Rhabdoviridae* and the genus *Alphavirus* of the *Togaviridae*, because each genus contains both terrestrial viruses and aquatic viruses. As supporting evidence of this hypothesis, replication of Chikungunya virus can be studied in zebrafish, while that of fish alphavirus can be studied in mosquito cells (Palha *et al.*, 2013; Hikke *et al.*, 2014).

As for virus transfer between freshwater and saltwater environments, fish viruses in the genus *Novirhabdovirus* (family *Rhabdoviridae*), the genus *Aquabirnavirus* (family *Birnaviridae*), the genus *Isavirus* (family *Orthomyxoviridae*), the genus *Alphavirus* (family *Togaviridae*), and the genus *Betanodavirus* (family *Nodaviridae*) are well known for their infectivity in both freshwater and marine fishes (Crane and Hyatt, 2011).

Table 14.1 Non-arboviruses, viruses located in the boundaries between arboviruses and non-arboviruses, and other viruses with intriguing relationships with arboviruses

Virus	Description
Family *Arenaviridae*	A novel arenavirus from a boid snake replicates not only in the snake but in mammalian and tick cells, when grown at 30°C (Hepojoki *et al*., 2015). Tacaribe virus, which had been isolated from bats and mosquito earlier, was recently isolated from a tick (*Amblyomma americanum*) (Sayler *et al*., 2014). These reports necessitate reassessment of the vector-borne status of these arenaviruses
Family *Birnaviridae*	The genus *Entomobirnavirus* includes *Espirito Santo virus*, *mosquito X virus*, and *Culex virus*. The first virus was isolated from the blood of a dengue patient, while the latter two viruses were isolated from mosquitoes. Host ranges of these viruses outside the class Insecta are unknown. Long before these recent discoveries, birnaviruses, such as infectious bursal disease virus had been isolated from mosquitoes (Howie and Thorsen, 1981)
Family *Bornaviridae*	The discovery of bornavirus integration in the genomes of insects and spiders revealed an unexpected host association of the bornaviruses at least in the past (Horie *et al*., 2011). The biological properties of tick-borne Nyamanini virus, including intranuclear replication, suggest close evolutionary relationships with bornaviruses and filoviruses (Herrel *et al*., 2012)
Family *Bunyaviridae*	Many early reports of transmission of rodent-borne viruses (genus *Hantavirus*) by gamasid mites, chiggers, and fleas were interpreted as artefacts without a proof of biological transmission. The dismissal has not dented the publication of similar reports, however. In one of more recent reports, increase in virus titre in chiggers was found (Zhang *et al*., 2002). Similarly, Bayou virus-specific RNA was detected in ticks and mites, raising a possibility of these ectoparasites serving in the maintenance or transmission of the virus (Houck *et al*., 2001). Whether or not these examples represent true biological or merely mechanical transmission is still unclear; and more independent research by multiple, unaffiliated groups is necessary before confirmation (Yu and Tesh, 2014). Nonetheless, it is of interest to note that Hantaan virus is the only rodent-borne bunyavirus still listed as 'possible' arbovirus in the International Catalogue of Arboviruses online. On the other hand, the isolation of a natural reassortant of a plant virus and an arbovirus presented a classification problem of reassortant viruses (Webster *et al*., 2011). It was also proposed that retrovirus genome fragments integrated in the genome of a nematode derived from a phlebovirus genome (Malik *et al*., 2000). Although this report appeared odd at first, a recent discovery of a new phlebovirus genome in a nematode (Bekal *et al*., 2011) has made it more difficult to ignore. The capsid protein gene of newly recognized dsRNA plant virus family, *Amalgaviridae*, is homologous to the nucleocapsid gene of phleboviruses (Krupovic *et al*., 2015)
Family *Caliciviridae*	The discovery of the similarity of genome organization between caliciviruses and cricket paralysis virus (a member of the family Dicistroviridae) generated an interest in the evolutionary histories of the two seemingly disparate animal virus families (Koonin and Gorbalenya, 1992). The most likely reservoir of human norovirus (Genogroup I) has been shellfish. A recent study, however, revealed far more concentration of this virus in micro- and mesozooplankton (in which a variety of copepods and other crustaceans are the dominant group) than in oysters. This raised a question if aquatic arthropods are reservoirs, vectors, or bystander hosts (Gentry *et al*., 2013)
Family *Chuviridae* (proposed)	This new RNA virus family isolated from many groups of arthropods and containing bisegmented and circular RNA viruses is reported to occupy a phylogenetic position 'intermediate' or 'basal' between segmented and non-segmented RNA virus families and proposed to be ancestral to all negative-strand RNA virus families (Li *et al*., 2015)
Family *Coronaviridae*	The viruses (called Runde virus for one of them) with virion morphology indistinguishable from that of typical coronaviruses in electron micrographs were isolated from ixodid ticks associated with seabirds. One of the viruses is even pathogenic to chicken (Traavik *et al*., 1977; Traavik, 1978; Saikku *et al*., 1980). This virus was later identified to be related to avian coronavirus (Traavik, 1979). Human enteric coronavirus could be adapted to replicate in mosquito cells (Luby *et al*., 1999). More recently, putative origin of CoV ancestors in insects has been speculated (Drexler *et al*., 2013)
Family *Dicistroviridae*	Detection of IgM antibody specific to Cricket paralysis virus in humans and several species of mammals has remained inexplicable (Longsworth *et al*., 1973). Taura syndrome virus of shrimps readily replicates in mosquito cell cultures (Arunrut *et al*., 2011)
Family *Flaviviridae*	One of the major interests in the genus *Flavivirus* has been the phylogenetic position of Tamana bat virus (TABV), the most distantly related flavivirus in the genus. Interestingly, the NS3 gene sequence was found in the genome of a fish (Belyi *et al*., 2010a). The reports of Hepatitis C virus (genus *Hepacivirus*) replication in mosquitoes or transmission by ticks (Germi *et al*., 2001; Wurzel *et al*., 2002) have been controversial. Recent discoveries of hepaciviruses from bats and rodents and of a new flavivirus from a nematode (Bekal *et al*., 2014) and the establishment of a new genus (*Pegivirus*) for formerly GB-C group viruses within this family (Kapoor *et al*., 2013; Quan *et al*., 2013) have significantly impacted on our understanding of the evolution of the flaviviruses as well as on the phylogenetic tree of the family *Flaviviridae*. The new data presented in the text of this chapter provide a support to the hypothesis that the genus *Flavivirus* derived from the other genera of this family, with TABV at an intermediate stage between those genera and the insect-specific subgroup flaviviruses
Family *Hepadnaviridae*	The members of this family have a narrow host range, infecting mammals and birds. Early claims of persistence of hepatitis B antigen in mosquitoes attracted attention, but no convincing evidence of transmission by arthropods has been found. A recent discovery of integration of hepadnavirus genome fragments in the genome of ticks raised a possibility of the existence of an undiscovered arthropod hepadnavirus (Katzourakis and Gifford, 2010)
Family *Herpesviridae*	Murid herpes virus 4 (strain 68), which is naturally found infecting rodents (genera *Myodes* and *Apodemus*) in Slovakia, was isolated repeatedly from *Ixodes ricinus* ticks and green lizard, suggesting apossible transmission of the virus between the tick and the lizard (Ficova *et al*., 2011). No definitive proof of the viral replication in ticks or transmission between the two has been presented yet
Family *Iridoviridae*	Two iridoviruses (invertebrate iridescent virus 6 of insects and a virus isolated from a chameleon) replicate in viper spleen cells and crickets, respectively (McIntosh and Kimura, 1974; Weinmann *et al*., 2007). A possibility of mosquito-borne transmission of a ranavirus to terrestrial turtles was raised (Kimble *et al*., 2014)

Table 14.1 Continued

Virus	Description
Family *Mesoniviridae* (proposed)	More than 15 viruses of this family (i.e. Nam Dinh virus, Dak Nong virus, and Cavally virus) were isolated from mosquitoes (Nga *et al.*, 2011; Zirkel *et al.*, 2011; Lauber *et al.*, 2012; Kuwata, *et al.*, 2013; Vasilakis *et al.*, 2014). Thus far, they are tentatively classified as mosquito-specific non-arboviruses. However, more complete host range studies are desirable because of replication in two phyla of hosts of the aforementioned corona-like virus isolated from ticks, as described earlier in this table
Family *Nimaviridae*	*Infectious hypodermic and haematopoietic necrosis virus*, a dsDNA virus, was found to be related to mosquito densoviruses (family Iridoviridae) (Shike *et al.*, 2000)
Family *Nodaviridae*	Flock House virus, a nodavirus isolated from a species of beetle, replicates in three kingdoms of hosts, insects (including mosquitoes and flies), mammals, yeasts, nematodes, and plants (Dasgupta *et al.*, 2003). An aquatic nodavirus of a freshwater prawn, *Macrobrachium rosenbergii* nodavirus, readily replicates in mosquito cells (Sudhakaran *et al.*, 2007)
Family *Nyamiviridae* (proposed)	The genome fragments of Nyamanini virus, a tick-borne negative strand RNA virus, were found integrated in the genome of zebra fish (Horie and Tomonaga, 2011). The close relationships of this virus with bornaviruses and filoviruses were mentioned earlier for the Bornaviridae
Family *Orthomyxoviridae*	Insect transmission of influenza A virus was reported multiple times in the past, but without proof. Recent rise of concern over the spread of avian influenza prompted a renewed interest in the role of mosquitoes and flies in mechanical transmission (Barbazan *et al.*, 2008; Sawabe *et al.*, 2009)
Family *Parvoviridae*	Aleutian disease virus of mink was reported to persist in mosquitoes as long as 35 days, suggesting replication in the vectors (Shen *et al.*, 1973). Hepatopancreatic parvovirus of shrimps and parvoviruses of mosquitoes were found to be very close phylogenetically, suggesting a possible genetic transfer (Roekring *et al.*, 2002)
Family *Picornaviridae*	Mechanical transmission of foot-and-mouth disease virus by flies and ticks has been documented many times (Hyslop, 1970). Accordingly, isolation of this virus from haematophagous arthropods, such as ticks, was not surprising (Sang *et al.*, 2006). The significance of replication of human enterovirus A and human enterovirus B in mosquito cells (White, 1987) is unknown. In a case of aseptic meningitis associated with an episode of tick bite, sequences of Human enterovirus A (Coxsackie A9) were detected in the cerebrospinal fluid of the patient and in ticks; and the CF antibody to the enterovirus was elevated. This raised a possibility of the existence of a tick-borne enterovirus (Freundt *et al.*, 2005). A picornavirus in the genus *Cardiovirus* was isolated from patients and from soft and hard ticks in Central Asia (L'Vov *et al.*, 2014). This virus is phylogenetically close to *Theiler's murine encephalomyelitis virus*
Family *Poxviridae*	Lumpy skin disease virus was found to be transmitted to cattle not only mechanically by ticks but also by bite of emergent larvae vertically infected, suggesting long persistence in ticks (Lubinga *et al.*, 2013; Tuppurainen *et al.*, 2013)
Prion	Although prions are not viruses, they are covered by the ICTV. The mechanical transmission of prions by insects, mites, and ticks (Lupi, 2003) has been a medical concern. Sarcophagous flies and grass mites, which fed on scrapie-infected hamsters, were reported to harbour the prion and mechanically transmit scrapie (Post *et al.*, 1999). Several findings suggested involvement of hay mites or maintenance of scrapie to sheep and goats, but these reports could not be definitively confirmed (Carp *et al.*, 2000)
Family *Reoviridae*	Genus *Coltivirus*: A genomic study of reoviruses revealed an evolutionary link between an aquatic virus of carp (genus *Aquareovirus*) and members of the genus *Coltivirus* transmitted by ticks (Mohd Jaafar *et al.*, 2008). Genus *Orbivirus*: An orbivirus, JKT-7400, isolated from *Culex* mosquitoes in Indonesia replicates in insect cells and causes lethal infection in rabbits (but without replication) (Vazeille *et al.*, 1988)
Family *Retroviridae*	HIV remains infectious in ticks adapted to human dwelling in Africa, *Ornithodoros moubata*, for as long as 10 days. Since these ticks ingest as much as 100 times more blood than mosquitoes, it was proposed that the role in HIV transmission of this tick is possible under an intense HIV transmission (Humphrey-Smith *et al.*, 1993)
Family *Rhabdoviridae*	Genus *Vesiculovirus*: Fish rhabdoviruses, Spring viraemia of carp virus and Pike fry virus, have been shown to replicate in *Drosophila* flies (Bussereau *et al.*, 1975). In turn, vesicular stomatitis virus replicates in a fish cell line (Kelly and Loh, 1972). Genus *Novirhabdovirus*: Infectious haematopoietic necrosis virus replicates in *Drosophila* flies and mosquito cells in culture (Bussereau *et al.*, 1975; Scott *et al.*, 1980). A crustacean ectoparasite of salmon, *Lepeophtheirus salmonis*, probably acts as mechanical rather than biological vector to salmon (Jakob *et al.*, 2011); however, this virus could be isolated from mayflies collected from streams and fish hatcheries (Shors and Winston, 1989). Similarly, Viral haemorrhagic septicaemia virus replicates in a moth cell line (Lorenzen and Olesen, 1995). Genus *Lyssavirus*: Mokola virus replicates in mosquito cells, which strongly suggested to the researchers its intermediary stage in the evolution of vertebrate Rabies virus complex from mosquito-associated rhabdoviruses (Buckley, 1975; Aitken *et al.*, 1984). Persistent antigen synthesis in mosquito cell cultures (Seganti *et al.*, 1990) and replication in mosquito and fish cell lines (Solis and Mora, 1970; Reagan and Wunner, 1985) of Rabies virus have been reported. It was not certain if these unusual results reflected mere sharing of virus receptors among animal groups and thus were artificial or were of evolutionary significance
Family *Roniviridae*	Yellow head virus of shrimp replicates in insect cells (Gangnonngiw *et al.*, 2010)
Family *Togaviridae*	As described in the section 'Virologic attributes of selected environments and hosts', replication of some alphaviruses in fish cell cultures has been known. However, little attention has been paid to learn of the significance until recently. Recent proposal of the origin of alphaviruses in a fish virus (Forrester, *et al.*, 2012) revived the interests in the older records because of isolation of aquatic alphavirus from fishes (Weston *et al.*, 2002). The role of an ectoparasitic copepod (*Lepeophtheirus salmonis*) in the virus transmission is currently understood to be mechanical vector (Petterson *et al.*, 2009). The role of *Lepidophthirus macrorhini*, an ectoparasite of elephant seal, may be a vector of an alphavirus (La Linn *et al.*, 2001) is unknown. The relevance of the reports of isolation of togaviruses from mites, replication in tick cells and of vertical transmission in ticks (Syverton and Berry, 1941; Sulkin, 1945; Reeves *et al.*, 1947; Howitt *et al.*, 1948; Hurlbut and Thomas, 1960; Pudney *et al.*, 1982) has received little attention despite their potential importance. The discovery of Eilat virus, an alphavirus which replicates only in mosquitoes but not in vertebrates (Nasar *et al.*, 2012), further presented intriguing questions about its position in alphavirus lineage and the determinants of host range

Table 14.1 Continued

Virus	Description
Family *Totiviridae*	Two dsRNA viruses were isolated from *Armigeres* and *Culex* mosquitoes (Zhai *et al*., 2010; Isawa *et al*., 2011). The *Armigeres* isolate replicates also in vertebrates (Zhai *et al*., 2010)
Unassigned virus	A close genetic relationship between Chronic bee paralysis virus and members of the genus *Alphavirus* of the family *Togaviridae* was demonstrated in multiple homologues through a protein analysis (Kuchibhatla *et al*., 2013). A new tick-borne virus is unusual in that the genome is composed of four RNA segments, two of which are related to the NS genes of the genus *Flavivirus* and the remainder to a nematode (*Toxocara canis*) (Qin *et al*., 2014)

For the virus transfer from terrestrial to aquatic environments, it is recognized that certain fishes feed on larvae and even adults of vectors (Murray, 1885). In fact, many small fishes have been used for biological control of mosquito larvae. These fishes, in turn, serve as preys for other aquatic vertebrates. It has been known for many years that several alphaviruses [such as EEEV, WEEV, and Venezuelan equine encephalomyelitis virus (VEEV)] replicate well in the cell cultures derived from fishes (Sorbet and Sanders, 1954; Officer, 1964; Middlebrooks *et al*., 1979; Wolf and Mann, 1980). Southern Elephant Seal virus, another alphavirus, was isolated from an ectoparasitic louse of the seal (La Linn *et al*., 2001). Infection of copepods by fish rhabdoviruses is significant, since some copepod species feed on mosquito larvae and many aquatic vertebrates depend on zooplankton as a major source of nutrients (Mulcahy *et al*., 1990).

Conversely, in virus transfer from aquatic to terrestrial environments, fishes and aquatic invertebrates serve as an important source of viral infection of many birds and other terrestrial animals through food chain. In fact, some fish viruses have been known to replicate in terrestrial and aquatic insects or insect cell cultures, such as several rhabdoviruses of the genus *Vesiculovirus* (Kelly and Loh, 1972; Bussereau *et al*., 1975; Scott *et al*., 1980; Shors and Winston, 1989; Lorenzen and Olessen, 1995). There are bats which specialize in eating fishes (i.e. minnows). Some species of mosquitoes feed almost exclusively on fishes (Sloof and Marks, 1965; Okudo *et al*., 2004). Larvae of vectors, in particular mosquitoes, are known to be orally infected in water contaminated with any of multiple arboviruses released from infected larvae or by cannibalism (such as YFV, DENV, EEEV, JEV, LACV, RVFV, SINV, SLEV, and WNV) (Whitman and Antunes, 1938; Burgdorfer and Varma, 1967; Miller *et al*., 1978; Turell *et al*., 1990; Bara *et al*., 2013). In dambos, an aquatic environment in Africa shared by congregations of many species of wildlife, transfer of vertebrate viruses to mosquito larvae through ingestion of viruses released from sick, injured, or dead wildlife is possible. Virus transfer from one group of mosquitoes to another is also possible because not only predatory *Toxorhynchites* mosquitoes but certain species of *Culex*, *Ochlerotatus*, and other mosquito genera, prey on other mosquito larvae. Thus, aquatic stage of vectors, like ticks and lice parasitizing amphibious mammals, birds, and reptiles, can play a role in virus transfer in both directions. Other examples of direct transmission of arboviruses (including Omsk haemorrhagic fever virus) in aquatic environments were compiled earlier (Kuno and Chang, 2005).

The following examples of viruses provide further supports to interbiome virus transfer. The genome of a reovirus of grass carp revealed an evolutionary link between aquareoviruses and coltiviruses (Mohd Jaafar *et al*., 2008). A similar conclusion was obtained between unclassified Infectious hypodermic haematopoietic necrosis virus and mosquito brevidensoviruses (family *Parvoviridae*), between parvoviruses infecting shrimps and insects (Shike *et al*., 2000; Roekring *et al*., 2002) and between Cricket paralysis virus and Taura syndrome virus (family *Dicistroviridae*) of shrimps (Mari *et al*., 2002). Aquatic nodavirus of shrimps also replicates in mosquito cells (Sudhakaran *et al*., 2007) as well as in a variety of aquatic insects, such as dragonfly nymphs, water bugs, diving beetles, and back swimmers (Sudhakaran *et al*., 2008). Similarly, Yellow head virus of shrimp (a ronivirus) replicates in moth cells *in vitro* (Gangnonngiw *et al*., 2010). Thus, it was not surprising to find evidence of gene transfer of viruses between shrimp and ixodid tick and between sand fly and daphnia (Liu *et al*., 2010; Ballinger *et al*., 2013). A recent discovery of a new member of the genus *Seadornavirus* closely related to Banna virus (BANV) from the intestinal contents of a freshwater carp (Reuter *et al*., 2013) is also of interest. While a report of Midway virus [family *Nyamiviridae* (proposed)] genome integration in fishes is compatible with the fact that this is a virus of ticks infesting seabirds, it is more difficult to understand the similar integration of Tamana bat virus (TABV) sequence in the genome of fish (Belyi *et al*., 2010a).

Interbiome transfer of arthropod-associated DNA viruses is well illustrated by the detection of a polyhedrin gene of a baculovirus of a lepidopteran insect in planktons in freshwater and marine environments as well, strongly suggesting viral gene transport through water from terrestrial to aquatic environments (Hewson *et al*., 2011). Although parvoviruses are also not arboviruses, the fact that the genus *Brevidensovirus* contains members in both terrestrial and aquatic environments suggests gene flow in the past. Thus, a report of a close relationship between *African swine fever virus* (ASFV) (family *Asfarviridae*) and a DNA virus infecting dinoflagellates (Ogata *et al*., 2009) may not be entirely enigmatic. The significance of the discovery of ASFV-like sequence in a metagenomic study of ponds in North America (Wan *et al*., 2013) is unknown, but may be a potential concern in agriculture.

Peridomestic and domestic environments

In this section, arboviral transmission in the environments modified by human habitation is examined with an emphasis on the evolutionary trends of the biological traits of selected vectors and vector–virus interactions.

Aedes aegypti

The domesticated anthropophilic subpopulations of this mosquito are believed to have evolved through multiple events of domestication from the sylvan subpopulation feeding on subhuman primates (but not humans) in Africa (Christophers, 1960; Tabachnick, 1991; Brown *et al*., 2011). The remarkable adaptability of this mosquito, which normally prefers to breed in clean water with diurnal feeding behaviour in urban settings, is illustrated in many unusual observations in various parts of the tropics, such as preference to breed in feral environments with much less anthropophilicity, breeding in septic tanks, and nocturnal feeding pattern (Salvan and Mouchet, 1994; Diarrasouba and Dosson-Yovo, 1997; Chadee and Martinez, 2000; Burke *et al.*, 2010). This is one of the mosquitoes known for multiple feeding, which enhances virus transmission. The physiological bases of preferential attraction to humans of this mosquito were identified as high concentration of isoleucine which is thought to promote higher egg production (Scott *et al.*, 1997; Harrington *et al.*, 2001) and higher expression of an odorant receptor specific to human (McBride *et al.*, 2014).

Aedes albopictus

Generally, this species is a vector in perisylvan, feral and peridomestic environments, with a wider range of vertebrates as sources of bloodmeal than *Ae. aegypti*. Although this mosquito, unlike *Ae. aegypti*, does not so much depend on human for a bloodmeal, its dependence gradually increases from perisylvan towards urban environments. Even in the perisylvan and feral areas, this mosquito still needs primate blood at least partially in its bloodmeal (Rudnick, 1983). The cause of the increasing number of reports of indoor breeding and urban adaptation in parts of tropical Asia (Dieng *et al.*, 2010; Kuno, 2012; Li, Y. *et al.*, 2014) has not been definitively determined yet but may suggest behavioural change associated with further adaptation to urban environments. The peculiar attraction of this mosquito to rubber, the cause of recent global dispersal by means of used tire trade, was observed much earlier in rubber plantations in Africa and Asia, but the significance remained unknown until mid-1980s, when the infestation through international used tire trade was discovered in North America (Hawley *et al.*, 1987; Kuno, 2012).

Culex pipiens

Two biotypes of this mosquito are recognized, 'pipiens ecotype' and 'molestus ecotype' (Chevillon *et al.*, 1995). The former is peridomestic and ornithophilic, while the latter has adapted more to man-made environments and is anthropophilic. *Culex pipiens quinquefasciatus*, on the other hand, feeds on both birds and mammals and serves as a bridge vector (Hamer *et al.*, 2008). The shift in host preference of *Culex pipiens* in North America from birds to humans was reported to have coincided with the dispersal of its preferred hosts (American robins) and rise of human cases of West Nile fever (Kilpatrick *et al.*, 2006a).

Interspecific competition

Ae. albopictus is an indigenous mosquito in Asia, where infestation of urban areas by introduced *Ae. aegypti* began to spread in the 19th century (Smith, 1956). As the infestation of *Ae. aegypti* spread in tropical Asia, a concomitant 'decline' of *Ae. albopictus* population was recognized in multiple urban centres (such as Kolkata and Bangkok). This phenomenon was interpreted 'displacement' through interspecific competition between the two species of mosquito (Senior-White, 1934; Gilotra *et al.*, 1967). In contrast, in the Americas, where infestation by *Ae. aegypti* had been established at least a few centuries earlier, it was reported that newly introduced *Ae. albopictus* 'displaced' *Ae. aegypti* instead (Lounibos, 2007); and the phenomenon was understood as competition between ecological homologues according to the principle established by Paul DeBach (1966). An obvious interest that arose from these contrasting observations across the Pacific was if the dominant mosquito is reversed depending on the order of establishment between the two species of mosquito or the real cause of contrasting observations is something else.

The early claims in Asia were quickly disputed by Chan *et al.* (1971). The critics stressed the importance of progressive increase of the habitats favoured by *Ae. aegypti* associated with rapid urbanization, which was accompanied by the inevitable loss of the peridomestic habitats favoured by *Ae. albopictus*, as suggested by Macdonald (1956), rather than true competition between the two species over the same breeding sites under a condition favourable to both. Furthermore, a closer re-examination of all early accounts revealed that most early reports were anecdotal accounts or gloss impressions without a scientifically valid baseline mosquito density data absolutely necessary for comparison. Also, the decline of *Ae. albopictus* population was not observed uniformly in all large population centres including Singapore (Chan *et al.*, 1971). In more recent surveys, with an only exception of Bangkok, in all other major urban centres including Kolkata, resurgence of *Ae. albopictus* has been reported; and in parts of Asia, actually *Ae. albopictus* became indoor breeder and displaced *Ae. aegypti* (Kuno, 2012). In the Americas, despite early claims of displacement of *Ae. aegypti* by introduced *Ae. albopictus*, persistent co-existence was observed (O'Neil and Juliano, 2012).

A confounding factor that complicates 'displacement' is the fact that spontaneous extinction of *Ae. aegypti* populations before the arrival of *Ae. albopictus* without a vector control program has been documented in such semitropical areas as the countries along the Mediterranean coast (Holstein, 1967) and in southern states of the United States (excepting southern Florida) which are considered marginal for *Ae. aegypti* because of suboptimal winter temperature (Tinker and Hayes, 1959). Another difficulty for assessing the difference in dominant species was that while densities of the two mosquito species were measured in both peridomestic and domestic environments in Asia, only outdoor densities were measured in the Americas. As for the DeBach's principle, because it was based on interspecific competition between two insect species under natural conditions, its direct application to urban mosquitoes is questionable because human-modified environments are not natural and complicated by numerous man-made confounding factors that vary constantly in qualities depending on place and time. Accordingly, it was concluded that in many cases of the reports of competitive displacement, actually more than one competitive mechanism and other non-competitive factors were involved (Reitz and Trumble, 2002).

Arboviral transmission in urban areas

Among the arboviruses transmitted by *Ae. aegypti* in urban areas [Chikungunya virus (CHIKV), four serotypes of DENV, and YFV], DENV is exceptional in that it is the only arbovirus perpetually transmitted in tropical regions between one species of mosquito vector and human, so long as the size of susceptible human population is sufficiently large and other factors are favourable for the vectors. Furthermore, vertical transmission of DENV in the urban vector has been documented repeatedly. Although the flight range of *Ae. aegypti* is limited, movement of infected humans in vector-infested urban areas contributes to rapid and extensive dispersal of the virus (Stoddard *et al.*, 2013; Reiner *et al.*, 2014).

YFV transmission in small- or medium-size urban areas has been recorded in the past, but these urban areas were characterized by location in or near the 'zone of emergence' (Mondet, 2001) and availability of open corridors that permit constant flow of human activities with surrounding YF-endemic feral or sylvan environments. As revealed in multiple studies in South America and Africa, YF outbreaks have occurred by the incursion of the sylvan or feral virus population into these smaller urban centres located not too far. Excepting for the island of Trinidad with a limited land mass, within large YFV-endemic sylvan regions in Africa and South America, the geographic foci of activity are thought to shift continuously, incursion to urban areas occurring at irregular intervals most often not at the same location year after year (Mondet, 2001). It is proposed that, contrary to the expectation of some epidemiologists, the probability of perpetual transmission of YFV in large modern urban centres is extremely small, because, unlike DENV, a subpopulation of YFV completely adapted to urban *Ae. aegypti* has not evolved yet. It is emphasized that this viral trait is distinct from vector competence measured by the magnitude of viral replication alone. In fact, confirmation of vertical transmission of YFV in urban *Ae. aegypti* under natural conditions has been rare, while in the sylvan vectors (such as *Haemagogus* spp.) it has been far more frequently recorded (Fontenille *et al.*, 1997; Mondet *et al.*, 2002).

CHIKV is transmitted in urban areas by *Ae. aegypti*, but unlike dengue and YF, it is far more difficult to identify its endemic foci, except that they are somewhere in the vast region of tropical Africa and Asia. As for its wide spread to the New World, as of 2014, it is still too early to confirm definitive establishment of endemicity in the region. Typically, an epidemic occurs unpredictably and suddenly but does not last more than 2 years in any small or medium-size urban centre and quickly disappears, the consecutive 3-year outbreak in Bangkok in 1962–64 being the longest (Halstead *et al.*, 1969). The constantly shifting foci of outbreaks, a unique transmission pattern of this virus, renders long-term CHIK surveillance in a fixed location difficult. As expected, confirming vertical transmission of CHIKV in this urban species of mosquito has been difficult (Vazeille *et al.*, 2009). Concurrent outbreaks of chikungunya and dengue in the same area in the tropics and frequent findings of mosquitoes coinfected by the two viruses have been a unique feature.

Persistent infection and pathology of viral infection

Pathology in persistent infection

The data in the previous sections of viral associations with arthropods, plants and vertebrates demonstrate a strong indication that the evolutionary process of a viral lineage entails repeated contacts of virus and its host over a long period which encompass an enormous number of generations of host before a stable virus–host relationship is established. To sustain such a long association with host, persistent infection with a minimal level of pathology to the host is essential, since severe pathology with a high mortality typically observed in acute viral infections renders establishment of persistent infection difficult.

Reviews of pathologies of arboviral infections in vectors revealed that cytopathic effects, mortality and/or reduced fecundity by certain viruses have been recognized (Tesh *et al.*, 1986; Faran *et al.*, 1987; Endris *et al.*, 1992; Bowers *et al.*, 2003). The mortality of *Ornithodoros* ticks infected by African swine fever virus is as high as 40% during first few gonotrophic cycles (Hess *et al.*, 1989). Still, overall, these pathologies are considered mild when assessed not at individual but at population level, since a combination of large population size of a vector species and high reproductive rate compensates the loss of vectors. On the other hand, the negative impacts of pathologies and loss of vertebrate hosts on virus transmission are far greater, given much smaller sizes of vertebrate populations and lower rates of reproduction (Kuno and Chang, 2005). For example, VSNJV transmission in an undisturbed island (an ideal location for the observation of natural transmission) ceased to exist after a decline of susceptible wild swine, deer, cattle, and equine populations (Killmaster *et al.*, 2011).

Generally, adaptation of a virus to a new vertebrate host is understood to proceed in three stages: (i) sporadic acute infection often with severe pathologies in a small proportion of the host for a short period without establishment of a stable virus–host relationship; (ii) occurrence of a greater magnitude of and more frequent outbreak of acute infection for a longer period; and (iii) establishment of a more stable relationship characterized by viral persistence and reduced levels of pathology in the host (Parrish *et al.*, 2008). All arboviral infections in vertebrates fall in the first two categories, because these viruses cause acute infections; and no example of persistent arboviral infection in vertebrates in nature comparable to the chronic/persistent infections of other animal RNA viruses has ever been scientifically proven (Kuno, 2001a). Because arboviral infections in vertebrates essentially represent an acute infection, the conclusion that the host-specific viral agents involved in persistent infection are the origins of acute infections typically demonstrated in new hosts (Villareal *et al.*, 2000) is useful for the identification of the progenitors or their close siblings of arbovirus groups (or lineages).

Virulence

According to the traditional concept, virulence of a pathogen would be reduced over a long period until a stable equilibrium is established between the pathogen and its hosts for a better mutual species survival; and the relationships characterized by severe pathologies (measured by mortality) to the hosts are manifestations of short histories of pathogen–host interactions (Smith, 1921; Smith, 1934; Chamberlain, 1982). The concept was based on observations of pathogen–host relationship over years, assuming little change in the composition of host and its geographic location. This concept is still used by many virologists to understand the evolution of more pathogenic animal viruses (Rethwilm and Bodem, 2013). However, this concept (termed 'dogma' by some critics) was challenged early by Ball (1943), who argued that high virulence actually represented longer relationships rather than more recent events. The renewed criticisms (Ewald, 1983; Johnson, 1986; Nuttall *et al.*,

1991; Read, 1994) became a subject of heated debate. Ewald (1983) even proposed that the diseases transmitted by vectors are more severe than those directly transmitted. Their critical comments were, in turn, challenged by others (Lipsitch *et al.*, 1996; Weiss, 2002; Ebert and Bull, 2003).

The sources of the absence of any consensus above are as follows. First, lack of consensus over the definition of the term 'virulence' has been one of the primary causes of controversy for many years (Casadevall and Pirofski, 2000). Second, although 'virulence' has been considered a property of pathogens at all times, in the very early period, the dominant yardstick for measuring arboviral virulence was the pathological impact on hosts, and transmissibility was not found acceptable then (Gordon Smith, 1960), because transmissibility inevitably entails constantly changing composition of host and its location. In modern times, the dominant yardstick used for arboviruses has been epidemic/epizootic transmission potential. Thus, the former measured virulence from the standpoint of host, while the latter measures it from the standpoint of virus. It should be noted that 'competent' vertebrates (characterized by a high level of viral replication by the current practice) are not necessarily effective in increasing transmission because of a higher mortality and shorter life (WHO, 1985). For example, because of 79% mortality of red grouse infected by Louping ill virus, the bird ('competent host') demonstrating a high and sustained level of viraemia was ruled out as an effective host for the maintenance of viral transmission (Reid, 1984). The impact of strong virulence on vertebrates is most clearly elucidated in the viruses with a very limited vertebrate host range, such as sylvan YF and DEN viruses with only subhuman primates as hosts. The pattern of YFV epizootic in sylvan environments is typically cyclical. The long intervals (on average > 15 years) between major epizootics characterized by absence or low incidence in the red howler monkey population in Trinidad (Downs, 1991) represent the periods necessary for recovery of the monkey population which suffers from a very high mortality. Similarly, it was reported that sylvan DENV-2 outbreak in Senegal occurred at intervals of 5–8 years (Diallo *et al.*, 2003); while epizootics of African horse sickness occurred in South Africa in cycles of about 20–25 years, until the 20th century. Conversely, 'less competent' hosts may be more effective in transmission, as shown among rodent species infected by TBEV (Labuda and Randolph, 1999). Interestingly, it was found experimentally that lower dose of virus would lead to a lower level of viraemia of longer duration than higher dose which would lead to a higher level of viraemia of shorter duration (Althouse and Hanley, 2015). Similarly, 'incompetent' vectors are not necessarily ineffective, as demonstrated in YF outbreak in Africa by an 'incompetent' *Ae. aegypti* (Miller *et al.*, 1989).

Third, it was also proposed that vertical transmission in vectors leads to more benign virulence compared with horizontal transmission (Lambrechts and Scott, 2009). Even this view generally accepted among many arbovirologists, however, had been criticized by some that horizontal transmission was correlated better with the evolution of more reduced virulence (Lipsitch *et al.*, 1996). Fourth, given the fact that virulence is a sum total of multifactorial expression, many proposals of the modern critics based on experimental data or mathematical model have been, by necessity, designed with assumptions and unnatural modes of transmission or methods employed (such as intrathoracic inoculation of mosquitoes or use of artificially high viral concentration in bloodmeal), rendering interpretation of the conclusions difficult. However statistically significant correlational analyses between cause and outcome (i.e. change in 'virulence') may be, they do not always establish a direct cause–effect relationship due to difficulties in integrating the impacts of miscellaneous confounding factors. Some of these factors include involvement of multiple hosts, changing herd immunity or size of susceptible population, and dispersal of pathogen (either naturally or artificially). As an example, a renewed outbreak of severe WNV infection in North America could not be explained by genetic change in virus alone, most likely other unidentified factor(s) playing an equally important role as well (Duggal *et al.*, 2013). Also, in the past, compared with Lineage 1, Lineage 2 of WNV had been considered benign, based on limited epidemic/epizootic activities and dispersal. The recent activities of the Lineage 2 in Africa and Europe seriously questioned this impression (Venter *et al.*, 2009; Bakonyi *et al.*, 2013). When wildlife are the hosts, too often accurate baseline data of their spatial/temporal distribution and population size were unavailable for the measurement of virulence change. Differences in vectorial capacity in the broader context of environmental conditions among reports for a particular virus–host relation constitute another source of discrepant conclusions (Christofferson *et al.*, 2014).

Epidemic and epizootic records are typically depicted in patterns of repeated cycles of rise and fall over limited periods. When epidemic/epizootic potential is the definition of virulence, sometimes, the intervals between the cycles are very long. While extinction of viral lineage is known phylogenetically for some flaviviruses, far less attention has been paid to the causes of the long outbreak inactivity. It is unclear if significant decline of symptomatic cases can be directly correlated with increased herd immunity or decline (or extinction) of virulent viral subpopulation (genotype). Studies to unravel the mechanisms contributing to sharp decline in transmission (such as Western equine encephalitis in North America) or sudden resurgence of O'nyong nyong fever in Africa after a long inactivity (Forrester *et al.*, 2008; Bergren *et al.*, 2014; Lanciotti *et al.*, 1998) deserve attention. If the hypothesis of CHIK outbreaks in Asia and the Americas in the 19th century (Carey, 1971) was correct, it strongly suggests extinction of old lineages of CHIKV there. As for sporadic or rare isolations of arboviruses or their genotypes (such as California encephalitis virus, Cacipacore virus, and Japanese encephalitis virus) (Eldridge *et al.*, 2001; Batista *et al.*, 2011; Kim *et al.*, 2015), it is uncertain if they suggest a sign towards viral extinction, insufficient field research or normal state of endemicity; but it is at least compatible with the concept that vectors are the reservoirs of arboviruses (Philip and Burgdorfer, 1961).

According to the early concept, 'reduced virulence' meant increase in the proportion of innately resistant subpopulation in the hosts after years of indigenous transmission rather than increased herd immunity or reduced pathogenicity of pathogen. Increase of an innately resistant subpopulation of host over years that has been revealed in Myxoma virus infection of European rabbits introduced to Australia (Best and Kerr, 2000) was one of the classic examples. This rationale was the reason for the importation of Indian monkeys to West Africa for YFV isolation by the Rockefeller Foundation researchers (Barrie, 1997; Kuno, 2014). The innate resistance to certain viral infections has been known to be manifested in lower levels of immunological and cytopathological responses. In a study of antibody (IgY) response to Buggy Creek virus (a cimicid-borne alphavirus) infection between the indigenous cliff swallow and exogenous house sparrow, the antibody level of the former was much less than that of more susceptible latter birds (Fassbinder-Orth *et al.*, 2013). With the consideration of host resistance, virulence status at any given time may be better understood to be the overall balance between the two opposing forces, rise of resistance in vertebrate host and emergence of a new viral variant with a higher virulence (Duggal *et al.*, 2014).

When one follows the early interpretation viewed from the standpoint of host rather than of pathogen, the available arbovirus data address more clearly the difference in disease severity between closely related indigenous and exogenous vertebrates for a virus of interest. If a particular virus and its vertebrate hosts have both existed for a long time in the same geographic region, it is termed an indigenous-indigenous relation. On the other hand, at a given location, if either partner of this pair is exogenous, it is an indigenous-exogenous relation. High levels of virulence have been observed in a large number of indigenous–exogenous relations compiled (Kuno and Chang, 2005). Thus, 'virulence' of African horse sickness virus in Africa is more severe in imported (exogenous) horses than in indigenous zebras. Bluetongue virus (BTV), which has apparently existed since antiquity in Africa, rarely produces demonstrable signs of illness in indigenous sheep breeds, cattle, and wild ruminants; but in European breeds of sheep and cattle imported (exogenous hosts), the pathologies of the viral infection are severe. In Omsk haemorrhagic fever virus infection in rodents, the pathologies in imported North American species of muskrat are severe, while those in indigenous voles in Russia are mild.

Similarly, YFV being exogenous in the New World, monkeys there suffer a far higher mortality in infection by the virus, compared with the monkeys in Africa where the virus–monkey relationship is indigenous-indigenous. In North America, a lineage of Buggy Creek virus is more cytopathic in a newly arrived (imported) house sparrow than in indigenous cliff swallow (Brown *et al.*, 2010b). In a more recent example, the mortality of crows in North America by WNV (exogenous virus) was initially devastatingly high but began to decline nearly a decade later (Reed *et al.*, 2009), as expected.

Phylogeny and empirical data in support of phylogenetic tree topology

Basics of phylogeny

The qualities of phylogenetic data illustrating genetic relations among a group of related viruses first depend on a number of parameters and subjective decisions based on available data, personal assumption or belief. At a minimum, they include rooting or not rooting a tree, what outgroup to select (if rooting), identification of informative site, exclusion of a problematic genomic segment(s) or taxa not considered orthologous, synonymous/nonsynonymous substitution and level of saturation, statistic method of choice, preferred phylogenetic program, and mode of presenting the pattern of evolution (such as cladogenic, reticulate or other modes) most appropriate for a particular group of viruses.

The qualities also depend on sampling of viruses. Incomplete sampling of viruses, in particular missing links, is a serious problem (Wiens, 2003). For studying the evolutionary history by molecular phylogeny, only appropriate coding sequences are useful for alignment. This partly explains the paucity of fully sequenced, untranslated regions (UTR) for too many arboviruses or strains in the depositories of sequence data. However, UTRs of some viruses do have significant markers of evolution, as described later. Here lies the necessity of stressing the importance of UTR because of the markers of evolutionary history found there, even though they cannot be easily aligned.

As sampling has improved because of discoveries of missing links and increase in the number of available viruses with more complete genome sequences, it has become more evident that vector-borne viruses cluster more clearly according to vector group. Thus, at the generic level, such a clear segregation is found in the genus *Flavivirus* of the *Flaviviridae*, the genus *Orbivirus* of the *Reoviridae* (Gaunt *et al.*, 2001; Belaganahalli *et al.*, 2011; Mohd Jaafar *et al.*, 2014), and *Tibrogargan* group [genus *Tibrovirus* (proposed)] of the *Rhabdoviridae* (Gubala *et al.*, 2011). Within the mosquito-borne subgroup of flaviviruses, lineages are segregated by the mosquito group associated (i.e. *Aedes* or *Culex*) (Gaunt *et al.*, 2001). In the *Uukuniemi* group of phleboviruses, clades are segregated into insect-transmitted and tick-transmitted viruses (Palacios *et al.*, 2013b); similarly, within the genus *Nairovirus* of the *Bunyaviridae*, viruses are even segregated into distinct clusters depending on the subgroup of tick vectors (soft ticks versus hard ticks) (Honig *et al.*, 2004). At the level of virus family, viruses are clustered to a particular genus according to host group (Labuda, 1991; Bourhy, 2005; Attoui *et al.*, 2005). When the patterns of clustering in all vector-borne virus families are examined, it is clear that vector-borne viruses cluster according to vector group but not to vertebrate host group, which strongly suggests that vertebrate host range is determined secondarily by the vectors.

Directly correlating arboviruses and vertebrate hosts, which essentially ignores the traits of vectors and of ecological determinants of host range, is difficult because different groups of vertebrates (i.e. birds, rodents, ruminants, primates) serve as hosts of a particular virus by a vector (or a group of closely related species of vectors, such as culicine mosquitoes) depending on vector species, promiscuity of vector, location, and ecosystem. Bridge vectors of arboviruses facilitating infection of distinct groups of vertebrates in feral and peridomestic environments are such good examples. Accordingly, all analyses or techniques to directly correlate the properties of arboviruses with vertebrate hosts by bypassing the role of vectors, such as codon usage bias, phylogeny (Kitchen *et al.*, 2011) and principal component analysis (Aguas and Ferguson, 2013) are problematic. As revealed in recent studies, GC contents of vector-borne orbiviruses correspond to a specific vector group (midge, mosquito, tick); and the pattern of codon usage of VSV genotypes correspond to vectors rather than to the livestock hosts (Mohd Jaafar *et al.*, 2014; Liang *et al.*, 2014). In over 400 plant viruses transmitted by arthropod vectors, too, viral traits were found to relate specifically to vector groups rather than to plant groups (Power, 2000).

Evaluation of phylogenetic tree topology and empirical data, with a focus on the genus *Flavivirus*

All molecular phylogenetic relations proposed for viruses are deduced hypotheses based on a number of assumptions (Kelchner and Thomas, 2006). It is emphasized that molecular phylogeny, when used alone, is devoid of biological characteristics or other genomic traits which are integral components of the definition of virus species by the ICTV. Thus, questionable assumptions have negative impacts on the accuracy of molecular phylogeny. Until recently, in contrast to palaeontologists who can independently assess the qualities of molecular phylogeny by using functional anatomy, dating of fossil records and geological events, most virologists have suffered from the absence of analogous markers of evolution (Hipsley and Müller, 2014). Sometimes, Ice Age (12,000 years ago), historical records of human or animal movement and geographic distribution, emergence of human population centres sufficiently large to support viral transmission, rise of agriculture, and passive transportation of viruses by migratory animals have been used to offset the weakness. Viral 'fossil' sequences integrated in the genomes of eukaryotic hosts, however, may prove to be a useful resource for arboviral research in the near future, if more such fossil sequences are discovered. In fact, dating to Miocene Period of the extant Ebola virus based on an orthology established with the fossil sequences has been reported already (Taylor *et al.*, 2014).

Today, given improved sampling and availability of vastly enriched viral traits, for a proponent of any molecular phylogeny, it is highly desirable to present a set of additional data to corroborate it. In the following discussion, all additional or supplemental data in support of proposed phylogenies are called 'empirical data' (Parker *et al.*, 2008). As far as flaviviruses of genus *Flavivirus* are concerned, the history of host range shift is useful as an empirical marker, because the viruses in this genus sharing the identical genome organization are clearly segregated into subgroups according to host range, as mentioned above. Furthermore, complete genome sequence data are available from many members of this genus. Accordingly, this group is selected for evaluating the qualities of proposed phylogenies.

NS3/ORF tree

In both NS3 gene and open reading frame (ORF) trees proposed, the basic tree topologies (Fig. 14.1A) are identical (Cook and Holmes, 2005). For simplicity, in the following passages the tree is called ORF tree. Tamana bat virus (TABV) sequence is customarily excluded, because of the considerable genetic distance from all other flaviviruses. The proposed history of host range shift based on this tree topology (Gritsun and Gould, 2007a) is schematically shown in Fig. 14.1B. From the unidentified origin, two major branches bifurcated, one branch leading to the evolution of a group of insect-specific subgroup viruses and the other branch leading to the evolution of the first intermediate, composite group. Subsequently, from the first intermediate, composite group diverged the mosquito-borne subgroup and the second intermediate group. Finally, from the second intermediate group diverged no-known-vector (NKV) subgroup and tick-borne subgroup, as shown in Fig. 14.1B.

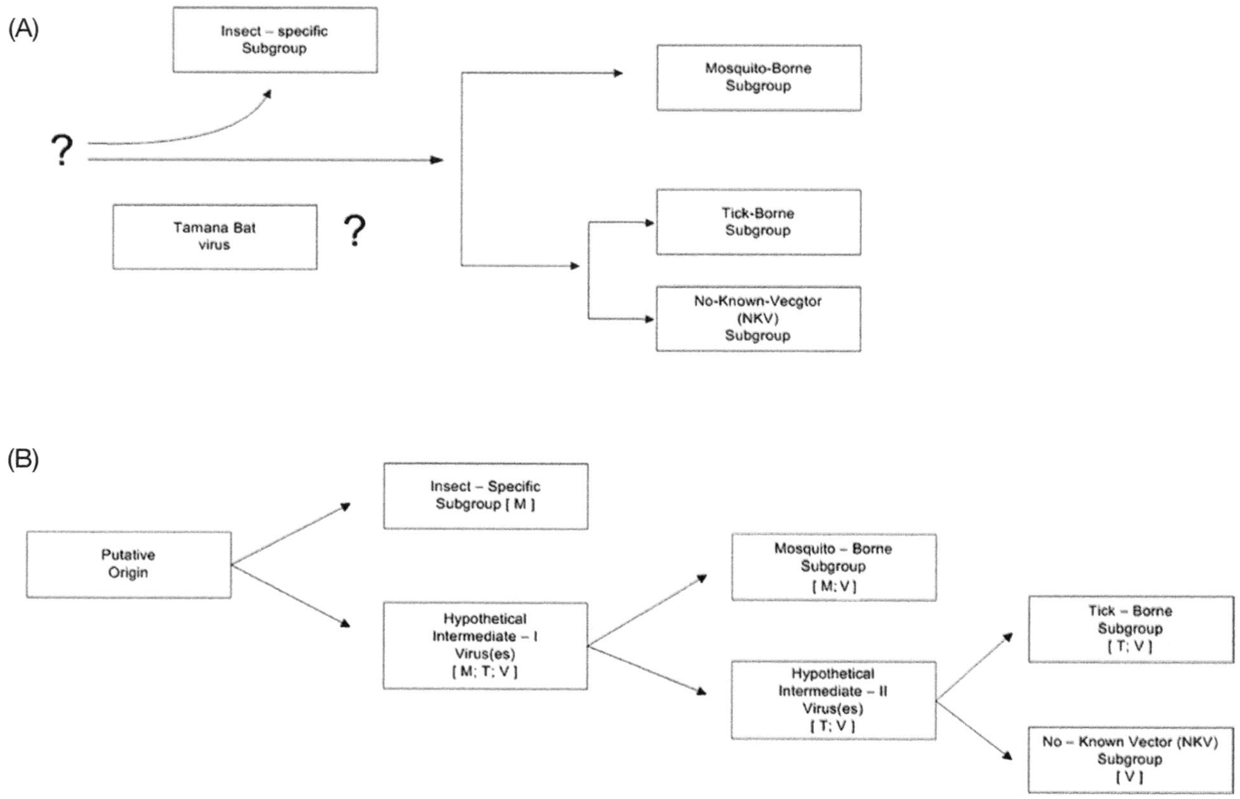

Figure 14.1 Basic topology of flaviviruses of the genus *Flavivirus* and deduced host range shift according to the ORF/NS3 tree. A: Schematic, basic topology (Source: Cook and Holmes, 2005). The phylogenetic position of *Tamana bat virus* (TABV) was not included in the phylogenetic studies due to its considerable genetic distance from all other flaviviruses. B: The history of host range shift (Source: Gritsun and Gould, 2007a). The letters in brackets refer to host groups (M, mosquitoes; T, ticks; V, vertebrates). Viruses of each subgroup: See Fig. 14.3 legend for the member viruses.

The theoretical advantages for the ORF tree are reduced probability of sampling errors, that full length ORF encompasses all genes, thus representing the entire sum of virus traits, and that thus far no evidence of interspecific recombination among flaviviruses has been found. According to the proposal (Gritsun and Gould, 2007a), it was hypothesized that the putative viruses in the first intermediate stage are presumed to have replicated in three groups of hosts: ticks, mosquitoes, and vertebrates; while the putative viruses in the second intermediate stage are the precursors of extant tick-borne and NKV subgroups. As for support to the putative viruses in the first intermediate stage, sharing of a stem loop located immediately upstream of the terminal long stable hairpin in the 3′ UTRs among mosquito-borne, tick-borne, and NKV subgroups was proposed (Gritsun and Gould, 2007b).

However, no extant first intermediate stage virus that possesses this unique host range has been found, while the viruses, which had emerged earlier in this proposal (insect-specific subgroup), have been abundantly isolated. The difficulty of finding such a putative first intermediate virus is not surprising given the fact that the host range specificity among the four subgroups is so strict that no flavivirus (or for that matter arbovirus) with efficient replication and biological transmission in nature by two classes (*Arachnida* and *Insecta*) of vectors has ever been isolated (Kuno, 2007). Regarding the precursor viruses in the second intermediate stage, from which tick-borne and NKV subgroups evolved, lack of empirical data for the sharing of traits between the two subgroups is problematic. The only shared trait used in support of this tree (Cook and Holmes, 2005), antigenic cross reaction between tick-borne and NKV subgroups, is invalid, because the original reference cited actually revealed no cross reaction at all (Calisher *et al.*, 1989; Calisher and Gould, 2003). Furthermore, one of the conserved sequences (CS2) in the 3′ UTR, which is found in the NKV subgroup, is not shared by the tick-borne subgroup; instead, it is shared by all members of the mosquito-borne subgroup (Table 14.2).

The utility of full length ORF (or genome) for phylogenetic studies of viruses depends on the group of viruses used. Because full length ORF of flaviviruses is a concatenated chain of genes, possible distortion of tree topology by the inclusion of highly variable genomic segment(s) needs to be considered. It was pointed out that full-length genomes or very long sequences with more genes would not necessarily improve phylogenetic accuracy, since most phylogenomic techniques poorly accommodate incongruence (Phillips *et al.*, 2004; Leigh *et al.*, 2011). Furthermore, unlike our understanding only 2 decades ago, the

Table 14.2 Presence of additional, possible evolutionary markers in the 3′ UTR of the flaviviral subgroups according to the NS5 tree. The subgroups (except for the two vector-borne subgroups) are arranged from top to bottom in branching order based on the NS5 tree. The vertical arrangement of the two vector-borne subgroups is arbitrary because the two subgroups diverged from the NKV subgroup, according to this tree. This analysis is based on the sequences available in GenBank as of 2014

					Motif	
Subgroup	Virus	Repeat	CS2	CS1	Hexa-nucleotide	Penta-nucleotide
Insect-specific	AEFV (Narita-21, SPFLD-MQ-2011-MP6)	RS	*			
	CXFV (NIID-21–2, Tokyo, RP-2011, Mex07, Iowa07)	RS				Yes
	KRV	RS			Yes	Yes
	CFAV	RS			Aatggc	Yes
	QBV	RS			Cttggc	Yes
	Mosquito Flavivirus (LSFlaviV-A20–09)	RS			Cttggc	Yes
NKV	MODV	DR	Yes		Yes	Yes
	MMLV	DR	Yes		Yes	cTcag
	APOIV	DR	RCS2 and CS2		Yes	cCcag
	RBV	DR	Yes		Yes	cCcag
Tick-borne	AHFV/DTV/KFDV/KSIV/LGTV/LIV/OHFV/POWV/TBEV	**			Yes	Yes***
Mosquito-borne	Mosquito-specific lineage – see Fig. 14.3 legend					
	CHAOV/DGV/LAMV	See Fig. 14.4	Yes	Yes+		Yes
	NHUV		Yes	Yes+		Yes
	Barkedji virus/NOUV	See Fig. 14.4	Yes	?	?	?
	Mosquito-borne lineages					
	YFV and DENV-JEV Lineages (23 viruses in Fig. 14.3)	See Fig. 14.4	Yes	Yes		Yes****

*Blank indicates none.
**Multiple repeat sequences found uniquely in the tick-borne subgroup (which are not RS, DR, CS3, CS2, or CS1; Wallner, *et al.*, 1995) are not shown.
***Present in all tick-borne viruses except that it is absent in AHFV and is cccag in KSIV.
****This pentanucleotide is located posterior to CS1 in all mosquito-borne viruses except for YOKV in which it is located upstream of CS2. The location in ENTV is unknown because its sequence downstream of CS1 is incomplete.
+ An imperfect sequence is found.
? The sequences downstream of CS2 in *Barkedji virus* and NOUV deposited in GenBank are incomplete.
Motifs: CS1: aScatattgacRccWgggaWaagac (source: Hahn *et al.*, 1987); CS2: ggWctagaggttagWggagaccc (source: Hahn *et al.*, 1987); Hexanucleotide: attggc (source: Kuno, unpublished); Pentanucleotide: cacag (source: Wengler and Castle, 1986); DR: ttgtaaata of the NKV subgroup (source: Leyssen *et al.*, 2002); RS: tgac__a__cgctcc__ccc_agtcccc of the insect-specific subgroup (source: Hoshino *et al.*, 2009).

modes of lateral gene transfer without involvement of either recombination or reassortment are now recognized to be many and varied.

NS5 tree

The basic topology of the NS5 tree (Kuno, *et al.*, 1998; 2009) is schematically illustrated in Fig. 14.2A. NS5 gene was selected because RdRp domain of this gene is the most conserved in the genomes of flaviviruses as well as in many RNA virus families. In this tree, if TABV is included, the virus would be located at the root of the genus *Flavivirus* tree (Kuno *et al.*, 2009; Qin *et al.*, 2014).

The history of host range shift shown in Fig. 14.2B is a slight modification of the earlier NS5 tree. According to this scheme, from the unknown origin emerged insect-specific subgroup (which includes AEFV, CFAV, KRV, CXFV, and others – Fig. 14.3 legend). Subsequently, the host range shifted to vertebrates (NKV subgroup). It is emphasized that the NKV viruses are not mere vertebrate viruses but have derived from insect-specific (or arthropod-specific) flaviviruses, as shown in the pattern of sharing of evolutionary markers along the branching order (Table 14.2). This agrees with a speculation that the NKV subgroup viruses 'may have had an arthropod stage earlier in their evolutionary history' (Porterfield, 1980). It is proposed that both mosquito-borne and tick-borne subgroups emerged from respective progenitors which had derived from the NKV subgroup. The progenitors (or their close relatives) for the former subgroup are proposed to be such mosquito-specific lineage viruses as Barkedji virus, Chaoyang virus (CHAOV), Nounané virus (NOUV), Nhumirim virus (NHUV) and others (Fig. 14.3 legend), which replicate only in mosquitoes despite their phylogenetic clustering within the mosquito-borne subgroup. The imperfect CS1 found in CHAOV, Donggang virus (DGV), Lammi virus (LAMV), and NHUV (Table 14.2; Fig. 14.4) may be considered a primordial stage before transition to CS1 in mosquito-borne subgroup. Thus far, one possible extant progenitor virus for the tick-borne subgroup based on phylogenetic position is Ngoye virus isolated from ticks, which does not replicate in either vertebrates or mosquitoes (Grard *et al.*, 2006). The viruses in both insect-specific subgroup and mosquito-specific lineage (within the mosquito-borne subgroup) replicate only in mosquitoes. However, the former are distinguished from the latter by much shorter ORF lengths or absence of CS1 and CS2 (Fig. 14.3; Table 14.2).

Existence of similar putative progenitors (or their siblings) preceding the emergence of multiple lineages of vector group-specific arboviruses within a viral group is also strongly suspected in the genus *Orbivirus* of the family *Reoviridae* and two genera (*Orthobunyavirus* and *Phlebovirus*) of the family *Bunyaviridae* because each of these genera comprises both tick-borne and insect-borne members. Even among the arthropod-associated members of the family *Rhabdoviridae* dominated by insect-borne viruses, a small number of viruses (such as Connecticut virus and Sawgrass virus) are tick-borne; and some members of these insect-borne rhabdoviruses demonstrate limited replication in tick cell cultures (Pudney *et al.*, 1982).

When the lengths (measured in the number of amino acids) of ORF are examined in relation to the tree topology based on NS5, a pattern of progressive increase along the branching order in the phylogram is revealed. As shown in Fig. 14.3, overlap in ORF length range among four subgroups, is either

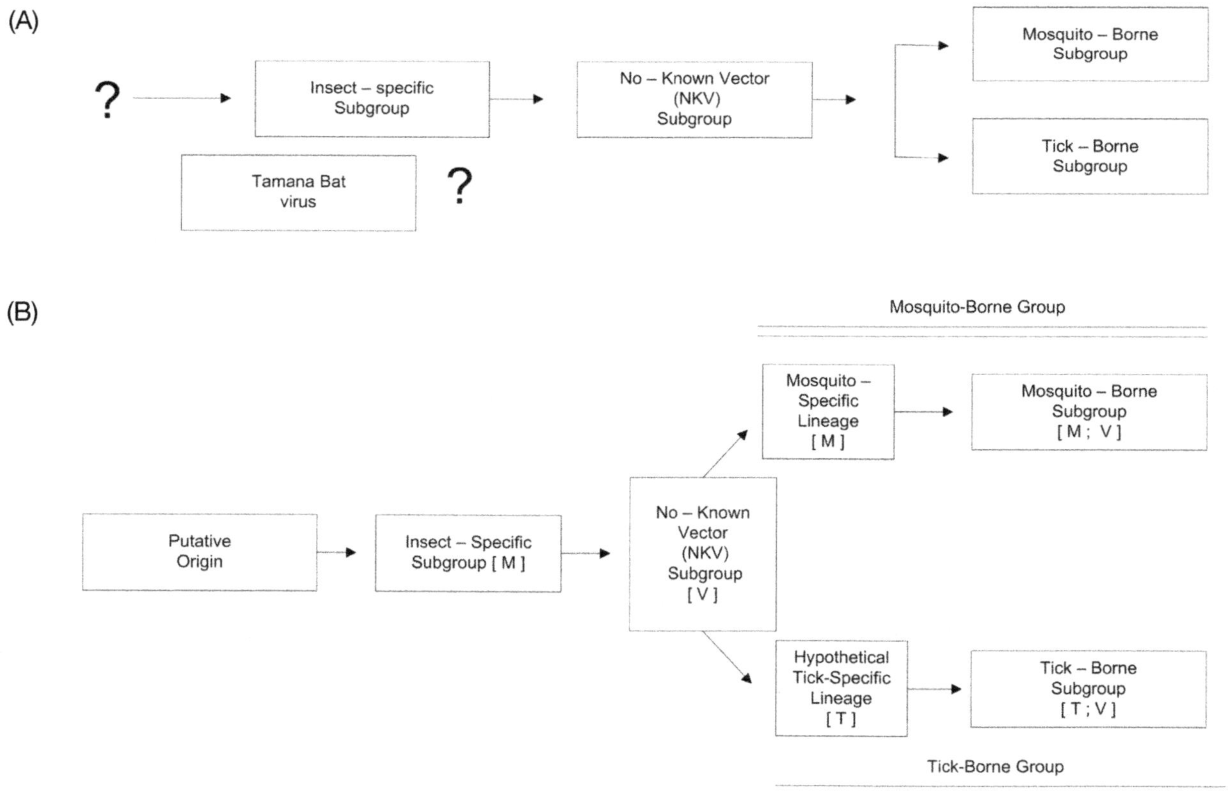

Figure 14.2 Basic topology of flaviviruses of the genus *Flavivirus* and deduced history of host range shift according to the NS5 tree. A: Schematic, basic topology (Source: Kuno *et al.*, 1998). B: The history of host range shift (Source: Kuno *et al.*, 2009). The letters in brackets refer to host groups (M, mosquitoes; T, ticks; V, vertebrates).

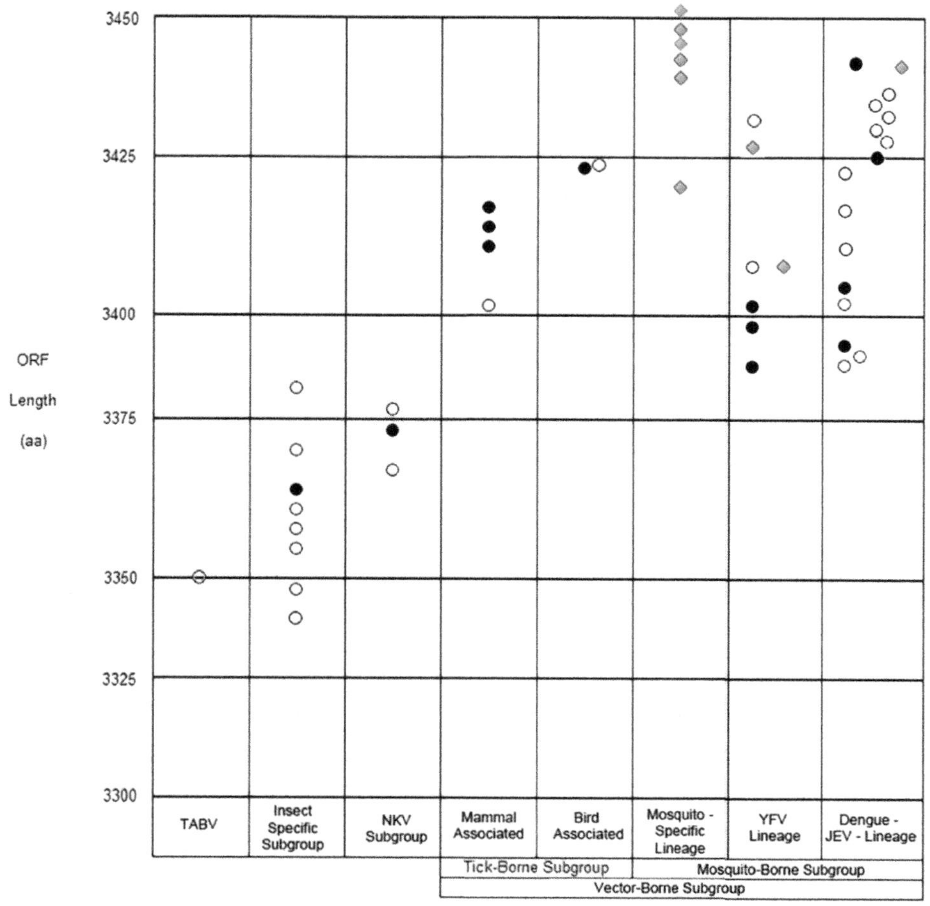

Figure 14.3 Relationship between the ORF Length and Branching Order among the Members of the Genus *Flavivirus*. Ordinate: Length of ORF (number of amino acids). The scales in 3560–3375 and 3400–3425 aa are artificially made longer to accommodate many viruses falling in the respective range; Abscissa: The branching order of *Tamana bat virus* (TABV), insect-specific subgroup, NKV subgroup, and vector-borne subgroups is arranged from left to right according to the NS5 tree (Sources: Vázquez *et al.*, 2012; Lee *et al.*, 2013). The order of tick-borne and mosquito-borne branches on the horizontal scale is arbitrary. For some viruses in which ORF length difference has been found among strains, only the length of the earliest report is used. Symbols: Open circle, ORF length of one virus; Solid circle, ORF length shared by two or more viruses; Diamond, viruses in mosquito-borne branch which have host range in only one of the two phyla. Viruses (ORF length in number of amino acids) TABV (*Tamana bat virus*) (3350). Insect-specific subgroup: AEFV (3341); CFAV (3341); KRV (3357); CXFV (3362); NAKV (3373); QBV (3359); CXFV-Uganda (3364); Hanko Virus (3385); Palm Creek Virus (3364). NKV subgroup: APOIV (3371); MMLV (3374); MODV (3374); RBV (3379). Tick-borne subgroup: mammal-associated lineage: KADV (3405); GGYV (3416); RFV (3416); POWV (3415); DTV (3415); KFDV (3416); KSIV (3416); AHFV (3416); RFV (3416); LGTV (3414); OHFV (3414); TBEV-FE (3414); TBEV-Vas (3414); LIV (3414). Tick-borne subgroup: seabird-associated lineage: TYUV (3422); MEAV (3423); SREV (3422). Mosquito-borne subgroup: mosquito-specific lineage: Barkedji Virus (3420); CHAOV (3435); DGV (3444); LAMV (3434); NOUV (3442); ILOV (3451); NHUV (3446). Mosquito-borne subgroup: YFV lineage: SEPV (3405); WSLV (3405); YFV (3411); POTV (3390); EHV (3401); BOUV (3390); UGSV (3393); BANV (3393); JUGV (3390); SABV (3390); SPOV (3429). ENTV (3411) and YOKV (3425) are both vertebrate viruses without a known vector but replicate in mosquito cell cultures. Mosquito-borne subgroup: dengue-JEV lineage: DENV-1 (3391); DENV-2 (3391); DENV-3 (3390); DENV-4 (3387); KEDV (3408); ZIKV (3419); TMUV (3410); IGUV (3416); NMV (3409); KOKV (3410); BSQV (3429); Duck Egg-Drop Syndrome Virus (DESV) (3425); NTAV(3427); ROCV (3425); ILHV (3424); BAGV (3426); SLEV (3430); ALFV (3434); USUV (3434); JEV (3432); MVEV (3434); WNV (3433). *Rabensburg virus* (3433), a variant of WNV, replicates in mosquitoes, but its vertebrate host range is temperature-sensitive.

minimal or none. In particular, no overlap in length range is observed between the NKV and the two vector-borne subgroups, strongly suggesting abrupt genetic changes during this host range transition. Among the members of the tick-borne subgroup, ORF length range is very narrow (only 11 amino acids difference between the shortest and the longest), compared with 47 amino acids difference in the mosquito-borne subgroup. This conservation in the tick-borne viruses reflects the mean substitution rate of this subgroup being only less than one half of the rate for the mosquito-borne subgroup. Still, even within the highly conserved tick-borne subgroup it is of interest to note that the lengths of the mammal-associated viruses are shorter than the lengths of sea bird-associated viruses.

In the mosquito-borne subgroup, it is noted in Fig. 14.3 that the ORF lengths of the seven viruses in the mosquito-specific lineage (proposed as progenitors or their siblings) and two viruses in the YFV lineage [*Entebbe bat virus* (ENTV) and *Yokose virus* (YOKV)] are at or near the longest in each lineage. One notable distinction between the two lineages is that while the former replicate only in mosquitoes, the latter bat viruses replicate in vertebrate and mosquito cells *in vitro*, despite the absence of known vectors (Varelas-Wesley and Calisher, 1982; Kuno, 2007). It is also noted that *Rabensburg virus* (RabV) in the JEV complex (which has the length close to the highest in the DENV-JEV lineage; Fig. 14.3) was reported to be restricted for replication in vertebrates (Aliota *et al.*, 2012). Thus, the

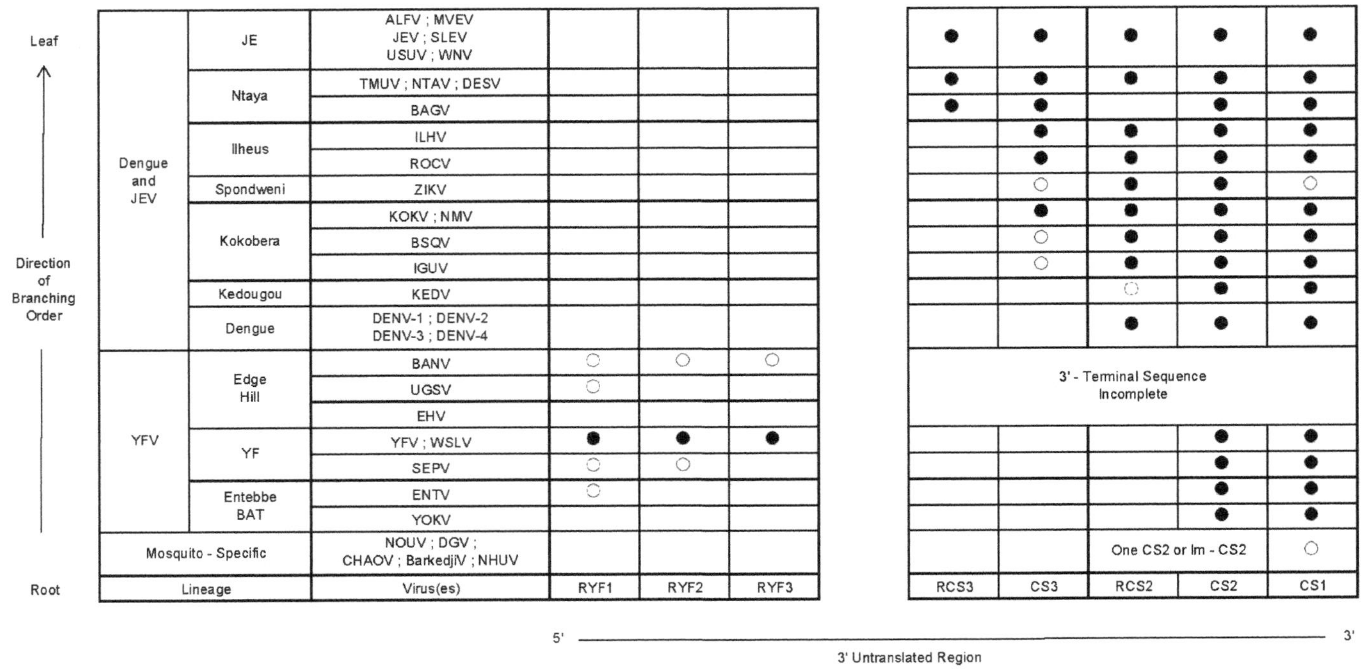

Figure 14.4 Correlation between the branching order among lineages of the mosquito-borne subgroup of the genus *Flavivirus* and conserved sequence organization in the 3′ untranslated region. Repeat or conserved sequences are arranged on the horizontal scale in the 5′→3′ direction. Branching order is plotted on the vertical scale from bottom to top, based on a NS5 gene tree topology (Sources: Vázquez *et al*., 2012; Lee *et al*., 2013). Abbreviations: RYF, Repeat sequence unique to the YF lineage; CS, Conserved sequence; RCS, Repeat of conserved sequence; Im, Imperfect. Symbols: Empty circle, Imperfect sequence; Solid circle, Perfect sequence. For YFV, the organization of the strains from Africa is used, although some of the New World strains have more than one CS2 or lack RYF (Bryant *et al*., 2005). In the mosquito-specific lineage, the sequences downstream of CS2 for NOUV and *Barkedji virus* in GenBank are unavailable, as of 2014.

shared trait of all viruses in the mosquito-borne subgroup with a host range in one phylum is long ORF length.

Thus, it is of interest to further investigate if the ORF lengths of the progenitor viruses in the mosquito-borne subgroup were too long in the first place for them to directly become arboviruses; and the lengths of ENTV, YOKV and RabV exceeded the lineage-specific functional limit genetically preset, resulting in secondary loss of host range in one phylum due to limited capacity. In this regard, it has been recognized that limited number of genes is a feature common to RNA viruses because of short lengths of genomes imposed for maintaining viral species stability, given much higher rates of mutation than in DNA viruses (Holmes, 2003; Moya *et al*., 2004).

Also, an intriguing relationship between ORF length and vertebrate host group is observed in the mosquito-borne subgroup. Like in the tick-borne subgroup, the viruses with the longest mean ORF length in the mosquito-borne subgroup (JEV complex) have birds as primary hosts; and, like in tick-borne subgroup, mammal-associated viruses (in particular, primate-associated YFV lineage and DENV) have ORF length range near the shortest. Even within the strains of one virus, such as SLEV, a strain identified phylogenetically to be most ancestral replicates only moderately in certain bats but does not in other mammals and birds which are typically infected by most SLEV strains assumed to have evolved later (Kopp *et al*., 2013).

When flaviviral ORF is broken down to individual genes and the length relationship analysed gene by gene, progressive increase in length along the branching order was again observed for prM, E, NS3, and NS5; while, progressive reduction was observed for NS1 and NS4B. For four other genes, no consistent pattern was observed (Kuno *et al*., 2009).

A survey of the 3′UTR of the mosquito-borne subgroup reveals an interesting trend of progressive increase in complexity of the organization of conserved sequences (CSs) in the direction of branching order of viral lineages, as shown in Fig. 14.4. This pattern is just the opposite of the evolutionary hypothesis based on progressive loss of CSs from the putative panflavivirus progenitor (Gritsun and Gould, 2006).

Besides CSs, additional, possible evolutionary markers in the 3′ UTR in combination further support the branching order depicted in the NS5 tree. As shown in Table 14.2, from the insect-specific to vector-borne, all subgroups share the canonical pentanucleotide (cacag). Although it is found in only one of 4 viruses of the NKV subgroup, almost identical sequences are found in other viruses in that subgroup. The insect-specific and NKV subgroups are also linked by the shared hexanucleotide (attggc). The fact that this motif is not found in all members of the insect-specific subgroup most likely reflects geographic separation for long periods and the resultant genetic diversity among the members of this subgroup because of frequent duplication or deletion observed in the 3′UTRs among flaviviruses. Regarding the emergence of the two vector-borne subgroups from the NKV subgroup, the following three sets of data support the branching order of the NS5 tree: (i) the structural elements in the 5′ and 3′ UTRs of Modoc virus (an NKV virus) are shared with both vector-borne flavivirus subgroups (Leyssen, *et al*., 2002); (ii) the two vector-borne subgroups are linked with the NKV subgroup by the sharing of the canonical pentanucleotide in the 3′ UTR (Table 14.2) and the low level, transient replication of tick-borne viruses in mosquito cells and of mosquito-borne viruses in tick cells, which is possibly a vestigial evidence of the past sharing of host range by the two vector-borne subgroups (Kuno, 2007); and (iii) CS2 and the aforementioned hexanucleotide (attggc) in the 3′ UTR of the NKV subgroup are shared by the mosquito-borne subgroup in one direction and by tick-borne subgroup in other direction, respectively (Table 14.2).

As for the significance of the insect-specific subgroup flaviviruses isolated from nearly all continents (Fig. 14.3 legend), they may be the 'seed viruses' predicted by Albert Sabin as the primordial lineages of flaviviruses which had disseminated to multiple continents where different lineages of extant vector-borne flaviviruses subsequently evolved (Sabin, 1959). In this chapter, however, these insect-specific subgroup viruses are interpreted to be their descendants but not the direct progenitors, which are instead proposed to be the siblings of the mosquito-specific and tick-specific lineages within the mosquito-borne and tick-borne subgroups, respectively, according to the proposed scheme presented in Fig. 14.2B.

The significance of the motifs found in the 3′ UTRs of the members of genus *Flavivirus* is also studied in the context of evolutionary relationship of this genus with three other genera of the family *Flaviviridae*. As shown in Table 14.3 (Kuno, unpublished), the motifs in the NKV subgroup, insect-specific subgroup, and TABV are shared by some members of other genera, revealing another set of footprints of this relationship. In particular, the sharing of the motifs of the genus *Flavivirus* (including TABV, insect-specific subgroup and NKV subgroup) in pegiviruses and hepaciviruses (Table 14.3) is more extensive than in pestiviruses. Thus, a close evolutionary relationship between TABV with these two genera is revealed for the first time. Furthermore, sharing of the flaviviral pentanucleotide by the unaffiliated viruses (one isolated from a nematode and another from a plant) provides additional data to support their classification as members of the family *Flaviviridae*. Although the nematode flavivirus was found phylogenetically affiliated with pestiviruses (Bekal *et al.*, 2014), its sharing of the canine hepacivirus-specific sequence (gaggccgcatgactgcagaga) suggests a relation with hepaciviruses as well (Kuno, unpublished).

Table 14.3 Shared motifs in the 3′UTRs and coding sequence lengths among four genera and unclassified members of the family *Flaviviridae*. TABV of genus *Flavivirus*, three genera (*Pestivirus*, *Pegivirus*, and *Hepacivirus*), and unclassified members are listed in a random order without a consideration of evolutionary direction. It should be noted that, because of the difference in genome organization among genera, the relative positions of these motifs are not necessarily identical among viruses.

Genus	Virus (strain)	Penta nucleotide	Hexa nucleotide	DR (NKV)	RS (insect)	TABV repeat	Coding length (nt)
Flavivirus	TABV					+	10,053
Pestivirus	*Classical swine fever virus* (CSFV) (Eystrup)	+		+*			11,697
	CSFV (Riems)	+		+*			11,697
	Bovine viral diarrhoea virus (BVDV)-1	+		+			11,967
	BVDV-2 (SH-28)			+*			11,688
	Border Disease virus (BDV) (BD 31)	+		+*			11,691
	Giraffe-1 pestivirus	+*		+*			11,970
	Porcine pestivirus (Bungowannah)	+	+*				11,757
	Pronghorn antelope Pestivirus	+*	+*	+			11,694
Pegivirus	*Rodent pegivirus* (CC-61)	+	+		+		10,455
	Equine pegivirus (C0035)		+			+	10,035
	Bat pegivirus (PDB-1698)	+				+*	9978
	Human pegivirus (GBV-C)	+					8622
	Simian pegivirus (SPgV-OB23)	+				+*	8652
Hepacivirus	*Canine hepacivirus* (AAK-2011)		+	+*			8829
	Rodent hepacivirus (RHV-339)	+	+*			+*	8247
	Rodent hepacivirus (SAR-46)		+				8346
	Rodent hepacivirus (NLR07)					+*	8568
	Bat hepacivirus (PDB-829)			+			9075
	Hepatitis C virus Genotype 2	+*	+*				9102
Unclassified	*Soybean cyst nematode virus 5*	+	+*				17,718
	Gentian kobu-sho associated virus	+					22,176

The traits of the following viruses (or strains) not listed in this table are nonetheless included in the discussion.
Genus *Pestivirus*: *Classical swine fever virus* (Alford/187); *Border disease virus* (X818; H2121; Gifhorn); *Sheep pestivirus* (Aydin/04-TR); *Bovine viral diarrhoea virus-3*
Genus *Pegivirus*: *Bat pegiviruses* (PDB-34.1; PDB-76.1; PDB-620; PDB-694; PDB-1715; GBV-D; PDB-303; PDB-491.2]); *Simian pegiviruses* (GBV-A; SPgVkrc_RC32; SPgVkrc_RC08; SPgVkrc_RT11); *Theiler's Disease-associated pegivirus* (Horse Serum A)
Genus *Hepacivirus*: *Hepatitis C virus* (genotype 1; genotype 3; genotype 4; genotype 5; genotype 6; GBV-B); *Bat hepacivirus* (PDB-452); *Equine hepacivirus* (JPN3)
Motifs: Pentanucleotide, Hexanucleotide, DR of NTV subgroup, and RS of insect-specific subgroup: same as in Table 14.2; TABV Repeat: ggggcttga__aaccccccc; +: Presence of motif without a substitution; +*: Motif with a substitution or substitutions.

Based on multiple sources of information obtained thus far and if the assumption of the origin of the genus *Flavivirus* in other flaviviral genera was correct, the history of the evolutionary change of the vector-borne lineages would be depicted as a changing pattern of the mosaic of sequence elements integrated in the 3′ UTR at each stage of major host range shift, in contrast to phylogenetic tree depicted on the basis of the properties of the coding regions.

In addition to the sharing of pentanucleotide shown in Tables 14.2 and 14.3, it was reported that the evolutionary link between the insect-specific subgroup of genus *Flavivirus* and genus *Pestivirus* is also revealed in the presence of duplicated RNA structures at the enhancer region of the 3′ UTR (Gritsun *et al.*, 2014). Because many sequence elements are repeated, it is highly likely that subsequent duplication and deletion contributed to further variation in the composition of the sequence elements.

Also, with respect to the pattern of increasing ORF length along the evolutionary direction depicted in the NS5 tree of genus *Flavivirus* (Fig. 14.3), the length ranges of the coding sequences of hepaciviruses (8247–9102 nt) and pegiviruses (8622–10,455 nt) (Table 14.3) either overlap with or are shorter compared with the corresponding range (10,023–10,155 nt) of the insect-specific subgroup or the length (10,053 nt) of TABV. Thus, the data from those two genera fall in the length ascending pattern quite well, assuming that the viruses of these two genera were ancestral to the genus *Flavivirus*. It is noted that many members of the genera *Pegivirus* and *Hepacivirus* have been only recently isolated and that variation in motif sharing among viruses within a given genus is noted, as in insect-specific subgroup (Table 14.2). Given a strong possibility of the discovery of more new viruses in the future, the current knowledge of motif sharing among the four genera is clearly incomplete. Still, the utility of the motifs in 3′ UTR for an evolutionary study is evident. Because currently 3′UTR sequence data are available from only 4 of 12 known viruses of the NKV subgroup and from a small number of insect-specific flaviviruses, more 3′ UTR sequencing of these viruses is desirable.

When scientists attempt to succinctly explain a seemingly complicated set of intertwined data, Occam's Razor is constantly in their minds. For unravelling the long history of host range shift of flaviviruses, however, the available data presented above do not favour the application of such a parsimony-based logic. Thus, the above data contradict a theory that DENV and YFV are the viruses that had originated in nonhuman primates (Holmes and Drummond, 2007), although it is possible that they could have derived very indirectly from some members of the genera *Hepacivirus* and/or *Pegivirus* associated with primates, as shown in Table 14.3.

Utilities or significance of genome size in other virus groups

Although the significance of the relationships between genome traits and host range found in the genus *Flavivirus* as empirical data may be limited to this genus, still potential utilities of similar data are found in other virus groups. In the genus *Alphavirus* of the *Togaviridae*, in contrast to the genus *Flavivirus*, the member viruses are not clearly demarcated into host range-based subgroups. Still, an analysis of the lengths of the non-structural polyprotein sequences (Table 14.4) reveals a classification into three distinct subgroups without an overlap in length: 7782–7805 nt among 2 fish viruses without a vector; 7296–7545 nt among 20 arboviruses transmitted by mosquitoes or cimicid bugs to mammals or birds; 5385–5802 nt among eight viruses (BEBV, BFV, BCRV, *Eilat virus*, NDUV, SESV, TROV, and WHAV) each with a peculiarity of host range, as described below. This clear segregation is in sharp contrast to the very homogeneous lengths (3702–3777 nt) of the structural polyprotein (or precursor) sequences of alphaviruses, with the only exceptions of a longer length (3960–3965 nt) of two fish viruses (Table 14.4; Kuno, unpublished).

The uniqueness of the host range data among the eight alphaviruses with a short length is as follows: BCRV is a recombinant virus and, unlike most alphaviruses vectored by mosquitoes, uses a sedentary ectoparasitic cimicid bug. Although BCRV and FMV are sympatric and transmitted by the same bug, the length of ORF of non-structural protein sequence of the former is much shorter than the latter (Table 14.4). Furthermore, one of the two lineages of BCRV, like a geographic variant of FMV(Stone Lakes virus), poorly replicates in avian hosts (Brault *et al.*, 2009; Brown *et al.*, 2010b). Thus, FMV and BCRV provide a unique opportunity to study the impacts of the change in nsORF length on arboviral transmission. SESV is a virus associated with an ectoparasitic louse, but its biological transmission to Southern Elephant Seal is still unconfirmed despite induction of neutralizing antibody. *Eilat virus* is a mosquito-specific virus incapable of replication in vertebrate cells (Nasar *et al.*, 2012). For the four viruses (BEBV, BFV, NDUV and TROV), although marsupials and swine have been suspected for BFV and NDUV, respectively, vertebrate hosts have not been identified despite many years of field research. As for WHAV, because of the absence of conclusive evidence of natural biological transmission, strictly speaking, the arbovirus status is still uncertain, despite proof of artificial transmission to laboratory rodents by mosquitoes, detection of viral RNA and seropositivity in some birds (Tomplins *et al.*, 2010). Assuming that Eilat virus is one of the precursors (or their siblings) of the vector-borne alphaviruses, it is possible that many of the other viruses in this short-length group, like the aforementioned ancestral SLEV strain (Kopp *et al.*, 2013), represent an early primitive stage after evolving to become arboviruses characterized by the limited capacity of replication in vertebrates.

In the family *Bunyaviridae*, the length ranges of L-segment among five genera are host range specific and do not overlap: genus *Orthobunyavirus* (6875–6980 nt); genus *Hantavirus* (6530–6562 nt); genus *Nairovirus* (11,840–12,255 nt); genus *Phlebovirus* (6341–6423 nt); and genus *Tospovirus* (8876–8897 nt). The corresponding lengths of the three new mosquito-associated bunyaviruses (Herbert virus, Tai virus and Kibale virus) range from 7428 to 7454 nt (Marklewitz *et al.*, 2013). Thus, for a proper classification of these new bunyaviruses to a new genus, selected genome (or ORF) length serves as a useful, supplemental criterion.

In the family *Reoviridae*, the three arboviral genera are similarly segregated by non-overlapping range of genome length: genus *Orbivirus* (17.96–19.8 kb); genus *Seadornavirus* (20.6–20.9 kb); genus *Coltivirus* (29.1–29.2 kb). The genus *Orbivirus* is subclassified on the bases of vector group and size of outer

Table 14.4 The lengths (nt) of non-structural and structural (or precursor) polyprotein sequences of alphaviruses

	Non-structural Polyprotein (nt)			
Virus	Long	Intermediate	Short	Structural Polyprotein (nt)
AURAV		7500		3735
BEBV			5517	3747
BFV			5385	3720
BCRV			5529	3717
CHIKV		7425		3747
EEEV		7482		3726*
Eilat Virus			5406	3702
FMV		7374		3717
GETV		7404		3762*
HJV		7353		3711
Igbo Ora Virus		7542		3744
MAYV		7314		3729
MIDV		7236		3777
MUCV		7368		3765
NDUV			5448	3723
Ockelbo Virus		7548		3738
ONNV		7545		3744
RRV		7443		3765
SAGV		7404		3762
SESV			5478	3756
SFV		7296		3762
SINV		7542		3738
Sleeping Dis. Virus	7782			3965
SPDV	7805			3960
TROV			5802	3732
UNAV		7530		3750
VEEV		7476		3768*
WEEV		7404		3760
WHAV			5565	3741
Range	7782–7805	7236–7548	5385–5802	3702–3965

*Precursor.
Only one length is arbitrarily selected per virus, when a variation in length exists among strains.

capsid protein 1′ unique to each vector group (Mohd Jaafar *et al.*, 2014); the genus is also segregated into tick- or insect (midge, mosquito, and sand fly)-associated viruses based on the length of the segment 3, since the non-overlapping length ranges are 1.93–2.02 kb and 2.28–2.98 kb, respectively.

In the family *Rhabdoviridae*, genome plasticity similar to the one illustrated by ORF expansion and contraction in the family *Flaviviridae* was recently revealed (Walker *et al.*, 2015).

Motifs in the 3′ UTR

The 3′ UTRs among the members of the genus *Alphavirus* have been recognized for a greater diversity of repeat sequence elements (RSEs) unique to a particular virus or to multiple members of a virus complex. Generally, with a few exceptions, the RSEs found in the viruses of the New World are not shared by the viruses of the Old World, and vice versa (Pfeffer *et al.*, 1998). A closer re-examination of available sequences, however, reveals a more extensive sharing of markers or motifs. As an example, a sequence (acccctgaatagtaacaaaa) is shared among AURAV, BFV, CHIKV, Eilat virus, GETV, RRV, SAGV, and TROV. A motif (attaacc) is shared between MAYV and MIDV; while another (aaactcgatgta_ttccgagga) is shared among HJV, Ockelbo virus, SINV, WEEV, and WHAV. Similarly, a shorter motif (ttttat or tttatt) is found in 16 viruses in Table 14.4 as well as in four pestiviruses (Border disease virus, Bovine viral diarrhoea virus-3, Classical swine fever virus, and Pronghorn antelope pestivirus). These examples represent motif sharing in alphaviruses in two worlds, while sharing of another marker (cccttag) is limited to the viruses in the Old World, Eilat virus and ONNV. Sharing of multiple markers [including the canonical 19-nt 3′ terminal pan-alphavirus conserved sequence element (CSE)] with many viruses by Eilat virus is intriguing, given a possibility of this virus being a progenitor (or its sibling) of the extant vector-borne alphaviruses.

The aforementioned flaviviral hexanucleotide (attggc) is found in CHIKV, MIDV, MUCV, SFV and VEEV. Furthermore, the canonical flavivirus pentanucleotide (cacag) is found in CHIKV, Eilat virus, MIDV, and SPDV. Although this pentanucleotide, like the 19-nt canonical alphavirus CSE, is apparently not absolutely necessary for replication of all vector-borne flaviviruses, sharing of flaviviral penta- and hexanucleotides may suggest a marker of evolutionary link between the families *Flaviviridae* and *Togaviridae*. Alternatively, these motifs may be interpreted to be the past or present markers of arthropod association independently adopted in multiple virus families. As an example, a survey of 3′UTRs in 11 other positive, non-segmented ssRNA virus families and an unclassified group revealed that this pentanucleotide is found, besides in *Flaviviridae* and *Togaviridae*, predominantly in some members (but not all) of the order *Nidovirales* (*Coronaviridae, Mesoniviridae, Roniviridae, Negevirus*). Except for a dozen coronaviruses, all were isolated from arthropods in terrestrial and aquatic environments. Even among coronaviruses, an avian coronavirus possessing this pentanucleotide was repeatedly isolated from ticks (Traavik, 1979). As for the hexanucleotide, it is also found in the selected members of nidoviruses in three families (*Arteriviridae, Coronaviridae, Roniviridae*), *Negevirus*, and several members of the family *Picornaviridae*. It is noted that the relationships of some enteroviruses of the family *Picornaviridae* and of the *Arteriviridae* with haematophagous insects have been investigated in the past because of occasional, intriguing observations including those in the Table 14.1. The significance of the sharing of such markers in alphaviruses might be also sought in the context of unravelling the complicated history of multiple intercontinental crossings in both directions in the evolution of lineages (Powers *et al.*, 2001; Forrester *et al.*, 2012).

Genetic determinants of host range among arboviruses

Host range of any arboviral lineage is determined by a combination of four sets of determinants: viral determinants; host determinants (such as virus receptor, antivirus defence mechanism, and others); biological factors of host (such as bionomics, behaviour, geographic distribution and movement); and environmental factors in ecosystem (such as flora, fauna, and host population dynamics). In this chapter, only the first set is examined, since the others are the subjects in other chapters. Regrettably, despite many studies, we still have few answers to such questions as to why few DNA arboviruses evolved (with the sole exception of ASFV), why alphaviruses are generally more promiscuous than flaviviruses in vector selection, why certain arboviruses are transmitted by ticks but not by midges or mosquitoes, and many other important issues.

Functional specification of viral genes

One of the basic assumptions on biological transmission is that this mode of transmission evolved in multiple virus families by convergence, through utilization of a set of genetic mechanisms uniquely available to a given virus group under particular ecological conditions. Because of the diversity of those underlying conditions leading to the evolution of biological transmission, generalization of the mechanisms involved for all arbovirus groups is difficult. Accordingly, the genomic markers responsible for adaptation to two phyla of hosts are thought to vary among arbovirus groups.

Still, some of the generalizations obtained thus far are useful. According to the 'evolutionary logics' of Dolja and Koonin (2011), viral genomes are basically classified into two categories, the genes essential for replication and expression and the genes involved in virus–host interactions which are host-specific. The genes in the first category include RdRp in the flavivirus-like superfamily, RdRp containing L-protein genes in the family *Bunyaviridae* and RT in the retroviruses. The genes in the second category include host-specific interactive genes generating capsid and envelope proteins that react with the cellular receptors of vertebrates (Javier and Rice, 2011).

The other important generalization is that because viral replication and host selection are inseparable, theoretically all genes involved in adsorption-replication-maturation-release process are involved in host range determination, even though some are critically important while the contributions of others may be negligible. Thus, it was found that both structural and non-structural protein genes are involved in host range determination of YFV and Eilat virus (Beasley *et al.*, 2013; Nasar et al., 2015).

Another important lesson learned is that, as described later, the genetic defects associated with host range restriction identified by *in vitro* genetic experiments using cell cultures are not necessarily reproducible *in vivo* (Stollar and Hardy, 1984; Bryant *et al.*, 2007). Also, gain in host range in in vitro studies often results in loss in fitness in the natural, primary hosts (Ciota *et al.*, 2015). Naturally, the relevance of *in vitro* studies regarding fitness trade-off associated with arboviral host alternation in the context of natural transmission was seriously questioned (Coffey *et al.*, 2013).

Basics of viral RNAs as they relate to arboviruses

Mutation rate

Theoretically, because of their higher rates of mutation, the more genetically variable viral RNAs are, more likely change the viruses significantly or assume a broader host range (Woolhouse *et al.*, 2001). It was also theorized that, in analysing the emergence of new hosts for a virus, the more phylogenetically distant potential hosts are from the current hosts, the less likely occurs host range shift between these hosts without a significant changes in genomes (DeFilippis and Villareal, 2000). Large genome population size, a common characteristic of RNA viruses, also favours emergence of the subpopulations that are better adaptable to a new host.

The low mutation rates of arboviruses have been traditionally interpreted to be the consequence of a combination of the constraints imposed by the necessity of replication in two phyla of hosts and purifying effect. Because of the traditional, oversimplified depiction of arbovirus transmission in a cycle alternating between vector and vertebrate, opportunities for mutation in two phyla of hosts have been assumed to be roughly equal in many laboratory studies. However, in reality, the proportions of time arboviruses spend in vectors in their natural transmission cycles are far greater than those in vertebrates. Thus, the prevalent practice of placing an equal weight in vertebrates in many trade-off experiments has been problematic. Most importantly, dynamic quantitative and qualitative analysis of re-integration of the arbovirus subpopulation replicating in vertebrate hosts back to the viral subpopulation in the vector(s) in each cycle of transmission (or more realistically at the end of each transmission season) has rarely been, if ever, performed under natural conditions, despite the fact that these information are crucially important in validating the biological transmission of arboviruses. Thus, when genome sequences of the viruses isolated from blood-engorged vectors during a natural transmission are examined, it is first necessary to determine if they represent the subpopulation which had earlier proliferated in the vertebrates and which are on the way to re-integration to the viral subpopulation in the vectors or the original subpopulation in the vectors which merely proliferated in response to the intake of bloodmeal.

Genome length

Unlike DNA viruses, longer the viral RNA genome, the more unstable is the RNA population as a species. Accordingly, typical positive-strand ssRNA viral genome lengths are around 9–12 kb, their small number of genes demonstrating functional pleiotropy and epistasis by necessity. The arboviruses with a longer total genome length typically belong to negative-strand ssRNA viruses (such as genus *Ephemerovirus* of the family *Rhabdoviridae*) and segmented RNA viruses. Thus, the probability of evolution of arboviruses in the positive ssRNA virus families with a much longer (>15 kb) genome [such as the families *Coronaviridae* and *Mesoniviridae* (Table 14.1) and unclassified plant/nematode flaviviruses] is very small. This may explain why an arbovirus equally efficient in replication in and biological transmission by two classes of arthropods (i.e. ticks and mosquitoes) has never evolved. As described earlier in the section 'Phylogeny and empirical data in support of phylogenetic tree topology', the host ranges of some members of the genus *Flavivirus* [such as Nounané, Barkedji, Lammi, Chaoyang, Nhumirim, and Ilomantsi viruses in the mosquito-specific lineage (Fig. 14.3) and four viruses in the mosquito-borne subgroup (ENTV, SOKV, and YOKV in YFV lineage and Rabensburg virus in DENV-JEV lineage)] are limited to only one phylum. One common denominator in all these viruses is that their ORF lengths are at or near the longest in each respective lineage (Fig. 14.3). These and the data from other virus families covered in the section 'Phylogeny and empirical data in support of phylogenetic tree topology' collectively support a significant ORF length change associated with major shift of host range in arthropod-associated viruses.

Gene transfer

Intraspecific recombination among genotypes of arboviruses has been frequently reported. However, confirmed cases of interspecific recombination have not been reported, with the exceptions of WEEV complex of alphaviruses. In contrast, many natural or experimental cases of interspecific reassortment have been documented in all segmented RNA arbovirus families. Furthermore, reassortment between ss RNA virus and segmented RNA entity (or virus) was found possible (Qin *et al.*, 2014). However, surprisingly, drastic host range shift as a result of either recombination or reassortment is yet to be documented.

Ubiquitous presence of RdRp in many RNA virus families has been speculated to be the result of modular transfer, although the origins of RdRp in positive-strand and negative-strand viruses are believed to be different. Another transfer mechanisms include gene transfer mediated by a retrovirus with or without integration in the genome of different host prior to second integration in other virus, as described in the section of plant virus-animal virus associations. The possible examples of viral genome transfer include *Sindbis virus* nsP1 and nsP2 from plant viruses (Ahlquist *et al.*, 1985), a tymovirus (isolated from *Culex* mosquitoes) from plant tymovirus (Wang *et al.*, 2012), circovirus from nanoviruses (plant viruses) (Gibbs and Weiller, 1999), and *Thogotovirus* glycoprotein (GP64) gene from baculovirus (Pearson and Rohrmann, 2002).

For many other animal viruses strongly speculated to have acquired certain genes from distantly related viruses, the exact mechanisms remain unknown, but horizontal gene transfer mediated by arthropods is clearly one of them (Dolja and Koonin, 2011). On the other hand, other gene transfer between microbial inhabitants (including viruses) and their hosts (in particular, arthropods) in the evolution of many animal virus lineages has been either neglected or underestimated.

Specific examples of host range genetic determinants in arboviruses

Presented below are some of the determinants identified. It should be noted that most of the studies were performed only *in vitro*.

Genus *Flavivirus*

Regarding the vector group determinants of flaviviruses, when C-prM-E genes of a dengue infectious clone were replaced with the corresponding genes from Langat virus (LGTV), new chimera replicated only in mosquito cells but not in tick cells, suggesting that the major vector group (mosquito versus tick) determinants reside in non-structural protein genes but not in structural protein genes or in 3′ UTR (Mandl *et al.*, 1993; Pletnev *et al.*, 2001; Engel *et al.*, 2011; Tumban *et al.*, 2011). Also, it was shown that multiple genes (E, M, NS3, NS4A, and NS4B) are involved in the tick-specific replication of LGTV (Mitzel *et al.*, 2008). A similar conclusion of the non-structural protein genes being host class determinants was obtained in studies of JEV, TBEV, and a NKV virus (Yoshii *et al.*, 2008; Charlier *et al.*, 2010). Although flaviviral 3′ UTR may not contain a major determinant, it may still act in combination with the determinants in the coding region. As mentioned earlier, a unique trait of the mosquito-specific lineage flaviviruses within the mosquito-borne subgroup (Fig. 14.3 legend), is that they, unlike the viruses within the insect-specific subgroup, possess CS1 and CS2 as a shared trait as members of the mosquito-borne subgroup (Table 14.2 and Fig. 14.4). The functions of CS1 and CS2 are, however, interpreted as markers of insect association rather than as genetic host range determinants.

Dengue virus

The domain III of the E protein has been identified repeatedly *in vitro* as the determinant regulating replication in two phyla of hosts (Wang *et al.*, 2000; Moncayo *et al.*, 2004), while mutations in the NS4B and the hairpin structure of the 3′UTR were similarly identified by others (Hanley *et al.*, 2003; Villordo and Gamarnik, 2013).

NKV subgroup

The factors responsible for the absence of replication of Modoc virus (MODV) in mosquito cells were identified as intracellular host factors rather than viral envelope protein, which pointed out the importance of non-structural genes in host range determination (Charlier *et al.*, 2010). This was also confirmed in a reverse genetic experiment between a mosquito-borne virus and MODV (Tumban *et al.*, 2013).

Genus *Alphavirus*

Chikungunya virus (CHIKV)

Amino acid substitutions in E1 and E2 proteins and organization of direct repeats in the 3′UTR of this virus were found to be important in vector switch and adaptation to mosquitoes, respectively (Tsetsarkin and Weaver, 2011; Chen *et al.*, 2013).

Eilat virus

Involvement of multiple non-structural and structural genes was found responsible for the inability of replication of this mosquito-associated virus in vertebrate cells (Nasar *et al.*, 2014).

O'nyong nyong virus

The determinant of this virus uniquely vectored by *Anopheles* mosquitoes among alphaviruses was identified to be NS3 gene (Saxton-Shaw *et al.*, 2013).

Sindbis virus (SINV)

Early research focused on the difference in composition of N-linked oligosaccharides in glycoproteins between vertebrates and insects. The vertebrate-restricted mutants were also temperature-sensitive, producing small plaques at 34.5°C but none at 40°C. It was concluded that it is the attachment of the oligosaccharide at residue 275 on the pE1/E2 protein, which interferes with the maturation of the mutants in vertebrate cells; while in mosquito cells this additional glycosylation of E2 was neither favoured nor selected against. For other mutants which replicate normally in vertebrate cells but which are host range-restricted in mosquito cells, it was found that viral replication is blocked at a stage later than adsorption-penetration. Surprisingly, it was found that host range-defective mutants identified *in vitro* could replicate normally *in vivo* (Stollar and Hardy, 1984). This prompted a need of exercising caution in interpreting the results of host-range mutants studied *in vitro*. In other study, the mutants with a defect in replication in mosquito cells were found unable to cleave pE2 glycoprotein into E3 and E2 (Boehme *et al.*, 2000). It was also proposed that SINV employs 2 kinds of receptors (one specific to vertebrate cells and the other broadly reactive across animal kingdom) with which the virus can replicate in two phyla of hosts (Strauss *et al.*, 1994).

Venezuelan equine encephalomyelitis virus (VEEV)

The gene regulating vector range of this virus was found to be E2. In contrast to the epizootic strains (subtypes IAB and IC), which are transmitted by an epizootic vector (*Aedes taeniorhynchus*), the E2 gene of the enzootic VEEV (subtype IE), which is maintained in sylvan environments by *Culex* (*Melaniconion*) spp., is vector range-restricted (Brault *et al.*, 2002).

Genus *Asfavirus*

African swine fever virus (ASFV)

The genes regulating replication and dissemination in ticks by this sole DNA arbovirus were identified to be Multigene Family 360 (Burrage *et al.*, 2004). It is of interest to note that ASFV replicates in soft ticks but not in hard ticks. This specificity was also found in some tick-borne RNA arboviruses as well.

Other issues concerning host range

Although vertebrates constitute one of the pillars of the arboviral triad (arthropod-virus-vertebrate), ironically, identification of the true natural vertebrate hosts has proved to be an elusive goal for too many viruses. The available data are incomplete for those viruses for a variety of reasons. Because host range is one of the criteria used in species demarcation for many virus groups, accuracy (or difference) of host range for justification in the establishment of a virus species is of critical importance in viral systematics. This issue significantly impacts on such subjects as virus–host cospeciation, tree topology of phylogram, strategy for disease surveillance, and arbovirus status of viruses.

Difficulties of host range identification and sampling bias

Generally, identification of true natural hosts for new animal viruses is a difficult task. For example, although fruit bats are now strongly suspected as reservoirs of Ebola virus based on sero- and viral RNA positivities, after many years of arduous research, including investigations of vertebrates, invertebrates, and plants, the reservoir of this virus has not been determined yet (Swanepoel *et al.*, 1996). Similarly, the hosts of a recently isolated Tibrovirus [proposed]-related rhabdovirus causing haemorrhagic fever have not been identified (Grard *et al.*, 2012). The other problem is mysterious virus disappearance in the same location over years in the population of a particular host from which a rare virus had been originally isolated. This renders confirmation of the host many years later difficult. Often, hitherto unknown host range is revealed only by accident or under an unusual circumstance, such as consequences of importation of exotic animals through animal trade, invasion of exotic viruses to new territories through an unidentified mechanism, and zoonotic disease outbreaks related to human encroachment upon habitats of wildlife.

Some of the viruses isolated from arthropods (i.e. *Netivot virus* isolated from *Culex* mosquito) reveal replication in vertebrate cells only after multiple blind passages (Tesh *et al.*, 1986). For a few ephemeroviruses (Kotonkan virus and Obodhiang virus), replication in vertebrate cell cultures could be confirmed only with co-cultivation with mosquito cells as helper cells (Buckley, 1973).

Currently, mosquito-borne viruses account for an overwhelming majority of known arboviruses. Although mosquitoes are unquestionably the most important group of haematophagous arthropods in terms of number of species and richness in ecosystem they occupy, the high proportion of mosquito-borne arboviruses is partly attributed to sampling bias in the past, as attention has been focused only on mosquito vectors by too many arbovirologists.

Promiscuity of vectors

Host survey only in one place in one season of year may generate incomplete data, if the available hosts in another season are quite different, because of promiscuous feeding of some vectors (Tesh *et al.*, 1972). In a study of *Culex fatigans* in Pakistan, the mosquitoes collected in houses fed on humans more often than on cattle in sheds, but the same mosquitoes resting in cattle sheds in winter fed more often on birds and bovids. However, the mosquitoes changed the hosts to humans and bovids during the spring and then back to humans and birds in the summer (Reisen and Boreham, 1979). Similarly, regarding opportunistic feeding behaviour of vectors, virus isolation from a particular vector alone does not necessarily identify the natural vector(s), as demonstrated in multiple reports of isolation of tick-borne viruses from mosquitoes and vice versa (Kuno, 2007; Lwande *et al.*, 2013). Broad host range of many ticks is well known. *Ixodes scapularis*, as a larva and a nymph, feeds on more than 52 species of mammals, 60 species of birds, and 8 species of reptiles; while *Ixodes persulcatus*, as an adult, feeds on at least 212 species (mammals, birds, and lizards).

Principal host

When numerous vectors or vertebrates are identified as hosts, it is often difficult to determine which is (or are) the principal host(s). Quite often, only a small number of them are critically important for sustaining natural transmission. As an example, in a study of Tick-borne encephalitis virus (TBEV) transmission, it was found that 20% of the vertebrates were involved in almost three-quarters of transmission by ticks, all others playing only a marginal role (Randolph *et al.*, 1999). A similar conclusion was arrived at for mosquito-borne WNV as well (Kilpatrick *et al.*, 2006b).

Epidemiologic/epizootic data

Seropositivity is often a useful indicator for a presumptive identification of vertebrate hosts. Strictly speaking, however, when little is known about actual viral replication and its significance in the seropositive vertebrates, seropositivity alone only confirms virus-vertebrate contact manifested in immune response to the virus by the bite of infected haematophagous arthropods. Whether or not the vertebrates in question are true natural hosts is another matter. As examples, a patient in whom a plant virus was detected developed antibody to that virus; while in other case, specific antibodies to an insect-specific virus were detected in mammals (Scotti and Longworth, 1980; Colson *et al.*, 2010). Antibody to Triatoma virus (a dicistrovirus) is frequently detected in Chagas disease patients, even though the virus does not replicate in human or other vertebrates (Querido *et al.*, 2015). Some mosquito-specific flaviviruses (i.e. Lammi virus and Ilomantsi virus) are known to serologically react with antiflaviviral antibodies in human serums (Huhtamo *et al.*, 2014). However, even if serological reaction was found specific to the plant- or insect-specific viruses, still one would not attempt to identify human as host of these viruses. Mere molecular detection of a viral genome in a vector or vertebrate (as in many metagenomic studies) does not establish a host relationship of the detected virus either.

When a zoonotic viral disease outbreak occurs, sometimes vertebrates previously unknown as hosts of the virus are found infected for the first time, most often because the physical (i.e. distant geographic separation) or ecological barrier that had prevented the contact between the virus and the new hosts was broken. In these cases, 'new hosts' observed in non-indigenous areas immediately after virus invasion (such as birds of North America after invasion of WNV) are not the hosts naturally selected. The cause of infection in these new hosts (or host crossover) is occasionally attributed to viral mutations detected in association with this outbreak. However, often overlooked in some of these reports is a simpler explanation that infection would have occurred in the new hosts regardless of the difference in disease severity and viral mutation because of innate genetic susceptibility of the new hosts to the virus. This innate susceptibility, which had been unknown before, was revealed only recently for the first time after the removal of barrier. Except for medical (for human) or veterinary importance, most spillover hosts transiently observed in major outbreaks, such as alligator, whale, and zoo animals in captivity infected by WNV and African penguins by EEEV, are generally not seriously considered in the natural host range.

The methods employed in host range studies in relation to virus isolation and laboratory experiments

In vivo test

Whenever possible, inclusion of this test is desirable for determining the host range, provided that the vectors or vertebrates selected are appropriate for the virus of interest. In reality, too often the hosts for laboratory tests (typically widely available laboratory rodents and colonized arthropods) are used arbitrarily in the absence of the information about the true natural hosts.

Prior to the advent of cell culture technique, for arthropods, intrathoracic (or intrahemocoelic) inoculation was developed to determine potential vector range (Hurlbut, 1951). Although this method is advantageous if the objective is to determine the maximal range of replication among multiple groups of arthropods, results need to be interpreted with caution, because this technique artificially bypasses natural barrier (midgut barrier) in vectors. The disadvantage of pledget feeding is insensitivity, as explained elsewhere in this chapter. The disadvantages of *in vivo* tests, in particular for vertebrate experiments, are labour-intensiveness and high cost. Because of these problems, these tests have been much less frequently used in modern times and largely replaced by *in vitro* system in many laboratories. Another problem is a need of a high-security facility, if the viruses must be so handled according to biosafety level classification and governmental regulations. Still, as demonstrated earlier, *in vivo* tests have a critical advantage over *in vitro* system, as pointed out earlier (Stollar and Hardy, 1984).

In vitro test

Viral replication *in vitro* is a highly useful trait for either quickly determining the group of hosts or roughly estimating the maximum range at or above host's family level, as demonstrated in the specificity of the 4 flavivirus subgroups which are clearly demarcated by the difference in host range (Kuno, 2007). In some arboviruses, replicative specificity of a virus is manifested even at vector's generic level in cell culture system. For examples, ONNV, which is transmitted by *Anopheles* mosquitoes, replicates in *Anopheles* cell cultures but not in *Aedes* cell cultures (Varma and Pudney, 1971); and some mosquito-borne arboviruses replicate in *Ae. albopictus* but not in *Ae. aegypti* or sand fly cells. However, such a strict *in vitro* specificity is generally less common. Many viruses transmitted by dipteran vectors (such as mosquitoes and sand flies) cross-infect the cells derived from other orders of insects (such as *Lepidoptera*), although cross-infectivity between two classes of arthropods (insects and ticks) is rare. Similarly, replication of many arboviruses is not limited to the cell cultures derived from mammalian species and extends to the cultures derived from poikilothermic vertebrates as well (Leake *et al.*, 1977).

In addition to difference in sensitivity among cell lines used, which is often a major source of conflicting results, passage history of virus or strain difference also affects the outcome, as revealed in a rare case of persistent infection of a tick cell line by a particular strain of JEV or WEV (Pudney *et al.*, 1982; Ciota *et al.*, 2015). Selecting an inappropriate cell culture generates data difficult to interpret, as in the use of only mosquito cells rather than tick cells for a presumably tick-borne virus (Qin *et al.*, 2014).

Arthropod cell cultures are often used for simulating genetic changes in arboviruses associated with alternation of host between two phyla. It is important to note that for transmission by vectors in nature, viral replication in the salivary gland is critical but that, unlike vertebrate cell cultures, tissue or organ origin of the cells in culture is totally unknown for nearly all arthropod vector cell lines available, including the C6/36 clone of *Aedes albopictus.*

Host of original virus isolation as a guide

As far as vector identification is concerned, with the exceptions mentioned elsewhere in mind, generally this traditional method has proven useful for many arboviruses. When a virus is isolated from a species of mosquito, however, too often its arbovirus status has been routinely studied only with one or two species of commonly used mosquitoes and rodents (or respective cell cultures), either because it has been an established tradition or because other vertebrates, arthropod hosts (such as ticks, sand flies, cimicid bugs and midges) or their cell cultures have not been readily available in most laboratories. Similarly, when viruses were isolated from ticks, they were tested only in ticks and vertebrates (or their cell cultures). With this prevalent practice, one may miss an opportunity to discover true vectors, as when a virus isolated from a tick turns out much later to be a mosquito-borne virus or even a vertebrate-specific virus. In other example, the real vectors of Orungo virus (an orbivirus originally isolated from *Anopheles* mosquitoes) are now identified as *Culicoides* midges (Mohd Jaafar *et al.*, 2014). Also, when a virus is isolated from an ectoparasitic arthropod (such as ticks and lice) on a vertebrate, the virus could be merely a virus of the ectoparasite (despite induction of antibody to the virus in the vertebrate), a virus of the vertebrate accidentally ingested by the ectoparasite (if it does not replicate in the vector), a 'non-arbovirus of biphylum host range' (defined later in the section of biological transmission) or a true arbovirus.

Absence of biological transmission in the field

The following examples demonstrate that determination of host range solely based on laboratory tests in the absence of evidence of vector-borne transmission in the field is difficult, misleading or equivocal.

Family *Bunyaviridae*

Ganjam virus, a variant of tick-borne Nairobi sheep disease virus, was isolated from and replicates experimentally in mosquitoes or mosquito cell cultures (Pudney *et al.*, 1982). However, no evidence of mosquito-borne transmission in the field has been found and attempts to reproduce the laboratory results failed (Dandewate *et al.*, 1981; Sudeep *et al.*, 2009).

Family *Flaviviridae*

Including SOKV (not in Fig. 14.3), at least 10 members of the mosquito-borne subgroup replicate only in one of two phyla of hosts, despite the fact that they cluster phylogenetically in the mosquito-borne branch (Fig. 14.3; Kuno *et al.*, 1998).

Family *Reoviridae*

Similar to the report of Ganjam virus, replication of tick-borne Colorado tick fever virus in mosquito cells was reported (Yunker and Cory, 1969). However, when tested *in vivo* using adult mosquitoes, replication of this virus was found to be negligible at best (Hurlbut and Thomas, 1960); and the *in vitro* result could not be reproduced (Varelas-Wesley and Calisher, 1982).

Family *Rhabdoviridae*

Mokola virus, a member of the genus *Lyssavirus*, replicates in adult mosquitoes and mosquito cell cultures and has been speculated to be the precursor of Rabies virus (Table 14.1); but there exists no evidence of vector-borne transmission.

Family *Togaviridae*

Several terrestrial alphaviruses replicate in the cell cultures derived from fishes, but they are not naturally transmitted to fishes (see Table 14.1; 'Virologic attributes of selected environments and hosts' above). The significance of alphavirus isolation from ticks and mites has remained unclear in the absence of a tick-borne alphavirus. It is, however, of interest to further investigate if such tick-borne alphaviruses exist but have not been discovered yet or if such alphaviruses once existed but subsequently became extinct.

Molecular methods

As a benefit of the applications of molecular phylogeny, if sampling is adequate and database rich, unknown vector (but not vertebrate) group of a suspected arbovirus can be identified reasonably accurately by examining the viral clustering with the viruses of known host range. Still, it needs to be determined if the virus is an arbovirus or a non-arbovirus of biphylum host range (to be defined later). Host range analysis based on branching pattern in a phylogenetic tree may be significantly affected when discovery of missing, important viruses alters the pattern of virus clustering. For examples, recent discoveries of a phlebovirus related to *Uukuniemi* group and a new member of the *Flaviviridae* (both from a species of nematode) and of negative-strand RNA viruses including the family *Chuviridae* will significantly impact on our interpretation of the evolutionary history of host range shift in multiple RNA virus families (Bekal *et al.*, 2011; Bekal *et al.*, 2014; Li *et al.*, 2015). When the viruses of biphylum host range (including arboviruses) are excluded, 'nucleotide composition analysis' employing sequences (>3,000 nt) was found to identify accurately the host range at the phylum level (Kapoor *et al.*, 2010). As for host range specificity for dsRNA viruses at family level, interpretation may be complicated by a probability that the genera constituting a family are not monophyletic (Gibrat *et al.*, 2013).

Width and change of host range

Generalist vs. specialist

Genetic control of vertebrate host range is well illustrated among closely related viruses. As an example, among the members of neurotropic JEV complex viruses, JEV replicates in pigs, horses and birds; WNV replicates in horses and birds but not in pigs; and SLEV replicates in birds but rarely in equines or pigs. For this reason, recent reports of SLEV isolation from a horse in South America (Rosa *et al.*, 2013) and seropositivity in horses deserve further investigations.

Many viruses are classified either as specialists or generalists for convenience; and the advantages and disadvantages of each have been debated (Woolhouse *et al.*, 2001). According to a theory, viruses with a high rate of mutation (such as many RNA viruses) tend to adopt a broader host range as a matter of survival, particularly if the population sizes of all major vertebrate hosts are insufficient or fluctuate over time considerably; while the viruses adapted to vertebrate hosts, whose population sizes are large and remain stable, do not need to depend on many species of hosts. Most arboviruses belong to the former category. For example, WNV, which occurs in all continents except Antarctica, has been isolated from well over 350 species of birds representing 14 orders and more than 40 species of non-avian vertebrates (equines, felines, canines, ungulates, rodents, bats, and reptiles). As for vectors, it was interpreted that the transmission patterns of Ross River virus by a broader range of vectors including both salt- and freshwater mosquitoes in Australia and of WNV or SLEV by multiple *Culex* species in North America reflect the advance in viral survival, persistent transmission or overwintering (Glass, 2005; Lord, 2010). On the other hand, DENV is exceptional in that in tropical urban environments it persists for years between *Ae. aegypti* and only one vertebrate (human) without replenishment of virus from external sources, so long as the size of susceptible human population is sufficiently large and conditions for vectors are favourable. Similarly, increasing population size of the cimicid bug infesting bird nests was found favourable for the persistence of BCRV which has only cliff swallow and house sparrow as its vertebrate hosts (Brown *et al.*, 2001).

Broad host range

According to one of the traditional thoughts, arboviruses with animal hosts in two phyla evolved independently in multiple genera or families of animal viruses originally with host range in one phylum, by secondarily acquiring replicative trait in the hosts of another phylum. Although no arbovirus has hosts in more than 2 phyla, among animal viruses Nodamura virus and Flock House virus of the *Nodaviridae* (non-arboviruses) are unique in that either one infects insects (including mosquitoes), vertebrates, yeasts, nematodes, and/or plants. The hosts of the members of the family *Birnaviridae* encompass multiple phyla, including vertebrates, arthropods, mollusks, rotifers, nematodes, bryozoans, and echiderms. Members of the *Totiviridae* infect protozoa, fungi, mosquitoes, salmon, and mammalian cells. It is of interest to note that the genome length ranges of these virus families with an exceptionally broad host range are only 4.32–5.54, 5.71–6.18 kb, and 4.6–7.6 kb for the *Nodaviridae*, *Birnaviridae*, and *Totiviridae*, respectively. Short length as a shared trait for broad host range, however, is applicable only at generic level within a given family rather than at family level of viruses. As an example, the coding length ranges of the non-arbovirus groups (genus *Pestivirus* and of the unclassified viruses) are longer than the range of the arbovirus-rich genus *Flavivirus* even though they all belong to the same virus family.

For certain viruses, cross-kingdom IRES has been proposed to be the viral trait endowing infectivity in a broader range of hosts (Woolaway *et al.*, 2001; Terenin *et al.*, 2005; Ventoso, 2012). Although this hypothesis may apply to insect-specific viruses of the *Dicistroviridae*, retroviruses, hepaciviruses, pestiviruses and others, however, it does not apply to most arboviruses because of the absence of IRES.

Lateral transfer of factor(s) is another mechanism considered. In an experiment of Tomato spotted wilt virus (a vector-transmitted plant bunyavirus), if a transcription factor from the vector (which binds to RdRp of viral RNA) was expressed in unnatural hosts (such as human cells), even the plant virus was found to replicate in mammalian cells (De Madeiros *et al.*, 2005). Further, the mechanism of gene transfer of small circular ssDNA viruses, such as between nanoviruses (plant viruses) and circoviruses (vertebrate viruses), is different.

It is noted that for nonviral pathogens (such as bacteria and fungi) with a host range in multi-kingdoms the shared mechanism for the evolution of pathogens with such a broad host range was identified as close physical proximity between pathogen and new hosts which ensures survival of the pathogen in the pathogen's original habitat (van Baarlen *et al.*, 2007). For viruses, a similar conclusion was obtained that, within genomic constraints described elsewhere in this chapter, emergence of very distantly related new spillover hosts is a result of stochastic outcome of repeated contacts with a very small proportion of genetic variants (endowed with a unique fitness) of a virus population. This is facilitated through co-existence in close proximity in a given environment over a long period rather than following a particular rule, genetic or otherwise (Parrish *et al.*, 2008).

Reduction in host range

According to a proposal, the three 'host-restricted' viruses in the mosquito-borne subgroup (ENTV, SOKV, YOKV), which have hosts in only one phylum, are hypothesized to have originally had hosts in two phyla but subsequently lost host range in one phylum (Kuno *et al.*, 1998). This proposal is partially similar to the hypothesis that predicted eventual regression of vector-borne viruses to vertebrate viruses (Baker, 1943). Theoretically, however, regression to arthropod viruses is also possible. Poor *in vivo* replication of Stone Lakes virus and of

Buggy Creek virus in cliff swallow and house sparrow (Brault *et al.*, 2009; Brown *et al.*, 2010b) may be interpreted either as a sign of regressed vertebrate dependence or an intermediate stage before acquiring satisfactory biological transmission cycle. On the other hand, the inability of Rabensburg virus to replicate in vertebrates was interpreted to be in an intermediate stage between arthropod-specific and vector-borne (Aliota and Kramer, 2012) rather than a result of regression. As for the mosquito-specific lineage flaviviruses within the mosquito-borne subgroup (Fig. 14.3 legend), it is proposed that more likely these viruses are the progenitors (or their close siblings) of the mosquito-borne subgroup and have never adapted to vertebrates.

Arthropod-specific viruses closely related to arboviruses

After the discovery of CFAV in 1975, the second insect-specific subgroup flavivirus was isolated from *Culex* mosquitoes in 1981 (Okuno *et al.*, 1984); but, unfortunately further characterization of this virus was not pursued. The increasing number of discoveries of new flaviviruses belonging to the insect-specific subgroup and to the mosquito-specific lineage viruses (within the mosquito-borne subgroup; Fig. 14.3 legend) is not surprising, since the number of undiscovered viruses associated with arthropods is thought to be large for the reasons described in the first several sections of this chapter. Arthropod-specific viruses have been isolated not only from the family *Flaviviridae* but also from the families *Bunyaviridae* (i.e. Ferak virus, Gouléako virus, Jonchet virus), *Reoviridae* (i.e. Fako virus), *Rhabdoviridae* (i.e. Spodoptera frugiperda rhabdovirus), and *Togaviridae* (i.e. Eilat virus).

While the majority of these viruses are expected to be either the pathogens or commensalistic inhabitants of arthropods, the following 2 groups present a classification problem for arbovirologists. First, in the group of new viruses isolated from arthropods, too often initial host range results obtained *in vitro* alone are still incomplete. Thus, until more comprehensive data are obtained, these viruses cannot be readily classifiable either to arthropod-specific non-arboviruses or to possible arboviruses. The second problematic groups are the viruses whose genomic traits are nearly or completely identical to the extant arboviruses. *Eilat virus*, an alphavirus (Nasar *et al.*, 2012) and mosquito-specific lineage flaviviruses within the mosquito-borne subgroup are such examples. Because of the suspected role as progenitor (or its sibling) in the evolution of extant terrestrial alphaviruses, it is highly possible that *Eilat virus* too, like the mosquito-specific lineage in the mosquito-borne subgroup of flaviviruses, has never adapted to vertebrates. Other candidates in this category include Moussa virus, a rhabdovirus (Quan *et al.*, 2010), *Negevirus* (a proposed new group isolated from sand flies and mosquitoes) (Vasilakis *et al.*, 2013), Herbert virus, Tai virus, Kibale virus, Kigluaik virus (family *Bunyaviridae*) (Marklewitz *et al.*, 2013; Ballinger *et al.*, 2014) and St. Croix River virus of the genus *Orbivirus* (Attoui *et al.*, 2001). Still, for these viruses it is worthwhile to examine viral replication in vertebrate cells co-cultivated with arthropod cells.

Vertebrate viruses without a known vector

Among many viruses in tentative arbovirus status (probable or possible) are found the vertebrate viruses which replicate at least in arthropod cells *in vitro* but whose natural vectors are unknown despite repeated attempts to identify. Three possibilities are considered to explain the absence of vectors for those viruses. First, for the viruses with proven replication in arthropods at least under laboratory conditions, vectors will be eventually found if investigated in depth in the field, as it was the case of Gabek Forest virus. Second, some of those viruses with a biphylum host range may be, like nodaviruses and aquatic viruses, non-arboviruses. Third, some of the viruses were once arboviruses but subsequently lost host range in vectors, as hypothesized for three flavivirus members (ENTV, SOKV and YOKV) in the mosquito-borne subgroup.

Host replacement

That change in host range represents host replacement rather than broadening or reduction was exemplified in the change of vectors of VEEV. According to a report, the epizootic strains of the virus (subtypes IAB and IC) do not replicate in several species of enzootic vectors including *Culex* (*Melanoconion*) spp. Apparently, the epizootic strains lost fitness in enzootic vectors when they adapted to the epizootic vector, *Ae. taeniorhynchus* (Brault *et al.*, 2002). This may be an example of pleiotropy.

Re-examination of the tenets of arboviruses

Defining the boundaries of arboviruses was not a simple task for early arbovirologists, as documented in their reports (WHO, 1967, 1985). Given the enormous amount of new data accumulated since, it is worthwhile to re-examine the traits of arboviruses, to find out where we stand now.

Biological transmission

Arbovirus was defined by the WHO on a combination of several traits including biological transmission. This unique mode of transmission is one of the salient features on which the concept of arboviruses as a distinct entity was established in the first place. And yet, according to the definition, 'arboviruses are viruses that are maintained in nature principally, or to an important extent, through biological transmission ....' One of the reasons for not making it an absolute requirement for all viruses in arbovirus classification was to make room for a future accommodation of many viruses (in particular, rodent-borne viruses) then still in a questionable status due to lack of all necessary data (WHO, 1985). Further, partial reproduction of biological transmission under laboratory conditions was accepted to fulfil the requirement for an arbovirus, given difficulties of securing necessary data under natural conditions (WHO, 1967). These unique considerations have become major sources of multiple problems described in this chapter.

The difficulty of confirming biological transmission is well demonstrated by the fact that among over 550 registered viruses listed in the International Catalogue of Arboviruses 1985 (ASTMH, 1985), only approximately 20% were classified as arboviruses; and more than 70% were in a tentative status of either 'probable' or 'possible' arboviruses, primarily for incompleteness of the available data. An examination of the status in the online version of the Catalogue (wwwn.cdc.gov/arbocat/VirusBrowser.aspx) in 2015 revealed little change in the status of the viruses in tentative categories since 1985. Many researchers are certainly surprised to learn of the very familiar 'arboviruses' (such as BFV, VSNJV and Usutu virus) still being in 'possible arbovirus' category and others (i.e. Oropouche virus, Thogoto virus, and Toscana virus) in 'probable arbovirus' category.

In contrast to the difficulty of confirming biological transmission, full genome sequencing within a day or two has become far popular, generating a large amount of molecular information, thanks to recent technological innovations. Thus, a number of the viruses in the 'probable' or 'possible' status have been fully genome-characterized, while confirmation of biological transmission of these viruses lagged far behind. This unbalance of knowledge is a problem arbovirology faces today and is one of the unintended consequences derived from a combination of enormous popularity and interest in molecular research and unfortunately reduced interest in, emphasis on or support to basic biological or field-oriented research (Johnson, 1991; ACAV, 2001).

Furthermore, despite the advancement in sequencing technology, as described earlier in the section of mutation-related issues, unlike under laboratory conditions, confirming biological transmission under natural conditions based on viral sequence data has been difficult because of enormous complication in sampling virus sequences and assigning them to the type of host exactly according to the order in the transmission cycle. Thus, unless a simpler, reliable and more economical method to determine biological transmission be developed, required proof of biological transmission be modified, or other acceptable alternative be established, the number of the viruses fully characterized at molecular levels but still in a tentative status will keep rising.

Despite existence of the ICTV, the absence of any international body that can impact significantly and globally by stressing standardization of the arboviral terms and their usage is another concern, because 'arbovirus' is not a term applicable in viral systematics. In the meanwhile, because the complicated definition of 'arbovirus' was either unknown to, not clearly understood by or not accepted by some virologists as too unrealistic to fulfil, it has become a convenient practice for many of them to liberally apply it to the viruses in a tentative or even questionable status as well. Troublingly, because of the deeply established tradition of holding vertebrates as reservoirs of arboviruses, 'arbovirus' has been defined as 'any virus of vertebrates which is transmitted by an arthropod' (Mahy, 2001). Terminological confusion is already evident in the literature as shown by the use of 'vector' for a haematophagous arthropod in reference to the viruses associated with it even when the viruses are arthropod-specific, by the inclusion of the family *Parvoviridae* in a list of arbovirus families (Clements, 2012), and by defining arboviruses as viruses that are transmitted only to mammalian hosts by arthropod vectors (Conway *et al.*, 2014).

While the definition of arbovirus stressed the importance of biological transmission between haematophagous arthropods and vertebrates, it made no reference to the ultimate fate (including survival) of arbovirus, if all hosts in one of the two phyla become unavailable or fitness to one phylum of hosts is completely lost for any reason. Johnson (1967), based on his strong belief that arboviruses are basically vertebrate viruses, wondered if vectors are absolutely necessary for all arboviruses for the maintenance in wildlife. Others, based on a mathematical model, reported absolute requirement of the two phyla of hosts for arboviral survival (Fine and LeDuc, 1978; Korenberg and Kovalevskii, 1994). The necessity of dependence on two phyla was also supported by the observations of very low rates of vertical transmission in vectors, which were judged to be insufficient for the survival of the arboviruses in arthropods alone (Tesh, 1984). Yet for many others, two phyla of hosts were necessary for genetic/fitness stability or maintenance of virulence (Moutailler *et al.*, 2011).

Although not much is known about the outcome of the maintenance of unselected wild strains of most arboviruses *in vivo* only in one phylum of hosts for many years, as far as T1 ribonuclease fingerprint and electrophoretic mobility of genomic segments are concerned, Toscana virus RNA was found unchanged after 12 transovarial transmission cycles in sand fly for nearly 2 years (Bilsel *et al.*, 1988). Reduced infectivity of VEEV strains serially passaged only in mosquito or in mouse was reported, but 10 passages (Coffey *et al.*, 2008) was too short to provide an answer as to the ultimate survival of the virus in either host. On the other hand, in the colonies of *Ornithodoros* ticks infected with African swine fever virus, if the ticks were continuously fed swine blood free of the virus and maintained without a contact with vertebrate hosts, the colonies eventually became virus-free over time, despite transstadial and transovarial transmissions (Hess *et al.*, 1989). In contrast, soft ticks, once infected with Karshi virus, remained infective to a vertebrate host for at least 8 years (Turell, 2015).

As for vector-borne plant viruses, the following two examples reveal interesting results. Wound tumour virus (a phytoreovirus transmitted by leafhoppers), when maintained in sweet clover for up to 24 years without a passage in insect vectors, lost two genome segments and became incapable of replicating in its vectors *in vivo*, through regression (Reddy and Black, 1977). In contrast, Rice dwarf virus, when transmitted transovarially only in the leafhopper vectors for 6 years, still retained capacity to replicate in rice plants (Honda *et al.*, 2007), suggesting importance of the kind of host selected for this type of studies.

Viruses of arthropod-vertebrate biphylum host range

As an inseparable interest in this discussion, it is worthwhile to view both arboviruses and any other viruses on the borderline (between arthropod-specific and vertebrate-specific viruses) from a slightly different angle and recognize all of them as 'viruses with an arthropod-vertebrate biphylum host range' (hereafter called 'biphylum viruses,' for convenience). Here, the underlying yardstick for designation as an arbovirus (which is unstated in the WHO definition) is level of dependence (or independence) on two phyla of hosts for a viral species survival. While the basis for classification is clarified with this yardstick, the system based on the criterion still does not solve the ambiguity of arbovirus classification because the yardstick is highly subjective and realistically difficult to measure qualitatively and/or quantitatively.

Characteristics of biphylum viruses

It is proposed that many viruses in a tentative arbovirus status and other viruses on the borderline do not need to depend on biological transmission for an absolute viral species survival, despite available proof of transmission obtained under laboratory conditions (including generation of viraemia), serological evidence of neutralizing antibody detected in and/or virus isolation from vertebrates in the field. Those viruses, which are considered essentially the viruses of haematophagous arthropods, may not necessarily depend on bloodmeal but persist replicating in the arthropods while still retaining replicative capacity in the secondary phylum of hosts (vertebrates) even at a low level. For these viruses, the outcomes of sporadic contacts with the secondary phylum of hosts (such as incidental viral replication and even symptomatic infection) associated with blood-feeding activity of the vectors are inconsequential as far as genetic composition and survival of the virus population are concerned. Similarly, for the viruses considered essentially vertebrate viruses, limited replication in haematophagous arthropods is considered an incidental expression of a trait not absolutely essential for viral survival.

Non-essential traits in RNA viruses

How such a non-essential trait is retained is a question that needs to be answered, since evolutionary process usually filters out nonessentials and because RNA viruses can hardly afford to waste genome space due to limited number of genes permissible for a virus species stability (Holmes, 2003). However, contrary to this theoretical limit, actually non-essential functions have been found in RNA viruses. Unusual nonessential traits are often discovered under experimental conditions. As an example, a mutant of Rabies virus adapted to *Drosophila* fly could be isolated (Plus and Atanasiu, 1966); and this viral antigen synthesis and limited replication have been observed in mosquito or fish cells (Table 14.1). Further, many arboviruses replicate in the cells derived from moths (order *Lepidoptera*) (Pudney *et al.*, 1982) and some (DENV, YFV,SLEV, VSIV) replicate in the cells of non-haematophagous mosquito unrelated to transmission of these viruses (Kuno, 1981). In another example, the exact functional significance of carbon dioxide sensitivity conferred by variants of Sigma virus (a rhabdovirus) in dipteran insects (*Drosophila* flies, mosquitoes, and muscoid flies) is still unknown despite numerous studies since the first discovery in 1930s (Longdon *et al.*, 2012). Also, limited levels of replication of many vertebrate-specific viruses in mosquito cells have been revealed (White, 1987). Aquatic vertebrate viruses replicating in the cells of terrestrial, haematophagous arthropods and conversely terrestrial viruses of arthropods replicating in aquatic vertebrates (see 'Virologic attributes of selected environments and hosts' above; and Table 14.1) are other examples. Furthermore, viral replication in an unnatural phylum of hosts has been documented not only in RNA but in DNA viruses (White, 1987; Table 14.1).

At present, the exact non-essential host range determinants of most of these biphylum non-arboviruses have not been investigated genetically; and the functions of non-essential elements of RNA viruses other than virulence and membrane fusion have not been investigated in depth either. For similar observations in vertebrate-specific viruses, some explained persistent retention of an atypical, nonessential phenotype in a minority viral subpopulation even under unfavourable conditions, by introducing a concept of 'memory' (Ruiz-Jarabo *et al.*, 2000). Although in arboviruses expression of an infrequent host range phenotype is generally understood in terms of genetic heterogeneity in viral population (Novella *et al.*, 2007), obviously the genetic mechanisms of stochastic infection by biphylum non-arboviruses are most likely different. For the emergence of epidemic/epizootic strains at irregular intervals from enzootic strains, in addition to mutation, atypical translation (Firth and Brierley, 2012) may be considered one of the possible mechanisms. Similarly, for biphylum non-arboviruses, expression of non-essential element as a result of pleiotropy needs to be considered. Thus, for all biphylum non-arboviruses, the significance of limited (or incidental) replication may be investigated in the contexts of sum total of atypical translation mechanism, cellular environments favourable (including virus receptors) and unfavourable (including innate resistance as well as of virus interference mechanisms) for viral replication in the secondary hosts.

Biphylum viruses most likely evolved independently in many different viral lineages by convergence, each lineage deploying a unique set of genetic and environmental factors. Accordingly, variation in the degree on dependence on biological transmission naturally developed among biphylum viruses, some becoming fully dependent (arboviruses), others becoming only partially dependent (viruses in ambiguous or tentative status), and the rest remaining basically independent (non-arboviruses).

Possible examples of biphylum viruses

It is highly possible that re-examination of all viruses in the Arbovirus Catalogue, Table 14.1 and other documents worldwide in the contexts of host specificity and viral persistence in both arthropod and vertebrate would yield a considerable number of biphylum viruses with little dependence on the secondary phylum of hosts (thus, non-arboviruses). For examples, the wild strains of viruses which reveal replication in the secondary phylum of hosts only after multiple blind passages or by co-cultivation with helper cells (such as Netivot virus, Kotonkan virus and Obodhiang virus), some bunya- and rhabdoviruses (such as Turlock virus and Flanders virus, respectively), some sand fly-borne viruses, VSVI, and other viruses in Table 14.1 (such as Invertebrate iridescent virus 6) may be good candidates for a possible re-classification in the group of basically arthropod-associated biphylum viruses.

What complicates application of this concept, however, is a group of primarily arthropod-associated viruses which minimally replicate in vertebrates but which still depend on them as source of bloodmeal necessary not so much to complete biological transmission cycle but to ensure viral replication and vertical transmission in the arthropods (such as Stone Lakes virus and one of two lineages of Buggy Creek virus). Similar dependence on ingested bloodmeal was also revealed in arboviruses (African swine fever virus and Crimean–Congo haemorrhagic fever virus infecting ticks and Kaeng Khoi virus infecting cimicid bugs). In these ticks and bugs, increased viral replication in the vectors was synchronized with ingestion of bloodmeal. In ticks, this enhanced viral replication was attributed to stimulated proliferation of susceptible new cells in the salivary gland and ovary rather than to increased viral replication in existing cells (Rennie *et al.*, 2001; Dickson and Turell, 1992). For some viruses, virus clearance experiment similar to that used for African swine fever virus in tick colonies (Hess *et al.*, 1989) may be useful to determine vertebrate dependence of vector-associated viruses, even though the method is laborious, time-consuming and of safety concern (especially for flying infected vectors).

Transovarial transmission (TOT)

Interestingly, California encephalitis virus is one of the candidates for reassessment of the status. Although this virus has been classified as an arbovirus for many years, according to the conclusion of a group of arbovirologists including the scientist who coined the term 'arbovirus' in 1958, this virus 'is probably a mosquito virus that is partially adapted to a narrow range of vertebrate hosts but is not reliant on vertebrates for its continuous existence' (Reeves *et al.*, 1983). The authors reached this conclusion primarily because of the unusually high rate (up to 99%, depending on several variables among early studies) of transovarial transmission (TOT) of this virus in the vectors. A survey of three virus families (*Asfarviridae*, *Bunyaviridae*, and *Rhabdoviridae*) revealed more viruses with a similarly high rate of TOT (Tesh, 1984). Indeed, in accordance with this hypothesis, the TOT rate for an insect-specific flavivirus in one study was nearly 100% (Saiyasombat *et al.*, 2011), while the TOT rates for many arboviruses have been much less than 1%.

But, application of TOT rate as a guide for determination of the level of dependence (or independence) on two phyla of hosts, like the determination of viraemia threshold, poses an assortment of complicated issues difficult to resolve, given a considerable variation in the rate depending on virus genotype, vector species selected, location, sampling methods, seasonal variation, number of ovarian cycle that needs to be examined, relevance of laboratory data, and others. The major vector of EEEV in the U.S., *Culiseta melanura*, is such an example. The overwintering mechanism of this virus and TOT in this mosquito are still unresolved since conflicting reports were presented nearly 35 years ago (Morris and Srihongse, 1978). In a recent study, a subpopulation of a principal vector (*Aedes triseriatus*) naturally infected by California encephalitis virus (La Crosse genotype) and demonstrating nearly 100% TOT in an endemic area of North America accounted for only 0.09% of the total surveyed, while the mosquito subpopulations with no TOT accounted for 96.2% (Reese *et al.*, 2010). However, as shown below, the concept of 'stabilised infection' (>90% TOT rate) observed in some mosquito-borne viruses (Turell *et al.*, 1992) was found not applicable to tick-borne viruses whose TOT rates in ticks are much lower (Labuda and Nuttall, 2004).

Low TOT rate, however, may not necessarily pose a problem for viral maintenance in vectors because their high fecundities compensate (Randolph, 1998). Also, lower TOTs in ticks are compensated by longer viral persistence due to longer longevity of the vectors, as shown in ASFV and Karshi virus persisting in soft ticks for at least 2 and 8 years, respectively (Endris *et al.*, 1992; Turell, 2015).

The definition of TOT rate is also an issue. According to Turell *et al.* (1982), a TOT rate (for example, 20%) can be interpreted in 3 ways: (i) all females transmit virus to 20% of their progeny; (ii) 20% of the females transmit virus to 100% of their progeny; and (iii) a combination of (i) and (ii) by which most females transmit at a very low rate but a few females transmit at nearly 100%. The third interpretation fits best the report by Reese *et al.* (2010). On the other hand, it was reported that an *Ornithodoros* tick colony, despite TOT, lost African swine fever virus when ticks were reared without a contact with the vertebrate hosts (Hess *et al.*, 1989). Regardless of conflicting reports on TOT rate, however, it is still prudent to re-evaluate all arboviruses with respect to dependence on two phyla of hosts, using multiple criteria.

Vertebrate viruses with a stochastic biphylum host range

Compared with the number of arthropod-associated viruses which are likely to fall in the group of biphylum non-arboviruses, the basically vertebrate viruses with biphylum trait are smaller in number but include at least the viruses with limited replication in vectors, such as Mokola virus (family *Rhabdoviridae*; genus *Lyssavirus*), three flaviviruses (ENTV, SOKV and YOKV) in the mosquito-borne subgroup, some viruses of fishes, other vertebrate-specific viruses (White, 1987), rodent-associated bunyaviruses and an arenavirus of snake (Table 14.1), and some bat viruses (rhabdoviruses) of the newly proposed genus *Ledantevirus* (Blasdell *et al.*, 2014). Persistent infection of these viruses in vertebrates must be confirmed *in vivo* in one or two suitable species. In reality, however, such a persistence test has been rarely performed for most of these viruses, except for a limited number of known arboviruses, due to a variety of daunting reasons. First, for many of these viruses, true vertebrate hosts are still unknown. Second, colonies of desired vertebrate species are rarely available in most laboratories (i.e. bats) or too expensive (i.e. subhuman primates). Third, the results with laboratory rodents arbitrarily selected and most commonly employed may not always be relevant.

Significance of biphylum host range

If firmly confirmed, the existence of a group of biphylum non-arboviruses which can persist only in one phylum, rather than insufficient amount of field study, may explain why the identities of natural vertebrate hosts of many arthropod-associated viruses and vectors of vertebrate viruses have remained unknown for so many years and why clear segregation of true arboviruses from non-arboviruses is difficult.

If this proposition is acceptable, then, the next question which needs to be answered is whether these 'non-arboviruses' represent one of the three intermediate stages of host range transition, vertebrate viruses transitioning to become arthropod viruses, arthropod viruses evolving to become full-fledged arboviruses, and arboviruses which partially lost host range through regression (as proposed for ENTV, SOKV and YOKV). In contrast to this somewhat teleological reasoning, the other alternative possibility is that for some (but not all) of these viruses replication in the secondarily acquired phylum of hosts reflects a purely incidental sharing of virus receptors and/or cellular environments (including weak virus inhibitory mechanism) favourable for opportunistic viral replication; hence it does not have a significance in evolutionary direction of the host range for these viruses.

Actually, it is highly probable that all the above possibilities are necessary to explain the heterogeneous observations better. As summarized later, heterogeneity is the trademark of arboviruses, the products of variable mechanisms which independently evolved and contributed to the evolution of biological transmission as a shared trait among diverse groups of virus lineages. Because the mechanisms involved in the evolution of these viruses were heterogeneous, so is the significance behind the replication in secondary phylum of hosts.

Viraemia and simulated transmission experiment

The requirement of viraemia in infected vertebrates that defines arboviruses has been disputed by the discoveries of nonviraemic transmission of arboviruses by ticks, black flies, and mosquitoes (Jones *et al.*, 1987; Mead *et al.*, 2000; Higgs *et al.*, 2005; Reisen *et al.*, 2007; Van den Hurk *et al.*, 2009). The possibility of technical insensitivity of the virus titration methods used for viraemia in the past as the primary reason for negative result, however, has not been entirely ruled out yet. Remarkably, this is another example of convergence, because nearly identical viral transmission mechanism between infective and uninfected vectors cofeeding the same plant without viral replication in the plant evolved in some vector-borne plant virus lineages (families *Dicistroviridae* and *Reoviridae*).

Considerable variation in viraemia level depending on passage history or strain of virus used has also been recognized. Viraemia may occur irregularly or intermittently in some reptiles and birds, as in the infections by some alphaviruses (Kuno, 2001a). Also, not all vertebrates exhibiting viraemia serve in biological transmission. For example, experimental equine infections with neurotropic arboviruses (such as WEEV, EEEV, JEV, and WNV) may demonstrate very short and low-titred viraemia. Realistically, however, in nature horses do not serve for further transmission, because they are dead-end hosts. Similarly, in humans infected by neurotropic arboviruses (with some exceptions, such as WNV) viraemia rarely develops. For examples, none of the California serogroup bunyaviruses have been isolated from blood or cerebrospinal fluid; and isolation of Japanese encephalitis virus from human blood is very infrequent, although several such cases have been documented among more than a few million JE cases reported in Asia over nearly 80 years.

In laboratory experiments to confirm viraemia to satisfy arbovirus definition, very narrow window of opportunity to detect viraemia and low concentration of infectious virus in blood of vertebrates frequently encountered with the use of much-preferred low-passage strains of virus, as well as other variables, render determination of viraemia level and threshold difficult. Also, some non-neurotropic viruses repeatedly neuroadapted only in the brain of laboratory rodents (following a traditional method), when subcutaneously inoculated into vertebrates, often generate viraemia of substantially low titre because of selection of an irrelevant phenotype.

The validity of the threshold of infection determined in artificial experiments is another concern because of uncertainty of and variation in minimal level (threshold) of viraemia necessary for transmission for a particular virus (Lord *et al.*, 2006). Although many researchers arbitrarily consider viraemia thresholds near 2–3 log PFU/ml blood to be acceptable for transmission, in an evaluation of a DENV-2 vaccine candidate, a small number of female mosquitoes could acquire infection by biting a volunteer with a viraemia titre less than 0.1 PFU/ml (Bancroft *et al.*, 1982). In other case, the threshold obtained by means of pledget feeding was often found more than a few orders of magnitude higher than that by feeding on viraemic mammals with a similar titre and thus unreliable (Jupp *et al.*, 1976; Turell, 1988). In a study of Louping ill virus infection of three kinds of vertebrates (laboratory rodent, 1-day old domestic chick and sheep) demonstrating a similar level of viraemia, larvae of *Ixodes ricinus* were infected by feeding on the chick and sheep but not on rodent; between the two 'competent' vertebrates, only the threshold in sheep was relevant because chick is not a natural host. Furthermore, the threshold of viraemia necessary for infecting nymphs was 10-fold less than for larvae (Reid, 1984).

Simulation experiments are often technically demanding. The difficulty and need of technical improvisation for the confirmation of biological transmission by small vectors (such as sand flies and midges) or in small vertebrates are well known (Watts *et al.*, 1988; Kading *et al.*, 2014). For infecting cattle with VSNJV by the bite of black fly, selection of bodily site is important, because clinical sign develops in mouth, nostril, and coronary bands of the foot but not in the neck (Mead *et al.*, 2009). Also, it was reported that mosquitoes alternatively inoculate virus extravascularly rather than directly into the vascular system (Turell and Spielman, 1992; Styer *et al.*, 2007), which further complicates the optimal design of laboratory experiments. Regarding difficultly of standardizing experimental procedure, for transmitting Qalyub virus to suckling mice through bite of infective tick nymphs, it was found necessary to feed not individually but in groups of 11–20 nymphs per mouse (Miller *et al.*, 1985). Similarly, a few mosquitoes rather than one mosquito infected by Chandipura virus per mouse were absolutely necessary for transmitting the virus to mouse (Ramachandra Rao *et al.*, 1967).

Although not explicitly stated in the WHO definition, in laboratory experiments for biological transmission it should be possible to re-isolate virus from experimentally inoculated animals, following Koch's Postulates. In the cattle infected by VSNJV, however, no infectious virus could be recovered from any of the infected tissues despite development of pathological signs and detection of viral RNA (Mead *et al.*, 2009).

Extrinsic incubation period (EIP)

It was pointed out that the early definition of arboviruses was based primarily on the transmission by mosquitoes. It was criticized that direct application of the EIP concept to tick-borne transmission is difficult because of significant differences in tick's haematophagous life stage, life cycle, and much longer longevity (Nuttall *et al.*, 1994). It is stressed that the patterns of replication of the arboviruses associated with ectoparasitic arthropods (such as ticks and lice) in relation to vertebrate hosts are different from those in mosquitoes, which is why direct applications of some of the arbovirology concepts based on mosquito vectors including EIP are difficult. As an example, *Ornithodoros* ticks infected with Karshi virus, once they became infective, could continuously transmit the virus to rodents for longer than 1 year; but the timing of encounter with vertebrate hosts is unpredictable, which rendered EIP determination difficult (Turell *et al.*, 2004). In other example, the EIP of Fort Morgan virus (FMV) in the sedentary cimicid bugs in bird nests is different depending on the biology of the avian occupant of the nest (resident house sparrow vs. migratory cliff swallow) and ambient temperature. Thus, with resident house sparrow as host, the EIP of FMV at 22° C was as short as 18 days; but in a period during which temperature was dropped to near 4° C, it was as long as 311 days (Rush *et al.*, 1980). EIP determination is also complicated when vertebrate hosts are aquatic.

Dead-end host

The vertebrates which recovered from RNA arboviral infection and which are no longer essential in transmission have been traditionally considered dead-end hosts. However, over-emphasis of this concept leads to under-estimation of the role of infected vertebrates as amplifying hosts as well, particularly under an intense rate of transmission. Cattle infected by VSNJV are good examples (Smith *et al.*, 2011). Also, many humans infected by CHIKV, DENV, or YFV in urban areas during epidemics serve as amplifying hosts.

In rodents infected by TBEV, despite persistence of neutralizing antibody, viral transmission to uninfected vertebrates still occurs (Nuttall and Labuda, 2003). In the sheep previously infected by Crimean–Congo haemorrhagic fever virus but recovered, under certain conditions, viral transmission by ticks in second infection was found possible (Wilson *et al.*, 1991). Similarly, in the rats first immunized with RVFV and later challenged intranasally, infectious virus could be isolated 28 days after challenge, despite high levels of neutralizing antibody (Anderson *et al.*, 1991). It should be noted that incompletely neutralized arboviruses in virus-antibody immune complex are still infectious in vectors, facilitating another rounds of transmission (Johnson and Varma, 1976; Wheeler *et al.*, 2012). To further complicate this concept, African swine fever virus infection in swine rarely results in induction of Nt antibodies (Viñuela, 1985).

Furthermore, determination of immune status by plaque reduction neutralization test (PRNT) is not a simple matter. Some arboviruses do not produce clearly readable plaques even after many passages for adaptation in any commonly available vertebrate cell lines. Also, accuracy of serological test depends on multiple factors including the criteria selected for the measurement of Nt antibody titre. Induction of cross-reactive Nt antibodies in flaviviral infections presents another kind of problem. Some of the dengue immune individuals with pre-existing Nt antibodies to a specific serotype (determined either with 50% or with 90% plaque reduction criterion) were nonetheless later infected naturally by the same serotype (Endy *et al.*, 2004; Sirivichayakul *et al.*, 2014). Application of immunofluorescence test, such as fluorescent focus inhibition, for some viruses is also difficult because of a difficulty of obtaining an antiserum with a sufficient antibody titre in experimental animals. If PRNT is better substituted by a molecular technique (i.e. RT-PCR), reliable criteria for neutralization broadly applicable to most arboviruses are yet to be established, because distinction of infectious genome from PCR-reactive noninfectious genome and/or its fragments must be established first.

Direct transmission and mechanical transmission

Direct transmission of arboviruses was incorporated in the revised definition of arbovirus (WHO, 1985). A compilation of multiple modes of direct transmission illustrated in a large number of the examples (Kuno, 2001b) testifies its common occurrence. Just like human infections as a result of drinking raw milk from TBEV-infected goats or cows, red grouse becomes infected by eating ticks infected by LIV (Gilbert *et al.*, 2004). Also, intranasal infection and oral infection have long been of strategic interest for the non-invasive administration of arbovirus vaccines (i.e. dengue vaccines) since 1940s. Even sexual transmission of flaviviruses in human and animals was found possible (Habu *et al.*, 1977; Foy *et al.*, 2011). Recent cases of 'person-to-person' transmission of a new tick-borne phlebovirus in China (Bao *et al.*, 2011) represent not an emergence of a new mode of transmission but the classic contact transmission reported multiple times in the fatal YF infections among researchers in the early part of the 20th century.

Direct transmission of arboviruses between vectors is also known, as shown in the aforementioned examples of *per os* larval infection in aquatic environments. Both intra- and inter-generational transfers of TBEV between infected (nymphs or adults) and uninfected (larva or nymph) occur through a vertebrate host on which multiple generations of ticks cofeed (Korenberg, 1976). Venereal transmission is also known for some arboviruses. Cannibalism of weakened or dead larvae (infected by a densovirus) of a moth (*Galleria mellonella*) infesting beehives or of dying shrimps (infected by Taura syndrome virus) by healthy caterpillars or shrimps, respectively, is another mode of direct transmission in arthropods. Horizontal transmission of iridoviruses among isopods, tipulid (crane fly) larvae, caterpillars, mosquitoes and crickets is also known. Although little has been studied for arboviruses, this mode of transmission by predatory mosquito larvae is highly possible.

Some midges in Asia in the genera of *Trithecoides* and *Culicoides* are known to engage in blood feeding on blood-engorged female mosquitoes (*Aedes*, *Anopheles*, *Armigeres*, and *Culex*). BTV has been isolated from the aforementioned two genera of midges in Asia. Little is known at present if such an unusual feeding behaviour of insects has any significance in disseminating arboviruses between haematophagous insects (James, 1969; Ma *et al.*, 2013). Also, faecal–oral route of transmission of insect-specific viruses well recognized in nonhaematophagous insects (i.e, lepidopteran caterpillars) and haematophagous Triatoma bugs (Marti *et al.*, 2014) has not been investigated much for arbovirus vectors, but is possible in ectoparasitic haematophagous arthropods living in congregation. As an example, it was reported that faeces of ixodid tick larvae feeding on TBEV-infected mice was infectious (Benda, 1958).

Mechanical transmission has been traditionally excluded from the definition of arbovirus but has been recognized to be crucially important in veterinary virology (Carn, 1996). This exclusion is puzzling, because, as described above, another mode of transmission without a need of viral replication in vectors for transmission to vertebrates, direct transmission, was accepted for arboviral transmission. It is important even for the viruses vertically and/or directly transmitted. As an example, spread of Equine infectious anaemia virus (a lentivirus and a non-arbovirus) heavily depends on this mode of transmission by blood sucking flies. Lumpy skin disease virus (a poxvirus) is not only mechanically transmitted by but transstadially or vertically in ticks and is even detectable in tick saliva (Lubinga *et al.*, 2013; Tuppurainen *et al.*, 2013). In vector-borne plant viruses, following four modes of transmission were interpreted to suggest an evolutionary trend towards biological transmission in the following order: non-persistent and non-circulative; semipersistent but noncirculative; persistent and circulative; and persistent and propagative. The first three types and the last type are analogous to mechanical transmission and biological transmission of arboviruses, respectively. Little is known if mechanical transmission of animal viruses by arthropods requires a helper (non-structural, virus-encoded protein), as in some aphid-borne plant viruses (Gray and Banerjee, 1999).

The distinction of mechanical from biological transmission is blurred when 'unnatural' vectors support not only transient arboviral replication but engage in promiscuous blood-feeding activities, such as WNV-infected ticks and other examples presented in this chapter. Records of mechanical transmission have been accumulating for such arboviruses as CHIKV, RVFV, and VSV, WEEV, and WNV (Rao *et al.*, 1968; Barnett, 1956; Hoch *et al.*, 1985; Smith *et al.*, 2009; Doyle *et al.*, 2011). Mechanical transmission was found epizootically significant for VEEV in the Americas and RVFV in Africa. High virus titre in viraemia, multiple feeding, and interrupted feeding collectively favour this mechanism (Davies, 1990).

Reservoir and amplifying host

'Reservoir' has traditionally referred to a population of hosts which are persistently infected with a causative agent of a disease and which can also serve as a source in the transmission of the agent to the populations of other species. Historically, it is most likely that the concept of regarding vertebrates as reservoirs of arboviruses had its origin in the discoveries of the modes of mosquito-borne transmission of filarial and malarial parasites by Patrick Manson and Ronald Ross, respectively in late 19th century. The word 'reservoir host' was used as early as in 1913. According to a historical study, in the 1920s, rodents, domestic animals, and birds were clearly linked to the causation of a variety of human illnesses (Hardy, 2003). The concept of zoonotic origin of the causes of human diseases (such as bubonic plague, dysentery, malaria, and influenza) was further popularized in the publications by a number of microbiologists in early 20th century, in particular Theobald Smith (1934) and Hans Zinsser (1935). For arboviruses, from the YF investigation by Harold Wolferstan Thomas in Brazil in 1905 evolved a suspicion of the involvement of sylvan animals in the disease transmission. The intriguing report of the coincidence of YF outbreak and die-off of monkeys in a forest in Trinidad by Andrew Balfour (1914), YFV isolation in monkeys in West Africa by the Rockefeller Foundation researchers in 1927, discovery of jungle YF in Colombia by Roberto Franco in 1907 and by Fred Soper in Brazil in 1937, as well as the early discoveries of flaviviruses without a vector (NKV subgroup flaviviruses) in rodents and bats in North America collectively solidified the concept of reservoirs of arboviruses in wildlife (Theiler and Downs, 1973; Johnson, 1991; Kuno, 2014). Initially, lower vertebrates were suspected to be responsible for the maintenance of arboviruses in nature (Theiler and Downs, 1973). With the prevalent notion of vertebrates as reservoirs of arboviruses in the background, culling livestock or wildlife population was sometimes proposed as a means of mitigating the spread of some arboviral diseases.

Although the term 'reservoir' in the causation of diseases has been used to identify the biotic origins of pathogens in nature, it has been a source of controversy because of lack of or ambiguity of the definition, disagreement over definition, contradiction, dispute over which components (including habitat and minimal population size of reservoir as well as length of reservoir persistence) of the definition to emphasize and/or misunderstanding (Haydon *et al.*, 2002; Ashford, 2003; Higgs and Beaty, 2005; Clements, 2012).

Long-term persistence of infectious virions or RNA of arboviruses in experimental animals (such as laboratory rodents and subhuman primates) and rare human cases have been reported. Their scientific values aside, however, the relevance of those laboratory or exceptional data in the context of reservoir has been questionable, because these animals are not natural hosts for the viruses used and human victims in rare examples would not serve as reservoir. Although the term 'reservoir' is still widely used to refer to vertebrate hosts today, an exhaustive survey of literature did not uncover a reliable evidence of vertebrate reservoir for any RNA arbovirus, leaving only vectors as reservoirs (Kuno, 2001a). Thus, the vertebrate hosts of RNA arboviruses should be regarded as 'amplifying hosts' but not reservoirs. It was also concluded by others that, by the current definition, reservoir could not be distinguished from amplifying host (Clements, 2012). Accordingly, with an exception of African swine fever virus (a DNA virus) with both hogs and ticks as reservoirs, the earlier proposal of vectors as true reservoirs (Philip and Burgdorfer, 1961) still stands. As mentioned earlier, soft ticks infected with Karshi virus remained infective to a vertebrate for nearly 8 years (Turell, 2015).

Currently, a virus of interest is BTV which has been reported to have 'overwintered' in Europe (Wilson *et al.*, 2008). A recent report of persistence of viral RNA of serotype 25 of BTV in the blood of goats for 2 years (Vögtlin *et al.*, 2013) needs to be interpreted with caution, as described below. Previously, reports of persistence of this virus for more than 3 years in a bull or in the skin (Luedke *et al.*, 1977; Takamatsu *et al.*, 2003) were later found to be not reproducible (Walton, 2004; Melville *et al.*, 2004). For other similar reports of long-term virus persistence, it was strongly speculated that multiple, temporally overlapping episodes of contact (direct) transmission of the virus between infected and healthy animals congregated in small spaces over a long period could be an alternative explanation of such reports of 'overwintering' (Batten *et al.*, 2014). Residual, localized vector activity in unusually warm winter is another possibility.

As for long persistence of viral RNA (or its fragments) in infected overwintering animals, it has been observed in more than several arboviruses, such as JEV, WNV, BTV and VSV. In many of these reports, viral RNA detectable by RT-PCR persisted, for examples for 111–222 days in the ruminants infected by BTV compared with 21 days of viraemia (Bonneau *et al.*, 2002) or up to 2452 days in the urine of humans infected with WNV (Murray *et al.*, 2010). In all similar reports of natural infections, when spring arrived after several months, infectious virus has never been isolated from those animals. The mechanism of long-term shedding of viral RNA (or fragments) without detection of infectious virions has never been elucidated, despite its importance.

Proper understanding of the required trait of true reservoir is also important for more clearly analysing why persistent dengue endemicity (but not sporadic outbreaks of short durations) in temperate regions infested by *Ae. albopictus* has never been established or unravelling a century-old medical enigma, 'no YF in Asia despite geographic distribution of Ae. aegypti '(Dudley, 1934; Frederiksen, 1955; WHO, 1985). The latter phenomenon is akin to 'anophelism without malaria' in parts of Europe and has been recognized since the publication of the first accurate geographic distribution maps of YF occurrence and of the vector in the world (Boyce, 1911). For this puzzle, a clear understanding of the true reservoir (vector) of this virus and of the ecological conditions in sylvatic areas of Asia conducive to stable maintenance of viral endemicity in the broader context of vectorial capacity rather than only high replicative potential [vector competence] in a given ecological setting is more important.

This is because establishment of sylvatic transmission is the prerequisite for YF anywhere in the tropics.

The above YF puzzle also has a historic importance and deserves a brief mention. After the first (and the only) documented case of YF in the Pacific in 1911 (Morris, 1995), on an advice of Patrick Manson, prevalence of *Ae. aegypti* in tropical Asia was surveyed. The fear of introduction of YF from the Americas to Asia through opening of the Panama Canal was then communicated by British officials to Wickliffe Rose [director of the International Health Commission of the Rockefeller Foundation (RF) and who was on an Asian tour at the time]. This prompted Rose to adopt YF research as a major philanthropic mission of the RF in 1915, leading not only to the historic isolation of the first human-pathogenic virus (YFV) and development of its vaccine but to the emergence of a distinguished group of pioneering virologists in that institution whose contributions, as well as contributions from other virologists elsewhere in the world, were indispensable for the later birth of arbovirology as a unique branch in virology.

Classification of the viruses with a single phylum host range and disease causation of arboviruses

Interpretation of the following virus–host associations with respect to arbovirus status is complicated. The first problem concerns the so-called viruses of 'restricted host range.' In this chapter, the word was applied only to the viruses which phylogenetically cluster with known arboviruses but which can replicate in only one phylum of hosts. However, in the absence of consensus of the term, other researchers have broadly included the viruses with host range in only one phylum, such as 'insect-specific viruses' and NKV subgroup of flaviviruses as well. The term 'NKV' (for no known vector) is used in this chapter only because it has been extensively used in the past and familiar to most readers. Confusingly, the term excludes other flaviviruses without a known vector (such as ENTV, SOKV, and YOKV). Also, the term implies a remote possibility of finding a vector in the future, however small the possibility may be. In reality, however, it may not be realistic, because inability of replication in arthropod cells was confirmed *in vivo* (Hurlbut and Thomas, 1969) and *in vitro* genetic studies (see 'Genetic determinants of host range among arboviruses' above). Furthermore, none of the early intense field investigations (Theiler and Downs, 1973) and subsequent research yielded a vector in nearly 70 years. Such terms as 'NKV' and 'host restriction' were most likely coined to refer to the exceptions to the traditional notion that the viruses studied in the context of arbovirus research are supposed to have hosts in arthropods and vertebrates, an assumption that is abundantly found problematic in this chapter. The usage of these terms will almost certainly become a source of complication in the future unless the terms are better defined or new terms coined, since these groups already exist in multiple arbovirus families.

Disease causation has never been an absolute requirement in the definition of arbovirus. And yet, a large proportion of the WHO guideline for arboviruses is devoted to clinical syndrome, epidemiology, pathogenesis, diagnosis, treatment and control (WHO, 1985). Because of the long tradition of regarding vertebrates as reservoirs of arboviruses, human infections by these viruses have been traditionally termed zoonoses. However, if one follows the system of disease nomenclature, given absence of vertebrates serving as true reservoirs of RNA arboviruses, the scientifically accurate term would be either 'arthroponosis' or 'arthropozoonosis' (Hubálek, 2003), the terms probably nobody has ever used. Although many arboviruses isolated early fortunately produced an overt disease more readily in laboratory animals, for many other viruses numerous blind passages were necessary before manifestation of a disease.

Naturally, as a consequence of over-emphasis on disease occurrence, when researchers discovered in the course of arbovirus investigation vertebrate viruses without a known vector and arthropod viruses without a known vertebrate host (and hence no disease), they placed many of these 'orphan viruses' low in research priority for uncertainty of arbovirus status or lack of medical or economic importance. With this practice, except for fortuitous isolations of a small number of viruses with a unique significance, discovery of arthropod-associated viruses without a disease association has been largely neglected. Ironically, many of these orphan viruses have turned out to be important viruses in viral genetics, evolution, ecology and systematics. In reality, the true diseases in natural vertebrate hosts for a large number of viruses listed in the International Catalogue of Arboviruses (ASTMH, 1985) also remain unknown, because vertebrate pathology has been studied artificially only in unnatural laboratory animals (i.e. rodents). To make the matter further complicated, some arthropod-associated viruses (such as an orbivirus, JKT-7400; Table 14.1) cause a disease in vertebrates without replication. Toxic or immunopathological effects of viral antigens could not be excluded.

Heterogeneity

Although a set of shared traits was used to establish arboviruses as a distinct group, as more data accumulated, it became evident that many of the viral traits are far more heterogeneous than homogenous among virus lineages, making application of a unified principle, mechanism, definition, rule, or understanding difficult in many aspects of arbovirology, as described below.

First, haematophagy evolved not only in the arthropods but also in other disparate groups of organisms, such as nematodes, leeches, fishes, birds, and mammals. In insects, even some moths are haematophagous; and some midges blood-feed on engorged female mosquitoes. And, the mechanisms responsible for this evolution among haematophagous groups are not the same. Second, the virus lineages which independently adapted to or evolved within haematophagous arthropods for biological transmission derived from diverse groups, differing in origin, genome organization, mode of replication, and interactions with host. Third, the bionomics, haematophagous life stage and blood feeding behaviour of vectors as well as the factors involved in vertebrate host selection in terms of favourable ecological conditions are diverse. The mechanisms of blood digestion between insects and ticks are fundamentally different. Thus, presence and efficacy of barriers (such as midgut barrier and salivary gland barrier) that affect vector competence are variable among vector groups.

The factors in viral persistence, including vertical transmission and trans-stadial transmission in vectors, are also heterogeneous. For example, in contrast to the pattern in mosquitoes, virus persistence is disrupted and titre sharply drops or may even disappear at moulting in ticks, while it lasts far longer in ectoparasitic lice. Fourth, vertebrate factors facilitating tissue specificity and determination of host range for viruses are also heterogeneous. For example, among tick-borne viruses, the initial sites of replication (before subsequent dissemination in body) of ASFV, Colorado tick fever virus, CCHFV, and TBEV are lymphocytes, erythrocytes, mononuclear phagocytes, and Langerhans cells, respectively; and many neurotropic viruses demonstrate a proclivity of longer persistence in nervous tissues. On the other hand, the viscerotropic viruses which replicate in other tissues generally do not persist long, with an exception of Colorado tick fever virus. Naturally, on the part of vectors, dependence on viraemia for transmission is variable, because viraemia may or may not develop, depending on virus and vertebrate. Fifth, the mechanisms involved in the evolution of the viruses with hosts in two phyla (vectors and vertebrates) are most likely multiple, the one responsible for biological transmission being only one of them. Sixth, as a consequence of multiple sources of heterogeneity above, the mechanisms and determinants (genetic as well as ecological) involved in virulence expression, virus survival strategy, and evolutionary history of host range shift are similarly heterogeneous. In sum, for many traits of arboviruses heterogeneity is the common denominator.

Expanding scope of research on invertebrate-associated viruses

Counterproductiveness of the traditional practice of subdividing virology on the basis of host group (such as plant virology, veterinary virology, insect virology, arbovirology, etc.) and the necessity of pooling the ideas, data, and resources among virologists across subdivisions had been keenly recognized by early 1970s. Regrettably, the progress has been very slow in this respect, as reflected in a critical comment addressing the ambiguity of the boundaries of arbovirology (Junglen and Drosten, 2013).

As for the expanding field of invertebrate-associated viruses in aquatic environments, it is important not to underestimate the significance of the accumulating amount of evidence that the viral gene pool in the aquatic environments is far greater than that in terrestrial environments and that many viruses in the two classes of environments are related (Brum and Sullivan, 2015). In fact, the concept of tracing the origins of terrestrial viruses to the ocean reservoirs evolved in the studies of caliciviruses (Smith *et al.*, 1998). Thus far, the role of copepods, shrimps, and other aquatic arthropod vectors involved in the transmission of viruses to fishes (Table 14.1) has been mostly mechanical and/or food-chain-based transmission. But, as the data confirming viral replication in aquatic arthropods accumulated, the possible role of virus-infected parasitic copepods in biological transmission of Taura syndrome virus (TSV) and White spot syndrome virus in aquatic environments has been more strongly suspected (Overstreet *et al.*, 2009; Mendoza-Cano *et al.*, 2014). The proposal of aquatic origin of alphaviruses (Forrester *et al.*, 2012) has further added an importance to studying aquatic environments in arbovirology.

Vectors involved in transmission are most likely not limited to the Phylum *Arthropoda*. In fact, replication of Infectious haematopoietic necrosis virus and Spring viraemia of carp virus in biological transmission and of *Viral haemorrhagic septicaemia virus* in mechanical transmission by leeches (Phylum *Annelida*) have been reported (Mulcahy *et al.*, 1990; Faisal and Schultz, 2009). Marine leeches are also suspected to play a role at least as mechanical vectors of the herpes virus responsible for fibropapillomatosis of marine turtle (Greenblatt *et al.*, 2004). More recently, a possible role of leech in the transmission of mimiviruses to humans was speculated because of the isolation of such a giant dsDNA virus from leech used for medical treatment and suspected aetiological association with the cases of human pneumonia (Boughalmi *et al.*, 2013b). Accordingly, the earlier interest in leeches as reservoirs of animal viruses (Shope, 1957) has been rekindled.

In terrestrial environments, like the TSV transmission from copepods to shrimps in aquatic environments, highly complicated transmission patterns of polydnaviruses from parasitic wasps to lepidopterous hosts and of Kashmir bee virus (*Dicistroviridae*) or Sacbrood virus (*Iflaviridae*) by parasitic varroa mites to honeybees have been recognized (Whitfield and Asgari, 2003; Shen *et al.*, 2005). It has been proven that non-haematophagous insects nonetheless become infected with arboviruses by predation on virus-infected arthropods, and then themselves ingested by vertebrates, thus infecting vertebrates indirectly (Gillett, 1958). Horizontal transmission of vertically transmitted Sigma virus and related viruses that cause CO_2 sensitivity could have occurred between *Drosophila* flies and other dipteran families, because the viruses causing the sensitivity are also found in muscoid fly (*Muscina stabalans*) and *Culex* mosquitoes. Although in the past host range shift of animal viruses has been almost exclusively focused on vertebrates, these represent examples of arthropod-arthropod transmission of viruses and hence arthropod host range shift of viruses, a field poorly researched in arbovirology. To further complicate the mode of vector-borne virus transmission, aggravation of parasitic infection in humans by the endosymbiotic viruses of protozoan parasites (such as *Leishmania* RNA virus 1 within *Leishmania* parasites) has been reported (Zangger *et al.*, 2013). Coupled with highly prevalent modes of direct transmission and indirect transmission through food chain, these examples of transmission mechanisms constitute an important source of information to unravel the histories of host range shift in arboviruses.

Regarding beneficial applications of arthropod-associated viruses, besides the prevalent applications of baculovirus vectors in virological or molecular genetic studies, other research applications ranging from recombinant vaccine development based on insect-specific viruses, alphaviruses, VSV and others to therapeutic usage of viruses, such as M1 (closely related to

Getah virus and Ross River virus) and a genetically modified oncolytic *Maraba virus* (a vesiculovirus from sandflies), have been in progress (Brun *et al.*, 2010; Lin *et al.*, 2014). Discoveries during research in some of the insect- or vector-associated viruses significantly impacted on other branches of virology. 'Insect cypoviruses were the first to reveal 7-methyl guanosine cap structure on eukaryotic mRNAs; Cricket paralysis virus has provided insight into novel mechanisms for initiation of protein synthesis; nodaviruses have revealed the first example of host RNA silencing that is suppressed by animal virus; and polydnaviruses of insects disclosed a remarkable symbiotic relationship between virus and host that involves virus-mediated immune suppression' (Friesen, 2007).

As a result of recent focusing on the viruses of hitherto poorly investigated groups of organisms, remarkable discoveries have been reported almost monthly. As an example, the genome of an isopod is now recognized to have endogenous viral equivalents representing 4 viral families (*Bunyaviridae*, *Circoviridae*, *Parvoviridae*, *Totiviridae*) and order *Mononegavirales* (Thése *et al.*, 2014). From a nematode a new member of the family *Flaviviridae* and a phlebovirus in the tick-borne *Uukuniemi* group were discovered (Bekal *et al.*, 2011; Bekal *et al.*, 2014). Thus, the family *Flaviviridae* now has two new members, one associated with a nematode and another with a plant (Atsugi *et al.*, 2013). Sequences similar to a few NS genes of the genus *Flavivirus* have been found in an unusual 4-segmented RNA virus isolated from ticks (Qin *et al.*, 2014). Ever-increasing number of discoveries of new viruses with unusual traits will continuously demand a constant revision of virus systematics and of the definition of virus species.

Deeply impressed by the close relationships of viruses infecting arthropods, vertebrates and plants, Kenneth Manley Smith quoted the words of Aristotle who had recognized an infinite variety of living things, 'the boundaries between them are indistinct and doubtful' (Smith, 1967). The miscellaneous issues raised and opening of many new frontiers described in this chapter remind us of the necessities of revising arbovirus definition, of reassessing the scope of arbovirology and the boundaries of arboviruses with a broader vision and of being prepared to face many surprises and challenges that await us.

References

ACAV (American Committee on Arthropod-borne Viruses) (2001). Identification of arboviruses and certain rodent-borne viruses: reevaluation of the paradigm. Emerg. Infect. Dis. *7*, 756–758.

Adelman, Z.N., Miller, D.M., and Myles, K.M. (2013). Bed bugs and infectious diseases: a case for the arboviruses. PloS Pathogens *9*, e1003462.

Aguas, R., and Ferguson, N.M. (2013). Feature selection methods for identifying genetic determinants of host species in RNA viruses. PloS Comput. Biol. *9*, e1003254.

Ahlquist, P., Strauss, E.G., Rice, C.M., Strauss, J.H., Haseloff, J., and Zimmern, D. (1985). Sindbis virus proteins nsP1 and nsP2 contain homology to non-structural proteins from several RNA plant viruses. J. Virol. *53*, 536–542.

Aitken, T.H., Kowalski, R.W., Beaty, B.J., Buckley, S.M., Wright, J.D., Shope, R.E., and Miller, B.R. (1984). Arthropod studies with rabies-related Mokola virus. Am. J. Trop. Med. Hyg. *33*, 945–952.

Aliota, M.T., and Kramer. L.D. (2012). Replication of West Nile virus, Rabensburg lineage in a mammalian cells is restricted by temperature. Parasit. Vectors *5(293)*, doi: 10.1186/1756-3305-5-293.

Althouse, B.M., and Hanley, K.A. (2015). The tortoise or the hare? Impacts of within-host dynamics on transmission success of arthropod-borne viruses. Phil. Tr. R. Soc. B. Biol. Sci. *370*, pii: 20140299.

Anderson, G.W., Lee, J.O., and Anderson, A.O. (1991). Efficacy of Rift Valley fever virus vaccine against an aerosol infection in rats. Vaccine *9*, 710–714.

Andrewes, C.H. (1957). Factors in virus evolution. Adv. Virus Res. *4*, 1–24.

Arunrut, N., Phromjai, J., Gangnnongiw, W., Kanthong, N., Sriurairatana, S., and Kaitpathomchai, W. (2011). *In vitro* cultivation of shrimp Taura syndrome virus (TSV) in a C6/36 mosquito cell line. J. Fish Dis. *34*, 805–810.

Asher, D.M. (1979). Persistent tick-borne encephalitis infection in man and monkeys: relation to chronic neurologic disease. In Arctic and Tropical Arboviruses, E. Kurstak, ed. (Academic Press, New York), pp. 179–195.

Ashford, R.W. (2003). When is a reservoir not a reservoir? Emerg. Infect. Dis. *9*, 1495–1496.

ASTMH. (1985). International Catalogue of Arboviruses 1985, N. Karabatsos, ed. (American Society of Tropical Medicine and Hygiene, San Antonio, TX).

Atsugi, G., Tomita, R., Kobayashi, K., and Seikine, K.T. (2013). Prevalence and genetic diversity of an unusual viruses associated with kobusho disease of gentian in Japan. J. Gen. Virol. *94*, 2360–2365.

Attoui, H., Stirling, J.M., Munderloh, U.G., Billoir, F., Brookes, S.M., Burroughs, J.N., de Micco, P., Mertens, P.P., and de Lamballerie, X. (2001). Complete sequence characterization of the genome of the *St. Croix River virus*, a new orbivirus isolated from cells of *Ixodes scapularis*. J. Gen. Virol. *82*, 795–804.

Attoui, H., Jaafar, F.M., Belhouchet, M., Biagini, P., Cantaloube, J.F., de Micco, P., and de Lamballerie, X. (2005). Expansion of family Reoviridae to include nine-segmented ds RNA viruses: isolation and characterization of a new virus designated *Aedes pseudoscutellaris* reovirus assigned to a proposed genus (*Dinovernavirus*). Virology *343*, 212–223.

Baer, G.M., and Woodall, D.F. (1966). Bat salivary gland virus carrier state in a naturally infected Mexican free tail bats. Am. J. Trop. Med. Hyg. *15*, 769–771.

Baker, A.C. (1943). The typical epidemic series. Am. J. Trop. Med. *23*, 559–566.

Baker, M.L., Schountz, T., and Wang, L.-F. (2012). Antiviral immune responses of bats: a review. Zoon. Publ. Hlth. *60*, 104–116.

Bakonyi, T., Ferenczi, E., Erdélyi, K., Kutasi, O., Csogo, T., Seidel, B., Weissenbock, H., Brugger, K., Ban, E., and Nowotny, N. (2013). Explosive spread of a neuroinvasive lineage 2 West Nile virus in Central Europe, 2008/2009. Vet. Microbiol. *165*, 61–70.

Balfour, A. (1914). Wild monkeys as reservoir for virus of yellow fever. Lancet *1*, 1176–1178.

Ball, G.H. (1943). Parasitism and evolution. Am. Naturalist *77*, 345–364.

Ballinger, M.J., Bruenn, J.A., Kotov, A.A., and Taylor, D.J. (2013). Selectively maintained paleoviruses in Holoarctic water fleas reveal an ancient origin for phleboviruses.Virology *446*, 276–282.

Ballinger, M.J., Buenn, J.A., Hay, J., Czechowski, D., and Taylor, D.J. (2014). Discovery and evolution of bunyavirids in arctic phantom midges and ancient bunyavirid-like sequences in insect genomes. J. Virol. *88*, 8783–8794.

Bancroft, W.H., Scott, R.M., Brandt, W.E., McCown, J.M., Eckels, K.H., Hayes, D.E., Gould, D.J., and Russell, P.K. (1982). Dengue-2 vaccine infection of *Aedes aegypti* mosquitoes by feeding on viremic patients. Am. J. Trop. Med. Hyg. *31*, 1229–1231.

Bao, C.-J., Guo, X.-L., Qum, X., Hu, J.-L., Zhou, M.-H., Varma, J.K., Cui, L.-B., Yang, H.-T., Jiao, Y.-J., Klena, J.D., *et al.* (2011). A family cluster of infections by a newly recognized bunyavirus in Eastern China, 2007: further evidence of person-to-person transmission. Clin. Infect. Dis. *53*, 1208–1214.

Bara, J.J., Clark, T.M., and Remold, S.K. (2013). Susceptibility of larval *Aedes aegypti* and *Aedes albopictus* (Diptera: Culicidae) to dengue virus. J. Med. Entomol. *50*, 179-184.

Barbazan, P.,Thitifhanyanont, A., Misse, D., Dubot, A., Bosc, P., Luangsri, N., Gonzalez, J.P., and Kittayapong, P. (2008). Detection of H5N1 avian influenza virus from mosquitoes collected in an infected poultry farm in Thailand. Vector-Borne Zoon. Dis. *8*, 105–109.

Barnett, H.C. (1956).The transmission of western equine encephalitis virus by the mosquito *Culex tarsalis* Coq. Am. J. Trop. Med. Hyg. *5*, 86–98.

Barrie, H.J. (1997). Artifacts and archives. Diary notes on a trip to West Africa in relation to a yellow fever expedition under the auspices of the Rockefeller Foundation, 1926, by Oskar Klotz. Canad. Bull. Med. Hist. *14*, 133–163.

Batista, W.C., Tavares, G.S.B., Vieira, D.S., Honda, E.R., Pereira, S.S., and Tada, M.S. (2011). Notification of the first isolation of Cacipacore virus in a human in the State of Rondonia, Brazil. Rev. Soc. Bras. Med. Trop. *44*, S28–S30.

Batten, C., Darpel, K., Henstock, M., Fay, P., Veronesi, E., Gubbins, S., Graves, S., Frost, L., and Oura, C. (2014). Evidence for transmission of bluetongue virus serotype 26 through direct contact. PloS One *9*, e96049.

Beasley, D.W., Morin, M., Lamb, A.R., Hayman, E., Watts, D.M., Lee, C.K., Trent, D.W., and Monath, T.P. (2013). Adaptation of yellow fever virus 17DD to Vero cells is associated with mutations in structural and non-structural protein genes. Virus Res. *176*, 280–284.

Bekal, S., Domier, L.L., Niblack, T.L., and Lambert, K.N. (2011). Discovery and initial analysis of novel viral genomes in the soybean cyst nematode. J. Gen. Virol. *92*, 1870–1879.

Bekal, S., Domier, L.L., Gonfa, B., McCoppin, N.K., Lambert, K.N., and Bhalerao, K. (2014). A novel flavivirus in the soybean cyst nematode. J. Gen. Virol. *95*, 1272–1280.

Belaganahalli, M.N., Maan, S., Maan, S., Tesh, R.B., Attoui, H., and Mertens, P.P.C. (2011). Umatilla virus genome sequencing and phylogenetic analysis: identification of Stretch Lagoon orbivirus as a new member of the *Umatilla* virus group. PloS One *6*, e23605.

Belyi, V.A., Levine, A.J., and Shalka, A.M. (2010a). Unexpected inheritance: multiple integrations of ancestral bornavirus and Ebola virus/Marburg sequences in vertebrate genomes. PloS One *6*, e1001030.

Belyi, V.A., Levine, A.J., and Shalka, A.M. (2010b). Sequences from ancestral single-stranded DNA viruses in vertebrate genomes: the *Parvoviridae* and *Circoviridae* are more than 40 to 50 million years old. J. Virol. *84*, 12458–12462.

Benda, R. (1958). The common tick, *Ixodes ricinus* L., as a reservoir and vector of tick-borne encephalitis. II. J. Hyg. Epidemiol. Microbiol. Immunol. 2, 331–244.

Bergren, N.A., Auguste, A.J., Forrester, N.L., Negi, S.S., Braun, W.A., and Weaver, S.C. (2014). Western equine encephalitis virus: evolutionary analysis of a declining alphavirus based on complete genome sequences. J. Virol. *88*, 9260–9267.

Best, S.M., and Kerr, P.J. (2000). Coevolution of host and virus: The pathogenesis of virulent and attenuated strains of myxoma virus in resistant and susceptible European rabbits. Virology *267*, 36–48.

Bilsel, P.A., Tesh, R.B., and Nichol, S.T. (1988). RNA genome stability of Toscana virus during serial transovarial transmission in the sandfly, *Phlebotomus perniciosus.* Virus Res. *11*, 87–94.

Blasdell, K.R., Guzman, H., Widen, S.G., Firth, C., Wood, T.G., Holmes, E.C., Tesh, R.B., Vasilakis, N., and Walker, P.J. (2014). Ledantevirus: A proposed new genus in the Rhabdoviridae has a strong ecological association with bats. Am. J. Trop. Med. Hyg. *92*, 405–410.

Boehme, K.W., Williams, J.C., Johnston, R.E., and Heidner, H.W. (2000). Linkage of an alphavirus host-range restriction to the carbohydrate-processing phenotypes of the host cells. J. Gen. Virol. *81*, 161–170.

Bonneau, K.R., C.D. Demaula, B.A. Mullens, *et al.* (2002). Duration of viraemia infectious to *Culicoides sonorensis* in bluetongue virus-infected cattle and sheep. Vet. Microbiol. *88*, 115–125.

Boughalmi, M., Pagnier, I., Asherfi, S., Colson, P., Raoult, D., and La Scola, B. (2013a). First isolation of a *Marseillevirus* in the Diptera *Syrphidae eristalis tenax.* Intervirology *56*, 386–394.

Boughalmi, M., Pagnier, I., Aherfi, S., Colson, P., Raoult, D., and La Scola, B. (2013b). First isolation of a giant virus from wild *Hirudo medicinalis.* Viruses *5*, 2920–2930.

Bourhy, H., Cowley, J.A., Larrous, F., Holmes, E.C., and Walker, P.J. (2005). Phylogenetic relationships among rhabdoviruses inferred using the L-polymerase gene. J. Gen. Virol. *86*, 2849–2858.

Bowers, D.F., Coleman, C.G., and Brown, D.T. (2003). Sindbis virus-associated pathology in *Aedes albopictus* (Diptera: Culicidae). J. Med. Entomol. *40*, 698–705.

Boyce, R.W. (1911).Yellow Fever and Its Prevention (John Murray, London).

Brault, A.C., Powers, A.M., and Weaver, S.C. (2002). Vector infection determinants of Venezuelan encephalitis virus reside within the E2 envelope glycoprotein. J. Virol. *76*, 6387–6392.

Brault, A.C., Armijos, M.V., Wheeler, S., Wright, S., Fang, Y., Langevin, S., and Reisen, W.K. (2009). Stone Lakes virus (Family Togaviridae, Genus Alphavirus), a variant of Fort Morgan virus isolated from swallow bugs (Hemiptera: Cimicidae) west of the Continental Divide. J. Med. Entomol. *46*, 1203–1309.

Braverman, Y. (1994). *Nematocera* (*Ceratopogonidae, Psychodidae, Simuliidae*, and *Culicidae*) and control methods. Res. Sci. Tech. Off. Int. Epiz. *13*, 1175–1199.

Brinton, M.A. (1980). Non-arbo togaviruses. In The Togaviruses-Biology, Structure, Reproduction, R.W. Schlesinger, ed. (Academic Press, New York and London), pp. 623–666.

Brinton, M.A. (1981). Genetically controlled resistance to flaviviruses and lactate-dehydrogenase-elevating virus-induced disease. Curr. Top. Microbiol. Immunol. *92*, 1–14.

Brown, C.R., Komar, N., Quick, S.B., Sethi, R.A., Panella, N.A., Brown, M.B., and Pfeffer, M. (2001). Arbovirus infection increases with group size. Proc. R. Soc. Lond. B *268*, 1833–1840.

Brown, C.R., Moore, A.T., Young, G.R., and Komar, N. (2010a). Persistence of Buggy Creek virus (Togaviridae, Alphavirus) for two years in unfed swallow bugs (Hemiptera; Cimicidae: *Oeciacus vicarious*). J. Med. Entomol. *47*, 436–441.

Brown, C.R., Moore, A.T., O'Brien, V.A., Padhi, A., Knutie, S.A., Young, G.R., and Komar, N. (2010b). Natural infection of vertebrate hosts by different lineages of Buggy Creek virus (family Togaviridae, genus Alphavirus). Arch. Virol. *155*, 745–749.

Brown, J.E., McBride, C.S., Johnson, P., Ritchie, S., Paupy, C., Bossin, H., Lutomiah, J., Fernandez-Salas, I., Ponlawat, A., Cornel, A.J., *et al.* (2011). Worldwide patterns of genetic differentiation imply multiple 'domestications' of *Aedes aegypti*, a major vector of human diseases. Proc. Biol. Sci. *278*, 2446–2454.

Brum, J.R., and M.B. Sullivan. (2015). Rising to the challenge: accelerated pace of discovery transforms marine virology. Nat. Rev. Microbiol. *13*, 147–159.

Brun, I., McManus, D., Lefvre, C., Hu, K., Falls, T., Atkins, H., Bell, J.C., McCart, J.A., Mahoney, D., and Stojdl, D.F. (2010). Identification of genetically modified *Maraba virus* as an oncolytic rhabdovirus. Mol. Therapy *18*, 1440–1449.

Bryant, J.E., Vasconcelos, P.F.C., Rijnbrand, R.C.A., Mutebi, J.P., Higgs, S., and Barrett, A.D. (2005). Size heterogeneity in the 3′ noncoding region of South American isolates of yellow fever virus. J. Virol. *79*, 3807–3821.

Bryant, J.E., Calvert, A.E., Mesesan, K., Crabtree, M.B., Volpe, K.E., Silengo, S., Kinney, R.M., Huang, C.Y., Miller, B.R., and Roehrig, J.T. (2007). Glycosylation of the dengue 2 virus E protein at N67 is critical for virus growth *in vitro* but not for growth in intrathoracically inoculated *Aedes aegypti* mosquitoes. Virology *366*, 415–423.

Buckley, S.M. (1973). Singh's *Aedes albopictus* cell cultures as helper cells for the adaptation of *Obodhiang* and *Kotonkan viruses* of the rabies serogroup to some vertebrate cell cultures. Appl. Microbiol. *25*, 695–696.

Buckley, S.M. (1975). Arbovirus infection of vertebrate and insect cell cultures, with special emphasis on Mokola, Obodhiang, and Kotonkan viruses of rabies serogroup. Ann. N.Y. Acad. Sci. *266*, 241–250.

Burgdorfer, W., and Varma, M.G.R. (1967). Trans-stadial and transovarial development of disease agents in arthropods. Annu. Rev. Entomol. *12*, 347–376.

Burke, R., R. Barrera, R., M. Lewis, M., Kluchinsky, T., and Claborn, D. (2010). Septic tanks as larval habitats for the mosquitoes *Aedes aegypti* and *Culex quinquefasciatus* in Playa-Playita, Puerto Rico. Med. Vet. Entomol. *24*, 117–123.

Burrage, T.G., Lu, Z., Neilan, J.G., Rock, D.L., and Zsak, L. (2004). *African swine fever virus* multigene family 360 genes affect virus replication and generalization of infection in *Ornithodoros porcinus* ticks. J. Virol. *78*, 2445–2453.

Bussereau, F., Kinkelin, P., and Le Berre, M. (1975). Infectivity of fish rhabdoviruses for *Drosophila melanogaster.* Ann. Microbiol. *126A*, 389–395.

Calisher, C.H., and Gould, E. (2003). Taxonomy of the virus family *Flaviviridae.* Adv. Virus Res. *59*, 1–19.

Calisher, C.H., Karabatsos, N., Dalrymple, J.M., Shope, R.E., Porterfield, J.S., Westaway, E.G., and Brandt, W.E. (1989). Antigenic relationships between flaviviruses as determined by cross-neutralization tests with polyclonal antisera. J. Gen. Virol. *70*, 37-43.

Calisher, C.H., Childs, J.E., Field, H.E., Holmes, K.V., and Schountz, T. (2006). Bats: important reservoir hosts of emerging viruses. Clin. Microbiol. Rev. *19*, 531–545.

Carey, D.E. (1971). Chikungunya and dengue: A case of mistaken identity? J. Hist. Med. *26*, 243–262.

Carn, V.M. (1996). The role of dipteran insects in the mechanical transmission of animal viruses. Br. Vet. J. *152*, 377–393.

Carp, R.I., Meeker, H.C., Rubenstein, R., Siqurdarsn, S., Papini, M., Kascsak, R.J., Kozlowski, P.B., and Wisniewski, H.M. (2000). Characteristics of scrapie isolates derived from hay mites. J. Neurol. *6*, 137–144.

Carroll, S.A., Bird, B.H., Rollin, P.E., and Nichol, S.T. (2010). Ancient common ancestry of Crimian-Congo hemorrhagic fever virus. Mol. Phylogenet. Evol. *55*, 1103–1110.

Casadevall, A., and Pirofski, L.A. (2000). Host–pathogen interactions: Basic concepts of microbial commensalism, colonization, infection, and disease. Infect. Immun. *68*, 6511–6518.

Casals, J. (1971). Arboviruses: Incorporation in a general system of virus classification. In Comparative Virology, Maramorosch and E. Kurstak, eds. (Academic Press, New York and London), pp. 307–333.

Chadee, D.D., and Martinez. R. (2000). Landing periodicity of *Aedes aegypti* with implications for dengue transmission in Trinidad, West Indies. J. Vector Ecol. *25*, 158–163.

Chamberlain, R.W. (1982). Arbovirology-Then and Now. Am. J. Trop. Med. Hyg. *31*, 430–437.

Chan, K.L., Chan, Y.C., and Ho, B.C. (1971). *Aedes aegypti* (L.) and *Aedes albopictus* (Skuse) in Singapore City. 4. Competition between species. Bull. Wld. Hlth. Org. *44*, 643–649.

Charlier, N., Davidson, A., Dallmeier, K., Molenkamp, R., De Clercq, E., and Neyts, J. (2010). Replication of not known vector flavivirus in mosquito cells is restricted by intracellular host factors rather than by the viral envelope proteins. J. Gen. Virol. *91*, 1693–1697.

Chen, R., Wang, E., Tsetsarkin, A.K., and Weaver, S.C. (2013). Chikungunya virus 3' untranslated region: Adaptation to mosquitoes and a population bottleneck as major evolutionary forces. PloS Path. *9*, e1003591.

Cheng, L. (1976). Marine insects (North-Holland Publishing Co., Amsterdam, Oxford, New York).

Cheng, R.L., Y. Xi, Y.H. Lou, *et al.* (2014). The brown planthopper nudivirus DNA integrated in its host genome. J. Virol. *88*, 5310–5318.

Chevillon, C., Eritja, R., Pasteur, N., and Raymond, M. (1995). Communalism, adaptation, and gene flow: Mosquitoes of the *Culex pipiens* Complex in different habitats. Genet. Res. *66*, 147–157.

Christie, J. (1872). 'Kidinga pepo': A peculiar form of exanthematous disease. Br. Med. J. *June 1*, 577–579.

Christie, J. (1881). On epidemics of dengue fever: Their diffusion and etiology. Glasgow Med. J. *16*, 161–176.

Christofferson, R.C., Chisenhall, D.M., Wearing, H.J., and Mores, C.N. (2014). Chikungunya viral fitness measures within the vector and subsequent transmission potential. PloS One *9*, e110538.

Christophers, S.R. (1960). *Aedes aegypti* (L.) The Yellow Fever Mosquito. (Cambridge University Press, London).

Ciota, A.T., Payne, A.F., and Kramer, L.D. (2015). West Nile virus adaptation to ixodid tick cells is associated with phenotypic trade-offs in primary hosts. Virology *482*, 128–132.

Clements, A.N. (2012). The Biology of Mosquitoes: Transmission of Viruses and Interactions with Bacteria, vol. 3 (CABI, Wallingford, UK).

Coffey, L.L., Vasilakis, N., Brault, A.C., Powers, A.M., Tripet, F., and Weaver, S.C. (2008). Arbovirus evolution *in vivo* is constrained by host alteration. Proc. Natl. Acad. Sci. U.S.A. *105*, 6970–6975.

Coffey, L.L., Forrester, N., Tsetsarkin, K., Vasilakis, N., and Weaver, S.C. (2013). Factors shaping the adaptive landscape for arboviruses: implications for the emergence of disease. Rev. Future Microbiol. *8*, 155–176.

Colson, P., Richet, H., Desnues, C., Balique, F., Moal, V., Grob, J.J., Berbis, P., Lecoq, H., Harle, J.R., Berland, Y., *et al.* (2010). Pepper mild mottle virus, a plant associated with specific immune responses, fever, abdominal pains, and pruritus in humans. PloS One *5*, e10041.

Conway, M.J., Colpitts, J.M., and Fikrig, E. (2014). Role of the vector in arbovirus transmission. Annu. Rev. Virol. *1*, 71–88.

Cook, S., and Holmes, E.C. (2005). A multigene analysis of the phylogenetic relationships among the flaviviruses (Family Flaviviridae) and the evolution of vector transmission. Arch. Virol. *151*, 309–325.

Corbett, M.K. (1964). Introduction. In Plant Virology, M.K. Corbett and H.D. Sisler, eds. (University of Florida Press, Gainesville, Florida), pp. 1–16.

Councilman, W.T. (1890). Etiology and prevention of yellow fever. A pathological and anatomical histology. US Marine Hosp. Serv. Publ. Hlth. Bull. No. 2, 151–159.

Crane, M., and Hyatt, A. (2011). Viruses of fish: an overview of significant pathogens. Viruses 3, 2025–2046.

Crochu, S., Cook, S., Attoui, H., Charrel, R.N., De Chesse, R., Belhouchet, M., Lemasson, J.J., de Micco, P., and de Lamballerie, X. (2004). Sequences of flavivirus-related RNA viruses persist in DNA form integrated in the genome of *Aedes* spp. mosquitoes. J. Gen. Virol. *85*: 1971–1980.

Cui, J., and Holmes, E.C. (2012). Endogenous RNA viruses of plants in insect genomes. Virology *427*, 77–79.

Dandawate, C.N., Singh, K.R., and Dhanda, V. (1981). Experimental infection of certain mosquitoes and ticks with Ganjam virus. Ind. J. Exp. Biol. *19*, 185–186.

Davies, C.R. (1990). Interrupted feeding of blood-sucking insects: causes and effects. Parasitology Today *6*, 19–22.

DeBach, P. (1966). The competitive displacement and coexistence principles. Annu. Rev. Entomol. *11*, 183–212.

De Filippis, V.R., and Villareal, L.P. (2000). An introduction to the evolutionary ecology of viruses. In Viral Ecology, C.J. Hurst, ed. (Academic Press, San Diego, London, Boston, and New York), pp. 125–208.

De Medeiros, R.B., Figuerido, J., de O. Resende, R., and Avila, A.C. (2005). Expression of a viral polymerase-bound host factor turns human cell lines permissive to a plant-and insect-infecting virus. Proc. Natl. Acad. Sci. U.S.A. *102*, 1175–1180.

Diallo, M., Ba, Y., Sall, A.A., Diop, O.M., Ndione, J.A., Mondo, M., Girault, L., and Mathiot, C. (2003). Amplification of the sylvatic cycle of dengue virus type 2, Senegal, 1999–2000: entomologic findings and epidemiologic considerations. Emerg. Infect. Dis. *9*, 362–367.

Diarrassouba, S., and Dossou-Yovo, J. (1997). Rhthme d'activité atypique chez *Aedes aegypti* en zone de savane sub-soudonienne de Côte d'Ivore. Bull. Soc. Pathol. Exot. *90*, 361–363.

Dickson, D.L., and Turell, M.J. (1992). Replication and tissue tropisms of Crimean–Congo hemorrhagic fever virus in experimentally infected adult *Hyalomma truncatum* (Acari: *Ixodidae*). J. Med. Entomol. *29*, 767–773.

Dieng, H., Saifur, R.G.M., Hassan, A.B., Salamah, M.R.C., Boots, M., Satho, T., Jaal, Z., and Abubakar, S. (2010). Indoor breeding of *Aedes albopictus* in northern Peninsular Malaysia and its potential epidemiological implications. PLoS ONE *5*, e11790.

Djikeng, A., Kuzmickas, R., Anderson, N.G., and Spiro, D.J. (2009). Metagenomic analysis of RNA viruses in a fresh water lake. PloS ONE *4*, e7264.

Dolja, V.V., and Koonin, E.V. (2011). Common origins and host-dependent diversity of plant and animal viromes. Curr. Opin. Virol. *1*, 322–321.

Downs, W.G. (1991). History of yellow fever in Trinidad. In Studies on the Natural History of Yellow Fever in Trinidad, E.S. Tikasingh, ed. (Port of Spain, Trinidad: CAREC), Monograph Series *1*, 2–5.

Doyle, M.S., Swope, B.N., Hogette, J.A., Burkhalter, K.L., Savage, H.M., and Nasci, R.S. (2011). Vector competence of the stable fly (Diptera: Muscidae) for West Nile virus. J. Med. Entomol. *48*, 656–668.

Drexler, J.F., Corman, V.M., and Drosten, C. (2013). Ecology, evolution and classification of bat coronaviruses in the aftermath of SARS. Antiviral Res. *101*, 45–56.

Duchêne, S., Holmes, E.C., and Ho, S.Y. (2014). Analysis of evolutionary dynamics in viruses are hindered by a time-dependent bias in rate estimation. Proc. Biol. Sci. *281*(1786), doi: 10.1098/rspb.2014.0732.

Dudley, S.F. (1934). Can yellow fever spread into Asia? An essay on the ecology of mosquito-borne disease. J. Trop. Med. Hyg. 37, 273–278.

Duggal, N.K., D'Anton, M., Xiang, J., Seiferth, R., Day, J., Nasci, R., and Brault, A.C. (2013). Sequence analyses of 2012 West Nile virus isolates from Texas fail to associate viral genetic factors with outbreak magnitude. Am. J. Trop. Med. Hyg. *89*, 205–210.

Duggal, N.K., Bosco-Lauth, A., Bowen, R.A., Wheeler, S.S., Reisen, W.K., Felix, T.A., Mann, B.R., Romo, H., Swetnam, D.M., *et al.* (2014). Evidence for co-evolution of West Nile virus and house sparrows in North America. PloS Negl. Trop. Dis. *8*, e3262.

Ebert, D., and Bull, J.J. (2003). Challenging the trade-off model for the evolution of virulence: its virulence management feasible? Trends Microbiol. *11*, 15–20.

Eisler, R. (2003). Health risks of gold miners: a synoptic review. Env. Geochem. Health 25, 325–345.

Eldridge, B.F., Glaser, C., Pedrin, R.E., and Chiles, R.E. (2001). The first reported case of California encephalitis in more than 50 years. Emerg. Infect. Dis. *7*, 451–452.

Endris, R.G., and Hess, W.R. (1992). Experimental transmission of African swine fever virus by the soft tick *Ornithodoros* (*Pavloskeyella*) *macrocanus* (Acari: *Ixodoidea*: *Argasidae*). J. Med. Entomol. *29*, 652–656.

Endris, R.G., Hess, W.R., and Caiado, J.M. (1992). African swine fever virus infection in the Iberian soft tick, *Ornithodoros* (*Pavloskeyella*) *macrocanus* (Acari: Argasidae). J. Med. Entomol. *29*, 874–878.

Endy, T.P., Nisalak, A., Chunsuttiwat, S., Vaughn, D.W., Green, S., Ennis, F.A., Rothman, A.L., and Libraty, D.H. (2004). Relationship of pre-existing dengue virus (DV) neutralizing antibody levels to viremia and severity of disease in a protective cohort study of DV infection in Thailand. J. Infect. Dis. *189*, 990–1000.

Engel, A.R., Mitzel, D.N., Hanson, C.T., Wolfinbarger, J.B., Bloom, M.E., and Pletnev, A.G. (2011). Chimeric tick-borne encephalitis virus is attenuated in *Ixodes scapularis* ticks and *Aedes aegypti* mosquitoes. Vector Borne Zoon. Dis. *11*, 665–674.

Ewald, P.W. (1983). Host–parasite relations, vectors, and the evolution of disease severity. Annu. Rev. Evol. System. *14*, 465–485.

Faisal, M., and Schulz. C.A. (2009). Detection of viral hemorrhagic septicemia virus (VHSV) from the leech *Myzobdella lugubris* Leidy, 1851. Parasit. Vectors *2*, 45.

Fan, G., and Li, J. (2011). Regions identity between the genome of vertebrates and non-retroviral families of insect viruses. Virol. J. *8*, 511.

Faran, M.E., Turell, M.J., Romoser, W.S., Routier, R.G., Gibbs, P.H., Cannon, T.L., and Bailey, C.L. (1987). Reduced survival of adult *Culex pipiens* infected with Rift Valley fever virus. Am. J. Trop. Med. Hyg. *37*, 403–409.

Fassbinder-Orth, C.A., Barak, V.A., and Brown, C.R. (2013). Immune responses of a native and an invasive bird to Buggy Creek virus (Togaviridae: Alphavirus) and its arthropod vector, the swallow bug (*Oeciacus vicarious*). PloS One *8*, e58045.

Fereres, A. (2015). Insect vectors as drivers of plant virus emergence. Curr. Opin. Virol. *10C*, 42–46.

Ficova, M., Batakova, T., Pancik, P., Vaclav, P., Prokov, P., Halasova, Z., and Kudelova, M. (2011). Molecular detection of murine herpesvirus 68 in ticks feeding on free-living reptiles. Microb. Ecol. *62*, 862–867.

Fine, P.E.M., and LeDuc, J.W. (1978). Towards a quantitative understanding of the epidemiology of Keystone virus in the eastern United States. Am. J. Trop. Med. Hyg. *27*, 322–338.

Firth, A.E., and Brierley, I. (2012). Non-canonical translation in RNA viruses. J. Gen. Virol. *93*, 1385–1409.

Fontenille, D., Diallo, M., Mondo, M., Ndiaye, M., and Thonnon, J. (1997). First evidence of natural vertical transmission of yellow fever virus in *Aedes aegypti*, its epidemic vector. Tr. R. Soc. Trop. Med. Hyg. *91*, 533–535.

Forrester, N.L., Kenny, J.L., Deardoff, E., Wang, E., and Weaver, S.,C. (2008). Western equine encephalitis submergence: lack of evidence for a decline in virus virulence. Virology *380*, 170–172.

Forrester, N.L., Palacios, G., Tesh, R.B., Savji, N., Guzman, H., Sherman, M., Weaver, S.C., and Lipkin, W.I. (2012). Genome-scale phylogeny of the *Alphavirus* genus suggests a marine origin. J. Virol. *86*, 2729–2738.

Fort, P., Albertini, A., Van-Hua, A., Berthomieu, A., Roche, S., Delsuc,F., Pasteur, N., Capy, P., Gaudin, Y., and Weill, M. (2012). Fossil rhabdoviral sequences integrated into arthropod genomes: ontogeny, evolution, and potential functionality. Mol. Biol. Evol. *29*, 381–390.

Foy, B.D., Kobylinski, K.C., Chilson Foy, J.L., Blitvich, B.J., Travassos da Rosa, A., Haddow, A.D., Lanciotti, R.S., and Tesh, R.B.(2011). Probable non-vector-borne transmission of Zika virus, Colorado, USA. Emerg. Infect. Dis. *17*, 880–882.

Frederiksen, H. (1955). Historical evidence for interference between dengue and yellow fever. Am. J. Trop. Med. Hyg. *4*, 483–491.

Freundt, E.C., Beatty, D.C., Stegall-Faulk, T., and Wright, S.M. (2005). Possible tick-borne human enterovirus resulting in aseptic meningitis. J. Clin. Microbiol. *43*, 3471–3473.

Friesen, P.D. (2007). Insect viruses. In Fields Virology, Fifth Edition, D.M. Knipe and P.M. Howley, eds. (Lippincott Williams and Wilkins, Philadelphia), vol. 1, pp. 707–736.

Gangnonngiw, W., Kanthong, N., and Flegel, T.W. (2010). Successful propagation of shrimp yellow head virus in immortal mosquito cells. Dis. Aquat. Org. *90*, 77–83.

Gaunt, M.W., Sall, A.A., de Lamballerie, X., Falconar, A.K.I., Dzhivanian, T.I., and Gould, E.A. (2001). Phylogenetic relationships of flaviviruses correlate with their epidemiology, disease association and biogeography. J. Gen. Virol. *82*, 1867–1876.

Geevarghese, G., and Banerjee, K. (1990). Role of bats in the natural cycle of arboviruses. Curr. Sci. *59*, 26–31.

Gentry, J., Vinje, I., Guadagnoli, D., and Lipp, E.K. (2013). Norovirus distribution within an estuarine environments. Appl. Env. Microbiol. *75*, 5474–5480.

Germi, R., Crance, J.M., Garin, D., Guimet, J., Thelu, M.A., Jouan, A., Zarski, J.P., and Drouet, E. (2001). Mosquito cells bind and replicate hepatitis C virus. J. Med. Virol. *64*, 6–12.

Gibbs, A.J. (ed.) (1973). Viruses and invertebrates (Amsterdam and London: North-Holland Publishing Co.).

Gibbs, M.J., and Weiller. G.F. (1999). Evidence that a plant virus switched hosts to infect a vertebrate and then recombined with a vertebrate-infecting virus. Proc. Natl. Acad. Sci. U.S.A. *96*, 8022–8027.

Gibrat, J.-F., Mariadassou, M., Boudinot, P., and Delmas, B. (2013). Analyses of the radiation of birnaviruses from diverse host phyla and of their evolutionary affinities with other double-stranded RNA and positive strand RNA viruses using robust structure-based multiple sequence alignments and advanced phylogenetic methods. BMC Evol. Biol. *13*, 154.

Gilbert, L., Jones, L.D., Laurenson, M.K., Gould, E.A., Reid, H.W., and Hudson, P.J. (2004). Ticks need not bite their red grouse hosts to infect them with louping ill virus. Proc. R. Soc. Lond. B (Suppl. 4) *271*, S202–S205.

Gillett, J.D. (1958). Laboratory tests on the maintenance of yellow fever virus in certain arthropods. Tr. R. Soc. Trop. Med. *52*, 269–271.

Gilotra, S.K., Rozeboom, L.E., and Bhattacharya, N.C. (1967). Observations on possible competitive displacement between populations of *Aedes aegypti* Linnaeus and *Aedes albopictus* Skuse in Calcutta. Bull. Wld. Hlth. Org. *37*, 437–446.

Glass, K. (2005). Ecological mechanisms that promote arbovirus survival: a mathematical model of Ross River virus transmission. Trans. R. Soc. Trop. Med. Hyg. *99*, 252–260.

Gordon, K.H.J., Williams, M.R., Baker, J.S., Gibson, J.M., Bawden, A.L., Milgae, A.G., Larkin, P.J., and Hanzlik, T.N. (2001). Replication-independent assembly of an insect virus (Tetraviridae) in plant cells. Virology *288*, 36–50.

Gordon Smith, C.E. (1960). Factors in the past and future evolution of the arboviruses. Trans. R. Soc. Trop. Med. Hyg. *54*, 113–129.

Gorman, B.M. (1990). The bluetongue virus. Curr. Top. Microbiol. Immunol. *162*, 1–19.

Grard, G., Lemasson, J.J., Sylla, M., Dubot, A., Cook, S., Molez, J.F., Pourrut, X., Charrel, R., Gonzalez, J.P., Munderloh, U., *et al.* (2006). *Ngoye virus*: a novel evolutionary lineage within the genus *Flavivirus*. J. Gen. Virol. *87*, 3273–3277.

Grard, G., Fair, J.N., Lee, D., Silkas, E., Steffen, I., Muyembe, J.J., Sittler, T., Veeraraghavan, N., Ruby, J.G., Wang, C., *et al.* (2012). A novel rhabdovirus associated with acute hemorrhagic fever in Central Africa. PloS Pathogens *8*, e1002924.

Gray, S.M., and Banerjee, N. (1999). Mechanisms of arthropod transmission of plant and animal viruses. Microbiol. Mol. Biol. Rev. *63*, 128–148.

Greenblatt, R.J., Work, T.M., Balazs, G.H., Sutton, C.A., Casey, R.N., and Casey, J.W. (2004). The *Ozobranchus* leech is a candidate mechanical vector for the fibropapilloma-associated turtle herpesvirus found latently infecting skin tumors on Hawaiian green turtles (*Chelonia mydas*). Virology *321*, 101–110.

Gritsun, T.S., and Gould, E.A. (2006). Direct repeats in the 3′ untranslated regions of mosquito-borne flaviviruses: possible implications for virus transmission. J. gen. Virol. *87*, 3297–3305.

Gritsun, T.S., and Gould, E.A. (2007a). Direct repeats in the flavivirus 3′ untranslated region; a strategy for survival in the environment? Virology *358*, 258–265.

Gritsun, T.S., and E.A. Gould (2007b). Origin and evolution of 3′ UTR of flaviviruses: long direct repeats as a basis for the formation of secondary structures and their significance for virus transmission. Adv. Virus Res. *69*, 203–248.

Gritsun, D.J., Jones, I.M., Gould, E.A., and Gritsun, T.S. (2014). Molecular archaeology of *Flaviviridae* untranslated regions: duplicated RNA structures in the replication enhancers of flaviviruses and pestiviruses emerged via convergent evolution. PloS One *9*, e92056.

Gu, S.H., Lim, B.K., Arai, S., Kim, J.A., Nicolas, V., Lalis, A., Denys, C., Cook, J.A., Dominguez, S.R., Holmes, K.V., *et al.* (2014). Molecular

phylogeny of hantaviruses harbored by insectivorous bats in Côte d'Ivoire and Vietnam. Viruses *6*, 1897–1910.
Gubala, A., Davis, S., Weir, S., Melville, L., Cowled, C., and Boyle, D. (2011). Tibrogargan and Coastal Plains rhabdoviruses: genomic characterization, evolution of novel genes and seroprevalence in Australian livestock. J. Gen. Virol. *92*, 2160–2170.
Gundersen-Rindal, D.E., and Lynn, D.E. (2003). Polydnavirus integration in lepidopteran host cells *in vitro*. J. Insect Physiol. *49*, 453–462.
Habu, A., Y. Murakami, A. Ogata, *et al.* (1977). Dysfunction of spermatogenesis by Japanese encephalitis viral infection of swine and viral shedding into semen (in Japanese). Uirusu *27*, 21–26.
Hahn,C.S., Hahn, Y.S., Rice, C.M., Lee, E., Dalgarno, L., Strauss, E.G., and Strauss, J.H. (1987). Conserved elements in the 3′-untranslated region of flavivirus RNAs and potential cyclization sequences. J. Mol. Biol. *198*, 33–41.
Halstead, S.B., Scanlon, J.E., Umpaivit, P., and Udomsakdi, S. (1969). Dengue and chikungunya virus infection in man in Thailand, 1962–64. IV. Epidemiologic studies in the Bangkok Metropolitan area. Am. J. Trop. Med. Hyg. *18*, 997–1021.
Hamer, G.L., Kitron, U.D., Brown, J.D., Loss, S.R., Ruiz, M.O., Goldberg, T.L., and Walker, E.D. (2008). *Culex pipiens* (Culicidae): a bridge vector of West Nile virus to humans. J. Med. Entomol. *45*, 125–128.
Hanley, K.A., Manlucu, L.R., Gilmore, L.E., Blaney, J.E., Hanson, C.T., Murphy, B.R., and Whitehead, S.S. (2003). A trade-off in replication in mosquito versus mammalian systems conferred by a point mutation in the NS4B protein of dengue virus type 4. Virology *312*, 222–232.
Hardy, A. (2003). Animals, disease, and man: making connections. Persp. Biol. Med. *46*, 200–215.
Hare, F.E. (1898). The 1897 epidemic of dengue in Northern Queensland. Austr. Med. Gaz. March *21*, 98–107.
Harrington, L.C., Edman, J.D., and Scott, T.W. (2001). Why do female *Aedes aegypti* (Diptera: Culicidae) feed preferentially and frequently on human blood? J. Med. Entomol. *38*, 411–422.
Hawley, W.A., Reiter, P., Copeland, R.S., Pumpuni, C.B., and Craig, G.B. (1987). *Aedes albopictus* in North America: Probable introduction in used tires from northern Asia. Science *236*, 1114–1116.
Haydon, D.T., Cleaveland, S., Taylor, L.H., and Laurenson, M.K. (2002). Identifying reservoirs of infection: A conceptual and practical challenge. Emerg. Infect. Dis. *8*, 1468–1473.
Hepojoki, J., Kipar, A., Korzykov, Y., Bell-Sakyi, L., Vapalahti, O., and Hetzel, U. (2015). Replication of boid inclusion body disease associated arenaviruses is temperature sensitive in both boid and mammalian cells. J. Virol. *89*, 1119–1128.
Herrel, M., Hoefs, N., Staeheli, P., and Schneider, U. (2012). Tick-borne Nyamanini virus replicates in the nucleus and exhibits unusual genome and matrix protein properties. J. Virol. *86*, 10739–10747.
Hess, W.R., Endris, R.G., Lousa, A., and Caiado, J.M. (1989). Clearance of African swine fever virus from infected tick (Acari) colonies. J. Med. Entomol. *26*, 314–317.
Hewson, I., Brown, J.M., Gitlin, S.A., and Doud, D.F. (2011). Nucleopolyhedrovirus detection and distribution in terrestrial, freshwater, and marine habitats of Appledore Island, Gulf of Maine. Microb. Ecol. *62*, 48–57.
Higgs, S., and Beaty, B.J. (2005). Natural cycles of vector-borne pathogens. In Biology of Disease Vectors, W.C. Marquardt, ed. (Elsevier Academic Press, Burlington, San Diego) second edition, pp. 167–185.
Higgs, S., Schneider, B.S., Vanlandingham, D.L., Klingler, K.A., and Gould, E.A. (2005). Nonviremic transmission of West Nile virus. Proc. Natl. Acad. Sci. U.S.A. *102*, 8871-8874.
Hikke, M.C., Verest, M., Vlak, J.M., and Pijlman, G.P. (2014). Salmonid alphavirus replication in mosquito cells: towards a novel vaccine production system. Microbiol. Biotechnol. *7*, 480–484.
Hipsley, C.A., and Müller, J. (2014). Beyond fossil calibrations: realities of molecular clock practices in evolutionary biology. Front. Genet. *5*, 138.
Hirsch, A. (1883). Dengue. In Handbook of Geographical and Historical Pathology (The New Sydenham Society, London), vol. 1, pp. 55–81.
Hoch, A.L., Gargan, T.P., and Bailey, C.L. (1985). Mechanical transmission of Rift Valley fever virus by hematophagous diptera. Am. J. Trop. Med. Hyg. *34*, 188–193.
Holmes, E.C. (2003). Error thresholds and the constraints to RNA virus evolution. Trends Microbiol. *11*, 543–546.
Holmes, E.C., and Drummond, A.J. (2007). The evolutionary genetics of viral emergence. Curr. Top. Microbiol. Immunol. *315*, 51–66.
Holstein, M. (1967). Dynamics of *Aedes aegypti* distribution, density, and seasonal prevalence in the Mediterranean area. Bull. Wld. Hlth. Org. *36*, 541–543.
Honda, K., Wei, T., Hagiwara, K., Higashi, T., Kimura, I., Akutsu, K., and Omura, T. (2007). Retention of *Rice dwarf virus* by descendans of pairs of viruliferous vector insects after rearing for 6 years. Phytopathology *97*, 712–716,
Honig, J.E., Osborne, J.C., and Nichol, S.T. (2004). The high genetic variation of viruses of the Genus *Nairovirus* reflects the diversity of their predominant tick hosts. Virology *318*, 10–16.
Hoogstraal, H. (1979). The epidemiology of tick-borne Crimean–Congo hemorrhagic fever in Asia, Europe, and Africa. J. Med. Entomol. *15*, 307–417.
Hoogstraal, H., and Aeschlmann, A. (1982). Tick-host specificity. Mitt. Schweiz. Entomol. Ges. Bull. Soc. Entomol. Suisse *55*, 5–32.
Horie, M., and Tomonaga, K. (2011). Non-retroviral fossils in vertebrate genomes. Viruses 3, 1836–1848.
Horie, M., Kobayashi, Y., Suzuki, Y., and Tomonaga, K. (2013). Comprehensive analysis of endogenous bornavirus-like elements in eukaryotic genomes. Phil. Tr. R. Soc. London B Biol. Sci. *368(1626)*, 20120499.
Horzinek, M.C. (1981). Nonarbo togavirus infections of animals: comparative aspects and diagnosis. In Comparative Diagnosis of Viral Diseases, E. Kurstak and C. Kurstak, eds. (Academic Press, New York and London), vol. 4, pp. 441–476.
Hoshino, K., Isawa, H., Tsuda,Y., Sawabe, K., and Kobayashi, M. (2009). Isolation and characterization of a new insect flavivirus from *Aedes albopictus* and *Aedes flavopictus* mosquitoes in Japan. Virology *391*, 119–129.
Houck, M.A., Qin, H., and Roberts, H.R. (2001). Hantavirus transmission: potential role of ectoparasites. Vector Borne Zoon. Dis. *1*, 75–79.
Howie, R.I., and Thorsen, J. (1981). Identification of a strain of infectious bursal disease virus isolated from mosquitoes. Canad. J. Comp. Med. *45*, 315–320.
Howitt, B.F., Doge, H.R., Bishop, L.K., and Gorrie, R.H. (1948). Virus of eastern equine encephalomyelitis isolated from chicken mites (*Dermanyssus gallinae*) and chicken lice (*Eomenachanthu stramineus*). Proc. Soc. Exp. Biol. Med. *68*, 622–625.
Hubálek, Z. (2003). Emerging human infectious diseases: Anthroponoses, zoonoses, and sapronoses. Emerg. Infect. Dis. *9*, 403–405.
Huff, C.G. (1938). Studies on the evolution of some disease-producing organisms. Quart. Rev. Biol. *13*, 196–206.
Huhtamo, E., Cook, S., Moureau, G., Uzcategui, N.Y., Sironen, T., Kuivanen, S., Putkuri, N., Kurkela, S., Harbach, R.E., Firth, A.E., *et al.* (2014). Novel flaviviruses from mosquitoes: mosquito-specific evolutionary lineages within the phylogenetic group of mosquito-borne flaviviruses. Virology *464–465*, 320–329.
Hull, R. (2002). Mathews' Plant Virology, Fourth Edition (Academic Press, San Diego), pp. 805–807.
Humphrey-Smith, I., Donker, G., Turzo, A., Chastel, C., and Schmidt-Mayerova, H. (1993). Evaluation of mechanical transmission of HIV by the African soft tick, *Ornithodoros moubata*. AIDS *7*, 341–347.
Hurlbut, H.S. (1951). The propagation of Japanese encephalitis virus in the mosquito by parenteral introduction and serial passage. Am. J. Trop. Med. Hyg. *31*, 448–451.
Hurlbut, H.S., and Thomas, J.I. (1960). The experimental host range of the arthropod-borne animal viruses in arthropods. Virology *12*, 391–407.
Hurlbut, H.S., and Thomas, J.I. (1969). Further studies on the arthropod host range of arboviruses. J. Med. Entomol. *6*, 423–427.
Hyslop, N.S.G. (1970). The epizootiology and epidemiology of foot and mouth disease. Adv. Vet. Sci. Comp. Med. *14*, 261–307.
ICTV (International Committee on Taxonomy of Viruses). (2012). Virus Taxonomy: Classification and Nomenclature of Viruses, Ninth Report, A.M.Q. King, M.J. Adams, E.B. Carstens, E.J. Lefkowitz, eds. (Elsevier Academic Press, Amsterdam).
Isawa,H., Kuwata, R., Hoshino, K., Tsuda, Y., Sakai, K., Watanabe, S., Nishimura, M., Satho, T., Kataoka, M., Nagata, N., *et al.* (2011). Identification and molecular characterization of a new nonsegmented double-stranded RNA virus isolated from *Culex* mosquitoes in Japan. Virus Res. *155*, 147–155.
Ishii, A., Ueno, K., Orba, Y., Sakai, M., Moonga, L., Hang'ombe, B.M., Mweene, A.S., Uemura, T., Ito, K., Hall, W.W., *et al.* (2014). A nairovirus isolated from African bays causes haemorrhagic gastroenteritis

and severe hepatic disease in mice. Nature Commun. *5(5651)*, doi: 10.1038/ncomms6651.

Jakob, E., Barker, D.E., and Garver, K.A. (2011). Vector potential of the salmon louse *Lepeophtheirus salmonis* in the transmission of infectious haematopoietic necrosis virus (IHNV). Dis. Aquat. Org. *97*, 155–165.

James, M.T. (1969). A study in the origin of parasitism. Bull. Entomol. Soc. Am. *15*, 251–253.

Javier, R.T., and Rice, A.P. (2011). Emerging theme: Cellular PDZ proteins as common targets of pathogenic viruses. J. Virol. *85*, 11544–11556.

Johnson, B.K., and Varma, M.G.R. (1976). Infection of *Aedes aegypti* cell line with infectious arbovirus–antibody complexes. Trans. R. Soc. Trop. Med. Hyg. *70*, 230–234.

Johnson, H.N. (1967). Biological implications of antigenically related mammalian viruses for which arthropod vectors are unknown and avian associated soft tick viruses. Jpn. J. Med. Res. *20*, 160–166.

Johnson, H.N. (1991). Virologist and Naturalist with the Rockefeller Foundation and the California Department of Public Health-Oral History Transcript/1991. The Brancroft Library of the University of California, Berkeley, CA. (Reprinted by Forgotton Books, San Bernandino, CA; 2012).

Johnson, K.M., Tesh, R.B., and Peralta, P.H. (1969). Epidemiology of Vesicular stomatitis virus: some new data and a hypothesis for transmission of the Indiana serotype. J. Am. Vet. Med. Assoc. *155*, 2133–2140.

Johnson, R.B. (1986). Human disease and the evolution of pathogen virulence. J. Theoret. Biol. *122*, 19–24.

Jones, L.D., Davies, C.R., Steels, G.M., and Nuttall, P.A.C. (1987). A novel mode of arbovirus transmission involving a nonviremic host. Science *237*, 775–777.

Junglen, S., and Drosten, C. (2013). Virus discovery and recent insights into virus diversity in arthropods. Curr. Opin. Microbiol. *16*, 507–513.

Jupp, P.G. (1976). The susceptibility of four South African species of *Culex* to West Nile and Sindbis viruses by two different infecting methods. Mosq. News *36*, 166–173.

Kading, R.C., Biggerstaff, B.J., Young, G., and Komar, N. (2014). Mosquitoes used to draw blood for arbovirus viremia determinations in small vertebrates. PloS One *9*, e99342.

Kapoor, A., Simmonds, P., Lipkin, W.I., Zaidi, S., and Delwart, E. (2010). Use of nucleotide composition analysis to infer hosts for three novel picornavirus-like viruses. J. Virol. *84*, 10322–10328.

Kapoor, A., Simmonds, P., Scheel, T.K.H., Hjelle, B., Cullen, J.M., Burbelo, P.D., Chauhan, L.V., Duraisamy, R., Sanchez Leon, M., Jain, K., *et al.* (2013). Identification of rodent homologs of hepatitis C virus and pegiviruses. mBio *4*, e00216–e00213.

Katzourakis, A., and Gifford, R.J. (2010). Endogenous viral elements in animal genomes. PLoS Genet. *6*, e1001191.

Kelchner, S.A., and Thomas, M.A. (2006). Model use in phylogenetics: nine key questions. Trends Ecol. Evol. *22*, 87–94.

Kelly, R.K., and Loh, P.C. (1972). Some properties of an established fish cell line from *Xiphophorus helleri* (red sword tail). In Vitro *9*, 73–80.

Killmaster, L.F., Stallknecht, D.E., Howerth, E.W., Moulton, J.K., Smith, P.F., and Mead, D.G. (2011). Apparent disappearance of vesicular stomatitis New Jersey virus from Ossabaw Island, Georgia. Vector Borne Zoon. Dis. *11*, 559–565.

Kilpatrick, A.M., Kramer, L.D., Jones, M.J., Marra, P.P., and Daszak, P. (2006a). West Nile virus epidemic in North America are driven by shifts in mosquito feeding behavior. PloS Biol. *4*, e82.

Kilpatrick, A.M., Daszak, P., Jones, M.J., Marra, P.P., and Kramer, L.D. (2006b). Host heterogeneity dominates West Nile virus transmission. Proc. Biol. Sci. *273*, 2327–2333.

Kim, H.-W., Cha, G.-W., Jeong, Y.-E., Lee, W.-G., Chang, K.-S., Roh, J.-Y., Yang, S.-C., Park, M.-Y., and Shin, E.-H. (2015). Detection of Japanese encephalitis virus genotype V in *Culex orientalis* and *Culex pipiens (Diptera: Culicidae)* in Korea. PLOS One, *Feb. 6*, doi: 10.137/journal. Pone.0116547.

Kim, K.C. (1985). Coevolution of Parasitic Arthropods and Mammals (John Wiley & Sons, New York).

Kimble, S.J., Karna, A.K., Johnson, A.J., Hoverman, J.T., and Williams, R.N. (2014). Mosquitoes as a potential vector of ranavirus transmission in terrestrial tutles. Ecohealth, *Sept. 12*, PMID: 25212726.

Kitchen, A., Shackelton, L.A., and Holmes, E.C. (2011). Family level phylogenies reveal modes of macroevolution in RNA viruses. Proc. Natl. Acad. Sci. U.S.A. *108*, 238–243.

Klenerman, P., Hengartner, H., and Zinkernagel, R.M. (1997). A nonretroviral RNA virus persists in DNA form. Nature *390*, 298–301.

Koblet, H. (1993). Viral evolution and insects as a possible virologic turning table. In Vitro Cell. Dev. Biol. *29A*, 274–283.

Koonin, E.V., and Gorbalenya. A.E. (1992). An insect picornavirus may have genome organization similar to that of caliciviruses. FEBS *297*, 81–86.

Kopp, A., Gillesie, T.R., Hobelsberger, D., Estrada, A., Harper, J.M., Miller, R.A., Eckerle, I., Muller, M.A., Podsiadlowski, L., Leendertz, F.H., *et al.* (2013). Provenance and geographic spread of St. Louis encephalitis virus. mBio *4*(3), e00322–13.

Korenberg, E.I. (1976). Some contemporary aspects of natural focality and epidemiology of tick-borne encephalitis. Folia Parasitol. *23*, 357–366.

Korenberg, E.I., and Kovalevskii, Y.V. (1994). A model for relatiohnships among the tick-borne encephalitis virus, its main vectors, and hosts. Adv. Dis. Vector Res. *10*: 65-92.

Krupovic, M., Dolia, V.V., and Koonin, E.V. (2015). Plant viruses of the *Amalgaviridae* family evolved via recombination between viruses with double-stranded and negative-strand RNA genomes. Biol. Direct *10*, 12.

Kuchibhatla, D.B., Sherman, W.A., Chung, B.Y., Cook, S., Schneider, G., Eisenhaber, B., and Karlin, D.G. (2013). Powerful sequence similarity search methods and in-depth manual analysis can identify remote homologs in many apparently 'orphan' viral proteins. J. Virol. *88*, 10–20.

Kuno, G. (1981). Replication of dengue, yellow fever, St. Louis encephalitis and vesicular stomatitis viruses in a cell line (TRA-171) derived from *Toxorhynchites amboinensis.* In Vitro *17*, 1011–1015.

Kuno, G. (2001a). Persistence of arboviruses and antiviral antibodies in vertebrate hosts: its occurrence and impacts. Rev. Med. Virol. *11*, 165–190.

Kuno, G. (2001b). Transmission of arboviruses without involvement of arthropod vectors. Acta Virol. *45*, 139–150.

Kuno, G. (2007). Host range specificity of flaviviruss: correlation with *in vitro* replication. J. Med. Entomol. *44*, 93–101.

Kuno, G. (2012). Revisiting Houston and Memphis: the background histories behind the discovery of the infestation by *Aedes albopictus* (Diptera: Culicidae) in the United States and their significance in the contemporary research. J. Med. Entomol. *49*, 1163–1176.

Kuno, G. (2014). Revisiting the Tumultuous Yellow Fever Investigations in the First Three Decades of the Twentieth Century (Small Batch Book, Amherst, MA, USA), pp. 27–31, 41–42.

Kuno, G., and Chang, G.J. (2005). Biological transmission of arboviruses: reexamination of and new insights into components, mechanisms, and unique traits as well as their evolutionary trends. Clin. Microbiol. Rev. *18*, 608–637.

Kuno, G., Chang, G.J., Tsuchiya, R., Karabatsos, N., and Cropp, C.B. (1998). Phylogeny of the Genus *Flavivirus.* J. Virol. *72*, 73–83.

Kuno, G., Chang, G.J., and Chien, L.J. (2009). Correlations of phylogenetic relation with host range, length of ORF or genes, organization of conserved sequences in the 3' noncoding region, and viral classification among the members of the Genus *Flavivirus.* In Viral Genes: Diversity, Properties and Parameters, Z. Feng and M. Long, eds. (Nova Biomedical Books, New York), pp. 1–33.

Kuwata, R., Satho, T., Isawa, H., Yen, N.T., Phong, T.V., Nga, P.T., Kurashige, T., Hiramatsu, Y., Fukumitsu, Y., Hoshino, K., *et al.* (2013). Characterization of Dak Nong virus, an insect nidovirus isolated from *Culex* mosquitoes in Viet Nam. Arch. Virol. *158*, 2273–2284.

Labuda, M. (1991). Arthropod vectors in the evolution of bunyaviruses. Acta Virol. *35*, 98–105.

Labuda, M., and Nuttall, P.A. (2004). Tick-borne viruses. Parasitology *129*, S221–S245.

Labuda, M., and Randolph, S.E. (1999). Survival strategy of tick-borne encephalitis virus: cellular basis and environmental determinants. Zentbl. Bakteriol. *289*, 512–524.

La Linn, M., Gardner, J., Warrilow, D., Darnell, G.A., McMahon, C.R., Field, I., Hyatt, A.D., Slade, R.W., and Suhrbier, A. (2001). Arbovirus of marine mammals: a new alphavirus isolated from the elephant seal louse, *Lepidophthirus macrorhini.* J. Virol. *75*, 4103–4109.

Lambrechts, L., and Scott, T.W. (2014). Mode of transmission and the evolution of arbovirus virulence in mosquito vectors. Proc. R. Soc. B. *276*, 1369–1378.

Lanciotti, R.S., Ludwig, M.L., Rwaguma, E.B., Lutwama, J.J., Kram, T.M., Karabatsos, N., Cropp, B.C., and Miller, B.R. (1998). Emergence of epidemic O'nyong nyong fever in Uganda after a 35-year absence: genetic characterization of the virus. Virology *252*, 258–268.

Lastra, J.R., and Esparza, J. (1976). Multiplication of vesicular stomatitis virus in the leafhopper *Peregrinus maidis* (Ashm.), a vector of a plant rhabdovirus. J. Gen. Virol. *32*, 139–142.

Lauber, C., Ziebuhr, J., Junglen, S., Drosten, C., Zirkel, F., Nga, P.T., Morita, K., Snijder, E.J., and Gorbalenya, A.E. (2012). Mesoniviridae: a proposed new family in the Order Nidovirales formed by a single species of mosquito-borne viruses. Arch. Virol. *157*, 1623–1628.

Leake, C.J., Varma, M.G., and Pudney, M. (1977). Cytopathic effect and plaque formation by arboviruses in a continuous cell line (XTC-2) from the toad *Xenopus laevis*. J. Gen.Virol. *35*, 335–339.

Lee, J.S., N. Grubaugh, J.P. Kondig, *et al.* (2013). Isolation and genomic characterization of Chaoyang virus strain ROK 144 from *Aedes vexans nipponi* from the Republic of Korea. Virology *435*, 220–224.

Leigh, J.W., Lapointe, F.-J., Lopez, P., and Bapteste, E. (2011). Evaluating phylogenetic congruence in the post-genomic era. Genome Biol. Evol. *3*, 571–587.

Lewis, E.A. (1946). Nairobi sheep disease: the survival of the virus in the ticks, *Rhipicephalus appendiculatus*. Parasitology *37*, 55–59.

Leyssen, P., Charlier, N., Lemey, P., Billoir, F., Vandamme, A.M., De Clercq, E., de Lamballerie, X., and Neyts, J. (2002). Complete genome sequence, taxonomic assignment, and comparative analysis of the untranslated regions of the Modoc virus, a flavivirus with no known vector. Virology *293*, 125–140.

Li, C.X., Shi, M., Tian, J.H., Lin, X.D., Kang, Y.J., Chen, L.J., Qin, X.C., Xu, J., Holmes, E.C., and Zhang, Y.Z. (2015). Unprecedented genomic diversity of RNA viruses in arthropods reveals the ancestry of negative-sense RNA viruses. Elife *4*, doi: 10.7554/eLife.05378.

Li, J.-L., Cornman, R.S., Evans, J.D., Pettis, J.S., Zhao, Y., Murphy, C., Peng, W.J., Wu, J., Hamilton, M., Boncristiani, H.F., *et al.* (2014). Systemic spread and propagation of a plant pathogenic virus in European honeybee, *Apis mellifera*. mBio *5(1)*, e00898–13.

Li, Y., Kamara, F., Zhou, G., Puthiyakunnon, S., Li, C., Liu, Y., Zhou, Y., Yao, L., Yan, G., and Chen, X.G. (2014). Urbanization increases *Aedes albopictus* larval habitats and accelerates mosquito development and survivorship. PloS Negl. Trop. Dis. *8*, e3301.

Liang, Y., He, M., and Teng, C.B. (2014). Evolution of the vesticular stomatitis viruses: divergence and codon usage bias. Virus Res. SO168–1702(14)00341–4.

Lin, Y., Zhang, H., Liang, J., Li, K., Zhu, W., Fu, L., Wang, F., Zheng, X., Shi, H., Wu, S., *et al.* (2014). Identification and characterization of alphavirus M1 as a selective oncolytic virus targeting ZAP-defective human cancers. Proc. Natl. Acad. Sci. U.S.A. *111*, E4504–E4512.

Lipsitch, M., Siller, S., and Novak, M.A. (1996). The evolution of viulence in pathogens with vertical and horizontal transmission. Evolution *50*, 1729–1741.

Liu, H., Fu, Y., Jiang, D., Li, G., Xie, J., Cheng, J., Peng, Y., Ghabrial, S.A., and Yi, X. (2010). Widespsread horizontal gene transfer from double-stranded RNA viruses to eukaryotic nuclear genomes. J. Virol. *84*, 11876–11887.

Longdon, B., Wilfert, L., and Jiggins, F.M. (2012). The Sigma viruses of *Drosophila*. In Rhabdoviruses: Molecular Taxonomy, Evolution, Genomics, Ecology, Cytopathology, and Control, R.G. Dietzgen and I.V. Kuzmin, eds. (Norfolk, UK, Caister Academic Press), pp. 117–132.

Longsworth, J.F., Robertson, J.S., Tinsley, T.W., Rowlands, D.J., and Brown, F. (1973). Reactions between an insect picornavirus and naturally occurring IgM antibodies in several mammalian species. Nature *242*, 314–316.

Lord, C.C. (2010). The effect of multiple vectors on arbovirus transmission. Israel J. Ecol. Evol. *56*, 371–392.

Lord, C.C., Rutledge, C.R., and Tabachnick, W.J. (2006). Relationships between host viremia and vector susceptibility for arboviruses. J. Med. Entomol. *43*, 623–630.

Lorenzen, N.L., and Olesen, N.J. (1995). Multiplication of VHS virus in insect cells. Vet. Res. *26*, 428–432.

Lounibos, L.P. (2007). Competitive displacement and reduction. AMCA Bull. *23* (Suppl.), 276–282.

Lovisolo, O., and Rösler, O. (2003). Searching for palaentological evidence of viruses that multiply in Insecta and Acarina. Acta Zool. Cracoviensia *46* (Suppl. Fossil Insects), 37–50.

Lubinga, J.C., Tuppurainen, E.S., Stoltsz, W.H., Ebersohn, K., Coetzer, J.A., and Venter, E.H. (2013). Detection of lumpy skin disease virus in saliva of ticks fed on lumpy skin disease virus-infected cattle. Exp. Appl. Acarol. *61*, 129–138.

Luby, J.P., Clinton, R., and Kurtz, S. (1999). Adaptation of human enteric coronavirus to growth in cell lines. J. Clin. Virol. *12*, 43–51.

Luedke, A.J., Jones, R.H., and Walton, T.E. (1977). Overwintering mechanism for bluetongue virus: biological recovery of latent virus from a bovine by bites of *Culicoides variipennis*. Am. J. Trop. Med. Hyg. *26*, 313–325.

Luis, A.D., Hayman, D.T., O'Shea, T.J., Cryan, P.M., Gilbert, A.T., Pulliam, J.R., Mills, J.N., Timonin, M.E., Willis, C.K., Cunningham, A.A., *et al.* (2013). A comparison of bats and rodents as reservoirs of zoonotic viruses: are bats special? Proc. R. Soc. B *280*, 20122753.

Lupi, O. (2003). Could ectoparasites act as vectors for prion diseases? Int. J. Dermatol. *42*, 425–429.

L'Vov, D.K., Al'Khovskii, S.V., Shchelkanov, M., *et al.* (2014). Genetic characterization of the Syr-Darya Valley fever virus (SDVFV) (*Picornaviridae, Cardiovirus*) isolated from the blood of patients and ticks in Kazakhstan and Turkmenistan (an abridged English translation of the Russian title). Vopr. Virusol., *59*, 15–19.

Lwande, O.W., Lutomiah, J., Obanda, V., Gakuya, F., Mitisya, J., Mulwa, F., Michuki, G.,Chepkorir, E.,Fisher, A., Venter, M., *et al.* (2013). Isolation of tick- and mosquito-borne arboviruses from ticks sampled from livestock and wild animal hosts in Ijara District, Kenya. Vector Borne Zoon. Dis. *13*, 637–642.

Ma, Y., Xu, J., Yang, Z., Wang, X., Lin, Z., Zhao, W.,Wang, Y., Li, X., and Shi, H. (2013). A video clip of the biting midge *Culicoides anopheles* ingesting blood from an engorged *Anopheles* mosquito in Hainan, China. Parasites Vectors *6*, 326.

McBride, C.S., Baier, F., Omondi, A.B., Spitzer, S.A., Lutomiah, J., Sang, R., Ignell, R., and Vosshall, L.B. (2014). Evolution of mosquito preference for humans linked to an odorant receptor. Nature *515(7526)*, 222–227.

Macdonald, W.W. (1956). *Aedes aegypti* in Malaya I – Distribution and dispersal. Ann. Trop. Med. Parasitol. *50*, 385–398.

McIntosh, A.H., and Kimura, M. (1974). Replication of the insect Chilo iridescent virus (CIV) in a poikilothermic vertebrate cell line. Intervirology *4*, 257–267.

Mahy, B.W.J. (2001). A Dictionary of Virology, Third Edition (Academic Press, San Diego, London, New York), p. 21.

Major, L., La Linn, M., Slade, R.W., Schroder, W.A., Hyatt, A.D., Gardner, J., Cowley, J., and Suhrbier, A. (2009). Ticks associated with Macquarie Island penguins carry arboviruses from four genera. PLoS One *4*, e4375.

Malik, H.S., Henikoff, S., and Eickbush, T.H. (2000). Poised for contagion: evolutionary origins of the infectious abilities of invertebrate retroviruses. Genome Res. *10*, 1307–1318.

Mandl, C.W., Holzmann, H., Kunz, C.,and Heinz, F.X. (1993). Complete genome sequence of Powassan virus: evaluation of genetic elements in tick-borne versus mosquito-borne flaviviruses. Virology *194*, 173–184.

Mari, J., Poulos, B.T., Lightner, D.V., and Bonami, J.R. (2002). Shrimp Taura syndrome virus: genomic characterization and similarity with members of the genus cricket paralysis-like viruses. J. Gen. Virol. *83*, 915–926.

Marklewitz, M., Zirkel, F., Rwego, I.B., Heidemann, H., Trippner, P., Kurth, A., Kallies, R., Briese, T., Lipkin, W.I., Drosten,C., *et al.* (2013). Discovery of a unique novel clade of mosquito-associated bunyaviruses. J. Virol. *87*, 12850–12865.

Marklewitz, M., Zirkel, F., Kurth, A., Drosten, C., and Junglen, S. (2015). Evolutionary and phenotypic analysis of live virus isolates suggests arthropod origin of a pathogenic RNA virus family. Proc. Natl. Acad. Sci. U.S.A. pii:201502036.

Marti, G.A., Balsalobre, A., Susevich, M.L., Rabinovich, J.E., and Echeverria, M.G. (2014). Detection of triatomine infection by Triatoma virus and horizontal transmission: protecting insectaries and prospects for biological control. J. Invertebr. Pathol. *124*, 57–60.

Mead, D.G., Ramberg, F.B., Besselsen, D.G., and Mare, C.J. (2000). Transmission of vesicular stomatitis virus from infected to noninfected black flies co-feeding on nonviremic deer mice. Science *287*, 485–487.

Mead, D.G., Lovett, K.R., Murphy, M.D., Pauszek, S.J., Smoliga, G., Gray, E.W., Noblet, R., Overmeyer, J., and Rodriguez, L.L. (2009). Experimental transmission of Vesicular stomatitis New Jersey virus from *Simulium vittatum* to cattle: clinical outcome is influenced by site of insect feeding. J. Med. Entomol. *46*, 866–872.

Melville, L.F., Hunt, N.T., Davis, S.S., and Weir, R.P. (2004). Bluetongue virus does not persist in naturally infected cattle. Vet. Ital. *40*, 502–507.

Mendoza-Cano, F., Sánchez-Paz, A., Terán-Díaz, B., Galvan-Alvarez, D., Encinas-Garcia, T., Enriquez-Espinosa, T., and Hernandez-Lopez, J. (2014). The endemic copepod *Calanus pacificus californicus* as a potential vector of white spot syndrome virus. J. Aquat. Anim. Health *26*, 113–117.

Middlebrooks, B.L., Ellender, R.D., and Wharton, J.H. (1979). Fish cell culture: a new cell line from *Cynocion nebulosus*. In Vitro *15*, 109–111.

Miller, B.R., DeFoliart, G.R., Hansen, W.R., and Yuill, T.M. (1978). Infection rates of *Aedes triseriatus* following ingestion of La Crosse virus by the larvae. Am. J. Trop. Med. Hyg. *27*, 605–608.

Miller, B.R., Loomis, R., Dejean, A., and Hoogstraal, H. (1985). Experimental studies on the replication and dissemination of Qalyub virus (*Bunyaviridae*: *Nairovirus*) in the putative vector, *Ornithodoros* (Pavlovskeyella) *erraticus*. Am. J. Trop. Med. Hyg. *34*, 180–187.

Miller, B.R., Monath, T.P., Tabachnick, W.J., and Ezike, V.I. (1989). Epidemic yellow fever caused by an incompetent vector. Trop. Med. Parasitol. *40*, 396–399.

Mitzel, D.N., Best, S.M., Masnick, M.F., Porcella, S.F., Wolfinbarger, J.B., and Bloom, M.E. (2008). Identification of genetic determinants of a tick-borne flavivirus associated with hsot-specific adaptation and pathogenicity. Virology *381*, 268–276.

Mohd Jaafar, F., Goodwin, A.E., Belhouchet, M., Merry, G., Fang, Q., Cantaloube, J.F., Biagini, P., de Micco, P., Mertens, P.P., and Attoui, H. (2008). Complete characterization of the American grass carp reovirus genome (Genus *Aquareovirus*: Family Reoviridae) reveals an evolutionary link between aquareoviruses and coltiviruses. Virology *373*, 310–321.

Mohd Jaafar, F., Belhouchet, M., Belaganahalli, M., Belaganahalli, M., Tesh, R.B., Mertens, P.P., and Attoui, H. (2014). Full-genome characterization of *Orungo*, *Lebombo*, and *Changuinola* viruses provides evidence for co-evolution of orbiviruses with their arthropod vectors. PLoS One 9, e86392.

Monath, T.P. (1988). The Arboviruses: Epidemiology and Ecology (CRC Press, Boca Raton, Florida).

Moncayo, A.C., Fernandez, Z., Ortiz, D., Diallo, M., Sall, A., Hartman, S., Davis, C.T., Coffey, L., Mathiot, C.C., Tesh, R.B., *et al.* (2004). Dengue emergence and adaptation to peridomestic mosquitoes. Emerg. Infect. Dis. *10*, 1790–1796.

Mondet, B. (2001). Yellow fever epidemiology in Brazil. Bull. Soc. Pathol. Ex. *94*, 260-267.

Mondet, B., Vasconcelos, P.F., Travassos da Rosa, A.P., Travassos da Rosa, E.S., Rodrigues, S.G., Travassos Rosa, J.F., and Bicout, D.J. (2002). Isolation of yellow fever virus from nulliparous *Haemagogus* (*Haemagogus*) *janthinomys* in eastern Amazon. Vector Borne Zoon. Dis. *2*, 47–50.

Moreno-Delafuente, A., Garzo, E., Moreno, A., and Fereres, A. (2013). A plant virus manipulates the behavior of its whitefly vector to enhance its transmission efficiency and spread. PLoS One *8(4)*, e61543.

Morris, A.D. (1995). The epidemic that never was: yellow fever in Hawaii. Hawaii Med. J. *54*, 781–784.

Morris, C.D., and Srihongse, S. (1978). An evaluation of the hypothesis of transovarial transmission of eastern equine encephalomyelitis virus by *Culiseta melanura*. Am. J. Trop. Med. Hyg. *27*, 1246–1250.

Mourya, D.T., Yadav, P.D., Basu, A., Shete, A., Patil, D.Y., Zawar, D., Majumdar, T.D., Kokate, P., Sarkale, P., Raut, C.G., *et al.* (2014). Maisoor virus, a novel bat Phlebovirus is closely related to STFS and Heartland viruses. J. Virol. *88*, 3605–3609.

Moutailler, S., Roche, B., Thiberge, J.-M., Caro, V., Rougeon, F., and Failloux, A.B. (2011). Host alternation is necessary to maintain the genome stability of *Rift Valley fever virus*. PloS Negl. Trop. Dis. *5(5)*, e1156.

Moya, A., E.C. Holmes, and F. González-Candelas. (2004). The population genetics and evolutionary epidemiology of RNA viruses. Nature Rev. *2*, 279–288.

Mulcahy, D., Klaybor, D., and Batts, W.N. (1990). Isolation of infectious hematopoietic necrosis virus from leech (*Piscicola salmositica*) and copepod (*Salmincola* sp.), ectoparasites of sockeye salmon *Onchorhynchus nerka*. Dis. Aquat. Org. *8*, 29–34.

Murray, C.H. (1885). Young trout destroyed by mosquitoes. Bull. US Fish. Comm. *5*, 243.

Murray, K., Walker, C., Herrington, E., Lewis, J.A., McCormick, J., Beasley, D.W., Tesh, R.B., and Fisher-Hoch, S. (2010). Persistent infection with West Nile virus years after initial infection. J. Infect. Dis. *201*, 2–4.

Nakamura, J. (1967). Japanese encephalitis in animals (in Japanese). Shinkei-Shimpo (Advances in Neurology) *11*, 223–233.

Nasar, F., Palacios, G., Gorchakov, R.V., Guzman, H., Da Rosa, A.P., Savji, N., Popov, V.L., Sherman, M.B., Lipkin, W.I., Tesh, R.B., *et al.* (2012). Eilat virus, a unique alphavirus with host range restricted to insects by RNA replication. Proc. Natl. Acad. Sci. U.S.A. *109*, 14622–14627.

Nasar, F., Gorchakov, R.V., Tesh, R.B., and Weaver, S.C. (2014). Eilat virus host range restriction is present at multiple levels of virus life cycle. J. Virol. *89*, 1404–1418.

Ng, T.F.F., Driscoll, C., Carlos, M.P., Prileau, A., Schmieder, R., Dwivedi, B., Wong, J., Cha, Y., Head, S., Breitbart, M., *et al.* (2013). Distinct lineage of *Vesiculovirus* from big brown bats, United States. Emerg. Infect. Dis. *19*, 1978–1980.

Nga, P.T., Parquet, M.C., Lauber, C., Parida, M., Nabeshima, T., Yu, F., Thuy, N.T., Ito, T., Okamoto, K., *et al.* (2011). Discovery of the first insect nidovirus, a missing evolutionary link in the emergence of the largest RNA virus genomes. PLoS Path. 7, e1002215.

Novella, I.S., Ebendick-Corpus, B.E., Zárate, S., and Miller, E.L. (2007). Emergence of mammalian cell-adapted Vesicular stomatitis virus from persistent infections of insect vector cells. J. Virol. *81*, 6664–6668.

Nuttall, P.A., and Labuda, M. (2003). Dynamics of infection in tick vectors and at the tick–host interface. Adv. Virus Res. *60*, 233–272.

Nuttall, P.A., Jones, L.D., and Davies, C.R. (1991). The role of arthropod vectors in arbovirus evolution. Adv. Dis. Vector Res. *8*, 15–44.

Nuttall, P.A., L.D. Jones, M. Labuda, *et al.* (1994). Adaptation of arthropods to ticks. J. Med. Entomol. *31*, 1–9.

Officer, J.E. (1964). Ability of a fish cell line to support the growth of mammalian viruses. Proc. Soc. Exp. Biol. Med. *116*, 190–194.

Ogata, H., Toyoda, K., Tomaru, Y., Nakayama, Y., Shirai, Y., Claverie, J.M., and Nagasaki, K. (2009). Remarkable sequence similarity between the dinoflagellate-infecting marine girus and the terrestrial pathogen African swine fever virus. Virol. J. *6*, 178.

Okudo, H., Toma, T., Sasaki, Y., Higa, Y., Fujikawa, M., Miyagi, I., and Okazawa, T. (2004). A crab-hole mosquito, *Ochlerotatus baisasi*, feeding on mudskipper (Gobiidae: Oxudercinae) in the Ryukyu Islnds. J. Am. Mosq. Contr. Assoc. *20*, 134–137.

Okuno, Y., Inagaki, A., Fukunaga, T., Tadano, M., and Fukai, K. (1984). Electron microscopic observation of a newly isolated flavivirus-like virus from field-caught mosquitoes. J. Gen. Virol. *65*, 803–807.

O'Neal, P.A., and Juliano, S.A. (2012). Seasonal variation in compatibility and coexistence of *Aedes* mosquitoes: stabilizing effects of egg mortalty or equalizing effects of resources? J. Anim. Ecol. *82*, 256–265.

Overstreet, R.O., Jovonovich, J., and M. Hongwei, M. (2009). Parasitic crustaceans as vectors of viruses, with an emphasis on three panaeid viruses. Integr. Comp. Biol. *49*, 127–141.

Palacios, G., Forrester, N.L., Savji, N., Travassos da Rosa, A.P., Guzman, H., Detoy, K., Popov, V.L., Walker, P.J., Lipkin, W.I., Vasilakis, N., *et al.* (2013a). Characterization of Farmington virus, a novel virus from birds that is distantly related to members of the Family Rhabdoviridae. Virol. J. *10(1)*, 219.

Palacios, G., Savji, N., Travassos da Rosa, A., Guzman, H., Yu, X., Desai, A., Rosen, G.E., Hutchison, S., Lipkin, W.I., and Tesh, R. (2013b). Characterization of the *Uukuniemi virus* group (*Phlebovirus*: *Bunyaviridae*): evidence for seven distinct species. J. Virol. *87*, 3187–3195.

Palha, N., Guivel-Benhassine, F., Briolat, V., Lutfalla, G., Sourisseau, M., Ellett, F., Wang, C.H., Leischke, G.J., Herbomel, P., Schwartz, O., *et al.* (2013). Real-time whole-body visualization of chikungunya virus infection and host interferon response in zebrafish. PloS Path. *9(9)*, e1003619.

Parker, J., Rambaut, A., and Pybus, O.G. (2008). Correlating viral phenotypes with phylogeny: accounting for phylogenetic uncertainty. Infect. Genet. Evol. *8*, 239–246.

Parrish, C.R., Holmes, E.C., Morens, D.M., Park, E.C., Burke, D.S., Calisher, C.H., Laughlin, C.A., Saif, L.J., and Daszak, P. (2008). Cross-species virus transmission and the emergence of new epidemic diseases. Microbiol. Mol. Biol Rev. *72*, 457–476.

Pattersson, J.H., and Fiz-Palacios, O. (2014). Dating the origin of the genus *Flavivirus* in the light of Beringian biogeography. J. Gen. Virol. doi: 10.1099/vir.0.065227-0.

Pearson, M.N., and Rohrmann, G.F. (2002). Transfer, incorporation and substitution of envelope fusion proteins among members of the *Baculoviridae*, *Orthomyxoviridae*, and *Metaviridae* (insect retroviruses) families. J. Virol. *76*, 5301–5304.

Petterson, E., Sandberg, M., and Santi, N. (2009). Salmonid alphavirus associated with *Lepeophtheirus salmonis* (Copepoda: Caligidae) from Atlantic salmon, *Salmo salar* L. J. Fish Dis. *32*, 477–479.

Philip, C.B., and Burgdorfer, W. (1961). Arthropod vectors as reservoirs of microbial disease agents. Annu. Rev. Entomol. *6*, 391–412.

Phillips, M.J., Delsuc, F., and Penny, D. (2004). Genome-scale phylogeny and the detection of systematic biases. Mol. Biol. Evol. *21*, 1455–1458.

Pletnev, A.G., Bray, M., Hanley, K.A., *et al.* (2001). Tick-borne Langat/mosquito-borne dengue flavivirus chimera, a candidate live attenuated vaccine for protection against disease caused by members of the tick-borne encephalitis virus complex: evaluation in rhesus monkeys and in mosquitoes. J. Virol. *75*, 8259–8267.

Plus, N., and Atanasiu, P. (1966). Sélection d'un mutant du virus rabique adapté á un mutant du virus rabique adapté à un insecte. Compt. R. Acad. Sci. Paris *263*, 89–92.

Plyusnin, A., and Sironen, T. (2014). Evolution of hantaviruses: co-speciation with reservoir hosts for more than 100 myr. Virus Res. *187*, 22–26.

Poinar, G., and Poinar, R. (1998). Parasites and pathogens of mites. Annu. Rev. Entomol. *43*, 449–469.

Poinar, G., and Poinar, R. (2005). Fossil evidence of insect pathogens. J. Invertebr. Pathol. *89*, 243–250.

Pool, W.A., Brownlee, A., and Wilson, D.R. (1930). The etiology of 'Louping ill.' J. Comp. Pathol. Ther. *43*, 253–290.

Porterfield, J.S. (1980). Antigenic characteristics and classification of *Togaviridae*. In The Togaviruses- Biology, Structure, Replication, R.W. Schlesinger, ed. (Academic Press, New York and London), pp. 13–46.

Post, K., Riesner, D., Walldorf, V., and Mehlhom, H. (1999). Fly larvae and pupae as vectors for scrapie. Lancet *354*, 1969–1970.

Power, A.G. (2000). Insect transmission of plant viruses: a constraint on virus variability. Curr. Opin. Plant Biol. *3*, 336–340.

Powers, A.M., Brault, A.C., Shirako, Y., Strauss, E.G. Kang, W., Strauss, J.H., and Weaver, S.C. (2001). Evolutionary relationships and systematics of the alphaviruses. J. Virol. *75*, 10118–10131.

Price, E.R. (1950). Western equine encephalomyelitis. Commun. Dis. Ctr. Bull. *9*, 5–6.

Pudney, M., Leake, C.J., and Buckley, S.M. (1982). Replication of arboviruses in arthropod *in vitro* systems: an overview. In Invertebrate Cell Culture Applications, K. Maramorosch, J. Mitsuhashi, eds. (Academic Press, San Francisco, Sydney, Toronto), pp. 159–194.

Qin, X.-C., Shi, M., Tian, J.-H., Lin, X.D., Gao, D.Y., He, J.R., Wang, J.B., Li, C.X., Kang, Y.J., Yu, B., *et al.* (2014). A tick-borne segmented RNA virus contains genome segments derived from unsegmented viral ancestors. Proc. Natl. Acad. Sci. U.S.A. *111*, 6744–6749.

Quan, P.L., Junglen, S., Tashmukhamedova, A., Hutchison, S.K., Kurth, A., Ellerbrok, H., Egholm, M., Briese, T., Leendertz, F.H., *et al.* (2010). Moussa virus: a new member of the Rhabdoviridae isolated from *Culex decens* mosquitoes in Côte d'Ivoire. Virus Res. *147*, 17–24.

Quan, P.L., Firth, C., Conte, J.M., Williams, S.H., Zambrana-Torrelio, C.M., Anthony, S.J., Ellison, J.A., Gilbert, A.T., Kuzmin, I.V., Niezgoda, M., *et al.* (2013). Bats are a major natural reservoir for hepaciviruses and pegiviruses. Proc. Natl. Acad. Sci. U.S.A. *110*, 8194–8199.

Querido, J., Echeverria, M.G., Marti, G.A., Costa, R.M., Susevich, M.L., Rabinovich, J.E., Copa, A., Montano, N.A., Garcia, L., Cordova, M., *et al.* (2015). Seroprevalence of Triatoma virus (Dicistroviridae: Cripaviridae [sic]) antibodies in Chagas disease patients. Parasit. Vectors *8*, 29.

Ramachandra Rao, T., Singh, K.R.P., Dhanda, V., and Bhatt, P.N. (1967). Experimental transmission of Chandipura virus by mosquitoes. Ind. J. Med. Res. *55*, 1306–1310.

Randolph, S.E. (1998). Ticks are not insects: consequences of contrasting vector biology for transmission potential. Parasitol. Today *14*, 186–192.

Randolph, S.E., Miklisová, D., Lysy, J., Rogers, D.J., and Labuda, M. (1999). Incidence from coincidence: patterns of tick infestations on rodents facilitates transmission of tick-borne encephalitis virus. Parasitology *118*, 177–186.

Rao, T.R., Paul, S.D., and Singh, K.R.P. (1968). Experimental studies on the mechanical transmission of chikungunya virus by *Aedes aegypti*. Mosq. News *28*, 406–408.

Read, A.F. (1994). The evolution of virulence. Trends Microbiol. *2*, 73–76.

Reagan, K.J., and Wunner, W. (1985). Rabies virus interaction with various cell lines is independent of the acetylcholine receptor. Arch. Virol. *84*, 277–282.

Reanney, D.C. (1982). The evolution of RNA viruses. Annu. Rev. Microbiol. *36*, 47–73.

Reddy, D.V.R., and Black, L.M. (1977). Deletion mutations of the genome segments of Wound tumor virus. Virology *61*, 458–473.

Reed, L.M., Johansson, M.A., Panella, N., McLean, R., Creekmore, T., Puelle, R., and Komar, N. (2009). Declining mortality in American crow (*Corvus brachyrhynchus*) following natural West Nile virus infection. Avian Dis. *53*, 458–461.

Reese, S.M., Mossel, E.C., Beaty, M.K., Beck, E.T., Geske, D., Blair, C.D., Beaty, B.J., and Black, W.C. (2010). Identification of super-infected *Aedes triseriatus* mosquitoes collected as eggs from the field and partial characterization of the infecting La Crosse viruses. Virol. J. *7(76)*, doi10.1186/1743-422X-7-76.

Reeves, W.C., Hammon, W.M., Furman, D.P., McClure, H.E., and Brookman, B. (1947). Recovery of western equine encephalomyocarditis virus from wild bird mites (*Liponyssus sylviarum*) in Kern Country, California. Science *105*, 411–412.

Reeves, W.C., Emmons, R.W., and Hardy, J.L. (1983). Historical perspectives on California encephalitis virus in California. Prog. Clin. Biol. Res. *123*, 19–29.

Reid, H.W. (1984). Epidemiology of louping ill. In Vectors in Virus Biology, M.A. Mayo and K.A. Harrap, eds. (London and New York: Academic Press), pp. 161–178.

Reiner, R.C., Stoddard, S.T., and Scott, T.W. (2014). Socially structured human movement shapes dengue transmission despite the diffusive effect of mosquito dispersal. Epidemics *6*, 30–36.

Reisen, W.K., and Boreham, P.F.L. (1979). Host selection patterns of some Pakistan mosquitoes. Am. J. Trop. Med. Hyg. *28*, 408–421.

Reisen, W.K., Fang, Y., and Martinez, V. (2007). Is nonviremic transmission of West Nile virus by *Culex* mosquitoes (Diptera: Culicidae) nonviremic? J. Med. Entomol. *44*, 299–302.

Reitz, S.R., and Trumble, J.T. (2002). Competitive replacement among insects and arachnids. Annu. Rev. Entomol. *47*, 435–465.

Rennie, L., Wilkinson, P.J., and Mellor, P.S. (2001). Transovarial transmission of African swine fever virus in the argasid tick *Ornithodoros moubata*. Med. Vet. Entomol. *15*, 140–146.

Rethwilm, A., and Bodem, J. (2013). Evolution of foamy viruses: the most ancient of all retroviruses. Viruses *5*, 2349–2374.

Reuter, G., Boros, A., Delwad, E., and Pankovics, P. (2013). Novel seadornavirus (Family Reoviridae) related to Banna virus in Europe. Arch. Virol. *158*, 2163–2167.

Ribeiro, J.M.C. (1995). Blood feeding arthropods: live syringes or invertebrate pharmacologists? Infect. Agents Dis. *4*, 143–152.

Roekring, S., Nielsen, L., Owens, L., Pattanakitsakul, S.N., Malasit, P., and Flegel, T.W. (2002). Comparison of penaeid and insect parvoviruses suggests that viral transfer may occur between two distantly related arthropod groups. Virus Res. *87*, 79–87.

Roiz, D., Vázquez, A., Sánchez-Seco, M.P., Tenorio, A., and Rizzoli, A. (2009). Detection of novel insect flavivirus sequences integrated in *Aedes albopictus* (Diptera: Culicidae) in Northern Italy. Virol. J. *6*, 93.

Roossinck, M.J. (2011). The good viruses: viral mutualistic symbioses. Nat. Rev. Microbiol. *9*, 99–108.

Rosa, R., Costa, E.A., Marques, R.E., Oliveira, T.S., Furtini, R., Bomfim, M.R., Teixeira, M.M., Paixao, T.A., and Sabntos, R.L. (2013). Isolation of Saint Louis encephalitis virus from a horse with neurological disease in Brazil. PloS Negl. Trop. Dis. *7*, e2537.

Rudnick, A. (1983). The ecology of the dengue virus complex in Peninsular Malaysia. Proc. Int. Conf. Dengue and Dengue Hemorrhagic Fever, 7–15.

Ruiz-Jarabo, C.M., Arias, A., Baranowski, E., Escarmis, C., and Domingo, E. (2000). Memory in viral quasispecies. J. Virol. *74*, 3543–3547.

Rush, W.A., Francy, D.B., Smith, G.C., and Cropp, C.B. (1980). Transmission of an arbovirus by a member of the family *Cimicidae*. Ann. Entomol. Soc. Am. 73, 315–318.

Sabin, A.B. (1959). Survey of knowledge and problems in field of arthropod-borne virus infection. Arch. Ges. Virusforsch. *9*, 1–10.

Saikku, P., Main, A.J., Ulmanen, I., and Brummer-Korvenkonito, M. (1980). Viruses in *Ixodes uriae* (Acari: Ixodidae) from seabird colonies at Røst Islands, Lofoten, Norway. J. Med. Entomol. *17*, 360–366.

Saiyasombat, R., Bolling, B.G., Brault, A.C., Bartholomay, L.C., and Blitvich, B.J. (2011). Evidence of efficient transovarial transmission of Culex flavivirus by *Culex pipiens* (Diptera: Culicidae). J. Med. Entomol. *48*, 1031–1038.

Saiyasombat, R., Carrillo-Tripp, J., Miller, W.A., Bredenbeek, P.J., and Blitvich, B.J. (2014). Substitution of the premembrane and envelope protein genes of Modoc virus with the homologous sequences of West Nile virus generate a chimeric virus that replicates in vertebrates but not mosquito cells. Virol. J. *11*, 150.

Salvan, M., and Mouchet, J. (1994). *Aedes albopictus* et *Aedes aegypti* a l'Ile de La Réunion. Ann. Soc. Belge. Med. Trop. *74*, 323–326.

Sang, R., Onyango, C., Gachoya, J., Mabinda, E., Konongoi, S., Ofula, V., Dunster, L., Okoth, F., Coldren, R., Tesh, R., *et al.* (2006). Tick-borne arbovirus surveillance in market livestock, Nairobi, Kenya. Emerg. Infect. Dis. *12*, 1074–1080.

Sawabe, K., Tanabayashi, K., Hotta, A., Hoshino, K., Isawa, H., Sasaki, T., Yamada, A., Kurahashi, H., Shudo, C., and Kobayashi, M. (2009). Survival of avian H5N1 A viruses in *Calliphora nigribarbis* (Diptera: Calliphoridae). J. Med. Entomol. *46*, 852–855.

Saxton-Shaw, K.D., Ledermann, J.P., Borland, E.M., Stovall, J.L., Mossel, E.C., Singh, A.J., Wilusz, J., and Powers, A.M. (2013). O'nyong nyong virus molecular determinants of unique vector specificity reside in non-structural protein 3. PloS Negl. Trop. Dis. *7(1)*, e1931.

Sayler, K.A., Barbet, A.F., Chamberlain, C., Clapp, W.L., Alleman, R., Loeb, J.C., and Lednicky, J.A. (2014). Isolation of *Tacaribe virus*, a Caribbean arenavirus, from host-seeking *Amblyomma americanum* ticks in Florida. PloS One *9(12)*, e115769.

Scott, J.L., Fendrick, J.L., and Leong, J.C. (1980). Growth of infectious hematopoietic necrosis virus in mosquito and fish cell lines. Wasmann J. Biol. *38*, 21–29.

Scott, T.W., Naksathit, A., Day, J.F., Kittayapong, P., and Edman, J.D. (1997). A fitness advantage for *Aedes aegypti* and the viruses it transmits when females feeding only on human blood. Am. J. Trop. Med. Hyg. *57*, 235–239.

Scotti, P.D., and Longworth, J.F. (1980). Naturally occurring IgM antibodies to a small RNA insect vius in some mammalian sera in New Zealand. Intervirology *13*, 186–191.

Seganti, L., Superti, F., Bianchi, S., Orsi, N., Divizia, M., and Pana, A. (1990). Susceptibility of mammalian, avian, fish, and mosquito cell lines to rabies viruses. Acta Virol. *34*, 155–163.

Selling, B.H., Allison, R.F., and P. Kaesberg, P. (1990). Genomic RNA of an insect virus directs synthesis of infectious virions in plants. Proc. Nat. Acad. Sci. U.S.A. *87*, 434–438.

Senior-White, R. (1934). Three years mosquito control work in Calcutta. Bull. Entomol. Res. *25*, 551–593.

Shen, D.T., Gorham, J.R., Harwood, R.F., and Padgett, G.A. (1973). The persistence of Aleutian disease virus in the mosquito *Aedes fitehii*. Arch. Ges. Virusforsch. *40*, 375–381.

Shen, M., Cui, L., Ostiguy, N., and Cox-Foster, D. *et al.* (2005). Intricate transmission routes and interactions between picorna-like viruses (*Kashmir bee virus* and *Sacbrood virus*) with the honeybee host and parasitic varroa mites. J. gen. Virol. *86*, 2281–2289.

Shike, H., Dhar, A.K., Burns, J.C., Shimizu, C., Jousset, F.X., Klimpel, K.R., and Bergoin, M. (2000). Infectious hypodermal and hematopoietic necrosis virus of shrimp is related to mosquito brevidensoviruses. Virology *277*, 167–177.

Shope, R.E. (1957). The leech as a potential virus reservoir. J. Exp. Med. *105*, 373–382.

Shors, S.T., and Winston, V. (1989). Detection of infectious hematopoietic necrosis virus in an invertebrate (*Callibaetis* sp.). Am. J. Vet. Res. *50*, 1307–1309.

Sirivichayakul, C., Sabchareon, A., Limkittikul, K., and Yoksan, S. (2014). Plaque reduction neutralization antibody test does not accurately predict protection against dengue infection in Ratchaburi Cohort, Thailand. Virol. J. *11*, 48.

Sloof, R., and Marks, E.N. (1965). Mosquitoes (Culicidae) biting a fish (*Periophthalmidae*). J. Med. Entomol. *2*, 16.

Smith, A.W., Skilling, D.E., Cherry, N., Mead, J.H., and Watson, D.O. (1998). Calicivirus emergence from ocean reservoirs: zoonotic and interspecies movements. Emerg. Infect. Dis. *4*, 13–20.

Smith, C.E.G. (1956). The history of dengue in tropical Asia and its probable relationship to the mosquito *Aedes aegypti*. J. Trop. Med. Hyg. *59*, 243–251.

Smith, K.M. (1967). Insect virology (New York and London: Academic Press), viii, 1–7.

Smith, P.F., Howorth, E.W., Carter, D., Gray, E.W., Noblet, R., and Mead, D.G. (2009). Mechanical transmission of vesicular stomatitis New Jersey virus by *Simulium vittatum* (Diptera: Simuliidae) to domestic swine (*Sus scrofa*). J. Med. Entomol. *46*, 1537–1540.

Smith, P.F., Howerth, E.W., Carter, D., Gray, E.W., Noble, R., Smoliga, G., Rodriguez, L.L., and Mead, D.G. (2011). Domestic cattle as a non-conventional amplifying host of vesicular stomatitis New Jersey virus. Med. Vet. Entomol. *25*, 184–191.

Smith, T. (1921). Parasitism as a factor in disease. Science *54*, 99–108.

Smith, T. (1934). Parasitism and Disease (Princeton University Press, Princeton, NJ).

Sojka, D., Franta, Z., Horn, M., Caffrey, C.R., Mares, M., and Kopacek, P. (2013). New insights into the machinery of blood digestion by ticks. Trends Parasitol. *29*, 276–285.

Solis, J., and Mora. E.C. (1970). Virus susceptibility range of the fathead minnow (*Pimephales promelas*) poikilothermic cell line. Appl. Microbiol. *19*, 1–4.

Soret, M.G., and Sanders, M. (1954). *In vitro* method for cultivating eastern equine encephalomyelitis virus in teleost embryos. Proc. Soc. Exp. Biol. Med. *87*, 526–529.

Stamm, D.D. (1966). Relationships of birds and arboviruses. Auk *83*, 84–97.

Stoddard, S.T., Forshey, B.M., Morrison, A.C., Paz-Soldan, V.A., Vazquez-Prokopec, G.M., Astete, H., Reiner, R.C., Vilcarromero, S., Elder, J.P., Halsey, E.S., *et al.* (2013). House-to-house human movement drives dengue virus transmission. Proc. Natl. Acad. Sci. U.S.A. *110*, 994–999.

Stollar, V., and Hardy, J.L. (1984). Host-dependent mutants of Sindbis virus whose growth is restricted in cultured *Aedes albopictus* cells produce normal yield of virus in intact mosquitoes. Virology *134*, 177–183.

Strauss, J.H., Wang, K.S., Schmaljohn, A.L., Kuhn, R.J., and Strauss, E.G. (1994). Host-cell receptors for Sindbis virus. Arch. Virol. *9*(Suppl.), 473–484.

Studzinski, G.P. (1999). Overview of apoptosis. In Overview of Apoptosis: A Practical Approach, G.P. Studzinski, ed. (Oxford University Press, New York and Oxford), pp. 1–17.

Styer, L.M., Kent, K.A., Albright, R.G., Bennett, C.J., Kramer, L.D., and Bernard, K.A. (2007). Mosquito inoculate high doses of West Nile virus as they probe and feed on live hosts. PLoS Path. *3*, 1262–1270.

Sudeep, A.B., Jadi, R.S., and Mishra, A.C. (2009). Ganjam virus. Ind. J. Med. Res. *130*, 514–519.

Sudhakaran, R., Parameswaran, V., and Hameed, A.S.S. (2007). *In vitro* replication of *Macrobrachium rosenbergii* nodavirus and extra small virus in C6/36 mosquito cell line. J. Virol. Methods *116*, 112–118.

Sudhakaran, R., Hanibabu, P., Kumar, S.R., Saratthi, M., Babu, V.S., Venkatesan, C., and Hameedl, A.C. (2008). Natural aquatic insect carriers of *Macrobrachium rosenbergii* nodavirus (Mr NV) and extra small virus (XSV). Dis. Aquat. Org. *79*, 141–145.

Sulkin, S.E. (1945). Recovery of equine encephalomyelitis virus (western type) from chicken mites. Science *101*, 381–383.

Sulkin, S.E. (1962). The bats as a reservoir of viruses in nature. Progr. Med. Virol. *4*, 157–207.

Sulkin, S.E., and Allen. A.R. (1974). Virus infections in bats. Monogr. Virol. *8*, 1–103.

Swanepoel, R., Leman, P.A., Burt, F.J., Zachariades, N.A., Braack, L.E., Ksiazek, T.G., Rollin, P.E., Zaki, S.R., and Peters, C.J. (1996). Experimental inoculation of plants and animals with Ebola virus. Emerg. Infect. Dis. *2*, 321–325.

Sylverton, J.T., and Berry, G.P. (1941). Hereditary transmission of the western type of equine encephalomyelitis virus in the wood tick, *Dermacentor andersoni* Stiles. J. Exp. Med. *73*, 507–530.

Tabachnick, W.J. (1991). Evolutionary genetics and arthropod-borne diseases. The yellow fever mosquito. Am. Entomol. *37*, 14–24.

Takamatsu, H., Mellor, P.S., Mertens, P.P., Kirkam, P.A., Burroughs, J.N., and Parkhouse, R.M. (2003). A possible overwintering mechanism for bluetongue virus in the absence of the insect vector. J. Gen. Virol. *84*, 227–235.

Taylor, D.J., Ballinger, M.J., Zhan, J.J., Hanzly, L.E., and Bruenn, J.A. (2014). Evidence that ebolaviruses and cuevaviruses have been diverging from Marburg viruses since the Miocene. Peer J. *2*, *e556*.

Terenin, I.M., Dmitriev, S.E., Andreev, D.,E., Royall, E., Belsham, G.J., Roberts, L.O., and Shatsky, I.N. (2005). A cross-kingdom internal ribosome entry site reveals a simplified mode of internal ribosome entry. Mol. Cell. Biol. *25*, 7879–7888.

Tesh, R.B. (1984). Transovarial transmission of arboviruses in their invertebrate vectors. In Current Topics in Vector Research, K.F. Harris, ed. (Praeger Publisher, New York), pp. 57–76.

Tesh, R.B., Chaniotis, B.N., Carrera, B.R., and Johnson, K.M. (1972). Further studies on the natural host preferences of Panamanian phlebotomine sand flies. Am. J. Epidemiol. *95*, 88–93.

Tesh, R.B., Peleg, J., Samina, I., Margalit, J., Bodkin, D.K., Shope, R.E., and Knudson, D. (1986). Biological and antigenic characterization of Netivot virus, an unusual new orbivirus recovered from mosquitoes in Israel. Am. J. Trop. Med. Hyg. *35*, 418–428.

Theiler, M., and Downs, W.G. (1973). The arthropod-borne viruses of vertebrates (Yale University Press, New Haven and London).

Thèse, J., Bézier, A., Periquet, G., Drezen, J.M., and Herniou, E.A. (2011). Paleozoic origin of insect large dsDNA viruses. Proc. Natl. Acad. Sci. U.S.A. *108*, 15931–15935.

Thèse, J., Leclercq, S., Moumen, B., Cordaux, R., and Gilbert, C. (2014). Remarkable diversity of endogenous viruses in a crustacean genome. Genome Biol. Evol. *6(8)*, 2129–2140.

Tinker, M.E., and Hayes, G.R. (1959). The 1958 *Aedes aegypti* distribution in the United States. Mosq. News *19*, 73–78.

Tompkins, D.M., Paterson, R., Massey, B., and Gleeson, D.M. (2010). Whataroa virus four decades on: emerging, persisting, or fading out? J. R. Soc. New Zealand *40*, 1–9.

Traavik, T. (1978). Experimental 'Runde' virus infections in embryonated eggs and chickens. Acta Path. Microbiol. Scand. Sect. B *86*, 299–301.

Traavik, T. (1979). Arboviruses in Norway. In Arctic and Tropical Arboviruses, E. Kurstak, ed. (Academic Press, New York, San Francisco, London), pp. 67–81.

Traavik, T., Mehl, R., and Kjeldberg, E. (1977). 'Runde' virus, a coronavirus-like agent associated with seabirds and ticks. Arch. Virol. *55*, 25–38.

Tsetsarkin, K.A., and Weaver, S.C. (2011). Sequential adaptive mutations enhance efficient vector switching by chikungunya virus and its epidemic emergence. PloS Path. *7*, e1002412.

Tumban, E., Mitzel, D.N., Maes, N.E., Hanson, C.T., Whitehead, S.S., and Hanley, K.A. (2011). Replacement of the 3′ untranslated variable region of mosquito-borne dengue virus with that of tick-borne Langat virus does not alter vector specificity. J. Gen. Virol. *92*, 841–848.

Tumban, E., Maes, N.E., Schirtzinger, E.E.,Young, K.I., Hanson, C.T., Whitehead, S.S., and Hanley, K.A. (2013). Replacement of conserved or variable sequences of the mosquito-borne dengue virus 3′ UTR with homologous sequences from Modoc virus does not change infectivity for mosquitoes. J. Gen. Virol. *94*, 783–788.

Tuppurainen, E.S., Lubinga, J.C., Stoltsz, W.H., Troskie, M., Carpenter, S.T., Coetzer, J.A., Venter, E.H., and Oura, C.A. (2013). Evidence of vertical transmission of lumpy skin disese virus in *Rhipicephalus decoloraus* ticks. Ticks Tick Borne Dis. *4*, 329–333.

Turell, M.J. (1988). Reduced Rift Valley fever virus infection rates in mosquitoes associated with pledget feedings. Am. J. Trop. Med. Hyg. *39*, 597–602.

Turell, M.J. (2015). Experimental transmission of Karshi (mammalian tick-borne flavivirus group) virus by *Ornithodoros* ticks > 2,900 days after initial virus exposure supports the role of soft ticks as a long-term maintenance mechanism for certain flaviviruses. PLoS Negl. Trop. Dis. *9*, e0004012.

Turell, M.J., and Spielman, A. (1992). Nonvascular delivery of Rift Valley fever virus by infected mosquitoes. Am. J. Trop. Med. Hyg. *47*, 190–194.

Turell, M.J., Hardy, J.L., and Reeves, W.C. (1982). Stabilized infection of California encephalitis virus in *Aedes dorsalis*, and its implications for viral maintenance in nature. Am. J. Trop. Med. Hyg. *31*, 1252–1259.

Turell, M.J., Linthicum, K.J., and Beaman, J.R. (1990). Transmission of Rift Valley fever virus by adult mosquitoes after ingestion of virus as larvae. Am. J. Trop. Med. Hyg. *43*, 677–690.

Turell, M.J., Mores, C.N., Lee, J.S., Paragas, J.J., Shermuhemedova, D., Endy, T.P., and Khodjaev, S. (2004). Experimental transmission of Karshi and Langat (tick-borne encephalitis virus complex) viruses by *Ornithodoros* ticks (Acari: Argasidae). J. Med. Entomol. *41*, 973–997.

Valiente Moro, C., Chauve, C., and Zenner, L. (2005). Vectorial role of some Dermanyssoid mites (Acari, *Mesostigmata, Dermanyssoidea*). Parasite *12*, 99–109.

Van Baarlen, P., van Belkum, A., Summerbell, R.C., Crous, P.W., and Thomma, B.P. (2007). Molecular mechanisms of pathogenicity: how do pathogenic microorganisms develop cross-kingdom host jumps? FEMS Microbiol. Rev. *31*, 239–277.

Van den Hurk, A.F., Smith, C.S., Field, H.E., Smith, I.L., Northill, J.A., Taylor, C.T., Jansen, C.C., Smith, G.A., and MacKenzie, J.S. (2009). Transmission of Japanese encephalitis virus from the black flying fox, *Pteropus alectro* to *Culex annulirostris* mosquitoes, despite the absence of detectable viremia. Am. J. Trop. Med. Hyg. *81*, 457–462.

Varelas-Wesley, I., and Calisher, C.H. (1982). Antigenic relationships of flaviviruses with undetermined arthropod-borne status. Am. J. Trop. Med. Hyg. *31*, 1273–1284.

Varma, M.G.R., and Pudney, M. (1971). Infection of *Anopheles stephensi* and *Aedes aegypti* cell lines with arboviruses isolated from *Anopheles* mosquitoes. Tr. R. Soc. Trop. Med. Hyg. *65*, 102–103.

Vasilakis, N., Deardorff, E.R., Kenney, J.L., Rossi, S.L., Hanley, K.A., and Weaver, S.C. (2009). Mosquitoes put the brake on arbovirus evolution: experimental evolution reveals slower mutation accumulation in mosquito than vertebrate cells. PloS Path. *5*, e1000467.

Vasilakis, N., Forrester, N.L., Palacios, G., Nasar, F., Savji, N., Rossi, S.L., Guzman, H., Wood, T.G., Popov, V., Gorchakov, R., *et al.* (2013). Negevirus-a proposed new taxon of insect-specific viruses with wide geographic distribution. J. Virol. *87*, 2475–2488.

Vasilakis, N., Guzman, H., Firth, C., Forrester, N.L., Widen, S.G., Wood, T.G., Rossi, S.L., Ghedin, E., Popov, V., Blasdell, K.R., *et al.* (2014). Mesoniviruses are mosquito-specific viruses with extensive geographic distribution and host range. Virol. J. *11*, 97.

Vazeille, M.C., Rosen, L., and Guillon, J.-C. (1988). An orbivirus of mosquitoes which induces CO_2 sensitivity in mosquitoes and is lethal for rabbits. J. Virol. *62*, 484–487.

Vazeille, M., Mousson, L., and Failloux, A.-B. (2009). Failure to demonstrate experimental vertical transmission of the epidemic strain of chikungunya virus in *Aedes albopictus* from La Réunion Islnd, Indian Ocean. Mem. Inst. Oswaldo Cruz *104*, 632–635.

Vázquez, A., Sánchez-Seco, M.-P., Palacios, G., Molero, F., Reyes, N., Ruiz, S., Aranda, C., Marques, E., Esosa, R., Moreno, J., *et al.* (2012). Novel flaviviruses detected in different species of mosquitoes in Spain. Vector-Borne Zoon. Dis. *12*, 223–229.

Venter, M., Human, S., Zaayman, D., Gerdes, G.H., Williams, J., Steyl, J., Leman, P.A., Paweska, J.T., Setzkorn, H., Rous, G., *et al.* (2009). Lineage 2 West Nile virus as cause of fatal neurologic disease in horses, South Africa. Emerg. Infect. Dis. *15*, 877–884.

Ventoso, I. (2012). Adaptive changes in alphavirus mRNA translation allowed colonization of vertebrate hosts. J. Virol. *86*, 9484–9494.

Verwoerd, D.W. (2012). History of orbivirus research in South Africa. J. S. Afr. Vet. Assoc. *83*, E1–E6.

Villareal, L.P., Defilippis,V.R., and Gottlieb, K.A. (2000). Acute and persistent viral life strategies and their relationship to emerging diseases. Virology *272*, 1–6.

Villordo, S.M., and Gamarik, A.V. (2013). Differential RNA sequence requirement for dengue virus replication in mosquito and mammalian cells. J. Virol. *87*, 9365–9372.

Viñuela, E. (1985). African swine fever virus. Curr. Topics Microbiol. Immunol. *116*, 151–170.

Vögtlin, A., Hofmann, M.A., Nenninger, C., Renzuello, S., Steinrigl, A., Loitsch, A., Schwermer, H., Kaufmann, C., Thur, B., *et al.* (2013). Long-term infection of goats with bluetongue virus serotype 25. Vet. Microbiol. *166(1–2)*, 165–173.

Walker, P.J. (2005). Bovine ephemeral fever in Australia and the World. Curr. Top. Microbiol. Immunol. *292*, 57–80.

Walker, P.J., Firth, C., Widen, S.G., Blasdell, K.R., Guzman, H., Wood, T.G., Paradkar, P.N., Holmes, E.C., Tesh, R.B., and Vasilakis, N. (2015). Evolution of genome size and complexity in the *Rhabdoviridae*. PLoS Pathogens *11(2)*, e1004664.

Wallner, G., Mandl, C.W., Kunz, C., and Heinz, F.X. *et al.* (1995). The flavivirus 3′-noncoding region: extensive size heterogeneity independent of evolutionary relationships among strains of tick-borne encephalitis virus. Virology *213*, 169–178.

Walton, T.E. (2004). The history of bluetongue and a current global overview. Vet. Ital. *40*, 31–38.

Wan, X.F., Barnett, J.L., Cunningham, F., Cheng, S., Yang, G., Nash, S., Long, L.P., Ford, L., Blackmon, S., Zhang, Y., *et al.* (2013). Detection of African swine fever virus-like sequences in ponds in the Mississippi Delta through metagenomic sequencing. Virus Genes *46*, 441–446.

Wang, E., Ni, H., Barrett, A.D.T., Watowich, S.J., Gubler, D.J., and Weaver, S.C. (2000). Evolutionary relationships of endemic/epidemic and sylvatic viruses. J. Virol. *74*, 3227-3234.

Wang, L., Lv, X., Zhai, Y., Fu, S., Wang, D., Rayner, S., Tang, Q., and Liang, G. (2012). Genomic characterization of a novel virus of the family Tymoviridae isolated from mosquitoes. PLoS ONE *7*, e39845.

Watts, D.M., MacDonald, C., Bailey, C.L., Meegan, J.M., Peters, C.J., and Mckee, K.T. (1988). Experimental infection of *Phlebotomus papatasi* with sandfly fever Sicilian virus. Am. J. Trop. Med. Hyg. *39*, 611–616.

Webster, C.G., Reitz, S.R., Perry, K.L., and Adkins, S. (2011). A natural M RNA reassortant arising from two species of plant-and insect-infecting bunyaviruses and comparison of its sequence and biological properties to parental species. Virology *413*, 216–225.

Weinmann, N., Papp, T., Alves de Mattos, A.P., Teifke, J.P., and Marschang, R.E. (2007). Experimental infection of crickets (*Gryllus bimaculatus*) with an invertebrate iridovirus isolated from a high-casgued chameleon (*Chamaeleo hoehnelii*). J. Vet. Diagn. Invest. *19*, 674–679.

Weiss, R.A. (2002). Virulence and pathogenesis. Trends Microbiol. *10*, 314–317.

Wengler, G., and Castle, E. (1986). Analysis of structural properties which possibly are characteristics for the 3′-terminal sequence of the genome RNA of flaviviruses. J. Gen. Virol. *67*, 1183–1188.

Weston, J., Villoing, S., Brémont, M., Castric, J., Pfeffer, M., Jewhurst, V., McLoughlin, M., Rodseth, O., Christie, K.E., Koumans, J., *et al.* (2002). Comparison of two aquatic alphaviruses, salmon pancreas disease virus and sleeping disease virus, by using genome sequence analysis, monoclonal reactivity, and cross-infection. J. Virol. *76*, 6155–6163.

Wheeler, S.S., Vineyard, M.P., Barker, C.M., and Reisen, W.K. (2012). Importance of recrudescent avian infection in West Nile virus overwintering: incomplete antibody neutralization of virus allows infrequent vector infection. J. Med. Entomol. *49*, 895-902.

White, L.A. (1987). Susceptibility of *Aedes albopictus* C6/36 cells to viral infection. J. Clin. Microbiol. *25*, 1221–1224.

Whitfield, J.B., and Asgari, S. (2003). Vector or not? Phylogenetics of polydnaviruses and their wasp carriers. J. Insect Physiol. *49*, 397–405.

Whitman, L., and Antunes, P.C.A. (1938). Studies on *Aedes aegypti* infected in the larval stage with the virus of yellow fever. Proc. Soc. Exp. Biol. Med. *37*, 664–666.

WHO (World Health Organization Scientific Group) (1967). Arboviruses and human diseases. WHO Tech. Rep. Ser. 369: 5–84.

WHO (World Health Organization Scientific Group). (1985). Arthropod-borne and rodent-borne viral diseases. WHO Tech. Rept. Series *719*.

Wiens, J.J. (2003). Missing data, incomplete taxa, and phylogenetic accuracy. System. Biol. *52*, 528–538.

Wilson, A., Darpel, K., and Mellor, P.S. (2008). Where does bluetongue virus sleep in the winter? PloS Biol. *6*, e210.

Wilson, M.L., Gonzalez, J.-P., Cornet, J.-P., and Camicas, J.L. (1991). Transmission of *Crimean–Congo haemorrhagic fever virus* from experimentally infected sheep to *Hyaloma truncatum* ticks. Res. Virol. *142*, 395–404.

Wolf, K., and Mann, J.A. (1980). Poikilotherm vertebrate cell lines and viruses: current listing for fishes. In Vitro *16*, 168–179.

Woolaway, K.E., Lazaridis, K., Belsham, G.J., Carter, M.J., and Roberts, L.O. (2001). The 5′ untranslated region of *Rhopalosiphum padi* virus contains an internal ribosome entry site which functions efficiently in mammalian, plant, and insect transmission systems. J. Virol. *75*, 10244–10249.

Woolhouse, M.E., Taylor, L.H., and Haydon, D.T. (2001). Population biology multihost pathogens. Science *292*, 1109–1112.

Wurzel, L.G., Cable, R.G., and Leiby, D.A. (2002). Can ticks be vectors for hepatitis C virus? N. Engl. J. Med. *347*, 1724–1725.

Yoshii, K., Goto, A., Kawakami, K., Kariwa, H., and Takashima, I. (2008). Construction and application of chimeric virus-like particles of tick-borne encephalitis virus and mosquito-borne Japanese encephalitis virus. J. gen. Virol. *89*, 200–211.

Yu, X.-J., and Tesh, R.B. (2014). The role of mites in the transmission and maintenance of *Hantaan virus* (*Hantavirus*: *Bunyaviridae*). J. Infect. Dis. *210*, 1693–1699.

Yunker, C.E., and Cory, J. (1969). Colorado tick fever virus: growth in a mosquito cell line. J. Virol. *3*, 631–632.

Yuval, B. (1992). The other habit: sugar feeding by mosquitoes. Bull. Soc. Vector Ecol. *17*, 150–156.

Zangger, H., Ronet, C., Desponds, C., Kuhlmann, F.M., Robinson, J., Hartley, M.A., Prevel, F., Castiglioni, P., Pratlong, F., Bastien, P., *et al.* (2013). Detection of *Leishmania RNA virus* in *Leishmania* parasites. PLoS Negl. Trop. Dis. *7*, e2006.

Zhai, Y., Attoui, H., Mohd Jaafar, F., Wang, H.Q., Cao, Y.X., Fan, S.P., Sun, Y.X., Liu, L.D., Mertens, P.P., Meng, W.S., *et al.* (2010). Isolation and full-length sequence analysis of *Armigeres subalbatus* totivirus, the first totivirus isolate from mosquitoes representing a proposed novel genus (*Arlivirus*) of the Family *Totiviridae*. J. Gen. Virol. *91*, 2836–2845.

Zhang, Y., Zhu, J., Tao, K., Wu, G.,Guo, H., Wang, J., Zhang, J., and Xing, A. (2002). Proliferation and location of Hantaan virus in gamasid mites and chigger mites; a molecular biological study (in Chinese). Zhonghua Yi Xue Za Zhi *82*, 1415–1419.

Zinsser, H. (1935). Rats, Lice and History (Little Brown & Co, Boston).

Zirkel, F., Kurth, A., Quan, P.-L., Briese, T., Ellerbrok, H., Pauli, G., Leendertz, F.H., Lipkin, W.I., Ziebuhr, J., Drosten, C., *et al.* (2011). An insect nidovirus emerging from a primary rain forest. mBio *2*, 1–10.

Part III

Arbovirus Diagnosis and Control

Laboratory Diagnosis of Arboviruses

15

Amy J. Lambert and Robert S. Lanciotti

Abstract

In recent years, there have been significant advances in the international capability for the diagnosis of arboviral infections. These advances have predominantly occurred in the field of nucleic acid-based diagnosis; driven by a growing diversity of highly transferable and relatively affordable molecular assays. These comparatively newly described molecular assays complement standard serological assays that continue to be applied preferentially in many laboratories for the detection of arboviral antibodies. In addition to nucleic acid-based and serological efforts, virus isolation remains the standard for arboviral diagnosis and is undertaken most often in cell culture. Here, we summarize a variety of classical and newly developed methods for arboviral diagnosis, including both serological and virus detection techniques. The utility of these approaches for application to human clinical and ecological samples in both diagnostic and research settings is discussed.

Overview

The incredible diversity of arboviruses presents a daunting challenge to their detection at the broad taxonomic level, requiring a combination of tools, reagents and understanding that is available only in specialized reference laboratories. As such, the efforts of the majority of investigators are focused on a comparatively small number of arboviruses, relevant to their specific research or diagnostic interests. In fact, the most widely used laboratory tests for the detection of arboviruses and their infections are limited to those that detect key, medically significant pathogens such as WN, DEN, and CHIK viruses. Here, we discuss the current state of arbovirus laboratory detection and diagnosis, including methods that are applicable to both high impact and relatively obscure pathogens. Defining the discussion, a hallmark of many arboviral infections is a short-lived viraemia in the human host which dictates that the most frequently used methods are antibody-based in their capability. Accordingly, we will outline the two main areas of arbovirus laboratory diagnosis: (i) the identification of infection through predominantly antibody based serological methods and (ii) direct virus detection. Much of what is known about the laboratory diagnosis of arboviruses has been determined by a relative abundance of research that has been conducted on select few, emerging pathogens, mainly flaviviruses, such as WN and DEN viruses. Therefore, examples given frequently involve these agents by default, not design, as their diagnoses have been the most rigorously investigated among arboviruses.

Serology

In most cases, the viraemia associated with an arboviral infection peaks and declines prior to the onset of symptoms, as shown for WNV (Fig. 15.1). As such, the diagnosis of arboviral infections is most often conducted by the detection of antibodies directed against a given virus in serum and CSF samples. At the onset of illness and in the weeks after the development of symptoms, IgM rises rapidly, followed by a relatively delayed, albeit sustained IgG and neutralization antibody response (Fig. 15.1). Ideally, the most accurate serological diagnosis is facilitated by the availability of both acute and convalescent serum samples, drawn during the first week after the onset of symptoms of illness and two or three weeks thereafter, respectively. The evaluation of such paired samples allows for the identification of a rise in antibody titres that definitively identifies a recent infection. Classical methods of serology include haemagglutination inhibition (HI) and complement fixation (CF) tests that have been largely replaced by enzyme-linked immunosorbent assays (ELISA) and immunofluorescence assays (IFA) in most clinical laboratories. However, it should be noted that HI and CF tests continue to be used quite successfully in some research laboratories. There are commercially available kits for the detection of antibodies directed against a limited number of arboviruses including DEN, JE, WN, SLE, EEE and LAC viruses, among others. These commercial kits primarily utilize ELISA and IFA formats and are available from a number of companies that are not discussed here. Historically, the plaque reduction neutralization test (PRNT) has been considered the 'gold standard' for the serological diagnosis of arbovirus infections. The PRNT continues to be the method of choice for serological confirmation and differentiation of infections and is an irreplaceable tool for accurate diagnosis in many laboratories.

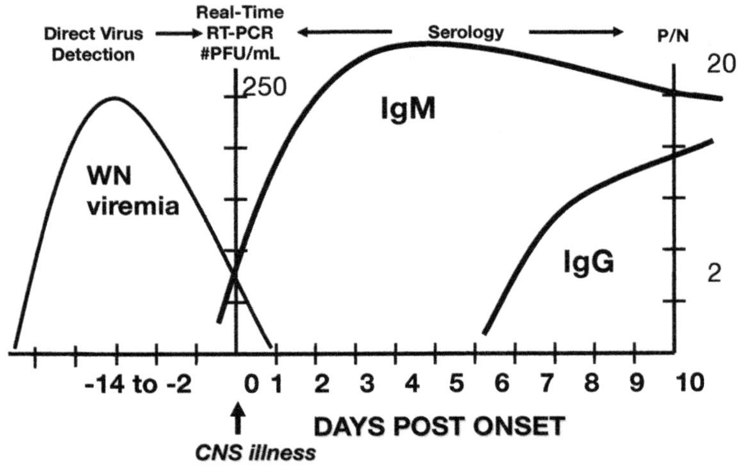

Figure 15.1 Theoretical depiction of WNV viremia and immune response as determined through empirical observation and generalization of thousands of results generated by WNV real-time RT-PCR, IgM and IgG ELISAs. PFU, plaque forming units; P/N, positive/negative value as determined by ELISA. Neutralization antibodies follow a similar pattern of response to IgG antibodies.

While the majority of arboviral serological diagnoses are straightforward, there are perplexing issues that can prevent the absolute distinction of a causative agent in some instances. Cross-reactivity, particularly among flaviviruses, can confound (Martin *et al.*, 2000) and occasionally prohibit species level identification even when the PRNT is applied using a comparative battery of related viruses for differentiation. Original antigenic sin, which causes an overwhelming immune response to a primary antigen during a secondary flavivirus infection, is also problematic (Kuno *et al.*, 1993). Finally, serological based determination of bunyavirus infections should be regarded with some caution as the ability of these viruses to reassort genomic segments could, in rare instances, cause the misidentification of a novel reassortant as a known agent from which the reassortant virus M segment, which encodes the immunoreactive envelope proteins, was derived.

ELISA

The antibody capture ELISA format for the detection of IgM (IgM antibody capture or MAC-ELISA) is the preferred method for frontline detection of recent arboviral infections. The assay is built upon a microtitre plate well, onto which an anti-IgM antibody is bound and a test specimen (a dilution of serum or CSF) is then added, capturing IgM in the sample. After IgM capture, a specific antigen is then applied, followed by the addition of an enzyme-linked antibody and a chromogenic substrate that changes colour in the presence of the enzyme. The amount of colour that develops is generally proportionate to the amount of antigen-specific IgM present in the test specimen and is objectively measured as optical density (OD), or absorbance. In most cases, IgM directed against arboviruses appears at levels that are detectable by the ELISA within the first week after the onset of symptoms (Fig. 15.1). IgM levels continue to rise during the first several weeks after the appearance of clinical illness, declining thereafter (Fig. 15.1). Therefore, the presence of IgM in a test specimen can be considered presumptive evidence of a recent infection. However, IgM has been shown to persist for a period of years in certain instances (Roehrig *et al.*, 2003; Kapoor *et al.*, 2004; Prince *et al.*, 2005) which prevents the confirmation of a recent infection based upon the presence of IgM alone. For DEN virus infections, IgM is detectable in 90–95% of patients within the first 6 days after the onset of symptoms (Vaughn *et al.*, 1997). Although, given its relatively protracted viraemia, DEN NS1 antigen is detectable sooner and tests utilizing this target are more sensitive than the MAC-ELISA in the early days after the development of DEN-related symptoms (Hang *et al.*, 2009). For Ross River virus, there is an IgA assay that is analogous to the IgM ELISA and has been shown to be beneficial in determining a recent infection due to the fact that IgA declines more rapidly than IgM in infected patients (Carter *et al.*, 1987). There is also an IgA assay that has been successfully used for the detection of WNV CNS infections (Nixon and Prince, 2006). IgG is less commonly used as a target for detecting recent arbovirus infections due to its relatively delayed and long lived response however, IgG avidity tests have been used to discriminate between primary and secondary flavivirus infections (Levett *et al.*, 2005). Low avidity IgG antibodies are characteristic of recent infections whereas high avidity IgG antibodies are seen after six months from the onset of symptoms (Fox *et al.*, 2006). Of clinical interest, the detection of IgM in a CSF sample is considered direct evidence of intrathecal antibody production and indicates a recent CNS infection, a condition associated with a number of arboviral agents including WN, JE and EEE viruses. Arboviral IgG antibodies are usually detected by an indirect sandwich method and have a relatively delayed response, making the detection of a significant rise in IgG titre in a convalescent sample of utility in confirming a recent diagnosis. Do to the extended presence of IgG, assays detecting this isotype can also be used to infer past exposure. As such, IgG ELISAs are useful as screening tools in serological studies of populations where identifying a history of exposure is more important than determining a recent infection.

To support interpretation, there are a number of schemes that are used to qualify a threshold value for the definition of a positive ELISA result. Some laboratories use a multiple of the OD from a weak positive that is run along with samples of interest, in doing so controlling for run to run variability. Other laboratories calculate a threshold value by using some calculation based upon the mean absorbance of the population (i.e.

more than five standard deviations above the mean). A conservative method defines the threshold at 2 × the absorbance of the negative control. Using this measure, the resultant value is commonly communicated as a positive/negative, or P/N ratio.

In recent years, the microsphere immunoassay (MIA) has been proposed as an alternative to the ELISA for arboviral antibody detection (Johnson *et al.*, 2005; Basile *et al.*, 2013). MIAs involve the detection of positive antibody-antigen reactions through the identification of fluorescence on a microsphere or bead in a flow cytometer. These assays are relatively fast (they take less than 5 hours to complete, compared with 2 days for the MAC-ELISA) and have a multiplexing capability that makes them attractive for the simultaneous targeting of broad taxonomical groupings of arboviral antibodies (Basile *et al.*, 2013). However, MIAs can be reagent intensive and less transferable than the ELISA, which is particularly limiting to their utility in global laboratories of restricted resources.

Indirect IFA

Some laboratories use indirect IFA for the detection and differentiation of IgM and IgG which can be useful in the determination of recent infections. Indirect IFAs have been developed for a number of arboviruses including WN, LAC, DEN, JE, EEE and YF (Malan *et al.*, 2003; Beaty *et al.*, 1982; Nagarkatti and Nagarkatti, 1980; Hildreth and Beaty, 1984; Niedrig *et al.*, 2008) and utilize slides that possess multiple wells onto which inactivated, infected cells are transfixed as the antigen substrate. Dilutions of patient samples are then added to the slide wells, fostering the antigen–antibody reaction in positive samples. Following the addition of patient serum, a fluorescein labelled 'conjugated' antibody (anti-IgM or anti-IgG) is then added to the well. Positive results are determined by fluorescence microscopy. A quantitative value can be assigned through the evaluation of multiple dilutions of a sample, which allows for the identification of an endpoint titre. In paired samples, a fourfold rise in titre is considered to be diagnostic of a recent infection. In some cases, a single elevated titre of > 1:128 is considered to be diagnostic. IgM can be detected by immunofluorescence within the first several days after the onset of symptoms. IgG can be detected by IFA shortly after IgM and is relatively long-lasting, paralleling the sustained longevity of neutralization antibodies.

HI assay

Many arboviruses, including members of the families *Bunyaviridae*, *Flaviviridae* and *Togaviridae* haemagglutinate erythrocytes. Exploiting this property, the HI assay is a useful tool for the quick screening of group-reactive antibodies (that act to inhibit haemagglutination) in patient samples (Hsuing *et al.*, 1994; Beaty *et al.*, 1995). HI antibodies rise quickly, before the end of the first week of onset of illness, and are long lived; consistent with a mixture of IgM and IgG isotypes. The test is performed by adding a fixed and predetermined amount of viral antigen to the wells of a 96 well plate. Following this, dilutions of patient samples are added to the plate in rows and red blood cells (preferably gander or chick erythrocytes) are then applied to all wells. After a short incubation, results are read visually. In the absence of HI antibodies, viruses that possess the haemagglutination capability act to complex red blood cells forming a lattice that appears as a homogenous, light pink solution in a negative well. In the presence of HI-antibodies, complexing of the antibodies with the viral antigen prevents haemagglutination, causing the red blood cells to pool together and form a 'button' at the bottom of a positive well. The highest dilution at which haemagglutination is inhibited defines that sample's HI titre. For each virus, there is a restricted optimal pH range for HI activity that needs to be established prior to testing. Serum samples are often tested at 1:10 dilutions, followed by twofold dilutions to an endpoint titre, thereafter. The HI test is particularly sensitive when applied to paired sera and a fourfold change in titre between acute and convalescent samples is diagnostic of a recent infection. A titre of > 1:80 in a single specimen is also considered to be presumptive evidence of a recent infection. Furthermore, a titre of > 1:2560 can be evidence of a secondary antibody response to a flavivirus infection.

HI antibodies are long lasting, broadly reactive and recognize common epitopes among members of the same antigenic groups. Therefore, the HI assay is useful in screening unknowns and demonstrating group relationships among arboviruses. The broad-reactivity and longevity of HI antibodies make the HI assay of great utility to epidemiological studies. Furthermore, the HI assay does not require species-specific reagents, and is very rapid, rendering it useful in wildlife surveys conducted on any number and type of animals.

CF test

The CF test exploits a biological factor that is common to immunologically normal mammals and is known as 'complement'. When free in solution, complement acts upon sensitized red blood cells, causing them to lyse. In the presence of antibody–antigen complexes, complement interacts with these complexes preferentially and does not act on red blood cells, preventing lysis and indicating a CF antibody positive sample. Results are read visually and, as with the HI test, the presence of a red blood cell button at the bottom of a well indicates a positive sample. Negative wells appear as a light pink solution that has been generated by the lysis of red blood cells in the absence of antibody–antigen complexes. As with other serological assays, dilutions of patients samples can be made to assign a quantitative value or titre to a sample. The highest dilution for which a positive reaction is visualized is that sample's CF antibody titre. For paired sera, a fourfold difference in antibody titre is considered diagnostic. In some cases, a CF titre of 1:32 is considered to be a presumptive positive for a recent infection. CF tests are relatively specific and can be used as a complement to HI assays to narrow the diagnosis from group reactive to species specific (Stark and Lewis, 2000). CF antibodies rise more slowly than HI antibodies, peaking at around six weeks after illness in some cases. For this reason, convalescent samples should be taken two to three months after the onset of symptoms. Finally, it is important to note that some people fail to develop CF antibodies and advanced age may contribute to the development of a stunted or delayed CF antibody response which can limit the ability to determine a negative diagnosis based on the results of a CF test alone. For this reason, CF tests should be used in conjunction with other serological assays.

PRNT

In the arbovirus diagnostic laboratory, assays that evaluate the ability of a sample to neutralize virus are of invaluable importance for the confirmation of virus-specific antibodies and the differentiation of infection with cross-reacting viruses. For the PRNT, the investigator is looking for evidence of neutralizing activity as determined by a reduction in plaque numbers from a virus preparation of known titre. The assay is most often conducted in a six-well plate that has confluent, susceptible cells, often Vero cells, lining the bottom of each well. Dilutions of patient serum are then combined with an equal volume of a known quantity of virus and incubated at 37°C for 1 hour or at 5°C overnight. The incubated dilutions are then inoculated onto the cell sheet at the bottom of each well, incubated for a short time (commonly 1 h at 37°C) to allow for adsorption, and overlayed for the first time with a nutritive agar. Following a multi-day period of incubation that is required for virus replication, the cells are then overlayed for a second time with an agar medium possessing a stain that assists in the visualization and counting of viral plaques. A conservative measure of positivity requires 90% neutralization in a well, i.e. 10 PFU from a 100 PFU dilution of virus. The highest dilution at which 90% neutralization is achieved is that sample's PRNT titre. Due to the biological nature of the test, the detection of neutralizing antibodies is considered to be reflective of protective immunity. Therefore, the assay is of supreme utility for evaluating a protective response to vaccination. In addition to using plaque reduction, neutralization can also be measured by fluorescent focusing techniques and in laboratory animals (i.e. newborn mice). Microneutralization tests that utilize a convenient 96-well format and can be automated have also been developed for a number of agents (Vorndam and Beltran, 2002; Taketa-Graham *et al.*, 2010; Simões *et al.*, 2012). Due to the use of live virus in the PRNT, a laboratorian should first familiarize themselves with relevant biosafety guidelines as described in the *Biosafety in Microbiological and Biomedical Laboratories* (BMBL), 5th edition, available from the US government at www.cdc.gov/biosafety/publications/bmbl5/.

Direct detection

Serological laboratory diagnosis is complemented, confirmed and refined to the individual species and strain levels by direct virus detection and identification. Direct methods can be applied to acute body fluids as well as to other samples, including post-mortem brain, kidney, liver, lung and lymph node tissues that are ideally harvested according to an understanding of the pathology of the suspected agent. In addition to human samples, these methods can be applied to ecological samples, including mosquito pools as well as the supernatants from tick and animal tissue homogenates. Utilized less frequently than serology due to the absence of a long lasting viraemia in the human host for many arboviruses, direct identification is nonetheless very important to the arboviral diagnostic laboratory. Direct approaches are particularly useful in investigations of the emergence of an unknown or novel virus for which serological methods are limited due to the indirect nature of those methods and cross-reactivity. Historically, for arbovirus diagnosis, surveillance, and research, virus isolation has been the standard assay against which other direct detection methods are evaluated. Other methods include antigen-capture ELISA and relatively widely used molecular assays, including RT-PCR. Less commonly used methods of direct detection include electron microscopy and antigen detection through immunohistochemistry (IHC) and IFA that generally require more specialized expertise, equipment and facilities.

Electron microscopy

Examination of viral particles by electron microscopy conducted on thin sections of tissues or fixed cell culture preparations is an excellent way of identifying the larger taxonomical grouping of an unknown arbovirus. With rare exception, the virion morphology of all members of a virus family is shared, allowing for relatively quick identification of a viral pathogen at the family level under the management of a skilled expert in microscopy. Recently, microscopy has been important for the discovery of a number of emergent arboviruses, including SFTS and Heartland viruses of the family *Bunyaviridae* (Yu *et al.*, 2011; Goldsmith *et al.*, 2013; McMullan *et al.*, 2012). Electron microscopy is particularly effective in narrowing the focus of outbreak investigations when combined with techniques that are sensitive at more refined taxonomic levels, such as nucleic-acid based methods.

Detection of viral antigens by immunohistochemical techniques

Immunohistochemical (IHC) staining techniques have been successfully applied for the detection of arboviral antigens in peripheral blood mononuclear cells and tissues for the identification of a number of arboviral CNS, viscerotropic and arthritic infections (De Brito *et al.*, 1992; Hall *et al.*, 1991). Colorado tick fever virus-infected erythrocytes have also been detected using IHC staining of a smear of clotted blood. Formalin-fixed and paraffin-embedded tissues can be used for post-mortem diagnoses and are logistically advantageous in their ability to be stored for months at a time prior to IHC examination. For many arboviruses, monoclonal antibodies can be used for direct IFA and for more obscure agents, polyclonal antibody preparations are of utility. Preparations of antibodies as well as other reagents to support arboviral testing are available at the Centers for Disease Control and Prevention, Division of Vector-Borne Diseases, Fort Collins, CO, USA.

Antigen-capture ELISA

Antigen-capture ELISAs have been developed and used for the detection of a number of agents including JE, CHIK, WN, and DEN viruses (Mei *et al.*, 2012; Shukla *et al.*, 2009; Saxena *et al.*, 2013; Hunt *et al.*, 2002). The generalized detection limit of this format (~$10^{3.5}$ PFU/ml) is relatively useful for the surveillance of arboviruses in infected arthropods and somewhat less useful for human clinical testing when compared to molecular methods of virus detection. As an exception, the application of an antigen capture ELISA for the detection of DEN virus NS1 antigen to acute phase human clinical samples has been helpful in the identification of DEN infections (Hang *et al.*, 2009) and is particularly useful in laboratories with serological, but not molecular, capabilities. As previously mentioned, for DEN virus, the NS1 antigen ELISA is preferred to the IgM ELISA for acute samples owing to a relatively sustained and high titre viraemia in the early stages of clinical illness (Hang *et al.*, 2009). However, NS1 assays are less sensitive for DEN detection than

RT-PCR based assays. Furthermore, detection of the NS1 antigen is reduced in the presence of IgM and IgG antibodies. As a result, the sensitivity of detection for secondary DEN virus infections is diminished.

In addition to antigen-capture ELISAs, there are now immunochromatographic antigen-capture assays for the detection of arboviruses, including WNV, EEE, WEE and RVFV (Nasci *et al.*, 2002, 2003; Turell *et al.*, 2011) in ecological samples, primarily mosquito pools. These assays are easy to use and highly portable. However, they require high titres of virus for detection (~$10^{5.0}$ PFU/ml) and, as such, are not recommended for use on human clinical samples.

Nucleic acid detection techniques

Nucleic acid sequence-based diagnostics is the fastest growing field in the laboratory detection of arboviruses. These methods use a variety of different formats that support the identification of arboviruses through the detection and description of genomic RNA in human acute phase fluid samples, tissues and ecological samples. Molecular assays are also quite useful for the identification and characterization of virus isolates. The most commonly used molecular methods are virus-specific, RT-PCR based in their capability and utilize real-time and agarose gel-based detection formats. Additional methods include nucleic acid sequence-based amplification (NASBA) and reverse transcription-loop-mediated isothermal amplification (RT-LAMP) methods. In general, these virus-specific assays are highly transferable across public health and research laboratories and have become widely used on a global scale. In contrast, some of the most useful nucleic acid-based methods for the identification of unknown and novel arboviruses are group reactive, consensus-style assays that are less transferable, but of extreme utility when used in conjunction with nucleic acid sequencing for the detection of groups of viruses and the discovery and description of emerging agents (Kuno *et al.*, 1998; Lambert and Lanciotti, 2009; Kay *et al.*, 2010).

RNA isolation

Virtually all nucleic acid-based detection methods require that nucleic acid is isolated and purified from complex samples prior to amplification and detection. Presently, there are a number of highly effective methods that are commercially available for this purpose. The most commonly used RNA purification procedures begin with the addition of a sample of interest (i.e. serum or supernatant of a tissue homogenate) to a lysis buffer that acts to lyse the viral particle, releasing viral genomic RNA into solution. This is followed by the targeted binding of the RNA to a substrate (often silica), followed by the elimination of unwanted molecules (fats and proteins) through washing of the bound nucleic acid. Finally, the desired, purified nucleic acid is eluted from the substrate in water (or elution buffer) and is ready for use in nucleic acid-based detection assays. It is important to carefully follow good RNA handling techniques (use of gloves, prevention of aerosols of human origin) during RNA isolation to reduce the likelihood of RNase based degradation of RNAs. In addition, it can be beneficial to the sensitivity of detection and diagnosis to enhance the concentration of eluted, purified RNAs by maximizing the input of the original sample at the lysis reaction step, followed by minimizing the elution volume of RNA at the elution step.

RT-PCR

RT-PCR is one of the most sensitive and widely used technologies in laboratory diagnosis and research. For the detection of arboviruses, RT-PCR assays range in capability from virus strain specific to serogroup and genus specific (Lanciotti *et al.*, 2000; Kuno *et al.*, 1998, Lambert and Lanciotti, 2009). Generally, RT-PCR begins with a reaction mixture that is made to contain all critical components for amplification and is put together on ice to protect those components and prevent premature enzymatic activity. At a minimum, this mixture contains nucleotides, buffer, enzymes for RT and PCR, as well oligonucleotide primers that are designed to bind to unique RNA sequences, allowing for the targeted detection of a specific virus or group of viruses. Following the addition of sample RNA, the reaction mixture is subjected to thermocycling for amplification of cDNAs. Historically, the majority of researchers have used agarose gel-based visualization and size discrimination of the amplicons that are generated by these assays for the preliminary identification of a positive. This method is useful, especially when combined with a nucleotide sequencing capability to verify the origin of cDNAs at the sequence level. Other methods of verifying target nucleic acid sequences within amplicons include nested-PCR, Southern blot and microarray oligonucleotide hybridization, as well as endonuclease digestion. To enhance the sensitivity of RT-PCR based detection, a laboratorian can maximize the input volume of sample RNA into a reaction and reduce the volume of water that is added to maintain the overall volume of that reaction.

Real-time 5′-exonuclease fluorogenic assays

Real-time RT-PCR assays, that exploit the 5′exonuclease function of Taq Polymerase to degrade internal, sequence-specific probes, resulting in the emission of a fluorescent signal and allowing for the simultaneous amplification and detection of target sequences, are of exceptional utility for virus-specific diagnosis. These assays have been developed and applied for the detection of WN, SLE, TBE, DEN, EEE, WEE, LAC, CTF, RVF, CHIK and Zika viruses, among others (Lanciotti *et al.*, 2000; Lanciotti and Kerst, 2001; Lambert *et al.*, 2003, 2005, 2007; Johnson *et al.*, 2005; Achazi *et al.*, 2011; Smith *et al.*, 2009; Bird *et al.*, 2007; Faye *et al.*, 2013). Advantages to the real-time RT-PCR format, relative to the traditional gel-based detection format, include enhanced sensitivity and specificity, a reduced likelihood of contamination, the ability to quantitate targets, and the generation of rapid results with a high throughput capability. Furthermore, the sequence-specific detection format of real-time assays eliminates the need for post-amplification characterization of amplicons. For these reasons, real-time assays are some of the most widely used in the field of arboviral molecular diagnosis. It is important to note that for some arboviral pathogens, including DEN, YF, CHIK, RVF, CTF, and sandfly fever group viruses, RT-PCR based methods, most often real-time in their capability, can be very useful when applied to acute-phase human clinical samples due to the presence of detectable levels of viraemia after the onset of symptoms (Johnson *et al.*, 2005; Lanciotti *et al.*, 2007; Bird *et al.*, 2007; Lambert *et al.*, 2007; Kay *et al.*, 2010; Nunes *et al.*, 2011). In this capacity, these assays are most often applied to acute phase serum and CSF samples as well as to post-mortem tissues. However, for blood cell-associated agents such as Heartland and CTF viruses, whole blood has also been empirically observed to be a preferred sample for molecular diagnoses.

NASBA

NASBA is a unique amplification format that has been successfully used for the detection of WN, SLE, CHIK, and DEN viruses, among others (Lanciotti and Kerst, 2001; Telles *et al.*, 2009; Wu *et al.*, 2001). Like RT-PCR, NASBA exploits oligonucleotide primer based genomic detection and reverse transcription that acts, through a modified protocol, to generate a dsDNA product from a target RNA. One of the two primers used in reverse transcription is customized to generate a T7 promoter sequence on the dsDNA product of reverse transcription. As a result, during the subsequent isothermal amplification step, T7 RNA polymerase acts to continuously produce complementary RNA strands from this template. These RNA copies are the measured products of amplification and are detected either through the use of an electrochemiluminescent probe, or a molecular beacon in a real-time format. Theoretically, this isothermal strategy is somewhat advantageous for sensitive detection of targets due to the continuous nature of NASBA amplification. In fact, the system has demonstrated sensitivities that are roughly equivalent to those afforded by real-time RT-PCR based assays.

RT-LAMP

In recent years, there have been a number of reports describing the use of RT-LAMP for the detection of arboviral RNAs including WN, DEN, CHIK, RVF, JE and YF viruses (Parida *et al.*, 2004; Teoh *et al.*, 2013; Lu *et al.*, 2012, Peyrefitte *et al.*, 2008; Li *et al.*, 2011; Kwallah *et al.*, 2013). Basically, the RT-LAMP assay uses strand displacement DNA polymerization for amplification. This reaction is facilitated by the use of six primers that recognize distinct regions of target nucleic acid, making the assay highly specific. RT-LAMP occurs in an isothermal state (63°C) in the presence of a modified DNA polymerase (Bst DNA polymerase) and reverse transcriptase. During the reaction, loop structures are formed by priming activity to enable subsequent rounds of amplification. This procedure is rapid, with results available in less than one hour. Products of amplification can be detected by agarose gel electrophoresis. Alternatively, real-time monitoring can be conducted by measuring turbidity (that changes during the course of a positive reaction due to an accumulation of the amplification by-product, magnesium pyrophosphate) with a turbidimeter. Finally, dye-based detection can also be used for the visual determination of a positive result in a reaction tube. In general, the sensitivity of the RT-LAMP method approximates that of standard RT-PCR. Due to the lack of extensive technological requirements, the assay is attractive for use in laboratories of varied and limited resources.

Virus isolation

Despite the advent of highly sensitive, specific and rapid molecular technologies, virus isolation remains the definitive method for the diagnosis of arboviral infections. To this day, novel viruses continue to be derived from both ecological and human clinical samples utilizing this broadly reactive, albeit time intensive, classical method. In addition to providing unequivocal evidence of infection, a virus isolate is an important resource for future research and diagnosis. If considering employment of this method, a laboratorian should first familiarize themselves with relevant biosafety guidelines as described in the *Biosafety in Microbiological and Biomedical Laboratories (BMBL)*, 5th edition, available from the US government at www.cdc.gov/biosafety/publications/bmbl5/.

For arboviruses, isolation is most often performed through the inoculation of either susceptible cells or suckling mice with sera or supernatants of homogenates derived from tissues of infected individuals, animal sources or mosquito pools. Although less common, direct inoculation of live, colony-reared mosquitoes can also be used to isolate arboviruses (Pervin *et al.*, 2002). Generally, undiluted serum, as well as 10^{-1} and 10^{-2} dilutions should be used to mitigate effects from viral autointerference and antibodies that might diminish the likelihood of deriving an isolate from an undiluted sample. Tissues and ecological samples should also be homogenized and diluted in a similar fashion, using a 10–20% weight to volume ratio, with a protein and antibiotic containing medium. Arboviruses can replicate in a variety of cell lines including chicken embryo, primary monkey and hamster kidney cells, C6/36 *Aedes albopictus* and AP61 *Aedes pseudoscutelaris* mosquito cell lines, Vero, LLC-MK2 and BHK-21 cell lines. After inoculation and adsorption onto a drained cell monolayer, cell cultures are fed with medium and incubated at a recommended temperature (most commonly 37°C for mammalian cells and 28°C for mosquito cell lines). Upon incubation, the time it takes for CPE to develop and the appearance of that CPE depend on the virus that is being isolated as well as the cell system that is being used. As an example, BHK-21 cells are rapidly destroyed by most arboviruses within days, while the development of arbovirus-related CPE in monkey kidney cell lines occurs more slowly. C636 and AP61 cell lines are susceptible to nearly all arboviruses, including DEN viruses that grow poorly in monkey kidney cells. However, mosquito cell lines show little evidence of viral CPE for most arboviruses, requiring that another method, such as IFA or RT-PCR be used to verify replication and isolation. Most often, the isolation of virus in cell culture is undertaken in cell culture flasks. Recently, there have been protocols developed for the isolation of viruses in shell vials that have been shown to have comparable or enhanced sensitivities relative to traditional methods of isolation (Caceda and Kochel *et al.*, 2007; Jason O. Velez, personal communication).

Most arboviruses are lethal to suckling mice (2–3 days old) that exhibit signs of illness, including paralysis and death within days and up to 2 weeks after intracerebral inoculation. Mice should be carefully regarded for signs of illness twice daily after inoculation. Following euthanization due to illness or death due to viral infection, mice should be frozen at −70°C until infected organs can be harvested for testing and passage in cell culture. For most arboviruses, brains are harvested to derive virus isolates due to the high level of replication that occurs within the CNS after intracerebral inoculation. For orthomyxoviruses and many orthobunyaviruses, livers should be harvested, in addition to brains, because these agents grow to comparatively high titres in the liver.

For quantitating isolates, plaque assays can be used to determine the number of plaque forming units (PFUs) in a given dilution of virus, which can then be used to calculate the overall titre of a preparation. These assays should be performed in susceptible cells that demonstrate CPE through the formation of plaques. Briefly, the method involves the inoculation of dilutions of a viral isolate onto a cell monolayer at the bottom of a six-well plate. Following a brief incubation that allows for adsorption, the cells are then overlayed with a nutritive agar and then incubated again for a period of days to allow for virus replication and plaque formation. Following this incubation period, the cells are then overlayed again with an agar that possesses a stain to assist in the visualization and counting of viral plaques in each well. The titre of a virus preparation can then be calculated with a knowledge of the number of plaques in a given well and the dilution from which that number of plaques was derived. Once a titre has been determined for a given virus preparation, that value can be of additional utility in the determination of quantitated RNA standards that can be extracted from virus dilutions and for which a value of 'PFU equivalents' is derived.

Methods for the identification and characterization of arboviral isolates

Following isolation, identification and characterization of newly derived arbovirus isolates has been historically provided through the use of predominantly antibody based methods. These methods include IFAs that utilize grouping fluids of antibodies developed against multiple serologically related viruses, to classify an isolate at the serogroup level. Following serogroup level classification, PRNTs, that use antibodies directed against individual species, are then performed for species level identification. Together, these methods are valuable for the identification of arboviruses in a group specific manner. However, they are time-consuming and can be limited in their ability to generate unequivocal, species identifying results due to antibody cross reactivity and/or a limited diversity of available antibodies.

The advent of molecular technologies, such as RT-PCR, and the application of these technologies in diagnostic and reference laboratories, provides a time-efficient alternative to traditional serological methods for the identification and characterization of virus isolates. The application of RT-PCR based molecular consensus assays, designed to detect a group of viruses of interest, followed by nucleotide sequencing for result confirmation and virus speciation has proven a powerful tool for emergent virus identification and discovery when applied to virus isolates (Kuno *et al.*,1998; Lanciotti *et al.*, 2008; Kay *et al.*, 2010). However, molecular identification of any agent is only possible with a priori knowledge of species-identifying genomic sequence data, a limiting factor in the usefulness of nucleic acid based technologies for arbovirus identification and characterization.

Next-generation sequencing and the future of arbovirus detection and description

Next-generation (NGEN) sequencing technologies promise to revolutionize the detection and identification of emergent viruses. In general, NGEN sequencing methods begin with the random amplification of a cDNA library from a viral genome. Following this, a variety of different NGEN sequencing methodologies, including semiconductor chip based methods, can be used for the generation of millions of sequencing reads from a single cDNA library. The sequence-independent nature of NGEN sequencing, combined with its phenomenal data output, allow for the rapid, comprehensive description of a viral genome without prior knowledge of identifying nucleic acid sequence data, a requirement for traditional methods of amplification and sequencing. This non-selective quality of NGEN sequencing offers a great advantage for the description of unknowns for which identifying molecular data are not available (McMullan *et al.*, 2012; Lanciotti *et al.*, 2013). In addition, the continued use of NGEN sequencing for the description of known viruses of previously undescribed nucleic acid sequence character is invaluable to the relatively rapid development of a comprehensive database of arboviral sequences. The data that lie therein can be used for future assay development and the comparative identification of a novel virus. For bunyaviruses, the segmented nature of their genome and their ability to reassort segments have complicated their molecular detection and description using standard methods of amplification and sequencing. NGEN methods allow for the simultaneous description of all three segment of the bunyavirus genome, a distinct advantage for the identification of a reassortant virus. This simultaneous description of all three bunyavirus segments has been recently used for the discovery of Heartland and Brazoran viruses (McMullan *et al.*, 2012; Lanciotti *et al.*, 2013) among others. An additional advantage, given its non-selective and comprehensive nature, NGEN can be used to generate previously intractable data in support of a heightened understanding of the possible contribution of arboviral quasispecies and minor variants to the emergence of human illness.

A recommendation for cautious use of point-of-care tests for arboviral diagnosis

As a final note, in recent years there has been a strong push in the public health community for point-of-care diagnostic tests, including immunochromatographic lateral flow or 'dipstick' assays for the detection of tropical infectious diseases (Peeling and Mabey, 2010). For arboviruses, there is great interest in point-of-care testing for a number of emerging agents, including YF virus for which vaccination campaigns could be initiated upon detection of an outbreak of YF-related illness. Given the inherent complexities of arboviral diagnosis, it is our recommendation that point-of-care testing always be used in the context of an algorithm that includes confirmation of results using standard measures of diagnosis, such as the PRNT, to verify the identity of a causative agent prior to the initiation of a public health intervention based on that diagnosis.

References

Achazi, K., Nitsche, A., Patel, P., Radonić, A., Donoso Mantke, O., and Niedrig, M. (2011). Detection and differentiation of tick-borne encephalitis virus subtypes by a reverse transcription quantitative real-time PCR and pyrosequencing. J. Virol. Methods *171*, 34–39.

Basile, A.J., Horiuchi, K., Panella, A.J., Laven, J., Kosoy, O., Lanciotti, R.S., Venkateswaran, N., and Biggerstaff, B.J. (2013). Multiplex Microsphere Immunoassays for the Detection of IgM and IgG to Arboviral Diseases. PLoS One *8*, e75670.

Beaty, B.J., Casals, J., Brown, K.L., Gundersen, C.B., Nelson, D., McPherson, J.T., and Thompson, W.H. (1982). Indirect fluorescent-antibody technique for serological diagnosis of La Crosse (California) virus infections. J. Clin. Microbiol. *15*, 429–434.

Beaty, B.J., Calisher, C.H., and Shope, R.E. (1995). Arboviruses. In Diagnostic Procedures for Viral, Rickettsial and Chlamydial Infections, 7th edn., E.H. Lennette, P.A. Lennette, and E.T. Lennette, eds. (American Public Health Association, Washington, DC), pp. 189–212.

Bird, B.H., Bawiec, D.A., Ksiazek, T.G., Shoemaker, T.R., and Nichol, S.T. (2007). Highly sensitive and broadly reactive quantitative reverse transcription-PCR assay for high-throughput detection of Rift Valley fever virus. J. Clin. Microbiol. *45*, 3506–3513.

Caceda, E.R., and Kochel, T.J. (2007). Application of modified shell vial culture procedure for arbovirus detection. PLoS One *2*, e1034.

Carter, I.W., Fraser, J.R., and Cloonan, M.J. (1987). Specific IgA antibody response in Ross River virus infection. Immunol. Cell. Biol. *65*, 511–513.

De Brito,T., Siqueira, S.A., Santos, R.T., Nassar, E.S., Coimbra, T.L., and Alves, V.A. (1992). Human fatal yellow fever. Immunohistochemical detection of viral antigens in the liver, kidney and heart. Pathol. Res. Pract. *188*, 177–181.

Faye, O., Faye, O., Diallo, D., Diallo, M., Weidmann, M., and, Sall, A.A. (2013). Quantitative real-time PCR detection of Zika virus and evaluation with field-caught Mosquitoes. Virol. J. *22*, 311.

Fox, J.L., Hazell, S.L., Tobler, L.H., and Busch, M.P. (2006). Immunoglobulin G avidity in differentiation between early and late antibody responses to West Nile virus. Clin. Vaccine Immunol. *13*, 33–36.

Goldsmith, C.S., Ksiazek, T.G., Rollin, P.E., Comer, J.A., Nicholson, W.L., Peret, T.C., Erdman, D.D., Bellini, W.J., Harcourt, B.H., Rota, P.A., *et al.* (2013). Cell culture and electron microscopy for identifying viruses in diseases of unknown cause. Emerg. Infect. Dis. *19*, 886–891.

Hall, W.C., Crowell, T.P., Watts, D.M., Barros, V.L., Kruger, H., Pinheiro, F., and Peters, C.J. (1991). Demonstration of yellow fever and dengue antigens in formalin-fixed paraffin-embedded human liver by immunohistochemical analysis. Am. J. Trop. Med. Hyg. *45*, 408–417.

Hang, V.T., Nguyet, N.M., Trung, D.T., Tricou, V., Yoksan, S., Dung, N.M., Van Ngoc, T., Hien, T.T., Farrar, J., Wills, B., and Simmons, C.P. (2009). Diagnostic accuracy of NS1 ELISA and lateral flow rapid tests for dengue sensitivity, specificity and relationship to viraemia and antibody responses. PLoS Negl. Trop. Dis. *3*, e360.

Hildreth, S.W., and Beaty, B.J. (1984). Detection of eastern equine encephalomyelitis virus and Highlands J virus antigens within mosquito pools by enzyme immunoassay (EIA). I. A laboratory study. Am. J. Trop. Med. Hyg. *33*, 965–972.

Hsuing, G.D. (1994). Hemagglutination and hemagglytination-inhibition test. In Hsuing's diagnostic virology, 4th edn., G.D. Hsuing, C.K.Y. Fong, and M.L. Landry, eds. (Yale University Press, New Haven, CT), pp. 69–75.

Hunt, A.R., Hall, R.A., Kerst, A.J., Nasci, R.S., Savage, H.M., Panella, N.A., Gottfried, K.L., Burkhalter, K.L., and Roehrig, J.T. (2002). Detection of West Nile virus antigen in mosquitoes and avian tissues by a monoclonal antibody-based capture enzyme immunoassay. J. Clin. Microbiol. *40*, 2023–2030.

Johnson, A.J., Noga, A.J., Kosoy, O., Lanciotti, R.S., Johnson, A.A., and B.J Biggerstaff. (2005). Duplex microsphere-based immunoassay for detection of anti-West Nile virus and anti-St. Louis encephalitis virus immunoglobulin m antibodies. Clin. Diagn. Lab. Immunol. *12*, 566–574.

Johnson, B.W., Russell, B.J., and Lanciotti, R.S. (2005). Serotype-specific detection of dengue viruses in a fourplex real-time reverse transcriptase PCR assay. J. Clin. Microbiol. *43*, 4977–4983.

Kapoor, H., Signs, K., Somsel. P., Downes, F.P., Clark, P.A., and Massey, J.P. (2004). Persistence of West Nile Virus (WNV) IgM antibodies in cerebrospinal fluid from patients with CNS disease. J. Clin. Virol. *31*, 289–291.

Kay, M.K., Gibney, K.B., Riedo, F.X., Kosoy, O.L., Lanciotti, R.S., and Lambert, A.J. (2010). Toscana virus infection in American traveler returning from Sicily, 2009. Emerg. Infect. Dis. *16*, 1498–1500.

Kuno, G., Gubler, D.J., and Oliver, A. (1993). Use of 'original antigenic sin' theory to determine the serotypes of previous dengue infections. Trans. R. Soc. Trop. Med. Hyg. *87*, 103–105.

Kuno, G., Chang, G.J., Tsuchiya, K.R., Karabatsos, N., and Cropp, C.B. (1998). Phylogeny of the genus Flavivirus. J. Virol. *72*, 73–83.

Kwallah, A., Inoue, S., Muigai, A.W., Kubo, T., Sang, R., Morita, K., and Mwau, M. (2013). A real-time reverse transcription loop-mediated isothermal amplification assay for the rapid detection of yellow fever virus. J. Virol. Methods *193*, 23–27.

Lambert, A.J., and Lanciotti, R.S. (2009). Consensus amplification and novel multiplex sequencing method for S segment species identification of 47 viruses of the *Orthobunyavirus*, *Phlebovirus*, and *Nairovirus* genera of the family *Bunyaviridae*. J. Clin. Microbiol. *47*, 2398–2404.

Lambert, A.J., Martin, D.A., and Lanciotti, R.S. (2003). Detection of North American eastern and western equine encephalitis viruses by nucleic acid amplification assays. J. Clin. Microbiol. *41*, 379–385.

Lambert, A.J., Nasci, R.S., Cropp, B.C., Martin, D.A., Rose, B.C., Russell, B.J., and Lanciotti, R.S. (2005). Nucleic acid amplification assays for detection of La Crosse virus RNA. J. Clin. Microbiol. *43*, 1885–1889.

Lambert, A.J., Kosoy, O., Velez, J.O., Russell, B.J., and Lanciotti, R.S. (2007). Detection of Colorado Tick Fever viral RNA in acute human serum samples by a quantitative real-time RT-PCR assay. J. Virol. Methods *140*, 43–48.

Lanciotti, R.S., and Kerst, A.J. (2001). Nucleic acid sequence-based amplification assays for rapid detection of West Nile and St. Louis encephalitis viruses. J. Clin. Microbiol. *39*, 4506–4513.

Lanciotti, R.S., Kerst, A.J., Nasci, R.S., Godsey, M.S., Mitchell, C., Savage, H., Komar, N., Panella, N.A., Allen, B.C., Volpe, K.E., Davis, B.S., and Roehrig, J.T. (2000). Rapid detection of West Nile virus from human clinical specimens, field-collected mosquitoes, and avian samples by a TaqMan reverse transcriptase-PCR assay. J. Clin. Microbiol. *38*, 4066–4071.

Lanciotti R.S., Kosoy, O.L., Laven, J.J., Panella, A.J., Velez, J.O., Lambert, A.J., and Campbell, G.L. (2007). Chikungunya virus in US travelers returning from India, 2006. Emerg. Infect. Dis. *13*, 764–767.

Lanciotti, R.S., Kosoy, O.L., Laven, J.J., Velez, J.O., Lambert, A.J., Johnson, A.J., Stanfield, S.M., and Duffy, M.R. (2008). Genetic and serologic properties of Zika virus associated with an epidemic, Yap State, Micronesia, 2007. Emerg. Infect. Dis. *14*, 1232–1239.

Lanciotti, R.S., Kosoy, O.I., Bosco-Lauth, A.M., Pohl, J., Stuchlik, O., Reed, M., and Lambert, A.J. (2013). Isolation of a novel orthobunyavirus (Brazoran virus) with a 1.7 kb S segment that encodes a unique nucleocapsid protein possessing two putative functional domains. Virology *444*, 55–63.

Levett, P.N., Sonnenberg, K., Sidaway, F., Shead, S., Niedrig, M., Steinhagen, K., Horsman, G.B., and Drebot, M.A. (2005). Use of immunoglobulin G avidity assays for differentiation of primary from previous infections with West Nile virus. J. Clin. Microbiol. *43*, 5873–5875.

Li, S., Fang, M., Zhou, B., Ni, H., Shen, Q., Zhang, H., Han, Y., Yin, J., Chang, W., Xu, G., and Cao, G. (2011). Simultaneous detection and differentiation of dengue virus serotypes 1–4, Japanese encephalitis virus, and West Nile virus by a combined reverse-transcription loop-mediated isothermal amplification assay. Virol. J. *21*, 360.

Lu, X., Li, X., Mo, Z., Jin, F., Wang, B., Zhao, H., Shan, X., and Shi, L. (2012). Rapid identification of Chikungunya and Dengue virus by a real-time reverse transcription-loop-mediated isothermal amplification method. Am. J. Trop. Med. Hyg. *87*, 947–953.

McMullan L.K., Folk, S.M., Kelly, A.J., MacNeil, A., Goldsmith, C.S., Metcalfe, M.G., Batten, B.C., Albariño, C.G., Zaki, S.R., Rollin, P.E., Nicholson, W.L., and Nichol, S.T. (2012). A new phlebovirus associated with severe febrile illness in Missouri. N. Engl. J. Med. *367*, 834–841.

Malan, A.K., Stipanovich, P.J., Martins, T.B., Hill, H.R., and Litwin, C.M. (2003). Detection of IgG and IgM to West Nile virus. Development of an immunofluorescence assay. Am. J. Clin. Pathol. *119*, 508–515.

Martin, D.A., Biggerstaff, B.J., Allen, B., Johnson, A.J., Lanciotti, R.S., and Roehrig, J.T. (2002). Use of immunoglobulin m cross-reactions in differential diagnosis of human flaviviral encephalitis infections in the United States. Clin. Diagn, Lab. Immunol. *9*, 544–549.

Mei, L., Wu, P., Ye, J., Gao, G., Shao, L., Huang, S., Li, Y., Yang, X., Chen, H., and Cao, S. (2012). Development and application of an antigen capture ELISA assay for diagnosis of Japanese encephalitis virus in swine, human and mosquito. Virol. J. *9*, 4.

Nagarkatti, P.S., and Nagarkatti, M. (1980). Comparison of haemagglutination inhibition (HI) and indirect fluorescent antibody (IFA) techniques for the serological diagnosis of certain flavivirus infections. J. Trop. Med. Hyg. *83*, 115–117.

Nasci, R.S., Gottfried, K.L., Burkhalter, K.L., Kulasekera, V.L., Lambert, A.J., Lanciotti, R.S., Hunt, A.R., and Ryan, J.R. (2002). Comparison of vero cell plaque assay, TaqMan reverse transcriptase polymerase chain reaction RNA assay, and VecTest antigen assay for detection of West Nile virus in field-collected mosquitoes. J. Am Mosq. Control Assoc. *18*, 294–300.

Nasci, R.S., Gottfried, K.L., Burkhalter, K.L., Ryan, J.R., Emmerich, E., and Davé, K. (2003). Sensitivity of the VecTest antigen assay for eastern equine encephalitis and western equine encephalitis viruses. J. Am. Mosq. Control Assoc. *19*, 440–444.

Niedrig, M., Kürsteiner, O., Herzog, C., and Sonnenberg, K. (2008). Evaluation of an indirect immunofluorescence assay for detection of immunoglobulin M (IgM) and IgG antibodies against yellow fever virus. Clin. Vaccine Immunol. *15*, 177–181.

Nixon, M.L., and Prince, H.E. (2006). West Nile virus immunoglobulin A (WNV IgA) detection in cerebrospinal fluid in relation to WNV IgG and IgM reactivity. J. Clin. Virol. *37*, 174–178.

Nunes, M.R., Palacios, G., Nunes, K.N., Casseb, S.M., Martins, L.C., Quaresma, J.A., Savji, N., Lipkin, W.I., and Vasconcelos, P.F. (2011). Evaluation of two molecular methods for the detection of Yellow fever virus genome. J. Virol. Methods *174*, 29–34.

Parida, M., Posadas, G., Inoue, S., Hasebe, F., and Morita, K. (2004). Real-time reverse transcription loop-mediated isothermal amplification for rapid detection of West Nile virus. J. Clin. Microbiol. *42*, 257–263.

Peeling, R.W., and Mabey, D. (2010). Point-of-care tests for diagnosing infections in the developing world. Clin. Microbiol. Infect. *16*, 1062–1069.

Pervin, M., Tabassum, S., and Islam, M.N. (2002). Isolation and serotyping of dengue viruses by mosquito inoculation technique from clinically suspected cases of dengue fever. Bangladesh Med. Res. Counc. Bull. *28*, 104–111.

Peyrefitte, C.N., Boubis, L., Coudrier, D., Bouloy, M., Grandadam, M., Tolou, H.J., and Plumet, S. (2008). Real-time reverse-transcription loop-mediated isothermal amplification for rapid detection of rift valley Fever virus. J. Clin. Microbiol. *46*, 3653–3659.

Pfeffer, M., B. Proebster, R.M. Kinney, and O.R. Kaaden. 1997. Genus-specific detection of alphaviruses by a seminested reverse transcription-polymerase chain reaction. Am. J. Trop. Med. Hyg. *57*, 709–718.

Prince, H.E., Tobler, L.H., Lapé-Nixon, M., Foster, G.A., Stramer, S.L., and Busch, M.P. (2005). Development and persistence of West Nile virus-specific immunoglobulin M (IgM), IgA, and IgG in viremic blood donors. J. Clin. Microbiol. *43*, 4316–4320.

Roehrig, J.T., Nash, D., Maldin, B., Labowitz, A., Martin, D.A., Lanciotti, R.S., and Campbell, G.L. (2003). Persistence of virus-reactive serum immunoglobulin M antibody in confirmed west nile virus encephalitis cases. Emerg. Infect. Dis. *9*, 376–379.

Saxena, D., Kumar, J.S., Parida, M., Sivakumar, R.R., and Patro, I.K. (2013). Development and evaluation of NS1 specific monoclonal antibody based antigen capture ELISA and its implications in clinical diagnosis of West Nile virus infection. J. Clin. Virol. *58*, 528–534.

Shukla, J., Khan, M., Tiwari, M., Sannarangaiah, S., Sharma, S., Rao, P.V., and Parida, M. (2009). Development and evaluation of antigen capture ELISA for early clinical diagnosis of chikungunya. Diagn. Microbiol. Infect. Dis. *65*, 142–149.

Simões, M., Camacho, L.A., Yamamura, A.M., Miranda, E.H., Cajaraville, A.C., da Silva Freire, M. (2012). Evaluation of accuracy and reliability of the plaque reduction neutralization test (micro-PRNT) in detection of yellow fever virus antibodies. Biologicals *40*, 399–404.

Smith, D.R., Lee, J.S., Jahrling, J., Kulesh, D.A., Turell, M.J., Groebner, J.L., and O'Guinn, M.L. (2009).Development of field-based real-time reverse transcription-polymerase chain reaction assays for detection of Chikungunya and O'nyong-nyong viruses in mosquitoes. Am. J. Trop. Med. Hyg. *81*, 679–684.

Stark, L.M., and Lewis, A.L. (2000). Complement fixation test. In Clinical Virology Manual, 3rd edn., S. Specter, R.L. Hodinka, and S.A. Young, eds. (ASM Press, Washington DC), pp. 112–126.

Taketa-Graham, M., Powell Pereira, J.L., Baylis, E., Cossen, C., Oceguera, L., Patiris, P., Chiles, R., Hanson, C.V., and Forghani, B. (2010). High throughput quantitative colorimetric microneutralization assay for the confirmation and differentiation of West Nile Virus and St. Louis encephalitis virus. Am. J. Trop. Med. Hyg. *82*, 501–504.

Telles, J.N., Le Roux, K., Grivard, P., Vernet, G., and Michault, A. (2009). Evaluation of real-time nucleic acid sequence-based amplification for detection of Chikungunya virus in clinical samples. J. Med. Microbiol. *58*, 1168–1172.

Teoh, B.T., Sam, S.S., Tan, K.K., Johari, J., Danlami, M.B., Hooi, P.S., Md-Esa, R., and Abubakar, S. (2013). Detection of dengue viruses using reverse transcription-loop-mediated isothermal amplification. BMC Infect. Dis. *13*, 387.

Turell. M., Davé, K., Mayda, M., Parker, Z., Coleman, R., Davé, S., and Strickman, D. (2011). Wicking assay for the rapid detection of Rift Valley fever viral antigens in mosquitoes (Diptera: Culicidae). J. Med. Entomol. *48*, 628–633.

Vaughn, D.W., Green, S., Kalayanarooj, S., Innis, B.L., Nimmannitya, S., Suntayakorn, S., Rothman, A.L., Ennis, F.A., and Nisalak, A. (1997). Dengue in the early febrile phase: viremia and antibody responses. J. Infect. Dis. *176*, 322–330.

Vorndam, V., and Beltran, M. (2002). Enzyme-linked immunosorbent assay-format microneutralization test for dengue viruses. Am. J. Trop. Med. Hyg. *66*, 208–212.

Wu, S.J., Lee, E.M., Putvatana, R., Shurtliff, R.N., Porter, K.R., Suharyono, W., Watts, D.M., King, C.C., Murphy, G.S., Hayes, C.G., and Romano, J.W. (2001). Detection of dengue viral RNA using a nucleic acid sequence-based amplification assay. J. Clin. Microbiol. *39*, 2794–2798.

Yu, X.J., Liang, M.F., Zhang, S.Y., Liu, Y., Li, J.D., Sun, Y.L., Zhang, L., Zhang, Q.F., Popov, V.L., Li, C., *et al.* (2011). Fever with thrombocytopenia associated with a novel bunyavirus in China. N. Engl. J. Med. *364*, 1523–1532.

Conventional Vector Control: Evidence it Controls Arboviruses

16

Scott Ritchie and Gregor Devine

Abstract

Arboviruses cause significant human morbidity and mortality. Most arboviral infections for which a vaccine is not registered are controlled by managing the mosquito vector. Unfortunately, 'evidence based' studies that provide proof that this impacts infection or disease are almost nonexistent. Many arboviral diseases, especially encephalitides such as St. Louis encephalitis and Murray Valley encephalitis, are so rare in time and space that sentinel animals and mosquito infection rates must be used as proxies for disease activity. Thus, the impact of vector control interventions on mosquito populations is often the only measure of success we have. That effect is assessed using a continuum of investigations that range from simple bioassays and small plot/semi-field trials (does the intervention kill the target vector?) to far larger and rarer field trials (does the intervention reduce populations of the target vector?). The logistics and cost of these larger trials restricts their uptake but they are a basic requirement for demonstrating operationally useful impacts. These evaluations (including those that measure impacts on the aquatic habitat) must demonstrate population level effects on adult vectors over significant scales. They must compare treated and untreated control areas under statistically robust designs. Of all the arboviral diseases, dengue is by far the most prevalent and widespread. Its transmission patterns are complex and driven by climate, herd immunity, a superbly adapted mosquito, and modern patterns of human movement. Despite its importance, just as for rarer arboviruses, myriad studies show the impact of dengue control strategies on mosquitoes but few demonstrate an effect on the transmission of the virus. This chapter highlights the difficulties of dengue management in modern urban environments and considers why it is so hard to emulate the modern day successes of many malaria control programmes. It discusses whether an emphasis on managing the aquatic habitat is a historical hangover from the well-funded, vertically managed eradication programmes of the 1900s and notes the paucity of trials that support investment in any existing vector control tool. The limited control options and operational funds available suggest that the old paradigms of dengue prevention and eradication are no longer practicable and need to be augmented by more targeted but less ambitious outbreak responses that focus on the few tools that might justify the expense of deployment.

Introduction

Arboviruses carried by mosquitoes continue to cause significant mortality and morbidity throughout the world. Dengue results in an estimated 96 million clinical cases per year (Bhatt *et al.* 2013), a value that seems to increase annually. Other epizootic arboviruses such as West Nile Virus (WNV) continue to spread into new areas, causing severe outbreaks and significant human mortality in the USA (Petersen *et al.* 2013). While vaccines have been successfully developed and used to prevent arboviral disease in man [notably Yellow Fever (YF) and Japanese encephalitis (JE)], many arboviruses can only be controlled using vector control. Mosquito control methods aimed to prevent arbovirus outbreaks in man have been in motion for decades, and modern synthetic pesticides have been used since the end of the Second World War. One would think that the evidence that these work to prevent disease in man is similarly well developed. Unfortunately, most vector control methods have only been trialled to entomological endpoints, lacking proof of principal that they prevent disease.

Cluster, randomized trials (CRTs, sometimes referred to as randomized controlled trials, RCTs) are the gold standard to measure the impact of an intervention on disease in communities (Wolbers *et al.*, 2012; Hayes and Moulton, 2009). Confidence in the impact of insecticide-treated bednets to control malaria was achieved through several CRTs (Lengeler, 2004). These studies typically consist of several villages/communities ('clusters') randomly assigned to control and treatment arms of the study; disease endpoint is typically assessed by surveys of human infection. As such, the studies are logistically difficult and require significant resources to set up and conduct. Thus, smaller 'neglected' diseases may not have the industry or government support to fund large CRTs. Indeed, almost all vector control interventions are limited to entomological endpoints that answer 'does it kill mosquitoes?', 'does it significantly reduce adult mosquito populations?', etc. There is a spectrum of intervention trials that, on one hand, increase in complexity from simple bioassays to large CRTs and

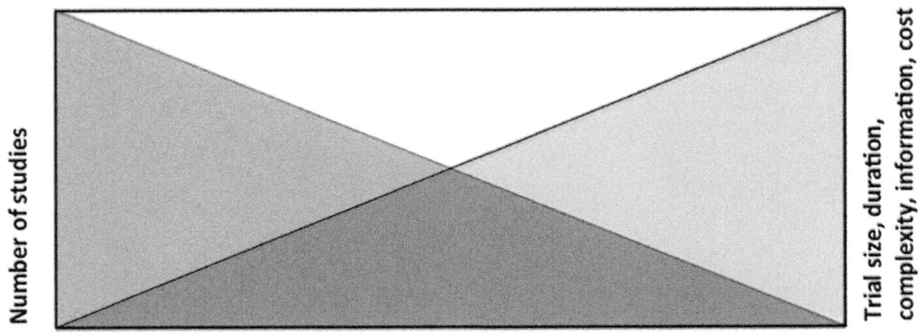

Product kills target insect	Efficacious under field conditions	Operationally effective	Impact on populations	Impact on disease
Lab bioassay	Semi-field cage, small outdoor trial, mesocosm trial	Outdoor field trial with caged insects, operational procedures applied, costings	Large field trial with control and treatment areas, CRT, entomological endpoints	Large CRT field trial with entomological and clinical/serological endpoints

Figure 16.1 Schematic of the relative attributes of trials of mosquito control interventions. Intervention goal and type is provided on the *x*-axis, with attributes on both vertical axes.

provide increasing confidence of the impact of the intervention on disease (Fig. 16.1). But on the other hand, these trials are increasingly costly and thus rare. Thus, while a plethora of publications exist on bioassays and small plot trials that confirm the intervention kills mosquitoes, little exists that confirm these interventions prevent disease.

This chapter will discuss the evidence that vector control does provide significant control of human disease, and briefly describe methods used to control arbovirus vectors. The review will focus on two areas: (i) control of urban dengue vectored by *Aedes aegypti*, a container-breeding mosquito that is the primary urban vector of dengue, and (ii) control of other arboviruses, especially epizootic ones such as encephalitids WNV and JEV, and alphaviruses such as Ross River virus (RRV) in Australia.

Evidence that mosquito control prevents transmission of epizootic arboviruses

Encephalitis outbreaks are extremely dynamic in space and time. Outbreaks of human disease can also be quite rare and isolated. For example, outbreaks of Murray Valley encephalitis (MVE) in Australia are generally limited to northern Australia, and human cases are less than five per year (Spencer *et al.*, 2001). Thus, disease endpoint trials are untenable, and proxies such as sentinel animal seroconversion are employed. The emergence of WNV in the USA has been characterized by widespread outbreaks encompassing hundreds or even thousands of clinical human cases. Thus, there is potential for clinical endpoint trials to be conducted in areas with a history of active WNV transmission. To date, the most comprehensive and convincing studies have been conducted in California where established mosquito control programmes employing sentinel animals and mosquito virus detection create a rich data source. In arguably the only study that measures the impact of vector control on WNV disease endpoint, Carney *et al.* (2008) compared WNV human cases (passively captured via the health system) in two areas treated by aerial ultra-low-volume (ULV) application of pyrethrin vs. an untreated control area during the course of an outbreak in central California in 2005. Adjusting for WNV incubation period, no new human cases occurred in either treatment area, while 18 occurred in the control area. The odds ratio for WNV infection (or actually human clinical disease) was six times higher in the untreated area than the sprayed area. This is the most thorough human disease endpoint study of encephalitis we could locate. Even then, the power of the study is quite limited compared to a CRT, with only two and one treatment and control clusters, respectively. The lack of definitive studies confirming the impact of vector control on disease has been of concern (Nasci *et al.*, 2001a) (Shapiro and Micucci, 2003). Tomerini *et al.* (2011) used an alternative approach to measure the impact of mosquito control on disease by comparing RRV incidence to the type and size of mosquito control programme in Queensland, Australia. Disease incidence was found to significantly decrease as mosquito control programmes increased in size.

Typically, the impact of vector control of mosquitoes is measured using longitudinal studies utilizing entomological indices such as mosquito collections, mosquito infection rates or sentinel animal seroconversion. For example, Lothrop *et al.* (2008) describe the impact of early and late season truck applied ULV on *Culex tarsalis* and *Culex quinquefasciatus* populations and WNV infection rates from 2004–2006 in the Coachilla Valley in southern California. Sentinel chickens were also used as a measure for WNV activity. The thorough network of traps and sentinel flocks allowed the movement of WNV through different ecosystems to be examined. From this, they were able to determine that early season vector control following initial detection of WNV helped reduce vector populations and WNV activity. Other examples of studies that measure the impact of vector control on adult mosquito populations include Reddy *et al.* (2006), Farajollahi *et al.* (2012) and Elnaiem *et al.* (2008). However, some studies are limited by not including an untreated control area in the analysis and simply relying upon changes in populations before and after control in a treated area; for example Breidenbaugh *et al.* (2009). Considering the transient and dynamic nature of wild mosquito populations, populations from untreated control areas should be included as a comparator with treated sites.

We will now provide a brief survey of traditional methods employed by health workers to control arbovirus outbreaks. This is meant as an overview, and readers should explore the primary literature for further detail.

Surveillance for arbovirus activity or risk

Most control agencies employ surveillance for arbovirus activity. Systems used include sentinel chickens for MVEV in Australia (Mackenzie *et al.*, 1998) and St. Louis encephalitis in the USA (Day and Stark, 2000), sentinel pigs for JEV in Australia (Hanna *et al.*, 1999), and even dead birds (esp. crows) for WNV in the USA (Mostashari *et al.*, 2003). Sentinel animals are bled and the presence of antibodies to the virus indicate recent infection; dead animals are submitted for PCR detection of the virus.

Many of these programmes also collect wild mosquitoes that are submitted for virus detection (formally virus culture but increasingly PCR or antigen detection Rapid Diagnostic Test (RDTs) with high specificity and sensitivity that can be administered in most laboratories (Padgett *et al.*, 2006), with the WNV RAMP test being an example (Sutherland and Nasci, 2007). Specific mosquito traps are used to target different vectors. For example, the gravid *Culex* trap (Reiter, 1987) uses hay infusion lure to attract gravid *Culex* vectors of WNV, and CO_2-baited light traps have been historically used to capture a wide range of mosquitoes that are processed for virus detection (van-den-Hurk *et al.*, 2008; Turell *et al.*, 2005). Recently, new traps have been designed that utilize PCR detection to process honey-treated cards that captured mosquitoes have expectorated on (Hall-Mendelin *et al.*, 2010; Ritchie *et al.*, 2013b). This system, coupled with a passive trap that captures mosquitoes without fans and lights, eliminates the need for batteries and processing of large numbers of mosquitoes. These sentinel mosquito traps have been used to detect several arboviruses in Australia, including Kunjin virus (KUNV), Barmah Forest virus (BMFV) and Ross River virus (RRV) (van den Hurk *et al.*, 2014).

The detection of virus activity will then trigger increased mosquito control and public health advisories warning people to avoid being bitten by mosquitoes. Of course, in many instances simply the presence of human cases of disease is a trigger for vector control and public health advisories. In some areas, especially remote areas lacking in organized vector control, weather events can be used to warn of potential arbovirus outbreaks, or at least a higher risk than normal. Satellite imaging of lush vegetation (normalized difference vegetation index; NDVI) resulting from high rainfall has been linked to epidemics of Rift Valley Fever virus (RVFV) in Africa (Linthicum *et al.*, 1999; Anyamba *et al.*, 2012). In Australia, indices based upon heavy rains linked to El-nino and La-nina cycles have been developed to highlight high risk scenarios for MVEV transmission (Spencer *et al.*, 2001). And periods of high temperature and low rainfall that lead to stagnation of water, especially in creeks subject to sewage outflow (Vazquez-Prokopec *et al.*, 2010a) and storm water catch basin has been associated with high populations of *Culex* and outbreaks of SLEV and WNV in the USA (Reisen, 2013) and Europe (Sirbu *et al.*, 2011), including the original outbreak of SLEV in St. Louis, Missouri in 1933 (Leake *et al.*, 1934).

Conventional mosquito control: source reduction

Source reduction (habitat modification) has also been used to eliminate larval habitat. Constructed man-made wetlands, sewage ponds, irrigation systems and even swimming pools can produce huge populations of a variety of *Culex* (Walton, 2012; Whelan, 1998; Eisen *et al.*, 2010; Reisen *et al.*, 2008). But with care, they can be designed, managed and even mitigated to minimize vegetation and organic content and associated production of mosquitoes. Jacups *et al.* (2011) report on modification of a sewage effluent fed marsh in a desert area near Alice Springs, central Australia that had produced high populations of *Culex annulirostris* with associated MVEV activity in sentinel chickens. After installing a gravity drain, water levels decreased and mosquito populations crashed, with MVEV and KUNV seroconversions in chickens ceasing. Human cases of KUNV and MVEV, totalling five in the 6 years before the intervention, also ceased. Drainage schemes, including runnelling and Open Marsh Water Management, are also employed to control saltmarsh mosquitoes (Dale and Knight, 2006). Animal husbandry can also reduce the threat from encephalitis. Pigs are the amplifying host for JEV, and the removal of piggeries in close association with houses and farmsteads has been advocated for prevention of JEV infection in man (Halstead and Jacobson, 2003; Solomon, 2006; van den Hurk *et al.*, 2008).

Conventional mosquito control: larvicides

A wide variety of vector control measures have been used to prevent, contain and eliminate arbovirus transmission. Evidence that larval control alone significantly reduces encephalitis transmission is lacking, and even studies showing a strong link to adult population impact are rare. That said, it would be unwise not to employ larval control in many situations. For example, drain sumps and storm water catch basin are an important source of *Culex* vectors of WNV in urban areas (Nasci *et al.*, 2001b; Ruiz *et al.*, 2004), and excellent formulations for sustained control of larvae and pupae in these habitats exist (Li *et al.*, 2011; Anderson *et al.*, 2011). These include biologicals such as *Bacillus thuringiensis* var. *israelensis* (Bti), *Bacillus sphaericus*, spinosad and insect growth regulators such as methoprene and pyriproxyfen. Wide acreage applications of various larvicides are also commonly used to control vectors, employing trucks, quad bikes, airplanes and helicopters. While much of this is conducted to prevent outbreaks of nuisance mosquitoes, in many instances reduction of arbovirus transmission risk is a tangible goal [for instance, to control RRV vectors in Australian saltmarsh (Kurucz *et al.*, 2009; Russell and Kay, 2004)].

Conventional mosquito control: adulticiding

Adulticiding using insecticidal aerosols is the primary form of mosquito control used to control encephalitis vectors. As stated earlier, only the Carney *et al.* (2008) study of aerial ULV in California clearly demonstrates that aerial applied insecticides provide protection from WNV infection in humans. Several studies document that aerial and ground ULV applications of insecticides (primarily synthetic pyrethoids, but also organophosphates such as naled and malathion) provide significant kill of exposed mosquitoes, typically held within mesh sentinel cages (for a review see Mount, 1998; Bonds, 2012). For example, Elnaiem *et al.* (2008) documented that aerial application of pyrethrin + piperonyl butoxide mix significantly reduced adult light trap captures of the primary WNV vector *Culex pipiens*, with a subsequent decline in sentinel chicken seroconversions in central California. Ground ULV, usually applied from truck mounted foggers, also show significant declines in *Culex* populations although Reddy *et al.* (2006) reported no impact on gravid *Cx. pipiens* and *Cx restuans* in urban Massachusetts, and Reiter *et al.* (1990) found transient reduction in populations.

In his thorough review of ULV fogging for vector control, Bonds (2012) highlights that successful control of mosquitoes using ULV fogging requires getting adequate droplet size and numbers to the target mosquitoes. Meteorological conditions can have a profound influence on the behaviour of the aerosol cloud, mosquito populations may only be present in the air for a short period of time (say, crepuscular period), and the insecticides used are also only active for a short period. Thus, control is localized and transient, requiring repeat applications.

There is public concern about the health impacts of insecticidal aerosols (Peterson *et al.*, 2006). The level of concern can become political, and can even result in cessation of vector control activities. Studies of the amount of insecticide and its potential toxicity to humans exposed to truck mounted ULV insecticides indicates that acute and subchronic exposure human-health risks from residential exposure to mosquito insecticides are low and are not likely to exceed levels of concern (Macedo *et al.*, 2010). Public concerns extend to the impact of insecticide on nontarget organisms, making frequent use of ULV fogging politically difficult.

Treatment of lower foliage and leaf litter under heavy vegetation with residual insecticides is termed barrier or harbourage sprays. These have been popular for control of nuisance mosquitoes, and provide control for several weeks (Hurst *et al.*, 2012), often much longer than traditional ULV fogging (Qualls *et al.*, 2012; Muzari *et al.* 2014). Barrier and harbourage sprays can be used to provide sustained control of vector populations in urban areas, and is currently being used for control of *Aedes albopictus* in the Torres Strait of Australia (Muzari *et al.*, 2014).

Elimination, the ultimate control

Local, regional and even country and continental wide elimination of vector species provides certain freedom from the risk of virus transmission. The results of the Fred Soper led campaigns resulted in the virtual elimination of *Ae. aegypti* from much of south and central America. In order to remove the risk of large scale outbreaks of the exotic RRV in New Zealand, the New Zealand government funded a large campaign to eliminate the saltmarsh mosquito *Ae. camptorhynchus* (Kay and Russell, 2013). This Australian mosquito, an important vector of RRV, became established in isolated coastal areas of New Zealand in 1998. With the high probability that RRV could be introduced from Australia into New Zealand, and that large populations of brush tail possums could serve as a reservoir, elimination of this competent vector was undertaken to prevent establishment of RRV. Using a comprehensive surveillance system and repeated treatment of affected coastal habitat with residual formulations of s-methoprene granules, *Ae. camptorhynchus* populations collapsed and were declared eliminated from New Zealand in 2010.

Final considerations: how does the evidence stack up?

It is obvious that clinical evidence that mosquito control interventions prevent transmission of human disease is limited. Indeed, properly controlled studies that show unequivocally that vector control significantly reduces the target adult mosquitoes are also limited. Does this mean that we should park our foggers and cast aside our dippers? This is analogous to the metaphor of the parachute and human health (Smith and Pell, 2006). We all accept that parachutes save lives for those jumping from airplanes. Yet there are no CRTs showing parachutes provide a health benefit. The authors of this seemingly tongue in cheek article published in the *British Medical Journal* go on to propose that 'everyone might benefit if the most radical protagonists of evidence based medicine organised and participated in a double blind, randomised, placebo controlled, crossover trial of the parachute'.

We think that we are at a similar crossroads with vector-borne diseases. While diseases such as malaria that have a large burden on society, and thus a large source of research funding, have successfully employed CRTs, most mosquito-borne diseases have a smaller footprint and are 'neglected'. Furthermore, the extremely limited and sporadic distribution of cases of diseases such as MVE and SLE make disease endpoint CRTs risky if not downright impossible to consider.

That said, we should demand solid evidence that mosquito control does significantly impact adult populations of the target species. Larval control interventions must do more than just dip for larvae – they must also measure the impact on adult mosquitoes that actually vector the pathogen. Proxies of disease transmission such as sentinel animals or mosquito virus detections are also very useful. These trials should also be controlled, with before and after intervention measures in treated and untreated areas, with suitable replication. Even these studies will require significant funding, collaboration and cooperation to complete.

Dengue

Background

Conventional vector control is the only option available for combating many arboviruses. For diseases such as dengue, emerging solutions including vaccines, sterile insect techniques and the release of disease-refractory mosquitoes are years from operational roll-out and will be unlikely to provide universal protection from transmission (Halstead, 2012; Hughes and Britton, 2013; Scott and Morrison, 2010).

Despite our total reliance on anti-vector measures for dengue management there have been few advances in methodology or efficiency over the past half a century. Even more surprisingly, although myriad publications demonstrate that vector control measures reduce populations of *Ae. aegypti* and *Ae. albopictus* mosquitoes there are very few papers that detail the impact of these efforts on the prevalence or incidence of disease. Carefully designed and replicated quantitative proofs that compare entomological parameters and disease incidence across matched areas remain largely absent.

Many of the most widely cited examples of successful arboviral disease control relate to campaigns that were based upon the tenets of eradication and that were unusually rigorously enforced. These include operations against yellow fever in the Americas during the 1950s and 1960s (Soper, 1967), dengue in Singapore during the 60s and 70s (Ooi *et al.*, 2006) and dengue in Cuba during the 1980s (Guzman Tirado, 2012). These campaigns were vertically and even militarily driven, utilized huge amounts of human and financial resources, and combined environmental management and source reduction with larviciding and adulticiding. They insisted on the cooperation of the wider community and enforced punitive legislation on those who did not comply.

The impact of these eradication programmes on disease is universally acknowledged, but much of the evidence for their efficacy remains circumstantial – dramatic impacts on mosquitoes over very large geographic scales coincided with the end of disease epidemics or periods of low transmission. These kinds of approaches were abandoned some decades ago in favour of more pragmatic, locally managed campaigns driven by the simplistic assumption that a reduction in adult mosquito numbers reduces transmission risk (Gubler, 1989). Unfortunately, there remains little understanding of the target entomological thresholds of that approach. Despite acclaimed vector control campaigns and consistently low *Aedes* indices, Singapore and Cuba remain vulnerable to major outbreaks of dengue (Egger *et al.*, 2008; Koh *et al.*, 2008). Focks *et al.* (2000) suggested that under specific conditions of herd immunity and host density, even low rates of virus introduction needed only 0.1 mosquito pupae per human to spark an epidemic: low entomological indices do not guarantee low incidence of infection.

The tools that we use in dengue vector control are blunt and the coverage that is achieved by them is hard to standardize. Well-funded, vertically managed malaria control programmes can reasonably aim to deliver a bed net to every person, or attain universal coverage of insecticides on all walls within a village. Vector control programmes for dengue tend to be locally managed, urban, poorly resourced and unevaluated. As a result the major approaches of source reduction, fumigation and targeted IRS can achieve very different levels of premise and mosquito coverage depending on the complexity of the urban environment, the cryptic nature of breeding sites, the skills, experience and motivation of the response teams, and the available legislation that might ensure community compliance and access to properties. In Brazil, more than 0.5 billion USD are invested in vector control every year and yet the spread of dengue fever continues (Barreto *et al.*, 2011).

In short, the impacts of vector control on entomological indicators and on disease incidence continue to be inferred by comparisons with historical pre-treatment or pre-campaign data or from unmatched non-intervention areas (Gurtler *et al.*, 2009; Suaya *et al.*, 2007). Their interpretation is confounded by the extremely heterogeneous nature of mosquito density, disease incidence and local resources: while one campaign claims success, another laments failure. Cluster randomized trials or case control comparisons for the impact of specific vector control tools on disease are lacking (Ballenger-Browning and Elder, 2009) and the available literature does not deliver the quantitative proof required to give operators some universal certainty that specific measures, applied in specific ways will impact disease. This may encourage beleaguered public health officials to simply go through the motions of control rather than optimize, evaluate and test their programmes.

The challenges of measuring the impact of entomological indices on dengue

Dengue outbreaks occur in irregular cycles driven by cross-protection, herd immunity and myriad biological and environmental variables. This spatial and temporal heterogeneity, against a variable background of routine control measures, makes robust demonstrations of vector control impacts difficult to plan, design and conduct. The parallel comparison of both entomological and virological indices during dengue outbreaks is complicated by the fact that the majority of viraemias can be subclinical or asymptomatic (Burke *et al.*, 1988) and that many cases may originate from outside the intervention zones. This latter observation emphasizes the importance of human movement in patterns of dengue transmission – a minor consideration in malaria transmission systems whose vectors are vagile, nocturnal and consequently target humans in and around their own homes (Stoddard *et al.*, 2009). During outbreaks, the xenodiagnostic surveillance of dengue in mosquitoes may provide a useful alternative or complement to human virological surveys (Garcia-Rejon *et al.*, 2008, 2011; Pinheiro *et al.*, 2005) but it will be of limited utility in hypo-endemic settings when infection rates may be very low (Chen *et al.*, 2010).

Intervention studies that are area-wide and largely preventative in nature (such as source reduction or the release of *Wolbachia*-infected mosquitoes) will require a choice of treatment and control areas, a roll-out of the intervention and the subsequent presence of dengue transmission in both control and intervention zones. Many believe that the only robust approach to this kind of analysis is in the use of cluster randomized trials (Wolbers *et al.*, 2012). Other interventions such as the focal treatment of houses with IRS or volatile emanators (Ritchie and Devine, 2013; Vazquez-Prokopec *et al.*, 2010b) in response to active transmission may be better suited to more manageable case control studies similar to those used to examine the impact of bed nets on malaria (D'Alessandro *et al.*, 1997; Rowland *et al.*, 1997; Schellenberg *et al.*, 2001). Small groups of houses surrounding cases identified in real time by passive surveillance programmes could be assigned to treatment or control arms. Their subsequent impact on virus transmission between neighbours and other householders could then be monitored.

Conventional entomological measures of *Ae. aegypti* density have, until very recently, focused on a variety of larval indices (Ballenger-Browning and Elder, 2009). These methods were developed during the yellow fever eradication campaign and exploited the relative ease with which the breeding sites of this low density, cryptic mosquito could be identified (Breteau, 1954; Conner and Monroe, 1923). Unfortunately, they correlate poorly with adult mosquito densities (Scott and Morrison, 2003) and, therefore, with dengue incidence (Focks and Chadee, 1997). The number of adult mosquitoes required for transmission can be diminishingly small. This is because they feed exclusively and frequently on human blood. One mosquito can be responsible for multiple human infections and densities of just 0.1 female adult per person may be sufficient to initiate an outbreak (Focks *et al.*, 2000).

In comparison with measures of malaria control, dengue researchers lack some key descriptive metrics that link mosquitoes to transmission. The most notable is the entomological inoculation rate (EIR), which is the rate at which people are bitten by infectious mosquitoes. Measures of EIR have been useful in describing declining malaria transmission in relation to vector control interventions (Cuzin-Ouattara *et al.*, 1999; Martins-Campos *et al.*, 2012; Osse *et al.*, 2012; Robert and Carnevale, 1991). They have not been applied to assessments of dengue transmission because of low mosquito densities (in comparison to *Anopheles*), low viral infection rates in *Aedes* mosquitoes, a lack of high-throughput cheap xenodiagnostics and poor tools for the collection of large numbers of dengue vectors. The last few years have seen significant developments in xenodiagnosis (Voge *et al.*, 2013) and adult mosquito trapping techniques (Ritchie *et al.*, 2013a) that may make EIRs a more accessible parameter in future dengue studies.

Larviciding versus adulticiding

Generally, there are two conventional vector control paradigms for dengue: (i) to maintain vectors at a level where they cannot transmit the disease and (ii) to break transmission once outbreaks occur. The former is heavily reliant on source reduction (often cited as the only sustainable way to control mosquito densities over the longer term; Gubler, 1998) and the latter focuses on killing viraemic mosquitoes by fumigation or IRS. The entomological thresholds that need to be achieved by either approach are extremely variable and poorly documented. As a consequence, campaigns commonly fail to prevent transmission although they may reduce or contain it (Vazquez-Prokopec *et al.*, 2010b). The entomological impact required will vary according to mosquito density, herd immunity, virus genotype, the urban environment, human density, and weather. Under these conditions, dengue incidence cannot be used as a reasonable gauge of control programme efficacy: a moderately successful control programme may realize modest decreases in transmission whose significance remains unknown because disease trends in the absence of vector control cannot be predicted.

Adulticiding and larviciding measures are usually conducted in tandem during epidemics, and larviciding often continues after transmission has slowed or ended (Burattini *et al.*, 2008; Oki *et al.*, 2011). This is a very different approach to malaria parasite control, where adulticiding is the central tenet of control. This is driven by the knowledge that decreases in the longevity of adult mosquitoes have an amplified effect on transmission while the linear impact of larviciding on overall abundance is far less certain. It is rare to see this difference in approach reflected in discussions on the most cost-effective dengue measures (Luz *et al.*, 2011).

Although dengue control is often said to be at its most sustainable when campaigns focus on larval source reduction (Gubler, 1998), the practical complexities of this approach are seldom discussed. Under many scenarios at least 90% of containers would have to be identified and treated in order to prevent transmission (Focks *et al.*, 2000). This can only be guaranteed where the most productive containers are limited in number and easily identifiable and accessible (i.e. water storage jars in rural Vietnam; Kay and Nam, 2005). At the opposite end of the spectrum, larval control can be almost impossible when there is no community participation, no guaranteed access to yards and gardens, myriad hidden breeding sites, known but inaccessible sites such as roof gutters and a rapid proliferation of non-degradable containers that negate the efforts of any specific treatment or cleanup campaign. Although most dengue management plans urge the delegation of responsibility for source reduction to the community, most homeowners are plainly unable or unwilling to take effective action. There is ample evidence that source reduction methods are often implemented poorly and consequently have little impact (Chadee *et al.*, 2005; Hayes *et al.*, 2003).

During times of active dengue transmission, killing adult mosquitoes has a theoretically greater effect than targeting larvae. Space-spraying from outside the household (i.e. truck-mounted fogging) has long been discredited (Reiter and Gubler, 1997) but indoor fumigation and residual spraying have direct impacts on viraemic, endophilic mosquitoes (Camargo, 1967; Giglioli, 1948; Ordonez Gonzalez *et al.*, 2011; Vazquez-Prokopec *et al.*, 2010b). IRS in particular exploits the traits shared by *Ae. aegypti* and many Anophelines: high coverage in transmission settings can potentially reduce parasite and pathogen transmission because of the vulnerability of largely endophagic and endophilic mosquitoes. The use of IRS by vector control teams might be supplemented or fully implemented by home-owners where the costs of appropriate insecticides and applicators are not prohibitive (Eisen *et al.*, 2009).

So what works?

Many papers demonstrate the link between larval or adult mosquito indices and increased disease transmission (Chadee, 2009; Pham *et al.*, 2011; Sanchez *et al.*, 2006) and many others document the impact of vector control tools on entomological indices (Ballenger-Browning and Elder, 2009; Erlanger *et al.*, 2008; Tsunoda *et al.*, 2013; Vanlerberghe *et al.*, 2011). However, there is shockingly little empirical evidence for the impact of specific components of control programmes on dengue prevalence or incidence.

In the post-eradication era, the most effective larviciding and source reduction campaigns appear to have taken place in Singapore and Cuba where vast human resources, universal access to property, punitive measures against non-compliant householders and a wealth of operational experience conspired to maximize the coverage of their control measures. The evidence for the impact of these extensive programmes comes from the correlation between the implementation of effective vector control campaigns and the subsequent long term decrease in the severity of dengue outbreaks.

In response to a large epidemic in 1981, Cuba reduced the percentage of houses positive for larva from 35 to <0.01 and appears to have been largely successful at preventing further serious outbreaks for a 20-year period. A serious outbreak in 2001 was attributed to a relaxation in vector control (Guzman *et al.*, 2006; Pelaez *et al.*, 2004). Before control programmes were instituted in Singapore, the most severe cases of dengue were associated with areas where ca. 16% of houses were positive for larvae (Ooi *et al.*, 2006). With source reduction, this house index was reduced to < 1% by the mid-1980s. Despite some smaller epidemics in the interim, dengue in Singapore was considered to have remained generally contained until 2005 when an unprecedented outbreak caused > 14,000 confirmed cases (subclinical and asymptomatic illness not included). This was surpassed in 2013 with > 22,000 cases. The percentage of positive houses has changed little but the loss of herd immunity, increasing temperature, human density and international travel have all been implicated in these control failures (Koh *et al.*, 2008). It is, of course impossible to know how much worse these epidemics would have been in the absence of Singapore's control programme.

On more limited experimental scales, there are more robustly quantitative assessments of impact. In Thailand, a serological survey of primary school children was carried out and dengue cases were mapped in order to inform a vector control campaign that included source reduction, screen lids, treatment with biocontrol agents and the deployment of lethal ovitraps. A very large number of lethal ovitraps were also deployed (5–10 per household). As these tools target blood-fed, ovipositing mosquitoes they are likely to have had a major impact upon virus transmission (Lee *et al.*, 2013). Results of the trials showed a reduction in entomological indices (per cent positive containers) and sero-positive children in the intervention areas (Kittayapong *et al.*, 2008).

That example shows that careful, real-time mapping of disease notifications can deliver a framework for evaluating the impact of spatially and temporally targeted approaches to dengue control. Subsequent analysis of those data sets can yield strong evidence for the effect of specific tools. The best illustration of this is from the hypo-endemic city of Cairns, Australia where, in 2003–2004, the application of IRS in the immediate vicinity of dengue case houses, in tandem with active tracing and treatment of contact addresses, had a significant impact on transmission once particular coverage thresholds (60% of target houses) were achieved (Vazquez-Prokopec *et al.*, 2010b).

Conclusion

Most campaigns against dengue are fought by locally resourced teams that lack the partnerships to implement comprehensive research and evaluation programmes. Resultantly, reviews of the impacts of vector control programmes on dengue tend to observe enormous variation between sites, an inability to pull apart the different components of integrated programmes, and the use of non-standard, insensitive entomological and virological indicators. Sophisticated trials of vector control interventions that use standard methodologies to analyse entomological and disease endpoints in parallel are largely unavailable (Ballenger-Browning and Elder, 2009; Erlanger *et al.*, 2008).

We would urge that more time and effort is spent in developing spatially focused operational programmes driven by real-time patterns of transmission. These will demand incidence and response recording systems of the kind that have been developed in Australia (Vazquez-Prokopec *et al.*, 2010b). Spatial and temporal analyses of these datasets can then test for evidence of the impact of focally applied measures such as IRS or indoor fumigation. Perhaps hypo-endemic areas, without much herd immunity and with manageable numbers of cases to track, may play a major role in providing initial evidence of operational impact. The effects of more diffusely applied source reduction and larval management tools will be less well served by this approach but the increasing adoption of adult indices as the entomological measure for control programmes (Pepin *et al.*, 2013) will at least guarantee a more relevant measure of insect density to correlate with disease.

It should be noted that, despite far greater resources and strong academic partnerships, malaria programmes share some of the dengue world's problems in creating an evidence base for control tools: in 2010, indoor residual spraying (IRS) was used to protect > 180 million people against malaria (Anon, 2011) despite the fact that the same year saw the publication of an efficacy review that concluded that there were too few robust IRS trials to allow malaria control programmes to quantify or predict its efficacy across different transmission settings (Pluess *et al.*, 2010).

The profound impacts of adulticiding and larval source management on the transmission of parasites and pathogens are largely accepted thanks to the weight of circumstantial evidence (Anon, 2012). The problem lies in being able to identify the components of campaigns that are most effective and to guarantee their effect across different programmes.

A final rant

Many reviews of dengue control focus on the historical success of the *Ae. aegypti* eradication programme implemented in the Americas during the 1950s and 1960s. This programme was driven by the threat of urban yellow fever transmission although, interestingly, the last outbreak of *Ae. aegypti* transmitted yellow fever in Brazil was in 1942 (Soper, 1963) and the PAHO-sponsored, nationwide eradication plan took place between 1947 and 1952 (Severo, 1955). This sustained action, conducted largely in the absence of disease, resulted in the purported elimination of *Ae. aegypti* from 22 countries in the region. These tremendous successes are attributed to international cooperation, adequate funding, good training, a universal political will and a military type campaign. Many reviews note the loss of commitment to the programme during the 1960s, the rise in insecticide resistance, and the fact that by the 1970s *Ae. aegypti* was re-establishing throughout the continent (e.g. PAHO, 1996). Few acknowledge that even if the political fervor had been retained, or adopted by the wider community, those vector control programmes would have been impossible to sustain. The PAHO campaign's successes were largely reliant on the 'perifocal treatment' of all potential breeding sites with DDT: one of the most persistent insecticides that the world has ever seen. In Brazil, most of the target population was rural, and households relied on a limited number of containers to store water, many of which were large, earthenware and easily identifiable (Camargo, 1967). Increasing urbanization, its associated detritus and the move to less environmentally stable chemistries means that those kinds of strategies cannot now work in most cities. During the PAHO eradication campaign in Brazil the city of Sao Paulo was home to ca. 2 million people (http://www.britannica.com/EBchecked/topic/701311/Sao-Paulo/261979/The-city-after-independence) and disposable, non-degradable containers were uncommon. Now it is a sprawling megacity of 20 million people and it contains the biggest concentration of slum housing in the Americas (http://en.wikipedia.org/wiki/S%C3%A3o_Paulo). As a measure of the potential numbers of artificial breeding habitats available, its population disposes of 6.5M tons of solid waste and 25 billion plastic bags per year (http://issuu.com/analiseeditorial/docs/sao_paulo_outlook_2012). Under those conditions, the city-wide control of *Ae. aegypti* is an impossible ideal. Perhaps only focal campaigns, targeted at real-time clusters, that aim simply to manage rather than prevent disease are operationally feasible.

References

Anderson, J.F., Ferrandino, F.J., Dingman, D.W., Main, A.J., Andreadis, T.G., and Becnel, J.J. (2011). Control of mosquitoes in catch basins in Connecticut with *Bacillus thuringiensis israelensis, Bacillus sphaericus,* [corrected] and spinosad. J. Am. Mosq. Control Assoc. *27*, 45–55.

Anon. (2011). World Malaria Report (World Health Organization, Geneva).

Anon. (2012). Global Strategy for Dengue Prevention and Control (World Health Organization, Geneva).

Anyamba, A., Linthicum, K.J., Small, J., Britch, S.C., and Tucker, C.J. (2012). Remote sensing contributions to prediction and risk assessment of natural disasters caused by large-scale Rift Valley Fever outbreaks. Proc. IEEE *99*, 1–11.

Ballenger-Browning, K.K., and Elder, J.P. (2009). Multi-modal *Aedes aegypti* mosquito reduction interventions and dengue fever prevention. Trop. Med. Int. Health *14*, 1542–1551.

Barreto, M.L., Teixeira, M.G., Bastos, F.I., Ximenes, R.A.A., Barata, R.B., and Rodrigues, L.C. (2011). Health in Brazil 3 Successes and failures in the control of infectious diseases in Brazil: social and environmental context, policies, interventions, and research needs. Lancet *377*, 1877–1889.

Bhatt, S., Gething, P.W., Brady, O.J., Messina, J.P., Farlow, A.W., Moyes, C.L., Darke, J.M., Brownstein, J.S., Hoen, A.G., Sankoh, O., *et al.* (2013). The global distribution and burden of dengue. Nature *496*, 504–507.

Bonds, J.A. (2012). Ultra-low-volume space sprays in mosquito control: a critical review. Med. Vet. Entomol. *26*, 121–130.

Breidenbaugh, M.S., Haagsma, K.A., Wojcik, G.M., and De Szalay, F.A. (2009). Efficacy of aerial spray applications using fuselage booms on Air Force C-130H aircraft against mosquitoes and biting midges. J. Am. Mosq. Control Assoc. *25*, 467–473.

Breteau, H. (1954). Yellow fever in French West Africa; an aspect of mass-preventive medicine. Bull. WHO *11*, 453–481.

Burattini, M.N., Chen, M., Chow, A., Coutinho, F.A.B., Goh, K.T., Lopez, L.F., Ma, S., and Massad, E. (2008). Modelling the control strategies against dengue in Singapore. Epidemiol. Infect. *136*, 309–319.

Burke, D.S., Nisalak, A., Johnson, D.E., and Scott, R.M. (1988). A prospective study of dengue infections in Bangkok. Am. J. Trop. Med. Hyg. *38*, 172–180.

Camargo, S. (1967). History of *Aedes aegypti* eradication in the Americas. Bull. WHO *36*, 602–603.

Carney, R.M., Husted, S., Jean, C., Glaser, C., and Kramer, V. (2008). Efficacy of aerial spraying of mosquito adulticide in reducing incidence of West Nile Virus, California, 2005. Emerg. Infect. Dis. *14*, 747–754.

Chadee, D.D. (2009). Dengue cases and *Aedes aegypti* indices in Trinidad, West Indies. Acta Tropica *112*, 174–180.

Chadee, D.D., Williams, F.L.R., and Kitron, U.D. (2005). Impact of vector control on a dengue fever outbreak in Trinidad, West Indies, in 1998. Trop. Med. Intl. Hlth *10*, 748–754.

Chen, C.-F., Shu, P.-Y., Teng, H.-J., Su, C.-L., Wu, J.-W., Wang, J.-H., Lin, T.-H., Huang, J.-H., and Wu, H.-S. (2010). Screening of dengue virus in field-caught *Aedes aegypti* and *Aedes albopictus* (Diptera: Culicidae) by one-step SYBR Green-based reverse transcriptase-polymerase chain reaction assay during 2004–2007 in Southern Taiwan. Vector-Borne Zoonotic Dis. *10*, 1017–1025.

Conner, M.E., and Monroe, W.M. (1923). Stegomyia indices and their value in yellow fever control. Am. J. Trop. Med. Hyg. *4*, 9–19.

Cuzin-Ouattara, N., Van den Broek, A.H.A., Habluetzel, A., Diabate, A., Sanogo-Ilboudo, E., Diallo, D.A., Cousens, S.N., and Esposito, F. (1999). Wide-scale installation of insecticide-treated curtains confers high levels of protection against malaria transmission in a hyperendemic area of Burkina Faso. Trans. R. Soc. Trop. Med. Hyg. *93*, 473–479.

Dale, P.E., and Knight, J.M. (2006). Managing salt marshes for mosquito control: impacts of runnelling, open marsh water management and grid-ditching in sub-tropical Australia. Wetl. Ecol. Manag. *14*, 211–220.

D'Alessandro, U., Olaleye, B., Langerock, P., Bennett, S., Cham, K., Cham, B., and Greenwood, B.M. (1997). The Gambian national impregnated bed net programme: evaluation of effectiveness by means of case–control studies. Trans. R. Soc. Trop. Med. Hyg. *91*, 638–642.

Day, J.F., and Stark, L.M. (2000). Frequency of Saint Louis encephalitis virus in humans from Florida, USA: 1990–1999. J. Med. Entomol. *37*, 626–633.

Egger, J.R., Ooi, E.E., Kelly, D.W., Woolhouse, M.E., Davies, C.R., and Coleman, P.G. (2008). Reconstructing historical changes in the force of infection of dengue fever in Singapore: implications for surveillance and control. Bull. WHO *86*, 187–196.

Eisen, L., Beaty, B.J., Morrison, A.C., and Scott, T.W. (2009). Proactive vector control strategies and improved monitoring and evaluation practices for dengue prevention. J. Med. Entomol. *46*, 1245–1255.

Eisen, L., Barker, C.M., Moore, C.G., Pape, W.J., Winters, A.M., and Cheronis, N. (2010). Irrigated agriculture is an important risk factor for West Nile virus disease in the hyperendemic Larimer-Boulder-Weld area of north central Colorado. J. Med. Entomol. *47*, 939–951.

Elnaiem, D.-E.A., Kelley, K., Wright, S., Laffey, R., Yoshimura, G., Reed, M., Goodman, G., Thiemann, T., Reimer, L., and Reisen, W.K. (2008). Impact of aerial spraying of pyrethrin insecticide on *Culex pipiens* and *Culex tarsalis* (Diptera: Culicidae) abundance and West Nile virus infection rates in an urban/suburban area of Sacramento County, California. J. Med. Entomol. *45*, 751–757.

Erlanger, T.E., Keiser, J., and Utzinger, J. (2008). Effect of dengue vector control interventions on entomological parameters in developing countries: a systematic review and meta-analysis. Med. Vet. Entomol. *22*, 203–221.

Farajollahi, A., Healy, S.P., Unlu, I., Gaugler, R., and Fonseca, D.M. (2012). Effectiveness of ultra-low volume nighttime applications of an adulticide against diurnal *Aedes albopictus*, a critical vector of dengue and chikungunya viruses. PLoS One *7*, e49181.

Focks, D.A., and Chadee, D.D. (1997). Pupal survey: an epidemiologically significant surveillance method for *Aedes aegypti*: an example using data from Trinidad. Am. J. Trop. Med. Hyg. *56*, 159–167.

Focks, D.A., Brenner, R.J., Hayes, J., and Daniels, E. (2000). Transmission thresholds for dengue in terms of *Aedes aegypti* pupae per person with discussion of their utility in source reduction efforts. Am. J. Trop. Med. Hyg. *62*, 11–18.

Garcia-Rejon, J., Alba Lorono-Pino, M., Arturo Farfan-Ale, J., Flores-Flores, L., Del Pilar Rosado-Paredes, E., Rivero-Cardenas, N., Najera-Vazquez, R., Gomez-Carro, S., Lira-Zumbardo, V., Gonzalez-Martinez, P., *et al.* (2008). Dengue virus-infected *Aedes aegypti* in the home environment. Am. J. Trop. Med. Hyg. *79*, 940–950.

Garcia-Rejon, J.E., Alba Lorono-Pino, M., Arturo Farfan-Ale, J., Flores-Flores, L.F., Lopez-Uribe, M.P., del Rosario Najera-Vazquez, M., Nunez-Ayala, G., Beaty, B.J., and Eisen, L. (2011). Mosquito infestation and dengue virus infection in *Aedes aegypti* females in schools in Merida, Mexico. Am. J. Trop. Med. Hyg. *84*, 489–496.

Giglioli, G. (1948). An investigation of the house-frequenting habits of mosquitoes of the British Guiana coastland in relation to the use of DDT. Am. J. Trop. Med. Hyg. *28*, 43–70.

Gubler, D.J. (1989). *Aedes aegypti* and *Aedes aegypti*-borne disease control in the 1990s: top down or bottom up. Charles Franklin Craig Lecture. Am. J. Trop. Med. Hyg. *40*, 571–578.

Gubler, D.J. (1998). Dengue and dengue hemorrhagic fever. Clin. Microbiol. Rev. *11*, 480–496.

Gurtler, R.E., Garelli, F.M., and Coto, H.D. (2009). Effects of a five-year citywide intervention program to control *Aedes aegypti* and prevent dengue outbreaks in Northern Argentina. PLoS Negl. Trop. Dis. *3*, e427.

Guzman, M.G., Pelaez, O., Kouri, G., Quintana, I., Vazquez, S., Penton, M., Avila, L.C., and Grp Multidisciplinario Control, E. (2006). Final characterization of and lessons learned from the dengue 3 epidemic in Cuba, 2001–2002. Rev. Panam. Salud. Publica. *19*, 282–289.

Guzman Tirado, C.M.G. (2012). Thirty years after the Cuban dengue hemorrhagic epidemic occurred in 1981. Revista Cubana de Med. Trop. *64*, 5–14.

Hall-Mendelin, S., Ritchie, S.A., Johansen, C.A., Zborowski, P., Cortis, G., Dandridge, S., Hall, R.A., and van den Hurk, A.F. (2010). Exploiting mosquito sugar feeding to detect mosquito-borne pathogens. Proc. Natl. Acad. Sci. U.S.A. *107*, 11255–11259.

Halstead, S.B. (2012). Dengue vaccine development: a 75% solution? Lancet *12*, 61510–61514.

Halstead, S.B., and Jacobson, J. (2003). Japanese encephalitis. Adv. Virus Res. *61*, 103–138.

Hanna, J.N., Ritchie, S.A., Phillips, D.A., Lee, J.M., Hills, S.L., van den Hurk, A.F., Pyke, A.T., Johansen, C.A., and Mackenzie, J.S. (1999). Japanese encephalitis in north Queensland, Australia, 1998. Med. J. Aust. *170*, 533–536.

Hayes, R.C., and Moulton, L.H. (2009). Cluster Randomised Trials (CRC Press, Boca Raton, Florida).

Hayes, J.M., Garcia-Rivera, E., Flores-Reyna, R., Suarez-Rangel, G., Rodriguez-Mata, T., Coto-Portillo, R., Baltrons-Orellana, R., Mendoza-Rodriguez, E., De Garay, B.F., Jubis-Estrada, J., *et al.* (2003). Risk factors for infection during a severe dengue outbreak in El Salvador in 2000. Am. J. Trop. Med. Hyg. *69*, 629–633.

Hughes, H., and Britton, N.F. (2013). Modelling the use of Wolbachia to control dengue fever transmission. Bull. Math. Biol. *75*, 796–818.

van den Hurk, A.F., Ritchie, S.A., Johansen, C.A., Mackenzie, J.S., and Smith, G.A. (2008). Domestic pigs and Japanese encephalitis virus infection, Australia. Emerg. Infect. Dis. *14*, 1736–1738.

van den Hurk, A.F., Hall-Mendelin, S., Townsend, M., Kurucz, N., Edwards, J., Ehlers, G., Rodwell, C., Moore, F.A., McMahon, J.L., *et al.* (2014). Applications of a sugar-based surveillance system to track arboviruses in wild mosquito populations. Vector Borne Zoonotic Dis. *14*, 66–73.

Hurst, T.P., Ryan, P.A., and Kay, B.H. (2012). Efficacy of residual insecticide biflex AquaMax applied as barrier treatments for managing mosquito populations in suburban residential properties in Southeast Queensland. J. Med. Entomol. *49*, 1021–1026.

Jacups, S., Kurucz, N., Whitters, R., and Whelan, P. (2011). Habitat modification for mosquito control in the Ilparpa Swamp, Northern Territory, Australia. J. Vector. Ecol., *36*, 292–299.

Kay, B., and Nam, V.S. (2005). New strategy against *Aedes aegypti* in Vietnam. Lancet *365*, 613–617.

Kay, B., and Russell, R. (2013). Mosquito Eradication: The Story of Killing Campto (Csiro Publishing, Clayton, Victoria, Australia).

Kittayapong, P., Yoksan, S., Chansang, U., Chansang, C., and Bhumiratana, A. (2008). Suppression of dengue transmission by application of integrated vector control strategies at sero-positive GIS-based foci. Am. J. Trop. Med. Hyg. *78*, 70–76.

Koh, B.K.W., Ng, L.C., Kita, Y., Tang, C.S., Ang, L.W., Wong, K.Y., James, L., and Goh, K.T. (2008). The 2005 dengue epidemic in Singapore: Epidemiology, prevention and control. Ann. Acad. Med. Singapore *37*, 538–545.

Kurucz, N., Whelan, P.I., Carter, J.M., and Jacups, S.P. (2009). A geospatial evaluation of *Aedes vigilax* larval control efforts across a coastal wetland, Northern Territory, Australia. J. Vector Ecol. *34*, 317–323.

Leake, J.P., Musson, E.K., and Chope, H.D. (1934). Epidemiology of epidemic encephalitis, St. Louis type. J. Am. Med. Assoc. *103*, 728–731.

Lee, C., Vythilingam, I., Chong, C.-S., Razak, M.A.A., Tan, C.-H., Liew, C., Pok, K.-Y., and Ng, L.-C. (2013). Gravitraps for management of dengue clusters in Singapore. Am. J. Trop. Med. Hyg. *88*, 888–892.

Lengeler, C. (2004). Insecticide-treated bed nets and curtains for preventing malaria. Cochrane Database Syst. Rev. *2*, CD000363.

Li, J., Sze, K., and P'ng, J. (2011). Evaluation of storm-water impacts on larviciding chemicals at catch basins for West Nile virus control. J. Environ. Engin. *138*, 182–189.

Linthicum, K.J., Anyamba, A., Tucker, C.J., Kelley, P.W., Myers, M.F., and Peters, C.J. (1999). Climate and satellite indicators to forecast Rift Valley fever epidemics in Kenya. Science *285*, 397–400.

Lothrop, H.D., Lothrop, B.B., Gomsi, D.E., and Reisen, W.K. (2008). Intensive early season adulticide applications decrease arbovirus transmission throughout the Coachella Valley, Riverside County, California. Vector Borne Zoonotic Dis. *8*, 475–489.

Luz, P.M., Vanni, T., Medlock, J., Paltiel, A.D., and Galvani, A.P. (2011). Dengue vector control strategies in an urban setting: an economic modelling assessment. Lancet *377*, 1673–1680.

Macedo, P.A., Schleier, J.J., Reed, M., Kelley, K., Goodman, G.W., Brown, D.A., and Peterson, R.K. (2010). Evaluation of efficacy and human health risk of aerial ultra-low volume applications of pyrethrins and piperonyl butoxide for adult mosquito management in response to West Nile virus activity in Sacramento County, California. J. Am. Mosq. Control Assoc. *26*, 57–66.

Mackenzie, J.S., Broom, A.K., Hall, R.A., Johansen, C.A., Lindsay, M.D., Phillips, D.A., Ritchie, S.A., Russell, R.C., and Smith, D.W. (1998). Arboviruses in the Australian region, 1990 to 1998. Commun. Dis. Intell. *22*, 93–100.

Martins-Campos, K.M., Pinheiro, W.D., Vitor-Silva, S., Siqueira, A.M., Melo, G.C., Rodrigues, I.C., Fe, N.F., Barbosa, M.D.G.V., Tadei, W.P., Guinovart, C., *et al.* (2012). Integrated vector management targeting *Anopheles darlingi* populations decreases malaria incidence in an unstable transmission area, in the rural Brazilian Amazon. Malaria J. *11*, 351.

Mostashari, F., Kulldorff, M., Hartman, J.J., Miller, J.R., and Kulasekera, V. (2003). Dead bird clusters as an early warning system for West Nile virus activity. Emerg. Infect. Dis. *9*, 641–646.

Mount, G.A. (1998). A critical review of ultralow-volume aerosols of insecticide applied with vehicle-mounted generators for adult mosquito control. J. Am. Mosq. Control Assoc. *14*, 305–334.

Muzari, O., Silcock, R., Davis, J., Ritchie, S.A., and Devine, G. (2014). Residual effectiveness of lambda-cyhalothrin harborage sprays against foliage-resting mosquitoes in north Queensland. J. Med. Entomol. *51*, 444–449.

Nasci, R.S., Newton, N.H., Terrillion, G.F., Parsons, R.E., Dame, D.A., Miller, J.R., Ninivaggi, D.V., and Kent, R. (2001a). Interventions: vector control and public education: panel discussion. Ann. N. Y. Acad. Sci. *951*, 235–254.

Nasci, R.S., Savage, H.M., White, D.J., Miller, J.R., Cropp, B.C., Godsey, M.S., Kerst, A.J., Bennett, P., Gottfried, K., and Lanciotti, R.S. (2001b). West Nile virus in overwintering *Culex* mosquitoes, New York City, 2000. Emerg. Infect. Dis. *7*, 742–744.

Oki, M., Sunahara, T., Hashizume, M., and Yamamoto, T. (2011). Optimal timing of insecticide fogging to minimize dengue cases: modeling dengue transmission among various seasonalities and transmission intensities. PLoS Negl. Trop. Dis. *5*, e1367.

Ooi, E.E., Goh, K.T., and Gubler, D.J. (2006). Denque prevention and 35 years of vector control in Singapore. Emerg. Infect. Dis. *12*, 887–893.

Ordonez Gonzalez, J.G., Thirion, J., Garcia Orozco, A., and Rodriguez, A.D. (2011). Effectiveness of indoor ultra-low volume application of Aqua Reslin Super during an emergency. J. Am. Mosq. Control Assoc. *27*, 162–164.

Osse, R., Aikpon, R., Padonou, G.G., Oussou, O., Yadouleton, A., and Akogbeto, M. (2012). Evaluation of the efficacy of bendiocarb in indoor residual spraying against pyrethroid resistant malaria vectors in Benin: results of the third campaign. Parasites Vectors *5*, 163.

Padgett, K.A., Cahoon-Young, B., Carney, R., Woods, L., Read, D., Husted, S., and Kramer, V. (2006). Field and laboratory evaluation of diagnostic assays for detecting West Nile virus in oropharyngeal swabs from California wild birds. Vector Borne Zoonotic Dis. *6*, 183–191.

Petersen, L.R., Brault, A.C., and Nasci, R.S. (2013). West Nile virus. Rev. J. Am. Med. Assoc. *310*, 308–315.

Peterson, R.K., Macedo, P.A., and Davis, R.S. (2006). A human-health risk assessment for West Nile virus and insecticides used in mosquito management. Environ. Health Perspect. *114*, 366–372.

Pelaez, O., Guzman, M.G., Kouri, G., Perez, R., San Martin, J.L., Vazquez, S., Rosario, D., Mora, R., Quintana, I., Bisset, J., *et al.* (2004). Dengue 3 epidemic, Havana, 2001. Emerg. Infect. Dis. *10*, 719–722.

Pepin, K.M., Marques-Toledo, C., Scherer, L., Morais, M.M., Ellis, B., and Eiras, A.E. (2013). Cost-effectiveness of novel system of mosquito surveillance and control, Brazil. Emerg. Infect. Dis. *19*, 542–550.

Pham, H.V., Doan, H.T.M., Phan, T.T.T., and Minh, N.N.T. (2011). Ecological factors associated with dengue fever in a central highlands Province, Vietnam. BMC Infect. Dis. *11*, 172.

Pinheiro, V.C.S., Tadei, W.P., Barros, P., Vasconcelos, P.F.C., and Cruz, A.C.R. (2005). Detection of dengue virus serotype 3 by reverse transcription-polymerase chain reaction in *Aedes aegypti* (Diptera, Culicidae) captured in Manaus, Amazonas. Memorias Do Instituto Oswaldo Cruz *100*, 833–839.

Pluess, B., Tanser, F.C., Lengeler, C., and Sharp, B.L. (2010). Indoor residual spraying for preventing malaria. Cochrane Database Syst. Rev. *4*, CD006657.

Qualls, W.A., Smith, M.L., Muller, G.C., Zhao, T.Y., and Xue, R.D. (2012). Field evaluation of a large-scale barrier application of bifenthrin on a golf course to control floodwater mosquitoes. J. Am. Mosq. Control Assoc. *28*, 219–224.

Reddy, M.R., Spielman, A., Lepore, T.J., Henley, D., Kiszewski, A.E., and Reiter, P. (2006). Efficacy of resmethrin aerosols applied from the road for suppressing *Culex* vectors of West Nile virus. Vector Borne Zoonotic Dis. *6*, 117–127.

Reisen, W.K. (2013). Ecology of West Nile virus in North America. Viruses *5*, 2079–2105.

Reisen, W.K., Takahashi, R.M., Carroll, B.D., and Quiring, R. (2008). Delinquent mortgages, neglected swimming pools, and West Nile virus, California. Emerg. Infect. Dis. *14*, 1747–1749.

Reiter, P. (1987). A revised version of the CDC Gravid mosquito trap. J. Am. Mosq. Control Assoc. *3*, 325–327.

Reiter, P., and Gubler, D.J. (1997). Surveillance and control of urban dengue vectors. In Dengue and Dengue Hemorrhagic Fever, D.J. Gubler, and G. Kuno, eds. (CAB International, Oxford), pp. 425–462.

Reiter, P., Eliason, D.A., Francy, D.B., Moore, C.G., and Campos, E.G. (1990). Apparent influence of the stage of bloodmeal digestion on the efficacy of ground applied ULV aerosols for the control of urban *Culex* mosquitoes. I. Field evidence. J. Am. Mosq. Control Assoc. *6*, 366–370.

Ritchie, S.A., and Devine, G.J. (2013). Confusion, knock-down and kill of *Aedes aegypti* using metofluthrin in domestic settings: a powerful tool to prevent dengue transmission? Parasites Vectors *6*, 262.

Ritchie, S.A., Cortis, G., Paton, C., Townsend, M., Shroyer, D., Zborowski, P., Hall-Mendelin, S., and van den Hurk, A. (2013). A simple non-powered passive trap for the collection of mosquitoes for arbovirus surveillance. J. Med. Entomol. *50*, 185–194.

Ritchie, S.A., Buhagiar, T.S., Townsend, M., Hoffmann, A., van den Hurk, A.F., McMahon, J.L., and Eiras, A. (2014) Field validation of the Gravid *Aedes* Trap (GAT) for collection of *Aedes aegypti* (Diptera: Culicidae). J. Med. Entomol. *51*, 210–219.

Robert, V., and Carnevale, P. (1991). Influence of deltamethrin treatment of bed nets on malaria transmission in the Kou valley, Burkina Faso. Bull. WHO *69*, 735–740.

Rowland, M., Hewitt, S., Durrani, N., Saleh, P., Bouma, M., and Sondorp, E. (1997). Sustainability of pyrethroid-impregnated bednets for malaria control in Afghan communities. Bull. WHO *75*, 23–29.

Ruiz, M.O., Tedesco, C., McTighe, T.J., Austin, C., and Kitron, U. (2004). Environmental and social determinants of human risk during a West Nile virus outbreak in the greater Chicago area, 2002. Int. J. Health Geogr. *3*, 8.

Russell, R.C., and Kay, B.H. (2004). Medical entomology: changes in the spectrum of mosquito-borne disease in Australia and other vector threats and risks, 1972--2004. Austr. J. Entomol. *43*, 271–282.

Sanchez, L., Vanlerberghe, V., Alfonso, L., Marquetti, M.D., Guzman, M.G., Bisset, J., and van der Stuyft, P. (2006). *Aedes aegypti* larval indices and risk for dengue epidemics. Emerg. Infect. Dis. *12*, 800–806.

Schellenberg, J., Abdulla, S., Nathan, R., Mukasa, O., Marchant, T.J., Kikumbih, N., Mushi, A.K., Mponda, H., Minja, H., Mshinda, H., *et al.* (2001). Effect of large-scale social marketing of insecticide-treated nets on child survival in rural Tanzania. Lancet *357*, 1241–1247.

Scott, T.W., and Morrison, A.C. (2003). *Aedes aegypti* density and the risk of dengue-virus transmission. In Ecological aspects of application of genetically modified mosquitoes, W. Takken, and T.W. Scott, eds. (Kluwer Academic, Norwell, MA), pp. 187–202.

Scott, T.W., and Morrison, A.C. (2010). Vector dynamics and transmission of dengue virus: implications for dengue surveillance and prevention strategies: vector dynamics and dengue prevention. Curr. Topics Microbiol. Immunol. *338*, 115–128.

Severo, O.P. (1955). Eradication of the *Aedes aegypti* mosquito from the Americas. Yellow fever, a symposium in commemoration of Carlos Juan Finlay, 1955. Available online: http://jdc.jefferson.edu/yellow_fever_symposium/6. Accessed 3 August 2015.

Shapiro, H., and Micucci, S. (2003). Pesticide use for West Nile virus. Can. Med. Assoc. J. *168*, 1427–1430.

Sirbu, A., Ceianu, C.S., Panculescu-Gatej, R.I., Vazquez, A., Tenorio, A., Rebreanu, R., Niedrig, M., Nicolescu, G., and Pistol, A. (2011). Outbreak of West Nile virus infection in humans, Romania, July to October 2010. Eur. Surveill. *16*, 19762.

Smith, G.C., and Pell, J.P. (2006). Parachute use to prevent death and major trauma related to gravitational challenge: systematic review of randomised controlled trials. Int. J. Prosthodont. *19*, 126–128.

Solomon, T. (2006). Control of Japanese encephalitis – within our grasp? N. Engl. J. Med. *355*, 869–871.

Soper, F.L. (1967). *Aedes aegypti* and yellow fever. Bull. WHO *36*, 521–527.

Spencer, J.D., Azoulas, J., Broom, A.K., Buick, T.D., Currie, B., Daniels, P.W., Doggett, S.L., Hapgood, G.D., Jarrett, P.J., *et al.* (2001). Murray Valley encephalitis virus surveillance and control initiatives in Australia. National Arbovirus Advisory Committee of the Communicable Diseases Network Australia. Commun. Dis. Intell. Q. Rep. *25*, 33–47.

Stoddard, S.T., Morrison, A.C., Vazquez-Prokopec, G.M., Soldan, V.P., Kochel, T.J., Kitron, U., Elder, J.P., and Scott, T.W. (2009). The role of human movement in the transmission of vector-borne pathogens. PLoS Negl. Trop. Dis. *3*, e481.

Suaya, J.A., Shepard, D.S., Chang, M.-S., Caram, M., Hoyer, S., Socheat, D., Chantha, N., and Nathan, M.B. (2007). Cost-effectiveness of annual targeted larviciding campaigns in Cambodia against the dengue vector *Aedes aegypti*. Trop. Med. Int. Health *12*, 1026–1036.

Sutherland, G.L., and Nasci, R.S. (2007). Detection of West Nile virus in large pools of mosquitoes. J. Am. Mosq. Control Assoc. *23*, 389–395.

Tomerini, D.M., Dale, P.E., and Sipe, N. (2011). Does mosquito control have an effect on mosquito-borne disease? The case of Ross River Virus disease and mosquito management in Queensland, Australia. J. Am. Mosq. Control Assoc. *27*, 39–44.

Tsunoda, T., Kawada, H., Huynh, T.T.T., Loan Le, L., San Hoang, L., Huu Ngoc, T., Huong Thi Que, V., Hieu Minh, L., Hasebe, F., Tsuzuki, A., *et al.* (2013). Field trial on a novel control method for the dengue vector, *Aedes aegypti* by the systematic use of Olyset (R) Net and pyriproxyfen in Southern Vietnam. Parasites Vectors *6*, 6.

Turell, M.J., Dohm, D.J., Sardelis, M.R., Oguinn, M.L., Andreadis, T.G., and Blow, J.A. (2005). An update on the potential of north American mosquitoes (Diptera: Culicidae) to transmit West Nile Virus. J. Med. Entomol. *42*, 57–62.

Vanlerberghe, V., Villegas, E., Oviedo, M., Baly, A., Lenhart, A., McCall, P.J., and Van der Stuyft, P. (2011). Evaluation of the effectiveness of insecticide treated materials for household level dengue vector control. PLoS Neglect. Trop. Dis. *5*, e994.

Vazquez-Prokopec, G.M., Vanden Eng, J.L., Kelly, R., Mead, D.G., Kolhe, P., Howgate, J., Kitron, U., and Burkot, T.R. (2010a). The risk of West Nile Virus infection is associated with combined sewer overflow streams in urban Atlanta, Georgia, USA. Environ. Health Perspect. *118*, 1382–1388.

Vazquez-Prokopec, G.M., Kitron, U., Montgomery, B., Horne, P., and Ritchie, S.A. (2010b). Quantifying the spatial dimension of dengue virus epidemic spread within a tropical urban environment. PLoS Neglect. Trop. Dis. *4*, e920.

Voge, N.V., Sanchez-Vargas, I., Blair, C.D., Eisen, L., and Beaty, B.J. (2013). Detection of dengue virus NS1 antigen in infected *Aedes aegypti* using a commercially available kit. Am. J. Trop. Med. Hyg. *88*, 260–266.

Walton, W.E. (2012). Design and management of free water surface constructed wetlands to minimize mosquito production. Wetl. Ecol. Manag. *20*, 173–195.

Whelan, P.I. (1998). The prevention of mosquito breeding in sewage treatment facilities. Bull. Mosquito Control Assoc. Austr. *10*, 19–28.

Wolbers, M., Kleinschmidt, I., Simmons, C.P., and Donnelly, C.A. (2012). Considerations in the design of clinical trials to test novel entomological approaches to dengue control. PLoS Neglect. Trop. Dis. *6*, e1937.

Biological Control of Arbovirus Vectors

17

Thomas Walker and Steven P. Sinkins

Abstract

Biological control methods have been used against arbovirus vectors, in particular mosquitoes, to help prevent the transmission of diseases such as dengue and chikungunya. Biocontrol is can be an attractive alternative to more conventional vector control strategies that involve insecticides due to its potential to have minimal impact on the environment. Numerous methods have been employed against field mosquito populations including natural mosquito predators such as *Toxorhynchites* mosquitoes, copepods and fish. Pathogenic bacteria, viruses and fungi have been proposed or used as mechanisms to control vector mosquitoes by having species-specific lethal effects on target populations. Several extracellular mosquito symbionts – midgut *Asaia* bacteria, midgut bacteria and yeast also have the potential to be used to inhibit arboviral replication in mosquitoes. Significant advances have recently been made in using the endosymbiotic bacterium *Wolbachia* to reduce the vector competence of mosquitoes that transmit arboviruses. In this chapter, we outline the various biocontrol strategies that have been proposed for insect vectors and the current status of research examining their implementation.

Introduction

For most arboviruses the lack of commercially available vaccines or effective drug treatments has meant that control has long focused on the vector. Mosquito control has relied on the use of insecticides targeting the adult or larval stages, in addition to removal of suitable larval sites. The usual response to dengue epidemics, where it is logistically feasible, is space spraying of insecticides; this can be very effective at immediately reducing the adult *Aedes* (*Ae.*) vector population when outbreaks occur, but often has negative effects on non-target organisms and can have damaging effects on the environment. In addition, prolonged use or exposure to additional pesticides often leads to insecticide-resistant mosquitoes. In many developing countries, the cost of implementing an insecticide-based vector control program is also problematic. Thus, there has been a clear stimulus for research into novel, environmentally sustainable biological systems that could reduce the use of, and reliance on, insecticides – both for environmental reasons and to reduce the selection pressure for insecticide resistance.

A number of biological control methods currently exist, or are under development, that could be used in integrated control programmes. Research on efficacy for disease vectors has primarily focused on mosquitoes, given their great importance in human a transmission of arboviruses that cause human disease, but several of these systems could also be applied to the other arthropod vectors of livestock and human arboviruses (sandflies, midges and ticks). The strategies under development can be broadly divided into two categories: population control, seeking to reduce vector numbers just as insecticides do; and population replacement, aiming to reduce or remove their competence to transmit viruses. There are overlaps between these two aims, since some systems could be used in both ways (either depending on the strain/circumstances, or both aims could even be achieved at the same time). Population control agents include natural predators of mosquito larvae such as *Toxorhynchites* mosquitoes, copepods and fish; and pathogenic microbes such as the bacterium *Bacillus thuringiensis israelensis* (Bti), which have been effectively used as 'bio-insecticides'; and the release of incompatible (in effect sterile) males using cytoplasmic incompatibility induced by the endosymbiont *Wolbachia*. More recently, the transinfection of *Aedes* mosquitoes that transmit dengue with certain strains of *Wolbachia* has emerged as a new replacement method to inhibit arbovirus transmission, and the use of extracellular symbionts to reduce vector competence has also been under investigation.

Predators

Species that are natural predators of mosquito larvae and adults have been utilized for the biological control of vectors of dengue and other arboviral diseases. An introduced mosquito predator should have the ability to adapt to the local environment and be able to reproduce and sustain a wild population in order to provide continuous control. Ideally, the species should have a preference for consumption of the target mosquito vector population, since it is very important that any released alien species have minimal ecological impacts on non-target species. Predatory mosquitoes, beetles, copepods, geckos and fish have all been shown to consume larvae or adult mosquitoes that transmit arboviruses.

Toxorhynchites and predatory aquatic insects

Toxorhynchites are large mosquitoes that do not blood fed but whose larvae are predacious on other mosquito larvae (Focks 2007). Predatory species that predate on mosquito species that transmit arboviral diseases include *Toxorhynchites* (*Tx.*) *splendens*, *Toxorhynchites amboinensisis* and *Toxorhynchites rutilus* (Focks *et al.*, 1985; Dominic *et al.*, 1998). The use of *Toxorhynchites* for biocontrol would require an assessment of how effective predation is on particular target mosquito species. The 'Asian tiger' mosquito *Ae. albopictus*, which can transmit DENV, Chikungunya (CHIKV), Yellow Fever Virus (YFV) and various encephalitis viruses such as Eastern Equine Encephalitis (EEE), appears to be particularly susceptible as larvae to predation by *Toxorhynchites* larvae. A study examining predation risk cues in the presence of *Tx. rutilus* revealed *Ae. albopictus* are more vulnerable to predation that other mosquitoes (Kesavaraju *et al.*, 2011). A field study also showed that there was a negative correlation between *Ae. albopictus* and *Tx. splendens* larval populations in cemetery ovitraps in Malaysia (Nyamah *et al.*, 2011), suggesting *Toxorhynchites* could be an effective predator of *Ae. albopictus* field populations. However, field trials of the practicalities of rearing and releasing *Toxorhynchites* for *Aedes* biocontrol are yet to be conducted.

Predatory water beetle species have also been proposed for biocontrol of mosquito larvae. *Acilius sulcatus* were shown to be very efficient predators of *Culex* (*Cx.*) *quinquefasciatus* larvae under field conditions (Chandra *et al.*, 2008). *Rhantus sikkimensis* also showed a high level of predatory consumption of mosquito larvae (Aditya *et al.*, 2006). The aquatic nymphs of the dragonfly *Brachytron pratense* were shown to efficiently consume *Anopheles subpictus* larvae highlighting the mosquito biocontrol potential of dragonflies under field conditions (Chatterjee *et al.*, 2007). The biocontrol potential of the nepidae bug, *Nepa cinerea*, against immature stages of mosquito species including *Aedes aegypti* was also demonstrated under laboratory conditions (Singh and Singh, 2004). Again though, field trials are still needed to assess whether artificially increasing the population densities of these predators could make significant contributions to mosquito control.

Fish

Mosquito species that breed in temporary pools in remote areas that dry up seasonally are problematic to control. Fish have been considered a potential biocontrol method and have been employed for controlling mosquito larvae since the 1930s. The most widely used fish for mosquito control has been *Gambusia affinis* (the mosquitofish), which is closely related to the common guppy. Mosquitofish are surface feeders and efficiently consume mosquito larvae and pupae. They are capable of inhabiting a wide variety of permanent and semi-permanent fresh water habitats including rice fields and wetlands. Additional species of fish that prey on mosquito larvae include *Ambassis nama*, *Parambassis ranga*, *Colisa fasciatus*, *Esomus danricus* and *Aplocheilus panchax*. A large number of studies have investigated the potential use of fish to control vectors of malaria (Ghosh and Dash, 2007). The annual killifish, *Nothobranchius guentheri*, has been investigated as a potential species to eradicate mosquito larval populations as this species can undergo suspended animation or diapause during the embryonic stages to survive periodic drought. Semi-field experiments with *Culex quinquefasciastus* larvae revealed that complete elimination of mosquito larvae could be achieved in the presence of 3 fish per m^2 of pond surface area (Matias and Adrias, 2010).

The use of fish to target the most important arboviral disease vector mosquitoes has been more limited. *Poecilia reticulata* and *Gambusia affinis* were shown to be a potential effective intervention strategy when used in indoor cement tanks to control *Ae. aegypti* larvae following severe outbreaks of Chikungunya is South India in 2006 (Ghosh *et al.*, 2011). Integrated dengue control programmes in Brazil have also implemented larvivorous fish to reduce *Ae. aegypti* populations (Regis *et al.*, 2013). The presence of predatory fish in water has also been shown to modify the oviposition behaviour of gravid *Ae. aegypti* females. In the presence of the larvivorous fish *Betta splendens*, *Ae. aegypti* oviposition was significantly reduced (Pamplona *et al.*, 2004). However, no significant difference was observed for *Ae. aegypti* oviposition with or without the larvivavrous fish *Poecilia reticulata*. Approaches to use natural fish extracts have also been explored for mosquito biocontrol. For example, toxins isolated from puffer fish such as *Arothron hispidus* have recently been shown to be effective in killing the larvae and eggs of mosquitoes including *Ae. aegypti* (Samidurai and Mathew, 2013).

Copepods

Mesocylops copepods are small freshwater crustaceans that are highly predatory on mosquito larvae (Marten and Reid, 2007). Copepods are relatively easy and inexpensive to mass-produce and have minimal effects on the environment. Early trials in the 1980s using *Mesocylops aspericornis* copepods failed to significantly reduce mosquito larval populations (Riviere *et al.*, 1987). However, *Ae. albopictus* was successful eradicated from Tahiti using copepods (Marten, 1990). The largest and most successful application of copepods for mosquito control was carried out in Vietnam to target *Ae. aegypti* (Vu *et al.*, 1998, 2005). From an initial introduction of copepods into a village in northern Vietnam in 1993, *Ae. aegypti* was eradicated from large surrounding areas by 2000, and dengue transmission was not being documented (Vu *et al.*, 2005). Copepod biocontrol was still being actively undertaken by communities in Vietnam even after the official intervention had ceased (Kay *et al.*, 2010; Sinh Nam *et al.*, 2012). However, further trials in Honduras using copepods failed to replicate the success seen in Tahiti and Vietnam (Marten and Reid, 2007), suggesting copepod biocontrol requires effective integration into existing health service programmes and extensive community engagement. There are also limitations in terms of mosquitoes to which it can be efficiently applied, since not all species make use as larval sites of the containers or water storage tanks that provide ideal breeding grounds for copepods (Hales and van Panhuis, 2005). Another limitation is that any biocontrol strategy that targets larval source reduction will not be immediately effective during arboviral outbreaks, as it requires time to significantly impact the adult mosquito population.

Pathogens

Bacteria, viruses and fungi that are pathogenic to mosquitoes have been used or considered to have potential for biocontrol purposes. An important consideration for biocontrol using these pathogenic agents is their species-specificity and potential effects on non-target organisms. The entomopathogenic bacterium *Bacillus thuringiensis israelensis* (Bti) has a long history of being successfully used in integrated mosquito vector control programmes. More recently research has been undertaken with both pathogenic fungi and pathogenic viruses that could lead to potentially safe and effective mechanisms to control insect populations that transmit arboviruses (again the emphasis of this research to date has very much been focused on mosquitoes).

Bti

Bacillus thuringiensis israelensis (Bti) are Gram-positive, spore-forming bacteria that release insecticidal toxins and virulence factors that selectively target the larval stages of insects, and Bti is considered a safe alternative to chemical insecticides for the control of mosquito larvae (Becker, 1997; Lacey, 2007). The four main insecticidal toxic proteins, Cry4A, Cry4B, Cry11A and Cyt1A, are secreted as water-soluble proteins but are then inserted into or translocate across host cell membranes (Crickmore *et al.*, 1998; Schnepf *et al.*, 1998). Binding of Cry toxins to the insect larval midgut membrane after ingestion leads to cell lysis and death (de Maagd *et al.*, 2001; Bravo *et al.*, 2007). Bti is formulated in a variety of ways including powders, granules, pellets and tablets. The length of time Bti persists in the environment after application varies (Lacey, 2007) but residual efficacy 16 weeks after application has been documented (Mulla *et al.*, 2004). The use of Bti against the most important arbovirus vector species has been relatively limited but tablets and granular Bti have been used to reduce numbers of *Ae. aegypti* larvae (Novak *et al.*, 1985; Armengol *et al.*, 2006; Ritchie *et al.*, 2010). Bti has also been applied synergistically in combination with predatory *Mesocyclops* copepods (Tietze *et al.*, 1994; Chansang *et al.*, 2004; Kittayapong *et al.*, 2006) and with insecticides (Seleena *et al.*, 2001). The density of *Ae. albopictus* mosquitoes was also significantly decreased by application of Bti (Lam *et al.*, 2010).

Bti use can have limitations in the longer term since resistance to Bti toxins can be selected (Georghiou and Wirth, 1997) and environmental factors such as water temperature can have an influence on efficacy (Elcin, 1995). In addition, the use of Bti in large mosquito breeding sites in urban environments is logistically demanding (Gomez-Dantes and Willoquet, 2009), and therefore slow-release, long-lasting formulations are needed to minimize the need for frequent application. Bti, like any larval source reduction strategy, also requires time to significantly impact the adult mosquito population. This delay between implementation and impact on disease transmission makes larval control interventions less suitable for bringing arboviral epidemics under control than methods that have lethal effects on adult mosquitoes (Gratz, 1991).

Densoviruses

Mosquito densoviruses (MDVs) are pathogenic DNA parvoviruses that replicate in the nuclei of mosquito cells. MDVs have been proposed as potential biocontrol agents due to their stability in the environment and being highly specific to target mosquito species (Carlson *et al.*, 2006). These non-enveloped viruses infect and kill mosquito larvae in a dose-dependent manner; any MDV-infected larvae that do pupate and eclose will have a reduced mean adult lifespan. Therefore, many infected individuals do not survive longer than the extrinsic incubation period for arboviruses, which reduces their vectorial capacity for disease transmission. Female mosquitoes infected with MDVs also have the capacity to transmit the virus vertically to their progeny suggesting MDVs could persist and spread through wild mosquito populations. A laboratory study compared the effect of three genetically distinct MDVs on *Ae. aegypti* (Ledermann *et al.*, 2004). Exposure to *Aedes aegypti* densovirus (AeDNV) resulted in larval mortality rates of up to 75%. Lower larval mortality (~30%) was observed for viruses isolated from *Haemagogus equinus* virus (HeDNV) and Peruvian *Aedes albopictus* cells (APeDNV) suggesting there are differences in MDV pathogenicity. An additional study showed that under laboratory conditions four MDVs were shown to result in more than 80% mortality of *Ae. aegypti* larvae (Hirunkanokpun *et al.*, 2008). The few surviving adult female mosquitoes produced up to 50% vertical transmission of MDV to their progeny. The efficacy and sustainability of AeDNV as a biocontrol agent was tested in and among oviposition sites in large laboratory cages but AeDNV was shown to not significantly reduce *Ae. aegypti* egg densities (Wise de Valdez *et al.*, 2010). Densoviruses have also been isolated from *Cx. pipiens* mosquitoes, named *Cp*DNV, and were shown to produce significant larval mortality (Jousset *et al.*, 2000; Baquerizo-Audiot *et al.*, 2009). Attempts to replicate CpDNV in various insect cell lines revealed specificity to Culicidae mosquito cells (Jousset *et al.*, 2000).

A direct inhibitory effect of MDVs on arboviral replication has also been demonstrated which could work synergistically with pathogenic effects. In *Ae. albopictus* C6/36 cells, DENV infection levels were significantly reduced in cell lines infected with MDVs (Burivong *et al.*, 2004; Wei *et al.*, 2006). Brevidensoviruses are single-stranded DNA MDVs that have been shown to inhibit the growth of dengue virus in cell culture (Mosimann *et al.*, 2011). However, evidence of co-infections between DENV, Japanese encephalitis virus (JEV) and MDV was observed in cell culture (Kanthong *et al.*, 2010). Co-infection of MDV and CHIKV in adult *Ae. aegypti* mosquitoes (Sivaram *et al.*, 2010) also suggests that MDVs may not be effective against all arboviruses. An alternative strategy of using MDVs for mosquito biocontrol proposes to utilize the potential of recombinant MDVs to be an effective delivery system to induce RNA interference (RNAi) in *Ae. albopictus* larvae (Gu *et al.*, 2011).

Fungi

Entomopathogenic fungi produce infective spores known as conidia that can attach and penetrate the cuticle of mosquitoes. Spores proliferate within mosquitoes and release toxins resulting in death. A large number of entomopathogenic fungi have been identified (Scholte *et al.*, 2004; Roy *et al.*, 2006) and have been used successfully in the control of agricultural insect pests (Shah and Pell, 2003). Fungi have been proposed for the control of malaria mosquito vectors (Blanford *et al.*, 2005; Knols *et al.*, 2010) and species such as *Beauveria bassiana* are also highly pathogenic to *Ae. aegypti* (Paula *et al.*, 2011; Paula *et al.*, 2011). Recently the effect of *B. bassiana* on *Ae. aegypti* under semi-field conditions was tested and revealed that mosquito survival was reduced by ~60–90% (Darbro *et al.*, 2012). Mosquitoes infected with *B. bassiana* also had 30% less contact with human blood feeding volunteers, suggesting this species could be used for biocontrol of *Ae. aegypti*. *Metarhizium anisopliae* causes high rates of mortality in adults of both *Ae. aegypti* and *Ae. albopictus* (Scholte *et al.*, 2007). A further study demonstrated that application of *B. bassiana* caused a reduction in the life span of *Ae. aegypti* and inhibited dengue virus replication in the mosquito midgut (Dong *et al.*, 2012). These fungal infections are thought to activate the Toll and JAK-STAT mosquito immune pathways, which will produce antimicrobial effector molecules such as cecropins that have anti-viral activity. Fungal infections can also significantly reduce the fecundity of *Ae. aegypti* under lab conditions (Reyes-Villanueva *et al.*, 2011) and kill *Ae. aegypti* eggs (Luz *et al.*, 2008). Furthermore, insecticide-resistant malaria vectors infected with fungi are also more susceptible to insecticides (Howard *et al.*, 2010).

Since entomopathogenic fungi predominantly target adult mosquitoes, they represent a promising biocontrol strategy for arboviral diseases. *M. anisopliae* and *B. bassiana* fungi can infect mosquitoes early in life but mortality occurs in adulthood. As adult mosquitoes can reproduce, selection pressure for resistance is likely to be less intense when compared to rapid-killing insecticides, so the evolution of fungus resistance is predicted to be much slower than the evolution of insecticide resistance (Knols *et al.*, 2010). There may also be a lower risk of resistance developing because several different toxins kill mosquitoes (Scholte *et al.*, 2007). As *B. bassiana* and *M. anisopliae* are commercially produced for biocontrol of agricultural pests, regulatory approval for the use against mosquito arboviral vectors will likely be relatively straightforward.

Despite great optimism for the use of fungi for arboviral control, there are limited studies describing the effect on wild mosquito populations. Further research is needed to determine the viability, infectivity and persistence of fungal spores in field settings (Knols *et al.*, 2010; Mnyone *et al.*, 2010). There are logistical issues that need to be solved and in particular research is needed on the optimal methods of delivery of fungal spores (Darbro and Thomas, 2009). As with any vector control intervention for arboviral diseases, a large-scale field trial is required to determine the efficacy of entomopathogenic fungi on virus transmission and disease suppression during outbreaks.

Symbionts

Midgut and other extracellular microbiota

Naturally acquired, resident microbiota that inhabit insect guts and other tissues can have major effects on vector competence for important pathogens. A general mechanism by which gut microbes can influence pathogen infection is thought to be through a basal up-regulation of immune genes (Dong *et al.*, 2009). *Plasmodium* (malaria parasite) transmission was shown to be directly inhibited by an *Enterobacter* bacterium isolated from wild *An. gambiae* mosquito populations in Zambia (Cirimotich *et al.*, 2011). This pathogen inhibition was shown to occur prior to invasion of the midgut epithelium and was mediated through a mosquito-independent interaction caused by bacterial generation of reactive oxygen species. As with malarial parasites, a critical stage in arboviral transmission is when the virus reaches the midgut after ingestion of an infectious bloodmeal. The presence of midgut bacteria has been demonstrated to influence viral dissemination and have an inhibitory effect on arboviral transmission. For example, a significant decrease in DENV infection was observed in *Ae. aegypti* that had field-derived bacterial symbionts in their midgut (Ramirez *et al.*, 2012). Reintroduction of some of these bacteria, either through the bloodmeal or sugar meal, led to a significant decrease in viral replication in the case of one bacterial isolate in particular, a *Proteus* species (Prsp_P). The mechanistic basis of the anti-dengue effect is likely to be through basal immune activation, but they could also potentially be directly influencing viral replication/infectivity by producing inhibitory metabolites; research is needed to examine this hypothesis. The activation of the mosquito immune response by DENV infection also regulates the density of microbiota in the midgut. Resident bacteria in mosquitoes do not always provide inhibitory effects on arboviruses. The midgut of *Ae. aegypti* adults is dominated by Gram-negative proteobacteria such as *Serratia odorifera* which can actually enhance susceptibility to DENV (Apte-Deshpande *et al.*, 2012).

Asaia are acetic acid alpha-proteobacteria that have been shown to be stably associated with the malaria vectors *Anopheles* (*An.*) *stephensi* (Favia *et al.*, 2007) and *An. gambiae* (Damiani *et al.*, 2010). *Asaia* bacteria have been isolated from *An. stephensi*, cultured in cell-free media and then transformed with foreign DNA. As high infection densities can occur the midgut and salivary glands of adult female mosquitoes, *Asaia* has been proposed as a potential biocontrol agent (Favia *et al.*, 2008). *Asaia* infections can be vertically transmitted to progeny or horizontally transmitted through oral feeding routes or during mating, which would aid the dissemination and maintenance of modified or selected virus-inhibiting isolates in wild populations. Recently it has been shown that larval development is delayed in *Anopheles* mosquitoes deprived of *Asaia* (Chouaia *et al.*, 2012), suggesting these bacteria may play beneficial roles in larvae. *Asaia* has also been shown to be capable of cross-colonizing other sugar-feeding insects including *Ae. aegypti* (Crotti *et al.*, 2009). The tissue tropism resulting from sugar feeding indicated that *Asaia* can colonize the midgut and salivary glands of *Ae. aegypti*. Furthermore, *Asaia* was identified in the ovaries and eggs of *Ae. aegypti* (Gusmao *et al.*, 2010) and was shown to infect *Ae. albopictus* (Chouaia *et al.*, 2010; Minard *et al.*, 2013). The occurrence of *Asaia* in mosquito midgut cells (Capone *et al.*, 2013) suggests the possibility of reducing vectorial competence, if isolates of *Asaia* isolates can be identified or selected that inhibit viruses such as DENV. Alternatively it could be modified to express and secrete inhibitory effectors, for example using dsRNA to induce RNAi, although such a strategy crosses into paratransgenic rather than purely biological control.

Mosquitoes also carry yeast including *Candida, Rodotorula* and *Cryptococcus* species. The yeast *Wickerhamomyces anomalus*, previously known as *Pichia anomala*, is stably associated with *An. stephensi* and infections are found in all life stages including the female midgut (Ricci *et al.*, 2011). Additional mosquito species have been shown to be infected with *W. anomalus* (Ricci *et al.*, 2011) and two yeast species were identified in *Ae. aegypti* with *Pichia* in the midgut and *Candida* in the midgut and ovaries (Gusmao *et al.*, 2010). Although the presence of symbiotic yeast in mosquito tissues means that a natural influence on vector competence is possible, further research is needed to examine the hypothesis, and whether yeast can be manipulated to act as arboviral biocontrol agents.

For all extracellular mosquito symbionts, research is needed on how they could be optimally delivered in the field – in particular the effectiveness and practicality (cost and longevity) of using sugar bait solutions loaded with virus-inhibiting microbiota. If modification to express heterologous anti-viral effectors is to be employed, it will also be essential to ensure that the expressed and secreted effectors are not unduly harmful to the bacteria themselves or to the host insect.

Wolbachia and cytoplasmic incompatibility

Wolbachia pipientis is an intracellular maternally inherited bacterium found in a range of arthropods, including many mosquito species, where it manipulates host reproduction by inducing crossing sterilities known as cytoplasmic incompatibility (CI). In its basic unidirectional form, CI is expressed as offspring early embryonic arrest when uninfected females mate with infected males – in other words *Wolbachia* modifies the sperm of infected males (Turelli, 2010). However a rescue function when *Wolbachia* is present in the egg allows the offspring of infected females to develop normally, regardless of the infection status of the males with which they mate. Thus in a mixed population infected females will produce a greater mean number of offspring. Once a threshold population frequency has been exceeded, which depends on the maternal transmission efficiency, penetrance of CI, and fitness costs (if any), the reproductive advantage of infection will steadily increase over time, and infection frequency will rise concordantly.

Wolbachia-induced cytoplasmic incompatibility can be used for population suppression by release of males that are incompatible with wild females, as an alternative to the sterile insect technique (SIT). SIT requires the use of irradiation to produce male sterility, which is usually detrimental to host fitness and mating competitiveness, while the incompatible insect technique (IIT) utilizing *Wolbachia* does not require irradiation and thus males are expected to be more competitive in the field. IIT has been examined for several Culicine mosquito species that naturally carry *Wolbachia*. In mosquitoes of the *Cx. pipiens* complex there are complex patterns of incompatibility when populations from different localities are crossed with each other, which can be unidirectional or bidirectional, and this was successfully exploited in a first field trial on *Cx. quinquefasciatus* mosquitoes, with a village population in Burma being successfully eliminated, at least temporarily (Laven, 1967). More recently its applicability to control of the same species in islands of the Indian Ocean such as La Réunion was demonstrated in cage suppression trials (Atyame *et al.*, 2011), following introgression of the incompatible *Wolbachia* into the target genetic background of the target population in order to maximize fitness and competitiveness.

Incompatible lines have been generated by microinjection of *Wolbachia* strains from other species, and fitness/male competiveness characterized, in *Ae. albopictus* (Xi *et al.*, 2006; Blagrove *et al.*, 2012, 2013; Moretti and Calvitti, 2013), as shown in Table 17.1. In *Ae. polynesiensis* small-scale pilot releases have been conducted using an incompatible line strain generating by introgression of a different *Wolbachia* strain from a sibling species, *Ae. riversi* (Brelsfoard *et al.*, 2008; Chambers *et al.*, 2011; O'Connor *et al.*, 2012). Studies of abundance patterns and dispersal using mark–release–recapture have been conducted for both species (Bellini *et al.*, 2010; Mercer *et al.*, 2012; Hapairai *et al.*, 2013) and are vital for estimating the suitability of a target and the scale of release that will be required for elimination. In general, IIT should be regarded as a strategy that is well suited to specific situations, particularly ecologically or physically isolated populations, such as those on oceanic islands, or to combat recent invasions. However it would likely be difficult to scale up the technique for use on a very large scale, since sex separation is required (using that fact that males eclose earlier on average than females and are smaller than female pupae, allowing physical sorting by size). In theory the technique could be combined with a genetic or transgenic sexing strain to allow sex separation on a large scale, although it would then no longer fall strictly into the biological control category, and would be subject to the additional restrictions on the use of genetic modification that are in place in many countries.

Wolbachia-mediated arboviral inhibition

In the presence of certain non-native strains of *Wolbachia* (transinfections), a strong inhibition of the development or dissemination of arboviruses has been observed in *Aedes* mosquitoes (Moreira *et al.*, 2009; Walker *et al.*, 2011; Blagrove *et al.*, 2012; van den Hurke, 2012; Blagrove *et al.*, 2013), as shown in Table 17.2. In *Ae. aegypti*, which is not naturally infected with *Wolbachia*, the *w*AlbB strain was established using embryo cytoplasm transfer from the closely related species *Ae. albopictus* (Xi *et al.*, 2005). Creating transinfections is more difficult when the donor and recipient species are not closely related (Braig *et al.*, 1994), so for the transfer of *Wolbachia* strains from *Drosophila* into *Ae. aegypti*, pre-adaptation of these *Wolbachia* in *Aedes* cell lines was used, followed by purification from the cell lines and microinjection into embryos (McMeniman *et al.*,

Table 17.1 Crossing relationships produced by various *Wolbachia* strains in *Ae. albopictus*

	Uninfected ♂	*w*AlbAandB ♂	*w*Mel ♂	*w*Pip ♂
Uninfected ♀ (cured)	Uninfected progeny	No progeny (unidirectional CI)	No progeny (unidirectional CI)	No progeny (unidirectional CI)
*w*AlbAandB ♀ (wildtype)	*w*AlbAandB progeny	*w*AlbAandB progeny (rescue cross)	No progeny (bidirectional CI)	No progeny (bidirectional CI)
*w*Mel ♀	*w*Mel progeny	No progeny (bidirectional CI)	*w*Mel progeny (rescue cross)	*Not done*
*w*Pip ♀	*w*Pip progeny	No progeny (bidirectional CI)	*Not done*	*w*Pip progeny (rescue cross)

Table 17.2 Summary of effects of native and introduced (transinfecting) *Wolbachia* strains in the *Aedes* mosquito vectors of dengue and chikungunya: Phenotypic effects of various *Wolbachia* strains in *Aedes aegypti* and *Aedes albopictus*

	Aedes aegypti	*Aedes albopictus*
*w*MelPop-CLA (origin *Drosophila melanogaster*)	Strongly inhibits DENV, CHIKV[a] Halves lifespan Unidirectional CI with wildtypes[a]	Very high fitness cost, low egg hatch within colony[d]
*w*Mel (origin *D. melanogaster*)	Strongly inhibits DENV, CHIKV[b] Some fitness costs[b] Unidirectional CI with wildtypes[b]	Strongly inhibits DENV, CHIKV[e] No major fitness costs detectable[e] Bidirectional CI with wildtypes[e]
*w*AlbB (origin *Ae. albopictus*)	Inhibits DENV[c] Some fitness costs[c] Unidirectional CI with wildtypes	Wildtype strain (with *w*AlbA); does not prevent DENV, CHIKV transmission
*w*Pip (origin *Cx. pipiens*)	*Not done*	Some fitness costs; males competitive[f] Bidirectional CI with wildtypes[f]
*w*Ri (origin *Drosophila simulans*)	*Not done*	Bidirectional CI with wildtypes[g]

[a]McMenimen *et al*., 2008; Moreira *et al*., 2009.
[b]Walker *et al*., 2011.
[c]Xi *et al*., 2005.
[d]Suh *et al*., 2009.
[e]Blagrove *et al*., 2012; Blagrove *et al*., 2013.
[f]Calvitti *et al*. 2010; Moretti and Calvitti, 2013.
[g]Xi *et al*., 2006.

2008, 2009; Walker *et al.*, 2011). Stable transinfections with two *Drosophila Wolbachia* strains, *w*MelPop-CLA (cell line -adapted) and *w*Mel, have been created in this manner. The vector competence of *Ae. aegypti* for DENV – dissemination of the virus and the presence of infectious DENV in mosquito saliva – was reduced in the presence of all three transinfecting *Wolbachia* strains (Moreira *et al.*, 2009; Bian *et al.*, 2010; Walker *et al.*, 2011). The degree of inhibition of DENV in adult females does however vary, with the strongest effects seen for the virulent *w*MelPop-CLA (Moreira *et al.*, 2009; Walker *et al.*, 2011). Furthermore, transmission of the DENV-2 strain was completely blocked under laboratory conditions for *w*Mel and *w*MelPop-CLA infected mosquitoes (Walker *et al.*, 2011).

Ae. aegypti mosquitoes are also competent vectors of additional arboviruses including the alphavirus CHIKV and flavivirus YFV (van den Hurk *et al.*, 2010). The *w*Mel and *w*MelPop-CLA strains appear to have variable effects on the vector competence to these arboviruses. Overall, infection and dissemination rates of CHIKV were significantly reduced by the *w*Mel strain following exposure to viraemic bloodmeals (van den Hurk *et al.*, 2012). Intrathoracic injections, which bypass the midgut barrier, revealed contrasting results with no significant effect of the *w*Mel strain on both CHIKV and YFV (van den Hurk *et al.*, 2012). However, the virulent *w*MelPop-CLA strain did inhibit YFV infection four-fold, highlighting the complex relationship between arboviruses, *Wolbachia* strains and the method of virus exposure. An inhibitory effect of naturally occurring *Wolbachia* on viral transmission has also been observed, for West Nile Virus (WNV) in *Cx. quinquefasciatus* and *Cx. pipiens*, which carry *w*Pip *Wolbachia*. Removal of *w*Pip resulted in a 2- to 3-fold increase in WNV transmission and higher viral titres (Glaser and Meola, 2010).

A chronic, large-scale host immune gene up-regulation occurs in the presence of *w*MelPop-CLA in *Ae. aegypti* (Kambris *et al.*, 2009; Moreira *et al.*, 2009). This up-regulation includes effector genes such as cecropins that are known to inhibit pathogens such as DENV. Gene knockdown studies have demonstrated that upregulated immune genes can contribute to the pathogen inhibition phenotype observed in naturally *Wolbachia*-uninfected species (Kambris *et al.*, 2010; Pan *et al.*, 2012), and also implicated elevated levels of Reactive Oxygen Species (Pan *et al.*, 2012). However, no such immune up-regulation was observed when *w*Mel was stably transferred from *Drosophila* into *Ae. albopictus*, but complete blockage of transmission of both DENV and CHIKV occurred in lab challenges as measured by number of infectious virions detectable in the saliva (Blagrove *et al.*, 2012, 2013). Furthermore, the *w*Mel and *w*MelPop strains in their native *Drosophila melanogaster* host do not induce immune gene up-regulation (Rances *et al.*, 2012; Wong *et al.*, 2011), but do protect against pathogenic *Drosophila* viruses (Hedges *et al.*, 2008; Texeira *et al.*, 2008, Osborne *et al.*, 2009). This suggests that other mechanisms beyond immune up-regulation are contributing to *Wolbachia*-mediated viral inhibition, especially where the host species is naturally adapted to the presence of *Wolbachia.*

In vitro experiments have shown a strong correlation between *Wolbachia* density and DENV inhibition (Frentiu *et al.*, 2010). *Ae. albopictus* mosquitoes naturally carry two strains of *Wolbachia* (Sinkins *et al.*, 1995), *w*AlbA and *w*AlbB, and are competent vectors of DENV (Mitchell *et al.*, 1987). The introduced *w*Mel strain from *Drosophila*, which blocked DENV and CHIKV transmission in the lab, is maintained at much higher densities in *Ae. albopictus* than the native *Wolbachia* (Blagrove *et al.*, 2012). Those *Wolbachia* strains that grow to high densities in transinfected mosquito species are likely to have the greatest impact on arboviral transmission. The absence of DENV in tissues containing high *w*MelPop-CLA infection densities in *Ae. aegypti* (Moreira *et al.*, 2009) also supports the correlation between *Wolbachia* density and arboviral inhibition. Likewise, heavy infections of *w*Mel in *Ae. aegypti* salivary glands may explain the complete blockage of DENV transmission observed in laboratory experiments (Walker *et al.*, 2011). The correlation between higher *Wolbachia* density and increased viral inhibition supports the hypothesis that host resource competition between *Wolbachia* and arboviruses may play an important role in the phenotype (Moreira *et al.*, 2009). For example, cholesterol is known to be an important component required for replication of flaviruses such as DENV (Carro and Damonte, 2013), as well as by *Wolbachia* itself for membrane integrity, and is a limited resource since it must be obtained by insects from their diet. In *Drosophila* it was recently demonstrated that dietary cholesterol levels can have a modulating effect on *Wolbachia* inhibition of viruses (Caragata *et al.*, 2013).

Wolbachia population replacement to reduce vector competence

High maternal transmission rates and high penetrance of CI also occur for transinfected *Wolbachia* strains in *Ae. aegypti* (McMeniman *et al.*, 2009; Walker *et al.*, 2011; Xi *et al.*, 2005) (Table 17.3). These phenotypic effects, which would allow *Wolbachia* strains to spread through mosquito populations, are maintained in genetically outbred diverse populations (Walker *et al.*, 2011; Yeap *et al.*, 2011). Therefore, rapid invasion of uninfected *Ae. aegypti* field populations would occur in the absence of major mosquito fitness costs. However, the *w*MelPop-CLA strain imparts much greater fitness costs than does *w*Mel. Adult mosquito lifespan is reduced by around 50% by *w*MelPop-CLA (McMeniman *et al.*, 2009; Yeap *et al.*, 2011), compared to around 10% for the *w*Mel strain (Walker *et al.*, 2011). Lifespan reduction is an advantage in terms of reducing disease transmission, given the viral extrinsic incubation period will not be achieved in short-lived females, but significantly raises the threshold frequency and thus makes it much more difficult to introduce such a strain into wild populations. The fecundity of *w*MelPop-CLA infected female mosquitoes is also reduced by around 50% relative to uninfected wildtype and *w*Mel-infected mosquitoes under semi-field conditions (Walker *et al.*, 2011). Thirdly, the viability of *w*MelPop-CLA infected eggs is significantly reduced during periods of embryonic quiescence (McMeniman and O'Neill, 2010). As the *w*Mel strain does not significantly affect *Ae. aegypti* egg viability (Walker *et al.*, 2011), each *Wolbachia* strain could be used effectively under different environmental conditions. For example, *Ae. aegypti* field populations that are present in areas with a long dry season could be suppressed by releasing the *w*MelPop-CLA strain, which would prevent hatching of the next generation after the dry season (McMeniman and O'Neill, 2010).

Wolbachia invasion of mosquito populations is dependent on an unstable threshold infection level that must be exceeded to allow the infection to spread to fixation (Turelli, 2010). CI induction will allow the infection to reach this unstable point unless significant fitness costs imposed by *Wolbachia* strains raise the unstable point. The *w*MelPop-CLA strain has significant associated fitness costs, and this has raised doubts whether this strain could invade wild *Ae. aegypti* populations (Turelli, 2010). Purpose built semi-field cages were used to determine the invasive potential of *w*MelPop-CLA and *w*Mel (Walker *et al.*, 2011). The *w*Mel strain increased rapidly from an initial starting frequency of 0.65 and reached fixation in one cage within 30 days (Walker *et al.*, 2011). In contrast, the *w*MelPop-CLA strain increased at a slower rate, reaching fixation in one cage after 40 days. Deterministic age-structured models predicted a similar pattern with faster invasion occurring for *w*Mel due to the smaller associated fitness costs (Walker *et al.*, 2011).

The first open releases of *w*Mel-infected mosquitoes into wild *Ae. aegypti* populations occurred in 2011 in two locations near Cairns in North-eastern Australia (Hoffmann *et al.*, 2011). Public support was achieved through community engagement and informing the public about this biocontrol approach to controlling dengue (McNaughton, 2012). Safety concerns were also addressed through research showing that *Wolbachia* is unlikely to infect non-target insects or infect mosquito predators (Popovici *et al.*, 2010). There was also no evidence that *Wolbachia* can induce an immune response in humans upon being bitten by *Wolbachia*-infected *Ae. aegypti* mosquitoes (Popovici *et al.*, 2010). A formal risk assessment for the release of *Wolbachia*-infected mosquitoes was followed by regulatory approval (Hoffmann *et al.*, 2011; Murphy *et al.*, 2010) prior to release in two suburbs of Cairns, Australia.

Mosquitoes were released for 10 weeks during the wet season and monitored using PCR-based screening to determine the change in *Wolbachia* frequencies in the release sites. In both locations, the *w*Mel strain successfully invaded *Ae. aegypti* populations reaching fixation within a few months following releases (Hoffmann *et al.*, 2011). The *w*Mel infection resulted in small fitness costs allowing an unstable equilibrium frequency of around 0.3 (30%) for successful invasion (Hoffmann *et al.*, 2011). This first open release demonstrated that *Wolbachia* is able to invade and persist in wild *Ae. aegypti* populations. Wild *w*Mel-infected *Ae. aegypti* populations could also be utilized for collection of material to be released in additional locations near the release sites. The two release site locations were also screened in September 2011 after the dry season to determine the *Wolbachia* infection frequency and over 98% of *Ae. aegypti* mosquitoes collected were infected with the *w*Mel strain (www.eliminatedengue.com). Additional releases of *w*Mel-infected mosquitoes have now occurred in surrounding suburbs and mosquitoes infected with the *w*MelPop-CLA strain have recently been released in Tri Nguyen Island, Vietnam to determine if this virulent *Wolbachia* strain can establish in wild populations. Laboratory experiments to determine the vector competence of wild, *w*Mel-infected mosquitoes collected from suburbs in Cairns have also been undertaken and a strong inhibitory effect on DENV is maintained in these populations (unpublished data).

The *Wolbachia* *w*Mel transinfection in *Ae. albopictus* induces complete bidirectional incompatibility with wildtype *w*AlbA+*w*AlbB-carrying *Ae. albopictus*, with no significant effects on egg hatch, fecundity, longevity, and relative male mating competitiveness in cage assays (Blagrove *et al.*, 2012,

Table 17.3 Crossing relationships produced by various *Wolbachia* strains in *Ae. aegypti*

	Uninfected ♂	*w*AlbB ♂	*w*Mel ♂
Uninfected ♀ (wildtype)	Uninfected progeny	No progeny (unidirectional CI)	No progeny (unidirectional CI)
wAlbB ♀	*w*AlbB progeny	*w*AlbB progeny (rescue cross)	Not done
*w*Mel ♀	*w*Mel progeny	Not done	*w*Mel progeny (rescue cross)

2013). Like unidirectional CI, bidirectional CI also provides a method to stably introduce *Wolbachia* strains into populations, since bidirectionally incompatible crossing types cannot stably co-exist. If incompatibility is complete and comparative fitness equal, whichever strain is at a local majority would be at a reproductive advantage, because its females would more frequently encounter and mate with males with which they are compatible. Once the *w*Mel strain reaches a population majority, it would be expected to reach local fixation, and to be stable to low-level immigration of wildtype individuals with which they are incompatible. The envisaged strategy for taking the *w*Mel strain to population fixation would consist of heavily male-biased releases (using pupal size sorting), such that as for IIT the release of males incompatible with wild females would also significantly reduce the overall numbers of biting females for the duration of the release period. Since *Ae. albopictus* in common with most mosquitoes shows marked seasonal variations in abundance, releases timed to begin around the start of the rainy season, as populations are just starting to expand, could achieve local *Wolbachia* replacement quickly and efficiently (Hancock *et al.*, 2011). The *w*Mel transinfection would first need to be introgressed into local genetic backgrounds of target populations, to maximize fitness and local adaptation, and ensure that no undesirable traits are introduced into those populations; this will require curing or partial curing of target wildtype lines using antibiotics. Bidirectional CI, and unidirectional CI with strains that impose significant fitness costs and thus higher threshold frequencies, require a larger scale of releases to be effective but do provide the advantage of enhanced control over the geographical extent to which population replacement would occur.

Several other viruses could also be targeted using *Wolbachia*-based strategies, such as WNV and Rift Valley Fever virus (RVFV), a *Phlebovirus* that primarily infects domestic livestock and more rarely humans, has increasing importance due to the influence of climate on the range of vector species (Weaver and Reisen, 2010). Transinfections of higher density *Drosophila Wolbachia* strains in *Culex pipiens* group mosquitoes are likely to provide greater inhibitory effects on viruses such as WNV and RVFV than do their native *Wolbachia*. However, a large number of mosquito species that have been shown to transmit WNV or be potential vectors under laboratory conditions and regional differences in primary vector (Andreadis, 2012), and RVFV has been shown to be experimentally transmitted by a range of mosquito species (Turell *et al.*, 2010) and Phlebotomine sandflies (Dohm *et al.*, 2000; Hoch *et al.*, 1984). Thus it is likely that a range vector species would need to be targeted if *Wolbachia* was to be used for effective biocontrol of these diseases.

Summary and future prospects

Biocontrol strategies to reduce the transmission of arboviral diseases are required to reduce the prolonged use of insecticides currently used as the primary vector control mechanism and to combat the development of resistance. Given that the large majority of arboviral disease transmission occurs in developing countries, biocontrol strategies should aim to be inexpensive and self-sustaining. Effective, safe and environmentally friendly options can include the use of mosquito predators, pathogens that kill mosquitoes and symbionts that inhibit arboviral replication. Predatory copepods have shown the greatest effect to date on natural mosquito populations, having been deployed very successfully to target dengue through the elimination of *Ae. aegypti* populations in rural Vietnam. Efforts to increase the uptake of this strategy, and integrate it effectively into existing control programmes, are now required. Bti has been extensively used due to its ability to selectively kill mosquito larvae but, like copepods, is ineffective as a means to combat rapid arboviral outbreaks. Several other predators and pathogens show promise but much more research is needed before their efficacy as a component of integrated control strategies can be readily determined.

Biocontrol strategies using *Wolbachia* have the potential to fulfil the critical requirement to be cost effective in the long term, since once it has been introduced into a particular population *Wolbachia* will be expected to remain at high frequency and continue to exert disease transmission control with no additional expenditure or effort required. Open releases of *Wolbachia*-infected *Ae. aegypti* mosquitoes are now under way with the aim or reducing DENV transmission in countries such as Vietnam; community engagement and support remains vital component of this roll-out. The long-term stability of the viral inhibition phenotype is an unknown – it might for example decrease with the passage of time as *Wolbachia* and host co-adapt, or alternatively viral escape mutations could be selected. Despite major advances, there are several aspects that will need to be better understood in order to be able to use *Wolbachia* optimally for disease control. A critical question is which *Wolbachia* strains are most effective at balancing the inhibitory effects on arboviral transmission with potential negative effects on host fitness. Different *Wolbachia* strains will have variable densities and tissue tropism in target mosquito vectors, which are likely key factors in determining these phenotypic effects. Mean bacterial density might however be reduced under particular ecological conditions in the wild, which could potentially reduce the effectiveness of viral inhibition. As with all the strategies that have been discussed, the focus of research to date has very much been on mosquitoes, and research to examine their applicability to other arboviral vectors of disease is also much needed.

References

Aditya, G., Ash, A., and Saha, G.K. (2006). Predatory activity of *Rhantus sikkimensis* and larvae of *Toxorhynchites splendens* on mosquito larvae in Darjeeling, India. J. Vector Borne Dis. *43*, 66–72.

Andreadis, T.G. (2012). The contribution of *Culex pipiens* complex mosquitoes to transmission and persistence of West Nile virus in North America. J. Am. Mosq. Control Assoc. *28*, 137–151.

Apte-Deshpande, A., Paingankar, M., Gokhale, M.D., and Deobagkar, D.N. (2012). *Serratia odorifera* a midgut inhabitant of *Aedes aegypti* mosquito enhances its susceptibility to Dengue-2 Virus. PLoS One *7*, e40401.

Armengol, G., Hernandez, J., Velez, J.G., and Orduz, S. (2006). Long-lasting effects of a *Bacillus thuringiensis* serovar israelensis experimental tablet formulation for *Aedes aegypti* (Diptera: Culicidae) control. J. Econ. Entomol. *99*, 1590–1595.

Atyame, C.M., Pasteur, N., Dumas, E., Tortosa, P., Tantely, M.L., Pocquet, N., Licciardi, S., Bheecarry, A., Zumbo, B., Weill, M., and Duron, O. (2011). Cytoplasmic incompatibility as a means of controlling *Culex pipiens quinquefasciatus* mosquito in the islands of the south-western Indian Ocean. PLoS Negl. Trop. Dis. *5*, e1440.

Baquerizo-Audiot, E., Abd-Alla, A., Jousset, F.X., Cousserans, F., Tijssen, P., and Bergoin, M. (2009). Structure and expression strategy of the genome of *Culex pipiens* densovirus, a mosquito densovirus with an ambisense organization. J. Virol. *83*, 6863–6873.

Becker, N. (1997). Microbial control of mosquitoes: management of the upper rhine mosquito population as a model programme. Parasitol. Today *13*, 485–487.

Bellini, R., Albieri, A., Balestrino, F., Carrieri, M., Porretta, D., Urbanelli, S., Calvitti, M., Moretti, R., and Maini, S. (2010). Dispersal and survival of *Aedes albopictus* (Diptera: Culicidae) males in Italian urban areas and significance for sterile insect technique application. J. Med. Entomol. *47*, 1082–1091.

Bian, G., Xu, Y., Lu, P., Xie, Y., and Xi, Z. (2010). The endosymbiotic bacterium *Wolbachia* induces resistance to dengue virus in *Aedes aegypti*. PLoS Pathog. *6*, e1000833.

Blagrove, M.S., Arias-Goeta, C., Failloux, A.B., and Sinkins, S.P. (2012). *Wolbachia* strain *w*Mel induces cytoplasmic incompatibility and blocks dengue transmission in *Aedes albopictus*. Proc. Natl. Acad. Sci. U.S.A. *109*, 255–260.

Blagrove, M.S., Arias-Goeta, C., Di Genua, C., Failloux, A.B., and Sinkins, S.P. (2013). A *Wolbachia w*Mel transinfection in *Aedes albopictus* is not detrimental to host fitness and inhibits Chikungunya virus. PLoS Negl. Trop. Dis. *7*, e2152.

Blanford, S., Chan, B.H., Jenkins, N., Sim, D., Turner, R.J., Read, A.F., and Thomas, M.B. (2005). Fungal pathogen reduces potential for malaria transmission. Science *308*, 1638–1641.

Braig, H.R., Guzman, H., Tesh, R.B., and O'Neill, S.L. (1994). Replacement of the natural *Wolbachia* symbiont of *Drosophila simulans* with a mosquito counterpart. Nature *367*, 453–455.

Brandler, S., and Tangy, F. (2013). Vaccines in development against West Nile virus. Viruses *5*, 2384–2409.

Brandler, S., Ruffie, C., Combredet, C., Brault, J.B., Najburg, V., Prevost, M.C., Habel, A., Tauber, E., Despres, P., and Tangy, F. (2013). A recombinant measles vaccine expressing chikungunya virus-like particles is strongly immunogenic and protects mice from lethal challenge with chikungunya virus. Vaccine *31*, 3718–3725.

Bravo, A., Gill, S.S., and Soberon, M. (2007). Mode of action of *Bacillus thuringiensis* Cry and Cyt toxins and their potential for insect control. Toxicon *49*, 423–435.

Burivong, P., Pattanakitsakul, S.N., Thongrungkiat, S., Malasit, P., and Flegel, T.W. (2004). Markedly reduced severity of Dengue virus infection in mosquito cell cultures persistently infected with *Aedes albopictus* densovirus (AalDNV). Virology *329*, 261–269.

Calvitti, M., Moretti, R., Lampazzi, E., Bellini, R., and Dobson, S.L. (2010). Characterization of a new *Aedes albopictus* (Diptera: Culicidae)-*Wolbachia pipientis* (Rickettsiales: Rickettsiaceae) symbiotic association generated by artificial transfer of the *w*Pip strain from *Culex pipiens* (Diptera: Culicidae). J. Med. Entomol. *47*, 179–187.

Capone, A., Ricci, I., Damiani, C., Mosca, M., Rossi, P., Scuppa, P., Crotti, E., Epis, S., Angeletti, M., Valzano, M., *et al.* (2013). Interactions between *Asaia*, *Plasmodium* and *Anopheles*: new insights into mosquito symbiosis and implications in Malaria Symbiotic Control. Parasit. Vectors *6*, 182.

Caragata, E.P., Rancès, E., Hedges, L.M., Gofton, A.W., Johnson, K.N., O'Neill, S.L., and McGraw, E.A. (2013). Dietary cholesterol modulates pathogen blocking by *Wolbachia*. PLoS Pathog. *9*, e1003459.

Carlson, J., Suchman, E., and Buchatsky, L. (2006). Densoviruses for control and genetic manipulation of mosquitoes. Adv. Virus Res. *68*, 361–392.

Carro, A.C., and Damonte, E.B. (2013). Requirement of cholesterol in the viral envelope for dengue virus infection. Virus Res. *174*, 78–87.

Chambers, E.W., Hapairai, L., Peel, B.A., Bossin, H., and Dobson, S.L. (2011). Male mating competitiveness of a *Wolbachia*-introgressed *Aedes polynesiensis* strain under semi-field conditions. PLoS Negl. Trop. Dis. *5*, e1271.

Chandra, G., Mandal, S.K., Ghosh, A.K., Das, D., Banerjee, S.S., and Chakraborty, S. (2008). Biocontrol of larval mosquitoes by *Acilius sulcatus* (Coleoptera: Dytiscidae). BMC Infect. Dis. *8*, 138.

Chansang, U.R., Bhumiratana, A., and Kittayapong, P. (2004). Combination of Mesocyclops thermocyclopoides and Bacillus thuringiensis var. israelensis: a better approach for the control of *Aedes aegypti* larvae in water containers. J. Vector Ecol. *29*, 218–226.

Chatterjee, S.N., Ghosh, A., and Chandra, G. (2007). Eco-friendly control of mosquito larvae by *Brachytron pratense* nymph. J. Environ. Health *69*, 44–48.

Chouaia, B., Rossi, P., Montagna, M., Ricci, I., Crotti, E., Damiani, C., Epis, S., Faye, I., Sagnon, N., Alma, A., *et al.* (2010). Molecular evidence for multiple infections as revealed by typing of *Asaia* bacterial symbionts of four mosquito species. Appl. Environ. Microbiol. *76*, 7444–7450.

Chouaia, B., Rossi, P., Epis, S., Mosca, M., Ricci, I., Damiani, C., Ulissi, U., Crotti, E., Daffonchio, D., Bandi, C., *et al.* (2012). Delayed larval development in *Anopheles mosquitoes* deprived of *Asaia* bacterial symbionts. BMC Microbiol. *12*(Suppl 1), S2.

Cirimotich, C.M., Dong, Y., Clayton, A.M., Sandiford, S.L., Souza-Neto, J.A., Mulenga, M., and Dimopoulos, G. (2011). Natural microbe-mediated refractoriness to *Plasmodium* infection in *Anopheles gambiae*. Science *332*, 855–858.

Crickmore, N., Zeigler, D.R., Feitelson, J., Schnepf, E., Van Rie, J., Lereclus, D., Baum, J., and Dean, D.H. (1998). Revision of the nomenclature for the Bacillus thuringiensis pesticidal crystal proteins. Microbiol. Mol. Biol. Rev. *62*, 807–813.

Crotti, E., Damiani, C., Pajoro, M., Gonella, E., Rizzi, A., Ricci, I., Negri, I., Scuppa, P., Rossi, P., Ballarini, P., *et al.* (2009). *Asaia*, a versatile acetic acid bacterial symbiont, capable of cross-colonizing insects of phylogenetically distant genera and orders. Environ. Microbiol. *11*, 3252–3264.

Damiani, C., Ricci, I., Crotti, E., Rossi, P., Rizzi, A., Scuppa, P., Capone, A., Ulissi, U., Epis, S., Genchi, M., *et al.* (2010). Mosquito–bacteria symbiosis: the case of *Anopheles gambiae* and *Asaia*. Microb. Ecol. *60*, 644–654.

Darbro, J.M., and Thomas, M.B. (2009). Spore persistence and likelihood of aeroallergenicity of entomopathogenic fungi used for mosquito control. Am. J. Trop. Med. Hyg. *80*, 992–997.

Darbro, J.M., Johnson, P.H., Thomas, M.B., Ritchie, S.A., Kay, B.H., and Ryan, P.A. (2012). Effects of *Beauveria bassiana* on survival, blood-feeding success, and fecundity of *Aedes aegypti* in laboratory and semi-field conditions. Am. J. Trop. Med. Hyg. *86*, 656–664.

Dohm, D.J., Rowton, E.D., Lawyer, P.G., O'Guinn, M., and Turell, M.J. (2000). Laboratory transmission of Rift Valley fever virus by *Phlebotomus duboscqi*, *Phlebotomus papatasi*, *Phlebotomus sergenti*, and *Sergentomyia schwetzi* (Diptera: Psychodidae). J. Med. Entomol. *37*, 435–438.

Dominic Amalraj, D., and Das, P.K. (1998). Estimation of predation by the larvae of *Toxorhynchites splendens* on the aquatic stages of *Aedes aegypti*. Southeast Asian J. Trop. Med. Public Health *29*, 177–183.

Dong, Y., Manfredini, F., and Dimopoulos, G. (2009). Implication of the mosquito midgut microbiota in the defense against malaria parasites. PLoS Pathog. *5*, e1000423.

Dong, Y., Morton, J.C. Jr, Ramirez, J.L., Souza-Neto, J.A., and Dimopoulos, G. (2012). The entomopathogenic fungus *Beauveria bassiana* activate toll and JAK-STAT pathway-controlled effector genes and anti-dengue activity in *Aedes aegypti*. Insect Biochem. Mol. Biol. *42*, 126–132.

Elcin, Y.M. (1995). Control of mosquito larvae by encapsulated pathogen *Bacillus thuringiensis* var. israelensis. J. Microencapsul. *12*, 515–523.

Favia, G., Ricci, I., Damiani, C., Raddadi, N., Crotti, E., Marzorati, M., Rizzi, A., Urso, R., Brusetti, L., Borin, S., *et al.* (2007). Bacteria of the genus *Asaia* stably associate with *Anopheles stephensi*, an Asian malarial mosquito vector. Proc. Natl. Acad. Sci. U.S.A. *104*, 9047–9051.

Favia, G., Ricci, I., Marzorati, M., Negri, I., Alma, A., Sacchi, L., Bandi, C., and Daffonchio, D. (2008). Bacteria of the genus *Asaia*: a potential paratransgenic weapon against malaria. Adv. Exp. Med. Biol. *627*, 49–59.

Focks, D.A. (2007). *Toxorhynchites* as biocontrol agents. J. Am. Mosq. Control Assoc. *23*, 118–127.

Focks, D.A., Sackett, S.R., Dame, D.A., and Bailey, D.L. (1985). Effect of weekly releases of *Toxorhynchites amboinensis* (Doleschall) on *Aedes aegypti* (L.) (Diptera: Culicidae) in New Orleans, Louisiana. J. Econ. Entomol. *78*, 622–626.

Frentiu, F.D., Robinson, J., Young, P.R., McGraw, E.A., and O'Neill, S.L. (2010). *Wolbachia*-mediated resistance to dengue virus infection and death at the cellular level. PLoS One *5*, e13398.

Georghiou, G.P., and Wirth, M.C. (1997). Influence of exposure to single versus multiple toxins of *Bacillus thuringiensis* subsp. *israelensis* on development of resistance in the mosquito *Culex quinquefasciatus* (Diptera: Culicidae). Appl. Environ. Microbiol. *63*, 1095–1101.

Ghosh, S.K., and Dash, A.P. (2007). Larvivorous fish against malaria vectors: a new outlook. Trans. R. Soc. Trop. Med. Hyg. *101*, 1063–1064.

Ghosh, S.K., Chakaravarthy, P., Panch, S.R., Krishnappa, P., Tiwari, S., Ojha, V.P., Manjushree, R., and Dash, A.P. (2011). Comparative efficacy of two poeciliid fish in indoor cement tanks against chikungunya vector *Aedes aegypti* in villages in Karnataka, India. BMC Public Health *11*, 599.

Glaser, R.L., and Meola, M.A. (2010). The native *Wolbachia* endosymbionts of *Drosophila melanogaster* and *Culex quinquefasciatus* increase host resistance to West Nile virus infection. PLoS One *5*, e11977.

Gomez-Dantes, H., and Willoquet, J.R. (2009). Dengue in the Americas: challenges for prevention and control. Cad. Saude Publica *25*, S19–S31.

Gratz, N.G. (1991). Emergency control of *Aedes aegypti* as a disease vector in urban areas. J. Am. Mosq. Control Assoc. *7*, 353–365.

Gu, J., Liu, M., Deng, Y., Peng, H., and Chen, X. (2011). Development of an efficient recombinant mosquito densovirus-mediated RNA interference system and its preliminary application in mosquito control. PLoS One *6*, e21329.

Gusmao, D.S., Santos, A.V., Marini, D.C., Bacci, M., Jr., Berbert-Molina, M.A., and Lemos, F.J. (2010). Culture-dependent and culture-independent characterization of microorganisms associated with *Aedes aegypti* (Diptera: Culicidae) (L.) and dynamics of bacterial colonization in the midgut. Acta Trop. *115*, 275–281.

Hales, S., and van Panhuis, W. (2005). A new strategy for dengue control. Lancet *365*, 551–552.

Hancock, P.A., Sinkins, S.P., and Godfray, H.C.J. (2011). Strategies for introducing *Wolbachia* to reduce transmission of mosquito-borne diseases. PLoS Negl. Trop. Dis. *5*, e1024.

Hapairai, L.K., Sang,M.A., Sinkins, S.P., and Bossin, H.C. (2013). Population studies of the filarial vector *Aedes polynesiensis* (Diptera: Culicidae) in two island settings of French Polynesia. J. Med. Entomol. *50*, 965–976.

Hedges, L.M., Brownlie, J.C., O'Neill, S.L., and Johnson, K.N. (2008). *Wolbachia* and virus protection in insects. Science *322*, 702.

Hirunkanokpun, S., Carlson, J.O., and Kittayapong, P. (2008). Evaluation of mosquito densoviruses for controlling *Aedes aegypti* (Diptera: Culicidae): variation in efficiency due to virus strain and geographic origin of mosquitoes. Am. J. Trop. Med. Hyg. *78*, 784–790.

Hoch, A.L., Turell, M.J., and Bailey, C.L. (1984). Replication of Rift Valley fever virus in the sand fly *Lutzomyia longipalpis*. Am. J. Trop. Med. Hyg. *33*, 295–299.

Hoffmann, A.A., Montgomery, B.L., Popovici, J., Iturbe-Ormaetxe, I., Johnson, P.H., Muzzi, F., Greenfield, M., Durkan, M., Leong, Y.S., Dong, Y., *et al.* (2011). Successful establishment of *Wolbachia* in *Aedes* populations to suppress dengue transmission. Nature *476*, 454–457.

Howard, A.F., Koenraadt, C.J., Farenhorst, M., Knols, B.G., and Takken, W. (2010). Pyrethroid resistance in *Anopheles gambiae* leads to increased susceptibility to the entomopathogenic fungi *Metarhizium anisopliae* and *Beauveria bassiana*. Malar. J. *9*, 168.

van den Hurk, A.F., Hall-Mendelin, S., Pyke, A.T., Frentiu, F.D., McElroy, K., Day, A., Higgs, S., and O'Neill, S.L. (2012). Impact of *Wolbachia* on infection with chikungunya and yellow fever viruses in the mosquito vector *Aedes aegypti*. PLoS Negl. Trop. Dis. *6*, e1892.

Jousset, F.X., Baquerizo, E., and Bergoin, M. (2000). A new densovirus isolated from the mosquito *Culex pipiens* (Diptera: culicidae). Virus Res. *67*, 11–16.

Kambris, Z., Cook, P.E., Phuc, H.K., and Sinkins, S.P. (2009). Immune activation by life-shortening *Wolbachia* and reduced filarial competence in mosquitoes. Science *326*, 134–136.

Kambris, Z., Blagborough, A.M., Pinto, S.B., Blagrove, M.S., Godfray, H.C., Sinden, R.E., and Sinkins, S.P. (2010). *Wolbachia* stimulates immune gene expression and inhibits *Plasmodium* development in *Anopheles gambiae*. PLoS Pathog. *6*, e1001143.

Kanthong, N., Khemnu, N., Pattanakitsakul, S.N., Malasit, P., and Flegel, T.W. (2010). Persistent, triple-virus co-infections in mosquito cells. BMC Microbiol. *10*, 14.

Kay, B.H., Tuyet Hanh, T.T., Le, N.H., Quy, T.M., Nam, V.S., Hang, P.V., Yen, N.T., Hill, P.S., Vos, T., and Ryan, P.A. (2010). Sustainability and cost of a community-based strategy against *Aedes aegypti* in northern and central Vietnam. Am. J. Trop. Med. Hyg. *82*, 822–830.

Kesavaraju, B., Khan, D.F., and Gaugler, R. (2011). Behavioral differences of invasive container-dwelling mosquitoes to a native predator. J. Med. Entomol. *48*, 526–532.

Kittayapong, P., Chansang, U., Chansang, C., and Bhumiratana, A. (2006). Community participation and appropriate technologies for dengue vector control at transmission foci in Thailand. J. Am. Mosq. Control Assoc. *22*, 538–546.

Knols, B.G., Bukhari, T., and Farenhorst, M. (2010). Entomopathogenic fungi as the next-generation control agents against malaria mosquitoes. Future Microbiol. *5*, 339–341.

Lacey, L.A. (2007). *Bacillus thuringiensis* serovariety *israelensis* and *Bacillus sphaericus* for mosquito control. J. Am. Mosq. Control Assoc. *23*, 133–163.

Lam, P.H., Boon, C.S., Yng, N.Y., and Benjamin, S. (2010). *Aedes albopictus* control with spray application of *Bacillus thuringiensis israelensis*, strain AM 65-52. Southeast Asian J. Trop. Med. Public Health *41*, 1071–1081.

Laven, H. (1967). Eradication of *Culex pipiens fatigans* through cytoplasmic incompatibility. Nature *216*, 383–384.

Ledermann, J.P., Suchman, E.L., Black, W.C.t., and Carlson, J.O. (2004). Infection and pathogenicity of the mosquito densoviruses AeDNV, HeDNV, and APeDNV in *Aedes aegypti* mosquitoes (Diptera: Culicidae). J. Econ. Entomol. *97*, 1828–1835.

Luz, C., Tai, M.H., Santos, A.H., and Silva, H.H. (2008). Impact of moisture on survival of *Aedes aegypti* eggs and ovicidal activity of *Metarhizium anisopliae* under laboratory conditions. Mem. Inst. Oswaldo Cruz *103*, 214–215.

de Maagd, R.A., Bravo, A., and Crickmore, N. (2001). How *Bacillus thuringiensis* has evolved specific toxins to colonize the insect world. Trends Genet. *17*, 193–199.

McMeniman, C.J., and O'Neill, S.L. (2010). A virulent *Wolbachia* infection decreases the viability of the dengue vector *Aedes aegypti* during periods of embryonic quiescence. PLoS Negl. Trop. Dis. *4*, e748.

McMeniman, C.J., Lane, A.M., Fong, A.W., Voronin, D.A., Iturbe-Ormaetxe, I., Yamada, R., McGraw, E.A., and O'Neill, S.L. (2008). Host adaptation of a *Wolbachia* strain after long-term serial passage in mosquito cell lines. Appl. Environ. Microbiol. *74*, 6963–6969.

McMeniman, C.J., Lane, R.V., Cass, B.N., Fong, A.W., Sidhu, M., Wang, Y.F., and O'Neill, S.L. (2009). Stable introduction of a life-shortening *Wolbachia* infection into the mosquito *Aedes aegypti*. Science *323*, 141–144.

McNaughton, D. (2012). The importance of long-term social research in enabling participation and developing engagement strategies for new dengue control technologies. PLoS Negl. Trop. Dis. *6*, e1785.

Marten, G.G. (1990). Evaluation of cyclopoid copepods for *Aedes albopictus* control in tires. J. Am. Mosq. Control Assoc. *6*, 681–688.

Marten, G.G., and Reid, J.W. (2007). Cyclopoid copepods. J. Am. Mosq. Control Assoc. *23*, 65–92.

Matias, J.R., and Adrias, A.Q. (2010). The use of annual killifish in the biocontrol of the aquatic stages of mosquitoes in temporary bodies of fresh water; a potential new tool in vector control. Parasit. Vectors *3*, 46.

Mercer, D.R., Marie, J., Bossin, H., Faaruia, M., Tetuanui, A., Sang, M.C., and Dobson, S.L. (2012). Estimation of population size and dispersal of *Aedes polynesiensis* on Toamaro motu, French Polynesia. J. Med. Entomol. *49*, 971–980.

Minard, G., Mavingui, P., and Moro, C.V. (2013). Diversity and function of bacterial microbiota in the mosquito holobiont. Parasit. Vectors *6*, 146.

Mitchell, C., Miller, B., and Gubler, D. (1987). Vector competence of *Aedes albopictus* from Houston, Texas, for dengue serotypes 1 to 4, yellow fever and Ross River viruses. J. Am. Mosq. Control Assoc. *3*, 460–465.

Mnyone, L.L., Kirby, M.J., Lwetoijera, D.W., Mpingwa, M.W., Simfukwe, E.T., Knols, B.G., Takken, W., and Russell, T.L. (2010). Tools for delivering entomopathogenic fungi to malaria mosquitoes: effects of delivery surfaces on fungal efficacy and persistence. Malar. J. *9*, 246.

Moreira, L.A., Iturbe-Ormaetxe, I., Jeffery, J.A., Lu, G., Pyke, A.T., Hedges, L.M., Rocha, B.C., Hall-Mendelin, S., Day, A., Riegler, M., *et al.* (2009). A *Wolbachia* symbiont in *Aedes aegypti* limits infection with dengue, Chikungunya, and *Plasmodium*. Cell *139*, 1268–1278.

Moretti, R., and Calvitti, M. (2013). Male mating performance and cytoplasmic incompatibility in a *w*Pip *Wolbachia* trans-infected line of *Aedes albopictus* (Stegomyia albopicta). Med. Vet. Entomol. *27*, 377–386.

Mosimann, A.L., Bordignon, J., Mazzarotto, G.C., Motta, M.C., Hoffmann, F., and Santos, C.N. (2011). Genetic and biological characterization of a densovirus isolate that affects dengue virus infection. Mem. Inst. Oswaldo Cruz *106*, 285–292.

Mulla, M.S., Thavara, U., Tawatsin, A., and Chompoosri, J. (2004). Procedures for the evaluation of field efficacy of slow-release formulations of larvicides against *Aedes aegypti* in water-storage containers. J. Am. Mosq. Control Assoc. *20*, 64–73.

Murphy, B., Jansen, C., Murray, J., and De Barro, P. (2010). Risk analysis on the Australian release of *Aedes aegypti* (L.) (Diptera: Culicidae) containing *Wolbachia* (CSIRO, Brisbane, Australia).

Novak, R.J., Gubler, D.J., and Underwood, D. (1985). Evaluation of slow-release formulations of temephos (Abate) and *Bacillus thuringiensis* var. *israelensis* for the control of *Aedes aegypti* in Puerto Rico. J. Am. Mosq. Control Assoc. *1*, 449–453.

Nyamah, M.A., Sulaiman, S., and Omar, B. (2011). Field observation on the efficacy of *Toxorhynchites splendens* (Wiedemann) as a biocontrol agent against *Aedes albopictus* (Skuse) larvae in a cemetery. Trop. Biomed. *28*, 312–319.

O'Connor, L., Plichart, C., Sang, A.C., Brelsfoard, C.L., Bossin, H.C., and Dobson, S.L. (2012). Open release of male mosquitoes infected with a *Wolbachia* biopesticide: field performance and infection containment. PLoS Negl. Trop. Dis. *6*, e1797.

Osborne, S.E., Leong, Y.S., O'Neill S.L., and Johnson, K.N. (2009). Variation in antiviral protection mediated by different *Wolbachia* strains in *Drosophila simulans*. PLoS Pathog. *5*, e1000656.

Pamplona Lde, G., Lima, J.W., Cunha, J.C., and Santana, E.W. (2004). [Evaluation of the impact on *Aedes aegypti* infestation in cement tanks of the municipal district of Caninde, Ceara, Brazil after using the *Betta splendens* fish as an alternative biological control]. Rev. Soc. Bras. Med. Trop. *37*, 400–404.

Pan, X., Zhou, G., Wu, J., Bian, G., Lu, P., Raikhel, A.S., and Xi, Z. (2012). *Wolbachia* induces reactive oxygen species (ROS)-dependent activation of the Toll pathway to control dengue virus in the mosquito *Aedes aegypti*. Proc. Natl. Acad. Sci. U.S.A. *109*, E23–E31.

Paula, A.R., Carolino, A.T., Paula, C.O., and Samuels, R.I. (2011a). The combination of the entomopathogenic fungus *Metarhizium anisopliae* with the insecticide Imidacloprid increases virulence against the dengue vector *Aedes aegypti* (Diptera: Culicidae). Parasit. Vectors *4*, 8.

Paula, A.R., Carolino, A.T., Silva, C.P., and Samuels, R.I. (2011b). Susceptibility of adult female *Aedes aegypti* (Diptera: Culicidae) to the entomopathogenic fungus *Metarhizium anisopliae* is modified following blood feeding. Parasit. Vectors *4*, 91.

Popovici, J., Moreira, L.A., Poinsignon, A., Iturbe-Ormaetxe, I., McNaughton, D., and O'Neill, S.L. (2010). Assessing key safety concerns of a *Wolbachia*-based strategy to control dengue transmission by *Aedes* mosquitoes. Mem. Inst. Oswaldo Cruz *105*, 957–964.

Ramirez, J.L., Souza-Neto, J., Torres Cosme, R., Rovira, J., Ortiz, A., Pascale, J.M., and Dimopoulos, G. (2012). Reciprocal tripartite interactions between the *Aedes aegypti* midgut microbiota, innate immune system and dengue virus influences vector competence. PLoS Negl. Trop. Dis. *6*, e1561.

Rances, E., Ye, Y.H., Woolfit, M., McGraw, E.A., and O'Neill, S.L. (2012). The relative importance of innate immune priming in *Wolbachia*-mediated dengue interference. PLoS Pathog. *8*, e1002548.

Regis, L.N., Acioli, R.V., Silveira, J.C., Jr., Melo-Santos, M.A., Souza, W.V., Ribeiro, C.M., da Silva, J.C., Monteiro, A.M., Oliveira, C.M., Barbosa, R.M., *et al.* (2013). Sustained reduction of the dengue vector population resulting from an integrated control strategy applied in two Brazilian cities. PLoS One *8*, e67682.

Reyes-Villanueva, F., Garza-Hernandez, J.A., Garcia-Munguia, A.M., Tamez-Guerra, P., Howard, A.F., and Rodriguez-Perez, M.A. (2011). Dissemination of *Metarhizium anisopliae* of low and high virulence by mating behavior in *Aedes aegypti*. Parasit. Vectors *4*, 171.

Ricci, I., Damiani, C., Scuppa, P., Mosca, M., Crotti, E., Rossi, P., Rizzi, A., Capone, A., Gonella, E., Ballarini, P., *et al.* (2011a). The yeast *Wickerhamomyces anomalus* (*Pichia anomala*) inhabits the midgut and reproductive system of the Asian malaria vector *Anopheles stephensi*. Environ. Microbiol. *13*, 911–921.

Ricci, I., Mosca, M., Valzano, M., Damiani, C., Scuppa, P., Rossi, P., Crotti, E., Cappelli, A., Ulissi, U., Capone, A., *et al.* (2011b). Different mosquito species host *Wickerhamomyces anomalus* (*Pichia anomala*): perspectives on vector-borne diseases symbiotic control. Antonie Van Leeuwenhoek *99*, 43–50.

Ritchie, S.A., Rapley, L.P., and Benjamin, S. (2010). *Bacillus thuringiensis* var. *israelensis* (Bti) provides residual control of *Aedes aegypti* in small containers. Am. J. Trop. Med. Hyg. *82*, 1053–1059.

Riviere, F., Kay, B.H., Klein, J.M., and Sechan, Y. (1987). *Mesocyclops aspericornis* (Copepoda) and *Bacillus thuringiensis* var. *israelensis* for the biological control of *Aedes* and *Culex* vectors (Diptera: Culicidae) breeding in crab holes, tree holes, and artificial containers. J. Med. Entomol. *24*, 425–430.

Roy, H.E., Steinkraus, D.C., Eilenberg, J., Hajek, A.E., and Pell, J.K. (2006). Bizarre interactions and endgames: entomopathogenic fungi and their arthropod hosts. Annu. Rev. Entomol. *51*, 331–357.

Samidurai, K., and Mathew, N. (2013). Mosquito larvicidal and ovicidal activity of puffer fish extracts against *Anopheles stephensi*, *Culex quinquefasciatus* and *Aedes aegypti* (Diptera: Culicidae). Trop. Biomed. *30*, 27–35.

Schnepf, E., Crickmore, N., Van Rie, J., Lereclus, D., Baum, J., Feitelson, J., Zeigler, D.R., and Dean, D.H. (1998). *Bacillus thuringiensis* and its pesticidal crystal proteins. Microbiol. Mol. Biol. Rev. *62*, 775–806.

Scholte, E.J., Knols, B.G., Samson, R.A., and Takken, W. (2004). Entomopathogenic fungi for mosquito control: a review. J. Insect Sci. *4*, 19.

Scholte, E.J., Takken, W., and Knols, B.G. (2007). Infection of adult *Aedes aegypti* and *Ae. albopictus* mosquitoes with the entomopathogenic fungus *Metarhizium anisopliae*. Acta Trop. *102*, 151–158.

Seleena, P., Lee, H.L., and Chiang, Y.F. (2001). Thermal application of *Bacillus thuringiensis* serovar *israelensis* for dengue vector control. J. Vector Ecol. *26*, 110–113.

Shah, P.A., and Pell, J.K. (2003). Entomopathogenic fungi as biological control agents. Appl. Microbiol. Biotechnol. *61*, 413–423.

Singh, R.K., and Singh, S.P. (2004). Predatory potential of *Nepa cinerea* against mosquito larvae in laboratory conditions. J. Commun. Dis. *36*, 105–110.

Sinh Nam, V., Thi Yen, N., Minh Duc, H., Cong Tu, T., Trong Thang, V., Hoang Le, N., Hoang San, L., Le Loan, L., Que Huong, V.T., Kim Khanh, L.H., *et al.* (2012). Community-based control of *Aedes aegypti* by using *Mesocyclops* in southern Vietnam. Am. J. Trop. Med. Hyg. *86*, 850–859.

Sinkins, S.P., Braig, H.R., and O'Neill, S.L. (1995). *Wolbachia pipientis*: bacterial density and unidirectional cytoplasmic incompatibility between infected populations of *Aedes albopictus*. Exp. Parasitol. *81*, 284–291.

Sivaram, A., Barde, P.V., Gokhale, M.D., Singh, D.K., and Mourya, D.T. (2010). Evidence of co-infection of chikungunya and densonucleosis viruses in C6/36 cell lines and laboratory infected *Aedes aegypti* (L.) mosquitoes. Parasit. Vectors *3*, 95.

Suh, E., Mercer, D.R., Fu, Y., and Dobson, S.L. (2009). Pathogenicity of life-shortening *Wolbachia* in *Aedes albopictus* after transfer from *Drosophila melanogaster*. Appl. Environ. Microbiol. *75*, 7783–7788.

Teixeira, L., Ferreira, A., and Ashburner, M. (2008). The bacterial symbiont *Wolbachia* induces resistance to RNA viral infections in *Drosophila melanogaster*. PLoS Biol. *6*, e2.

Tietze, N.S., Hester, P.G., Shaffer, K.R., Prescott, S.J., and Schreiber, E.T. (1994). Integrated management of waste tire mosquitoes utilizing *Mesocyclops longisetus* (Copepoda: Cyclopidae), *Bacillus thuringiensis* var. *israelensis*, *Bacillus sphaericus*, and methoprene. J. Am. Mosq. Control Assoc. *10*, 363–373.

Turell, M.J., Wilson, W.C., and Bennett, K.E. (2010). Potential for North American mosquitoes (Diptera: Culicidae) to transmit rift valley fever virus. J. Med. Entomol. *47*, 884–889.

Turelli, M. (2010). Cytoplasmic incompatibility in populations with overlapping generations. Evolution *64*, 232–241.

Vu, S.N., Nguyen, T.Y., Kay, B.H., Marten, G.G., and Reid, J.W. (1998). Eradication of *Aedes aegypti* from a village in Vietnam, using copepods and community participation. Am. J. Trop. Med. Hyg. *59*, 657–660.

Vu, S.N., Nguyen, T.Y., Tran, V.P., Truong, U.N., Le, Q.M., Le, V.L., Le, T.N., Bektas, A., Briscombe, A., Aaskov, J.G., *et al.* (2005). Elimination of dengue by community programs using *Mesocyclops* (Copepoda) against *Aedes aegypti* in central Vietnam. Am. J. Trop. Med. Hyg. *72*, 67–73.

Walker, T., Johnson, P.H., Moreira, L.A., Iturbe-Ormaetxe, I., Frentiu, F.D., McMeniman, C.J., Leong, Y.S., Dong, Y., Axford, J., Kriesner, P., *et al.* (2011). The wMel *Wolbachia* strain blocks dengue and invades caged *Aedes aegypti* populations. Nature *476*, 450–453.

Weaver, S.C., and Reisen, W.K. (2010). Present and future arboviral threats. Antiviral Res. *85*, 328–345.

Wei, W., Shao, D., Huang, X., Li, J., Chen, H., Zhang, Q., and Zhang, J. (2006). The pathogenicity of mosquito densovirus (C6/36DNV) and its interaction with dengue virus type II in *Aedes albopictus*. Am. J. Trop. Med. Hyg. *75*, 1118–1126.

Wilder-Smith, A., Ooi, E.E., Vasudevan, S.G., and Gubler, D.J. (2010). Update on dengue: epidemiology, virus evolution, antiviral drugs, and vaccine development. Curr. Infect. Dis. Rep. *12*, 157–164.

Wise de Valdez, M.R., Suchman, E.L., Carlson, J.O., and Black, W.C. (2010). A large scale laboratory cage trial of *Aedes* densonucleosis virus (AeDNV). J. Med. Entomol. *47*, 392–399.

Wong, Z.S., Hedges, L.M., Brownlie, J.C., and Johnson, K.N. (2011). *Wolbachia*-mediated antibacterial protection and immune gene regulation in *Drosophila*. PLoS One *6*, e25430.

Xi, Z., Khoo, C.C., and Dobson, S.L. (2005). *Wolbachia* establishment and invasion in an *Aedes aegypti* laboratory population. Science *310*, 326–328.

Xi Z., Khoo, C.C., and Dobson, S.L. (2006). Interspecific transfer of *Wolbachia* into the mosquito disease vector *Aedes albopictus*. Proc. Biol. Sci. *273*, 1317–1322.

Yeap, H.L., Mee, P., Walker, T., Weeks, A.R., O'Neill, S.L., Johnson, P., Ritchie, S.A., Richardson, K.M., Doig, C., Endersby, N.M., *et al.* (2011). Dynamics of the 'popcorn' *Wolbachia* infection in outbred *Aedes aegypti* informs prospects for mosquito vector control. Genetics *187*, 583–595.

RNA Interference: a Pathway to Arbovirus Control

18

Kathryn A. Hanley and Christy C. Andrade

Abstract

Arboviruses are responsible for a high burden of established and emerging disease worldwide, but specific antiviral therapies are lacking for the vast majority of these pathogens. RNA interference (RNAi) plays a central role in controlling arbovirus infections in arthropod vectors, and although the role of RNAi in vertebrate immunity remains controversial, it is clear that exogenous small RNAs can be used to stimulate antiviral effectors in vertebrate hosts. Thus a robust effort has been made to design and test RNAi-based strategies for control of arbovirus infections and prevention of arbovirus transmission. This chapter reviews these efforts, which group into three general categories: (i) use of exogenous small RNAs to prevent or treat arbovirus infections in humans and domestic animals, (ii) generation of genetically modified arthropod vectors that express antiviral small RNAs and thereby block transmission, and (iii) creation of recombinant arboviruses that possess target sites for host microRNAs (miRNAs) as live-attenuated vaccines. While these approaches all hold promise, there remain barriers to the safe and effective delivery of small RNAs and critical knowledge gaps regarding arbovirus–RNAi interactions that must be overcome before any of these approaches come to fruition.

Introduction

RNA interference (RNAi) is an ancient and ubiquitous defence against RNA virus infection in eukaryotes (Ding and Voinnet, 2007; Zhou and Rana, 2013). The small interfering RNA (siRNA) branch of the RNAi pathway is initiated when the endonuclease Dicer cleaves long double-stranded RNA (dsRNA) within the cytoplasm to produce small RNA fragments (siRNAs). Then siRNAs, coupled to the RNA-induced silencing complex (RISC), mediate cleavage of single-stranded RNA (ssRNA) with perfect complementarity to the siRNA. In the microRNA (miRNA) branch of the pathway, dsRNA is produced in the nucleus and exported to the cytoplasm where it is processed by Dicer to produce miRNAs; miRNAs coupled to RISC bind to complementary sequences in a ssRNA target. Unlike siRNAs, miRNAs tolerate multiple mismatches with the ssRNA target. The bound miRNA can inhibit translation of the target RNA or mediate cleavage. A third branch of RNAi, the piwi-interacting RNA (piRNA) pathway, is thought to act primarily in the germline and is not discussed further here (Siomi *et al.*, 2011).

Among arthropods, particularly mosquitoes (Blair, 2011), RNAi is the primary mechanism controlling arbovirus infections (Kingsolver *et al.*, 2013). Indeed, the arboviruses Sindbis and O'nyong nyong, which are normally benign for their mosquito vectors, become quite lethal when they are engineered to express viral suppressors of RNAi derived from insect viruses (Cirimotich *et al.*, 2009; Myles *et al.*, 2008). Chapter 8 describes insect immune responses, including RNAi, in detail.

The role of RNAi in vertebrate antiviral immunity is considerably more controversial (Svoboda, 2014). Until recently, it was largely accepted that the interferon pathway had superseded RNAi in innate defence against virus infection over the course of vertebrate evolution. Although the RNAi pathway persists in vertebrate cells and responds to stimulation by exogenous dsRNA, miRNA, or siRNA, it was not thought to play a meaningful role in controlling authentic virus replication. However in 2013 two studies demonstrated that Nodamuravirus (Li *et al.*, 2013) and Flock House virus (Maillard *et al.*, 2013), from which the viral suppressor of RNAi had been deleted, stimulated and were susceptible to RNAi-mediated control in mammalian cells. Additionally, Shapiro *et al.* (2014) have shown that knocking out Drosha, a key nuclease in the microRNA (miRNA) pathway, in mammalian fibroblasts resulted in significant increases in Sindbis virus replication. Together these findings suggest that RNAi does enact antiviral immunity in mammals.

On the other hand, there is also substantial evidence against a role of RNAi in mammalian defence against viruses. For example, Shapiro *et al.* found that knocking out Dicer, the lynchpin of both the siRNA and miRNA branches of RNAi, in fibroblasts did not enhance Sindbis virus replication (Shapiro *et al.*, 2014). Similarly Bogerd *et al.* (2014) demonstrated that human embryonic kidney (HEK) cells and HEK cells in which the Dicer gene had been ablated (noDice cells) supported similar levels of replication in a broad panel of different viruses. These findings prompted Bogerd *et al.* to conclude that viruses are resistant to the human RNAi response (Bogerd *et al.*, 2014). Thus the question of whether RNAi participates in the vertebrate immune response to virus infection under natural conditions remains unresolved at this time.

In this chapter, we summarize the approaches that have been taken to leverage RNAi, via exogenous small RNAs, genetically modified mosquitoes, and recombinant viruses, to treat arbovirus disease, block arbovirus transmission, and prevent arbovirus infections, respectively. Additionally we identify critical gaps in current knowledge about these approaches and fruitful directions for future research.

Use of exogenous small RNAs to treat arbovirus infection

For the vast majority of arboviruses, there are no specific therapies available to treat infection or alleviate disease (Botting and Kuhn, 2012; Lani *et al.*, 2014; Lim *et al.*, 2013). Thus intense effort has been devoted to discovery and development of small RNA therapies for arboviruses. To review this topic, we searched Pubmed using the name of each viral genus that contains arboviruses (e.g. 'flavivirus', 'alphavirus', 'orbivirus', etc.) combined with each of the terms 'RNA interference', 'RNAi', 'siRNA', and 'miRNA'. The results are summarized in Tables 18.1 to 18.4 and discussed in the subsections below.

Flaviviridae

The genus *Flavivirus* of the family *Flaviviridae* contains all of the known vector-borne flaviviruses. The flaviviruses fall into four major clades: the mosquito-borne flaviviruses, the tick-borne flaviviruses, the flaviviruses directly transmitted between vertebrates and the flaviviruses directly transmitted between arthropods (Cook *et al.*, 2012). The former two clades include many of the emerging and established arboviruses responsible for human disease, such as dengue, West Nile, Japanese encephalitis, yellow fever, tick-borne encephalitis, and Zika viruses. Moreover the flaviviruses are responsible for the majority of cases of arboviral disease. Dengue virus alone infects almost 400 million people annually (Bhatt *et al.*, 2013) while, despite the availability of vaccines for both, Japanese encephalitis virus and yellow fever virus respectively cause tens of thousands (Le Flohic *et al.*, 2013) to hundreds of thousands (Gubler, 2004) of cases of clinical disease each year. While most of the flaviviruses are zoonotic, the four serotypes of dengue virus have established human-endemic cycles that are ecologically distinct from their enzootic progenitors (Hanley *et al.*, 2013).

The flavivirus genome comprises a single segment of positive-sense RNA. Genome organization is conserved throughout the genus and consists of genes for three structural proteins [capsid (C), pre-membrane (prM) and envelope (E)] and seven non-structural genes (NS1, 2A, 2B, 3, 4A, 4B, 5) flanked by a 5′ and 3′ untranslated region (UTR).

A large number of studies have characterized the efficacy of small RNAs for inhibition of the replication of a broad array of vector-borne flaviviruses (Table 18.1). In most of these studies, mammalian cells in culture have been treated with siRNAs, short hairpin RNAs (shRNAs), or artificial miRNAs (amiRNAs) targeting the virus genome, either through transient transfection or expression from a plasmid or viral vector, hours to days prior to infection with the target virus. Nine of the ten flavivirus genes (excluding NS4A) and both UTRs were targeted in at least one of these studies. At least one of the small RNAs utilized in each of these studies significantly inhibited virus replication. Although small RNAs varied in their efficacy and some small RNAs in some studies did not affect virus replication, there was no obvious trend in efficacy relative to targeting of different genes or regions of the flavivirus genome. Notably, however, both Stein *et al.* (2011) and Xie *et al.* (2013) found that targeting the highly conserved 5′ cyclization sequence (5′ CS) was particularly effective. Moreover Stein *et al.* (2011) showed that a single siRNA targeting the 5′ CS could inhibit replication of all four serotypes of dengue virus. Such broad activity is critical for a successful anti-dengue therapy, given current lack of readily available diagnostics capable of identifying specific serotypes (Lim *et al.*, 2013).

A single study tested the impact of small RNAs on flavivirus replication in arthropod cells and found that a siRNA inhibited dengue virus replication in *Aedes albopictus* C6/36 cells (Wu *et al.*, 2010). Alhoot and colleagues (2011, 2012) have also shown that knockdown of multiple host genes in cultured mammalian cells inhibited dengue virus entry, replication, or production of infectious virions. These latter studies open the door to development of host-targeted therapies for flaviviral infections.

A smaller number of the studies summarized in Table 18.1 tested the effect of small RNAs applied after infection in cultured cells. The findings for post-infection efficacy were mixed. For example Geiss *et al.* (2005) found that when siRNAs were introduced 10 h post infection, replication of West Nile virus in Huh-7.5 cells was unaffected, whereas Ye *et al.* (2011) found that siRNAs did significantly inhibit West Nile virus replication when introduced 24 h post infection in BHK cells.

Although studies in cultured cells are a necessary foundation, studies *in vivo* offer substantially more insight into the potential utility of small RNA therapies for flaviviral disease. To date, all *in vivo* trials of small RNA therapies for flaviviruses have been conducted in mice. As in the studies in culture, treatment of animals prior to infection generally suppressed virus replication and improved survival, whereas results of treating after infection were mixed. For example, multiple studies have tested the impact of exogenous small RNAs on the survival of mice infected with Japanese encephalitis virus: Shen *et al.* (2014a) reported that pre-treatment of mice with antiviral shRNAs improved mouse survival, Murakami *et al.* (2005) reported that concurrent injection of mice with virus and antiviral shRNAs improved survival as well, but Kumar *et al.* (2006) reported that treatment of mice with antiviral siRNAs six hours after infection did not improve survival. Thus results from pre-infection treatments with antiviral small RNAs *in vivo* may provide an overly optimistic view of their potential utility as therapies for established infections.

Togaviridae

The *Togaviridae* family includes two genera: the monospecific genus *Rubivirus* and the genus *Alphavirus*. The latter contains 29 recognized species, 27 of which are transmitted by mosquitoes (Forrester *et al.*, 2012). The alphaviruses are second only to the flaviviruses in their importance as emerging human pathogens (Weaver *et al.*, 2012). Of particular note, chikungunya virus (CHIKV), which was already established in Africa and Asia, made its modern debut in the Americas in 2013. The virus was first detected in St. Martin but had spread to 31 countries and territories in North and South America, including the USA, at the time of this writing (Staples and Fischer, 2014). CHIKV is transmitted by the domestic mosquito *Ae. aegypti* and the peridomestic mosquito *Ae. albopictus*; consequently it has a tendency to generate explosive outbreaks. Infection typically results in debilitating joint pain, with occasional cardiac or neurological manifestations (Morrison, 2014). Other alphaviruses, such as Venezuelan equine encephalitis virus (VEEV) and eastern equine encephalitis virus (EEEV), are maintained in zoonotic reservoirs but occasionally infect equids and humans (Weaver *et al.*, 2012). EEEV has the lower incidence of the two viruses, but the higher case fatality rate, which ranges from 50% to 70% in humans (Hollidge *et al.*, 2010).

The positive-sense, single-stranded RNA genome of the alphaviruses encodes four non-structural proteins (nsP1–4) and five structural proteins (C, E3, E2, 6K, E1) (Forrester *et al.*, 2012). At present there are no licensed vaccines or specific treatments to prevent alphavirus infection or disease, thus there has been great interest in developing anti-alphavirus small RNAs (Table 18.2). Six of the nine studies summarized in Table 18.2 assess the effect of treating mammalian cells with small RNAs, including siRNAs, shRNAs and miRNAs, prior to infection. All six studies report that such treatment effectively inhibits viral replication. Interestingly, O'Brien *et al.* (2007) also passaged two strains of VEEV in the presence of an effective pool of siRNAs 10 times but found that only one of the two strains evolved resistance. Only one study, by Parashar *et al.* (2013), applied a small RNA treatment post infection; they found that treatment of mice with siRNAs against CHIKV up to 72 h post infection significantly inhibited virus replication. Three studies reported that combination of either multiple siRNAs or multiple miRNAs had, at best, an additive effect – no synergy was detected.

Two of the studies in Table 18.2 investigated the impact of small RNA treatment on replication of alphaviruses in arthropod vector cells and showed that that pre-treatment of a variety of tick cell lines in culture with siRNA or dsRNA suppressed Semliki forest virus replication (Barry *et al.*, 2013) while dsRNA treatment concurrent with O'nyong nyong virus infection inhibited virus replication in *Anopheles gambiae in vivo* (Keene *et al.*, 2004). In combination with similar studies in flaviviruses (see '*Flaviviridae*', above), this work suggests that exogenous, antiviral small RNAs may offer novel tools to block arbovirus transmission.

Rhabdoviridae

The six genera in the family *Rhabdoviridae* include arthropod-borne pathogens of plants and vertebrates as well directly transmitted pathogens of mammals (Bourhy *et al.*, 2005). The single-stranded, negative-sense RNA genome encodes three structural (N, M and G) and two non-structural (L and either NS or P, which together form the RNA-dependent RNA polymerase) proteins. To date, efforts to utilize exogenous small RNAs to control rhabdovirus infection have focused on the widely studied vesicular stomatitis virus (VSV), although there have also been single studies on bovine ephemeral fever virus (BEFV) (Walker, 2005) and Chandipura virus (CHPV) (Menghani *et al.*, 2012) (Table 18.3). BEFV infects cattle while the host range of VSV is exceptionally broad (Hubalek *et al.*, 2014). Both VSV and BEFV are primarily of veterinary concern, whereas CHPV is known to cause outbreaks of acute encephalitis among children in south Asia (Menghani *et al.*, 2012; Rajasekharan *et al.*, 2014).

Two studies that tested the effect of pre-treatment of cells with small RNAs targeting the viral genome found that both transfection and constitutive expression of siRNAs and shRNAs diminished viral replication by ten- to one hundred-fold in cultured mammalian cells. Of particular interest, Cureton *et al.* (Cureton *et al.*, 2012), following on the results of an siRNA screen for host cell proteins required for VSV replication, found that knocking down the human COPI co-aptamer complex before infection decreased intercellular spread of VSV by up to 90%, suggesting that host-targeted siRNA therapies for this virus could also be successful.

In contrast, Kumar and Arankalle (Kumar and Arankalle, 2010) tested the effect of siRNAs against CHPV virus introduced *in vivo after* infection. They found that intracerebral injection of mice with siRNAs targeting CHPV did not stimulate an interferon response but did decrease virus replication by four logs and significantly improved survival. These results suggest that CHPV siRNAs have potential as therapy for CHPV encephalitis.

Other arboviruses

We identified two studies that investigated the ability of small RNAs to inhibit replication of vector-borne bunyaviruses (Table 18.4). The negative-sense (or ambi-sense) RNA genome of the *Bunyaviridae* is divided into three segments that encode the viral RNA-dependent RNA polymerase and three structural proteins. Four genera of bunyaviruses infect animals; three of these are transmitted by arthropod vectors. The two studies tested the efficacy of siRNAs and shRNAs, respectively, to restrict replication of Hazara virus (genus *Nairovirus*) and Rift Valley fever virus (genus *Phlebovirus*) in cultured primate cells. Hazara virus is a member of the Crimean–Congo haemorrhagic fever virus (CCHFV) serogroup of nairoviruses and serves as a BSL-2 model for CCHFV. Rift Valley fever virus is a phlebovirus that causes severe disease in domestic ruminants and explosive outbreaks in humans in endemic regions in Africa and the Middle East (Hubalek *et al.*, 2014). The paucity of studies on small RNA therapies for Rift Valley fever virus is particularly surprising, given the threat posed by this virus to human health and economic prosperity in its endemic range, as well as its potential for global expansion (Hartley *et al.*, 2011).

While the study on Rift Valley fever virus tested only the effect of treating cells with siRNAs prior to infection, the study on Hazara virus tested the effect of treatment both pre- and post infection. Small RNA treatment was effective in inhibiting replication of both viruses irrespective of the timing of treatment, though there was substantial variation in the efficacy of RNA constructs targeting different regions of the genome. Flusin *et al.* (2011) also reported that siRNAs against the S segment of Hazara virus showed synergy with ribavirin in suppression of viral replication; this is the only study of which we are aware to test the potential synergy of small RNAs with other small molecule therapies on arbovirus replication.

The genera *Asfavirus* (family *Asfarviridae*) and *Orbivirus* (family *Reoviridae*) also contain vector-borne viruses that cause

Table 18.1 Studies testing the efficacy of exogenous small RNAs to inhibit replication of vector-borne flaviviruses

Species (strain)	Hosts tested	In culture or *in vivo*	Cell line or tissue	Small RNA species	Target in viral or host genome	Treated pre- or post infection	Small RNA concentration	Efficacy	Additional treatments	Reference
Dengue-2 (NGC)	Human	In culture	PBMC	siRNA	Host CD14, CLTC, DNM2	24 h pre	25–50 nM	Inhibition of DENV entry and replication; reduction in extracellular DENV		Alhoot *et al.* (2011)
Dengue-2 (NGC)	Human	In culture	HepG2	siRNA	Host *GRP78* gene	24 h pre	50 nM	Suppression of extracellular viral RNA and infectious virions		Alhoot *et al.* (2012)
Dengue-1 (Nauru Island) Dengue-2 (NGC) Dengue-3 (H87) Dengue-4 (Dominica)	Monkey	In culture	Vero	shRNA	Viral 5′ UTR and 3′ UTR	24 h pre from recombinant adenovirus	nr	One of eight shRNAs, targeting 5′ UTR, suppressed replication of all four DENV serotypes >48%		Korrapati *et al.* (2012)
Dengue-1 (West Pac) Dengue-2 (NGC, S210) Dengue-3 (H87) Dengue-4 (H241)	Human Mouse	Both	HuH-7	siRNA	Viral 5′ and 3′ UTR and C, E, NS3, NS5 genes	48 h pre in culture; 24 h pre and 24 and 72 h post *in vivo*	100 nM in culture; 10 mg/kg *in vivo*	Two of seven siRNAs inhibited replication of all four serotypes; siRNA targeting 5′ cyclization sequence was most effective in suppressing all four serotypes in culture and antibody-enhanced disease *in vivo*	siRNAs targeting homologous regions in yellow fever and West Nile virus did not inhibit dengue virus	Stein *et al.* (2011)
Dengue-2 (NGC)	Hamster Human	In culture	BHK-21 MDDC MDM	siRNA	Viral E gene	24 h pre and 24 h post in BHK-21 cells; 24 h post in MDM	300 pmol in culture	siRNAs suppressed virus replication in all cell types and reduced TNF-alpha production by MDDCs		Subramanya *et al.* (2010)
Dengue-1 (GZ02-218)	Mosquito	In culture	C6/36	siRNA	Viral prM gene	4 h pre	1.0 μg	One of four siRNAs protected cells from virus-induced CPE and reduced virus replication		Wu *et al.* (2010)
Dengue-2 (NGC)	Hamster	In culture	BHK-21	Artificial miRNAs (amiRNA)	Viral C, E, NS1, NS3, NS5 genes and 3′ UTR	6 h pre for transient transfection or stable expression from lentivirus	0.125 mg for transient transfection	6 of 21 amiRNAs, targeting C, E, NS1 and NS5, significantly increased survival of DENV-infected cells	Combination of 2 amiRNAs was not more effective than the single most effective amiRNA	Xie *et al.* (2013)
Japanese encephalitis (JaOAr-S982)	Human Pig Monkey Mouse	Both	HEK-293 PS Vero Neuro-2a	shRNA	Viral C E NS5	48 h pre in culture; 7 days pre *in vivo*	500 ng in culture; 1–5 μg/g *in vivo*	All shRNAs significantly reduced virus replication in cultured cells; shRNA targeting NS5 significantly improved mouse survival when delivered by retrovirus but not adenovirus vector		Anantpadma and Vrati (2012)
Japanese encephalitis (Nakayama)	Human Mouse	Both	Vero Neuro- 2a	siRNA shRNA	Viral E gene	24 h pre for transfection of siRNAs; 30 min pre or 30 min post or 6 h post *in vivo*	200 nM siRNAs	Significant inhibition of virus replication in culture and significant protection from mortality *in vivo* when treated 30 min pre or post-infection but not when treated 6 h post-infection	siRNAs also protected mice from mortality due to West Nile virus strain B956 when treated 30 min pre or post-infection but not when treated 6 h post-infection	Kumar *et al.* (2006)

Japanese encephalitis (nr)	Human Monkey Mouse	Both	HepG2 Vero	shRNA	Viral C, M, NS3 genes	5 and 24 h pre and up to 24 h post in culture; concurrent *in vivo*	50–250 nM shRNA or 1.25–5 mg/ml of plasmid in culture; 1.0–5.0 µg/g *in vivo*	Significant inhibition of virus replication in culture and *in vivo*; substantially better survival in mice treated with 5.0 µg/g		Murakami *et al*. (2005)
Japanese encephalitis (NJ 2008)	Human Mouse	Both	SK-N-SH	shRNA	Viral C, E, NS1, NS3, NS4B, NS5 genes	24 h pre in culture and *in vivo*; concurrent in culture	0.8 µg in culture; 5 µg/g *in vivo*	All 10 siRNAs significantly suppressed virus replication in culture, suppressed virus titre in mouse brain *in vivo*, and enhanced mouse survival *in vivo*		Shen *et al*. (2014a)
Japanese encephalitis genotype III (NJ2008, GQ918133; SA14, U14163); genotype I (HEN0701, FJ495189)	Mouse Human	Both	N2a SK-N-SH	siRNA	Viral E, NS3, NS4B genes	24 h pre in culture; 24 h pre or 72 and 24 h pre *in vivo*	na	siRNAs significantly inhibited virus replication in culture and protected against mortality *in vivo*	Lentivirus delivery of siRNAs did not stimulate an interferon response	Shen *et al*. (2014b)
Langat (TP21)	Human Mouse	In culture	HeLa Organotypic hippocampal cultures (OHCs)	siRNA	Viral 5′ and 3′ UTR and C, prM, E, NS1, NS3, NS5 genes	4 h pre or 1 h post in HeLa cells; 24 h pre and 1 h post in OHCs	200 nM in HeLa; 800 nN in OHCs	14 of 19 siRNAs significantly inhibited virus replication in pre-treated HeLa cells; one siRNA tested significantly inhibited virus replication when added post-infection in HeLa cells and in treated OHCs		Maffioli *et al*. (2012)
Tick-borne enceph-alitis (K23, Aina, Sofjin)	Human	In culture	HEK-293	shRNA	Viral E gene	48 h pre	1 µg plasmid	All three shRNAs inhibited replication K23 strain	Only one of three shRNAs inhibited across the three virus subtypes	Achazi *et al*. (2012)
West Nile (2471)	Monkey Mouse	Both	Vero	siRNA	Viral E gene	24 h pre in culture and *in vivo*	100 nM in culture; 180 µg *in vivo*	Significant inhibition of virus replication in culture but only partial protection from virus replication and mortality *in vivo*	siRNA treatment post-infection did not improve survival *in vivo*	Bai *et al*. (2005)
West Nile (NY99)	Human	In culture	Huh-7.5	siRNA	Viral C NS1 NS2A NS2B NS3 NS4B NS5	18 h pre or 10 h post	1 mg	Transfection of siRNAs prior to infection but not post-infection significantly inhibited virus replication	Delivery of siRNA via electroporation slightly increased siRNA efficacy, as did inhibition of virus translation	Geiss *et al*. (2005)
West Nile (Sarafend)	Monkey	In culture	Vero	shRNA	Viral C, NS2B, NS4B genes	Pre via stably transfected cells	na	shRNAs targeting C and NS2B inhibited viral replication but shRNA targeting NS4B did not		(Ong *et al*., 2008)
West Nile (Sarafend)	Monkey	In culture	Vero	siRNA	Viral NS5 gene	24 h pre	2 µg plasmid expressing siRNA	Significant inhibition of virus replication		Ong *et al*. (2006)

Table 18.1 Studies

Species (strain)	Hosts tested	In culture or *in vivo*	Cell line or tissue	Small RNA species	Target in viral or host genome	Treated pre- or post infection	Small RNA concentration	Efficacy	Additional treatments	Reference
West Nile (NY 99)	Human	In culture	HTB-11	siRNA	Viral C, prM, E, NS5 genes	Pre via cells stably expressing a retroviral vector	na	All four siRNAs inhibited virus replication, some significantly		Yang *et al*. (2008)
West Nile (B956)	Hamster Mouse	Both	BHK-21	siRNA	Viral C, prM, E, NS1, NS5 genes	24 h pre or 24 h post in culture; 15 h pre and 6 and 24 h post *in vivo*	150–300 pmol in culture; 50 µg/mouse *in vivo*	5 of 21 siRNAs significantly inhibited replication of both West Nile and St. Louis encephalitis virus (SLEV) in culture; one of two siRNAs protected against death from West Nile and SLEV *in vivo* when administered before infection but offered no protection post-infection	Transfer of macrophages from antiviral siRNA mice prevented encephalitis in new host	Ye *et al*. (2011)
Yellow fever (17DD)	Monkey Mouse	Both	Vero	shRNA	Viral E, NS1, NS5 genes	Pre via stably transfected cell lines in culture; 12 h pre *in vivo*	nr in culture; 50 µg plasmid *in vivo*	shRNAs against E and NS1 significantly suppressed viral replication in culture while NS5 shRNAs did not; E and NS1 shRNAs significantly extended survival in infected mice		Pacca *et al*. (2009)

na, not applicable; nr, not reported.

Blocking arbovirus transmission with small RNAs

The studies discussed in the section on 'Use of exogenous small RNAs to treat arbovirus infection' utilized exogenous, antiviral small RNAs to inhibit arbovirus replication in arthropod vector cells both in culture and *in vivo*. Franz *et al.* (2006) have taken this approach to the next level by generating transgenic *Aedes aegypti* expressing endogenous, inverted-repeat dsRNA targeting the dengue virus serotype 2 genome. *Ae. aegypti* is the major vector of dengue virus and also transmits yellow fever, chikungunya and Zika virus. The inserted dsRNA was coupled to the carboxypeptidase A promoter, and hence was expressed only in the midgut following a bloodmeal, the point at which dengue virus infection is established. Dengue virus serotype 2 replication in this line was severely restricted in early generations (Franz *et al.*, 2006), but by generation 17 expression of the transgene had waned and susceptibility to dengue virus was restored (Franz *et al.*, 2009). Subsequently, Franz *et al.* (2014) generated a new transgenic line expressing the same dsRNA; in this line transgene expression has been stable for 33 generations. Although stability is much improved, the transgene does impose a fitness cost on its carrier and only confers resistance to one of the four serotypes of dengue virus. The progress in this avenue of research is exciting but much work remains before this strategy for transmission control comes to fruition.

As an alternative to targeting the virus genome, it may be possible to target vector genes with siRNAs to disrupt arbovirus transmission. For example, Mysore *et al.* (2013) utilized chitosan/siRNA nanoparticles to knock down semaphorin-1a in developing larvae of *Ae. aegypti*. Later developmental stages showed defects in structures necessary for odour perception. As mosquitoes use olfactory cues for host location, disruption of odour perception offers a route by which to block arbovirus transmission.

Attenuating arboviruses using small RNAs

Stimulation of RNAi can also provide a platform for the rational development of live attenuated arbovirus vaccines and improve performance of arboviruses used for cancer therapy. In particular, insertion of miRNA target sites into viral genomes can suppress virus replication and restrict virus tropism. For example, Pletnev and colleagues generated a live attenuated vaccine candidate for tick-borne encephalitis virus by creating a chimeric virus in which the prM and E genes of dengue virus were replaced with that of tick-borne encephalitis virus (TBEV/DEN4) (Pletnev *et al.*, 1992). However TBEV/DEN4 manifests unacceptably high levels of neuroinvasiveness and neurovirulence for use as a vaccine. Insertion of a single target sequence for a miRNA expressed in brain tissue significantly attenuated both the neuroinvasiveness and neurovirulence of TBEV/DEN4 (Heiss *et al.*, 2011). Insertion of multiple target sequences for brain-expressed miRNAs also retarded evolution of resistance to miRNA via accumulation of mutation in the target sequences (Heiss *et al.*, 2012; Teterina *et al.*, 2014). Similarly, addition of neuron-specific miRNA target sequences into the genome of Semliki forest virus significantly attenuated neuroinvasiveness and neurovirulence in mice (Ylosmaki *et al.*, 2013), although other studies have shown that Semliki forest viruses carrying miRNA target sites are genetically unstable (Ratnik *et al.*, 2013). Interestingly, eastern equine encephalitis virus has achieved through evolution what the studies above seek to engineer, although the outcome is enhanced virulence rather than attenuation. The 3′ UTR of eastern equine encephalitis virus contains a target site for the haematopoietic-cell-specific miRNA miR-142-3p (Trobaugh *et al.*, 2014). The action of miR-142-3p prevents eastern equine encephalitis virus from infecting myeloid cells, but this restriction mutes the innate immune response to the virus, which in turn enhances neurological disease. Clearly it is possible to shift the tropism and virulence of arboviruses by addition of miRNA targets, but it is equally possible for arboviruses to realign these phenotypes via evolution at the target sites.

Challenges, gaps and future research directions

Barriers to delivery of exogenous small RNAs

As discussed above, exogenous small RNAs, including siRNAs, shRNAs, miRNAs and artificial miRNAs (amiRNAs) offer promise as anti-arboviral therapies. However, safe and effective delivery of exogenous small RNAs to target cells within infected hosts remains a major challenge. Barriers to delivery include unintended silencing of off-target genes, which could exacerbate toxicity, RNA-induced stimulation of the immune system, degradation of small RNAs prior to reaching target cells, and failure of siRNA uptake into target cells (Draz *et al.*, 2014). However these challenges are not insurmountable, and several antiviral small RNAs have gone to clinical trial including the anti hepatitis C virus drug SPC3649 (LNA) (Draz *et al.*, 2014).

While 'naked' siRNAs can be delivered to brain and lung tissues, delivery vehicles are generally needed to get small RNAs to their intended target (Whitehead *et al.*, 2009). In addition to peptides and conventional lipid-based carriers, various nanoparticles have been tested for use in small RNA delivery, including silica-based, metal-based, lipid-based and carbon-based nanoparticles, dendrimeres, hydrogels, and semiconductor nanocrystals (Draz *et al.*, 2014). Synthetic lipids have been a particularly popular approach for delivering small RNAs to combat virus infections *in vivo* (Whitehead *et al.*, 2009). For example, siRNAs complexed with a stable nucleic acid–lipid particles (siRNA-SNALP) vehicle have been developed against hepatitis B virus and Ebola virus and synthetic shRNA complexed with lipid nanoparticle vehicle have been developed against hepatitis C virus (Draz *et al.*, 2014). Both lipid-based and peptide-based vehicles have been used to deliver small RNAs for treatment of arboviruses. Kumar *et al.* (Kumar *et al.*, 2006) utilized siRNA-iFECT as a vehicle to deliver siRNAs against West Nile and Japanese encephalitis virus to the brains of mice, while Subramanya *et al.* (2010) used siRNA–peptide complexes to deliver anti-dengue small RNAs to human monocyte-derived dendritic cells.

Design and dosage of small RNA therapies

As noted in 'Use of exogenous small RNAs to treat arbovirus infection', there is relatively little need for prophylactic small RNAs but urgent need for therapeutic small RNAs. To date, however, most studies have demonstrated that pre-treatment of cells in culture or *in vivo* can inhibit subsequent arbovirus infection. Future research efforts must focus on optimizing the design and dosage of small RNAs for treatment of established arbovirus infections. In this context, it is notable that in several studies of alphaviruses (Table 18.2) combining multiple small RNAs did not generate a synergistic increase in virus inhibition. However another potential barrier to successful use of small RNA therapies is the evolution of virus resistance. O'Brien *et al.* (2007) showed that one of two strains of Venezuelan equine encephalitis virus evolved resistance to a pool of siRNAs in as few as 10 passages. Similarly tick-borne encephalitis virus into which a single miRNA target sequences had been inserted rapidly evolved resistance to miRNA-based control (Heiss *et al.*, 2012; Teterina *et al.*, 2014). However insertion of multiple miRNA target sites retarded the evolution of resistance (Heiss *et al.*, 2012; Teterina *et al.*, 2014). Thus, combination of multiple exogenous small RNAs targeting conserved sites in arbovirus genomes may not be necessary to optimize efficacy of small RNA therapies but may avert the evolution of resistant viruses.

Suppressors of RNA interference

Over the course of the host–pathogen arms race (Meyerson and Sawyer, 2011; Obbard and Dudas, 2014), arboviruses have evolved viral suppressors of RNAi (VSRs), including both viral proteins and non-coding RNAs. Kakumani *et al.* (2013) reported that the dengue virus NS4B protein acts as a viral suppressor of RNAi (VSR), while Schnettler *et al.* (2012) found that a subgenomic flavivirus RNAs (sfRNA) derived from the 3′ UTR of both dengue and West Nile virus served as a VSR. In addition, Hussain and Asgari (2014a) described a functional miRNA-like small RNA (DENV-vsRNA-5) encoded by dengue virus serotype 2 that regulates dengue virus replication in *Ae. aegypti*, although a debate is currently under way whether DENV-vsRNA-5 is actually an miRNA-like molecule encoded by dengue virus or whether it is an antiviral siRNA (Hussain and Asgari, 2014b; Skalsky *et al.*, 2014). VSRs and viral dependence on RNAi for regulation of replication may subvert attempts to implement small RNA therapies, particularly if such mechanisms function differently in humans relative to small animal model species.

Another reason that data from small animal models may be misleading in the pursuit of small RNA therapies for arboviruses is that the degree to which RNAi controls replication of a particular virus may be species- or even genotype-specific. Lambrechts *et al.* (2013) generated isofemale lines of *Ae. aegypti* that differed in their Dicer genotype. They showed that these lines differed significantly in susceptibility to genetically distinct isolates of dengue virus, revealing a genotype by genotype interaction between the vector and the virus. Similarly, van Mierlo *et al.* (2014) found that the VP1 protein from Nora virus isolated from *Drosophila melanogaster* acts as a VSR in *D. melanogaster* by antagonizing Argonaute-2 (AGO-2). However VP1 from Nora virus isolated from *Drosophila immigrans* was unable to antagonize AGO-2 in *D. melanogaster*, although it was able to do so in *D. immigrans*. Deeper understanding of host-specificity in viral suppression of RNAi will be necessary for designing effective therapeutic small RNAs.

Broad-spectrum small RNA therapies

Many arboviral infections result in similar symptoms, and the availability of broad-spectrum small RNAs would provide a therapeutic option for patients awaiting (or unable to obtain) specific arboviral diagnosis. Broad-spectrum small RNAs would allow for treatment of either multiple serotypes of a virus such as dengue or multiple species within a genus or even family. Several studies summarized in Tables 18.1 to 18.4 have identified small RNAs with efficacy *in vivo* or in culture against multiple strains of the same virus species (Korrapati *et al.*, 2012; O'Brien, 2007; Stein *et al.*, 2011), while two studies have identified small RNAs with *in vivo* efficacy against pairs of species, namely West Nile and St. Louis encephalitis virus (Ye *et al.*, 2011) and West Nile and Japanese encephalitis virus (Kumar *et al.*, 2006). Further progress in this direction will be critical in the development of small RNA therapies with broad utility for the control of arboviral diseases.

References

Achazi, K., Patel, P., Paliwal, R., Radonic, A., Niedrig, M., and Donoso-Mantke, O. (2012). RNA interference inhibits replication of tick-borne encephalitis virus in vitro. Antiviral Res. *93*, 94–100.

Alhoot, M.A., Wang, S.M., and Sekaran, S.D. (2011). Inhibition of dengue virus entry and multiplication into monocytes using RNA interference. PLoS Negl. Trop. Dis. *5*, e1410.

Alhoot, M.A., Wang, S.M., and Sekaran, S.D. (2012). RNA interference mediated inhibition of dengue virus multiplication and entry in HepG2 cells. PLoS One *7*, e34060.

Anantpadma, M., and Vrati, S. (2012). siRNA-mediated suppression of Japanese encephalitis virus replication in cultured cells and mice. J. Antimicrob. Chemother. *67*, 444–451.

Bai, F., Wang, T., Pal, U., Bao, F., Gould, L.H., and Fikrig, E. (2005). Use of RNA interference to prevent lethal murine west nile virus infection. J. Infect. Dis. *191*, 1148–1154.

Barry, G., Alberdi, P., Schnettler, E., Weisheit, S., Kohl, A., Fazakerley, J.K., and Bell-Sakyi, L. (2013). Gene silencing in tick cell lines using small interfering or long double-stranded RNA. Exp. Appl. Acarol. *59*, 319–338.

Bhatt, S., Gething, P.W., Brady, O.J., Messina, J.P., Farlow, A.W., Moyes, C.L., Drake, J.M., Brownstein, J.S., Hoen, A.G., Sankoh, O., *et al.* (2013). The global distribution and burden of dengue. Nature *496*, 504–507.

Bhomia, M., Sharma, A., Gayen, M., Gupta, P., and Maheshwari, R.K. (2013). Artificial microRNAs can effectively inhibit replication of Venezuelan equine encephalitis virus. Antiviral. Res. *100*, 429–434.

Blair, C.D. (2011). Mosquito RNAi is the major innate immune pathway controlling arbovirus infection and transmission. Future Microbiol. *6*, 265–277.

Bogerd, H.P., Skalsky, R.L., Kennedy, E.M., Furuse, Y., Whisnant, A.W., Flores, O., Schultz, K.L., Putnam, N., Barrows, N.J., Sherry, B., *et al.* (2014). Replication of many human viruses is refractory to inhibition by endogenous cellular microRNAs. J. Virol. *88*, 8065–8076.

Botting, C., and Kuhn, R.J. (2012). Novel approaches to flavivirus drug discovery. Expert. Opin. Drug. Discov. *7*, 417–428.

Bourhy, H., Cowley, J.A., Larrous, F., Holmes, E.C., and Walker, P.J. (2005). Phylogenetic relationships among rhabdoviruses inferred using the L polymerase gene. J. Gen. Virol. *86*, 2849–2858.

Chuang, S.T., Ji, W.T., Chen, Y.T., Lin, C.H., Hsieh, Y.C., and Liu, H.J. (2007). Suppression of bovine ephemeral fever virus by RNA interference. J. Virol. Methods *145*, 84–87.

Cirimotich, C.M., Scott, J.C., Phillips, A.T., Geiss, B.J., and Olson, K.E. (2009). Suppression of RNA interference increases alphavirus replication and virus-associated mortality in *Aedes aegypti* mosquitoes. BMC Microbiol. *9*, 49.

Cook, S., Moureau, G., Kitchen, A., Gould, E.A., de Lamballerie, X., Holmes, E.C., and Harbach, R.E. (2012). Molecular evolution of the insect-specific flaviviruses. J. Gen. Virol. *93*, 223–234.

Cureton, D.K., Burdeinick-Kerr, R., and Whelan, S.P. (2012). Genetic inactivation of COPI coatomer separately inhibits vesicular stomatitis virus entry and gene expression. J. Virol. *86*, 655–666.

Dash, P.K., Tiwari, M., Santhosh, S.R., Parida, M., and Lakshmana Rao, P.V. (2008). RNA interference mediated inhibition of Chikungunya virus replication in mammalian cells. Biochem. Biophys. Res. Commun. *376*, 718–722.

Ding, S.W., and Voinnet, O. (2007). Antiviral immunity directed by small RNAs. Cell *130*, 413–426.

Draz, M.S., Fang, B.A., Zhang, P., Hu, Z., Gu, S., Weng, K.C., Gray, J.W., and Chen, F.F. (2014). Nanoparticle-mediated systemic delivery of siRNA for treatment of cancers and viral infections. Theranostics *4*, 872–892.

Flusin, O., Vigne, S., Peyrefitte, C.N., Bouloy, M., Crance, J.M., and Iseni, F. (2011). Inhibition of Hazara nairovirus replication by small interfering RNAs and their combination with ribavirin. Virol. J. *8*, 249.

Forrester, N.L., Palacios, G., Tesh, R.B., Savji, N., Guzman, H., Sherman, M., Weaver, S.C., and Lipkin, W.I. (2012). Genome-scale phylogeny of the alphavirus genus suggests a marine origin. J. Virol. *86*, 2729–2738.

Franz, A.W., Sanchez-Vargas, I., Adelman, Z.N., Blair, C.D., Beaty, B.J., James, A.A., and Olson, K.E. (2006). Engineering RNA interference-based resistance to dengue virus type 2 in genetically modified *Aedes aegypti*. Proc. Natl. Acad. Sci. U.S.A. *103*, 4198–4203.

Franz, A.W., Sanchez-Vargas, I., Piper, J., Smith, M.R., Khoo, C.C., James, A.A., and Olson, K.E. (2009). Stability and loss of a virus resistance phenotype over time in transgenic mosquitoes harbouring an antiviral effector gene. Insect. Mol. Biol. *18*, 661–672.

Franz, A.W., Sanchez-Vargas, I., Raban, R.R., Black, W.C.T., James, A.A., and Olson, K.E. (2014). Fitness impact and stability of a transgene conferring resistance to dengue-2 virus following introgression into a genetically diverse *Aedes aegypti* strain. PLoS Negl. Trop. Dis. *8*, e2833.

Geiss, B.J., Pierson, T.C., and Diamond, M.S. (2005). Actively replicating West Nile virus is resistant to cytoplasmic delivery of siRNA. Virol. J. *2*, 53.

Gubler, D.J. (2004). The changing epidemiology of yellow fever and dengue, 1900 to 2003: full circle? Comp. Immunol. Microbiol. Infect. Dis. *27*, 319–330.

Hanley, K.A., Monath, T.P., Weaver, S.C., Rossi, S.L., Richman, R.L., and Vasilakis, N. (2013). Fever versus fever: the role of host and vector susceptibility and interspecific competition in shaping the current and future distributions of the sylvatic cycles of dengue virus and yellow fever virus. Infect. Genet. Evol. *19*, 292–311.

Hartley, D.M., Rinderknecht, J.L., Nipp, T.L., Clarke, N.P., and Snowder, G.D. (2011). Potential effects of Rift Valley fever in the United States. Emerg. Infect. Dis. *17*, e1.

Heiss, B.L., Maximova, O.A., and Pletnev, A.G. (2011). Insertion of microRNA targets into the flavivirus genome alters its highly neurovirulent phenotype. J. Virol. *85*, 1464–1472.

Heiss, B.L., Maximova, O.A., Thach, D.C., Speicher, J.M., and Pletnev, A.G. (2012). MicroRNA targeting of neurotropic flavivirus: effective control of virus escape and reversion to neurovirulent phenotype. J. Virol. *86*, 5647–5659.

Hollidge, B.S., Gonzalez-Scarano, F., and Soldan, S.S. (2010). Arboviral encephalitides: transmission, emergence, and pathogenesis. J. Neuroimmune Pharmacol. *5*, 428–442.

Hubalek, Z., Rudolf, I., and Nowotny, N. (2014). Arboviruses pathogenic for domestic and wild animals. Adv. Virus. Res. *89*, 201–275.

Hussain, M., and Asgari, S. (2014a). MicroRNA-like viral small RNA from Dengue virus 2 autoregulates its replication in mosquito cells. Proc. Natl. Acad. Sci. U.S.A. *111*, 2746–2751.

Hussain, M., and Asgari, S. (2014b). Reply to Skalsky *et al.*: A microRNA-like small RNA from Dengue virus. Proc. Natl. Acad. Sci. U.S.A. *111*, E2360.

Kakumani, P.K., Ponia, S.S., Rajgokul, K.S., Sood, V., Chinnappan, M., Banerjea, A.C., Medigeshi, G.R., Malhotra, P., Mukherjee, S.K., and Bhatnagar, R.K. (2013). Role of RNA interference (RNAi) in dengue virus replication and identification of NS4B as an RNAi suppressor. J. Virol. *87*, 8870–8883.

Keene, K.M., Foy, B.D., Sanchez-Vargas, I., Beaty, B.J., Blair, C.D., and Olson, K.E. (2004). RNA interference acts as a natural antiviral response to O'nyong-nyong virus (Alphavirus; Togaviridae) infection of Anopheles gambiae. Proc. Natl. Acad. Sci. U.S.A. *101*, 17240–17245.

Keita, D., Heath, L., and Albina, E. (2010). Control of African swine fever virus replication by small interfering RNA targeting the *A151R* and *VP72* genes. Antivir. Ther. *15*, 727–736.

Kingsolver, M.B., Huang, Z., and Hardy, R.W. (2013). Insect antiviral innate immunity: pathways, effectors, and connections. J. Mol. Biol. *425*, 4921–4936.

Korrapati, A.B., Swaminathan, G., Singh, A., Khanna, N., and Swaminathan, S. (2012). Adenovirus delivered short hairpin RNA targeting a conserved site in the 5′ non-translated region inhibits all four serotypes of dengue viruses. PLoS Negl. Trop. Dis. *6*, e1735.

Kumar, P., Lee, S.K., Shankar, P., and Manjunath, N. (2006). A single siRNA suppresses fatal encephalitis induced by two different flaviviruses. PLoS Med. 3, e96.

Kumar, S., and Arankalle, V.A. (2010). Intracranial administration of P gene siRNA protects mice from lethal Chandipura virus encephalitis. PLoS One *5*, e8615.

Lam, S., Chen, K.C., Ng, M.M., and Chu, J.J. (2012). Expression of plasmid-based shRNA against the E1 and nsP1 genes effectively silenced Chikungunya virus replication. PLoS One *7*, e46396.

Lambrechts, L., Quillery, E., Noel, V., Richardson, J.H., Jarman, R.G., Scott, T.W., and Chevillon, C. (2013). Specificity of resistance to dengue virus isolates is associated with genotypes of the mosquito antiviral gene Dicer-2. Proc. Biol. Sci. *280*, 20122437.

Lani, R., Moghaddam, E., Haghani, A., Chang, L.Y., AbuBakar, S., and Zandi, K. (2014). Tick-borne viruses: a review from the perspective of therapeutic approaches. Ticks Tick Borne Dis. *5*, 457–465.

Le Flohic, G., Porphyre, V., Barbazan, P., and Gonzalez, J.P. (2013). Review of climate, landscape, and viral genetics as drivers of the Japanese encephalitis virus ecology. PLoS Negl. Trop. Dis. *7*, e2208.

Li, Y., Lu, J., Han, Y., Fan, X., and Ding, S.W. (2013). RNA interference functions as an antiviral immunity mechanism in mammals. Science *342*, 231–234.

Lim, S.P., Wang, Q.Y., Noble, C.G., Chen, Y.L., Dong, H., Zou, B., Yokokawa, F., Nilar, S., Smith, P., Beer, D., *et al.* (2013). Ten years of dengue drug discovery: progress and prospects. Antiviral Res. *100*, 500–519.

Maffioli, C., Grandgirard, D., Leib, S.L., and Engler, O. (2012). SiRNA inhibits replication of Langat virus, a member of the tick-borne encephalitis virus complex in organotypic rat brain slices. PLoS One *7*, e44703.

Maillard, P.V., Ciaudo, C., Marchais, A., Li, Y., Jay, F., Ding, S.W., and Voinnet, O. (2013). Antiviral RNA interference in mammalian cells. Science *342*, 235–238.

Menghani, S., Chikhale, R., Raval, A., Wadibhasme, P., and Khedekar, P. (2012). Chandipura Virus: an emerging tropical pathogen. Acta Trop. *124*, 1–14.

Meyerson, N.R., and Sawyer, S.L. (2011). Two-stepping through time: mammals and viruses. Trends Microbiol. *19*, 286–294.

van Mierlo, J.T., Overheul, G.J., Obadia, B., van Cleef, K.W., Webster, C.L., Saleh, M.C., Obbard, D.J., and van Rij, R.P. (2014). Novel *Drosophila* viruses encode host-specific suppressors of RNAi. PLoS Pathog. *10*, e1004256.

Morrison, T.E. (2014). Re-emergence of chikungunya virus. J Virol. *88*, 11644–11647.

Murakami, M., Ota, T., Nukuzuma, S., and Takegami, T. (2005). Inhibitory effect of RNAi on Japanese encephalitis virus replication in vitro and *in vivo*. Microbiol. Immunol. *49*, 1047–1056.

Myles, K.M., Wiley, M.R., Morazzani, E.M., and Adelman, Z.N. (2008). Alphavirus-derived small RNAs modulate pathogenesis in disease vector mosquitoes. Proc. Natl. Acad. Sci. U.S.A. *105*, 19938–19943.

Mysore, K., Flannery, E.M., Tomchaney, M., Severson, D.W., and Duman-Scheel, M. (2013). Disruption of *Aedes aegypti* olfactory system development through chitosan/siRNA nanoparticle targeting of semaphorin-1a. PLoS Negl. Trop. Dis. *7*, e2215.

Obbard, D.J., and Dudas, G. (2014). The genetics of host-virus coevolution in invertebrates. Curr. Opin. Virol. *8C*, 73–78.

O'Brien, L. (2007). Inhibition of multiple strains of Venezuelan equine encephalitis virus by a pool of four short interfering RNAs. Antiviral Res. *75*, 20–29.

Ong, S.P., Choo, B.G., Chu, J.J., and Ng, M.L. (2006). Expression of vector-based small interfering RNA against West Nile virus effectively inhibits virus replication. Antiviral Res. 72, 216–223.

Ong, S.P., Chu, J.J., and Ng, M.L. (2008). Inhibition of West Nile virus replication in cells stably transfected with vector-based shRNA expression system. Virus Res. *135*, 292–297.

Pacca, C.C., Severino, A.A., Mondini, A., Rahal, P., D'Avila S.G., Cordeiro, J.A., Nogueira, M.C., Bronzoni, R.V., and Nogueira, M.L. (2009).

RNA interference inhibits yellow fever virus replication in vitro and *in vivo*. Virus Genes *38*, 224–231.
Parashar, D., Paingankar, M.S., Kumar, S., Gokhale, M.D., Sudeep, A.B., Shinde, S.B., and Arankalle, V.A. (2013). Administration of E2 and NS1 siRNAs inhibit chikungunya virus replication in vitro and protects mice infected with the virus. PLoS Negl. Trop. Dis. *7*, e2405.
Pletnev, A.G., Bray, M., Huggins, J., and Lai, C.J. (1992). Construction and characterization of chimeric tick-borne encephalitis/dengue type 4 viruses. Proc. Natl. Acad. Sci. U.S.A. *89*, 10532–10536.
Rajasekharan, S., Rana, J., Gulati, S., Gupta, V., and Gupta, S. (2014). Neuroinvasion by Chandipura virus. Acta Trop. *135*, 122–126.
Ramirez-Carvajal, L., and Long, C.R. (2012). Down-regulation of viral replication by lentiviral-mediated expression of short-hairpin RNAs against vesicular stomatitis virus ribonuclear complex genes. Antiviral Res. *95*, 150–158.
Ratnik, K., Viru, L., and Merits, A. (2013). Control of the rescue and replication of Semliki Forest virus recombinants by the insertion of miRNA target sequences. PLoS One *8*, e75802.
Schnettler, E., Sterken, M.G., Leung, J.Y., Metz, S.W., Geertsema, C., Goldbach, R.W., Vlak, J.M., Kohl, A., Khromykh, A.A., and Pijlman, G.P. (2012). Noncoding flavivirus RNA displays RNA interference suppressor activity in insect and Mammalian cells. J. Virol. *86*, 13486–13500.
Scott, T., Paweska, J.T., Arbuthnot, P., and Weinberg, M.S. (2012). Pathogenic effects of Rift Valley fever virus NSs gene are alleviated in cultured cells by expressed antiviral short hairpin RNAs. Antivir. Ther. *17*, 643–656.
Seyhan, A.A., Alizadeh, B.N., Lundstrom, K., and Johnston, B.H. (2007). RNA interference-mediated inhibition of Semliki Forest virus replication in mammalian cells. Oligonucleotides *17*, 473–484.
Shapiro, J.S., Schmid, S., Aguado, L.C., Sabin, L.R., Yasunaga, A., Shim, J.V., Sachs, D., Cherry, S., and tenOever, B.R. (2014). Drosha as an interferon-independent antiviral factor. Proc. Natl. Acad. Sci. U.S.A. *111*, 7108–7113.
Shen, T., Liu, K., Miao, D., Cao, R., and Chen, P. (2014a). Effective inhibition of Japanese encephalitis virus replication by shRNAs targeting various viral genes in vitro and in vivo. Virology *454–455*, 48–59.
Shen, T., Liu, K., Miao, D., Cao, R., Zhou, B., and Chen, P. (2014b). Lentivirus-mediated RNA interference against Japanese encephalitis virus infection in vitro and in vivo. Antiviral Res. *108*, 56–64.
Siomi, M.C., Sato, K., Pezic, D., and Aravin, A.A. (2011). PIWI-interacting small RNAs: the vanguard of genome defence. Nat. Rev. Mol. Cell Biol. *12*, 246–258.
Skalsky, R.L., Olson, K.E., Blair, C.D., Garcia-Blanco, M.A., and Cullen, B.R. (2014). A 'microRNA-like' small RNA expressed by Dengue virus? Proc. Natl. Acad. Sci. U.S.A. *111*, E2359.
Staples, J.E., and Fischer, M. (2014). Chikungunya virus in the Americas – what a vectorborne pathogen can do. N. Engl. J. Med. *371*, 887–889.
Stassen, L., Huismans, H., and Theron, J. (2007). Silencing of African horse sickness virus VP7 protein expression in cultured cells by RNA interference. Virus Genes *35*, 777–783.
Stein, D.A., Perry, S.T., Buck, M.D., Oehmen, C.S., Fischer, M.A., Poore, E., Smith, J.L., Lancaster, A.M., Hirsch, A.J., Slifka, M.K., *et al.* (2011). Inhibition of dengue virus infections in cell cultures and in AG129 mice by a small interfering RNA targeting a highly conserved sequence. J. Virol. *85*, 10154–10166.
Subramanya, S., Kim, S.S., Abraham, S., Yao, J., Kumar, M., Kumar, P., Haridas, V., Lee, S.K., Shultz, L.D., Greiner, D., *et al.* (2010). Targeted delivery of small interfering RNA to human dendritic cells to suppress dengue virus infection and associated proinflammatory cytokine production. J. Virol. *84*, 2490–2501.
Svoboda, P. (2014). Renaissance of mammalian endogenous RNAi. FEBS Lett. *588*, 2550–2556.
Teterina, N.L., Liu, G., Maximova, O.A., and Pletnev, A.G. (2014). Silencing of neurotropic flavivirus replication in the central nervous system by combining multiple microRNA target insertions in two distinct viral genome regions. Virology *456–457*, 247–258.
Trobaugh, D.W., Gardner, C.L., Sun, C., Haddow, A.D., Wang, E., Chapnik, E., Mildner, A., Weaver, S.C., Ryman, K.D., and Klimstra, W.B. (2014). RNA viruses can hijack vertebrate microRNAs to suppress innate immunity. Nature *506*, 245–248.
Walker, P.J. (2005). Bovine ephemeral fever in Australia and the world. Curr. Top. Microbiol. Immunol. *292*, 57–80.
Weaver, S.C., Winegar, R., Manger, I.D., and Forrester, N.L. (2012). Alphaviruses: population genetics and determinants of emergence. Antiviral Res. *94*, 242–257.
Whitehead, K.A., Langer, R., and Anderson, D.G. (2009). Knocking down barriers: advances in siRNA delivery. Nat. Rev. Drug Discov. *8*, 129–138.
Wu, X., Hong, H., Yue, J., Wu, Y., Li, X., Jiang, L., Li, L., Li, Q., Gao, G., and Yang, X. (2010). Inhibitory effect of small interfering RNA on dengue virus replication in mosquito cells. Virol. J. *7*, 270.
Xie, P.W., Xie, Y., Zhang, X.J., Huang, H., He, L.N., Wang, X.J., and Wang, S.Q. (2013). Inhibition of Dengue virus 2 replication by artificial micrornas targeting the conserved regions. Nucleic Acid Ther. *23*, 244–252.
Yang, Y., Wu, C., Wu, J., Nerurkar, V.R., Yanagihara, R., and Lu, Y. (2008). Inhibition of West Nile Virus replication by retrovirus-delivered small interfering RNA in human neuroblastoma cells. J. Med. Virol. *80*, 930–936.
Ye, C., Abraham, S., Wu, H., Shankar, P., and Manjunath, N. (2011). Silencing early viral replication in macrophages and dendritic cells effectively suppresses flavivirus encephalitis. PLoS One *6*, e17889.
Ylosmaki, E., Martikainen, M., Hinkkanen, A., and Saksela, K. (2013). Attenuation of Semliki Forest virus neurovirulence by microRNA-mediated detargeting. J. Virol. *87*, 335–344.
Zhang, K., Chen, Y., Pan, J., Ahola, T., and Guo, D. (2009). Lentiviral vector-derived shRNAs confer enhanced suppression of Semliki forest virus replication in BHK-21 cells compared to shRNAs expressed from plasmids. Biotechnol. Lett. *31*, 501–508.
Zhou, R., and Rana, T.M. (2013). RNA-based mechanisms regulating host–virus interactions. Immunol. Rev. *253*, 97–111.

Genetically Modified Vectors for Control of Arboviruses

19

Ken E. Olson and Alexander W.E. Franz

Abstract

Arthropod-borne viruses (arboviruses) are maintained in nature by cycling between haematophagus arthropod vectors and vertebrate hosts. Medically important mosquito vectors (*Aedes* and *Culex* spp.) and the arboviruses they transmit have increased their geographic range through human travel, trade and climate change. There are no vaccines or therapeutic drugs readily available to control most arboviruses leaving large segments of the world's human population at risk for disease. Arboviral disease prevention relies on vector control by eliminating breeding sites and using outdoor or indoor insecticides to reduce vector-human contact. These approaches are costly, they impact the environment and they are difficult to sustain. Despite vector control efforts, arboviral diseases continue to emerge in new geographic regions (eg. chikungunya in western hemisphere) or pose increasing threats to human health in urban areas (eg. dengue). New approaches are needed to impede arbovirus transmission and control the emergence, prevalence and spread of arboviruses. In this chapter, we will discuss genetically modified vectors (GMVs) as a potentially important advancement in vector control. This chapter will focus mainly on dengue viruses (DENVs) and the mosquito vector, *Ae. aegypti*, but will discuss other arboviruses of the families *Flaviviridae* and *Togaviridae* and their vectors where appropriate.

Introduction

Arthropod-borne viruses (arboviruses) are maintained in nature by cycling between haematophagus arthropod vectors and susceptible vertebrate hosts (Weaver and Barrett, 2004; Weaver and Reisen, 2010). Mosquito vectors are essential components of the transmission cycle of a number of clinically important arboviruses. The female mosquito requires a bloodmeal for egg development and if the vector is competent to support infection it can acquire virus from the blood of a viraemic host. In the mosquito, the arbovirus must infect the midgut, replicate, amplify and disseminate to salivary glands before the virus can be transmitted to the next vertebrate host. There are over 500 catalogued arboviruses mainly from four virus families (*Bunyaviridae, Flaviviridae, Reoviridae* and *Togaviridae*) and over 3000 described mosquito species (Drebot *et al.*, 2002) (www.mosquito.org/mosquito-info). Only a relatively small number of vector species are competent to transmit arboviruses to humans. For example, a single mosquito species, *Aedes aegypti* (*Aae*), accounts for millions of transmission events each year as the primary vector of dengue viruses (DENV) to humans (Bhatt *et al.*, 2013). There are few vaccines or antiviral therapies to prevent or control arbovirus infections in humans and conventional vector control approaches have failed to prevent outbreaks of DENV in urban settings. New tools for disease control are sorely needed.

Recent advances in knowledge and technologies associated with arbovirus structure, replication and genetics have enhanced our understanding of infection in vertebrate hosts and arthropods. In parallel, major developments in arthropod genetics and genomics over the last several decades have allowed new insights into the molecular biology of arbovirus–vector interactions and a better appreciation for the innate immune responses arboviruses encounter in the vector. A complete genome sequence of *Aae* (1.376 gigabase base pairs; (Nene *et al.*, 2007) is now available through VectorBase at www.vectorbase.org/organisms/aedes-aegypti. Additionally, transcriptome analysis of *Aae* infected with DENV compared with uninfected mosquitoes has revealed complex changes in vector gene transcription and has led to the discovery of genes differentially expressed during infection (Behura *et al.*, 2011; Bonizzoni *et al.*, 2012). Coupled with these advances, genetic manipulation of genomes of *Aae* and several other medically important vectors is now possible and tools and techniques are available for heritable expression of genes that negatively impact vector fitness or the vector's competence to transmit an arbovirus. Considerable effort has been expended to develop genetically modified vectors (GMVs) as tools for arbovirus disease control (Terenius *et al.*, 2008). GMVs are developed by introducing one or more genes into the nuclear genome of the vector depending on the GMV's desired effect on the vector population. GMVs can be engineered to reduce vector populations or replace existing populations with vector populations that are refractory to the pathogen, making them incompetent for transmitting arboviruses. This chapter will focus mainly on DENVs (serotypes 1–4; *Flavivirus; Flaviviridae*) and *Aae* because of their role in transmitting the most clinically important arbovirus affecting humans. *Aae* is probably the best studied vector of arboviruses and is easily reared and maintained in the laboratory for genetic manipulation. References to other arboviruses [flaviviruses and alphaviruses (family: *Togaviridae*)] and their vectors are made where additional studies provide unique information. Finally, tools such as Skeeter Buster (Magori *et al.*, 2009; Okamoto *et al.*, 2014), have been developed to provide a biologically detailed model of *Aae* population dynamics parameterized for a tropical city to quantitatively assess whether a GMV-based strategy for disease control can impact vector populations and their ability to transmit virus.

Aedes aegypti

Developing GMVs based on *Aae* for arbovirus control strategies is a 'no brainer'. *Aae* is an anthropophilic, peridomestic, day-biting mosquito vector having a limited flight range with most of its daily activities occurring indoors (Delatte *et al.*, 2010; Ooi *et al.*, 2006). Adult, female *Aae* take multiple bloodmeals during their lifetime increasing the risk of arbovirus transmission. Although *Aae* originated in Africa where it adapted to human habitats, the vector is now widespread in the Americas, Asia, Oceania, and parts of Australia mirroring the rise of human commerce and trade and the rise of tropical urban centres (Moore *et al.*, 2013; Powell and Tabachnick, 2013; Tabachnick, 1991). *Aae* is responsible for explosive urban DENV epidemics and is the vector species most often associated with the emergence of severe forms of DEN disease [DEN haemorrhagic fever (DHF) and DEN shock syndrome (DSS)] (Gubler, 2002; Weaver and Reisen, 2010). The incidence of DHF has dramatically risen since the 1950s. Greater than 2 billion people in 100 countries are at risk for DENV infection and 100–390 million are infected annually with tens of millions of cases manifesting some level of disease (Bhatt *et al.*, 2013; Guzman and Isturiz, 2010). All four DENV serotypes are readily maintained in an urban transmission cycle between *Aae* and humans (Mackenzie *et al.*, 2004). The DENV serotypes are hyperendemic in many regions contributing to the increased frequency of DHF and DSS (Gubler, 2002). *Aae* is also an important vector of yellow fever virus (YFV; *Flavivirus*) (Gubler, 2002; Weaver and Reisen, 2010) and chikungunya virus (CHIKV; *Alphavirus*) (Weaver, 2014). Historically, *Aae* has transmitted YFV in Africa and the Americas leading to major outbreaks in urban settings with significant mortality. Despite an efficacious vaccine for YFV, WHO estimates 200,000 human cases and 30,000 deaths, mostly in Africa, occur yearly making YFV a continuing health concern (Barrett and Higgs, 2007). The continuing fear is that YFV will spread anew into urban centres through its association with *Aae*. *Aae* also transmits various genotypes of CHIKV in Africa and Asia, islands in the Caribbean, and CHIKV is now emerging in the USA (Weaver, 2014) with potential significant impacts on human health and our medical infrastructure. The lack of licensed vaccines for DENV and CHIKV, and the absence of specific therapeutic treatments for DENV, YFV and CHIKV make *Aae* and the arboviruses it transmits a global public health threat. Conventional vector control efforts have largely failed to contain DENV transmission. *Aae* is a logical, initial choice for developing GMV-based control strategies due to the vectors central role in transmitting medically important arboviruses and relative ease of rearing and maintaining the vector in the laboratory. Finally, significant advances have been made towards our understanding of arbovirus–*Aae* interactions that lead to vector competence (Behura *et al.*, 2011; Bonizzoni *et al.*, 2012; Sanchez-Vargas *et al.*, 2009).

Vector competence and vectorial capacity

What makes *Aae* or any haematophagus arthropod vector competent for transmitting arboviruses? A few terms are helpful in this regard when considering GMV's as a disease control tool. The vectorial capacity (C) of an haematophagus arthropod includes not only the concepts of vector competence which refers to the intrinsic permissiveness of a vector to infection, replication, and transmission of an arbovirus but also vector abundance, host specificity, vector longevity, and the extrinsic incubation period (EIP) or length of time between the vector's initial infection and ability to transmit virus. C represents the number of infective bites generated by a single case of a particular viraemic individual, on a daily basis. The mathematical concept of C is defined as $C=(ma^2)(pn)(b)/{-\log(p)}$ (Ciota and Kramer, 2013; Delatte *et al.*, 2010). C is calculated by accounting for the influence of several important epidemiological factors that predict the likelihood of vector-borne disease (Anderson and Rico-Hesse, 2006; Bosio *et al.*, 1998; Delatte *et al.*, 2010). GMV's that negatively impact one or more components of this equation may significantly reduce vector-borne disease. One key variable explaining C is the 'infectability' of vectors and specifically, refers to the proportion of vectors that acquire an infectious bloodmeal and successfully become infected and transmit an arbovirus (b in the above equation). The b variable is also known as vector competence (Ciota and Kramer, 2013; Delatte *et al.*, 2010). The development of GMV strategies that block vector competence by reducing arbovirus entry, clearing vectors of the virus or eliminating virus transmission could bring the b value to zero, eliminating C and potentially disrupting the transmission cycle. While a determination of b is an important indicator of a mosquito population's ability to transmit DENV, other components of C play a large role in determining DENV transmission. For instance, C is heavily influenced by (p) the daily probability of vector survival, (n) the EIP and (ma^2) which combines blood feeding frequency with human biting rate. Anderson and Rico-Hesse (2006) have also shown that DENV genetics plays a role in determining C by showing that Southeast Asia (SEA) genotypes of DENV2 are more efficient in infecting field strains of *Aae* than the American (AM) genotype due to SEA genotypes having a shorter EIP (n) than AM genotypes (Anderson and Rico-Hesse, 2006). Additionally, given *Aae*'s preference for humans as targets for feeding in an urban transmission cycle, human biting rates will be high affecting C (Delatte *et al.*, 2010). Arboviral disease control strategies that decrease (b, ma^2), or daily survival (p) or increase (n) should adversely impact C and reduce the probability that mosquitoes will complete EIP for a given arbovirus. Modern molecular tools, genetics, sequencing, virus–vector interaction studies and transgenesis approaches are allowing researchers to identify and test gene-based strategies that significantly modulate the mathematical variables associated with C.

Aedes aegypti control

Global eradication of most arboviruses may be impossible since they are zoonotic viruses maintained in transmission cycles between animals (often birds or mammals) and arthropods. While DENVs are maintained in distinct sylvan transmission cycles between mosquito vectors and non-human primates, most DENV outbreaks are the result of urban transmission cycles between *Aae* and humans (Vasilakis *et al.*, 2007). Substantial control of some arboviral diseases, particularly DEN, requires improved vector control in combination with efficacious vaccines and drugs (Garcia *et al.*, 2009). Conventional vector control approaches rely on population reduction of the vector through the use of insecticides and reducing available sites needed for insect development (Gubler, 2002). In the past, *Aae* eradication campaigns combining source reduction of larval development sites with insecticides (e.g. organochlorine, dichloro-diphenyl-trichloroethane or DDT) were successful in eliminating mosquitoes that transmit arboviruses, but these programmes could not be sustained, because they were environmentally unsound and insecticide resistance occurred among vector populations (Lumjuan *et al.*, 2011). In recent decades, pyrethroid insecticides have played a major global role in the control of adult *Aae*, often in combination with the organophosphate insecticide (temephos) to control immatures (Bisset *et al.*, 2013). However, the evolution of resistance in *Aae* and other vectors to insecticides has compromised the effectiveness of control programmes (Bisset *et al.*, 2013; Saavedra-Rodriguez *et al.*, 2012, 2014). Unfortunately, the insecticide armamentarium for vector control is limited; few new chemical insecticides have been produced in the last several decades. Development of new chemical insecticides with new modes of actions is daunting; the low hanging fruit of chemical insecticides for vector control has been picked. Other control approaches such as bio-insecticides are promising with the use of *Bacillus thuringiensis israeliensis* (Bti) to target larvae (Mittal, 2003). *Aae* have developed little to no resistance to Bti toxin and Bti is environmentally safe, however, this approach to control is logistically challenging due to the wide range of containers *Aae* can utilize as breeding sites (Medlock *et al.*, 2012). *Wolbachia pipientis* is an obligate intracellular bacterium that lives inside insects and is transmitted vertically from mother to offspring (McGraw *et al.*, 2002; McGraw and O'Neill, 2013). In seminal studies, researchers forced *Wolbachia* (MelPop strain) infection in previously uninfected *Drosophila simulans* and *Aae* (McGraw *et al.*, 2002; McMeniman *et al.*, 2009). *Wolbachia* prevents successful reproduction between infected males and uninfected females constituting a genetic system to favour *Wolbachia* infection of a vector population (McGraw and O'Neill, 2013). Significantly, *Wolbachia* infected mosquitoes are resistant to arbovirus infections. *Wolbachia* (wMel and wMelpop strains) negatively impacts vectorial capacity (C) by interfering with arbovirus (DENV, YFV, CHIKV) replication and vector competence (b) although the mechanism of resistance is not fully understood (Moreira *et al.*, 2009a; Walker *et al.*, 2011). *Wolbachia* strain (wMelPoP) is associated with shortening vector life spans which would affect (C) by significantly reducing daily survival (p) (Moreira *et al.*, 2009b). The effectiveness of *Wolbachia* as a biocontrol agent is currently under intense study in Australia and Southeast Asia. Nevertheless, the use of GMVs to control transmission of arboviruses is a promising approach to disease control and at least one GMV approach is being tested in the field.

Genetically modified vectors for arbovirus control

In principle, arthropod vectors can be genetically modified to conditionally express genes that lower vector fitness (vector population reduction) or immunize vectors by expressing pathogen-specific molecules that impair vector competence (vector population replacement). To test these ideas three hurdles had to be overcome, (i) transgenesis systems had to be developed for efficient transformation of targeted vector species, (ii) effector molecules needed to be identified that either lowered vector survivorship or induced an anti-pathogen phenotype in the vector, and (iii) mechanisms needed to be developed to drive the effector system into wild-type, virus competent vector populations particularly for population replacement strategies (Chen *et al.*, 2007; James, 2005; Robert *et al.*, 2013; Windbichler *et al.*, 2007). These were not trivial undertakings. During the last several decades there has been significant progress in (i) and (ii). Development of new technologies for gene drive (iii) is currently being undertaken and new mathematical models have been generated for introgressing effector genes into vector populations. Recent studies using a mathematical model suggest that if GMVs are released in the right proportions complex drive systems may not be necessary to attain limited drive of an anti-pathogen gene (Okamoto *et al.*, 2014).

GMV, a short history

Stable, heritable germline transformation of insects was demonstrated over 30 years ago in *D. melanogaster* (*Dm*) using *P* (class II) transposable element (TE) as insertion vector (Rubin and Spradling, 1982). Class II TEs possess short inverted terminal repeats (ITRs) and express a transposase to generate DNA intermediates that invade host DNA by a cut-and-paste mechanism leading to target site duplication. Embryos of eye colour mutant flies were co-injected with plasmid DNA of the autonomous *P*-element encoding the *rosy* (*ry*) eye colour gene. Twenty-50% of injected *ry* mutant embryos developed into transformed flies having wild-type like eye colour (Rubin and Spradling, 1982). However, in subsequent generations researchers observed *P*-element excision and transposition rates of about 2.4% which could affect stability of the transgene. The *P*-element TE was then developed into a binary donor-helper transformation system in which the transposase of *P* was expressed *in trans* from a separate helper plasmid and the *P* sequence of the original TE encoding plasmid (donor) was disrupted by insertion of a selection marker. Using this non-autonomous *P*-element system had the advantage that one could achieve stable TE integration in the initial transformed generation but the transgene could no longer be excised (transposition) in subsequent generations because the non-integrated helper plasmid would be lost.

The binary *P* element transformation system became the design principle for other TEs, which were later used for the germline transformation of mosquitoes (Catteruccia *et al.*, 2000b; Coates *et al.*, 1998; Grossman *et al.*, 2001; Jasinskiene *et al.*, 1998). Initially, researchers tried to use the *P*-element to transform mosquito species such as *Anopheles gambiae* and

Ae. triseriatus (McGrane *et al.*, 1988; Miller *et al.*, 1987). While transformation was reported in each species the transformation events failed to show true transposition, transformation was inefficient and transformants were lost in subsequent generations. Additional attempts were made to use the *P*-element for the transformation of other non-drosophilid insects, but these were unsuccessful making it clear that the functionality of *P* was largely restricted to drosophilids. Thus, prospects for efficient genetic manipulation of mosquitoes depended heavily on the discovery of a new functional TE system in mosquitoes. The medfly *Ceratitis capitata* was the first non-drosophilid to be heritably transformed using the *Tc1 Minos* TE originating from *D. hydei* (Franz and Savakis, 1991; Loukeris *et al.*, 1995). As in all of these pioneering proof-of-principle experiments, successful genetic manipulation was confirmed by reporter gene expression from the integrated transgene. A second TE used for the transformation of non-drosophilid insects was *piggyBac*. This TE was discovered in a cabbage looper moth (*Trichoplusia ni;* Lepidoptera) cell line and further developed as a TE for genetic manipulation of insects (Cary *et al.*, 1989). *piggyBac* was successfully used to transform *C. capitata* (Handler *et al.*, 1998) and a range of other insect species, including *Bombyx mori* (Tamura *et al.*, 2000), *Tribolium castaneum* (Berghammer *et al.*, 1999), and the pink bollworm, *Pectinophora gossypiella* (Thibault *et al.*, 1999).

Significantly, in 1998 *Aae* was transformed using both *Hermes* TE and *Mos1 mariner* TE to achieve stable germline transformation (Coates *et al.*, 1998; Jasinskiene *et al.*, 1998). The *Hermes* TE originated from the house fly, *Musca domestica* and *Mos1 mariner* TE from *D. mauritiana*. Other mosquito vectors that have been genetically transformed by TE-mediated genetic manipulation include the malaria vectors *An. stephensi* and *An. gambiae*, and the arbovirus vectors *Cx. quinquefasciatus* and *Ae. albopictus* (Allen *et al.*, 2001; Catteruccia *et al.*, 2000b; Grossman *et al.*, 2001; Labbe *et al.*, 2010).

Thus, decades after the first TE-mediated germline transformation of *Dm*, TE-mediated transformation is now possible in a number of medically important mosquito vectors. Significantly, since the late 1990s we have developed molecular-genetic tools that can be readily applied to germline transformation of mosquito vectors. While overall maintenance and preparation of mosquitoes for germline transformation is more challenging than for *Dm*, procedures and molecular tools are available to facilitate mosquito transgenesis. *Aae*, as previously stated, is a relatively easy arthropod vector to manipulate and maintain in laboratory culture because this mosquito species has less stringent requirements in terms of water quality, larval food composition, and light conditions than other mosquito species and *Aae* eggs can be desiccated and stored for up to 3–4 months, an important advantage when establishing and maintaining transgenic lines.

Developing GMVs: procedures, equipment, and reagents

Microinjection of pre-blastoderm embryos

Germline transformation of mosquitoes requires microinjection of eggs in pre-blastoderm stage. When using TEs for transgene insertion, as discussed previously, the transforming DNA is typically provided in the form of two plasmid DNAs. The donor plasmid contains the left and right ITR sequences of the TE flanking the effector gene expression cassette and the eye tissue-specific selection marker expression cassette (Franz *et al.*, 2006) (Fig. 19.1). The original transposase of the TE has been inserted into a separate plasmid vector, the helper plasmid, under control of a minimal promoter such as *hsp70*. Both helper and donor are co-injected into mosquito embryos at defined concentrations (i.e. 0.5 μg/ml donor plasmid and 0.3 μg/ml helper plasmid).

Microinjection of embryos has been extensively described (Franz *et al.*, 2006; Jasinskiene *et al.*, 2007). Briefly, pre-blastoderm embryos (eggs) of *Aae* are collected from 10–15 females that had received a bloodmeal 72–96 h earlier. The eggs (light grey in colour) are aligned on a wet filter paper with their posterior poles containing the embryonic tissue all pointing towards the same direction. Approximately 100 eggs are mounted onto a cover slip prepared with a strip of double-stick Scotch tape at one of its edges. Eggs are desiccated for a few minutes and then covered with halocarbon 27 oil to stop further desiccation. Microinjection occurs with pulled 1 mm boro-silicate needles (P97 Flaming/Brown micropipette puller; Sutter Co., Novato, CA, USA). Pulled needles are briefly bevelled using the BV-10 microelectrode beveller (Sutter). Pulled needles are loaded with helper/donor plasmid DNA diluted in 5 mM KCl and 0.1 mM NaH_2PO_4 (pH 6.8) injection buffer and inserted into a needle holder, which is connected with an air tube to a Femtojet air compressor (Eppendorf, Germany) (Jasinskiene *et al.*, 1998). The needle holder is attached to a micro-manipulator (Leica, Germany). The aligned eggs are mounted onto the stage of a standard microscope. Each mosquito egg is injected with nanolitre quantities of DNA. Following injection, eggs are carefully detached and kept under moist condition at 28°C and 80% relative humidity for 4–5 days before hatching. Surviving females are pooled in numbers of 10–20 and outcrossed to a male of the recipient (non-transgenic) laboratory strain such as *Aae* Higgs white eye (HWE). Surviving males are individually mated with 15–20 HWE females each. After 2–3 days, individual crosses are pooled in numbers of three to reduce the number of required bloodfeeds. Mosquito families obtain artificial bloodmeals consisting of citrated sheep blood. Ideally, three egg papers per family are produced in five day intervals. Egg papers are hatched one week later and G_1 larvae are screened for eye-tissue-specific selection marker expression. Transgenic individuals of each family are outcrossed to the recipient strain for another generation. Resulting G_2 mosquitoes are then tested for gene-of-interest expression.

Transposable elements

The central component for mosquito germline transformation is a TE, which can integrate itself into the recipient genome in a predictable manner. TEs, which have been isolated and successfully used for the genetic modification of mosquitoes include *Minos* and *piggyBac* for *Anopheles* and *Hermes*, *piggyBac*, *mariner Mos1* for *Cx. quinquefasciatus*, *Aae*, and/or *Ae. albopictus*. Interestingly, none of these TE when autonomous, i.e. containing their functional transposase, show sufficient levels of mobility in most mosquito species tested. It has been speculated that TEs belonging to the *mariner* family might require highly specific physical and biochemical conditions, which determine their transposition efficiency (Sinzelle *et al.*,

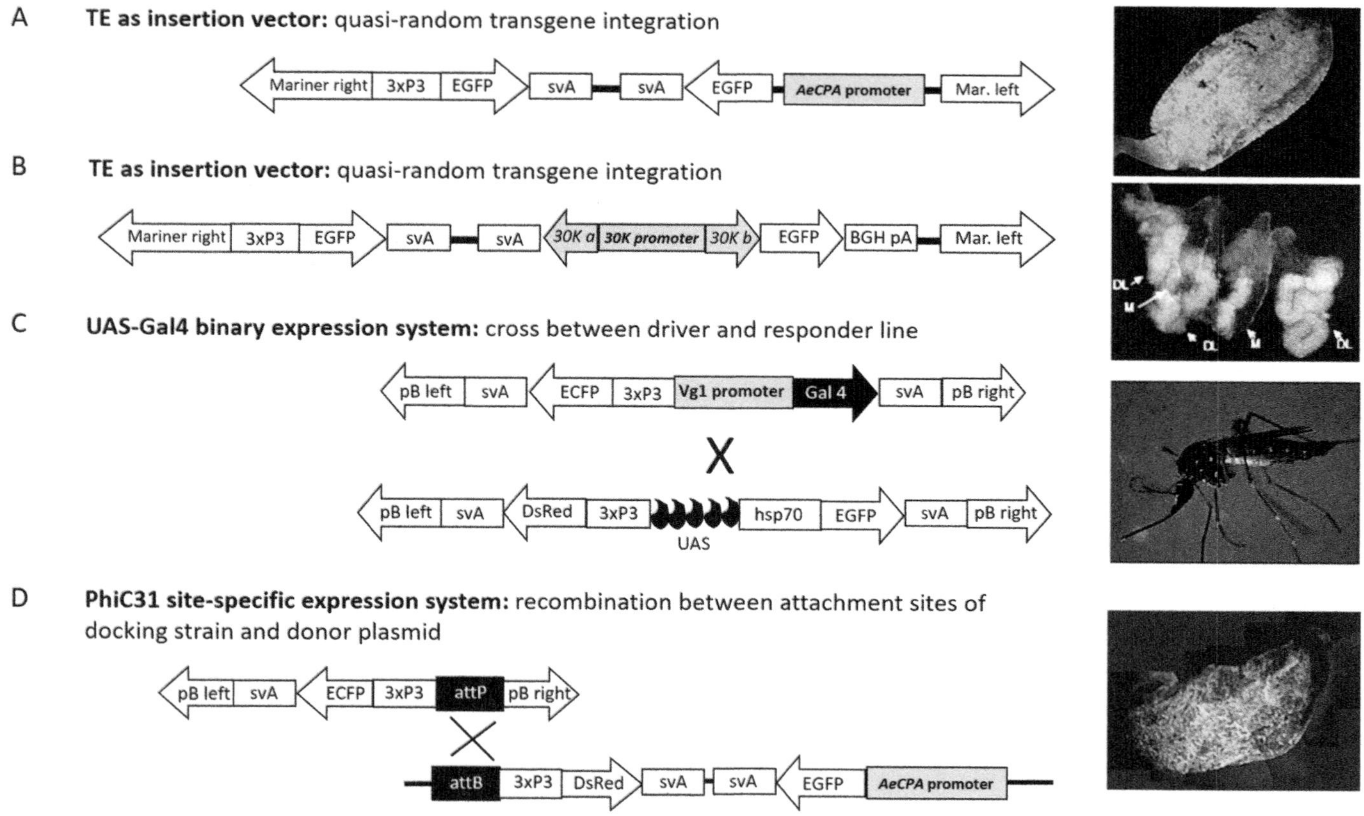

Figure 19.1 Selected transformation systems and tissue-specific expression of the EGFP reporter in *Aedes aegypti*. A) *Mariner Mos1* transformation of *Aae* generating GMV line Carb/egfp105, which expresses EGFP from *AeCPA* promoter in midguts after bloodmeal activation (Franz *et al*., 2011). B) *Mariner Mos1* transformation of *Aae* generating GMV line 30KExGM which expresses EGFP from *30K* promoter in salivary glands (Mathur *et al*., 2010). C) Gal4 driver and UAS donor lines. Fat body-specific EGFP expression from the *Vg1* promoter in a GMV *Aae* Vg–Gal4/UAS-EGFP hybrid female mosquito (Kokoza and Raikhel, 2011). D) PhiC31-mediated recombination of docking strain attP26 with the attB/DsRed2-carb/EGFP/svA donor plasmid. Midgut-specific EGFP expression from *AeCPA* promoter at 48 h after bloodmeal activation (Franz *et al*., 2011).

2008). However, TEs such as *Minos, piggyBac, Hermes,* and *mariner Mos1* are typically used as non-autonomous versions for mosquito germline transformation experiments.

Hermes belongs to the family of the *hobo, Ac, Tam3* (*hAT*) class II TE containing 17 bp imperfect ITRs (Warren *et al.*, 1994). Initial studies suggested that *Hermes* was able to transpose in *Aae* embryos (Sarkar *et al.*, 1997). The TE showed a preference for a GTNCAGAC sequence motif as insertion site. This TE was used for the first successful germline transformation attempts of *Aae* and *Cx. quinquefasciatus*. However, it was soon discovered that *Hermes* integration into mosquito genome does not follow the canonical cut-and-paste mechanism typical for class II TE (Jasinskiene *et al.*, 2000). Instead, partial integration of the donor plasmid flanking the ITR motifs of the TE was observed, without the typical target site duplication. In *Aae, Hermes* remobilization frequencies were <1%, in presence of transposase expressed from another TE in male reproductive tissue (Smith and Atkinson, 2011). After these observations were made, researchers favoured *piggyBac* and *mariner Mos1* as transgene insertion vectors over *Hermes* because their integration patterns appeared to be more predictable.

PiggyBac is arguably the most widely used TE for germline transformation of arthropods. The *piggyBac* TE was shown to be capable of transposition in embryos of *Aae, Dm,* and *Trichoplusia ni* (Lobo *et al.*, 1999). Meanwhile many *piggyBac*-like TEs have been detected in the genomes of a phylogenetically diverse range of organisms including fungi, plants, insects, crustaceans, amphibians, fishes and mammals (Sarkar *et al.*, 2003). The characteristic *piggyBac* insertion sequence motif is TTAA. In mosquitoes, *piggyBac* typically exhibits a precise cut-and-paste mechanism. However, in *Aae,* atypical patterns of *piggyBac*-mediated transgene integration have been observed (Adelman *et al.*, 2004). In several transformed lines, high unit-length copy numbers of donor and helper plasmid integrations occurred, which were arranged in large tandem arrays. These massive TE clusters were easily lost over several generations. Thus, using *piggyBac* in *Aae* can result in unexpected recombination events leading to loss of the transgene.

Another class II *Tc1*-like TE frequently used for insect germline transformation is *mariner Mos1* from *D. mauritiana* (Bryan *et al.*, 1990; Coates *et al.*, 1995; Garza *et al.*, 1991; Moreira *et al.*, 2000; Wang *et al.*, 2000; Wilson *et al.*, 2003). The TE only requires a TA motif as recognition sequence for its integration. Over the years, *mariner Mos1* has become an important TE for the generation of transgenic *Aae* (Adelman *et al.*, 2007; Franz *et al.*, 2006, 2014; Khoo *et al.*, 2010, 2013; Mathur *et al.*, 2010; Wilson *et al.*, 2003). *Mos1* transposase expression (*in trans*) generates rare germline remobilization events, demonstrating *Mos1* stability as an insertion system (Wilson *et al.*, 2003). Previous findings suggest that *mariner Mos1* transposition rates increase with increasing transposase concentrations (Lohe and Hartl, 1996).

Selection markers

Easy visual identification of transformants is an important factor when generating genetically modified organisms, because a high number of individuals (>1000) need to be screened for the desired phenotype/genotype. Thus, the design of scorable selection marker systems was of high priority when developing germline transformation systems for mosquitoes. The relatively large insect eye was identified as an ideal tissue to express a marker gene for selecting transformants.

Initially, the *Dm cinnarbar* (*cn*) eye pigment gene was used as a selection marker for selection of transgenic *Aae* based on transforming the white-eye *kh*^{w} strain (Jasinskiene *et al.*, 1998; Kokoza *et al.*, 2001). *Dm cn* was inserted into *Hermes* or *piggyBac* TE resulting in germline transformed individuals showing dark eye pigmentation as opposed to non-transformed individuals having white eyes. However, this system was limited to *Aae* and *Dm*. A more useful, universal insect-specific selection marker system became available with the design of the artificial, photoreceptor cell-specific *3xP3* promoter, containing three Pax-6 homodimer binding sites linked to a TATA box (Berghammer *et al.*, 1999). Pax-6 is involved in gene regulation during eye development (Sheng *et al.*, 1997). The *3xP3* promoter gave researchers the ability to express powerful fluorescent markers such as EGFP, ECFP, DsRed, mCherry, etc., in eye tissue of larvae, pupae, and adults of transgenic insects belonging to the orders Coleoptera, Lepidoptera, and Diptera. *3xP3* is now used routinely to drive expression of fluorescent markers in *An. stephensi*, *An. gambiae*, *Aae*, and *Ae. albopictus* (Grossman *et al.*, 2001; Kokoza *et al.*, 2001; Labbe *et al.*, 2010; Nolan *et al.*, 2002). In the white-eye background of HWE or *kh*^{w} mosquitoes, *3xP3* driven fluorescent marker expression is easily detectable using an epifluorescent microscope equipped with appropriate filter sets.

DNA promoters for expression of effector genes

Precise spatial and temporal control of effector gene expression is essential and having a toolbox of endogenous stage-, sex- or tissue-specific DNA promoters is desirable. Early attempts to express genes-of-interest in mosquitoes such as *Aae* and Anopheles spp. were hampered by the limited choice of functional, species-specific promoters. In the beginning, *Dm* promoters such as *hsp70*, *poly-ubiquitin*, or *actin 5C* were used to drive reporter gene expression in *Aedes*, *Culex*, and *Anopheles* species (Catteruccia *et al.*, 2000a; Pinkerton *et al.*, 2000; Sakai and Miller, 1992) with mixed success. However, based on anecdotal reports, it became clear that *Dm* derived promoters did not always lead to robust, predictable gene expression in mosquitoes.

Based on pathogen infection and transmission patterns in vectors, several key tissues are important target sites for effector gene expression in mosquitoes: midgut, fat body, salivary glands, ovaries, testis, and germline. A number of midgut-specific promoters have been isolated from *Aae* and *Anopheles* species. Initially, promoters of two *Anopheles* trypsin genes were tested for their ability to drive reporter gene expression in midgut tissue of *Dm* (Skavdis *et al.*, 1996). Moreira *et al.* (2000) then isolated and tested the promoter of the *carboxypeptidase A* (*CPA*) genes of *Aae* and *An. gambiae*. The promoters were linked to the firefly luciferase reporter gene and introduced into the *Aae* germline. The *AaCPA* promoter initiated luciferase expression in midguts of bloodfed mosquitoes between 4 and 48 h post bloodmeal (pbm), whereas the *An. gambiae CPA* enabled reporter gene expression in the same tissue between 0 and 24 h pbm. No luciferase expression was detected in tissues other than the midgut or in midguts of sugarfed mosquitoes. Thus, strong midgut-specific, bloodmeal inducible promoters were identified for tissue-specific expression in *An. gambiae* and *Aae* (Fig. 19.1A and D). The midgut is an ideal tissue for expression of anti-pathogen effector genes to inhibit bloodmeal acquired arboviruses at their initial site of infection (Franz *et al.*, 2006; Franz *et al.*, 2014; Ito *et al.*, 2002). Another bloodmeal inducible promoter is *vitellogenin 1* (*Vg1*), which was characterized and isolated from *Aae* and *An. stephensi* (Kokoza *et al.*, 2000; Nirmala *et al.*, 2006). In *An. stephensi*, the *Vg1* promoter regulates strong gene-of-interest expression in fat body between 24 and 48 h pbm and in *Aae* the homologous *Vg1* promoter led to peak expression at 24 h pbm (Fig. 19.1C). Several salivary gland-specific promoters have been characterized for *Aae* and *Anopheles* sp. Initially, promoters of the *Aae Maltase*-like (*Mal1*) and *apyrase* (*Apy*) genes were tested in transgenic mosquitoes for their ability to drive gene-of-interest expression (Coates *et al.*, 1999). However, *Apy* and *Mal1* promoters led to weak expression levels of reporters in salivary glands of transformed *Aae*, indicating weak promoter activity. Eventually, the promoter of the *An. stephensi* anti-platelet protein (AAPP), a member of the *30K* gene family, proved to be a strong, tissue-specific, constitutive promoter of reporter gene expression in salivary glands (Yoshida and Watanabe, 2006). Importantly, expression was strongest in the distal-lateral lobes of female salivary glands, which are relevant regions of the salivary glands tissues for arbovirus infection and transmission. In *Aae*, the *30K* promoter homologue is a bi-directional promoter driving expression of 30Ka and 30Kb proteins (Mathur *et al.*, 2010). The bi-directionality should be valuable in expressing two effector genes in salivary gland tissues. Germline transformation of *Aae* with a construct having a truncated version of 30Kb fused to the EGFP reporter resulted in strong reporter expression in distal-lateral lobes of the female salivary glands (Fig. 19.1B).

Recently, strong constitutive promoters from two *Aae ubiquitin* genes, *Ub(L40)* and *PUb*, have been characterized and tested in transgenic mosquitoes (Anderson *et al.*, 2010). The *Ub(L40)* promoter drives gene expression predominantly in larvae and ovaries, whereas the *PUb* promoter drives gene expression in embryos, larvae, pupae, males, and constitutively in the female midgut. Specifically, the *PUb* promoter has been shown to be highly valuable for constitutive effector gene expression in *Aae* (Khoo *et al.*, 2013). Soon after, two heat-shock protein 70 (*hsp70*)-like promoters of *Aae* were described, *AaHsp70Aa* and *AaHsp70Bb*. Both promoters significantly activate gene expression in various mosquito tissues such as salivary glands, midguts, and ovaries following a 1 h exposure at an elevated temperature (39°C) (Carpenetti *et al.*, 2012). Furthermore, specific promoters for female (*nanos*) and male-only (b_2-tubulin) germline expression in *Aae* and *An. gambiae* are available (Adelman *et al.*, 2007; Smith *et al.*, 2007). These promoters can be used for sex/germline-specific gene expression of enzymes that facilitate DNA integration/excision or recombination events. For a population reduction strategy described later, the female, flight muscle-specific *Actin-4* promoter of *Aae*,

Table 19.1 Mosquito promoters which have been used to drive expression of effector genes and/or reporters in transgenic mosquitoes

Promoter	Mosquito species	Tissue of expression	Inducible/constitutive	Heterologous gene expression
Carboxypeptidase A	*Ae. aegypti*, *An. gambiae*	Midgut	Bloodmeal	Effector gene
Aper1	*An. gambiae*, *An. stephensi*	Midgut	Before bloodmeal	Effector gene
vitellogenin 1	*Ae. aegypti*, *An. stephensi*	Fat body	Bloodmeal	Effector gene
30K/AAPP	*Ae. aegypti*, *An. stephensi*	Salivary glands	Constitutive	Effector gene
Poly-ubiquitin (PUb)	*Ae. aegypti*	All life stages, midgut	Constitutive	Effector gene
Ub_{L40}	*Ae. aegypti*	Early larvae, ovaries	Constitutive	Reporter only
AaHsp70Aa	*Ae. aegypti*	No tissue specificity	Heat-shock >37°C	Reporter only
AaHsp70Bb	*Ae. aegypt*	No tissue specificity	Heat-shock >37°C	Reporter only
nanos (nos)	*Ae. aegypti*, *An. gambiae*	Female germline	Constitutive	Effector gene
vasa	*An. gambiae*	Male/female germline	Constitutive	Reporter only
b2-tubulin	*Ae. aegypti*, *An. gambiae*	Head, ventral nerve cord, testes	Constitutive	Effector gene
Actin-4	*Ae aegypti*, *Ae. albopictus*, *An. stephensi*	Indirect flight muscle	Constitutive	Effector gene

Ae. albopictus, and *An. stephensi* has been characterized (Fu *et al.*, 2010; Labbe *et al.*, 2012; Marinotti *et al.*, 2013). Promoters which have been tested to drive heterologous gene expression in mosquitoes are listed in Table 19.1.

Transgenic systems to improve transgene expression in mosquitoes

PhiC31

The use of Class II TEs as transgene insertion vectors bears one major disadvantage, which is the nature of their quasi-random integration patterns into the host genome. As outlined above, TEs such as *piggyBac* and *mariner Mos1* require only very short recognition motifs, which are abundantly dispersed throughout the genome. Random transgene integration easily leads to position effect variegation, since not every integration locus allows robust gene expression. Instead, heterochromatin structures including promoters and enhancer elements from surrounding host genes can interfere with expression levels of newly integrated transgenes (Sabl and Henikoff, 1996; Wallrath and Elgin, 1995). A way to avoid position effects is to place docking (landing) sites into favourable loci of the genome, which act as predetermined recombination sites for the transgene.

The site-directed recombination system of the *Streptomyces* temperate phage, PhiC31 (*Caudovirales*, *Siphoviridae*) was analysed over 15 years ago (Thorpe and Smith, 1998). In presence of PhiC31 integrase, a serine recombinase, a short (39 bp) phage attachment (*attP*) site of the phage genome recombines via double-strand DNA breaks and crossover with the corresponding 34 bp bacterial attachment (*attB*) site of the host. As a result, the entire phage genome (~41,000 bp) is integrated into the *Streptomyces* genome. A great advantage of the PhiC31 serine recombinase is that recombination functions in unidirectional fashion. The PhiC31 recombinase does not require co-factors for recombination and integration, but does need co-factors to catalyse DNA excision from the host genome (Fish *et al.*, 2007). This leads to high stability of the transgene in the docking site. In heterologous organisms the system is more efficient when the *attP* site is anchored in the host genome and the *attB* site linked to the donor plasmid encoding the gene-of-interest. The PhiC31 system has been extensively used in a range of eukaryotic organisms such as plants (tobacco, wheat, *Arabidopsis*), insects (*Dm*, mosquitoes, *B. mori*), zebrafish, and mammals (mouse, rat, hamster, cow and human cell lines). Mosquito species in which the PhiC31 system has been successfully used are *An. gambiae* (Meredith *et al.*, 2013), *An. stephensi* (Isaacs *et al.*, 2012), *Aae* (Franz *et al.*, 2011; Khoo *et al.*, 2013; Nimmo *et al.*, 2006) and *Ae. albopictus* (Labbe *et al.*, 2010). The efficiency of the PhiC31 system largely depends on availability of mosquitoes containing the *attP* site in an optimal locus in the genome. Because the *attP* site is inserted into the mosquito genome via (quasi-random) TE-mediated integration the optimal docking strain needs to be identified based on gene-of-interest expression levels from the *attB* site-containing donor (Fig. 19.1D). Thus, it remains essential to generate a panel of docking strains and test them individually by 'supertransformation' with a donor plasmid containing the *attB* site, an eye marker, and a gene-of-interest expression cassette. It needs to be emphasized that in *Aae* at least, *3xP3* promoter driven eye marker expression level alone is not a sufficient criteria for identifying an optimal *attP* docking strain (Franz *et al.*, 2011; Nimmo *et al.*, 2006). Eye marker expression is rather insensitive to variegation (position effect) whereas reporter gene expression under control of an inducible, tissue-specific promoter does respond sensitively to position effects. (Franz *et al.*, 2011; Khoo *et al.*, 2013). The PhiC31 recombinase is typically provided as *in vitro* transcribed mRNA and co-injected with the donor plasmid into embryos of the docking strain. The donor plasmid encodes an eye marker different from that of the docking strain to easily screen for recombinants showing expression of both eye markers (Franz *et al.*, 2011). Thus, the PhiC31 site-specific integration system can facilitate reproducible, robust transgene expression in mosquito vectors provided optimal *attP* site containing docking strains have been identified and maintained in the insectary.

Insulators

Another approach to mitigate position effects of TE-based transgenes is the use of chromatin insulators. Two major systems have been described so far: the chicken b-globin *5′HS4* (hypersensitive site 4) insulators consisting of multiple elements and the *scs/scs′* (specialized chromatin structures) insulator originating from a portion of the *gypsy* retro-transposon from *Dm* (Kuhn and Geyer, 2003; Sarkar *et al.*, 2006). The two protective properties of chromatin insulators are their anti-enhancer activity blocking enhancer–promoter interactions and their anti-silencer activity, which prevents the spread of repressive heterochromatin by affecting histone modifications. This is an important aspect for *Aae*, whose genome contains vast amounts of heterochromatin.

The *scs/scs'* chromatin insulator has been tested in *An. stephensi* (Carballar-Lejarazu *et al.*, 2013). The insulator consists of ~350 bp of DNA adjacent to the long ITRs of the *gypsy* retrotransposon and several host-derived DNA binding proteins. Flanking an eye marker expression cassette on both sites by *scs/scs'* in inverted-repeat (IR) formation resulted in a significantly increased abundance of marker gene mRNA in comparison to mosquitoes expressing a marker gene without *scs/scs'* sequences. It needs to be emphasized that in both mosquito strains, the transgene had been inserted in the identical locus using the PhiC31 system. Furthermore, transgene expression levels among a range of transgenic lines with a gene-of-interest flanked by *scs/scs¢* were more consistent than in control lines lacking the insulator sequences. These results indicate that the exogenously derived *gypsy* insulator-like DNA is functional in *An. stephensi* and leads to higher and more consistent transgene mRNA abundance. It remains to be seen whether the *scs/scs'* chromatin insulators would have the same positive effects in other mosquito species such as *Aae*.

UAS-Gal4 binary expression system

Originally developed as a tool for targeted gene expression in *Dm* (Brand and Perrimon, 1993), the UAS-Gal4 system has been widely used in a variety of organisms including mosquitoes of the genera *Anopheles* and *Aedes* (Kokoza and Raikhel, 2011; Lynd and Lycett, 2012; O'Brochta *et al.*, 2012). The bipartite system is based on two independent transgenic lines: (i) the driver line containing the yeast *Gal4* activator gene under control of a promoter-of-choice and (ii) the responder line containing the binding sites (UAS, upstream activating sequences) for the Gal4 protein upstream of the gene-of-interest expression cassette. This system allows for very fine-tuned temporal and spatial gene-of-interest expression. Once a panel of UAS effector lines and Gal4 promoter lines have been generated, expression of a particular effector gene in various tissues can be compared or alternatively, expression of different effectors in the same tissue depending on the choice of driver line-responder line combinations, which can be generated through simple crosses. However, the UAS-Gal4 system *per se* does not mitigate position effects. Thus, Lynd and Lycett shielded the UAS responder transgene with chromatin insulators derived from the *Dm gypsy* retrotransposon to achieve comparable expression levels in different responder lines (Lynd and Lycett, 2012).

O'Brochta and colleagues successfully applied the UAS-Gal4 system for transposon-based enhancer trapping in *An. stephensi* (O'Brochta et al., 2012). This was possible because the *piggyBac* TE is highly mobile in *An. stephensi*. The *piggyBac* transposase promoter located near the terminal region of the TE was used to provide basal promoter functions for Gal4 expression. Enhancer detection depended on the remobilization of *piggyBac*-Gal4 transposons and incidental integration nearby genomic enhancer elements. Crosses were performed between six *piggyBac*-Gal4 lines and six UAS-fluorescent reporter lines, which all showed unique integrations sites in the mosquito genome. As a result of *piggyBac*-Gal4 remobilization, ~300 out of 24,000 transgenic progeny larvae showed novel transgene expression patterns in tissues such as midgut, fat body, and/or salivary glands. This work demonstrated the utility of this system for functional genomics to identify novel regulatory elements resulting in unique spatial and temporal transgene expression patterns.

For use in *Aae*, the essential DNA-binding and activation domains of Gal4 were fused directly, resulting in the chimeric Gal4Δ activator (Kokoza and Raikhel, 2011). In a proof-of-concept experiment, Gal4Δ of the transgenic driver line was placed under control of the bloodmeal inducible, fat body-specific *Vg* promoter (Fig. 19.1C). The responder line contained the EGFP reporter fused to UAS. Crossing the two transgenic lines resulted in hybrids strongly expressing EGFP in fatbody of bloodfed females, confirming that the system is functional in *Aae*.

GMVs for control strategies aimed at vector population reduction

Principle of sterile insect technology (SIT)

One possible strategy to control arbovirus transmission in the field is to reduce or eliminate the population of competent mosquito vectors in an endemic region. A well-established, non-transgenic approach for population reduction in some insect populations is sterile insect technology (SIT), which requires periodic mass releases of irradiated sterile insects (usually males) in target areas to outnumber the fertile wild-type male mating partners (Bushland *et al.*, 1955). Male releases of mosquitoes are desired since males do not bloodfeed or transmit arboviruses. The irradiated sterile males carry a significant fitness load, making them unfavourable mating partners. However, in mass release scenarios, the target area is 'flooded' with the sterile males; hence the probability increases for sterile males being chosen as mating partners among the targeted wild-type population. As a consequence, significantly fewer offspring will be generated in the target population within a single generation. Subsequent mass releases of sterile insects can eventually lead to population collapse, though the efficacy of SIT is highly dependent on the density of the original wild-type population. Another critically important factor is the likelihood and time period required for novel invasion of the cleared area by wild-type mosquitoes from other regions. Nevertheless, SIT has been highly successful in the Americas for eradicating the screw worm fly (*Cochliomyia hominivorax*) and to minimize medfly populations (*C. capitata*). Other potential targets for SIT programmes are the Mexican fruit fly (*Anastrepha ludens*) and the pink bollworm (*P. gossypiella*) (Franz and Robinson, 2011; Knipling *et al.*, 1968; Robinson, 2002; Robinson *et al.*, 2009). In Zanzibar, Africa, tsetse fly (*Glossina morsitans*) and in Okinawa, Japan, Melon fly (*Bactrocera cucurbitae*) populations have been successfully controlled using SIT (Koyama *et al.*, 2004; Vreysen, 2001). In 'classical' SIT, the production of sterile insects is a large-scale operation, requiring factory-size facilities in which billions of sterile insects are generated on a weekly basis (Alphey *et al.*, 2010). Therefore, fertile insects need to be mass-reared and fed, followed by sex separation and sterilization of males using ionizing radiation such as X-ray. Mass-rearing facilities initially produce equal numbers of the two sexes, but females are typically separated and discarded before release. The use of 'classical' SIT for mosquito population control has been examined decades ago (Asman *et al.*, 1981; Krishnamurthy *et al.*, 1977; Rai *et al.*, 1970; Weidhaas, 1962). In a few field tests, successful, temporal control has been reported for *Cx. qinquefasciatus*, on a small island off the coast of Florida and for *Cx. tarsalis* in California (Asman *et al.*, 1981; Benedict and Robinson, 2003). Nevertheless, today there is no large-scale SIT program in operation targeting any mosquito species (Asman *et al.*, 1981; Phuc *et al.*, 2007). Whereas sex separation based on pupa size is not a fundamental problem when applying SIT to mosquitoes, sterilization of the insects proved to be more difficult. Irradiation of pupae caused severe fitness damage to the vector, whereas irradiation of adult males was insufficient (Phuc *et al.*, 2007). An additional problem is that mosquito populations are far more responsive to density-dependent effects such as resource limitation (availability of larval food, oviposition sites) than agricultural pests such as the medfly. Thus, a mere reduction in female fecundity might not always result in a significant impact on the overall target population size because fewer female offspring need to compete with each other for limited food resources and survive as fit, competitive individuals (Phuc *et al.*, 2007).

Design and application of RIDL

To better match the principle of SIT-mediated population reduction with the characteristics of mosquito biology and the specific dynamics of mosquito populations, a novel concept termed RIDL (Release of Insects carrying a Dominant Lethal) was developed based on GMV technology (Alphey and Andreasen, 2002; Thomas *et al.*, 2000). The underlying principle of RIDL is that the *Aae*-based GMV expresses a dominant lethal gene in a conditional manner once the larva stage has been completed. The GMV would have the following requirements for efficient mosquito population reduction: (i) male RIDL GMVs are competitive for mating with wild-type *Aae* females; (ii) the lethal gene can be 'switched off' during mosquito rearing/propagation of the RIDL insects prior to their release; (iii) once 'switched on' in progeny, the progeny die due to expression of the dominant lethality trait; (iv) resource-limited, larva competition is still ensured since the gene is only expressed and effective at the pupal or adult stage (late-acting lethal). As a proof-of-concept, the RIDL system was successfully developed in *Dm* (Thomas *et al.*, 2000). Control of lethality was accomplished using the TET on/off system based on expression of a tetracycline-repressible transactivator (tTAV), a protein containing the tet repressor (tTA) fused to the activating domain of virion protein 16 (VP16) of herpes simplex virus (Gossen and Bujard, 1992; Phuc *et al.*, 2007). In absence of tetracycline, tTAV binds to its tetracycline responsive element (tetO7) allowing expression of the lethality gene. Overexpression of tTAV in transgenic *Dm* was lethal due to a positive feedback loop. In the presence of tetracycline, however, the substrate binds to the tTAV transactivator, thereby changing its conformation preventing tTAV from binding to tetO7 (Thomas *et al.*, 2000).

A transgenic *Aae* RIDL strain (OX513A) was genetically engineered to carry a dominant, non-sex-specific, late-acting lethal genetic system that is repressed in the presence of tetracycline (Phuc *et al.*, 2007). OX513A is a homozygous RIDL line of *Aae* (ROCK strain), transformed with the tTAV lethal positive-feedback system. During mass rearing of RIDL mosquitoes when both sexes are needed, tetracycline is added to the food diet, allowing mosquitoes to survive. After release of male OX513A RIDL mosquitoes in the field, they can mate with wild-type *Aae* females and any offspring generated by the mating will exist in a tetracycline-free environment. As a consequence, tTAV expression is activated and offspring die at the pupal stage, in theory leading to population collapse. Laboratory studies strongly indicate that the RIDL system has great potential for mosquito population control in the field. Wise de Valdez and colleagues went a step further and showed that periodic releases of *Aae* RIDL mosquitoes in controlled large laboratory cage experiments could eliminate wild-type-like mosquito populations (Wise de Valdez *et al.*, 2011). Transgenic RIDL males, which were introduced on a weekly basis into the cages containing a population of a genetically diverse laboratory strain (GDLS) of *Aae* at an initial ratio of 8.5–10:1 (RIDL:GDLS) significantly reduced the target population within 10–20 weeks.

Aae RIDL OX513A strain has been field-tested in several tropical regions following regulatory compliances of the host countries. Published results can be found for field experiments conducted in 2010/2011 in Malaysia and in 2010 in the Cayman Islands (Harris *et al.*, 2012; Lacroix *et al.*, 2012). In Malaysia, the release site was located in uninhabited government forested land in Pahang, Malaysia, which is an atypical habitat for *Aae*. Transgenic RIDL mosquitoes and non-transgenic controls were allowed to emerge in separate cages. Before their release, RIDL males and non-transgenic males were labelled with two different fluorescent dyes. Mosquitoes of the release area were then trapped (re-captured) at different time points using ovitraps. Presence of the RIDL transgene in progeny eggs was assessed by PCR. Major findings of this study were that longevity of RIDL males was similar to that of wild-types. However, overall dispersal of the transgenics was significantly lower (0–25 metres for the majority of males) than that of the wild-types (75–100 metres for the majority of males). In another experiment in the Cayman Islands, ~3 million *Aae* RIDL males were released during a 23-week period (Harris *et al.*, 2012). The experiment showed that the RIDL males were able to mate with wild-type *Aae* females, leading to population suppression. To induce population collapse in absence of immigration, it was necessary to release three OX513A males per hectare per week, for an approximate 10:1 (RIDL: wild-type males) ratio. Ovitrap indices in release areas of RIDL males became significantly lower (up to 80% reduction) over an extended period of time than ovitrap indices of non-RIDL-treated areas.

In a different tetracycline-repressible *Aae* RIDL strain (OX3604C), the transactivator, tTAV, is under control of a strong female-specific promoter to ensure conditional late-acting female lethality (Fu *et al.*, 2010). Expression of the effector gene occurs in L4 larvae and leads to an adult female that has a flightless phenotype. A critical aspect of this strategy is the *Aae Actin-4* control sequence, which contains a sex-specific splicing motif driving gene expression in the female indirect flight muscles. The *Actin-4* promoter has a translation initiation codon and Kozak sequence, CCACCATG, engineered into the *AeActin-4* 5′ UTR before the 5′ donor site of the intron to ensure RNA splicing. This leads to the expression of the tTAV effector in a stage-, tissue-, and sex-specific manner resulting in female-specific RIDL strains for population suppression (Fu *et al.*, 2010). Following mating with wild-type mosquitoes, the adult female-offspring are flightless and cannot mate or act as efficient vectors of arboviruses. The male-offspring pass on the trait to the next generation. OX3604C RIDL males, which have been introduced weekly into large laboratory cages containing stable target mosquito populations at initial ratios of 8.5–10:1 OX3604C:target eliminated the vector populations within 10–20 weeks (Wise de Valdez *et al.*, 2011). However, in five large field-cage experiments OX3604C was effective at suppressing but not eliminating vector populations due to mating disadvantage with wild-type populations (Facchinelli *et al.*, 2013). Insights from large outdoor cage experiments may provide an important part of the progressive, stepwise evaluation of GMVs. The female-killing RIDL system has also been successfully adapted for *Ae. albopictus*, and *An. stephensi* (Labbe *et al.*, 2012; Marinotti *et al.*, 2013).

In summary, the RIDL-based population reduction strategy is a highly promising approach to control mosquitoes and indirectly arbovirus transmission. However, as experimental data and observations suggest, RIDL alone will likely not constitute the magic bullet in arbovirus disease control. So far, the impact of RIDL-mediated mosquito population reduction on arbovirus (i.e. DENV) prevalence in an endemic area has not been assessed. It also needs to be kept in mind that RIDL is not a one-time-treatment solution. Instead, transgenic sterile males need to be periodically released over extended periods of time to achieve long-term success. Wild-type population densities and immigration potential strongly influence the efficacy of the RIDL strategy. Facilities and protocols have been developed to generate millions of RIDL males for repeated treatments of endemic areas to reduce *Aae* populations.

Homing endonuclease strategy for vector population reduction

Homing endonuclease genes (HEGs) are endonucleases encoding selfish genetic elements, which are common among a wide range of microorganisms such as fungi, plants, algae, bacteria and bacteriophages (Deredec *et al.*, 2011) and typically occur in genomes of chloroplasts and mitochondria.

A HEG sequence-specifically cleaves genomic DNA of a homologous chromosome, which does not already contain the same HEG (Burt and Koufopanou, 2004). During the recombinational DNA repair process of the cellular DNA repair machinery, the homologous HEG containing chromosome is used as template resulting in insertion of the HEG into the cleaved copy of the chromosome ('homing'). The unique, HEG-specific recognition sequence motifs can be 18–30 bp in length with the HEG being inserted in the middle of its recognition sequence and thus being protected from cleavage. HEGs suggest another GMV approach for generating sterile males for vector population reduction by inducing extreme reproductive sex ratios that suppress or eliminate vector populations. Naturally occurring sex distorters (male bias) have been described in *Aedes* and *Culex* genera although the mechanism of action is not clear (Galizi *et al.*, 2014; Hickey and Craig, 1966a; Sweeny and Barr, 1978). Galizi, Doyle and colleagues recently described a synthetic male-bias, sex distortion system in *An. gambiae* based on an HEG (Galizi *et al.*, 2014). By chance they discovered that the recognition sequence for HEG I-*Ppo*I (originating from the slime mould *Physarum polycephalum*) occurs in ribosomal gene repeat sequences located exclusively on the mosquito's X chromosome. GMVs were designed to place HEG expression under control of the β_2-tubulin promoter that allows I-*Ppo*I expression during spermatogenesis. The observed result is that transgenics generate only Y-sperm since the paternal X chromosome is shredded and heterozygotes produced from fertilization of the transgenic male sperm and maternally derived X chromosome of non-transgenic females show developmental arrest due to shredding of the X chromosome. Sterile transgenic males were propagated by backcrossing transgenic females with non-transgenic males. It was also demonstrated that distorter male mosquitoes can efficiently suppress caged wild-type mosquito populations, providing the foundation for a new class of genetic vector control strategies. So far, this approach has not been tested in the field. While the X-shredder approach has promise in *Anopheles* which have fully differentiated heteromorphic sex chromosomes where the male-determining locus resides on the non-recombining Y chromosome (Marin and Baker, 1998); in *Aedes* and *Culex* mosquitoes, male development is initiated by a dominant male-determining locus (M-locus) located on a homomorphic sex-determining chromosome (Newton *et al.*, 1974). This may make the X-shredder strategy questionable for other vector species relevant to arbovirus transmission.

GMVs for vector population replacement to control arbovirus transmission

A different concept for using GMVs to control arbovirus transmission in the field is based on vector population replacement. Population replacement strategies are inherently more complex than population reduction strategies because an effector gene (e.g. anti-pathogen gene) must be driven into the target population. Here, the goal is to replace highly competent wild-type vector populations with GMVs refractory to medically important pathogens such as DENVs. The idea behind population replacement is not new for suppressing disease transmission in the field and leaves the ecological niche for the vector intact (Crampton *et al.*, 1994; Curtis, 1968, 1976). Two important proof-of-principle studies demonstrated that *Aae* and *An. stephensi* could be genetically modified to express an anti-pathogen or effector gene and become resistant to model pathogens (Ito *et al.*, 2002; Kokoza *et al.*, 2000). In *Aae*, the coding sequence of the pre-pro-defensin (DefA) was placed under control of the bloodmeal-inducible, fat body-specific *Vg* promoter (Kokoza *et al.*, 2000). Transgenic females, which had received a bloodmeal, overexpressed DefA in the fat body, leading to accumulation of the antimicrobial peptide for 20–22 days in the mosquito haemolymph. It was then confirmed in an *in vitro* assay that transgenically produced DefA had similar levels of antibacterial activity against Gram-positive *Micrococcus luteus* bacteria as DefA isolated from wild-type mosquitoes. *An. stephensi* were also genetically engineered to be refractory to *Plasmodium berghei* by overexpressing a synthetic effector gene, SM1 (salivary gland- and midgut-binding peptide 1) (Ito *et al.*,

2002). SM1 consists of four units of a synthetic peptide that were expressed in midguts of bloodfed females in the context of the midgut-specific, bloodmeal-inducible *CPA* promoter (Ito *et al.*, 2002; Warren *et al.*, 1994). The SM1 peptide was identified by *in vivo* selection from a library of bacteriophages displaying random 12-amino-acid peptides. SM1 having the amino acid sequence PCQRAIFQSICN bound specifically to two epithelia in the mosquito that are traversed by the parasite: the distal lobes of the salivary glands and the lumenal surface of the midgut. SM1 expression inhibited *P. berghei* from crossing the epithelia, blocking further development of the malaria parasite from ookinetes into the oocyst stage.

Arbovirus infection of *Aedes aegypti*

An overview and understanding of arbovirus infection patterns in *Aae* leading to virus transmission suggests GMV-based approaches for generating pathogen refractory phenotypes to control arbovirus transmission. DENV infection of *Aae* is used here as an example. *Aae* adult females seek a human bloodmeal within days after emerging from the pupal stage. Females are attracted to humans by a complex array of cues that include humidity, visual cues, CO_2, lactic acid and other chemical odorants on or near the skin surface (Bernier *et al.*, 2000; Smallegange *et al.*, 2011). The bloodmeal is rich in amino acids that are required for egg production (Telang and Wells, 2004). *Aae* females can ingest bloodmeals every several days during their adult life, leading to multiple gonotrophic cycles (Scott *et al.*, 2000; Scott *et al.*, 1993). A possible way of altering *Aae* transmission potential is to develop GMVs that affect mating or feeding preferences. However, accomplishing this control strategy requires a much deeper understanding of vector genes involved with host seeking, host feeding preference and behaviour. Others have noted that accessory glands of male mosquitoes of *Aae* produce substances (presumably proteins or peptides) that are transferred to females during mating, and alter female physiology and behaviour by inhibiting subsequent female mating behaviour, stimulating oviposition and preoviposition behaviours, and host-seeking behaviour (Klowden, 1999; Lee and Klowden, 1999). Molecular and genetic analyses of male accessory glands may identify gene products that govern essential behaviours of vectors.

Aae and other bloodfeeding vectors acquire virus from a viraemic human or animal. In the case of DENV, during the acute febrile period of disease in humans greater than 10^7 $TCID_{50}$/ml of virus can circulate in the peripheral blood (Halstead, 2008). Blood enters the posterior midgut where virus replicates in midgut epithelial cells. About six to seven days later the virus escapes from the midgut and disseminates to secondary tissues such as fat body, nerve tissue, hemocytes and the salivary glands (Linthicum *et al.*, 1996; Salazar *et al.*, 2007). Once in the salivary glands, virus replicates and enters saliva for DENV transmission to a new host when the female mosquito takes a second bloodmeal. The extrinsic incubation period (EIP) for DENVs in *Aae* is 7–14 days under laboratory conditions (~28°C, 80% RH), although some mosquitoes can transmit as soon as 4 days post infection (Salazar *et al.*, 2007). In nature, vector competence varies among *Aae* populations for the same DENV2 strain (Bennett *et al.*, 2002). A significant part of the observed intraspecies variation can be attributed to genetic differences among vector populations that may involve differences in the mosquito's efficiency to support virus infection and may in part be associated with differences in the vector's innate immune responses that target the virus (Sanchez-Vargas *et al.*, 2009; Souza-Neto *et al.*, 2009; Waterhouse *et al.*, 2007; Xi *et al.*, 2008). The midgut appears to be the major organ that determines vector competence for DENV in *Aae* (Black *et al.*, 2002). Depending on the particular vector strain and virus strain combination, quantitative trait loci (QTL) that determine midgut infection barrier (MIB) and midgut escape barrier (MEB) have been described for DENVs (Bennett *et al.*, 2002, 2005; Bosio *et al.*, 2000; Lozano-Fuentes *et al.*, 2009). In presence of a MIB, the virus is not able to efficiently infect the midgut epithelium. In the presence of a MEB, the virus readily infects the midgut epithelium but its dissemination to secondary tissues is blocked or inhibited. DENV infections are clearly modulated in *Aae*'s midgut. DENV2 replication in the mosquito midgut is certainly affected by antiviral responses causing either transcriptional or translational repression (Black *et al.*, 2002). For instance, impairing the innate antiviral immune RNA interference pathway (RNAi), a known antiviral response, in the midgut can significantly increase arbovirus infection and dissemination from that organ (Campbell *et al.*, 2008; Cirimotich *et al.*, 2009; Khoo *et al.*, 2010; Sanchez-Vargas *et al.*, 2009).

RNAi consists of three distinct pathways, the small interfering RNA (siRNA), microRNA (miRNA), and piwi-interacting RNA (piRNA) pathways (Blair, 2011). The siRNA pathway constitutes the major anti-arboviral RNAi response. The siRNA pathway is triggered by double-stranded RNA (dsRNA), a pathogen-associated molecular pattern or PAMP. Since almost all arboviruses are RNA viruses, they will generate dsRNA during replication. The essential genetic components (*dicer2*, *argonaute2*, *r2d2*, and others) of the siRNA pathway have been identified in mosquitoes (Blair, 2011). Recent research has also implicated the piRNA pathway in mosquito anti-arboviral immunity, but components of the *Piwi* gene-family are not well defined (Morazzani *et al.*, 2012). Arboviruses must evade or suppress RNAi without causing pathogenesis in the vector to maintain their transmission cycle, but little is known about mechanisms of arbovirus modulation of RNAi. Genetic manipulation of mosquitoes to enhance their RNAi response can limit arbovirus infection and replication and has been used in novel strategies for interruption of arbovirus transmission. As we will discuss in a later section, expression of dsRNA derived from DENV sequence profoundly reduces *Aae* competence for DENV transmission (Franz *et al.*, 2006, 2014; Mathur *et al.*, 2010). We have shown that expression of DENV2 derived dsRNA in midgut tissues co-incident with intake of an infected bloodmeal virtually eliminates vector competence in GMVs (Franz *et al.*, 2006, 2014). In summary, the *Aae* midgut is the initial site of arbovirus replication in the vector and acts as an important tissue for determining vector competence. GMV strategies that prevent arbovirus replication in the midgut would be critical for reducing the prevalence of an arbovirus in vector populations.

The spatio-temporal tissue tropism of DENV2 strain Jamaica 1409 has been determined by examining non-midgut tissues of mosquitoes exposed to an infectious bloodmeal. Viral antigen is readily detected in fat body (abdomen, thorax or head) following dissemination from the midgut (Salazar *et al.*, 2007). Fat body appears to be a major site for DENV2 replication in the vector. DENV2 dissemination to abdominal fat body or thoracic fat body can be detected as early as 3 days post infection (dpi). Other infected organs and tissues include trachea, malphigian tubules, hemocytes, ommatidia of the compound eye, nerve tissue, and salivary glands (Salazar *et al.*, 2007). DENV2 Jamaica 1409 displays a significant tropism for tissue of the central nervous system and salivary glands. Oesophagus, cardia, hindgut and especially muscle rarely contain viral antigen (Salazar *et al.*, 2007). The importance of virus amplification in secondary tissues other than salivary glands, which is critical for virus transmission, is unclear, but as new female-, adult- and tissue-specific transcriptional promoters become known, researchers should be able to express effector genes that inhibit virus replication in specific organs to assess whether virus amplification in these tissues affects virus transmission (Franz *et al.*, 2006, 2009).

Aae females have a pair of salivary glands and each gland consists of three lobes attached to a common salivary duct (James, 2003). Each lobe comprises a secretory epithelium with a basal and apical surface. The basal ends of the epithelial cells form the outside surface of the glands and are bound to the basal lamina. How the virus crosses the basal lamina and the mechanism of virus entry into salivary gland epithelial cells is poorly understood. Female salivary glands differentiate into two lateral and one medial lobe, which produce distinctive secretions (James, 2003). The proximal region of the lateral lobes secretes enzymes involved in sugar feeding such as amylases and α-1-4 glycosidase and proteins from the medial and the distal lateral lobes participate in haematophagy and include apyrases, esterase, anticoagulants and vasodilators (Argentine and James, 1995; Beerntsen *et al.*, 1999; Smartt *et al.*, 1995). In *Aae* (strain: Chetumal) mosquitoes, DENV2 infection often began in the distal lateral lobes, spread to the proximal section of the lateral lobes and then throughout the entire organ later. Virus antigen concentration in salivary glands remained high from 14 to 21 days post infection, suggesting that older females in the population are excellent vectors for transmitting DENVs (Salazar *et al.*, 2007). GMVs have been engineered to inhibit virus replication in salivary glands as a way of preventing horizontal transmission of the virus to the next susceptible vertebrate host (Mathur *et al.*, 2010).

Expression of effector genes targeting DENV in *Aedes aegypti*

In 2006, we published a report describing the generation of *Aae* GMVs engineered to be refractory to DENV2 (Franz *et al.*, 2006). The GMVs were developed by transforming *Aae* HWE embryos with the *mariner Mos1* TE containing a 578 bp IR sequence based on a sequence region from the DENV2 genome encoding mostly prM/M protein. The IR sequence was designed to utilize the vector's siRNA pathway by expressing a dsRNA trigger to silence DENV2. The effector dsRNA was expressed co-incidentally with the bloodmeal by placing its expression under control of the *Aae CPA (AeCPA)* promoter. To increase efficiency of dsRNA formation following transcription in the midgut of bloodfed females, the sense and antisense fragments of the IR sequence were separated by the small intron of the *Aae sialokinin I* gene (Beerntsen *et al.*, 1999). In an initial transformation experiment to generate a GMV, one line, Carb77, was obtained which stably expressed the IR effector in midguts of bloodfed females between 27 and 48 h pbm (Franz *et al.*, 2006). The Carb77 transgene was inserted into a single locus within a non-protein encoding region of the genomic DNA. We confirmed that the IR effector was transcribed into long dsRNA, which served as a PAMP recognized by the endogenous RNAi pathway in the mosquito midgut. We then detected degradation of the dsRNA into 21 nt duplexes, the hallmark of siRNA activity. Carb77 females were highly resistant to DENV2 infection when virus was obtained through a bloodmeal. However, when virus was intrathoracically injected into female Carb77 to bypass the midgut tissue, DENV2 replicated well in secondary tissues of the mosquito including salivary glands. In contrast, bloodmeal acquired DENV2 was not detected in saliva of Carb77 mosquitoes. Additionally, the siRNA-mediated resistance phenotype to DENV2 was reversible in Carb77 mosquitoes: transient silencing of key genes of the RNAi pathway, such as *dcr2*, *r2d2*, or *ago2*, broke the GMV's DENV2 resistance phenotype. This confirmed that the anti-DENV2 effector set up a midgut infection barrier in the GMVs to prevent DENV strains of the same serotype. However, due to the homology-dependent nature of the siRNA pathway, Carb77 was not resistant to other DENV serotypes (DENV1, 3, 4) or other arboviruses. This shortcoming in the approach can be rectified by generating a GMV that expresses a dsRNA having sequence elements of all serotypes.

Carb77 mosquitoes exhibited a strong DENV2 resistance phenotype until generation 17 (G_{17}) (Franz *et al.*, 2009). Sequence analysis of the transgene revealed that the transgene was still intact at the DNA level. However, the IR effector was no longer expressed, even though the selection marker (EGFP) was still expressed in Carb77 mosquitoes after G_{17}. The underlying cause for silencing of the IR effector was never revealed; however, we speculated that heterochromatin re-arrangements at the transgene integration site might have inhibited IR effector expression possibly highlighting the need for gene insulators.

In a further experiment we investigated whether expression of the identical IR effector in another tissue such as the fat body would also cause profound resistance to DENV2. Placing the IR molecule under control of the bloodmeal-inducible fat body specific *Vg1* promoter (Kokoza *et al.*, 2000) led to expression of the dsRNA trigger for up to ~24 h pbm in an *Aae* GMV line (Vg40) (Franz *et al.*, 2009). DENV2 replicated well in the midgut epithelium of Vg40 mosquitoes. Typically, the virus started to disseminate from the midgut to secondary tissues such as the fat body at ~4 days pbm, a time point when the bloodmeal in the midgut was digested and the *Vg1* promoter no longer induced gene expression. To re-induce *Vg1*-driven anti-DENV2 effector expression at a time point when the virus was escaping the midgut, it was necessary to give mosquitoes a second, non-infectious bloodmeal. However, resistance to DENV2 was never observed in the Vg40 mosquitoes. DENV2 was able to infect the salivary glands of the Vg40 transgenic females, which then released the virus into their saliva. Thus, it seems to be ineffective to block the virus in the fat body because silencing DENV2 in this tissue apparently does not prevent the virus from being able to infect the salivary glands. Regardless, these observations emphasize the requirement that a siRNA-based blocking strategy must occur in the correct context (space and time) to generate DENV refractory mosquitoes.

The salivary glands are another critically important target tissue to antagonize DENV replication in *Aae*. With the discovery of the *30K* promoter we reasoned that it should be possible to generate GMVs expressing the anti-DENV2 IR effector in the mosquito salivary glands (Mathur *et al.*, 2010). The bi-directional *Aae 30K* promoter is constitutive in the adult female and allows simultaneous reporter gene and anti-DENV effector gene expression in salivary glands. In the wild-type, two gene products (30Ka and 30Kb) are controlled by the single promoter in opposite directions. 30Ka and 30Kb both consist of two exons each. The exon 2 of 30Ka was replaced with the coding sequence of EGFP whereas the exon 2 of 30Kb was replaced with the intron-containing anti-DENV2 IR sequence used in Carb77 and Vg40 mosquitoes. Five GMV lines, 1, 15, 22, 27 and P4 harboured this construct. Another transgenic line, P6 harboured a similar transgene with the exception that the intron 1 of 30Kb had been deleted. All transgenic lines had 2–3 transgene integration events. Transgenic mosquitoes of all lines expressed the transgene in the distal-lateral lobes of the female salivary glands. In four out of five lines tested, DENV2 titres and infection rates (range 27–43%) were significantly reduced in salivary glands when compared with the HWE recipient control (infection rate 62%). In other tissues outside the salivary glands, DENV2 titres in the transgenic lines were similar to the virus titre in the HWE control. Furthermore, in two of the lines, 1 and 15, no DENV2 was detectable in saliva, indicating that an RNAi response in salivary glands was able to block DENV2 transmission.

Since the Carb77 line lost the resistance phenotype in G_{17}, another group of transgenic *Aae* was generated having the identical transgene as Carb77. New GMVs were developed to identify lines having a strong but stable DENV2 refractory phenotype and to use these new lines for further tests and experiments aimed at assessing their eventual use in a field-based population replacement strategy. We identified a GMV line, Carb109 that strongly expressed the anti-DENV2 IR effector in the midguts of bloodfed females at 20 h pbm (Franz *et al.*, 2014). Carb109 mosquitoes were maintained in two separate colonies, Carb109M (male) and Carb109F (female) based on male or female G_1 founders. Carb109 mosquitoes had two transgene integration events. Genetic analyses and physical chromosome mapping identified two closely linked transgene integration sites in Carb109M mosquitoes associated with chromosome 3. Carb109M has been completely refractory to DENV2 bloodmeal infections for at least 33 generations in laboratory culture (Fig. 19.2). As with Carb77, occasionally single females produced elevated DENV2 titres. However, when DENV2 was recovered from some of these isolated high-titre samples and used for subsequent infection studies, Carb109M females did not show any virus titres suggesting that the few high DENV2 titre samples did not represent selection of viral mutants able to evade the siRNA pathway. NextGen sequencing confirmed that the Carb109M dsRNA trigger was recognized by the RNAi machinery of the mosquito and processed into 21 bp siRNAs. Similar to Carb77, Carb109M mosquitoes were highly resistant to different strains of DENV2 representing four different

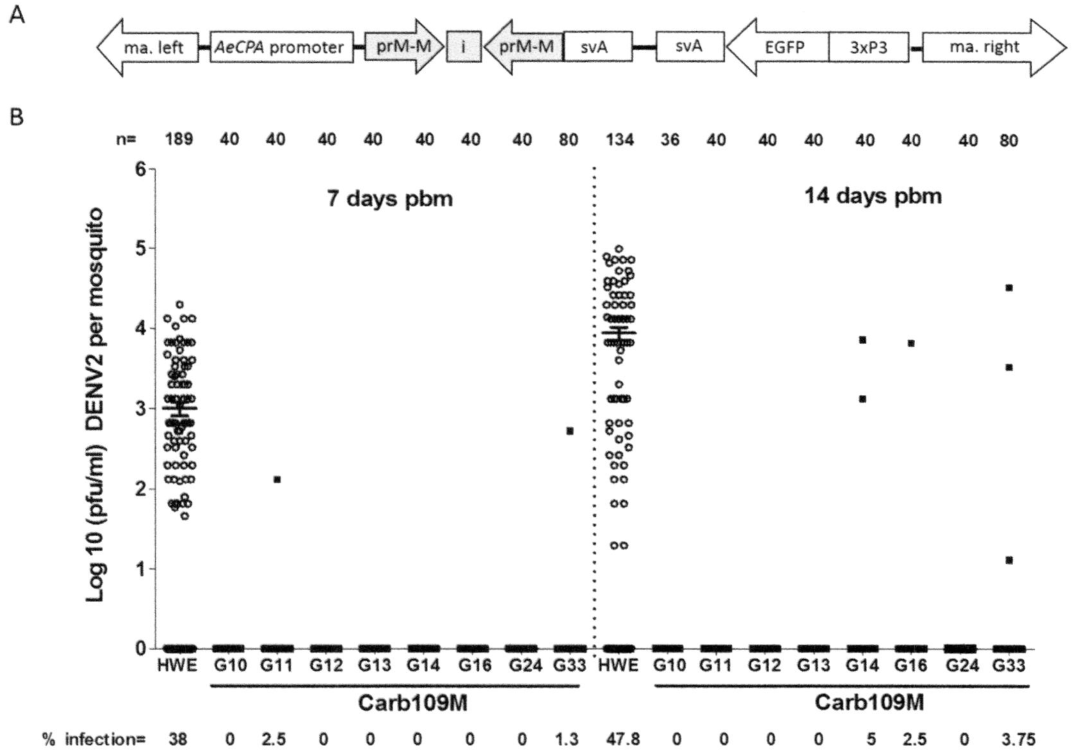

Figure 19.2 Analysis of Carb109M DENV2-resistance phenotype over 33 intercrossed generations. (A) Design of the transgene harboured by Carb109 mosquitoes based on the *mariner Mos1* TE containing the eye tissue-specific selection marker and an IR DNA (shaded) derived from the prM-M sequence of DENV2. A functional intron is inserted between the sense and antisense prM-M fragments. Expression of the IR sequence from the *AeCPA* promoter leads to dsRNA formation in midgut epithelium of bloodfed females, triggering RNAi. (B) HWE (control) and Carb109M mosquitoes received bloodmeals containing 10^6 PFU/ml DENV2-Jamaica1409. Virus titres in the mosquitoes were assessed at 7 and 14 days post bloodmeal (dpi). Each data point represents the virus titre of a single female. Mean values and standard errors are indicated.

genotypes (American, Asian 2, American-Asian, and Cosmopolitan). We estimated that the 578 nucleotide sequence IR RNA shared 86% minimal nucleotide identity among the different DENV2 genotypes.

A population replacement strategy can be effective in the field if the anti-pathogen effector is stably inherited and expressed by the wild-type target population without imposing substantial fitness burdens. To test this prerequisite for the DENV2-refractory GMVs, the Carb109M transgene was introgressed into a genetically diverse laboratory strain (GDLS) of *Aae* composed of a mixture of 10 wild-type populations collected from DENV endemic areas of Chiapas State in Southern Mexico (Wise de Valdez *et al.*, 2011). Introgression was performed via five consecutive backcrosses of Carb109M × GDLS and Carb109F × GDLS resulting in Carb109M/GDLS.BC5 and Carb109F/GDLS.BC5, respectively (Franz *et al.*, 2014). In theory, 97% of alleles in each BC5 strain were expected to have originated from the GDLS strain except for those HWE alleles linked tightly to the Carb109 transgene (Falconer, 1989). The number of backcrosses had a strong effect on transgene stability in the GDLS population over subsequent generations. When Carb109M/GDLS.BC1 or Carb109F/GDLS.BC1 heterozygotes were crossed with GDLS parents, the Carb109 transgene was lost in the cage population after five generations with no selection for transgenic individuals. This showed that either the transgene itself had a dominant fitness load or that a deleterious allele occurred near the Carb109 transgene insertion site. However, when a similar experiment was conducted with Carb109F/GDLS.BC5 mosquitoes, the transgene frequency reached stability among heterozygotes, supporting the interpretation that the wild-type alleles had a dominant positive effect on fitness. Thus, five generations of backcrossing did not improve the low fitness of transgenic homozygotes but did improve the fitness of transgenic heterozygotes. When selection for the transgene was applied to Carb109F/GDLS.BC5 and Carb109M/GDLS.BC5 heterozygotes after each subsequent generation of intercrossing, transgenic allele frequencies reached the predicted values in two out of six replicates according to Fisher's Model of Natural Selection (Falconer, 1989; Fisher, 1950; Franz *et al.*, 2014). The initial rate and pattern of decline in the frequency of EGFP-expressing BC_1 and BC_5 GMVs was consistent with a fitness load linked to the transgene or with the HWE background in which the transgene was inserted. Careful interpretation of the data eventually led to the conclusion that a deleterious allele was located at or near the Carb109 insertion site(s) in the HWE recipient strain (Sinzelle *et al.*, 2008). From these efforts we were able to develop a stable homozygous GMV line Carb109M/GDLS.BC5.HZ. This line displays a strong, almost completely refractory phenotype to DENV2 infection and should be an important anti-pathogen GMV for future studies (Fig. 19.3).

In summary, the results of the backcross and selection experiments clearly emphasize the importance of outcrossing transgenes into genetically diverse, recently colonized

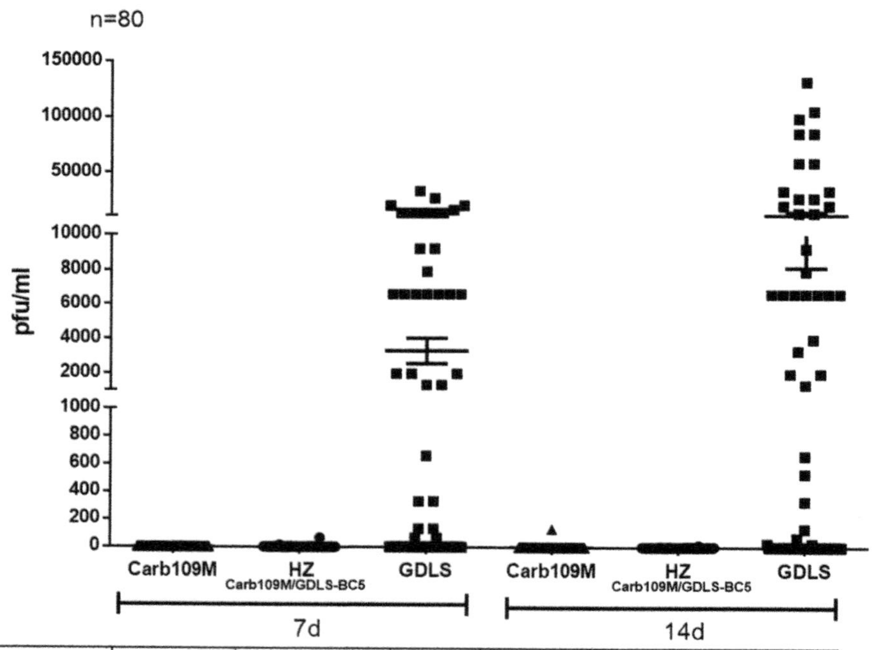

Mosquito strain	Carb 109	HZ Carb109M/GDLS-BC5	GDLS	Carb 109	HZ Carb109M/GDLS-BC5	GDLS
Days post-infection		7d			14d	
% infection	**0**	**2.5**	**40**	**1.25**	**1.25**	**45**
Mean	0	0.99	3297	1.65	0.165	11118
Std. Deviation	0	7.507	6538	14.76	1.476	26496
Std. Error	0	0.8393	731	1.65	0.165	2962
Virus titer in blood	$3x10^6$ to $7 x 10^6$ pfu/ml					

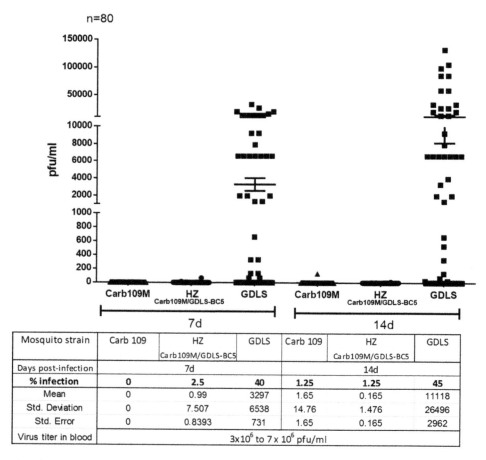

Figure 19.3 Analysis of resistance to DENV2 infection of homozygous Carb109M/GDLS.BC5.HZ mosquitoes. GDLS, Carb109M, Carb109M/GDLS.BC5.HZ mosquitoes were challenged with DENV2-Jamaica1409 (titre in the bloodmeal: >10^6 PFU/ml). Virus titres of mosquitoes were assessed at 7 and 14 dpi. Each data point represents the virus titre of a single female. Mean values and standard errors are indicated.

mosquito strains before releasing transgenes into population cages or natural populations. To date, Carb77 and Carb109M represent the only GMV lines that have been successfully engineered to be refractory to an arbovirus. Carb109M showed that the transgene could be introgressed into a DENV2 competent vector population (GDLS) rendering the population incompetent for the virus. One would expect that following a similar, RNAi-based concept, effector genes targeting other *Aae*-transmitted arboviruses could be readily designed. However, constructing effector genes to target multiple arboviruses has been difficult. We are currently designing a chimeric IR effector gene incorporating the conserved NS5 (RdRP) coding regions of DENV1–4. In support of this approach several examples have been described in which RNAi-based antiviral approaches have been used in transgenic plants to develop resistance to multiple tospoviruses (Bucher *et al.*, 2006; Peng *et al.*, 2014). Other effector genes are also possible such as single-chain antibodies to specifically capture and immobilize virions of DENV1–4 or an engineered hammerhead ribozyme (based on *Tetrahymena thermophila* group I intron catalytic RNA), which would cleave the highly conserved cyclization sequence motif in the viral genome of all four DENV serotypes (Carter *et al.*, 2010). None of these approaches has resulted in complete, simultaneous resistance in *Aae* to DENV1–4. Any DENV control based on a GMV approach will require suppression of all four serotypes due to their hyperendemic occurrence in many tropical regions of the world.

Gene drive systems

An important requirement for a successful population replacement strategy is to tightly link the antiviral effector construct to an efficient, species-specific gene drive system to drive the transgene into a wild-type population at an elevated rate until fixation (Hay *et al.*, 2010; James, 2005). Important criteria for a gene drive system are (i) the gene drive must be powerful enough to spread an anti-pathogen effector gene to fixation in a wild-type population even if the effector is associated with fitness costs; (ii) in relation to (i), the driver should be able to completely transform a wild-type population because even a small percentage of wild-type individuals could still promote disease transmission; (iii) it has to be highly specific for the target population to avoid uncontrollable spread of the transgene in nature; (iv) the function of the driver has to be insensitive to point mutations, which could easily occur within a few generations; (v) it should be perpetual, meaning that re-introduction of engineered, pathogen-resistant vectors in the same target area would not be necessary.

Currently, there is no gene drive system readily available for *Aae*. In the past, several genetic systems have been considered potential candidates for the development of gene drive systems. These included autonomous TE, which would be able to spread through wild-type populations. A well-documented example is the *P*-element of *Drosophila*, which has invaded *Dm* populations all over the world following likely horizontal acquisition from *Drosophila willistoni* (Engels, 1997; Kidwell, 1992). However, as described above, class II TEs which are currently in use for germline transformation of arthropods typically do not exhibit any 'driver' capability in heterologous hosts and have limitations on size of the effector gene. In theory, HEG could be used for population reduction or population replacement in the field. Introduction of a HEG into a naive population is expected to rapidly lead towards fixation of the HEG from a low starting frequency. An HEG could be linked to a strong anti-pathogen effector gene to be spread through a wild-type population to reduce vector competence for the target pathogen (Sinkins and Gould, 2006). Previously, it has been shown that commercially available HEG are able to excise transgenes flanked by their specific recognition sequences in the soma and germline of *Aae* (Aryan *et al.*, 2013; Traver *et al.*, 2009). Crossing experiments of a transgenic *An. gambiae* donor line expressing the HEG I-*Sce*I in the male germline with target lines containing the I-*Sce*I recognition site showed that the HEG was able to rapidly invade the target line populations under laboratory conditions (Windbichler *et al.*, 2011). Furthermore, the HEG induced high rates of recognition site-specific cleavage in genomic DNA of the target strains, resulting in a relative increase in HEG gene frequency among the cage population. Ultimately, these experiments demonstrate strong gene drive capabilities of HEG in mosquito vectors. However, a major issue remaining before HEG can be generally considered as mosquito species-specific gene drivers is the development of techniques to allow easy engineering and manipulation of HEGs so they can recognize genome-specific sequences specifically adapted to targeted vector populations.

Sex ratio distortion caused by meiotic drive is another phenomenon, with promise as a gene driver. Naturally occurring meiotic drive has been observed decades ago in the mosquitoes *Cx. pipiens* and *Aae* (Craig *et al.*, 1960; Hickey and Craig, 1966b; Sweeny and Barr, 1978). Meiotic drive-based sex ratio distortion typically results in a large excess of males as an aberration from the expected Mendelian 0.5 ratio. In all *Culicinae*, males are heterozygous (Mm) at the sex locus located on chromosome I, whereas females are homozygous (mm) (Gilchrist and Haldane, 1946). The meiotic driver MD is tightly linked to the dominant male determining locus and blocks maturation of female-determining sperm through breakage of the homologous chromosome carrying the recessive female-determining m allele and a sensitive responder locus (*ms*) (Cha *et al.*, 2006; Owusu-Daaku *et al.*, 2007). However, MD is ineffective when the responder locus of the *m* allele is insensitive (*mi*). Since the MD carrying chromosomes selectively destroy their sensitive (*ms*) homologue, they have the potential to rapidly increase in frequency in natural populations. Thus, an anti-pathogen effector gene could be linked to *mi* and would be expected to spread if released into an *ms* bearing mosquito population together with MD (Sinkins and Gould, 2006). Despite recent efforts to precisely locate and identify gene candidates responsible for naturally occurring meiotic drive in *Aae*, its molecular mechanism remains elusive to date (Shin *et al.*, 2012).

A different meiotic drive mechanism termed *Medea* (maternal-effect, dominant embryonic arrest) has been observed in *Tribolium* beetles. Here, a selfish DNA element causes the death of all offspring of heterozygous females that do not inherit a copy of the *Medea* gene(s) (Beeman and Friesen, 1999; Beeman *et al.*, 1992; Wade and Beeman, 1994). *Medea* spreads by killing non-*Medea*-bearing homologous chromosomes, thereby causing a relative increase in the frequency of the *Medea*-bearing homologue. The molecular nature underlying *Medea* in *Tribolium* is still under investigation (Lorenzen *et al.*, 2008). Meanwhile, a synthetic selfish DNA element following the *Medea* principle has been successfully developed for *Dm* (Chen *et al.*, 2007; Hay *et al.*, 2010). The underlying mechanism which enables *Medea* to drive itself through wild-type populations involves death of any offspring not inheriting a maternal or paternal copy of the element when present in females. *Medea* bearing males on the other hand produce viable progeny inheriting the *Medea* element. In *Dm*, the synthetic element comprises two microRNAs expressed from the bicoid promoter that silence a maternally expressed gene: MYD88, essential for embryogenesis ('loss of function'), and a tightly linked antidote such as a microRNA-insensitive copy of the MYD88 gene under control of the early zygotic bnk promoter ('restoring function'). Importantly, the silencing of MYD88 must not lead to embryonic arrest prior to zygotic transcription otherwise the antidote will not be able to restore loss of MYD88 function. At the molecular level, toxin and antidote encoding cDNAs can be arranged in such a way that allows additional linkage of 'cargo', which would be the anti-pathogen effector gene. In theory, when assuming that a synthetic *Medea* bears no fitness cost for the organism (100% survival in presence of the element is always better than 0% survival in absence of the element), the element would spread to fixation within a population even when introduced at a very low frequency. According to mathematical prediction models, however, *Medea*'s drive capability is heavily dependent on a minimal threshold introduction frequency, indicating that the element causes fitness costs to the host organism. Despite its great success in *Dm*, it has been challenging to develop a synthetic *Medea* for mosquitoes such as *Aae*. A major obstacle has been a lack of strong, maternal effect and/or zygotic promoters that would allow for species-specific, tightly controlled embryonic arrest and its rescue. Engineered HEG and a synthetic *Medea* are the emerging candidates for genetic drive of antiviral effectors. However, currently no gene drivers for *Aae* are available.

Self-limiting gene drive: reduce and replace strategy

Another genetic arbovirus/vector control strategy would combine population reduction with population replacement (Robert *et al.*, 2013). Recent mathematical modelling suggests that 'reduce and replace' (R&R) could be a powerful alternative to only population replacement and would not require a gene driver since the anti-pathogen effector gene is linked to a transgene inflicting population reduction (female killing) such as described previously for RIDL. Importantly, RIDL transgenes and an anti-DENV2 effector transgene are readily available and could be easily linked to each other to further test their efficacy under laboratory conditions. Gould and colleagues have shown through mathematical modelling that releases of a hypothetical strain of *Aae* carrying a conditional/female mortality gene (RIDL) and an anti-pathogen gene could reduce the number of arbovirus competent vectors and virus transmission (Legros *et al.*, 2013; Okamoto *et al.*, 2013; Robert *et al.*, 2013). They found that the R&R strategy can decrease the frequency of competent vectors below 50% 2 years after releases end. Therefore, this combined approach appears preferable to releasing a strain carrying just a dominant lethal gene (RIDL), which leads to only temporary population suppression. However, the fixation of an anti-pathogen gene in the population also is unlikely when considering genetic drift at small population sizes combined with the dynamics of vector population recovery. Additionally, immigration by wild-type mosquitoes could be impacted as population reduction would increase the long-term frequency of the anti-pathogen gene. These modelling results indicate that driving an anti-pathogen construct to fixation through an R&R strategy will be self-limiting but they also suggest that R&R may be useful in some dengue-endemic environments leading to a spatially heterogeneous decrease in competent vectors to decrease arbovirus transmission.

We have shown that introgressing an anti-pathogen gene into a virus competent *Aae* population in cage experiments can significantly impact mosquito infection and virus transmission (Franz *et al.* 2014). As stated previously, it is assumed that in most release scenarios introgression of an anti-pathogen gene will require linkage with a selfish genetic element or similar technologies to achieve fixation. Unfortunately, GMVs having anti-pathogen gene-drive mechanisms will not be available in the near future as a public health tool for disease control. Thus, attention is now focused on transgenic strategies aimed at mosquito population suppression, an approach generally perceived to be practical. By contrast, aiming at replacing vector competent mosquito populations with vector incompetent populations by releasing mosquitoes carrying a single anti-pathogen gene without a gene-drive mechanism is widely considered impractical. Okamoto and colleagues have recently reported the use of Skeeter Buster, a previously published stochastic, spatially explicit model of *Aae* to investigate whether a number of approaches for releasing mosquitoes with only an anti-pathogen construct would be efficient and effective in the tropical city of Iquitos, Peru (Magori *et al.*, 2009; Okamoto *et al.*, 2014). To assess the performance of such releases using realistic release numbers, the authors compared the transient and long-term effects of this strategy with two other genetic control strategies: R&R and female killing RIDL. They found that releasing mosquitoes carrying only an anti-pathogen construct can substantially decrease vector competence of a natural population, even at release ratios well below that required for R&R and RIDL. Finally, although current genetic control strategies based on population reduction are compromised by immigration of wild-type mosquitoes, releasing mosquitoes carrying only an anti-pathogen gene was considerably more robust to such immigration. Obviously, these conclusions by modelling need to be tested in cage experiments and limited field trials; however, *Aae* GMVs are available to test both the self-limiting R&R drive strategy and the introgression strategy based on an anti-pathogen transgene moving into a target vector population in the absence of gene drive.

Regulatory considerations for using GMVs to control arbovirus transmission

As previously discussed, novel GMV laboratory-based strategies are now available, which have the potential to control arbovirus transmission by reducing or replacing vector populations. GMV technology is not a 'magic bullet' for eliminating arboviral disease threats, but ideally will be used as part of an overall integrated approach to arboviral disease management. Researchers acknowledge that with GMV technologies there are minimal background, experience, or specific regulatory pathways to guide the transition of GMV products from the laboratory to the field. However, considerable thought and debate has occurred to ensure the release of safe and efficacious GMVs that are consistent with the highest technical and ethical standards (Alphey and Beech, 2012; Ramsey *et al.*, 2014; Reeves *et al.*, 2012). Researchers must move GMV-based products beyond the laboratory by rigorously selecting appropriate field sites for field-cage and later release studies to evaluate a GMV's ability to reduce or replace vector populations (Facchinelli *et al.*, 2011). Essential to the success of this technology is national, state, and community engagement and approval (Lavery *et al.*, 2008; Lavery *et al.*, 2010). Therefore proponents of GMV technology must simultaneously develop the science and contribute meaningfully to community dialogue regarding if, when, where, and how GMV technology could and should be applied. Part of this process is to meet the legal, technical, and community regulatory requirements for GMV testing and use.

Acknowledgements

This work was funded, in part, by a grant from the Foundation for the NIH through the Grand Challenges in Global Health Initiative and by NIH-NIAID grants RO1AI48470 and R21AI112782.

References

Adelman, Z.N., Jasinskiene, N., Vally, K.J., Peek, C., Travanty, E.A., Olson, K.E., Brown, S.E., Stephens, J.L., Knudson, D.L., Coates, C.J., *et al.* (2004). Formation and loss of large, unstable tandem arrays of the piggyBac transposable element in the yellow fever mosquito, *Aedes aegypti*. Transgen. Res. *13*, 411–425.

Adelman, Z.N., Jasinskiene, N., Onal, S., Juhn, J., Ashikyan, A., Salampessy, M., MacCauley, T., and James, A.A. (2007). nanos gene control DNA mediates developmentally regulated transposition in the yellow fever mosquito *Aedes aegypti*. Proc. Natl. Acad. Sci. U.S.A. *104*, 9970–9975.

Allen, M.L., O'Brochta, D.A., Atkinson, P.W., and Levesque, C.S. (2001). Stable, germ-line transformation of *Culex quinquefasciatus* (Diptera: Culicidae). J. Med. Entomol. *38*, 701–710.

Alphey, L., and Andreasen, M. (2002). Dominant lethality and insect population control. Mol. Biochem. Parasitol. *121*, 173–178.

Alphey, L., and Beech, C. (2012). Appropriate regulation of GM insects. PLoS Neglect. Trop. Dis. *6*, e1496.

Alphey, L., Benedict, M., Bellini, R., Clark, G.G., Dame, D.A., Service, M.W., and Dobson, S.L. (2010). Sterile-insect methods for control of mosquito-borne diseases: an analysis. Vector Borne Zoonotic Dis. *10*, 295–311.

Anderson, J.R., and Rico-Hesse, R. (2006). *Aedes aegypti* vectorial capacity is determined by the infecting genotype of dengue virus. Am. J. Trop. Med. Hyg. *75*, 886–892.

Anderson, M.A., Gross, T.L., Myles, K.M., and Adelman, Z.N. (2010). Validation of novel promoter sequences derived from two endogenous ubiquitin genes in transgenic *Aedes aegypti*. Insect Mol. Biol. *19*, 441–449.

Argentine, J.A., and James, A.A. (1995). Characterization of a salivary gland-specific esterase in the vector mosquito, *Aedes aegypti*. Insect Biochem. Mol. Biol. *25*, 621–630.

Aryan, A., Anderson, M.A., Myles, K.M., and Adelman, Z.N. (2013). TALEN-based gene disruption in the dengue vector *Aedes aegypti*. PloS One *8*, e60082.

Asman, S.M., McDonald, P.T., and Prout, T. (1981). Field studies of genetic control systems for mosquitoes. Annu. Rev. Entomol. *26*, 289–318.

Barrett, A.D., and Higgs, S. (2007). Yellow fever: a disease that has yet to be conquered. Annu. Rev. Entomol. *52*, 209–229.

Beeman, R.W., and Friesen, K.S. (1999). Properties and natural occurrence of maternal-effect selfish genes ('Medea' factors) in the red flour beetle, *Tribolium castaneum*. Heredity *82*, 529–534.

Beeman, R.W., Friesen, K.S., and Denell, R.E. (1992). Maternal-effect selfish genes in flour beetles. Science *256*, 89–92.

Beerntsen, B.T., Champagne, D.E., Coleman, J.L., Campos, Y.A., and James, A.A. (1999). Characterization of the Sialokinin I gene encoding the salivary vasodilator of the yellow fever mosquito, *Aedes aegypti*. Insect Mol. Biol. *8*, 459–467.

Behura, S.K., Gomez-Machorro, C., Harker, B.W., deBruyn, B., Lovin, D.D., Hemme, R.R., Mori, A., Romero-Severson, J., and Severson, D.W. (2011). Global cross-talk of genes of the mosquito *Aedes aegypti* in response to dengue virus infection. PLoS Neglect. Trop. Dis. *5*, e1385.

Benedict, M.Q., and Robinson, A.S. (2003). The first releases of transgenic mosquitoes: an argument for the sterile insect technique. Trends Parasitol. *19*, 349–355.

Bennett, K.E., Olson, K.E., Munoz Mde, L., Fernandez-Salas, I., Farfan-Ale, J.A., Higgs, S., Black, W.C.T., and Beaty, B.J. (2002). Variation in vector competence for dengue 2 virus among 24 collections of *Aedes aegypti* from Mexico and the United States. Am. J. Trop. Med. Hyg. *67*, 85–92.

Bennett, K.E., Flick, D., Fleming, K.H., Jochim, R., Beaty, B.J., and Black, W.C.t. (2005). Quantitative trait loci that control dengue-2 virus dissemination in the mosquito *Aedes aegypti*. Genetics *170*, 185–194.

Berghammer, A.J., Klingler, M., and Wimmer, E.A. (1999). A universal marker for transgenic insects. Nature *402*, 370–371.

Bernier, U.R., Kline, D.L., Barnard, D.R., Schreck, C.E., and Yost, R.A. (2000). Analysis of human skin emanations by gas chromatography/mass spectrometry. 2. Identification of volatile compounds that are candidate attractants for the yellow fever mosquito (*Aedes aegypti*). Anal. Chem. *72*, 747–756.

Bhatt, S., Gething, P.W., Brady, O.J., Messina, J.P., Farlow, A.W., Moyes, C.L., Drake, J.M., Brownstein, J.S., Hoen, A.G., Sankoh, O., *et al.* (2013). The global distribution and burden of dengue. Nature *496*, 504–507.

Bisset, J.A., Marin, R., Rodriguez, M.M., Severson, D.W., Ricardo, Y., French, L., Diaz, M., and Perez, O. (2013). Insecticide resistance in two *Aedes aegypti* (Diptera: Culicidae) strains from Costa Rica. J. Med. Entomol. *50*, 352–361.

Black, W.C.T., Bennett, K.E., Gorrochotegui-Escalante, N., Barillas-Mury, C.V., Fernandez-Salas, I., de Lourdes Munoz, M., Farfan-Ale, J.A., Olson, K.E., and Beaty, B.J. (2002). Flavivirus susceptibility in *Aedes aegypti*. Arch. Med. Res. *33*, 379–388.

Blair, C.D. (2011). Mosquito RNAi is the major innate immune pathway controlling arbovirus infection and transmission. Future Microbiol. *6*, 265–277.

Bonizzoni, M., Dunn, W.A., Campbell, C.L., Olson, K.E., Marinotti, O., and James, A.A. (2012). Complex modulation of the *Aedes aegypti* transcriptome in response to dengue virus infection. PloS One *7*, e50512.

Bosio, C.F., Beaty, B.J., and Black, W.C.t. (1998). Quantitative genetics of vector competence for dengue-2 virus in *Aedes aegypti*. Am. J. Trop. Med. Hyg. *59*, 965–970.

Bosio, C.F., Fulton, R.E., Salasek, M.L., Beaty, B.J., and Black, W.C.t. (2000). Quantitative trait loci that control vector competence for dengue-2 virus in the mosquito *Aedes aegypti*. Genetics *156*, 687–698.

Brand, A.H., and Perrimon, N. (1993). Targeted gene expression as a means of altering cell fates and generating dominant phenotypes. Development *118*, 401–415.

Bryan, G., Garza, D., and Hartl, D. (1990). Insertion and excision of the transposable element mariner in Drosophila. Genetics *125*, 103–114.

Bucher, E., Lohuis, D., van Poppel, P.M., Geerts-Dimitriadou, C., Goldbach, R., and Prins, M. (2006). Multiple virus resistance at a high frequency using a single transgene construct. J. Gen. Virol. *87*, 3697–3701.

Burt, A., and Koufopanou, V. (2004). Homing endonuclease genes: the rise and fall and rise again of a selfish element. Curr. Opin. Genet. Dev. *14*, 609–615.

Bushland, R.C., Lindquist, A.W., and Knipling, E.F. (1955). Eradication of screw-worms through release of sterilized males. Science *122*, 287–288.

Campbell, C.L., Keene, K.M., Brackney, D.E., Olson, K.E., Blair, C.D., Wilusz, J., and Foy, B.D. (2008). *Aedes aegypti* uses RNA interference in defense against Sindbis virus infection. BMC Microbiol. *8*, 47.

Carballar-Lejarazu, R., Jasinskiene, N., and James, A.A. (2013). Exogenous gypsy insulator sequences modulate transgene expression in the malaria vector mosquito, Anopheles stephensi. Proc. Natl. Acad. Sci. U.S.A. *110*, 7176–7181.

Carpenetti, T.L., Aryan, A., Myles, K.M., and Adelman, Z.N. (2012). Robust heat-inducible gene expression by two endogenous hsp70-derived promoters in transgenic *Aedes aegypti*. Insect Mol. Biol. *21*, 97–106.

Carter, J.R., Keith, J.H., Barde, P.V., Fraser, T.S., and Fraser, M.J., Jr. (2010). Targeting of highly conserved Dengue virus sequences with anti-Dengue virus trans-splicing group I introns. BMC Mol. Biol. *11*, 84.

Cary, L.C., Goebel, M., Corsaro, B.G., Wang, H.G., Rosen, E., and Fraser, M.J. (1989). Transposon mutagenesis of baculoviruses: analysis of Trichoplusia ni transposon IFP2 insertions within the FP-locus of nuclear polyhedrosis viruses. Virology *172*, 156–169.

Catteruccia, F., Nolan, T., Blass, C., Muller, H.M., Crisanti, A., Kafatos, F.C., and Loukeris, T.G. (2000a). Toward Anopheles transformation: Minos element activity in anopheline cells and embryos. Proc. Natl. Acad. Sci. U.S.A. *97*, 2157–2162.

Catteruccia, F., Nolan, T., Loukeris, T.G., Blass, C., Savakis, C., Kafatos, F.C., and Crisanti, A. (2000b). Stable germline transformation of the malaria mosquito *Anopheles stephensi*. Nature *405*, 959–962.

Cha, S.J., Chadee, D.D., and Severson, D.W. (2006). Population dynamics of an endogenous meiotic drive system in *Aedes aegypti* in Trinidad. Am. J. Trop. Med. Hyg. *75*, 70–77.

Chen, C.H., Huang, H., Ward, C.M., Su, J.T., Schaeffer, L.V., Guo, M., and Hay, B.A. (2007). A synthetic maternal-effect selfish genetic element drives population replacement in Drosophila. Science *316*, 597–600.

Ciota, A.T., and Kramer, L.D. (2013). Vector–virus interactions and transmission dynamics of West Nile virus. Viruses *5*, 3021–3047.

Cirimotich, C.M., Scott, J.C., Phillips, A.T., Geiss, B.J., and Olson, K.E. (2009). Suppression of RNA interference increases alphavirus replication and virus-associated mortality in *Aedes aegypti* mosquitoes. BMC Microbiol. *9*, 49.

Coates, C.J., Turney, C.L., Frommer, M., O'Brochta, D.A., Warren, W.D., and Atkinson, P.W. (1995). The transposable element mariner can excise in non-drosophilid insects. Mol. Gen. Genet. *249*, 246–252.

Coates, C.J., Jasinskiene, N., Miyashiro, L., and James, A.A. (1998). Mariner transposition and transformation of the yellow fever mosquito, *Aedes aegypti*. Proc. Natl. Acad. Sci. U.S.A. *95*, 3748–3751.

Coates, C.J., Jasinskiene, N., Pott, G.B., and James, A.A. (1999). Promoter-directed expression of recombinant fire-fly luciferase in the salivary glands of Hermes-transformed *Aedes aegypti*. Gene *226*, 317–325.

Craig, G.B., Jr., Hickey, W.A., and Vandehey, R.C. (1960). An inherited male-producing factor in *Aedes aegypti*. Science *132*, 1887–1889.

Crampton, J.M., Warren, A., Lycett, G.J., Hughes, M.A., Comley, I.P., and Eggleston, P. (1994). Genetic manipulation of insect vectors as a strategy for the control of vector-borne disease. Ann. Trop. Med. Parasitol. *88*, 3–12.

Curtis, C.F. (1968). Possible use of translocations to fix desirable genes in insect pest populations. Nature *218*, 368–369.

Curtis, C.F. (1976). Population replacement in *Culex fatigans* by means of cytoplasmic incompatibility. 2. Field cage experiments with overlapping generations. Bull. World Health Organization *53*, 107–119.

Delatte, H., Desvars, A., Bouetard, A., Bord, S., Gimonneau, G., Vourc'h, G., and Fontenille, D. (2010). Blood-feeding behavior of *Aedes albopictus*, a vector of Chikungunya on La Réunion. Vector Borne Zoonotic Dis. *10*, 249–258.

Deredec, A., Godfray, H.C., and Burt, A. (2011). Requirements for effective malaria control with homing endonuclease genes. Proc. Natl. Acad. Sci. U.S.A. *108*, E874–E880.

Drebot, M.A., Henchal, E., Hjelle, B., LeDuc, J.W., Repik, P.M., Roehrig, J.T., Schmaljohn, C.S., Shope, R.E., Tesh, R.B., Weaver, S.C., *et al.* (2002). Improved clarity of meaning from the use of both formal species names and common (vernacular) virus names in virological literature. Arch. Virol. *147*, 2465–2472.

Engels, W.R. (1997). Invasions of P elements. Genetics *145*, 11–15.

Facchinelli, L., Valerio, L., Bond, J.G., Wise de Valdez, M.R., Harrington, L.C., Ramsey, J.M., Casas-Martinez, M., and Scott, T.W. (2011). Development of a semi-field system for contained field trials with *Aedes aegypti* in southern Mexico. Am. J. Trop. Med. Hyg. *85*, 248–256.

Facchinelli, L., Valerio, L., Ramsey, J.M., Gould, F., Walsh, R.K., Bond, G., Robert, M.A., Lloyd, A.L., James, A.A., Alphey, L., *et al.* (2013). Field cage studies and progressive evaluation of genetically-engineered mosquitoes. PLoS Neglect. Trop. Dis. *7*, e2001.

Falconer, D.S. (1989). Introduction to Quantitative Genetics, 3rd edn (Longman Wiley, New York).

Fish, M.P., Groth, A.C., Calos, M.P., and Nusse, R. (2007). Creating transgenic Drosophila by microinjecting the site-specific phiC31 integrase mRNA and a transgene-containing donor plasmid. Nature protocols *2*, 2325–2331.

Fisher, R.A. (1950). Gene frequencies in a cline determined by selection and diffusion. Biometrics *6*, 353–361.

Franz, A.W., Sanchez-Vargas, I., Adelman, Z.N., Blair, C.D., Beaty, B.J., James, A.A., and Olson, K.E. (2006). Engineering RNA interference-based resistance to dengue virus type 2 in genetically modified *Aedes aegypti*. Proc. Natl. Acad. Sci. U.S.A. *103*, 4198–4203.

Franz, A.W., Sanchez-Vargas, I., Piper, J., Smith, M.R., Khoo, C.C., James, A.A., and Olson, K.E. (2009). Stability and loss of a virus resistance phenotype over time in transgenic mosquitoes harbouring an antiviral effector gene. Insect Mol. Biol. *18*, 661–672.

Franz, A.W., Jasinskiene, N., Sanchez-Vargas, I., Isaacs, A.T., Smith, M.R., Khoo, C.C., Heersink, M.S., James, A.A., and Olson, K.E. (2011). Comparison of transgene expression in *Aedes aegypti* generated by mariner Mos1 transposition and PhiC31 site-directed recombination. Insect Mol. Biol. *20*, 587–598.

Franz, A.W., Sanchez-Vargas, I., Raban, R.R., Black, W.C.t., James, A.A., and Olson, K.E. (2014). Fitness impact and stability of a transgene conferring resistance to dengue-2 virus following introgression into a genetically diverse *Aedes aegypti* strain. PLoS Neglect. Trop. Dis. *8*, e2833.

Franz, G., and Robinson, A.S. (2011). Molecular technologies to improve the effectiveness of the sterile insect technique. Genetica *139*, 1–5.

Franz, G., and Savakis, C. (1991). Minos, a new transposable element from *Drosophila hydei*, is a member of the Tc1-like family of transposons. Nucleic Acids Res. *19*, 6646.

Fu, G., Lees, R.S., Nimmo, D., Aw, D., Jin, L., Gray, P., Berendonk, T.U., White-Cooper, H., Scaife, S., Kim Phuc, H., *et al.* (2010). Female-specific flightless phenotype for mosquito control. Proc. Natl. Acad. Sci. U.S.A. *107*, 4550–4554.

Galizi, R., Doyle, L.A., Menichelli, M., Bernardini, F., Deredec, A., Burt, A., Stoddard, B.L., Windbichler, N., and Crisanti, A. (2014). A synthetic sex ratio distortion system for the control of the human malaria mosquito. Nature Commun. *5*, 3977.

Garcia, G.P., Flores, A.E., Fernandez-Salas, I., Saavedra-Rodriguez, K., Reyes-Solis, G., Lozano-Fuentes, S., Guillermo Bond, J., Casas-Martinez, M., Ramsey, J.M., Garcia-Rejon, J., *et al.* (2009). Recent rapid rise of a permethrin knock down resistance allele in *Aedes aegypti* in Mexico. PLoS Neglect. Trop. Dis. *3*, e531.

Garza, D., Medhora, M., Koga, A., and Hartl, D.L. (1991). Introduction of the transposable element mariner into the germline of Drosophila melanogaster. Genetics *128*, 303–310.

Gilchrist, B.M., and Haldane, J.B. (1946). Sex-linkage in *Culex molestus*. Experientia *2*, 372.

Gossen, M., and Bujard, H. (1992). Tight control of gene expression in mammalian cells by tetracycline-responsive promoters. Proc. Natl. Acad. Sci. U.S.A. *89*, 5547–5551.

Grossman, G.L., Rafferty, C.S., Clayton, J.R., Stevens, T.K., Mukabayire, O., and Benedict, M.Q. (2001). Germline transformation of the malaria vector, *Anopheles gambiae*, with the piggyBac transposable element. Insect Mol. Biol. *10*, 597–604.

Gubler, D.J. (2002). The global emergence/resurgence of arboviral diseases as public health problems. Arch. Med. Res. *33*, 330–342.

Guzman, A., and Isturiz, R.E. (2010). Update on the global spread of dengue. Int. J. Antimicrobial Agents *36(Suppl 1)*, S40–S42.

Halstead, S.B. (2008). Dengue virus–mosquito interactions. Annu. Rev. Entomol. *53*, 273–291.

Handler, A.M., McCombs, S.D., Fraser, M.J., and Saul, S.H. (1998). The lepidopteran transposon vector, piggyBac, mediates germ-line transformation in the Mediterranean fruit fly. Proc. Natl. Acad. Sci. U.S.A. *95*, 7520–7525.

Harris, A.F., McKemey, A.R., Nimmo, D., Curtis, Z., Black, I., Morgan, S.A., Oviedo, M.N., Lacroix, R., Naish, N., Morrison, N.I., *et al.* (2012). Successful suppression of a field mosquito population by sustained release of engineered male mosquitoes. Nat. Biotechnol. *30*, 828–830.

Hay, B.A., Chen, C.H., Ward, C.M., Huang, H., Su, J.T., and Guo, M. (2010). Engineering the genomes of wild insect populations: challenges, and opportunities provided by synthetic Medea selfish genetic elements. J. Insect Physiol. *56*, 1402–1413.

Hickey, W.A., and Craig, G.B., Jr. (1966a). Distortion of sex ratio in populations of *Aedes aegypti*. Can. J. Genet. Cytol. *8*, 260–278.

Hickey, W.A., and Craig, G.B., Jr. (1966b). Genetic distortion of sex ratio in a mosquito, *Aedes aegypti*. Genetics *53*, 1177–1196.

Isaacs, A.T., Jasinskiene, N., Tretiakov, M., Thiery, I., Zettor, A., Bourgouin, C., and James, A.A. (2012). Transgenic *Anopheles stephensi* coexpressing single-chain antibodies resist *Plasmodium falciparum* development. Proc. Natl. Acad. Sci. U.S.A. *109*, E1922–E1930.

Ito, J., Ghosh, A., Moreira, L.A., Wimmer, E.A., and Jacobs-Lorena, M. (2002). Transgenic anopheline mosquitoes impaired in transmission of a malaria parasite. Nature *417*, 452–455.

James, A.A. (2003). Blocking malaria parasite invasion of mosquito salivary glands. J. Exp. Biol. *206*, 3817–3821.

James, A.A. (2005). Gene drive systems in mosquitoes: rules of the road. Trends Parasitol. *21*, 64–67.

Jasinskiene, N., Coates, C.J., Benedict, M.Q., Cornel, A.J., Rafferty, C.S., James, A.A., and Collins, F.H. (1998). Stable transformation of the yellow fever mosquito, *Aedes aegypti*, with the Hermes element from the housefly. Proc. Natl. Acad. Sci. U.S.A. *95*, 3743–3747.

Jasinskiene, N., Coates, C.J., and James, A.A. (2000). Structure of hermes integrations in the germline of the yellow fever mosquito, *Aedes aegypti*. Insect Mol. Biol. *9*, 11–18.

Jasinskiene, N., Juhn, J., and James, A.A. (2007). Microinjection of *A. aegypti* embryos to obtain transgenic mosquitoes. J. Visual. Exp. *5*, 219.

Khoo, C.C., Piper, J., Sanchez-Vargas, I., Olson, K.E., and Franz, A.W. (2010). The RNA interference pathway affects midgut infection- and escape barriers for Sindbis virus in *Aedes aegypti*. BMC Microbiol. *10*, 130.

Khoo, C.C., Doty, J.B., Heersink, M.S., Olson, K.E., and Franz, A.W. (2013). Transgene-mediated suppression of the RNA interference pathway in *Aedes aegypti* interferes with gene silencing and enhances Sindbis virus and dengue virus type 2 replication. Insect Mol. Biol. *22*, 104–114.

Kidwell, M.G. (1992). Horizontal transfer of P elements and other short inverted repeat transposons. Genetica *86*, 275–286.

Klowden, M.J. (1999). The check is in the male: male mosquitoes affect female physiology and behavior. J. Am. Mosq. Control Assoc. *15*, 213–220.

Knipling, E.F., Laven, H., Craig, G.B., Pal, R., Kitzmiller, J.B., Smith, C.N., and Brown, A.W. (1968). Genetic control of insects of public health importance. Bull. World Health Org. *38*, 421–438.

Kokoza, V.A., and Raikhel, A.S. (2011). Targeted gene expression in the transgenic *Aedes aegypti* using the binary Gal4-UAS system. Insect Biochem. Mol. Biol. *41*, 637–644.

Kokoza, V., Ahmed, A., Cho, W.L., Jasinskiene, N., James, A.A., and Raikhel, A. (2000). Engineering bloodmeal-activated systemic immunity in the yellow fever mosquito, *Aedes aegypti*. Proc. Natl. Acad. Sci. U.S.A. *97*, 9144–9149.

Kokoza, V., Ahmed, A., Wimmer, E.A., and Raikhel, A.S. (2001). Efficient transformation of the yellow fever mosquito *Aedes aegypti* using the piggyBac transposable element vector pBac[3xP3-EGFP afm]. Insect Biochem. Mol. Biol. *31*, 1137–1143.

Koyama, J., Kakinohana, H., and Miyatake, T. (2004). Eradication of the melon fly, Bactrocera cucurbitae, in Japan: importance of behavior, ecology, genetics, and evolution. Annu. Rev. Entomol. *49*, 331–349.

Krishnamurthy, B.S., Curtis, C.F., Subbarao, S.K., Chandrahas, R.K., and Adak, T. (1977). Studies on the induction of high sterility male linked translocations in *Culex* p. *fatigans*. Indian J. Med. Res. *65(Suppl)*, 1–12.

Kuhn, E.J., and Geyer, P.K. (2003). Genomic insulators: connecting properties to mechanism. Curr. Opin. Cell Biol. *15*, 259–265.

Labbe, G.M., Nimmo, D.D., and Alphey, L. (2010). piggybac- and PhiC31-mediated genetic transformation of the Asian tiger mosquito, *Aedes albopictus* (Skuse). PLoS Neglect. Trop. Dis. *4*, e788.

Labbe, G.M., Scaife, S., Morgan, S.A., Curtis, Z.H., and Alphey, L. (2012). Female-specific flightless (fsRIDL) phenotype for control of *Aedes albopictus*. PLoS Neglect. Trop. Dis. *6*, e1724.

Lacroix, R., McKemey, A.R., Raduan, N., Kwee Wee, L., Hong Ming, W., Guat Ney, T., Rahidah, A.A.S., Salman, S., Subramaniam, S., Nordin, O., *et al.* (2012). Open field release of genetically engineered sterile male *Aedes aegypti* in Malaysia. PloS One 7, e42771.

Lavery, J.V., Harrington, L.C., and Scott, T.W. (2008). Ethical, social, and cultural considerations for site selection for research with genetically modified mosquitoes. Am. J. Trop. Med. Hyg. *79*, 312–318.

Lavery, J.V., Tinadana, P.O., Scott, T.W., Harrington, L.C., Ramsey, J.M., Ytuarte-Nunez, C., and James, A.A. (2010). Towards a framework for community engagement in global health research. Trends Parasitol. *26*, 279–283.

Lee, J.J., and Klowden, M.J. (1999). A male accessory gland protein that modulates female mosquito (Diptera: Culicidae) host-seeking behavior. J. Am. Mosq. Control Assoc. *15*, 4–7.

Legros, M., Xu, C., Morrison, A., Scott, T.W., Lloyd, A.L., and Gould, F. (2013). Modeling the dynamics of a non-limited and a self-limited gene drive system in structured *Aedes aegypti* populations. PloS One *8*, e83354.

Linthicum, K.J., Platt, K., Myint, K.S., Lerdthusnee, K., Innis, B.L., and Vaughn, D.W. (1996). Dengue 3 virus distribution in the mosquito *Aedes aegypti*: an immunocytochemical study. Med. Vet. Entomol. *10*, 87–92.

Lobo, N., Li, X., and Fraser, M.J., Jr. (1999). Transposition of the piggyBac element in embryos of *Drosophila melanogaster*, *Aedes aegypti* and *Trichoplusia* ni. Mol. Gen. Genet. *261*, 803–810.

Lohe, A.R., and Hartl, D.L. (1996). Autoregulation of mariner transposase activity by overproduction and dominant-negative complementation. Mol. Biol. Evol. *13*, 549–555.

Lorenzen, M.D., Gnirke, A., Margolis, J., Garnes, J., Campbell, M., Stuart, J.J., Aggarwal, R., Richards, S., Park, Y., and Beeman, R.W. (2008). The maternal-effect, selfish genetic element Medea is associated with a composite Tc1 transposon. Proc. Natl. Acad. Sci. U.S.A. *105*, 10085–10089.

Loukeris, T.G., Livadaras, I., Arca, B., Zabalou, S., and Savakis, C. (1995). Gene transfer into the medfly, Ceratitis capitata, with a Drosophila hydei transposable element. Science *270*, 2002–2005.

Lozano-Fuentes, S., Fernandez-Salas, I., de Lourdes Munoz, M., Garcia-Rejon, J., Olson, K.E., Beaty, B.J., and Black, W.C.t. (2009). The neovolcanic axis is a barrier to gene flow among *Aedes aegypti* populations in Mexico that differ in vector competence for Dengue 2 virus. PLoS Neglect. Trop. Dis. *3*, e468.

Lumjuan, N., Rajatileka, S., Changsom, D., Wicheer, J., Leelapat, P., Prapanthadara, L.A., Somboon, P., Lycett, G., and Ranson, H. (2011). The role of the *Aedes aegypti* Epsilon glutathione transferases in conferring resistance to DDT and pyrethroid insecticides. Insect Biochem. Mol. Biol. *41*, 203–209.

Lynd, A., and Lycett, G.J. (2012). Development of the bi-partite Gal4-UAS system in the African malaria mosquito, *Anopheles gambiae*. PloS One 7, e31552.

McGrane, V., Carlson, J.O., Miller, B.R., and Beaty, B.J. (1988). Microinjection of DNA into *Aedes triseriatus* ova and detection of integration. The Am. J. Trop. Med. Hyg. *39*, 502–510.

McGraw, E.A., and O'Neill, S.L. (2013). Beyond insecticides: new thinking on an ancient problem. Nat. Rev. Microbiol. *11*, 181–193.

McGraw, E.A., Merritt, D.J., Droller, J.N., and O'Neill, S.L. (2002). *Wolbachia* density and virulence attenuation after transfer into a novel host. Proc. Natl. Acad. Sci. U.S.A. *99*, 2918–2923.

Mackenzie, J.S., Gubler, D.J., and Petersen, L.R. (2004). Emerging flaviviruses: the spread and resurgence of Japanese encephalitis, West Nile and dengue viruses. Nat. Med. *10*, S98–S109.

McMeniman, C.J., Lane, R.V., Cass, B.N., Fong, A.W., Sidhu, M., Wang, Y.F., and O'Neill, S.L. (2009). Stable introduction of a life-shortening *Wolbachia* infection into the mosquito *Aedes aegypti*. Science *323*, 141–144.

Magori, K., Legros, M., Puente, M.E., Focks, D.A., Scott, T.W., Lloyd, A.L., and Gould, F. (2009). Skeeter Buster: a stochastic, spatially explicit modeling tool for studying *Aedes aegypti* population replacement and population suppression strategies. PLoS Neglect. Trop. Dis. *3*, e508.

Marin, I., and Baker, B.S. (1998). The evolutionary dynamics of sex determination. Science *281*, 1990–1994.

Marinotti, O., Jasinskiene, N., Fazekas, A., Scaife, S., Fu, G., Mattingly, S.T., Chow, K., Brown, D.M., Alphey, L., and James, A.A. (2013). Development of a population suppression strain of the human malaria vector mosquito, *Anopheles stephensi*. Malaria J. *12*, 142.

Mathur, G., Sanchez-Vargas, I., Alvarez, D., Olson, K.E., Marinotti, O., and James, A.A. (2010). Transgene-mediated suppression of dengue viruses in the salivary glands of the yellow fever mosquito, *Aedes aegypti*. Insect Mol. Biol. *19*, 753–763.

Medlock, J.M., Hansford, K.M., Schaffner, F., Versteirt, V., Hendrickx, G., Zeller, H., and Van Bortel, W. (2012). A review of the invasive mosquitoes in Europe: ecology, public health risks, and control options. Vector Borne Zoonotic Dis. *12*, 435–447.

Meredith, J.M., Underhill, A., McArthur, C.C., and Eggleston, P. (2013). Next-generation site-directed transgenesis in the malaria vector mosquito *Anopheles gambiae*: self-docking strains expressing germline-specific phiC31 integrase. PloS One *8*, e59264.

Miller, L.H., Sakai, R.K., Romans, P., Gwadz, R.W., Kantoff, P., and Coon, H.G. (1987). Stable integration and expression of a bacterial gene in the mosquito *Anopheles gambiae*. Science *237*, 779–781.

Mittal, P.K. (2003). Biolarvicides in vector control: challenges and prospects. J. Vector Borne Dis. *40*, 20–32.

Moore, M., Sylla, M., Goss, L., Burugu, M.W., Sang, R., Kamau, L.W., Kenya, E.U., Bosio, C., Munoz Mde, L., Sharakova, M., *et al.* (2013). Dual African origins of global *Aedes aegypti* s.l. populations revealed by mitochondrial DNA. PLoS Neglect. Trop. Dis. 7, e2175.

Morazzani, E.M., Wiley, M.R., Murreddu, M.G., Adelman, Z.N., and Myles, K.M. (2012). Production of virus-derived ping-pong-dependent piRNA-like small RNAs in the mosquito soma. PLoS Path. *8*, e1002470.

Moreira, L.A., Edwards, M.J., Adhami, F., Jasinskiene, N., James, A.A., and Jacobs-Lorena, M. (2000). Robust gut-specific gene expression in transgenic *Aedes aegypti* mosquitoes. Proc. Natl. Acad. Sci. U.S.A. *97*, 10895–10898.

Moreira, L.A., Iturbe-Ormaetxe, I., Jeffery, J.A., Lu, G., Pyke, A.T., Hedges, L.M., Rocha, B.C., Hall-Mendelin, S., Day, A., Riegler, M., *et al.* (2009a). A Wolbachia symbiont in *Aedes aegypti* limits infection with dengue, Chikungunya, and Plasmodium. Cell *139*, 1268–1278.

Moreira, L.A., Saig, E., Turley, A.P., Ribeiro, J.M., O'Neill, S.L., and McGraw, E.A. (2009b). Human probing behavior of *Aedes aegypti* when infected with a life-shortening strain of *Wolbachia*. PLoS Neglect. Trop. Dis. 3, e568.

Nene, V., Wortman, J.R., Lawson, D., Haas, B., Kodira, C., Tu, Z.J., Loftus, B., Xi, Z., Megy, K., Grabherr, M., *et al.* (2007). Genome sequence of *Aedes aegypti*, a major arbovirus vector. Science *316*, 1718–1723.

Newton, M.E., Southern, D.I., and Wood, R.J. (1974). X and Y chromosomes of *Aedes aegypti* (L.) distinguished by Giemsa C-banding. Chromosoma *49*, 41–49.

Nimmo, D.D., Alphey, L., Meredith, J.M., and Eggleston, P. (2006). High efficiency site-specific genetic engineering of the mosquito genome. Insect Mol. Biol. *15*, 129–136.

Nirmala, X., Marinotti, O., Sandoval, J.M., Phin, S., Gakhar, S., Jasinskiene, N., and James, A.A. (2006). Functional characterization of the promoter of the vitellogenin gene, AsVg1, of the malaria vector, *Anopheles stephensi*. Insect Biochem. Mol. Biol. *36*, 694–700.

Nolan, T., Bower, T.M., Brown, A.E., Crisanti, A., and Catteruccia, F. (2002). piggyBac-mediated germline transformation of the malaria mosquito Anopheles stephensi using the red fluorescent protein dsRED as a selectable marker. J. Biol. Chem. *277*, 8759–8762.

O'Brochta, D.A., Pilitt, K.L., Harrell, R.A., 2nd, Aluvihare, C., and Alford, R.T. (2012). Gal4-based enhancer-trapping in the malaria mosquito *Anopheles stephensi*. G3 (Bethesda) *2*, 1305–1315.

Okamoto, K.W., Robert, M.A., Lloyd, A.L., and Gould, F. (2013). A reduce and replace strategy for suppressing vector-borne diseases: insights from a stochastic, spatial model. PloS One *8*, e81860.

Okamoto, K.W., Robert, M.A., Gould, F., and Lloyd, A.L. (2014). Feasible introgression of an anti-pathogen transgene into an urban mosquito population without using Gene-Drive. PLoS Neglect. Trop. Dis. *8*, e2827.

Ooi, E.E., Goh, K.T., and Gubler, D.J. (2006). Dengue prevention and 35 years of vector control in Singapore. Emerg. Infect. Dis. *12*, 887–893.

Owusu-Daaku, K.O., Butler, R.D., and Wood, R.J. (2007). Meiotic drive by the Y-linked D gene in *Aedes aegypti* (L.) (Diptera: Culicidae) is associated with disruption of spermiogenesis, leading to premature senescence of spermatozoa. Arthropod Struct. Dev. *36*, 233–243.

Peng, J.C., Chen, T.C., Raja, J.A., Yang, C.F., Chien, W.C., Lin, C.H., Liu, F.L., Wu, H.W., and Yeh, S.D. (2014). Broad-spectrum transgenic resistance against distinct tospovirus species at the genus level. PloS One 9, e96073.

Phuc, H.K., Andreasen, M.H., Burton, R.S., Vass, C., Epton, M.J., Pape, G., Fu, G., Condon, K.C., Scaife, S., Donnelly, C.A., *et al.* (2007). Late-acting dominant lethal genetic systems and mosquito control. BMC Biol. *5*, 11.

Pinkerton, A.C., Michel, K., O'Brochta, D.A., and Atkinson, P.W. (2000). Green fluorescent protein as a genetic marker in transgenic *Aedes aegypti*. Insect Mol. Biol. *9*, 1–10.

Powell, J.R., and Tabachnick, W.J. (2013). History of domestication and spread of *Aedes aegypti*--a review. Memorias do Instituto Oswaldo Cruz *108(Suppl. 1)*, 11–17.

Rai, K.S., McDonald, P.T., and Asman, S.M. (1970). Cytogenetics of two radiation-induced, sex-linked translocations in the yellow-fever mosquito, *Aedes aegypti*. Genetics *66*, 635–651.

Ramsey, J.M., Bond, J.G., Macotela, M.E., Facchinelli, L., Valerio, L., Brown, D.M., Scott, T.W., and James, A.A. (2014). A regulatory structure for working with genetically modified mosquitoes: lessons from Mexico. PLoS Neglect. Trop. Dis. *8*, e2623.

Reeves, R.G., Denton, J.A., Santucci, F., Bryk, J., and Reed, F.A. (2012). Scientific standards and the regulation of genetically modified insects. PLoS Neglect. Trop. Dis. *6*, e1502.

Robert, M.A., Okamoto, K., Lloyd, A.L., and Gould, F. (2013). A reduce and replace strategy for suppressing vector-borne diseases: insights from a deterministic model. PloS One *8*, e73233.

Robinson, A.S. (2002). Genetic sexing strains in medfly, Ceratitis capitata, sterile insect technique programmes. Genetica *116*, 5–13.

Robinson, A.S., Vreysen, M.J., Hendrichs, J., and Feldmann, U. (2009). Enabling technologies to improve area-wide integrated pest management programmes for the control of screwworms. Med. Vet. Entomol. *23(Suppl. 1)*, 1–7.

Rubin, G.M., and Spradling, A.C. (1982). Genetic transformation of *Drosophila* with transposable element vectors. Science *218*, 348–353.

Saavedra-Rodriguez, K., Suarez, A.F., Salas, I.F., Strode, C., Ranson, H., Hemingway, J., and Black, W.C.t. (2012). Transcription of detoxification genes after permethrin selection in the mosquito *Aedes aegypti*. Insect Mol. Biol. *21*, 61–77.

Saavedra-Rodriguez, K., Strode, C., Flores, A.E., Garcia-Luna, S., Reyes-Solis, G., Ranson, H., Hemingway, J., and Black, W.C.t. (2014). Differential transcription profiles in *Aedes aegypti* detoxification genes after temephos selection. Insect Mol. Biol. 23, 199–215.

Sabl, J.F., and Henikoff, S. (1996). Copy number and orientation determine the susceptibility of a gene to silencing by nearby heterochromatin in *Drosophila*. Genetics *142*, 447–458.

Sakai, R.K., and Miller, L.H. (1992). Effects of heat shock on the survival of transgenic *Anopheles gambiae* (Diptera: Culicidae) under antibiotic selection. J. Med. Entomol. *29*, 374–375.

Salazar, M.I., Richardson, J.H., Sanchez-Vargas, I., Olson, K.E., and Beaty, B.J. (2007). Dengue virus type 2: replication and tropisms in orally infected *Aedes aegypti* mosquitoes. BMC Microbiol. *7*, 9.

Sanchez-Vargas, I., Scott, J.C., Poole-Smith, B.K., Franz, A.W., Barbosa-Solomieu, V., Wilusz, J., Olson, K.E., and Blair, C.D. (2009). Dengue virus type 2 infections of *Aedes aegypti* are modulated by the mosquito's RNA interference pathway. PLoS Path. *5*, e1000299.

Sarkar, A., Yardley, K., Atkinson, P.W., James, A.A., and O'Brochta, D.A. (1997). Transposition of the Hermes element in embryos of the vector mosquito, *Aedes aegypti*. Insect Biochem. Mol. Biol. *27*, 359–363.

Sarkar, A., Sim, C., Hong, Y.S., Hogan, J.R., Fraser, M.J., Robertson, H.M., and Collins, F.H. (2003). Molecular evolutionary analysis of the widespread piggyBac transposon family and related 'domesticated' sequences. Mol. Genet. Genom. *270*, 173–180.

Sarkar, A., Atapattu, A., Belikoff, E.J., Heinrich, J.C., Li, X., Horn, C., Wimmer, E.A., and Scott, M.J. (2006). Insulated piggyBac vectors for insect transgenesis. BMC Biotechnol. *6*, 27.

Scott, T.W., Chow, E., Strickman, D., Kittayapong, P., Wirtz, R.A., Lorenz, L.H., and Edman, J.D. (1993). Blood-feeding patterns of *Aedes aegypti* (Diptera: Culicidae) collected in a rural Thai village. J. Med. Entomol. *30*, 922–927.

Scott, T.W., Amerasinghe, P.H., Morrison, A.C., Lorenz, L.H., Clark, G.G., Strickman, D., Kittayapong, P., and Edman, J.D. (2000). Longitudinal studies of *Aedes aegypti* (Diptera: Culicidae) in Thailand and Puerto Rico: blood feeding frequency. J. Med. Entomol. 37, 89–101.

Sheng, G., Thouvenot, E., Schmucker, D., Wilson, D.S., and Desplan, C. (1997). Direct regulation of rhodopsin 1 by Pax-6/eyeless in Drosophila: evidence for a conserved function in photoreceptors. Genes Dev. *11*, 1122–1131.

Shin, D., Mori, A., and Severson, D.W. (2012). Genetic mapping a meiotic driver that causes sex ratio distortion in the mosquito *Aedes aegypti*. J. Heredity *103*, 303–307.

Sinkins, S.P., and Gould, F. (2006). Gene drive systems for insect disease vectors. Nat. Rev. Genet. *7*, 427–435.

Sinzelle, L., Jegot, G., Brillet, B. Rouleaux-Bonnin, Bigot, Y., and Auge-Gouillou, C. (2008). Factors acting on Mos1 transposition efficiency. BMC Mol. Biol. *9*, 106.

Skavdis, G., Siden-Kiamos, I., Muller, H.M., Crisanti, A., and Louis, C. (1996). Conserved function of anopheles gambiae midgut-specific promoters in the fruitfly. EMBO J. *15*, 344–350.

Smallegange, R.C., Verhulst, N.O., and Takken, W. (2011). Sweaty skin: an invitation to bite? Trends Parasitol. 27, 143–148.

Smartt, C.T., Kim, A.P., Grossman, G.L., and James, A.A. (1995). The Apyrase gene of the vector mosquito, *Aedes aegypti*, is expressed specifically in the adult female salivary glands. Exp. Parasitol. *81*, 239–248.

Smith, R.C., and Atkinson, P.W. (2011). Mobility properties of the Hermes transposable element in transgenic lines of *Aedes aegypti*. Genetica *139*, 7–22.

Smith, R.C., Walter, M.F., Hice, R.H., O'Brochta, D.A., and Atkinson, P.W. (2007). Testis-specific expression of the beta2 tubulin promoter of *Aedes aegypti* and its application as a genetic sex-separation marker. Insect Mol. Biol. *16*, 61–71.

Souza-Neto, J.A., Sim, S., and Dimopoulos, G. (2009). An evolutionary conserved function of the JAK-STAT pathway in anti-dengue defense. Proc. Natl. Acad. Sci. U.S.A. *106*, 17841–17846.

Sweeny, T.L., and Barr, A.R. (1978). Sex ratio distortion caused by meiotic drive in a mosquito, *Culex pipiens* L. Genetics *88*, 427–446.

Tabachnick, W.J. (1991). The evolutionary relationships among arboviruses and the evolutionary relationships of their vectors provides a method for understanding vector–host interactions. J. Med. Entomol. *28*, 297–298.

Tamura, T., Thibert, C., Royer, C., Kanda, T., Abraham, E., Kamba, M., Komoto, N., Thomas, J.L., Mauchamp, B., Chavancy, G., *et al.* (2000). Germline transformation of the silkworm *Bombyx mori* L. using a piggyBac transposon-derived vector. Nat. Biotechnol. *18*, 81–84.

Telang, A., and Wells, M.A. (2004). The effect of larval and adult nutrition on successful autogenous egg production by a mosquito. J. Insect Physiol. *50*, 677–685.

Terenius, O., Marinotti, O., Sieglaff, D., and James, A.A. (2008). Molecular genetic manipulation of vector mosquitoes. Cell Host Microbe *4*, 417–423.

Thibault, S.T., Luu, H.T., Vann, N., and Miller, T.A. (1999). Precise excision and transposition of piggyBac in pink bollworm embryos. Insect Mol. Biol. *8*, 119–123.

Thomas, D.D., Donnelly, C.A., Wood, R.J., and Alphey, L.S. (2000). Insect population control using a dominant, repressible, lethal genetic system. Science *287*, 2474–2476.

Thorpe, H.M., and Smith, M.C. (1998). In vitro site-specific integration of bacteriophage DNA catalyzed by a recombinase of the resolvase/invertase family. Proc. Natl. Acad. Sci. U.S.A. *95*, 5505–5510.

Traver, B.E., Anderson, M.A., and Adelman, Z.N. (2009). Homing endonucleases catalyze double-stranded DNA breaks and somatic transgene excision in *Aedes aegypti*. Insect Mol. Biol. *18*, 623–633.

Vasilakis, N., Holmes, E.C., Fokam, E.B., Faye, O., Diallo, M., Sall, A.A., and Weaver, S.C. (2007). Evolutionary processes among sylvatic dengue type 2 viruses. J. Virol. *81*, 9591–9595.

Vreysen, M.J. (2001). Principles of area-wide integrated tsetse fly control using the sterile insect technique. Medecine tropicale: revue du Corps de sante colonial *61*, 397–411.

Wade, M.J., and Beeman, R.W. (1994). The population dynamics of maternal-effect selfish genes. Genetics *138*, 1309–1314.

Walker, T., Johnson, P.H., Moreira, L.A., Iturbe-Ormaetxe, I., Frentiu, F.D., McMeniman, C.J., Leong, Y.S., Dong, Y., Axford, J., Kriesner, P., *et al.* (2011). The wMel Wolbachia strain blocks dengue and invades caged *Aedes aegypti* populations. Nature *476*, 450–453.

Wallrath, L.L., and Elgin, S.C. (1995). Position effect variegation in Drosophila is associated with an altered chromatin structure. Genes Dev. *9*, 1263–1277.

Wang, W., Swevers, L., and Iatrou, K. (2000). Mariner (Mos1) transposase and genomic integration of foreign gene sequences in Bombyx mori cells. Insect Mol. Biol. *9*, 145–155.

Warren, W.D., Atkinson, P.W., and O'Brochta, D.A. (1994). The Hermes transposable element from the house fly, Musca domestica, is a short inverted repeat-type element of the hobo, Ac, and Tam3 (hAT) element family. Genet. Res. *64*, 87–97.

Waterhouse, R.M., Kriventseva, E.V., Meister, S., Xi, Z., Alvarez, K.S., Bartholomay, L.C., Barillas-Mury, C., Bian, G., Blandin, S., Christensen, B.M., *et al.* (2007). Evolutionary dynamics of immune-related genes and pathways in disease-vector mosquitoes. Science *316*, 1738–1743.

Weaver, S.C. (2014). Arrival of chikungunya virus in the new world: prospects for spread and impact on public health. PLoS Neglect. Trop. Dis. *8*, e2921.

Weaver, S.C., and Barrett, A.D. (2004). Transmission cycles, host range, evolution and emergence of arboviral disease. Nat. Rev. Microbiol. *2*, 789–801.

Weaver, S.C., and Reisen, W.K. (2010). Present and future arboviral threats. Antiviral Res. *85*, 328–345.

Weidhaas, D.E. (1962). Chemical sterilization of mosquitoes. Nature *195*, 786–787.

Wilson, R., Orsetti, J., Klocko, A.D., Aluvihare, C., Peckham, E., Atkinson, P.W., Lehane, M.J., and O'Brochta, D.A. (2003). Post-integration behavior of a Mos1 mariner gene vector in *Aedes aegypti*. Insect Biochem. Mol. Biol. *33*, 853–863.

Windbichler, N., Papathanos, P.A., Catteruccia, F., Ranson, H., Burt, A., and Crisanti, A. (2007). Homing endonuclease mediated gene targeting in Anopheles gambiae cells and embryos. Nucleic Acids Res. *35*, 5922–5933.

Windbichler, N., Menichelli, M., Papathanos, P.A., Thyme, S.B., Li, H., Ulge, U.Y., Hovde, B.T., Baker, D., Monnat, R.J., Jr., Burt, A., *et al.* (2011). A synthetic homing endonuclease-based gene drive system in the human malaria mosquito. Nature *473*, 212–215.

Wise de Valdez, M.R., Nimmo, D., Betz, J., Gong, H.F., James, A.A., Alphey, L., and Black, W.C.t. (2011). Genetic elimination of dengue vector mosquitoes. Proc. Natl. Acad. Sci. U.S.A. *108*, 4772–4775.

Xi, Z., Ramirez, J.L., and Dimopoulos, G. (2008). The *Aedes aegypti* toll pathway controls dengue virus infection. PLoS Path. *4*, e1000098.

Yoshida, S., and Watanabe, H. (2006). Robust salivary gland-specific transgene expression in Anopheles stephensi mosquito. Insect Mol. Biol. *15*, 403–410.

Arbovirus Vaccines

Scott B. Halstead

20

Abstract

Arthropod-borne (Arbo) viruses are a broad group of human and veterinary pathogens that are transmitted by the bite of a member of the arthropod phylum, generally mosquitoes or ticks. This review provides basic information on the nature of diseases with a general description of ecology and global disease prevalence together with information on licensed vaccines or vaccines in development for 20 arboviral species belonging to four families – *Bunyaviridae, Flaviviridae, Reoviridae* and *Togaviridae*.

Introduction

Arboviruses are a group of viruses with functionally similar mechanisms of transmission. These viruses replicate in and are transmitted between vertebrate hosts and arthropod vectors. The author has selected 20 arboviruses that comprise a sufficient threat to human health to warrant the use of vaccines. When licensed vaccines are not available, the chapter indicates the status of vaccine development for potential use in humans. The threats posed by arboviruses to domestic and wild animals and the status of veterinary vaccine development are not covered. Vaccines designed for use in animals are considered only if such vaccines have been accompanied by the development of vaccines for use in humans. For each human arboviral disease a brief perspective is provided on its epidemiology, geographical distribution and vaccine status.

Bunyaviridae

Nairovirus

Crimean–Congo haemorrhagic fever (CCHF)

Disease perspective

This is a widespread viral zoonosis involving domestic and wild animals that in large series of cases in humans has a 5% mortality rate. The causative virus is found across Asia, in Eastern Europe, the Middle East and in a belt across central Africa, South Africa and Madagascar. Across this wide area, the virus circulates as many geographic genetic variants that are maintained in cycles involving vertebrate hosts and hard ticks whose species vary with their preferred hosts. Genetic sequencing suggests that birds are responsible for long distance distribution of infected ticks (Bente *et al.*, 2013). Common vertebrate hosts include the European hare, Middle-African hedgehogs and multimammate rats. In addition, CCHF virus is maintained in ticks by transovarial and transstadial transmission. Virus can also be spread between infected and uninfected ticks when both are feeding on a vertebrate (Bente *et al.*, 2013). Human cases occur at a time of tick biting activity and abundance. Recent outbreaks or areas of high prevalence have been reported in Turkey, Iran, Kosovo, India and Pakistan. Virus has been isolated from at least 31 species of ticks of the genera *Haemaphysalis*, and *Hyalomma*. Sporadic infection of people is usually the result of bite by Hyalomma ticks because both the immature and adult forms seek bloodmeals. Clusters of cases occur when people butcher or eat livestock infected with the virus. The virus in blood is highly infectious and often spreads in hospital settings from acutely ill patients to hospital staff (Oncu, 2013). Disease begins with the acute onset of flu-like symptoms 5–6 days after a tick bite or exposure to infected blood or tissues. Signs of haemorrhage appear 3–5 days later accompanied by agitation, mental confusion, soft palate petechiae, epistaxis, haematuria and bleeding in the g.i. tract. The disease progresses to a stage of disseminated intravascular coagulation, kidney failure, shock and acute respiratory distress syndrome (Akinci *et al.*, 2013). Persons suspected of CCHF should be immediately isolated, treated and maintained using strict containment precautions (Leblebicioglu *et al.*, 2012). Ribavirin has been shown to be antiviral *in vitro* and is used empirically. Despite many published clinical studies, there is as yet no consensus that appropriate double-blinded, randomized clinical trials have demonstrated unequivocal efficacy of ribavirin to reduce mortality of CCHF (CDC, 2014; Ascioglu *et al.*, 2011; Soares-Weiser *et al.*, 2010). There is some evidence that hyperimmune globulin may favourably affect the outcome of severe disease (Kubar *et al.*, 2011).

Vaccines

As the result of sustained transmission to humans, work in Bulgaria resulted in the development of a mouse brain-derived CCHF vaccine which has been in use since the early 1970s and the preparation of specific immune globulin from hyperimmunized volunteers, which is banked and utilized for treatment of patients (Papa *et al.*, 2011). Sequencing shows that the Bulgarian vaccine strain, isolated prior to 1970, is still closely related to viruses circulating in Bulgaria, Kosovo and Turkey. So far, individuals given this vaccine have not developed CCHF and vaccination of high-risk groups has reduced the incidence of CCHF. The vaccine elicits both humoral and cellular immunity with low neutralizing antibody titres (Mousavi-Jazi *et al.*, 2012). A successful vaccination procedure has recently been described in which the genes for modified M segment glycoprotein of CCHFVare inserted into Modified Ankara strain of vaccinia. Using peripheral inoculation of CCHFV into interferon receptor deficient mice is a good animal model of human disease (Buttigieg *et al.*, 2014), when 5- to 8-week-old type I interferon receptor deficient mice were given two doses of vaccine intramuscularly, they successfully resisted intradermal infection challenge using 200 TCID 50 of CCHFV.

Orthobunyavirus

La Crosse, California encephalitis

Disease perspective

La Crosse virus (LACV) is a member of the California serogroup of viruses. The prototype member of this group was California virus, however, LACV is the most geographically distributed member of the group and causes the most human disease in the upper Midwestern, mid-Atlantic and southeastern United States, particularly in Appalachia. LACV causes 70–130 recognized cases of encephalitis in the USA (up to 30% of total encephalitis cases), with attack rates in children as high as 7.2/100,000 (Bennett *et al.*, 2012). California virus was linked to human encephalitis in 1945. La Crosse was recovered from a Wisconsin child with encephalitis in 1960. Other members of the subgroup that cause minimal human morbidity include snowshoe hare, Jamestown Canyon and trivitattus viruses. *Aedes triseriatus*, a treehole-breeding mosquito is both the host and the vector. This mosquito breeds in hardwood deciduous forests but also in small artificial containers such as discarded cans, bottles and tyres. The mosquito bites actively during daylight. The virus is maintained by vertical transmission from female mosquitoes to their eggs and can even be transmitted venereally between adults. In addition, the virus is maintained and amplified in complex vertebrate-mosquito cycles involving chipmunks, the grey squirrel, foxes and woodchucks. *Aedes albopictus* may serve as an auxillary vector establishing brief human to human transmissions. There is a high ratio of inapparent to apparent infections. Risk to infection varies with the time of exposure to wooded areas, and specifically to areas with treeholes (Hollidge *et al.*, 2010). The clinical spectra of infections with LACV and other members of the group varies from inapparent infections, mild febrile illness, to aseptic meningitis and encephalitis. It is believed that Jamestown Canyon virus may contribute as the cause of many encephalitis cases attributed to La Crosse (Bennett *et al.*, 2012). Neurological signs occur after a 3- to 4-day prodrome of fever, headache and vomiting. Fortunately, the mortality rate is around 1.9% (Miller *et al.*, 2012). Although, post-illness sequelae, including diminished cerebral performance and seizures are moderately common (Miller *et al.*, 2012).

Approximately 80–100 La Crosse neuroinvasive disease cases are reported each year in the USA (http://www.cdc.gov/lac/tech/epi.html, accessed 3 June 2014).

Vaccine development status

Viruses of the California serogroup are composed of three single-stranded, negative-sense RNA genome segments, designated small (S), medium (M) and large (L). The M segment encodes a single polyprotein that is processed into two glycoproteins (G_N and G_C) and a non-structural protein of unknown function (NS_M). Proceeding from evidence that infection and challenge of subhuman primates revealed that infection with Jamestown Canyon virus (JCV) protected against LACV, but, not visa versa, genes for the G_N and G_C proteins of JCV were inserted into the replicative machinery of LACV. When rhesus monkeys were infected with the JCV/LCV chimera, they were viraemic and resisted challenge with JCV. In other experiments, mice immunized with JCV/LACV resisted challenge with JCV, LACV and Tahyna virus (a European member of the subgroup) (Bennett *et al.*, 2012). In earlier work, when mice lacking interferon-alpha/beta receptors were immunized using a single injection of plasmid DNA encoding the viral envelope proteins G_N and G_C, they were fully protected against lethal challenge (Pavlovic *et al.*, 2000).

Oropouche

Disease perspective

The virus was isolated from a febrile patient in Vega de Oropouche, Trinidad, in 1955 (Anderson *et al.*, 1961). Oropouche viruses exist as a monophylogenetic group with four lineages. Lineage I contains most Brazilian strains, lineage II Peruvian strains, III strains from Panama and lineage IV a single strain from Manaus, Brazil in 1981 (Pinheiro *et al.*, 2014). The virus is maintained in two different transmission cycles, an ill-defined jungle cycle involving sloths as vertebrate hosts and presumably transmitted by vector haematophagous insects. It is thought that humans contract the virus infection in jungle habitats introducing it into urban settlements where biting midges appear to serve as vectors (Baisley *et al.*, 1998; Pinheiro *et al.*, 1982). Urban epidemics are explosive and of short duration. An epidemic in the city of Belem involved at least 11,000 patients in 1961. Since then, there have been many urban epidemics in Brazil and also in Argentina, Panama, Peru and Trinidad (Aguilar *et al.*, 2011a; Baisley *et al.*, 1998). Most cases are limited to the Brazilian Amazon region. The disease is characterized by the sudden onset of fever, headache, chills, dizziness, muscular pain, arthralgia and photophobia (dengue-line) 4–8 days after bite of an infected insect. Arthralgia and prostration may be incapacitating. The acute phase lasts 2–5 days, but arthralgia may persist for weeks. Recurrence of symptoms a week or two later have been reported. In most cases, a maculopapular rash on the thorax, back, arms and legs is observed accompanying defervescence. Occasional patients experience aseptic meningitis (Bastos Mde *et al.*, 2012).

Oropouche is the second most common arboviral disease in Brazil, surpassed only by dengue. Remarkably, for a disease involving such large numbers, mortality is rare (Pinheiro *et al.*, 2014).

Vaccine development status
No vaccines are under construction.

Phlebovirus

Rift Valley fever (RVF)

Disease perspective
RVF is primarily a disease of cattle and sheep transmitted by mosquitoes with scattered outbreaks occurring in Egypt, sub-Saharan Africa, Saudi Arabia, Yemen and Madagascar. Viral infections produce particularly severe disease in pregnant cattle and sheep and in newborn lambs. Humans are infected from bites of infected mosquitoes or more commonly from direct contact with body fluids or tissues of animals with the disease (Hubalek *et al.*, 2014). The virus passes transovarially in *Aedes neomelanoconion* that breed in water-filled 'dambos' scattered across Kenyan flood plains (Peters, 1997). *Culex pipiens* was implicated as a vector of human infection during the Egypt outbreak. Viral infection produces a broad spectrum of responses from inapparent to mild to severe. After an incubation period of 3–7 days there is a flu-like febrile illness accompanied by nausea and vomiting, flushed face, icterus in some cases followed by hallucinations and delirium in severe cases. During early convalescence blurred vision may be noted leading to transient central blindness. Meningoencephalitis and severe haemorrhage occur independently in around 1% of cases resulting in an elevated mortality rate (Peters and Harrison, 2014).

High levels of rainfall that increase mosquito populations are pre-conditions for outbreaks. There have been multiple outbreaks in Africa over the past 70 years, the largest being in Egypt in 1977–78 with 200,000 cases and nearly 600 deaths. The most recent outbreak was in South Africa in 2010, with nearly 100 cases occurring in livestock and abattoir workers.

Vaccines
For laboratory workers and individuals at high risk of acquiring RVF, a formalin-inactivated Salk Institute-Government Services Division (TSI-GSD)-200 vaccine using virus grown in diploid rhesus lung cell line (FRhL-2, DBS 103 cells) is available and has been shown to raise and maintain high titres of neutralizing antibodies.(Pittman and Plotkin, 2013) The main strategy for preventing human cases of RVF is to vaccinate cattle and sheep in the face of epizootics (Ikegami and Makino, 2009; Monath, 2013). Several attenuated strains or replicating constructs of RVFVs, have been developed and some have been safely used to vaccinate domestic animals: MP-12 derived by passage of virus in the presence of a mutagen (Caplen *et al.*, 1985), a natural plaque isolate, clone 13 (Indran and Ikegami, 2012), M segment subunits, Gn and Gc have been used to raise protective antibodies, using reverse genetics RVFV has been engineered into a replicon that is immunogenic but does not spread from cell to cell (Oreshkova *et al.*, 2013), adenovirus, alphavirus (VEE or SIN) replicons, vaccinia or Newcastle's disease virus expressing Gn and Gc genes of RVFV have been developed and are protective experimentally (Indran and Ikegami, 2012; Murakami *et al.*, 2014; Warimwe *et al.*, 2013).

Severe fever with thrombocytopenia syndrome (SFTS)

Disease perspective
Between late March and July 2009, middle-aged farmers, male and female living in Hubei and Henan Provinces were admitted to hospital with a high fever accompanied by thrombocytopenia. Among 171 confirmed patients in this early report, most presented with an acute high fever accompanied by thrombocytopenia, gastrointestinal symptoms, lymphadenopathy and elevations of alanine aminotransferase, aspartate aminotransferase, creatine kinase and lactate dehydrogenase, proteinuria and in some instances, haematuria. Death occurred in 21 cases. A new Bunyavirus, genus *Phlebovirus*, was isolated from these patients and from *Haemaphysalis longicornis* ticks that are distributed widely in Asia and are known to feed on many domestic farm animals and household pests (Yu *et al.*, 2011). Since the initial outbreak report more patients with SFTS have been identified in six Chinese provinces predominantly in rural hilly areas (Ding *et al.*, 2013). There has been a report of person to person transmission by contact with patient blood (Bao *et al.*, 2013; Cui *et al.*, 2013; Liu *et al.*, 2012). Epidemiological studies have found very low prevalence of SFTS antibodies in the general population with concentrations suggesting that grazing and goat raising are risk factors (Liang *et al.*, 2014). Goats in China have been found to circulate SFTS antibodies (Zhao *et al.*, 2012).

Vaccine development status
Reports describing the development of vaccines against this agent have not been published.

Sandfly fever

Disease perspective
Sandfly fever, or more correctly Phlebotomus (Pappataci) fever is caused by phleboviruses, Sicilian, and those of the Naples species complex including the Toscana clade (Palacios *et al.*, 2014). Sicilian and Naples viruses are transmitted by *Phlebotomus papatasi* prevalent in Southern Europe, North Africa, the Balkans, Eastern Mediterranean, Iraq, Iran, Pakistan, Afghanistan and India. Toscana virus is transmitted by *Phlebotomus perniciousus* and *Phlebotomus perfiliewi* in Portugal, Spain France, Italy, Turkey and Cyprus (Alkan *et al.*, 2013; Charrel *et al.*, 2012). Among endemic populations, infections mainly occur during childhood, are inapparent or mild and result in life-long immunity. Accordingly, sandfly fever is a bigger health problem among susceptible adults who enter endemic areas. These include military personnel and tourists with occasional epidemics among indigenous adults. Sandflies are active during late spring and summer months, feed at night and are found within a short distance of breeding sites. Viruses are maintained in nature by transovarial transmission resulting in characteristic infected sandfly 'patches.' The virus can also be transmitted from viraemic humans to biting female sandflies.

The incubation period for sandfly fever is 3–5 days. The illness is acute febrile dengue-like disease without a generalized maculopapular rash. The disease generally abates within 3 days leaving individuals with severe prostration and post-illness depression. These symptoms wane within a week. Meningitis and meningoencephalitis have been observed during Toscana infections and this virus may be an important cause of acute neurological disease in southern Europe (Colomba *et al.*, 2012). Fatalities are rare. Immunity is type specific with infection by heterologous viruses possible. The incidence of human disease caused by this group of viruses is not well known, but, clearly because of its widespread prevalence, this is a significant cause of human morbidity.

Vaccine development status

While components of sandfly fever viruses have been incorporated into Rift Valley fever vaccines, the development of vaccines against any sandfly fever viruses has not been reported (Lihoradova *et al.*, 2013).

Flaviviridae

Flavivirus

Dengue

Disease perspective

There are four dengue viruses. As described in greater detail elsewhere, the following scenario best describes the evolution of these viruses to become human pathogens and their subsequent global dispersal (Halstead, 2008). The dengue viruses evolved from a flavivirus ancestor in subhuman primates most probably in the Indonesia archipelago. This surmise is based on (i) the observation that neutralizing antibodies to each of the four dengue viruses circulate in the blood of wild caught monkeys from every country in Southeast Asia, (ii) Indonesian primate populations and their viruses have been separated for prolonged periods over the past million years. During the Quaternary glaciations period peak ice accumulation occurred 10,000, 60,000 and 150,000 years ago, resulting in low sea levels that served to connect the Indonesian island chain. The islands separated again during interglacial periods. Indeed, as might be predicted from this geological history, Indonesian primate species differ from island to island. When the Dutch, Portuguese and English established colonies in SE Asia, they brought with them *Aedes aegypti* from Africa. Soon, one by one, each of the four dengue viruses emerged into newly established urban transmission cycles (Wang *et al.*, 2000). *Aedes aegypti* was also introduced into the American tropics resulting in its present global distribution. In the modern era the growth of cities plus steady improvements in ground and air transportation distributed *Aedes aegypti* and the four dengue viruses into virtually every tropical geographical niche. In this way dengue has become a global scourge. Currently dengue viruses are transmitted in 128 countries with an estimated 4 billion persons at risk to infection (Brady *et al.*, 2012). Of these, 290 million may be infected yearly, resulting in 90 million clinically overt febrile cases (Bhatt *et al.*, 2013).

Two other African zoonotic viruses emerged into the urban transmission cycle: yellow fever in West Africa and subsequently transported across the Atlantic and chikungunya, whose East African urban epidemics have spread to Asia on numerous occasions (Carey, 1971). One time only (until 2013) chikungunya made the journey from East Africa to the West Indies. There, in 1827–8 African slaves who would have known the Swahili term for this disease, 'ki pepo denga' the term that soon emerged was the Spanish homonym, 'dengue' (Carey, 1971; Christie, 1872, 1881; Halstead, 2008). Because the 1828 chikungunya epidemic self-extinguished, with the passage of time, the diagnostic term 'dengue' was transferred to another febrile exanthem, also transmitted by *Aedes aegypti*, caused possibly by the American genotype DENV 2 that had made the transit from its African sylvatic cycle at a time as yet unknown (Vasilakis *et al.*, 2007).

Dengue disease outcomes in humans are controlled by a unique immunological phenomenon. Dengue antibodies, circulating at the time of infection, may increase disease severity across a spectrum that at the severe end included the dengue acute vascular permeability syndrome dengue haemorrhagic fever/dengue shock syndrome (DHF/DSS). This syndrome is characterized by thrombocytopenia, altered haemostasis, activation of complement, increased vascular permeability to fluids and smaller macromolecules and elevated liver enzymes all occurring late in illness at around the time of defervescence. The role of antibodies in altering the outcome of human dengue infections *in vivo* is clearly established by a unique clinical finding – the occurrence of the vascular permeability syndrome in infants born to dengue-immune mothers (Halstead, 1970; Halstead *et al.*, 2002). However, the larger fraction of all severe dengue disease accompanies second heterotypic dengue infections in each of the twelve different sequences of a second dengue infection that are possible (Gibbons *et al.*, 2007; Halstead, 2003).

Vaccine development status

There are no licensed dengue vaccines. This chapter will briefly describe dengue vaccine development from a historical perspective including those vaccines that have been inoculated or tested in humans. For a more complete treatment of this subject, please refer to recent reviews (Blaney *et al.*, 2010; Chavez *et al.*, 2010; Durbin and Whitehead, 2010; Faheem *et al.*, 2011; Guzman *et al.*, 2010; Halstead, 2014b; Halstead and Thomas, 2013; Konishi, 2012; Malabadi *et al.*, 2012; Murphy and Whitehead, 2011; Schmitz *et al.*, 2011; Swaminathan *et al.*, 2010; Thomas, 2011; Thomas and Endy, 2011a,b; Webster *et al.*, 2009).

Serially passaged, live-attenuated vaccines

As soon as the dengue viruses were recovered in the laboratory, a successful dengue vaccine was developed. Sabin reported that DENV 1 could be adapted to grow in the brains of young adult Swiss Webster mice following serial passage. When mouse brain DENV-1 was inoculated into susceptible volunteers dengue symptoms were somewhat reduced. At the seventh passage, volunteers no longer developed fever or malaise, but continued to experience a maculopapular total body rash (Sabin and Schlesinger, 1945). Vaccines were made of 15th and 19th mouse passage virus. Of 16 susceptible adults given passage 15 virus stabilized with human albumin, none developed fever, all developed rash and all resisted challenge with wild-type DENV-1 (Sabin, 1952). The 19th passage virus was stabilized using bovine albumin, lyophilized and a dilution of 1:10,000 successfully immunized volunteers (Sabin, 1952). This work was extended by Wisseman who took the 18th passage virus seven more times in suckling mouse brain (Wisseman *et al.*, 1963). This was followed by five serial passages at 'limiting' dilutions. The virus recovered was designated as the MD-1 strain and 31, 32 and 33 mouse brain-passaged viruses used as vaccine master seed, production seed and vaccine, respectively. Ten volunteers each were given 20,000 and 200,000 suckling mouse LD 50 doses of MD-1 (Wisseman, 1966). These volunteers developed neutralizing antibodies and resisted challenge with live-DENV 1. A phase II study was conducted in 300 flavivirus susceptible adult volunteers, 12% of whom developed a dengue rash. All responded with neutralizing antibodies and all four volunteers tested resisted challenge with wild-type DENV 1 (Wisseman, 1966). During the 1963 Puerto Rican DENV-3 outbreak, this vaccine was given to 561 persons with 552 controls. No enhanced disease was observed in vaccinees, but the incidence of disease was reduced by 39% (Bellanti *et al.*, 1966). No further studies are reported on this vaccine.

Japanese workers obtained essentially identical results as Sabin and colleagues. The Mochizuki strain of DENV-1 was isolated in 2 week old adult white mice inoculated with acute phase blood from patients with dengue fever from the 1943 Nagasaki outbreak. The Japanese were able to adapt DENV-1 to mice more quickly than the Americans as they harvested first passage mouse brains blindly on day 5, passing brain suspensions into a second group of mice. All these mice developed mild neurological signs while half of the next passage mice became paralysed (Hotta, 1969; Kimura and Hotta, 1944). At the fifth mouse passage, DENV-1 Mochizuki continued to produce classical dengue fever in susceptible human adult volunteers. At the 16th passage human disease response was modified; susceptibles inoculated with the 35th mouse passage had no dengue symptoms or rash (Hotta, 1952, 1969). In 1944, Japanese workers inoculated two and three susceptible adult volunteers with 37th mouse brain passaged and 40th mouse brain passaged DENV-1, respectively. Within two weeks, all developed neutralizing antibodies and each individual resisted challenge with wild-type DENV-1 (Hotta, 1969). Hotta also found that two our of four volunteers inoculated with ox bile treated DENV-1 developed neutralizing antibodies and resisted homotypic challenge (Hotta, 1954).

DENV-2 vaccine development was attempted but with not the success of DENV-1. Three strains of DENV-2 were recovered from soldiers with dengue fever on New Guinea in 1944. When inoculated into susceptible adult volunteers these strains consistently produced milder febrile exanthema than did DENV-1. However, adapting DENV-2 to mice proved difficult. Serial intracerebral passages of DENV-2 into adult white mice failed to result in lethal infection (Sabin, 1952; Sabin and Schlesinger, 1945). Following the finding by Meiklejohn and co-workers that DENV-2 could be rapidly adapted to and grow to higher titre in baby mouse than in adult mouse brains, Schlesinger and colleagues were able to establish lethal infections by passing DENV-2 in baby mouse brain (Meiklejohn *et al.*, 1952; Schlesinger and Frankel, 1952). Continued passage of DENV-2 in suckling mice gradually increased the pathogenicity of virus for adult mice. DENV-2 New Guinea C seed viruses in general circulation usually have more than 100 mouse brain passages (Schlesinger and Frankel, 1952). Five adult volunteers were inoculated with DENV-2 virus passed 40 times in suckling mice and an additional 12 volunteers inoculated with virus passed 46 times (Schlesinger *et al.*, 1956). Each of these vaccines raised neutralizing antibody responses in susceptibles but challenge experiments designed to demonstrate solid immunity were not successful (Schlesinger *et al.*, 1956).

WRAIR DENV-2 vaccine

In 1971, recognizing the severity of the world-wide dengue problem, the Dengue Task Force of the Armed Forces Epidemiology Board's Virus Commission initiated a cooperative scientific effort to develop DENV 1–4 vaccines. A temperature sensitive candidate DENV-2 vaccine was obtained by clonal selection of virus that produced small plaques in primary African green monkey kidney cells at the Walter Reed Army Institute of Research (WRAIR) (Eckels *et al.*, 1976, 1980; Harrison *et al.*, 1977). Immunogenicity of this vaccine, PR-159 S-1, was tested in yellow fever immune or susceptible adult human volunteers (Bancroft *et al.*, 1981, 1984). In 98 adult volunteers, 90% of 70 yellow fever-immunes seroconverted with DENV-2 neutralizing antibodies after a single dose containing 4.5 logs of virus but the same dose produced only a 61% seroconversion rate in susceptible donors (Bancroft *et al.*, 1984). The vaccine was mildly symptomatic.

Hawaii/Mahidol tetravalent vaccines

Workers at the Department of Tropical Medicine and Medical Microbiology of the University of Hawaii Medical School screened tissue culture-passaged wild-type DENV-1, -2 and -4 and mouse-passaged DENV-3 in cell systems that had been used to propagate viral vaccines licensed for use in the USA. These included WI-38 continuous human embryo lung, primary chick and duck embryo fibroblasts, primary rabbit kidney cells, primary dog kidney cells (PDK) and primary African green monkey kidney cells (GMK). Only PDK and GMK cells supported replication of all four DENV(Halstead, 1978). Serial passage in PDK at low dilution of seven different wild-type DENV (two DENV-1, two DENV-2 and three DENV-4) resulted in phenotypic changes at passage levels unique but reproducible for each DENV (Halstead *et al.*, 1984a; Halstead and Marchette, 2003). The phenotypic changes observed were small plaques in LLC-MK2 cells, absent plaque formation on GMK, lack of cytopathic effect in LLC-MK2 cells, shut-off virus replication at temperatures above 38°C, reduced viraemia in susceptible rhesus monkeys and progressive increase in mean day of death in suckling mice inoculated intracerebrally. With DENV-1, phenotypic changes emerged rapidly at low PDK passage and a candidate vaccine could be prepared at passage 13. For DENV-2 phenotypic changes emerged gradually and were fully stable at passages above 50; gradual changes were also observed on passage of three DENV-4 strains with useable variants obtained between passages 20 and 48. These phenotypic changes were used successfully to identify human dengue vaccine candidates at Mahidol University (see below). PDK passaged DENV-2 16803 was transferred to the WRAIR and incorporated in a live-attenuated DENV development programme.

Of importance, when passaged dengue-4 viruses were cloned, strains with abruptly different phenotypic properties were obtained. Only uncloned viruses demonstrated gradual passage-related phenotypic changes (Halstead *et al.*, 1984a; Halstead and Marchette, 2003). A cloned virus, DENV 4 H-241 at 35 PDK passages was subsequently passaged three times in FRhL cells. This virus was inoculated into five susceptible adults, two of whom developed mild dengue illness with fever and rash caused by large plaque-size revertant viruses, while three other volunteers failed to seroconvert (Eckels *et al.*, 1984). In the laboratory, it was noted that inoculation of cloned PDK-passaged viruses on FRhL monolayers were genetically unstable as evidenced by plaque size changes (Halstead *et al.*, 1984b). Of possible relevance to this *in vivo* observation was the subsequent discovery that passage of dengue viruses in FRhL cells changed infectivity of viruses for rhesus monkeys (Anez *et al.*, 2009).

After 15 passages in PDK or GMK cells in Hawaii, strains of each of the four DENV were transferred to Mahidol University for further passage, pre-clinical and clinical characterization with support from the Rockefeller Foundation and the South-east Asia Regional Office of the World Health Organization (Halstead and Marchette, 2003). In 1988, attenuated DENV 1–4 viruses were licensed by Mahidol University to Aventis Pasteur Merieux Vaccines. Clinical development after that date was organized and supported by Aventis (now Sanofipasteur). PDK-passaged viral strains DEN-1 16007, DEN-2 16681 and DEN-4 1036 were tested in small groups of flavivirus-susceptible adult Thai volunteers. Acceptable reactogenic and immunogenic strains were identified for DEN-1, -2 and -4 viruses at passage levels PDK13, PDK53, and PDK48, respectively (Bhamarapravati and Sutee, 2000; Bhamarapravati and Yoksan, 1997; Bhamarapravati *et al.*, 1987). In the hands of these investigators, DENV-3 viruses did not grow in PDK cells. Attenuation was attempted by serial passage in primary GMK. Based upon studies in flavivirus susceptible adult volunteers, an acceptable vaccine candidate was found for DENV-3 16562 at the 30th GMK cell passage. A vaccine candidate was made with three passages in FRhL cells (Bhamarapravati and Yoksan, 1997). Serial passage was shown to select for amino acid changes predominantly in non-structural regions. DENV-1 16007, PDK 13 differed from parental virus by 14 amino acids, while DENV-2 16681, PDK 53 vaccine differed by nine amino acids (Huang *et al.*, 2000, 2003).

DENV-2 16681 PDK53 was tested in 10 US Army soldiers with acceptable reactogenicity and homotypic neutralizing antibodies in all subjects (Vaughn *et al.*, 1996). Administration of 10^3–10^4 plaque-forming units (PFU) of each of four monovalent candidates elicited neutralizing antibody seroconversion in 3/5, 5/5, 5/5 and 5/5 American volunteers, respectively (Kanesa-thasan *et al.*, 2001). PDK vaccine strains when cloned and passaged in Vero cells retained phenotypic attributes including immunogenic attenuation and immunogenicity for human beings (Kinney *et al.*, 2010a,b).

Bivalent and trivalent formulations using DENV-1, 2 and 4 PDK-passaged vaccine candidates elicited nearly 100% respective seroconversions in susceptible Thai adults (Bhamarapravati and Yoksan, 1989). With candidate vaccine made in Bangkok successful tetravalent neutralizing antibody seroconversion to administration of tetravalent dengue vaccine was observed in four volunteers (Bhamarapravati and Yoksan, 2000). However, when vaccine viruses were produced in the facilities of Pasteur Meriuex Vaccines and a mixture of DENV-1, -2 and -4 grown in PDK cells and a DENV-3 grown in GMK were given to partially dengue-immune children and susceptible adults the response was a DENV-3 viraemia and DENV-3-specific neutralizing antibodies (Kanesa-thasan *et al.*, 2001; Kitchener *et al.*, 2006; Sabchareon *et al.*, 2004; Sun *et al.*, 2003). Further, when a cloned, mutagenized and Vero-adapted strain of DENV-3 16562 GMK 30 FRhL 3 was injected in 15 susceptible Asian adults all developed classical clinical dengue fever (Sanchez *et al.*, 2006). This result led to the abandonment of further development of the Mahidol live-attenuated vaccine.

WRAIR tetravalent vaccine

The WRAIR also developed a tetravalent live-attenuated dengue vaccine based upon serial passage of wild-type DENV in PDK cells (Eckels *et al.*, 2003). Specific virus strains and PDK cell passage levels were selected based on a series of Phase 1 safety and immunogenicity studies conducted at the WRAIR and the University of Maryland School of Medicine Center for Vaccine Development (Kanesa-Thasan *et al.*, 2003). The following candidate vaccines were produced with terminal passage in FRhL cells – DEN-1 45AZ5 PDK20/FRhL3, DEN-2 16803 PDK50/FRhL3, DEN-3 CH53489 PDK20/FRhL3 and DEN-4 341750 PDK20/FRhL4. Candidate DENV-2 and-4 vaccine strains were fully passaged at the University of Hawaii (Halstead and Marchette, 2003; Marchette *et al.*, 1990), but tested for clinical responses at WRAIR (Hoke *et al.*, 1990; Kanesa-Thasan *et al.*, 2003). Among the selected passage levels, the seroconversion rates were 100%, 92%, 46% and 58% for a single dose of DENV-1, -2, -3 and -4 respectively. The WRAIR DENV-2, -3 and -4 vaccine viruses were well tolerated by volunteers. The DENV-1 PDK20/FRhL3 monovalent candidate was associated with high reactogenicity with 40% developing fever and generalized rash. Vaccine related reactions consisted of modified symptoms of dengue fever to include headache, myalgia, and rash (Edelman *et al.*, 1994, 2003; Mackowiak *et al.*, 1994; Sun *et al.*, 2003).

Sixteen different high and low dosage formulations of the WRAIR tetravalent vaccine were tested in 64 adult volunteers (Edelman *et al.*, 2003; Sun *et al.*, 2009). The formulations were derived by using undiluted vaccine (five to six logs of virus) or a 1:30 dilution for each virus serotype. Seroconversion rates after a single dose of tetravalent vaccine were 83%, 65%, 57% and 25% to DENV-1, -2, -3, -4, respectively, similar to that seen with monovalent vaccines. Few additional seroconversions were seen following a booster dose 1 month after the first dose. Though the sample size was small, a trend towards increased reactogenicity was observed when a full dose of the DENV-1 component was combined with lower doses of DENV-2 and/ or DENV-4. With these viral strains, at the doses evaluated, viral interference did not affect antibody response but may have modified reactogenicity. In an attempt to reduce reactogenicity due to the DENV-1 component and to increase the immunogenicity of the DENV-4 component, a 17th formulation using DENV-1 45AZ5 PDK 27 and DENV-4341750 PDK 6 was evaluated. Expanded testing of vaccine formulations along with booster doses at an interval of 6 months yielded higher seroconversion rates along with acceptable reactogenicity in adults (Sun *et al.*, 2009).

Although hospital-based studies in Thailand have shown that most primary dengue virus infections in young children are clinically silent DHF/DSS is regularly seen during secondary infections in children 2 years of age and older (Green *et al.*, 1999; Vaughn *et al.*, 2000). For this reason an evaluation of tetravalent vaccines in young children was required. The new formulation WRAIR vaccine was found to be well tolerated and immunogenic in a phase 1 study of seven flavivirus-naive children, aged 6–9 years, in Thailand (Simasathien *et al.*, 2008). Subsequently, two doses of this vaccine were given, 6 months apart, to 34 flavivirus-susceptible infants aged 12–15 months (Watanaveeradej *et al.*, 2011). After the second dose, 85.7% developed trivalent DENV neutralizing antibodies and 53.6% tetravalent neutralizing antibody responses. None developed clinical symptoms or signs of dengue fever.

Further development of LAV dengue vaccine at WRAIR has been put on hold pending further development by GSK of a tetravalent inactivated whole virus vaccine under co-development with the Institute Oswaldo Cruz in Rio de Janeiro, Brazil (S. Thomas, personal communication, 2008). It is not clear what actions the developers may take in response to the report that passage of dengue viruses in FRhL cells resulted in a Glu237 to Gly substitution that enhanced interaction with heparan sulfate and reduced infectivity of both vaccine and wild-type DENV-4 strains for susceptible rhesus monkeys (Anez *et al.*, 2009).

NIH tetravalent vaccine

The cloning of DENV-4 (WRAIR 814669, Dominica, 1981) by C.J. Lai and colleagues at the National Institutes of Health permitted the introduction or deletion of selected genes in the dengue genome opening a new era of dengue vaccine research (Lai *et al.*, 1991). A DENV-4 mutant transcribed from recombinant cDNA with a 30-nucleotide deletion in the 3′ untranslated region (10478–10507; DENV-4 2AD30) produced lower viraemia and slightly decreased neutralizing antibody responses in rhesus monkeys compared with wild type DENV 4 (Men *et al.*, 1996). This virus, grown in VERO cells, was evaluated in 20 volunteers (Durbin *et al.*, 2001). Following inoculation of 10^5 PFU in 0.5 ml subcutaneously there were minimal local or systemic symptoms. Ten volunteers (50%) developed a mild rash (unnoticed by the volunteers themselves. A mild increase in serum ALT levels was seen in five volunteers (25%) with a transient increase to five times the upper limit of normal (238 IU/L) in one volunteer. Seventy per cent of volunteers had low-level viraemia and all developed neutralizing antibodies 28 days following a single dose of vaccine. This virus could not be propagated by *Aedes aegypti* or *Aedes albopictus* after feeding on infected human blood (Troyer *et al.*, 2001). To lessen ALT responses, additional mutations were introduced into rDENV4 Δ30 to further attenuate the virus in animal models and ultimately human vaccinees. Based on the elevated liver enzymes associated with the 10^5 PFU dose of rDENV4 Δ30 and the known involvement of liver infection in dengue virus pathogenesis, a large panel of mutant viruses was screened for level of replication in the HuH-7 human hepatoma cell line, a surrogate for human liver cells and selected viruses were further analysed for level of viraemia in SCID-HuH-7 mice. It was hypothesized that rDENV4 Δ30 derivatives with restricted replication *in vitro* and *in vivo* in HuH-7 human liver cells would be restricted in replication in the liver of vaccinees. Two mutations identified by this screen, NS3 4995 and NS5 200,201, were separately introduced into rDENV4 Δ30 and found to further attenuate the vaccine candidate for SCID-HuH-7 mice and rhesus monkeys while retaining sufficient immunogenicity in rhesus monkeys to confer protection (Blaney *et al.*, 2010). Clinical studies of rDENV4 Δ30–4995 revealed a high incidence of an erythematous rash at the injection site, a low level of ALT elevation, no viraemia and adequate neutralizing antibody response (Wright *et al.*, 2009). In humans, the rDENV4 Δ30–200,201 vaccine candidate administered at 10^5 PFU exhibited greatly reduced viraemia, high infectivity and lacked liver toxicity while inducing serum neutralizing antibody at a level comparable to that observed in volunteers immunized with rDENV4 Δ30 (McArthur *et al.*, 2008). The virologic, safety, and immunogenicity profiles of this vaccine at 10^3 and 10^5 PFU were similar. In tetravalent vaccine trials, rDEN4D30 and rDEN4D30–200,201 were both evaluated at a dose of 10^3 PFU (Durbin *et al.*, 2013a).

A similar rDENV1 Δ30 construct was made and evaluated in animal models and found to be attenuated and immunogenic. rDENV1 Δ30 was given at a dose of 10^3 PFU to 20 healthy adult human volunteers. Neutropenia and a transient asymptomatic rash were observed in 40% of vaccinees while 95% developed homotypic neutralizing antibodies that persisted throughout 6 months of observation. This candidate was selected for inclusion in a tetravalent dengue vaccine formulation (Durbin *et al.*, 2006a, 2013a).

DENV 2Δ30 viruses proved to be overattenuated in susceptible adult volunteers. A candidate vaccine was made from inserting the prM and E structural proteins of DENV2 NGC into the non-structural proteins and non-translated region of rDENV4 Δ30. This resulted in the chimera rDEN2/4 Δ30(ME) (Durbin *et al.*, 2006b). This virus was evaluated at a dose of 1000 PFU in 20 healthy dengue-naive adult volunteers. Low level vaccine viraemia, transient asymptomatic rashes and mild neutropenia were noted in some vaccinees. All developed and maintained significant neutralizing antibody titres over 6 months observation. This virus was selected as a future component of a tetravalent formulation (Durbin *et al.*, 2013a).

Vaccine candidates, rDENV3Δ30/31 and rDENV3/4 Δ30(ME) were made which contain the membrane (M) precursor and envelope (E) genes of DENV3 inserted into DENV-4 with a 30 nucleotide deletion in the 3′ untranslated region (Blaney *et al.*, 2008; Durbin *et al.*, 2011b).

Eight monovalent DENV vaccine candidates have been evaluated in 13 separate NIAID-sponsored clinical trials conducted at the Centre for Immunization Research, Johns Hopkins Bloomberg School of Public Health and the University of Vermont School of Medicine (Durbin *et al.*, 2011a; Durbin and Whitehead, 2011). The final mixture (TV 003) contains 10^3 PFU each of DENV-1 D30, DENV 2/4Δ30, DENV3Δ30/31 and DENV 4 D30. Sixty flavivirus naive adult volunteers were given 3 3 3 3 vaccine; 92% developed tetravalent neutralizing antibody responses following a single dose (Durbin *et al.*, 2013a). When a second dose was administered 6 months later, 100% circulated neutralizing antibodies to all four DENV (Durbin *et al.*, 2013a). FDA has waived the requirement for monkey neurovirulence studies. These viruses have been licensed for vaccine production to VA Biotech, Vietnam, Biological E and Panacea of India and Butantan, Sao Paulo, Brazil.

A phase II clinical study of Butantan tetravalent dengue vaccine is in progress.

Takeda/Inviragen/CDC tetravalent vaccine

The US Centers for Disease Control and Prevention (CDC) developed a tetravalent chimeric dengue vaccine by splicing the prM and E genes of attenuated DENV-1, -3 and -4 into non-structural RNA of the successfully attenuated DENV-2 16681 PDK-53 (see Hawaii/Mahidol vaccine). All of these viruses had been derived from DENV passed in PDK or GMK at the University of Hawaii with further passage of DENV-2, 3 and 4 in Bangkok. (Halstead and Marchette, 2003; Kinney *et al.*, 1997; Kinney and Huang, 2001; Vaughn *et al.*, 1996) Attenuating mutations for the DENV-2 vaccine reside outside the structural gene region justifying the use of the virus backbone to receive *PrM-E* genes from the other dengue virus types (Butrapet *et al.*, 2000; Huang *et al.*, 2003). Mutations at three loci were found to contribute individually, and synergistically, to DENV-2 PDK-53 attenuation, specifically, a C to T transition at nucleotide 57 in the 5¢ non-coding region (5¢NC-57), a Gly-to-Asp substitution in non-structural protein 1, amino acid 53 (NS1–53) and a Glu-to-Val substitution in non-structural protein 3, amino acid 250 (NS3–250) (Butrapet *et al.*, 2000). It is possible that the PDK-53 virus specific mutations in prM (DEN-2 PDK-53-V virus only), NS2A, and/or NS4A, might further modulate attenuation to an undefined extent. Additionally, the juxtaposition of heterologous genes in the chimeric DEN viruses, particularly in DENV-2/3 and DENV-2/4 chimeras may also contribute to an attenuated phenotype, relative to wild-type virus (Huang *et al.*, 2003). Chimeras for DENV-1 have been produced using structural genes from both the Mahidol DENV-1 PDK-13 vaccine virus and the near wild-type DENV-1 16007 virus (Butrapet *et al.*, 2002). The structural genes of the DENV-1 16007 virus appear to be more immunogenic in mice than those of the PDK-13 vaccine virus (Huang *et al.*, 2000). Chimeras for DENV-2/3 and DENV-2/4 have been constructed using the wild-type virus structural genes, and have characterized for biological markers (Osorio *et al.*, 2011a,b). Monkeys inoculated with high dose tetravalent chimeric vaccine developed neutralizing antibodies to all four DENV and resisted challenge with DENV-1 to -4. Those inoculated with a lower dose formulation were only partially protected against challenge with DENV-1 or -2. The chimeric viruses all had small plaque size, were temperature sensitive and reduced replication in C6/36 cells and mice (Osorio *et al.*, 2011b). All four DENV were rederived under Good Manufacturing Practices and assembled as a mixture of DENV-2 PDK 53 and a mixture of DENV-2 PDK 53 DENV-2 chimeras using genes from wild-type DENV-1, -3 and -4. These are referred to as DENVax 1, 2, 3 and 4.

This vaccine was licensed to Inviragen and then sold to Takeda Pharmaceuticals. The tetravalent vaccine candidate has undergone pre-clinical testing in mice and non-human primates comparing intradermal and subcutaneous routes of delivery (Huang *et al.*, 2003; Osorio *et al.*, 2011a,b) The ID delivery of a tetravalent preparation of DENVax (10^5 PFU per DENV type) produced superior immune responses in cynamolgous monkeys (Osorio *et al.*, 2011a,b) In the ID group, no viraemia was detected following challenge with 10^{-5} PFU of DENV-1 West Pacific or DENV-2 New Guinea C viruses. Inviragen has initiated two phase I clinical trials in flavivirus naive adults, one at the St. Louis University Center for Vaccine Development and other in Medellin, Columbia. (Osorio *et al.*, 2011b) Study designs are randomized, placebo controlled, assessing two dosage levels of tetravalent DENVax administered SC or ID. Two doses were administered at time 0 and 90 days. Low dose formulations contain DENV-2/1 1.8×10^4 PFU, DENV-2/3 1×10^4 PFU, DENV-2/4 2×10^5 PFU and DENV-2 PDK 53 5×10^3 PFU while high-dose formulations contain DENV-1 2×10^4, DENV-2 5×10^4, DENV-3 1×10^5 and DENV-4 3×10^5. Adverse events observed were acceptable and differed little from placebo. After two doses, both by ID and SC routes a high percentage of susceptible adult vaccinees developed neutralizing antibody responses to all four DENV. DENV-4 antibody responses were consistently lower than responses to the three other viruses (D. Stinchcomb, personal communication, June 2012).

Acambis/sanofipasteur vaccine (CYD)

Chimeric DENV vaccines were derived by coupling genes encoding wild-type dengue virus prM and E structural proteins to the capsid and non-structural genes from yellow fever vaccine (CYD) (Chambers *et al.*, 1999; Guirakhoo *et al.*, 2001; Guirakhoo *et al.*, 2000; Rice *et al.*, 1989). The operating hypothesis was that the attenuation characteristics of 17D YF vaccine virus would be imparted to the dengue chimera (Monath *et al.*, 2002). Developed by Acambis, the vaccine was licensed to sanofipasteur. Intracellular replication of the construct is performed by YF RNA polymerase. Vero cells serve as the substrate for vaccine virus production.

Dengue/yellow fever virus chimeras for each of the four dengue virus serotypes consist of *PrM-E*-genes from the PUO-359/TVP-1140 Thai strain for serotype 1, the PUO-218 Thai strain for serotype 2, the PaH881/88 Thai strain for serotype 3, and the 1228 (TVP-980) Indonesian strain for serotype 4. The CYD tetravalent vaccine (TDV) combined the four CYD viruses as a single vaccine containing five logs of each serotype ('5555 formulation'). The vaccine is freeze-dried and contains no adjuvant or preservative. It is presented in a single-dose vial or in a five-dose multi-dose vial (Guirakhoo *et al.*, 2000, 2001; Guy *et al.*, 2011). Neurovirulence in mice inoculated with dengue/YF chimeras was reduced compared to the yellow fever 17D vaccine virus (Monath *et al.*, 2005). Viraemia in rhesus macaques is similar to that of yellow fever 17D virus and greatly reduced compared to wild-type dengue viruses. High titred DENV 1–4 neutralizing antibody titres were raised following inoculation of five logs of each component; somewhat lower titres with three logs (Guirakhoo *et al.*, 2001, 2004). Rhesus monkeys given five logs of the DENV-2 vaccine were protected from viraemia when challenged with homotypic wild-type dengue virus but, anamnestic neutralizing antibody responses suggested incomplete protection (Halstead, 2013). When monkeys were vaccinated using virus mixtures breakthrough viraemias were observed with DENV-1 and DENV-4 wild-type virus challenge (Guirakhoo *et al.*, 2001, 2004; Guirakhoo and Pugachev, 2002). Humans demonstrated high rates of seroconversion with a DENV-2/YF chimera (CYD2) with higher levels of DENV-2 neutralizing antibodies produced in individuals previously vaccinated with 17D YF (Guirakhoo *et al.*, 2006). Antibody responses in susceptible adults and children to the standard three dose series (0, 6 and 12 months) were nearly identical to responses to a two dose schedule in 17D yellow fever immunes (Poo *et al.*, 2011). Even broader responses were observed to one or two doses of CYD in children and adults who were residents of dengue-endemic countries (Capeding *et al.*, 2011).

In susceptible humans, low levels of viraemia were observed following administration of CYD TDV, with DENV-4 being recovered most frequently (Guy, 2009; Guy and Almond, 2008). Viraemia was noted less frequently following administration of a second or third dose of vaccine. After receiving a single dose of CYD TDV formulated to contain 10 5 PFU/ml each, approximately 12% of susceptible adults develop neutralizing antibodies to all four DENV (Guy, 2009; Morrison *et al.*, 2010). It was observed that administration of a second dose of live vaccine was not able to boost antibody response unless delivered 4–6 months after the first dose; further, a third dose, also administered 6 months later, was require to achieve 100% tetravalent neutralizing antibody response (Guy *et al.*, 2010). Limited data suggest that by extending the interval between first and second dose of tetravalent vaccine to 12–15 months, nearly 100% tetravalent seroconversions can be achieved (Guy *et al.*, 2010; Morrison *et al.*, 2010). In individuals circulating YF antibodies or children circulating dengue antibodies, administration of two doses of CYD was sufficient to evoke full tetravalent neutralizing antibody responses (Guy *et al.*, 2010).

Clinical studies of CYD TDV have progressively involved larger and larger groups of susceptibles, yellow fever-immune and dengue-immune human volunteers (Guy, 2009; Guy *et al.*, 2010; Lang, 2009; Morrison *et al.*, 2010). In 2009, a Phase IIb proof of concept efficacy and large-scale immunization safety trial was initiated among 4000 Thai children AGED 4–11 years (CYD23). Participants received three subcutaneous injections of either CYD TDV or a placebo at months 0, 6 and12. In the CYD 23 trial, from studies of a small random sample it was estimated that 90% of enrolled Thai children circulated neutralizing antibodies from previous Japanese encephalitis (JE) vaccination or wild-type DENV or JE infections. Despite the background flavivirus immunity and documented boosts in DENV-1 to -4 neutralizing antibodies following two and three doses of vaccine neither the initial, second nor third dose of vaccine protected against DENV-2 disease (Halstead, 2012; Sabchareon *et al.*, 2012). Although numbers are small, successive doses of vaccine resulted in no obvious trend of increased protection against disease with DENV -1, -3 or -4. When cases from all post-immunization periods were combined (intention to treat) individual protective efficacy for DENV-1 was 61.3, for DENV-3 was 82.0 and for DENV-4 was 90.0, all judged statistically significant by the study's authors. An additional phase III clinical trial was initiated Australian adults in 2010, the objectives of this placebo-controlled trial being to assess lot-to-lot consistency, safety and immunogenicity (Guy *et al.*, 2011). Other phase III trials involving 10,275 Asian children ages 2–14 years resulted in a combined efficacy of 56% (press release, 28 April 2014, Sanofipasteur, Lyon, France).

Dengue vaccines in pre-clinical development

A very large number of dengue vaccines are in pre-clinical development. Data from selected constructs developed over the past decade are summarized in Table 20.1.

Japanese encephalitis (JE)

Disease perspective

Japanese encephalitis (JE) is the leading recognized cause of viral encephalitis in Asia. JE viruses are transmitted by *Culex*

Table 20.1 Selected dengue vaccines in pre-clinical development

		Dengue Virus Monkey Challenge		
Vaccine type	Composition	Viraemia	Anamnestic Aby	Reference
Measles vector	DIII Den1–4	ND	ND	Brandler *et al.* (2010)
Complex Adeno vector	D III Den1,2 + D III Den3,4	Den 2,4	Yes	Holman *et al.* (2007), Raviprakash *et al.* (2008)
Adenovirus 5 vector	DIII Den1–4	ND	ND	Khanam *et al.* (2009)
Adeno 5 vector	Hairpin si RNA 5′ NTR Den1–4	ND	ND	Korrapati *et al.* (2012)
Pichia pastoris expression	Virus-like particle.395 aa DIII Den2	ND	ND	Mani *et al.* (2013)
Hepatitis B core	Virus-like particles, D III Den2	ND	ND	Arora *et al.* (2012)
VEE replicon	85% DIII Den1–4	Yes, Den 1,2	yes	White *et al.* (2013)
C deficient WNV chimera	Den 2 prM E	ND	ND	Suzuki *et al.* (2009)
Inactivated, formalin	Den2 alum	1/3	Yes	Putnak *et al.* (2005)
Inactivated, prime	LAV boost Den1–4	No Den 1–4	Yes to Den 1, 3, 4	Simmons *et al.* (2010)
Fusion partner *Neisseria* meningitidis P64k	Den2, alum	Yes, Den 1,2	Yes, Den 2	Valdes *et al.* (2009)
Prime-boost	Den2/P64k	ND	ND	Valdes *et al.* (2010)
Fusion partner, P64k	DIII Den1,3,4 + c Den2	ND	ND	Izquierdo *et al.* (2014)
Drosophila expression	80% E Den2	low dose no Den2 viraemia	Yes	Clements *et al.* (2010)
Drosophila expression	80% E Den1–4	Low dose, no viraemia	No Aby reported	Clements *et al.* (2010)
DNA+ Vaxfectin	Den1–4 E prM	low Den2 viraemia	No Aby reported	Porter *et al.* (2012)
DNA shuffling	Den1–4	Den 1,2	yes	Raviprakash *et al.* (2006)

spp. mosquitoes throughout Asia, a rice-growing region with an indigenous population of more than 3 billion people, also a major tourist destination. The rice paddy-breeding *Culex tritaeniorhynchus summarosus*, a night-biting mosquito that feeds preferentially on large domestic animals and birds and infrequently on humans, is the principal vector of zoonotic and human JE in northern Asia. A more complex ecology prevails in southern Asia, from Taiwan to India, where *Cx. tritaeniorhynchus* and members of the closely related *Cx. vishnui* group are vectors. Seasonal rains are accompanied by increases in mosquito populations and increased transmission. Pigs serve as amplifying hosts. In contrast, humans are likely dead end hosts because they experience short duration and low level viraemia (Halstead and Jacobson, 2008). There is no direct human to human transmission. In endemic areas, the incidence of JE disease is greater in the young; attack rates in the 3–15 years age group are 5–10 times higher than in older persons (Solomon, 2004). While the higher disease rates in younger persons reflect high immunity rates in adults there is some evidence that young children are intrinsically at greater risk to develop encephalitis than are susceptible adults (Hammon *et al.*, 1958). Numerous epidemiological observations document a weak protective effect of prior dengue virus infection on subsequent overt JE disease (Edelman *et al.*, 1975; Grossman *et al.*, 1973; Hammon *et al.*, 1958; Libraty *et al.*, 2000; Tarr and Hammon, 1974). From the standpoint of risk, it is important to understand that reported cases vastly underestimate the infectious burden. The ratio of infections to symptomatic JE cases has been estimated to vary between 1:25–1:300; the lower rates (1:200–1:300) were observed in northern Asians indigenous to the zoonotic heartland of JE; higher rates have been measured in non-indigenous military personnel (Benenson *et al.*, 1975; Gajanana *et al.*, 1995; Halstead and Grosz, 1962). Determinants for occurrence of overt neurological disease following infection are not well understood (Myint *et al.*, 2007). JE resembles West Nile infections in this respect (Gyure, 2009; Solomon and Vaughn, 2002).

Typically, symptoms start suddenly following a variable incubation period of 2 days to 2 weeks. Neurological involvement is sometimes heralded by lethargy, fever, headache, abdominal pain, nausea and vomiting (Tsai and Solomon, 2004). These symptoms may be followed by a combination of nuchal rigidity, photophobia, altered consciousness, hyperexcitability, masked facies, muscle rigidity, cranial nerve palsies, tremulous eye movements, tremors and involuntary movement of the extremities, paresis, incoordination and pathological reflexes (Solomon and Vaughn, 2002). Sensory deficits are rare. In children, 50–85% develop focal or general seizures compared with 10% of adult cases (Halstead *et al.*, 2013; Tsai and Solomon, 2004). Seizures have been associated with poor clinical outcome (Solomon *et al.*, 2002).

Opening pressure of cerebral spinal fluid (CSF) may be elevated in up to 50% of cases; protein levels are usually normal or mildly elevated (Solomon and Vaughn, 2002). CSF pleocytosis ranges from 10 to a few thousand cells per cubic millimetre (median of several hundred), and are predominantly of lymphocytic origin (Halstead and Tsai, 2004). Electroencephalogram demonstrates diffuse delta wave activity and, rarely, spike and seizure patterns (Tsai and Solomon, 2004). Imaging studies demonstrate diffuse white matter oedema and abnormal signals in the thalamus – often bilateral and haemorrhagic – basal ganglia, cerebellum, midbrain, pons and spinal cord (Kimura *et al.*, 1997; Kumar *et al.*, 1992; Misra *et al.*, 1994). In non-fatal cases, clinical improvement begins after approximately one week, paralleling defervescence. Recovery of neurological function may take weeks to years. Seizure disorders, motor and cranial nerve paresis, and movement disorders may persist in up to one third of patients. Persistent behavioural and/or psychological abnormalities occur in 45–75% of survivors and are more severe in children (Kumar *et al.*, 1993). There is no specific therapy for JE; supportive care focuses on controlling seizures, ventilator support of respiratory failure, and monitoring and reducing cerebral oedema (Tsai and Solomon, 2004). Anecdotal use of interferon-alpha and ribavirin has been reported, but when studied more carefully yielded negative results (Harinasuta *et al.*, 1985; Kumar *et al.*, 1993). Fatality rates vary between 5–40% often reflecting the standard of medical care available.

In economically advanced Asian countries, such as Japan, Korea and Taiwan nationwide implementation of JE vaccination has almost eliminated human disease while in moderate and low income countries, such as Thailand, Sri Lanka and Nepal, the recent integration of JE vaccine into routine immunization programmes has led to a marked reduction of JE (Dumre *et al.*, 2013; Halstead *et al.*, 2013; Upreti *et al.*, 2013). Before the introduction of JE vaccine, summer outbreaks of JE occurred regularly in Japan, Korea, China, Okinawa, and Taiwan. Over the past decade, there has been a pattern of steadily enlarging recurrent seasonal outbreaks in Nepal and India, with small outbreaks in the Philippines, Indonesia, and the northern tip of Queensland, Australia (Bhattachan *et al.*, 2009; Halstead *et al.*, 2013; Kari *et al.*, 2006; Rayamajhi *et al.*, 2011). Despite current vaccination efforts, a consensus estimate is that annually there are 67,900 cases of JE throughout its enzootic region (Campbell *et al.*, 2011).

Vaccines

Formalin-inactivated

Three vaccines are in use: (i) an inactivated mouse brain-derived vaccine produced in Thailand, Vietnam, Taiwan and Korea; (ii) an inactivated vaccine cultivated on primary hamster kidney (PHK) cells produced in China; and (iii) inactivated vaccines cultivated on Vero cells produced in Europe, India, China and Japan.

Mouse brain vaccine

An inactivated mouse brain vaccine using either Nakayama or Beijing-1 virus strains was developed in Japan and produced beginning in the 1950s. Local production of this vaccine contributed to control of JE in Thailand, India, Korea, Taiwan, Viet Nam, Malaysia and Sri Lanka (Halstead *et al.*, 2013) For several decades this vaccine was available for use by tourists or military personnel in developed countries (BIKEN, distributed by Sanofi Pasteur as JE-Vax®). The Korean Green Cross Vaccine Co. also produces a mouse brain JE vaccine. This was licensed for use in the United Kingdom and available in Europe on a named patient basis. Seroconversion rates, quantitative neutralizing antibody titres following vaccination, and efficacy rates varied according to the population studied (indigenous versus non-indigenous) and number of doses administered (one, two or three doses in the primary immunization series) (Halstead *et al.*, 2013). A single efficacy trial showed equivalent protection afforded by either Beijing-1 or Nakayama strains (Hoke *et al.*, 1988). For travellers, a three dose immunization series has been recommended (Immunization Practices Advisory Committee (ACIP), 1993). Based on data from the UK Australia, Canada, and the USA the rate of hypersensitivity reactions is estimated to be between 0.7 and 104 per 10,000 vaccinees (Halstead *et al.*, 2013). The cause of temporally associated neurological or hypersensitivity reactions are not clearly understood; the presence of murine neural proteins, gelatin and/or thimerosal in vaccine preparations have all been implicated but none proven as causative. The occurrence of a single case of acute disseminated encephalomyelitis (ADEM) temporally related to vaccination in Japan prompted the Japanese government (May 2005) to suspend routine childhood JE vaccination. BIKEN ceased production of JE-VAX® in 2005.

Hamster kidney cell vaccine

A second inactivated vaccine, manufactured from virus grown in primary hamster kidney (PHK) cells, has been in wide use, from 1968 to 1990, as China's principal JE vaccine (Gu and Ding, 1987). Approximately 70 million doses of the PHK cell culture inactivated JE vaccine (Beijing-3, P-3 strain) were administered in China yearly until 2005. Urticarial allergic reactions were reported in approximately 1 in 15,000 vaccine recipients (Hoke *et al.*, 1988). Field trials in China demonstrated vaccine efficacy ranging from 76% to 95% (Halstead and Tsai, 2004; Oya, 1988).

Vero cell vaccines

IC51 or JE-VC In the USA and Europe this licensed vaccine is available as IXIARO®, in Australia and New Zealand as JESPECT®. In India, this vaccine is manufactured by Biological E as JEEV. The vaccine was developed at the Walter Reed Army Institute of Research (WRAIR, Silver Spring, MD) from JE SA14-14-2 virus strain passaged eight times in primary dog kidney (PDK) cells, then cultivated in Vero cells in serum-free medium, formalin-inactivated and formulated with 0.1% aluminium hydroxide (Eckels *et al.*, 1988). The absence of serum allows for a simplified purification process and potentially, superior safety profile (Montagnon and Vincent-Falquet, 1998; Srivastava *et al.*, 2001). Kollaritsch and colleagues have published a useful review of the product's development lifecycle (Kollaritsch *et al.*, 2009).

Several early phase clinical studies established the safety and immunogenicity of the IC51 candidate (Kollaritsch *et al.*, 2009; Lyons *et al.*, 2007). Multinational phase 3 immunogenicity trials demonstrated the IC51 vaccine was well tolerated and elicited non-inferior immune responses compared to control (JE-VAX®) (Tauber *et al.*, 2007). There was no evidence of an increased incidence of rare adverse events (including anaphylaxis or anaphylactoid reactions) compared to placebo (Tauber *et al.*, 2008).

Additional studies demonstrated high (83%) seroconversion rates 1 year following vaccination, superiority of the standard dosing regimen (2 × 6 μg) compared to using low- or high-dose strategies and the observation that the presence of TBE antibodies had no safety impact but heightened the JE immune response following a single dose. There was no adverse impact on safety or immunogenicity of either vaccine when IC51 was given together with hepatitis A vaccine, HAVRIX® 1440 (Dubischar-Kastner *et al.*, 2010; Kaltenbock *et al.*, 2009; Schuller *et al.*, 2008, 2009).

A pivotal noninferiority immunogenicity study compared two doses of JE-IC51given on days 0 and 28 to three doses of JE-MB given on days 0, 7 and 28 to adults aged ≥ 18 years in the USA, Austria and Germany (Tauber *et al.*, 2007). In the 'per protocol' analysis, 352 (96%) of 365 JE-IC51 recipients developed a PRNT50 ≥ 10 compared with 347 (94%) of 370 JE-MB recipients at 28 days after the last dose (CDC, 2010). Two doses of vaccine are necessary as 28 days after receiving one dose of the standard 6-g regimen, only 95 (41%) of 230 JE-IC51 recipients had seroconverted with a PRNT50 ≥ 10. By contrast, 97% (110/113) of the subjects who had received two doses had a PRNT50 ≥10 (Tauber *et al.*, 2008). Antibodies raised to JE-CV protected mice against infection with four genotypes of JE (Erra *et al.*, 2013).

Safety for children was established in an open-label trial in the Philippines in which 195 infants aged 2–11 months were randomly assigned to receive JE-VC ($n = 131$) or 7-valent pneumococcal conjugate vaccine ($n = 64$). An additional 1674 children, aged 1–17 years, were randomly assigned to receive JE-VC ($n = 1280$) or hepatitis A vaccine ($n = 394$) (CDC, 2013). Among the 1411 children who received JE-VC, 23 (2%) reported a serious adverse event within 7 months of the first dose. The most common serious adverse events were pneumonia ($n = 6$) and febrile seizures ($n = 5$). Only three serious adverse events were reported within 2 weeks after a dose of JE-VC, including one report each of a febrile convulsion, cellulitis, and gastroenteritis. One death resulted from suspected bacterial meningitis and pneumonia in a male aged 12 years at 4 months after the second dose of JE-VC. No other neurological or hypersensitivity events were reported as serious adverse events. Among 48 children aged 1 and 2 years who were randomly assigned to receive JE-VC in India, five (10%) reported injection site tenderness, and one (2%) reported fever within 7 days after either dose (CDC, 2013). The only unsolicited adverse events were one report each of skin lesion and skin rash. No serious adverse events or deaths were reported.

In 2010, the ACIP approved JE-VC for use in adults ≥ 17 years and in 2013, approval was issued for the use of JE-VC in children, ages 2 months to 17 years (CDC, 2010, 2013).

JENVAX This Vero cell-derived purified inactivated JE vaccine is derived from an Indian JE virus, manufactured by Bharat Biotech received marketing approval from the Drug Controller General of India in 2013, and was developed as a public–private partnership with the Indian Council of Medical Research. In unpublished clinical trials reported in the press, JENVAC is described as showing safety and immunogenicity equivalent to that of SA-14-14-2. In an unstated number of individuals 1–50 years, seroconversions were reported at 98.7% 28 days after one dose and 99.8% 28 days after two doses.

Similar formalin-inactivated Vero cell-grown JE vaccines have been developed by two vaccine manufacturers in Japan and one in China. The development and licensure of second generation, non-mouse brain derived JE vaccines provides improved safety profile and lower dosage requirements and opportunities for greatly expanded use.

Live-attenuated vaccines

SA-14-14-2 This vaccine was developed from a genotype III JEV was isolated from *C. pipiens* in Xian in 1954. To achieve attenuation, it was passaged in mice and then over 100 times in primary hamster kidney cells (PHK), chick embryo tissues and suckling mice (Halstead *et al.*, 2013; Yu, 2010). The parental and vaccine viruses differ by four amino acids in the envelope and one each in NS2, NS3 and NS4B (Ni *et al.*, 1997). Studies suggest that a heparin sensitive mutation at Glu-306 leads to a loss of neuroinvasiveness (Lee *et al.*, 2004). This remarkably non-neurovirulent live-attenuated vaccine, has gradually been introduced into China, where it has demonstrated an excellent safety profile and very high efficacy (88–96%) in large scale trials ($n > 200{,}000$ children) and effectiveness (Hennessy *et al.*, 1996; Kumar *et al.*, 2009; Liu *et al.*, 1997; Tsai *et al.*, 1998). In China, SA 14-14-2, currently administered as a two- or three-dose vaccine, replacing the inactivated PHK vaccine (Yu Yong Xin, personal communication, 17 December 2009). Since its licensure in China in 1988, more than 300 million doses have been produced and administered to over 120 million children. Numerous large scale evaluations of vaccine safety demonstrate low rates (0.2–6%) of short-lived local and systemic (i.e. fever) reactogenicity and even lower (2.3 per 10,000) predicted rates of neurotoxicity (Halstead *et al.*, 2013; Halstead and Thomas, 2010). Case–control studies of a large vaccine trial in Nepal showed rapid onset of protection followed by a 5-year efficacy of 96% after a single dose of vaccine (Bista *et al.*, 2001; Ohrr *et al.*, 2005; Tandan *et al.*, 2007). A similar result was observed in India (Kumar *et al.*, 2009). In a small, single study, SA 14-14-2 vaccine was co-administered with live measles vaccine in children; normal immune responses were retained to each vaccine (Gatchalian *et al.*, 2008). Of interest, over the past decade genotype I JE viruses have largely replaced genotype III viruses throughout much of Asia (Schuh *et al.*, 2014). Fortunately, in experimental *in vivo* models, SA-14-14-2 provides excellent protection against genotypes I and III JEV (Liu *et al.*, 2010). Recently, the vaccine has been licensed for use and millions of doses administered in Nepal, India, Sri Lanka, and South Korea (Elias *et al.*, 2009; Halstead and Jacobson, 2008; Yaich, 2009). The Program for Advanced Technology in Health (PATH) negotiated concessional prices for the use of SA 14-14-2 in India, Sri Lanka and Nepal for public health prevention programmes (http://www.path.org/projects/japanese_encephalitis_project.php). SA 14-14-2 was incorporated into the national extended programme for immunizations (EPI) in Nepal and China (personal communications Dr Jay Tandan and Dr Yu Yong Xin, respectively, 17 December 2009).

On 9 October 2013, the WHO prequalified vaccine produced by the Chinese manufacturer, Chengdu Institute of Biological Products (http://www.who.int/mediacentre/news/releases/2013/japanese_encephalitis). On 22 November 2013, the Global Alliance for Vaccines and Immunization (GAVI) opened a funding window for purchase and distribution of WHO prequalified JE vaccines.

The dosage of this vaccine is 0.5 ml in children less than 6 years, 1.0 ml in older children and adults. In China, doses are given to children at 1 and 2 years of age and at school entry; large-scale immunization studies suggest that a single dose can be administered effectively at any time after 12 months of age (Tandan *et al.*, 2007). In the Nepal trial, SA-14-14-2 was immediately effective in preventing disease when an outbreak was in progress. (Bista *et al.*, 2001).

ChimeriVaxTM- JE, JE-CV or IMOJEV Sanofi Pasteur produces a JE vaccine developed by Acambis, Inc., Cambridge, Massachusetts using the chimeric virus technology generated at the St. Louis University Health Sciences Center, St. Louis, Missouri. IMOJEV is produced using infectious clone technology based on insertion of prM and E genes from JE SA 14-14-2 virus into the non-structural genes of YF 17D viral strain as viral 'backbone' (Arroyo *et al.*, 2001a; Chambers *et al.*, 1999; Monath *et al.*, 2000, 2003). The resulting chimeric RNA was electroporated into Vero cells. Progeny virus particles contain JE-specific antigenic determinants that elicit neutralizing antibodies as well as cytotoxic T lymphocytes (Konishi *et al.*, 1998). YF 17D was chosen as backbone to the chimera because of its proven record of safety and efficacy.

A single dose of JE-CV was shown to protect against a virulent wt JEV challenge in mice (Guirakhoo *et al.*, 1999; Monath *et al.*, 2000). Passive protection experiments also demonstrated that mouse serum raised against IMOJEV was protective against all four JEV genotypes (Beasley *et al.*, 2004; Bonaparte *et al.*, 2014). Moreover, a computer analysis of human leucocyte antigen (HLA) class II restricted T-cell epitopes in JEV E protein revealed a high degree of conservation among putative T helper epitopes in IMOJEV and circulating JEV representing all four genotypes (De Groot *et al.*, 2007). Because it is a living agent, IMOJEV has been evaluated for its ability to replicate in and to be transmitted by vector mosquitoes. Individual *Cx. tritaeniorhynchus*, *Aedes albopictus*, and *Ae aegypti* mosquitoes ingested a virus-laden bloodmeal or were inoculated intrathoracically (IT). IMOJEV did not replicate following oral feeding in any of the three mosquito species. In *Cx. Tritaeniorhynchus* replication was not detected after IT inoculation of IMOJEV. No genetic changes were associated with replication of vaccine virus in mosquitoes (Bhatt *et al.*, 2000). None of three Australian mosquitoes (*Cx. annulirostris*, *Cx. gelidus*, and *Ae. vigilax*) became infected after being fed orally with 6.1 log10 PFU/ml of IMOJEV (Reid *et al.*, 2006).

Inoculation of IMOJEV in non-human primates resulted in no illness, transient, low viraemia, followed by high titres of anti-JE neutralizing antibodies. A WHO monkey neurovirulence test scored only minimal brain and spinal cord lesions. Vaccinated monkeys were protected against IC or intranasal (IN) virulent JE virus challenge (Monath *et al.*, 2000). Safety and immunogenicity of IMOJEV (high and low dose) was established in a Phase 1 trial comparing immune responses in YF immune and flavivirus-naive volunteer. Low level viraemias were observed in both groups and the frequency of all adverse events was similar between groups. Overall, 96% of subjects who received IMOJEV developed neutralizing antibodies to one or more of the wild-type JE strains tested (Beijing, P3, Nakayama) (Monath *et al.*, 2002). In a Phase 2 study, IMOJEV was well tolerated with viraemias of short duration and low titre; 94% of subjects administered graded doses (1.8–5.8 log10) of IMOJEV developed neutralizing antibodies. A second dose, administered 30 days later, had no booster effect. There is evidence that IMOJEV induces sterilizing immunity in mice (Monath *et al.*, 2003).

Two randomized controlled phase III studies were conducted: 410 participants received one IMOJEV injection, 410 received three injections of JE-VAX (licensed killed mouse brain product). Seroconversion after a single IMOJEV vaccination (99.1%) was statistically non-inferior to that after three doses of JE-VAX (95.1%). 1601 susceptible adults and children received JE-CV, 403 received placebo. Adverse reaction rates were significantly lower with IMOJEV (67.6%) than with MBD-JE (82.2%) ($P < 0.001$), and the reactogenicity profile of IMOJEV was comparable with that of placebo (Guy *et al.*, 2010; Torresi *et al.*, 2010). After an early trial in 200 naive 12- to 24 month-old Thai infants had shown good antibody responses to IMOJEV (Chokephaibulkit *et al.*, 2010), a trial in 1200 naive infants, ages 12–18 months, in Thailand and the Philippines demonstrated consistency of three successive industrial scale manufacturing lots of live attenuated IMOJEV. A single dose of IMOJEV was well tolerated and elicited a protective immune response with seroconversions occurring in 95% (Feroldi *et al.*, 2012). In addition, in a phase III, observer-blinded trial, 300 Thai children, ages 9–18 months, were randomized 1:1 to receive one dose of IMOJEV or SA14-14-2. Neutralizing antibody seroconversion rates in both groups exceeded 99%. Thus, IMOJEV achieved non-inferiority with a 'gold-standard' vaccine known to protect humans against JE (Feroldi *et al.*, 2014). It has been possible to give IMOJEV with measles, mumps and rubella vaccines without evidence of interference with any component (Huang *et al.*, 2014). Vaccine is licensed for use in Australia and Thailand.

Two doses of IMOJEV, 0.5 ml SC, are given to children (aged ≥ 12 months) and adults at 0 and 28 days. For individuals aged between 12 and 24 months, the preferred injection site is the anterolateral thigh. For individuals aged >2 years, the deltoid is the recommended injection site. Immune responses to IMOJEV have been followed for 5 years, with 65% receiving a single dose and 75% receiving two doses retaining detectable neutralizing antibodies (Nasveld *et al.*, 2010).

Tick-borne encephalitis

Disease perspective

TBEV circulates in nature as three subtypes, Western, Siberian and Far Eastern. The disease is also referred to as Central European encephalitis, Far Eastern encephalitis, Russian spring–summer encephalitis and diphasic milk fever. TBEV are genetically and antigenically related to a large group of zoonotic tick-borne flaviviruses some of which affect humans, producing encephalitis (Langat, Powassan, Louping ill) or haemorrhagic fever (Kyasanur Forest Disease, Omsk haemorrhagic fever). The Far Eastern and Siberian strains of TBEV are closely related, while Western TBEV is actually more closely related genetically to Louping Ill virus, than to Central European and Siberian strains of TBEV. (Mansfield *et al.*, 2009)Western TBEV is transmitted primarily by *Ixodes ricinus*, whereas the vector for the Siberian and far-Eastern subtypes is *I. Persulcatus*. *I. ricinus* is a three-host tick. Each parasitic stage (larva, nymph, adult) feed for a period of a few days on a different host. Ticks become active when the mean outdoor temperatures are at or above 11°C. As each stage in the life cycle takes approximately 1 year to moult to the next stage, the entire life cycle may be completed in 2–6 years, depending on geographical location. Once infected with TBEV, ticks remain infected throughout their life cycle. Furthermore, TBEV is passed from adult transovarially to larvae. Larvae and nymphs feed predominantly on ground dwelling rodents, particularly, yellow-necked field mice and voles. Rodents can be chronically infected. In addition, uninfected larvae or nymphs can become infected when feeding on a host at the same time that infected nymphs or larvae are feeding (co-transmission) (Amicizia *et al.*, 2013; Barrett *et al.*, 2013; Mansfield *et al.*, 2009).

The average incubation period between tick bite and onset of symptoms is 11 days (4–28 days, range). Infection with the Far Eastern subtype of TBEV results in a disease with a monophasic illness course while infection with the Western subtype usually produces a biphasic course in nearly 75% of patients with the first phase lasting between 1 and 7 days. The incubation period is generally 7–14 days. During a biphasic infection symptoms initially include fever, fatigue, headache, vomiting and pain in the neck, shoulders or lower back. This may be followed by an asymptomatic period lasting 2–10 days and then progress to neurological involvement of the meninges, brain parenchyma or spinal cord. Acute TBE is characterized by encephalitic symptoms in 45–56% of patients. Symptoms range from mild meningitis to severe meningoencephalomyelitis, which is characterized by muscular weakness which develops 5–10 days after fever subsides. Severely affected patients may demonstrate altered consciousness and a poliomyelitis-like syndrome that may result in long-term disability (Amicizia *et al.*, 2013; Dumpis *et al.*, 1999; Gritsun *et al.*, 2003). Severity of illness increases with the age of the patient. Pareses and lasting sequelae appear to occur less frequently in young patients (Kunze *et al.*, 2004). After approximately age 40, TBE patients increasingly develop the encephalitis form of the disease (Kunze *et al.*, 2005). In patients older than 60 years, TBE takes a severe course, often leading to paralysis and death. In Siberia, approximately 80% of cases present with fever but without neurological sequellae.

The case fatality rate of TBE is approximately 1–2% in infections with the Western subtype, but as high as 20–40% with infections by the Far-Eastern subtype. Infection with the Siberian subtype produces a mortality rate of 2–3% (Amicizia *et al.*, 2013). A chronic form of TBE is observed in patients from Siberia or Far-Eastern Russia and is thought to be associated with infections by the Siberian subtype of TBEV (Gritsun *et al.*, 2003). This occurs in two forms: (i) neurological symptoms that occur for years following bite of infected tick, including Kozhevnikov's epilepsy, progressive neuritis of the shoulder plexus, lateral and dispersed sclerosis and a Parkinson's-like disease with progressive muscle atrophy, often accompanied by mental deterioration; abd (ii) hyperkinesis or epileptoid syndrome.

An increase in reported cases of TBE has occurred over the past two decades. This has been attributed by some to warming temperatures in Europe and Asia, but a wide range of political, ecological, economic and demographic factors appear to play role (Kunze, 2010). This includes both increases and decreases in forestation. Newly planted fields may bring farmers into contact with infected ticks. There is a growing participation by residents and tourists in outdoor pursuits such as hiking and fishing. Also, the recent economic downturn may have resulted in increasing foraging for food in forests by persons who are unvaccinated. In central Europe, two seasonal peaks of TBE occur, one in June/July and the second in September/October, corresponding to two waves of feeding by larvae and nymphs. All three known subtypes of TBEV are capable of co-circulating in the same area, as is currently the situation in Estonia. The disease is widespread across Europe and Asia with recent extension to areas not previously affected, such as Norway (Mansfield *et al.*, 2009). About 3000 hospitalized cases of TBE are recorded annually in Europe (Heinz *et al.*, 2007). Russia and Western Siberia contribute the largest number of cases, 10–15,000 in some years. In Western Europe, formerly Austria and now the Czech Republic has the highest incidence rates, with 400–1000 cases annually (Mansfield *et al.*, 2009).

Vaccines

Four licensed vaccines are available to prevent TBE disease. All are formalin-inactivated virus grown in tissue culture cells: (i) Austrian (Baxter) FSME-Immun 0.5ml and FSME-Immun 0.25ml Junior, (ii) Encepur (Novartis) AdultS and Children, (iii) MoscowR TBE vaccine (Federal State Enterprise of Chumakov Institute of Poliomyelitis and Viral Encephalitides, Russian Academy of Medical Sciences, Russia) and (iv) EnceVirR(Scientific Production Association Microgen, Russia) TBE vaccine (Amicizia *et al.*, 2013; Barrett *et al.*, 2013). FSME-ImmunR and EncepurR are available in the European Union and have been authorized by the European Medicines Agency (EMA) (Amicizia *et al.*, 2013). A single-dose of a chimeric West Nile/TBE vaccine mutagenized by amino acid alterations to core protein has been shown to be protective in a subhuman primate model (Rumyantsev *et al.*, 2013).

In Europe over 80 million doses of TBE vaccines have been administered. Austrian vaccine is prepared growing the Neudorfl virus strain in specific pathogen-free chick embryo cells with aluminium hydroxide as an adjuvant. Human albumin is added as a stabilizer to reduce tumour necrosis factor alpha and interleukin 1beta production in vaccinees. The Novartis vaccine is produced using the K23 strain of TBEV grown on chick embryo cells. Following inactivation by formaldehyde, vaccine is purified by continuous flow density gradient ultracentrifugation. Alum is used as an adjuvant. The vaccine was initially stabilized using processed bovine gelatin. Reactions thought to be allergic phenomena to gelatin resulted in a reformulation to use sucrose as stabilizer. Of the four vaccines that are produced, that in Austria has been in use for the longest period. No formal prospective efficacy trial was conducted prior to licensing. This vaccine has been used since 1976 with upwards of 88% of the Austrian population now fully vaccinated. A prospective, randomized, phase II double-blind dose finding study (0.6, 1.2 and 2.4 mg antigen) was carried out in volunteers, 16–65 years using the Baxter vaccine ($n=504$) who were bled 21 days following administration of two doses at an interval of 30 days. Similar dose finding studies (0.6, 1.2 and 2.4 mg antigen) were carried out in children, ages 1–5 years ($n=639$) and 6–15 years ($n=639$). Seroconversion rates from 85–97% of cases. A dose of 1.2 mg was selected as optimal. Subsequently, this dose was given to over 3000 adults in a large safety trial (Ehrlich *et al.*, 2003; Amicizia *et al.*, 2013). Using data on the incidence of vaccination among cases of serologically proven TBEV infections, a field effectiveness of 99% has been calculated among fully immunized individuals (Heinz *et al.*, 2007). Efficacy correlates with the long duration persistence of neutralizing antibodies following administration of two doses of the Baxter vaccine (Loew-Baselli *et al.*, 2011; Prymula *et al.*, 2012).

The standard immunization schedule consists of three immunizations administered intramuscularly on day 0, a second dose administered 1–3 months later and a third dose administered 5–12 months after the second dose. It is now recommended that booster doses be given at 5-year intervals indefinitely (Amicizia *et al.*, 2013; Barrett *et al.*, 2013). Rapid immunization schedules have been introduced, three immunizations to be administered on days 0, 7 and 21. The Enceptor rapid immunization schedule includes a fourth vaccination after 12–18 months. For those under 60 years of age, booster doses are recommended at 5-year intervals; for those older than 60 years, booster doses are recommended at three year intervals. A retrospective study found a high level of protection following two doses of the Baxter vaccine (Heinz *et al.*, 2007). Vaccination is recommended for persons living in TBEV enzootic areas, woodcutters, farmers, military personnel, laboratory workers and tourists who plan to hike, bicycle or camp in enzootic areas (Barrett *et al.*, 2008). Based upon limited data TBE vaccines are not recommended for use as post-exposure prophylaxis or as a therapeutic vaccine (Broker and Kollaritsch, 2008). This leaves a significant gap in the management of unvaccinated individuals who experience tick bites. An immune globulin concentrate containing IgG antibodies against TBEV was in use as a post-exposure treatment. However, reports of severe clinical outcomes in patients given TBE immune globulins plus evidence of *in vitro* ADE led to the abandonment of post-exposure treatment (Aebi and Schaad, 1994; Ozherelkov *et al.*, 2008; Waldvogel *et al.*, 1996). In the absence of evidence of a central role for macrophages as TBEV infection target cells, the phenomenon observed is more likely to be an example of antibody-mediated pathogenicity of viral infections in the brain (Gould and Buckley, 1989; Halstead, 1982; Webb *et al.*, 1967). Recent studies in a mouse model of TBE failed to demonstrate antibody enhanced infection (Kreil *et al.*, 1998). Some workers continue to argue that during critical time windows administration of high levels of neutralizing antibodies may be protective (Broker and Kollaritsch, 2008; Pen'evskaia and Rudakov, 2010).

West Nile

Disease perspective

West Nile virus (WNV) was isolated from a Ugandan patient with a mild fever in 1937. The virus was not isolated again until the 1950s, when the first WNV epidemics were described in Israel and Egypt. A large-scale epidemiological study in the 1950s identified seroprevalence rates as high as 61% in the Nile Delta of Egypt (Gray and Webb, 2014). For many years WNV, a zoonotic pathogen, was thought to cause a dengue fever-like disease in humans and encephalitis in horses in South Asia, the Middle East and Africa. WNV appeared in the Western Hemisphere in 1999 in New York City, before spreading rapidly across North America, causing disease in wild birds, horses, and humans (Petersen *et al.*, 2013). The virus has also been isolated in the West Indies and neutralizing antibody-positive birds and horses have been identified in Mexico, Jamaica, and the Dominican Republic; West Nile virus also appeared recently in South America, including Brazil and Argentina. The spectrum of disease extends from a mild febrile exanthem to fatal encephalitis;

neuroinvasive disease occurs in approximately 30% of reported WNV cases (Petersen *et al.*, 2013). In Europe, since 2008, there has been an unprecedented increased WNV activity, including the sustained emergence of a lineage 2 WNV, with a rapid rise in the number of cases of neuroinvasive disease of animals and humans (Gray and Webb, 2014).

Based upon a study of 576 blood donors experiencing early stage West Nile infection, 26% became symptomatic of whom half sought medical attention (Zou *et al.*, 2010). In symptomatic cases that evolve to a neurological outcome, after an incubation period of around 2–14 days, patients develop a high fever, headache, neck stiffness which may progress to stupor, disorientation, coma, tremors, convulsions, muscle weakness, vision loss, numbness and paralysis. Patients with neuroinvasive disease often have a flu-like prodrome followed by development of neurological signs and symptoms. It is estimated that approximately 35–40% of patients with WNV neuroinvasive disease have meningitis, 55–60% have encephalitis, and 5–10% acute flaccid paralysis, although numbers vary in different case series. Patients with encephalitis typically have fever (85–100%), headache (47–90%), and an altered mental status (46–74%) (Davis *et al.*, 2008). Following infection, many patients develop movement disorders which include tremors of various types, myoclonus, and Parkinsonism. Cerebellar signs and symptoms occur in a variable number. Weakness is also common and may be generalized and non-specific or a lower motor neuron pattern of flaccid paralysis associated with absent or reduced reflexes and preserved sensation. Cranial nerve palsies have been reported in 10% or more of cases and most commonly involve the ophthalmic (second) and facial (seventh) nerves. Visual problems in patients with WNND have become increasingly recognized. Patients often complain of blurry vision, trouble seeing and photophobia. Clinically, reduced vision, active non-granulomatous uveitis, vitreitis, multifocal chorioretinitis and optic neuritis have been described. Immunosuppressed patients, such as patients with organ transplants or AIDS, often develop a prolonged serious encephalitis during WNV infections (Davis *et al.*, 2008). Severe outcomes of WNV infection are more frequently observed in persons of advanced age or those with diabetes. A genetic factor, CCR5, leads to more serious complications.

Phylogenetic analyses describe five distinct WNV lineages. Strains of lineage 1, responsible for fatal neuroinvasive disease in humans and animals, include the NY-99 strain introduced in North America and the closely related Israel 1998 strain (Brandler and Tangy, 2013). Lineage 1 strains are also found in Africa, Asia, the Middle East, Europe and Australia (Kunjin subtype). Lineage 2, considered as a zoonosis in sub-Saharan Africa and Madagascar, has recently been introduced in Greece, Hungary and Italy and is associated with severe neuroinvasive infections in birds, horses and humans. Lineage 3 has been isolated in the Czech Republic, lineage 4 strains, in Russia, and linage 5 strains have only been described in India (Brandler and Tangy, 2013). Until the virus was imported into the United States in 1999, it had been only associated with isolated outbreaks of encephalitis in elderly patients in Israel and Romania. In the USA WNV is transmitted by *Culex pipiens* (Eastern USA), *Culex tarsalis* (Midwest and West) and *Culex quinquefasciatus* (Southeast). Infections in humans are predominantly (80%) asymptomatic. The virus is transmitted in nature by various species of *Culex* mosquitoes predominantly between various species of birds. Mammals, reptiles and amphibians of many species are infected, sometimes with fatal outcome. Crows, blue jays and hawks are particularly at risk to fatal WNV infection. The number of dying birds in a locale during early summer months is used to monitor wildlife infection and predict possible human or equine cases (Diamond, 2009). Ecology studies suggest that *Culex pipiens*, the dominant enzootic (bird-to-bird) and bridge (bird-to-human) vector of WNV in urbanized areas in the north-east and north-central USA, shifts its feeding preferences from birds to humans during the late summer and early fall, coincident with the dispersal of its avian hosts. Most (~85%) human infections in the USA occur in the late summer with a peak number of cases in August and September. In warmer parts of the country, virtually year-round transmission has been observed.

Following its introduction in 1999, WNV has spread to cause clinical disease in 47 US states (all except Maine, Alaska and Hawaii), Canada, Caribbean islands and Central America. In the USA, from 1999 until 2013, the CDC reported 39,557 cases and 1668 deaths are attributed to WNV infection. Recent WNV outbreaks have been reported in Russia, Israel, Greece and the Czech Republic. Perhaps 20% of infected persons will have mild symptoms including fever, headache, body aches, nausea, vomiting and sometimes a skin rash on the trunk. About 1 in 150 infected humans, higher in individuals 60 years or older, will develop encephalitis. There is no effective drug treatment. Recent epidemiological data show a sharp decline in the number WNV cases in the past years. The outbreak in the USA revealed that WNV can be spread through blood transfusions, organ transplants, intrauterine exposure and breast feeding (Nett *et al.*, 2012; Pealer *et al.*, 2003; Petersen *et al.*, 2013). Since 2003, blood banks in the USA have routinely screened plasma for virus.

Vaccine development status

While the introduction of several different kinds of vaccines for horses has proved to be a potent tool for control of veterinary disease there is no licensed human vaccine against WNV. Vaccines in clinical or late pre-clinical development are reviewed briefly.

Chimeric vaccines

Two ChimeriVax West Nile (WN-CV) viruses have been constructed (Monath, 2002; Monath *et al.*, 2001). In the first, WN01, the complete sequences of the prM/E genes of the YFV 17D virus were replaced with those of the WNV strain NY99 introduced in the USA, without any modification (Arroyo *et al.*, 2001b; Monath *et al.*, 2001). WN01was developed by Intervet as a single dose vaccine for horses (PreveNileTM, Schering-Plough Animal Health/Merck) was commercially available since 2006 but withdrawn because of a number of reports of severe adverse events (De Filette *et al.*, 2012). For aWNV vaccine for human use (WN02), three mutations were introduced into the WN01 E gene [E107 (L→F), E316 (A→V) and E440 (K→R)]. These mutations are known to attenuate the closely related JEV. In the human WN02 vaccine, these mutations were shown to independently enhance neuroattenuation of the chimera, so that reversion at one residue or even two residues would maintain the attenuated phenotype (Arroyo *et al.*, 2001b). Even in a very unlikely event that all three mutations simultaneously reverted to wt codons, the chimera, would be similar to the attenuated WN01 which is not neuroinvasive and is even less neurovirulent than YFV 17D vaccine by direct intracerebral (i.c.) inoculation of mice and monkeys (Arroyo *et al.*, 2004). A passage 5 WN02 vaccine was safe and highly immunogenic in a human phase 1 study in healthy adults (Monath *et al.*, 2006). However, this vaccine produced heterogeneous plaque populations: large plaque (LP) and small plaque (SP) were identified in the vaccine lot at a ratio of 50:50. The mutation at M66 (L→P) that is responsible for the SP phenotypic change, appears to be an adaptation to propagation in serum free (SF) Vero cells. The SP clone was selected for production of vaccine lots to be used in phase 2 clinical trials in humans. US adults aged ≥ 50 years were randomized to one of four treatment groups: ChimeriVax-WN02 10^3 plaque-forming units (PFU) ($n = 121$), 10^4 PFU ($n = 122$), 10^5 PFU ($n = 110$) and placebo ($n = 120$). Neutralizing antibodies were observed in 92.1%, 93.2%, and 95.4%, respectively, of subjects by day 28. No significant adverse events were observed (Dayan *et al.*, 2012).

Replication-incompetent vaccines

RepliVAX West Nile vaccine is propagated in C-expressing cells (or as a unique two-component virus) using methods similar to those used to produce today's economical and potent LAVs. Due to deletion of most of the gene for the C protein, RepliVAX cannot spread between normal cells, and is unable to cause disease in vaccinated animals. RepliVAX provides an efficacious WN vaccine in animal models that demonstrate that it is potent, economical to produce, and safe for immunization of the immunocompromised host (Suzuki *et al.*, 2009; Widman *et al.*, 2008).

Measles vaccine as vector

Genes for WNV IS-98-ST1 envelop protein were inserted into the genome of live-attenuated Schwarz measles vaccine (MV). Inoculation of this construct into MV-susceptible mice induced high levels of specific anti-WNV neutralizing antibodies and protected mice from a lethal challenge with WNV. In addition, antibodies raised in mice were given to BALB/c mice which were protected against lethal outcome following a challenge with a high dose of WNV (Despres *et al.*, 2005; Lorin *et al.*, 2005).

West Nile/Dengue 4

Several WNV and Dengue virus type 4 chimeras have been constructed and evaluated as vaccines. In one chimera, genes for the WNVNY99 membrane precursor and envelope were cloned on a Dengue 4 (WN/DEN4) background and the other had a 30 nucleotide deletion in the 3′ non-coding region of DEN4 (WN/DEN4-3′Δ30). Both these vaccines were attenuated in Rhesus macaques and prevented viraemia in the macaques upon challenge. A follow-up study with the WN/DEN4-3′Δ30 virus showed that it was unable to infect geese, and that it was safe in immunocompromised mice and attenuated in monkeys (Pletnev *et al.*, 2003, 2006). This vaccine was evaluated in healthy flavivirus-naive adult volunteers age 18–50 years in single and two dose studies. The first study evaluated a range of doses of vaccine given as a single dose with higher seroconversions being observed at lower viral concentrations (75% vs. 55%). A booster dose at 6 months resulted in a seroconversion rate of 89% (Durbin *et al.*, 2013b).

DNA

A plasmid DNA encoding the WNV membrane and envelope proteins injected intramuscularly in mice and horses provided protection against a WNV challenge.(Davis *et al.*, 2001) This study paved the way for the licensing of the first DNA vaccine for animal use, i.e. West Nile- Innovator® DNA. Phase I clinical trials for a DNA vaccine expressing WNV NY99 preM and E genes were conducted in 2007 (Martin *et al.*, 2007). Three doses of this vaccine under control of a cytomegalovirus promoter raised similar neutralizing antibodies and T-cell responses and was without adverse events in young and older adults (Ledgerwood *et al.*, 2011). More recently, a capsid deleted Kunjin virus DNA vaccine was developed with the capsid being provided in trans. These single-cycle viruses replicate once to generate VLPs, which were highly immunogenic in mice and horses (Chang *et al.*, 2008).

Other vaccines

A very large number of constructs have been developed, many using approaches reviewed in the section 'Dengue vaccines in pre-clinical development' and in Table 20.1. Most have been subjected to a variety of pre-clinical tests resulting in useful immune responses and in some instances protection in a challenge virus model. For further details, several excellent recent reviews should be consulted (Brandler and Tangy, 2013; De Filette *et al.*, 2012; Iyer and Kousoulas, 2013).

Yellow fever (YF)

Disease perspective

Yellow fever is the prototype of the genus flavivirus. The virus is enzootic in three cycles, two involving African subhuman primates and one cycle in the American tropics. West African YFV has escaped from its zoonotic reservoir on numerous occasions into an urban transmission cycle involving the domesticated African variant of *Aedes aegypti*. This YFV was imported to the American hemisphere in the 1600s, when large urban outbreaks took an enormous toll. Sometime after its introduction to the New World, the virus escaped and established a complex sylvatic cycle. YFV was identified as the viral cause of urban yellow fever and its vector shown to be *Aedes aegypti* by Walter Reed in Cuba in 1900.

Disease severity is quite variable; a high proportion of cases are mild or moderate and cannot be distinguished from other viral syndromes. After the bite of an infected mosquito there is an incubation period of 3–6 days or more. Classical disease begins with abrupt onset of fever accompanied by headache, photophobia, dizziness, back pain, malaise, anorexia, gastrointestinal disturbances and irritability (Kerr, 1951). Patients often are unable to sleep and may be hyper-alert. A macular blanching rash, conjunctival congestion and a facies suggesting alcohol intoxication are premonitory signs. On physical examination there may be mid-epigastric tenderness, an enlarged liver and a white coated tongue, described as 'small' with red margins and tip (Beeuwkes, 1936; Berry and Kitchen, 1931). The patient may have bradycardia relative to degree of fever (Faget's sign). Laboratory findings include a marked leukopenia with relative neutropenia, elevation of serum SGOT and SGPT. On day 4–6 of fever there may be a brief remission. Patients may progress to a period of intoxication accompanying a rebound in fever (saddle back fever curve) accompanied by vomiting, jaundice, haematemesis, melena, haematuria, metrorrhagia, petechiae, ecchymoses, epistaxis, oozing of blood from gums or from needle puncture sites and anuria. Alkaline phosphatase values are normal, while bilirubin values vary from 3 to 5 g/l. Most yellow fever patients exhibit albuminuria starting during the early acute stage and worsening in proportion to overall disease severity. Fatal outcome maybe heralded by delirium and intense agitation (Kerr, 1951).

In Africa, yellow fever virus causes a wider spectrum of illness than in the Americas. Yellow fever in African adults is often a relatively mild disease with case fatality rates relatively lower than the terrifyingly high rates common in American outbreaks (Monath, 1997). These differences were sufficient to motivate the Rockefeller Foundation in 1926 to study YF in laboratories sited in Lagos, Nigeria and Accra, Ghana (Strode, 1951). The African patient, Asibi, who had a mild, self-limiting disease, yielded the prototype yellow fever virus strain and parent to the 17D vaccine. Almost immediately, three Rockefeller researchers infected with the Asibi strain acquired yellow fever and died (Strode, 1951). Over several centuries there are extensive anecdotal reports in the medical literature of disease attack and severity rates being lower in individuals of African ancestry than in whites or Asians. In the Caribbean and southern United States yellow fever outbreaks in Africans who had lived in the Americas for generations were frequently mild with low attack rates (Carter, 1931; Kiple, 1984; Kiple and Kiple, 1977; Kiple and Ornelas, 1996). There is also a lower incidence of overt YF disease and death rates in children than in adults. Beeuwkes reported that of 91 individuals under the age of 30 the death rate was 8% while of 31 persons 30 years and older, 32% died (Beeuwkes, 1936).

Seven genotypes, based on greater than 7% differences in nucleotides, have been identified from full-length YFV sequences, two in the Americas, three in Angola, East and Central Africa and two in West Africa (Mutebi *et al.*, 2001; von Lindern *et al.*, 2006). African viruses are maintained in two sylvatic cycles. Genotype 1 is in West Africa, where virus is transmitted among a wide range of subhuman primate species by *Aedes furcifur-taylori* or *Aedes luteocephalus* and other species of *Stegomyia*. The relative stability of the yellow fever genome in Africa suggests that virus is also maintained by transovarial transmission (Sall *et al.*, 2010). Virus in this cycle is transmitted by these vector species to humans and then from human to human in urban as well as in rural areas by *A aegypti*. This species in Africa occupies a wider range of habitats and has a broader host-feeding preference than outside Africa. Outbreaks occur in mid-rainy season (August) to early dry season (October). A relatively close genetic relationship exists between American and West African genotypes, consistent with the hypothesized West African origin of American yellow fever viruses which then adapted to a new group of vertebrate hosts and mosquito vectors. Following prolonged separation phenotypic differences emerged between American sylvatic and West African urban YF viruses (Deubel *et al.*, 1987; Fitzgeorge and Bradish, 1980; Henderson *et al.*, 1970).

The viruses with the greatest genetic distance from West African and American viruses are transmitted in the Central/East African sylvatic cycles among subhuman primates by *Aedes africanus* and from monkeys to humans by *Aedes simpsoni* (Chang *et al.*, 1995; Lepiniec *et al.*, 1994; Mutebi *et al.*, 2001; von Lindern *et al.*, 2006). Virus from this cycle produced the large 1960 Ethiopian outbreak, smaller 1990 and 1996 outbreaks in Kenya and the 2012 outbreak in Darfur, Sudan (http://www.who.int/csr/don/2012_12_06/en/) (Monath, 1989, 1994, 1997; Soghaier *et al.*, 2013) Despite the presence of *Aedes aegypti* in urban centres of East Africa, no Central African YF virus has ever escaped to the urban cycle (Strode, 1951). African subhuman primates support high enough viraemia to sustain transmission, but infections in these animals are clinically inapparent. The resistance to fatal yellow fever in African as compared with Asian subhuman primates is consistent with the evolution of the virus in Africa as a zoonosis. Little work has been done to define phenotypic differences between West and Central/East African viruses. In fact, there is no formal evidence that the pylogenetically older Central/East African viruses are capable of being transmitted by urbanized *Ae. aegypti*. The invasion of Asia by yellow fever has been the emergence event most widely predicted and feared during the 20th century (Dudley, 1934; Strode, 1951). The most reasonable explanation for the failure of this emergence event is the fact that transmission of YF in urban cycles on the East Coast of Africa has not occurred. The Central/East African YF zoonotic cycle provides an effective barrier to the spread of West African YF urban strains.

Based upon abundant evidence, historical and genetic, yellow fever virus moved out of the West African sylvatic cycle resulting in African and American urban outbreaks (Sall *et al.*, 2010). In the New World, yellow fever was recognized as one of the classic epidemic diseases. This emergence event occurred at a date unknown but likely the early 1600s (Strode, 1951). Domesticated East and West African *A aegypti* undoubtedly infested many of the sailing vessels which plied the trade in human beings supplying sugar plantation workers to the British, French, Spanish, Dutch and Portuguese colonies in Americas and Asia (Moore *et al.*, 2013). On occasion, the human cargo and crew must have offered a combustible mixture of yellow fever-infecteds, -immunes and -susceptibles that supported endemic transmission during the 2- to 3-month journey across the Atlantic Ocean. It is equally plausible that virus was transported in living female *A. aegypti*, infected either in Africa or on-board. World trade soon introduced yellow fever to many of the towns and cities on the coasts of North and South America and to the banks of the Mississippi and Amazon rivers.

Another remarkable event occurred silently in the Americas; yellow fever virus escaped from the urban cycle to a forest cycle, ultimately infecting a wide range of South American subhuman primates as well as marsupials, transmitted by several species of *Haemagogus* mosquitoes producing YF epizootics that travel up and down the vast tropical forests in the river basins of South America east of the Andean Cordillera. As evidence that the New World zoonotic cycle began relatively recently, in nature and in the laboratory, American subhuman primates experience high fatality rates when infected with wild-type yellow fever virus (Strode, 1951; Waddell and Taylor, 1945, 1946, 1947, 1948).

Early in the 20th century the urban yellow fever mosquito vector was identified and effective mosquito control methods designed and implemented (Strode, 1951). Using these methods, as early as 1934, urban yellow fever was eradicated from the Western Hemisphere and, by 1960, *A. aegypti* had been eradicated from 13 large countries occupying 85% of the Central and South American land mass (Monath, 1994). Despite this show of competence, by the 1970s, *A. aegypti* began to reinvade and by 1990 had virtually attained its previous range. As early as the late 1800s, there was evidence that humans were infected with yellow fever virus from the American sylvatic cycle, principally adults with occupational exposure to forest or forest fringe areas of Bolivia, Brazil, Venezuela, Colombia, Ecuador and Peru. At present, epizootics and human cases occur during months of high rainfall and humidity, January–May (Monath *et al.*, 2008). Over the past 50 years, in these six countries, there have been between 50 and 500 cases reported each year with as many as 200–250 deaths (PAHO).

The reintroduction of yellow fever into urban areas of the Americas from the sylvatic cycle has been repeatedly predicted. Several putative occurrences have been reported, but continuing urban outbreaks have not occurred. The reason for the failure of this much heralded event is not clear. One must ask whether jungle *Haemagogus*-adapted IIB viruses have lost their ability to be transmitted by *A aegypti*? In the 1940s, Waddell and Taylor demonstrated that South American yellow fever strains were capable of being serially transmitted among subhuman primates by *A aegypti* (Waddell and Taylor, 1945, 1946, 1947, 1948). However the proper experiment has not been done. This would be to compare the competence of colonies of American *A aegypti* to transmit type IIA and IIB viruses. Fortunately for the world, but unfortunately for experimental design, American urban yellow fever virus strains do not exist and only the transmissability of sylvatic YF strains have been studied (Aitken *et al.*, 1977; Lourenco de Oliveira *et al.*, 2002, 2004; Miller and Ballinger, 1988; Miller *et al.*, 1989; Mutebi *et al.*, 2004; Tabachnik *et al.*, 1985).

Vaccines

YF virus was attenuated in 1937 by workers at the Rockefeller Foundation, New York. YF vaccine is manufactured by major pharmaceutical manufacturers as well as developing world producers. At present, only strains of 17D yellow fever virus are licensed as human vaccines. The prototype Asibi virus was passaged 18 times in cultures of minced mouse embryo tissues then in minced whole chick embryo. After 58 passages, the virus, now designated as 17D, was grown in chick embryo cultures from which brain and spinal cord had been removed. Final passage of virus for use as a vaccine was (and is) made in embryonated hens' eggs (Lloyd *et al.*, 1936; Monath *et al.*, 2008; Theiler and Smith, 1937a,b). All live attenuated YF vaccines are made from strains designated YF 17DD or 17D 204 (Monath *et al.*, 1983). There are 20 amino acid changes in the genome and four nucleotide differences in the 3′ non-coding region between wild-type Asibi strain and 17D vaccine viruses (dos Santos *et al.*, 1995; Monath *et al.*, 2008; Rice *et al.*, 1985). Wild-type YF viruses share amino acids at envelope protein positions 52, 200, 305 and 380, suggesting that amino acid differences observed at these sites between wild and vaccine viruses may be responsible for attenuation. Interestingly, a single or limited passage of wild-type YF virus in HeLa cells resulted in loss of neurotropism and viscerotropism accompanied by a completely different set of amino acid changes (Dunster *et al.*, 1990, 1999). It is possible that mutations in NS proteins convey attenuation. NS amino acid changes and nucleotide changes in the 3′ NCR in YF are present in attenuated chimeric vaccines made when dengue prM and E genes are inserted on the YF 17D backbone (Guirakhoo *et al.*, 2006; Monath *et al.*, 2002). There are also eight E protein amino acid changes, eight in NS proteins and four nucleotide differences in the 3′ NCR between 17DD (Brazil) and 17 D 204 (USA, UK, France, Senegal) vaccines (Dupuy *et al.*, 1989; Galler *et al.*, 1998; Jennings *et al.*, 1993; Rice *et al.*, 1985).

17D vaccine is given by the subcutaneous route in a volume of 0.5 ml, generally in the upper arm. The minimum dose requirement is 1000 mouse LD_{50} or the equivalent of 10,000 PFU. Vaccines vary with respect to stabilizer additives and salt content. Some contain sodium chloride and buffer salts. All vaccines are lyophilized. As YF vaccines do not contain antibiotics, when reconstituted, vaccines should be administered within 1 h. When maintained on ice, vaccines in multi-dose vials given under epidemic conditions are used for a period of 8 h. Vaccines are supplied in single and multiple dose vials, up to 20 doses.

One dose of vaccine is known to protect more than 90% of vaccinees raising antibodies that circulate essentially for a lifetime (Monath *et al.*, 2008). Neutralizing antibodies are thought to be the primary correlate of protective immunity (Monath *et al.*, 2008; Pugachev *et al.*, 2002). Because of its safety and efficacy, innate and acquired immune responses to 17 D vaccine have been studied in detail in human volunteers (Pulendran, 2009). Administration of YF 17D results in a massive expansion of CD8+ T cells, peaking at 15 days (Miller *et al.*, 2008). Vaccination with YF 17D induces polyvalent adaptive immune responses, including the production of cytotoxic T cells, a mixed T helper 1 (T_H1) and T_H2 cell profile and robust neutralizing antibodies that can persist for up to 40 years after vaccination (Barrett and Teuwen, 2009). YF 17D signals through RIG1 and MDA5, activating transcription factors that regulate the expression of type I IFNs. The solute carrier family two genes, *SLC2A6* and *EIF2AK4*, that mediate an integrated stress response in innate immune cells, predicted high CD8+ T cells and antibody responses (Querec *et al.*, 2009). In addition, with respect to antibody responses, TNF receptor superfamily, receptor 17, a receptor for B cell-activating factor (BAFF) was a key gene in predictive signatures (Querec *et al.*, 2009).

YF vaccine is made in 10 facilities world-wide. Five of these are principal suppliers of vaccine – sanofipasteur (USA), Bio-Manguinhos (Brazil), Chiron/Novartis (UK), sanofipasteur, Marcy l'Etoile (France) and the Pasteur Institute of Dakar, Senegal (Africa). In 1966, yellow fever 17D vaccine was discovered to be contaminated with avian leukosis virus. (ALV) All vaccines produced at that time contained the agent, due to the high prevalence of ALV infection in flocks used for egg production. New seeds free of avian leukosis virus were developed in the 1970s and all manufacturers now employ leukosis-free YF virus seeds and leukosis-free flocks as stipulated by WHO standards.

17D vaccine strains retain a degree of neurovirulence demonstrated on inoculation of mice or monkeys intracerebrally (Monath *et al.*, 2005). This appears to have a counterpart in young infants and the elderly (Vaccine-associated neurological disease – YEL-AND). Encephalitis in adults is rare. Between 1945 and 1991 only 21 adult cases of YF post-vaccination encephalitis were reported worldwide (Monath *et al.*, 2008). Between 1991 and March 2006, after institution of a formal adverse events reporting system, there were 29 post-vaccination encephalitis cases reported, mostly in individuals over the age of 60 years (Marfin *et al.*, 2005; McMahon *et al.*, 2007). In 1952–53, five cases of encephalitis were reported among 1800 infants who were vaccinated under the age of 1 year (Stuart, 1956). Since then restrictions were imposed on giving 17D vaccine to infants under 9 months of age. As a result, the incidence of encephalitis has been very low.

In 2001, multi-organ failure was described in seven individuals receiving YF 17D vaccine, six were fatal (vaccine-associated viscerotropic disease – YEL-AVD) (Martin *et al.*, 2001; Vasconcelos *et al.*, 2001). These were observed in individuals immunized both with 17DD and 17D 204 vaccine strains. The signs and symptoms of the syndrome resembled wild-type yellow fever disease, including rapid onset of fever, malaise and myalgia within 2–5 days of vaccination followed by jaundice, oliguria, cardiovascular signs and haemorrhage. Large amounts of YF antigen were found in liver, heart. Virus recovered did not differ significantly from dominant genome present in vaccines (Engel *et al.*, 2006; Jennings *et al.*, 1994). A partial genome sequence from a individual who died after receiving vaccine in 1975 revealed that only 17 DD vaccine virus was recovered from organs (Engel *et al.*, 2006). For a time, it was thought that individuals who had thymus resections were uniquely at risk to YEL-AVD, but as more cases have been described no important risk factors have been established. Indeed, several large scale prospective and retrospective studies have found no increased risk to younger, healthy individuals given YF vaccine and equivocal risk among elderly vaccinees (Breugelmans *et al.*, 2013; Nordin *et al.*, 2013; Rafferty *et al.*, 2013; Thomas *et al.*, 2011). Because jungle yellow fever remains a travel hazard, the prudent course is to carefully evaluate specific risk associated with proposed travel before vaccinating elderly persons (Monath, 2012; Rafferty *et al.*, 2013). Individuals with YEL-AVD appear to mount unusual cytokine responses along with brisk and fairly normal acquire immune responses (Silva *et al.*, 2010).

Reoviridae

Spinareovirinae

Coltivirus

Colorado tick fever (CTF)

Disease perspective

Colorado tick fever is caused by a *Coltivirus*, a genus in the subfamily *Spinovirinae* of the *Reovirus* family (Attoui *et al.*, 2005). The virus, transmitted to a wide variety of vertebrates by the Rocky Mountain wood tick, *Dermacentor andersoni*, is acquired by larvae that feed on infected rodents and is maintained through the nymphal and adult stages. The transmission season is from March to November in the geographic distribution of the tick vector, the high plains and mountainous regions of the western US at elevations of 4000–10,000 feet (Brackney *et al.*, 2010; Klasco, 2002; McLean *et al.*, 1981). Humans become infected by the bite of adult ticks. Unusually, for pathogenic viruses of humans, the virus grows in haematopoietic cells, particularly erythrocytes (Philipp *et al.*, 1993). Virus may circulate in blood for several months after onset of infection. While person to person transmission has not been reported, the virus can be transmitted by blood transfusion (Leiby and Gill, 2004).

The incubation period is 3–4 days after exposure to tick bite. The onset is abrupt with fever, chills, malaise, headache, retro-orbital pain, myalgia, and hyperaesthesia. Nausea, vomiting and abdominal pain may occur. Some patients exhibit a maculopapular and petechial eruptions. The disease classically is biphasic. An initial attack lasts 2–3 days followed by defervescence and then a second bout of fever. Leukopenia is the hallmark of the disease. Examination of bone marrow aspirates shows maturation arrest of myeloid and erythroid elements with depletion of megakaryocytes. A reduced platelet count is found in most patients with haemorrhage in some cases (Klasco, 2002).

It is important to differentiate CTF from Rocky Mountain spotted fever (RMSF) quickly. RMSF, a potentially fatal acute febrile illness that is successfully treated with antibiotics, presents with a leukocytosis and often with a haemorrhagic rash. RMSF is caused by rickettsia transmitted by *Dermacentor andersoni*, but also by other tick species. RMSF occurs throughout Canada, central America, parts of South America and the USA (except Hawaii, Maine, Vermont and Alaska).

Vaccine development status

A vaccine was prepared from virus grown in suckling mouse brain. Virus was precipitated using calcium phosphate, eluted, filtered and formalin-inactivated (Thomas *et al.*, 1963). Half of 30 vaccinees given two doses of vaccine 1 month apart circulated high titres of neutralizing antibodies 6 months later (Thomas *et al.*, 1967).

Togaviridae

Alphaviridae

Chikungunya

Disease perspective

Chikungunya infections in humans results in a benign, dengue-like syndrome characterized in adults by the abrupt onset of fever, arthralgia, maculopapular rash, leukopenia and often followed by a prolonged period of post-febrile arthralgia/arthritis. The virus was isolated in 1952 during an outbreak in Tanzania where in the Makonde language the term means 'that which bends up' referring to the characteristic symptom of arthralgia (Robinson, 1955). Chikungunya is a well-established zoonosis involving many sub-human primate species in South, Central and West Africa and transmitted by several mosquito vectors. The virus circulates as three genotypes, Asian, East, Central and South African and West African. Although human illnesses are frequently reported in the area of the zoonosis, virus from zoonotic foci appears to spill over into an East African human transmission cycle (vectors are *Aedes aegypti or Aedes* albopictus) at intervals of approximately 40–50 years (Halstead, 2014a). Historically, this East African urban epidemic seeded pandemics throughout the Asian as well as the American tropics. Chikungunya-like disease swept India in 1824, 1871, 1902, 1923, 1963, 1964 and 2006 (Carey, 1971; Pulmanausahakul *et al.*, 2011; Renault *et al.*, 2012; Weaver, 2013). In 1779, recognized by its distinctive clinical features, chikungunya was epidemic in the Indonesian archipelago (Bylon, 1780). During the 1870 epidemic on Zanzibar the Swahili term for chikungunya was *Ki denga pepo* or *dyenga*. Christie, physician to the Sultan of Zanzibar, recognized that this virus had been transported to the West Indies, producing the 1827–28 epidemic, where the Spanish homonym *dengue* was adopted (Carey, 1971; Christie, 1872, 1881).

During or before the Second World World War, chikungunya appears to have become endemically established in the urban cycle throughout much of South-East Asia. Serological evidence of chikungunya virus infection has been found throughout the Philippines, possibly during the Second World War; since then, localized outbreaks have occurred in Manila, Philippines, in 1967 and Negros, Philippines, in 1968 (Basaca-Sevilla and Halstead, 1966; Campos *et al.*, 1969; Macasaet *et al.*, 1969) and as recently as 2012–13 (Yoan *et al.*, 2015). An extensive neutralizing antibody survey documented chikungunya activity possibly starting in World War II in Kalimantan and Sulawesi, Indonesia. During the late 1950s and the 1960s, chikungunya was endemic in urban populations of Thailand, Cambodia, and South Vietnam, continuing into the 1970s (Chastel, 1964; Halstead *et al.*, 1969; Vu-Qui *et al.*, 1967). In Burma, intermittent outbreaks recorded in 1963 and from 1970 to 1973 (Ming *et al.*, 1974). A large chikungunya epidemic affected much of Indonesia in 1983 and 1984 (R. D. Slemons, personal communication) and Burma in 1984 and 1985 (Thein *et al.*, 1992). Chikungunya virus was isolated in Australia in 1989 (Harnett and Bucens, 1990). From 1990 to 1995, chikungunya remained endemic at low levels in Thailand, Myanmar (formerly Burma), and Indonesia (Thaikreuea *et al.*, 1997). Little or no chikungunya virus infection has been reported in the 20th century in Papua New Guinea, the Solomon Islands, Vanuatu (formerly New Hebrides), the Caroline Islands, the Pacific Islands (Tesh *et al.*, 1975).

In 2005–6, a notable epidemic occurred on Réunion, Mauritius, Madagascar, Mayotte and the Seychelles (Renault *et al.*, 2012). The genotype responsible for this outbreak soon spread to India and to South-East Asia (Lahariya and Pradhan, 2006; Pulmanausahakul *et al.*, 2011). The remarkable attribute of the Réunion outbreak was that the virus was transmitted by *Aedes albopictus* (Reiter *et al.*, 2006; Tsetsarkin *et al.*, 2011). This virus was brought by tourists to Europe, resulting in modest autochthonous outbreaks in south-eastern France and north-eastern Italy (Grandadam *et al.*, 2011; Moro *et al.*, 2010; Poletti *et al.*, 2011; Rezza *et al.*, 2007). Chikungunya was introduced into Saint Martin in December 2013. As of 7 May 2014, the CDC reported chikungunya cases from 14 Caribbean countries (http//wwnc.cdc.gov/travel/notices/watch/chikungunya-caribbe).

During pan-Asian pandemics chikungunya resulted in many millions of clinical illnesses and some deaths. Due to the panic that accompanies these outbreaks and the very real post-illness disabilities, chikungunya exacts major economic and political burdens. For these reasons a major effort for dengue vaccine development has emerged.

Vaccine development status

A large number of experimental chikungunya vaccines have been developed and one carried into phase I/II clinical trials in humans (Table 20.2). In view of the low mortality associated with chikungunya virus infection, the sporadic nature of

Table 20.2 Chikungunya vaccines in pre-clinical and clinical development

Description	Candidate	Development stage	Reference
Grown in AGMK cells, formalin inactivated	Ross strain, killed	Phase I	Harrison *et al.* (1971)
Grown in CEK cells, formalin-inactivated	Ross strain, killed	Pre-clinical	Harrison (1967), White *et al.* (1972)
Grown in BHK21 cells, UV inactivated	Ross strain, 177th smb passage, killed	Pre-clinical	Nakao and Hotta (1973)
Virus serially passaged in MRC5 cells	181/clone 25, live-attenuated	Phase I/II	Levitt *et al.* (1986), McClain *et al.* (1998), Edelman *et al.* (2000)
EMC IRES replaces subgenomic promoter	CHIK-IRES live-attenuated	Pre-clinical	Plante *et al.* (2011), Roy *et al.* (2014)
CHIK structural genes inserted into VEE non-structural genes	VEE/EEE/CHIK, live-attenuated chimeric	Pre-clinical	Wang, E. *et al.* (2011), Wang *et al.* (2008)
CHIK structural genes inserted into measles vaccine	CHIK/measles	Pre-clinical	Brandler *et al.* (2013)
Structural gene plasmid	DNA, E 1+2+ 3	Pre-clinical	Mallilankaraman *et al.* (2011), Muthumani *et al.* (2008)
E1, E2 and C genes inserted into adenovirus	Adenovirus chikungunya vaccine	Pre-clinical	Wang, D. *et al.* (2011)
E1,2,3,capsid and 6K proteins expressed from DNA	Virus-like particles	Pre-clinical	Akahata *et al.* (2010)
Proteins expressed from recombinant baculovirus system	Virus-like particles	Pre-clinical	Metz *et al.* (2013a)
Baculovirus expression system	E 2 subunit	Pre-clinical	Metz *et al.* (2013b)

pandemics and correspondingly, the difficulty of designing and conducting phase III clinical trials, the commercial production of a chikungunya vaccine is likely to have a low priority.

Eastern equine encephalitis (EEE)

Disease perspective

EEE is a zoonosis maintained in nature in bird–mosquito cycles. The virus circulates as North and South American types. The North American virus is predominantly circulated by *Culiseta melanura*, breeding in coastal swamps and feeding on numerous species of birds. The virus may produce severe encephalitis with high mortality rate in humans and horses. Epizootics occur unpredictably resulting in periodic outbreaks of EEE in horses and less often in humans. In North America virus activity occurs in fresh water swamps of the Eastern Seaboard, Gulf Coast with Florida having the most human cases. *C. Melanura* serves as the zoonotic vector with bridging mosquito vectors bringing virus to humans and horses, such as *Coquillettidia perturbans*, *Aedes sollicitans*, *Aedes vexans* and *Aedes canadensis*. Virus activity varies markedly from year to year in response to still unknown ecological factors. Most infections in birds are silent, but infections in pheasants are often fatal, and epizootics in these species are used as sentinels for periods of increased viral activity (Hubalek *et al.*, 2014; Zacks and Paessler, 2010).

In the USA, EEE is a very low incidence disease, with a median of eight cases occurring annually in the Atlantic and Gulf States from 1964–2007 (Tsai, 2014). Transmission is often observed in focal endemic areas, the coast of Massachusetts, the six southern counties of New Jersey, and north-eastern Florida. Cases have been recognized on Caribbean islands (Zacks and Paessler, 2010). The case–infection ratio is lowest in children (1:8) and somewhat higher in adults (1:29). EEE occurs rarely in the Caribbean, Central or South America (Carrera *et al.*, 2013).

Vaccine development status

An effective killed vaccine is licensed for horses but no EEE vaccine is available for humans. For laboratory personnel, simple formalin-inactivated vaccines are in use (Tsai, 2014). Other vaccine candidates such as a Sindbis/EEE chimeric vaccine and a recombinant EEE virus attenuated by internal ribosomal entry site control are in pre-clinical development (Pandya *et al.*, 2012; Roy *et al.*, 2013).

Mayaro

Disease perspective

The disease is a non-fatal, dengue-like disease characterized by fever, chills, headache, eye pain, generalized myalgia, arthralgia, diarrhoea, vomiting and rash of 3–5 days' duration. Severe joint pain is a prominent feature of this illness. Arthralgias sometimes persist for months and can be incapacitating (Halsey *et al.*, 2013; Tesh *et al.*, 1999).

The virus was isolated in Trinidad and is enzootic in forested areas of Central and South America. The virus or cases have been identified in French Guiana, Suriname, Venezuela, Peru, Bolivia and Brazil (Halsey *et al.*, 2013; Munoz and Navarro, 2012). The principal cycle involves *Haemagogus* mosquitoes, marsupials and small mammals, a cycle that is somewhat analogous to the jungle cycle of yellow fever virus. Cases have been reported in travellers returning to the USA, Netherlands, France and Germany. Small epidemics have been reported several times in Manaus, Brazil (Mourao *et al.*, 2012). Experimentally, the virus has been transmitted by *Aedes aegypti*, although at present no outbreaks have been attributed to this vector. As humans increasingly work in or visit the American tropical forests, the incidence of Mayaro will continue to rise.

At present, annual morbidity burden can be estimated in the thousands.

Vaccine development status

Weaver and colleagues have described molecular strategies for producing generic alphavirus vaccines that could be used as a template to produce a Mayaro vaccine (Chattopadhyay *et al.*, 2013; Plante *et al.*, 2011). An inactivated Mayaro vaccine has been produced in human diploid cell cultures (Robinson *et al.*, 1976).

O'nyong nyong

Disease perspective

O'nyong nyong virus circulates in an ill-defined zoonotic cycle in South, Central and West Africa. O'nyong nyong infection produces an acute fever accompanied by rash and polyarthritis. The virus was isolated from humans and anopheles mosquitoes from northern Uganda in 1959. It is the only virus whose primary vectors are anopheline mosquitoes (*Anopheles funestus and Anopheles gambiae*) (Vanlandingham *et al.*, 2005). There have been two epidemics of o'nyong nyong fever. The first occurred from 1959–62 in Uganda, Kenya, Tanzania Zaire, Malawi and Mozambique affecting over two million people. The second occurred in 1996–97 and was confined to Uganda (Kiwanuka *et al.*, 1999; Lanciotti *et al.*, 1998; Lutwama *et al.*, 1999; Powers *et al.*, 2000; Sanders *et al.*, 1999). A small outbreak was reported in a refugee camp in Cote d'Ivoire in 2003 (Posey *et al.*, 2005). Historical evidence of an epidemic in 1904–05 suggests that o'nyong nyong, like chikungunya, emerges from its zoonotic cycle at intervals of 30–50 years.

Vaccine development status

The chikungunya candidate vaccine, CHIKV/IRES (Table 20.1), that in preclinical studies was demonstrated to be safe, immunogenic and efficacious against CHIK virus also induced cross-protective immunity against ONNV. A single dose of CHIKV/IRES elicited a strong cross-neutralizing antibody response and conferred protection against ONNV challenge in the A129 mouse model. CHIKV/IRES-immune A129 mothers transferred antibodies to their infants that were protective (Partidos *et al.*, 2012).

Ross River fever (RRF)

Disease perspective

RRF (epidemic polyarthritis) was first recognized in Australia as a clinical entity in 1928. The virus was isolated in 1963 from mosquitoes. The disease is characterized by the sudden onset of fever. About 95% of these cases report joint pain. This is typically symmetrical with acute onset, affecting the fingers, toes, ankles, wrists, back, knees and elbows. Fatigue, myalgia and headache are common during the acute illness. A maculopapular rash occurs in 50% of patients. The acute fever is followed frequently by post-illness arthralgia. In most patients there is steady improvement over a period of months to a year or more. Chronic fatigue is reported in some patients (Suhrbier and La Linn, 2004).

Most cases occur in Queensland, Western Australia and the Northern Territory. Cases commonly occur in adults. Areas near the breeding sites of vector mosquitoes, *Culex annulirostritis*, *Aedes vigilax* and *Ae camptorynchus* are at risk of infection (Jacups *et al.*, 2008). The main reservoir hosts are kangaroos and wallabies, although horses, opossums, birds and flying foxes may play a role. Cases occur during the wet season in different parts of Australia. Epidemic polyarthritis is limited to Australia, Fiji and isoland of the South Pacific including American Samoa and the Cook Islands. The largest outbreak occurred in 1979–1980 in the Western Pacific and affected more than 60,000 people. Recent RRF outbreaks have occurred at 3- to 4-year intervals, averaging 400 notifications per epidemic and predominantly affecting individuals 30–50 years of age with equal numbers of males and females (http://www.health.gov.au/internet/main/publishing.nsf/content.cc/cda-cdi2903i.htm, accessed 3 June 2014).

Vaccine development status

Binary ethylenimine (BEI) was used to destroy the infectivity of RRV without abolishing the antigenicity or immunogenicity of the virion. Mice immunized intramuscularly with BEI-inactivated virus, with or without alhydrogel adjuvant, produced antibody which neutralized Ross River virus *in vitro*, and the mice also failed to develop viraemia when challenged intravenously with live virus. Serum neutralization and *in vivo* protection were greatest when BEI-inactivated virus was administered without adjuvant (Yu and Aaskov, 1994). Antibody from BALB/c mice immunized with this vaccine neutralized all strains of Ross River virus tested, *in vitro*, albeit to different degrees (Aaskov *et al.*, 1997).

RRV grown in protein-free cell culture and inactivated using formalin plus UV, purified and concentrated by ultracentrifugation produced an immunogenic and protective vaccine (Holzer *et al.*, 2011). After active immunizations, the vaccine protected adult mice from viraemia while interferon alpha/beta receptor knock-out mice were protected from death and disease. In passive transfer studies, humans given vaccine raised antibodies that protected adult mice from viraemia and young mice from development of arthritic signs similar to human RRV-induced disease. Based on the good correlation between antibody titres in human sera and protection of animals, a correlate of protection was defined. This correlate is of particular importance for the evaluation of the vaccine because of the low incidence of RRV disease in Australia renders a classical efficacy trial impractical. Because RRV grows in human monocytes/macrophages it was important that antibody-dependent enhancement of infection was not observed in mice even at low to undetectable concentrations of vaccine-induced antibodies.

Ross River virus (RRV) is endemic in Australia and several South Pacific Islands. This vaccine was tested in 382 healthy, RRV-naive adults in a phase 1/2 dose-escalation study at 10 sites in Austria, Belgium and the Netherlands. Subjects were randomized to receive 1.25 mg, 2.5 mg, 5 mg or 10 mg aluminium hydroxide-adjuvanted or non-adjuvanted RRV vaccine, a second dose after 3 weeks and a booster at 6 months. The optimal vaccine formulation was the adjuvanted 2.5 mg dose that produced neutralizing antibodies in 92.7% (82.2–98.0%) of vaccinees (Aichinger *et al.*, 2011).

Sindbis

Disease perspective

Sindbis infections of human adults produce a febrile viral exanthem, often including arthralgia. Infections of children are milder. The joints, especially the ankles, wrists, knees, fingers and toes, less often the hips, shoulders, elbows, neck and back are involved symmetrically. Achilles and wrist tendons may be inflamed. Joint symptoms resolve in 3–4 months in 60% of cases, but in others may persevere for 3–4 years. The virus was first isolated in 1952 in Cairo, Egypt, and is maintained in nature by transmission between birds and various species of *Culex* mosquitoes. The virus is distributed widely over four continents, Africa, Europe, Asia and Australia. Areas of Scandinavia are prone to epidemic outbreaks. Disease occurs in individuals who participate in outdoor activities during the summer months (Suhrbier *et al.*, 2012). Outbreaks in Scandinavia may result in hundreds to thousands of cases.

Vaccine development status

The technology to produce a Sindbis vaccine is available. Sindbis virus, itself, has been used as a vector or as a chimera to build vaccines against other pathogens (Center *et al.*, 2013; Dar *et al.*, 2012, 2013; Maruggi *et al.*, 2013; Roy *et al.*, 2013; Sun *et al.*, 2014; Zhu *et al.*, 2013).

Venezuelan equine encephalitis (VEE)

Disease perspective

VEE virus was isolated from an epizootic in Venezuelan horses in 1938 (Beck and Wyckoff, 1938; Kubes and Rios, 1939). VEE circulates in nature as six subtypes. Virus types I and III have multiple antigenic variants. Types IAB and IC have caused epizootics and human epidemics. Human cases were first identified in 1943. Hundreds of thousands of equine and human cases have occurred over the past 70 years. In humans, the incubation period is 2–5 days, followed by the abrupt onset of fever, chills, headache, sore throat, myalgia, malaise, prostration, photophobia, nausea, vomiting, and diarrhoea. In 5–10% of cases, there is a biphasic illness; the second phase is heralded by seizures, projectile vomiting, ataxia, confusion, agitation, and mild disturbances in consciousness. There is cervical lymphadenopathy and conjunctival suffusion. Cases of meningoencephalitis may demonstrate cranial nerve palsy, motor weakness, paralysis, seizures, and coma. Microscopic examination of tissues reveals inflammatory infiltrates in lymph nodes, spleen, lung, liver and brain. Lymph nodes show cellular depletion, necrosis of germinal centres, and lymphophagocytosis. The liver shows patchy hepatocellular degeneration, the lungs demonstrate a diffuse interstitial pneumonia with intra-alveolar haemorrhages, and the brain shows patchy cellular infiltrates.

During 1971, 1AB epizootics moved through Central America and Mexico to southern Texas. This and earlier epizootics in the 1960s are now attributed to the circulation of incompletely inactivated VEE vaccine (Aguilar *et al.*, 2011b; Weaver, 1997; Weaver *et al.*, 1996; Zacks and Paessler, 2010). After two decades of quiescence, epizootic disease emerged again in Venezuela and Colombia in 1995.This occurrence of 1C virus has been attributed to escape of virulent VEE from a laboratory (Aguilar *et al.*, 2011b; Brault *et al.*, 2001). VEEV resides in ill-defined sylvatic reservoirs in the South American rainforests. Known hosts include rodents and aquatic birds with transmission by *Culex melaconion* species. Vectors for horse-to-horse and horse-to-human transmission include *Aedes taeniorhynchus* and *Psorophora confinnis*. Epizootics, most introduced by humans, can move rapidly, up to several miles per day. Human cases are proportional to and follow epizootic occurrences. Viraemia levels in human blood are high enough to infect mosquitoes. Because virus can be recovered from human pharyngeal swabs, and household attack rates are often as high as 50%, it is widely believed that person-to-person transmission occurs, although direct evidence is lacking. VEE type 1E continues to circulate and infect horses and humans at low levels in the Gulf Coast of Mexico approaching the Texas border (Adams *et al.*, 2012; Aguilar *et al.*, 2011b). Virus types II–VI are restricted to relatively small foci; each has a unique vector–host relationship and rarely results in human infections.

Vaccines

Alarmed by VEE epizootics in Latin America, whose population relied heavily on horses and mules for agriculture and transportation, as soon as virus was isolated in 1938, 'inactivated' vaccines were prepared (Beck and Wyckoff, 1938; Kubes and Rios, 1939). Formalin-inactivated preparations were initially made from mouse brain and other animal tissues following infection with wild-type, subtype IAB strains isolated during epizootics (Paessler and Weaver, 2009). These vaccines actually caused illness and death in some animals suggesting that residual live virus remained in vaccine lots (Tigertt and Downs, 1962). This problem continued as 'inactivated' VEEV vaccines were produced and deployed separately in many enzootic countries. Sequencing studies showing conservation among all IAB strains isolated from 1938–1973 (Kinney *et al.*, 1992; Weaver *et al.*, 1999), as well as the isolation of live virus from at least one human vaccine preparation, indicated that the 'escape' of VEEV from incompletely inactivated vaccines likely occurred repeatedly resulting in repeated epizootics (Paessler and Weaver, 2009). The disappearance of subtype IAB VEEV since 1973, when inactivated vaccines were discontinued and replaced by the live-attenuated TC-83 strain, supports this conclusion.

The prototype VEEV, Trinidad Donkey 1, was isolated in 1944 by intracerebral inoculation of a guinea pig with brain tissues from a donkey with fatal encephalitis in Trinidad (Tigertt and Downs, 1962). The virus, subtype 1AB, was passaged 12 times in embryonated eggs and then further attenuated by serial passages in guinea pig heart (GPH) cells resulting in the TC-83 strain (Berge *et al.*, 1961). The virus was plaque purified in chick embryo fibroblasts (CEF) and passaged four additional times in FGPH cultures (Pittman and Plotkin, 2013). Because of frequent serious infections of laboratory workers with VEEV, many by the respiratory route, TC-83 was used to vaccinate humans as an investigational new drug product. TC-83 resulted in seroconversions in about 80% accompanied by mild to moderate flu-like symptoms in around 20% of human vaccinees. TC-83 was first tested extensively in equids during the 1971 Texas VEE epizootic where it may have contributed to limiting the spread of the epizootic northward. Although in horses the vaccine produces viraemia, fever and leucopenia, robust neutralizing antibodies are raised and horses are protected against challenge with VEEV (Paessler and Weaver, 2009).

The reactogenicity and immunogenicity of TC-83 may result from only two attenuating mutations among the 12 mutations observed following prolonged passage in GPH cultures of the Trinidad donkey strain (Kinney *et al.*, 1989). These mutations are subject to reversion in horses or humans receiving vaccine, a fact that has contributed to the reintroduction of vaccine strains into human and horse populations. TC-83 was isolated from mosquitoes collected during the 1971 VEE epizootic/epidemic in Louisiana in 1971 (Pedersen *et al.*, 1972). Efforts have been made to improve the safety of TC-83 by making viral replication dependent on an internal ribosome entry site of encephalomyelocarditis virus and inactivating the subgenomic promoter with 13 synonymous mutations (Volkova *et al.*, 2008). This recombinant virus further attenuated TC-83, but was completely unable to infect mosquito cells or mosquitoes *in vivo*.

The only other VEE vaccine administered to humans is TC-84, formalin-inactivated and made from the TC-83 production seed, TC-82, with an additional passage in CEFs. TC-84 has a local reaction rate of 5% but no systemic reactions. Currently, commercial equine vaccines marketed in the USA consist of inactivated TC-83, often combined with the other principal alphaviral encephalitides, eastern and western equine encephalitis viruses.

A large number of new VEEV vaccine constructs have been developed and have shown promise in pre-clinical testing. The genome of Sindbis virus has been used to prepare a Sindbis/VEE chimera. The vaccine is constructed to express VEE structural proteins but contain cis-acting RNA elements. SIN83 chimeric virus proved to be protective but, as opposed to TC83, did not cause any detectable clinical disease in animal models (Paessler and Weaver, 2009). Sindbis and Semlilki forest disease viruses have been used to prepare alphavirus replicon vaccines. Here the structural genes are packaged in non-VEEV viral non-structural proteins and cis-acting RNA producing a replicating particle that presents VEEV structural protein antigens. Vaccinia virus recombinants and DNA alphavirus vaccines are also in pre-clinical development (Dupuy *et al.*, 2011; Paessler and Weaver, 2009). There has been recent progress in developing a trivalent vaccine against EEE, WEE and VEE (Wolfe *et al.*, 2014).

Western equine encephalitis

Disease perspective

WEE infections occur principally in the USA and Canada west of the Mississippi River, mainly in rural areas, where water impoundments, irrigated farmland, and naturally flooded land provide breeding sites for *Culex tarsalis* (Zacks and Paessler, 2010). The virus is transmitted in a cycle involving mosquitoes, birds, and other vertebrate hosts. Humans and horses are susceptible to encephalitis. The case–infection ratio varies by age, having been estimated at 1:58 in children younger than 4 years and 1:1150 in adults. Infections are most severe at the extremes of life; a third of cases occur in children younger than 1 year. Recurrent human epidemics have been reported from the Yakima Valley in Washington State and the Central Valley of California; the largest outbreak on record resulted in 3400 cases and occurred in Minnesota, North and South Dakota, Nebraska, and Montana as well as Alberta, Manitoba and Saskatchewan, Canada. Epizootics in horses precede human epidemics by several weeks. For the past 20 years, only three cases of WEE have been reported in the USA, presumably reflecting successful mosquito abatement.

Vaccine development status

Two types of chimeric Sindbis/WEE chimeras have been developed and used as live virus vaccines. The first was derived from the structural proteins of WEEV strain CO92-1356, while a second generation vaccine included genes from EEE and was made from WEEV strain McMillan. Each of these two constructs was attenuated when inoculated intracerebrally in 6-week-old mice. Four weeks later, vaccinated mice developed no detectable disease following intranasal challenge of five logs of WEEV (Atasheva *et al.*, 2009).

References

Aaskov, J., Williams, L., and Yu, S. (1997). A candidate Ross River virus vaccine: preclinical evaluation. Vaccine *15*, 1396–1404.

Adams, A.P., Navarro-Lopez, R., Ramirez-Aguilar, F.J., Lopez-Gonzalez, I., Leal, G., Flores-Mayorga, J.M., Travassos da Rosa, A.P., Saxton-Shaw, K.D., Singh, A.J., Borland, E.M., *et al.* (2012). Venezuelan equine encephalitis virus activity in the Gulf Coast region of Mexico, 2003–2010. PLoS Negl. Trop. Dis. *6*, e1875.

Aebi, C., and Schaad, U.B. (1994). [TBE-immunoglobulins – a critical assessment of efficacy]. Schweiz. Med. Wochenschr. *124*, 1837–1840.

Aguilar, P.V., Barrett, A.D., Saeed, M.F., Watts, D.M., Russell, K., Guevara, C., Ampuero, J.S., Suarez, L., Cespedes, M., Montgomery, J.M., *et al.* (2011a). Iquitos virus: a novel reassortant Orthobunyavirus associated with human illness in Peru. PLoS Negl. Trop. Dis. *5*, e1315.

Aguilar, P.V., Estrada-Franco, J.G., Navarro-Lopez, R., Ferro, C., Haddow, A.D., and Weaver, S.C. (2011b). Endemic Venezuelan equine encephalitis in the Americas: hidden under the dengue umbrella. Future Virol *6*, 721–740.

Aichinger, G., Ehrlich, H.J., Aaskov, J.G., Fritsch, S., Thomasser, C., Draxler, W., Wolzt, M., Muller, M., Pinl, F., Van Damme, P., *et al.* (2011). Safety and immunogenicity of an inactivated whole virus Vero cell-derived Ross River virus vaccine: a randomized trial. Vaccine *29*, 9376–9384.

Aitken, T.H.G., Downs, W.G., and Shope, R.E. (1977). *Aedes aegypti* strain fitness for yellow fever virus transmission. Am. J. Trop. Med. Hyg. *26*, 985–989.

Akahata, W., Yang, Z.Y., Andersen, H., Sun, S., Holdaway, H.A., Kong, W.P., Lewis, M.G., Higgs, S., Rossmann, M.G., Rao, S., and Nabel, G.J. (2010). A virus-like particle vaccine for epidemic Chikungunya virus protects nonhuman primates against infection. Nat. Med. *16*, 334–338.

Akinci, E., Bodur, H., and Leblebicioglu, H. (2013). Pathogenesis of Crimean–Congo hemorrhagic Fever. Vector Borne Zoonotic Dis. *13*, 429–437.

Alkan, C., Bichaud, L., de Lamballerie, X., Alten, B., Gould, E.A., and Charrel, R.N. (2013). Sandfly-borne phleboviruses of Eurasia and Africa: epidemiology, genetic diversity, geographic range, control measures. Antiviral Res. *100*, 54–74.

Amicizia, D., Domnich, A., Panatto, D., Lai, P.L., Cristina, M.L., Avio, U., and Gasparini, R. (2013). Epidemiology of tick-borne encephalitis (TBE) in Europe and its prevention by available vaccines. Hum. Vaccine Immunother. *9*, 1163–1171.

Anderson, C.R., Spence, L., Downs, W.G., and Aitken, T.H. (1961). Oropouche virus: a new human disease agent from Trinidad, West Indies. Am. J. Trop. Med. Hyg. *10*, 574–578.

Anez, G., Men, R., Eckels, K.H., and Lai, C.J. (2009). Passage of dengue virus type 4 vaccine candidates in fetal rhesus lung cells selects heparin-sensitive variants that result in loss of infectivity and immunogenicity in rhesus macaques. J. Virol. *83*, 10384–10394.

Arora, U., Tyagi, P., Swaminathan, S., and Khanna, N. (2012). Chimeric Hepatitis B core antigen virus-like particles displaying the envelope domain III of dengue virus type 2. J. Nanobiotechnol. *10*, 30.

Arroyo, J., Guirakhoo, F., Fenner, S., Zhang, Z.X., Monath, T.P., and Chambers, T.J. (2001a). Molecular basis for attenuation of neurovirulence of a yellow fever Virus/Japanese encephalitis virus chimera vaccine (ChimeriVax-JE). J. Virol. *75*, 934–942.

Arroyo, J., Miller, C.A., Catalan, J., and Monath, T.P. (2001b). Yellow fever vector live-virus vaccines: West Nile virus vaccine development. Trends Mol. Med. *7*, 350–354.

Arroyo, J., Miller, C., Catalan, J., Myers, G.A., Ratterree, M.S., Trent, D.W., and Monath, T.P. (2004). ChimeriVax-West Nile virus live-attenuated vaccine: preclinical evaluation of safety, immunogenicity, and efficacy. J. Virol. *78*, 12497–12507.

Ascioglu, S., Leblebicioglu, H., Vahaboglu, H., and Chan, K.A. (2011). Ribavirin for patients with Crimean–Congo haemorrhagic fever: a systematic review and meta-analysis. J. Antimicrob. Chemother. *66*, 1215–1222.

Atasheva, S., Wang, E., Adams, A.P., Plante, K.S., Ni, S., Taylor, K., Miller, M.E., Frolov, I., and Weaver, S.C. (2009). Chimeric alphavirus vaccine candidates protect mice from intranasal challenge with western equine encephalitis virus. Vaccine *27*, 4309–4319.

Attoui, H., Mohd Jaafar, F., de Micco, P., and de Lamballerie, X. (2005). Coltiviruses and seadornaviruses in North America, Europe, and Asia. Emerg. Infect. Dis. *11*, 1673–1679.

Baisley, K.J., Watts, D.M., Munstermann, L.E., and Wilson, M.L. (1998). Epidemiology of endemic Oropouche virus transmission in upper Amazonian Peru. Am. J. Trop. Med. Hyg. *59*, 710–716.

Bancroft, W.H., Top, F.H., Jr., Eckels, K.H., Anderson, J.H., Jr., McCown, J.M., and Russell, P.K. (1981). Dengue-2 vaccine: virological, immunological, and clinical responses of six yellow fever-immune recipients. Infect. Immun. *31*, 698–703.

Bancroft, W.H., Scott, R.M., Eckels, K.H., Hoke, C.H., Jr., Simms, T.E., Jesrani, K.D., Summers, P.L., Dubois, D.R., Tsoulos, D., and Russell, P.K. (1984). Dengue virus type 2 vaccine: reactogenicity and immunogenicity in soldiers. J. Infect. Dis. *149*, 1005–1010.

Bao, H., Ramanathan, A.A., Kawalakar, O., Sundaram, S.G., Tingey, C., Bian, C.B., Muruganandam, N., Vijayachari, P., Sardesai, N.Y., Weiner, D.B., *et al.* (2013). Non-structural protein 2 (nsP2) of Chikungunya virus (CHIKV) enhances protective immunity mediated by a CHIKV envelope protein expressing DNA Vaccine. Viral. Immunol. *26*, 75–83.

Barrett, A.D., and Teuwen, D.E. (2009). Yellow fever vaccine – how does it work and why do rare cases of serious adverse events take place? Curr. Opin. Immunol. *21*, 308–313.

Barrett, P.N., Plotkin, S.A., and Ehrlich, H.J. (2008). Tick-borne encephalitis virus vaccines. In Vaccines, S.A. Plotkin, W.A. Orenstein, and P.A. Offit, eds. (Elsevier, Philadelphia), pp. 841–856.

Barrett, P.N., Portsmouth, D., and Ehrlich, H.J. (2013). Tick-borne encephalitis virus vaccines., In Vaccines, S.A. Plotkin, W.A. Orenstein, P.A. Offit, and P.J. Hotez, eds. (Elsevier Saunders, Philadelphia), pp. 773–788.

Basaca-Sevilla, V., and Halstead, S.B. (1966). Recent virological studies on haemorrhagic fever and other arthropod- borne virus infections in the Philippines. J. Trop. Med. Hyg. *69*, 203–208.

Bastos Mde, S., Figueiredo, L.T., Naveca, F.G., Monte, R.L., Lessa, N., Pinto de Figueiredo, R.M., Gimaque, J.B., Pivoto Joao, G., Ramasawmy, R., and Mourao, M.P. (2012). Identification of Oropouche Orthobunyavirus in the cerebrospinal fluid of three patients in the Amazonas, Brazil. Am. J. Trop. Med. Hyg. *86*, 732–735.

Beasley, D.W., Li, L., Suderman, M.T., Guirakhoo, F., Trent, D.W., Monath, T.P., Shope, R.E., and Barrett, A.D. (2004). Protection against Japanese encephalitis virus strains representing four genotypes by passive transfer of sera raised against ChimeriVax-JE experimental vaccine. Vaccine *22*, 3722–3726.

Beck, C.E., and Wyckoff, R.W. (1938). Venezuelan equine encephalomyelitis. Science *88*, 530.

Beeuwkes, H. (1936). Clinical manifestations of yellow fever in the West African native as observed during four extensive epidemics of the disease in the gold Coast and Nigeria. Trans. R. Soc. Trop. Med. Hyg. *30*, 61–86.

Bellanti, J.A., Bourke, A.T.C., Buescher, E.L., Cadigan, F.C., Cole, G.A., El Batawi, Y., Hatgi, J.N., McCown, J.M., Negron, H., Ordonez, J.V., *et al.* (1966). Report of dengue vaccine trial in the Caribbean, 1963: a collaborative study. Bull. World Health Org. 35, 93.

Benenson, M.W., Top, F.H., Jr., Gresso, W., Ames, C.W., and Altstatt, L.B. (1975). The virulence to man of Japanese encephalitis virus in Thailand. Am. J. Trop. Med. Hyg. *24*, 974–980.

Bennett, R.S., Gresko, A.K., Nelson, J.T., Murphy, B.R., and Whitehead, S.S. (2012). A recombinant chimeric La Crosse virus expressing the surface glycoproteins of Jamestown Canyon virus is immunogenic and protective against challenge with either parental virus in mice or monkeys. J.Virol. *86*, 420–426.

Bente, D.A., Forrester, N.L., Watts, D.M., McAuley, A.J., Whitehouse, C.A., and Bray, M. (2013). Crimean–Congo hemorrhagic fever: history, epidemiology, pathogenesis, clinical syndrome and genetic diversity. Antiviral Res. *100*, 159–189.

Berge, T.O., Banks, I.S., and Tigertt, W.D. (1961). Attenuation of Venezuelan equine encephalomyelitis virus by in vitro cultivation in guinea pig heart cells. Am. J. Hyg. *73*, 209–218.

Berry, G.P., and Kitchen, S.F. (1931). Yellow fever accidentally contracted in the laboratory: a study of seven cases. Am. J. Trop. Med. *11*, 365–434.

Bhamarapravati, N., and Yoksan, S. (1989). Study of bivalent dengue vaccine in volunteers. Lancet *1*, 1077.

Bhamarapravati, N., and Yoksan, S. (1997). Live attenuated tetravalent dengue vaccine, In Dengue and Dengue Hemorrhagic Fever, D.J. Gubler, and G. Kuno, eds. (CAB International, New York), pp. 367–377.

Bhamarapravati, N., and Yoksan, S. (2000). Live attenuated tetravalent dengue vaccine. Vaccine *18*(Suppl.), 44–47.

Bhamarapravati, N., Yoksan, S., Chayaniyayothin, T., Angsubphakorn, S., and Bunyaratvej, A. (1987). Immunization with a live attenuated dengue-2-virus candidate vaccine (16681-PDK 53): clinical, immunological and biological responses in adult volunteers. Bull. World Health Org. *65*, 189–195.

Bhatt, S., Gething, P.W., Brady, O.J., Messina, J.P., Farlow, A.W., Moyes, C.L., Drake, J.M., Brownstein, J.S., Hoen, A.G., Sankoh, O., *et al.* (2013). The global distribution and burden of dengue. Nature *496*, 504–507.

Bhatt, T.R., Crabtree, M.B., Guirakhoo, F., Monath, T.P., and Miller, B.R. (2000). Growth characteristics of the chimeric Japanese encephalitis virus vaccine candidate, ChimeriVax-JE (YF/JE SA14-14-2), in *Culex tritaeniorhynchus*, *Aedes albopictus*, and *Aedes aegypti* mosquitoes. Am. J. Trop. Med. Hyg. *62*, 480–484.

Bhattachan, A., Amatya, S., Sedai, T.R., Upreti, S.R., and Partridge, J. (2009). Japanese encephalitis in hill and mountain districts, Nepal. Emerg. Infect. Dis. *15*, 1691–1692.

Bista, M.B., Banerjee, M.K., Shin, S.H., Tandan, J.B., Kim, M.H., Sohn, Y.M., Ohhr, H.C., and Halstead, S.B. (2001). Efficacy of single dose SA 14-14-2 vaccine against Japanese encephalitis: a case–control study. Lancet *358*, 791–795.

Blaney, J.E., Jr., Sathe, N.S., Goddard, L., Hanson, C.T., Romero, T.A., Hanley, K.A., Murphy, B.R., and Whitehead, S.S. (2008). Dengue virus type 3 vaccine candidates generated by introduction of deletions in the 3′ untranslated region (3′-UTR) or by exchange of the DENV-3 3′-UTR with that of DENV-4. Vaccine *26*, 817–828.

Blaney, J.E., Jr., Durbin, A.P., Murphy, B.R., and Whitehead, S.S. (2010). Targeted mutagenesis as a rational approach to dengue virus vaccine development. Curr. Top. Microbiol. Immunol. *338*, 145–158.

Bonaparte, M., Dweik, B., Feroldi, E., Meric, C., Bouckenooghe, A., Hildreth, S., Hu, B., Yoksan, S., and Boaz, M. (2014). Immune response to live-attenuated Japanese encephalitis vaccine (JE-CV) neutralizes Japanese encephalitis virus isolates from south-east Asia and India. BMC Infect. Dis. *14*, 156.

Brackney, M.M., Marfin, A.A., Staples, J.E., Stallones, L., Keefe, T., Black, W.C., and Campbell, G.L. (2010). Epidemiology of Colorado tick fever in Montana, Utah, and Wyoming, 1995–2003. Vector Borne Zoonotic Dis. *10*, 381–385.

Brady, O.J., Gething, P.W., Bhatt, S., Messina, J.P., Brownstein, J.S., Hoen, A.G., Moyes, C.L., Farlow, A.W., Scott, T.W., and Hay, S.I. (2012). Refining the global spatial limits of dengue virus transmission by evidence-based consensus. PLoS Negl. Trop. Dis. *6*, e1760.

Brandler, S., and Tangy, F. (2013). Vaccines in development against West Nile virus. Viruses *5*, 2384–2409.

Brandler, S., Ruffie, C., Najburg, V., Frenkiel, M.P., Bedouelle, H., Despres, P., and Tangy, F. (2010). Pediatric measles vaccine expressing a dengue tetravalent antigen elicits neutralizing antibodies against all four dengue viruses. Vaccine *28*, 6730–6739.

Brandler, S., Ruffie, C., Combredet, C., Brault, J.B., Najburg, V., Prevost, M.C., Habel, A., Tauber, E., Despres, P., and Tangy, F. (2013). A recombinant measles vaccine expressing chikungunya virus-like particles is strongly immunogenic and protects mice from lethal challenge with chikungunya virus. Vaccine *31*, 3718–3725.

Brault, A.C., Powers, A.M., Medina, G., Wang, E., Kang, W., Salas, R.A., De Siger, J., and Weaver, S.C. (2001). Potential sources of the 1995 Venezuelan equine encephalitis subtype IC epidemic. J. Virol. *75*, 5823–5832.

Breugelmans, J.G., Lewis, R.F., Agbenu, E., Veit, O., Jackson, D., Domingo, C., Bothe, M., Perea, W., Niedrig, M., Gessner, B.D., and Yactayo, S. (2013). Adverse events following yellow fever preventive vaccination campaigns in eight African countries from 2007 to 2010. Vaccine *31*, 1819–1829.

Broker, M., and Kollaritsch, H. (2008). After a tick bite in a tick-borne encephalitis virus endemic area: current positions about post-exposure treatment. Vaccine *26*, 863–868.

Butrapet, S., Huang, C.Y., Pierro, D.J., Bhamarapravati, N., Gubler, D.J., and Kinney, R.M. (2000). Attenuation markers of a candidate dengue type 2 vaccine virus, strain 16681 (PDK-53), are defined by mutations in the 5′ noncoding region and non-structural proteins 1 and 3. J. Virol. *74*, 3011–3019.

Butrapet, S., Rabablert, J., Angsubhakorn, S., Wiriyarat, W., Huang, C., Kinney, R., Punyim, S., and Bhamarapravati, N. (2002). Chimeric dengue type 2/type 1 viruses induce immune responses in cynomolgus monkeys. Southeast Asian J. Trop. Med. Public Health *33*, 589–599.

Buttigieg, K.R., Dowall, S.D., Findlay-Wilson, S., Miloszewska, A., Rayner, E., Hewson, R., and Carroll, M.W. (2014). A novel vaccine against Crimean–Congo Haemorrhagic Fever protects 100% of animals against lethal challenge in a mouse model. PLoS One *9*, e91516.

Bylon, D. (1780). Korte aatekening, wegens eene algemeene ziekte, doorgans genaamd de knokkel-koorts. Verhandelungen van het Bataviaasch Genootschop der Konsten in Wetenschappen *2*, 17–30.

Campbell, G.L., Hills, S.L., Fischer, M., Jacobson, J.A., Hoke, C.H., Hombach, J.M., Marfin, A.A., Solomon, T., Tsai, T.F., Tsu, V.D., and Ginsburg, A.S. (2011). Estimated global incidence of Japanese encephalitis: a systematic review. Bull. World Health Org. *89*, 766–774, 774A–774E.

Campos, L.E., San Juan, A., and Cenabre, L.C. (1969). Isolation of chikungunya virus in the Philippines. Acta Med. Philippina *5*, 152–155.

Capeding, R.Z., Luna, I.A., Bomasang, E., Lupisan, S., Lang, J., Forrat, R., Wartel, A., and Crevat, D. (2011). Live-attenuated, tetravalent dengue vaccine in children, adolescents and adults in a dengue endemic country: randomized controlled phase I trial in the Philippines. Vaccine *29*, 3863–3872.

Caplen, H., Peters, C.J., and Bishop, D.H. (1985). Mutagen-directed attenuation of Rift Valley fever virus as a method for vaccine development. J. Gen. Virol. *66*, 2271–2277.

Carey, D.E. (1971). Chikungunya and dengue: A case of mistaken identity? J. Hist. Med. Allied Sci. *26*, 243–262.

Carrera, J.P., Forrester, N., Wang, E., Vittor, A.Y., Haddow, A.D., Lopez-Verges, S., Abadia, I., Castano, E., Sosa, N., Baez, C., *et al.* (2013). Eastern equine encephalitis in Latin America. N. Engl. J. Med. *369*, 732–744.

Carter, H.R. (1931). Yellow Fever, An Epidemiological and Historical Study of its Place of Origin (Williams and Wilkins, Baltimore).

CDC (2010). Japanese encephalitis vaccines: Recommendations of the Advisory Committee on Immunization Practices (ACIP). MMWR *59*, RR-1–27.

CDC (2013). Use of Japanese Encephalitis Vaccine in Children: Recommendations of the Advisory Committee on Immunization Practices, 2013. MMWR *62*, 898–900.

CDC (2014). Crimean-Congo hemorrhagic fever (CCHF). Available online: http://www.cdc.gov/vhf/crimean-congo/. Accessed 28 September 2015.

Center, R.J., Miller, A., Wheatley, A.K., Campbell, S.M., Siebentritt, C., and Purcell, D.F. (2013). Utility of the Sindbis replicon system as an Env-targeted HIV vaccine. Vaccine *31*, 2260–2266.

Chambers, T.J., Nestorowicz, A., Mason, P.W., and Rice, C.M. (1999). Yellow fever/Japanese encephalitis chimeric viruses: construction and biological properties. J. Virol. *73*, 3095–3101.

Chang, D.C., Liu, W.J., Anraku, I., Clark, D.C., Pollitt, C.C., Suhrbier, A., Hall, R.A., and Khromykh, A.A. (2008). Single-round infectious particles enhance immunogenicity of a DNA vaccine against West Nile virus. Nat. Biotechnol. *26*, 571–577.

Chang, G.J., Cropp, B.C., Kinney, R.M., Trent, D.W., and Gubler, D.J. (1995). Nucletide sequence variation of the envelope protein gene identifies two distinct genotypes of the yellow fever virus. J. Virol. *69*, 5773–5780.

Charrel, R.N., Bichaud, L., and de Lamballerie, X. (2012). Emergence of Toscana virus in the Mediterranean area. World J Virol *1*, 135–141.

Chastel, C. (1964). Human infections in Cambodia with chikungunya or a closely allied virus. III. Epidemiology. Bull. Soc. Pathol. Exot. *57*, 65–82.

Chattopadhyay, A., Wang, E., Seymour, R., Weaver, S.C., and Rose, J.K. (2013). A chimeric vesiculo/alphavirus is an effective alphavirus vaccine. J. Virol. *87*, 395–402.

Chavez, J.H., Silva, J.R., Amarilla, A.A., and Moraes Figueiredo, L.T. (2010). Domain III peptides from flavivirus envelope protein are useful antigens for serologic diagnosis and targets for immunization. Biologicals *38*, 613–618.

Chokephaibulkit, K., Sirivichayakul, C., Thisyakorn, U., Sabchareon, A., Pancharoen, C., Bouckenooghe, A., Gailhardou, S., Boaz, M., and Feroldi, E. (2010). Safety and Immunogenicity of a single administration of live-attenuated Japanese encephalitis vaccine in previously primed 2- to 5-year-olds and naive 12- to 24-month-olds: Multicenter randomized controlled trial. Pediatr. Infect. Dis. J. *29*, 1111–1117.

Christie, J. (1872). Remarks on 'kidinga Pepo': a peculiar form of exantematous disease. BMJ *1*, 577–579.

Christie, J. (1881). On epidemics of dengue fever: their diffusion and etiology. Glasgow Med. J. 3, 161–176.

Clements, D.E., Coller, B.A., Lieberman, M.M., Ogata, S., Wang, G., Harada, K.E., Putnak, J.R., Ivy, J.M., McDonell, M., Bignami, G.S., *et al.* (2010). Development of a recombinant tetravalent dengue virus vaccine: immunogenicity and efficacy studies in mice and monkeys. Vaccine *28*, 2705–2715.

Colomba, C., Saporito, L., Ciufolini, M.G., Marchi, A., Rotolo, V., De Grazia, S., Titone, L., and Giammanco, G.M. (2012). Prevalence of Toscana sandfly fever virus antibodies in neurological patients and control subjects in Sicily. N. Microbiol. *35*, 161–165.

Cui, F., Cao, H.X., Wang, L., Zhang, S.F., Ding, S.J., Yu, X.J., and Yu, H. (2013). Clinical and epidemiological study on severe fever with thrombocytopenia syndrome in Yiyuan County, Shandong Province, China. Am. J. Trop. Med. Hyg. *88*, 510–512.

Dar, P.A., Ganesh, K., Nagarajan, G., Sarika, S., Reddy, G.R., and Suryanarayana, V.V. (2012). Sindbis virus replicase-based DNA vaccine construct encoding FMDV-specific multivalent epitope gene: studies on its immune responses in guinea pigs. Scand. J. Immunol. *76*, 345–353.

Dar, P.A., Suryanaryana, V.S., Nagarajan, G., Reddy, G.R., Dechamma, H.J., and Kondabattula, G. (2013). DNA prime-protein boost strategy with replicase-based DNA vaccine against foot-and-mouth disease in bovine calves. Vet. Microbiol. *163*, 62–70.

Davis, B.S., Chang, G.J., Cropp, B., Roehrig, J.T., Martin, D.A., Mitchell, C.J., Bowen, R., and Bunning, M.L. (2001). West Nile virus recombinant DNA vaccine protects mouse and horse from virus challenge and expresses in vitro a noninfectious recombinant antigen that can be used in enzyme-linked immunosorbent assays. J. Virol. *75*, 4040–4047.

Davis, L.E., Beckham, J.D., and Tyler, K.L. (2008). North American encephalitic arboviruses. Neurol. Clin. *26*, 727–757, ix.

Dayan, G.H., Bevilacqua, J., Coleman, D., Buldo, A., and Risi, G. (2012). Phase II, dose ranging study of the safety and immunogenicity of single dose West Nile vaccine in healthy adults ≥ 50 years of age. Vaccine *30*, 6656–6664.

De Filette, M., Ulbert, S., Diamond, M., and Sanders, N.N. (2012). Recent progress in West Nile virus diagnosis and vaccination. Vet. Res. *43*, 16.

De Groot, A.S., Martin, W., Moise, L., Guirakhoo, F., and Monath, T. (2007). Analysis of ChimeriVax Japanese Encephalitis Virus envelope for T-cell epitopes and comparison to circulating strain sequences. Vaccine *25*, 8077–8084.

Despres, P., Combredet, C., Frenkiel, M.P., Lorin, C., Brahic, M., and Tangy, F. (2005). Live measles vaccine expressing the secreted form of the West Nile virus envelope glycoprotein protects against West Nile virus encephalitis. J. Infect. Dis. *191*, 207–214.

Deubel, V., Schlesinger, J.J., and Digoutte, J.P. (1987). Comparative immunochemical and biological analysis of African and South American yellow fever viruses. Arch. Virol. *94*, 331–338.

Diamond, M.S. (2009). Progress on the development of therapeutics against West Nile virus. Antiviral Res. *83*, 214–227.

Ding, F., Zhang, W., Wang, L., Hu, W., Soares Magalhaes, R.J., Sun, H., Zhou, H., Sha, S., Li, S., Liu, Q., *et al.* (2013). Epidemiologic features of severe fever with thrombocytopenia syndrome in China, 2011–2012. Clin. Infect. Dis. *56*, 1682–1683.

Dubischar-Kastner, K., Eder, S., Buerger, V., Gartner-Woelfl, G., Kaltenboeck, A., Schuller, E., Tauber, E., and Klade, C. (2010). Long-term immunity and immune response to a booster dose following vaccination with the inactivated Japanese encephalitis vaccine IXIARO, IC51. Vaccine *28*, 5197–5202.

Dudley, S.F. (1934). Can yellow fever spread into Asia? J. Trop. Med. Hyg. *37*, 273–278.

Dumpis, U., Crook, D., and Oksi, J. (1999). Tick-borne encephalitis. Clin. Infect. Dis. *28*, 882–890.

Dumre, S.P., Shakya, G., Na-Bangchang, K., Eursitthichai, V., Rudi Grams, H., Upreti, S.R., Ghimire, P., Kc, K., Nisalak, A., Gibbons, R.V., and Fernandez, S. (2013). Dengue virus and Japanese encephalitis virus epidemiological shifts in Nepal: a case of opposing trends. Am. J. Trop. Med. Hyg. *88*, 677–680.

Dunster, L.M., Gibson, C.A., Stephenson, J.R., Minor, P.D., and Barrett, A.D. (1990). Attenuation of virulence of flaviviruses following passage in HeLa cells. J. Gen. Virol. *71*, 601–607.

Dunster, L.M., Wang, H., Ryman, K.D., Miller, B.R., Watowich, S.J., Minor, P.D., and Barrett, A.D. (1999). Molecular and biological changes associated with HeLa cell attenuation of wild-type yellow fever virus. Virology *261*, 309–318.

Dupuy, A., Despres, P., Cahour, A., Girard, M., and Bouloy, M. (1989). Nucleotide sequence comparison of the genome of two 17D-204 yellow fever vaccines. Nucleic Acids Res. *17*, 3989.

Dupuy, L.C., and Reed, D.S. (2012). Nonhuman primate models of encephalitic alphavirus infection: historical review and future perspectives. Curr. Opin. Virol. *2*, 363–367.

Dupuy, L.C., Richards, M.J., Ellefsen, B., Chau, L., Luxembourg, A., Hannaman, D., Livingston, B.D., and Schmaljohn, C.S. (2011). A DNA vaccine for venezuelan equine encephalitis virus delivered by intramuscular electroporation elicits high levels of neutralizing antibodies in multiple animal models and provides protective immunity to mice and nonhuman primates. Clin. Vaccine Immunol. *18*, 707–716.

Durbin, A.P., and Whitehead, S.S. (2010). Dengue vaccine candidates in development. Curr. Top. Microbiol. Immunol. 338, 129–143.

Durbin, A.P., and Whitehead, S.S. (2011). Next-generation dengue vaccines: novel strategies currently under development. Viruses 3, 1800–1814.

Durbin, A.P., Karron, R.A., Sun, W., Vaughn, D.W., Reynolds, M.J., Perreault, J.R., Thumar, B., Men, R., Lai, C.-J., Elkins, W.R., *et al.* (2001). Attenuation and immunogenicity in humans of a live dengue virus type-4 vaccine candidate with a 30 nucleotide deletion in its 3′-untranslated region. Am. J. Trop. Med. Hyg. *65*, 405–413.

Durbin, A.P., McArthur, J., Marron, J.A., Blaney, J.E., Jr., Thumar, B., Wanionek, K., Murphy, B.R., and Whitehead, S.S. (2006a). The live attenuated dengue serotype 1 vaccine rDEN1Delta30 is safe and highly immunogenic in healthy adult volunteers. Hum. Vaccine *2*, 167–173.

Durbin, A.P., McArthur, J.H., Marron, J.A., Blaney, J.E., Thumar, B., Wanionek, K., Murphy, B.R., and Whitehead, S.S. (2006b). rDEN2/4Delta30(ME), a live attenuated chimeric dengue serotype 2 vaccine is safe and highly immunogenic in healthy dengue-naive adults. Hum. Vaccine *2*, 255–260.

Durbin, A.P., Kirkpatrick, B.D., Pierce, K.K., Schmidt, A.C., and Whitehead, S.S. (2011a). Development and clinical evaluation of multiple investigational monovalent DENV vaccines to identify components for inclusion in a live attenuated tetravalent DENV vaccine. Vaccine *29*, 7242–7250.

Durbin, A.P., Whitehead, S.S., Shaffer, D., Elwood, D., Wanionek, K., Thumar, B., Blaney, J.E., Murphy, B.R., and Schmidt, A.C. (2011b). A single dose of the DENV-1 candidate vaccine rDEN1Delta30 is strongly immunogenic and induces resistance to a second dose in a randomized trial. PLoS Negl. Trop. Dis. *5*, e1267.

Durbin, A.P., Kirkpatrick, B.D., Pierce, K.K., Elwood, D., Larsson, C.J., Lindow, J.C., Tibery, C., Sabundayo, B.P., Shaffer, D., Talaat, K.R., *et al.* (2013a). A single dose of any of four different live attenuated tetravalent dengue vaccines is safe and immunogenic in flavivirus-naive adults: a randomized, double-blind clinical trial. J. Infect. Dis. *207*, 957–965.

Durbin, A.P., Wright, P.F., Cox, A., Kagucia, W., Elwood, D., Henderson, S., Wanionek, K., Speicher, J., Whitehead, S.S., and Pletnev, A.G. (2013b). The live attenuated chimeric vaccine rWN/DEN4Delta30 is well-tolerated and immunogenic in healthy flavivirus-naive adult volunteers. Vaccine *31*, 5772–5777.

Eckels, K.H., Brandt, W.E., Harrison, V.R., McCown, J.M., and Russell, P.K. (1976). Isolation of a temperature-sensitive dengue-2 virus under conditions suitable for vaccine development. Infect. Immun. *14*, 1221–1227.

Eckels, K.H., Harrison, V.R., Summers, P.L., and Russell, P.K. (1980). Dengue-2 vaccine: preparation from a small-plaque virus clone. Infect. Immun. *27*, 175–180.

Eckels, K.H., Scott, R.M., Bancroft, W.H., Brown, J., Dubois, D.R., Summers, P.L., Russell, P.K., and Halstead, S.B. (1984). Selection of attenuated dengue 4 viruses by serial passage in primary kidney cells. V. Human response to immunization with a candidate vaccine prepared in fetal rhesus lung cells. Am. J. Trop. Med. Hyg. *33*, 684–689.

Eckels, K.H., Yu, Y.X., Dubois, D.R., Marchette, N.J., Trent, D.W., and Johnson, A.J. (1988). Japanese encephalitis virus live-attenuated vaccine, Chinese strain SA14-14-2; adaptation to primary canine kidney cell cultures and preparation of a vaccine for human use. Vaccine *6*, 513–518.

Eckels, K.H., Dubois, D.R., Putnak, R., Vaughn, D.W., Innis, B.L., Henchal, E.A., and Hoke, C.H., Jr. (2003). Modification of dengue virus strains by passage in primary dog kidney cells: preparation of candidate vaccines and immunization of monkeys. Am. J. Trop. Med. Hyg. *69*, 12–16.

Edelman, R., Schneider, R.J., Chieowanich, P., Pornpibul, R., and Voodhikul, P. (1975). The effect of dengue virus infection on the clinical sequelae of Japanese encephalitis: a one year follow-up study in Thailand. Southeast Asian J. Trop. Med. Public Health *6*, 308–315.

Edelman, R., Tacket, C.O., Wasserman, S.S., Vaughn, D.W., Eckels, K.H., Dubois, D.R., Summers, P.L., and Hoke, C.H., Jr. (1994). A live attenuated dengue-1 vaccine candidate (45AZ5) passaged in primary dog kidney cell culture is attenuated and immunogenic for humans. J. Infect. Dis. *170*, 1448–1455.

Edelman, R., Tacket, C.O., Wasserman, S.S., Bodison, S.A., Perry, J.G., and Mangiafico, J.A. (2000). Phase II safety and immunogenicity study of live chikungunya virus vaccine TSI-GSD-218. Am. J. Trop. Med. Hyg. *62*, 681–685.

Edelman, R., Wasserman, S.S., Bodison, S.A., Putnak, R.J., Eckels, K.H., Tang, D., Kanesa-Thasan, N., Vaughn, D.W., Innis, B.L., and Sun, W. (2003). Phase I trial of 16 formulations of a tetravalent live-attenuated dengue vaccine. Am. J. Trop. Med. Hyg. *69*, 48–60.

Ehrlich, H.J., Pavlova, B.G., Fritsch, S., Poellabauer, E.M., Loew-Baselli, A., Obermann-Slupetzky, O., Maritsch, F., Cil, I., Dorner, F., and Barrett, P.N. (2003). Randomized, phase II dose-finding studies of a modified tick-borne encephalitis vaccine: evaluation of safety and immunogenicity. Vaccine *22*, 217–223.

Elias, C., Okwo-Bele, J.M., and Fischer, M. (2009). A strategic plan for Japanese encephalitis control by 2015. Lancet Infect. Dis. *9*, 7.

Engel, A.R., Vasconcelos, P.F., McArthur, M.A., and Barrett, A.D. (2006). Characterization of a viscerotropic yellow fever vaccine variant from a patient in Brazil. Vaccine *24*, 2803–2809.

Erra, E.O., Askling, H.H., Yoksan, S., Rombo, L., Riutta, J., Vene, S., Lindquist, L., Vapalahti, O., and Kantele, A. (2013). Cross-protective capacity of Japanese encephalitis (JE) vaccines against circulating heterologous JE virus genotypes. Clin. Infect. Dis. *56*, 267–270.

Faheem, M., Raheel, U., Riaz, M.N., Kanwal, N., Javed, F., us Sahar Sadaf Zaidi, N., and Qadri, I. (2011). A molecular evaluation of dengue virus pathogenesis and its latest vaccine strategies. Mol. Biol. Rep. *38*, 3731–3740.

Feroldi, E., Pancharoen, C., Kosalaraksa, P., Watanaveeradej, V., Phirangkul, K., Capeding, M.R., Boaz, M., Gailhardou, S., and Bouckenooghe, A. (2012). Single-dose, live-attenuated Japanese encephalitis vaccine in children aged 12–18 months: randomized, controlled phase 3 immunogenicity and safety trial. Hum. Vaccine Immunother. *8*, 929–937.

Feroldi, E., Pancharoen, C., Kosalaraksa, P., Chokephaibulkit, K., Boaz, M., Meric, C., Hutagalung, Y., and Bouckenooghe, A. (2014). Primary immunization of infants and toddlers in Thailand with Japanese encephalitis chimeric virus vaccine in comparison with SA14-14-2: a randomized study of immunogenicity and safety. Pediatr. Infect. Dis. J. *33*, 643–649.

Fitzgeorge, R., and Bradish, C.J. (1980). The in vivo differentiation of strains of yellow fever virus in mice. J. Gen. Virol. *46*, 1–13.

Gajanana, A., Thenmozhi, V., Samuel, P.P., and Reuben, R. (1995). A community-based study of subclinical flavivirus infections in children in an area of Tamil Nadu, India, where Japanese encephalitis is endemic. Bull. World Health Org. *73*, 237–244.

Galler, R., Post, P.R., Santos, C.N., and Ferreira, I.I. (1998). Genetic variability among yellow fever virus 17D substrains. Vaccine *16*, 1024–1028.

Gatchalian, S., Yao, Y., Zhou, B., Zhang, L., Yoksan, S., Kelly, K., Neuzil, K.M., Yaich, M., and Jacobson, J. (2008). Comparison of the immunogenicity and safety of measles vaccine administered alone or with live, attenuated Japanese encephalitis SA 14-14-2 vaccine in Philippine infants. Vaccine *26*, 2234–2241.

Gibbons, R.V., Kalanarooj, S., Jarman, R.G., Nisalak, A., Vaughn, D.W., Endy, T.P., Mammen, M.P., Jr., and Srikiatkhachorn, A. (2007). Analysis of repeat hospital admissions for dengue to estimate the frequency of third or fourth dengue infections resulting in admissions and dengue hemorrhagic fever, and serotype sequences. Am. J. Trop. Med. Hyg. *77*, 910–913.

Gould, E.A., and Buckley, A. (1989). Antibody-dependent enhancement of yellow fever and Japanese encephalitis virus neurovirulence. J. Gen. Virol. *70*, 1605–1608.

Grandadam, M., Caro, V., Plumet, S., Thiberge, J.M., Souares, Y., Failloux, A.B., Tolou, H.J., Budelot, M., Cosserat, D., Leparc-Goffart, I., and Despres, P. (2011). Chikungunya virus, southeastern France. Emerg. Infect. Dis. *17*, 910–913.

Gray, T.J., and Webb, C.E. (2014). A review of the epidemiological and clinical aspects of West Nile virus. Int. J. Gen. Med. *7*, 193–203.

Green, S., Vaughn, D.W., Kalayanarooj, S., Nimmannitya, S., Suntayakorn, S., Nisalak, A., Lew, R., Innis, B.L., Kurane, I., Rothman, A.L., and Ennis, F.A. (1999). Early immune activation in acute dengue illness is related to development of plasma leakage and disease severity. J. Infect. Dis. *179*, 755–762.

Gritsun, T.S., Lashkevich, V.A., and Gould, E.A. (2003). Tick-borne encephalitis. Antiviral Res. *57*, 129–146.

Grossman, R.A., Edelman, R., Chieowanich, P., Voodhikul, P., and Siriwan, C. (1973). Study of Japanese encephalitis virus in Chiangmai valley, Thailand. II.Human clinical infections. Am. J. Epidemiol. *98*, 121–132.

Gu, P.W., and Ding, Z.F. (1987). Inactivated Japanese encephalitis (JE) vaccine made from hamster kidney cell culture (a review). JE and Hers Bull. *2*, 15–26.

Guirakhoo, F., and Pugachev, K.V. (2002). Viremia and immunogenicity in nonhuman primates of a tetravalent yellow fever-dengue chimeric vaccine: genetic reconstructions, dose adjustment, and antibody responses against wild-type dengue virus isolates. Virology *298*, 146–159.

Guirakhoo, F., Zhang, Z.X., Chambers, T.J., Delagrave, S., Arroyo, J., Barrett, A.D., and Monath, T.P. (1999). Immunogenicity, genetic stability, and protective efficacy of a recombinant, chimeric yellow fever-Japanese encephalitis virus (ChimeriVax-JE) as a live, attenuated vaccine candidate against Japanese encephalitis. Virology *257*, 363–372.

Guirakhoo, F., Weltzin, R., Chambers, T.J., Zhang, Z.X., Soike, K., Ratterree, M., Arroyo, J., Georgakopoulos, K., Catalan, J., and Monath, T.P. (2000). Recombinant chimeric yellow fever-dengue type 2 virus is immunogenic and protective in nonhuman primates. J. Virol. *74*, 5477–5485.

Guirakhoo, F., Arroyo, J., Pugachev, K.V., Miller, C., Zhang, Z.X., Weltzin, R., Georgakopoulos, K., Catalan, J., Ocran, S., Soike, K., *et al.* (2001). Construction, safety, and immunogenicity in nonhuman primates of a chimeric yellow fever-dengue virus tetravalent vaccine. J. Virol. *75*, 7290–7304.

Guirakhoo, F., Pugachev, K., Zhang, Z., Myers, G., Levenbook, I., Draper, K., Lang, J., Ocran, S., Mitchell, F., Parsons, M., *et al.* (2004). Safety

and efficacy of chimeric yellow Fever-dengue virus tetravalent vaccine formulations in nonhuman primates. J. Virol. 78, 4761–4775.
Guirakhoo, F., Kitchener, S., Morrison, D., Forat, R., McCarthy, K., Nichols, R., and Yoksan, S. (2006). Live attenuated chimeric yellow fever dengue type 2 (ChimeriVax-DEN2) vaccine: Phase I clinical trial for safety and immunogenicity: effect of yellow fever pre-immunity in induction of cross neutralizing antibody responses to all 4 dengue serotypes. Hum. Vaccine 2, 60–67.Guy, B. (2009). Immunogenicity of sanofi pasteur tetravalent dengue vaccine. J. Clin. Virol. 46(Suppl 2), S16–S19.
Guy, B., and Almond, J.W. (2008). Towards a dengue vaccine: progress to date and remaining challenges. Comp. Immunol. Microbiol. Infect. Dis. 31, 239–252.
Guy, B., Guirakhoo, F., Barban, V., Higgs, S., Monath, T.P., and Lang, J. (2010). Preclinical and clinical development of YFV 17D-based chimeric vaccines against dengue, West Nile and Japanese encephalitis viruses. Vaccine 28, 632–649.
Guy, B., Barrere, B., Malinowski, C., Saville, M., Teyssou, R., and Lang, J. (2011). From research to phase III: Preclinical, industrial and clinical development of the Sanofi Pasteur tetravalent dengue vaccine. Vaccine 29, 7229–7241.
Guzman, M.G., Hermida, L., Bernardo, L., Ramirez, R., and Guillen, G. (2010). Domain III of the envelope protein as a dengue vaccine target. Exp. Rev. Vaccines 9, 137–147.
Gyure, K.A. (2009). West Nile virus infections. J. Neuropathol. Exp. Neurol. 68, 1053–1060.
Halsey, E.S., Siles, C., Guevara, C., Vilcarromero, S., Jhonston, E.J., Ramal, C., Aguilar, P.V., and Ampuero, J.S. (2013). Mayaro virus infection, Amazon Basin region, Peru, 2010–2013. Emerg. Infect. Dis. 19, 1839–1842.
Halstead, S.B. (1970). Observations related to pathogenesis of dengue hemorrhagic fever. VI. Hypotheses and discussion. Yale J. Biol. Med. 42, 350–362.
Halstead, S.B. (1978). Studies on the attenuation of dengue 4. Asian J. Infect. Dis. 2, 112–117.
Halstead, S.B. (1982). Immune enhancement of viral infection. Prog. Allergy 31, 301–364.
Halstead, S.B. (2003). Neutralization and antibody dependent enhancement of dengue viruses. Adv. Virus. Res. 60, 421–467.
Halstead, S.B. (2008). Dengue: overview and history. In Dengue, S.B. Halstead, ed. (Imperial College Press, London), pp. 1–28.
Halstead, S.B. (2012). Dengue vaccine development: a 75% solution? Lancet 380, 1535–1536.
Halstead, S.B. (2013). Identifying protective dengue vaccines: Guide to mastering an empirical process. Vaccine 31, 4501–4507.
Halstead, S.B. (2014a). Chikungunya. In Feigin and Cherry's Textbook of Pediatric Infectious Diseases, J.D. Cherry, G.J. Harrison, S.L. Kaplan, W.J. Steinbach, P.J. Hotez, eds. (Elsevier Saunders, Philadelphia), pp. 2241–2248.
Halstead, S.B. (2014b). Dengue vaccines. In Dengue and Dengue Hemorrhagic Fever, D.J. Gubler, ed. (CAB International, Wallingford, UK), pp. 551–579.
Halstead, S.B., and Grosz, C.R. (1962). Subclinical Japanese encephalitis I. Infection of Americans with limited residence in Korea. Am. J. Hyg. 75, 190–201.
Halstead, S.B., and Jacobson, J. (2008). Japanese encephalitis vaccines. In Vaccines, 5th ed., S.A. Plotkin, W.A. Orenstein, and P.A. Offit, eds. (Elsevier, Philadelphia), pp. 311–352.
Halstead, S.B., and Marchette, N. (2003). Biological properties of dengue viruses following serial passage in primary dog kidney cells: studies at the University of Hawaii. Am. J. Trop. Med. Hyg. 69, 5–11.
Halstead, S.B., and Thomas, S.J. (2010). Japanese encephalitis: new options for active immunization. Clin. Infect. Dis. 50, 1155–1164.
Halstead, S.B., and Thomas, S.J. (2013). Dengue vaccines. In Vaccines, S.A. Plotkin, W. Orenstein, and P.A. Offit, eds. (Elsevier Saunders, Philadelphia), pp. 1042–1051.
Halstead, S.B., and Tsai, T.F. (2004). Japanese encephalitis vaccines. In Vaccines, S.A. Plotkin, and W.A. Orenstein, eds. (Elsevier, Philadelphia), pp. 919–958.
Halstead, S.B., Scanlon, J., Umpaivit, P., and Udomsakdi, S. (1969). Dengue and chikungunya virus infection in man in Thailand, 1962–1964: IV. Epidemiologic studies in the Bangkok metropolitan area. Am. J. Trop. Med. Hyg. 18, 997–1021.
Halstead, S.B., Diwan, A., Marchette, N.J., Palumbo, N.E., and Srisukonth, L. (1984a). I. Attributes of uncloned virus at different passage levels. Am. J. Trop. Med. Hyg. 33, 654–665.
Halstead, S.B., Eckels, K.H., Putvatana, R., Larsen, L.K., and Marchette, N.J. (1984b). Selection of attenuated dengue 4 viruses by serial passage in primary kidney cells. IV. Characterization of a vaccine candidate in fetal rhesus lung cells. Am. J. Trop. Med. Hyg. 33, 679–683.
Halstead, S.B., Lan, N.T., Myint, T.T., Shwe, T.N., Nisalak, A., Soegijanto, S., Vaughn, D.W., and Endy, T. (2002). Infant dengue hemorrhagic fever: research opportunities ignored. Emerg. Infect. Dis. 12, 1474–1479.
Halstead, S.B., Jacobson, J., and Dubischar-Kastner, K. (2013). Japanese encephalitis vaccines. In Vaccines, 6th edn, S.A. Plotkin, W.Orenstein, and P.A. Offit, eds. (Elsevier Saunders, Philadelphia), pp. 312–351.
Hammon, W.M., Tigertt, W.D., Sather, G.E., Berge, T.O., and Meiklejohn, G. (1958). Epidemiologic studies of concurrent 'virgin' epidemics of Japanese B encephalitis and of mumps on Guam, 1947–1948, with subsequent observations including dengue, through 1957. Am. J. Trop. Med. Hyg. 7, 441–468.
Harinasuta, C., Nimmanitya, S., and Titsyakorn, U. (1985). The effect of interferon-alpha A on two cases of Japanese encephalitis in Thailand. Southeast Asian J. Trop. Med. Public Health 16, 332–336.
Harnett, G.B., and Bucens, M.R. (1990). Isolation of chikungunya virus in Australia. Med. J. Aust. 152, 328–329.
Harrison, V.R., Binn, L.N., and Randall, R. (1967). Comparative immunogenicities of chikungunya vaccines prepared in avian a mammalian tissues. Am. J. Trop. Med. Hyg. 16, 786–791.
Harrison, V.R., Eckels, K.H., Bartelloni, P.J., and Hampton, C. (1971). Production and evaluation of a formalin-killed chikungunya vaccine. J. Immunol. 107, 643–647.
Harrison, V.R., Eckels, K.H., Sagartz, J.W., and Russell, P.K. (1977). Virulence and immunogenicity of a temperature-sensitive dengue-2 virus in lower primates. Infect. Immun. 18, 151–156.
Heinz, F.X., Holzmann, H., Essl, A., and Kundi, M. (2007). Field effectiveness of vaccination against tick-borne encephalitis. Vaccine 25, 7559–7567.
Henderson, B.E., Cheshire, P.P., Kirya, G.B., and Lule, M. (1970). Immunological studies with yellow fever and selected African group B arboviruses in rhesus and vervet monkeys. Am. J. Trop. Med. Hyg. 19, 110–119.
Hennessy, S., Zhengle, L., Tsai, T.F., Strom, B.L., Wan, C.M., Liu, H.L., Wu, T.X., Yu, H.J., Liu, Q.M., Karabatsos, N., *et al.* (1996). Effectiveness of live-attenuated Japanese encephalitits vaccine (SA14-14-2): a case control study. Lancet 347, 1583–1571.
Hoke, C.H., Jr., Malinoski, F.J., Eckels, K.H., Scott, R.M., Dubois, D.R., Summers, P.L., Simms, T., Burrous, J., Hasty, S.E., and Bancroft, W.H. (1990). Preparation of an attenuated dengue 4 (341750 Carib) virus vaccine. II. Safety and immunogenicity in humans. AmJTropMedHyg 43, 219–226.
Hoke, C.H., Jr., Nisalak, A., Sangawhipa, N., Jatanasen, S., Laorakpongse, T., Innis, B.L., Kotchasenee, S., Gingrich, J.B., Latendresse, J., Fukai, K., and Burke, D.S. (1988). Protection against Japanese encephalitis by inactivated vaccines. N. Engl. J. Med. 319, 608–614.
Hollidge, B.S., Gonzalez-Scarano, F., and Soldan, S.S. (2010). Arboviral encephalitides: transmission, emergence, and pathogenesis. J. Neuroimmune Pharmacol. 5, 428–442.
Holman, D.H., Wang, D., Raviprakash, K., Raja, N.U., Luo, M., Zhang, J., Porter, K.R., and Dong, J.Y. (2007). Two complex, adenovirus-based vaccines that together induce immune responses to all four dengue virus serotypes. Clin. Vaccine Immunol. 14, 182–189.
Holzer, G.W., Coulibaly, S., Aichinger, G., Savidis-Dacho, H., Mayrhofer, J., Brunner, S., Schmid, K., Kistner, O., Aaskov, J.G., Falkner, F.G., *et al.* (2011). Evaluation of an inactivated Ross River virus vaccine in active and passive mouse immunization models and establishment of a correlate of protection. Vaccine 29, 4132–4141.
Hotta, S. (1952). Experimental Studies in Dengue I. Isolation, identification and modification of the virus. J. Infect. Dis. 90, 1–9.
Hotta, S. (1954). Experiments of active immunization against dengue with mouse-passaged unmodified virus. Acta Trop. 11, 97–104.
Hotta, S. (1969). Dengue and Related Hemorrhagic Diseases (Warren H. Green, Inc., St. Louis).
Huang, C.Y., Butrapet, S., Pierro, D.J., Chang, G.J., Hunt, A.R., Bhamarapravati, N., Gubler, D.J., and Kinney, R.M. (2000). Chimeric dengue type 2 (vaccine strain PDK-53)/dengue type 1 virus as a

potential candidate dengue type 1 virus vaccine. J. Virol. *74*, 3020–3028.

Huang, C.Y., Butrapet, S., Tsuchiya, K.R., Bhamarapravati, N., Gubler, D.J., and Kinney, R.M. (2003). Dengue 2 PDK-53 virus as a chimeric carrier for tetravalent dengue vaccine development. J. Virol. *77*, 11436–11447.

Huang, L.M., Lin, T.Y., Chiu, C.H., Chiu, N.C., Chen, P.Y., Yeh, S.J., Boaz, M., Hutagalung, Y., Bouckenooghe, A., and Feroldi, E. (2014). Concomitant administration of live attenuated Japanese encephalitis chimeric virus vaccine (JE-CV) and measles, mumps, rubella (MMR) vaccine: Randomized study in toddlers in Taiwan. Vaccine *32*, 5363–5369.

Hubalek, Z., Rudolf, I., and Nowotny, N. (2014). Arboviruses pathogenic for domestic and wild animals. Adv. Virus Res. *89*, 201–275.

Ikegami, T., and Makino, S. (2009). Rift valley fever vaccines. Vaccine *27(Suppl 4)*, D69–D72.

Immunization Practices Advisory Committee (ACIP) (1993). Inactivated Japanese encepahltis virus vaccine: recommendations of the ACIP. MMWR Morb. Mortal. Wkly. Rep. *42*, 1–15.

Indran, S.V., and Ikegami, T. (2012). Novel approaches to develop Rift Valley fever vaccines. Front Cell Infect. Microbiol. *2*, 131.

Iyer, A.V., and Kousoulas, K.G. (2013). A review of vaccine approaches for West Nile virus. Int. J. Environ. Res. Public Health *10*, 4200–4223.

Izquierdo, A., Garcia, A., Lazo, L., Gil, L., Marcos, E., Alvarez, M., Valdes, I., Hermida, L., Guillen, G., and Guzman, M.G. (2014). A tetravalent dengue vaccine containing a mix of domain III-P64k and domain III-capsid proteins induces a protective response in mice. Arch Virol. *159*, 2597–2604.

Jacups, S.P., Whelan, P.I., and Currie, B.J. (2008). Ross River virus and Barmah Forest virus infections: a review of history, ecology, and predictive models, with implications for tropical northern Australia. Vector Borne Zoonotic Dis. *8*, 283–297.

Jennings, A.D., Whitby, J.E., Minor, P.D., and Barrett, A.D.T. (1993). Comparison of the nucleotide and deduced amino acid sequences of the structural protein genes of the yellow fever 17DD vaccine strain from Senegal with those of other yellow fever vaccine viruses. Vaccine *11*, 679–681.

Jennings, A.D., Gibson, C.A., Miller, B.R., Mathews, J.H., Mitchell, C.J., Roehrig, J.T., Wood, D.J., Taffs, F., Sil, B.K., Whitby, S.N., *et al.* (1994). Analysis of a yellow fever virus isolated from a fatal case of vaccine-associated human encephalitis. J. Infect. Dis. *169*, 512–518.

Kaltenbock, A., Dubischar-Kastner, K., Eder, G., Jilg, W., Klade, C., Kollaritsch, H., Paulke-Korinek, M., von Sonnenburg, F., Spruth, M., Tauber, E., *et al.* (2009). Safety and immunogenicity of concomitant vaccination with the cell-culture based Japanese encephalitis vaccine IC51 and the hepatitis A vaccine HAVRIX1440 in healthy subjects: A single-blind, randomized, controlled Phase 3 study. Vaccine *27*, 4483–4489.

Kanesa-Thasan, N., Sun, W., Kim-Ahn, G., Van Albert, S., Putnak, J.R., King, A., Raengsakulsrach, B., Christ-Schmidt, H., Gilson, K., Zahradnik, J.M., *et al.* (2001). Safety and immunogenicity of attenuated dengue virus vaccines (Aventis Pasteur) in human volunteers. Vaccine *19*, 3179–3188.

Kanesa-Thasan, N., Edelman, R., Tacket, C.O., Wasserman, S.S., Vaughn, D.W., Coster, T.S., Kim-Ahn, G.J., Dubois, D.R., Putnak, J.R., King, A., *et al.* (2003). Phase 1 studies of Walter Reed Army Institute of Research candidate attenuated dengue vaccines: selection of safe and immunogenic monovalent vaccines. Am. J. Trop. Med. Hyg. *69*, 17–23.

Kari, K., Liu, W., Gautama, K., Mammen, M.P., Jr., Clemens, J.D., Nisalak, A., Subrata, K., Kim, H.K., and Xu, Z.Y. (2006). A hospital-based surveillance for Japanese encephalitis in Bali, Indonesia. BMC Med. *4*, 8.

Kerr, J.A. (1951). The clinical aspects and diagnosis of yellow fever. In Yellow Fever, G.K. Strode, ed. (McGraw-Hill Book Company, Inc., New York), pp. 389–425.

Khanam, S., Pilankatta, R., Khanna, N., and Swaminathan, S. (2009). An adenovirus type 5 (AdV5) vector encoding an envelope domain III-based tetravalent antigen elicits immune responses against all four dengue viruses in the presence of prior AdV5 immunity. Vaccine *27*, 6011–6021.

Kimura, K., Dosaka, A., Hashimoto, Y., Yasunaga, T., Uchino, M., and Ando, M. (1997). Single-photon emission CT findings in acute Japanese encephalitis. Am. J. Neuroradiol. *18*, 465–469.

Kimura, R., and Hotta, S. (1944). On the inoculation of dengue virus into mice. Nippon Igaku *3379*, 629–633.

Kinney, R.M., and Huang, C.Y. (2001). Development of new vaccines against dengue fever and Japanese encephalitis. Intervirology *44*, 176–197.

Kinney, R.M., Johnson, B.J., Welch, J.B., Tsuchiya, K.R., and Trent, D.W. (1989). The full-length nucleotide sequences of the virulent Trinidad donkey strain of Venezuelan equine encephalitis virus and its attenuated vaccine derivative, strain TC-83. Virology *170*, 19–30.

Kinney, R.M., Tsuchiya, K.R., Sneider, J.M., and Trent, D.W. (1992). Molecular evidence for the origin of the widespread Venezuelan equine encephalitis epizootic of 1969 to 1972. J. Gen. Virol. *73*, 3301–3305.

Kinney, R.M., Butrapet, S., Chang, G.J., Tsuchiya, K.R., Roehrig, J.T., Bhamarapravati, N., and Gubler, D.J. (1997). Construction of infectious cDNA clones for dengue 2 virus: strain 16681 and its attenuated vaccine derivative, strain PDK-53. Virology *230*, 300–308.

Kinney, R., Kinney, C.Y.H., Barban, V., Lang, J., and Guy, B. (2010a). Dengue serotype 1 attenuated strain, US P. Office, ed. (US Sanofi Pasteur Center for Disease Control and Prevention), pp. 1–70.

Kinney, R., Kinney, C.Y.H., Barban, V., Lang, J., and Guy, B. (2010b). Dengue serotype 2 attenuated strain, U.P. Office, ed. (US: Sanofi Pasteur Center for Disease Control and Prevention), pp. 1–77.

Kiple, K.F. (1984). The Caribbean Slave: A biological History (Cambridge University Press, Cambridge).

Kiple, K.F., and Kiple, V.H. (1977). Black yellow fever immunities, innate and acquired. Soc. Sci. Hist. *2*, 419–436.

Kiple, K.F., and Ornelas, K.C. (1996). Race, war and tropical medicine in the eighteenth-century Caribbean. Clio. Med. *35*, 65–79.

Kitchener, S., Nissen, M., Nasveld, P., Forrat, R., Yoksan, S., Lang, J., and Saluzzo, J.F. (2006). Immunogenicity and safety of two live-attenuated tetravalent dengue vaccine formulations in healthy Australian adults. Vaccine *24*, 1238–1241.

Kiwanuka, N., Sanders, E.J., Rwaguma, E.B., Kawamata, J., Ssengooba, F.P., Najjemba, R., Were, W.A., Lamunu, M., Bagambisa, G., Burkot, T.R., *et al.* (1999). O'nyong-nyong fever in south-central Uganda, 1996–1997: clinical features and validation of a clinical case definition for surveillance purposes. Clin. Infect. Dis. *29*, 1243–1250.

Klasco, R. (2002). Colorado tick fever. Med. Clin. North Am. *86*, 435–440, ix.

Kollaritsch, H., Paulke-Korinek, M., and Dubischar-Kastner, K. (2009). IC51 Japanese encephalitis vaccine. Exp. Opin. Biol. Ther. *9*, 921–931.

Konishi, E. (2012). Issues related to recent dengue vaccine development. Trop. Med. Health *39*, 63–71.

Konishi, E., Yamaoka, M., Khin Sane, W., Kurane, I., and Mason, P.W. (1998). Induction of protective immunity against Japanese encephalitis in mice by immunization with a plasmid encoding Japanese encephalitis virus premembrane and envelope genes. J. Virol. *72*, 4925–4930.

Korrapati, A.B., Swaminathan, G., Singh, A., Khanna, N., and Swaminathan, S. (2012). Adenovirus delivered short hairpin RNA targeting a conserved site in the 5′ non-translated region inhibits all four serotypes of dengue viruses. PLoS Negl. Trop. Dis. *6*, e1735.

Kreil, T.R., Burger, I., Attakpah, E., Olas, K., and Eibl, M.M. (1998). Passive immunization reduces immunity that results from simultaneous active immunization against tick-borne encephalitis virus in a mouse model. Vaccine *16*, 955–959.

Kubar, A., Haciomeroglu, M., Ozkul, A., Bagriacik, U., Akinci, E., Sener, K., and Bodur, H. (2011). Prompt administration of Crimean–Congo hemorrhagic fever (CCHF) virus hyperimmunoglobulin in patients diagnosed with CCHF and viral load monitorization by reverse transcriptase-PCR. Jpn J. Infect. Dis. *64*, 439–443.

Kubes, V., and Rios, F.A. (1939). The causative agent of infectious equine encephalomyelitis in Venezuela. Science *90*, 20–21.

Kumar, R., Kohli, N., Mathur, A., and Wakhlu, I. (1992). Use of the computed tomographic scan in Japanese encephalitis. Ann. Trop. Med. Parasitol. *86*, 77–81.

Kumar, R., Mathur, A., and Singh, K.B. (1993). Clinical sequelae of Japanese encephalitis in children. Indian J. Med. Res. *97*, 9–13.

Kumar, R., Tripathi, P., and Rizvi, A. (2009). Effectiveness of one dose of live-attenuated SA 14–14–2 vaccine against Japanense encephalitis. N. Engl. J. Med. *360*, 1465–1466.

Kunze, U. (2010). TBE--awareness and protection: the impact of epidemiology, changing lifestyle, and environmental factors. Wien. Med. Wochenschr. *160*, 252–255.

Kunze, U., Asokliene, L., Bektimirov, T., Busse, A., Chmelik, V., Heinz, F.X., Hingst, V., Kadar, F., Kaiser, R., Kimmig, P., *et al.* (2004). Tick-borne encephalitis in childhood – consensus 2004. Wien. Med. Wochenschr. *154*, 242–245.

Kunze, U., Baumhackl, U., Bretschneider, R., Chmelik, V., Grubeck-Loebenstein, B., Haglund, M., Heinz, F., Kaiser, R., Kimmig, P., Kunz, C., *et al.* (2005). The Golden Agers and Tick-borne encephalitis. Conference report and position paper of the International Scientific Working Group on Tick-borne encephalitis. Wien. Med. Wochenschr. *155*, 289–294.

Lahariya, C., and Pradhan, S.K. (2006). Emergence of chikungunya virus in Indian subcontinent after 32 years: A review. J. Vector Borne Dis. *43*, 151–160.

Lai, C.J., Zhao, B.T., Hori, H., and Bray, M. (1991). Infectious RNA transcribed from stably cloned full-length cDNA of dengue type 4 virus. Proc. Natl. Acad. Sci. U.S.A. *88*, 5139–5143.

Lanciotti, R.S., Ludwig, M.L., Rwaguma, E.B., Lutwama, J.J., Kram, T.M., Karabatsos, N., Cropp, B.C., and Miller, B.R. (1998). Emergence of epidemic O'nyong-nyong fever in Uganda after a 35-year absence: genetic characterization of the virus. Virology *252*, 258–268.

Lang, J. (2009). Recent progress on Sanofi Pasteur's dengue vaccine candidate. J. Clin. Virol. *46(Suppl. 2)*, S20–S24.

Leblebicioglu, H., Bodur, H., Dokuzoguz, B., Elaldi, N., Guner, R., Koksal, I., Kurt, H., and Senturk, G.C. (2012). Case management and supportive treatment for patients with Crimean–Congo hemorrhagic fever. Vector Borne Zoonotic Dis. *12*, 805–811.

Ledgerwood, J.E., Pierson, T.C., Hubka, S.A., Desai, N., Rucker, S., Gordon, I.J., Enama, M.E., Nelson, S., Nason, M., Gu, W., *et al.* (2011). A West Nile virus DNA vaccine utilizing a modified promoter induces neutralizing antibody in younger and older healthy adults in a phase I clinical trial. J. Infect. Dis. *203*, 1396–1404.

Lee, E., Hall, R.A., and Lobigs, M. (2004). Common E protein determinants for attenuation of glycosaminoglycan-binding variants of Japanese encephalitis and West Nile viruses. J. Virol. *78*, 8271–8280.

Leiby, D.A., and Gill, J.E. (2004). Transfusion-transmitted tick-borne infections: a cornucopia of threats. Transfus. Med. Rev. *18*, 293–306.

Lepiniec, L., Dalgarno, L., Huong, V.T., Monath, T.P., Digoutte, J.P., and Deubel, V. (1994). Geographic distribution and evolution of yellow fever viruses based on direct sequencing of genomic cDNA fragments. J. Gen. Virol. *75*, 417–423.

Levitt, N.H., Ramsburg, H.H., Hasty, S.E., Repik, P.M., Cole, F.E., Jr., and Lupton, H.W. (1986). Development of an attenuated strain of chikungunya virus for use in vaccine production. Vaccine *4*, 157–162.

Liang, S., Bao, C., Zhou, M., Hu, J., Tang, F., Guo, X., Jiao, Y., Zhang, W., Luo, P., Li, L., *et al.* (2014). Seroprevalence and risk factors for severe fever with thrombocytopenia syndrome virus infection in Jiangsu Province, China, 2011. Am. J. Trop. Med. Hyg. *90*, 256–259.

Libraty, D.H., Nisalak, A., Endy, T.P., Suntayakorn, S., Vaughn, D.W., and Innis, B.L. (2000). Risk factors for severe disease in Japanese encephalitis: the effect of prior dengue virus infection. Paper presented at Annual Meeting of the American Society of Tropical Medicine and Hygiene (Houston, TX).

Lihoradova, O.A., Indran, S.V., Kalveram, B., Lokugamage, N., Head, J.A., Gong, B., Tigabu, B., Juelich, T.L., Freiberg, A.N., and Ikegami, T. (2013). Characterization of Rift Valley fever virus MP-12 strain encoding NSs of Punta Toro virus or sandfly fever Sicilian virus. PLoS Negl. Trop. Dis. *7*, e2181.

von Lindern, J.J., Aroner, S., Barrett, N.D., Wicker, J.A., Davis, C.T., and Barrett, A.D. (2006). Genome analysis and phylogenetic relationships between east, central and west African isolates of Yellow fever virus. J. Gen. Virol. *87*, 895–907.

Liu, X., Yu, Y., Li, M., Liang, G., Wang, H., Jia, L., and Dong, G. (2011). Study on the protective efficacy of SA14-14-2 attenuated Japanese encephalitis against different JE virus isolates circulating in China. Vaccine *29*, 2127–2130.

Liu, Y., Li, Q., Hu, W., Wu, J., Wang, Y., Mei, L., Walker, D.H., Ren, J., Wang, Y., and Yu, X.J. (2012). Person-to-person transmission of severe fever with thrombocytopenia syndrome virus. Vector Borne Zoonotic Dis. *12*, 156–160.

Liu, Z.L., Hennessy, S., Strom, B.L., Tsai, T.F., Wan, C.M., Tang, C.S., Xiang, C.F., Bilker, W.B., Pan, X.P., Yao, Y.J., *et al.* (1997). Short-term safety of live attenuated Japanese encephalitis vaccine (SA14–14–2): Results of a randomized trial with 26,239 subjects. J. Infect. Dis. *176*, 1366–1369.

Lloyd, W., Theiler, M., and Ricci, N.I. (1936). Modification of the virulence of yellow fever virus by cultivation in tissues in vitro. Trans. R. Soc. Trop. Med. Hyg. *29*, 481–529.

Loew-Baselli, A., Poellabauer, E.M., Pavlova, B.G., Fritsch, S., Firth, C., Petermann, R., Barrett, P.N., and Ehrlich, H.J. (2011). Prevention of tick-borne encephalitis by FSME-IMMUN vaccines: review of a clinical development programme. Vaccine *29*, 7307–7319.

Lorin, C., Combredet, C., Labrousse, V., Mollet, L., Despres, P., and Tangy, F. (2005). A paediatric vaccination vector based on live attenuated measles vaccine. Therapie *60*, 227–233.

Lourenco de Oliveira, R., Vazeille, M., Bispo de Filippis, A.M., and Failloux, A.B. (2002). Oral suasceptibility to yellow fever virus of *Aedes aegypti* from Brazil. Mem. Inst. Oswaldo Cruz *97*, 437–439.

Lourenco de Oliveira, R., vazeille, M., de Filippis, A.M., and Failloux, A.B. (2004). *Aedes aegypti* in Brazil: genetically differentiated populations with high susceptibility to dengue and yellow fever viruses. Trans. R. Soc. Trop. Med. Hyg. *98*, 43–54.

Lutwama, J.J., Kayondo, J., Savage, H.M., Burkot, T.R., and Miller, B.R. (1999). Epidemic O'Nyong-Nyong fever in southcentral Uganda, 1996–1997: entomologic studies in Bbaale village, Rakai District. Am. J. Trop. Med. Hyg. *61*, 158–162.

Lyons, A., Kanesa-thasan, N., Kuschner, R.A., Eckels, K.H., Putnak, R., Sun, W., Burge, R., Towle, A.C., Wilson, P., Tauber, E., and Vaughn, D.W. (2007). A Phase 2 study of a purified, inactivated virus vaccine to prevent Japanese encephalitis. Vaccine *25*, 3445–3453.

McArthur, J.H., Durbin, A.P., Marron, J.A., Wanionek, K.A., Thumar, B., Pierro, D.J., Schmidt, A.C., Blaney, J.E., Jr., Murphy, B.R., and Whitehead, S.S. (2008). Phase I clinical evaluation of rDEN4Delta30-200,201: a live attenuated dengue 4 vaccine candidate designed for decreased hepatotoxicity. Am. J. Trop. Med. Hyg. *79*, 678–684.

Macasaet, F.F., Villami, P.T., and Wexler, S. (1969). Epidemiology of arbovirus infections in Negros Oriental. II.Serologic findings of the epidemic in Amlan. Philippine Med. Assoc. *45*, 311–317.

McClain, D.J., Pittman, P.R., Ramsburg, H.H., Nelson, G.O., Rossi, C.A., Mangiafico, J.A., Schmaljohn, A.L., and Malinoski, F.J. (1998). Immunologic interference from sequential administration of live attenuated alphavirus vaccines. J. Infect. Dis. *177*, 634–641.

Mackowiak, P.A., Wasserman, S.S., Tacket, C.O., Vaughn, D.W., Eckels, K.H., Dubois, D.R., Hoke, C.H., Jr., and Edelman, R. (1994). Quantitative relationship between oral temperature and severity of illness following inoculation with candidate attenuated dengue virus vaccines. Clin. Infect. Dis. *19*, 948–950.

McLean, R.G., Francy, D.B., Bowen, G.S., Bailey, R.E., Calisher, C.H., and Barnes, A.M. (1981). The ecology of Colorado tick fever in Rocky Mountain National Park in 1974. I.Objectives, study design, and summary of principal findings. Am. J. Trop. Med. Hyg. *30*, 483–489.

McMahon, A.W., Eidex, R.B., Marfin, A.A., Russell, M., Sejvar, J.J., Markoff, L., Hayes, E.B., Chen, R.T., Ball, R., Braun, M.M., and Cetron, M. (2007). Neurologic disease associated with 17D-204 yellow fever vaccination: a report of 15 cases. Vaccine *25*, 1727–1734.

Malabadi, R.B., Ganguly, A., Silva, J.A., Parashar, A., Suresh, M.R., and Sunwoo, H. (2012). Overview of plant-derived vaccine antigens: Dengue virus. J. Pharm. Pharm. Sci. *14*, 400–413.

Mallilankaraman, K., Shedlock, D.J., Bao, H., Kawalekar, O.U., Fagone, P., Ramanathan, A.A., Ferraro, B., Stabenow, J., Vijayachari, P., Sundaram, S.G., *et al.* (2011). A DNA vaccine against chikungunya virus is protective in mice and induces neutralizing antibodies in mice and nonhuman primates. PLoS Negl. Trop. Dis. *5*, e928.

Mani, S., Tripathi, L., Raut, R., Tyagi, P., Arora, U., Barman, T., Sood, R., Galav, A., Wahala, W., de Silva, A., *et al.* (2013). Pichia pastoris-expressed dengue 2 envelope forms virus-like particles without pre-membrane protein and induces high titer neutralizing antibodies. PLoS One *8*, e64595.

Mansfield, K.L., Johnson, N., Phipps, L.P., Stephenson, J.R., Fooks, A.R., and Solomon, T. (2009). Tick-borne encephalitis virus – a review of an emerging zoonosis. J. Gen. Virol. *90*, 1781–1794.

Marchette, N.J., Dubois, D.R., Larsen, L.K., Summers, P.L., Kraiselburd, E.G., Gubler, D.J., and Eckels, K.H. (1990). Preparation of an attenuated dengue 4 (341750 Carib) virus vaccine. I. Pre-clinical studies. Am. J. Trop. Med. Hyg. *43*, 212–218.

Marfin, A.A., Eides, R.S., Kozarsky, P.E., and Cetron, M.S. (2005). Yellow fever and Japanese encephalitis vaccines: indications and complications. Infect. Dis. Clin. North Am. *19*, 151–168.

Martin, J.E., Pierson, T.C., Hubka, S., Rucker, S., Gordon, I.J., Enama, M.E., Andrews, C.A., Xu, Q., Davis, B.S., Nason, M., *et al.* (2007). A West Nile virus DNA vaccine induces neutralizing antibody in healthy adults during a phase 1 clinical trial. J. Infect. Dis. *196*, 1732–1740.

Martin, M., Tsai, T.F., Cropp, B., Chang, G.J., Holmes, D.A., Tseng, J., Shieh, W., Zaki, S.R., Al-Sanouri, I., Cutrona, A.F., *et al.* (2001). Fever and multisystem organ failure associated with 17D-204 yellow fever vaccination: a report of four cases. Lancet *358*, 98–104.

Maruggi, G., Shaw, C.A., Otten, G.R., Mason, P.W., and Beard, C.W. (2013). Engineered alphavirus replicon vaccines based on known attenuated viral mutants show limited effects on immunogenicity. Virology *447*, 254–264.

Meiklejohn, G., England, B., and Lennette (1952). Propagation of dengue virus strains in unweaned mice. Am. J. Trop. Med. Hyg. *1*, 51–58.

Men, R., Bray, M., Clark, D., Chanock, R.M., and Lai, C.J. (1996). Dengue type 4 virus mutants containing deletions in the 3′ noncoding region of the RNA genome: analysis of growth restriction in cell culture and altered viremia pattern and immunogenicity in rhesus monkeys. J. Virol. *70*, 3930–3937.

Metz, S.W., Gardner, J., Geertsema, C., Le, T.T., Goh, L., Vlak, J.M., Suhrbier, A., and Pijlman, G.P. (2013a). Effective chikungunya virus-like particle vaccine produced in insect cells. PLoS Negl. Trop. Dis. *7*, e2124.

Metz, S.W., Martina, B.E., van den Doel, P., Geertsema, C., Osterhaus, A.D., Vlak, J.M., and Pijlman, G.P. (2013b). Chikungunya virus-like particles are more immunogenic in a lethal AG129 mouse model compared to glycoprotein E1 or E2 subunits. Vaccine *31*, 6092–6096.

Miller, A., Carchman, R., Long, R., and Denslow, S.A. (2012). La Crosse viral infection in hospitalized pediatric patients in Western North Carolina. Hosp. Pediatr. *2*, 235–242.

Miller, B.R., and Ballinger, M.E. (1988). *Aedes albopictus* mosquitoes introduced into Brazil: vector competence for yellow fever and dengue viruses. Trans. R. Soc. Trop. Med. Hyg. *82*, 476–477.

Miller, B.R., Monath, T.P., Tabachnik, W.J., and Ezike, V.I. (1989). Epidemic yellow fever caused by an incompetent mosquito vector. Trop. Med. Parasitol. *40*, 396–399.

Miller, J.D., van der Most, R.G., Akondy, R.S., Glidewell, J.T., Albott, S., Masopust, D., Murali-Krishna, K., Mahar, P.L., Edupuganti, S., Lalor, S., *et al.* (2008). Human effector and memory CD8+ T cell responses to smallpox and yellow fever vaccines. Immunity *28*, 710–722.

Ming, C.K., Thein, S., Thaung, U., Tin, U., Myint, K.S., Swe, T., Halstead, S.B., and Diwan, A.R. (1974). Clinical and laboratory studies on haemorrhagic fever in Burma, 1970–72. Bull. World Health Org. *51*, 227–235.

Misra, U.K., Kalita, J., Jain, S.K., and Mathur, A. (1994). Radiological and neurophysiological changes in Japanese encephalitis. J. Neurol. Neurosurg. Psychiatry *57*, 1484–1487.

Monath, T.P. (1989). Recent epidemics of yellow fever in Africa and the risk of future urbanization and spread., In Arbovirus Research in Australia., M.F. Uren, J. Blok, and L.H. Manderson, eds. (Queensland Institute of Medical Research, Brisbane).

Monath, T.P. (1994). Yellow fever and dengue – the interactions of virus, vector and host in the re-emergence of epidemic disease. Semin. Virol. *5*, 133–145.

Monath, T.P. (1997). Epidemiology of yellow fever: current status and speculations on future trends. In Factors in the Emergence of Arbovirus Diseases, J.F. Saluzzo, and B. Dodet, eds. (Elsevier, Paris), pp. 143–156.

Monath, T.P. (2002). Editorial: jennerian vaccination against West Nile virus. Am. J. Trop. Med. Hyg. *66*, 113–114.

Monath, T.P. (2012). Review of the risks and benefits of yellow fever vaccination including some new analyses. Exp. Rev. Vaccines *11*, 427–448.

Monath, T.P. (2013). Vaccines against diseases transmitted from animals to humans: a one health paradigm. Vaccine *31*, 5321–5338.

Monath, T.P., Kinney, R.M., Schlesinger, J.J., Brandriss, M.W., and Bres, P. (1983). Ontogeny of yellow fever 17D vaccine: RNA oligonucliotide fingerprint and monoclonal antibody analyses of vaccines produced world-wide. J. Gen. Virol. *64*, 627–637.

Monath, T.P., Levenbook, I., Soike, K., Zhang, Z.X., Ratterree, M., Draper, K., Barrett, A.D., Nichols, R., Weltzin, R., Arroyo, J., and Guirakhoo, F. (2000). Chimeric yellow fever virus 17D-Japanese encephalitis virus vaccine: dose–response effectiveness and extended safety testing in rhesus monkeys. J. Virol. *74*, 1742–1751.

Monath, T.P., Arroyo, J., Miller, C., and Guirakhoo, F. (2001). West Nile virus vaccine. Curr. Drug Targets Infect. Dis. *1*, 37–50.

Monath, T.P., McCarthy, K., Bedford, P., Johnson, C.T., Nichols, R., Yoksan, S., Marchesani, R., Knauber, M., Wells, K.H., Arrroyo, J., and Guirakhoo, F. (2002). Clinical proof of principle for ChimeriVax: recombinant live, attenuated vaccines against flavivirus infections. Vaccine *20*, 1004–1018.

Monath, T.P., Guirakhoo, F., Nichols, R., Yoksan, S., Schrader, R., Murphy, C., Blum, P., Woodward, S., McCarthy, K., Mathis, D., *et al.* (2003). Chimeric live, attenuated vaccine against Japanese encephalitis (ChimeriVax-JE): phase 2 clinical trials for safety and immunogenicity, effect of vaccine dose and schedule, and memory response to challenge with inactivated Japanese encephalitis antigen. J. Infect. Dis. *188*, 1213–1230.

Monath, T.P., Myers, G.A., Beck, R.A., Knauber, M., Scappaticci, K., Pullano, T., Archambault, W.T., Catalan, J., Miller, C., Zhang, Z.X., *et al.* (2005). Safety testing for neurovirulence of novel live, attenuated flavivirus vaccines: infant mice provide an accurate surrogate for the test in monkeys. Biologicals *33*, 131–144.

Monath, T.P., Liu, J., Kanesa-Thasan, N., Myers, G.A., Nichols, R., Deary, A., McCarthy, K., Johnson, C., Ermak, T., Shin, S., *et al.* (2006). A live, attenuated recombinant West Nile virus vaccine. Proc. Natl. Acad. Sci. U.S.A. *103*, 6694–6699.

Monath, T.P., M.S., C., and Teuwen, D.E. (2008). Yellow fever vaccine. In Vaccines, S.A. Plotkin, W.A. Orenstein, and P.A. Offit, eds. (Saunders Elsevier, Philadelphia), pp. 959–1055.

Montagnon, B.J., and Vincent-Falquet, J.C. (1998). Experience with the Vero cell line. Dev. Biol. Stand. *93*, 119–123.

Moore, M., Sylla, M., Goss, L., Burugu, M.W., Sang, R., Kamau, L.W., Kenya, E.U., Bosio, C., Munoz Mde, L., Sharakova, M., and Black, W.C. (2013). Dual African origins of global *Aedes aegypti* s.l. populations revealed by mitochondrial DNA. PLoS Negl. Trop. Dis. *7*, e2175.

Moro, M.L., Gagliotti, C., Silvi, G., Angelini, R., Sambri, V., Rezza, G., Massimiliani, E., Mattivi, A., Grilli, E., Finarelli, A.C., *et al.* (2010). Chikungunya virus in North-Eastern Italy: a seroprevalence survey. Am. J. Trop. Med. Hyg. *82*, 508–511.

Morrison, D., Legg, T.J., Billings, C.W., Forrat, R., Yoksan, S., and Lang, J. (2010). A novel tetravalent dengue vaccine is well tolerated and immunogenic against all 4 serotypes in flavivirus-naive adults. J. Infect. Dis. *201*, 370–377.

Mourao, M.P., Bastos Mde, S., de Figueiredo, R.P., Gimaque, J.B., Galusso Edos, S., Kramer, V.M., de Oliveira, C.M., Naveca, F.G., and Figueiredo, L.T. (2012). Mayaro fever in the city of Manaus, Brazil, 2007–2008. Vector Borne Zoonotic Dis. *12*, 42–46.

Mousavi-Jazi, M., Karlberg, H., Papa, A., Christova, I., and Mirazimi, A. (2012). Healthy individuals' immune response to the Bulgarian Crimean–Congo hemorrhagic fever virus vaccine. Vaccine *30*, 6225–6229.

Munoz, M., and Navarro, J.C. (2012). [Mayaro: a re-emerging Arbovirus in Venezuela and Latin America]. Biomedica *32*, 286–302.

Murakami, S., Terasaki, K., Ramirez, S.I., Morrill, J.C., and Makino, S. (2014). Development of a novel, single-cycle replicable rift valley Fever vaccine. PLoS Negl. Trop. Dis. *8*, e2746.

Murphy, B.R., and Whitehead, S.S. (2011). Immune Response to Dengue Virus and Prospects for a Vaccine. Annu. Rev. Immunol. *29*, 587–619.

Mutebi, J.P., Wang, H., Li, L., Bryant, J.E., and Barrett, A.D. (2001). Phylogenetic and evolutionary relationships among yellow fever virus isolates in Africa. J. Virol. *75*, 6999–7008.

Mutebi, J.P., Gianella, A., and Travassos da Rosa, A. (2004). Yellow fever virus infectivity for Bolivian *Aedes aegypti* mosquitoes. Emerg. Infect. Dis. *10*, 1657–1660.

Muthumani, K., Lankaraman, K.M., Laddy, D.J., Sundaram, S.G., Chung, C.W., Sako, E., Wu, L., Khan, A., Sardesai, N., Kim, J.J., *et al.* (2008). Immunogenicity of novel consensus-based DNA vaccines against Chikungunya virus. Vaccine *26*, 5128–5134.

Myint, K.S., Gibbons, R.V., Perng, G.C., and Solomon, T. (2007). Unravelling the neuropathogenesis of Japanese encephalitis. Trans. R. Soc. Trop. Med. Hyg. *101*, 955–956.

Nakao, E., and Hotta, S. (1973). Immunogenicity of purifed, inactivated chikungunya virus in monkeys. Bull. World Health Org. *48*, 559–562.

Nasveld, P.E., Ebringer, A., Elmes, N., Bennett, S., Yoksan, S., Aaskov, J., McCarthy, K., Kanesa-Thasan, N., Meric, C., and Reid, M. (2010). Long term immunity to live attenuated Japanese encephalitis chimeric virus vaccine: Randomized, double-blind, 5-year phase II study in healthy adults. Hum. Vaccine *6*, 1–9.

Nett, R.J., Kuehnert, M.J., Ison, M.G., Orlowski, J.P., Fischer, M., and Staples, J.E. (2012). Current practices and evaluation of screening solid organ donors for West Nile virus. Transpl. Infect. Dis. *14*, 268–277.

Ni, H., Watowich, S.J., and Barrett, A.D.T. (1997). Molecular basis of attenuation and virulence of Japanese encephalitis virus. In Factors in the Emergence of Arbovirus Diseases, J.F. Saluzzo, and B. Dodet, eds. (Elsevier, Paris), pp. 203–211.

Nordin, J.D., Parker, E.D., Vazquez-Benitez, G., Kharbanda, E.O., Naleway, A., Marcy, S.M., Molitor, B., Kuckler, L., and Baggs, J. (2013). Safety of the yellow Fever vaccine: a retrospective study. J. Travel Med. *20*, 368–373.

Ohrr, H.C., Tandan, J.B., Sohn, Y.M., Shin, S.H., Pradhan, D.P., and Halstead, S.B. (2005). Effect of a single dose of SA 14–14–2 vaccine 1 year after immunization in Nepalese children with Japanese encephalitis: a case–control study. Lancet *366*, 1375–1378.

Oncu, S. (2013). Crimean–Congo hemorrhagic fever: an overview. Virol. Sin. *28*, 193–201.

Oreshkova, N., van Keulen, L., Kant, J., Moormann, R.J., and Kortekaas, J. (2013). A single vaccination with an improved nonspreading Rift Valley fever virus vaccine provides sterile immunity in lambs. PLoS One *8*, e77461.

Osorio, J.E., Brewoo, J.N., Silengo, S.J., Arguello, J., Moldovan, I.R., Tary-Lehmann, M., Powell, T.D., Livengood, J.A., Kinney, R.M., Huang, C.Y., and Stinchcomb, D.T. (2011a). Efficacy of a tetravalent chimeric dengue vaccine (DENVax) in Cynomolgus macaques. Am. J. Trop. Med. Hyg. *84*, 978–987.

Osorio, J.E., Huang, C.Y., Kinney, R.M., and Stinchcomb, D.T. (2011b). Development of DENVax: A chimeric dengue-2 PDK-53-based tetravalent vaccine for protection against dengue fever. Vaccine *29*, 7251–7260.

Oya, A. (1988). Japanese encephalitis vaccine. Acta Paediatr. Jpn *30*, 175–184.

Ozherelkov, S.V., Kalinina, E.S., Kozhevnikova, T.N., Sanin, A.V., Timofeeva, T., Timofeev, A.V., and Stivenson, D.R. (2008). [Experimental study of the phenomenon of antibody dependent tick-borne encephalitis virus infectivity enhancement in vitro]. Zh. Mikrobiol. Epidemiol. Immunobiol. *Nov-Dec*, 39–43.

Paessler, S., and Weaver, S.C. (2009). Vaccines for Venezuelan equine encephalitis. Vaccine *27 Suppl 4*, D80–85.

Palacios, G., Tesh, R.B., Savji, N., Travassos da Rosa, A.P., Guzman, H., Bussetti, A.V., Desai, A., Ladner, J., Sanchez-Seco, M., and Lipkin, W.I. (2014). Characterization of the Sandfly fever Naples species complex and description of a new Karimabad species complex (genus Phlebovirus, family Bunyaviridae). J. Gen. Virol. *95*, 292–300.

Pandya, J., Gorchakov, R., Wang, E., Leal, G., and Weaver, S.C. (2012). A vaccine candidate for eastern equine encephalitis virus based on IRES-mediated attenuation. Vaccine *30*, 1276–1282.

Papa, A., Papadimitriou, E., and Christova, I. (2011). The Bulgarian vaccine Crimean–Congo haemorrhagic fever virus strain. Scand. J. Infect. Dis. *43*, 225–229.

Partidos, C.D., Paykel, J., Weger, J., Borland, E.M., Powers, A.M., Seymour, R., Weaver, S.C., Stinchcomb, D.T., and Osorio, J.E. (2012). Cross-protective immunity against o'nyong-nyong virus afforded by a novel recombinant chikungunya vaccine. Vaccine *30*, 4638–4643.

Pavlovic, J., Schultz, J., Hefti, H.P., Schuh, T., and Molling, K. (2000). DNA vaccination against La Crosse virus. Intervirology *43*, 312–321.

Pealer, L.N., Marfin, A.A., Petersen, L.R., Lanciotti, R.S., Page, P.L., Stramer, S.L., Stobierski, M.G., Signs, K., Newman, B., Kapoor, H., *et al.* (2003). Transmission of West Nile virus through blood transfusion in the United States in 2002. N. Engl. J. Med. *349*, 1236–1245.

Pedersen, C.E., Jr., Robinson, D.M., and Cole, F.E., Jr. (1972). Isolation of the vaccine strain of Venezuelan equine encephalomyelitis virus from mosquitoes in Louisiana. Am. J. Epidemiol. *95*, 490–496.

Pen'evskaia, N.A., and Rudakov, N.V. (2010). [Efficiency of use of immunoglobulin preparations for the postexposure prevention of tick-borne encephalitis in Russia (a review of semi-centennial experience)]. Med. Parazitol. (Mosk) *Jan-Mar*, 53–59.

Peters, C.J. (1997). Emergence of Rift Valley fever. In Factors in the Emergence of Arbovirus Diseases, J.F. Saluzzo, and B. Dodet, eds. (Elsevier, Paris), pp. 253–264.

Peters, C.J., and Harrison, G.J. (2014). Rift Valley fever. In Feigin and Cherry's Textbook of Pediatric Infectious Diseases, J.D. Cherry, G.J. Harrison, S.L. Kaplan, W.J. Steinbach, and P.J. Hotez, eds. (Elsevier Saunders, Philadelphia), pp. 2528–2531.

Petersen, L.R., Brault, A.C., and Nasci, R.S. (2013). West Nile virus: review of the literature. JAMA *310*, 308–315.

Philipp, C.S., Callaway, C., Chu, M.C., Huang, G.H., Monath, T.P., Trent, D., and Evatt, B.L. (1993). Replication of Colorado tick fever virus within human hematopoietic progenitor cells. J. Virol. *67*, 2389–2395.

Pinheiro, F.P., Travassos da Rosa, A.P., Gomes, M.L., LeDuc, J.W., and Hoch, A.L. (1982). Transmission of Oropouche virus from man to hamster by the midge *Culicoides paraensis*. Science *215*, 1251–1253.

Pinheiro, F.P., Travassos da Rosa, A.P.A., and Vasconcelos, P.F. (2014). Oropouche fever. In Feigin and Cherry's Textbook of Pediatric Infectious Diseases, J.D. Cherry, G.J. Harrison, S.L. Kaplan, W.J. Steinbach, and P.J. Hotez, eds. (Elsevier Saunders, Philadelphia), pp. 2537–2544.

Pittman, P.R., and Plotkin, S.A. (2013). Biodefense and special pathogen vaccines. In Vaccines, S.A. Plotkin, Orenstein, W.A, Offit, P.A., eds. (Elsevier Saunders, Philadelphia), pp. 1008–1017.

Plante, K., Wang, E., Partidos, C.D., Weger, J., Gorchakov, R., Tsetsarkin, K., Borland, E.M., Powers, A.M., Seymour, R., Stinchcomb, D.T., *et al.* (2011). Novel chikungunya vaccine candidate with an IRES-based attenuation and host range alteration mechanism. PLoS Pathog. *7*, e1002142.

Pletnev, A.G., Claire, M.S., Elkins, R., Speicher, J., Murphy, B.R., and Chanock, R.M. (2003). Molecularly engineered live-attenuated chimeric West Nile/dengue virus vaccines protect rhesus monkeys from West Nile virus. Virology *314*, 190–195.

Pletnev, A.G., Swayne, D.E., Speicher, J., Rumyantsev, A.A., and Murphy, B.R. (2006). Chimeric West Nile/dengue virus vaccine candidate: preclinical evaluation in mice, geese and monkeys for safety and immunogenicity. Vaccine *24*, 6392–6404.

Poletti, P., Messeri, G., Ajelli, M., Vallorani, R., Rizzo, C., and Merler, S. (2011). Transmission potential of chikungunya virus and control measures: the case of Italy. PLoS One *6*, e18860.

Poo, J., Galan, F., Forrat, R., Zambrano, B., Lang, J., and Dayan, G.H. (2011). Live-attenuated tetravalent dengue vaccine in dengue-naive children, adolescents, and adults in Mexico City: randomized controlled phase 1 trial of safety and immunogenicity. Pediatr. Infect. Dis. J. *30*, e9-e17.

Porter, K.R., Ewing, D., Chen, L., Wu, S.J., Hayes, C.G., Ferrari, M., Teneza-Mora, N., and Raviprakash, K. (2012). Immunogenicity and protective efficacy of a vaxfectin-adjuvanted tetravalent dengue DNA vaccine. Vaccine *30*, 336–341.

Posey, D.L., O'Rourke, T., Roehrig, J.T., Lanciotti, R.S., Weinberg, M., and Maloney, S. (2005). O'Nyong-nyong fever in West Africa. Am. J. Trop. Med. Hyg. 73, 32.

Powers, A.M., Brault, A.C., Tesh, R.B., and Weaver, S.C. (2000). Re-emergence of Chikungunya and O'nyong-nyong viruses: evidence for distinct geographical lineages and distant evolutionary relationships. J. Gen. Virol. *81*, 471–479.

Prymula, R., Pollabauer, E.M., Pavlova, B.G., Low-Baselli, A., Fritsch, S., Angermayr, R., Geisberger, A., Barrett, P.N., and Ehrlich, H.J. (2012). Antibody persistence after two vaccinations with either FSME-IMMUN(R) Junior or ENCEPUR(R) Children followed by third vaccination with FSME-IMMUN(R) Junior. Hum. Vaccin. Immunother. *8*, 736–742.

Pugachev, K.V., Ocran, S.W., Guirakhoo, F., Furby, D., and Monath, T.P. (2002). Heterogeneous nature of the genome of the ARILVAX yellow fever 17D vaccine revealed by consensus sequencing. Vaccine *20*, 996–999.

Pulendran, B. (2009). Learning immunology from the yellow fever vaccine: innate immunity to systems vaccinology. Nat. Rev. Immunol. *9*, 741–747.

Pulmanausahakul, R., Roytrakul, S., Auewarakul, P., and Smith, D.R. (2011). Chikungunya in Southeast Asia: understanding the emergence and finding solutions. Int. J. Infect. Dis. *15*, e671–676.

Putnak, R.J., Coller, B.A., Voss, G., Vaughn, D.W., Clements, D., Peters, I., Bignami, G., Houng, H.S., Chen, R.C., Barvir, D.A., *et al.* (2005). An evaluation of dengue type-2 inactivated, recombinant subunit, and live-attenuated vaccine candidates in the rhesus macaque model. Vaccine 23, 4442–4452.

Querec, T.D., Akondy, R.S., Lee, E.K., Cao, W., Nakaya, H.I., Teuwen, D., Pirani, A., Gernert, K., Deng, J., Marzolf, B., *et al.* (2009). Systems biology approach predicts immunogenicity of the yellow fever vaccine in humans. Nat. Immunol. *10*, 116–125.

Rafferty, E., Duclos, P., Yactayo, S., and Schuster, M. (2013). Risk of yellow fever vaccine-associated viscerotropic disease among the elderly: a systematic review. Vaccine *31*, 5798–5805.

Raviprakash, K., Apt, D., Brinkman, A., Skinner, C., Yang, S., Dawes, G., Ewing, D., Wu, S.J., Bass, S., Punnonen, J., and Porter, K. (2006). A chimeric tetravalent dengue DNA vaccine elicits neutralizing antibody to all four virus serotypes in rhesus macaques. Virology *353*, 166–173.

Raviprakash, K., Wang, D., Ewing, D., Holman, D.H., Block, K., Woraratanadharm, J., Chen, L., Hayes, C., Dong, J.Y., and Porter, K. (2008). A tetravalent dengue vaccine based on a complex adenovirus vector provides significant protection in rhesus monkeys against all four serotypes of dengue viruses. J. Virol. *82*, 6927–6934.

Rayamajhi, A., Ansari, I., Ledger, E., Bista, K.P., Impoinvil, D.E., Nightingale, S., Kumar, R., Mahaseth, C., Solomon, T., and Griffiths, M.J. (2011). Clinical and prognostic features among children with acute encephalitis syndrome in Nepal; a retrospective study. BMC Infect. Dis. *11*, 294.

Reid, M., Mackenzie, D., Baron, A., Lehmann, N., Lowry, K., Aaskov, J., Guirakhoo, F., and Monath, T.P. (2006). Experimental infection of *Culex annulirostris*, *Culex gelidus*, and *Aedes vigilax* with a yellow fever/Japanese encephalitis virus vaccine chimera (ChimeriVax-JE). Am. J. Trop. Med. Hyg. *75*, 659–663.

Reiter, P., Fontenille, D., and Paupy, C. (2006). *Aedes albopictus* as an epidemic vector of chikungunya virus: another emerging problem? Lancet Infect. Dis. *6*, 463–464.

Renault, P., Balleydier, E., D'Ortenzio, E., Baville, M., and Filleul, L. (2012). Epidemiology of chikungunya infection on Réunion Island, Mayotte, and neighboring countries. Med. Mal. Infect. *42*, 93–101.

Rezza, G., Nicoletti, L., Angelini, R., Romi, R., Finarelli, A.C., Panning, M., Cordioli, P., Fortuna, C., Boros, S., Magurano, F., *et al.* (2007). Infection with chikungunya virus in Italy: an outbreak in a temperate region. Lancet *370*, 1840–1846.

Rice, C.M., Lenches, E.M., Eddy, S.R., Shin, S.J., Sheets, R.L., and Strauss, J.H. (1985). Nucleotide sequence of yellow fever viruses: Implications for flavivirus gene expression and evolution. Science *229*, 726–733.

Rice, C.M., Grakoui, A., Galler, R., and Chambers, T.J. (1989). Transcription of infectious yellow fever RNA from full-length cDNA templates produced by in vitro ligation. N. Biologist *1*, 285–296.

Robinson, D.M., Cole, F.E., Jr., McManus, A.T., and Pedersen, C.E., Jr. (1976). Inactivated Mayaro vaccine produced in human diploid cell cultures. Mil. Med. *141*, 163–166.

Robinson, M.C. (1955). An epidemic of virus disease in Southern Province, Tanganyika Territory in 1952–1953. I.Clinical features. Trans. R. Soc. Trop. Med. Hyg. *49*, 28–32.

Roy, C.J., Adams, A.P., Wang, E., Leal, G., Seymour, R.L., Sivasubramani, S.K., Mega, W., Frolov, I., Didier, P.J., and Weaver, S.C. (2013). A chimeric Sindbis-based vaccine protects cynomolgus macaques against a lethal aerosol challenge of eastern equine encephalitis virus. Vaccine *31*, 1464–1470.

Roy, C.J., Adams, A.P., Wang, E., Plante, K., Gorchakov, R., Seymour, R.L., Vinet-Oliphant, H., and Weaver, S.C. (2014). Chikungunya vaccine candidate is highly attenuated and protects nonhuman primates against telemetrically monitored disease following a single dose. J. Infect. Dis. *209*, 1891–1899.

Rumyantsev, A.A., Goncalvez, A.P., Giel-Moloney, M., Catalan, J., Liu, Y., Gao, Q.S., Almond, J., Kleanthous, H., and Pugachev, K.V. (2013). Single-dose vaccine against tick-borne encephalitis. Proc. Natl. Acad. Sci. U.S.A. *110*, 13103–13108.

Sabchareon, A., Lang, J., Chanthavanich, P., Yoksan, S., Forrat, R., Attanath, P., Sirivichayakul, C., Pengsaa, K., Pojjaroen-Anant, C., Chambonneau, L., *et al.* (2004). Safety and immunogenicity of two tetravalent live attenuated dengue vaccines in 5–12 year old Thai children. Pediatr. Infect. Dis. J. *23*, 99–109.

Sabchareon, A., Wallace, D., Sirivichayakul, C., Limkittikul, K., Chanthavanich, P., Suvannadabba, S., Jiwariyavej, V., Dulyachai, W., Pengsaa, K., Wartel, T.A., *et al.* (2012). Protective efficacy of the recombinant, live-attenuated, CYD tetravalent dengue vaccine in Thai schoolchildren: a randomised, controlled phase 2b trial. Lancet *380*, 1559–1567.

Sabin, A.B. (1952). Research on dengue during World War II. Am. J. Trop. Med. Hyg. *1*, 30–50.

Sabin, A.B., and Schlesinger, R.W. (1945). Production of immunity to dengue with virus modified by propagation in mice. Science *101*, 640–642.

Sall, A.A., Faye, O., Diallo, M., Firth, C., Kitchen, A., and Holmes, E.C. (2010). Yellow fever virus exhibits slower evolutionary dynamics than dengue virus. J. Virol. *84*, 765–772.

Sanchez, V., Gimenez, S., Tomlinson, B., Chan, P.K., Thomas, G.N., Forrat, R., Chambonneau, L., Deauvieau, F., Lang, J., and Guy, B. (2006). Innate and adaptive cellular immunity in flavivirus-naive human recipients of a live-attenuated dengue serotype 3 vaccine produced in Vero cells (VDV3). Vaccine *24*, 4914–4926.

Sanders, E.J., Rwaguma, E.B., Kawamata, J., Kiwanuka, N., Lutwama, J.J., Ssengooba, F.P., Lamunu, M., Najjemba, R., Were, W.A., Bagambisa, G., and Campbell, G.L. (1999). O'nyong-nyong fever in south-central Uganda, 1996–1997: description of the epidemic and results of a household-based seroprevalence survey. J. Infect. Dis. *180*, 1436–1443.

dos Santos, C.N., Post, P.R., Carvalho, R., Ferreira, I.I., Rice, C.M., and Galler, R. (1995). Complete nucleotide sequence of yellow fever virus vaccine strains 17DD and 17D-213. Virus Res. *35*, 35–41.

Schlesinger, R.W., Gordon, I., Frankel, J.W., Winter, J.W., Patterson, P.R., and Dorrance, W.R. (1956). Clinical and serologic response of man to immunization with attenuated dengue and yellow fever viruses. J. Immunol. *77*, 352–364.

Schlesinger, W., and Frankel, J.W. (1952). Adaptation of the New Guinea B strain of dengue virus to suckling and to adult swiss mice; a study in viral variation. Am. J. Trop. Med. Hyg. *1*, 66–77.

Schmitz, J., Roehrig, J., Barrett, A., and Hombach, J. (2011). Next generation dengue vaccines: a review of candidates in preclinical development. Vaccine *29*, 7276–7284.

Schuh, A.J., Ward, M.J., Leigh Brown, A.J., and Barrett, A.D. (2014). Dynamics of the emergence and establishment of a newly dominant genotype of Japanese encephalitis virus throughout Asia. J. Virol. *88*, 4522–4532.

Schuller, E., Jilma, B., Voicu, V., Golor, G., Kollaritsch, H., Kaltenbock, A., Klade, C., and Tauber, E. (2008). Long-term immunogenicity of the new Vero cell-derived, inactivated Japanese encephalitis virus vaccine IC51 Six and 12 month results of a multicenter follow-up phase 3 study. Vaccine *26*, 4382–4386.

Schuller, E., Klade, C.S., Wolfl, G., Kaltenbock, A., Dewasthaly, S., and Tauber, E. (2009). Comparison of a single, high-dose vaccination regimen to the standard regimen for the investigational Japanese encephalitis vaccine, IC51: a randomized, observer-blind, controlled Phase 3 study. Vaccine *27*, 2188–2193.

Silva, M.L., Espirito-Santo, L.R., Martins, M.A., Silveira-Lemos, D., Peruhype-Magalhaes, V., Caminha, R.C., de Andrade Maranhao-Filho, P., Auxiliadora-Martins, M., de Menezes Martins, R., Galler, R., *et al.* (2010). Clinical and immunological insights on severe, adverse neurotropic and viscerotropic disease following 17D yellow fever vaccination. Clin. Vaccine Immunol. *17*, 118–126.

Simasathien, S., Thomas, S.J., Watanaveeradej, V., Nisalak, A., Barberousse, C., Innis, B.L., Sun, W., Putnak, J.R., Eckels, K.H., Hutagalung, Y., *et al.* (2008). Safety and immunogenicity of a tetravalent live-attenuated dengue vaccine in flavivirus naive children. Am. J. Trop. Med. Hyg. *78*, 426–433.

Simmons, M., Burgess, T., Lynch, J., and Putnak, R. (2010). Protection against dengue virus by non-replicating and live attenuated vaccines used together in a prime boost vaccination strategy. Virology *396*, 280–288.

Soares-Weiser, K., Thomas, S., Thomson, G., and Garner, P. (2010). Ribavirin for Crimean–Congo hemorrhagic fever: systematic review and meta-analysis. BMC Infect. Dis. *10*, 207.

Soghaier, M.A., Hagar, A., Abbas, M.A., Elmangory, M.M., Eltahir, K.M., and Sall, A.A. (2013). Yellow Fever outbreak in Darfur, Sudan in October 2012; the initial outbreak investigation report. J. Infect. Public Health *6*, 370–376.

Solomon, T. (2004). Flavivirus encephalitis. N. Engl. J. Med. *351*, 370–378.

Solomon, T., and Vaughn, D.W. (2002). Pathogenesis and clinical features of Japanese encephalitis and West Nile virus infections. Curr. Top. Microbiol. Immunol. *267*, 171–194.

Solomon, T., Dung, N.M., Kneen, R., Thao le, T.T., Gainsborough, M., Nisalak, A., Day, N.P.J., Kirkham, F.J., Vaughn, D.W., Smith, S.M., and White, N.J. (2002). Seizures and raised intracranial pressure in Vietnamese patients with Japanese encephalitis. Brain *125*, 1084–1093.

Srivastava, A.K., Putnak, J.R., Lee, S.H., Hong, S.P., Moon, S.B., Barvir, D.A., Zhao, B., Olson, R.A., Kim, S.O., Yoo, W.D., *et al.* (2001). A purified inactivated Japanese encephalitis virus vaccine made in Vero cells. Vaccine *19*, 4557–4565.

Strode, G.K., ed. (1951). Yellow Fever (McGraw-Hill Book Company, New York).

Stuart, G. (1956). Reactions following vaccination against yellow fever. In Yellow Fever Vaccinations, K.C. Smithburn, ed. (WHO, Geneva), pp. 143–156.

Suhrbier, A., and La Linn, M. (2004). Clinical and pathologic aspects of arthritis due to Ross River virus and other alphaviruses. Curr. Opin. Rheumatol. *16*, 374–379.

Suhrbier, A., Jaffar-Bandjee, M.C., and Gasque, P. (2012). Arthritogenic alphaviruses – an overview. Nat. Rev. Rheumatol. *8*, 420–429.

Sun, C., Gardner, C.L., Watson, A.M., Ryman, K.D., and Klimstra, W.B. (2014). Stable, high-level expression of reporter proteins from improved alphavirus expression vectors to track replication and dissemination during encephalitic and arthritogenic disease. J. Virol. *88*, 2035–2046.

Sun, W., Edelman, R., Kanesa-Thasan, N., Eckels, K.H., Putnak, J.R., King, A.D., Houng, H.S., Tang, D., Scherer, J.M., Hoke, C.H., Jr., and Innis, B.L. (2003). Vaccination of human volunteers with monovalent and tetravalent live-attenuated dengue vaccine candidates. Am. J. Trop. Med. Hyg. *69*, 24–31.

Sun, W., Cunningham, D., Wasserman, S.S., Perry, J., Putnak, J.R., Eckels, K.H., Vaughn, D.W., Thomas, S.J., Kanesa-Thasan, N., Innis, B.L., and Edelman, R. (2009). Phase 2 clinical trial of three formulations of tetravalent live-attenuated dengue vaccine in flavivirus-naive adults. Hum. Vaccin. *5*, 33–40.

Suzuki, R., Winkelmann, E.R., and Mason, P.W. (2009). Construction and characterization of a single-cycle chimeric flavivirus vaccine candidate that protects mice against lethal challenge with dengue virus type 2. J. Virol. *83*, 1870–1880.

Swaminathan, S., Batra, G., and Khanna, N. (2010). Dengue vaccines: state of the art. Exp. Opin. Ther. Pat. *20*, 819–835.

Tabachnik, W.J., Wallis, G.P., Aitkin, T.H.G., Miller, B.R., Amato, G.D., Lorenz, L., Powell, J.R., and Beatty, B.J. (1985). Oral infection of *Aedes aegypti* with yellow fever virus: geographic variation and genetic considerations. Am. J. Trop. Med. Hyg. *34*, 1219–1224.

Tandan, J.B., Ohrr, H., Sohn, Y.M., Yoksan, S., Ji, M., Nam, C.M., and Halstead, S.B. (2007). Single dose of SA 14–14–2 vaccine provides long-term protection against Japanese encephalitis: a case–control study in Nepalese children 5 years after immunization. Vaccine *25*, 5041–5045.

Tarr, G.C., and Hammon, W.M. (1974). Cross-protection between group B arboviruses: Resistance in mice ot Japanese B encephalitis and St. Louis encephalitis viruses induce by dengue virus immunization. Infect. Immun. *9*, 909–915.

Tauber, E., Kollaritsch, H., Korinek, M., Rendi-Wagner, P., Jilma, B., Firbas, C., Schranz, S., Jong, E., Klingler, A., Dewasthaly, S., and Klade, C.S. (2007). Safety and immunogenicity of a Vero-cell-derived, inactivated Japanese encephalitis vaccine: a non-inferiority, phase III, randomised controlled trial. Lancet *370*, 1847–1853.

Tauber, E., Kollaritsch, H., von Sonnenburg, F., Lademann, M., Jilma, B., Firbas, C., Jelinek, T., Beckett, C., Knobloch, J., McBride, W.J., *et al.* (2008). Randomized, double-blind, placebo-controlled phase 3 trial of the safety and tolerability of IC51, an inactivated Japanese encephalitis vaccine. J. Infect. Dis. *198*, 493–499.

Tesh, R.B., Gadjusek, D.C., and Garruto, R.M. (1975). The distribution and prevalence of group A arbovirus neutralizing antibodies among human populations in Southeast Asia and the Pacific Islands. Am. J. Trop. Med. Hyg. *24*, 664–675.

Tesh, R.B., Watts, D.M., Russell, K.L., Damodaran, C., Calampa, C., Cabezas, C., Ramirez, G., Vasquez, B., Hayes, C.G., Rossi, C.A., *et al.* (1999). Mayaro virus disease: an emerging mosquito-borne zoonosis in tropical South America. Clin. Infect. Dis. *28*, 67–73.

Thaikreuea, L., Charearnsook, O., Reanphumkarnkit, S., Dissomboon, P., Phonjan, R., Ratchbud, S., Kounsang, Y., and Buranapiyawong, D. (1997). Chikungunya in Thailand: A re-emerging disease? Southeast Asian J. Trop. Med. Public Health *28*, 359–364.

Theiler, M., and Smith, H.H. (1937a). The effect of prolonged cultivation in vitro upon the pathogenicity of yellow fever virus. J. Exp. Med. *65*, 767–786.

Theiler, M., and Smith, H.H. (1937b). Use of yellow fever virus modified by in vitro cultivation for human immunization. J. Exp. Med. *65*, 787–800.

Thein, S., Linn, M.L., Aaskov, J., Aung, M.M., Aye, M., Zaw, A., and Myint, A. (1992). Development of a simple indirect enzyme-linked immunosorbent assay for the detection of immunoglobulin M antibody in serum from patients following an outbreak of chikungunya virus infection in Yangon, Myanmar. Trans. R. Soc. Trop. Med. Hyg. *86*, 438–442.

Thomas, L.A., Eklund, C.M., Philip, R.N., and Casey, M. (1963). Development of a vaccine against Colorado tick fever for use in man. Am. J. Trop. Med. Hyg. *18*, 678–685.

Thomas, L.A., Philip, R.N., Patzer, E., and Casper, E. (1967). Long duration of neutralizing-antibody response after immunization of man with a formalinized Colorado tick fever vaccine. Am. J. Trop. Med. Hyg. *16*, 60–62.

Thomas, R.E., Lorenzetti, D.L., Spragins, W., Jackson, D., and Williamson, T. (2011). Reporting rates of yellow fever vaccine 17D or 17DD-associated serious adverse events in pharmacovigilance data bases: systematic review. Curr. Drug. Saf. *6*, 145–154.

Thomas, S.J. (2011). The necessity and quandaries of dengue vaccine development. J. Infect. Dis. *203*, 299–303.

Thomas, S.J., and Endy, T.P. (2011a). Critical issues in dengue vaccine development. Curr. Opin. Infect. Dis. *24*, 442–450.

Thomas, S.J., and Endy, T.P. (2011b). Vaccines for the prevention of dengue: Development update. Hum. Vaccine *7*, 674–684.

Tigertt, W.D., and Downs, W.G. (1962). Studies on the virus of Venezuelan equine encephalomyelitis in Trinidad, W.I.I. The 1943–1944 epizootic. Am. J. Trop. Med. Hyg. *11*, 822–834.

Torresi, J., McCarthy, K., Feroldi, E., and Meric, C. (2010). Immunogenicity, safety and tolerability in adults of a new single-dose, live-attenuated vaccine against Japanese encephalitis: Randomised controlled phase 3 trials. Vaccine *28*, 7993–8000.

Troyer, J.M., Hanley, K.A., Whitehead, S.S., Strickman, D., Karron, R.A., Durbin, A.P., and Murphy, B.R. (2001). A live attenuated recombinant dengue-4 virus vaccine candidate with restricted capacity for dissemination in mosquitoes and lack of transmission from vaccinees to mosquitoes. Am. J. Trop. Med. Hyg. *65*, 414–419.

Tsai, T.F. (2014). Alphaviruses: Eastern Equine Encephalitis, In Feigin and Cherry's Textbook of Pediatric Infectious Diseases, J.D. Cherry, Harrison, G.J., Kaplan, S.L., Steinbach, W.J., Hotez, P.J., eds. (Elsevier Saunders, Philadelphia), pp. 2226–2230.

Tsai, T.F., Yu, Y.X., Jia, L.L., Putvatana, R., Zhang, R., Wang, S., and Halstead, S.B. (1998). Immunogenicity of live attenuated SA14-14-2 Japanese encephalitis vaccine – a comparison of 1- and 3-month immunization schedules. J. Infect. Dis. *177*, 221–223.

Tsai, T.R., and Solomon, T. (2004). Flaviviruses (yellow fever, dengue, dengue hemorrhagic fever, Japanese encephalitis, St. Louis encephalitis tick-borne encephalitis). In Principals and Practice of Infectious Diseases, G.L. Mandell, J.E. Bennett, and R. Dolin, eds. (Elsevier, Philadelphia), pp. 1926–1950.

Tsetsarkin, K.A., Chen, R., Leal, G., Forrester, N., Higgs, S., Huang, J., and Weaver, S.C. (2011). Chikungunya virus emergence is constrained in Asia by lineage-specific adaptive landscapes. Proc. Natl. Acad. Sci. U.S.A. *108*, 7872–7877.

Upreti, S.R., Janusz, K.B., Schluter, W.W., Bichha, R.P., Shakya, G., Biggerstaff, B.J., Shrestha, M.M., Sedai, T.R., Fischer, M., Gibbons, R.V., *et al.* (2013). Estimation of the impact of a Japanese encephalitis immunization program with live, attenuated SA 14-14-2 vaccine in Nepal. Am. J. Trop. Med. Hyg. *88*, 464–468.

Valdes, I., Hermida, L., Martin, J., Menendez, T., Gil, L., Lazo, L., Castro, J., Niebla, O., Lopez, C., Bernardo, L., *et al.* (2009). Immunological evaluation in nonhuman primates of formulations based on the chimeric protein P64k-domain III of dengue 2 and two components of Neisseria meningitidis. Vaccine *27*, 995–1001.

Valdes, I., Hermida, L., Gil, L., Lazo, L., Castro, J., Martin, J., Bernardo, L., Lopez, C., Niebla, O., Menendez, T., *et al.* (2010). Heterologous prime-boost strategy in non-human primates combining the infective dengue virus and a recombinant protein in a formulation suitable for human use. Int. J. Infect. Dis. *14*, e377–383.

Vanlandingham, D.L., Hong, C., Klingler, K., Tsetsarkin, K., McElroy, K.L., Powers, A.M., Lehane, M.J., and Higgs, S. (2005). Differential infectivities of o'nyong-nyong and chikungunya virus isolates in *Anopheles gambiae* and *Aedes aegypti* mosquitoes. Am. J. Trop. Med. Hyg. *72*, 616–621.

Vasconcelos, P.F., Luna, E.J., Galler, R., Silva, L.J., Coimbra, T.L., Barros, V.L., Monath, T.P., Rodigues, S.G., Laval, C., Costa, Z.G., *et al.* (2001).

Serious adverse events associated with yellow fever 17DD vaccine in Brazil: a report of two cases. Lancet *358*, 91–97.

Vasilakis, N., Shell, E.J., Fokam, E.B., Mason, P.W., Hanley, K.A., Estes, D.M., and Weaver, S.C. (2007). Potential of ancestral sylvatic dengue-2 viruses to re-emerge. Virology *358*, 402–412.

Vaughn, D.W., Hoke, C.H., Jr., Yoksan, S., LaChance, R., Innis, B.L., Rice, R.M., and Bhamarapravati, N. (1996). Testing of a dengue 2 live-attenuated vaccine (strain 16681 PDK 53) in ten American volunteers. Vaccine *14*, 329–336.

Vaughn, D.W., Green, S., Kalayanarooj, S., Innis, B.L., Nimmannitya, S., Suntayakorn, S., Endy, T.P., Raengsakulrach, B., Rothman, A.L., Ennis, F.A., and Nisalak, A. (2000). Dengue viremia titer, antibody response pattern, and virus serotype correlate with disease severity. J. Infect. Dis. *181*, 2–9.

Volkova, E., Frolova, E., Darwin, J.R., Forrester, N.L., Weaver, S.C., and Frolov, I. (2008). IRES-dependent replication of Venezuelan equine encephalitis virus makes it highly attenuated and incapable of replicating in mosquito cells. Virology 377, 160–169.

Vu-Qui, D., Nguyen-thi, K.T., and Ly, Q.B. (1967). Antibodies to chikungunya virus in Vietnamese children in Saigon. Bull. Soc. Pathol. Exot. *60*, 335–359.

Waddell, M.B., and Taylor, R.M. (1945). Studies on cyclic passage of yellow fever virus in South American mammals and mosquitoes. I Marmosets (*Calithrix autira*) and Cebus monkeys (*Cebus versutus*) in combination with *Aedes aegypti* an *Haemagogus equinus*. Am. J. Trop. Med. 25, 225–230.

Waddell, M.B., and Taylor, R.M. (1946). Studies on the cyclic passage of yellow fever virus to South American mammals and mosquitoes. II.Marmosets (*Callithrix penicilliata* and *Leontocebus crysomela*) in combination with *Aedes aegypti*. Am. J. Trop. Med. *26*, 455–465.

Waddell, M.B., and Taylor, R.M. (1947). Studies on the cyclic passage of yellow fever virus in South American mammals and mosquitoes. III. Further observations on Haemagogus equinus as a vector of the virus. Am. J. Trop. Med. *27*, 471–476.

Waddell, M.B., and Taylor, R.M. (1948). Studies on the cyclic passage of yellow fever virus in South American mammals and mosquitoes. IV.Marsupials (*Metachirus nudicaudatus* and *Marmosa*) in combination with *Aedes aegypti* as vector. Am. J. Trop. Med. *28*, 87–100.

Waldvogel, K., Bossart, W., Huisman, T., Boltshauser, E., and Nadal, D. (1996). Severe tick-borne encephalitis following passive immunization. Eur. J. Pediatr. *155*, 775–779.

Wang, D., Suhrbier, A., Penn-Nicholson, A., Woraratanadharm, J., Gardner, J., Luo, M., Le, T.T., Anraku, I., Sakalian, M., Einfeld, D., and Dong, J.Y. (2011). A complex adenovirus vaccine against chikungunya virus provides complete protection against viraemia and arthritis. Vaccine *29*, 2803–2809.

Wang, E., Ni, H., Xu, R., Barrett, A.D., Watowich, S.J., Gubler, D.J., and Weaver, S.C. (2000). Evolutionary relationships of endemic/epidemic and sylvatic dengue viruses. J. Virol. *74*, 3227–3234.

Wang, E., Volkova, E., Adams, A.P., Forrester, N., Xiao, S.Y., Frolov, I., and Weaver, S.C. (2008). Chimeric alphavirus vaccine candidates for chikungunya. Vaccine *26*, 5030–5039.

Wang, E., Kim, D.Y., Weaver, S.C., and Frolov, I. (2011). Chimeric Chikungunya viruses are nonpathogenic in highly sensitive mouse models but efficiently induce a protective immune response. J. Virol. *85*, 9249–9252.

Warimwe, G.M., Lorenzo, G., Lopez-Gil, E., Reyes-Sandoval, A., Cottingham, M.G., Spencer, A.J., Collins, K.A., Dicks, M.D., Milicic, A., Lall, A., *et al.* (2013). Immunogenicity and efficacy of a chimpanzee adenovirus-vectored Rift Valley fever vaccine in mice. Virol. J. *10*, 349.

Watanaveeradej, V., Simasathien, S., Nisalak, A., Endy, T.P., Jarman, R.G., Innis, B.L., Thomas, S.J., Gibbons, R.V., Hengprasert, S., Samakoses, R., *et al.* (2011). Safety and immunogenicity of a tetravalent live-attenuated dengue vaccine in flavivirus-naive infants. Am. J. Trop. Med. Hyg. *85*, 341–351.

Weaver, S.C. (1997). Convergent evolution of epidemic Venezuelan equine encephalitis viruses. In Factors in the Emergence of Arbovirus Diseases, J.F. Saluzzo, and B. Dodet, eds. (Elsevier, Paris), pp. 241–249.

Weaver, S.C. (2013). Urbanization and geographic expansion of zoonotic arboviral diseases: mechanisms and potential strategies for prevention. Trends Microbiol. *21*, 360–363.

Weaver, S.C., Salas, R., Rico-Hesse, R., Ludwig, G.V., Oberste, M.S., Boshell, J., and Tesh, R.B. (1996). Re-emergence of epidemic Venezuelan equine encephalomyelitis in South America. VEE Study Group. Lancet *348*, 436–440.

Weaver, S.C., Pfeffer, M., Marriott, K., Kang, W., and Kinney, R.M. (1999). Genetic evidence for the origins of Venezuelan equine encephalitis virus subtype IAB outbreaks. Am. J. Trop. Med. Hyg. *60*, 441–448.

Webb, H.E., Wight, D.G.D., Platt, G.S., and Smith, C.E.G. (1967). Langat virus encephalitis in mice. 1. The effect of administration of specific antiserum. J. Hyg. *66*, 343–354.

Webster, D.P., Farrar, J., and Rowland-Jones, S. (2009). Progress towards a dengue vaccine. Lancet Infect. Dis. *9*, 678–687.

White, A., Berman, S., and Lowenthal, J.P. (1972). Comparative immunogenicities of Chikungunya vaccines propagated in monkey kidney monolayers and chick embryo suspension cultures. Appl. Microbiol. *23*, 951–952.

White, L.J., Sariol, C.A., Mattocks, M.D., Wahala, M.P.B.W., Yingsiwaphat, V., Collier, M.L., Whitley, J., Mikkelsen, R., Rodriguez, I.V., Martinez, M.I., *et al.* (2013). An alphavirus vector-based tetravalent dengue vaccine induces a rapid and protective immune response in macaques that differs qualitatively from immunity induced by live virus infection. J. Virol. *87*, 3409–3424.

Widman, D.G., Frolov, I., and Mason, P.W. (2008). Third-generation flavivirus vaccines based on single-cycle, encapsidation-defective viruses. Adv. Virus Res. *72*, 77–126.

Wisseman, C.L., Jr. (1966). Prophylaxis of dengue, with special reference to live virus vaccine. Proc. Japan. Soc. Trop. Med. *7*, 51–56.

Wisseman, C.L., Jr., Sweet, B.H., Rosenzweig, E.C., and Eylar, O.R. (1963). Attenuated Living Type 1 Dengue Vaccines. Am. J. Trop. Med. Hyg. *12*, 620–623.

Wolfe, D.N., Heppner, D.G., Gardner, S.N., Jaing, C., Dupuy, L.C., Schmaljohn, C.S., and Carlton, K. (2014). Current strategic thinking for the development of a trivalent alphavirus vaccine for human use. Am. J. Trop. Med. Hyg. *91*, 442–450.

Wright, P.F., Durbin, A.P., Whitehead, S.S., Ikizler, M.R., Henderson, S., Blaney, J.E., Thumar, B., Ankrah, S., Rock, M.T., McKinney, B.A., *et al.* (2009). Phase 1 trial of the dengue virus type 4 vaccine candidate rDEN4{Delta}30-4995 in healthy adult volunteers. Am. J. Trop. Med. Hyg. *81*, 834–841.

Yaich, M. (2009). Investing in vaccines for developing countries: How public–private partnerships can confront neglected diseases. Hum. Vaccin. *5*, 368–369.

Yoon, I.K., Alera, M.T., Lago, C.B., Tac-An, I.A., Villa, D., Fernandez, S., Thaisomboonsuk, B., Klungthong, C., Levy, J.W., Velasco, J.M., *et al.* (2015). High rate of subclinical chikungunya virus infection and association of neutralizing antibody with protection in a prospective cohort in the Philippines. PLoS Negl. Trop. Dis. *9*, e0003764.

Yu, S., and Aaskov, J.G. (1994). Development of a candidate vaccine against Ross River virus infection. Vaccine *12*, 1118–1124.

Yu, X.J., Liang, M.F., Zhang, S.Y., Liu, Y., Li, J.D., Sun, Y.L., Zhang, L., Zhang, Q.F., Popov, V.L., Li, C., *et al.* (2011). Fever with thrombocytopenia associated with a novel bunyavirus in China. N. Engl. J. Med. *364*, 1523–1532.

Yu, Y. (2010). Phenotypic and genotypic characteristics of Japanese encephalitis attenuated live vaccine virus SA14-14-2 and their stabilities. Vaccine *28*, 3635–3641.

Zacks, M.A., and Paessler, S. (2010). Encephalitic alphaviruses. Vet. Microbiol. *140*, 281–286.

Zhao, L., Zhai, S., Wen, H., Cui, F., Chi, Y., Wang, L., Xue, F., Wang, Q., Wang, Z., Zhang, S., *et al.* (2012). Severe fever with thrombocytopenia syndrome virus, Shandong Province, China. Emerg. Infect. Dis. *18*, 963–965.

Zhu, W., Fu, J., Lu, J., Deng, Y., Wang, H., Wei, Y., Deng, L., Tan, W., and Liang, G. (2013). Induction of humoral and cellular immune responses against hepatitis C virus by vaccination with replicon particles derived from Sindbis-like virus XJ-160. Arch. Virol. *158*, 1013–1019.

Zou, S., Foster, G.A., Dodd, R.Y., Petersen, L.R., and Stramer, S.L. (2010). West Nile fever characteristics among viremic persons identified through blood donor screening. J. Infect. Dis. *202*, 1354–1361.

Small Molecule Drug Development for Dengue Virus

21

Qing-Yin Wang and Pei-Yong Shi

Abstract

Dengue virus imposes one of the largest social and economic burdens of any mosquito borne viral pathogen. Currently, there is no clinically approved vaccine or effective antiviral therapy available. Tremendous progresses have been made in the past decade towards development of antivirals for dengue virus. In this chapter, we summarize the current approaches in antivirals development and report the progress towards discovery of direct acting antivirals or host inhibitors against dengue virus infection. Concerted efforts have to be continuously undertaken and will eventually lead to an effective dengue therapy.

Introduction

Dengue is the most common arboviral illness in humans. It is caused by the infection of dengue virus (DENV). The spectrum of illness ranges from a mild, non-specific febrile syndrome to classic dengue fever (DF), to the severe forms of the disease, dengue haemorrhagic fever (DHF) and dengue shock syndrome (DSS). Owing to many factors (including explosive population growth, poorly planned urbanization, and increased international travel), DENV is now spreading in many settings in the world where it was not previously endemic. A recent study estimated 390 million dengue infections with 96 million infections exhibiting disease symptoms each year (Bhatt *et al.*, 2013). There are four distinct serotypes of DENV (DENV-1, -2, -3, and -4). Infection with one DENV serotype confers lifelong homotypic immunity to that serotype and a very brief period of partial heterotypic immunity to other serotypes, but a person can eventually be infected by all four serotypes. Several serotypes can be in co-circulation during an epidemic. No vaccine or a specific antiviral is currently available for DENV. Current care of dengue patients is supportive in nature, mainly through clinical monitoring and fluid balancing. Therefore, there is an obvious and urgent need to develop effective and tolerated treatments.

DENV belongs to genus *Flavivirus* within family *Flaviviridae*. Besides DENV, many flaviviruses are emerging or reemerging pathogens, including the West Nile virus (WNV), yellow fever virus (YFV), Japanese encephalitis virus (JEV), and tick-borne encephalitis virus (TBEV). The flavivirus genome is a single-strand, plus-sense RNA of approximately 11,000 nucleotides. It consists of a 5′ untranslated region (UTR), a long open reading frame (ORF), and a 3′ UTR. The ORF encodes a polyprotein that is processed into 10 mature proteins, including three structural proteins [capsid (C), premembrane (prM), and envelope (E)] and seven non-structural proteins (NS1, NS2A, NS2B, NS3, NS4A, NS4B, and NS5). Structural proteins are components of virus particles, while non-structural proteins are responsible for viral replication, virion assembly, and evasion of host immune response (Lindenbach, 2007).

A few approaches have been undertaken to develop safe and effective vaccines including attenuation of natural strains by serial passage in cell cultures or deletion of a small fragment within the 3′ UTR of viral genome, development of chimeric viruses by replacing the prM-E genes of the yellow fever virus-17D vaccine with the corresponding genes from DENV, and production of engineered viral proteins using recombinant virus expression systems to elicit specific immune responses to DENV (Rothman, 2011). However, a major challenge is to produce a tetravalent vaccine that is effective against all four serotypes of DENV. The concern is that vaccines effective against only single serotypes may increase the risk of antibody-dependent enhancement if immunized individuals are subsequently infected by a different serotype of DENV. The history of dengue vaccine development and its technical aspects are a vast subject and beyond the scope of this chapter. Here, we focus on the recent development of small molecule antivirals as an intervention strategy for DENV. Therapeutic antibodies have also been pursued with promising results, and are not included in this chapter; readers are encouraged to refer to a recent review on this topic (Chan *et al.*, 2013).

Developmental stages of antivirals

Antivirals are the molecules that block virus replication by targeting a particular step of a virus life cycle without causing unacceptable side effects to the host. The era of antiviral therapy is now in its sixth decade. Up to date, there are more than 60 clinically efficacious antiviral compounds have been developed and are available for treatment of HIV, influenza, hepatitis B virus (HBV), hepatitis C virus (HCV), and herpes simplex virus (HSV). Antiviral development, like all other drug discovery, is long, challenging, and risky. Development of a new medicine takes about 10–15 years on average. The overall process of drug discovery can be divided into four stages.

1 Target identification and validation. The viral proteins are the first and preferred choice of target. Almost all the efficacious antivirals available today work via targeting important virus enzymes, such as polymerases, proteases, or neuraminidase. Host factors are the next choice of target for therapeutic intervention. With the advent of genome-wide screening technologies, our understanding of the complex interaction between the virus and the host are expanding. The host pathways utilized by viral replication may eventually lead to potent, selective, and broad-spectrum antivirals.
2 Lead discovery and optimization. The lead discovery and generation is usually achieved by screening of chemical libraries via target-based or cell-based approach. In target-based approach, screening campaigns are performed against a validated target(s). The screening methodology can be either enzyme activity-based, structure-based *in silico* docking, or rational design depending on the target of choice. In cell-based approach, chemical libraries are assayed for antiviral activity in the context of viral replication. Once a 'hit' is identified, the 'hit' will be validated to ensure its genuine on-target antiviral activity. After validation, the 'hit' becomes a 'lead' for which medicinal chemistry will be performed to improve the potency, toxicity, solubility, and other drug-like properties of the compound. Such lead optimization involves multiple cycles of hypothesis-synthesis-activity process (Jones, 1998).
3 Preclinical development and clinical trials. Preclinical studies are performed in experimental animals to examine the pharmacological and toxicological profiles of a compound, including absorption, distribution, metabolism, excretion, and toxicity (aka ADMET). A compound that successfully emerged from preclinical testing becomes an antiviral candidate, and upon regulatory approval, clinical trials can then be initiated. This stage is a rate-limiting step of the whole antiviral development because translation of preclinical information to clinical output is a step with high attrition rate. Clinical trials include phase I safety test in healthy volunteers, phase II efficacy test in patients, and phase III safety and efficacy test in large number of patients.
4 Final registration and post-marketing studies. Following successful clinical trials, final approval and registration of a drug by health authorities has to be obtained prior to marketing. The antiviral drug then becomes available for clinicians and patients. After registration, the new drug usually enters phase IV clinical trial, a post-marketing study to delineate additional information about treatment's risks, benefits, and optimal use.

Rationale and target product profile

It has been reported that there are distinct differences (>10-fold) in plasma viraemia between severe (DHF/DSS) and non-severe (DF) patients (Libraty *et al.*, 2002a,b). This suggests that reducing viral loads by an antiviral drug in the early stage of viral infection could prevent or lessen the chances of patients from progressing to DHF/DSS. Both Wang *et al.* (2003, 2006) and Guilarde *et al.* (2008) observed the presence of DENV during post-defervescence in patients with severe dengue. These findings imply that the clinical manifestations of severe dengue may, in part, be virally driven and support the hypothesis that antivirals given at later stages may still be beneficial. The current rationale for a dengue antiviral is to rapidly reduce viraemia by >10-fold through treatment with direct-acting antivirals (DAAs) in the early phase of DENV infection; such reduction of viraemia is expected to translate into clinical benefits and prevent/decrease the incidence of DSH/DSS.

Since dengue epidemics occur mainly in developing countries, where children and pregnant women make up a significant portion of patient population, a cheap oral drug that encompasses an excellent safety profile with once daily dose would be ideal. Table 21.1 outlines the target product profile for an ideal antiviral drug for DENV.

Discovery of inhibitors of dengue virus

In this section, we review the progress towards discovery of DAAs and host factor inhibitors. We focus on compound classes that inhibit clinically validated targets (i.e. protease and polymerase), and those that have shown proof-of-concept of antiviral activities in animal models. Interested readers are encouraged to refer to recent reviews in this rapidly evolving area (Noble *et al.*, 2010; Green *et al.*, 2012; Lim *et al.*, 2013).

Classic targets for DAA development

Protease inhibitors

Protease inhibitors (PIs) are an important class of antiviral drugs. PIs prevent viral replication by selectively binding to viral

Table 21.1 Target product profile of a dengue drug

Properties	Minimal product profile	Added value
Route of administration	Oral	
Dosing regimen	Once a day for 3–5 days	
Clinical efficacy	Attenuates the duration and severity of symptoms Reduces incidence of severe dengue disease Active against all four serotypes of viruses	Effective in other flaviviral diseases
Safety	Good safety profile in adult and children with few or no secondary effects	Safe in pregnant women and children Safe for prophylactic use
Chemical manufacturing control	Long shelf-life in endemic countries (>3 years) Low/reasonable costs of goods	

proteases and blocking proteolytic processing of polyprotein precursors that are essential for the production of infectious viral particles. Given the success of PIs in treating HIV/AIDS and more recently HCV infection, DENV NS3 protease is an obvious target for therapeutic intervention. Similar to HCV protease, the serine protease activity of DENV NS3 is dependent on association with a hydrophilic region of 40 amino acids of NS2B. Crystal structures of DENV protease showed that it adopts two different conformations (referred to as 'open' and 'closed' conformation) depending on the absence or presence of a bound ligand (Erbel *et al.*, 2006; Noble *et al.*, 2012, 2013). In the 'open' conformation of the DENV 2 protease (without ligand bound), the C-terminal region of NS2B was positioned away from the protease active site. In the 'closed' conformation of the DENV-2 protease (with a ligand bound), NS2B wraps around the NS3 protease, with the C-terminal residues forming a β-hairpin that forms part of the S2 and S3 subsites and making direct interactions with the P2 and P3 residues of the substrate. A very similar 'closed' conformation was observed for the WNV protease. NMR studies also confirmed the existence of the 'open' and 'closed' conformation and it has been shown that low molecular weight inhibitors can shift the conformational equilibrium towards the 'closed' conformation (Su *et al.*, 2009; de la Cruz *et al.*, 2011; Kang *et al.*, 2013; Kim *et al.*, 2013).

There are two primary avenues for finding inhibitors of DENV protease. The first is to model inhibitors on the sequences of peptide substrate and then optimize pharmacokinetics by eliminating metabolic weak spots and changing the physical properties of the inhibitor so that it can be administered by the oral route. While this is a very demanding strategy, nine out of ten registered HIV protease inhibitors and the recently approved two HCV protease inhibitors have been designed using this approach (Tsantrizos, 2008). We and others have shown that it is possible to develop peptidomimetics with low micromolar inhibition of the DENV protease (Yin *et al.*, 2006a,b; Schuller *et al.*, 2011). By modifying an optimized substrate of DENV protease with different electrophilic functional groups that are designed to form a reversible, covalent bond with the active site serine, the activity of the inhibitor could be increased more than 1000-fold (Yin *et al.*, 2006b). This suggests that similar to the HCV protease inhibitors, such a warhead may be an important component of the DENV protease inhibitors. However, since a potent inhibition still requires at least two basic residues, a cell-permeable and orally bioavailable peptidomimetic drug for DENV protease will be challenging to develop. Cyclisation has been shown to affect the potency of peptide inhibitors (Xu *et al.*, 2012), suggesting that this approach may be viable to develop more potent and membrane-permeable peptide inhibitors. Larger peptides that are stabilized by disulfide bonds also inhibit DENV protease *in vitro* and in cell culture (Rothan *et al.*, 2012), further demonstrating that a peptide-based inhibitor is the most likely source of a DAA to inhibit DENV protease.

The second avenue for lead finding is through *in silico* docking or enzyme-based high-throughput screening (HTS). These activities have resulted in the discovery of several classes of nonpeptidic compounds that bind to the active site of enzyme, but all exhibited low micromolar inhibition *in vitro*; some of these compounds even have charged substituents that cause cell permeability an issue. One particular class of compounds,

Table 21.2 Chemical structures of selected DENV protease inhibitors

Compound	Structure	Reference
1	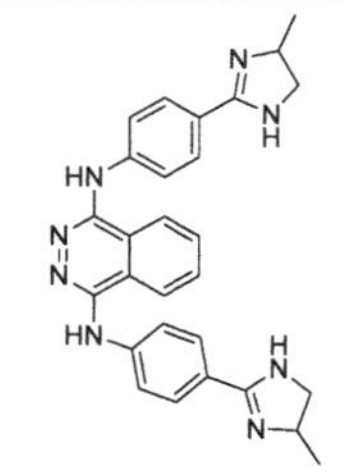	Bodenreider *et al.* (2009)
2	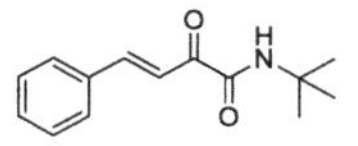	Steuer *et al.* (2011)
3	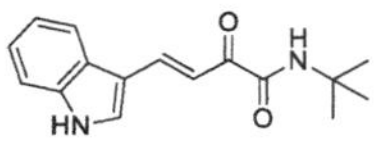	Steuer *et al.* (2011)

compound 1 (Table 21.2) was identified by an enzyme-based HTS campaign (Bodenreider *et al.*, 2009). Compound 1 exhibited an IC_{50} of 6 μM against DENV-2 protease that correlated well with the binding affinity measured by NMR spectroscopy (K_d = 9.4 μM), isothermal titration calorimetry (ITC; K_d = 7.3 μM), and surface plasmon resonance (SPR; K_D = 3.7 μM). To generate a more potent inhibitor, medicinal chemistry was initiated using compound 1 as the starting point. To rationalize the structure–activity relationship (SAR) and guide further compound syntheses, the authors predicated the binding mode of compound 1 by automatic docking using the structure of WNV protease bound with a tetra-peptide aldehyde inhibitor. Approximately 130 compounds were synthesized; however, none of the analogues exhibited submicromolar activity and no clear SAR emerged from these studies. If co-crystal structures of these compounds bound to the DENV protease can be generated, this will certainly guide lead optimization.

Another interesting class of compounds came from the screening of small-molecular aldehydes. By combining the styryl pharmacophore with a ketoamide function to serve as electrophilic trap for the catalytic serine, the authors reported the discovery of a fragment-like lead compound with reasonable target affinity and good ligand efficiency (Table 21.2, compound 2; Steuer *et al.*, 2011), Subsequent SAR study yielded compound 3 showing anti-DENV activity in cell culture, which might serve as a valuable starting point for further development of DENV protease inhibitors.

Proteases from the four serotypes of DENV share very similar substrate specificity (Li *et al.*, 2005). However, the amino acid variation is high among the four serotypes of DENV protease. Such amino acid variation is slightly lower than that of the HCV protease of different genotypes. In 1500 HCV NS3-protease sequences analysed from treatment-naive patient samples, only 47% of amino acids were found to be conserved (Cento *et al.*, 2012). Along the same line, sequence similarity across DENV serotypes for NS3 is between 63 to 74%, suggesting that the genetic barrier to develop resistance could also be low for DENV PI. Therefore, dengue PIs would likely have to be administered as a combination therapy with other DAAs.

Polymerase inhibitors

The RNA-dependent RNA polymerase (RdRp) is located within the C-terminal region of NS5 protein, and is responsible for synthesis of both the negative-strand RNA and positive-strand RNAs. Similar to other nucleic acid polymerases, DENV RdRp has the typical right-hand polymerase structure, consisting of a palm, a finger, and a thumb subdomain (Yap *et al.*, 2007). Since the RdRp is essential for viral replication and a host analogue is not found, it is considered of great value as a target for the antiviral development. Indeed, polymerase inhibitors have been widely used as antivirals against HSV and HBV; the polymerase inhibitors have become the cornerstone in the treatment of HIV/AIDS infection (Cihlar and Ray, 2010). HIV drug discovery has shown two different ways to inhibit viral polymerase, including nucleoside and non-nucleoside inhibitors. Zidovudine, the first drug licensed to treat HIV/AIDS patients, is a nucleoside inhibitor (NI) that is converted to its nucleoside triphosphate analogue by kinases present in the host cells, which then competes with endogenous nucleotide for incorporation into viral DNA. Once incorporated, it serves as a chain terminator of viral transcripts (Cihlar and Ray, 2010). Nevirapine, approved in 1996, is a non-nucleoside inhibitor (NNI) that relies on allosteric disruption of the enzymatic activity (de Bethune, 2010). While the above mentioned drugs target the reverse transcriptase in HIV, the two approaches should be applicable to develop inhibitors of DENV RdRp.

Nucleoside analogues with ribose 2′-C-methly or 4′-C-azido modifications are potent inhibitors of HCV NS5b polymerase (Carroll *et al.*, 2003; Klumpp *et al.*, 2006). Interestingly, 7-deaza-2′-C-methyl-adenosine (MK0608) is a potent inhibitor of DENV both in cell culture and in a DENV viraemia mouse model (Olsen *et al.*, 2004; Schul *et al.*, 2007). Therefore, a panel of adenosine nucleosides containing various 2′-C-alkyl and C-7 substitutions were synthesized to explore the SAR for DENV inhibition, yielding NITD008 (Table 21.3, compound 4) as a potent DENV inhibitor. Antiviral spectrum testing showed that NITD008 is specific to *Flaviviridae* family viruses; it similarly inhibited WNV, YFV, and TBEV (Yin *et al.*, 2009b). Using a DENV replicon, NITD008 suppressed viral RNA synthesis but not viral translation. The triphosphate form of NITD008 inhibited RNA elongation catalysed by a recombinant polymerase of DENV. Moreover, the therapeutic potential of NITD008 was assessed in the DENV mouse models; a single dose or multiple-dose delayed treatment suppressed peak viraemia, reduced cytokine elevation, and completely protected the infected mice from death. However, following 2-week toxicology studies in rats and dogs, a no observed adverse effect level (NOAEL) could not be established for NITD008 in either species, leading to the termination of NITD008 development. As exemplified by NITD008, unpredictable toxicity is a risk and challenge for nucleoside/nucleotide antivirals development; on the other hand, nucleoside/nucleotide analogue have the advantage over other classes of inhibitors because of their high genetic barrier to resistance. Indeed, continuous culturing of DENV or WNV

Table 21.3 Chemical structures of selected DENV RdRp inhibitors

Compound	Structure	Reference
4	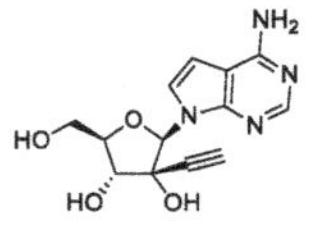	Yin *et al.* (2009b), Chen *et al.* (2010)
5	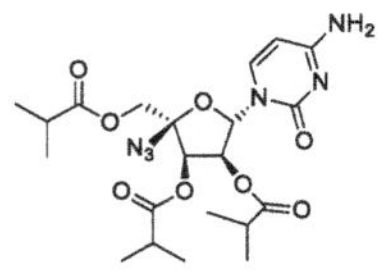	Nguyen *et al.* (2013)
6	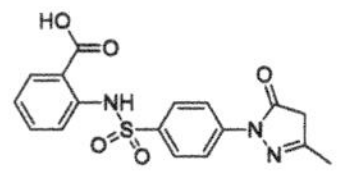	Yin *et al.* (2009a), Niyomrattanakit *et al.* (2010)
7	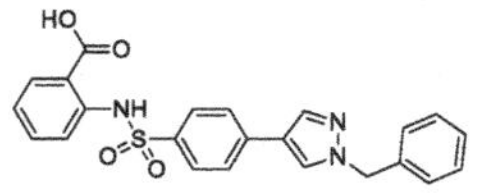	Yin *et al.* (2009a), Niyomrattanakit *et al.* (2010)
8	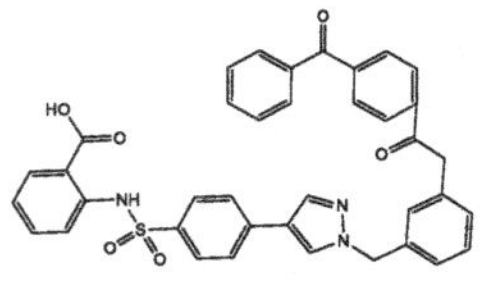	Yin *et al.* (2009a), Niyomrattanakit *et al.* (2010)
9	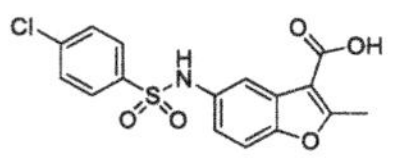	Noble *et al.* (2013)

in cell lines with NITD008 for up to 4 months did not lead to the emergence of resistant viruses (Yin *et al.*, 2009b).

Another interesting nucleoside analogue that has caught the attention of the dengue community is balapiravir (R1626; Table 21.3, compound 5), a tri-isobutyrate ester prodrug of 4-azidocytidine (R1479) that was initially developed for HCV treatment. Balapiravir was successfully tested in a phase IIa study in treatment of naive patients with HCV infections, but the development was later abandoned due to haematological toxicity (Klumpp and Smith, 2011). Since balapiravir has anti-DENV activity, it was repurposed for a phase II trial for the treatment of DENV infection (Nguyen *et al.*, 2013). Dengue patients received balapiravir at doses of 1500 mg ($n = 10$) or 3000 mg ($n = 22$) every 12 h for 5 days. The patients were recruited based on the following criteria: (i) fever with temperature ≥38°C; (ii) a positive NS1 rapid test; (iii) onset of symptoms <48 h prior to initial dosing. Surprisingly, no protective activity of balapiravir was observed. The reasons behind the lack of efficacy need investigation; however, this should not create any pessimism regarding the potential use of NIs for the treatment of DENV infection.

NNIs bind to allosteric pockets of polymerase to block enzymatic activity. The inhibition can be achieved through structural alteration of enzyme to an inactive conformation, blocking the conformational switch from polymerase initiation to elongation, or impeding the processivity of polymerase elongation. In search for DENV RdRp NNIs, we performed a HTS of more than 1 million compounds against a DENV-2 full-length recombinant NS5 protein, from which compound 6 (Table 21.3) was identified with an IC_{50} of 7.2 μM and was used as the starting point for further optimization (Yin *et al.*, 2009a). The optimization effort generated compound 7 (Table 21.3) with an improved IC_{50} of 0.7 μM against DENV RdRp, with no activity against human DNA polymerase or HCV RdRp. To map the site of association between the inhibitor and the polymerase, a photo-affinity experiment was performed with compound 8 (IC_{50} = 1.5 μM, Niyomrattanakit *et al.*, 2010), which irreversibly inhibited the enzyme upon UV irradiation. On the basis of the labelled residue (methionine at position 320) and computational study using the crystal structure of the DENV RdRp, a binding mode was proposed at a novel allosteric site between the finger and the thumb subdomains of the enzyme; binding of the compound to RdRp would occlude RNA template from entering the RNA tunnel of the enzyme. A similar binding mode was recently reported for another small molecule inhibitor, NITD107 (Table 21.3, compound 9, Noble *et al.*, 2013). The cocrystal structure showed that NITD107 binds to the RNA binding groove of the polymerase and thereby inducing conformational changes in the protein. However, cellular efficacy of these compounds was still lacking. The cocrystal structure of NITD107-RdRp has paved the way for lead optimization.

Emerging targets for DAA development

Capsid inhibitor

The DENV C protein is essential in virus assembly to ensure specific encapsidation of the viral genome. The mature form of DENV capsid is a highly basic protein of 12 kDa that forms homodimers in solution consisting of four alpha helices (a1–a4) (Jones *et al.*, 2003; Ma *et al.*, 2004). From a DENV infection-based HTS of approximate 200,000 compounds, Bryd and colleagues identified an inhibitor, ST-148 (Table 21.4, compound 10), which targets the C protein (Byrd *et al.*, 2013). This compound inhibited DENV-2 in a viral titre reduction assay with an EC_{50} of 16 nM and an EC_{90} of 125 nM. It displayed weaker activities against other serotypes, with EC_{50}s of 2.832, 0.512, and 1.150 μM for DENV-1, -3, and -4, respectively. ST-148 resistant DENV-2 harboured a single amino acid change of S34L in C protein. Infectious virus with the engineered mutation showed a 550-fold reduction in susceptibility to ST-148 treatment. Binding of ST-148 to DENV C protein was evaluated by measuring the change in the intrinsic fluorescence of the protein upon compound binding; surprisingly, ST-148 bound equally well to the wild-type and mutant C proteins, raising the question of its mechanism-of-inhibition. Ser34 is located in the α1-α1′ helices of the dimeric C structure. The authors speculated that ST-148 might block the interaction between the C protein and membranes. Following oral or intraperitoneal administration of ST-148 at 50 mg/kg/day in the AG129 mice, the plasma level of ST-148 peaked at 1 h after dosing; the plasma concentration of ST-148 respectively reached concentrations of 468- and 7750-fold above the *in vitro* EC_{50}. On average, twice daily (BID) treatment with ST-148 reduced peak plasma viraemia by 52-fold, and reduced viral load in the spleen and liver by 3- and 20-fold, respectively. The pharmacokinetic data indicate that ST-148 has a limited oral bioavailability, fairly rapid clearance, and good systemic availability following intraperitoneal administration. Overall, the results warrant further preclinical development of ST-148.

Table 21.4 Chemical structures of selected emerging targets inhibitors

Compound	Structure	Reference
10	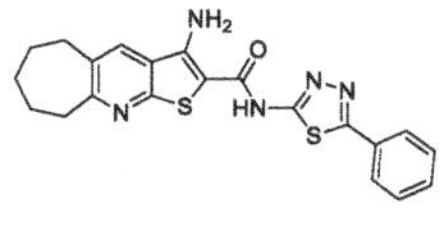	Byrd *et al.* (2013)
11	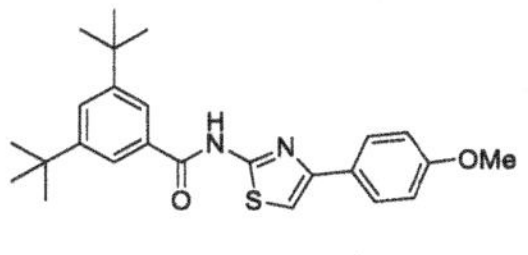	Xie *et al.* (2011)
12	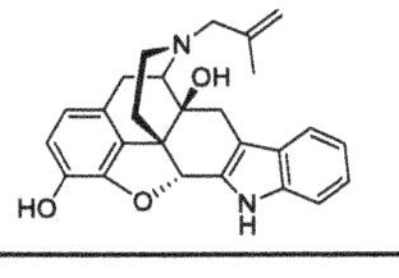	van Cleef *et al.* (2013)

NS4B inhibitors

The less well-characterized NS4B protein has emerged as attractive therapeutic target in a manner similar to targeting HCV NS5A (Gao *et al.*, 2010; Suk-Fong Lok, 2013). NS4B is a small hydrophobic protein of 27 kDa. Its N-terminal region is localized in the endoplasmic reticulum (ER) lumen, while the C-terminal region contains three membrane-spanning segments (Miller *et al.*, 2006). NS4B proteins of DENV serotypes share 78–85% amino acid sequence identity, whereas those of YFV, WNV, and DENV share 35% identity. Flavivirus NS4B is essential for viral replication and for suppression of interferon (IFN) JAK/STAT signalling (Munoz-Jordan *et al.*, 2003, 2005). A number of NS4B inhibitors have been reported for potential antiviral development, though the exact mechanism by which these molecules target NS4B and thereby inhibiting viral replication remains to be elucidated (Xie *et al.*, 2011; van Cleef *et al.*, 2013; Patkar *et al.*, 2009).

From an HTS campaign of 1.8 million compounds against DENV-2 replicon in A549 cells, we identified a potent and pan-serotype DENV inhibitor, NITD-618 (Table 21.4, compound 11), with EC_{50}s ranging from 1.0 to 4.1 μM. It was not active against other RNA viruses (EC_{50} >40 μM), including the two closely related flaviviruses (WNV and YFV), two plus-strand RNA alphaviruses [chikungunya virus, Western equine encephalitis virus (WEEV)], and a negative-strand RNA rhabdovirus [vesicular stomatitis virus (VSV)]. Mechanism of action studies indicated that NITD-618 acts by suppression of viral RNA synthesis, while resistance selection studies identified amino acid changes within the NS4B protein (P104L and A119T); when engineered into the wild-type replicon or DENV, both mutations conferred resistance to NITD-618. Notably, P104 and A119 are conserved among all four serotypes of DENV, but not in other flaviviruses, which might account for the selectivity of NITD-618 in inhibiting DENV. Since residue P104L was previously shown to be important for the NS3–NS4B interaction (Umareddy *et al.*, 2006), it is attempting to speculate that NITD-618 could interrupt such complex formation. Unfortunately, the high lipophilicity of NITD-618 resulted in poor pharmacokinetic properties which hindered testing its activity in the AG129 mouse; medicinal chemistry attempts to reduce its lipophilicity resulted in the loss of activity or the reduction in antiviral selectivity against DENV.

Using a similar DENV-2 replicon assay, van Cleef and coworkers screened the NIH Clinical Collection and reported the identification of a δ-opioid receptor antagonist SDM25N (Table 21.4, compound 12) that inhibits DENV at the step of viral RNA replication (van Cleef *et al.*, 2013). By culturing the DENV replicon cells for several passages under selection of SDM25N and antibiotics, a single amino acid substitution (F164L) in the NS4B protein that confers compound resistance was found. Interestingly, the P104L substitution in the NS4B protein that confers resistance to NITD-618 also provided resistance to SDM25N. Residue P104 is in the predicted transmembrane domain 3 of the NS4B protein, whereas residue F164 resides in the predicted cytoplasmic loop of the protein. It was postulated that substitutions at these two amino acids that are located in different subcellular compartments induce similar conformational changes of the NS4B protein, resulting in resistance to SDM25N. Another interesting observation of SDM25N is that the compound exhibits antiviral activity only in mammalian cells, but not in the C6/36 mosquito cells, suggesting that host environment plays a role in mediating the compound efficacy.

Other groups working on related flaviviruses have also found inhibitors that target NS4B protein. An HTS using pseudo-infectious YFV particles, which express Renilla luciferase in a replication-dependent manner, identified two classes of inhibitors targeting the NS4B protein (Patkar *et al.*, 2009). Virus resistant to compounds CCG-3394 and CCG-4088 harboured a K128R mutation in the cytoplasmic loop of NS4B. Taken together, these three studies establish NS4B as a target for flavivirus drug discovery.

Host target inhibitors

Besides identifying DAAs, looking for host-targeted agents offers another attractive avenue for therapeutic intervention. Successful host-targeted antiviral agents include the HCV drug ribavirin, the recently identified cyclophilin inhibitors, and the HIV entry inhibitor Maravoric that targets CCR5 co-receptor. DENV E protein, like the ones of many different viruses, is glycosylated with N-linked glycans. Since viruses do not encode their own carbohydrate-modifying enzymes, the glycan process and proper folding of the viral glycoproteins rely on host cellular ER α-glucosidase. Inhibition of α-glucosidase disturbs the maturation and function of viral E proteins, preventing the formation of mature, infectious viral particles. Indeed, the essential role of ER α-glucosidase II in the DENV life cycle was recently demonstrated in a genome-wide siRNA knockdown study (Sessions *et al.*, 2009). Besides affecting virion processing, cellular α-glucosidase is also important for the glycan modification of DENV NS1, a protein that is essential for viral replication. Therefore, treatment of infected cells with α-glucosidase inhibitors could also block viral replication (Rathore *et al.*, 2011). Furthermore, α-glucosidase inhibitors, such as castanospermine and celgosivir, have been shown to inhibit DENV in the AG129 mouse model (Rathore *et al.*, 2011; Whitby *et al.*, 2005; Watanabe *et al.*, 2012).

Table 21.5 Chemical structures of selected glucosidase inhibitors

Compound	Structure	Reference
13	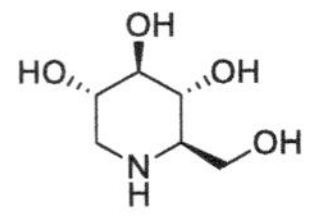	Courageot *et al.* (2000)
14	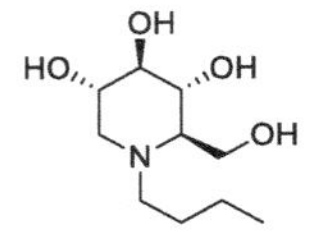	Miller *et al.* (2012)
15	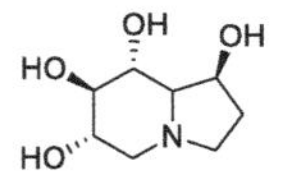	Whitby *et al.* (2005)
16	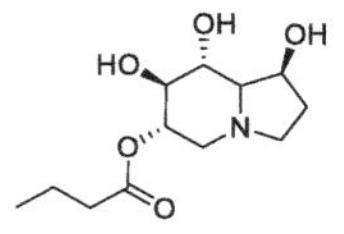	Schul *et al.* (2007), Rathore *et al.* (2011), Watanabe *et al.* (2012)

Inhibitors of ER α-glucosidase are essentially derivatives of two iminosugars, namely 1-deoxynojirimycin (DNJ, Table 21.5, compound 13, Courageot *et al.*, 2000) and castanospermine (CAST; Table 21.5, compound 15, Whitby *et al.*, 2005). Imino sugars are similar in structure to glucose, the substrate of α-glucosidases, so they competitively inhibit the enzyme as transition-state substrate analogues (Hempel *et al.*, 1993). DNJ and CAST, as well as their derivative, have been demonstrated in cell culture and/or animal models to inhibit the morphogenesis of enveloped viruses from at least 12 different families, including DENV, HCV, and HIV/AIDS (Chang *et al.*, 2013). Furthermore, clinical trials of N-butyl-deoxynojirimycin (NB-DNJ, Table 21.5, compound 14) and 6-*O*-butanoyl castanospermine (celgosivir/Bu-Cast, Table 21.5, compound 16) have been evaluated for the safety and antiviral efficacy against HIV/AIDS and HCV, respectively. In both cases, modest reduction in viral titre in the serum of some of the treated patients was observed (Durantel, 2009; Fischl *et al.*, 1994). More importantly, the celgosivir trial showed that this drug was well tolerated in the 12–24 week treatment, at a dosage of 400–600 mg/day, without serious adverse events. Mild to moderate, reversible gastrointestinal symptoms could be controlled with anti-diarrhoea agents and with a low-sucrose/starch and high-glucose diet. Encouraged by the above results, a clinical trial for celgosivir was launched in Singapore in 2012 for treatment of dengue patients (http://www.celaden.sg/). The efficacy results of celgosivir in dengue patients remain to be seen; the outcome of this trial will impact on the development of other alpha-glucosidase inhibitors.

As exemplified by α-glucosidase inhibitors, targeting host processes that are conservatively required by a range of viruses could lead to the development of an antiviral agent with a broad spectrum. In addition, inhibitors of host targets usually have higher genetic barrier. On the other hand, the intrinsic challenge associated with host target is the toxicity. The gastrointestinal side effects observed in the iminosugar trials are due to the non-selective inhibition of intestinal disaccharidases which also use glucose as the substrate. The crystal structure of yeast ER α-glucosidase I has recently been solved (Barker and Rose, 2013). The structure information could be useful for rational design of more potent and specific inhibitors of ER α-glucosidase.

Conclusions and challenges

Tremendous progresses have been made in the past decade towards development of antivirals for DENV. The new knowledge about DENV replication and pathogenesis has provided better rationale for the ongoing antiviral efforts. Despite these progresses, the following knowledge gap and the intrinsic feature of DENV replication complex will continue to be the obstacles for DENV drug development.

1 The target cells and organs of DENV infection in humans. Monocytes and dendritic cells have been well documented to be the main replication sites for DENV (Balsitis *et al.*, 2009). Other tissues and organs for viral replication are poorly defined, especially during the acute viraemia time (Rosen *et al.*, 1999; Jessie *et al.*, 2004). This information is critical to guide where the antiviral drug should pharmacologically target to in patients.
2 Challenges of DENV protease and RdRp. Although protease and polymerase are proven antiviral targets for HIV and HCV, the flavivirus protease and RdRp have their intrinsic challenges for antiviral development. Specifically, the active site of flavivirus protease is very flat (Noble *et al.*, 2012) and the P1/P3 residues of the protease substrate are highly charged Arg or Lys, making the design of peptidomimetics difficult (Noble and Shi, 2012). For RdRp, the stability and enzyme activity of the recombinant RdRp are low, preventing a robust enzyme-based HTS (Lim *et al.*, 2013).
3 Low chance to identify inhibitors of viral targets from phenotypic screening. The replicon- and viral infection-based HTS has rarely revealed inhibitors of viral targets. This is because a much greater number of host proteins than the 10 viral proteins are required to support viral replication. As mentioned before, DAA is a more straightforward approach than the host-targeting approach for antiviral development. In reality, cell-based phenotypic screening has hardly generated 'hits' that inhibits viral targets.
4 Lack of animal model that depictures human diseases. Although the AG129 mouse model can be used to examine antiviral activities by monitoring viraemia reduction, this mouse model lacks innate immune response (due to lack of interferon receptors) and does not depicture human diseases. Therefore, the current model cannot be used to evaluate the antiviral effects on the relief of disease symptoms. Future development of humanized mouse model could be useful for studies of dengue pathogenesis as well as evaluation of vaccine and antivirals.

Compared with DAA, inhibitors that can reverse the pathophysiological pathway of severe dengue disease have a clear advantage by offering longer window of therapeutic time. Future research should be encouraged to understand and define the molecular mechanism of dengue pathogenesis such as vascular leakage. Ideally, dengue therapy should ultimately combine one DAA drug with another drug to reverse the pathology of DHF/DSS. Though challenging, we are optimistic that the continuous, concerted effort will lead to an effective dengue therapy.

Finally, two critical questions are frequently asked about dengue drug discovery. Is there a realistic treatment window for such an acute viral infection? Will antiviral drug reduce viraemia and prevent severe diseases? The only way to answer these questions is to test safe compounds with demonstrated antiviral activity and safety window in human patients through clinical trial.

References

Balsitis, S.J., Coloma, J., Castro, G., Alava, A., Flores, D., McKerrow, J.H., Beatty, P.R., and Harris, E. (2009). Tropism of dengue virus in mice and humans defined by viral non-structural protein 3-specific immunostaining. Am. J. Trop. Med. Hyg. *80*, 416–424.

Barker, M.K., and Rose, D.R. (2013). Specificity of processing α-glucosidase I is guided by the substrate conformation: crystallographic and in silico studies. J. Biol. Chem. *288*, 13563–13574.

de Bethune, M.P. (2010). Non-nucleoside reverse transcriptase inhibitors (NNRTIs), their discovery, development, and use in the treatment of HIV-1 infection: a review of the last 20 years (1989–2009). Antiviral Res. *85*, 75–90.

Bhatt, S., Gething, P.W., Brady, O.J., Messina, J.P., Farlow, A.W., Moyes, C.L., Drake, J.M., Brownstein, J.S., Hoen, A.G., Sankoh, O., *et al.* (2013). The global distribution and burden of dengue. Nature *496*, 504–507.

Bodenreider, C., Beer, D., Keller, T.H., Sonntag, S., Wen, D., Yap, L., Yau, Y.H., Shochat, S.G., Huang, D., Zhou, T., *et al.* (2009). A fluorescence quenching assay to discriminate between specific and nonspecific inhibitors of dengue virus protease. Anal. Biochem. *395*, 195–204.

Byrd, C.M., Dai, D., Grosenbach, D.W., Berhanu, A., Jones, K.F., Cardwell, K.B., Schneider, C., Wineinger, K.A., Page, J.M., Harver, C., *et al.* (2013). A novel inhibitor of dengue virus replication that targets the capsid protein. Antimicrob. Agents Chemother. *57*, 15–25.

Carroll, S.S., Tomassini, J.E., Bosserman, M., Getty, K., Stahlhut, M.W., Eldrup, A.B., Bhat, B., Hall, D., Simcoe, A.L., LaFemina, R., *et al.* 2003. Inhibition of hepatitis C virus RNA replication by 2′-modified nucleoside analogs. J. Biol. Chem. *278*, 11979–11984.

Cento, V., Mirabelli, C., Salpini, R., Dimonte, S., Artese, A., Costa, G., Mercurio, F., Svicher, V., Parrotta, L., Bertoli, A., *et al.* (2012). HCV genotypes are differently prone to the development of resistance to linear and macrocyclic protease inhibitors. PLoS ONE *7*, e39652.

Chan, K.R., Ong, E.Z., and Ooi, E.E. (2013). Therapeutic antibodies as a treatment option for dengue fever. Exp. Rev. Anti-infective Ther. *11*, 1147–1157.

Chang, J., Block, T.M., and Guo, J.T. (2013). Antiviral therapies targeting host ER alpha-glucosidases: Current status and future directions. Antiviral Res. *99*, 251–260.

Chen, Y.L., Yin, Z., Duraiswamy, J., Schul, W., Lim, C.C., Liu, B., Xu, H.Y., Qing, M., Yip, A., Wang, G., *et al.* (2010). Inhibition of dengue virus RNA synthesis by an adenosine nucleoside. Antimicrobial Agents Chemother. *54*, 2932–2939.

Cihlar, T., and Ray, A.S. (2010). Nucleoside and nucleotide HIV reverse transcriptase inhibitors: 25 years after zidovudine. Antiviral Res. *85*, 39–58.

van Cleef, K.W.R., Overheul, G.J., Thomassen, M.C., Kaptein, S.J.F., Davidson, A.D., Jacobs, M., Neyts, J., van Kuppeveld, F.J.M., and van Rij, R.P. (2013). Identification of a new dengue virus inhibitor that targets the viral NS4B protein and restricts genomic RNA replication. Antiviral Res. *99*, 165–171.

Courageot, M.P., Frenkiel, M.P., Dos Santos, C.D., Deubel, V., and Despres, P. (2000). Alpha-glucosidase inhibitors reduce dengue virus production by affecting the initial steps of virion morphogenesis in the endoplasmic reticulum. J. Virol. *74*, 564–572.

de la Cruz, L., Nguyen, T.H. D., Ozawa, K., Shin, J., Graham, B., Huber, T., Otting, G., 2011. Binding of low molecular weight inhibitors promotes large conformational changes in the dengue virus NS2B-NS3 protease: fold analysis by pseudocontact shifts. J. Am. Chem. Soc. *133*, 19205–19215.

Durantel, D. (2009). Celgosivir, an alpha-glucosidase I inhibitor for the potential treatment of HCV infection. Curr. Opin. Investig. Drugs *10*, 860–870.

Erbel, P., Schiering, N., D'Arcy, A., Renatus, M., Kroemer, M., Lim, S.P., Yin, Z., Keller, T.H., Vasudevan, S.G., and Hommel, U. (2006). Structural basis for the activation of flaviviral NS3 proteases from dengue and West Nile virus. Nat. Struct. Mol. Biol. *13*, 372–373.

Fischl, M.A., Resnick, L., Coombs, R., Kremer, A.B., Pottage, J.C., Jr., Fass, R.J., Fife, K.H., Powderly, W.G., Collier, A.C., and Aspinall, R.L. (1994). The safety and efficacy of combination N-butyl-deoxynojirimycin (SC-48334) and zidovudine in patients with HIV-1 infection and 200–500 CD4 cells/mm3. J. Acquir. Immune Defic. Syndr. *7*, 139–147.

Gao, M., Nettles, R.E., Belema, M., Snyder, L.B., Nguyen, V.N., Fridell, R.A., Serrano-Wu, M.H., Langley, D.R., Sun, J.H., O'Boyle, D.R., *et al.* (2010). Chemical genetics strategy identifies an HCV NS5A inhibitor with a potent clinical effect. Nature *465*, 96–100.

Green, J., Bandarage, U., Luisi, K., and Rijnbrand, R. (2012). Recent advances in the discovery of dengue virus inhibitors. In Annual Reports in Medicinal Chemistry, vol. 47, C.D. Manoj, ed. (Academic Press, Burlington, MA, USA), pp. 297–317.

Guilarde, A.O., Turchi, M.D., Siqueira, J.B., Jr., Feres, V.C., Rocha, B., Levi, J.E., Souza, V.A., Boas, L.S., Pannuti, C.S., and Martelli, C.M. (2008). Dengue and dengue hemorrhagic fever among adults: clinical outcomes related to viremia, serotypes, and antibody response. J. Infect. Dis. *197*, 817–824.

Hempel, A., Camerman, N., Mastropaolo, D., and Camerman, A. (1993). Glucosidase inhibitors: structures of deoxynojirimycin and castanospermine. J. Med. Chem. *36*, 4082–4086.

Jessie, K., Fong, M.Y., Devi, S., Lam, S.K., and Wong, K.T. (2004). Localization of dengue virus in naturally infected human tissues, by immunohistochemistry and in situ hybridization. J. Infect. Dis. *189*, 1411–1418.

Jones, C.T., Ma, L., Burgner, J.W., Groesch, T.D., Post, C.B., and Kuhn, R.J. (2003). Flavivirus capsid is a dimeric alpha-helical protein. J. Virol. *77*, 7143–7149.

Jones, P.S. (1998). Strategies for antiviral drug discovery. Antivir. Chem. Chemother. *9*, 283–302.

Kang, C., Gayen, S., Wang, W., Severin, R., Chen, A.S., Lim, H.A., Chia, C.S.B., Schüller, A., Doan, D.N.P., Poulsen, A., Hill, J., Vasudevan, S.G., and Keller, T.H. (2013). Exploring the binding of peptidic West Nile virus NS2BGÇôNS3 protease inhibitors by NMR. Antiviral Res. *97*, 137–144.

Kim, Y.M., Gayen, S., Kang, C., Joy, J., Huang, Q., Chen, A.S., Wee, J.L.K., Ang, M.J.Y., Lim, H.A., Hung, A.W., *et al.* (2013). NMR analysis of a novel enzymatically active unlinked dengue NS2B-NS3 protease complex. J. Biol. Chem. *288*, 12891–12900.

Klumpp, K., Leveque, V., Le, P.S., Ma, H., Jiang, W.R., Kang, H., Granycome, C., Singer, M., Laxton, C., Hang, J.Q., *et al.* (2006). The novel nucleoside analog R1479 (4′-azidocytidine) is a potent inhibitor of NS5B-dependent RNA synthesis and hepatitis C virus replication in cell culture. J. Biol. Chem. *281*, 3793–3799.

Klumpp, K., and Smith, D.B. (2011). Discovery and clinical evaluation of the nucleoside analog Balapiravir (R1626) for the treatment of HCV infection. In Antiviral Drugs, Kazmierski, W.M., ed. (John Wiley & Sons, Inc., Hoboken, NJ, USA), pp. 287–304.

Li, J., Lim, S.P., Beer, D., Patel, V., Wen, D., Tumanut, C., Tully, D.C., Williams, J.A., Jiricek, J., Priestle, J.P., Harris, J.L., and Vasudevan, S.G. (2005). Functional profiling of recombinant NS3 proteases from all four serotypes of dengue virus using tetrapeptide and octapeptide substrate libraries. J. Biol. Chem. *280*, 28766–28774.

Libraty, D.H., Endy, T.P., Houng, H.S., Green, S., Kalayanarooj, S., Suntayakorn, S., Chansiriwongs, W., Vaughn, D.W., Nisalak, A., Ennis, F.A., and Rothman, A.L. (2002a). Differing influences of virus burden and immune activation on disease severity in secondary dengue-3 virus infections. J. Infect. Dis. *185*, 1213–1221.

Libraty, D.H., Young, P.R., Pickering, D., Endy, T.P., Kalayanarooj, S., Green, S., Vaughn, D.W., Nisalak, A., Ennis, F.A., and Rothman, A.L. (2002b). High circulating levels of the dengue virus non-structural protein NS1 early in dengue illness correlate with the development of dengue hemorrhagic fever. J. Infect. Dis. *186*, 1165–1168.

Lim, S.P., Koh, J.H.K., Seh, C.C., Liew, C.W., Davidson, A.D., Chua, L.S., Chandrasekaran, R., Cornvik, T.C., Shi, P.Y., and Lescar, J. (2013). A crystal structure of the Dengue virus NS5 polymerase delineates inter-domain amino acids residues that enhance its thermostability and de novo initiation activities. J. Biol. Chem. *288*, 31105–31114.

Lim, S.P., Wang, Q.Y., Noble, C.G., Chen, Y.L., Dong, H., Zou, B., Yokokawa, F., Nilar, S., Smith, P., Beer, D., Lescar, J., and Shi, P.-Y. (2013). Ten years of dengue drug discovery: Progress and prospects. Antiviral Res. *100*, 500–519.

Lindenbach, B.D., Thiel, H.J., and Rice, C.M. (2007). Flaviviridae: the viruses and their replication. In: Fields Virology, D.M. Knipe and P.M. Howley, eds. (Lippincott William & Wilkins, Philadelphia), pp. 1101–1152.

Ma, L., Jones, C.T., Groesch, T.D., Kuhn, R.J., and Post, C.B. (2004). Solution structure of dengue virus capsid protein reveals another fold. Proc. Natl. Acad. Sci. U.S.A. *101*, 3414–3419.

Miller, J.L., Lachica, R., Sayce, A.C., Williams, J.P., Bapat, M., Dwek, R., Beatty, P.R., Harris, E., and Zitzmann, N. (2012). Liposome-mediated delivery of iminosugars enhances efficacy against dengue virus in vivo. Antimicrob. Agents Chemother. *56*, 6379–6386.

Miller, S., Sparacio, S., and Bartenschlager, R. (2006). Subcellular localization and membrane topology of the Dengue virus type 2 Non-structural protein 4B. J. Biol. Chem. *281*, 8854–8863.

Munoz-Jordan, J.L., Sanchez-Burgos, G.G., Laurent-Rolle, M., and Garcia-Sastre, A. (2003). Inhibition of interferon signaling by dengue virus. Proc. Natl. Acad. Sci. U.S.A. *100*, 14333–14338.

Munoz-Jordan, J.L., Laurent-Rolle, M., Ashour, J., Martinez-Sobrido, L., Ashok, M., Lipkin, W.I., and Garcia-Sastre, A. (2005). Inhibition of alpha/beta interferon signaling by the NS4B protein of flaviviruses. J. Virol. *79*, 8004–8013.

Nguyen, N.M., Tran, C.N.B., Phung, L.K., Duong, K.T.H., Huynh, H.l.A., Farrar, J., Nguyen, Q.T.H., Tran, H.T., Nguyen, C.V.V., Merson, L., *et al.* (2013). A randomized, double-blind placebo controlled trial of Balapiravir, a polymerase inhibitor, in adult dengue patients. J. Infect. Dis. *207*, 1442–1450.

Niyomrattanakit, P., Chen, Y.L., Dong, H., Yin, Z., Qing, M., Glickman, J.F., Lin, K., Mueller, D., Voshol, H., Lim, J.Y.H., *et al.* (2010). Inhibition of dengue virus polymerase by blocking of the RNA tunnel. J. Virol. *84*, 5678–5686.

Noble, C.G., and Shi, P.Y. (2012). Structural biology of dengue virus enzymes: Towards rational design of therapeutics. Antiviral Res. *96*, 115–126.

Noble, C.G., Chen, Y.L., Dong, H., Gu, F., Lim, S.P., Schul, W., Wang, Q.Y., and Shi, P.Y. (2010). Strategies for development of dengue virus inhibitors. Antiviral Res. *85*, 450–462.

Noble, C.G., Seh, C.C., Chao, A.T., and Shi, P.Y. (2012). Ligand-bound structures of the dengue virus protease reveal the active conformation. J. Virol. *86*, 438–446.

Noble, C.G., Lim, S.P., Chen, Y.L., Liew, C.W., Yap, L., Lescar, J., and Shi, P.Y. (2013). Conformational flexibility of the dengue virus RNA-dependent RNA polymerase revealed by a complex with an inhibitor. J. Virol. *87*, 5291–5295.

Olsen, D.B., Eldrup, A.B., Bartholomew, L., Bhat, B., Bosserman, M.R., Ceccacci, A., Colwell, L.F., Fay, J.F., Flores, O.A., Getty, K.L., *et al.* (2004). A 7-deaza-adenosine analog is a potent and selective inhibitor of hepatitis C virus replication with excellent pharmacokinetic properties. Antimicrobial Agents Chemotherapy *48*, 3944–3953.

Patkar, C.G., Larsen, M., Owston, M., Smith, J.L., and Kuhn, R.J. (2009). Identification of inhibitors of yellow fever virus replication using a replicon-based high-throughput assay. Antimicrobial Agents Chemotherapy *53*, 4103–4114.

Rathore, A.P., Paradkar, P.N., Watanabe, S., Tan, K.H., Sung, C., Connolly, J.E., Low, J., Ooi, E.E., and Vasudevan, S.G. (2011). Celgosivir treatment misfolds dengue virus NS1 protein, induces cellular pro-survival genes and protects against lethal challenge mouse model. Antiviral Res. *92*, 453–460.

Rosen, L., Drouet, M.T., and Deubel, V. (1999). Detection of dengue virus RNA by reverse transcription-polymerase chain reaction in the liver and lymphoid organs but not in the brain in fatal human infection. Am. J. Trop. Med. Hyg. *61*, 720–724.

Rothan, H.A., Bdulrahman, A.Y., Asikumer, P.G., Thman, S., Ahman, N.A., and Yusof, R. (2012). Protegrin-1 inhibits dengue NS2B-NS3 serine protease and viral replication in MK2 cells. J. Biomed. Biotechnol. *2012*, 251482.

Rothman, A.L. (2011). Immunity to dengue virus: a tale of original antigenic sin and tropical cytokine storms. Nat. Rev. Immunol. *11*, 532–543.

Sayce, A.C., Miller, J.L., and Zitzmann, N. (2010). Targeting a host process as an antiviral approach against dengue virus. Trends Microbiol. *18*, 323–330.

Schul, W., Liu, W., Xu, H.Y., Flamand, M., and Vasudevan, S.G. (2007). A dengue fever viremia model in mice shows reduction in viral replication and suppression of the inflammatory response after treatment with antiviral drugs. J. Infect. Dis. *195*, 665–674.

Schuller, A., Yin, Z., Brian Chia, C.S., Doan, D.N., Kim, H.K., Shang, L., Loh, T.P., Hill, J., and Vasudevan, S.G. (2011). Tripeptide inhibitors of dengue and West Nile virus NS2B-NS3 protease. Antiviral Res. *92*, 96–101.

Sessions, O.M., Barrows, N.J., Souza-Neto, J.A., Robinson, T.J., Hershey, C.L., Rodgers, M.A., Ramirez, J.L., Dimopoulos, G., Yang, P.L., Pearson, J.L., and Garcia-Blanco, M.A. (2009). Discovery of insect and human dengue virus host factors. Nature *458*, 1047–1050.

Steuer, C., Gege, C., Fischl, W., Heinonen, K.H., Bartenschlager, R., and Klein, C.D. (2011). Synthesis and biological evaluation of +¦-ketoamides as inhibitors of the Dengue virus protease with antiviral activity in cell-culture. Bioorganic Medicinal Chem. *19*, 4067–4074.

Su, X.C., Ozawa, K., Qi, R., Vasudevan, S.G., Lim, S.P., and Otting, G. (2009). NMR analysis of the dynamic exchange of the NS2B cofactor between open and closed conformations of the West Nile virus NS2B-NS3 protease. PLoS Negl. Trop. Dis. 3, e561.

Suk-Fong Lok, A. (2013). HCV NS5A inhibitors in development. Clinics Liver Dis. *17*, 111–121.

Tsantrizos, Y.S. (2008). Peptidomimetic therapeutic agents targeting the protease enzyme of the human immunodeficiency virus and hepatitis C virus. Accounts Chem. Res. *41*, 1252–1263.

Umareddy, I., Chao, A., Sampath, A., Gu, F., and Vasudevan, S.G. (2006). Dengue virus NS4B interacts with NS3 and dissociates it from single-stranded RNA. J. Gen. Virol. *87*, 2605–2614.

Wang, W.K., Chao, D.Y., Kao, C.L., Wu, H.C., Liu, Y.C., Li, C.M., Lin, S.C., Ho, S.T., Huang, J.H., and King, C.C. (2003). High levels of plasma dengue viral load during defervescence in patients with dengue hemorrhagic fever: implications for pathogenesis. Virology *305*, 330–338.

Wang, W.K., Chen, H.L., Yang, C.F., Hsieh, S.C., Juan, C.C., Chang, S.M., Yu, C.C., Lin, L.H., Huang, J.H., and King, C.C. (2006). Slower rates of clearance of viral load and virus-containing immune complexes in patients with dengue hemorrhagic fever. Clin. Infect. Dis. *43*, 1023–1030.

Watanabe, S., Rathore, A.P., Sung, C., Lu, F., Khoo, Y.M., Connolly, J., Low, J., Ooi, E.E., Lee, H.S., and Vasudevan, S.G. (2012). Dose- and schedule-dependent protective efficacy of celgosivir in a lethal mouse model for dengue virus infection informs dosing regimen for a proof of concept clinical trial. Antiviral Res. *96*, 32–35.

Whitby, K., Pierson, T.C., Geiss, B., Lane, K., Engle, M., Zhou, Y., Doms, R.W., and Diamond, M.S. (2005). Castanospermine, a potent inhibitor of dengue virus infection in vitro and in vivo. J. Virol. *79*, 8698–8706.

Xie, X., Wang, Q.Y., Xu, H.Y., Qing, M., Kramer, L., Yuan, Z., and Shi, P.Y. (2011). Inhibition of dengue virus by targeting viral NS4B protein. J. Virol. *85*, 11183–11195.

Xu, S., Li, H., Shao, X., Fan, C., Ericksen, B., Liu, J., Chi, C., and Wang, C. (2012). Critical effect of peptide cyclization on the potency of peptide inhibitors against Dengue virus NS2B-NS3 protease. J. Med. Chem. *55*, 6881–6887.

Yap, T.L., Xu, T., Chen, Y.L., Malet, H., Egloff, M.P., Canard, B., Vasudevan, S.G., and Lescar, J. (2007). Crystal structure of the dengue virus RNA-dependent RNA polymerase catalytic domain at 1.85-angstrom resolution. J. Virol. *81*, 4753–4765.

Yin, Z., Patel, S.J., Wang, W.L., Chan, W.L., Ranga Rao, K.R., Wang, G., Ngew, X., Patel, V., Beer, D., Knox, J.E., *et al.* (2006a). Peptide inhibitors of dengue virus NS3 protease. Part 2: SAR study of tetrapeptide aldehyde inhibitors. Bioorg. Med. Chem. Lett. *16*, 40–43.

Yin, Z., Patel, S.J., Wang, W.L., Wang, G., Chan, W.L., Rao, K.R., Alam, J., Jeyaraj, D.A., Ngew, X., Patel, V., *et al.* (2006b). Peptide inhibitors of Dengue virus NS3 protease. Part 1: Warhead. Bioorg. Med. Chem. Lett. *16*, 36–39.

Yin, Z., Chen, Y.L., Kondreddi, R.R., Chan, W.L., Wang, G., Ng, R.H., Lim, J.Y.H., Lee, W.Y., Jeyaraj, D.A., Niyomrattanakit, P., *et al.* (2009a). N-Sulfonylanthranilic acid derivatives as allosteric inhibitors of dengue viral RNA-dependent RNA polymerase. J. Med. Chem. *52*, 7934–7937.

Yin, Z., Chen, Y.L., Schul, W., Wang, Q.Y., Gu, F., Duraiswamy, J., Reddy Kondreddi, R., Niyomrattanakit, P., Lakshminarayana, S.B., Goh, A., *et al.* (2009b). An adenosine nucleoside inhibitor of dengue virus. Proc. Natl. Acad. Sci. U.S.A. *106*, 20435–20439.

Part IV

Future Trends

Arbovirology: Back to the Future

22

Robert B. Tesh and Charles H. Calisher

Abstract

This chapter briefly reviews the origins and development of arbovirology as a discipline and the changes that have occurred in the focus, techniques and orientation of arbovirus research over time. Arbovirus research was initially patient- or disease-oriented and the focus was mainly on diagnosis, treatment and control. Currently, arbovirus research is primarily focused on the pathogenetic aspects of arboviral diseases and on the structural and genetic characteristics of their aetiological agents. The early emphasis on field work, virus isolation and shoe leather epidemiology has been largely replaced by laboratory-based molecular and genetic studies, non-cultural diagnostic techniques and computer modelling of arboviral cycles and outbreaks. Both types of research (field and laboratory-based) provide valuable insights into the emergence, prevention and control of arboviral diseases, but the older classical methods have recently fallen into disfavour. Some of the contributing factors leading to this situation are discussed and recommendations are provided on how to better balance the teaching, funding and practice of arbovirology.

The very early days of viral disease diagnosis

Well before the dawn of modern disease diagnosis, people who were sick recovered, remained sick, or died. Tribal shamans or elders might make pronouncements based on their previous experiences or observations, but no useful tools were available for specific diagnostic purposes. Later, as society and education developed and individuals were trained in medicine, some advances were made but specific, accurate, and precise diagnoses still were impossible. However, certain diseases had pathognomonic signs and so the diagnostic possibilities were narrowed. That is, medicine moved from 'guesses' to 'good guesses', sometimes even better than good guesses, as experts in various areas of treatment became more available or were taught by previous generations.

Accurate diagnoses of diseases are important. People want to know what is making them sick or killing them; farmers want to know what is reducing their herds or crops; public health officials want to know the cause and source of illness in order to prevent or even cure them. But these concerns reflect a more modern perspective. Epidemiology did not appear as a science until nearly 200 years ago, when Pierre Charles Alexandre Louis began accumulating data regarding infectious disease states. It was only then that odd, coincidental or deliberate observations (Anton van Leeuwenhoek's invention of the microscope; Dimitri Iosifovich Ivanovsky's and Martinus Beijerinck's independent discoveries that tobacco mosaic disease is caused by a virus; Ignaz Semmelweis' recognition of the relationship between puerperal ('childbed') fever, principally caused by streptococci, and the lack of hand washing by physicians examining women delivering babies; John Snow's study of cholera in London; and contributions by Agostino Bassi, Louis Pasteur, Robert Koch and Joseph Lister to our understanding of the 'germ theory') were assembled into an orderly overview did it become obvious that further advances were necessary (Murphy, 1989).

The aetiological agent of yellow fever, a plague recognized for hundreds of years but not adequately diagnosed until the turn of the last century, was the first virus identified as causing a human disease. Using humans as experimental hosts, Walter Reed, James Carroll, Aristides Agramonte, and Jesse Lazear proved Carlos Finlay's hypothesis regarding the existence of an arthropod vector and its relationship to transmission of yellow fever virus. At the same time, their work led to the validation and recognition of the significance of Henry Rose Carter's hypothesis that there is an 'extrinsic incubation period' of this agent in mosquitoes.

In 1927, Thomas M. Rivers published a paper describing the differentiation of bacteria and viruses, establishing virology as a field of study distinct from bacteriology, and establishing virology as a separate discipline of clinical interest (Murphy, 1989). The next year, Wilbur Sawyer, Frank Horsfall, Richard Shope, and others joined Rivers in founding the Rockefeller Foundation Virus Laboratory; activities of the laboratory began in earnest in 1928, with Sawyer as director (Calisher, 2013). Thirty-two yellow fever infections, including five deaths, were reported in workers in the New York, Brazil, and Africa laboratories of the Rockefeller Foundation and Sawyer rightly considered it high priority to devise a vaccine to protect those working with the virus. Additional work was necessary before this could come to fruition. Edward Jenner had prepared a crude but effective vaccine against smallpox more than 100 years previously and Arnold Theiler and others had discovered the first bluetongue virus and prepared a rudimentary vaccine against it (Murphy, 1989). Still, accurate and sophisticated viral diagnosis was a very long way from being an established discipline.

With Wray Lloyd and others, Sawyer showed that *Aedes aegypti* mosquitoes could transmit yellow fever virus to laboratory mice and to monkeys. They adapted yellow fever virus (YFV) to replicate in various cell cultures *in vitro* and showed that the adapted virus was at least partially attenuated. This talented group of Rockefeller Foundation scientists developed an intraperitoneal protection test in mice, showed that monkeys given this attenuated virus preparation became immune to challenge with fully virulent YFV, and showed that otherwise susceptible monkeys administered serum from immune monkeys could also withstand challenge with virulent strains of YFV – thus the neutralization test. Sawyer's vaccine was protective, but a better and less reactogenic vaccine still was needed. These pioneering efforts to develop not simply a vaccine but useful techniques for studying viruses and for conducting studies which provided information about YFV and the disease it causes were a huge series of leaps forward.

Max Theiler, then at the Department of Tropical Medicine of the Harvard Medical School, had begun working with YFV at a time when facilities were primitive, techniques generally undeveloped, and supplies and adequate safety protocols unavailable. At the same time, the Rockefeller Foundation was emphasizing studies of yellow fever in Africa and in the Americas. Its focus was principally on geographic distribution, impact on local populations, and disease pathology. When the neutralization test was applied, more specificity was introduced for diagnosis and when it was shown that the albino laboratory mouse could be used to amplify the virus, other advances were made. Prior to these discoveries, monkeys had been used as the preferred hosts for laboratory studies of YFV and for diagnostic assays, but the mouse provided a simpler, less expensive, and easier to handle host, and put test evaluations on a more statistically relevant basis. Subsequent studies using newborn mice provided even more advantages. Their immature (soft) skulls and less well developed immune systems allowed for uniform production of reagents, thus newborn mice provided a simpler system for virus isolations. Collectively, these studies, stemmed from not only an interest in YFV but, more importantly, in the disease that it caused, which was the principal focus of research at the time. Theiler joined the Rockefeller Foundation laboratories in 1930 and worked to standardize the virulence and pathogenicity of a French strain of the virus; concomitantly, he found that the incubation period (time from inoculation to signs of illness) decreased with passage (Calisher, 2013).

Eventually these and other discoveries led to the implementation of methods for control of the mosquito vector and to the development of a yellow fever vaccine by Theiler and others. These important steps taken towards combating vector-borne diseases provided the groundwork for the field of arbovirology, which developed and began to flourish soon after.

Virus identification and serological methods

While searching for YFV on two continents, Rockefeller Foundation personnel serendipitously isolated numerous other viruses but could not identify them. These were sent to the New York laboratories where their numbers soon overwhelmed the capacity of the laboratories. Using the complement fixation (CF) test developed for work with studies of *Mycobacterium tuberculosis*, Jordi Casals and others in the Rockefeller laboratories began studies of viruses affecting the central nervous system. Results of preliminary tests were confirmed by neutralization tests, enabling them to distinguish rabies virus from poliomyelitis viruses and from viruses isolated from arthropods, such as St. Louis encephalitis, Japanese encephalitis, louping ill, Russian spring–summer encephalitis (tick-borne encephalitis), eastern equine encephalitis, western equine encephalitis, and Venezuelan equine encephalitis viruses (Calisher, 2013; Downs, 1982). Casals and his collaborators also found that, whereas eastern equine encephalitis, western equine encephalitis, and Venezuelan equine encephalitis viruses were different, they were related to each other by degrees of cross-reactivity, and that St. Louis encephalitis, Japanese encephalitis, louping ill, and Russian spring–summer encephalitis viruses also were distinct but interrelated antigenically. The first three viruses were placed in an antigenic group designated the Group A arboviruses, and the second series of viruses were classified as the Group B arboviruses. These were the first data-based attempts at virus classification. Previously, viruses were grouped based on host type, disease manifestations, and mode of transmission. Subsequent investigations characterized the nucleic acid type of the viral genome (i.e. RNA or DNA), whether it was single- or double-stranded, whether the genome was segmented or not, virion size and shape, and other physical and chemical features.

Because of the large number and disparate characteristics (absence of antigenic relatedness, arthropod vector sources, differences in geographic range, varying clinical features and pathogenesis, etc.), it soon became clear that there would be insufficient letters for the many virus groups expected to be recognized, so alternate designations were developed based on the name of the first recognized member of the group (i.e. Bunyamwera group, California encephalitis group, rabies group, etc.). The findings of Hirst and his colleagues that erythrocytes from chickens agglutinated in the presence of influenza A virus and that antibody to that virus inhibited such agglutination formed the foundation for another useful laboratory diagnostic test, the haemagglutination inhibition (HI) test. Theiler suggested that Casals determine whether arboviruses agglutinated erythrocytes and when he showed that some of them did, the HI test was incorporated into the serological armamentarium for arbovirus identification and for diagnosis of arboviral infections. Application of these methods (virus isolation, HI, CF, and neutralization tests, as well as immunofluorescence assays) became the basis of arboviral diagnosis for many years and to these were added electron microscopy to visualize viruses in infected tissues and cell cultures. Later, enzyme-linked immunosorbent assays (ELISA) and application of monoclonal antibody detection systems revolutionized the field, as described below.

Classical HI, CF, and neutralization tests of single serum samples were incapable of determining whether antibody was the result of a recent or a remote infection. However, adaptation of the HI test using inhibition kinetics allowed the determination of antibody avidity and, therefore, speculation as to the extent of relatedness of very closely related viruses. It was still to be determined whether the antibody found in a previously infected individual could be attributed to a particular virus or not. Introduction of ELISA provided such temporal information by determining whether the antibody was immunoglobulin M (recent) or G (remote). In addition, ELISA usually is more sensitive and reveals higher titres, relative to HI tests. Commercial kits now have been developed to detect antibody to many viruses. The manufacturers advertise these kits as sensitive and specific but not all commercial kits meet these assertions and complete reliance on them can be diagnostically misleading.

In response to the growing interest and recognition of arboviruses as significant causes of human illnesses, the World Health Organization established a series of reference centres to serve both as confirmatory and research laboratories, relying mainly on national laboratories while only providing a minimum of funding and support. The centres functioned effectively for many decades but have more recently been in decline due to a lack of funding, difficulty and cost of international shipments of infectious agents, concerns about national patrimony and intellectual property, and the switch from laboratory-based testing of samples to the use of commercial kits and non-cultural diagnostic techniques. For these reasons as well as issues of human use and privacy, it has become very difficult to obtain paired acute- and convalescent-phase serum samples from individual patients who have had specific virus infections, the *sine qua non* of former days, for comparative studies and as control reagents. As virus discovery and laboratory diagnosis of arboviral infections moved from culture and basic serological assays to more sophisticated, molecular-driven non-cultural techniques, new epidemiological information has been accumulated that would not have been dreamed of even two decades ago.

For example, using polymerase chain reaction (PCR) and next generation sequencing, it is now possible to detect and to determine the probable host source of a virus, its genotype and probable geographic origin, the molecular determinants causing pathogenesis and whether the virus is a reassortant or recombinant. Furthermore, viral nucleic acid sequences often can be detected in tissue samples from vertebrates and invertebrates that no longer contain infectious virus, even from individuals that are no longer symptomatic. These are major improvements that have allowed the development of detailed epidemiological hypotheses and of more specific methods to protect susceptible individuals from infection. Current advances in the production of bedside-relevant virus and antibody detection systems will only advance patient care and treatment.

Research into clinical aspects of virus diseases

In its earliest days as a field of study or discipline, virology was focused on viral diseases and their aetiology, treatment, and prevention. Virologists used the tools they had available to identify the infecting agent and then fumbled to search for a cure or to construct a vaccine to prevent the virus and the disease it caused from spreading to other people, livestock, or wild animals. Elegant studies by patient-oriented physicians and veterinarians were done with relatively simple methods. For example, in 1917 Montgomery summarized for publication his extraordinary, innovative studies of the virus causing Nairobi sheep disease, work that was done in essential isolation, with few reliable tools and methods available, and essentially by himself (Montgomery, 1917). More sophisticated studies later by Nathanson, Gonzalez-Scarano and their colleagues showed that the L RNA segment of bunyaviruses, which encodes the viral polymerase, is the major determinant of neuroattenuation and that *in vitro* phenotypes (temperature sensitivity and altered plaque morphology) segregated with the L RNA segment (Endres *et al.*, 1991). In contrast, attenuation in mice and infectivity for mosquitoes of some virus clones mapped to the M RNA segment, suggesting that the bunyavirus glycoproteins, which are involved in virus entry, also play a role in virulence.

Virus taxonomy

Possibly the most profound effects on our understanding of viruses and their evolution and taxonomy have come through use of genome analyses. More importantly, contemporary methods can be used to engineer viral genomes by reverse genetics to determine gene functions by analysing the phenotypic effects of particular gene sequences. That is, reverse genetics can be used to connect a given genetic sequence in a virus with specific effects on the host organism.

These techniques are now done routinely and inexpensively, and the results can be used to place viruses in taxa or to create new taxa. Conversely, recognition of the nucleic acid composition and sequence of a virus does not identify its phenotypic properties and certainly does not identify a species. Until recently, a virus species was defined as 'a polythetic class of viruses that constitutes a replicating lineage and occupies a particular ecologic niche.' The definition recently was changed to replace 'polythetic' with 'monothetic'.

Species are ideals, not concrete entities; they are names on a list of names, not entities that have properties. As Van Regenmortel *et al.* insisted, 'a part of a thing is a thing and not a property' (Van Regenmortel *et al.*, 2013). A suitable analogy might be that a collection of neurons is part of a person but it is not the person. A virus may be pathogenic for one host but not for another, may replicate in arthropods of one species but not another, may haemagglutinate or not, be squat and bullet-shaped in one host early in the infection but elongated in another developmental stage or host late in the infection, may be stable at room temperature for many days or only for a few hours, may alter cell function in various manners, or be sensitive to different drugs or compounds. These are some of the many properties viruses may possess and which cannot at this time be predicted by genomic sequencing. In order to determine such biological properties, biological methods using intact infectious virus must be employed, such as inoculation of vertebrate or invertebrate hosts, characterization of pathological lesions, production of reassortants, studies of replication cycles and plaque morphology, determination of antigenic and other characteristics.

Such classical techniques are in disuse at this time. Few undergraduate or even graduate schools now teach the history and fundamentals of virological techniques. Likewise, funding agencies emphasize the use of costly and leading edge technologies. The US National Institutes of Health, Department of

Defense, and other organizations world-wide are funding 'magic bullet' research (platforms for rapid, field, or bedside-relevant, non-cultural diagnoses, broad spectrum or syndrome-specific antivirals, and sequence consensus vaccines). Their meagre budgets and funding priorities are driving arbovirus research in other, more applied directions. However, there is more to understanding arboviruses and the diseases they cause than merely determining sequences, gene products, and 'magic bullet' therapies.

Field studies of viruses

In the competition for recognition and research funding, field studies for some time have had low priority within funding agencies. It has long been a joke among older virologists that some naive students assume that viruses come from freezers. Some viruses come directly from patients in hospital settings or clinics; but others do not, particularly arboviruses and other emerging zoonotic agents. Many viruses in clinical samples now can be tested and at least provisionally identified directly by rapid and sensitive non-cultural techniques that can be done at the bedside or in a physician's office. This fulfils one goal of case management, namely to determine what disease is being managed. Rapid detection of a virus also provides information that may be useful in understanding the epidemiology of the agent as well as to alert and to assist public health officials in stemming its spread in a population. Nonetheless, there still are many human illnesses that are not laboratory-diagnosed and there are many viruses that are unrecognized or unknown. This may be due to inadequate laboratory facilities, misdiagnosis by physicians, incorrect specimen collection and preservation, and use of or dependence on molecular techniques that cannot detect the agent. Too often, what are actually being tested for are recognized viruses. Novel, uncommon, and emerging viruses generally are discovered in samples collected and cultured from vertebrate and invertebrate hosts in the field.

Much publicity and hype has been given to the potential of the new technology (RT-PCR, next generation sequencing, metagenomics, satellite imagery and GPS, wide availability of telecommunications) in detecting and identifying emerging viral diseases. But this is largely fantasy being propagated by technology enthusiasts, commercial interests, and well-meaning people who do not know any better. Newly emerging and novel zoonotic and arthropod-borne viral pathogens are not discovered by metagenomic studies of rodent faeces, computer-generated maps of 'hot spots' of biodiversity, or African villagers using their cell phones to report local disease outbreaks or wildlife deaths. Whereas these things might alert us to localities where viral diseases are occurring or might emerge, the actual aetiological agent generally is discovered by trained scientific teams on the ground in the field.

A recent analysis of new human pathogenic viruses, discovered from 1897 to 2010, identified a total of 213 disease causing agents (Rosenberg *et al.*, 2013). Of these 213 viruses, 68% were known or presumed to be zoonotic, and 39% were arthropod-borne. The period with the highest rate of virus discovery (seven or eight per year) was between 1950 and 1969, after which the annual total decreased. Many of the new viruses discovered during this 20-year interval were arboviruses (Rosenberg *et al.*, 2013). This period coincided with comprehensive programmes, supported by the Rockefeller Foundation (Downs, 1982), the Institute Pasteur, and other funding agencies, to isolate viruses from humans, other vertebrates, and arthropod vectors at field stations in tropical America, Africa, and India. The productivity of this strategy illustrates the importance of location, approach, long-term commitment, and sponsorship of pathogen discovery (Rosenberg *et al.*, 2013). In contrast to the arthropod-borne viruses, which showed a precipitous decline after the 1970s, non-arbovirus discovery remained constant at approximately two new viruses yearly from the mid-1950s through 2010, despite major advances in diagnostic methods and technology (Rosenberg *et al.*, 2013).

These data argue against the dogma that genome sequencing and non-cultural techniques have largely replaced the need for the more classical virological techniques. A recent publication from the US Centers for Disease Control and Prevention emphasized the importance of retaining the classical techniques (isolation of viruses in cell cultures, electron microscopy, serology, histopathology and immunochemistry assays) in diagnostic laboratories (Goldsmith *et al.*, 2013). The initial outbreaks caused by severe acute respiratory syndrome coronavirus, Middle East respiratory syndrome coronavirus, Nipah henipavirus, ebolaviruses, severe fever with thrombocytopenia syndrome and Heartland phleboviruses, and Sin Nombre and other hantaviruses as well as West Nile flavivirus and chikungunya alphavirus in the Americans were recognized after the viruses were cultured in diagnostic laboratories from clinical or field-collected samples. Subsequent full genome sequencing, phylogenetic studies, as well as biochemical and structural analyses provided valuable additional information as to the pathogens' taxonomic relationships, evolutionary origins, composition, and structure. But, with few exceptions, other important characteristics, such as pathogenicity, host range, antigenicity, host immunological response, vector or reservoir relationships, and drug or other chemical sensitivity, all depend on having infectious virus to work with. As noted above, the latter types of information cannot be obtained from a sequence or a phylogenetic tree.

Field studies of zoonotic and arthropod-borne viruses require a team of people with a variety of special skills, such as entomologists, mammalogists, ornithologists, epidemiologists, and microbiologists, as well as vehicle drivers, trappers, or field collectors. Field workers toil under diverse climate conditions, crawl through sewers, wade into swamps, tolerate insects, are aware of poisonous or otherwise potentially dangerous life forms, and miss meals, while at the same time maintaining good relations with landowners as well as with local and national authorities. Language skills and knowledge of local culture also are important. Field workers must be skilled in safety protocols, be vaccinated with necessary and available vaccines, be aware of dangerous political or terrorist situations, and be willing to rise early and go to bed late. Obviously, field work requires quite a different skill set from what is required in a molecular virology laboratory, sequencing facility, or bioinformatics core. Also, in order for field studies to be productive, long-term support is required; otherwise, the data accumulated from one brief field trip comprise point prevalence studies, which are useful only for that single location, at that specific time, under those specific ecological conditions. Extrapolation or modelling of such data to the overall epidemiological situation can be, and often is, misleading and, therefore, is an erroneous approach. Field studies are expensive; for academic researchers, financing them depends on competing for grants, contracts, or other awards.

Much lip service is given to biodiversity and to the importance of studying the gene pool of organisms in tropical forests and other pristine ecosystems before they are altered or disappear. Likewise, much publicity is given to the detection and identification of newly emerging viral pathogens, but the types of studies required to find them are not being funded. As noted before, such agents are not discovered by blogs, social media, or other sophisticated electronic systems, reporting wildlife deaths and sporadic disease outbreaks in remote human populations or by metagenomic studies of randomly collected faeces from vertebrates. Such activities and outcomes require experienced field teams in these localities, as there were in the past with the Rockefeller Foundation and Institute Pasteur overseas research facilities. The critical questions are: where do these viruses occur, what are their natural cycles, how do humans or domestic animals become infected, and what are their epidemic or pandemic potentials? How can we expect to control or end the spread of viruses about which we know only their sequences and not their biological characteristics?

Many young investigators working with arboviruses today are ignorant of or uninterested in these types of biological questions. The current fascination with the most recently developed tools and with gene function is understandable and undoubtedly will reveal many new aspects of viruses, but the natural history of viruses and their ecological relationships also are important. The ultimate control and prevention of arthropod-borne and other zoonotic viral diseases depends on a thorough understanding of their ecology and epidemiology. Students now may receive graduate degrees by sequencing a few viruses and analysing their phylogenetic relationships or by characterizing a gene expressed during virus replication and its effects on cellular function, without having any real knowledge of the ecology of the agent, its modes of transmission, or its pathogenesis in vertebrates. Providing a few brief sentences in the introduction of a manuscript, thesis, or grant proposal, giving superficial acknowledgement of its public health or veterinary importance and its pathogenic potential, is insufficient.

Suggestions for restoring balance to arbovirology and other fields of virus research

1. Universities and other virology training centres should teach potential virologists using a broader historical perspective of the various disciplines. Students also should learn basic laboratory and field techniques in addition to biochemistry, genetics, and cell biology. This should include growing and maintaining cell cultures, isolating viruses from clinical or field materials, participating in a field study, collecting clinical samples or arthropod vectors, and observing actual patients with viral infections. Such information in curricula would broaden the views and knowledge of students, and it would teach them to be professionals, rather than simply training them to be specialized technicians.
2. Virology training should place emphasis on the continued importance of virus isolation, when virus isolation is possible. Without a virus isolate, one cannot determine its virulence, pathogenesis, antigenicity, or drug sensitivity.
3. Despite the greater sensitivity and specificity of some of the newer molecular technology (next generation sequencing, bioinformatics, structural analysis, etc.) it is still far too expensive and complex for many diagnostic facilities, especially those in tropical regions where many of the arboviral pathogens and emerging zoonotics are endemic. For simple identification of many common arboviruses, the classical techniques (isolation and serology) provide the essential basic information at much less cost. But because of pressure from international agencies and commercial interests, many laboratories in developing countries have lost the capacity to do the classical diagnostic tests and have become dependent on RT-PCR and other molecular-based commercial tests (Inglis, 2013; Hall-Mendelin *et al.*, 2010). The end result has been that they no longer have personnel trained in the classical techniques or can obtain a regular supply of the very reagents or kits needed to conduct the molecular tests on which they depend.
4. Another, more difficult, problem is the selection criteria for university faculty who will teach virology students. One of the major criteria for selection of research faculty in most medical and graduate schools of biomedical sciences in the US is the ability to obtain external funding to support the individual's research. The National Institutes of Health (NIH) is the major source of support for biomedical research in the US Because of NIH funding priorities and the composition of its virology and microbiology study sections, field-based research has a low priority compared to the newer 'cutting edge' technologies. Consequently, bioinformatics, genomics, cell biology, biochemistry, and vaccinology have priority over field-based research or pathogen discovery. As a result, laboratory-based research in the former areas is favoured over more basic field or ecological research. Not surprisingly, persons trained or working in these better-funded subject areas usually are selected for new faculty positions. They, in turn, teach students what they know (i.e. molecular biology, genomics, cell biology, basic immunology, etc.) and the cycle gets repeated. The research priorities of the NIH are determined in large part by the political leaders who control the budget process and by pressure from academic and professional groups. NIH study sections are generally selected based on the types of research proposals submitted. So, if academic researchers are predominantly laboratory- or computer-based scientists, their research proposals naturally will reflect their interests. Consequently, the selection and representation of study section members will reflect the same bias. The only feasible way to alter this situation is to lobby political officials, funding agencies, and academic institutions to restore some balance between field and laboratory research. This will not be an easy task in a world where virtual reality, computer modelling, and *in vitro* experimentation are increasingly used to study and attempt to understand the natural world in which we live.

Recent advances in genomics, molecular biology, and technology have provided tools that imaginative investigators have applied to problems unsolvable only a decade or so ago. This chapter is in no way meant to either minimize these remarkable advances or to discourage anyone from employing them. Our intent here is to encourage inclusivity among virologists, so that they do not consider that 'classical' equals 'outdated'. We have attempted to point out that some classical techniques are still quite useful for attaining important biological information and diagnostic goals. If a technique has been superseded by a better (more rapid, more specific, less expensive) technique then the better one certainly should be adapted for use. Nonetheless, high quality research using simple techniques should be continued if the procedure provides information not otherwise available. Novel and known viruses will continue to emerge and the need to detect, characterize, and control them will remain (Keusch *et al.*, 2009; Woolhouse *et al.*, 2012). The first line of defence against emerging viruses is surveillance, so we should use whatever proven and effective techniques are available to achieve those goals.

References

Calisher, C.H. (2013). Lifting the Impenetrable Veil: from Yellow Fever to Ebola Hemorrhagic Fever and SARS (Rockpile Press, Red Feather Lakes, Colorado).

Downs, W.G. (1982). The Rockefeller Foundation Virus Program: 1951–1971 with update to 1981. Annu. Rev. Med. 33, 1–29.

Endres, M.J., Griot, C., Gonzalez-Scarano, F., and Nathanson, N.J. (1991). Neuroattenuation of an avirulent bunyavirus variant maps to the L RNA segment. Virology *65*, 5465–5470.

Goldsmith, C.S., Ksiazek, T.G., Rollin, P.E., Comer, J.A., Nicholson, W.L., Peret, T.C.T., Erdman, D.D., Bellini, W.J., Harcourt, B.H., Rota, P.A., *et al.* (2013). Cell culture and electron microscopy for identifying viruses in diseases of unknown cause. Emerg. Infect. Dis. *19*, 886–891.

Hall-Mendelin, S., Ritchie, S.A., Johansen, C.A., Zborowski, P., Cortis, G., Dandridge, S., Hall, R.A., and van den Hurk A.F. (2010). Exploiting mosquito sugar feeding to detect mosquito-borne pathogens. Proc. Natl. Acad. Sci. U.S.A. *107*, 11255–11259.

Inglis, T.J.J. (2013). Review article: the lab without walls; a deployable approach to tropical infectious diseases. Am. J. Trop. Med. Hyg. *88*, 614–618.

Keusch, G.T., Pappaioanou, M., Gonzalez, M.C., Scott, K.A., and Tsai, P., eds. (2009). Sustaining Global Surveillance and Response to Emerging Zoonotic Diseases (The National Academies Press, Washington, DC).

Montgomery, E. (1917). On a tick borne gastro-enteritis of sheep and goats occurring in East Africa. J. Comp. Path. *30*, 28–57.

Murphy, F.A. (2012). The Foundation of Virology: Discoverers and Discoveries, Inventors and Inventions, Developers and Technologies (Infinity Publishing, West Conshahocken, Pennsylvania).

Rosenberg, R., Johansson, M.A., Powers, A.M., and Miller, B.R. (2013). Search strategy has influenced the discovery rate of human viruses. Proc. Natl. Acad. Sci. U.S.A. *110*, 13961–13964.

Van Regenmortel, M.H.V., Ackermann, H.-W., Calisher, C.H., Dietzgen, R.G., Horzinek, M.C., Keil, G.M., Mahy, B.W.J., Martelli, G.P., Murphy, F.A., Pringle, C., *et al.* (2013). Virus species polemics: 14 senior virologists oppose a proposed change to the ICTV definition of virus species. Arch. Virol. *158*, 1115–1119.

Woolhouse, M., Scott, F., Hudson, Z., Howey, R., and Chase-Topping, M. (2012). Human viruses: discovery and emergence. Phil. Trans. R. Soc. B. *367*, 2864–2871.

Index

A

adaptation 12, 31, 72, 73, 90, 93, 99, 102, 103, 105, 128, 130, 135, 140, 141, 148, 156, 162, 164, 167–169, 171, 172, 216, 221, 224, 225, 230, 231, 242, 244, 253, 258, 259, 263, 264, 295, 298, 300, 352, 363, 369, 386
Adelaide River virus (ARV) 26, 78, 123, 125, 220
Adenoviridae 224
adjuvant 39, 344, 350, 358, 368
adulticiding 283, 285, 286, 287
Aedeomyia squamipennis 193, 196, 212
Aedes (Ae.) aegypti 4, 41, 59, 72, 77, 84–86, 91, 108–112, 114–118, 137, 138, 140, 142, 156, 164, 169, 193–199, 205, 210, 212–216, 220, 229–232, 246, 247, 255, 257–263, 265–268, 282, 284–290, 292–302, 305, 311–313, 315–317, 319, 321, 325, 326, 331–335, 340, 343, 348, 352–355, 357, 360, 361, 366, 367, 370, 371, 386
Aedes aegypti densovirus (AeDNV) 293, 300, 302
Aedes aegypti Higgs white eye (HWE) 315–330
Aedes (Ae.) albopictus 39, 56, 59, 60, 70, 83, 86, 91, 100, 112, 114, 116–118, 130, 131, 135, 140, 141, 146, 147, 150, 155, 158, 160–164, 167, 169, 193–199, 210, 213–216, 230, 246, 254, 257–263, 265–268, 276, 284, 288, 292–302, 304, 305, 318, 320, 321, 324, 332, 333, 338, 343, 348, 356, 361, 367, 369
Aedes (Ae.) canadensis 62, 357
Aedes flavivirus (AeFV) 28, 122, 124, 199, 213, 235, 236, 237
Aedes (Ae.) furcifur-taylori 353
Aedes (Ae.) lineatopennis 62, 193, 196
Aedes (Ae.) luteocephalus 353
Aedes (Ae.) neomelanoconion 339
Aedes (Ae.) polynesiensis 194, 295, 299, 300
Aedes (Ae.) pseudoscutellaris 194, 210, 257
Aedes (Ae.) riversi 295
Aedes (Ae.) sollicitans 146, 357
Aedes (Ae.) taeniorhynchus 62, 133, 146, 164, 172, 195, 244, 248, 359
Aedes (Ae.) triseriatus 39, 62, 83, 191–193, 195, 196, 200–216, 251, 264, 265, 318, 333, 338
Aedes (Ae.) vexans 62, 193, 194, 212, 263, 357
Aedes (Ae.) vigilax 80, 152, 196, 198, 213, 289, 348, 369
African horse sickness virus (AHSV) 23, 37, 127, 129, 148, 231, 232, 233, 310, 314
African swine fever virus (ASFV) 27, 127, 132, 175, 192, 213–215, 223, 229, 231, 242, 244, 249, 250, 251 253, 254, 256, 258, 260, 261, 264, 265, 267, 310, 313
Alphaviridae 152, 355
alphavirus 6, 10–13, 31–33, 40–43, 49, 51, 73, 89, 91–106, 121–123, 128, 130, 133, 134, 136, 143, 144, 150, 152–154, 164, 173, 175, 178, 196–198, 209, 211–214, 221, 224–226, 228, 229, 233, 240–244, 246, 248, 252, 256, 258, 260–265, 267, 268, 279, 282, 296, 304, 305, 309, 312–314, 316, 332, 339, 357, 359, 360, 362, 366, 367, 370, 371, 377, 388
aluminium hydroxide 347, 350, 358
Amalgaviridae 227, 262
amplifying hosts 160, 168, 192, 222, 225, 253, 254, 346
Anopheles (An.) gambiae 42, 108, 110, 115, 116, 133, 194, 213, 294, 299, 300, 305, 309, 313, 317, 318, 320, 321, 324, 329, 333–335, 357, 370
Anopheles (An.) stephensi 267, 294, 299, 301, 318, 320–322, 324, 331–335
animal husbandry 1, 4, 5, 283
Annelida 256
antibodies 5, 10, 34, 38–40, 43, 57, 60, 62, 68, 76, 78–82, 101, 103–105, 143, 151, 152, 155, 156, 159, 160, 245, 253, 262, 263, 265, 266, 271–283, 329, 333, 339–345, 347–350, 352, 354, 355, 357–359, 361, 362, 366, 370, 371, 373, 379
antigen-capture 274, 275
antigenicity 10, 163, 358, 388, 389
antiviral 5, 33, 42, 45, 48, 49, 54–56, 59, 64–70, 75, 82, 83, 86, 87, 92–106, 108, 110, 112, 115, 117, 118, 121, 159, 172, 202, 203, 207, 210- 214, 257, 262, 301, 302, 303, 304, 305, 308, 310, 311, 312, 313, 314, 315, 325, 329, 332, 337, 373, 374, 376–381, 388
apoptosis 35, 42, 43, 55, 57–59, 64, 66–70, 76, 100, 108, 109, 111, 112, 116, 118, 203, 204, 207, 210, 211, 214, 216, 221, 266
aquatic environments 219, 220, 223, 224, 226, 229, 242, 253, 256
Arachnida 235
Aransas Bay virus (ABV) 27, 188
Arboledas virus (ADSV) 22, 216
arboviral plasticity 168
arbovirology 9, 83, 208, 209, 219, 226, 249, 252, 255–257, 385, 386, 389
Arthropod-borne viruses (Arboviruses) 1, 3, 4, 9, 11, 13, 14, 31, 45, 75, 89, 121, 157, 159, 160, 167–169, 172, 175–178, 180, 187, 188, 191–193, 196–198, 200, 203, 204, 209, 210, 213, 215, 216, 219–221, 223, 225–230, 232, 233, 236, 240, 242–244, 246–268, 271–277, 281–284, 289, 291–296, 303–305, 310–313, 315–317, 320, 322, 324–326, 329, 335, 337, 362, 364, 365, 370, 386, 388, 389
Arenaviridae 1, 227
arenavirus 225–227, 251, 261, 266
Arteriviridae 226, 242
arthralgia 63, 91, 92, 104, 149, 221, 222, 338, 355, 357, 358
Arthropoda 12, 220, 256
Artificial miRNAs (amiRNAs) 304, 306, 311
Asaia 112, 115, 178, 291, 294, 299
Asfarviridae 12, 27, 121, 127, 175, 192, 196, 229, 251, 305
Aura virus (AURAV) 14, 32, 91, 122, 241
autophagy 51, 52, 54–60, 203, 204, 207, 208, 211–215
avian leukosis virus (ALV) 354

B

Bacillus sphaericus 283, 288, 300, 301
Bacillus thuringiensis 283, 288, 291, 293, 298–301, 317
Bacillus thuringiensis var. *israelensis* (Bti) 283, 291, 293, 298, 301, 317
Bactrocera cucurbitae 322, 333
Baculoviridae 220, 224, 226, 264
Banna virus (BAV) 26, 220, 229, 265

Barkedji virus (BJV) 213, 235–238
Barmah Forest virus (BFV) 14, 91, 153, 240, 241, 249, 283, 365
bats 12, 35, 62, 81, 83, 91, 135, 137, 144, 151, 154–158, 160–162, 164, 165, 225, 227, 229, 238, 244, 247, 251, 254, 257, 258, 260, 261, 263–266
Batu Cave virus (BCV) 16, 220
Beauveria bassiana 113, 115, 294, 299, 300
Bebaru virus (BEBV) 14, 240, 241
BHK-21 cells 60, 276, 306, 308, 309, 310, 314
Binary ethylenimine (BEI)-inactivated virus 358
binary P element transformation system 317
biocontrol 113, 114, 116, 287, 291–295, 297- 301, 317
biodiversity 132, 153, 388, 389
bioinformatics 11, 108, 116, 183, 184, 185, 187, 188, 388, 389
biological transmission 82, 85, 132, 136, 159, 191, 209, 219, 221, 225–227, 235, 240, 242, 243, 246, 248–256, 262
Biosafety in Microbiological and Biomedical Laboratories (BMBL) 274, 276
Birnaviridae 198, 223, 226, 227, 247
bloodmeal 1, 61, 107, 113, 200, 203, 204, 206, 207, 210, 223, 224, 230, 232, 243, 250, 290, 294, 296, 311, 315, 316, 318, 319–322, 324- 328, 333, 337, 348
Blue Crab virus (BCV) 27, 220
Bluetongue virus (BTV) 37, 38, 42, 43, 127, 128, 131–133, 135–137, 144, 146, 148, 155, 156, 158, 161, 164, 176, 198, 210, 215, 216, 221, 233, 253, 254, 258, 260, 263, 266, 267, 268, 386
Border disease virus 239, 241
Bornaviridae 71, 224, 227, 228
bornavirus 222, 227, 258, 261
bottlenecks 128, 147, 148, 156, 167, 168, 171
bovine ephemeral fever virus (BEFV) 26, 35, 77–80, 82–87, 123–125, 129, 130, 132, 152, 221, 305, 310, 312
Brazoran virus (BRAXV) 20, 126, 160, 277, 278
brevidensoviruses 229, 266, 293
Buggy Creek virus (BCRV) 14, 128, 196, 198 211, 224, 233, 240, 241, 247, 248, 250, 258, 260
Bukalasa bat virus (BBV) 16, 220
Bunyaviridae 1–3, 6, 12, 13, 17, 20, 22, 27, 28, 31, 36, 37, 39, 40, 42, 43, 61, 67–70, 121, 122, 124, 126–129, 131, 133, 145, 150, 151, 155, 160–162, 175, 176, 192, 193, 196, 198, 199, 207, 209–214, 216, 223, 224, 226, 227, 233, 236, 240, 242, 246, 248, 251, 257, 264, 268, 273, 274, 278, 305, 315, 337, 368
Bunyip Creek virus (BCV) 25

C

C6/36 cells 111, 268, 293, 304, 344
Cache Valley virus (CVV) 17, 62, 68, 126, 155, 211, 212
Cacipacore virus (CPCV) 15, 232, 258
Calbertado virus (CBTV) 28, 199
Caliciviridae 226, 227
California encephalitis virus (CEV) 3, 18, 41, 68, 70, 193, 202, 210, 212–214, 216, 232, 251, 260, 265, 267, 338, 386
carboxypeptidase A, (CPA) 311, 320, 325–327
Cavally virus (CAVV) 28, 199, 228
cell fusing agent virus (CFAV) 28, 122, 124, 150, 156, 192, 199, 211, 215, 216, 222, 235–237, 248
Centers for Disease Control and Prevention (CDC) 13, 90, 91, 101, 137, 156, 177, 188, 248, 274, 290, 337, 344, 347, 351, 356, 361, 388
Ceratitis capitata 318, 322, 333, 334
Chandipura virus (CHPV) 26, 76, 83, 85, 86, 125, 136, 161, 214, 215, 252, 265, 305, 313, 314
Chaoyang virus (CHAOV) 28, 122, 124,235–237, 243, 263
Chikungunya virus (CHIKV) 2–5, 9, 14, 40, 54, 85, 89, 91–93, 95–97, 99–105, 114, 122, 128, 130–133, 135–137, 139, 140, 142, 144, 147, 148, 150, 156, 160, 164, 167, 169–172, 175, 176, 188, 189, 198, 210, 214, 226, 230–232, 241, 244, 253, 254, 259, 261, 264, 265, 267, 278, 279, 288, 292, 293, 296, 299, 304, 305, 313–317, 335, 356, 357, 360, 361, 363, 364, 366–371, 377
cholesterol 51–53, 56–60, 74, 101, 102, 297, 299
Chuviridae 222, 227, 246
cimicid bugs 224, 240, 246, 250, 252
Circoviridae 224, 257, 258
Classical swine fever virus (CFSV) 226, 239, 241
climate change 4, 9, 101, 102, 131, 135, 146, 147, 157, 158, 160, 163, 315
Cluster, randomized trials (CRTs) 281, 282, 284
CO2-baited light traps 283
Cochliomyia hominivorax 322
Colorado tick fever virus (CTFV) 22, 29, 127, 129, 220, 246, 256, 268, 274, 355, 361, 365, 366, 368, 370
Coltivirus 38, 127, 129, 229, 264
community participation 286, 300, 301
competent vectors 75, 135, 137, 232, 296, 330
complementation 84, 169, 333
complement fixation (CF) 103, 279, 386
Connecticut virus (CNTV) 27, 81, 84, 236
copepods 220, 226, 227, 229, 256, 291–293, 298, 300, 301
Coquillettidia perturbans 357
Coronaviridae 227, 242, 243
Cricket paralysis virus 227, 229, 257
Crimean-Congo haemorrhagic fever virus (CCHFV) 20, 62–64, 66–68, 128, 137, 145, 154, 161, 197, 212, 216, 250, 253, 256, 268, 305, 338, 368
Culex (Cx.) annulirostris 78, 80, 146, 267, 283, 348, 369
Culex (Cx.) decens 80, 85, 163, 214, 265
Culex flavivirus (CxFV) 28, 122, 124, 150, 159, 162, 192, 199, 209, 211, 213–215, 265
Culex (Cx.) gelidus 348, 369
Culex (Cx.) pipiens quinquefasciatus 13, 83, 117, 151, 155, 169, 171, 172, 195, 198, 199, 211, 215, 230, 258, 262, 265, 284, 292, 293, 295, 296, 298, 300, 301, 318, 319, 329, 331, 351
Culex (Cx.) quinquefasciatus 195, 198, 199, 295, 296, 318, 319
Culex (Cx.) tarsalis 13, 62, 140, 141, 154, 167, 172, 195, 196, 199, 210, 220, 257, 282, 288, 323, 351, 359
Culex (Cx.) tritaeniorhynchus 81, 146, 195, 214, 215, 346, 348, 361
Culex (Cx.) vishnui 195, 214, 346
Culicoides 37, 61, 62, 78–83, 86, 87, 129, 132, 133, 135–137, 146, 151, 152, 155–158, 161, 164, 195, 198, 210, 215, 223, 246, 253, 258, 263, 368
Culiseta inornata 62, 68, 193, 212, 215
Culiseta (Cs.) Melanura 196, 197, 215, 251, 264, 357
cytoplasmic incompatibility 114, 115, 117, 118, 291, 295, 298–301, 332
cytopathic effect (CPE) 11, 55, 67, 76, 95, 152, 231, 263, 276, 277, 306, 341
cytotoxic T lymphocytes (CTL) 348

D

Dak Nong virus (DKNV) 28, 228, 262
DDT 221, 287, 288, 317, 333
de novo 33, 38, 51, 52, 58, 177, 184, 186, 189, 380
dead-end hosts 2, 89, 90, 252, 253
deforestation 1, 4, 5, 135, 145, 165
Dengue fever (DF) 91, 117, 138, 156, 162, 259, 279, 285, 288, 289, 341–343, 350, 362, 365, 368, 373, 374, 379, 381
dengue haemorrhagic fever (DHF) 137, 142, 143, 221, 316, 340, 343, 373, 374, 379
dengue shock syndrome (DSS) 316, 340, 343, 373, 374, 379
Dengue virus (DENV) 12, 15, 38–44, 56–60, 86, 107, 114–118, 122–124, 128, 133, 135, 137, 155, 156, 158–165, 167–169, 172, 175, 177–180, 187–189, 192, 209, 212, 213, 215, 216, 229–232, 235, 237, 238, 240, 243, 247, 250, 252, 253, 257, 259–261, 265–267, 278, 279, 288, 290, 292, 293, 294, 296–301, 304, 306, 311–316, 331–335, 340–346, 352, 360–371, 373–381
Densoviruses 228, 229, 266, 293, 299, 300
diapause 147, 191, 200, 203, 208, 214, 216, 292
diarrhoea 63, 64, 92, 151, 239, 357, 358, 378
Dicistroviridae 224–227, 229, 247, 252, 256, 265
dipstick assays 277
Diptera 12, 13, 69, 76, 82–86, 116, 117, 121, 129, 156–160, 162, 163, 165, 201, 204, 210–216, 257–259, 261, 262, 265, 266, 279, 288–290, 298, 299, 300, 301, 320, 331, 333, 334
disease incidence 282, 285
disease transmission 108, 112, 220, 254, 284, 286, 293, 297, 298, 324, 329
Donggang virus (DGV) 28, 122, 124, 236
double-stranded RNA (dsRNA) 44, 69, 115, 118, 121, 126, 153, 165, 175, 178, 210, 214, 227, 229, 247, 260, 261, 294, 303, 305, 309–312, 325–327
Drosopphila hydei 318, 332, 333

Drosophila melanogaster 86, 108, 115–117, 191, 192, 200–204, 207–210, 212–216, 258, 296, 300, 301, 312, 317, 332, 333
Durham virus (DURV) 27, 80, 82, 220

E

Eastern equine encephalitis (EEEV) 3, 10, 13, 14, 89, 90, 100–102, 105, 106, 122, 128, 130, 131, 137, 143, 168, 175, 187, 191, 196, 197, 215, 216, 225, 229, 241, 245, 251, 252, 279, 292, 305, 311, 357, 361, 368–370, 386
ebola virus (EBOV) 85, 234, 244, 258, 266, 311
ecosystem 2, 4, 5, 137, 147, 155, 161, 165, 219, 220, 223, 233, 242, 244, 282, 389
egg hatch 296, 297
Eilat virus (EILV) 6, 11–13, 28, 95, 104, 122, 150, 162, 209, 214, 228, 240, 241, 242, 244, 248, 264
electron microscopy 10, 11, 44, 47, 49–51, 59, 60, 274, 278, 386, 388, 390
embryogenesis 200, 201, 204, 208, 330
embryonic quiescence 297, 300
endoplasmic reticulum (ER) 41, 45, 49, 57, 59, 60, 65, 74, 94, 377, 379
entomological inoculation rate 286
Entebbe bat virus (ENTV) 16, 122, 124, 144, 220, 235, 237, 238, 243, 247, 248, 251, 255
enzootic vectors 12, 248
enzootic 4, 12, 90, 91, 98, 100, 104, 159, 164, 211, 215, 244, 248, 250, 304, 346, 350–352, 357, 359
enzyme-linked immunosorbent assays (ELISA) 271–274, 278, 279, 362, 386, 387
ephemerovirus 72–75, 78–82, 86, 129, 130, 198, 244
epidemic 1–5, 69, 90–92, 100, 103–105, 121, 133, 135, 137, 140, 144–150, 153, 154, 157, 158, 160, 162–165, 167, 172, 175, 189, 196, 198, 209, 212, 231, 232, 250, 253, 257, 259–262, 264, 265, 267, 278, 283, 285–291, 293, 316, 338, 339, 350, 353–362, 364, 366, 367, 369, 371, 373, 374, 389
epidemiology 4, 6, 42, 67, 78, 82–84, 86, 89, 106, 142, 148, 154–166, 168, 172, 175, 177, 191, 196, 197, 200, 201 209, 212, 216, 255, 260–262, 264, 265, 289, 302, 313, 337, 341, 360–362, 365–367, 369, 385, 388, 389
epileptoid syndrome 349
epizootic 69, 77, 85, 90, 98, 100, 104, 146, 156, 160, 166, 173, 214, 222, 232, 244, 245, 248, 250, 281, 282, 339, 353, 354, 357–360, 365, 370
equine infectious anaemia virus 253
erythematous rash 343
exogenous 66, 104, 110, 222, 233, 303–306, 309, 310, 311, 312,331
extrinsic incubation period (EIP) 108, 111, 117, 140, 146, 147, 210, 252, 253, 293, 297, 316, 325, 385
Eyach virus (EYAV) 22, 38, 127, 129, 177, 178

F

Farmington virus (FARV) 27, 224, 264
FASTA 184, 185
FASTQ 184, 185, 186
febrile illness 2, 6, 38, 63, 64, 69, 77, 79, 81, 90, 91, 99, 151, 155, 161, 178, 221, 278, 338, 339, 355
fecundity 200, 201, 231, 294, 297, 299, 323
field studies 197–199, 201, 202, 204, 331, 388, 389
filial infection rates 192, 214, 216
Filoviridae 71, 224
Fisher's Model of Natural Selection 328
fitness landscapes 167–169, 171, 187
Flanders virus (FLAV) 26, 80, 250
Flaviviridae 1–3, 10, 12–14, 28, 31, 33, 35, 41, 43, 45, 58, 92, 121–124, 129, 152, 167, 168, 176, 192, 194, 198, 198, 199, 209–211, 213, 223, 224, 226, 227, 233, 239, 241, 242, 246, 248, 257–260, 273, 304, 305, 315, 337, 340, 373, 376, 380
Flock House virus 211, 214, 228, 247, 303
Fort Morgan virus (FMV) 14, 122, 128, 220, 240, 241, 252, 258
fumigation 140, 285–287

G

Gabek Forest virus 21, 248
Gamboa virus (GAMV) 12, 18, 196, 212
Gan Gan virus (GGV) 22, 220
Ganjam virus 20, 62, 67, 68, 246, 259, 266
gene driver 329, 330
gene transfer 72, 229, 236, 243, 247, 263, 333
genetic determinants 133, 164, 219, 220, 242, 243, 255, 257, 264
genetic diversity 67, 121, 126–132, 147, 148, 154, 155, 159, 163, 164, 166–173, 205–207, 238, 257, 360, 361
genetic drift 205, 206, 207, 330
genetic polymorphisms 207
genetic shift 37, 127, 128, 205–207
genetically diverse laboratory strain (GDLS) 323, 328
genetically modified vectors (GMVs) 315–319, 322–331
genetics 5, 9, 11, 68, 70, 132, 133, 137, 142, 155, 160, 167, 172, 210, 216, 255, 261, 264, 266, 313, 314, 315, 316, 331, 333, 339, 380, 387, 389
genome size 6, 38, 43, 75, 86, 124, 129, 186, 216, 240, 267
genomic sequencing 11, 38, 80, 131, 187, 387
genotypes 9, 43, 128–131, 133, 144, 146, 148–150, 154, 172, 206, 207, 209, 210, 211, 213, 221, 232, 233, 243, 313, 316, 328, 347, 348, 353, 355, 360, 362, 363, 375, 379
genus *Alphavirus* 10–12, 14, 89, 121, 132, 226, 229, 240, 241, 244, 258, 304
genus *Aquareovirus* 37, 228, 264
genus *Asfavirus* 27, 196, 244, 305, 310
genus *Coltivirus* 12, 22, 37, 38, 126, 198, 228, 240, 355
genus *Cypovirus* 37, 220
genus *Dinovernavirus* 37, 210, 257
genus *Entomobirnavirus* 227
genus *Ephemerovirus* 12, 26, 35, 78, 123, 125, 152, 243, 310
genus *Flavivirus* 225, 227, 229, 233, 234, 236, 238, 239, 240, 243, 247, 260, 264, 278, 304, 373
genus *Hepacivirus* 33, 123, 227, 239, 240
genus *Isavirus* 38, 126, 226
genus *Ledantevirus* 36, 251, 258
genus *Lyssavirus* 35, 72, 123, 228, 256, 251
genus *Nairovirus* 12, 20, 36, 61, 124, 126, 193, 196, 197, 233, 240, 261, 264, 278, 305, 310, 337
genus Novirhabdovirus 35, 72, 123, 226, 228
genus *Nucleorhabdovirus* 35, 123, 125
genus *Orbivirus* 12, 23, 26, 37, 126, 148, 195, 198, 228, 233, 236, 240, 248, 305, 310
genus *Orthobunyavirus* 12, 17, 20, 36, 37, 61, 65, 66, 124, 126, 129, 150, 151, 193, 196, 236, 240, 278
genus *Pestivirus* 33, 123, 239, 240, 247
genus *Phlebovirus* 12, 20, 22, 36, 37, 61, 68, 69, 124, 126, 129, 145, 150, 151, 160, 162, 193, 196, 212, 214, 215, 216, 236, 240, 264, 278, 298, 305, 310, 339, 360, 388
genus *Seadornavirus* 12, 26, 37, 126
genus *Sigmavirus* 35, 72, 80
genus *Tibrovirus* 12, 26, 35, 79, 233
genus *Tospovirus* 12, 36, 37. 61, 124, 126, 129, 240
genus *Vesiculovirus* 12, 26, 35, 77, 123, 125, 196, 226, 228, 229, 310
germline transformation 317–320, 329, 332, 334, 335
Getah virus (GETV) 14, 122, 241, 257
Global Alliance for Vaccines and Immunization 348
globalization 1, 4–6, 130, 131
Glossina morsitans 322
Gouléako virus (GOUV) 20, 161, 199, 214, 248
Great Island virus (GIV) 24, 220

H

haemagglutination inhibition assay (HI) 271, 273, 278, 386, 387
Haemagogus spp. 231
Haemaphysalis 62, 145, 151, 195, 215, 337, 339
haematophagous 9, 12, 29, 37, 169, 197, 220–226, 228, 242, 244, 245, 249, 250, 252, 253, 255, 256, 338
haematophagy 221, 224, 255, 326
haemostasis 340
hantaviruses 12, 61, 196, 197, 216, 225, 261, 265, 268, 388
harbourage sprays 284
Hawaii/Mahidol tetravalent vaccines 341
Hazara virus (HAZV) 20, 62, 126, 305
Heartland virus (HRTV) 22, 63, 66, 69, 126, 151, 162, 264, 274, 275, 277, 388
hepaciviruses 225, 227, 239, 240, 247, 265
Hepadnaviridae 224, 227
hepatitis B virus (HBV) 111, 311, 373, 376

hepatitis C virus (HCV) 57, 86, 103, 179, 227, 239, 260, 262, 268, 311, 371, 373, 379, 380, 381
Herbert virus (HEBV) 28, 199, 240, 248
herd immunity 192, 200, 232, 281, 285–287
herpes simplex virus (HSV) 323, 373, 376
Herpesviridae 227
heterologous 10, 37, 43, 206, 207, 211, 295, 321, 329, 340, 344, 363, 370
heterotypic dengue infections 340
Highlands J virus (HJV) 14, 122, 128, 131, 178, 187, 188, 241, 278
high-throughput 11, 12, 135, 152, 155, 172, 177, 188, 278, 286, 375, 380
homeothermic 191
homing endonuclease genes (HEGs) 324, 331, 332
homologous 10, 11, 43, 55, 72, 131–133, 165, 206, 224, 227, 265, 267, 306, 320, 324, 329
horizontal transmission (HT) 12, 67, 77, 107, 197, 200, 214, 232, 253, 256, 263, 326
host metabolism 45
host range 2, 6, 10–13, 33–35, 73, 79, 83, 89, 90, 104, 129, 133, 147, 150, 162, 169, 209, 210, 214, 219–221, 223, 226–228, 232–240, 242–251, 255, 256, 258, 261, 262, 264, 267, 305, 335, 368, 388
host replacement 248
Hyalomma truncatum 193, 196, 212, 216, 259
hyperimmune globulin 337

I

Iflaviridae 225, 256
IgA assay 272
Igbo Ora Virus 241
IgG avidity 272
IgM 227, 263, 266, 271, 272, 273, 274, 275, 277, 278, 279
Illumina 170, 177, 181, 183, 184, 186, 187
immune evasion 39, 47, 48, 57, 61, 66, 127, 225
Immunization Practices Advisory Committee (ACIP) 347, 361, 365
immunization safety 345
Immunofluorescent assay (IFA) 154, 202, 271, 273, 274, 276, 277, 278
immunogenicity 67, 341–343, 345–348, 358–360, 362–365, 367–371
immunohistochemistry (IHC) 274, 380
incompatible insect technique 295
incubation period 63, 64, 77, 107, 108, 111, 117, 140, 146, 147, 162, 277, 282, 293, 297, 316, 325, 339, 346, 349, 351, 352, 355, 358, 385, 386
indoor fumigation 286, 287
indoor residual spraying (IRS) 285, 286, 287, 289
inhibitors 41, 60, 67, 103, 110, 205, 373–381
innate immunity 64, 69, 75, 86, 87, 104, 105, 116, 117, 123, 203, 213, 215, 313, 314, 368
insecticidal aerosols 283, 284
Insecta 220, 227, 235, 263
insecticide resistance 115, 117, 287, 291, 294, 317, 331
insecticides 107, 283–287, 289, 291, 293, 294, 298, 315, 317, 333
insecticide-treated bednets 281
insect-specific viruses 6, 13, 80, 135, 150, 191, 192, 198, 199, 209, 211, 216, 225, 247, 253, 256, 267
Insulators 321, 322, 326, 333
Interbiome transfer 226, 229
Interferon (IFN) 32, 33, 35, 37–43, 48, 57–61, 64–70, 75, 76, 78, 81–84, 86, 87, 93–106, 108, 110, 115, 116, 127, 131, 133, 205, 211, 264, 303, 305, 307, 314, 338, 346, 354, 358, 364, 377, 379, 380
interhost bottlenecks 167
Interhost diversity 148, 168, 189
International Committee for Taxonomy of Viruses (ICTV) 1, 9, 10, 13, 27, 35, 220, 222, 228, 234, 249, 261, 390
intra-host genetics 167
inverted-repeat (IR) sequence 311, 322, 326, 327, 328, 329
inverted terminal repeats (ITRs) 317, 319, 322
Ion Torrent 181–187
Iridoviridae 226–228
iridoviruses 227, 253
irradiation 295, 323, 377
irrigation systems 1, 4, 283
Israel turkey meningoencephalitis virus (ITV) 15, 223
Ixodes persulcatus 195, 245, 349
Ixodes ricinus 159, 195, 227, 252, 258, 349
Ixodes scapularis 72, 155, 171, 245, 257, 260
Ixodid ticks 61, 62, 158, 213, 222, 224, 227

J

JAK-STAT 32, 39, 40, 57, 102, 108, 115, 117, 205, 212, 294, 299, 335, 377
Jamestown Canyon virus (JCV) 18, 61, 62, 63, 67, 126, 131, 211, 213, 338, 361
Japanese encephalitis virus (JEV) 3, 4, 14, 15, 33, 39–45, 55, 58–60, 101, 122–124, 128, 131, 133, 137, 140, 143–146, 148, 149, 152, 153, 155–166, 171, 191, 194, 197, 210, 212, 214, 215, 222, 229, 232, 235, 237, 238, 243, 246, 247, 252, 254, 261, 262, 264, 267, 268, 278, 281–283, 289, 290, 293, 304, 307, 311–314, 333, 345, 348, 351, 360–371, 373, 386
JE-CV 347–349, 361, 365
JE-Vax 346
Johnston Atoll virus (JAV) 27, 189
Jurona virus (JURV) 26, 125, 220

K

Kaeng Khoi virus (KKV) 18, 250
Kamiti River virus (KRV) 26, 122, 124, 150, 192, 199, 212, 220, 235–237
Karshi virus (KSIV) 16, 122, 124, 224, 235, 237, 249, 251, 252, 254
Kasba virus (KASV) 25, 220
Kashmir bee virus (KBV) 20, 256, 266
Kasokero virus (KASOV) 20, 220
Keystone virus (KEYV) 18, 193, 201, 212, 213, 260
Kibale virus (KIBV) 28, 199, 240, 248
Kotonkan virus (KOTV) 26, 78, 86, 244, 250, 258
Koutango virus (KOUV) 15, 122, 124, 149, 212
Kozhevnikov's epilepsy 349
Kunjin virus (KUNV) 10, 15, 41, 44, 58, 142, 149, 156, 158, 162, 163, 283, 352
Kyasanur Forest disease 3, 155, 162, 165, 349
Kyasanur Forest disease virus (KFDV) 14, 122, 124, 137, 144, 145, 161, 195, 215, 235, 237

L

Laboratory Diagnosis 4, 271
La Crosse virus (LACV) 18, 61–63, 65, 66, 70, 126, 191, 192, 196, 200–216, 229, 251, 264, 265, 278, 338, 361, 368
Lammi virus (LAMV) 28, 235–237, 243, 245
Langat virus (LGTV) 14, 59, 122, 124, 224, 235, 237, 243, 267, 313, 371
La nina cycles 283
larval control 283, 284, 286, 289, 293
larval habitats 142, 258, 263
larvicides 283, 301
Le Dantec virus (LDV) 27, 35, 81
lentivirus 225, 253, 306, 307, 309, 310, 314
Lepidoptera 229, 246, 250, 253, 261, 318, 320, 332
Leptotrombidium scutellare 197
library preparation 183
Linker Amplified Shotgun Library (LASL) 177
live attenuated arbovirus vaccines 311
live-attenuated vaccines 93, 303, 340, 348, 367, 369
livestock 31, 42, 61–64, 69, 77–80, 83, 86, 135, 143–145, 158, 160, 161, 163, 164, 233, 254, 261, 263, 266, 291, 298, 337, 339, 387
longevity 251, 252, 273, 286, 295, 297, 316, 323
Long-term persistence 75, 137, 192, 254
loop-mediated isothermal amplification (LAMP) 275, 276, 278, 279
louping ill virus (LIV) 15, 27, 122, 124, 148, 159, 232, 252, 386
louping ill disease 144, 222, 265, 349
luciferase 309, 320, 332, 378
Lutzomyia spp 76, 77, 145, 193, 195, 196
Lymphocytic choriomeningitis virus (LCMV) 224, 226

M

MAC-ELISA 272, 273
macroevolution 220, 221, 262
maculopapular rash 338, 340, 355, 358
Maraba virus (MARAV) 26, 125, 216, 257, 258
marine environments 226, 229
mariner Mos1 318, 319, 321, 326, 327, 332
Mayaro virus (MAYV) 14, 91, 122, 145, 241, 357, 364, 367, 369, 370
measles vaccine 299, 348, 352, 356, 361, 362, 363, 366

mechanical transmission 85, 157, 227, 228, 253, 254, 256, 259, 261, 265, 266
Medea 329–333
Medea-bearing homologue 329
medfly 318, 322, 323, 333, 334
Melon fly 322
meningitis 63, 64, 91, 163, 228, 260, 338, 340, 347, 349, 351
Mesoniviridae 6, 13, 28, 198, 199, 222, 223, 228, 242, 243, 263
metagenomics 157, 175, 178, 187, 188, 226, 388
microarray oligonucleotide hybridization 275
microRNA (miRNA) 98–100, 105, 114, 118, 303, 304–306, 309, 311–314, 325, 330
microsphere immunoassay (MIA) 273, 277
Middle East respiratory syndrome coronavirus (MERS) 388
Middelburg virus (MIDV) 14, 122, 241
midges 31, 37, 62, 78, 79–86, 107, 121, 124, 136, 152, 156, 157, 161, 197, 210, 220, 223, 224, 242, 246, 252, 253, 255, 257, 288, 291, 310, 338
midgut bacteria 113, 114, 117, 291, 294
midgut epithelium 107, 294, 325, 326, 327
midgut escape barrier 56, 107, 325
midgut infection barrier 107, 325, 326
Midway virus (MIDWV) 224, 229
mode of transmission 10, 150, 220, 226, 242, 248, 253, 386
Modoc virus (MODV) 16, 122, 124, 144, 235, 237, 238, 244, 263, 265, 267
Mokola virus 84, 228, 246, 251, 257
molecular assays 271, 274, 275
molecular biology 5, 9, 10, 11, 38–40, 42, 69, 92, 107, 129, 132, 191, 204, 315, 389, 390
molecular clock 221, 261
mosquito density 230, 285, 286
mosquito infection 172, 281, 282, 330
mosquito larvae 229, 253, 291–293, 298, 299, 301
mosquito populations 146, 150, 159, 281–284, 289, 291, 293, 294, 297, 298, 323, 324, 330, 339, 346
mosquito-borne flaviviruses 15, 16, 29, 34, 42, 49, 159, 260, 263, 304
mouse brain-derived CCHF vaccine 338
Moussa virus (MOUV) 28, 80, 85, 152, 163, 214, 248, 265
MP-12 339, 366
Mucambo virus (MUCV) 14, 241
Muller's ratchet 169
Murray Valley encephalitis virus (MVEV) 15, 57, 58, 84, 122, 124, 143, 146, 149, 150, 152, 153, 158, 160, 164, 211, 213, 237, 283, 290
mutant swarms 169
mutation rates 132, 167, 169, 171, 173, 207, 242
myalgia 63, 64, 90, 92, 342, 355, 357, 358
MYD88 108, 109, 330

N

Nairobi sheep disease virus (NSDV) 20, 62, 67, 68, 126, 224, 246
Nairovirus 12, 20, 36, 62–64, 66–69, 126, 129, 225, 261, 313
Nakiwogo virus (NAKV) 28, 237
Nam Dinh virus (NDiV) 28, 228
nanoviruses 243, 247
Naples viruses 339
natural host range 10, 245
nausea 63, 64, 91, 339, 346, 351, 358
Ndumu virus (NDUV) 14, 122, 240, 241
Negevirus 6, 11, 13, 155, 198, 216, 223, 224, 242, 248, 267
Netivot virus (NETV) 25, 244, 250, 266
neurological disease 2, 99, 152, 265, 311, 340, 346, 355
neurovirulence 42, 86, 93, 102, 104, 130, 311, 314, 343, 344, 348, 354, 360, 363, 367
neutralization 39, 40, 42, 44, 47, 59, 78, 79, 80, 104, 106, 253, 258, 266, 268, 271–274, 279, 358, 364, 386
neutralizing antibody 41, 78, 79, 225, 240, 250, 253, 260, 338, 341–346, 349, 350, 356, 357, 364, 366, 367, 369, 370, 371
New Minto virus (NMV) 27, 81, 85, 237
Newcastle's disease virus 339
next generation sequencing (NGS) 2, 11, 14, 80, 153–155, 160, 161, 163, 165, 170, 172, 188, 201, 212, 277, 387, 388, 389
Ngoye virus (NGOV) 24, 28, 236, 260
Nhumirim virus (NHUV) 235, 236, 237, 243
Nidovirales 242, 263
Nimaviridae 226, 228
Nipah henipavirus 388
NITD008 376
NITD107 377
Nodamura virus 247, 303
Nodaviridae 223, 226, 228, 247
nodavirus 228, 229, 248, 257, 266
No known vector (NKV) 124, 144, 225, 234, 235–240, 243, 244, 254, 255, 263
non-nucleoside inhibitor (NNI) 376
Nora virus 312
Nothobranchius guentheri 292
Nounané virus (NOUV) 28, 122, 124, 235–238, 243
novirhabdovirus 71, 73, 198
nucleic acid sequence-based amplification (NASBA) 275, 276, 278, 279
nucleocapsid 36, 37, 42, 51, 60, 67, 69, 71, 73, 74, 85, 92–95, 105, 123, 124, 127, 160, 205, 227, 278
nucleoside inhibitor (NI) 376, 381
Nucleorhabdovirus 129, 224
Nyando virus (NDOV) 19, 220
nymphs 229, 252, 253, 292, 349, 350

O

O. moubata moubata 192
O'nyong nyong virus (ONNV) 14, 43, 91, 99, 122, 130, 133, 213, 232, 241, 244, 246, 262, 266, 303, 305, 309, 313, 357
Obodhiang virus (OBOV) 26, 78, 82, 244, 250
Ochlerotatus flavivirus (OcFV) 28, 199
Ockelbo virus (OCKV) 14, 241
Omsk haemorrhagic fever virus (OHFV) 14, 122, 124, 144, 229, 233
Orbivirus 25, 37, 42, 127–129, 132, 144, 154, 161, 188, 222, 228, 233, 246, 255, 257, 258, 264, 266, 267, 304, 310
Order *Mononegavirales* 26, 71, 257
original antigenic sin 272, 278, 380
Ornithodoros moubata 192, 196, 214, 215, 223, 228, 261, 265
Ornithodoros tick 231, 249, 251, 252, 267
Oropouche virus (OROV) 19, 62, 63, 66, 67, 69, 126, 136, 145, 146, 162, 249, 338, 360, 368
Orthobunyavirus 12, 36, 37, 39, 40, 42, 44, 61, 63–66, 68–70, 125, 129, 134, 136, 144, 148, 151, 154, 155, 158, 160, 165, 196, 210, 211, 213, 276, 278, 338, 360
Orthomyxoviridae 1, 12, 27, 31, 38, 121, 126, 175, 188, 189, 192, 222, 224, 226, 228, 264
Orungo virus (ORUV) 25, 198, 246
overwintering 70, 136, 147, 156, 157, 162, 163, 191, 192, 197, 198, 200, 201, 203, 209–212, 214–216, 247, 251, 254, 263, 266, 268, 289
oviposition 150, 192, 197, 199, 200, 207, 292, 293, 323, 325

P

Pacui virus (PACV) 22, 210
Paramyxoviridae 71, 226
Parvoviridae 224, 226, 228, 229, 249, 257, 258
passive trap 163, 283, 290
pathogen-associated molecular patterns (PAMPs) 64, 94, 95, 97, 108, 109
pathogenicity 10, 12, 37, 68, 77, 133, 140, 163, 175, 202, 232, 254, 267, 293, 300, 301, 341, 350, 370, 386, 388
Pectinophora gossypiella 318
pegiviruses 225, 239, 240, 262, 265
P-element 317, 318, 329
perifocal treatment 287
persistent infection 49, 68, 72, 75, 76, 100, 136, 150, 162, 191, 192, 198, 204, 210, 220, 231, 246, 251, 264
person-to-person transmission 253, 257, 359, 366
pesticides 281, 291
Pestivirus 39, 40, 60, 225, 226, 239, 240, 241, 247, 260
phenotypic plasticity 168
PhiC31 319, 321, 322, 332, 333
Phlebotomine sandflies 6, 61, 62, 124, 136, 157, 161, 189, 192, 196, 198, 215, 216, 298
Phlebotomus (Pappataci) fever 339
Phlebotomus papatasi 62, 70, 193, 194, 215, 267, 299, 339
Phlebotomus perfiliewi 151, 339
Phlebotomus perniciosus 70, 196, 211, 216, 258

Phlebovirus 12, 22, 36, 37, 61–67, 69, 70, 125, 144, 151, 154, 156, 161–163, 165, 166, 197, 211, 216, 221, 223, 225, 227, 233, 246, 253, 257, 264, 278, 305, 339, 360, 388
photophobia 338, 346, 351, 352, 358
phylogenetic tree topology 233, 234, 243
phylogeny 72, 75, 101, 132, 160, 163, 166, 219, 220, 233, 234, 243, 246, 260–262, 264, 265, 278, 313
phytophagous arthropods 224
Pichia anomala 113, 116, 117, 294, 301
Picornaviridae 169, 228, 242, 263
piggyBac 318–322, 331–335
pink bollworm 318, 322, 335
piperonyl butoxide 284, 289
Piwi gene-family 325
piwi-interacting RNA (piRNA) 303, 325, 334
plaque forming units (PFUs) 54, 73, 252, 272, 274–277, 327, 328, 342–345, 348, 352, 354
plaque reduction neutralization test (PRNT) 253, 271–274, 277, 279, 347
plasmid DNA 309, 317, 318, 338, 352
poikilothermic 191, 197, 223, 246, 263, 266
point mutation 97, 127, 128, 261, 329
point-of-care diagnostic tests 277
poly-A tail 32, 177, 178
polymerase chain reaction (PCR) 11, 13, 29, 79, 117, 153, 170, 176–178, 180–183, 187, 192, 198, 203, 213, 253, 275–279, 283, 288, 289, 297, 323, 365, 380, 387
polythetic 10, 11, 387
population control 291, 323, 331, 335
population growth 1, 4, 5, 13, 142, 373
Powassan virus (POWV) 14, 42, 122, 124, 148, 155, 168, 171, 235, 237, 263
Poxviridae 1, 13, 127, 226, 228
predators 291, 292, 297, 298
programmed cell death (PCD) 66, 111, 203
Pronghorn antelope pestivirus 239, 241
prostration 338, 340, 358
protease inhibitors 374, 375, 379, 380
Psorophora confinnis 359
Punique virus (PUNV) 21, 166
Punta Toro virus (PTV) 21, 63, 69, 366
pyrethrin 282, 284, 288, 289
pyrethroid insecticides 317, 333

Q

Qalyub virus (QYBV) 20, 252, 264
quantitative trait loci (QTL) 200, 212, 325, 331
Quaranfil virus (QRFV) 27, 38, 178
Quaternary glaciations period 340

R

Rabensburg virus (RaBV) 171, 220, 237, 243, 248
Rabies virus (RABV) 82, 85, 123, 220, 223–226, 228, 246, 250, 265, 266, 386
randomized controlled trials (RCTs) 281
rapid amplification of cDNA ends (RACE) 178
Rapid Diagnostic Test (RDTs) 283
rDEN2/4 Δ30(ME) 343
rDENV3/4 Δ30 343
rDENV3Δ30/31 343
reactive oxygen species (ROS) 59, 114, 117, 131, 294, 296, 301
reactogenicity 342, 348, 349, 359, 360
Real-time 5′-exonuclease fluorogenic assays 275
reassortment 10, 12, 37, 39, 41, 44, 121, 126–129, 131–134, 148, 151, 155, 161, 164, 165, 205–207, 209–212, 215, 236, 243
recombination 10, 40, 72, 73, 90, 105, 121, 128, 131–133, 148, 155, 156, 158, 162, 164, 165, 235, 236, 243, 262, 319, 320, 321, 324, 332
Red Queen hypothesis 170
reduced virulence 232
Reed Ranch virus (RERAV) 27, 81, 220
Release of Insects carrying a Dominant Lethal (RIDL) 323, 324, 330
relative male mating competitiveness 297
Reoviridae 1, 2, 3, 13, 23, 26, 28, 31, 37, 39, 41, 121, 122, 126–129, 148, 152, 176, 192, 195, 198, 210, 220, 223, 224, 226, 228, 233, 236, 240, 246, 248, 252, 257, 264, 265, 305, 315, 337, 355
Reoviruses 31, 37, 126, 153, 154, 175, 178, 198, 228
RepliVAX West Nile 352
reservoirs 37, 91, 107, 143, 144, 146, 147, 160, 165, 192, 224, 225, 227, 232, 244, 249, 254–256, 261, 263, 264, 266, 305, 359
retrotransposons 111, 116, 222
Retroviridae 228
Reverse transcription-loop-mediated isothermal amplification (RT-LAMP) 275, 276
Reverse transcription polymerase chain reaction (RT-PCR) 11, 153, 157, 170, 202, 204, 253, 254, 272, 274–278, 388, 389
Rhabdoviridae 1, 2, 6, 12, 13, 26–28, 31, 35, 40, 43, 71, 75, 80, 82–86, 121–123, 125, 127, 129, 151, 152, 163, 169, 175, 189, 191, 192, 196, 198, 201, 209, 210, 214–216, 223, 224, 226, 228, 233, 236, 241, 243, 246, 248, 251, 258, 264, 265, 267, 305
Rhabdoviruses 31, 35, 39, 43, 71–73, 75, 77, 79, 80–87, 124, 129, 132, 152, 153, 156, 158, 175, 198, 222, 224, 228, 229, 236, 250, 251, 258, 261, 310, 312
Rhopalosiphum padi virus 224, 268
ribavirin 305, 310, 313, 337, 346, 360, 369, 378
Rift Valley fever virus (RVFV) 21, 27, 40, 41, 61–63, 65–70, 111, 116, 124, 131, 137, 140, 144–146, 160, 162, 165, 176, 196, 211, 214–216, 229, 253, 254, 257, 260, 261, 264, 267, 275, 278, 279, 283, 298–301, 305, 314, 339, 340, 361, 366, 368
Rio Bravo virus (RBV) 16, 122, 124, 126, 137, 144, 220, 225, 235, 237
Rio Grande virus (RGV) 22, 212
RNA interference (RNAi) 108–112, 115–117, 124, 129, 171, 204, 205, 207, 212–214, 293, 294, 300, 303, 304, 312–314, 325, 331–334
RNA isolation 179, 188, 275
RNA replication 214, 264, 378, 379
RNA synthesis 376, 377, 379, 380
RNA-dependent RNA polymerase (RdRp) 31–33, 35, 36, 38, 40, 42, 43, 48, 49, 58, 59, 65, 66, 68–71, 73, 74, 81, 83, 94, 123, 124, 147, 167, 205, 206, 236, 242, 243, 247, 305, 329, 376, 377, 379–381
RNAi-based antiviral approaches 329
RNase 65, 101, 104, 148, 179–181, 275
RNA-induced silencing complex (RISC) 110, 303
Rockefeller Foundation 2, 91, 233, 254, 255, 258, 262, 342, 353, 354, 385, 386, 388–390
rodents 1, 12, 61, 62, 78, 81, 89–91, 121, 135, 144, 145, 160, 197, 221, 225, 227, 233, 240, 245–247, 251–255, 263, 265, 349, 355, 359
Roniviridae 223, 226, 228, 242
Ross River virus (RRV) 2, 14, 102, 122, 130, 132, 133, 156, 196, 198, 209, 213, 214, 220, 241, 247, 257, 260, 272, 278, 282, 283, 284, 290, 300, 358, 360, 364, 365, 370, 371
ruminants 37, 62, 63, 86, 144, 145, 154, 163, 166, 233, 254, 305

S

SA-14-14-2 347, 348
Saboya virus (SABV) 16, 122, 124, 220, 223, 237
Salmon pancrease disease virus (SPDV) 241
Sanger sequencing 170, 176–178, 180, 184, 186
Sawgrass virus (SAWV) 27, 35, 81, 85, 236
Seadornavirus 129, 152, 154, 155, 222, 265, 360
Secoviridae 224
selective pressure 89, 128, 131, 147, 156, 167–169, 171, 202
selfish DNA 329, 330
Semliki forest virus (SFV) 14, 33, 40–43, 91, 93, 95–101, 103–105, 112, 116, 117, 122, 130, 132, 305, 311, 314
Sena Madureira virus (SMV) 27, 220
sentinel 78, 79, 82, 85, 152, 153, 158, 159, 281–284, 357
sequelae 2, 90, 91, 338, 349, 363, 365
sequence-independent single primer amplification (SISPA) 163, 177, 178, 181–183, 189
seroconversion 78, 79, 282–284, 341, 342, 344–347, 349, 350, 352, 359
serological assays 151, 271, 273, 387
Serratia odorifera 114, 115, 294, 298
severe acute respiratory syndrome (SARS) 388
Severe fever with thrombocytopenia syndrome virus (SFTSV) 22, 61, 63, 66, 68- 70, 126, 129, 132, 151, 162, 366, 371
Sex ratio distortion 329, 332, 334, 335
sexual transmission 61, 137, 253
short hairpin RNAs (shRNAs) 304–311, 314
Sigma virus (SIGMAV) 26, 71, 75, 78, 84, 191, 192, 200, 201, 203, 207–210, 212, 214–216, 250, 256, 263
simulated transmission 252

Sindbis virus (SINV) 12, 14. 39, 41, 42, 90–93, 95–106, 109, 112, 114, 117, 122, 130, 136, 144, 152, 160, 163, 172, 198, 205, 212, 215, 224, 229, 241, 243, 244, 257, 258, 262, 266, 267, 303, 331, 333, 358, 359, 362, 369, 371
Skeeter Buster 316, 330, 333
small interfering RNA (siRNA) 97, 110–112, 205, 303–314, 325–327, 378
siRNA-iFECT 311
Small Molecule Drug 373
small molecule inhibitor 377
small RNA therapies 304, 305, 311, 312
snowshoe hare virus (SSHV) 18, 61, 62, 126, 129, 206, 214
Sokuluk virus (SOKV) 16, 243, 246–248, 251, 255
source reduction methods 286
Southern blot 275
southern elephant seal virus (SESV) 14, 89, 122, 152, 229, 240, 241
Spinareovirinae 23, 37, 38, 355
Spodoptera frugiperda rhabdovirus 248
St Croix River virus (SCRV) 28, 127, 220, 222, 248, 257
St. Louis encephalitis (SLEV) 2, 15, 122–124, 128, 131, 132, 137, 144, 146, 148, 149, 154, 155, 159, 161, 164, 168, 169, 171, 191, 192, 195, 197, 210–213, 215, 225, 229, 237, 238, 240, 247, 250, 262, 278, 279, 281, 283, 308, 312, 370, 386
stabilized infection 191, 192, 200–205, 207–210, 216, 267
sterile insect techniques (SIT) 284, 295, 299, 322, 323, 331, 332, 334, 335
Stone Lakes virus 240, 247, 250, 258
substitution rates 167, 237
suckling mice 80, 252, 276, 341, 348
sugar meal 113, 294
superinfection 39, 131, 206, 207, 211, 215
survival of the flattest 169, 173
sylvatic cycles 4, 313, 353
sylvatic transmission 255

T

Tamana bat virus (TABV) 16, 227, 229, 234, 236, 237, 239, 240
Taura syndrome virus 227, 229, 256, 257, 263
tetracycline-repressible transactivator (tTAV) 323, 324
tetravalent vaccine 341–345, 363, 364, 368, 373
therapeutic small RNAs 311, 312
Thogoto virus (THOV) 27, 43, 70, 126, 128, 131–133, 243, 249
Tibrovirus 72, 79, 244
Tick-borne encephalitis virus (TBEV) 14, 27, 39–43, 57, 59, 122–124, 128, 130, 133, 136, 137, 144, 148, 149, 157, 159, 160, 213, 215, 224, 232, 235, 237, 243, 245, 253, 256, 260, 262, 265, 267, 277, 311–313, 349, 350, 360, 365, 366, 368, 373, 376
ticks 1, 9, 12, 31, 33, 37, 38, 56, 61, 62, 64, 68, 69, 72, 81–83, 85, 107, 121, 124, 126, 132, 136, 144, 145, 147–149, 151, 154, 155, 157, 158, 160–163, 165, 187, 188, 192, 196, 210, 212, 213, 215, 216, 217, 220, 222–229, 231, 233–236, 242–246, 249–254, 256–260, 263–268, 291, 310, 313, 337, 339, 349, 355
tissue tropism 10, 73, 83, 84, 98, 104, 172, 259, 294, 298, 325
tissue-specific DNA promoters 320
Togaviridae 1–3, 5, 10–15, 28, 31, 41, 89, 121, 122, 129, 150, 158, 164, 167, 168, 175, 176, 192, 196, 198, 209, 211–213, 223, 224, 226, 228, 229, 240, 242, 246, 248, 258, 260, 265, 273, 304, 313, 315, 316, 355
Toll pathway 59, 108, 109, 111, 112, 114, 117, 118, 202, 205, 208, 210, 216, 301, 335
Toscana virus (TOSV) 21, 62, 63, 66, 68–70, 126, 144, 145, 154, 157, 163, 196, 211, 216, 249, 258, 278, 339, 362
Tospovirus 12, 36, 126, 129, 133, 329, 334
Totiviridae 223, 226, 229, 247, 257, 268
Toxorhynchites (Tx.) splendens 292, 298, 299, 301
Toxorhynchites amboinensisis 292
Toxorhynchites rutilus 292
Toxorhynchites 229, 262, 291, 292, 298, 299, 301
transcription 32, 35–38, 41, 42, 49, 52, 55, 59, 61, 64–69, 71–76, 78, 81–84, 86, 94–97, 102, 103, 108–110, 115, 116, 121, 123–125, 127, 129, 131, 133, 177, 180, 202–204, 208, 213, 247, 275–279, 289, 315, 325, 326, 330, 334, 354, 369, 380
transgene 311, 313, 317–319, 321–323, 326–332, 334, 335
transgenesis 316–318, 333, 334
transgenic mosquitoes 110, 111, 313, 320, 321, 327, 331–333
Transovarial transmission (TOT) 61, 62, 70, 192, 196, 197, 199–204, 206–216, 249, 251, 258, 264–266, 339, 353
transposable element (TE) 182, 183, 317–322, 326, 327, 329, 331–335
Tribolium castaneum 318, 329, 331
tritaeniorhynchus summarosus 346
tropism 10, 33, 34, 70, 73, 83, 84, 98, 99, 102, 104, 117, 121, 133, 150, 172, 207, 210, 259, 294, 298, 311, 325, 326, 334, 379
Trocara virus (TROV) 14, 152, 164, 240, 241
tsetse fly 112, 322, 335
Turlock virus (TURV) 19, 250
Tymoviridae 223, 224, 267

U

UAS-Gal4 binary expression system 322
Umatilla virus (UMAV) 25, 127, 258
Una virus (UNAV) 14, 241
untranslated regions (UTR) 33, 34, 39, 40, 129, 176, 208, 219, 233, 235, 238–244, 260, 263, 267, 304, 306, 307, 311, 312, 324, 361, 373
Upolu virus (UPOV) 27, 188
urbanization 1, 4–6, 107, 118, 142, 230, 287, 367, 373
Usutu virus (USUV) 15, 122, 124, 143, 144, 155, 156, 158, 164, 165, 237, 249
Uukuniemi virus (UUKV) 21, 66, 70, 126, 264

V

vaccine 5, 6, 39, 42, 57, 81, 84–86, 92, 93, 100, 102, 103, 105, 107, 121, 129, 131, 133, 137, 155, 159, 162, 164, 173, 175, 252, 253, 255–257, 261, 265, 281, 284, 289, 291, 299, 302–305, 311, 315–317, 337–371, 373, 379, 385- 388
vector control 5, 107, 111, 115, 154, 197, 281–291, 293, 294, 298, 300, 302, 315–317, 324, 330, 334
Venezuelan equine encephalitis virus (VEEV) 12, 14, 42, 100–106, 122, 133, 136, 156, 159, 162, 164, 167–169, 171–173, 175, 229, 241, 244, 248, 249, 254, 305, 312, 313, 359, 360, 365, 368, 371, 386
Vero cells 40, 55, 57, 178, 207, 258, 274, 342–344, 346–348, 352, 369, 370
vertebrate hosts 1, 2, 4, 9, 12, 62, 71, 72, 75, 76, 80, 89, 105, 107, 130, 131, 135–137, 143–147, 150, 159, 167, 191, 192, 197, 198, 200, 201, 209, 210, 225, 231, 233, 240, 242, 244, 245, 247, 249, 251–255, 258, 262, 267, 303, 315, 337, 338, 353, 359
vertical transmission 12, 70, 136, 146, 147, 150, 154, 155, 162, 165, 191–193, 196, 199, 210–216, 228, 230–232, 249, 250, 256, 260, 267, 293, 338
vesicular stomatitis virus (VSV) 42, 54, 71, 73–77, 79, 81–86, 96, 101, 128, 169, 172, 203, 211, 215, 216, 221, 224, 228, 233, 254, 256, 262–264, 305, 312, 314, 377
vesiculovirus 72, 73, 75,76, 83, 85, 86, 123, 129, 198, 215, 216, 257, 264
viraemia 12, 61, 62, 76–78, 93, 98–100, 136, 155, 164, 169, 223, 225, 228, 232, 250–252, 254, 256, 258, 271, 272, 274, 275, 278, 285, 341–346, 348, 352, 353, 358, 359, 371, 374, 376, 377, 379
Viral 'fossil' sequences 234
viral determinants 150, 242
Viral haemorrhagic septicaemia virus 82, 228, 256
viral persistence 142, 160, 198, 224, 225, 231, 250, 251, 256
viral quasispecies 128, 132, 170, 172, 173, 265
viral replication 34, 39, 45, 51, 53–55, 57, 59, 64, 66, 67, 77, 123, 127, 129, 131, 203, 215, 227, 231, 232, 242, 244–246, 248, 250, 251–253, 256, 294, 305, 307, 308, 314, 359, 373, 374, 376–381
viral suppressors of RNAi (VSRs) 303, 312
virogenesis 155, 171, 200
virulence 12, 33, 34, 41, 44, 61, 90–92, 98, 99, 101, 102, 105, 123, 129, 130, 135, 142, 144, 151, 152, 155, 163, 164, 167, 171, 187, 203, 207, 209, 210, 219, 231–233, 249, 250, 256, 259, 260, 262, 265, 268, 293, 301, 311, 333, 361, 362, 364, 366, 368, 386, 387, 389
virus discovery 156, 160, 175, 187, 188, 262, 387, 388
virus identification 13, 82, 86, 277, 386
virus isolation 11, 151–153, 156, 158, 162, 201, 245, 246, 250, 271, 274, 276, 385, 386, 389
virus transmission 40, 61, 75, 132, 145, 146, 156, 158, 159, 209, 223, 228, 230, 231, 256, 260, 264, 266, 284, 285, 287, 290, 294, 311, 316, 325, 326, 330, 360, 361
virus host interactions 34, 50, 56, 61, 71, 74, 77, 89, 92, 170, 171, 219, 226, 242
vitellogenin 320, 321, 334

W

Walter Reed Army Institute of Research (WRAIR) 341–343, 347
West Nile virus (WNV) 3, 6, 12, 13, 15, 39–43, 45, 47, 49–51, 53–60, 83, 101, 105, 107–109, 111, 112, 114, 116, 117, 122–124, 128, 130, 133, 135, 154–165, 167, 168, 171, 172, 175, 177, 180, 187, 188, 191, 192, 195, 197, 204, 205, 209–212, 214–216, 225, 229, 232, 233, 237, 245, 247, 252, 254, 257, 259, 261, 262, 264–268, 271, 272, 275, 278, 279, 281–284, 288, 289, 290, 296, 298–300, 304, 306, 312–314, 332, 345, 350–352, 360–369, 373, 376, 377, 380, 381
Western equine encephalitis virus (WEEV) 12, 14, 90, 91, 93, 95, 98, 100–102, 104, 122, 128, 130, 132, 137, 144, 175, 178, 191, 196–198, 216, 220, 222, 225, 229, 241, 243, 252, 254, 257, 258, 278, 359, 360, 377, 386
Whataroa virus (WHAV) 114, 122, 240, 241, 267
wMel 117, 295–299, 301, 317, 335
Wolbachia pipientis 117, 295, 299, 301, 317
Wolbachia 59, 112, 114–118, 178, 179, 285, 289, 291, 295–302, 317, 333–335
World Health Organization (WHO) 6, 137, 138, 143, 165, 268, 288, 332, 342, 387

Y

Yellow fever virus (YFV) 2, 4, 12, 14, 16, 39, 41–43, 45, 58, 59, 84, 114, 116, 122–124, 128, 130, 132, 133, 136–138, 140, 144, 148, 150, 153, 155, 157, 159, 164, 175, 178, 179, 210, 212, 220, 226, 229–233, 237, 238, 240, 242, 243, 250, 253–255, 258, 260, 264, 278, 279, 292, 296, 300, 304, 313, 314, 316, 317, 344, 351–354, 357, 360, 362–367, 369–371, 373, 376–378, 380, 385, 386
Yellow head virus 228, 229, 260
Yokose virus (YOKV) 16, 122, 124, 235, 237, 238, 243, 247, 248, 251, 255

Z

Zika virus (ZIKAV) 16, 122, 124, 137, 139, 140, 149, 156–158, 160, 162, 260, 275, 278, 304, 311
zone of emergence 231

CPSIA information can be obtained
at www.ICGtesting.com
Printed in the USA
LVOW05s1126230416
484885LV00039B/177/P

9 781910 190210